Computer Aided Analysis and Design for Electrical Engineers

Computer Aided Analysis and Design for Electrical Engineers

B. JAMES LEY
NEW YORK UNIVERSITY

HOLT, RINEHART AND WINSTON, INC.
NEW YORK CHICAGO SAN FRANCISCO ATLANTA DALLAS
MONTREAL TORONTO LONDON SYDNEY

TO MY WIFE

Library of Congress Catalog Card Number 71-116731
SBN: 03-082867-8
Printed in the United States of America
0 1 2 3 22 9 8 7 6 5 4 3 2 1

PREFACE

The President's Science Advisory Panel has estimated that in 1971–72, $400 million will be spent annually for undergraduate computer education. This will account for approximately 4 percent of the money that will be spent on higher education that year. This is probably a very conservative prediction.

In 1971–72 all engineering calculations or designs that can be programmed will be carried out with the aid of a computer. Since the use of a digital computer in solving a problem generally requires a completely different approach from the conventional methods that have been taught in various engineering curricula, it is obvious that radical changes in undergraduate as well as graduate engineering education must take place. For example, we may ask how important the knowledge of calculus is in the differentiation or integration of various functions, or, for that matter, how helpful the knowledge of Laplace transforms is in the solution of simultaneous integrodifferential equations if a digital computer is to be employed.

The reader should not conclude that the author is calling for the elimination of such topics from the engineering curriculum, but rather that a need exists for a reevaluation of such mathematical concepts in light of a computer oriented curriculum. Using the words of Professor Charles L. Miller of the Massachusetts Institute of Technology, "a complete restructuring of the practice of engineering and wholly new approaches to executing the engineering process" are required.

The purpose of this book is to make the use of a digital computer as natural a tool in solving electrical engineering problems as the slide rule was 20 years ago. The author believes that the best way to achieve this goal is by imitating existing programs and continuing this practice throughout the years of formal education. In other words, the solution of many different types of electrical engineering problems should be considered along with the programming and necessary numerical techniques so that the student will learn how to solve problems in his chosen field. The reader should be cautioned that the knowledge of electrical engineering, programming, and numerical methods gained by taking separate courses in these three areas is not sufficient; what is required is a marriage of all three. For example, the present day policy of a required computer course in the freshman year is wasteful: little if any real benefit is derived from a course that teaches only Fortran and its use in abstract examples. What is needed is continued practice in the use of the computer to solve meaningful problems in the student's chosen field throughout his entire education.

In order to achieve the goal stated above, only problems in or related to electrical engineering will be considered in this book. The problems are of various degrees of difficulty, ranging from the sophomore year through the graduate level. Although it is assumed that the reader possesses some electrical engineering background, the electrical aspect of each problem is nevertheless discussed to a degree sufficient for the reader to understand the solution to the problem.

No attempt is made to include all numerical techniques that might prove useful to the electrical engineer; likewise, no attempt is made to include examples from all areas of electrical engineering. The author has attempted to present a sufficient number of problems, along with the necessary numerical techniques, to permit most electrical engineers to become proficient in using a digital computer in solving their technical problems.

Although this book is intended primarily for a one-semester, senior/first-year-graduate course, some instructors may prefer to go at a slower pace (which would result in less outside reading) by using it in a two semester course. The book may also be used at a much lower level. For example, this book could be used in a one-semester sophomore course by using the following material:

1. Chapter 1.
2. Chapter 2.
3. Chapter 3: Sections 3.1, 3.2, and 3.4 (RC network and speed regulator).
4. Chapter 4: Sections 4.1, 4.2 (quadratic root only), and 4.5.
5. Chapter 5: Sections 5.1, 5.3, and 5.4.
6. Chapter 6: Sections 6.1, 6.3, and 6.4.
7. Chapter 10: Sections 10.1, 10.2, and 10.5 (voltage divider design).
8. Chapter 11: Sections 11.1, 11.2, and 11.3.
9. Chapter 13: Sections 13.1 and 13.2.

Similarly, this lower level course could start at the junior year, rather than at the sophomore year, by simply including more advanced topics. It is thus possible, with suitable selection, to use this book as an introductory text at either the sophomore or junior level and then continue its use as a supplementary reference in later years. The book may also be used for self study purposes by practicing electrical engineers who are interested in learning how to use the computer in the solution of their industrial problems.

For classroom use, whether it be at the sophomore or graduate level, the author strongly recommends requiring a computer term project (dealing with some phase of electrical engineering) that would be written during the last half of the course (see Section 9.4). Similarly, for the practicing engineer, the early use of the computer in connection with one of his current assignments is strongly advised. Following is a brief discussion of each chapter.

Chapter 1 discusses the general-purpose digital computer in terms of the computer's five functional blocks. The knowledge of numerical methods and coding are then brought in by considering the steps in solving a problem when the digital computer is employed. Flow charting and the Fortran language are then introduced by means of a simple example.

Chapter 2 is devoted to presenting the basic rules of the Fortran language. This particular algebraic language was chosen because of its widespread use. No attempt is made to use the machine language of a particular computer in solving a problem because it is much more difficult to learn and apply. It is also unnecessary for the majority of engineers to possess such knowledge.

Chapter 3 applies the knowledge gained in Chapter 2 to solve a variety of network problems. Problems such as the calculation of the dissipative insertion loss or the calculation of the frequency response from the time response should serve to show the reader the need for a computer solution.

Chapter 4 deals with algebraic as well as iterative methods that can be used for finding roots of equations. The discussion of the solution of the quadratic equation should be of particular interest to the beginner because it shows that there is usually a great deal more to any problem than one might first suspect.

Chapter 5 develops various numerical integration methods, emphasizing their similarities and differences. Error determination and error comparison are made by means of several electrical examples.

Chapter 6 deals with the solution of simultaneous linear algebraic equations. Multimesh circuit with d-c and sinusoidal excitation are solved. Section 6.6 will prove particularly useful to those readers who deal with the solution of complex coefficient equations.

Chapter 7 considers a number of active network examples. The worstcase design of a flip-flop illustrates the large number of calculations the computer can be called on to perform in the design of a fairly complicated circuit. The shunt-peaked transistor amplifier design for a specified bandwidth illustrates a typical trial-and-error type of problem that can easily be solved by the digital computer.

Chapter 8 is a discussion of how we may write different types of subroutines. This subject is useful to the reader who has already solved a number of problems of his own origin and is at the point of solving longer and more complicated problems. A discussion of how we may use several subroutines of the IBM 360 Scientific Subroutine Package is also included. These illustrative examples should encourage the reader to make fuller use of this package.

Chapter 9 deals with various matrix operations and their corresponding computer programs. This is followed by the use of matrix techniques in the discussion of various topological problems.

Chapter 10 is an introduction to Monte Carlo techniques. Several Fortran random-number generators are discussed along with their use in the solution of various electrical systems. The worst-case and Monte Carlo solution of the Digital-to-Analog Resistive Ladder Network once again shows the advantage of a computer solution to a difficult problem.

Chapter 11 covers the subject of graphs. This material is generally considered the most interesting subject by the majority of students because it permits them to express themselves artistically in the presentation of vast and complicated results.

Chapter 12 is a discussion of the general area of polynomial approximations and interpolation. Several different methods are discussed along with their Fortran programs. Error-determination and error-comparisons are considered in connection with several electrical examples.

Chapter 13 discusses several numerical methods for solving ordinary linear and nonlinear differential equations. An introduction to some of the numerical techniques that are used in solving partial differential equations is also included.

The author would like to acknowledge the many valuable suggestions and constructive criticism made by his students in the course of the preparation of this book. In particular, he wishes to thank Julio Galvez and Robert Guarino who gave permission to use their term projects as student examples (see Sections 11.5 and 9.4, respectively). The author is also indebted to his wife Dorothy for her help in the preparation of the manuscript and to Alexander Blumenfeld for his detailed reading of the entire manuscript from a technical point of view.

B. JAMES LEY

New York City
April 1970

CONTENTS

CHAPTER 1

The General Purpose Digital Computer

1.1 INTRODUCTION

This chapter is used to introduce the reader to the general purpose digital computer through its five functional blocks. A general discussion of how one uses a digital computer in the solution of an engineering problem follows. This is then illustrated by considering some of the problems involved in using the computer to find the roots of the quadratic equation.

The reader should note that the quadratic equation example allows for an early use of the computing facilities and in addition serves to give a deeper insight into the various phases involved in a digital computer solution and establish the necessary motivation to learn the general rules of Fortran that are discussed in Chapter 2.

The reader should also note that this book has been written so that it is essentially machine-independent, even though all the programs were written so that they would run on an IBM 360/30 computer. The use of other model computers, as well as other computing centers, will therefore require minor changes from time to time.

1.2 THE FUNCTIONAL BLOCKS OF A DIGITAL COMPUTER

At this point, it is appropriate to ask the question, "How important is it to know something about the inner workings of a digital computer before one begins a study of the use of the computer as an engineering tool?" The answer to this question is that thanks to the short but glorious history of the computing industry, the new user requires little if any detailed knowledge of the inner workings of a computer. In fact, the motto of most university computing centers is "Hide the computer from the student." In other words, the only way to turn out a large volume of student and research problems is to operate on a closed shop basis. In addition, it should be stressed that with teletype terminals and data-phone links, one could use a computer that was located thousands of miles away. The only thing that is required is a general knowledge of the structure of a computer.

Figure 1.1 is a picture of a typical third generation digital computing

Figure 1.1 IBM System/360 computer (Courtesy, IBM)

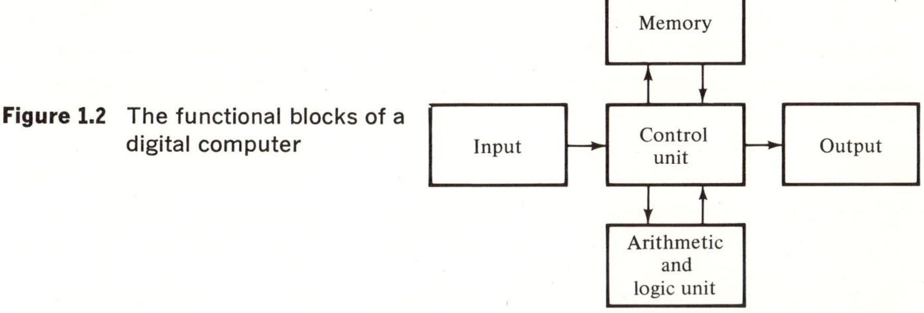

Figure 1.2 The functional blocks of a digital computer

center. In order to better understand how to communicate with such a computer, one should recognize that a general purpose digital computer is physically analogous to the ordinary desk calculator and differs mainly in only two respects; it is much faster, and it has a memory. The possession of a memory gives it the ability to store a program of instruction that can be modified during execution by the program itself; thus, it has the capability of carrying out very long and involved calculations at electronic speeds. The structure of any general purpose digital computer can be crudely represented by the block diagram shown in Figure 1.2.

Input

In attempting to solve a technical problem with the aid of a digital computer, it is first necessary to develop unambiguous step-by-step procedures, usually called algorithms, that correspond to the logical steps that must be performed in order to obtain the problem's solution. These instructions must be suitably coded before they can be read into and stored in the computer's memory.

In general, there is usually more than one way to read information into any given computer. The typewriter or console switches represent two such manual types of inputs. The magnetic tape, paper tape, and punched card represent additional examples of input devices.

Magnetic tape is erasable and therefore can be used repeatedly. It is therefore not only possible to use it as an input as well as an output device, but it may also be used as auxiliary memory storage.

Punched paper tape is probably one of the oldest input devices. Holes are punched across the tape at a density corresponding to approximately ten lines per inch. These holes can be read by electric contact means or photoelectrically. An example is shown in Figure 1.3. In this example, an eight-channel code is used on a one-inch-wide tape.

The small feed holes ride on the tape read sprocket and serve only in

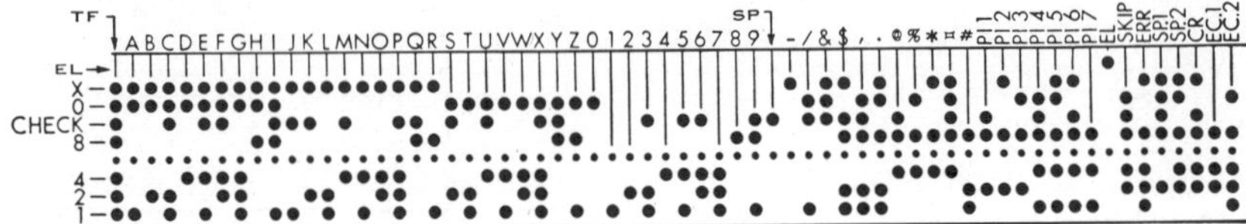

Figure 1.3 Paper tape (Courtesy, IBM)

the process of moving the tape. The lower 4 punched holes correspond to the weighted code, 1-2-4-8, as shown on the left. The number 1 can thus be represented by a single punch in the weight-1 channel, the number 2 by a single punch in the weight-2 channel, the number 3 by a punch in the weight-1 and the weight-2 channels, and so on.

The X and O channels are used with the 1-2-4-8 channels to represent alphabetic and other special characters. For example, the letter *A* is represented by punches in the X, O, and 1 channels.

The check (or parity) channel is used for error detection. Each line or column is punched so that it contains an odd number of holes. An examination will show that a check hole has been punched for any line that would have consisted of an even number of holes.

An E. L. (end-of-line) channel punch represents the end of a record and when sensed by the tape reader, an end-of-record signal is sent to the computer.

One of the most frequently used input devices is the 12 row, 80 column punch card shown in Figure 1.4. An examination of this card indicates that in each column there are 12 punching locations. Ten of these positions or rows are numbered from 0 to 9. The other two positions are referred to as the 11th and 12th rows, with the 12th row being

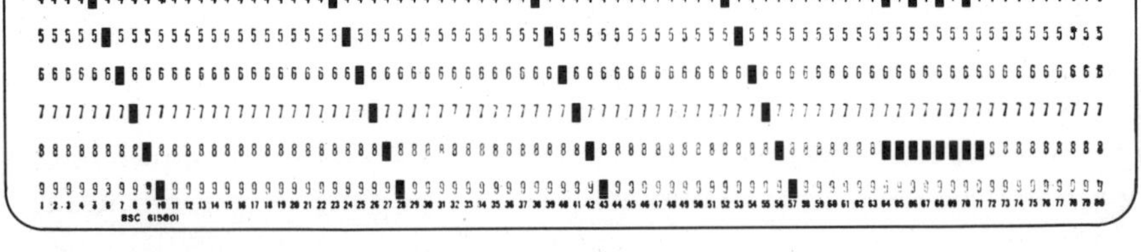

Figure 1.4 The 12-row, 80-column punch card

closest to the top edge. We thus refer to the 12th edge as the top edge and the 9th edge as the bottom edge. In a similar way, we refer to the left edge as the column 1 edge and the right edge as the column 80 edge.

Information can be put on these cards by means of a special key punch machine that will be described in Appendix A. This machine has a keyboard similar to that of a standard typewriter and is capable of punching a unique hole pattern as well as printing the character at the top of the column. A character may be coded in any one of the 80 columns by using one, two, or three punches per column, as shown in Figure 1.4. The ten decimal digits are shown in the first ten columns and correspond to single punches per column in rows 0 through 9 respectively. The 26 letters of the alphabet are coded by means of two punches per column. The first nine letters, A through I respectively, have a common 12th row punch along with a 1 through 9 row punch. The next nine letters, J through R respectively, have a common 11th row punch along with a 1 through 9 row punch. The last 8 letters, S through Z respectively, have a common 0 row punch along with a 2 through 9 row punch.

Since the combination of 12 things taken one at a time represent 12 distinct items and the ten decimal digits, 0 through 9, account for 10 of the 12 possibilities, only two other characters can be defined by a single column punch. These two characters are the + and − symbols as shown in columns 73 and 74. Since the + character is represented by a single punch in row 12, the 12th row is sometimes referred to as the + row and similarly the 11th row is sometimes called the − row. The other useful special characters are shown in columns 64 through 74 and are represented by two or three punches per column.

On input, the *card reader* senses the unique hole pattern and converts the information to electrical pulses so that they can be transmitted and eventually stored in the computer's memory. In general, each card in a deck of cards is read sequentially, 80 columns at a time.

Output

Most computers have more than one type of output device. Magnetic tape, paper tape, punched card, or oscilloscope picture display are some of the more common forms. When punched cards are the form of the output, it is necessary to process these cards by an *off-line* printer (not under computer control) in order to obtain printed results. An IBM 407 Off-Line Printer is shown in Figure 1.5. Many of the larger present-day computers have *on-line* printer output. Because of the multi-function operation of these computers and the use of a high-speed line printer, this can be a very efficient way of supplying the desired output. Since

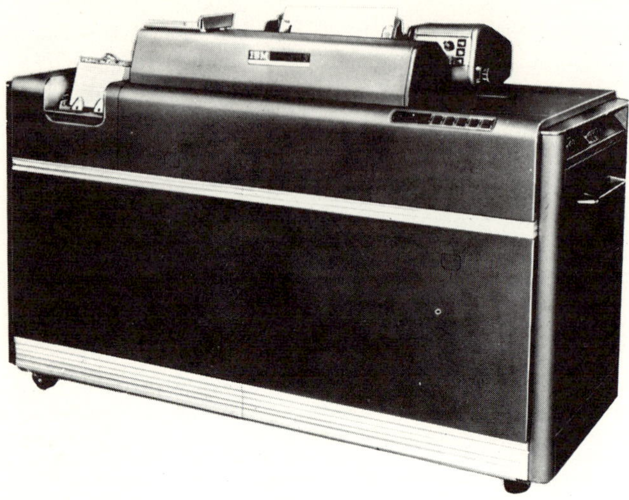

Figure 1.5 IBM 407 Printer (Courtesy, IBM)

the printing of one job is taking place at the same time that another job is being executed, the user is charged for only the execution time.

The Control Unit

As the name of the functional block implies, the control unit is the device that determines the sequence of the execution of the individual instructions stored in the computer memory. Normally the coded instructions that have been stored in memory are executed sequentially. If a given instruction is a jump or transfer instruction, however, the accumulator register is examined to see whether it is negative, zero, or positive. This will, in general, prevent the sequential execution of the next instruction and instead will cause a jump or transfer to an instruction that has been specified by another memory address.

The operation of the control unit is carried out in two distinct cycles. During the instruction cycle, an instruction is brought from a particular memory address specified by the instruction counter register, which is located in the control unit. The instruction counter is then incremented by one so that it corresponds to the address of the next instruction that normally would be obtained from the memory. Respecification of the instruction counter will later take place if the instruction should turn out to be a transfer instruction. During the second or execution cycle, the appropriate operands are brought from the memory to the arithmetic unit to carry out the required arithmetic or logic operations specified by the instruction. The result, in turn, is stored in the memory.

The Arithmetic Unit

The arithmetic unit performs the operations of addition, subtraction, multiplication, and division as ordered by the control unit. The control unit supplies the required operands from their memory storage locations, the necessary timing and sequence signals to control the sequence of the arithmetic operation that has been specified by the instruction, and then causes the computed result to be stored in memory.

The electronics of arithmetic units may differ greatly and depend on such things as whether decimal or binary number representation is used and, as in the case of addition, whether the digits of the two operands are added sequentially as in a "series" machine, or whether they are added all at once, as in a "parallel" machine. The reader is referred to Appendix B for a more detailed discussion of computer arithmetic operations.

After reading Appendix B, the reader should realize that it is not necessary to understand fully the arithmetic operations of addition, subtraction, multiplication, or division, or to understand other existing arithmetic algorithms in order to become a competent user of a digital computer. What is important is for the reader to realize that the arithmetic unit is capable of performing only addition, subtraction, multiplication, and division, and that these operations are basically just addition. As a consequence, the reader should be aware that his formal knowledge of algebra, geometry, calculus, and so on, is not directly applicable to the digital computer solution of a technical problem. Corresponding numerical methods that employ only addition must be found. The general subject of numerical methods will thus play an important role in learning how to use a digital computer in the solution of electrical engineering problems.

It should also be realized that in addition to the arithmetic operations, the arithmetic unit is also capable of performing the simple logical operation of determining whether a number is less than, equal to, or greater than zero.

The Memory Unit

One of the most important blocks of a digital computer is its memory. The possession of a memory and the ability to execute instructions at electronic speeds are the two major differences that exist between the digital computer and the ordinary electric desk calculating machine.

During the execution of a program, the instructions and data that are stored in the memory are transmitted to the control and arithmetic units and the computed results are stored back in memory. The time required to execute an instruction is not only dependent on the electronics of the arithmetic unit, but also on the time it takes to retrieve and store information in the memory.

Magnetic core, magnetic disk, magnetic drum, magnetic tape, delay line, plated wire, thin film, and cryogenic devices represent the principal types of memory system used today.

Coincident-current magnetic-core storage uses tiny toroidal cores that have a rectangular hysteresis loop. This particular method of storage is probably the most important one in the group. In this type of storage, the cores are wired in rows and columns to form a memory plane. Each core is capable of being magnetized clockwise or counterclockwise, and the direction of magnetization can be used to store a 0 or a 1 bit. An important property of this type of storage is that it permits random or direct access with all cells being equally accessible.

Magnetic disk, magnetic drum, magnetic tape, and delay line represent examples of cyclic storage systems. In this type of storage, memory locations circulate or cycle and periodically information may be read-in or read-out of a particular location. The access time to a given location depends on the proximity of a particular location to the reading and writing devices at the time of the execution of a read or write instruction. The main advantage of this type of storage over core storage is that simpler access electronics can be used, and obviously the main disadvantage is the greater access time. In general, most third generation computers use both types of storage. The main memory employs core storage for speed and the peripheral memory employs magnetic disks for storing large data files, such as subroutines, and so on.

Memory is subdivided into units called words and words into units called cells. Each word is assigned a number and is individually addressable. It is thus possible to fetch or store information from a given location in memory by using the appropriate memory address. A cell contains a single character and in binary form it is made up of several bits. The actual character is determined by the 1's and 0's stored by the individual bits. In the IBM 360 system the basic addressable unit is an 8-bit field called a *byte*. One byte can represent eight binary bits, two decimal digits, or one alphameric character.

A characteristic property that applies to all types of memories is destructive read-in and nondestructive read-out. That is, each time information is stored in a given location in memory, the contents of that location are erased as the new information is read in; and each time information is read out, the contents of that memory location remain undisturbed.

It is also important to realize that all information processed by the computer must be stored in numeric form. Both input data, as well as instructions (commands) that the computer will execute (obey), must be stored numerically. Since individual instructions are represented in terms of numbers like data, they can be operated upon arithmetically and thereby modified or changed into new instructions. This single observation sheds light on the real power of the digital computer as an engineering tool in solving technical problems. Not every instruction that is to be executed by the computer has to be coded and stored initially in the computer memory. Using the technique of iteration, only the basic algorithm instructions need be coded and stored. During the execution phase, each time an instruction is executed, it can be modified so that during the next iteration a different set of operands can be involved. For example, a basic set of 10 different instructions may be executed 1000 times and it therefore corresponds to the execution of 10,000 different instructions.

1.3 SOLVING PROBLEMS WITH THE AID OF A DIGITAL COMPUTER

Any discussion of how one uses a digital computer in the solution of an engineering problem would have to include the following:

1. Planning
2. Coding
3. Computing
4. Interpreting

Planning

This is obviously the most important phase in the solution of a problem, but unfortunately it is the one that is usually given the least attention. It is apparent that the engineer should have a clear idea of the goal that he is attempting to achieve, but it is only fair to point out that the problem might be very complex, the results very difficult to predict, or the methods of solution so numerous that it is difficult for him to find an exact method of attack. Nevertheless, the initial phase of planning should include the writing of a clear verbal description of the problem. As simple as this might seem, the newcomer will soon learn that this is indeed a very difficult task (except in the case of a trivial problem). For example, some of the questions that have to be answered are:

1. What initial information exists, how accurate is it, and in what form is it given?
2. What method or models should be used?
3. What relevant mathematics applies?

4. What type of results do I want, and how accurate should they be?
5. How will I know if the results are correct or if the problem has a solution?

After a verbal description of the problem has been written, a mathematical description is required. At this point it may be necessary to enlist the aid of some mathematician in an advanced analysis group. It is important to warn the reader, however, that he and not the mathematician is the expert as far as the physical problem is concerned. Much can be lost in the translation from engineering language to a language the mathematician can understand. Conversely, much can be lost by the engineer, particularly in terms of physical insight, if he delegates too much authority to the mathematician.

After a mathematical description has been written, a numerical description is required. One must never lose sight of the fact that addition is the only mathematical operation the computer can perform. Here again one may call on the help of an expert, but the same warning applies.

The fourth step in the planning phase is to supply the procedural steps that will be required in the problem solution. Most engineers prefer to do this graphically by using a block diagram or a flow chart which shows the sequence of the computational steps with each block representing a specific step.

Coding

The next major step is to rewrite the problem solution, as defined by the flow chart, into a language that can be understood by the computer. Numerical instructions are accepted and understood by the control unit only when they are written in machine language. Since machine language is precise and compact, it is also a very limited language and it is very difficult to write good machine language programs. As a consequence, many different types of languages have been developed to overcome these difficulties. Fortran (formula translator) is by far the most widely used.

The use of Fortran involves two phases. The first phase translates the instructions coded in the Fortran language (the source program) into machine language commands (the object program). The second phase executes the machine language program, along with the necessary input data, to produce the desired output results.

In summary, there are two steps required to produce the source program. The first step is to state the problem and define an unambiguous step-by-step procedure for solving it. The procedure is called an algorithm, and we frequently use a flow chart as a means of indicating the necessary steps. The flow chart is then used as a guide in writing the corresponding Fortran source program. The computer, in turn, translates this source program into machine language for subsquent execution.

Computing

After the problem has been coded in Fortran, it will have to be key-punched on cards so that it can be processed by the computer. In addition, job cards will have to be added to the source program to produce a deck that can be batch processed. The color, order, and number of cards will vary, depending upon the requirements of the user and the particular computing center that will be used to do the processing. An example of a typical deck is shown in Figure 1.6.

The first card is a green job card and requires the punching of the name of the user, the identification or charge number, and the name of the program. The second and third cards are prepunched control cards. The next set of cards is the Fortran source program and this is followed by a pink pre-punched control card that would be the last card in the deck if no input data were involved. If the problem requires the reading in of input data, a yellow pre-punched data control card followed by the required data cards and a pink pre-punched control card must be added.

The program could now be processed by the computing center operators

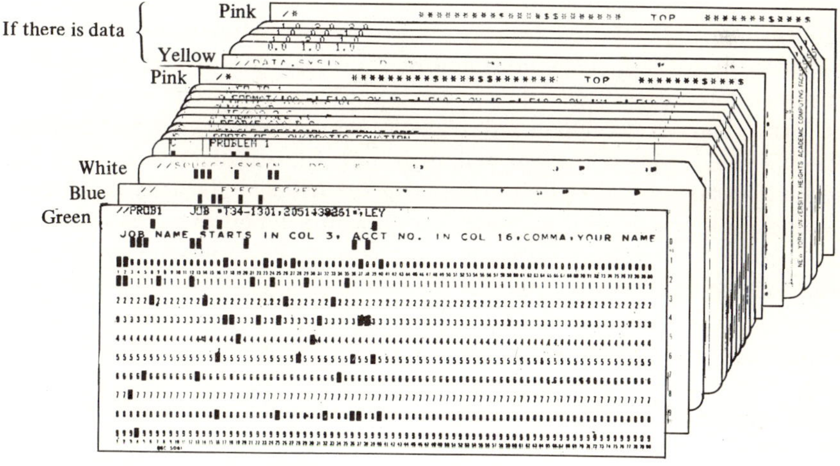

Figure 1.6 A Fortran source deck with control cards for batch processing

and the results returned. The turn-around-time would be a function of the type of computing equipment available, the number of users, and the complexity of the problem itself. Typical turn-around-times will run from 15 minutes to 24 hours and are obviously influenced by any computer down time.

The reader will soon learn that the desired results are rarely obtained the first time a new program is run. The amount of effort required to "debug" the program will depend on the type of errors that have been made. Typographical and Fortran language errors are usually the easiest to correct. This is primarily due to the printed diagnostics given by the computer. In processing the Fortran source deck, each executable Fortran statement or each punched card that is processed will be listed sequentially by number. If any of the Fortran rules have been violated, the corresponding list numbers (the number of a statement where an error occurs) will be printed along with a general description of what the error may be. With the aid of the printed error description and the number of the statement that the error occurs in, and a little practice, it is usually a simple matter to detect and repunch the faulty cards and resubmit the deck for further processing.

Executing errors are usually much more difficult to detect and much time may be spent in eliminating them. The error here, more often than not, will be a logical error that was committed in developing the solution algorithm. One type of execution error will cause the computer to "hang up" or stop. This type of error is usually caused by the attempted execution of an illegal operation. Another type of error will cause the computer to run for an excessively long time and will probably be halted automatically after a specified amount of computer time has been used. This type of error is usually due to the execution of a group of statements that form a loop and the computer has no way of getting out of the loop.

Even though the computer will print an alphameric code that will describe the type of execution error that has been made, it is difficult to detect exactly where and why the error occurs from an examination of the program listing. Rather than call for a dump (memory print out) so that an attempt can be made to trace the execution to the step where the error takes place, the following procedure is recommended. Modify the existing program by strategically adding additional WRITE statements. For example, if the error indicates that you are attempting to raise a negative argument to some power (take the logarithm of a negative argument), insert WRITE statements in the steps leading to any exponentiation operation, so that the numerical values of the variables involved can be printed out each step of the way. The numerical examination of the intermediate results will generally give the required clue as to why the program is not working.

Once the program is made operable, additional modifications can be made so that headings and other useful descriptive information can be printed out in the final run.

Interpreting

The computer user will soon learn that the computer is capable of printing a vast amount of information in a form that commands respect. The newcomer should establish a mode of operation based on the assumption that any results furnished by the computer are *incorrect.*

The execution errors of "hanging up" and getting into an "infinite loop," as described above, are obvious errors; however, errors that exist in a printed output are not necessarily obvious and are therefore very difficult to detect. The engineer has the responsibility of establishing the validity of all printed results. This may be done by hand calculating a particular set of conditions, running a similar problem where the results are known, or incorporating in the basic program itself automatic checking features.

Once the accuracy of the results have been established, they must be considered in light of the goals developed in the planning phase. Frequently it will be necessary to go back to the beginning and redefine the problem based on the insight gained from the initial attempt. It is thus obvious that the engineer must have a complete understanding of each step in the computer solution. The goal of this book is to build, from the knowledge of electrical engineering, the required skills of algorithm writing, numerical analysis, and Fortran coding that will make the use of the digital computer a formidable tool in the practice of the engineering profession.

The Solution of the Quadratic Equation

Before we begin our study of the basic Fortran language, let us examine the four steps just outlined and see what is involved by means of the example that follows.

In many of the numerical methods used in solving problems, the problem of finding the roots of a polynomial is encountered. Consider as a simple example, the case of a quadratic

$$AX^2 + BX + C = 0$$

for various sets of coefficients A, B, and C. Thus, as far as the planning stage is concerned, we are interested in writing a general computer program that is capable of finding the roots of a quadratic for various sets of A-B-C coefficients.

The mathematical description of the problem would correspond to letting $X1$ and $X2$ be the two roots, thus

$$X1 = \frac{-B + \sqrt{B^2 - 4AC}}{2A}$$

and
$$X2 = \frac{-B - \sqrt{B^2 - 4AC}}{2A}; \; A \neq 0$$

When $A = 0$, the solution would contain only the single root

$$X1 = -C/B$$

In this example, the mathematical description requires the use of the arithmetic operations of addition, subtraction, multiplication, division, and taking the square root. Since the first four are basic operations as far as a digital computer is concerned, only the square root operation must be considered in more detail. Thus, as far as a numerical description of the problem is concerned, we need only to find a method of finding the square root of a number that involves the basic arithmetic operations of the computer. This problem will be considered in Chapter 4, and it may be neglected at this point if we realize that the Fortran language possesses a built-in library routine that numerically finds the square root of a number.

The fourth step in the planning stage is to supply the procedural steps required in finding the roots of a quadratic. Figure 1.7 is a flow chart that shows the required steps. An examination of this figure shows that a flow chart is made up of a connected set of boxes whose shape indicates the operation that is to be performed. In general, all flow charts will begin with a box labeled *Start* and end with a box labeled *Stop*. In this example there is no *Stop* box because the number of coefficient data sets is not specified. In this particular flow chart, information flows from left to right and from line to line. The small numbered circles indicate common connection points and are used to reduce the number of connecting lines. Note that the numerical numbers used in the flow chart correspond respectively to the numerical labeled Fortran statements.

To follow this flow chart for finding the roots of a quadratic, begin at the *Start* box and move right to the circle numbered 1. We will see later that this number 1 stands for the numerical label of the next box. This next box has the shape of a punch card and corresponds to the instruction that will read in the first set of coefficients. The shape of the next box is the decision symbol. At this point we test to see whether A is equal to zero or if it is less than or greater than zero. If it is equal to zero, we move to the right and reach the box labeled *number 3*, which corresponds to the instruction that calculates the single root $X1$. We

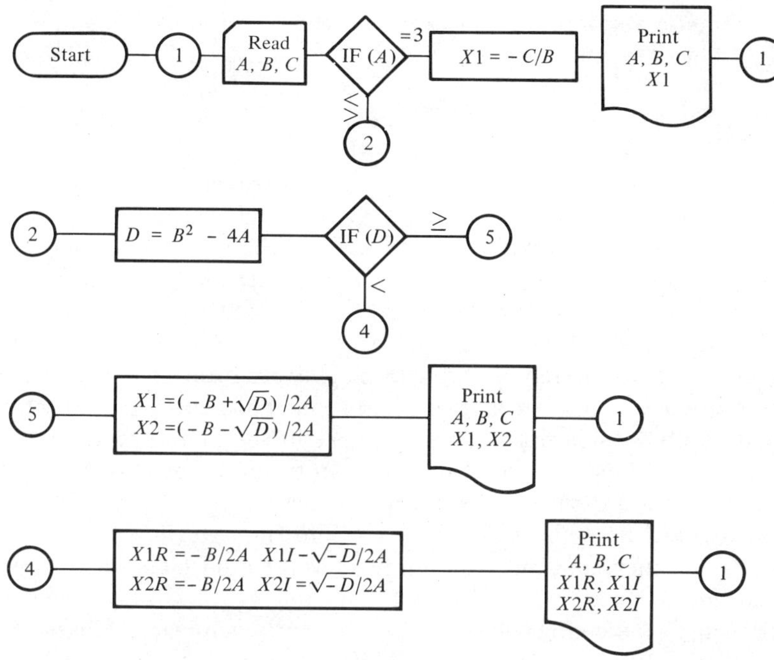

Figure 1.7 Quadratic root solution flowchart

then move to the right and print out the numerical values of A, B, and C as well as the value of the root $X1$. Moving further to the right, we reach a small circle numbered 1. Since this is the same point as the first circle we encountered, we are back at the read point and read the next set of A-B-C coefficients.

Alternately, if we had found A was not equal to zero, we would know that two roots existed and we would be directed to the circle numbered 2 on the second line. At this point we would calculate the value of the discriminant D. We would next test the discriminant to see if it was less than zero or equal to or greater than zero.

If the discriminant was equal to or greater than zero, we would know that the two roots were real, and we would go to the small circle numbered 5 on the third line. At this point we would calculate the two roots and then print out the numerical values of A, B, and C and the values of the two roots $X1$ and $X2$. Moving further to the right, we reach a small circle numbered 1. Since this is the same point as the first circle we encountered, we loop back and read the next set of coefficients.

Alternately, if the discriminant had been less than zero, we would know that the two roots were complex and we would go to the circle numbered 4 on the fourth line. Assuming that our computer cannot do complex arithmetic, we would separately find the real and the imaginary part of each root. The numerical values of A, B, and C and the real and imaginary values of each root would then be printed before we looped back for the next data set.

At this point it should be clear why there was no Stop box. The algorithm described by this flow chart would go on forever, always seeking the next set of coefficients and finding their roots. In actual computer execution, whenever the computer tries to execute the READ instruction and finds that no more data sets exist, it will automatically terminate the process and print out an error message that indicates there are no more data. One should regard this as an allowable error. Note the flexibility of a program that is written in this way. The program is capable of finding the roots of a single set of coefficients (one data card) or 100 sets of coefficients (100 data cards) with no changes in the main program.

At this time we should also point out that, in general, flow charts may be drawn with various degrees of detail. In a very large and complicated problem, an overall flow chart could be used to indicate how one part was related to another, and individual flow charts would be used to give the required detail.

We next turn to the coding phase. Since this is the subject of the next chapter, we avoid details and merely show the correspondence that exists between the Fortran program and the flow chart.

Figure 1.8 is the Fortran source program listing printed by the com-

```
C      ROOTS OF A QUADRATIC EQUATION
C      SINGLE PRECISION F FORMAT CODE
C
     1 READ(5,6)A,B,C
     6 FORMAT(3F5.1)
       IF(A)2,3,2
     3 X1=-C/B
       WRITE(6,9)A,B,C,X1
     9 FORMAT('0A =',F10.2,3X,'B =',F10.2,3X,'C =',F10.2,3X,'X1 =',F10.2,
      1//)
       GO TO 1
     2 D=B*B-4.*A*C
       IF(D)4,5,5
     5 X1=(-B+D**0.5)/(2.*A)
       X2=(-B-D**0.5)/(2.*A)
       WRITE(6,7)A,B,C
     7 FORMAT('0A =',F10.2,3X,'B =',F10.2,3X,'C =',F10.2)
       WRITE(6,10)X1,X2
    10 FORMAT(' X1 =',F10.2,3X,'X2 =',F10.2,//)
       GO TO 1
     4 X1R=-B/(2.*A)
       X1I=(-D)**0.5/(2.*A)
       X2R=-B/(2.*A)
       X2I=-((-D)**0.5/(2.*A))
       WRITE(6,7)A,B,C
       WRITE(6,8)X1R,X1I,X2R,X2I
     8 FORMAT(' X1R =',F10.2,3X,'X1I =',F10.2,/,' X2R =',F10.2,3X,'X2I ='
      1,F10.2,//)
       GO TO 1
       RETURN
       END
```

Figure 1.8 Fortran program for quadratic root solution

puter. Note that there is a one-to-one correspondence that exists between each line printed and each card that has been keypunched. To follow the general column alignment, consider the first printed line. The letter *C* would be found punched in the first column of the corresponding card and the letter *R* (Roots) would be found punched in the 7th column. In general, columns 1 through 5 are used for the statement label or number, and columns 7 through 72 contain the actual Fortran statement.

Note that a numeric statement label can be assigned arbitrarily to any executable statement and consists of from 1 through 5 decimal digits. Statement numbers may appear anywhere in columns 1 through 5, but must not contain embedded blanks. It is important to recognize that the numbers appearing on the left (S.0001, S.0002, . . . , S.0026) have been assigned by the computer (not the programmer) to each executable statement. These numbers will be referred to by the computer in indicating the location of a statement that contains a specific error.

The first three lines of Figure 1.8 are comments and are ignored by the Fortran complier. Comment cards may be used for headings, as they are in this example, or they may be used to explain what is taking place at a given point in the program. The next line (computer number S.0001) is the READ statement and is numerically labeled 1. This statement corresponds to the punch card box shown in the first line of the flow chart. The execution of this statement will cause the first data card to be read. The READ statement indicates that the respective numerical values of the three variables named *A*, *B*, and *C* will be found on the first data card. The number 6 that appears inside the parenthesis refers to the number of the format statement that describes how the numbers will appear on the data card.

The term 3F5.1 indicates that all three variables have been punched in F5.1 format code. An F5.1 number is a real number written with one digit after the decimal point, right justified, in a field five columns wide. Figure 1.9 shows how the first data card looks. In the first 5 columns, we find that the value of *A* is 0.0, in the next five columns (columns 6 through 10), we find that *B* has a value of 1.0, and in the next five columns (columns 11 through 15), we find *C* has a value of 1.0.

Computer statement S.0003 corresponds to the decision symbol used in the first line of the flow chart. The statement IF(A)2,3,2 means that if *A* is less than zero, control is transferred to the statement labeled 2 (the first 2 after IF(A)); if *A* is equal to zero, control is transferred to the statement labeled 3; while if *A* is greater than zero, control would be transferred to the third number, which is 2. Thus if *A* is equal to zero, we go to 3 and if *A* is not equal to zero (less than or greater than zero), we go to 2. Note that this is identical to the logic shown in the flow chart.

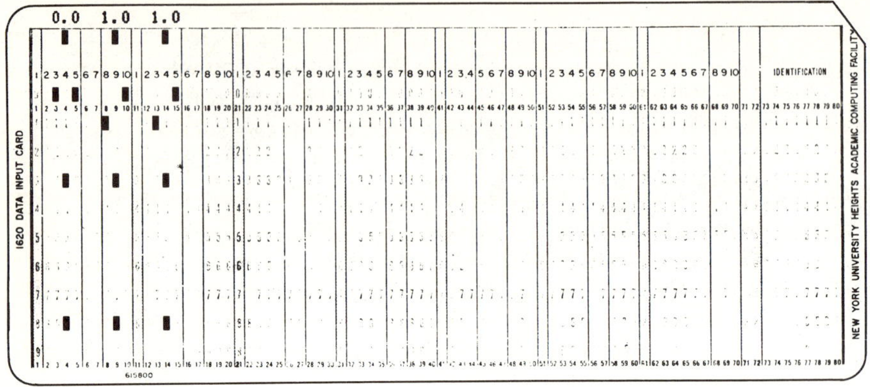

Figure 1.9 The first data card for the quadratic root problem

Since A is equal to zero, control is transferred to statement number 3 (S.0004) and $X1 = -C/B$ calculates the value of the single root. The next statement (S.0005) would be sequentially executed and this would cause the numerical values of A, B, C, and $X1$ to be printed in accordance with the format statement that has a numerical label of 9 (S.0006). The next statement is GO TO 1 and this causes an unconditional transfer to S.0001 and the execution of this instruction would read in the next data card.

Alternately, if A had not been equal to zero, control would be transferred to statement number 2 (S.0008). Here the value of the discriminant would be calculated and then tested (S.0009) to see if it was less than, equal to, or greater than zero. If it was equal to or greater than zero, control would be transferred to statement number 5 (S.0010). In this and the following statement, the values of $X1$ and $X2$ would be found. In the next statement (S.0012), the values of A, B, and C would be printed in accordance with the description specified by format statement number 7 (S.0013). The next statement (S.0014) would print out the numerical values of $X1$ and $X2$ in accordance with the description specified by format statement number 10 (S.0015). The next statement (S.0016) is GO TO 1 and causes control to be transferred to statement number 1 (S.0001) and the execution of this statement reads in the next data card.

Alternately, if the discriminant had been less than zero, control would be transferred to statement number 4 (S.0017). This would cause the sequential execution of statements S.0017 through S.0024 in which the real and imaginary values of the two roots would be calculated and printed along with the values of A, B, and C. Control is then returned to statement 1 so that the next set of data may be processed.

```
A =      0.0    B =      1.00   C =      1.00   X1 =      -1.00


A =      1.00   B =     -2.00   C =      1.00
X1 =      1.00   X2 =      1.00


A =      1.00   B =      0.0    C =     -1.00
X1 =      1.00   X2 =     -1.00


A =      1.00   B =      2.00   C =      2.00
X1R =     -1.00   X1I =      1.00
X2R =     -1.00   X2I =     -1.00


IHC217I
```

Figure 1.10 Quadratic equation output for four sets of input data

We next consider the computing phase. Figure 1.10 shows the computer results that were printed for the following four sets of input data:

	A	B	C
Set 1	0.0	1.0	1.0
Set 2	1.0	−2.0	1.0
Set 3	1.0	0.0	−1.0
Set 4	1.0	2.0	2.0

Set number 1 corresponds to the case of a single root with the value of A equal to zero. Sets 2 and 3 correspond to the case of two real roots and set 4 corresponds to the case of complex roots.

Note the error code that has been printed after the fourth data set, IHC217I. This code indicates the inability of the computer to find a fifth data card and as a consequence, the computing is automatically terminated.

The fourth and last phase is interpreting. Hand calculating any one of the results printed in Figure 1.10 will show that the computer results are correct. At this point the reader may be tempted to think that we have achieved our initial goal of finding the roots of

$$AX^2 + BX + C = 0$$

since four test problems have produced exact results, but in Chapter 4 we will return to this problem and show that we have just scratched the surface. Although this problem appears to be very simple and straightforward, we will find that considerable additional thinking and coding are required for an accurate solution of the quadratic. Nevertheless, the example, as it is considered above, should give us some insight into the various phases involved in a digital computer solution and establish the necessary motivation to learn the general rules of Fortran that are discussed in Chapter 2.

PROBLEMS

The following problems refer to the material presented in Appendix A.

1-1. Use the IBM 26 or 29 keypunch to punch a Fortran program deck for:
   a. Figure 2.3 (see also Figure 1.8)
   b. Figure 3.3
   c. Figure 3.7
   d. Figure 3.9
   e. Figure 3.12

1-2. Change the order of the statement numbers used in the program of Problem 1-1a so that statement number 1 is changed to statement number 10, statement number 2 is changed to statement number 9, . . . , and statement number 10 is changed to statement number 1. Now use the deck that was produced in Problem 1-1a and the duplicating and correcting ability of the keypunch to produce a new card deck with the appropriate statement number changes.

1-3. Assuming that the first 50 numbers of Figure 10.18 are to be used as numerical input data for a program that you will later run, punch up ten data cards so that each card will contain five numbers and each number will occupy ten columns (right justified).

1-4. Repeat Problem 1-3 using drum control and a blank drum card.

1-5. Develop an appropriate drum card and repeat Problem 1-4, but this time space the numbers so that they occupy 15 columns, right justified.

1-6. Repeat Problem 1-1a using drum control and an appropriate Fortran drum control card.

1-7. Prepare a drum card that has a 0 punched in each of its 80 columns. Now use this card and the keypunch operating under drum control to produce a duplicate of the deck punched in Problem 1-1a.

1-8. Use an appropriate drum card to number sequentially each of the cards (columns 79 and 80) produced in Problem 1-7.

1-9. Number 50 cards (columns 79 and 80) similar to those shown in Figure A.8. After shuffling these cards, use the IBM 82 sorter to reorder the sequence. Record each step of the procedure in a manner similar to that shown in Figures A.8 through A.12.

The following problems refer to the material presented in Appendix B.

1-10. Perform the following arithmetic operations using the 9's as well as the 10's complements:
a. 938 − 260 b. −260 + 938
c. −54.71 + 34.68 d. −34.68 − 54.71

1-11. Represent the following decimal numbers in terms of their binary equivalents:
a. 23 b. 49 3. 3.75 d. 3.76 e. −69.86

1-12. Repeat Problem 1-11 using:
a. Ternary numbers (base 3)
b. Quinary numbers (base 5)
c. Octal numbers (base 8)
d. Hexadecimal numbers (base 16). Hint: use values of a's equal to

$$a = 0, 1, 2, 3, 4, 5, 6, 7, 8, 9, \text{A, B, C, D, E, or F}$$

1-13. Perform the following arithmetic operations using both 1's and 2's complements:
a. 1011 − 111 b. 111 − 1011
c. 10011.101 − 11.100 d. −11.100 − 10011.101

1-14. Use floating-point notation to represent the following numbers in terms of five-significant digits and a two-digit exponent:
a. 3.14159 b. 99999999.9999
c. 0.0000368574 d. $-1.6892\ 10^{-36}$

1-15. Using the multiplication algorithm described in Appendix B, carry out the following arithmetic operations using a three-digit long distributor register and a six-digit long accumulator register:

	MULTIPLICAND	MULTIPLIER
a.	23	89
b.	367	32
c.	186	186
d.	1344	344

1-16. Using the division algorithm discussed in Appendix B, carry out the following arithmetic operations:
a. $38)_{10}/5)_{10}$ b. $268)_{10}/16)_{10}$
c. $234)_5/33)_5$ d. $1101001)_2/1011)_2$
e. $11.101)_2/0.11)_2$ f. $11.101)_2/1101001.01)_2$

The following problems refer to the material presented in Chapter 1.

1-17. Name the five functional blocks of a digital computer.

1-18. State the two major differences between a general purpose digital computer and the ordinary desk calculator.

1-19. What is an algorithm?

1-20. Name five ways of introducing information into a digital computer.

1-21. Name five operations the arithmetic unit is capable of performing.

1-22. Name five different means of storing computer information.

1-23. What is the purpose of job control cards?

1-24. Use the appropriate job cards required by your computer center and submit the Fortran deck punched in Problem 1-1a (use data shown in Figure 1.10).

1-25. Examine the computer output obtained for Problem 1-24. If any errors occurred, correct and resubmit.

1-26. Repeat Problem 1-24 using data of your own choosing. Explain any errors in the computed results in terms of the limitations of the program of Figure 1.10.

1-27. Draw a flow chart for each of the following problems:

a. Given three different numbers—A, B, and C—find the largest number.

b. Repeat step a and include the possibility of equal numbers.

c. Given N numbers $A1, A2, \ldots, AN$, find the smallest number(s).

d. Given N numbers, rearrange them in numerical descending order so that $A1$ will be the largest and AN the smallest.

e. Given the numerical value of N, determine factorial N ($N!$).

CHAPTER 2

The Fortran Language

2.1 INTRODUCTION

This chapter deals with the coding phase of the digital computer solution. No attempt is made to include either machine language or symbolic machine language coding, since both of these depend on the hardware of the particular computer being used. Each of these languages is relatively difficult to learn and the time and effort spent to become proficient in their use is questionable for electrical engineers since the average user is likely to be working with a computing center that replaces its computer about every three years.

Several problem-oriented languages, which are machine-independent, have been developed in the past ten years. Fortran (formula translator) is by far the most widely used, and even at computing centers that emphasize other problem-oriented languages, Fortran compliers are generally available. As a consequence, we will restrict our coding to the Fortran language with a main goal of obtaining proficiency rather than diversity. Since numerical commands are accepted and understood by the control unit only when they are written in machine language, a translation from Fortran must be provided. Thus, two languages will be involved in communicating with the computer: machine language and Fortran. Note that only the programs that translate Fortran into machine language are machine-dependent. Virtually every computer manufacturer today provides a Fortran compiler that transforms each Fortran statement into a series of machine language instructions and stores them at memory locations that minimize computation time.

In the past there have been several different versions of Fortran and no doubt the future will include additional changes. No attempt is made to point out all the variations that exist; instead we will restrict our discussion to Fortran IV as it generally applies to the IBM 360 system. The reader will learn that any differences that exist, as far as other machines are concerned, will be minor ones.

2.2 ELEMENTS OF THE FORTRAN LANGUAGE

A Fortran statement will generally contain constants, variables, and control words. The general rules for writing such statements may be mastered by studying the following four Fortran statement categories:

Arithmetic Statement. The arithmetic statement has the general form

$$\text{Variable} = \text{Arithmetic Expression.}$$

During the execution phase, a numerical value will be found for the expression on the right-hand side, and this result will be assigned to the variable on the left-hand side.

Control Statement. The control statements will determine the sequence of the execution and the termination of the program statements.

Specification Statement. Specification statements are used to specify the format of input data or output results. Another type of specification statement, the dimension statement, specifies the amount of storage that must be reserved for subscripted variables.

Input-Output Statement. An input or output statement is used to read data into the machine or to print or punch output results.

Using the above statements, and working from the flow chart of the problem, a Fortran source program is first written on a coding form similar to the one shown in Figure 2.1. After the programmer is satisfied with his initial effort, each line (each Fortran statement) is punched on a Fortran statement card similar to the one shown in Figure 2.2. The complete set of cards is referred to as the Fortran source deck. Figure 2.3 shows the Fortran source deck for the quadratic root problem discussed in Chapter 1.

Since it is often necessary to identify certain statements in a program, note that columns 1 through 5 in Figures 2.2 or 2.3 are reserved for the statement label or number. In general, statement numbers may be assigned arbitrarily to an executable statement (and format statement) and can consist of from one through five decimal digits. Statement numbers may appear anywhere in columns 1 through 5, but must not contain embedded blanks. Leading zeros or blanks are ignored.

Program ROOTS OF QUADRATIC EQUATION

FORTRAN CODING FORM

```
C    PROBLEM 1
C    ROOTS OF QUADRATIC EQUATION
C    SINGLE PRECISION F FORMAT CODE
C
    1 READ(5,6)A,B,C
    6 FORMAT(3F5.1)
      IF(A)2,3,2
    3 X1=C/B
      WRITE(6,9)A,B,C,X1
    9 FORMAT('0A =',F10.2,3X,'B =',F10.2,3X,'C =',F10.2,3X,'X
     11 =',F10.2,//)
      GO TO 1
    2 D=B*B-4.A*C
      IF(D)4,5,5
    5 X1=(-B+D**0.5)/(2.*A)
      X2=(-B-D**0.5)/(2.*A)
      WRITE(6,7)A,B,C
    7 FORMAT('0A =',F10.2,3X,'B =',F10.2,3X,'C =',F10.2)
      WRITE(6,10)X1,X2
   10 FORMAT(' X1 =',F10.2,3X,'X2 =',F10.2,//)
      GO TO 1
```

Figure 2.1 Fortran coding form for quadratic root problem

FORTRAN CODING FORM

Program ______

C FOR COMMENT

STATEMENT NUMBER	Cont	FORTRAN STATEMENT
4		X1R=-B/(2.*A)
		X1I=(-D)**0.5/(2.*A)
		X2R=-B/(2.*A)
		X2I-((-D)**0.5/(2.*A))
		WRITE(6,7)A,B,C
		WRITE(6,8)X1R,X1I,X2R,X2I
8		FORMAT(' X1R =',F10.2,3X,'X1I =',F10.2,/,' X2R =',F10.2
	1	,3X,'X2I =',F10.2,//)
		GO TO 1
		RETURN
		END

Figure 2.1 Continued

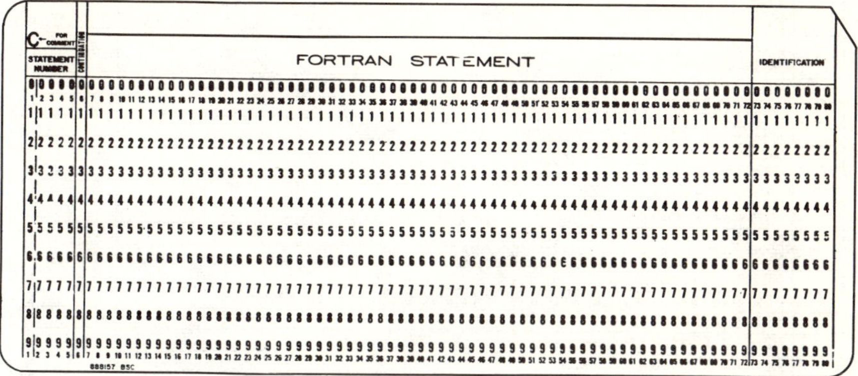

Figure 2.2 A Fortran statement card

Fortran statements are punched one to a card from columns 7 through 72. Columns 73 through 80 are not interpreted by the Fortran complier and may be used for identification purposes as described in Section 1.2. If a Fortran statement is too long and will exceed column 72, it may be continued on successive cards (columns 7 through 72) by punching on each continuation card any character other than zero in column 6 (the continuation column). It is usually good practice to punch a 1 in column 6 of the first continuation card, a 2 in column 6 of the second continuation card, and so on.

Comment cards may be used for headings, as shown in Figure 2.3, or they may be used to explain what is taking place at a given point in the program. Comment cards must have a C punched in column 1, and columns 2 through 80 may be used for the appropriate comment. Comment cards are ignored by the Fortran complier. The reader is urged to make liberal use of comment cards in writing his program so that it may be more easily understood at some later time.

A more detailed examination of Figure 2.3 will reveal that embedded blanks are not permitted within variable names of Fortran control words and that all Fortran statements require that each term must be terminated by one or more of the following delimiters: a blank, . , (,) , * , / , + , − , or column 73.

2.3 ARITHMETIC STATEMENTS

There are four basic types of constants: integer, real, complex, and double precision.

```
      END
      RETURN
      GO TO 1
     1,F10.2,//)
    8 FORMAT(' X1R =',F10.2,3X,'X1I =',F10.2,/,' X2R =',F10.2,3X,'X2I ='
      WRITE(6,8)X1R,X1I,X2R,X2I
      WRITE(6,7)A,B,C
      X2I=-((-D)**0.5/(2.*A))
      X2R=-B/(2.*A)
      X1I=(-D)**0.5/(2.*A)
    4 X1R=-B/(2.*A)
      GO TO 1
   10 FORMAT(' X1 =',F10.2,3X,'X2 =',F10.2,//)
      WRITE(6,10)X1,X2
    7 FORMAT('0A =',F10.2,3X,'B =',F10.2,3X,'C =',F10.2)
      WRITE(6,7)A,B,C
      X2=(-B-D**0.5)/(2.*A)
    5 X1=(-B+D**0.5)/(2.*A)
      IF(D)4,5,5
    2 D=B*B-4.*A*C
      GO TO 1
     1//)
    9 FORMAT('0A =',F10.2,3X,'B =',F10.2,3X,'C =',F10.2,3X,'X1 =',F10.2,
      WRITE(6,9)A,B,C,X1
    3 X1=-C/B
      IF(A)2,3,2
    6 FORMAT(3F5.1)
    1 READ(5,6)A,B,C
C
C     SINGLE PRECISION F FORMAT CODE
C     ROOTS OF A QUADRATIC EQUATION
```

Figure 2.3 Fortran source deck for the quadratic root problem of Section 1.3

Integer Constant (Fixed-Point)

An integer constant is any positive or negative whole number written without a decimal point. A negative integer constant must be preceded by a minus sign, but the sign may be omitted from a positive integer constant. The Fortran compiler will automatically assume all unsigned integer constants to be positive. The maximum magnitude of an integer constant is

$$2^{31} - 1 \text{ or } 2147483647$$

Note that the maximum magnitude of an integer constant is a function of the number of bits used in storage. This number will therefore vary for different machines. Examples of some valid integer constants are:

$$0, \ +10, \ -36, \ 2147483647$$

Note that the + sign is not necessary for the representation of the integer constant 10, and any output record would not include this sign. Examples of some invalid integer constants are:

0., 10.0, 2,696, 5168721344

The first and second examples are invalid integer constants because they contain a decimal point. The third example is invalid because of the comma and the fourth because of its magnitude (in the IBM 360 system). In general, an integer constant may contain any number of decimal digits, but the number of digits stored will be a function of the number of bits allocated for storage.

Real Constant (Floating-Point)

There are two basic types of real floating-point numbers: those without an exponent (F-type), and those with an exponent (E-type). Any real number must be written with a decimal point and would consist of from one through seven decimal digits. This maximum number of decimal digits preserved in single precision calculations will vary for different machines and is a function of the number of bits allocated for storage. Examples of some valid F-type constants are:

$$-1.7934, \ -9999.99, \ 0.0, \ 1.2356$$

An E-type number consists of a sign (a positive sign does not have to be written) followed by from one through seven decimal digits, including a decimal point embedded therein. This, in turn, is followed by E (power of 10) and a two-decimal positive or negative integer exponent. The range of an E-type number is

$$16^{-63} \text{ to } 16^{+63}$$

or the approximate magnitude is 10^{75}. This value will also vary with different machines, but in most cases will be at least 10^{38}. Examples of some valid E-type constants are:

5.64E−02, +.564E−1, 564.E−04, +0.0056400E+01

which are respectively equivalent to

$$5.64 \ 10^{-2}, \ 0.564 \ 10^{-1}, \ 564. \ 10^{-4}, \ 0.0056400 \ 10^{+1}$$

and are all equal to the same number 0.0564.

Double Precision Constant

A double precision constant must be written with a decimal point and would consist of from 1 through 16 decimal digits optionally followed by a D (power of 10) and a two-decimal positive or negative integer exponent. The maximum magnitude is

$$16^{63} \text{ or approximately } 10^{75}. \text{(in the 360 system)}.$$

Not all machines will be capable of double precision calculations and those that are will not necessarily be capable of preserving 16 decimal digits. Examples of some valid D-type constants are:

189.689721387, −56.4D−02, 0.0, 0.236589547660543D−51

Variables

Fortran variables are given symbolic or meaningful names and represent the memory location in which the numerical value of the variable is stored. The name of the variable may contain from one through six alphabetic or numeric characters and the first one must be alphabetic. In the predefined specification, the Fortran translator recognizes integer variables if they begin with the letters I, J, K, L, M, or N. All other letters correspond to real variables. Examples of some valid integer variables are:

NEXT, MARY22, ITEMP, I, LIMIT

Examples of some valid real variables are:

APOLLO, GEMINI, TEMP, A, X15

Examples of some invalid real variables are:

5STAR, TE MP, RESPONSE, ALPHA*

The first is invalid because the first character in the name is not alphabetic. The second is invalid because of the embedded blank. The third is invalid because it contains more than six characters and the fourth because it contains an illegal character.

Subscripted Variables

A subscripted variable is used to represent the elements of an array and is designated by an integer or real variable name followed by parentheses that enclose the subscript. Some examples of valid subscripted variables are:

BETA(I10), Y(7), X(J,5), N(4*K+3,L),
ITEM(3,7,2), IALPHA(7*I+4,4*J,K)

The first three are examples of real subscripted variables and the last three are integer subscripted variables. The first two are single subscripted variables, the middle two are double subscripted variables, and the last two are triple subscripted variables. The subscript consists of constants or variables that must be in the integer mode and restricted to the form of constant times variable plus or minus constant ($C*V \pm C'$). Thus a valid subscript may be one of the following seven forms:

Form	
C	Where V is an unsigned integer variable and C and C′ are unsigned integer constants
V	
C*V	
V+C′	
V−C′	
C*V+C′	
C*V−C′	

An array may have a maximum of seven subscripts, but this number may be different for other machines. Each subscript is separated by a comma, and the numerical value of a subscript must be greater than zero and less than 32767. Some examples of invalid subscripted variables are:

ALPHA(I*7), ALPHA(7/I), ALPHA(4*7*I), ALPHA(7*K+4,4*A,K)

The first example is invalid because the subscript is variable times constant; this could be corrected by writing 7*I. The second is invalid because division is not permitted. The third is not valid because the subscript is of the form constant times constant times variable, and the fourth is invalid because the second subscript contains the real variable A.

Arithmetic Expressions

An arithmetic expression consists of constants and variables separated by arithmetic operation symbols and parentheses. No two arithmetic operators may appear in sequence. The precedence of arithmetic operation symbols is listed below in descending order.

SYMBOL	OPERATION
**	Exponentiation (to the power of)
*,/	Multiplication, Division
+,−	Addition, Subtraction

In order to better understand the precedence rules, consider the following examples:

ARITHMETIC EXPRESSION	MEANING
$A + B/C$	Divide B by C and add that result to A.
$A + D{*}{*}E{*}F - G$	Raise D to the E power and then multiply by F. Add this result to A and subtract G.

Note that among operations of equal precedence, the grouping will be from left to right. For an example consider the following:

$$A{*}B - C/D + E{*}F$$

The highest precedence is multiplication or division. Thus first $A{*}B = X$ is found, then $C/D = Y$ is found, and then $E{*}F = Z$ is found. We now have

$$X - Y + Z$$

Since subtraction and addition are of equal precedence, the order is from left to right. We first subtract Y from X and then add Z.

Parentheses can be used to override the usual rules of precedence and the operations are carried out on the basis of inner parentheses first, thus

$$A{*}(B - C)/(D + E){*}F$$

means: first subtract C from B and then add D to E. Now multiply A by $B-C$ and in turn divide this by $D+E$, and finally multiply this result by F

$$\frac{A(B - C)}{D + E} F$$

Mathematical Function Subprograms

One of the important features of Fortran is that it provides a library of commonly used functions. These functions are numerically evaluated by built-in efficiently coded subroutine subprograms. It is therefore not necessary for the user to write a numerical routine that evaluates the sine or cosine of an angle, the square root of an expression, the absolute value of an expression, and so on. Since these function subprograms are permanently stored in the peripheral memory, they are automatically transferred to the main memory before the execution phase of a program that requires them. All the user need do in the main program is to supply the numerical value of the argument and call on the subprogram by name to evaluate the function for this value. Some of the more useful functions are listed in the table below:

FUNCTION	SINGLE PRECISION NAME	DOUBLE PRECISION NAME
Trigonometric sine	SIN(X)	DSIN(X)
Trigonometric cosine	COS(X)	DCOS(X)
Trigonometric arctangent	ATAN(X)	DATAN(X)
Trigonometric arctangent of (X/Y)	ATAN2(X,Y)	DATAN2(X,Y)
Hyperbolic tangent	TANH(X)	DTANH(X)
Exponential (ϵ^X)	EXP(X)	DEXP(X)
Natural logarithm	ALOG(X)	DLOG(X)
Logarithm to the base 10	ALOG10(X)	DLOG10(X)
Square root	SQRT(X)	DSQRT(X)
Absolute value	ABS(X)	DABS(X)

where the argument X (or X/Y) is any real expression. For an example of an arithmetic statement that uses several library functions, consider the following:

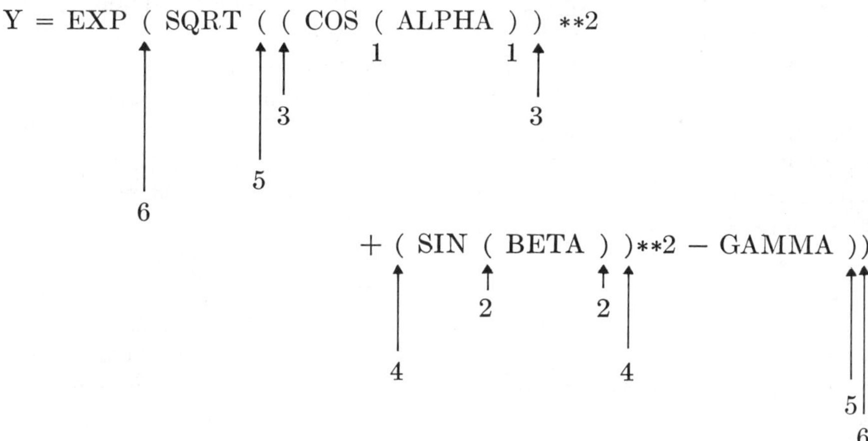

In order to interpret this expression, we must work from the innermost parenthesis and from left to right. We first evaluate ALPHA, which is the argument of the COS library function. We next evaluate BETA, which is the argument of the SIN library function. The contents within the third set of parentheses is now squared and this is followed by the squaring of the contents within the fourth set of parentheses. The latter is now added to the former and GAMMA is subtracted. The result at this point is the numerical value of the contents of the fifth set of parentheses and this result corresponds to the argument of the SQRT library function, which, in turn, is the argument of the exponential library function EXP; thus

$$Y = \epsilon^{\sqrt{\cos^2 \text{ALPHA} + \sin^2 \text{BETA} - \text{GAMMA}}}$$

In addition to the above predefined subprograms that are permanently stored in the computer, the programmer may wish to include some of his own subprograms. The general details of this subject, along with a discussion of the use of the scientific subroutines, will be covered in Chapter 8.

Restrictions

The only allowable form of an arithmetic statement is

Variable = Arithmetic Expression

The left-hand side may not be an expression. For example,

A − 3. = B**C + D/E

is not valid since A − 3. is an expression.

In general, most Fortran compilers do not permit mixed-mode expressions. During the translation phase, a mixed-mode diagnostic error will be printed for each invalid statement and execution of the program will be prevented. Mixed mode does not apply to arithmetic statements. The variable name on the left-hand side may be integer or real mode and the expression on the right-hand side may be of the opposite mode. The execution of this type of statement would be carried out in the mode of the arithmetic expression side and then converted and stored at the memory location corresponding to the mode and name of the left-hand variable.

For Fortran compilers that permit mixed-mode expressions, integer mode constants and variables are first floated or converted to real (floating-point) numbers, and the expression is then evaluated in the real mode. The programmer who writes mixed-mode expressions must be aware of the possible consequences. For example, if the variable I had the numerical value of 10 and the variable J had the numerical value of 4,

I/J + 2

would have the numerical value of 4, but

I/J + 2.

would have the numerical value of 4.5.

In raising a number to a power, a real number may have either a real or integer exponent; thus, either

R**I

or

R**FLTI

is permissible. Raising a floating-point or real number to an integer or real power results in a real number. If $I = 3$ and FLTI $= I$, then FLTI equals 3., and raising the real variable R to the respective powers corresponds to

R**3

and

R**3.

It is important for the programmer to recognize the difference between these two expressions as far as execution is concerned. The first expression is essentially equivalent to

R*R*R

but the equivalent of the second expression is based on the identity

R**FLTI ≡ EXP(FLTI*ALOG(R))

Since the evaluation of the second expression involves the numerical routines of the library functions EXP and ALOG, it should be apparent that in this example the first form would be preferable since it would require fewer calculations. In general, when the exponent is an integer and equal to or less than about 20, the first form is more efficient.

The user should also be aware that when the second form is used and R represents some complicated real expression that may have a negative value for some set of conditions, an execution error will result because the library function will involve taking the logarithm of a negative argument that is not permissible.

In contrast, an integer number can be raised to only an integer power; thus,

I**J

is valid, but

I**FLTJ

is not because this would result in an integer mode argument for the library function ALOG.

An additional restriction that would apply to exponentiation would be for the case where R itself would be an expression that corresponded to the arithmetical operation of raising a number to a power; thus,

A**B**I

is not valid and would not be acceptable by most Fortran compilers because it is ambiguous. Do we mean

$$(A^B)^I = A^{B*I}$$

or

$$(A^B)^I = A^{B^I}$$

Since it is not clear which operation should occur first, parentheses must be used; thus

(A**B)**I

and

A**(B**I)

would eliminate the ambiguity.

2.4 CONTROL STATEMENTS

Fortran statements are executed sequentially until a control statement is reached. The execution of the control statement will cause control to be transferred to the statement number that is specified. Since only executable statements may be numbered, except for the case of format statements, control can be transferred only to an executable statement.

Unconditional GO TO Statement

The simplest control statement is the GO TO statement which has the form

GO TO SN

where SN is the statement number of an executable statement in the source program. This GO TO statement will cause control to be transferred to the statement with the number SN and it will become the next executable statement. An example of a GO TO statement is

GO TO 54

The Computed GO TO Statement

The computed GO TO statement permits multiple branch transfer control. The actual statement number to which control is transferred depends on the value of an integer variable at the time the statement is executed. The computed GO TO statement has the form

GO TO (SN_1,SN_2, . . . ,SN_N), I

where SN_1, SN_2, . . . , SN_N are the numbers of various statements elsewhere in the source program and I stands for an unsigned nonsubscripted integer variable. Whenever this statement is executed, the value of I must not be less than 1 or greater than N. Control is transferred to statement SN_1, SN_2, . . . , SN_N depending on whether the current value of I is 1, 2, . . . , N, respectively. An example of a computed GO TO statement is

```
GO TO (30,42,50,9),IABLE
```

Meaning if IABLE were equal to 3 at the time this statement was executed, control would be transferred to the statement labeled number 50, which is the third statement number.

The Arithmetic IF Statement

The arithmetic IF statement is a conditional transfer statement having the form

$$\text{IF(Arithmetic Expression)}SN_1,SN_2,SN_3$$

The control word IF is followed by an arithmetic expression or argument enclosed in parentheses and this is followed by three statement numbers separated by commas. If the value of the argument is negative, control transfers to statement SN_1; if the value of the argument is equal to zero, control passes to statement SN_2; and if the value of the argument is positive, control is transferred to the statement SN_3. This may be graphically pictured by means of the flowgraph shown in Figure 2.4. An example of the statement that corresponds to a three branch type of decision is

```
IF(D)4,5,6
```

meaning that if the value of *D* is less than zero, control is transferred to statement number 4; if the value of the argument is equal to zero, control passes to statement number 5; while if the value of the argument is greater than zero, control is transferred to statement number 6. It is also possible to use the IF statement as a two-way transfer as illustrated by the examples below.

1. IF(D)4,4,6
2. IF(D)4,5,4
3. IF(D)4,6,6

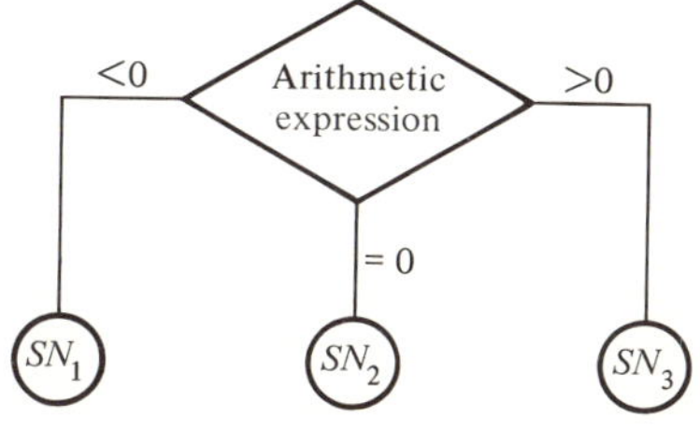

Figure 2.4 A graphical representation of the arithmetic IF statement

The first example causes a transfer to statement number 4, if the value of D is less than or equal to zero, and a transfer to statement number 6 if the value of D is greater than zero. The second causes a transfer to 4 if D is less than or greater than zero and a transfer to 5 if D is equal to zero. The third causes a transfer to 4 if D is less than zero and a transfer to 6 if D is equal or greater than zero.

The DO Statement

The DO or iteration statement is one of the most frequently used statements in the Fortran language. It permits a portion of the source program to be executed over and over. The form of the DO statement is

```
DO SN J=INIT,LIMIT,INCR
```

where SN is a statement number defining the last or terminal statement in the iteration loop, J is any nonsubscripted integer variable, and INIT, LIMIT, and INCR are unsigned integer constants or unsigned nonsubscripted integer variables having a numerical value greater than zero. Before the first execution of the statements that follow, down to and including the terminal statement, the variable J is given the initial value INIT. The variable J is then tested to see if it exceeds the limit value LIMIT. If it does, control passes to the one following the statement with the SN number, and the statements in the loop are not executed at all. If it does not, the iteration loop is executed. Each succeeding time, the integer variable J is incremented by the integer value INCR and again tested to see if it exceeds the limit value, and so on. The number of times the statements in the loop are executed is thus given by

$$(\text{LIMIT} - \text{INIT})/\text{INCR} + 1$$

If the increment value is 1, the Fortran translator will automatically recognize this fact if INCR, along with the preceding comma, is omitted. Some examples of valid DO statements are

```
DO 20 I=1,N,2
DO 22 M=2,MN,2
DO 55 N=1,NT
DO 55 N=INIT,5,2
```

Some examples of invalid DO statements are

```
DO 20 I=1,A,2
DO 22 M=-2,MN,2
DO MAX N=1,NT,-1
```

In the first example the error is the use of a real variable A for the LIMIT value. In the second example the use of a negative initial value is an error. In the third example there are two errors: the scope number must be a number and not a variable name, and the increment must be a positive quantity.

For an example that shows the advantage of a DO statement, consider the problem that requires a series of statements to be executed ten times. The following statements represent one way of doing this:

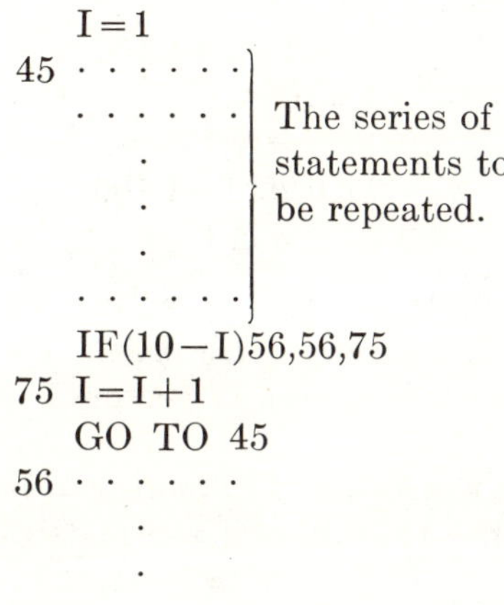

Here I is initially set equal to 1. The series of statements are then sequentially executed for the first time and then I is tested to see if it is equal to 10. Since in the initial pass I is equal to 1, control is transferred to statement number 75 and I is incremented by one so that the new value of I equals 2. The GO TO 45 statement causes the statements to be executed for a second time. This will continue until I equals 10 and the statements are executed the tenth time. At this point the test for I compared to 10 will cause control to be transferred to statement number 56. Now consider the same problem using a DO statement.

```
   DO 40 I=1,10
   ......  }
   ......  }  The series of statements
      .    }  to be repeated. The last
      .    }  one has been given the statement
      .    }  number 40.
40 ......  }
56 ......
      .
```

Note that in the second program the DO statement automatically includes the initializing step, the testing step, the incrementing step, and the GO TO step that were required as separate statements in the first program.

As an additional example that shows how the DO loop may be employed, consider the column matrix *A* that contains 100 elements. Let it be required to rearrange the elements so that the value of the first element will be equal to the value of the 100th element, the second the 99th, and so on. The following program will obtain the desired result:

```
   DO 10 I=1,100
   J=101-I
10 B(J)=A(I)
   DO 20 I=1,100
20 A(I)=B(I)
```

The first DO 10 loop employs a dummy subscripted variable *B* to store the element values of *A* in the reverse order. The second DO 20 loop destroys the original element values of vector *A* and replaces it by the reverse order element values.

For an example that shows how DO loops can be nested, consider the problem of obtaining the transpose of the matrix *A* which contains *M* rows and *N* columns.

```
   DO 10 I=1,M
   DO 10 J=1,N
10 AT(J,I)=A(I,J)
```

In the first DO 10 statements, *I* is initially set equal to 1 and tested to see whether it is greater than *M*. Since initially *I* will not be greater than *M*, the second DO 10 statement is executed sequentially. Here *J* is initially set equal to 1 and tested to see whether it is greater than *N*. Since it is not, the next statement that has the scope number 10 will be executed, yielding AT(1,1) = A(1,1). We now go from the scope number 10 to the lower DO 10 statement. At this point *J* is incremented by the implied increment of 1, so that *J* now has the value of 2, and it is then tested to see whether it is greater than *N*. If it is not, statement 10 is executed for the second time yielding AT(2,1) = A(1,2). The lower DO 10 loop will continue to be executed until *J* is incremented to the value of $N + 1$. At this point the test of whether *J* is greater than *N* is satisfied and it indicates that the lower DO 10 loop has been executed *N* times. Since the outer DO 10 loop has not been completed, we go from the scope number 10 to the upper DO 10 statement and *I* is incremented by 1, so that the value of *I* now equals 2. *I* is now tested to see whether it is greater than *M*. If it is not, the second DO 10 statement is executed sequentially. Since we are coming from the statement above and not from the scope statement, J is set to its initial value of 1. The iteration variable J is then tested to see whether it is greater than N. Since it is not, the next statement that has the scope number will be executed, yielding AT(1,2) = A(2,1). The lower DO 10 loop will now be executed for a total of N times before the iteration variable in the outer iteration loop is incremented to 3. This process will continue until the outer loop has been executed M times. At this point the variable AT will be the transpose of A, and control will pass to the statement following the scope statement, statement 10.

For an example of a more complicated set of nested DO loops, consider the problem of matrix multiplication. Matrix multiplication requires the first matrix to have as many columns as the second matrix has rows. Let the first matrix A have I rows and J columns and the second matrix B have J rows and K columns. The resulting matrix C will therefore have I rows and K columns. A general element of the C matrix would be given by

$$C_{MN} = \sum_{L=1}^{J} A_{ML} * B_{LN} \text{ where } M = 1, 2, \ldots, I \text{ and } N = 1, 2, \ldots, K$$

or

$$C_{MN} = \sum_{M=1}^{I} \sum_{N=1}^{K} \sum_{L=1}^{J} A_{ML} * B_{LN}$$

and a suitable Fortran program would be

```
   DO 10 M=1,I
   DO 10 N=1,K
   C(M,N)=0.0
   DO 10 L=1,J
10 C(M,N)=C(M,N)+A(M,L)*B(L,N)
```

Initially M would be set equal to 1 and then N would be set equal to 1. Sequentially executing the third statement would set C(1,1)=0.0. The lower DO 10 loop would then be executed J times. The initial execution would yield

C(1,1)=0.0+A(1,1)*B(1,1)

The second execution of the lower DO 10 loop would yield

C(1,1)=A(1,1)*B(1,1)+A(1,2)*B(2,1)

and the Jth execution would yield

C(1,1)=A(1,1)*B(1,1)+A(1,2)*B(2,1)+ . . . +A(1,J)*B(J,1)

Control would then be transferred from the scope statement to the middle

DO 10 statement and N would be incremented by 1 so that it would have a value of 2. Sequentially executing the third statement would set C(1,2)=0.0. The lower DO 10 loop would then be executed J times, yielding the value of C(1,2). Control would again be transferred from the scope statement to the middle DO 10 loop and N would once more be incremented by 1. This would continue until the middle DO 10 loop had been executed K times. At this point, control would be transferred to the outer DO 10 loop and M would be incremented by 1 so that it would have a value of 2. The second statement would then be executed sequentially, setting N to its initial value of 1. The execution of the third statement would then yield C(2,1) = 0.0. The execution of the fourth statement would set L to its initial value of 1 and the lower DO 10 loop would be executed J times. This process would continue until the outer loop had been executed I times and the resulting matrix C would be equal to the product of matrix A times B.

The CONTINUE Statement

The CONTINUE statement is a dummy step essentially equivalent to "do nothing go on." It may be put anywhere in the program without affecting the sequence of execution. It is used for the sole purpose of representing a labeled junction point, and it is often needed as the last statement in a DO loop to avoid ending the DO loop with a GO TO statement. Consider the following example that shows an invalid use of the DO statement:

```
 5 DO 10 J=1,N
   X
   IF(E)10,10,20
20 Y
10 GO TO 5
```

where X and Y are arithmetic statements and E is an arithmetic expression. The last statement in an iteration loop cannot be one of the transfer statements. The GO TO 5 statement shown above would have the effect of transferring control to statement number 5, which would initialize the variable J each time. Similarly, if statement 10 were omitted, the scope number in the DO loop changed to 20, and the numbers 10 in the IF statement changed to 5, the same effect would result.

```
 5 DO 20 J=1,N
   X
   IF(E)5,5,20
20 Y
```

Any branch taken within an iteration loop, unless it is to transfer out of the loop, must eventually lead to the statement with the scope number and may not lead directly to the DO statement itself. The CONTINUE statement may thus be used as the last statement in the loop to avoid ending the loop in a GO TO statement.

```
 5 DO 10 J=1,N
   X
   IF(E)10,10,20
20 Y
10 CONTINUE
```

One must return from the scope number statement in order to increment the variable *J*.

The END Statement

The END statement is a nonexecutable statement that physically is the last statement in the source program. It is needed to inform the Fortran compiler that no more FORTRAN statements follow.

2.5 SPECIFICATION STATEMENTS

The DIMENSION Statement

All subscripted variables or arrays must be dimensioned. The DIMENSION statement reserves the number of memory cells required to store a given array and must appear before the array name can be used in any subsequent arithmetic statement. In many Fortran compilers, a sequence error will result if the DIMENSION statement is not the first statement of the source program. The form of the DIMENSION statement is

$$\text{DIMENSION Variable } (S_1,S_2, \ldots ,S_N)$$

where Variable stands for the name of the array and S_1, S_2, . . . , S_N are unsigned nonzero integer constants that stand for the maximum value of each of the subscripts. Some examples of valid DIMENSION statements are

```
DIMENSION X(6),Y(12)
DIMENSION A(4,3),B(6,19,66),C(3,4,5,6)
```

Note that more than one array may be dimensioned by the same statement and that each variable is separated from the next variable by a comma. In the first example, 6 consecutive memory places are reserved for the variable X, and 12 places are reserved for Y. This representation allows any element value to be obtained from memory by using the appropriate number in parentheses, following the variable name. In the second example, the first array A is a double-subscripted variable containing four rows and three columns, and 12 places are reserved for the value of its elements; the second array B is a triple-subscripted variable having six rows, 19 columns, and 66 layers, while the third variable C is a quadruple-subscripted variable that reserves 360 (3*4*5*6) places for the value of its elements. In general, elements of an array are stored columnwise in consecutive memory locations; thus, the elements of $A(4,3)$ would be stored in the following order:

$$A_{11}, A_{21}, \ldots, A_{41}, A_{12}, A_{22}, \ldots, A_{24}, A_{13}, A_{23}, \ldots, A_{43}.$$

For higher number subscripted variables, the subscripts are incremented in order from left to right; thus, the elements of $B(6,19,66)$ would be stored in the following order:

$$B_{111}, B_{211}, \ldots, B_{6,19,1}, B_{112}, B_{212}, \ldots, B_{6,19,2}, \ldots, B_{6,19,66}.$$

An example of an invalid DIMENSION statement is

DIMENSION X(I,J)

Even though the numerical values of I and J may be known, the DIMENSION statement must give the dimension as a number and not as a variable name. In writing a general program, the programmer may realize that the value of the subscripts will be different for different sets of conditions; it is therefore the responsibility of the programmer to choose subscripts large enough so that for any set of conditions, no subscript will be larger than the maximum size that has been specified.

The I, F, E, D Numeric Format Codes

A format specification is required in reading numerical data into the computer and for printing or punching numerical output results. The four types of numeric format codes are specified by means of the following general forms:

aIw
aFw.d
aEw.d
aDw.d

where a is optional and is used to indicate the number of times the format code applies; I, F, E, D are the format codes; w is an integer constant that specifies the total column width of the data; and *d* is an integer constant that specifies the number of decimal places to the right of the decimal point. The I format code is used in the transferral of integer data, the F format code is used in the transferral of real data that does not contain an exponent, and the D and E format codes are used in the transferral of real data that contain a D or an E exponent. The following are valid and invalid examples of internal data to be printed in I4 format:

INTERNAL VALUE	PRINTED VALUE
−0	−0
−327	−327
+327	327
−6895	6895

If the number to be printed has less than four digits, the left print positions are blank. If the number has more than four digits, only the least significant four digits are printed. If the number possesses a minus sign, an additional position should be specified in the field width.

The following are valid and invalid examples of internal data to be printed in F5.1 format:

INTERNAL VALUE	PRINTED VALUE
−166.32	166.3
+10.40	10.4
−0.	−0.0
+2193.36	193.4

If an insufficient number has been specified for *d*, the number of digits in the fractional part of the number, the fractional part is rounded or truncated from the right (note the rounding that takes place in the last example). If an insufficient number has been specified for *w*, the total column width of the data, the left-most significant part (including sign) is lost. In general, *w* should be at least $d + 3$.

The following are some valid and invalid examples of internal data to be printed in E12.5 and D16.9 format:

INTERNAL VALUE	PRINTED VALUE E12.5	PRINTED VALUE D16.9
−166.32	−0.16632E 03	−0.166320000D 03
.000000000008	0.80000E−11	0.800000000D−11
−2345678.932	−0.23457E 07	−0.234567893D 07

Note the rounding off that takes place in the third example when E12.5 format code is used. In general, w should be at least equal to $d + 7$ for an E or D format code. Thus if we wish to print out the maximum number of significant digits that have been preserved for an E-type code, E14.7 format should be specified. In the case of a D-type number, D23.16 should be used. If the specified value of w is larger than that required, the number is printed right-justified and the excess left-most print positions are blank. We will see later that we may avoid erratic carriage control in printing by making the value of w at least equal to 15 and 24, respectively (E15.7 and D24.16).

When reading in input data from cards, the beginning of the exponent part must be punched with an *E* (or *D*), or if that is omitted, by a + or − sign. For example,

E+4, E+04, E 04, +4, +04

are permissible and each means 10^4. In addition, decimal points need not be punched on the data cards when F, E, or D format is used. The decimal point will be taken from the d portion of the format code. If a decimal point is punched on the data card, it takes precedence over the position indicated by the d portion of the format code.

The Numeric FORMAT Statement

The FORMAT statement is used with a READ or WRITE statement to specify the form of the data that will be read into the computer or the desired form of the results that will be punched or printed out of the computer. Since the FORMAT statement is referred to by either a READ or WRITE statement, all FORMAT statements must have a statement number or label. FORMAT statements are nonexecutable and may be placed anywhere in the source program. It is usually good practice to place the FORMAT statement immediately after the corresponding READ or WRITE statement. The form of the FORMAT statement is

SN FORMAT (aF_1 aF_2, . . . ,aF_N)

where SN is the statement number consisting of from one to five digits and aF_1, aF_2, . . . , aF_N specify the repetitive format code of the variables in the input or output list. Some examples of valid numeric FORMAT statements are

```
30 FORMAT(3F5.2,I3,2E14.7)
40 FORMAT(3D25.16)
```

In the first example the first three variables in the list are to be read in or printed out in F5.2 format, the fourth integer variable is to use I3 format, and the last two variables in the list are to use E14.7 format. In the second example all three variables use the same format, D25.16.

A slash (/) may be written within a FORMAT statement and it causes the items that follow the slash to be printed on a new line. The slash thus signals the beginning of a new record. A new record is written whenever a slash or right parenthesis is encountered. For example, assume that the following numeric FORMAT statement was referred to by a WRITE command:

```
10 FORMAT(I6,E15.7/I4,F10.4)
```

The first two variables in the WRITE statement list would be printed in I6 and E15.7 format, respectively. The next line would contain the last two variables in the list and would be printed in I4 and F10.4 format respectively.

Blank output records can be printed or input records skipped by using consecutive slashes. If N successive slashes appear in a FORMAT statement, $N - 1$ lines are skipped or $N - 1$ blank outputs result. For example,

```
10 FORMAT(I6,E15.7///I4,F10.4)
```

would produce the same output as the previous example, except there would be two blank lines between the first line output and the last line output. If the N slashes appear at the beginning or end of the FORMAT statement, N blank lines result. For example,

```
10 FORMAT(//I6,E15.7/I4,F10.4)
```

is identical to the first example except that two blank lines will first be printed before the two output lines are printed.

Additional parentheses are permitted within the main FORMAT statement parentheses. For example, the FORMAT statement

```
10 FORMAT(3(I4,E15.7))
```

indicates that the number 3 preceding the enclosed set of parentheses is the repetition factor of the enclosed format codes. The above statement would be equivalent to

```
10 FORMAT(I4,E15.7,I4,E15.7,I4,E15.7)
```

The Scale Factor P

It has already been pointed out that E (or D) format calls for a number to be printed out as a fraction between 0.1 and 1.0 and this, in turn, is multiplied by 10 to some decimal power. Scientific notation represents a number between 1.0 and 10.0 multiplied by 10 to some decimal power. In printing, scientific notation numbers may be obtained by writing 1P before the E (or D) specification. This will cause the decimal point to be between the first and second digits and will decrease the exponent by 1. The general form of the scale factor is

sPaEw.d

where *s* is the scale factor, *P* stands for place, and *a* is the repetition factor that indicates the number of times the format code Ew.d applies. For example,

INTERNAL VALUE	PRINTED VALUE IN 1PE12.5
—166.325	—1.66325E 02

The scale factor moves the decimal point *s* places to the right and reduces the exponent by *s*.

Double Precision

Some problems may require a greater accuracy than single precision is capable of yielding. In addition, the manipulation of a combination of very large and very small numbers in a long and involved routine can result in inaccurate results caused by rounding errors. Further, the subtraction of two nearly equal numbers, each having a certain number of significant figures of accuracy, will not yield a result of comparable accuracy.

Sixteen significant digits may be preserved in arithmetic operations simply by naming the double-precision variable in a declaration statement. The form is

DOUBLE PRECISION A,B,C

This statement declares *A*, *B*, and *C* as double precision variables.

It is important to realize that if an arithmetic expression involves single- as well as double-precision variables, the single-precision variables will be converted to double-precision numbers, and the resulting expression will be evaluated in double precision. Note that this will not necessarily yield a result that has the same accuracy that would be found for the corresponding arithmetic expression involving only double-precision constants and variables. A further discussion of this point will be illustrated in Chapter 4.

The A Format Code

The use of the A format code permits the storage of alphanumeric characters. That is, in addition to the ten decimal numbers, each of the remaining Fortran characters has a numerical equivalent and can be stored in memory. The form of the A FORMAT code is

aAw

where *a* is optional and is used to indicate the number of times the format code applies, *A* stands for the format code, and *w* is an integer constant that specifies the number of characters of data.

If a WRITE statement refers to a FORMAT statement that has a value of *w* smaller than the required number, only the left-most *w* characters will be printed, and if *w* is greater than the required number, the additional leading characters will be blanks.

Since all arithmetic operations can be performed on alphanumeric variables, it is important to recognize the consequence of mixed-mode expressions. Any alphanumeric variable beginning with the letters *I*, *J*, *K*, *L*, *M*, or *N* would be considered as an integer and this would result in a different number of bits being used in storage than would be used if the variable was considered as a real number. Converting integer alphanumeric variables to real numbers in a mixed-mode expression will result in error.

The H Format Code

The *H* or Hollerith Field code permits the printing of headings and other alphanumeric messages as part of the output. The general form is

wH

where *w* is an integer constant specifying the number of characters that follow the letter *H*. The format code *wH* is written in the FORMAT statement followed by *w* alphanumeric characters. On output these *w* characters are printed as part of the output record. An example of a FORMAT statement that uses the H code is

```
10 FORMAT(8H RESULTS)
```

Note that blanks are significant and must be included in the count. An error in the space count will generally prevent the execution of the program.

When the FORMAT statement is associated with a WRITE command, the first space after *H* is used for carriage control of the printer. The following line printer carriage control characters may be used:

Blank	Single-space printing.
0 (zero)	Double-space printing.
+	No space before printing.
1	Skip to the top of a new page before printing.

For example,

```
10 FORMAT(8H1RESULTS)
```

will cause RESULTS to be printed at the top of a new page.

The above rules apply in general to the line printer. Carriage control will be governed by the appearance of any one of the four characters occupying printer position 1. The character itself will not be printed and any other character could cause improper carriage operation.

Literal Data in FORMAT Statement

Literal data may be made up of alphanumeric and special characters enclosed in apostrophes and written within the FORMAT statement. On output, all characters, including any blanks, within the apostrophes are printed as part of the output. For example,

```
10 FORMAT('1RESULTS')
```

would produce the identical results of the last Hollerith example. Again the first character that is to be printed in the output record is used for carriage control and, as a consequence, it is not printed. One obvious advantage of using the enclosed apostrophes form rather than the Hollerith form is that it avoids the necessity of space counting and possible execution halting that could result from a miscount. One disadvantage of the use of the enclosed apostrophes form is that it prevents the use of words that require apostrophes (such as it's), but this objection can easily be removed by using two successive apostrophes (it''s).

The X Format Code

The X format or skip code has the form

wX

where *w* is an unsigned integer constant specifying the number of blanks to appear in the output or the number of characters to be skipped on input. If the FORMAT statement

```
10 FORMAT(I6,10X,E15.7)
```

was associated with a print statement, the first variable would be printed in integer mode, right-justified, in the first six columns. The next ten columns would be blank due to the 10 X and the second variable would appear in the next 15 columns in E15.7 format. An interesting extension of this example is the possibility of the first variable being large enough to spill into column 1. Not only could this cause improper printing of the numerical value of the first variable, but it could also result in improper carriage control. The latter is a common error and could be avoided by rewriting the above as follows:

```
10 FORMAT(1H ,I6,10X,E15.7)
```

or

```
10 FORMAT(1X,I6,10X,E15.7)
```

The added 1H blank or 1X term automatically ensures single-space carriage operation.

2.6 INPUT-OUTPUT STATEMENTS

Input-output statements cause data to be transmitted in and results to be printed or punched out. It is therefore necessary to specify the form that the data will have. The form of the data is described by a separate FORMAT statement referred to in the input or output statement and corresponds to the respective variables in the I-0 list.

The READ Statement

The form of the READ statement is

```
READ(5,SN)List
```

where SN is the statement number of the FORMAT statement that gives the format code for each of the variables to be read in. The number 5 is a data set reference number and may be different for different compilers. List corresponds to the names of the variables that are separated by commas and for which numerical values are to be read in from the data card. Note that a separate data card (cards) is needed for each READ statement. The execution of a READ statement causes the card reader to read data cards until the number of input items matches the number of variables in the List. Note that once a data card or a part of a data card has been read in, it cannot be read again. The following is an example of a READ statement and its corresponding FORMAT statement:

```
   READ(5,10)A,B,C
10 FORMAT(3F5.1)
```

The above READ statement causes input data from a data card to be read in and stored in memory locations corresponding to the variables A, B, and C. The 10 in the READ statement refers to statement number 10. This is the FORMAT statement that specifies the form that the input data will have. In this example we indicate that all three variables are written in the same F5.1 format code; that is, the numerical value of A is found in the first five columns of the data card and contains one digit to the right of the decimal point; similarly, the value of B is found in the next five columns, and the value of C is found in columns 11 through 15. Figure 2.5 shows the four data cards that were used in the quadratic root problem of Section 1.3.

The Punch Statement

The form of the statement used for punching output results is

WRITE(7,SN)List

where SN is the statement number of the FORMAT statement that gives the format code for each of the variables or arrays in the List. The num-

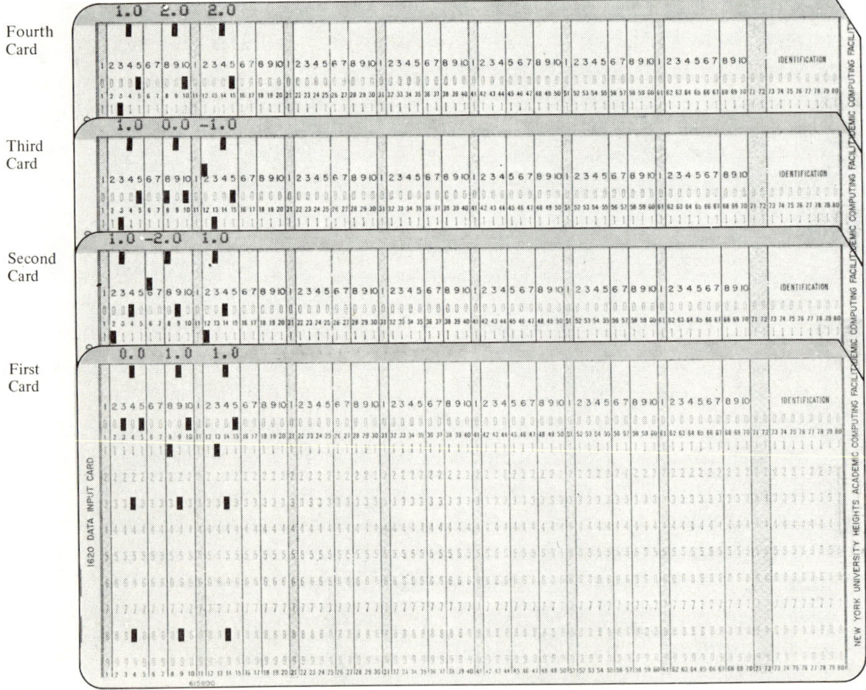

Figure 2.5 The four data cards of the quadratic root problem of Section 1.3

ber 7 is a data set reference number that refers to a peripheral punching device. The following is an example of a punch statement and its corresponding FORMAT statement:

```
   WRITE(7,20)A,B,C
20 FORMAT(F5.2,2E14.7)
```

The above WRITE statement causes the numerical values stored in memory for the three variables *A*, *B*, and *C* to be punched out. The 20 in the WRITE statement refers to statement number 20. This is the FORMAT statement that specifies the form of the output data. In this example, a single card will be punched and it will contain the current numerical values of the variables. The variable *A* will be punched in the first five columns in F5.2 format code, *B* will be punched in the next 14 columns in E14.7 format code, and *C* will be punched in the next 14 columns (columns 20 through 33) in E14.7 format code. Figure 2.6 shows the resulting four cards that would be punched for the four data sets of the quadratic root problem of Section 1.3. Note that these cards are punched and the printed character does not appear at the top of the punched columns. Rather than interpreting the punch code, it is possible to duplicate each card by using the IBM 026 or IBM 29 keypunch with the print switch thrown to the *on* position. It is also possible to obtain the printed column characters by using an IBM 82 Interpreter, but here the printed characters will not be in line with the corresponding punched column.

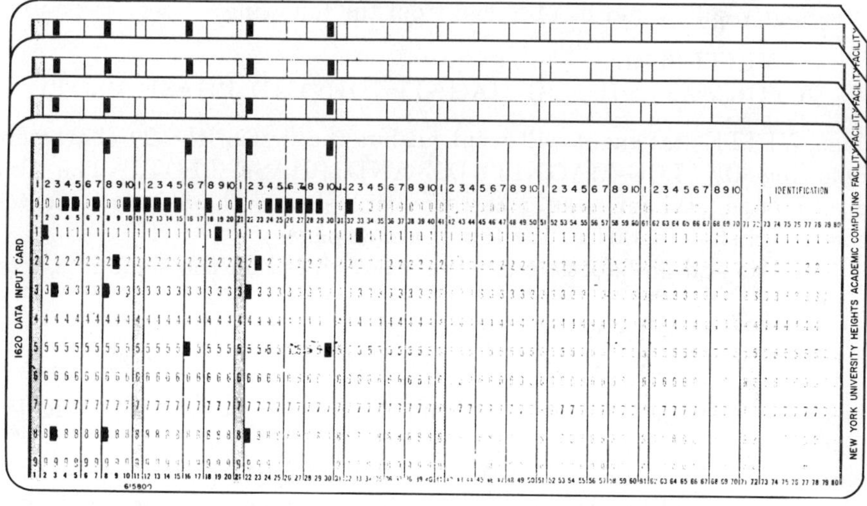

Figure 2.6 Punch card output for the four data sets of the quadratic root problem of Section 1.3

The Print Statement

The form of the statement used for printing output results is

```
WRITE(6,SN)List
```

where SN is the statement number of the FORMAT statement that gives the format code for each of the variables in the List. The number 6 is a data set reference number that refers to an output device. The following is an example of a print statement and its corresponding FORMAT statement:

```
   WRITE(6,30)A,B,C
30 FORMAT(2F10.4,E14.7)
```

The above WRITE statement causes the numerical values stored in memory for the variables *A*, *B*, and *C* to be printed out. The 30 is the FORMAT statement nnmber, and this specifies that the variable *A* will be printed in columns 1 through 10 and *B* in columns 11 through 20 (both in F10.4 format code). In addition, it specifies that *C* will appear in columns 21 through 34 and will be printed in E14.7 format.

In order to have the computer print output headings, the List may be omitted. For example, the following WRITE statement and its corresponding FORMAT statement causes the heading "LOG-MAGNITUDE AND PHASE PLOT" to be printed out.

```
   WRITE(6,40)
40 FORMAT(' LOG-MAGNITUDE AND PHASE PLOT')
```

A similar result would be obtained from the following:

```
   WRITE(6,50)
50 FORMAT(29H1LOG-MAGNITUDE AND PHASE PLOT)
```

This WRITE statement will print all but the first of the 29 characters following H, "LOG-MAGNITUDE AND PHASE PLOT." The first character is used for printer carriage control, and the 1 here indicates the record should start at the top of a new page.

Numeric and alphabetic output can be printed on the same line. For example,

```
   WRITE(6,60)CRMS
60 FORMAT(7H ANSWER,11X, 'I RMS =',F6.3)
```

has a WRITE statement that contains the single variable CRMS. The blank after 7H causes single spacing. The output line (columns 2 through 7) will contain "ANSWER" followed by 11 blank spaces, then "I RMS =" will be printed, and finally the numerical value of CRMS will be printed in F6.3 format code. Note that it was not necessary to consider carriage control in the space following the first apostrophe since I RMS = is printed on the same line as ANSWER and carriage control has already been established by the blank after 7H.

Indexing in Lists

The List of an input or output statement is a series of variable names separated by commas; thus, each variable that is to be read in or printed out must be specified by name in the List. Since this is not very practical for subscripted variables, the Fortran translator permits the value of the subscript to be defined within the List itself and iterated by an implied DO type loop. As an example, consider

```
  READ(5,3)(C(I),I = 1,41)
3 FORMAT(10F5.2)
```

This statement causes $C(1)$, $C(2)$, . . . , $C(41)$ to be read in from data cards. The first card must contain $C(1)$, $C(2)$, . . . , $C(10)$ in F5.2 format code. The second card must contain $C(11)$, $C(12)$, . . . , $C(20)$, and so on. The fifth card must contain $C(41)$ in columns 1 through 5.

It is also possible to combine indexed and nonindexed variables in the same WRITE statement; for example,

```
  READ(5,5)N,(C(I),I = 1,N)
5 FORMAT(I6/(10F5.2))
```

causes the value of N to be read from the first data card. The value of N then gives the LIMIT index that specifies the number of elements that will be read in for the array C. The value of $C(1)$, $C(2)$, . . . , $C(N)$ will then be read from as many subsequent data cards until the number of input items equals the value of N.

It is also possible to index double subscripted arrays. Consider

```
   DO 10 I = 1,5
10 READ(5,20)(C(I,J),J = 1,10)
20 FORMAT(10F5.2)
```

These statements will cause ten elements to be read from each data card in F5.2 format. The first card will contain $C(1,1)$, $C(1,2)$, . . . , $C(1,10)$; the second card $C(2,1)$,$C(2,2)$, . . . , $C(2,10)$; and so on. By using nested parentheses and implied DO loops, the same result can be obtained with a single READ statement.

```
   READ(5,20)((C(I,J),J = 1,10),I = 1,5)
20 FORMAT(10F5.2)
```

Note that the outermost set of parentheses is analogous to the outermost

DO, and the innermost parentheses is analogous to the innermost DO. The value of I is thus initially set equal to 1 and the value of J is incremented by 1 from 1 to 10. I is then incremented by 1 to a value of 2, and the inner loop is initialized and incremented from 1 to 10 again. This is repeated until I reaches a value of 5 and J reaches a value of 10; control then passes to the next executable statement.

The above example involves a two-dimensional array that is made up of five rows and ten columns. The execution of the READ statement was equivalent to reading in the elements of the array one row at a time. By interchanging the order of the iteration variables, the following WRITE statement will be equivalent to printing the transposition of the original matrix:

```
   WRITE(6,30)((C(I,J),I = 1,5),J = 1,10)
30 FORMAT(5F5.2)
```

The following are additional examples of nested Lists and their corresponding READ or WRITE sequence:

LIST	READ OR WRITE SEQUENCE
(C(I),I=1,41,2)	C(1), C(3), . . . , C(41)
(C(I),I=N,41)	C(N), C(N+1), . . . , C(41)
(C(I),D(I),I=1,41)	C(1), D(1), C(2), . . . , C(41), D(41)
((C(I,J),I=1,41),J=1,5)	C(1,1), C(2,1), . . . , C(41,1), C(1,2), . . . , C(41,5)
((C(I,J),J=1,5),I=1,41)	C(1,1), C(1,2), . . . , C(1,5) C(2,1), C(2,2), . . . , C(41,5)
(((C(I,J,K),I=1,41),J=1,5),K=1,3)	C(1,1,1), C(2,1,1), . . . , C(41,1,1), C(1,2,1), C(2,2,1), . . . , C(41,5,1), C(1,1,2), C(2,1,2), . . . , C(41,5,3)

PROBLEMS

2-1. What is Fortran?

2-2. Name the four types of Fortran statements.

2-3. What is a statement label and why is it needed?

2-4. What are comment cards and how are they identified?

2-5. What columns are used in writing a Fortran statement?

2-6. What columns are used in writing a comment?

2-7. Name five Fortran delimiters. How are they used?

2-8. What are the source and object decks? What languages are used?

2-9. Identify the type of each of the following constants and variables. If any are invalid indicate the reason why.

a. 12.E(J) b. 296

c. 40E+3 d. 0.707E−8.

e. ABLE(M(I,J)) f. MAX$

g. APOLLO8
h. AVERAGE
i. 5.0**2
j. X10(I + 5,J*4)
k. 2TIMES
l. 0
m. A8.6
n. 3.3D−06
o. BETA(5*A+6,3*4,238)

2-10. Indicate the order of the arithmetic operations in the following expressions:
a. ((A+6.2)**4/B+2.4)*F
b. (((3.5+5.7)+A**2/D)+6.4)
c. A1*((A2+A3)/A4+A5*(A6−A7)**A8/A9+A10)**A11

2-11. Identify the errors in the following arithmetic statements:
a. SQRT(Y)=ABLE*(BAKER+CHARLIE)
b. −J=I**2+K
c. A=((X+Y)A**+I**2.)
d. A*X=H+B*SIN(3*ABLE)*−C
e. CIR=2.*π*R
f. ALPHA(I,J)=I**(J+2M)
g. Y=SQRT(COS(3.*ABS(A+B−C)))
h. A(10)=(X+Y)**(Z/W)
i. A(5*I*3)=SQRT(ABS(X*X−Y*Y))
j. ZETA=2.*(A**I)/(B**J)+ATAN(I*J)

2-12. Evaluate the following arithmetic statements:

a. L=(I+J+K)/2	A=3.0
b. L=I*J−K**3	B=−2.0
c. L=I/J−I/K	C=−1.0
d. X=A*B**J−B*C	I=5
e. X=SQRT(A*A+(ABS(B+C))**2)	J=2
f. X=A*B/4.0+I	K=4

2-13. Write the appropriate Fortran statement for each of the following. Choose meaningful names for all variables. For mode purposes, simply change the first letter to retain the mnemonic meaning.

a. $$R_B = \frac{V_{BB}(1 - VT) - V_{BE(\mathrm{OFF})}}{(V_{CE(\mathrm{SAT})} - V_{BE(\mathrm{OFF})})/(R_{K2}(1 - RT))} + I_{CBO}$$

b. $$g(t) = 1 - (1 - t + 0.5\,t^2)\epsilon^{-t}$$

c. $$Y_{2,3} = -0.5Y_1 + j(0.75(Y_1{}^2 - 1))^{1/2}$$

d. $$P_{\mathrm{out}} = \frac{\left[\dfrac{(R_3{*}R_{I2})/(R_3 + R_{I2})}{R_1 + (R_3{*}R_{I2})/(R_3 + R_{I2})}\right]^2}{R_{I2}} V_{\mathrm{in}}{}^2$$

e. $$G(j\omega) = \frac{100(1 + j\omega/2.5)}{(1 + j\omega/10)^2(1 + j2\omega)(1 + j4\omega)}$$

2-14. Find and correct any errors in the following Fortran statements:

a. IF(A−I)1,2,2 b. GO TO 3,5
c. GO TO (N−14) d. GO TO 127
e. DO 10, J=1,50,K f. DO 10 PEN=1,10,+1
g. GO TO (4,7,2)ITEM
h. DO 10 IKE=NIXON,JOHN,LYNDON
i. GO TO 10 I=0,−10,−1
j. IF(W−X+Y),3,4,5
k. IF(A=3.14159)5,6,7

2-15. Given the elements of the M by N array A, write appropriate Fortran programs for each of the following:

a. Determine the sum of the outside elements.
b. Determine the sum of all interior elements.
c. Determine the sum of alternate elements starting with a_{11}.
d. Whenever $M = N$, determine the sum of the two diagonals.

2-16. Repeat Problem 2-15 for the elements of the L by M by N array B.

2-17. Express each of the following numbers in I9, F10.4, E14.2, and D16.9 format. Preserve the maximum number of significant figures in each case.

a. 3.141592653 b. 10^6 c. 10^{-4} d. −3156.02678

2-18. If $I = 510$, $J = -49$, and $K = -59648$, what column would each data set start for the following READ and FORMAT statements

```
   READ(5,10) I, J, K
10 FORMAT(3I15)
```

2-19. Three variables A, B, and C, appear on input data cards as follows (b represents a blank, S represents the sign of the number, and X represents the decimal digit):

a. bbSXX.XXXXbbSXX.XXXXbbSXX.XXXX
b. bSX.XXXXbSX.XXXXbbSX.XXXXESXX
c. bbSX.XXXXESXXbbSX.XXXXESXXbbbXXXXXXXX

Give the corresponding READ and FORMAT statements for reading in this data.

2-20. Write the Fortran print and FORMAT statements for obtaining the following output:

```
Blank line
bbAb=bSXXX.XXXX
Blank line
bbBb=bSXXX.XXX
```

2-21. Show the output that will result from the execution of the following statements:

a.
```
      WRITE(6,10)CRMS
   10 FORMAT(21H CALCULATION OF I RMS,//,
                              7H ANSWER,5X,7HI RMS =,F6.3)
```
b.
```
      WRITE(6,10)X1R,X1I,X2R,X2I
   10 FORMAT(' X1R =',E15.7,3X,'X1I =',E15.7,//,
                              'X2R =',E15.7,3X, 'X2I =', E15.7)
```

2-22. Given a two-dimensional array, A, having five rows and five columns, write the Fortran statements required to read in the element values.

a. Assume that the elements will be read in columnwise, five numbers to the card in E15.7 format code.

b. Assume that the elenents will be read in by rows, five numbers to the card in E15.7 format code.

2-23. Write a Fortran program that reads in the M by M elements ($M \leq 10$) of the array A and determines and prints out whether the array is symmetrical or not.

2-24. What does the following program do?

```
      I=1
      DO 10 J=1,50
   10 I=I*J
      WRITE(6,20)I
   20 FORMAT('1I =',I10)
      STOP
      END
```

2-25. What does the following program do? Are there any errors?

```
      X=3.
      DO 10 I=1,100
      XNEXT=(2.*X+37./X*X)/3.
      IF(ABS(X-XNEXT)-1.E-6)20,20,10
   10 X=XNEXT
   20 WRITE(6,30)XNEXT
   30 FORMAT('1ANSWER =',E15.7)
      STOP
      END
```

2-26. If we have no built-in Fortran function for evaluating the hyperbolic sine, we will have to use the identity

$$\sinh X = \frac{\epsilon^{X} - \epsilon^{-X}}{2}$$

Write a Fortran program for evaluating and printing the sinh X for $X = 0$ to $X = 10^{-8}$ in 10^{-10} unit steps.

2-27. Since the value of X used in Problem 2-26 is very small, the identity used in the calculation of sinh X will not yield very accurate results. Extend the program of Problem 2-26 to include the calculation of the sinh X using the approximation

$$\sinh X \approx X + \frac{X^3}{3!} + \frac{X^5}{5!} + \frac{X^7}{7!}$$

Print the values obtained by both methods and their difference for each of the argument values.

2-28. Write a Fortran program that will perform the following:

a. Read in a maximum of 100 numbers, five numbers to a data card and each number in E15.7 format code.

b. Compute and print the arithmetic mean value:

$$AM = \frac{\sum_{I=1}^{M} X_I}{M}$$

c. Compute and print the standard deviation:

$$SD = \sqrt{\frac{\sum_{I=1}^{M} X_I^2 - \left[\sum_{I=1}^{M} X_I\right]^2 / M}{M - 1}}$$

CHAPTER 3

Network Examples

3.1 INTRODUCTION

The main purpose of this chapter is to show the reader how some simple electrical network problems may be solved so that he will begin to develop a background and understanding of how the digital computer may be used as a super tool in solving problems in his chosen field. Most of the examples are restricted to the general area of network analysis and deal with relatively basic concepts that are well known by most electrical engineers. All the examples use the built-in library functions (such as SQRT, EXP, ATAN, SIN, COS, and so on) and thus avoid the need for lengthy numerical method discussions at this time.

3.2 MATCHING NETWORK DESIGN

Design of a Minimum-Loss Resistor Matching Pad

One of the problems encountered in electrical engineering is impedance matching. For a nonsymmetrical T- or Pi-section, the impedance seen at the input usually differs from the terminating impedance, but there exists a pair of impedances such that if one end is terminated in one of the impedances, the input impedance seen at the other end will be the other impedance and vice versa. These impedances are called the image impedances of the network and are labeled Z_{I1} and Z_{I2} depending on the end referred to. If we are interested in designing a matching network that will match over a broad band of frequencies, a network that uses

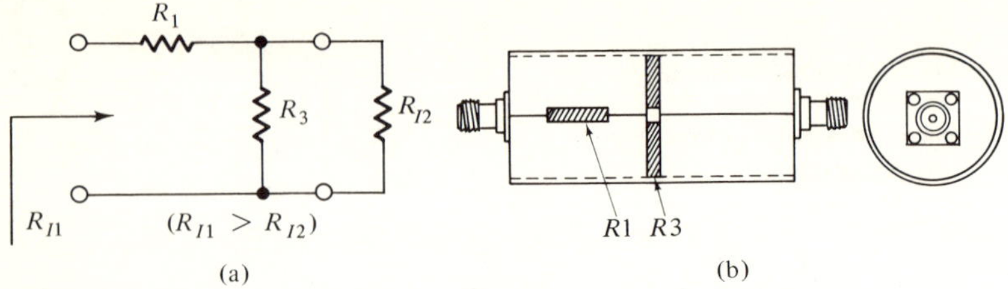

Figure 3.1 (a) Minimum-loss pad; (b) coaxial-type broad band attenuator (R_3 is a coaxial disk resistor)

only resistors is required. Since resistors are involved, the matching network must introduce a certain amount of loss. In certain applications a fairly large loss may be actually required, but in other applications where the signal level is low, the loss should be kept at a minimum.

If our primary interest is in designing a network to provide broad-band matching, the minimum-loss pad or attenuator shown in Figure 3.1a and the coaxial type of construction shown in Figure 3.1b should be used. When properly designed this network will provide the required matching. That is, if the resistance R_{I2} is connected to the right-hand terminals (the load side), the input resistance will be R_{I1}. Conversely, if the resistance R_{I1} is the load resistance and it is connected to the left-hand terminals, the input resistance seen at the right-hand terminals will be R_{I2}.

The following design equations are based on the load resistor R_{I2} being less than the input resistor R_{I1}.[1]

$$R_1 = R_{I1}\sqrt{1 - \frac{R_{I2}}{R_{I1}}}$$

$$R_3 = \frac{R_{I2}}{\sqrt{1 - \dfrac{R_{I2}}{R_{I1}}}}$$

$$\theta_{\min} = \cosh^{-1}\sqrt{\frac{R_{I1}}{R_{I2}}} \quad \text{nepers loss}$$

If the reverse is true, the same equations are used, but the network should be turned end-for-end so that the load resistor will be R_{I1}, as shown in Figure 3.2.

It is of interest to note that the design equations for R_1 and R_3 may

[1] See *Reference Data for Radio Engineers,* 4th ed. (New York: International Telephone and Telegraph Corp., 1962), p. 252; and W. L. Everitt and G. E. Anner, *Communication Engineering* (New York: McGraw Hill Book Company, 1956), p. 481.

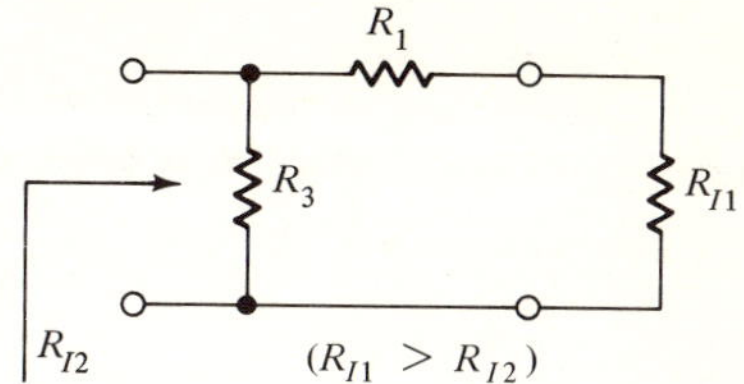

Figure 3.2 Minimum-loss pad connection for the case of the load resistor being the larger value

be obtained by a more direct approach, once the pad configuration is known. Thus from Figures 3.1a and 3.2, we have

$$R_{I1} = R_1 + \frac{R_3 R_{I2}}{R_3 + R_{I2}}$$

$$R_{I2} = \frac{R_3(R_1 + R_{I1})}{R_1 + R_3 + R_{I1}}$$

Solving these two equations for R_1 and R_3 yields the desired design equations.

Figure 3.3 is a Fortran program that may be used to calculate the minimum-loss resistor matching pad. The first four lines are comment statements that are used to identify the program. Computer statement S.0001 reads in values of R_{I1} and R_{I2} (where R_{I1} must be larger than R_{I2}). The dummy variable A is introduced in S.0003 and is equal to the numerical value of $\sqrt{1 - R_{I2}/R_{I1}}$. Computer statements S.0004 and S.0005 readily calculate the required values for R_1 and R_3.

Although it is possible, by using the library function EXP and a suitable iteration procedure, to find the minimum loss θ from

$$\theta_{\min} = \cosh^{-1}\sqrt{\frac{R_{I1}}{R_{I2}}}$$

the arccosh may be computed more easily by using its logarithmic

```
                C
                C       DESIGN OF MINIMUM LOSS
                C         RESISTOR MATCHING PAD
                C
S.0001                1 READ(5,2) RI1,RI2
S.0002                2 FORMAT(2F10.3)
S.0003                  A=SQRT(1.-RI2/RI1)
S.0004                  R1=RI1*A
S.0005                  R3=RI2/A
S.0006                  X=SQRT(RI1/RI2)
S.0007                  ATTDB=8.686*ALOG(X+SQRT(X*X-1.))
S.0008                  WRITE(6,3) RI1,RI2,R1,R3,ATTDB
S.0009                3 FORMAT('1DESIGN OF MINIMUM LOSS RESISTOR MATCHING PAD',////,' RI1
                       1=',F10.3,4X,'RI2 =',F10.3,////,' R1 =',F10.3,4X,'R3 =',F10.3,///,'
                       2 ATTENUATION =',F10.3,' DB')
S.0010                  GO TO 1
S.0011                  RETURN
S.0012                  END
```

Figure 3.3 Program for designing minimum loss pads

representation

$$\text{arccosh } X = Ln(X + (X^2 - 1)^{0.5})$$

where

$$X = \sqrt{\frac{R_{I1}}{R_{I2}}}$$

Computer statement S.0006 is used to find the numerical value of X and S.0007 determines the loss in decibels (1 neper = 8.686 db). Statements S.0008 and S.0009 are used to print out the results. Statement S.0010 returns control to statement 1 so that the next set of R_{I1} and R_{I2} values can be processed.

Figure 3.4 shows the computed results for two different sets of input data. Notice the execution error IHC217I shown at the bottom of the second set. This execution error occurs because the computer attempts to read a third data card. Since only two sets of data are supplied, an execution error must obviously occur when the computer attempts to read in the missing third set.

The above execution error should be considered as a permissible error because it allows the computer to terminate execution automatically after the last data card has been processed. In other words, the above program will handle 1, 2, or as many data cards as are supplied.

The reader is encouraged to verify the computer results. Using the values of R_1 and R_3 obtained from the computer solution and the value of R_{I2}, hand calculate the input resistance R_{I1}. Also check the pad loss by hand calculation from

$$P_{\text{loss}} = 10 \text{ Log}_{10}(P_{\text{in}}/P_{\text{out}})$$

where

$$P_{\text{in}} = V_{\text{in}}^2/R_{I1}$$

```
DESIGN OF MINIMUM LOSS RESISTOR MATCHING PAD

RI1 =    500.000     RI2 =    400.000

R1 =    223.607     R3 =    894.427

ATTENUATION =      4.180 DB

DESIGN OF MINIMUM LOSS RESISTOR MATCHING PAD

RI1 =     72.000     RI2 =     50.000

R1 =     39.799     R3 =     90.453

ATTENUATION =      5.406 DB

IHC217I
```

Figure 3.4 Two minimum-loss pads designed by the program of Figure 3.3

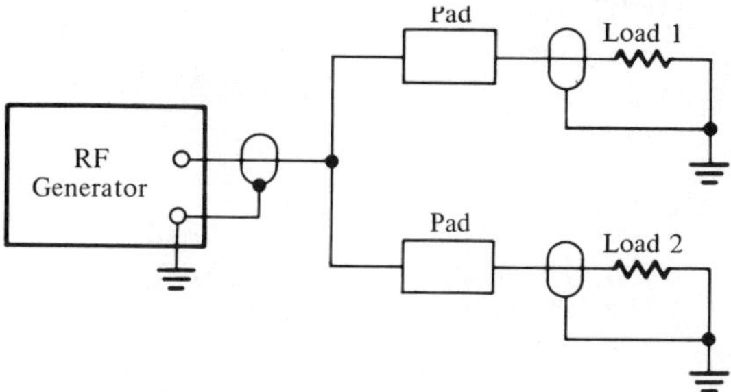

Figure 3.5 Generator load isolation

and

$$P_{\text{out}} = \frac{\left[\dfrac{(R_3 * R_{I2})/R_3 + R_{I2}}{R_1 + (R_3 * R_{I2})/(R_3 + R_{I2})}\right]^2 V_{\text{in}}^{\;2}}{R_{I2}}$$

Design of an Asymmetrical Resistive T-Section for a Prescribed dB Loss

In the above discussion we were interested in matching two different resistances over a broad band of frequencies with the minimum possible loss. We frequently require a specified loss in addition to the matching. Large loss sections may also be required when we are trying to isolate the load from the generator. For example, Figure 3.5 shows two different loads being driven by the signal generator. If the pads are sufficiently large (approximately 20 dB or more), the loads will have a negligible effect on each other. The importance of this fact may be realized by recognizing that the two loads could correspond to respective amplifier-trigger circuits, such as a tunnel diode or Schmitt. On firing, the effect would be reflected back to the input of the amplifier. The pads would thus provide the necessary isolation to prevent the triggering of one channel to influence the other.

Figure 3.6 shows the required matching section. The design equations

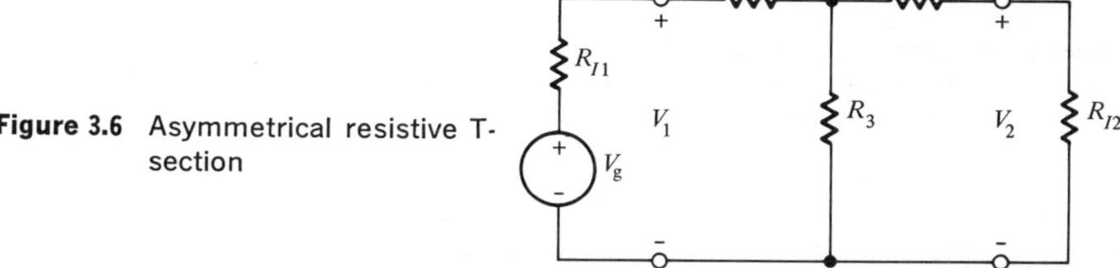

Figure 3.6 Asymmetrical resistive T-section

```
        C        DESIGN OF UNBALANCED RESISTIVE
        C         MATCHING NETWORK
        C             FOR A
        C        PRESCRIBED DB LOSS
        C
S.0001          1 READ (5,2) ATTDB,RI1,RI2
S.0002          2 FORMAT(3F10.3)
S.0003            THETA=.1151*ATTDB
S.0004            A=SQRT(RI1*RI2)
S.0005            R3=2.*A/(EXP(THETA)-EXP(-THETA))
S.0006            R2=RI2/TANH(THETA)-R3
S.0007            R1=RI1/TANH(THETA)-R3
S.0008            WRITE(6,3)RI1,RI2,ATTDB,R1,R2,R3
S.0009          3 FORMAT(////,' RI1 =',F8.3,4X,'RI2 =',F8.3,//,' TRANSMISSION LOSS =
                 1',F6.3,' DB',//,' R1 =',F8.3,' R2 =',F8.3,' R3 =',F8.3)
S.0010            GO TO 1
S.0011            RETURN
S.0012            END
```

Figure 3.7 Program for designing asymmetrical resistive T-sections

follow and are based on specifying R_{I1}, R_{I2}, and the required pad loss θ in nepers:[2]

$$R_3 = \frac{\sqrt{R_{I1} * R_{I2}}}{\sinh \theta}$$

$$R_2 = \frac{R_{I2}}{\tanh \theta} - R_3$$

$$R_1 = \frac{R_{I1}}{\tanh \theta} - R_3$$

Figure 3.7 is the corresponding Fortran design program. The first five statements are comment statements and are used to identify the program. Computer statements S.0001 and S.0002 read in the required attenuation in decibels and the two image resistances. Computer statements S.0004 to S.0007 calculate the required numerical values for R_1, R_2, and R_3. Computer statements S.0008 and S.0009 cause the corresponding information to be printed and S.0010 returns control to statement 1 so that the next data card can be processed.

[2] *Reference Data for Radio Engineers,* 4th ed., pp. 252, 480.

```
RI1 =   72.000        RI2 =   50.000

TRANSMISSION LOSS =  5.000 DB

R1 =   39.908 R2 =    -2.450 R3 =    98.717


RI1 =   72.000        RI2 =   50.000

TRANSMISSION LOSS =10.000 DB

R1 =   45.832 R2 =    18.940 R3 =    42.179


RI1 =   72.000        RI2 =   50.000

TRANSMISSION LOSS =20.000 DB

R1 =   61.328 R2 =    38.883 R3 =    12.128

IHC217I
```

Figure 3.8 Three asymmetrical T-sections designed by the program of Figure 3.7

Figure 3.8 shows the computer results for three different sets of input data. Note that the first set resulted in a negative value for R_2. This may be explained by referring to the second data set in Figure 3.4 and realizing that the minimum loss for this particular set of image resistances is 5.406 dB.

Design of an Asymmetrical T-Section for a Prescribed Transmission Gain

As an alternative to the above design, it is sometimes more desirable to be able to specify the pad requirements in terms of the image resistances and the network transmission or voltage transfer ratio (V_2/V_1).

Referring to Figure 3.6, the voltage transmission VG is given by

$$VG = \frac{V_2}{V_1} = \frac{(R_3(R_2 + R_{I2}))/(R_2 + R_3 + R_{I2})}{R_1 + (R_3(R_2 + R_{I2}))/(R_2 + R_3 + R_{I2})} * \frac{R_{I2}}{R_2 + R_{I2}}$$

Specifying the pad in terms of VG, R_{I1}, and R_{I2} permits the calculation of the pad elements from the following equations:

$$R_1 = \frac{2\,R_{I2}(1 - VG)}{VG^2\ \text{DENOM}} - R_{I1}$$

$$R_2 = R_{I2}\left(\frac{2(R_{I2} - R_{I1}VG)}{R_{I1}VG^2\ \text{DENOM}} - 1\right)$$

$$R_3 = \frac{2\,R_{I2}}{VG\ \text{DENOM}}$$

where

$$\text{DENOM} = \frac{R_{I2}}{VG^2 - R_{I1}} - 1$$

Figure 3.9 is the Fortran design program. Figure 3.10 shows the computer results for $R_{I1} = 50.0$ ohms, $R_{I2} = 72.0$ ohms, and the transmission gain equal to 0.30, 0.40, 0.50, 0.60, and 0.70. Note that a negative

```
C       DESIGN OF UNBALANCED RESISTIVE
C           MATCHING NETWORK
C              FOR A
C       PRESCRIBED VOLTAGE GAIN
C
      1 READ(5,2)RI2,RI1,VG
      2 FORMAT(3F10.3)
        DENOM=RI2/((VG)**2*RI1)-1.
        R1=2.*RI2*(1.-VG)/(VG**2*DENOM)-RI1
        R2=RI2*(2.*(RI2-RI1*VG)/(RI1*VG**2*DENOM)-1.)
        R3=2.*RI2/(VG*DENOM)
        WRITE(6,3)RI2,RI1,VG,R1,R2,R3
      3 FORMAT(////,'    RI2 =',F8.3,4X,'         RI1 =',F8.3,//,' TRANSMIS
     1SION GAIN =',F6.3,//,' R1 =',F8.3,' R2 =',F8.3,' R3 =',F8.3)
        GO TO 1
        RETURN
        END
```

Figure 3.9 Program for designing asymmetrical resistive T-sections for a prescribed voltage gain

```
    RI2 =  72.000                    RI1 =  50.000
TRANSMISSION GAIN = 0.300
R1 =  24.667 R2 =  49.600 R3 =  32.000


    RI2 =  72.000                    RI1 =  50.000
TRANSMISSION GAIN = 0.400
R1 =  17.500 R2 =  45.000 R3 =  45.000


    RI2 =  72.000                    RI1 =  50.000
TRANSMISSION GAIN = 0.500
R1 =  10.504 R2 =  41.748 R3 =  60.504


    RI2 =  72.000                    RI1 =  50.000
TRANSMISSION GAIN = 0.600
R1 =   3.333 R2 =  40.000 R3 =  80.000


    RI2 =  72.000                    RI1 =  50.000
TRANSMISSION GAIN = 0.700
R1 =  -4.526 R2 =  40.168 R3 = 106.105
IHC217I
```

Figure 3.10 Five asymmetrical T-sections designed by the program of Figure 3.9.

value for R_1 results when the transmission gain is specified as 0.70. This may be explained by recalling that in the minimum loss L-Pad of Figures 3.1a and 3.2, the series arm adjacent to the smaller of the two image resistances is zero. Since R_{I1} is less than R_{I2} in Figure 3.6, minimum loss conditions require R_1 to be zero.

Design of Symmetrical Resistive T-Attenuators

The asymmetrical T-section shown in Figure 3.6 can be made a symmetrical T-section by making $R_1 = R_2$. In general, for a symmetrical T-section, the impedance seen at the input terminals differs from the terminating impedance, but there exists an impedance, Z_0, called the *characteristic impedance,* such that if either end is terminated in this

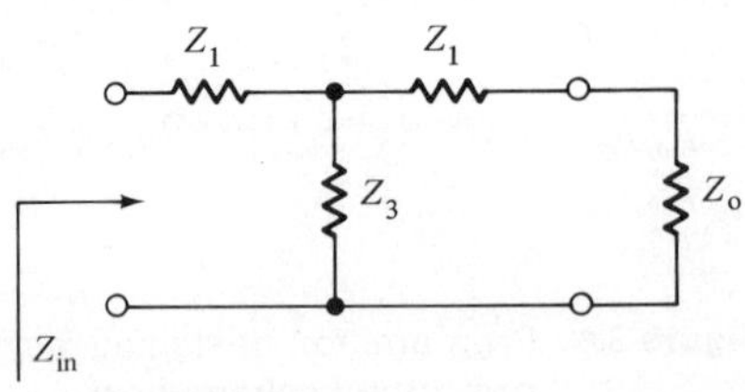

Figure 3.11 A symmetrical T-section

```
C
C     DESIGN OF SYMMETRICAL
C     RESISTIVE T-ATTENUATOR SECTIONS
C
      READ(5,1)R0
    1 FORMAT(F10.3)
      WRITE(6,2)R0
    2 FORMAT('1ATTENUATOR T-SECTION',//,' CHARACTERISTIC RESISTANCE =',F
     110.3,/////)
    3 READ(5,4)ATTDB
    4 FORMAT(F10.3)
      ALPHA=.1151*ATTDB
      R1=R0*TANH(ALPHA/2.)
      R3=2.*R0/(EXP(ALPHA)-EXP(-(ALPHA)))
      WRITE(6,5)R1,R3,ATTDB
    5 FORMAT('0R1 =',F10.3,4X,'R3 =',F10.3,4X,'ATTENUATION =',F 8.3,' DB
     1')
      GO TO 3
      RETURN
      END
```

Figure 3.12 Program for designing symmetrical resistive T-sections

impedance, the impedance seen at the other end is also Z_0. From the symmetrical T-section shown in Figure 3.11, we therefore find

$$Z_{\text{in}} = Z_0 = Z_1 + \frac{Z_3(Z_1 + Z_0)}{Z_1 + Z_3 + Z_0}$$

or

$$Z_0 = \sqrt{Z_1^2 + 2Z_1Z_3}$$

As an attenuator, a symmetrical T- or Pi-section may be used to introduce a specified loss between a generator and a matched load without changing the matching conditions. If the elements are resistive, the loss will be constant and the match will be preserved over a wide band of frequencies. Specifying the characteristic resistance R_0 and the desired

```
ATTENUATOR T-SECTION

CHARACTERISTIC RESISTANCE =    500.000



R1 =    28.743    R3 =  4334.473    ATTENUATION =   1.000 DB

R1 =    57.297    R3 =  2152.958    ATTENUATION =   2.000 DB

R1 =    85.477    R3 =  1419.636    ATTENUATION =   3.000 DB

R1 =   113.109    R3 =  1048.574    ATTENUATION =   4.000 DB

R1 =   140.031    R3 =   822.642    ATTENUATION =   5.000 DB

R1 =   166.100    R3 =   669.507    ATTENUATION =   6.000 DB

R1 =   191.192    R3 =   558.196    ATTENUATION =   7.000 DB

R1 =   215.205    R3 =   473.239    ATTENUATION =   8.000 DB

R1 =   238.058    R3 =   406.053    ATTENUATION =   9.000 DB

R1 =   259.693    R3 =   351.490    ATTENUATION =  10.000 DB

R1 =   363.138    R3 =   162.653    ATTENUATION =  16.000 DB

IHC217I
```

Figure 3.13 Attenuator T-sections designed by the program of Figure 3.12

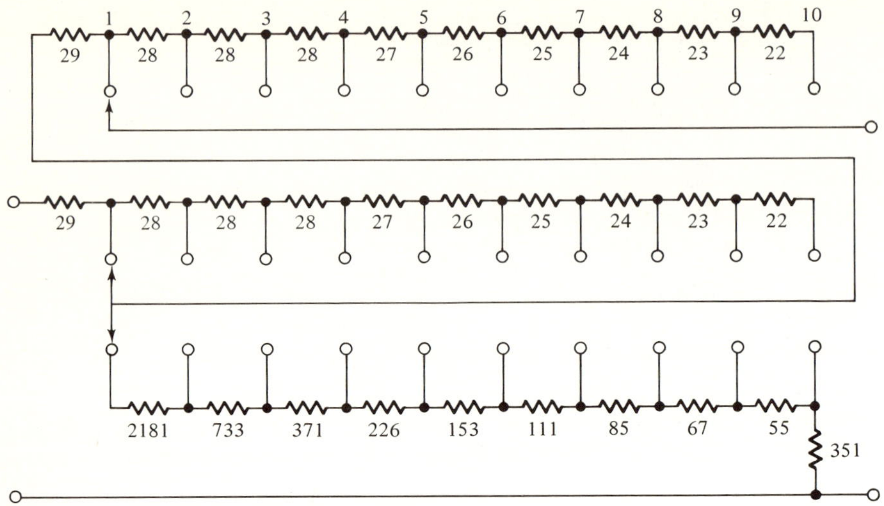

Figure 3.14 A 500-ohm, 10-dB, 1-dB step attenuator

dB loss ATTDB, the design equations are as follows:[3]

$$R_3 = \frac{R_0}{\sinh(0.1151\ \text{ATTDB})}$$

$$R_1 = R_0 \tanh \frac{0.1151\ \text{ATTDB}}{2}$$

Figure 3.12 is the Fortran design program and Figure 3.13 gives the computed T-sections based on a characteristic resistance of 500 ohms and attenuations of 1, 2, 3, 4, 5, 6, 7, 8, 9, 10, and 16 dB.

Using three single-pole 12-position wafer switches, the 10-dB (1-dB

[3] *Ibid.*, pp. 252, 480.

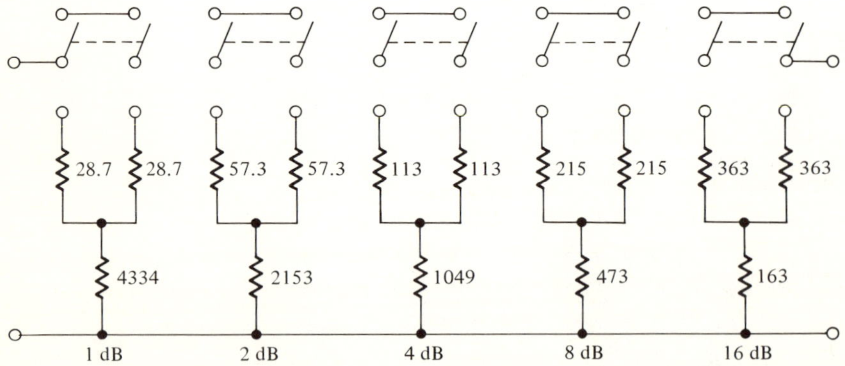

Figure 3.15 A 500-ohm, 31-dB, 1-dB step attenuator

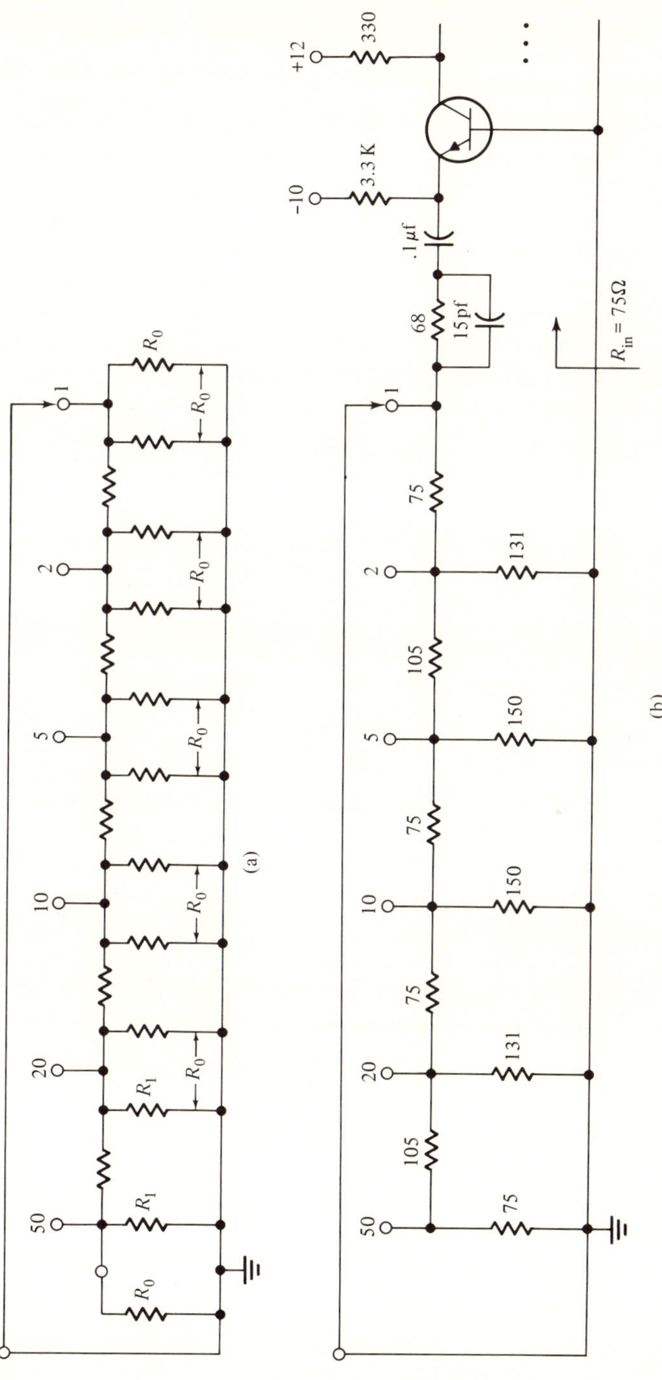

Figure 3.16 A 50-ohm input voltage divider attenuator

step) attenuator shown in Figure 3.14 can be constructed. Alternately, using five double-pole double-throw switches, the attenuator shown in Figure 3.15 can be constructed. This latter attenuator would provide a maximum attenuation of 31 dB in 1 dB steps.

Figure 3.16 shows a useful modification of the last ladder attenuator. Here symmetrical Pi-sections have been cascaded to provide voltage divisions of 2, 5, 10, 20, and 50. Note that the input impedance is equal to $R_0/2$ and is independent of the voltage tap position. Such a voltage divider is useful in limiting the peak-to-peak voltage swing. For example, if an amplifier-trigger circuit is designed to operate linearly about a nominal input voltage of 100 millivolts, the above attenuator would be set to the divide-by-1 position if the input signal were approximately 100 millivolts, and to the divide-by-10 position if the input voltage were approximately one volt.

Specifying the voltage divider ratio VR and the characteristic resistance R_0, the values of R_1 and R_3 can be found from the following equations:

$$R_1 = R_0 \frac{VR + 1}{VR - 1}$$

$$R_3 = R_0 \frac{VR - \dfrac{1}{VR}}{2}$$

Figure 3.17 is the Fortran design program for computing the values of R_1 and R_3.

Figure 3.18 shows the computer output for two Pi-sections that were designed for a characteristic resistance of 100 ohms (input impedance of 50 ohms) and a voltage division of 2 and 2.5 respectively. Figure 3.16b shows the final form of the voltage attenuator. Note that the use of Pi-sections permitted the combination of two resistors that were in parallel and thus saved a total of five precision resistors. With proper construction

```
C
C         VOLTAGE ATTENUATOR DESIGN
C
        1 READ(5,2)RO,VR
        2 FORMAT(2F10.3)
          R1=RO*(VR+1.)/(VR-1.)
          R3=RO*(VR-1./VR)/2.
          WRITE(6,3)RO,VR,R1,R3
        3 FORMAT(////.' THE CHARACTERISTIC RESISTANCE =',F8.3,//,' THE VOLTA
         1GE RATIO =',F8.3,///,' R1 =',F8.3,3x,'R3 =',F8.3)
          GO TO 1
          RETURN
          END
```

Figure 3.17 Program for designing symmetrical resistive Pi-section

```
THE CHARACTERISTIC RESISTANCE = 100.000
THE VOLTAGE RATIO =    2.000

R1 = 300.000   R3 =  75.000


THE CHARACTERISTIC RESISTANCE = 100.000
THE VOLTAGE RATIO =    2.500

R1 = 233.333   R3 = 105.000
IHC217I
```

Figure 3.18 Two Pi-sections designed by the program of Figure 3.17

(good ground returns), this attenuator may be connected to an input BNC connector by a 50-ohm cable that has a maximum length of six inches, and it may be used over a frequency range of 0 to 150 megahertz.

Design of a Lossless L-C Single Frequency Matching Network

Consider the case of a load resistance R_2 that is to be matched, at a single frequency, to a generator resistance R_1 so that maximum power will be transferred. Ideally, to deliver maximum power to the load, no power should be lost in the matching network. Clearly, if no power is to be lost in the matching section, it must contain only reactances.

Referring to Figure 3.19 we see that if R_2 is shunted by X_1 (a pure reactance), the resulting impedance Z will be

$$Z = \frac{R_2 jX_1}{R_2 + jX_1} = \frac{R_2 X_1^2 + jR_2^2 X_1}{R_2^2 + X_1^2}$$

and the

$$Re\{Z\} = \frac{R_2 X_1^2}{R_2^2 + X_1^2} = \frac{X_1^2}{R_2^2 + X_1^2} R_2$$

will be less than R_2. Thus if R_1 is less than R_2, we shunt the larger resis-

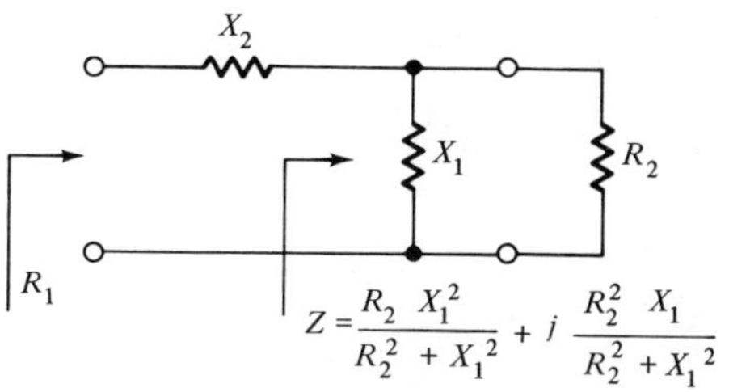

Figure 3.19 Reactive matching section

tance (R_2) with a pure reactance, X_1, so that the $Re\ \{Z\} = R_1$, and balance out the reactive component, $Im\ \{Z\}$, by adding a series reactance X_2 of proper magnitude and sign. The design equations are

$$X_1 = R_2 \sqrt{\frac{R_1}{R_2 - R_1}}$$

$$X_2 = \sqrt{R_1(R_2 - R_1)}$$

Since X_1, the shunt element, can be either an inductance or a capacitance, the series element X_2 must be a capacitance or an inductance. It also follows that if R_1 is greater than R_2, the network must be turned end-for-end and that if R_2 is not a pure resistance or if it is desired that Z_{in} have a reactive component (to provide a conjugate match), adding an additional reactance and/or modifying X_2 will produce the required match.

Figure 3.20 is the Fortran design program for computing the elements of the matching section. The program calculates the element values for either possibility; that is, the series capacitance and shunt inductance or the series inductance and shunt capacitance. Figure 3.21 shows the computed results for three cases:

1. $R_1 = 20.0$ ohms, $R_2 = 300.0$ ohms, and angular frequency $\omega = 5\ 10^6$ rps
2. $R_1 = 200$ ohms, $R_2 = 300$ ohms, and $\omega = 5\ 10^6$ rps
3. $R_1 = 200$ ohms, $R_2 = 300$ ohms, and $\omega = 5\ 10^7$ rps.

```
                 C
                 C       L-C SINGLE FREQUENCY
                 C       MATCHING NETWORK
                 C
S.0001                 1 READ(5,2)R1,R2,W
S.0002                 2 FORMAT(2F10.3,E10.4)
S.0003                   A=SQRT(R1*(R2-R1))
S.0004                   B=R2*SQRT(R1/(R2-R1))
                 C
                 C       SERIES CAPACITANCE AND SHUNT INDUCTANCE
                 C
S.0005                   CSER=1./(W*A)
S.0006                   SHUNTL=B/W
                 C
                 C       SERIES INDUCTANCE AND SHUNT CAPACITANCE
                 C
S.0007                   SERL=A/W
S.0008                   CSHUNT=1./(W*B)
S.0009                   WRITE(6,3)R1,R2,W,CSER,SHUNTL,SERL,CSHUNT
S.0010                 3 FORMAT('1L-C MATCHING SECTION',//,' R1 =',F10.3,' R2 =',F10.3,'  F
                        1REQUENCY =',E10.3,' RADIANS PER SECOND',/////' SERIES CAPACITANCE
                        2AND SHUNT INDUCTANCE L-SECTION',//,' C =',E10.3,4X,'L =',E10.3,///
                        3 / ,' SERIES INDUCTANCE AND SHUNT CAPACITANCE L-SECTION',//,' L ='
                        4,E10.3,4X,'C =',E10.3)
S.0011                   GO TO 1
S.0012                   RETURN
S.0013                   END
```

Figure 3.20 Program for designing L-C matching sections

```
L-C MATCHING SECTION

R1 =     20.000 R2 =     300.000   FREQUENCY = 0.500E 07 RADIANS PER SECOND


SERIES CAPACITANCE AND SHUNT INDUCTANCE L-SECTION

C = 0.267E-08        L = 0.160E-04


SERIES INDUCTANCE AND SHUNT CAPACITANCE L-SECTION

L = 0.150E-04        C = 0.249E-08

L-C MATCHING SECTION

R1 =    200.000 R2 =     300.000   FREQUENCY = 0.500E 07 RADIANS PER SECOND


SERIES CAPACITANCE AND SHUNT INDUCTANCE L-SECTION

C = 0.141E-08        L = 0.849E-04


SERIES INDUCTANCE AND SHUNT CAPACITANCE L-SECTION

L = 0.283E-04        C = 0.471E-09

L-C MATCHING SECTION

R1 =    200.000 R2 =     300.000   FREQUENCY = 0.500E 08 RADIANS PER SECOND


SERIES CAPACITANCE AND SHUNT INDUCTANCE L-SECTION

C = 0.141E-09        L = 0.849E-05


SERIES INDUCTANCE AND SHUNT CAPACITANCE L-SECTION

L = 0.283E-05        C = 0.471E-10
IHC217I
```

Figure 3.21 Three L-C matching sections computed by the program of Figure 3.20

Design of a Lossless Matching Section with a Specified Phase Shift

The L-C matching section discussed above provided a single frequency maximum power match. Another class of lossless matching sections also provides a specified phase shift.[4] One application of this type of section is the driving of the elements of an antenna array where, in addition to the matching feature, a specified phase shift is also required.

Figure 3.22 shows the general form of the desired matching section. Since the network is to be lossless, all the Z's must be purely reactive and may be specified in terms of the two image resistances R_{I1} and R_{I2}

[4] See W. L. Everitt and G. E. Anner, *Communication Engineering* (New York: McGraw-Hill Book Company, 1956, pp. 421–24).

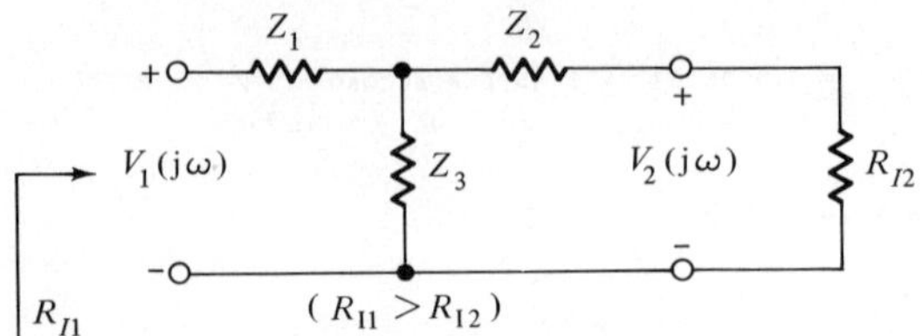

Figure 3.22 Matching lossless T-section with a prescribed phase shift

and the phase shift β as follows:

$$Z_1 = j\frac{R_{I1}}{\sqrt{R}}\left[\frac{1-\sqrt{R}\cos\beta}{\sin\beta}\right] = j\frac{R_{I1}}{C}$$

$$Z_2 = jR_{I1}\left[\frac{\sqrt{R}-\cos\beta}{R\sin\beta}\right] = j\frac{R_{I1}}{A}$$

$$Z_3 = -j\frac{R_{I1}}{\sqrt{R}\sin\beta} = \frac{R_{I1}}{jB}$$

where

$$R = \frac{R_{I1}}{R_{I2}}$$

β = the phase angle between $V_1(j\omega)$ and $V_2(j\omega)$ ("+" for $V_2(j\omega)$ lagging $V_1(j\omega)$ and "−" for $V_2(j\omega)$ leading $V_1(j\omega)$.)

and

$$A = \frac{R\sin\beta}{\sqrt{R}-\cos\beta}$$

$$B = \sqrt{R}\sin\beta$$

$$C = \frac{\sqrt{R}\sin\beta}{1-\sqrt{R}\cos\beta}$$

Since we are matching two resistances, R_{I1} and R_{I2}, one of the reactances must be of opposite sign from the other two. Due to the location of j in the above expression, it is apparent that A and B must be of the same sign. This condition (for $R > 1$) results in the four possible configurations shown in Figure 3.23.

If $V_2(j\omega)$ lags $V_1(j\omega)$, β and $\sin\beta$ will both be positive and this will result in B being positive. Thus Z_3 must be a capacitance for a lagging

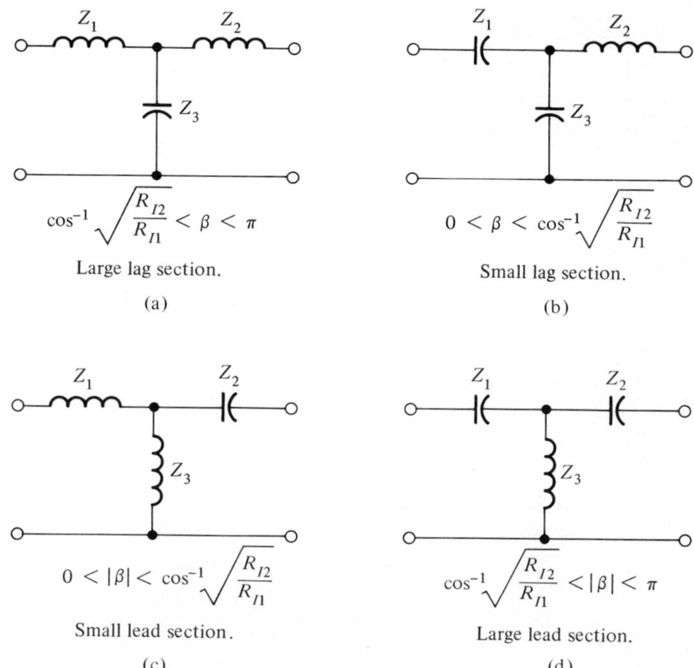

Figure 3.23 The four possible configurations of a lead-lag matching T-section

phase shift. Since $\sqrt{R}$ is greater than 1, A will also be positive and this will result in Z_2 being an inductance. C, however, will be positive or negative depending on the amount of phase lag required. For a large phase lag, C will be positive if $\cos^{-1}\sqrt{(R_{I2}/R_{I1})} < \beta \leq \pi$, and Z_1 will be an inductance as shown in Figure 3.23a. For a small phase lag, C will be negative if $0 < \beta < \cos^{-1}\sqrt{(R_{I2}/R_{I1})}$ and Z_1 will be a capacitance as shown in Figure 3.23b. When $\beta = \cos^{-1}\sqrt{(R_{I2}/R_{I1})}$, $C = \infty$ and Z_1 will be a short circuit so that the T-section will become an L-section.

If $V_2(j\omega)$ leads $V_1(j\omega)$, β and $\sin\beta$ will be negative and this will result in B being negative. Thus Z_3 must be an inductance for a leading phase shift. Since R is greater than 1, and the $\sin\beta$ is negative, A must be negative and this will result in Z_2 being a capacitance. If $0 < |\beta| < \cos^{-1}\sqrt{(R_{I2}/R_{I1})}$, a small phase lead will result and C will be positive. This will result in Z_1 being an inductance as shown in Figure 3.23c. For a large phase lead, C will be negative if $\cos^{-1}\sqrt{(R_{I2}/R_{I1})} < |\beta| \leq \pi$ and Z_1 will be a capacitance as shown in Figure 3.23d. When $|\beta| =$

```
C
C     DESIGN OF AN ASYMMETRICAL
C        MATCHING SECTION
C           FOR A
C     SPECIFIED PHASE SHIFT
C
    1 READ(5,2) RI1,RI2,BETA,W
    2 FORMAT(3F10.3,E10.3)
      WRITE(6,3) RI1,RI2,BETA,W
    3 FORMAT('1ASYMMETRICAL MATCHING SECTION FOR A SPECIFIED PHASE SHIFT
     1',////,' RI1 =',F10.3,4X,'RI2 =',F10.3,4X,'BETA =',F6.2,' DEGREES'
     2,4X,'FREQUENCY =',F10.3)
      R=RI1/RI2
      SR=SQRT(R)
      ARG=BETA/57.3
      A=R*SIN(ARG)/(SR-COS(ARG))
      B=SR*SIN(ARG)
      C=SR*SIN(ARG)/(1.-SR*COS(ARG))
      IF(ARG) 10,20,30
   20 WRITE(6,4)
    4 FORMAT('1BETA = 0')
      GO TO 100
   10 X3=RI1/ABS(B)
      AL3=X3/W
      X2=RI1/ABS(A)
      C2=1./(W*X2)
      IF(C) 40,50,60
   50 WRITE(6,5)
    5 FORMAT('1C = 0')
   40 X1=RI1/ABS(C)
      C1=1./(W*X1)
      WRITE(6,6)C1,C2,AL3
    6 FORMAT(//' C1 =',E10.3,' C2 =',E10.3,' L3 =',E10.3)
      GO TO 100
   60 X1=RI1/C
      AL1=X1/W
      WRITE(6,7)AL1,C2,AL3
    7 FORMAT(//,' L1 =',E10.3,' C2 =',E10.3,' L3 =',E10.3)
      GO TO 100
   30 X2=RI1/A
      AL2=X2/W
      X3=RI1/B
      C3=1./(W*X3)
      IF(C) 70,50,80
   70 X1=RI1/ABS(C)
      C1=1./(W*X1)
      WRITE(6,8)C1,AL2,C3
    8 FORMAT(//,' C1 =',E10.3,' L2 =',E10.3,' C3 =',E10.3)
      GO TO 100
   80 X1=RI1/C
      AL1=X1/W
      WRITE(6,9)AL1,AL2,C3
    9 FORMAT(//,' L1 =',E10.3,' L2 =',E10.3,' C3 =',E10.3)
  100 CONTINUE
      GO TO 1
      RETURN
      END
```

Figure 3.24 Program for designing asymmetrical lossless matching sections with a prescribed amount of phase shift

$\cos^{-1}\sqrt{(R_{I2}/R_{I1})}$, $C = \infty$ and Z_1 will be a short circuit and the T-section will degenerate into the L-section.

Figure 3.24 is the Fortran design program for computing the elements of the matching section. Specifying R_{I1}, R_{I2} (R_{I1} must be larger than R_{I2}), the required phase shift β, and the angular frequency ω, the computer calculates the numerical values of the three required elements. Figure 3.25 shows the computer output for $R_{I1} = 800$ ohms, $R_{I2} = 200$ ohms, $\omega = 5\ 10^6$ radians per second, and $\beta = -12°$, $-65°$, $30°$, and $110°$ respectively. The values of β have been chosen so that all four possible configurations result. Figure 3.26 shows the resulting network and the phasor diagram for the case of $\beta = -12°$.

```
ASYMMETRICAL MATCHING SECTION FOR A SPECIFIED PHASE SHIFT

RI1 =    800.000     RI2 =    200.000     BETA =-12.00 DEGREES     FREQUENCY = 0.500E 07

L1 = 0.368E-03 C2 = 0.203E-09 L3 = 0.385E-03


ASYMMETRICAL MATCHING SECTION FOR A SPECIFIED PHASE SHIFT

RI1 =    800.000     RI2 =    200.000     BETA =-65.00 DEGREES     FREQUENCY = 0.500E 07

C1 = 0.293E-08 C2 = 0.575E-09 L3 = 0.883E-04


ASYMMETRICAL MATCHING SECTION FOR A SPECIFIED PHASE SHIFT

RI1 =    800.000     RI2 =    200.000     BETA = 30.00 DEGREES     FREQUENCY = 0.500E 07

C1 = 0.341E-09 L2 = 0.907E-04 C3 = 0.250E-09


ASYMMETRICAL MATCHING SECTION FOR A SPECIFIED PHASE SHIFT

RI1 =    800.000     RI2 =    200.000     BETA =110.00 DEGREES     FREQUENCY = 0.500E 07

L1 = 0.143E-03 L2 = 0.997E-04 C3 = 0.470E-09
THC217I
```

Figure 3.25 Examples of the four possible configurations of Figure 3.23 computed by the program of Figure 3.24

3.3 LADDER NETWORK CALCULATIONS

The Input Resistance of a Ladder Network

In Section 3.2 we recalled that one of the properties of a symmetrical T- or Pi-section was that if the section was terminated in its characteristic impedance, Z_o, the impedance seen at the other end would also be equal to Z_0. Consider the symmetrical resistive T-section shown in Figure 3.27. For the element values shown, the characteristic resistance will be equal to

$$R_0 = \sqrt{R_1{}^2 + 2R_1R_3} = \sqrt{50^2 + 2(50)(4000)} = 634.43 \text{ ohms}$$

It is apparent that if this section is terminated in other than 634.43 ohms, the input resistance will not be equal to the characteristic resistance. We now investigate the input resistance of a cascade of identical sections that have been terminated in other than the characteristic resistance, as shown in Figure 3.28.

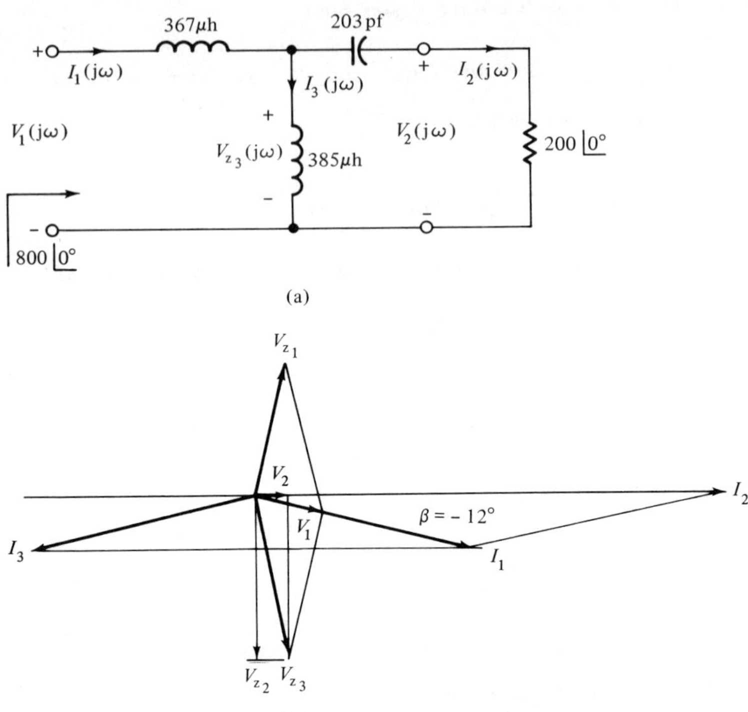

Figure 3.26 The resulting network and phasor diagram for case I of the computer solution shown in Figure 3.25; (a) network; (b) phasor diagram

Figure 3.29 is a Fortran program that calculates the input resistance for a given value of load resistance as a function of the number of cascaded sections. Computer statement S.0002 reads in the value of the load resistance RLOAD, the combined series arm resistance SERR, and the parallel or shunt arm resistance PARR. Computer statements S.0006 to S.0008 make each of the combined series arm resistances equal to SERR and each of the shunt arm resistances equal to PARR. Thus $R_1 = R_2 = RS(I)/2$ and $R_3 = RP(I)$. Computer statement S.0009 next calculates the input resistance seen at the first section, RIN(1). Once this

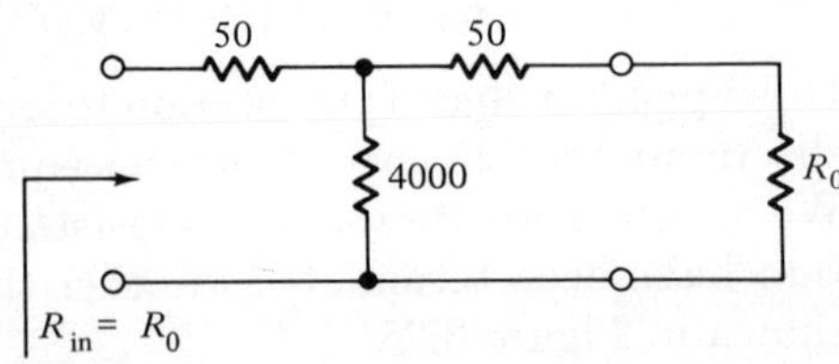

Figure 3.27 A symmetrical resistive T-section having a characteristic resistance of 634.43 ohms

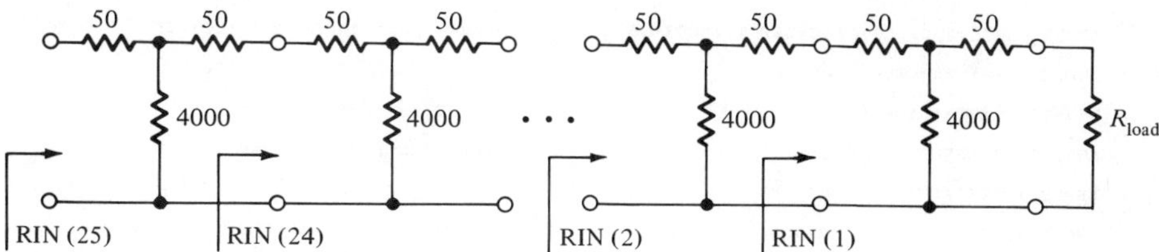

Figure 3.28 A cascade of 25 T-sections

resistance has been found, the DO 20 loop is able to calculate the input resistance of the remaining sections.

Figure 3.30 gives the input resistance at each section for load resistances of 200, 400, 630, 800, and 1000 ohms, respectively. An examination of this computer output reveals that in each case the input resistance seen at successive sections asymptotically approaches the characteristic resistance. It further shows that if the load resistance is less than the characteristic resistance, the input resistance will be less than R_0, while if the load resistance is greater than the characteristic resistance, the input resistance will be greater than the characteristic resistance. Note also that a maximum of 15 sections are required for any of the load resistances investigated before the input resistance is within 1 percent of the characteristic resistance (634.43 ohms).

A practical observation of the above discussion will reveal that if a 50-ohm 20-db pad is inserted between a line and its load, the input impedance seen at the pad will be essentially equal to 50 ohms and will be for all practical purposes independent of the actual load value.

```
C
C         INPUT RESISTANCE OF LADDER NETWORK
C
          DIMENSION RIN(25),RS(25),RP(25)
        1 READ(5,2)RLOAD,SERR,PARR
        2 FORMAT(3F10.2)
          WRITE(6,3)SERR,PARR,RLOAD
        3 FORMAT('1INPUT RESISTANCE OF CASCADED T SECTIONS',//,' SERIES RESI
         1STANCE =',F8.2,//,' SHUNT RESISTANCE =',F8.2,//,' LOAD RESISTANCE
         2=',F8.2)
          DO 10 I=1,25
          RS(I)=SERR
       10 RP(I)=PARR
          RIN(1)=RS(1)/2.+RP(1)*(RS(1)/2.+RLOAD)/(RP(1)+RS(1)/2.+RLOAD)
          WRITE(6,4)RIN(1)
        4 FORMAT('0INPUT RESISTANCE AT SECTION  1 =',F8.2)
          DO 20 I=2,25
          RIN(I)=RS(I)/2.+RP(I)*(RS(I)/2.+RIN(I-1))/(RP(I)+RS(I)/2.+RIN(I-1)
         1)
       20 WRITE(6,5)I,RIN(I)
        5 FORMAT('0INPUT RESISTANCE AT SECTION',I3,' =',F8.2)
          GO TO 1
          RETURN
          END
```

Figure 3.29 A Fortran program for calculating the input resistance of a ladder network

```
INPUT RESISTANCE OF CASCADED T SECTIONS

SERIES RESISTANCE =  100.00

SHUNT RESISTANCE = 4000.00

LOAD RESISTANCE =  200.00

INPUT RESISTANCE AT SECTION  1 =  285.29
INPUT RESISTANCE AT SECTION  2 =  359.36
INPUT RESISTANCE AT SECTION  3 =  421.36
INPUT RESISTANCE AT SECTION  4 =  471.67
INPUT RESISTANCE AT SECTION  5 =  511.48
INPUT RESISTANCE AT SECTION  6 =  542.37
INPUT RESISTANCE AT SECTION  7 =  565.96
INPUT RESISTANCE AT SECTION  8 =  583.76
INPUT RESISTANCE AT SECTION  9 =  597.08
INPUT RESISTANCE AT SECTION 10 =  606.98
INPUT RESISTANCE AT SECTION 11 =  614.30
INPUT RESISTANCE AT SECTION 12 =  619.69
INPUT RESISTANCE AT SECTION 13 =  623.65
INPUT RESISTANCE AT SECTION 14 =  626.55
INPUT RESISTANCE AT SECTION 15 =  628.67
INPUT RESISTANCE AT SECTION 16 =  630.23
INPUT RESISTANCE AT SECTION 17 =  631.36
INPUT RESISTANCE AT SECTION 18 =  632.19
INPUT RESISTANCE AT SECTION 19 =  632.80
INPUT RESISTANCE AT SECTION 20 =  633.24
INPUT RESISTANCE AT SECTION 21 =  633.56
INPUT RESISTANCE AT SECTION 22 =  633.80
INPUT RESISTANCE AT SECTION 23 =  633.97
INPUT RESISTANCE AT SECTION 24 =  634.09
INPUT RESISTANCE AT SECTION 25 =  634.18
```

```
INPUT RESISTANCE OF CASCADED T SECTIONS

SERIES RESISTANCE =  100.00

SHUNT RESISTANCE = 4000.00

LOAD RESISTANCE =  400.00

INPUT RESISTANCE AT SECTION  1 =  454.49
INPUT RESISTANCE AT SECTION  2 =  497.99
INPUT RESISTANCE AT SECTION  3 =  531.96
INPUT RESISTANCE AT SECTION  4 =  558.05
INPUT RESISTANCE AT SECTION  5 =  577.81
INPUT RESISTANCE AT SECTION  6 =  592.64
INPUT RESISTANCE AT SECTION  7 =  603.69
INPUT RESISTANCE AT SECTION  8 =  611.87
INPUT RESISTANCE AT SECTION  9 =  617.90
INPUT RESISTANCE AT SECTION 10 =  622.33
INPUT RESISTANCE AT SECTION 11 =  625.59
INPUT RESISTANCE AT SECTION 12 =  627.97
INPUT RESISTANCE AT SECTION 13 =  629.71
INPUT RESISTANCE AT SECTION 14 =  630.99
INPUT RESISTANCE AT SECTION 15 =  631.92
INPUT RESISTANCE AT SECTION 16 =  632.60
INPUT RESISTANCE AT SECTION 17 =  633.09
INPUT RESISTANCE AT SECTION 18 =  633.45
INPUT RESISTANCE AT SECTION 19 =  633.72
INPUT RESISTANCE AT SECTION 20 =  633.91
INPUT RESISTANCE AT SECTION 21 =  634.05
INPUT RESISTANCE AT SECTION 22 =  634.15
INPUT RESISTANCE AT SECTION 23 =  634.23
INPUT RESISTANCE AT SECTION 24 =  634.28
INPUT RESISTANCE AT SECTION 25 =  634.32
```

Figure 3.30 The input resistance seen at each section for load resistances of 200, 400, 630, 800 and 1000 ohms respectively

Digital-to-Analog Resistive Ladder

The basic accuracy of many digital-to-analog and analog-to-digital converters is based on the design of a binary resistive ladder network. Figure 3.31 shows an R-$2R$ ten-bit ladder network. Each of the ten input voltages $V_1, V_2, \ldots, V_{10}$ is restricted to either a logical "1" or "0" level and would correspond to say 10 volts or 0 volts respectively, depending on the state of the flip-flop the input is connected to. In practice, buffer or level amplifiers are used to connect the flip-flop register to the ladder network so that more precise voltage signals are supplied.

It may be shown that if any one of the inputs is equal to 10 volts

```
INPUT RESISTANCE OF CASCADED T SECTIONS

SERIES RESISTANCE =  100.00

SHUNT RESISTANCE = 4000.00

LOAD RESISTANCE =  630.00

INPUT RESISTANCE AT SECTION  1 =  631.20
INPUT RESISTANCE AT SECTION  2 =  632.07
INPUT RESISTANCE AT SECTION  3 =  632.71
INPUT RESISTANCE AT SECTION  4 =  633.17
INPUT RESISTANCE AT SECTION  5 =  633.51
INPUT RESISTANCE AT SECTION  6 =  633.76
INPUT RESISTANCE AT SECTION  7 =  633.94
INPUT RESISTANCE AT SECTION  8 =  634.07
INPUT RESISTANCE AT SECTION  9 =  634.17
INPUT RESISTANCE AT SECTION 10 =  634.24
INPUT RESISTANCE AT SECTION 11 =  634.29
INPUT RESISTANCE AT SECTION 12 =  634.33
INPUT RESISTANCE AT SECTION 13 =  634.36
INPUT RESISTANCE AT SECTION 14 =  634.38
INPUT RESISTANCE AT SECTION 15 =  634.39
INPUT RESISTANCE AT SECTION 16 =  634.40
INPUT RESISTANCE AT SECTION 17 =  634.41
INPUT RESISTANCE AT SECTION 18 =  634.41
INPUT RESISTANCE AT SECTION 19 =  634.42
INPUT RESISTANCE AT SECTION 20 =  634.42
INPUT RESISTANCE AT SECTION 21 =  634.42
INPUT RESISTANCE AT SECTION 22 =  634.42
INPUT RESISTANCE AT SECTION 23 =  634.43
INPUT RESISTANCE AT SECTION 24 =  634.43
INPUT RESISTANCE AT SECTION 25 =  634.43
```

```
INPUT RESISTANCE OF CASCADED T SECTIONS

SERIES RESISTANCE =  100.00

SHUNT RESISTANCE = 4000.00

LOAD RESISTANCE =  800.00

INPUT RESISTANCE AT SECTION  1 =  751.03
INPUT RESISTANCE AT SECTION  2 =  717.38
INPUT RESISTANCE AT SECTION  3 =  693.86
INPUT RESISTANCE AT SECTION  4 =  677.22
INPUT RESISTANCE AT SECTION  5 =  665.35
INPUT RESISTANCE AT SECTION  6 =  656.82
INPUT RESISTANCE AT SECTION  7 =  650.68
INPUT RESISTANCE AT SECTION  8 =  646.24
INPUT RESISTANCE AT SECTION  9 =  643.02
INPUT RESISTANCE AT SECTION 10 =  640.68
INPUT RESISTANCE AT SECTION 11 =  638.98
INPUT RESISTANCE AT SECTION 12 =  637.74
INPUT RESISTANCE AT SECTION 13 =  636.84
INPUT RESISTANCE AT SECTION 14 =  636.19
INPUT RESISTANCE AT SECTION 15 =  635.71
INPUT RESISTANCE AT SECTION 16 =  635.36
INPUT RESISTANCE AT SECTION 17 =  635.11
INPUT RESISTANCE AT SECTION 18 =  634.93
INPUT RESISTANCE AT SECTION 19 =  634.79
INPUT RESISTANCE AT SECTION 20 =  634.65
INPUT RESISTANCE AT SECTION 21 =  634.62
INPUT RESISTANCE AT SECTION 22 =  634.57
INPUT RESISTANCE AT SECTION 23 =  634.53
INPUT RESISTANCE AT SECTION 24 =  634.50
INPUT RESISTANCE AT SECTION 25 =  634.48
```

Figure 3.30 Continued

(the nth input) and all the other inputs are equal to 0 volts, Figure 3.31 is equivalent to Figure 3.32. The output voltage is therefore given by

$$V_{\text{out}} = \frac{(2^n R)/(2^n - 1)}{2^n R + (2^n R)/(2^n - 1)} 10$$

$$= \frac{1}{2^n} 10$$

and would thus correspond to $(1/2)10 = 5$ volts if the first input (V_1) had been equal to 10 volts and all of the other inputs equal to 0 volts. By using the superposition principle, it readily follows that the total output voltage is the algebraic sum of the individual contributions and

```
INPUT RESISTANCE OF CASCADED T SECTIONS

SERIES RESISTANCE =   100.00

SHUNT RESISTANCE = 4000.00

LOAD RESISTANCE = 1000.00

INPUT RESISTANCE AT SECTION  1 =  881.68

INPUT RESISTANCE AT SECTION  2 =  805.67

INPUT RESISTANCE AT SECTION  3 =  754.88

INPUT RESISTANCE AT SECTION  4 =  720.06

INPUT RESISTANCE AT SECTION  5 =  695.74

INPUT RESISTANCE AT SECTION  6 =  678.56

INPUT RESISTANCE AT SECTION  7 =  666.30

INPUT RESISTANCE AT SECTION  8 =  657.51

INPUT RESISTANCE AT SECTION  9 =  651.18

INPUT RESISTANCE AT SECTION 10 =  646.60

INPUT RESISTANCE AT SECTION 11 =  643.28

INPUT RESISTANCE AT SECTION 12 =  640.87

INPUT RESISTANCE AT SECTION 13 =  639.12

INPUT RESISTANCE AT SECTION 14 =  637.84

INPUT RESISTANCE AT SECTION 15 =  636.92

INPUT RESISTANCE AT SECTION 16 =  636.24

INPUT RESISTANCE AT SECTION 17 =  635.75

INPUT RESISTANCE AT SECTION 18 =  635.39

INPUT RESISTANCE AT SECTION 19 =  635.13

INPUT RESISTANCE AT SECTION 20 =  634.94

INPUT RESISTANCE AT SECTION 21 =  634.80

INPUT RESISTANCE AT SECTION 22 =  634.70

INPUT RESISTANCE AT SECTION 23 =  634.63

INPUT RESISTANCE AT SECTION 24 =  634.57

INPUT RESISTANCE AT SECTION 25 =  634.53

THC217I
```

Figure 3.30 Continued

would be given by

$$V_{\text{out}} = \frac{1}{2} V_1 + \frac{1}{4} V_2 + \frac{1}{8} V_3 + \frac{1}{16} V_4 + \frac{1}{32} V_5 + \frac{1}{64} V_6 + \frac{1}{128} V_7 + \frac{1}{256} V_8 + \frac{1}{512} V_9 + \frac{1}{1024} V_{10}$$

The output voltage therefore corresponds to the analog of the digital register and would have 2^{10} distinct levels ranging in value from 0.0 volts to 9.99023 volts (10.0 − 0.009766 volts) in increments of 0.009766 volts (10/1024 = 0.009766 volts).

If we wanted to know the actual output voltage for a given binary

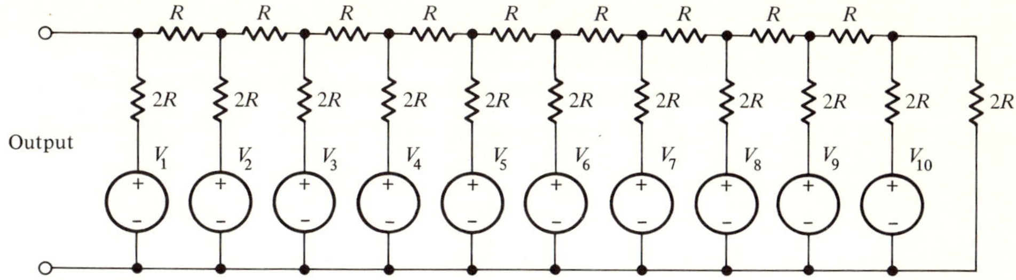

Figure 3.31 The *R-2R* binary resistive ladder

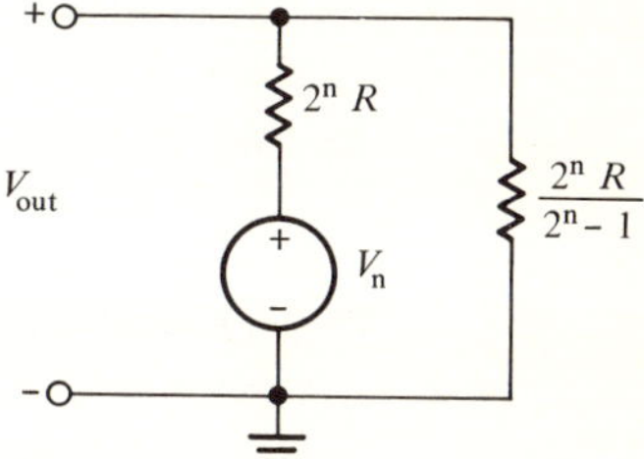

Figure 3.32 The single input equivalent of Figure 3.31

```
C       DIGITAL TO ANALOG
C       RESISTIVE LADDER
C         DESIGN
C
        DIMENSION R(20),V(10),RIN(10),ROUT(10),AN(10),AJ(10)
        READ(5,1) (R(I),I=1,20)
      1 FORMAT(5F10.2)
        WRITE(6,4) (R(I),I=1,20)
      4 FORMAT('1DIGITAL TO ANALOG RESISTIVE LADDER',////,'  R1 =',F7.1'
       1R2 =',F7.1,'  R3 =',F7.1,'  R4 =',F7.1,'  R5 =',F7.1,'  R6 =',F7.1
       2,'  R7 =',F7.1,'  R8 =',F7.1,'  R9 =',F7.1,' R10 =',F7.1,//,' R11
       3=',F7.1,' R12 =',F7.1,' R13 =',F7.1,' R14 =',F7.1,' R15 =',F7.1,'
       4R16 =',F7.1,' R17 =',F7.1,' R18 =',F7.1,' R19 =',F7.1,' R20 =',F7.
       51,////,3X,'V(1)',6X,'V(2)',6X,'V(3)',6X,'V(4)',6X,'V(5)',6X,'V(6)'
       6,6X,'V(7)',6X,'V(8)',6X,'V(9)',6X,'V(10)',4X,'V OUT',//)
      2 READ(5,3) (V(I),I=1,10)
      3 FORMAT(5F10.2)
        RIN(10)=R(20)
        DO 10 I=1,9
        K=11-I
        AN(K)=R(K)*RIN(K)/(R(K)+RIN(K))
     10 RIN(K-1)=R(K+9)+AN(K)
        AN(1)=R(1)*RIN(1)/(R(1)+RIN(1))
        AJ(1)=R(1)
        ROUT(1)=R(1)
        DO 20 I=2,10
        ROUT(I)=R(I+9)+AJ(I-1)
     20 AJ(I)=R(I)*ROUT(I)/(R(I)+ROUT(I))
        RATIO=1.
        VOUT=RIN(1)/(RIN(1)+R(1))*V(1)
        DO 30 I=2,10
        A=ROUT(I)*RIN(I)/(ROUT(I)+RIN(I))
        RATIO=RATIO*AJ(I-1)/(R(I+9)+AJ(I-1))
     30 VOUT=VOUT+A/(A+R(I))*V(I)*RATIO
        WRITE(6,5) (V(I),I=1,10),VOUT
      5 FORMAT(11F10.6)
        GO TO 2
        RETURN
        END
```

Figure 3.33 A Fortran program for computing the output voltage of the binary ladder shown in Figure 3.31

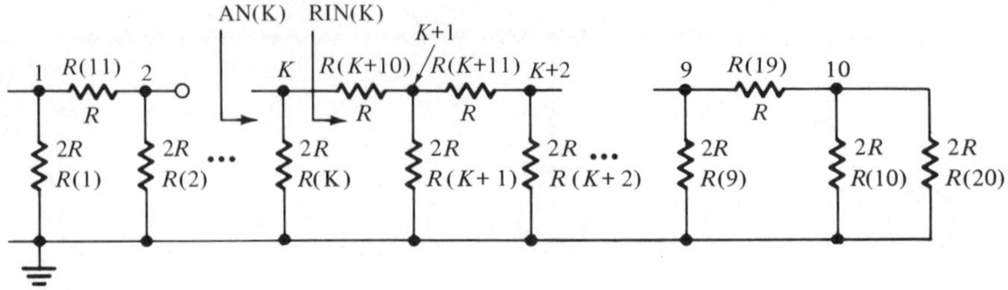

Figure 3.34 *AN(K)* and *R*!N*(K)* are defined

number stored in the flip-flop register, a simple Fortran program could be written that would contain the above output voltage equation and thus, for a given set of input voltages (flip-flop states), the output voltage could be easily calculated.

Figure 3.33 is a Fortran program that computes the output voltage for a given set of input voltages. Unlike the above described procedure, a much more general approach has been used so that we may later (Chapter 10) investigate the effect of ladder resistance variation (or input voltage differences from 0 or 10 volts) on the output voltage. The program first reads in and then prints out the 20 ladder resistances. The 10 input voltages are next read into memory. The DO 10 loop then calculates $AN(K)$ and $R\text{IN}(K-1)$ which are defined in Figure 3.34. Note that $AN(1)$ and $R\text{IN}(1)$ must be calculated by separate statements and that for $R = 1000$ ohms, $R\text{IN}(K)$ will be equal to 2000 ohms, independent of the value of K. The DO 20 loop then calculates $R\text{OUT}(I)$ and $AJ(I)$ which are defined in Figure 3.35. Note that $AJ(1)$ and $R\text{OUT}(1)$ must be calculated by other means.

The DO 30 loop and computer statement S.0020 are now used to compute the actual value of the output voltage. The contribution of $V(1)$ must be handled as a special case and we see from computer statement S.0020 that the contribution is given by "RIN(1)/(RIN(1) + R(1))*V(1)." Since, in this example (the value of the R's in the ladder),

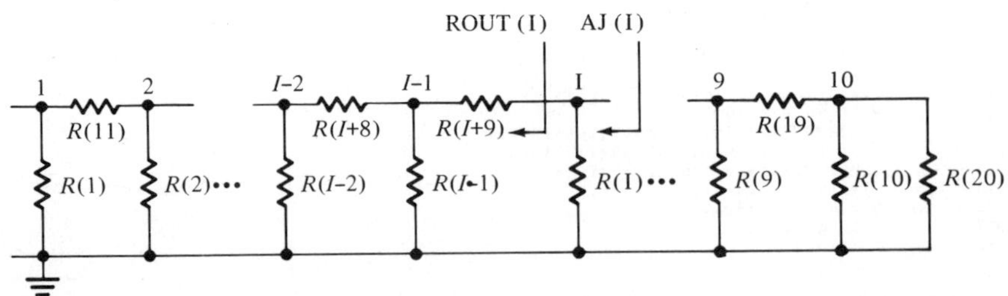

Figure 3.35 *AJ(I)* and *ROUT(I)* are defined

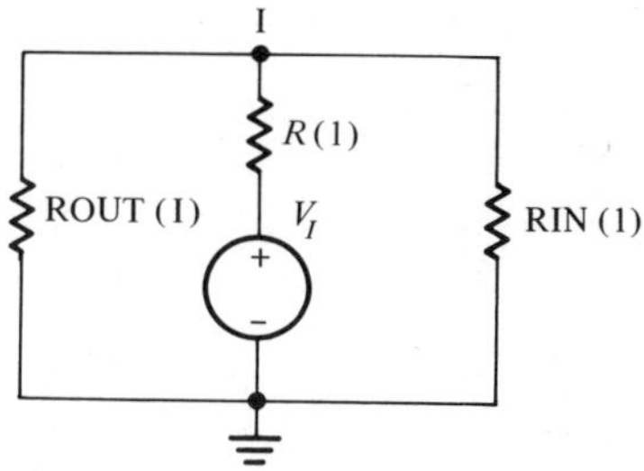

Figure 3.36 Calculation of the node voltage

RIN(1) and R(1) both equal 2000 ohms, the contribution would be equal to $\frac{1}{2}V(1)$. That is, if $V(1) = 10$ volts, the contribution to VOUT would be 5 volts, while if $V(1) = 0$ volts, the contribution to VOUT would be 0 volts.

Entering the DO 30 loop, computer statement S.0022 calculates the parallel combination of the two resistors (ROUT(I) and RIN(I)) looking to the left and right of node I, as shown in Figure 3.36. The voltage at node I is then given by

$$\frac{A}{A + R(I)}\, V(I)$$

In order to determine the contribution to VOUT, this voltage is multiplied by RATIO, as shown in computer statement S.0024. Note that RATIO is modified in each iteration. For example, initially when $I = 2$, RATIO $= AJ(1)/(R(11) + AJ(1))$. This is the contribution (the fraction) of the voltage appearing at node 2 to VOUT. While when $I = 3$, RATIO $= (AJ(1)/(R(11) + AJ(1)))*(AJ(2)/(R(12) + AJ(2))$ and this would be the contribution of the voltage appearing at node 3 to VOUT, and so on.

Figure 3.37 shows the computer output for several different flip-flop storage conditions. Note that this program would permit one to determine the change in the output voltage due to a change in any one of the ladder resistors from their nominal values. This problem will be considered in more detail in Chapter 10.

The Calculation of the Driving-Point Impedance

The calculation of the driving-point impedance of a ladder network may be readily organized for iterative computer computation. Consider the ladder network shown in Figure 3.38 in which the series arms have been expressed as impedances with even subscripts and the shunt arms as admittances with odd subscripts. In order to calculate the driving-point impedance, we may start at the load or right-hand port and take the

DIGITAL TO ANALOG RESISTIVE LADDER

R1 = 2000.0 R2 = 2000.0 R3 = 2000.0 R4 = 2000.0 R5 = 2000.0 R6 = 2000.0 R7 = 2000.0 R8 = 2000.0 R9 = 2000.0 R10 = 2000.0

R11 = 1000.0 R12 = 1000.0 R13 = 1000.0 R14 = 1000.0 R15 = 1000.0 R16 = 1000.0 R17 = 1000.0 R18 = 1000.0 R19 = 1000.0 R20 = 2000.0

V (1)	V (2)	V (3)	V (4)	V (5)	V (6)	V (7)	V (8)	V (9)	V (10)	V OUT
0.0	0.0	0.0	0.0	0.0	0.0	0.0	0.0	0.0	10.000000	0.009766
0.0	0.0	0.0	0.0	0.0	0.0	0.0	0.0	10.000000	0.0	0.019531
0.0	0.0	0.0	0.0	0.0	0.0	0.0	0.0	10.000000	10.000000	0.029297
0.0	0.0	0.0	0.0	0.0	0.0	0.0	10.000000	0.0	0.0	0.039062
0.0	0.0	0.0	0.0	0.0	0.0	0.0	10.000000	0.0	10.000000	0.048828
0.0	10.000000	0.0	0.0	0.0	0.0	0.0	0.0	0.0	0.0	2.499999
10.000000	0.0	0.0	0.0	0.0	0.0	0.0	0.0	0.0	0.0	5.000000

IHC217I

Figure 3.37 Output ladder voltage determined by the program of Figure 3.33

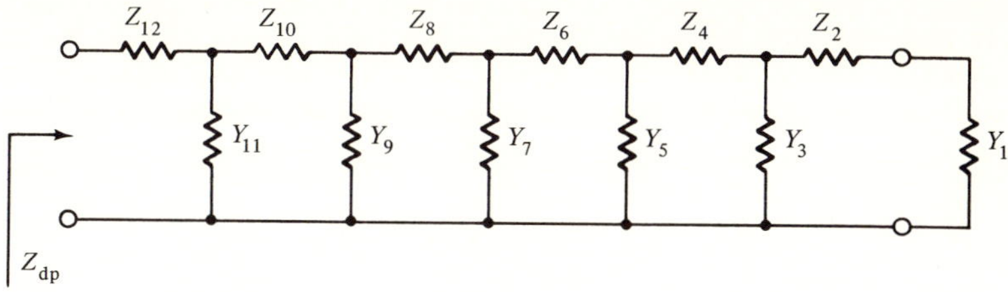

Figure 3.38 Calculation of the driving-point impedance of a ladder network

reciprocal of Y_1. This impedance may now be added to Z_2 and, in turn, the reciprocal of this sum can be added to Y_3. The reciprocal of the sum of these two admittances can now be found and added to Z_4. This procedure is continued until all elements of the network have been considered and at this point the resulting impedance will be equal to the driving-point impedance. Mathematically the above procedure can be expressed as a continued fraction expansion as follows:

$$Z_{dp} = Z_{12} + \cfrac{1}{Y_{11} + \cfrac{1}{Z_{10} + \cfrac{1}{Y_9 + \cfrac{1}{\ddots \; Z_4 + \cfrac{1}{Y_3 + \cfrac{1}{Z_2 + \cfrac{1}{Y_1}}}}}}}$$

In order to program the above procedure, we must realize that if our computer is not capable of performing complex arithmetic, the Z's and the Y's must be computed in such a way that only real numbers are involved. Recognizing this point and also recognizing the need to convert impedances to admittances and admittances to impedances, we can define these two conversions as follows:

$$Z = \frac{1}{Y} = \frac{1}{G + jB} = \frac{G}{G^2 + B^2} - j\frac{B}{G^2 + B^2}$$

$$Y = \frac{1}{Z} = \frac{1}{R + jX} = \frac{R}{R^2 + X^2} - j\frac{X}{R^2 + X^2}$$

We may now take advantage of the fact that we have used only even subscripts for the impedances and odd subscripts for the admittances and make use of the unused subscripts to store the temporary results of combining impedances that are in series and admittances that are in parallel.

Since the basic procedure requires the conversion of an admittance to an impedance and then the addition of this impedance to another impedance and then the conversion of this resulting impedance to an admittance and adding this to another admittance, these two steps can be written in terms of a general iterative procedure, and except for the starting and ending, this iterative loop will serve to do the bulk of the computation. Thus, to start the procedure, we may write:

```
ZR(1)=ZR(2)+YR(1)/(YR(1)**2+YJ(1)**2)
ZJ(1)=ZJ(2)-YJ(1)/(YR(1)**2+YJ(1)**2)
YR(2)=YR(3)+ZR(1)/(ZR(1)**2+ZJ(1)**2)
YJ(2) = YJ(3)-ZJ(1)/(ZR(1)**2+ZJ(1)**2)
```

where $ZR(1)$ and $ZJ(1)$ are the real and imaginary parts of the sum of the two impedances Z_2 and the reciprocal of Y_1. The terms $YR(2)$ and $YJ(2)$ are the real and imaginary parts of the sum of the two admittances Y_3 and the reciprocal of Z_1 ($Z_1 = ZR(1) + j\,ZJ(1)$). At this point we may enter the iteration loop and compute the two basic steps by executing the following computer statements

```
   DO 10 I=3,9,2
   ZR(I)=ZR(I+1)+YR(I-1)/(YR(I-1)**2+YJ(I-1)**2)
   ZJ(I)=ZJ(I+1)-YJ(I-1)/(YR(I-1)**2+YJ(I-1)**2)
   YR(I+1)=YR(I+2)+ZR(I)/(ZR(I)**2+ZJ(I)**2)
10 YJ(I+1)=YJ(I+2)-ZJ(I)/(ZR(I)**2+ZJ(I)**2)
```

In the last pass of this loop the values of $YR(10)$ and $YJ(10)$ are found. These two values are equal to the real and imaginary parts of the admittance looking to the right of the admittance Y_{11} in Figure 3.38. The following two statements may now be used to terminate the procedure and thus obtain the value of the driving-point impedance.

```
ZRDP=ZR(12)+YR(10)/(YR(10)**2+YJ(10)**2)
ZJDP=ZJ(12)-YJ(10)/(YR(10)**2+YJ(10)**2)
```

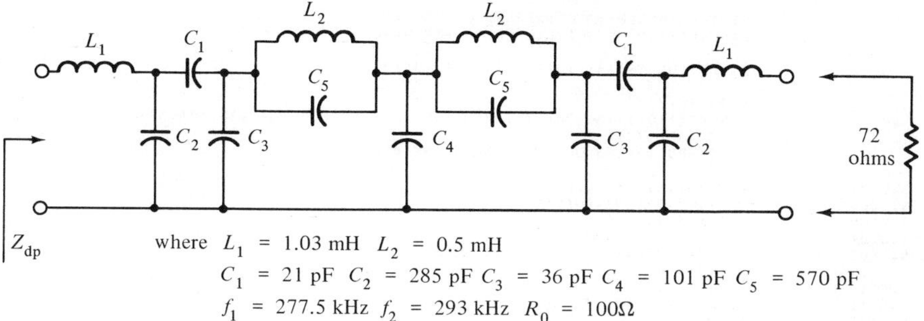

Figure 3.39 A 280–290 kHz bandpass filter

The reader is urged to go through the above statements step-by-step in order to see that this procedure may be readily changed so that it will be able to compute the driving-point impedance independent of the length of the actual ladder network under consideration.

Let us make use of the above procedure to find the driving-point impedance of the filter network shown in Figure 3.39.[5] This is a bandpass filter whose passband covers the frequency range of 280 kHz to 290 kHz, with a high rejection requirement in the vicinity of 296 kHz. The actual design of this filter is beyond the scope of this discussion, but it is helpful in terms of better understanding the following problem to realize that the individual sections of this filter were designed to take into consideration the problems of mismatch and noninfinite element Q's. For these reasons the infinite attenuation peak was made equal to 298 kHz, the lower cutoff frequency f_1 was made equal to 277.5 kHz, the upper cutoff frequency was made equal to 293.0 kHz, and the nominal or characteristic resistance R_0 was made equal to 100 ohms.

The final form of this filter was used in an actual communication system where it was driven by another bandpass filter whose passband was 280 kHz to 296 kHz. In reality then, this filter was the actual load of the 280–296 kHz filter. It was therefore necessary to determine the driving-point impedance of this filter before the effects of mismatch and the insertion loss characteristics of the 280–296 kHz filter could be determined.

Figure 3.40 shows a Fortran program for computing the driving-point impedance of this filter. Initially, computer statements S.0002 and S.0003 print the heading and subheadings. Computer statement S.0004 then sets ω equal to 2π times 269 kHz. The next statement specifies the frequency increment as 2π times 1 kHz. The computer statements S.0006

[5] For a more detailed discussion of this filter see B. James Ley, Samuel G. Lutz, and Charles F. Rehberg, *Linear Circuit Analysis* (New York: McGraw-Hill Book Company, 1959), pp. 555–58.

```
         C        FILTER PROBLEM   DISSIPATICNLESS CASE
         C        CALCULATION OF DRIVING PCINT IMPEDANCE
         C
S.0001            DIMENSION YR(13),YJ(13),ZR(13),ZJ(13)
         C
S.0002            WRITE(6,30)
S.0003         30 FORMAT('1DRIVING POINT IMPEDANCE'//6X,'FREQUENCY',1X,'MAGNITUDE',5
                 1X,'PHASE',//)
         C        ASSIGNED VALUES OF IMPEDANCES AND ADMITTANCES
         C
S.0004            W=2.*3.14159*.263E+6
S.0005            DW=2.*3.14159*.001E+6
S.0006            DO 40 K=1,53
S.0007            W=W+DW
S.0008            YR(1)=1./72.
S.0009            YJ(1)=0.0
S.0010            ZR(2)=0.0
S.0011            ZJ(2)=W*(1.03E-3)
S.0012            YR(3)=0.0
S.0013            YJ(3)=W*(285.E-12)
S.0014            ZR(4)=0.0
S.0015            ZJ(4)=-1.0/(W*21.E-12)
S.0016            YR(5)=0.0
S.0017            YJ(5)=W*(36.E-12)
S.0018            ZR(6)=0.0
S.0019            ZJ(6)=-(0.5E-3/570.E-12)/(W*0.5E-3-1./(W*570.E-12))
S.0020            YR(7)=0.0
S.0021            YJ(7)=W*(101.E-12)
S.0022            ZR(8)=0.0
S.0023            ZJ(8)=ZJ(6)
S.0024            YR(9)=0.0
S.0025            YJ(9)=YJ(5)
S.0026            ZR(10)=0.0
S.0027            ZJ(10)=ZJ(4)
S.0028            YR(11)=0.0
S.0029            YJ(11)=YJ(3)
S.0030            ZR(12)=0.0
S.0031            ZJ(12)=ZJ(2)
S.0032            YR(13)=YR(1)
S.0033            YJ(13)=0.0
          C
          C       CALCULATION OF DRIVING PCINT IMPEDANCE
          C
S.0034            ZR(1)=ZR(2)+YR(1)/(YR(1)**2+YJ(1)**2)
S.0035            ZJ(1)=ZJ(2)-YJ(1)/(YR(1)**2+YJ(1)**2)
S.0036            YR(2)=YR(3)+ZR(1)/(ZR(1)**2+ZJ(1)**2)
S.0037            YJ(2)=YJ(3)-ZJ(1)/(ZR(1)**2+ZJ(1)**2)
S.0038            DO 20 I=3,9,2
S.0039            ZR(I)=ZR(I+1)+YR(I-1)/(YR(I-1)**2+YJ(I-1)**2)
S.0040            ZJ(I)=ZJ(I+1)-YJ(I-1)/(YR(I-1)**2+YJ(I-1)**2)
S.0041            YR(I+1)=YR(I+2)+ZR(I)/(ZR(I)**2+ZJ(I)**2)
S.0042     20     YJ(I+1)=YJ(I+2)-ZJ(I)/(ZR(I)**2+ZJ(I)**2)
S.0043            ZRDP=ZR(12)+YR(10)/(YR(10)**2+YJ(10)**2)
S.0044            ZJDP=ZJ(12)-YJ(10)/(YR(10)**2+YJ(10)**2)
S.0045            ZDPM=(ZRDP**2+ZJDP**2)**.5
S.0046            ZDPP= ATAN(ZJDP/ZRDP)*180./3.14159
S.0047            F=W/(2.*3.14159)
S.0048         40 WRITE(6,50)F,ZDPM,ZDPP
S.0049         50 FORMAT(E15.3,2F10.2)
S.0050            RETURN
S.0051            END
```

Figure 3.40 A Fortran program for computing the driving-point impedance of the bandpass filter of Figure 3.39

to S.0048 make up the overall DO loop which computes the driving-point impedance for 53 different frequencies. In each iteration, ω is incremented by 2π times 1 kHz by the execution of computer statement S.0007 (in the first pass $\omega = 2\pi*270\ 10^3$). Computer statements S.0008 to S.0033 specify the actual Z's and Y's as they were defined in Figures 3.38 and 3.39. Computer statements S.0034–S.0037 start the continued fraction procedure discussed above, and S.0038–S.0042 are the iteration loop statements of the procedure. Computer statements S.0043 through S.0046 terminate the procedure yielding the magnitude and phase of the driving-point impedance.

Figure 3.41 indicates the variation of the magnitude and phase of the driving-point impedance as a function of frequency. Recalling that the passband of this filter was from 280 to 290 kHz, note the variation of the magnitude of the impedance in the 280–296 kHz range (this is the passband of the driving filter). Also note that the phase of the impedance far below the passband is essentially equal to $-90°$ and far above the passband is $+90°$, and thus the impedance in these two regions corresponds to one that is purely imaginary.

DRIVING POINT IMPEDANCE

FREQUENCY	MAGNITUDE	PHASE	FREQUENCY	MAGNITUDE	PHASE
0.264E 06	256.29	-89.86	0.291E 06	67.75	22.40
0.265E 06	241.26	-89.83	0.292E 06	67.71	-3.74
0.266E 06	226.17	-89.78	0.293E 06	25.61	77.03
0.267E 06	210.97	-89.72	0.294E 06	60.88	89.56
0.268E 06	195.64	-89.64	0.295E 06	84.60	89.97
0.269E 06	180.14	-89.51	0.296E 06	104.02	90.00
0.270E 06	164.42	-89.33	0.297E 06	121.38	90.00
0.271E 06	148.41	-89.06	0.298E 06	137.52	90.00
0.272E 06	132.01	-88.62	0.299E 06	152.85	90.00
0.273E 06	115.13	-87.88	0.300E 06	167.61	90.00
0.274E 06	97.63	-86.54	0.301E 06	181.93	90.00
0.275E 06	79.46	-83.89	0.302E 06	195.92	90.00
0.276E 06	60.89	-78.04	0.303E 06	209.63	90.00
0.277E 06	43.91	-63.50	0.304E 06	223.11	90.00
0.278E 06	36.79	-31.90	0.305E 06	236.40	90.00
0.279E 06	49.07	-4.70	0.306E 06	249.52	90.00
0.280E 06	68.63	0.50	0.307E 06	262.51	90.00
0.281E 06	79.45	-3.29	0.308E 06	275.36	90.00
0.282E 06	78.15	-5.39	0.309E 06	288.10	90.00
0.283E 06	72.59	-1.12	0.310E 06	300.74	90.00
0.284E 06	71.33	8.78	0.311E 06	313.28	90.00
0.285E 06	78.97	19.00	0.312E 06	325.74	90.00
0.286E 06	95.33	23.56	0.313E 06	338.11	90.00
0.287E 06	114.54	19.31	0.314E 06	350.42	90.00
0.288E 06	118.99	6.78	0.315E 06	362.65	90.00
0.289E 06	94.18	-3.56	0.316E 06	374.82	90.00
0.290E 06	66.62	4.31			

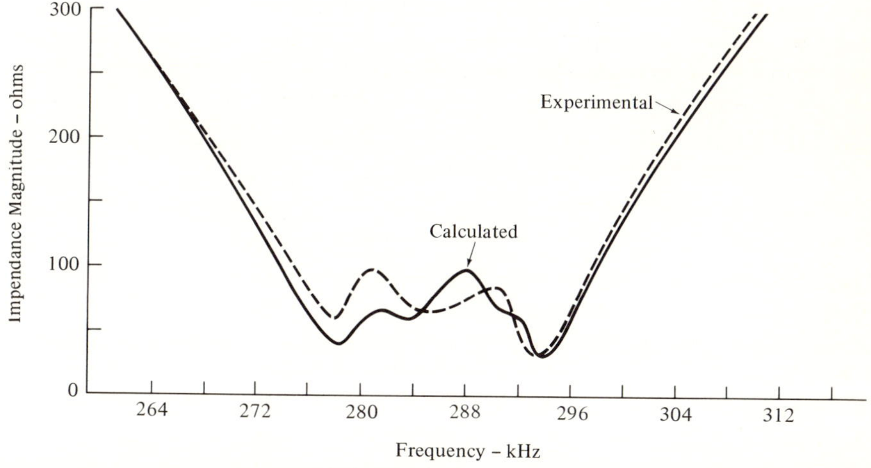

Figure 3.41 The magnitude and phase of the driving-point impedance of the bandpass filter of Figure 3.39

The Calculation of the Dissipationless Insertion Loss

In the last problem, we pointed out that the 280–290 kHz filter was actually the load or termination of another bandpass filter (280–296 kHz), and in order to determine the effects of mismatch and the insertion loss characteristics of the driving 280–296 kHz filter, the driving-point impedance (the load impedance of the 280–296 kHz filter) would first have to be determined as a function of frequency. Rather than turn to a new filter, and one in which the load impedance is the driving-point impedance of another filter, let us illustrate the technique of computing the insertion loss by using the 280–290 kHz filter with a constant load termination of 72 ohms.

Referring to Figure 3.42 we see that the insertion loss can be measured at a given frequency by first connecting the load to the generator and measuring the load voltage and then inserting the filter between the load and the generator and measuring the load voltage under this condition. The dB insertion loss is then calculated from:

$$\text{I.L.} = 20 \log_{10} \frac{V_1}{V_2}$$

This corresponds to the difference (loss) in power that would be delivered to the load if the load were connected directly to the generator as compared to the power delivered to the load with the filter inserted.

In order to organize a simple iterative procedure that can be used to compute the insertion loss, let us use an alternate method of defining the insertion loss as shown in Figure 3.43. Assuming a 1-volt load voltage, we first calculate the generator current with the load connected directly to the generator. This will be equal to a generator current of $2Y_0$ amperes. The filter is now inserted between the generator and the load, and again we compute the generator current required to produce a 1-volt load voltage. The dB insertion loss will then be given by

$$\text{I.L.} = 20 \log_{10} \frac{I_{G2}}{I_{G1}}$$

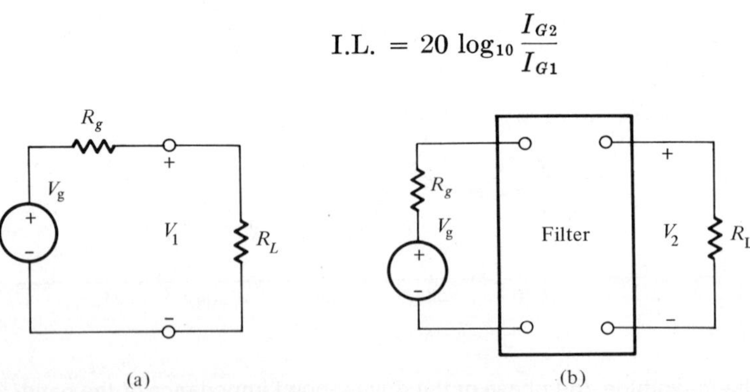

Figure 3.42 The measurement of insertion loss

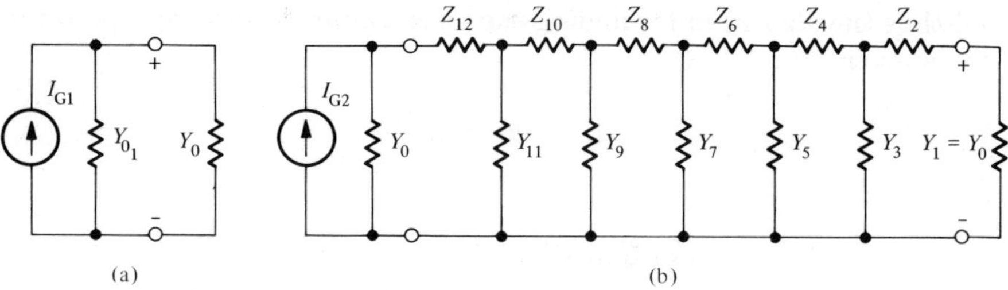

Figure 3.43 An alternate method for measuring insertion loss

The reader should verify that in a linear system this definition is equivalent to the voltage definition.

We now turn to the problem of developing an algorithm for calculating the insertion loss. Note that in the circuit of Figure 3.43b, the series arms have been expressed as impedances with even subscripts and the shunt arms as admittances with odd subscripts. Starting with a load voltage $V_1 = 1$ volt, we first calculate the load current $I_1 = Y_1V_1$. This current also flows through Z_2 and the voltage across Z_2 can now be computed from $V_2 = Z_2I_2$. At this point we can calculate the voltage across Y_3 from $V_3 = V_1 + V_2$ and in turn find the current $I_3 = Y_3V_3$. We are now in a position to calculate the current in the impedance Z_4 ($I_4 = I_2 + I_3$) and in turn the voltage across Z_4 ($V_4 = Z_4I_4$). This now permits us to calculate the voltage across Y_5 and the current I_5, and so on. The generator current can thus be found by the successive evaluation of the following equations:

$$
\begin{aligned}
I_1 &= Y_1V_1 = Y_0 \\
V_2 &= Z_2I_2 \\
V_3 &= V_1 + V_2 \\
I_3 &= Y_3V_3 \\
I_4 &= I_2 + I_3 \\
V_4 &= Z_4I_4 \\
V_5 &= V_3 + V_4 \\
&\;\;\vdots \\
V_{11} &= V_9 + V_{10} \\
I_{11} &= Y_{11}V_{11} \\
I_{12} &= I_{10} + I_{11} \\
V_{12} &= Z_{12}V_{12} \\
V_{in} &= V_{11} + V_{12} \\
I_{Y_0} &= Y_0V_{in} \\
I_{G2} &= I_{Y_0} + I_{12}
\end{aligned}
$$

and thus the algorithm for finding the insertion loss is developed and the loss is given by

$$\text{I.L.} = 20 \log_{10} \frac{I_{G2}}{I_{G1}}$$

```
C
C      FILTER PROBLEM   DISSIPATIONLESS CASE
C      CALCULATION OF INSERTION LOSS
C
       DIMENSION YR(13),YJ(13),ZR(13),ZJ(13),CR(13),CJ(13),VR(13),VJ(13)
       WRITE(6,10)
    10 FORMAT('1INSERTION LOSS CALCULATIONS'//6X,'FREQUENCY',3X,'DB LOSS'
      1//)
       W=2.*3.14159*.263E+6
       DW=2.*3.14159*.001E+6
C
C      ASSIGNED VALUES OF IMPEDANCES AND ADMITTANCES
C
       DO 20 K=1,53
       W=W+DW
       YR(1)=1./72.
       YJ(1)=0.0
       ZR(2)=0.0
       ZJ(2)=W*(1.03E-3)
       YR(3)=0.0
       YJ(3)=W*(285.E-12)
       ZR(4)=0.0
       ZJ(4)=-1.0/(W*21.E-12)
       YR(5)=0.0
       YJ(5)=W*(36.E-12)
       ZR(6)=0.0
       ZJ(6)=-(0.5E-3/570.E-12)/(W*0.5E-3-1./(W*570.E-12))
       YR(7)=0.0
       YJ(7)=W*(101.E-12)
       ZR(8)=0.0
       ZJ(8)=ZJ(6)
       YR(9)=0.0
       YJ(9)=YJ(5)
       ZR(10)=0.0
       ZJ(10)=ZJ(4)
       YR(11)=0.0
       YJ(11)=YJ(3)
       ZR(12)=0.0
       ZJ(12)=ZJ(2)
       YR(13)=YR(1)
       YJ(13)=0.0
C
C      CALCULATION OF INSERTION LOSS
C
       CR(2)=1./72.
       CJ(2)=0.0
       VJ(2)=(1./72.)*ZJ(2)
       VR(2)=(1./72.)*ZR(2)
       VR(3)=1.+VR(2)
       VJ(3)=VJ(2)
       CR(3)=VR(3)*YR(3)-VJ(3)*YJ(3)
       CJ(3)=VR(3)*YJ(3)+VJ(3)*YR(3)
       DO 30 J=4,12,2
       CR(J)=CR(J-2)+CR(J-1)
       CJ(J)=CJ(J-2)+CJ(J-1)
       VR(J)=CR(J)*ZR(J)-CJ(J)*ZJ(J)
       VJ(J)=CR(J)*ZJ(J)+CJ(J)*ZR(J)
       VR(J+1)=VR(J)+VR(J-1)
       VJ(J+1)=VJ(J)+VJ(J-1)
       CR(J+1)=VR(J+1)*YR(J+1)-VJ(J+1)*YJ(J+1)
 30    CJ(J+1)=VR(J+1)*YJ(J+1)+VJ(J+1)*YR(J+1)
       CRA=CR(13)+CR(12)
       CJA=CJ(13)+CJ(12)
       C=(CRA**2+CJA**2)**0.5
       DBL=(20./2.30259)*ALOG(C/(2./72.))
       F=W/(2.*3.14159)
    20 WRITE(6,40)F,DBL
    40 FORMAT(E15.3,F10.3)
       RETURN
       END
```

Figure 3.44 A Fortran program for computing the dissipationless insertion loss

INSERTION LOSS CALCULATIONS

FREQUENCY	DB LOSS	FREQUENCY	DB LOSS
0.264E 06	25.983	0.291E 06	0.178
0.265E 06	24.822	0.292E 06	0.009
0.266E 06	23.589	0.293E 06	6.050
0.267E 06	22.273	0.294E 06	18.199
0.268E 06	20.864	0.295E 06	29.588
0.269E 06	19.350	0.296E 06	41.513
0.270E 06	17.714	0.297E 06	56.933
0.271E 06	15.941	0.298E 06	98.760
0.272E 06	14.011	0.299E 06	68.490
0.273E 06	11.906	0.300E 06	58.261
0.274E 06	9.619	0.301E 06	53.584
0.275E 06	7.176	0.302E 06	50.927
0.276E 06	4.693	0.303E 06	49.281
0.277E 06	2.456	0.304E 06	48.222
0.278E 06	0.888	0.305E 06	47.537
0.279E 06	0.166	0.306E 06	47.103
0.280E 06	0.003	0.307E 06	46.844
0.281E 06	0.014	0.308E 06	46.713
0.282E 06	0.017	0.309E 06	46.676
0.283E 06	0.001	0.310E 06	46.709
0.284E 06	0.026	0.311E 06	46.797
0.285E 06	0.133	0.312E 06	46.926
0.286E 06	0.282	0.313E 06	47.088
0.287E 06	0.366	0.314E 06	47.275
0.288E 06	0.287	0.315E 06	47.481
0.289E 06	0.082	0.316E 06	47.703
0.290E 06	0.013		

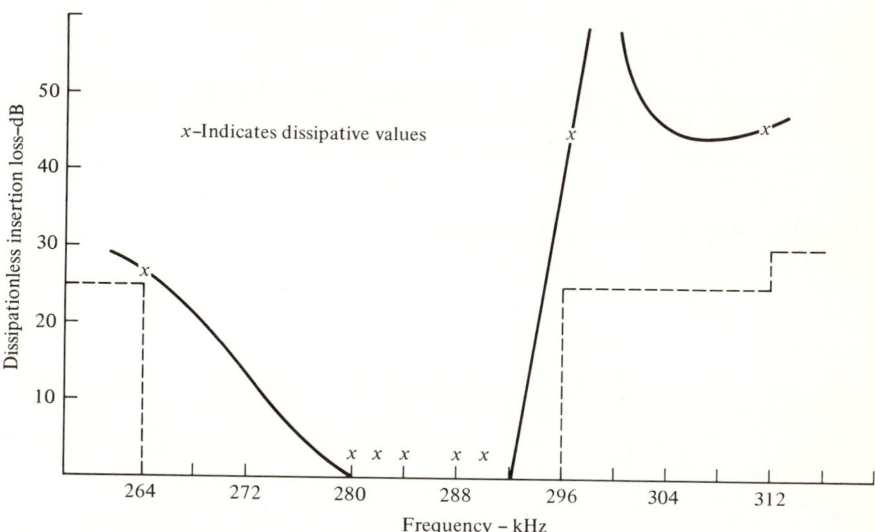

Figure 3.45 The dissipationless insertion loss of the filter of Figure 3.39

Figure 3.44 shows a Fortran program for computing the dissipationless insertion loss. Computer statements S.0002 and S.0003 print the heading and subheadings. Computer statement S.0004 initially sets ω equal to 2π times 263 kHz. The next statement specifies the frequency increment as 2π times 1 kHz. Computer statements S.0006–S.0056 make up the overall DO loop that computes the insertion loss from 264 kHz to 316 kHz. Each iteration ω is incremented by 2π times 1 kHz by the execution of computer statement S.0007. Computer statements S.0008 to S.0033 specify the actual Z's and Y's as they are defined in Figures 3.39 and 3.43b. Computer statements S.0034–S.0041 start the iteration procedure for calculating the generator current. Computer statements S.0042 to S.0050 are the iteration loop statements of the procedure and computer statements S.0051–S.0056 terminate the process, yielding the dB insertion loss.

Figure 3.45 indicates the variation of the insertion loss as a function of frequency. Note that the dissipationless insertion loss meets the general requirements of the reject band; a net loss of at least 25 dB below 264 kHz and a net loss of at least 25 dB from 296 to 312 kHz and at least 30 dB from 312 kHz to 8 MHz.

All of the previous discussion is based on lossless inductors and capacitors. Although capacitors essentially satisfy the lossless condition, inductors are usually found to be lossy as well as nonlinear devices. In many designs the lossless behavior will correspond very closely to the lossy behavior. For example, Figures 3.41 and 3.45 show the actual measured departure from the theoretical dissipationless curves when coil Q's of 300 or greater (over the frequency range) were used. In most designs, however, losses cannot be ignored since it is not always possible to obtain coils with high enough Q's. The next section will show how the minimum value of Q's can be found.

The Calculation of Dissipative Insertion Loss

The last section pointed out that the requirements of our 280–290 kHz bandpass filter in the reject band were a net loss of at least 25 dB below 264 kHz and between 296 kHz and 312 kHz and at least 30 dB from 312 kHz to 8 MHz. The results shown in Figure 3.45 reveal that the filter meets these requirements. We must realize, however, that although we have spent considerable time in designing the filter and computing the insertion loss, the design is far from completed because we have assumed the use of lossless inductors and capacitors. When we examine the passband requirements, it is possible that this design will have to be scrapped because of the inability of industry to build inductors with sufficiently high Q's.

In practice we can usually assume that the capacitors will satisfy the lossless condition and the inductors will not. It will therefore be necessary to determine the lowest value Q coils that can be used to satisfy the bandpass requirements. We should realize that the higher the Q required, the larger and more expensive the coil will be and the more difficult it will be to build. For example, the physical limit of building a 4 micro-

```
C         FILTER PROBLEM DISSIPATIVE CASE
C         CALCULATION OF INSERTION LOSS
C
          DIMENSION Q(6)
          DIMENSION YR(13),YJ(13),ZR(13),ZJ(13),CR(13),CJ(13),VR(13),VJ(13)
          READ(5,5) (Q(I),I=1,6)
        5 FORMAT(6F10.2)
          WRITE(6,10)
       10 FORMAT('1INSERTION LOSS CALCULATIONS'//6X,'FREQUENCY',3X,'DB LOSS'
         1//)
          DW=2.*3.14159*.001E+6
          DO 20 I=1,6
          W=2.*3.14159*.203E+6
          WRITE(6,15)Q(I)
       15 FORMAT(///,' Q =',F8.2,//)
C
C         ASSIGNED VALUES OF IMPEDANCES AND ADMITTANCES
C
          DO 20 K=1,53
          W=W+DW
          YR(1)=1./72.
          YJ(1)=0.0
          ZR(2)=W*1.03E-03/Q(I)
          ZJ(2)=W*(1.03E-3)
          YR(3)=0.0
          YJ(3)=W*(265.E-12)
          ZR(4)=0.0
          ZJ(4)=-1.0/(W*21.E-12)
          YR(5)=0.0
          YJ(5)=W*(36.E-12)
           R=W*.5E-3/Q(I)
           AL=.5E-3
          C=570.E-12
          DENOM=1.+((R*C)**2-2.*AL*C)*W**2+(AL*C*W**2)**2
          ZR(6)=R/DENOM
          ZJ(6)=((AL-R**2*C)*W-AL**2*C*W**3)/DENOM
          YR(7)=0.0
          YJ(7)=W*(101.E-12)
          ZR(8)=ZR(6)
          ZJ(8)=ZJ(6)
          YR(9)=0.0
          YJ(9)=YJ(5)
          ZR(10)=0.0
          ZJ(10)=ZJ(4)
          YR(11)=0.0
          YJ(11)=YJ(3)
          ZR(12)=ZR(2)
          ZJ(12)=ZJ(2)
          YR(13)=YR(1)
          YJ(13)=0.0
C
C         CALCULATION OF INSERTION LOSS
C
          CR(2)=1./72.
          CJ(2)=0.0
          VJ(2)=(1./72.)*ZJ(2)
          VR(2)=(1./72.)*ZR(2)
          VR(3)=1.+VR(2)
          VJ(3)=VJ(2)
          CR(3)=VR(3)*YR(3)-VJ(3)*YJ(3)
          CJ(3)=VR(3)*YJ(3)+VJ(3)*YR(3)
          DO 30 J=4,12,2
          CR(J)=CR(J-2)+CR(J-1)
          CJ(J)=CJ(J-2)+CJ(J-1)
          VR(J)=CR(J)*ZR(J)-CJ(J)*ZJ(J)
          VJ(J)=CR(J)*ZJ(J)+CJ(J)*ZR(J)
          VR(J+1)=VR(J)+VR(J-1)
          VJ(J+1)=VJ(J)+VJ(J-1)
          CR(J+1)=VR(J+1)*YR(J+1)-VJ(J+1)*YJ(J+1)
   30     CJ(J+1)=VR(J+1)*YJ(J+1)+VJ(J+1)*YR(J+1)
          CRA=CR(13)+CR(12)
          CJA=CJ(13)+CJ(12)
          C=(CRA**2+CJA**2)**0.5
          DBL=(20./2.30259)*ALOG(C/(2./72.))
          F=W/(2.*3.14159)
       20 WRITE(6,40) F,DBL
       40 FORMAT(E15.3,F10.3)
          RETURN
          END
```

Figure 3.46 A Fortran program for calculating the dissipative insertion loss of the filter of Figure 3.39

INSERTION LOSS CALCULATIONS

FREQUENCY DB LOSS

Q = 50.00		Q = 100.00		Q = 150.00	
0.264E 06	26.837	0.264E 06	26.362	0.264E 06	26.225
0.265E 06	25.787	0.265E 06	25.251	0.265E 06	25.096
0.266E 06	24.685	0.266E 06	24.077	0.266E 06	23.900
0.267E 06	23.529	0.267E 06	22.835	0.267E 06	22.631
0.268E 06	22.315	0.268E 06	21.516	0.268E 06	21.281
0.269E 06	21.042	0.269E 06	20.115	0.269E 06	19.840
0.270E 06	19.709	0.270E 06	18.625	0.270E 06	18.298
0.271E 06	18.320	0.271E 06	17.040	0.271E 06	16.649
0.272E 06	16.883	0.272E 06	15.359	0.272E 06	14.883
0.273E 06	15.414	0.273E 06	13.569	0.273E 06	13.004
0.274E 06	13.944	0.274E 06	11.754	0.274E 06	11.027
0.275E 06	12.516	0.275E 06	9.907	0.275E 06	9.004
0.276E 06	11.194	0.276E 06	8.147	0.276E 06	7.045
0.277E 06	10.044	0.277E 06	6.611	0.277E 06	5.328
0.278E 06	9.122	0.278E 06	5.432	0.278E 06	4.041
0.279E 06	8.450	0.279E 06	4.659	0.279E 06	3.255
0.280E 06	8.014	0.280E 06	4.226	0.280E 06	2.866
0.281E 06	7.777	0.281E 06	4.018	0.281E 06	2.703
0.282E 06	7.702	0.282E 06	3.947	0.282E 06	2.648
0.283E 06	7.765	0.283E 06	3.971	0.283E 06	2.662
0.284E 06	7.956	0.284E 06	4.085	0.284E 06	2.750
0.285E 06	8.276	0.285E 06	4.289	0.285E 06	2.918
0.286E 06	8.733	0.286E 06	4.576	0.286E 06	3.148
0.287E 06	9.353	0.287E 06	4.937	0.287E 06	3.411
0.288E 06	10.184	0.288E 06	5.396	0.288E 06	3.707
0.289E 06	11.315	0.289E 06	6.044	0.289E 06	4.115
0.290E 06	12.887	0.290E 06	7.058	0.290E 06	4.821
0.291E 06	15.094	0.291E 06	8.730	0.291E 06	6.082
0.292E 06	18.166	0.292E 06	11.645	0.292E 06	8.564
0.293E 06	22.247	0.293E 06	16.677	0.293E 06	13.935
0.294E 06	27.271	0.294E 06	23.731	0.294E 06	22.054
0.295E 06	32.904	0.295E 06	31.960	0.295E 06	31.357
0.296E 06	38.690	0.296E 06	40.988	0.296E 06	41.576
0.297E 06	43.937	0.297E 06	50.273	0.297E 06	53.288
0.298E 06	47.790	0.298E 06	57.358	0.298E 06	63.813
0.299E 06	49.724	0.299E 06	58.090	0.299E 06	62.318
0.300E 06	49.988	0.300E 06	55.194	0.300E 06	56.871
0.301E 06	49.503	0.301E 06	52.493	0.301E 06	53.227
0.302E 06	48.825	0.302E 06	50.547	0.302E 06	50.888
0.303E 06	48.196	0.303E 06	49.202	0.303E 06	49.360
0.304E 06	47.719	0.304E 06	48.302	0.304E 06	48.365
0.305E 06	47.367	0.305E 06	47.691	0.305E 06	47.702
0.306E 06	47.133	0.306E 06	47.296	0.306E 06	47.277
0.307E 06	46.994	0.307E 06	47.053	0.307E 06	47.017
0.308E 06	46.940	0.308E 06	46.929	0.308E 06	46.883
0.309E 06	46.949	0.309E 06	46.893	0.309E 06	46.840
0.310E 06	47.008	0.310E 06	46.921	0.310E 06	46.865
0.311E 06	47.109	0.311E 06	47.002	0.311E 06	46.945
0.312E 06	47.244	0.312E 06	47.123	0.312E 06	47.067
0.313E 06	47.408	0.313E 06	47.278	0.313E 06	47.222
0.314E 06	47.591	0.314E 06	47.457	0.314E 06	47.402
0.315E 06	47.792	0.315E 06	47.655	0.315E 06	47.602
0.316E 06	48.008	0.316E 06	47.871	0.316E 06	47.818

Figure 3.47 The dissipative insertion loss for six different Q values

henry coil at a frequency of 300 kHz is about a Q of 150, but this is a very expensive coil and any design requiring such a coil would probably be scrapped.

In order to show how we may find the lowest Q values, let us specify our bandpass requirements as less than 3 dB insertion loss with less than 2 dB distortion (ripple). Later we will see that higher Q's will give a smaller insertion loss and ripple in the passband, but the value of Q will

Q = 200.00		Q = 250.00		Q = 300.00	
0.264E 06	26.160	0.264E 06	26.122	0.264E 06	26.098
0.265E 06	25.023	0.265E 06	24.980	0.265E 06	24.953
0.266E 06	23.817	0.266E 06	23.769	0.266E 06	23.737
0.267E 06	22.536	0.267E 06	22.480	0.267E 06	22.444
0.268E 06	21.170	0.268E 06	21.105	0.268E 06	21.063
0.269E 06	19.709	0.269E 06	19.633	0.269E 06	19.584
0.270E 06	18.143	0.270E 06	18.053	0.270E 06	17.994
0.271E 06	16.462	0.271E 06	16.353	0.271E 06	16.281
0.272E 06	14.654	0.272E 06	14.521	0.272E 06	14.432
0.273E 06	12.719	0.273E 06	12.551	0.273E 06	12.440
0.274E 06	10.668	0.274E 06	10.454	0.274E 06	10.313
0.275E 06	8.548	0.275E 06	8.274	0.275E 06	8.091
0.276E 06	6.476	0.276E 06	6.128	0.276E 06	5.894
0.277E 06	4.652	0.277E 06	4.235	0.277E 06	3.951
0.278E 06	3.305	0.278E 06	2.849	0.278E 06	2.538
0.279E 06	2.522	0.279E 06	2.070	0.279E 06	1.763
0.280E 06	2.168	0.280E 06	1.743	0.280E 06	1.457
0.281E 06	2.036	0.281E 06	1.633	0.281E 06	1.364
0.282E 06	1.993	0.282E 06	1.599	0.282E 06	1.335
0.283E 06	2.002	0.283E 06	1.604	0.283E 06	1.338
0.284E 06	2.076	0.284E 06	1.669	0.284E 06	1.397
0.285E 06	2.226	0.285E 06	1.810	0.285E 06	1.532
0.286E 06	2.431	0.286E 06	2.001	0.286E 06	1.714
0.287E 06	2.646	0.287E 06	2.187	0.287E 06	1.881
0.288E 06	2.853	0.288E 06	2.339	0.288E 06	1.996
0.289E 06	3.126	0.289E 06	2.525	0.289E 06	2.122
0.290E 06	3.655	0.290E 06	2.943	0.290E 06	2.463
0.291E 06	4.664	0.291E 06	3.785	0.291E 06	3.192
0.292E 06	6.791	0.292E 06	5.622	0.292E 06	4.797
0.293E 06	12.315	0.293E 06	11.250	0.293E 06	10.497
0.294E 06	21.151	0.294E 06	20.574	0.294E 06	20.143
0.295E 06	31.002	0.295E 06	30.747	0.295E 06	30.574
0.296E 06	41.783	0.296E 06	41.821	0.296E 06	41.831
0.297E 06	54.725	0.297E 06	55.432	0.297E 06	55.871
0.298E 06	68.190	0.298E 06	72.226	0.298E 06	75.062
0.299E 06	64.728	0.299E 06	65.943	0.299E 06	67.061
0.300E 06	57.582	0.300E 06	57.947	0.300E 06	58.112
0.301E 06	53.482	0.301E 06	53.599	0.301E 06	53.642
0.302E 06	50.999	0.302E 06	51.035	0.302E 06	51.040
0.303E 06	49.395	0.303E 06	49.398	0.303E 06	49.399
0.304E 06	48.365	0.304E 06	48.353	0.304E 06	48.344
0.305E 06	47.684	0.305E 06	47.669	0.305E 06	47.651
0.306E 06	47.250	0.306E 06	47.230	0.306E 06	47.215
0.307E 06	46.985	0.307E 06	46.964	0.307E 06	46.948
0.308E 06	46.850	0.308E 06	46.828	0.308E 06	46.812
0.309E 06	46.807	0.309E 06	46.785	0.309E 06	46.769
0.310E 06	46.831	0.310E 06	46.811	0.310E 06	46.795
0.311E 06	46.913	0.311E 06	46.892	0.311E 06	46.877
0.312E 06	47.035	0.312E 06	47.015	0.312E 06	47.001
0.313E 06	47.192	0.313E 06	47.172	0.313E 06	47.159
0.314E 06	47.373	0.314E 06	47.355	0.314E 06	47.342
0.315E 06	47.573	0.315E 06	47.556	0.315E 06	47.544
0.316E 06	47.791	0.316E 06	47.774	0.316E 06	47.763

Figure 3.47 Continued

have little effect in the reject band except at frequencies very close to cutoff values.[6]

Figure 3.46 shows a Fortran program for computing the dissipative insertion loss. This program is similar to the dissipationless insertion loss program of Figure 3.44 and differs mainly in the addition of another DO 20 loop (computer statement S.0008), which permits the iteration of Q values and the inclusion of real element terms for Z_2, Z_6, Z_8, and Z_{12} (computer statements S.0016, S.0024–S.0028, S.0032, and S.0040).

Figure 3.47 shows the computed output for the six Q values that were

[6] As a consequence, the cutoff values of this design were taken as f_1 = 277.5 kHz and f_2 = 293 kHz, so that the effects of Q would show up less at the edges of the required passband of 280–290 kHz.

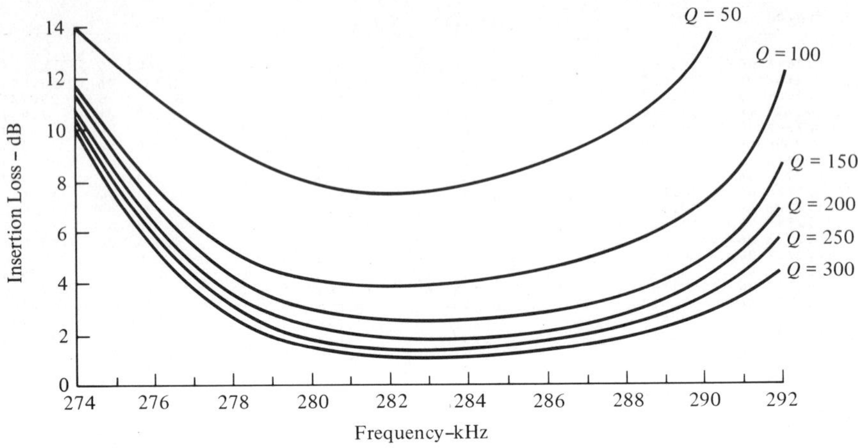

Figure 3.48 The dissipative insertion loss versus Q

specified and Figure 3.48 is a graph that shows the individual Q insertion loss in the frequency range of 275 kHz to 291 kHz. An examination of this figure will show that a minimum Q of 250 is needed to meet the design requirements. Note that the value of Q has little effect in the reject band but considerable effect in the passband. The design therefore requires that a minimum Q of 250 be maintained over the approximate frequency range of 275 to 295 kHz.

At this point, the reader may wonder why we have assumed the same value of Q for all four coils. The reason that this was done was because the value of L_1 (1.03 mhy) was essentially the same as L_2 (0.5 mhy, see Figure 3.39) and thus one could expect to obtain a given Q for one coil as easily as for the other. In designs where inductors that have a wider range in values are used, the Q values will have to be iterated separately and over ranges that represent the physical values of Q that can be obtained for a particular corresponding numerical inductor value.

It is also possible to modify the above program so that for each iteration of Q the maximum and minimum values of the insertion loss in the passband can be found, and whenever the conditions of maximum insertion loss and distortion (maximum insertion loss minus minimum insertion loss) are met, the program terminates by printing the individual Q value of each coil.

A Comparison of the Amplitude Response of Papoulis versus Butterworth Filters

The ideal normalized low-pass filter characteristic shown in Figure 3.49 indicates that signals that contain frequencies within the passband $(0 \leq \omega < 1)$ are transmitted with linear phase response and without loss,

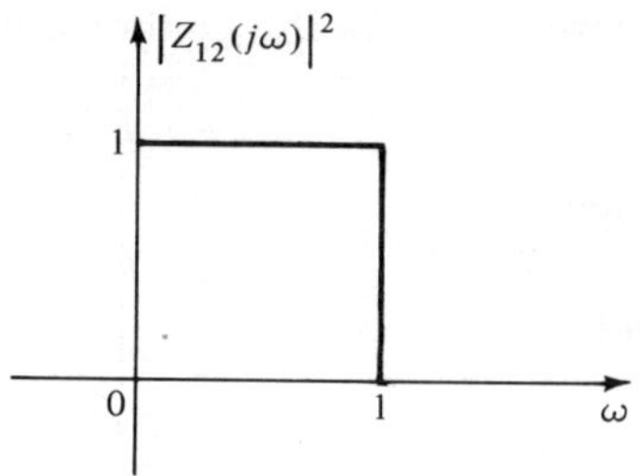

Figure 3.49 The ideal low-pass filter response

but zero transmission results for signal frequency components in the stop band ($\omega > 1$). Since it is physically impossible to build a network that has this type of transmission, the designer can only resort to approximating the ideal behavior.

In the early 1920s, O. J. Zobel and G. A. Campbell developed the so-called "classical" or image-parameter approach to filter design. This was a trial and error cascading of known amplitude and phase characteristics of symmetrical T-, Pi-, or L-sections. No attempt was made to design for both amplitude and phase response. In general, the designer would cascade the appropriate number of filter sections to satisfy the attenuation requirements in the stop band and accept whatever phase characteristic resulted.

"Modern" filter design, on the other hand, is based on approximating the filter requirements with a rational transfer function and using synthesis techniques to realize the physical filter elements. The Butterworth filter is one such type of filter, and its normalized magnitude function is given by

$$|AB(j\omega)| = \frac{1}{\sqrt{1 + \omega^{2n}}}$$

where n is a positive integer that corresponds to the number of poles. Such a transfer function results in the first $2n - 1$ derivatives of the amplitude function vanishing at $\omega = 0$. This produces a monotonic response that is extremely flat in the vicinity of the origin. For this reason a Butterworth response is also known as a *maximally flat* response.

In 1958, Papoulis[7] introduced his "optimum" filters. These filters had a monotonic, but not maximally flat response and possessed a faster falloff in the vicinity of cutoff (normalized $\omega = 1$). The normalized general filter equation for Papoulis filters is given by

$$|AP(j\omega)| = \frac{1}{\sqrt{1 + L_n(\omega^2)}}$$

[7] A. Papoulis, "Optimum Filters with Monotonic Response," *Proc. IRE*, 46, No. 3, March 1958.

where $L_n(\omega^2)$ is the generating polynomial and is found from

$$L_n(\omega^2) = \int_{-1}^{2\omega^2-1} v^2(x)\, dx$$

where

$$v(x) = a_0 + a_1P_1(x) + \cdots + a_kP_k(x)$$

and

$$a_0 = \frac{a_1}{3} = \frac{a_2}{5} = \cdots = \frac{a_k}{2k+1}$$

and $P_n(x)$ are Legendre polynomials of the first kind and are found from the following relation

$$P_n(x) = \frac{1}{2^n n!} \frac{d^n}{dx^n} (x^2 - 1)^n$$

For example, for a value of $n = 3$,

$$L_3(\omega^2) = 3\omega^6 - 3\omega^4 + \omega^2$$

Figure 3.50 is a Fortran program that determines the comparative response of one-to-eight-pole Papoulis and Butterworth filters. The heading is printed initially, and then the DO 20 loop is used to generate 40

```
C      PAPOULIS VERSUS BUTTERWORTH FILTERS
C
       DIMENSION W(40),P(16),B(16),AP(40),AB(40)
       WRITE(6,1)
     1 FORMAT('1A COMPARISON OF THE AMPLITUDE RESPONSE OF PAPOULIS VERSUS
      1 BUTTERWORTH FILTERS',///)
       DO 20 I=1,40
    20 W(I)=0.1*I
       N=0
    40 N=N+1
       IF (N-8)41,41,42
    41 WRITE(6,2)N
     2 FORMAT(//,10X,'POLYNOMINAL DEGREE =',I4)
       READ(5,3) (P(I),I=2,16,2)
     3 FORMAT(8F10.2)
       WRITE(6,4) (P(I),I=2,16,2)
     4 FORMAT(//,' THE COEFFICIENTS P(2), P(4), ..., P(16) ARE',//,8(3X,F
      18.2))
       READ(5,3) (B(I),I=2,16,2)
       WRITE(6,5) (B(I),I=2,16,2)
     5 FORMAT(//,' THE COEFFICIENTS B(2), B(4), ..., B(16) ARE',//,8(3X,F
      18.2))
C
C      CALCULATION OF THE AMPLITUDE FUNCTIONS
C
       WRITE(6,6)
     6 FORMAT(//,2X,' FREQUENCY W',3X,'AMPLITUDE AP(W)',2X,'AMPLITUDE AB(
      1W)',//,8X,'RADIANS',12X,'DB',15X,'DB',//)
       DO 50 I=1,40
       AP(I)=-10.0*ALOG10(1.+P(2)*W(I)**2+P(4)*W(I)**4+P(6)*W(I)**6+P(8)*
      1W(I)**8+P(10)*W(I)**10+P(12)*W(I)**12+P(14)*W(I)**14+
      2P(16)*W(I)**16)
       AB(I)=-10.0*ALOG10(1.+B(2)*W(I)**2+B(4)*W(I)**4+B(6)*W(I)**6+B(8)*
      1W(I)**8+B(10)*W(I)**10+B(12)*W(I)**12+B(14)*W(I)**14+
      2B(16)*W(I)**16)
    50 WRITE(6,7) W(I),AP(I),AB(I)
     7 FORMAT(3E17.5)
       GO TO 40
    42 STOP
       END
```

Figure 3.50 A program that computes the frequency response of one-to-eight-pole Butterworth and Papoulis filters

```
A COMPARISON OF THE AMPLITUDE RESPONSE OF PAPOULIS VERSUS BUTTERWORTH FILTERS


        POLYNOMINAL DEGREE =   1

THE COEFFICIENTS P(2), P(4), ...., P(16) ARE

     1.00      0.0       0.0       0.0       C.0       0.0       0.0       0.0

THE COEFFICIENTS B(2), B(4), ...., B(16) ARE

     1.00      0.0       0.0       r.0       0.0       0.0       0.0       0.0

  FREQUENCY W    AMPLITUDE AP(W)   AMPLITUDE AB(W)
    RADIANS            DB                DB

   0.10000E 00     -0.43211E-01      -0.43211E-01
   0.20000E 00     -0.17033E 00      -0.17033E 00
   0.30000E 00     -0.37426E 00      -0.37426E 00
   0.40000E 00     -0.64458E 00      -0.64458E 00
   0.50000E 00     -0.96910E 00      -0.96910E 00
   0.60000E 00     -0.13354E 01      -0.13354E 01
   0.70000E 00     -0.17319E 01      -0.17319E 01
   0.80000E 00     -0.21484E 01      -0.21484E 01
   0.90000E 00     -0.25768E 01      -0.25768E 01
   0.10000E 01     -0.30103E 01      -0.30103E 01
   0.11000E 01     -0.34439E 01      -0.34439E 01
   0.12000E 01     -0.38739E 01      -0.38739E 01
   0.13000E 01     -0.42975E 01      -0.42975E 01
   0.14000E 01     -0.47129E 01      -0.47129E 01
   0.15000E 01     -0.51188E 01      -0.51188E 01
   0.16000E 01     -0.55145E 01      -0.55145E 01
   0.17000E 01     -0.58995E 01      -0.58995E 01
   0.18000E 01     -0.62737E 01      -0.62737E 01
   0.19000E 01     -0.66370E 01      -0.66370E 01
   0.20000E 01     -0.69897E 01      -0.69897E 01
   0.21000E 01     -0.73320E 01      -0.73320E 01
   0.22000E 01     -0.76641E 01      -0.76641E 01
   0.23000E 01     -0.79865E 01      -0.79865E 01
   0.24000E 01     -0.82995E 01      -0.82995E 01
   0.25000E 01     -0.86034E 01      -0.86034E 01
   0.26000E 01     -0.88986E 01      -0.88986E 01
   0.27000E 01     -0.91855E 01      -0.91855E 01
   0.28000E 01     -0.94645E 01      -0.94645E 01
   0.29000E 01     -0.97359E 01      -0.97359E 01
   0.30000E 01     -0.10000E 02      -0.10000E 02
   0.31000E 01     -0.10257E 02      -0.10257E 02
   0.32000E 01     -0.10508E 02      -0.10508E 02
   0.33000E 01     -0.10752E 02      -0.10752E 02
   0.34000E 01     -0.10990E 02      -0.10990E 02
   0.35000E 01     -0.11222E 02      -0.11222E 02
   0.36000E 01     -0.11449E 02      -0.11449E 02
   0.37000E 01     -0.11670E 02      -0.11670E 02
   0.38000E 01     -0.11886E 02      -0.11886E 02
   0.39000E 01     -0.12098E 02      -0.12098E 02
   0.40000E 01     -0.12304E 02      -0.12304E 02
```

Figure 3.51 The output results of the program of Figure 3.50

normalized frequencies, ranging from 0.1 to 4.0 radians per second, in 0.1 rps steps. The dummy variable N is now set equal to zero and statement 40 then makes N equal to one. We next test to see whether the eight-pole filter response has been found. Since N is equal to one during the first iteration, we print the degree of the polynomial and read in and print the coefficients of both the Papoulis and Butterworth polynomials. For example, when reading in the coefficients for N equal to three, $P(2)$ would be equal to 1 and would correspond to the coefficient of the ω^2 term, $P(4)$ would be equal to -3 and would be equal to the coefficient

```
POLYNOMINAL DEGREE =   2

THE COEFFICIENTS P(2), P(4), ..., P(16) ARE

  0.0       1.00      0.0       0.0       0.0       0.0       0.0       0.0

THE COEFFICIENTS B(2), B(4), ..., B(16) ARE

  0.0       1.00      0.0       0.0       0.0       0.0       0.0       0.0

FREQUENCY W    AMPLITUDE AP(W)   AMPLITUDE AB(W)
  RADIANS           DB                DB

0.10000E 00     -0.43072E-03      -0.43072E-03
0.20000E 00     -0.69402E-02      -0.69402E-02
0.30000E 00     -0.35034E-01      -0.35034E-01
0.40000E 00     -0.10978E 00      -0.10978E 00
0.50000E 00     -0.26329E 00      -0.26329E 00
0.60000E 00     -0.52925E 00      -0.52925E 00
0.70000E 00     -0.93456E 00      -0.93456E 00
0.80000E 00     -0.14910E 01      -0.14910E 01
0.90000E 00     -0.21909E 01      -0.21909E 01
0.10000E 01     -0.30103E 01      -0.30103E 01
0.11000E 01     -0.39166E 01      -0.39166E 01
0.12000E 01     -0.48765E 01      -0.48765E 01
0.13000E 01     -0.58615E 01      -0.58615E 01
0.14000E 01     -0.68499E 01      -0.68499E 01
0.15000E 01     -0.78265E 01      -0.78265E 01
0.16000E 01     -0.87815E 01      -0.87815E 01
0.17000E 01     -0.97091E 01      -0.97091E 01
0.18000E 01     -0.10606E 02      -0.10606E 02
0.19000E 01     -0.11471E 02      -0.11471E 02
0.20000E 01     -0.12304E 02      -0.12304E 02
0.21000E 01     -0.13107E 02      -0.13107E 02
0.22000E 01     -0.13878E 02      -0.13878E 02
0.23000E 01     -0.14622E 02      -0.14622E 02
0.24000E 01     -0.15337E 02      -0.15337E 02
0.25000E 01     -0.16027E 02      -0.16027E 02
0.26000E 01     -0.16693E 02      -0.16693E 02
0.27000E 01     -0.17335E 02      -0.17335E 02
0.28000E 01     -0.17956E 02      -0.17956E 02
0.29000E 01     -0.18557E 02      -0.18557E 02
0.30000E 01     -0.19138E 02      -0.19138E 02
0.31000E 01     -0.19701E 02      -0.19701E 02
0.32000E 01     -0.20247E 02      -0.20247E 02
0.33000E 01     -0.20777E 02      -0.20777E 02
0.34000E 01     -0.21292E 02      -0.21292E 02
0.35000E 01     -0.21792E 02      -0.21792E 02
0.36000E 01     -0.22278E 02      -0.22278E 02
0.37000E 01     -0.22751E 02      -0.22751E 02
0.38000E 01     -0.23212E 02      -0.23212E 02
0.39000E 01     -0.23661E 02      -0.23661E 02
0.40000E 01     -0.24099E 02      -0.24099E 02

POLYNOMINAL DEGREE =   3

THE COEFFICIENTS P(2), P(4), ..., P(16) ARE

  1.00     -3.00      3.00      0.0       0.0       0.0       0.0       0.0

THE COEFFICIENTS B(2), B(4), ..., B(16) ARE

  0.0       0.0       1.00      0.0       0.0       0.0       0.0       0.0

FREQUENCY W    AMPLITUDE AP(W)   AMPLITUDE AB(W)
  RADIANS           DB                DB

0.10000E 00     -0.37191E-01      0.0
0.20000E 00     -0.93959E-01      0.0
0.30000E 00     -0.10771E 00      0.0
0.40000E 00     -0.12658E 00      -0.82835E-05
0.50000E 00     -0.20981E 00      -0.26092E-03
0.60000E 00     -0.27447E 00      -0.33991E-02
```

Figure 3.51 Continued

0.70000E 00	-0.50243E 00	-0.48305E 00
0.80000E 00	-0.78323E 00	-0.10111E 01
0.90000E 00	-0.15716E 01	-0.18510E 01
0.10000E 01	-0.30103E 01	-0.30103E 01
0.11000E 01	-0.49587E 01	-0.44272E 01
0.12000E 01	-0.71409E 01	-0.60053E 01
0.13000E 01	-0.93460E 01	-0.76543E 01
0.14000E 01	-0.11469E 02	-0.93092E 01
0.15000E 01	-0.13470E 02	-0.10931E 02
0.16000E 01	-0.15344E 02	-0.12499E 02
0.17000E 01	-0.17097E 02	-0.14003E 02
0.18000E 01	-0.18738E 02	-0.15442E 02
0.19000E 01	-0.20280E 02	-0.16817E 02
0.20000E 01	-0.21732E 02	-0.18129E 02
0.21000E 01	-0.23104E 02	-0.19383E 02
0.22000E 01	-0.24404E 02	-0.20583E 02
0.23000E 01	-0.25640E 02	-0.21733E 02
0.24000E 01	-0.26817E 02	-0.22835E 02
0.25000E 01	-0.27941E 02	-0.23894E 02
0.26000E 01	-0.29017E 02	-0.24912E 02
0.27000E 01	-0.30048E 02	-0.25893E 02
0.28000E 01	-0.31038E 02	-0.26838E 02
0.29000E 01	-0.31991E 02	-0.27751E 02
0.30000E 01	-0.32909E 02	-0.28633E 02
0.31000E 01	-0.33795E 02	-0.29487E 02
0.32000E 01	-0.34651E 02	-0.30313E 02
0.33000E 01	-0.35478E 02	-0.31114E 02
0.34000E 01	-0.36280E 02	-0.31892E 02
0.35000E 01	-0.37057E 02	-0.32646E 02
0.36000E 01	-0.37811E 02	-0.33380E 02
0.37000E 01	-0.38543E 02	-0.34094E 02
0.38000E 01	-0.39255E 02	-0.34788E 02
0.39000E 01	-0.39947E 02	-0.35465E 02
0.40000E 01	-0.40621E 02	-0.36125E 02

POLYNOMINAL DEGREE = 4

THE COEFFICIENTS P(2), P(4), ..., P(16) ARE

0.0	3.00	-8.00	6.00	0.0	0.0	0.0	0.0

THE COEFFICIENTS B(2), B(4), ..., B(16) ARE

0.0	0.0	0.0	1.00	0.0	0.0	0.0	0.0

FREQUENCY W RADIANS	AMPLITUDE AP(W) DB	AMPLITUDE AB(W) DB
0.10000E 00	-0.12631E-02	0.0
0.20000E 00	-0.18647E-01	-0.82835E-05
0.30000E 00	-0.81144E-01	-0.28163E-03
0.40000E 00	-0.20346E 00	-0.28445E-02
0.50000E 00	-0.35804E 00	-0.16927E-01
0.60000E 00	-0.47792E 00	-0.72335E-01
0.70000E 00	-0.51151E 00	-0.24341E 00
0.80000E 00	-0.56249E 00	-0.67358E 00
0.90000E 00	-0.11380E 01	-0.15548E 01
0.10000E 01	-0.30103E 01	-0.30103E 01
0.11000E 01	-0.61080E 01	-0.49742E 01
0.12000E 01	-0.96055E 01	-0.72426E 01
0.13000E 01	-0.12988E 02	-0.96177E 01
0.14000E 01	-0.16110E 02	-0.11975E 02
0.15000E 01	-0.18967E 02	-0.14254E 02
0.16000E 01	-0.21588E 02	-0.16430E 02
0.17000E 01	-0.24005E 02	-0.18498E 02
0.18000E 01	-0.26249E 02	-0.20461E 02
0.19000E 01	-0.28343E 02	-0.22326E 02
0.20000E 01	-0.30306E 02	-0.24099E 02
0.21000E 01	-0.32155E 02	-0.25789E 02
0.22000E 01	-0.33904E 02	-0.27402E 02
0.23000E 01	-0.35562E 02	-0.28944E 02
0.24000E 01	-0.37140E 02	-0.30421E 02
0.25000E 01	-0.38645E 02	-0.31838E 02
0.26000E 01	-0.40084E 02	-0.33200E 02

Figure 3.51 Continued

0.27000E 01	-0.41463E 02	-0.34511E 02
0.28000E 01	-0.42787E 02	-0.35774E 02
0.29000E 01	-0.44060E 02	-0.36993E 02
0.30000E 01	-0.45286E 02	-0.38170E 02
0.31000E 01	-0.46469E 02	-0.39309E 02
0.32000E 01	-0.47611E 02	-0.40412E 02
0.33000E 01	-0.48716E 02	-0.41481E 02
0.34000E 01	-0.49786E 02	-0.42519E 02
0.35000E 01	-0.50823E 02	-0.43526E 02
0.36000E 01	-0.51829E 02	-0.44504E 02
0.37000E 01	-0.52805E 02	-0.45456E 02
0.38000E 01	-0.53755E 02	-0.46383E 02
0.39000E 01	-0.54679E 02	-0.47285E 02
0.40000E 01	-0.55578E 02	-0.48165E 02

POLYNOMINAL DEGREE = 5

THE COEFFICIENTS P(2), P(4), ..., P(16) ARE

1.00	-8.00	28.00	-40.00	20.00	0.0	0.0

THE COEFFICIENTS B(2), B(4), ..., B(16) ARE

0.0	0.0	0.0	0.0	1.00	0.0	0.0

FREQUENCY W RADIANS	AMPLITUDE AP(W) DB	AMPLITUDE AB(W) DB
0.10000E 00	-0.39884E-01	0.0
0.20000E 00	-0.12369E 00	0.0
0.30000E 00	-0.18327E 00	-0.24850E-04
0.40000E 00	-0.19436E 00	-0.45143E-03
0.50000E 00	-0.21511E 00	-0.42349E-02
0.60000E 00	-0.32882E 00	-0.26180E-01
0.70000E 00	-0.50148E 00	-0.12097E 00
0.80000E 00	-0.56840E 00	-0.44294E 00
0.90000E 00	-0.77889E 00	-0.12991E 01
0.10000E 01	-0.30103E 01	-0.30103E 01
0.11000E 01	-0.79464E 01	-0.55554E 01
0.12000E 01	-0.13284E 02	-0.85683E 01
0.13000E 01	-0.18090E 02	-0.11698E 02
0.14000E 01	-0.22336E 02	-0.14760E 02
0.15000E 01	-0.26126E 02	-0.17684E 02
0.16000E 01	-0.29552E 02	-0.20451E 02
0.17000E 01	-0.32682E 02	-0.23066E 02
0.18000E 01	-0.35570E 02	-0.25539E 02
0.19000E 01	-0.38252E 02	-0.27882E 02
0.20000E 01	-0.40759E 02	-0.30107E 02
0.21000E 01	-0.43114E 02	-0.32224E 02
0.22000E 01	-0.45335E 02	-0.34244E 02
0.23000E 01	-0.47439E 02	-0.36174E 02
0.24000E 01	-0.49438E 02	-0.38022E 02
0.25000E 01	-0.51343E 02	-0.39794E 02
0.26000E 01	-0.53162E 02	-0.41498E 02
0.27000E 01	-0.54903E 02	-0.43137E 02
0.28000E 01	-0.56573E 02	-0.44716E 02
0.29000E 01	-0.58178E 02	-0.46240E 02
0.30000E 01	-0.59723E 02	-0.47712E 02
0.31000E 01	-0.61213E 02	-0.49136E 02
0.32000E 01	-0.62651E 02	-0.50515E 02
0.33000E 01	-0.64041E 02	-0.51851E 02
0.34000E 01	-0.65387E 02	-0.53148E 02
0.35000E 01	-0.66690E 02	-0.54407E 02
0.36000E 01	-0.67954E 02	-0.55630E 02
0.37000E 01	-0.69182E 02	-0.56820E 02
0.38000E 01	-0.70374E 02	-0.57978E 02
0.39000E 01	-0.71534E 02	-0.59106E 02
0.40000E 01	-0.72663E 02	-0.60206E 02

Figure 3.51 Continued

```
         POLYNOMINAL DEGREE =    6

THE COEFFICIENTS P(2), P(4), ..., P(16) ARE

     0.0        6.00      -40.00      105.00     -120.00       50.00       0.0        0.0

THE COEFFICIENTS B(2), B(4), ..., B(16) ARE

     0.0        0.0        0.0        0.0        0.0        1.00        0.0        0.0

  FREQUENCY W    AMPLITUDE AP(W)   AMPLITUDE AB(W)

      RADIANS             DB                 DB


    0.10000E 00      -0.24347E-02       0.0
    0.20000E 00      -0.31565E-01       0.0
    0.30000E 00      -0.10997E 00       0.0
    0.40000E 00      -0.19875E 00      -0.70409E-04
    0.50000E 00      -0.23324E 00      -0.10560E-02
    0.60000E 00      -0.24568E 00      -0.94412E-02
    0.70000E 00      -0.37443E 00      -0.59697E-01
    0.80000E 00      -0.56518E 00      -0.28863E 00
    0.90000E 00      -0.63743E 00      -0.10803E 01
    0.10000E 01      -0.30103E 01      -0.30103E 01
    0.11000E 01      -0.98548E 01      -0.61683E 01
    0.12000E 01      -0.16838E 02      -0.99634E 01
    0.13000E 01      -0.22837E 02      -0.13856E 02
    0.14000E 01      -0.28035E 02      -0.17611E 02
    0.15000E 01      -0.32635E 02      -0.21164E 02
    0.16000E 01      -0.36775E 02      -0.24510E 02
    0.17000E 01      -0.40550E 02      -0.27661E 02
    0.18000E 01      -0.44026E 02      -0.30636E 02
    0.19000E 01      -0.47253E 02      -0.33452E 02
    0.20000E 01      -0.50267E 02      -0.36125E 02
    0.21000E 01      -0.53097E 02      -0.38667E 02
    0.22000E 01      -0.55766E 02      -0.41091E 02
    0.23000E 01      -0.58294E 02      -0.43407E 02
    0.24000E 01      -0.60694E 02      -0.45625E 02
    0.25000E 01      -0.62981E 02      -0.47753E 02
    0.26000E 01      -0.65165E 02      -0.49797E 02
    0.27000E 01      -0.67255E 02      -0.51764E 02
    0.28000E 01      -0.69260E 02      -0.53659E 02
    0.29000E 01      -0.71187E 02      -0.55488E 02
    0.30000E 01      -0.73042E 02      -0.57255E 02
    0.31000E 01      -0.74830E 02      -0.58963E 02
    0.32000E 01      -0.76556E 02      -0.60618E 02
    0.33000E 01      -0.78224E 02      -0.62222E 02
    0.34000E 01      -0.79839E 02      -0.63777E 02
    0.35000E 01      -0.81403E 02      -0.65288E 02
    0.36000E 01      -0.82921E 02      -0.66756E 02
    0.37000E 01      -0.84394E 02      -0.68184E 02
    0.38000E 01      -0.85825E 02      -0.69574E 02
    0.39000E 01      -0.87217E 02      -0.70928E 02
    0.40000E 01      -0.88572E 02      -0.72247E 02

         POLYNOMINAL DEGREE =    7

THE COEFFICIENTS P(2), P(4), ..., P(16) ARE

     1.00      -15.00      105.00     -355.00      615.00     -525.00      175.00       0.0

THE COEFFICIENTS B(2), B(4), ..., B(16) ARE

     0.0        0.0        0.0        0.0        0.0        0.0        1.00        0.0

  FREQUENCY W    AMPLITUDE AP(W)   AMPLITUDE AB(W)

      RADIANS             DB                 DB


    0.10000E 00      -0.37191E-01       0.0
    0.20000E 00      -0.93959E-01       0.0
```

Figure 3.51 Continued

0.30000E 00	-0.10771E 00	0.0
0.40000E C0	-0.12658E 00	-0.82835E-05
0.50000E 00	-0.20981E 00	-0.26092E-03
0.60000E 00	-0.27447E 00	-0.33991E-02
0.70000E 00	-0.29002E 00	-0.29353E-01
0.80000E 00	-0.47635E 00	-0.18692E 00
0.90000E 00	-0.60719E 00	-0.89469E 00
0.10000E 01	-0.30110E 01	-0.30103E 01
0.11000E 01	-0.12387E 02	-0.68101E 01
0.12000E 01	-0.21214E 02	-0.11411E 02
0.13000E 01	-0.28500E 02	-0.16061E 02
0.14000E 01	-0.34724E 02	-0.20497E 02
0.15000E 01	-0.40195E 02	-0.24668E 02
0.16000E 01	-0.45101E 02	-0.28583E 02
0.17000E 01	-0.49562E 02	-0.32265E 02
0.18000E 01	-0.53664E 02	-0.35739E 02
0.19000E 01	-0.57465E 02	-0.39026E 02
0.20000E 01	-0.61011E 02	-0.42144E 02
0.21000E 01	-0.64338E 02	-0.45111E 02
0.22000E 01	-0.67474E 02	-0.47939E 02
0.23000E 01	-0.70440E 02	-0.50642E 02
0.24000E 01	-0.73257E 02	-0.53230E 02
0.25000E 01	-0.75938E 02	-0.55712E 02
0.26000E 01	-0.78498E 02	-0.58096E 02
0.27000E 01	-0.80948E 02	-0.60391E 02
0.28000E 01	-0.83296E 02	-0.62602E 02
0.29000E 01	-0.85552E 02	-0.64736E 02
0.30000E 01	-0.87724E 02	-0.66797E 02
0.31000E 01	-0.89816E 02	-0.68791E 02
0.32000E 01	-0.91836E 02	-0.70721E 02
0.33000E 01	-0.93788E 02	-0.72592E 02
0.34000E 01	-0.95677E 02	-0.74407E 02
0.35000E 01	-0.97507E 02	-0.76169E 02
0.36000E 01	-0.99281E 02	-0.77882E 02
0.37000E 01	-0.10100E 03	-0.79548E 02
0.38000E 01	-0.10268E 03	-0.81170E 02
0.39000E 01	-0.10430E 03	-0.82749E 02
0.40000E 01	-0.10589E 03	-0.84288E 02

POLYNOMINAL DEGREE = 8

THE COEFFICIENTS P(2), P(4), ..., P(16) ARE

0.0	10.00	-120.00	615.00	-1624.00	2310.00	-1680.00	490.00

THE COEFFICIENTS B(2), B(4), ..., B(16) ARE

0.0	0.0	0.0	0.0	0.0	0.0	0.0	1.00

FREQUENCY W RADIANS	AMPLITUDE AP(W) DB	AMPLITUDE AB(W) DB
0.10000E 00	-0.38377E-02	0.0
0.20000E 00	-0.42070E-01	0.0
0.30000E 00	-0.10905E 00	0.0
0.40000E 00	-0.13549E 00	0.0
0.50000E 00	-0.15069E 00	-0.62126E-04
0.60000E 00	-0.24196E 00	-0.12216E-02
0.70000E 00	-0.29499E 00	-0.14406E-01
0.80000E 00	-0.37460E 00	-0.12055E 00
0.90000E 00	-0.61317E 00	-0.73829E 00
0.10000E 01	-0.30098E 01	-0.30103E 01
0.11000E 01	-0.14861E 02	-0.74779E 01
0.12000E 01	-0.25291E 02	-0.12898E 02
0.13000E 01	-0.33720E 02	-0.18296E 02
0.14000E 01	-0.40877E 02	-0.23400E 02
0.15000E 01	-0.47154E 02	-0.28181E 02
0.16000E 01	-0.52775E 02	-0.32662E 02
0.17000E 01	-0.57883E 02	-0.36873E 02
0.18000E 01	-0.62577E 02	-0.40844E 02
0.19000E 01	-0.66925E 02	-0.44601E 02
0.20000E 01	-0.70982E 02	-0.48165E 02
0.21000E 01	-0.74786E 02	-0.51555E 02

Figure 3.51 Continued

0.22000E 01	-0.78372E 02	-0.54788E 02
0.23000E 01	-0.81764E 02	-0.57876E 02
0.24000E 01	-0.84983E 02	-0.60834E 02
0.25000E 01	-0.88049E 02	-0.63670E 02
0.26000E 01	-0.90975E 02	-0.66396E 02
0.27000E 01	-0.93775E 02	-0.69018E 02
0.28000E 01	-0.96460E 02	-0.71545E 02
0.29000E 01	-0.99039E 02	-0.73984E 02
0.30000E 01	-0.10152E 03	-0.76339E 02
0.31000E 01	-0.10391E 03	-0.78618E 02
0.32000E 01	-0.10622E 03	-0.80824E 02
0.33000E 01	-0.10845E 03	-0.82962E 02
0.34000E 01	-0.11061E 03	-0.85037E 02
0.35000E 01	-0.11270E 03	-0.87051E 02
0.36000E 01	-0.11473E 03	-0.89008E 02
0.37000E 01	-0.11670E 03	-0.90912E 02
0.38000E 01	-0.11861E 03	-0.92765E 02
0.39000E 01	-0.12047E 03	-0.94570E 02
0.40000E 01	-0.12228E 03	-0.96330E 02

Figure 3.51 Continued

of the ω^4 term, $P(6)$ would be equal to 3 and would correspond to the coefficient of the ω^6 term, and all other higher order coefficients would be equal to zero.

Computer statements S.0018 and S.0019 are now used to print the subheadings and the DO 50 loop computes and prints the dB magnitude of $|AP(j\omega)|$ and $|AB(j\omega)|$ for each of these 40 frequencies.

Figure 3.51 shows the computer output. In order to show how we might make use of this information, let us assume that we have already designed and put into production a six-pole Butterworth filter whose specifications required less than 0.25 dB ripple over a normalized bandwidth of from 0 to 0.6 radians per second and a minimum of 20 dB loss at $\omega = 1.5$ rps. Assume that due to other system changes we are now asked to see if we can maintain the same maximum loss in the 0 to 0.6 rps range, but increase the loss at 1.5 rps to 30 dB. Since the fabrication of a new filter circuit board would introduce other mechanical production line changes, it would be to our advantage to use some other type of six-pole filter. An investigation of Chebyshev six-pole filters would show that we could not satisfy the ripple requirements, but Figure 3.51 shows that the six-pole Papoulis filter will indeed satisfy the new requirements. For the actual synthesis of the new filter elements, the reader is referred to the previous reference.

3.4 LOG MAGNITUDE AND PHASE CALCULATIONS

The R-C Network

For a simple introductory example of a log magnitude and phase versus log frequency plot, consider the network shown in Figure 3.52. This example involves a single real pole and is representative of many simplified electrical systems. For example, the high frequency transfer function

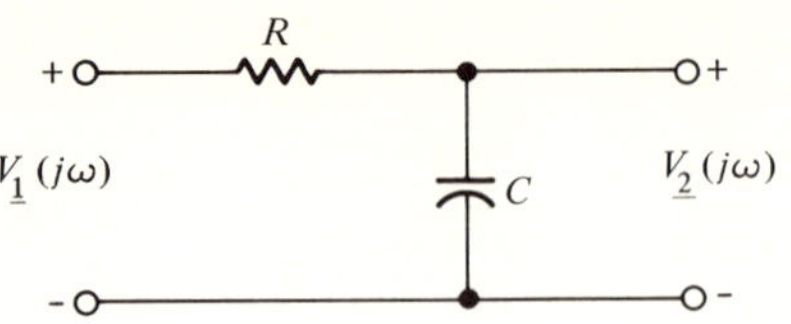

Figure 3.52 The R-C coupling network

of the R-C transistor or vacuum tube amplifier is described by a transfer function that is mathematically identical to the transfer function of this network.

The sinusoidal steady-state, open-circuit, positive-mode voltage transfer ratio is given by

$$G(j\omega) = \frac{V_2(j\omega)}{V_1(j\omega)} = \frac{1/(j\omega C)}{R + 1/(j\omega C)}$$

$$= \frac{1}{1 + j\omega RC} = \frac{1}{1 + j\omega/\omega_{p1}}$$

$$= \frac{1}{\sqrt{1 + (\omega/\omega_{p1})^2}} \underline{\big|\, - \tan^{-1}(\omega/\omega_{p1})}$$

where $\omega_{p1} = 1/RC$ = the corner frequency. Expressing the magnitude in decibels we have

$$|G(j\omega)|_{\text{dB}} = -20 \log_{10} \sqrt{1 + (\omega/\omega_{p1})^2}$$

and the phase is given by

$$\theta(j\omega) = -\tan^{-1}(\omega/\omega_{p1})$$

Let us assume that our problem is to obtain the necessary data to plot the magnitude and phase at one octave intervals for four octaves above and below the corner frequency, independent of the numerical values of *R* or *C*. In Chapter 11 we will discuss how the computer may be used to produce the plot.

Figure 3.53 shows a Fortran program that will compute the necessary data. Initially computer statements S.0001–S.0004 print the heading and subheadings. Computer statements S.0005 and S.0006 then read in the numerical values of *R* and *C*. Computer statement S.0007 now calculates the corner frequency ω_{p1} and computer statement S.0008 is used to set the initial value of ω to $\omega_{p1}/32$ which is five octaves below ω_{p1}. The DO 40 loop is now executed nine times since there are nine different frequency calculations to be made. During the first iteration, computer statement S.0010 sets ω equal to $2*\omega$ or $\omega_{p1}/16$ (four octaves below ω_{p1}). In each successive pass the frequency is doubled, thus providing the numerical values of the magnitude and phase at one octave intervals. Computer statements S.0011 and S.0012 are used to compute the dB

```
        C       LOG-MAGNITUDE AND PHASE PLOT
        C                    OF
        C            RC COUPLING NETWORK
        C
S.0001          WRITE(6,10)
S.0002       10 FORMAT('1LOG-MAGNITUDE AND PHASE PLOT'/,14X,'OF'/5X,'RC COUPLING N
               1ETWORK'//)
S.0003          WRITE(6,20)
S.0004       20 FORMAT(10H MAGNITUDE,5X,5HPHASE,10H FREQUENCY)
S.0005          READ(5,30)R,C
S.0006       30 FORMAT(2E10.2)
S.0007          WP1=1./(R*C)
S.0008          W=WP1/32.
S.0009          DO 40 I=1,9
S.0010          W=2.*W
S.0011          DB=-20.*ALOG10((1.+(W/WP1)**2)**.5)
S.0012          PHASE=-(ATAN(W/WP1))*57.3
S.0013       40 WRITE(6,50)DB,PHASE,W
S.0014       50 FORMAT(3F10.4)
S.0015          RETURN
S.0016          END
```

Figure 3.53 A Fortran program for computing the log magnitude and phase of the network of Figure 3.52

magnitude and phase and each iteration S.0013 and S.0014 prints the numerical values of the dB magnitude, phase, and the corresponding angular frequency ω.

Figure 3.54 shows the computer output and corresponding plots for a numerical value of $R = 1.00E5$ and $C = 1.00E-6$. It can be observed

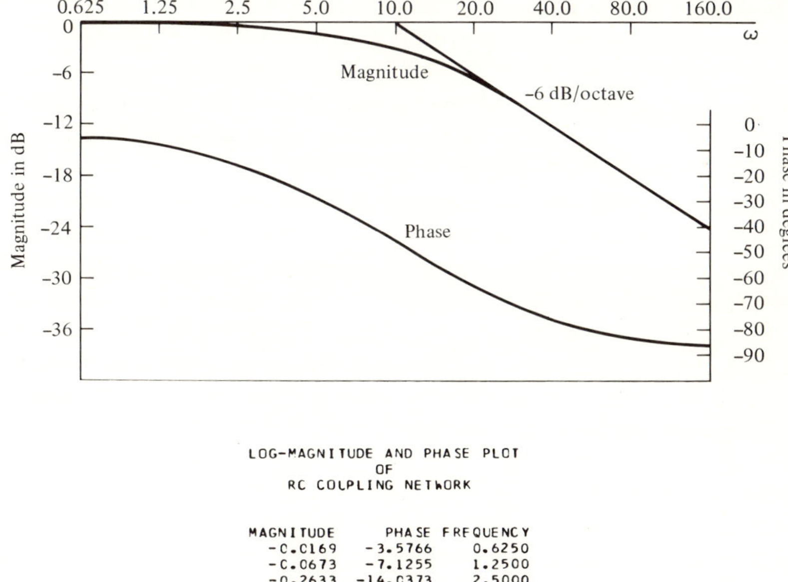


```
LOG-MAGNITUDE AND PHASE PLOT
              OF
     RC COUPLING NETWORK


MAGNITUDE      PHASE  FREQUENCY
  -0.0169    -3.5766     0.6250
  -0.0673    -7.1255     1.2500
  -0.2633   -14.0373     2.5000
  -0.9691   -26.5670     5.0000
  -3.0103   -45.0033    10.0000
  -6.9897   -63.4396    20.0000
 -12.3045   -75.9693    40.0000
 -18.1291   -82.8810    80.0001
 -24.0993   -86.4300   160.0001
```

Figure 3.54 The plotted results of the output of Figure 3.53

that for $\omega \ll \omega_{p1}$, the magnitude is asymptotic to 0 dB, while at high frequencies ($\omega \gg \omega_{p1}$), it is asymptotic to a straight line that has a -6 dB/octave (-20 dB/decade) slope that intersects the low-frequency asymptote at the corner frequency ω_{p1}. The actual departure of the curve from its asymptotes is limited to a range of approximately three octaves (or about one decade) above and below the corner frequency. At the corner frequency, the departure is 3 dB and at one octave above and below the corner frequency, the departure is approximately 1 dB.

It may also be observed that for a simple pole the phase lags approach 0° at low frequencies and $-90°$ at high frequencies. In addition we can see that the phase curve has odd symmetry about the corner frequency (phase shift $-45°$) and this corresponds to an 18.4° increment one octave either side of the corner frequency.

The Stability of a Speed Regulator

The following problem is used to illustrate the practical application of the above plots in determining the stability of a closed loop system. Consider the automatic speed regulator shown in Figure 3.55 in which the exciter and main generator are driven by a separate prime mover and the field of the controlled motor is excited by a constant voltage. An examination of this figure will reveal that the operation of this system is as follows: an input voltage v_1 is supplied to the dc amplifier that is equal to the difference between the reference voltage V and the dc tachometer generator voltage v_5. Since the voltage of the tachometer generator is directly proportional to the angular velocity of the controlled motor, under equilibrium conditions the input to the amplifier will be equal to the difference between the reference voltage V and the voltage of the

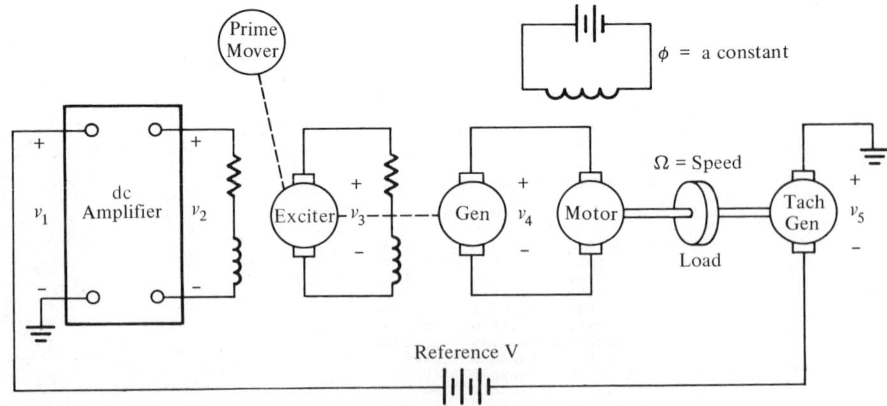

Figure 3.55 An automatic speed regulator

tachometer generator, which will be directly proportional to the equilibrium speed of the system. The output voltage of the amplifier is applied to the field of the exciter and it produces an exciter voltage that is proportional to its field current. The exciter voltage in turn produces a main generator field current that determines the magnitude of the voltage of the main generator, which, in turn, determines the magnitude of the armature current of the motor. The magnitude of this armature current, in the presence of the constant excitation field, will produce a developed torque that is equal and opposite to the load or retarding torque. Thus in equilibrium the developed torque is equal to and opposite to the retarding torque so that the net acceleration torque is zero and the system runs at a constant velocity.

Now if there is a sudden increase (decrease) in the load, the developed torque will be less (more) than the retarding torque and thus the net acceleration torque will be negative (positive) and the system will slow down (speed up). This will cause the tachometer generator voltage to decrease (increase), the input and output voltage of the amplifier to increase (decrease), the exciter field current and armature voltage to increase (decrease), the main field current and generator voltage to increase (decrease), and the motor armature current and developed torque to increase (decrease). A new equilibrium condition will be reached when the net acceleration torque is equal to zero. The actual reduction (increase) in speed that was needed to provide the increase (decrease) in the developed torque does not have to be large if the loop gain of the system is large. As a consequence, large changes in load can take place with relatively small (negligible) changes in motor speed.

In order to analyze this unit, let us assume that the voltage transfer ratio of each of the elements in the system is known. We may thus break the loop at the reference voltage and write the overall open loop voltage transfer ratio (v_5/v_1).

$$\frac{v_5}{v_1} = \frac{v_2}{v_1} \cdot \frac{v_3}{v_2} \cdot \frac{v_4}{v_3} \cdot \frac{\Omega}{v_4} \cdot \frac{v_5}{\Omega}$$

or

$$G(j\omega) = \frac{v_5(j\omega)}{v_1(j\omega)} = 100 \frac{1}{1 + j\omega/10} \frac{4}{1 + j2\omega} \frac{10}{1 + j4\omega} 0.1$$

$$= \frac{400}{(1 + j\omega/10)(1 + j2\omega)(1 + j4\omega)}$$

This transfer function can now be computed in a manner similar to the way it was calculated in the last problem and the resulting magnitude and phase can be plotted as a function of frequency.

Figure 3.56 shows a Fortran program for calculating the dB magnitude

```
C        LOG-MAGNITUDE AND PHASE PLOT
C                   OF AN
C        AUTOMATIC SPEED REGULATOR
C          UNCORRECTED
C
         WRITE(6,1)
       1 FORMAT('1LOG-MAGNITUDE AND PHASE PLOT',/13X,'OF',/,' AUTOMATIC SPE
        1ED REGULATOR',////,' THE UNCORRECTED SYSTEM',///,' MAGNITUDE',5X,'
        2PHASE',' FREQUENCY',/,5X,'DB',10X,'DEG',7X,'RPS')
         W=0.0
         DO 10 I=1,50
         W=W+.2
         DB=20.*(ALOG10(400.)-ALOG10((1.+(W/10.)**2)**.5)-ALOG10((1.+(2.*W)
        1**2)**.5)-ALOG10((1.+(4.*W)**2)**.5))
         PHASE=(-ATAN(W/10.)-ATAN(2.*W)-ATAN(4.*W))*180./3.14159
      10 WRITE(6,2) DB,PHASE,W
       2 FORMAT(3F10.2)
         RETURN
         END
```

Figure 3.56 A Fortran program for calculating the magnitude and phase of the open loop transfer function of Figure 3.55

and phase versus frequency and Figure 3.57 shows the corresponding results and plot. An examination of the data or Bode plot indicates that the system will be unstable when the loop is closed. This can be understood by recalling that threshold stability occurs when the 0-dB point on the magnitude curve occurs at the same frequency as the $-180°$ point on the phase curve. The closed loop system will thus be unstable when, at the frequency corresponding to a phase of $-180°$, the magnitude is greater than 0 dB. An examination of the data or plot of Figure 3.58 shows that this system will be unstable, since at the frequency corresponding to phase crossover ($\theta = -180°$), the magnitude is approximately 15.6 dB. It is therefore necessary to introduce a corrective network in order to stabilize the closed loop system.

A suitable corrective network is shown in Figure 3.58. The open circuit voltage transfer ratio of this corrective network is given by

$$\frac{v_1(j\omega)}{v(j\omega)} = \frac{R_1}{R_1 + R} \frac{1 + j\omega RC}{1 + j\omega \dfrac{R_1 RC}{R_1 + R}}$$

$$= \frac{1}{M} \frac{1 + j\omega/\omega_{01}}{1 + j\omega/(M\omega_{01})} = \frac{1}{M} \frac{1 + j\omega/\omega_{01}}{1 + j\omega/\omega_{p1}}$$

where

$$\omega_{01} = 1/RC$$

$$\omega_{p1} = \frac{R_1 + R}{R_1} \frac{1}{RC} = M\omega_{01}$$

Since it is beyond the scope of this problem to consider the actual design of this network, let us assume that we have arrived at the following element values:

$$R = 40 \text{ Kilohms}$$
$$R_1 = 13.3 \text{ Kilohms}$$
$$C = 10\ \mu\text{f}$$

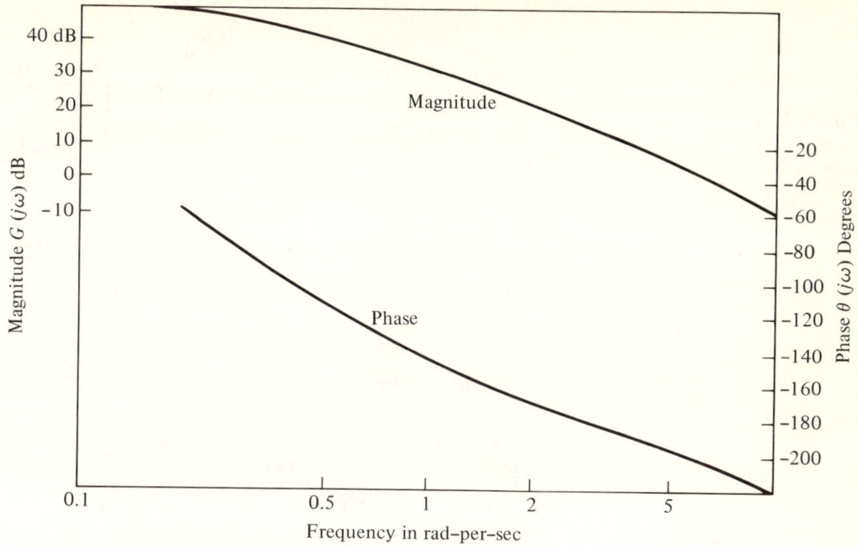

LOG-MAGNITUDE AND PHASE PLOT
OF
AUTOMATIC SPEED REGULATOR

THE UNCORRECTED SYSTEM

MAGNITUDE DB	PHASE DEG	FREQUENCY RPS			
49.25	-61.61	0.20	4.25	-199.23	5.20
44.37	-98.95	0.40	3.53	-200.43	5.40
39.85	-121.01	0.60	2.82	-201.59	5.60
35.99	-135.21	0.80	2.14	-202.72	5.80
32.70	-145.11	1.00	1.48	-203.81	6.00
29.87	-152.45	1.20	0.84	-204.88	6.20
27.39	-158.19	1.40	0.21	-205.92	6.40
25.20	-162.86	1.60	-0.40	-206.92	6.60
23.22	-166.77	1.80	-1.00	-207.90	6.80
21.44	-170.15	2.00	-1.58	-208.86	7.00
19.80	-173.12	2.20	-2.15	-209.79	7.20
18.30	-175.78	2.40	-2.71	-210.70	7.40
16.90	-178.20	2.60	-3.26	-211.59	7.60
15.59	-180.42	2.80	-3.79	-212.45	7.80
14.37	-182.47	3.00	-4.31	-213.29	8.00
13.22	-184.40	3.20	-4.83	-214.12	8.20
12.13	-186.21	3.40	-5.33	-214.92	8.40
11.09	-187.92	3.60	-5.82	-215.70	8.60
10.11	-189.55	3.80	-6.31	-216.47	8.80
9.17	-191.10	4.00	-6.78	-217.22	9.00
8.27	-192.59	4.20	-7.25	-217.95	9.20
7.40	-194.01	4.40	-7.71	-218.66	9.40
6.57	-195.39	4.60	-8.16	-219.36	9.60
5.77	-196.71	4.80	-8.61	-220.04	9.80
5.00	-197.99	5.00	-9.04	-220.71	10.00

Figure 3.57 Bode plot for the uncorrected system of an automatic speed regulator

Figure 3.58 A corrective network

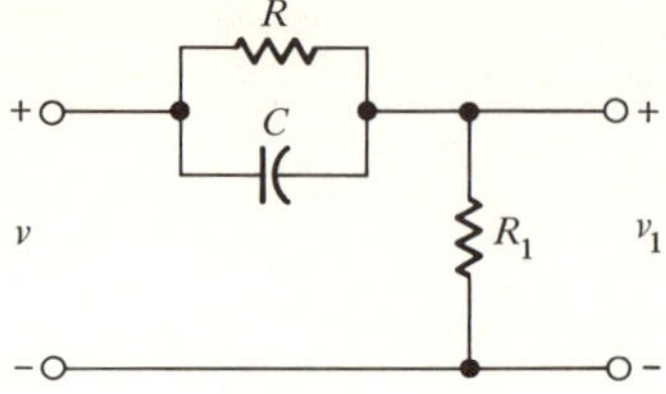

The voltage transfer ratio of the corrected system now becomes

$$\frac{v_5}{v} = \frac{v_1}{v}\frac{v_5}{v_1}$$

Thus

$$\begin{aligned} G(j\omega) &= \frac{1}{4}\frac{1 + j\omega/2.5}{1 + j\omega/10}\frac{400}{(1 + j\omega/10)(1 + j2\omega)(1 + j4\omega)} \\ &= \frac{100(1 + j\omega/2.5)}{(1 + j\omega/10)^2(1 + j2\omega)(1 + j4\omega)} \end{aligned}$$

and we may once again use the digital computer to compute the log magnitude and phase versus frequency for this corrected system. Figure 3.59 is the Fortran program and Figure 3.60 is the corresponding results and Bode plot. An examination of the open loop, transfer function response shows that the corresponding closed loop system will be stable because at the frequency corresponding to phase crossover (8.2 rps), the magnitude is approximately -8.4 dB. There is thus a gain margin of 8.4 dB. The phase margin can be similarly found by determining the phase shift at the frequency corresponding to 0-dB loop gain. Since, for this corrected system, the phase shift at 4.6 rps is $-158.69°$, we see that we will have a phase margin of 21.31° (180.0 − 158.69).

```
                C        LOG-MAGNITUDE AND PHASE PLOT
                C                     OF AN
                C        AUTOMATIC SPEED REGULATOR
                C           CORRECTED
                C
S.0001                   WRITE(6,1)
S.0002                 1 FORMAT('1LOG-MAGNITUDE AND PHASE PLOT',/13X,'OF',/,' AUTOMATIC SPE
                        1ED REGULATOR',////,' THE CORRECTED SYSTEM  ',///,' MAGNITUDE',5X,'
                        2PHASE',' FREQUENCY',/,5X,'DB',10X,'DEG',7X,'RPS')
S.0003                   W=0.0
S.0004                   DO 10 I=1,50
S.0005                   W=W+.2
S.0006                   DB=20.*(ALOG10(100.)+ALOG10((1.+(W/2.5)**2)**.5)-ALOG10(1.+(W/10.)
                        1**2)-ALOG10((1.+(2.*W)**2)**.5)-ALOG10((1.+(4.*W)**2)**.5))
S.0007                   PHASE=(ATAN(W/2.5)-2.*ATAN(W/10.)-ATAN(2.*W)-ATAN(4.*W))*180./3.14
S.0008                10 WRITE(6,2)DB,PHASE,W
S.0009                 2 FORMAT(3F10.2)
S.0010                   RETURN
S.0011                   END
```

Figure 3.59 A Fortran program for calculating the magnitude and phase of the corrected system

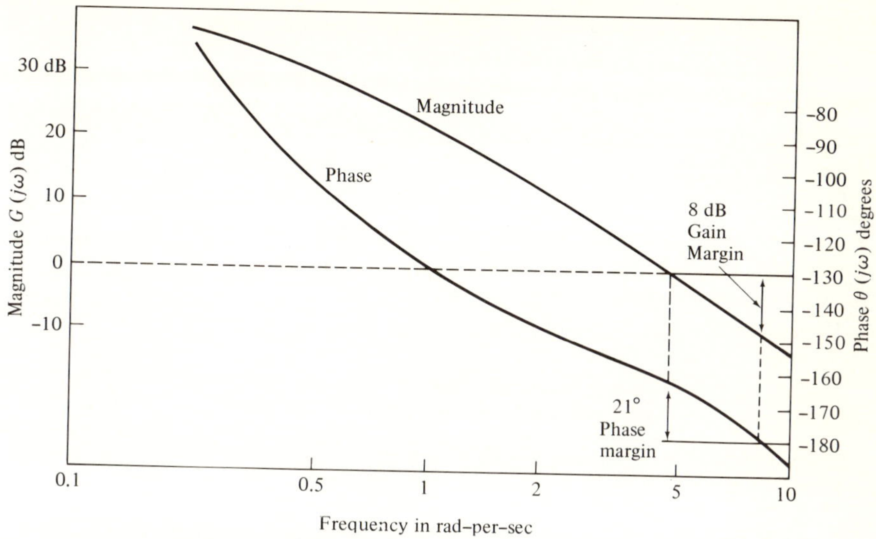

LOG-MAGNITUDE AND PHASE PLOT
OF
AUTOMATIC SPEED REGULATOR

THE CORRECTED SYSTEM

MAGNITUDE DB	PHASE DEG	FREQUENCY RPS			
37.23	-58.21	0.20	-1.57	-162.46	5.20
32.43	-92.19	0.40	-2.09	-163.72	5.40
28.04	-111.00	0.60	-2.61	-164.98	5.60
24.35	-122.11	0.80	-3.11	-166.23	5.80
21.26	-129.08	1.00	-3.60	-167.48	6.00
18.67	-133.72	1.20	-4.07	-168.72	6.20
16.45	-136.98	1.40	-4.54	-169.96	6.40
14.54	-139.40	1.60	-5.00	-171.18	6.60
12.86	-141.29	1.80	-5.45	-172.39	6.80
11.37	-142.87	2.00	-5.89	-173.59	7.00
10.05	-144.25	2.20	-6.33	-174.78	7.20
8.85	-145.52	2.40	-6.75	-175.96	7.40
7.76	-146.72	2.60	-7.17	-177.12	7.60
6.76	-147.89	2.80	-7.59	-178.27	7.80
5.83	-149.05	3.00	-8.00	-179.40	8.00
4.97	-150.22	3.20	-8.40	-180.51	8.20
4.16	-151.39	3.40	-8.79	-181.62	8.40
3.40	-152.57	3.60	-9.19	-182.70	8.60
2.68	-153.77	3.80	-9.57	-183.77	8.80
2.00	-154.99	4.00	-9.95	-184.82	9.00
1.34	-156.21	4.20	-10.33	-185.86	9.20
0.72	-157.45	4.40	-10.70	-186.88	9.40
0.12	-158.69	4.60	-11.07	-187.88	9.60
-0.46	-159.95	4.80	-11.43	-188.87	9.80
-1.02	-161.20	5.00	-11.79	-189.84	10.00

Figure 3.60 Bode plot for the corrected system of an automatic speed regulator

The Frequency Evaluation of a General Transfer Function

The first example of this section showed how a Fortran program could be written to evaluate the magnitude and phase versus frequency of a simple single-pole transfer function. We then extended this to the more complicated transfer functions that were encountered in the Speed Regulator problem and showed how these magnitude and phase curves could be used to determine the stability of the system. Since all these examples involved only real poles or real zeros, we now turn to the more general problem that would also include complex poles as well as complex zeros.

Consider the following general system transfer function:

$$G(jW(I)) = \frac{\prod_{J=1}^{\mathrm{NTERMS}} S(J)*(1.+jW(I)*T(J) - W(I)**2*TT(J))^{M(J)}}{(jW(I))^{N}}$$

The quantity $S(J)$ is the scale factor or gain of the individual factors that make up the transfer function. The $jW(I)$ term in the denominator corresponds to the net number of poles or zeros located at the origin. For example, if $N = +2$, this would correspond to two poles at the origin, while if $N = -3$, this would represent three zeros at the origin.

NTERMS is limited by the DIMENSION statement to 25 and thus the program permits the evaluation of transfer functions that have up to 25 other factors. These factors can be distinct or multiple and real or complex poles or zeros. If the exponent $M(J)$ is positive, the factor corresponds to a zero while if the exponent is negative, the factor represents the effects of a pole. The magnitude of $M(J)$ determines the multiplicity.

If $TT(J) = 0$, the factor corresponds to a real pole or zero. If $TT(J)$ does not equal zero, the factor can represent either complex poles or complex zeros. For example, if $TT(J) = 0$ and $T(J) = 5$, this would correspond to

$$1 + j\omega*5$$

or

$$1 + s5$$

which would be due to a real pole (or zero) located at $s = -1/5$. On the other hand, if $TT(J) = 2$ and $T(J) = 2$, this would correspond to

$$1 + j\omega*2 - \omega**2*2$$

or

$$1 + 2s + 2s^2$$

which would be due to a set of conjugate poles (or zeros) located at $s = -1/2 \pm j1/2$.

Figure 3.61 shows a Fortran program for evaluating a general transfer function. Initially the DIMENSION statement declares the variables S, T, TT, and M to have a maximum of 25 elements (this limits the number of factors to 25). It also limits the number of different frequency calculations to 100. The READ statement next reads in the value of N, which corresponds to the number of poles or zeros at the origin; the number of other factors, NTERMS; NUMW, the number of frequency calculations that will be made; the starting frequency, WLOW; and the frequency multiplying factor, WMULT.

Computer statements S.0004 and S.0005 now read in the individual factors that make up the overall transfer function. Computer statements S.0006–S.0019 are used to print out the heading and subheadings and the number of poles or zeros at the origin. The DO 20 loop is then used to calculate the effect of the individual scale factors.[8] Computer statement S.0023 sets the initial frequency $W(1)$ to WLOW, and the DO 30 loop evaluates the system transfer function for NUMW different frequencies, each one being WMULT times the value of the previous frequency value.

[8] Note that denominator scale factors must be given in terms of their reciprocal value.

```
                C      EVALUATION OF SYSTEM TRANSFER FUNCTION
                C
S.0001                 DIMENSION S(25),T(25),TT(25),M(25),W(100)
S.0002                 READ(5,1)N,NTERMS,NUMW,WLOW,WMULT
S.0003               1 FORMAT(3I5,E15.7,F10.4)
S.0004                 READ(5,2)(S(I),T(I),TT(I),M(I),I=1,NTERMS)
S.0005               2 FORMAT(3E15.7,I5)
S.0006                 WRITE(6,3)
S.0007               3 FORMAT('1EVALUATION OF SYSTEM TRANSFER FUNCTION',///)
S.0008                 IF(N)5,4,4
S.0009               4 WRITE(6,6)N
S.0010               6 FORMAT(10X,'THE NUMBER OF POLES AT THE ORIGIN =',I4,//)
S.0011                 GO TO 8
S.0012               5 WRITE(6,7)N
S.0013               7 FORMAT(10X,'THE NUMBER OF ZEROS AT THE ORIGIN =',I4,//)
S.0014               8 WRITE(6,9)
S.0015               9 FORMAT(9X,'SCALE',9X,'T',15X,'TT',15X,'M',/)
S.0016                 WRITE(6,10)(S(I),T(I),TT(I),M(I),I=1,NTERMS)
S.0017              10 FORMAT(5X,E12.5,3X,E12.5,5X,E12.5,5X,I3)
S.0018                 WRITE(6,11)
S.0019              11 FORMAT(///,12X,'FREQUENCY',13X,'MAGNITUDE',14X,'PHASE'/)
                C
                C      COMPUTE TRANSFER FUNCTION AS A FUNCTION OF FREQUENCY
                C
S.0020                 SCALE=1.0
S.0021                 DO 20 I=1,NTERMS
S.0022              20 SCALE=SCALE*S(I)
S.0023                 W(1)=WLOW
S.0024                 DO 30 I=1,NUMW
S.0025                 ZMAG=1./(W(I)**N)
S.0026                 FLTN=N
S.0027                 PHASE=-FLTN*90.
S.0028                 DO 40 J=1,NTERMS
S.0029                 ZMAG=ZMAG*(SQRT((1.-(W(I)**2)*TT(J))**2+(W(I)*T(J))**2))**M(J)
S.0030                 ANGLE=ATAN((W(I)*T(J))/(1.-(W(I)**2)*TT(J)))*57.29578
S.0031                 FLTM=M(J)
S.0032                 IF(1.-(W(I)**2)*TT(J)) 50,40,40
S.0033              50 ANGLE=ANGLE+180.
S.0034              40 PHASE=PHASE+ANGLE*FLTM
S.0035                 ZMAG=ZMAG*SCALE
S.0036                 WRITE(6,13)W(I),ZMAG,PHASE
S.0037              13 FORMAT(3(10X,E12.5))
S.0038              30 W(I+1)=W(I)*WMULT
S.0039                 RETURN
S.0040                 END
```

Figure 3.61 A Fortran program for evaluating a general transfer function

At the beginning of each iteration of the DO 30 loop, the magnitude (ZMAG) and phase contribution of the poles (or zeros) at the origin are computed. The DO 40 loop is then executed NTERMS, and each time the magnitude and angle contribution of the individual factors are multiplied and added to ZMAG and angle to determine the overall magnitude and phase for this particular frequency calculation.

Special attention should be called to computer statement S.0030. The library function ATAN is only a two-quadrant formula and this implies that phase shifts that are calculated by this function must be limited to $-\pi/2$ to $+\pi/2$ radians (or $-90°$ to $+90°$). Since the phase shift of a complex pole or a complex zero can vary from $-180°$ to $+180°$, computer statement S.0032 has been added to sense whenever the magnitude of the phase shift becomes more than 90°. Whenever 1. − (W(I)**2)*TT(J) is found to be negative, control is transferred to S.0033 and 180° are added to ANGLE to compensate for the two-quadrant limitation of ATAN. If 1. − (W(I)**2)*TT(J) is not found to be negative, no change is made in the evaluation of ANGLE.

After completing the DO 40 loop, computer statement S.0035 multiplies the scale factor contributions and ZMAG to find the overall magnitude value and S.0036–S.0037 print out the frequency, magnitude, and phase. Computer statement S.0038 then determines the new frequency for the next iteration of the DO 30 loop.

Figure 3.62 shows the computer results for the compensated speed-regulator transfer function

$$G(j\omega) = \frac{V_1(j\omega)}{V(j\omega)} = \frac{100(1 + j\omega/2.5)}{(1 + j\omega/10)^2(1 + j2\omega)(1 + j4\omega)}$$

By referring to Figure 3.60, we see that the results obtained here are identical to those found earlier, except that the former are expressed in dB.

Figure 3.63 shows the computer results for the following arbitrary and more complicated transfer function that involves complex poles and complex zeros:

$$G(j\omega) = \frac{(1 + j5\omega)(1 + j2\omega - \omega^2/10)^2}{(j\omega)^2(1 + j2\omega - 2\omega^2)(1 + j8\omega)^4}$$

The reader is encouraged to verify these results by hand-computing the value of the transfer function for one of the frequencies calculated in the computer program.

EVALUATION OF SYSTEM TRANSFER FUNCTION

THE NUMBER OF POLES AT THE ORIGIN = 0

SCALE	T	PT	M
0.10000E 03	0.40000E 00	0.0	1
0.10000E 01	0.10000E 00	0.0	-2
0.10000E 01	0.20000E 01	0.0	-1
0.10000E 01	0.40000E 01	0.0	-1

FREQUENCY	MAGNITUDE	PHASE
0.10000E 00	0.91108E 02	-0.31967E 02
0.11000E 00	0.89469E 02	-0.34898E 02
0.12100E 00	0.87576E 02	-0.38046E 02
0.13310E 00	0.85405E 02	-0.41415E 02
0.14641E 00	0.82938E 02	-0.45002E 02
0.16105E 00	0.80163E 02	-0.48803E 02
0.17716E 00	0.77076E 02	-0.52808E 02
0.19487E 00	0.73681E 02	-0.57004E 02
0.21436E 00	0.69997E 02	-0.61372E 02
0.23579E 00	0.66053E 02	-0.65886E 02
0.25937E 00	0.61892E 02	-0.70520E 02
0.28531E 00	0.57565E 02	-0.75242E 02
0.31384E 00	0.53134E 02	-0.80015E 02
0.34522E 00	0.48666E 02	-0.84804E 02
0.37975E 00	0.44228E 02	-0.89570E 02
0.41772E 00	0.39887E 02	-0.94275E 02
0.45949E 00	0.35704E 02	-0.98880E 02
0.50544E 00	0.31729E 02	-0.10335E 03
0.55599E 00	0.28006E 02	-0.10765E 03
0.61158E 00	0.24564E 02	-0.11175E 03
0.67274E 00	0.21421E 02	-0.11563E 03
0.74001E 00	0.18585E 02	-0.11926E 03
0.81402E 00	0.16054E 02	-0.12264E 03
0.89542E 00	0.13815E 02	-0.12575E 03
0.98496E 00	0.11854E 02	-0.12859E 03
0.10835E 01	0.10149E 02	-0.13117E 03
0.11918E 01	0.86755E 01	-0.13350E 03
0.13110E 01	0.74101E 01	-0.13559E 03
0.14421E 01	0.63283E 01	-0.13748E 03
0.15863E 01	0.54067E 01	-0.13918E 03
0.17449E 01	0.46235E 01	-0.14074E 03
0.19194E 01	0.39590E 01	-0.14219E 03
0.21113E 01	0.33954E 01	-0.14359E 03
0.23225E 01	0.29172E 01	-0.14496E 03
0.25547E 01	0.25108E 01	-0.14638E 03
0.28102E 01	0.21645E 01	-0.14788E 03
0.30912E 01	0.18684E 01	-0.14951E 03
0.34003E 01	0.16142E 01	-0.15131E 03
0.37403E 01	0.13950E 01	-0.15334E 03
0.41144E 01	0.12050E 01	-0.15561E 03
0.45258E 01	0.10396E 01	-0.15815E 03
0.49784E 01	0.89501E 00	-0.16099E 03
0.54762E 01	0.76813E 00	-0.16412E 03
0.60238E 01	0.65658E 00	-0.16755E 03
0.66262E 01	0.55845E 00	-0.17125E 03
0.72888E 01	0.47221E 00	-0.17522E 03
0.80177E 01	0.39666E 00	-0.17941E 03
0.88195E 01	0.33079E 00	-0.18378E 03
0.97014E 01	0.27372E 00	-0.18829E 03
0.10672E 02	0.22469E 00	-0.19288E 03

Figure 3.62 The evaluation of the transfer function

$$G(j\omega) = \frac{100(1 + j\omega/2.5)}{(1 + j\omega/10)^2(1 + j2\omega)(1 + j4\omega)}$$

by the computer program of Figure 3.61

```
EVALUATION OF SYSTEM TRANSFER FUNCTION


     THE NUMBER OF POLES AT THE ORIGIN =    2

    SCALE             T                TT               M

 0.10000E 01    0.50000E 01     0.0                    1
 0.10000E 01    0.20000E 01     0.20000E 01           -1
 0.10000E 01    0.80000E 01     0.0                   -4
 0.10000E 01    0.20000E 01     0.10000E 00            2


      FREQUENCY                MAGNITUDE              PHASE

   0.10000E 00              0.43140E 02           -0.29697E 03
   0.12500E 00              0.19979E 02           -0.31435E 03
   0.15625E 00              0.86399E 01           -0.33076E 03
   0.19531E 00              0.35321E 01           -0.34530E 03
   0.24414E 00              0.13909E 01           -0.35756E 03
   0.30518E 00              0.53820E 00           -0.36774E 03
   0.38147E 00              0.20750E 00           -0.37670E 03
   0.47684E 00              0.79795E-01           -0.38564E 03
   0.59605E 00              0.30099E-01           -0.39547E 03
   0.74506E 00              0.10848E-01           -0.40590E 03
   0.93132E 00              0.36956E-02           -0.41531E 03
   0.11642E 01              0.12087E-02           -0.42228E 03
   0.14552E 01              0.38876E-03           -0.42652E 03
   0.18190E 01              0.12490E-03           -0.42847E 03
   0.22737E 01              0.40350E-04           -0.42860E 03
   0.28422E 01              0.13136E-04           -0.42727E 03
   0.35527E 01              0.43123E-05           -0.42460E 03
   0.44409E 01              0.14292E-05           -0.42063E 03
   0.55511E 01              0.47922E-06           -0.41529E 03
   0.69389E 01              0.16319E-06           -0.40850E 03
```

Figure 3.63 The evaluation of the transfer function

$$G(j\omega) = \frac{(1 + j5\omega)(1 + j2\omega - \omega^2/10)^2}{(1 + j2\omega - 2\omega^2)(1 + j8\omega)^4(j\omega)^2}$$

by the computer program of Figure 3.61

The Calculation of the Frequency Response from the Time Response

In the automatic speed regulator example above, it was assumed that the transfer function of each of the elements was known. If a system is constructed of components whose transfer functions are not known analytically, or if the servomechanism is already built, experimental means must be employed to determine the transfer function so that the stability of the system can be studied. This is a laborious task and often leads to very inaccurate data. In addition, it is sometimes impossible to obtain the use of the system for measurement purposes if the system is already in operation.

It is therefore desirable to use the computer to determine the transfer function from a few rapid tests that can be made on the operating system. One such test is to analyze the output response that occurs from a step input. In order to show how this output time response will lead to the frequency response of the system, consider a typical output response such as that shown in Figure 3.64. This output response can be represented mathematically by a series of step functions that have been appropriately weighted and delayed in time. We thus may write

$$g(t) = A_1u(t) + A_2u(t - \Delta t) + A_3u(t - 2\Delta t) + \cdots$$
$$= \sum_{I=1}^{N} A_I u(t - I\Delta t)$$

and the Laplace transform of this output response would be given by

$$G(s) = \mathcal{L}\left[\sum_{I=1}^{N} A_I u(t - I\,\Delta t)\right]$$
$$= \sum_{I=1}^{N} A_I \frac{\epsilon^{-sI\,\Delta t}}{s}$$

Assuming, then, that we have a linear lumped system with an input excitation $f(t) = u(t)$ and an output response $g(t)$, we may solve for the transform $G(s)$ of the response from

$$G(s) = F(s)W(s)$$

where $W(s)$ is the transfer function of the system. We may thus write

$$G(s) = \frac{1}{s}\sum_{I=1}^{N} A_I\epsilon^{-sIt} = W(s)/s$$

where $F(s) = 1/s =$ the transform of the input step and

$$W(s) = \sum_{I=1}^{N} A_I\epsilon^{-sIt}$$

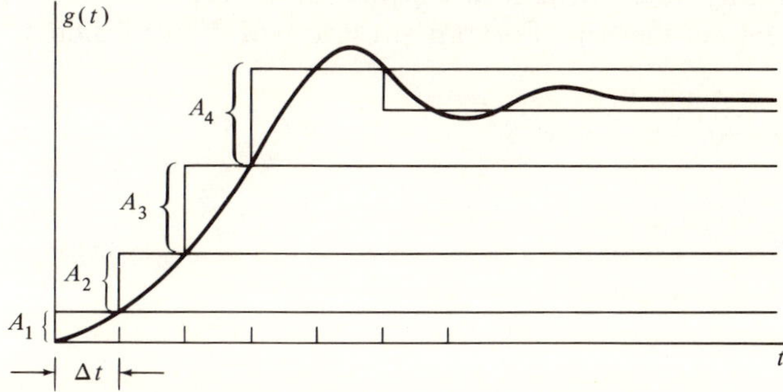

Figure 3.64 A typical system output response produced by a step input

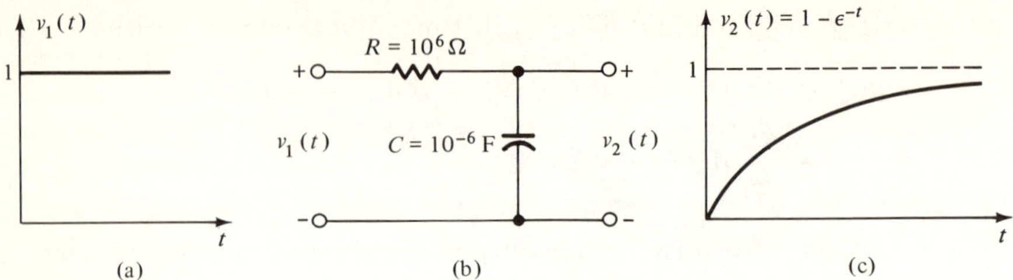

Figure 3.65 (a) The step input; (b) the system; (c) the output response

is the system transfer function. For the sinusoidal steady-state case $s = j\omega$ and $t = \infty$, so that

$$W(s) = \sum_{I=1}^{\infty} A_I \epsilon^{-jIt\omega}$$

$$= \sum_{I=1}^{\infty} A_I(\cos It\omega - j \sin It\omega)$$

and thus

$$|W(s)| = \left(\left(\sum_{I=1}^{\infty} A_I \cos It\omega\right)^2 + \left(\sum_{I=1}^{\infty} A_I \sin It\omega\right)^2\right)$$

and

$$\theta_o - \theta_i = \tan^{-1} \frac{-\sum_{I=1}^{\infty} A_I \sin It\omega}{\sum_{I=1}^{\infty} A_I \cos It\omega}$$

In order to test the Fortran program that we will write, let us first consider the step response of the simple RC network shown in Figure 3.65. Instead of experimentally measuring the output time response due to the unit input step, let us use the Fortran program of Figure 3.66 to

```
C        COMPUTER GENERATED INPUT DATA
C
         DIMENSION V(200)
         T=0.0
         T1=0.050
         DO 10 I=1,200
         T=T+T1
      10 V(I)=1.-EXP(-T)
         WRITE(7,1)(V(I),I=1,200)
         WRITE(6,1)(V(I),I=1,200)
       1 FORMAT(10F8.4)
         RETURN
         END
```

Figure 3.66 A Fortran program for generating the input data for the Fortran program of Figure 3.68

obtain accurate input data. Since the value of Δt must be small compared to the natural time constants of the system, but not too small to produce essentially nonzero values of A_I, let us use a value of $\Delta t = 0.05$ seconds and compute the output response every 0.05 seconds until $t = 10$ seconds (this assumes that the output time response has reached its steady-state value in ten times the time constant and that we have used a value of Δt corresponding to $\frac{1}{20}$ of the time constant). Figure 3.67 shows the input data that we will use. Note that the Fortran program of Figure 3.66 punches out as well as prints out the data; therefore, we avoid the need to hand-punch the data.

Let us now turn to the Fortran program of Figure 3.68. Initially the value of N, the number of data points; the time increment, TINC; the frequency multiplying factor, WMULT; and the maximum frequency, WMAX, are read in as data. Next the N data points are read in. Note that the first point $V(1)$ corresponds to the time TINC. The heading and input data are then printed, and this operation is followed by the printing of N, TINC, WMULT, WMAX, and the subheadings RADIAN FREQUENCY, DB MAGNITUDE, and ANGLE IN DEGREES.

Computer statement S.0010 then sets $A(1) = V(1)$, and the DO 10 loop, in turn, determines the remaining $N - 1$ step sizes (see Figure 3.64). Computer statement S.0013 is used to compute the starting frequency. The frequency is found by assuming that the transient has died out in N*TINC seconds. Using this value as the period T, the starting frequency is then taken as $\frac{1}{10}$ of $2\pi/T$.

Computer statement S.0014 is used to start the iteration loop and each time sets ω equal to WMULT times the previous value of ω. SUMC and SUMS are next set equal to zero, and the dummy variable BASE is given the value of TINC*W ($\Delta t * \omega$). The DO 20 loop now computes the sine and cosine summations. Note that mathematically we require I to run from 1 to ∞, but for all practical purposes we can let I run from

```
0.0488  0.0952  0.1393  0.1813  0.2212  0.2592  0.2953  0.3297  0.3624  0.3935
0.4230  0.4512  0.4780  0.5034  0.5276  0.5507  0.5726  0.5934  0.6133  0.6321
0.6501  0.6671  0.6834  0.6983  0.7135  0.7275  0.7408  0.7534  0.7654  0.7769
0.7878  0.7981  0.8079  0.8173  0.8262  0.8347  0.8428  0.8504  0.8577  0.8647
0.8713  0.8775  0.8835  0.8892  0.8946  0.8997  0.9046  0.9093  0.9137  0.9179
0.9219  0.9257  0.9293  0.9328  0.9361  0.9392  0.9422  0.9450  0.9477  0.9502
0.9526  0.9549  0.9571  0.9592  0.9612  0.9631  0.9649  0.9666  0.9683  0.9698
0.9713  0.9727  0.9740  0.9753  0.9765  0.9776  0.9787  0.9798  0.9807  0.9817
0.9826  0.9834  0.9842  0.9850  0.9857  0.9864  0.9871  0.9877  0.9883  0.9889
0.9894  0.9899  0.9904  0.9909  0.9913  0.9918  0.9922  0.9926  0.9929  0.9933
0.9936  0.9939  0.9942  0.9945  0.9948  0.9950  0.9953  0.9955  0.9957  0.9959
0.9961  0.9963  0.9965  0.9967  0.9968  0.9970  0.9971  0.9973  0.9974  0.9975
0.9976  0.9978  0.9979  0.9980  0.9981  0.9982  0.9983  0.9983  0.9984  0.9985
0.9986  0.9986  0.9987  0.9988  0.9988  0.9989  0.9989  0.9990  0.9990  0.9991
0.9991  0.9992  0.9992  0.9993  0.9993  0.9993  0.9994  0.9994  0.9994  0.9994
0.9995  0.9995  0.9995  0.9995  0.9996  0.9996  0.9996  0.9996  0.9996  0.9997
0.9997  0.9997  0.9997  0.9997  0.9997  0.9998  0.9998  0.9998  0.9998  0.9998
0.9998  0.9998  0.9998  0.9998  0.9998  0.9998  0.9999  0.9999  0.9999  0.9999
0.9999  0.9999  0.9999  0.9999  0.9999  0.9999  0.9999  0.9999  0.9999  0.9999
0.9999  0.9999  0.9999  0.9999  0.9999  0.9999  0.9999  0.9999  1.0000  1.0000
```

Figure 3.67 The input data for the program of Figure 3.68

```
        C      THE CALCULATION OF THE FREQUENCY RESPONSE
        C                         FROM
        C                   THE TIME RESPONSE
        C
S.0001         DIMENSION A(200),V(200)
S.0002         READ(5,1)N,TINC,WMULT,WMAX
S.0003       1 FORMAT(I5,3F10.3)
S.0004         READ(5,2) (V(I),I=1,N)
S.0005       2 FORMAT(10F8.4)
S.0006         WRITE(6,3) (V(I),I=1,N)
S.0007       3 FORMAT('1THE CALCULATION OF THE FREQUENCY RESPONSE FROM THE TIME R
              1SPONSE',//,(10F8.4))
S.0008         WRITE(6,4)N,TINC,WMULT,WMAX
S.0009       4 FORMAT(///,' N =',I5,'  T INCREMENT =',F6.3,/,'  FREQ MULTIPLIER =
              1',F6.3,'  MAXIMUM FREQUENCY =',F6.3,///,' RADIAN FREQUENCY',5X,'DB
              2 MAGNITUDE  ANGLE IN DEGREES',/)
S.0010         A(1)=V(1)
        C
        C      STEP INCREMENTS
        C
S.0011         DO 10 I=2,N
S.0012      10 A(I)=V(I)-V(I-1)
        C
        C      INITIALIZE FREQUENCY
        C
S.0013         W=2.*3.14159/(N*TINC*10.)
S.0014       5 W=W*WMULT
S.0015         SUMC=0.0
S.0016         SUMS=0.0
S.0017         BASE=TINC*W
        C
        C      SUM SINE AND COSINE
        C
S.0018         DO 20 I=1,N
S.0019         ARG=I*BASE
S.0020         SUMC=SUMC+A(I)*COS(ARG)
S.0021      20 SUMS=SUMS+A(I)*SIN(ARG)
S.0022         WMAG=SQRT(SUMC**2+SUMS**2)
S.0023         DB=20.*ALOG10(WMAG)
S.0024         ANGLE=ATAN(-SUMS/SUMC)*57.29578
S.0025         IF(SUMC) 30,40,40
S.0026      30 ANGLE=ANGLE-180.
S.0027      40 WRITE(6,6)W,DB,ANGLE
S.0028       6 FORMAT(3F17.7)
S.0029         IF(W-WMAX) 5,50,50
S.0030      50 CONTINUE
S.0031         RETURN
S.0032         END
```

Figure 3.68 A Fortran program that calculates the frequency response from the time response

1 to N (where N is the number of steps required for the output response to almost reach its final steady-state value). The dB magnitude and phase angle are then calculated and later printed out by computer statement S.0027. Note that as in the previous problem, we must test to see if the phase shift is less than $-90°$ (computer statement S.0025) and must make a correction if necessary (computer statement S.0026). Computer statement S.0029 finally tests to see whether the iteration should be repeated.

Figure 3.69 shows the computer output except for the data points that are deleted. Since $R = 10^6$ ohms and $C = 10^{-6}$ farads, the corner frequency equals $1/RC$ or 1 radian-per-second. We should thus have a -3 dB magnitude and a $-45°$ phase shift at this frequency. The computed results show that the magnitude is in close agreement, but the phase shift is slightly high. At one octave below the corner frequency (or 0.5 rps), the dB magnitude and phase shift should be approximately

```
THE CALCULATION OF THE FREQUENCY RESPONSE FROM THE TIME RESPONSE

N =  200   T INCREMENT = 0.050
 FREQ MULTIPLIER = 1.100  MAXIMUM FREQUENCY =10.000

RADIAN FREQUENCY       DB MAGNITUDE  ANGLE IN DEGREES

     0.6911486E-01    -0.2071845E-01    -0.4053310E 01
     0.7602626E-01    -0.2504607E-01    -0.4457176E 01
     0.8362883E-01    -0.3028703E-01    -0.4900968E 01
     0.9199166E-01    -0.3660677E-01    -0.5388534E 01
     0.1011907E 00    -0.4425277E-01    -0.5924030E 01
     0.1113097E 00    -0.5348061E-01    -0.6511950E 01
     0.1224406E 00    -0.6462079E-01    -0.7157201E 01
     0.1346846E 00    -0.7806456E-01    -0.7864996E 01
     0.1481529E 00    -0.9427905E-01    -0.8641015E 01
     0.1629680E 00    -0.1138158E 00    -0.9491250E 01
     0.1792647E 00    -0.1373417E 00    -0.1042200E 02
     0.1971911E 00    -0.1656408E 00    -0.1143991E 02
     0.2169100E 00    -0.1996381E 00    -0.1255181E 02
     0.2386009E 00    -0.2404280E 00    -0.1376469E 02
     0.2624608E 00    -0.2892805E 00    -0.1508554E 02
     0.2887067E 00    -0.3476694E 00    -0.1652109E 02
     0.3175771E 00    -0.4172929E 00    -0.1807773E 02
     0.3493346E 00    -0.5000765E 00    -0.1976100E 02
     0.3842678E 00    -0.5981662E 00    -0.2157544E 02
     0.4226943E 00    -0.7139614E 00    -0.2352405E 02
     0.4649635E 00    -0.8500517E 00    -0.2560796E 02
     0.5114595E 00    -0.1009174E 01    -0.2782587E 02
     0.5626051E 00    -0.1194186E 01    -0.3017374E 02
     0.6188653E 00    -0.1407968E 01    -0.3264444E 02
     0.6807514E 00    -0.1653306E 01    -0.3522726E 02
     0.7488261E 00    -0.1932799E 01    -0.3790807E 02
     0.8237083E 00    -0.2248688E 01    -0.4066890E 02
     0.9060786E 00    -0.2602665E 01    -0.4348862E 02
     0.9966860E 00    -0.2995669E 01    -0.4634378E 02
     0.1096354E 01    -0.3427770E 01    -0.4921097E 02
     0.1205988E 01    -0.3898365E 01    -0.5206947E 02
     0.1326586E 01    -0.4406700E 01    -0.5490089E 02
     0.1459243E 01    -0.4952028E 01    -0.5768515E 02
     0.1605165E 01    -0.5532467E 01    -0.6039771E 02
     0.1765680E 01    -0.6144006E 01    -0.6302187E 02
     0.1942246E 01    -0.6783468E 01    -0.6556100E 02
     0.2136470E 01    -0.7450822E 01    -0.6800710E 02
     0.2350115E 01    -0.8140757E 01    -0.7033148E 02
     0.2585124E 01    -0.8846622E 01    -0.7257768E 02
     0.2843635E 01    -0.9576872E 01    -0.7474603E 02
     0.3127995E 01    -0.1031870E 02    -0.7678175E 02
     0.3440793E 01    -0.1107696E 02    -0.7878250E 02
     0.3784870E 01    -0.1184036E 02    -0.8063748E 02
     0.4163355E 01    -0.1261952E 02    -0.8251682E 02
     0.4579687E 01    -0.1339705E 02    -0.8432480E 02
     0.5037653E 01    -0.1419528E 02    -0.8598891E 02
     0.5541415E 01    -0.1498320E 02    -0.8772076E 02
     0.6095553E 01    -0.1579414E 02    -0.8952415E 02
     0.6705105E 01    -0.1659129E 02    -0.9126274E 02
     0.7375611E 01    -0.1738258E 02    -0.9293155E 02
     0.8113168E 01    -0.1818581E 02    -0.9475848E 02
     0.8924479E 01    -0.1901416E 02    -0.9645947E 02
     0.9816922E 01    -0.1979546E 02    -0.9835133E 02
     0.1079861E 02    -0.2058443E 02    -0.1002460E 03
```

Figure 3.69 The output results for the computer program of Figure 3.68 for TINC = 0.05

equal to -1 dB and $-26.6°$, respectively, and at one octave above the corner frequency (or 2.0 rps), the dB magnitude and phase shift should be approximately equal to -7 dB and $-63.4°$, respectively. Again we see that in both cases the phase shift is slightly high. In fact, we can see that the phase shift becomes more negative than $-90°$ as the frequency approaches 10.0 rps. Since this is impossible for a simple pole, one of our assumptions must have caused this error. Should N have been larger?

```
THE CALCULATION OF THE FREQUENCY RESPONSE FROM THE TIME RSPONSE

0.0247  0.0488  0.0723  0.0952  0.1175  0.1393  0.1605  0.1813  0.2015  0.2212
0.2404  0.2592  0.2775  0.2953  0.3127  0.3297  0.3462  0.3624  0.3781  0.3935
0.4084  0.4230  0.4373  0.4512  0.4647  0.4780  0.4908  0.5034  0.5157  0.5276
0.5393  0.5507  0.5618  0.5726  0.5831  0.5934  0.6035  0.6133  0.6228  0.6321
0.6412  0.6501  0.6587  0.6671  0.6753  0.6834  0.6912  0.6988  0.7062  0.7135
0.7206  0.7275  0.7342  0.7408  0.7472  0.7534  0.7595  0.7654  0.7712  0.7769
0.7824  0.7878  0.7930  0.7981  0.8031  0.8079  0.8127  0.8173  0.8218  0.8262
0.8305  0.8347  0.8388  0.8428  0.8466  0.8504  0.8541  0.8577  0.8612  0.8647
0.8680  0.8713  0.8744  0.8775  0.8806  0.8835  0.8864  0.8892  0.8919  0.8946
0.8972  0.8997  0.9022  0.9046  0.9070  0.9093  0.9115  0.9137  0.9158  0.9179
0.9199  0.9219  0.9238  0.9257  0.9276  0.9293  0.9311  0.9328  0.9345  0.9361
0.9376  0.9392  0.9407  0.9422  0.9436  0.9450  0.9463  0.9477  0.9490  0.9502
0.9514  0.9526  0.9538  0.9549  0.9561  0.9571  0.9582  0.9592  0.9602  0.9612
0.9622  0.9631  0.9640  0.9649  0.9658  0.9666  0.9674  0.9683  0.9690  0.9698
0.9705  0.9713  0.9720  0.9727  0.9733  0.9740  0.9746  0.9753  0.9759  0.9765
0.9771  0.9776  0.9782  0.9787  0.9792  0.9798  0.9803  0.9807  0.9812  0.9817
0.9821  0.9826  0.9830  0.9834  0.9838  0.9842  0.9846  0.9850  0.9854  0.9857
0.9861  0.9864  0.9868  0.9871  0.9874  0.9877  0.9880  0.9883  0.9886  0.9889
0.9892  0.9894  0.9897  0.9899  0.9902  0.9904  0.9907  0.9909  0.9911  0.9913
0.9916  0.9918  0.9920  0.9922  0.9924  0.9926  0.9927  0.9929  0.9931  0.9933


N =  200  T INCREMENT = 0.025
 FREQ MULTIPLIER = 1.100  MAXIMUM FREQUENCY =10.000


RADIAN FREQUENCY     DB MAGNITUDE   ANGLE IN DEGREES

   0.1382298E 00   -0.1271066E 00   -0.7723259E 01
   0.1520527E 00   -0.1414726E 00   -0.8488975E 01
   0.1672578E 00   -0.1588300E 00   -0.9329082E 01
   0.1839835E 00   -0.1797986E 00   -0.1025027E 02
   0.2023817E 00   -0.2051143E 00   -0.1125974E 02
   0.2226198E 00   -0.2356716E 00   -0.1236498E 02
   0.2448816E 00   -0.2725282E 00   -0.1357393E 02
   0.2693695E 00   -0.3169526E 00   -0.1489468E 02
   0.2963063E 00   -0.3704570E 00   -0.1633540E 02
   0.3259367E 00   -0.4348305E 00   -0.1790414E 02
   0.3585302E 00   -0.5121772E 00   -0.1960846E 02
   0.3943830E 00   -0.6049530E 00   -0.2145497E 02
   0.4338210E 00   -0.7160220E 00   -0.2344881E 02
   0.4772028E 00   -0.8486477E 00   -0.2559267E 02
   0.5249228E 00   -0.1006482E 01   -0.2788589E 02
   0.5774148E 00   -0.1193539E 01   -0.3032320E 02
   0.6351559E 00   -0.1414035E 01   -0.3289297E 02
   0.6986710E 00   -0.1672153E 01   -0.3557602E 02
   0.7685377E 00   -0.1971584E 01   -0.3834366E 02
   0.8453910E 00   -0.2314926E 01   -0.4115762E 02
   0.9299296E 00   -0.2702778E 01   -0.4397176E 02
   0.1022922E 01   -0.3132818E 01   -0.4673694E 02
   0.1125213E 01   -0.3599096E 01   -0.4941335E 02
   0.1237733E 01   -0.4092607E 01   -0.5198636E 02
   0.1361505E 01   -0.4603996E 01   -0.5448485E 02
   0.1497654E 01   -0.5129137E 01   -0.5698343E 02
   0.1647418E 01   -0.5675332E 01   -0.5957170E 02
   0.1812159E 01   -0.6263165E 01   -0.6227972E 02
   0.1993373E 01   -0.6916358E 01   -0.6499339E 02
   0.2192709E 01   -0.7637992E 01   -0.6744543E 02
   0.2411978E 01   -0.8386615E 01   -0.6939810E 02
   0.2653173E 01   -0.9094589E 01   -0.7100238E 02
   0.2918489E 01   -0.9756451E 01   -0.7284532E 02
   0.3210335E 01   -0.1047640E 02   -0.7514813E 02
   0.3531366E 01   -0.1131422E 02   -0.7709108E 02
   0.3884501E 01   -0.1211167E 02   -0.7814413E 02
   0.4272948E 01   -0.1279554E 02   -0.7968742E 02
   0.4700240E 01   -0.1363047E 02   -0.8175359E 02
   0.5170261E 01   -0.1446891E 02   -0.8248572E 02
   0.5687284E 01   -0.1516369E 02   -0.8418103E 02
   0.6256009E 01   -0.1608781E 02   -0.8547150E 02
   0.6881606E 01   -0.1677841E 02   -0.8660390E 02
   0.7569762E 01   -0.1769823E 02   -0.8784398E 02
   0.8326734E 01   -0.1842186E 02   -0.8945662E 02
   0.9159402E 01   -0.1925925E 02   -0.8990337E 02
   0.1007534E 02   -0.2013782E 02   -0.9150038E 02
```

Figure 3.70 The output results for the computer program of Figure 3.68 for TINC = 0.025

Should TINC have been smaller? Or should we have supplied input data over a greater range of time (greater than 10 seconds)?

In order to answer the above questions in part, let us rerun the problem with TINC cut in half. Since N will be kept as 200, we will therefore be furnishing data to only five seconds (five times the time constant). Figure 3.70 shows that considerable improvement in the phase has resulted, but the phase shift still becomes more negative than $-90°$. The reader may experiment and find that by a combination of reducing TINC and specifying the output response over a greater period of time, very accurate phase results can be obtained. From a practical point of view, such accuracy is not required. The reader should also be aware that in a more complicated system the phase shift can become more negative than $-180°$. Since ATAN has been compensated to compute phase shifts up to only $-180°$, phase shifts less than $-180°$ will result in a jump in the phase value. Since the determination of the stability of a system involves phase shifts up to $-180°$, it is not necessary to compensate the ATAN function further.

PROBLEMS

3-1. Rewrite the program of Figure 3.3 so that θ_{min} is computed by using the library function EXP and a suitable iterative procedure.

3-2. Using the program of Figure 3.3, design a minimum-loss pad to match a 300-ohm television antenna to a 75-ohm coaxial line.

3-3. In certain locations in New York City, television signals must be attenuated in order to improve the picture quality.

   a. Using the appropriate computer program, design a 300-ohm attenuator ($R_{I1} = R_{I2} = 300$ ohms) to provide attenuation losses of 0, 1, 2, 4, 8, and 16 dB. Assuming that your attenuator will be connected between the 300-ohm twin lead-in wire and the receiver terminals, sketch the circuit diagram of your switching arrangement. Be sure you split the series arm resistor (use H-pad configuration) to maintain circuit symmetry.

   b. Repeat part a. using a 75-ohm coaxial line instead of the 300-ohm twin line.

3-4. Derive the design equations that were used in the program for obtaining the asymmetrical T-section for a prescribed transmission gain.

3-5. Modify the program of Figure 3.12 so that it will be able to determine the individual resistors required (as a function of R_0) for the 10-dB attenuator of Figure 3.14.

3-6. Repeat Problem 3-5 for the attenuator of Figure 3.16. Your program should be capable of determining the individual resistor values based on input data of the characteristic resistance, the number of attenuator positions, and the voltage ratio of successive attenuator positions.

3-7. Resistive T-sections may be converted to equivalent Pi-sections (wye to delta conversion) by means of the following relations:

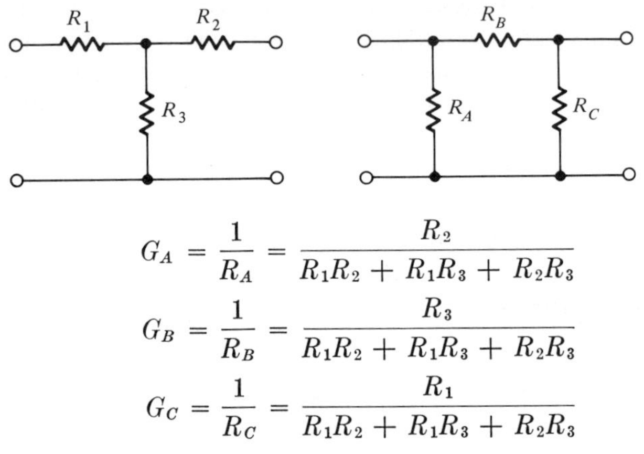

$$G_A = \frac{1}{R_A} = \frac{R_2}{R_1R_2 + R_1R_3 + R_2R_3}$$

$$G_B = \frac{1}{R_B} = \frac{R_3}{R_1R_2 + R_1R_3 + R_2R_3}$$

$$G_C = \frac{1}{R_C} = \frac{R_1}{R_1R_2 + R_1R_3 + R_2R_3}$$

Write a Fortran program that will read in the numerical values of R_1, R_2, and R_3 and compute the corresponding equivalent Pi-section values R_A, R_B, and R_C. Test your program with the following sets of input values

	R_1	R_2	R_3
a.	10 ohms	10	10
b.	10	10	5
c.	42.5	118.6	93.4

3-8. The dual of Problem 3-7 is the conversion of resistive Pi-sections to equivalent T-sections (delta to wye conversion). Repeat Problem 3-7 using the following design equations:

$$R_1 = \frac{G_C}{G_AB_B + G_AG_C + G_BG_C}$$

$$R_2 = \frac{G_A}{G_AG_B + G_AG_C + G_BG_C}$$

$$R_3 = \frac{G_B}{G_AG_B + G_AG_C + G_BG_C}$$

Check your program by using input data that correspond to the output Pi-section values obtained in Problem 3-7.

3-9. The design equations of the symmetrical resistive Pi-section

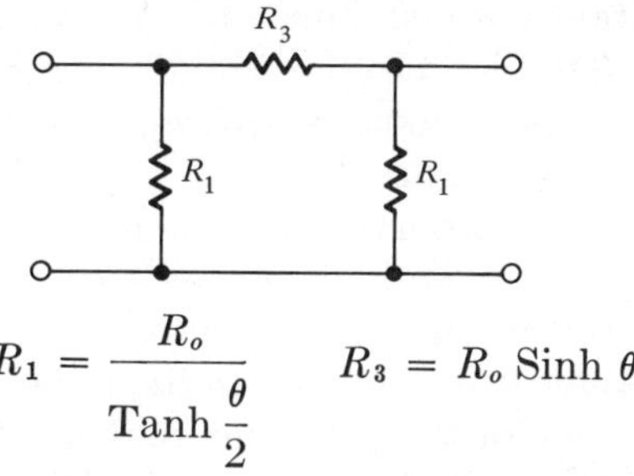

$$R_1 = \frac{R_o}{\text{Tanh}\,\dfrac{\theta}{2}} \qquad R_3 = R_o \text{ Sinh } \theta$$

are quite similar to those of the symmetrical resistive T-section. Develop an appropriate Fortran program for computing the values of the Pi-section elements, based on input data values of the characteristic resistance R_o and the decibel attenuation loss. Test your program by designing 50-ohm attenuators that have losses of 1, 2, 5, 10, and 20 decibels.

Using the values obtained for the 1-dB attenuator, terminate the Pi-section in 50 ohms and compute the input resistance and the decibel loss.

3-10. The bridged T attenuator

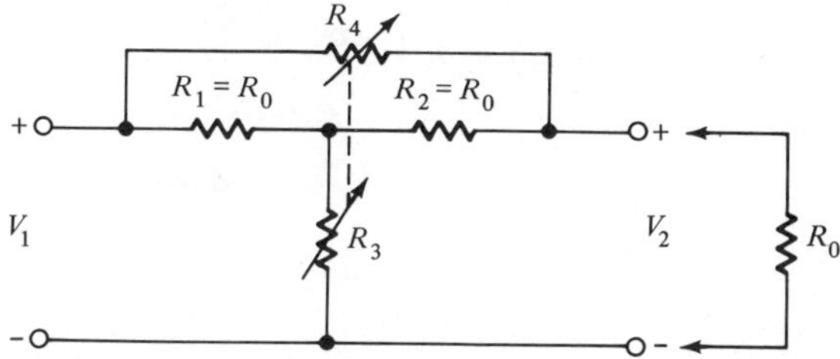

may be adjusted to give a specified loss by varying the values of R_3 and R_4. For example, if $R_3 = 0$ and $R_4 = \infty$, the input resistance will be equal to R_o (the characteristic resistance) and the attenuation will be infinite. Conversely, if $R_3 = \infty$ and $R_4 = 0$, the input resistance will also be equal to R_o, but the loss will be equal to zero. Between these two extremes the attenuation may be adjusted to any desirable value, with the input resistance still equal to R_o, by suitably adjusting the values of R_3 and R_4.

Develop a Fortran design program based on a specified characteristic resistance R_o, the desired voltage ratio V_2/V_1 (= VR), and the following design equations

$$R_3 = \frac{R_o}{\dfrac{1}{VR} - 1} \qquad R_4 = R_o\left(\frac{1}{VR} - 1\right)$$

Use this program to determine the values of R_3 and R_4 for a characteristic resistance of 600 ohms and voltage ratio values of 1.0, 0.9, . . . , and 0.0. Sketch the circuit diagram of your switching arrangement using one-pole twelve-position switch wafers similar to those shown in Figure 3.14.

3-11. The reader should observe that it is possible to use potentiometers for resistors R_3 and R_4 in Problem 3-10. Determine the problems involved in the mechanical coupling of the resistor shafts of R_3 and R_4 so that a continuous voltage ratio adjustment may be made. Hint: One of the problems that must be investigated is the change in the taper that will be required for the two potentiometers in order to track over a given voltage ratio range.

3-12. Problem 3-7 may be modified to convert complex impedance T-sections to complex admittance Pi-sections. The design equations are identical to those used in the resistive case except each term is complex

$$Y_A = G_A + jB_A = \frac{Z_2}{Z}$$

$$Y_B = G_B + jB_B = \frac{Z_3}{Z}$$

$$Y_C = G_C + jB_C = \frac{Z_1}{Z}$$

where

$$Z = Z_1Z_2 + Z_1Z_3 + Z_2Z_3$$

and

$$Z_1 = R_1 + jX_1$$
$$Z_2 = R_2 + jX_2$$
$$Z_3 = R_3 + jX_3$$

Write a Fortran program for this conversion assuming that the computer that will be used will not be capable of doing complex arithmetic. Test your program for the following sets of input

	Z_1	Z_2	Z_3
a.	1.0 + j1.0	1.0 + j1.0	1.0 + j1.0
b.	1.0 + j1.0	1.0 + j1.0	1.0 − j1.0
c.	3.2 + j5.6	6.8 + j9.2	3.7 − j13.9

3-13. Repeat Problem 3-12 for the case of the conversion of complex admittance Pi-sections to complex impedance T-sections.

3-14. Equivalent symmetrical T-networks may be obtained from measuring the short-circuit and open-circuit driving-point impedance of symmetrical two-port networks.

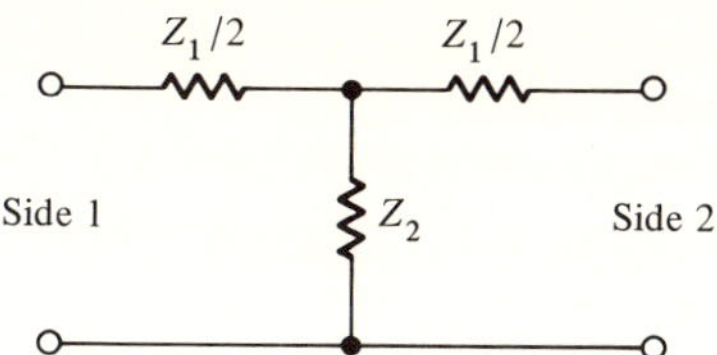

Referring to the symmetrical T-network shown, we see that the open-circuit driving-point impedance would be given by

$$Z_{oc} = R_{oc} + jX_{oc} = \frac{Z_1}{2} + Z_2$$

and the impedance seen at either side with the opposite side short circuited would be given by

$$Z_{sc} = R_{sc} + jX_{sc} = \frac{Z_1}{2} + \frac{(Z_1/2)Z_2}{(Z_1/2) + Z_2}$$

Solving these equations it thus follows that

$$Z_2 = \sqrt{Z_{oc}(Z_{oc} - Z_{sc})}$$

and

$$\frac{Z_1}{2} = Z_{oc} - Z_2$$

Using the above relations, develop a computer program for obtaining the equivalent T-network. Test your program using the following data:

$$Z_{oc} = 100 - j100 \qquad Z_{sc} = 80 + j15$$

3-15. Problem 3-14 can be extended to the asymmetrical T-case. Refering to the asymmetrical T-section shown below, we see that

$$Z_{o1} = Z_1 + Z_3$$

$$Z_{s1} = Z_1 + \frac{Z_2 Z_3}{Z_2 + Z_3}$$

$$Z_{o2} = Z_2 + Z_3$$

$$Z_{s2} = Z_2 + \frac{Z_1 Z_3}{Z_1 + Z_3}$$

Z_1 Z_2

Side 1 Z_3 Side 2

Solving these equations we find

$$Z_1 = Z_{o1} - Z_3$$
$$Z_2 = Z_{o2} - Z_3$$

and

$$Z_3 = \pm \sqrt{Z_{o1}(Z_{o2} - Z_{s2})}$$

Note that the alternate sign for Z_3 gives rise to two possible cases. In practice it is necessary to make an additional test to determine the phase condition due to a known load in order to establish the correct sign for Z_3.

Develop an appropriate computer program for obtaining both possible solutions. Test your program using the following data:

$$Z_{o1} = 0 \qquad Z_{s1} = 0.0 - j\,100.0$$
$$Z_{o2} = 0.0 + j\,100.0 \qquad Z_{s2} = \infty$$

3-16. The reactive matching program of Figure 3.20 may be used to obtain the L-C element values for a single-frequency resistive match. Assuming that both the source and load impedances may be complex, modify the resistive matching program to obtain a conjugate match. Hint: Adding an additional reactance and/or modifying X_2 will produce the required match.

3-17.

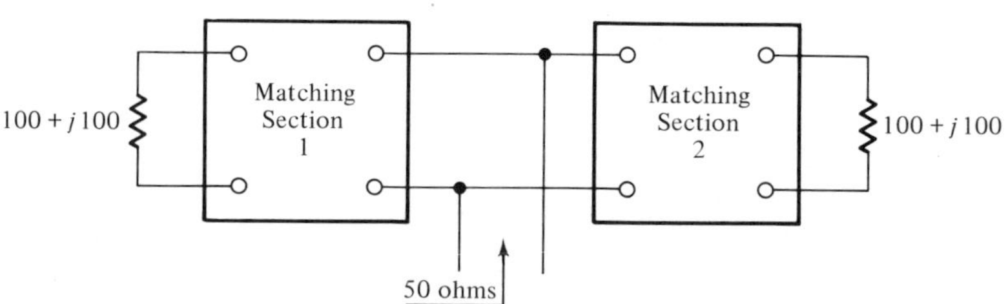

The current in load 1 is to have the same magnitude as the current in load 2, but its phase angle should lag by 90°. In addition, the impedance seen at the input of the paralleled matching sections should be 50 ohms. Using the appropriate computer programs, design the two matching sections.

3-18. Modify the program of Figure 3.29 so that it is capable of determining the input resistance of a general ladder network. Test your program by using the following ladder values:

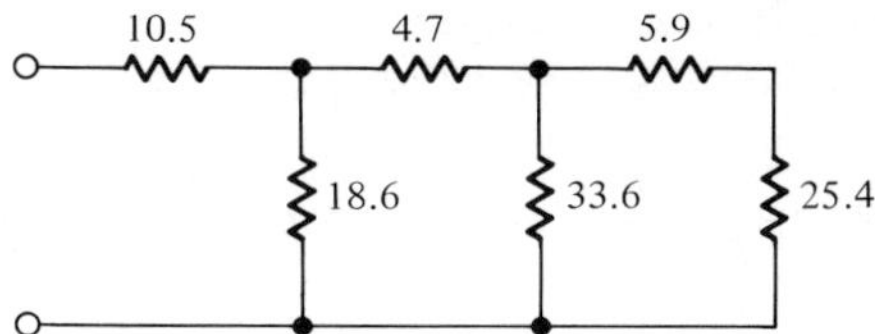

3-19. Assuming that the "0" logical level corresponds to a voltage of 0.10 volts (the saturation voltage of a transistor) instead of 0.0 volts, use the program of Figure 3.33 to determine the maximum error that will occur for any flip-flop register-storage condition. Hint: Is it necessary to investigate all 2^{10} register states?

3-20. Derive the equivalent network of Figure 3.32.

3-21. Modify the program of Figure 3.40 so that it is capable of finding the driving-point impedance of a ladder network, independent of the number of branches in the ladder.

3-22. Combine the computer programs of Figures 3.40 and 3.44 so that the insertion loss of the filter of Figure 3.39 can be computed when it is terminated with another filter. Note that this problem corresponds to the cascading of two identical filters, with the second filter terminated in 72 ohms.

3-23. Modify the program of Figure 3.46 so that the Q values for each coil are iterated separately. In addition, for each iteration of Q, the maximum and minimum value of the insertion loss in the passband should be found and whenever the condition of maximum insertion loss or maximum distortion is met, the program should go on to the next inductor. Hint: If the value of Q found for coil one is used in attempting to find the minimum Q that can be used for coil two, we will obviously find that coil two should have an infinite Q. A better approach might assume a small value of Q (say 50) for each coil and then increase the value of Q for coil one until a noticeable improvement results; then go on to coil two, and so on, until all coils have been considered. The entire process should then be repeated with smaller increments in the Q values until the passband requirements are met.

3-24. Modify the program of Figure 3.59 so that the gain and phase margins are automatically computed and printed.

3-25. Modify the program of Figure 3.59 so that a trial and error iteration of the choice of R_1, R, and C (Figure 3.58) may be made. Note that a change in C will not affect the low frequency gain. One investigation should therefore include the variation of C so that a minimum gain and phase margin is realized (say 10-dB gain margin and 30° phase margin).

3-26. Consider the one-section artificial delay line shown below

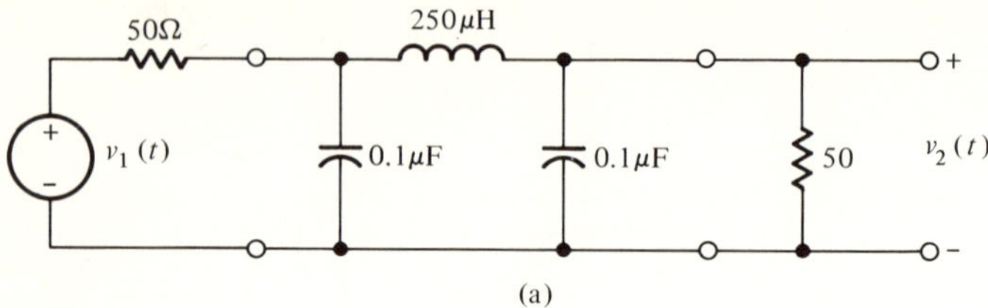

(a)

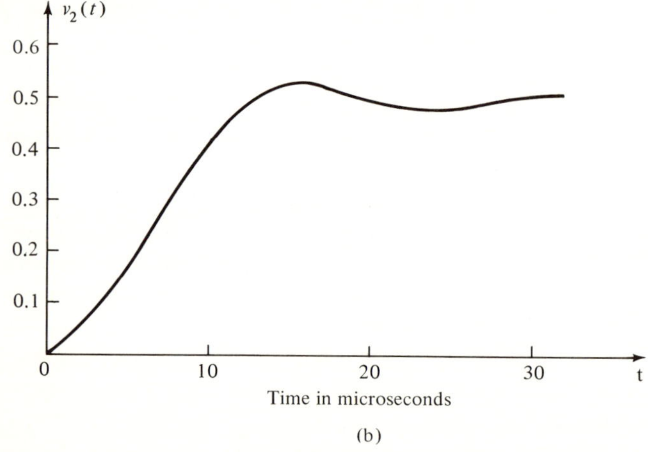

(b)

and the output time response due to a unit step input voltage.

Use the computer program of Figure 3.68 to determine the frequency response of this network $(V_2(j\omega)/V_1(j\omega))$. Verify your results by using the program of Figure 3.61 to evaluate the system transfer function.

CHAPTER 4

Numerical Methods for Finding Roots

4.1 INTRODUCTION

In Section 1.3 we discussed the four steps involved in the digital computer solution of an engineering problem, and we used the solution of the quadratic equation as a simple illustrative example. We pointed out that although the computer yielded exact solutions for four different sets of *A*-*B*-*C* coefficients, the program was far from perfect and, in reality, merely scratched the surface as far as its ability to find the roots of a second-degree polynomial for a general set of *A*-*B*-*C* coefficients.

In this chapter we continue with the solution of the quadratic equation and use it as an example to help point out the need for a deeper insight so that we may formulate better algorithms. We must always remember that the computer can only execute the instructions we give it; unfortunately, in many cases we are not aware of the full inplications of the instructions we write. Later we will also consider several other root-finding methods that are applicable in the solution of nonlinear algebraic and transcendental equations.

4.2 FINDING THE ROOTS OF THE QUADRATIC, CUBIC, AND QUARTIC EQUATIONS

THE QUADRATIC EQUATION

After studying Chapter 2, a closer examination of the program of Figure 1.8 will reveal that it is not capable of handling any set of *A*-*B*-*C* coefficients. For example, the use of F5.1 format code in the READ state-

```
C        ROOTS OF A QUADRATIC EQUATION
C        SINGLE PRECISION E15.7 FORMAT CODE
C
       1 READ(5,6)A,B,C
       6 FORMAT(3E15.7)
         IF(A)2,3,2
       3 X1=-C/B
         WRITE(6,9)A,B,C,X1
       9 FORMAT('0A =',E15.7,3X,'B =',E15.7,3X,'C =',E15.7,3X,'X1 =',E15.7,
        1//)
         GO TO 1
       2 D=B*B-4.*A*C
         IF(D)4,5,5
       5 X1=(-B+D**0.5)/(2.*A)
         X2=(-B-D**0.5)/(2.*A)
         WRITE(6,7)A,B,C
       7 FORMAT('0A =',E15.7,3X,'B =',E15.7,3X,'C =',E15.7)
         WRITE(6,10)X1,X2
      10 FORMAT(' X1=',E15.7,3X,'X2=',E15.7,//)
         GO TO 1
       4 X1R=-B/(2.*A)
         X1I=(-D)**0.5/(2.*A)
         X2R=-B/(2.*A)
         X2I=-((-D)**0.5/(2.*A))
         WRITE(6,7)A,B,C
         WRITE(6,8)X1R,X1I,X2R,X2I
       8 FORMAT(' X1R=',E15.7,3X,'X1I=',E15.7,/,' X2R=',E15.7,3X,'X2I=',E15
        1.7,//)
         GO TO 1
         STOP
         END
```

Figure 4.1 A program for finding the roots of a quadratic using E15.7 format code

ment automatically limits the range of the coefficients A, B, or C from -99.9 to 999.9. In a similar way, the use of F10.2 format code in the WRITE statement restricts the magnitude of the roots to values between 0.01 and 99,999.99. The upper limit here is not due to the specified width value w (Fw.d), but rather to this particular computer's limitation of preserving only seven significant digits in single-precision arithmetic.

Figure 4.1 shows a modification of the previous program in which E format is used instead of F format. The use of E format code permits the possibility of handling coefficients and roots that can range in value from approximately 10^{-75} to 10^{+75}. Since an IBM 360 computer has been used, only seven significant digits are preserved in single precision arithmetic and therefore E15.7 format has been used in both the READ and WRITE statements. The constant $d = 7$ permits the maximum number of significant figures to be read or printed. The constant $w = 15$ allows one more space than necessary and thus provides a separation of the coefficients on the data cards and also avoids possible erratic carriage control during printing.

Figure 4.2 shows that the output results are the same for the four data sets used in the original problem. A fifth data set has been added to show that even with the use of E15.7 format code, the combination of very large and very small numbers can lead to inaccurate results due to rounding errors. This may be seen by recalling that the quadratic

$$AX^2 + BX + C = 0$$

```
A =   0.0              B =  0.20C0000E 01    C =  0.1000000E 01   X1 = -0.50COCOOE CO


A =   0.1CCOCOOE 01    B =  0.2000000E 01    C =  0.1000000E 01
X1=  -0.1CCOCOCE 01    X2= -0.10C0000E 01


A =   0.10COCOCE 01    B =  0.0              C = -0.1000000E 01
X1=   0.10COCOCE 01    X2= -0.1C00000E 01


A =   0.10CCCOOE 01    B = -0.2000000E 01    C =  0.2000000E 01
X1R=   0.1CCOCOCE C1    X1I=   0.1CCOOCOE 01
X2R=   0.1CCCCCCE C1    X2I= -0.1000000E 01


A =   0.10COCOOE 01    B =  0.2000000E 07    C =  0.1000000E 03
X1=  -0.60COCOOE 01    X2= -0.1999994E 07


IHC217I
```

Figure 4.2 The output results of the program of Figure 4.1

has the following set of roots

$$X1 = -\frac{B}{2A} + \sqrt{\left(\frac{B}{2A}\right)^2 - \frac{C}{A}}$$

$$X2 = -\frac{B}{2A} - \sqrt{\left(\frac{B}{2A}\right)^2 - \frac{C}{A}}$$

so that for

$$X^2 + 2\ 10^6\, X + 100 = 0$$

we get

$$X1 = -10^6 + \sqrt{10^{12} - 100}$$
$$X2 = -10^6 - \sqrt{10^{12} - 100}$$

Recall now that in floating-point arithmetic each number is represented as a mantissa and an exponent, with the mantissa being less than one. For example, 10^{12} would be represented as

mantissa	exponent
.1000000	13

Thus, in adding two numbers such as 10^{12} and -100, the individual exponents are compared for equality, and the mantissa of the smaller number is shifted until the two exponents are equal. The mantisses are then added to give the sum. It should be clear that since only seven significant digits are preserved for each mantissa, considerable round-off error will result. An examination of the last two equations will show that since $\sqrt{10^{12} - 100}$ will be approximately equal to 10^6, we should expect

```
          C       ROOTS OF A QUADRATIC EQUATION
          C       DOUBLE PRECISION D25.16 FORMAT CODE
          C
S.0001            DOUBLE PRECISION A,B,C,D,X1,X2,X,X1R,X1I,X2R,X2I,ZX1,ZX2,ZRX1,ZIX1
                 1,ZRX2,ZIX2,F,G
S.0002          1 READ(5,6)A,B,C
S.0003          6 FORMAT(3E15.7)
S.0004            IF(A)2,3,2
S.0005          3 X1=-C/B
S.0006            ZX1=(A*X1+B)*X1+C
S.0007            WRITE(6,7)A,B,C
S.0008          7 FORMAT('0A =',D25.16,3X,'B =',D25.16,3X,'C =',D25.16)
S.0009            WRITE(6,11)X1,ZX1
S.0010         11 FORMAT(' X1 =',D25.16,3X,'ZX1 =',D25.16,//)
S.0011            GO TO 1
S.0012          2 D=B*B-4.*A*C
S.0013            IF(D)4,5,5
S.0014          5 X1=(-B+D**0.5)/(2.*A)
S.0015            X2=(-B-D**0.5)/(2.*A)
S.0016            ZX1=(A*X1+B)*X1+C
S.0017            ZX2=(A*X2+B)*X2+C
S.0018            WRITE(6,7)A,B,C
S.0019            WRITE(6,10)X1,X2,ZX1,ZX2
S.0020         10 FORMAT(' X1 =',D25.16,3X,'X2 =',D25.16,/,' ZX1 =',D25.16,2X,'ZX2 =
                 1',D25.16,//)
S.0021            GO TO 1
S.0022          4 X1R=-B/(2.*A)
S.0023            X1I=(-D)**0.5/(2.*A)
S.0024            ZRX1=(A*X1R+B)*X1R+C-A*X1I*X1I
S.0025            ZIX1=(2.*A*X1R+B)*X1I
S.0026            X2R=-B/(2.*A)
S.0027            X2I=-((-D)**0.5/(2.*A))
S.0028            ZRX2=(A*X2R+B)*X2R+C-A*X2I*X2I
S.0029            ZIX2=(2.*A*X2R+B)*X12
S.0030            WRITE(6,7)A,B,C
S.0031            WRITE(6,8)X1R,X1I,ZRX1,ZIX1
S.0032          8 FORMAT(' X1R =',D25.16,3X,'X1I =',D25.16,/,' ZRX1=',D25.16,3X,'ZIX
                 11=',D25.16)
S.0033            WRITE(6,9)X2R,X2I,ZRX2,ZIX2
S.0034          9 FORMAT(' X2R =',D25.16,3X,'X2I =',D25.16,/,' ZRX2=',D25.16,3X,'ZIX
                 12=',D25.16,//)
S.0035            GO TO 1
S.0036            STOP
S.0037            END
```

Figure 4.3 A Fortran program that uses D25.16 format code to find the roots of a quadratic

a very large error in $X1$ because we will be taking the difference of two large and nearly equal numbers.

One way to minimize the round-off error is to use DOUBLE PRECISION since in double precision the IBM 360 computer preserves 16 significant figures rather than seven. In adding 10^{12} and -100, the -100 number will affect the result since there are only ten orders of magnitude difference between the two numbers and 16 significant digits are being preserved. Figure 4.3 is a Fortran program that incorporates DOUBLE PRECISION. Note that initially we must declare (computer statement S.0001) all double-precision variables. In other words, any arithmetic statement that we wish the computer to execute in double precision must have all the variables involved declared as double-precision variables.[1] Note also that all WRITE statements use D25.16 format code so that 16 significant figures will be printed and $w = 25$ allows for two spaces more than necessary.

[1] For example, the reader will find it instructive to rerun the above program with the variable D omitted from the DOUBLE PRECISION list.

An accuracy check feature has also been added. This is accomplished by substituting each result back into the original quadratic to determine if the polynomial essentially has a zero value when evaluated for this root. Note that in the calculation of the value of the quadratic for a given root, the nested form has been used. The reason for this may be appreciated by first considering the evaluation of the quadratic

$$AX^2 + BX + C$$

using the form

$$A*X*X + B*X + C$$

in which three multiplications and two additions are required in contrast to the evaluation of the nested form of the quadratic

$$(A*X + B)*X + C$$

where only two multiplications and two additions are needed. Thus, a reduction in the number of arithmetic operations will not only reduce computer time, but will also reduce the effect of round-off error. Later in this chapter we will consider the nesting of polynomials in more general terms.

Note also that because of the inability of our computer to do complex arithmetic, the error check for complex roots must be handled in such a way that only real numbers are involved. This is done by realizing that if a complex root

$$X1 = X1R + jX1I$$

is substituted back into

$$ZX1 = AX^2 + BX + C$$

we have

$$\begin{aligned} ZX1 &= A(X1R + jX1I)^2 + B(X1R + jX1I) + C \\ &= (A*X1R + B)*X1R + C - A*X1I*X1I \\ &\qquad + j(2.*A*X1R + B)*X1I \\ &= ZRX1 + jZIX1 \end{aligned}$$

and we would expect both the real part ($ZRX1$) and the imaginary part ($ZIX1$) to be equal to zero if $X1 = X1R + jX1I$ is an exact root.

Figure 4.4 shows the results of double-precision calculations. Comparing the answers of the first four data sets to the previous computer results, we see that they are essentially the same. We should notice, however, that in double precision the coefficients read in are slightly different from the numerical values stored. This particular condition is peculiar to double-precision operation and is of small consequence since the coefficients are still accurate to 15 places. It should be pointed out, however,

```
A =    0.0                          B =    0.1999999999999999C 01    C =    0.SSSSSSSSSSSSSS9CD CC
X1 =  -0.5CCCCCCCOCOOCOCOD 00        ZX1 =    0.0

A =    0.99SSSSSSSS999990D 00       B =    0.1999999999999999C 01    C =    0.SSSSSSSSSSSSSS9CD CC
X1 =  -0.9SSSSSSSS9999990D 00        X2 =  -0.9999999999999990C OC
ZX1 =    0.0                         ZX2 =    0.0

A =    0.99SSSSSSSS999990D 00       B =    0.0                       C =  -0.SSSSSSSSSSSSSSS9CD CC
X1 =    0.SSSSSSSS99999998D 00       X2 =  -0.9999999999999998D 00
ZX1 =  -0.222C446C49250312D-15       ZX2 =   -0.2220446049250312C-15

A =    0.SSSSSSSSS9999990D 00       B =  -0.1999999999999999C 01    C =    0.199S9SSSSSSSS99D C1
X1R =    C.9SSSSSSS99999990D 00      X1I =    0.9999999999999998C CO
ZRX1=    C.222C446C49250312D-15      ZIX1=    0.0
X2R =    C.9SSSSSSS99999990D 00      X2I =  -0.999999999999S998C CO
ZRX2=    C.222C446C49250312D-15      ZIX2=    0.0

A =    0.9SSSSSSSSS999990D 00       B =    0.1999999999999999C 07    C =    0.SSSSSSSSSSSSSS9CD C2
X1 =  -0.5CCC2243369817720-04        X2 =  -0.1997999999949997C 07
ZX1 =  -C.4486737135241724D-02       ZX2 =  -0.4486737135241724C-02

A =    0.9SSSSSSSS9999990D 00       B =    0.4999999999999999C 10    C =    0.SSSSSSSSSSSSSSSCD C2
X1 =  -0.4768371582031249D-05        X2 =  -0.4999999999999994C 10
ZX1 =  -C.2374185791C15622D 05       ZX2 =  -0.2374185791015622C C5

IHC217I
```

Figure 4.4 The output results of the program of Figure 4.3

that decimal to binary (hexadecimal) conversion errors may take place whenever the decimal number contains a fractional part. The reason for this is that there may not be an exact binary equivalent of the decimal data word. For example, the decimal number $(0.8125)_{10}$ may be written in terms of four binary bits as $(0.1101)_2$, which is equivalent to

$$
\begin{aligned}
(0.1101)_2 &= 1*2^{-1} + 1*2^{-2} + 0*2^{-3} + 1*2^{-4} \\
&= .5000 + .2500 + .0000 + .0625 \\
&= .8125
\end{aligned}
$$

On the other hand, the decimal number $(0.8126)_{10}$ has no exact binary equivalent. Expressing it in terms of the first 25 binary bits, we have

$$(0.8126)_{10} = (0.1101000000000110100011011+)_2$$

so that in the case of a computer that has a word length of 24 bits, the number would be rounded and stored as

$$0.110100000000011010001110$$

If we now asked the computer to print the decimal equivalent of this stored number, we would get

$$0.8126002$$

Turning to the fifth data set, we see that the answers given by each program for $X2$ are roughly the same, but there are almost five orders of magnitude difference between the two $X1$ answers. Since we have already discussed why we expected the value of $X1$ obtained from the previous program to be inaccurate, it is timely to ask just how accurate the double-precision results are.

An examination shows that for the first two data sets the quadratic has a value of 0.0 when the roots are substituted, so the answers are exact. The third and fourth data sets show either a zero value or a value approximately equal to 10^{-15}. Since a value of 10^{-15} means 0.00000000000000x+, it also follows that the roots are virtually exact answers. The fifth data set, however, shows a remainder of

$$ZX1 = ZX2 = -0.4486737135241724\text{D}-02$$

when the respective roots are substituted. Since both of these remainders are not essentially zero, the question still remains as to the accuracy of the individual roots.

In order to determine the number of places to which these roots are accurate, let us define the roots calculated by the computer as

$$X1 = R_1 + \epsilon_1$$

and

$$X2 = R_2 + \epsilon_2$$

where R_1 and R_2 are the exact roots and ϵ_1 and ϵ_2 are the respective computer errors. Substituting each result into the quadratic, we find

$$\begin{aligned} ZX1 &= A(R_1 + \epsilon_1)^2 + B(R_1 + \epsilon_1) + C \\ &= \underbrace{AR_1^2 + BR_1 + C}_{=0} + (2AR_1 + B)\epsilon_1 + A\epsilon_1^2 \end{aligned}$$

and

$$\begin{aligned} ZX2 &= A(R_2 + \epsilon_2)^2 + B(R_2 + \epsilon_2) + C \\ &= (2AR_2 + B)\epsilon_2 + A\epsilon_2^2 \end{aligned}$$

In our example $A = 1$, $B = 2\ 10^6$, and $C = 100$. A is therefore small compared with B and since ϵ_1 and ϵ_2 are small, we may neglect the terms $A\epsilon_1^2$ and $A\epsilon_2^2$. It is thus possible to approximate the two errors by assuming $R_1 = X1$ and $R_2 = X2$. We therefore find

$$\begin{aligned} \epsilon_1 \approx \frac{ZX1}{2AR_1 + B} &= \frac{-0.4486737135241724\text{D}-02}{2(-0.5000224336981772\text{D}-04) + 0.1999999999999999\text{D}\ 07} \\ &\approx -0.45\ 10^{-8} \end{aligned}$$

$$\epsilon_2 \approx \frac{ZX2}{2AR_2 + B} = \frac{-0.4486737135241724\text{D}-02}{2(-0.1999999999999997\text{D }07) + 0.1999999999999999\text{D }07}$$

$$\approx 0.22\ 10^{-8}$$

or

$$\frac{\epsilon_1}{R_1} \approx 0.9\ 10^{-4} \qquad \text{and} \qquad \frac{\epsilon_2}{R_2} \approx 0.1\ 10^{-14}$$

Thus

$$R_1 \approx -0.5000224336981772\ 10^{-4} \quad \text{and} \quad \epsilon_1 \approx -0.000045\ 10^{-4}$$
$$R_2 \approx -0.1999999999949997\ 10^{7} \quad \text{and} \quad \epsilon_2 \approx 0.00000000000000022\ 10^{7}$$

Hence the first root is accurate to approximately four places and the second to 15 places.

If we now turn to the sixth data set we will see that the use of double precision has not avoided, but only delayed, the previous problem. Here we have

$$X^2 + 5\ 10^9\ X + 100 = 0$$

with

$$X1 = -2.5\ 10^9 + \sqrt{\frac{25}{4}\ 10^{18} - 100}$$

$$X2 = -2.5\ 10^9 - \sqrt{\frac{25}{4}\ 10^{18} - 100}$$

Since $(25/4)10^{18}$ is more than 16 orders of magnitude greater than -100, we must again expect a large error in the calculation of $X1$. In order to avoid this error we may recognize that the product of

$$X1 = \frac{-B + \sqrt{B^2 - 4AC}}{2A}$$

and

$$X2 = \frac{-B - \sqrt{B^2 - 4AC}}{2A}$$

is

$$X1*X2 = \frac{C}{A}$$

Therefore, whenever B^2 is large compared to $4AC$ and B is positive, we should first compute $X2$ (the larger magnitude root) and then compute $X1$ from

$$X1 = \frac{C}{(A*X2)}$$

while if B is negative, we first compute $X1$ (the larger root) and then compute $X2$ from

$$X2 = \frac{C}{(A*X1)}$$

Thus, for our sixth data set

$$X2 \approx -5.0\ 10^{9}$$

and since

$$X1*X2 = \frac{C}{A}$$

$$X1 \approx -0.2\ 10^{-7}$$

Figure 4.5 is a Fortran program that incorporates the above feature, and Figure 4.6 shows the results for the same six data sets. Using the

```
C       ROOTS OF A CUADRATIC ECUATICN
C       DOUBLE PRECISICN D25.16 FCRMAT CODE
C
        DOUBLE PRECISICN A,B,C,C,X1,X2,X,X1R,X1I,X2R,X2I,ZX1,ZX2,ZRX1,ZIX1
       1,ZRX2,ZIX2
      1 READ(5,6)A,B,C
      6 FORMAT(3E15.7)
        IF(A)2,3,2
      3 X1=-C/B
        ZX1=(A*X1+B)*X1+C
        WRITE(6,7)A,B,C
      7 FORMAT('0A =',D25.16,3X,'B =',C25.16,3X,'C =',C25.16)
        WRITE(6,11)X1,ZX1
     11 FORMAT(' X1 =',D25.16,3X,'ZX1 =',D25.16,//)
        GO TC 1
      2 D=B*B-4.*A*C
        IF(D)4,5,5
      5 IF(B)12,13,14
     12 X1=(-B+D**0.5)/(2.*A)
        X2=C/(A*X1)
        GO TC 15
     13 X1=(-4.*A*C)**0.5/(2.*A)
        X2=-X1
        GO TC 15
     14 X2=(-B-D**0.5)/(2.*A)
        X1=C/(A*X2)
     15 X=X1
        ZX1=(A*X+B)*X+C
        X=X2
        ZX2=(A*X+B)*X+C
        WRITE(6,7)A,B,C
        WRITE(6,10)X1,X2,ZX1,ZX2
     10 FORMAT(' X1 =',D25.16,3X,'X2 =',D25.16,/,' ZX1 =',D25.16,2X,'ZX2 =
       1',D25.16,//)
        GO TO 1
      4 X1R=-B/(2.*A)
        X1I=(-D)**0.5/(2.*A)
        ZRX1=(A*X1R+B)*X1R+C-A*X1I*X1I
        ZIX1=(2.*A*X1R+B)*X1I
        X2R=-B/(2.*A)
        X2I=-((-D)**0.5/(2.*A))
        ZRX2=(A*X2R+B)*X2R+C-A*X2I*X2I
        ZIX2=(2.*A*X2R+B)*X12
        WRITE(6,7)A,B,C
        WRITE(6,8)X1R,X1I,ZRX1,ZIX1
      8 FORMAT(' X1R =',D25.16,3X,'X1I =',C25.16,/,' ZRX1=',D25.16,3X,'ZIX
       11=',D25.16)
        WRITE(6,9)X2R,X2I,ZRX2,ZIX2
      9 FORMAT(' X2R =',D25.16,3X,'X2I =',C25.16,/,' ZRX2=',D25.16,3X,'ZIX
       12=',D25.16,//)
        GO TO 1
        STOP
        END
```

Figure 4.5 An improved version of the quadratic root program of Figure 4.3

```
A =     0.0                     B =    0.1999999999999999D 01   C =    0.9999999999999990D 00
X1 =  -0.5000000000000000D 00   ZX1 =   0.0


A =    0.9999999999999990D 00   B =    0.1999999999999999D 01   C =    0.9999999999999990D 00
X1 =  -0.9999999999999990D 00   X2 =  -0.9999999999999990D 00
ZX1 =   0.0                     ZX2 =   0.0


A =    0.9999999999999990D 00   B =    0.0                      C =   -0.9999999999999990D 00
X1 =    0.9999999999999998D 00  X2 =  -0.9999999999999998D 00
ZX1 =  -0.2220446049250312D-15  ZX2 =  -0.2220446049250312D-15


A =    0.9999999999999990D 00   B =   -0.1999999999999999D 01   C =    0.1999999999999999D 01
X1R =    0.9999999999999990D 00   X1I =    0.9999999999999998D 00
ZRX1=    0.2220446049250312D-15   ZIX1=    0.0
X2R =    0.9999999999999990D 00   X2I =   -0.9999999999999998D 00
ZRX2=    0.2220446049250312D-15   ZIX2=    0.0


A =    0.9999999999999990D 00   B =    0.1999999999999999D 07   C =    0.9999999999999990D 02
X1 =  -0.5000000000125004D-04   X2 =  -0.1999999999949997D 07
ZX1 =  -0.9947598300641401D-13  ZX2 =  -0.4486737135241724D-02


A =    0.9999999999999990D 00   B =    0.4999999999999999D 10   C =    0.9999999999999990D 02
X1 =  -0.2000000000000001D-07   X2 =  -0.4999999999999994D 10
ZX1 =  -0.9237055564881301D-13  ZX2 =  -0.2374185791015622D 05


IHC217I
```

Figure 4.6 The output results of the program of Figure 4.5

same assumptions that we used in the previous analysis, we see that for the sixth data set

$$\epsilon_1 \approx \frac{ZX1}{2AR_1 + B} = \frac{-0.9237055564881301\text{D}-13}{2(-0.2000000000000001\text{D}-07) + 0.4999999999999999\text{D}\ 10}$$

$$\approx -0.185\ 10^{-23}$$

$$\epsilon_2 \approx \frac{ZX2}{2AR_2 + B} = \frac{-0.2374185791015622\text{D}\ 05}{2(-0.4999999999999994\text{D}\ 10) + 0.4999999999999999\text{D}\ 10}$$

$$\approx 0.475\ 10^{-5}$$

or

$$\frac{\epsilon_1}{R_1} \approx 0.925\ 10^{-16} \quad \text{and} \quad \frac{\epsilon_2}{R_2} \approx -0.95\ 10^{-15}$$

Thus

$$R_1 \approx -0.2000000000000001\text{D}-07$$

and

$$\epsilon_1 \approx -0.000000000000000185\text{D}-07$$

$$R_2 \approx -0.4999999999999994\text{D}\ 10$$

and

$$\epsilon_2 \approx 0.000000000000000475\text{D}\ 10$$

Hence the first root is accurate to approximately 16 places and the second to 15 places.

At this point the reader may believe that the latest program is capable of accurately finding the roots for any set of $A-B-C$ coefficients. Further thought will show that we have come a long way, but we still have a distance to go. For example, the program would not be able to find the roots of

$$10^{50} X^2 + 10^{50} X + 10^{50} = 0$$

However, if we first divide by 10^{50}, there would be no problem in finding the roots of

$$X^2 + X + 1 = 0$$

This suggests that additional coding is needed to normalize the coefficients with respect to the largest coefficient or the arithmetic mean of the largest and smallest coefficients.

Another example of one of the shortcomings of this program is its inability to make a change in the variable when the change would improve the accuracy in finding the roots. For example,

$$10^{50} X^2 + 10^{25} X + 1 = 0$$

could be changed to

$$Y^2 + Y + 1 = 0$$

by letting $Y = 10^{25} X$. This latter equation could then be solved more accurately by most computers. The reader is urged to study this program further and find other shortcomings.

The Cubic Equation

The three roots of a cubic equation may also be found by the use of analytic formulas. The problem, however, is considerably more difficult than the solution of the quadratic equation and, in fact, has been investigated by such men as Euler, Cardano, Fontano, Lagrange, Descartes, and others. Many of the solutions presented by these men lead to irreducible cases and thus are not capable of solving all types of cubic equations.[2]

The common goal of most of the more recent investigations was to find a general solution that involved a relationship among the coefficients

[2] See *Handbook of Mathematical Functions*, ed. Milton Abramowitz and Irene Stegun (Washington, D.C.: National Bureau of Standards, Applied Mathematics Series, U.S. Government Printing Office, 1964), p. 17; and Louis Weisner, *Introduction to the Theory of Equations* (New York: The Macmillan Company, 1938), pp. 134–40.

of the cubic and thus provide a "table look-up" type of solution. In other words, finding the numerical value of some relation that is a function of the numerical values of the cubic coefficients allows one to consult a predetermined table and obtain accurate answers by linear interpolation. Our discussion will lead to a computer program that will provide the solution to any type of cubic. This program may later be easily modified to generate a table of solutions. We will follow the notation used by Neumark[3] who has presented perhaps the best recent discussion of this problem.

Starting with the general cubic equation

$$AX^3 + BX^2 + CX + D = 0$$

where A is positive, we factor

$$(3AX + B)^3 - 3(B^2 - 3AC)(3AX + B) + (27A^2D + 2B^3 - 9ABC) = 0$$

and define the variables determinant (DET) and Δ (DELTA)

$$\text{DET} = B^2 - 3AC$$

$$\Delta = \frac{27A^2D + 2B^3 - 9ABC}{2\,|(B^2 - 3AC)^{3/2}|}$$

The solution of the general cubic may now be obtained from the solution of two specific cubics that are both a function of only DELTA.

For the case of DET > 0, we change the variable by substituting

$$X = \frac{-B - 2Y(B^2 - 3AC)^{1/2}}{3A}$$

into the factored form of the cubic to obtain

$$4Y^3 - 3Y = \text{DELTA}$$

For the case of DET < 0, we substitute

$$X = \frac{-B - 2Z(3AC - B^2)^{1/2}}{3A}$$

and obtain

$$4Z^3 + 3Z = \text{DELTA}$$

Since both of these cubics are a function of DELTA, it is only necessary to find their solutions (Y_1, Y_2, and Y_3 or Z_1, Z_2, and Z_3) and then use the appropriate transformation to determine the solution to the general cubic (X_1, X_2, and X_3).

[3] S. Neumark, *Solution of Cubic and Quartic Equations* (London: Pergamon Press, 1965).

Let us first consider the solution of

$$4Z^3 + 3Z = \text{DELTA}$$

which is graphed in Figure 4.7a. An examination of this curve will show that we have one real root and two conjugate complex roots. We may also observe that it is necessary to consider only positive values of DELTA since the roots (Z_1, Z_2, and Z_3) will be only the negative of the roots we obtain when DELTA is negative.

Since the values of DELTA are continuous and may have any value, we let the real root be

$$Z_1 = \sinh \delta$$

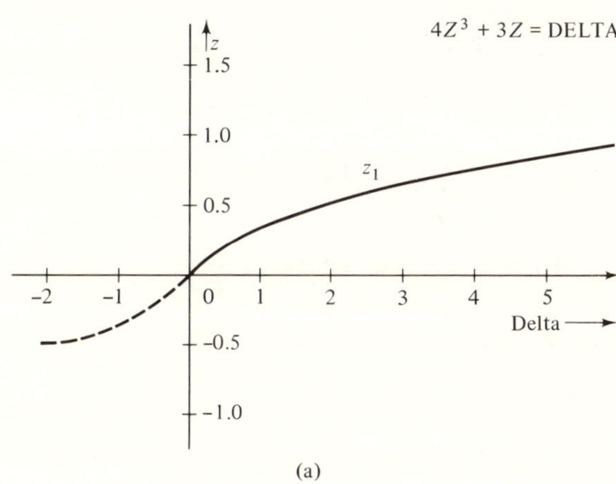

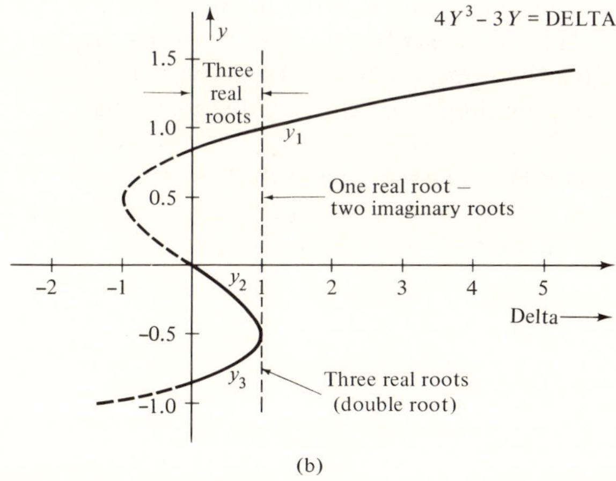

Figure 4.7 (a) The specific cubic $4Z^3 + 3Z = \text{DELTA}$; (b) the specific $4Y^3 - 3Y = \text{DELTA}$

and thus the specific cubic becomes

$$4 \sinh^3 \delta + 3 \sinh \delta = \sinh 3\delta = \text{DELTA}$$

Thus we find that

$$\delta = \frac{\sinh^{-1} (\text{DELTA})}{3}$$

and

$$Z_1 = \sinh \left(\frac{\sinh^{-1} (\text{DELTA})}{3} \right)$$

If we now factor the cubic we get

$$4Z^3 + 3Z - \text{DELTA} = (Z - Z_1)(4Z^2 + 4Z_1Z + (4Z_1^2 + 3)) = 0$$

and solving for the roots of the quadratic factor, we obtain the complex roots

$$Z_{2,3} = -0.5\ Z_1 \pm j(0.75(Z_1^2 + 1))^{1/2}$$

$$= \frac{1}{2} (-\sinh \delta \pm j \sqrt{3} \cosh \delta)$$

Let us now consider the solution of

$$4Y^3 - 3Y = \text{DELTA}$$

which is graphed in Figure 4.7b. An examination of this curve will show that it is necessary to consider only positive values of DELTA since the roots (Y_1, Y_2, and Y_3) will be only the negative of the roots we obtain when DELTA is negative. We may also observe that when the magnitude of DELTA is equal to or less than one, all the roots are real and when the magnitude of DELTA is greater than one, we have one real root and two conjugate complex roots.

For the case of three distinct real roots, we note that the magnitude of DELTA is less than one and if we choose Y_1 to be the only positive root, we see from Figure 4.7b that Y_1 must be restricted to the range

$$\frac{1}{2} \sqrt{3} < Y_1 < 1$$

If we now let

$$Y_1 = \cos \beta \quad \text{where} \quad 0° < \beta < 30°$$

the specific cubic becomes

$$4 \cos^3 \beta - 3 \cos \beta = \cos 3\beta = \text{DELTA}$$

We thus find that

$$\beta = \frac{\cos^{-1}(\text{DELTA})}{3}$$

and

$$Y_1 = \cos\left(\frac{\cos^{-1}(\text{DELTA})}{3}\right)$$

Factoring the cubic, we obtain

$$4Y^3 - 3Y - \text{DELTA} = (Y - Y_1)(4Y^2 + 4Y_1Y + (4Y_1^2 - 3)) = 0$$

and solving for the roots of the quadratic factor, we obtain

$$\begin{aligned} Y_{2,3} &= -\frac{1}{2}\cos\beta \pm \frac{1}{2}\sqrt{3}\sin\beta \\ &= -\cos(60° \pm \beta) \end{aligned}$$

For the case of DELTA $= \pm 1$, we have a double root and the three roots are given by

$$(\text{DELTA} = +1) \quad X_1 = \frac{-B - 2(B^2 - 3AC)^{1/2}}{3A},$$

$$X_2 = X_3 = \frac{-B + (B^2 - 3AC)^{1/2}}{3A}$$

or

$$(\text{DELTA} = -1) \quad X_1 = \frac{-B + 2(B^2 - 3AC)^{1/2}}{3A},$$

$$X_2 = X_3 = \frac{-B - (B^2 - 3AC)^{1/2}}{3A}$$

For the case where the magnitude of DELTA is greater than 1, we have one real root and two conjugate complex roots. If we let Y_1 be the real root and note from Figure 4.7b that Y_1 has a minimum value greater than 1, we may let

$$Y_1 = \cosh\alpha$$

and the specific cubic becomes

$$4\cosh^3\alpha - 3\cosh\alpha = \cosh 3\alpha = \text{DELTA}$$

We thus find that

$$\alpha = \frac{\cosh^{-1}(\text{DELTA})}{3}$$

and

$$Y_1 = \cosh\left(\frac{\cosh^{-1}(\text{DELTA})}{3}\right)$$

Factoring the cubic as before, we have

$$4Y^3 - 3Y - \text{DELTA} = (Y - Y_1)(4Y^2 + 4Y_1Y + (4Y_1^2 - 3)) = 0$$

and solving for the roots of the quadratic factor, we obtain the complex roots

$$Y_{2,3} = -0.5\ Y_1 \pm j(0.75(Y_1^2 - 1))^{1/2}$$
$$= \frac{1}{2}(-\cosh\alpha \pm j\sqrt{3}\sinh\alpha)$$

There are also two special cases that must be treated separately. They are the cases of DET = 0 and DELTA = 0. For the case of the determinant DET = 0, we have

$$\text{DET} = B^2 - 3AC = 0$$

and in this case the solution is obtained directly from the factored form of the general cubic, from which the solution for the real root is given by

$$X_1 = \frac{-B + (B^3 - 27A^2D)^{1/3}}{3A}$$

if $B^3 - 27\ A^2D$ is positive and by

$$X_1 = \frac{-B - |(B^3 - 27A^2D)|^{1/3}}{3A}$$

if $B^3 - 27\ A^2D$ is negative. The two complex roots, in turn, are found by multiplying the real root by

$$-0.5 \pm j0.5\sqrt{3}$$

If $B^3 - 27\ A^2D = 0$, we have a triple root and

$$X_1 = X_2 = X_3 = -\frac{B}{3A}$$

For the case of DELTA = 0, we have

$$27A^2D + 2B^3 - 9ABC = 0$$

and the solution by square roots is

$$X_1 = -\frac{B}{3A}$$

$$X_{2,3} = \frac{-B \pm (3B^2 - 9AC)^{1/2}}{3A}$$

Figure 4.8 is the corresponding Fortran program. Initially the heading is printed, and the four coefficients of the cubic are read into memory. The numerical values of DET and DELTA are then computed, and the dummy variable I is set equal to 1. DELTA is next tested to determine whether it is less than, equal to, or greater than zero. If it is less than zero, I is set equal to -1 and DELTA is made positive so that we work with the magnitude of DELTA. When the roots are determined, if I is negative, we change the sign of all the roots before printing the results. If DELTA is equal to zero, we transfer control to statement 11 (S.0093) and consider this special case. If DELTA is greater than zero, we test to see whether DET is less than, equal to, or greater than zero.

If DET is less than zero, we transfer control to statement 13 (S.0012) and find the solution of the specific cubic

$$4Z^3 + 3Z = \text{DELTA}$$

This is accomplished by first finding the value of δ from

$$\delta = \frac{\sinh^{-1}(\text{DELTA})}{3}$$

by using the identity[4]

$$\sinh^{-1} X = Ln(X + (X^2 + 1)^{0.5})$$

The value of $Z_1 = \sinh \delta$ is then computed by computer statement S.0013, and the values of $Z_{2,3}$ are found in turn. Computer statement S.0019 then tests to see if I is negative or positive. If I is negative, we change the sign of the roots Z_1, Z_2, and Z_3 before going on. Computer statements S.0025 through S.0030 are then used to compute and print the solutions X_1, X_2, X_3 from the relation

$$X = \frac{-B - 2Z(3AC - B^2)^{1/2}}{3A}$$

If, in computer statement S.0011, we had found DET = 0, control would have been transferred to statement 14 (S.0075) to handle this special case. On the other hand, if DET had been found to be positive, we would have transferred to statement 15 (S.0033) to find the solution to the

[4] See *Handbook of Mathematical Functions*, formula 4.6.20, p. 87.

```
                C      ROOTS OF CUBIC EQUATION
                C      A*S**3 + B*S**2 +C*S + D = 0
                C
S.0001                 WRITE(6,1)
S.0002               1 FORMAT('1THE ROOTS OF THE CUBIC EQUATION',//,' A*S**3 + B*S**2 + C
                      1*S + D = 0',//)
S.0003               5 READ(5,2) A,B,C,D
S.0004               2 FORMAT(4E15.7)
S.0005                 DET=B*B-3.*A*C
S.0006                 DELTA=(27.*A**2*D+2.*B**3-9.*A*B*C)/(2.*(ABS(DET))**1.5)
S.0007                 I=1
S.0008                 IF (DELTA) 10,11,12
S.0009              10 DELTA=-DELTA
S.0010                 I=-1
S.0011              12 IF(DET) 13,14,15
                C
                C      DETERMINANT LESS THAN ZERO
                C      SOLUTION FOUND FROM Z CUBIC = DELTA.
                C
S.0012              13 Y=ALOG(DELTA+((DELTA)**2+1.)**0.5)/3.
S.0013                 Z1=(EXP(Y)-EXP(-Y))/2.
S.0014                 Z2R=-0.5*Z1
S.0015                 Z3R=Z2R
S.0016                 Z2I=(0.75*Z1**2+1.)**0.5
S.0017                 Z3I=-Z2I
S.0018              16 DETABS=(ABS(DET))**0.5
S.0019                 IF(I) 17,18,18
S.0020              17 Z1=-Z1
S.0021                 Z2R=-Z2R
S.0022                 Z2I=-Z2I
S.0023                 Z3R=-Z3R
S.0024                 Z3I=-Z3I
S.0025              18 X1=(-B-2.*Z1*DETABS)/(3.*A)
S.0026                 X2R=(-B-2.*Z2R*DETABS)/(3.*A)
S.0027                 X3R=X2R
S.0028                 X2I=(-2.*Z2I*DETABS)/(3.*A)
S.0029                 X3I=-X2I
S.0030              19 WRITE(6,20)A,B,C,D,X1,X2R,X2I,X3R,X3I
S.0031              20 FORMAT(' ONE REAL ROOT AND TWO COMPLEX ROOTS',//,' A =',E15.7,'  B
                      1 =',E15.7,' C =',E15.7,'  D =',E15.7,//,'  X1 =',E15.7,/,' X2R =',
                      2E15.7,'  X2I =',E15.7,/,' X3R =',E15.7,'  X3I =',E15.7,/////)
S.0032                 GO TO 5
                C
                C      DETERMINANT GREATER THAN ZERO
                C      SOLUTION FOUND FROM Y CUBIC = DELTA.
                C
S.0033              15 IF(DELTA-1.) 21,22,23
                C
                C      TRIGNOMETRIC FUNCTION SOLUTION
                C
S.0034              21 A0=1.5707963050
S.0035                 A1=-0.2145988016
S.0036                 A2=0.0889789874
S.0037                 A3=-0.0501743046
S.0038                 A4=0.0308918810
S.0039                 A5=-0.0170881256
S.0040                 A6=0.0066700901
S.0041                 A7=-0.0012624911
S.0042                 X=DELTA
S.0043                 FX=((((((A7*X+A6)*X+A5)*X+A4)*X+A3)*X+A2)*X+A1)*X+A0
S.0044                 BETA=      ((1.-X)**0.5)*FX/3.
S.0045                 Y1=COS(BETA)
S.0046                 Y2=-0.5*COS(BETA)+0.5*3.**0.5*SIN(BETA)
S.0047                 Y3=-0.5*COS(BETA)-0.5*3.**0.5*SIN(BETA)
S.0048                 IF (I) 24,25,25
S.0049              24 Y1=-Y1
S.0050                 Y2=-Y2
S.0051                 Y3=-Y3
S.0052              25 X1=(-B-2.*Y1*DET**0.5)/(3.*A)
S.0053                 X2=(-B-2.*Y2*DET**0.5)/(3.*A)
S.0054                 X3=(-B-2.*Y3*DET**0.5)/(3.*A)
S.0055              26 WRITE(6,101) A,B,C,D,X1,X2,X3
S.0056             101 FORMAT(' THREE REAL ROOTS',//,' A =',E15.7,'  B =',E15.7,'  C =',E
                      115.7,'  D =',E15.7,//,' X1 =',E15.7,'  X2 =',E15.7,'  X3 =',E15.7,
                      2//)
S.0057                 GO TO 5
```

Figure 4.8 A program that uses a classical method to find the roots of a cubic

```
              C
              C       HYPERBOLIC FUNCTION SOLUTION
              C
S.0058          23 Y=ALOG(DELTA+((DELTA)**2+1.)**0.5)/3.
S.0059             Z1=(EXP(Y)+EXP(-Y))/2.
S.0060             Z2R=-0.5*Z1
S.0061             Z2I=(0.75*(Z1**2-1.))**0.5
S.0062             Z3R=Z2R
S.0063             Z3I=-Z2I
S.0064             GO TO 16
S.0065          22 DETABS=ABS(DET)**0.5
S.0066             IF(I)27,28,28
S.0067          27 X1=(-B+2.*DETABS)/(3.*A)
S.0068             X2=(-B-DETABS)/(3.*A)
S.0069             X3=X2
S.0070             GO TO 26
S.0071          28 X1=(-B-2.*DETABS)/(3.*A)
S.0072             X2=(-B+DETABS)/(3.*A)
S.0073             X3=X2
S.0074             GO TO 26
              C
              C       SPECIAL CASE  DETERMINANT = 0.
              C
S.0075          14 W=B**3-27.*A**2*D
S.0076             IF(W)29,30,31
S.0077          29 W=(-W)**(1./3.)
S.0078             W=-W
S.0079             GO TO 32
S.0080          31 W=W**(1./3.)
S.0081          32 X1=(-B+W)/(3.*A)
S.0082             X2R=-0.5*X1
S.0083             X2I=0.5*3.**0.5*X1
S.0084             X3R=X2R
S.0085             X3I=-X2I
S.0086             GO TO 19
S.0087          30 X1=-B/(3.*A)
S.0088             X2=X1
S.0089             X3=X1
S.0090             WRITE(6,103)
S.0091         103 FORMAT(' TRIPLE ROOT',//)
S.0092             GO TO 26
              C
              C       SPECIAL CASE  DELTA = 0.
              C
S.0093          11 IF(DET)33,34,34
S.0094          33 S=-3.*DET
S.0095             S=S**0.5
S.0096             X1=-B/(3.*A)
S.0097             X2R=X1
S.0098             X2I=S/(3.*A)
S.0099             X3R=X2R
S.0100             X3I=-X2I
S.0101             GO TO 19
S.0102          34 S=(3.*DET)**0.5
S.0103             X1=-B/(3.*A)
S.0104             X2=(-B+S)/(3.*A)
S.0105             X3=(-B-S)/(3.*A)
S.0106             GO TO 26
S.0107             RETURN
S.0108             END
```

Figure 4.8 Continued

specific cubic

$$4Y^3 - 3Y = \text{DELTA}$$

Here we would first determine if the magnitude of DELTA was less than, equal to, or greater than one. If the magnitude was less than one, there would be three real roots, and control would be transferred to statement 21 (S.0034) where the trigonometric function solution would be used to find the roots. This is accomplished by first finding the value of β from

$$\beta = \frac{\cos^{-1}(\text{DELTA})}{3}$$

by using the approximation[5]

$$\sin^{-1} X = \frac{\pi}{2} - (1 - X)^{1/2}(a_0 + a_1X + a_2X^2 + a_3X^3 + a_4X^4 + a_5X^5 + a_6X^6 + a_7X^7)$$

where

$$\begin{aligned} a_0 &= 1.57079\,63050 & a_4 &= .03089\,18810 \\ a_1 &= -.21459\,88016 & a_5 &= -.01708\,81256 \\ a_2 &= .08897\,89874 & a_6 &= .00667\,00901 \\ a_3 &= -.05017\,43046 & a_7 &= -.00126\,24911 \end{aligned}$$

which has an absolute error $\leq 2\ 10^{-8}$. The value of $Y_1 = \cos\beta$ is then computed by computer statement S.0045 and the values of $Y_{2,3}$ are found in turn. Computer statement S.0048 then tests to see whether I is negative or positive. If I is negative, we change the sign of the roots Y_1, Y_2, and Y_3 before going on. Computer statements S.0052 through S.0056 are then used to compute and print the solution X_1, X_2, X_3 from the relation

$$X = \frac{-B - 2Y(B^2 - 3AC)^{1/2}}{3A}$$

If, in computer statement S.0033, we had found $|\text{DELTA}| = 1$, control would have been transferred to statement 22 (S.0065) to find the roots for this case of a double real root. On the other hand, if $|\text{DELTA}| > 1$, control would be transferred to statement 23 (S.0058) where we would first find the value of α from

$$\alpha = \frac{\cosh^{-1}(\text{DELTA})}{3}$$

by using the identity[6]

$$\cosh^{-1} X = Ln(X + (X^2 - 1)^{1/2})$$

The value of $Y_1 = \cosh\alpha$ (note that Y_1 is called Z_1 in the computer program so that another section of the program may be used later) would then be computed by computer statement S.0059, and the values of $Y_{2,3}$ ($Z_{2,3}$ in program) would be found in turn. At this point, control would be transferred to statement 16 and that part of the program would be reused to compute and print the solution.

Figure 4.9 gives the computer results for several different cubics. Note that the examples chosen check the major portions of the program and the verification of these solutions is left as an exercise for the reader.

[5] See *Handbook of Mathematical Functions*, formula 4.4.46, p. 81.
[6] See *Handbook of Mathematical Functions*, formula 4.6.21, p. 87.

```
THE ROOTS OF THE CUBIC EQUATION

A*S**3 + B*S**2 + C*S + D = 0

THREE REAL ROOTS

A =  0.1000000E 01  B =  0.2000000E 01  C = -0.1000000E 01  D = -0.2000000E 01

X1 =  0.9999996E 00  X2 = -0.1000001E 01  X3 = -0.1999997E 01

THREE REAL ROOTS

A =  0.1000000E 01  B = -0.2000000E 01  C = -0.1000000E 01  D =  0.2000000E 01

X1 = -0.9999996E 00  X2 =  0.1000001E 01  X3 =  0.1999997E 01

THREE REAL ROOTS

A =  0.1000000E 01  B =  0.1000000E 01  C = -0.1000000E 01  D = -0.1000000E 01

X1 =  0.1000000E 01  X2 = -0.1000000E 01  X3 = -0.1000000E 01

THREE REAL ROOTS

A =  0.1000000E 01  B = -0.1000000E 01  C = -0.1000000E 01  D =  0.1000000E 01

X1 = -0.1000000E 01  X2 =  0.1000000E 01  X3 =  0.1000000E 01


IHC210I PROGRAM INTERRUPT(P) OLD PSW IS     FF15000F62007752
THREE REAL ROOTS

A =  0.1000000E 01  B =  0.3000000E 01  C =  0.3000000E 01  D =  0.1000000E 01

X1 = -0.1000000E 01  X2 = -0.1000000E 01  X3 = -0.1000000E 01

ONE REAL ROOT AND TWO COMPLEX ROOTS

A =  0.1000000E 01  B =  0.3000000E 01 C =  0.4000000E 01  D =  0.2000000E 01

 X1  = -0.1000000E 01
X2R  = -0.1000000E 01  X2I =  0.9999996E 00
X3R  = -0.1000000E 01  X3I = -0.9999996E 00



ONE REAL ROOT AND TWO COMPLEX ROOTS

A =  0.3000000E 01  B = -0.9000000E 01 C =  0.1300000E 02  D =  0.1000000E 01

 X1  = -0.7312965E-01
X2R =  0.1536565E 01  X2I = -0.1625263E 01
X3R =  0.1536565E 01  X3I =  0.1625263E 01



ONE REAL ROOT AND TWO COMPLEX ROOTS

A =  0.1000000E 01  B = -0.1000000E 01 C =  0.3000000E 01  D =  0.3000000E 01

 X1  = -0.7113453E 00
X2R =  0.8556725E 00  X2I = -0.2091427E 01
X3R =  0.8556725E 00  X3I =  0.2091427E 01



ONE REAL ROOT AND TWO COMPLEX ROOTS
```

Figure 4.9 The output results of the program of Figure 4.8

```
A =  0.1000000E 01  B =  0.3000000E 01 C =  0.1000000E 01  D =  0.3CCC000E 01

 X1 = -0.3014423E 01
X2R =  0.7211685E-02  X2I = -0.1021482E 01
X3R =  0.7211685E-02  X3I =  0.1021482E 01



ONE REAL ROOT AND TWO COMPLEX ROOTS

A =  0.1000000E 01  B = -0.1000000E 01 C =  0.0          D =  0.2000000E 01

 X1 = -0.1000284E 01
X2R =  0.1000141E 01  X2I = -0.1000284E 01
X3R =  0.1000141E 01  X3I =  0.1000284E 01



IHC217I
```

Figure 4.9 Continued

The Quartic Equation

The quartic equation is the highest-order polynomial that can be solved by direct formula. As in the case of the cubic equation, a number of different solutions have been proposed by various mathematicians. A common approach in many of these solutions is to factor the quartic so that a resolvent cubic is obtained. For example, Neumark[7] starts with the general quartic

$$AX^4 + BX^3 + CX^2 + DX + E = 0$$

and first factors this into two quadratic factors

$$(AX^2 + GX + H)(AX^2 + gX + h) = 0$$

where G, H, g, and h must be found from the four relations

$$\begin{aligned} G + g &= B \\ A(H + h + gG) &= AC \\ Gh + gH &= AD \\ Hh &= AE \end{aligned}$$

He next lets

$$Gg = AX$$

and assuming that X is known, he solves for G and g from this last equation and the first equation of the previous four, obtaining

$$G = \frac{B + \sqrt{B^2 - 4AX}}{2} \qquad \text{and} \qquad g = \frac{B - \sqrt{B^2 - 4AX}}{2}$$

[7] See Neumark, *Solution of Cubic and Quartic Equations*, Ch. 3.

Now using the middle two equations, he finds

$$H = \frac{C - X}{2} + \frac{B(C - X) - 2AD}{2\sqrt{B^2 - 4AX}}$$

and

$$h = \frac{C - X}{2} - \frac{B(C - X) - 2AD}{2\sqrt{B^2 - 4AX}}$$

When these four results are substituted into the last equation of the above four relations, he obtains the resolvent cubic

$$X^3 - 2CX^2 + (C^2 + BD - 4AE)X - (BCD - B^2E - AD^2) = 0$$

This cubic can now be solved by our previous program and the roots X_1, X_2, X_3 can be used to determine the values of G, g, H, and h. Note that this will result in three sets of values. In practice, the smallest real root of the resolvent cubic is used to obtain the required values of G, g, H, and h, and at this point the roots of the two quadratic factors of the quartic are found by using the quadratic formula.

In the next section we will begin our discussion of iterative methods that can be used to find the roots of polynomials. The reader will find that most of these methods are quite simple when compared with our classical solution of the quadratic, cubic, and quartic. The reader should be aware, however, that our discussion of the quadratic, cubic, and quartic was not just an academic exercise. The classical solution is more efficient and can yield more accurate results for every type of quadratic, cubic, or quartic than can any iterative scheme. In addition, we should also be aware that when iterative methods that are based on using a quadratic divisor are employed, the quadratic formula must ultimately be used in determining the roots.

4.3 THE METHOD OF HALF-INTERVAL SEARCH

In the last section we dealt with the algebraic solution of second, third, and fourth degree polynomials. In this and succeeding sections we deal with the more general problem of finding the zeros or roots of

$$f(t) = 0$$

where $f(t)$ may be an algebraic or transcendental equation. We will employ only iterative methods that proceed by successive approximations to obtain the roots.

The basic problem we face, therefore, has two parts. The first part is to isolate the roots or find their approximate locations. This may be

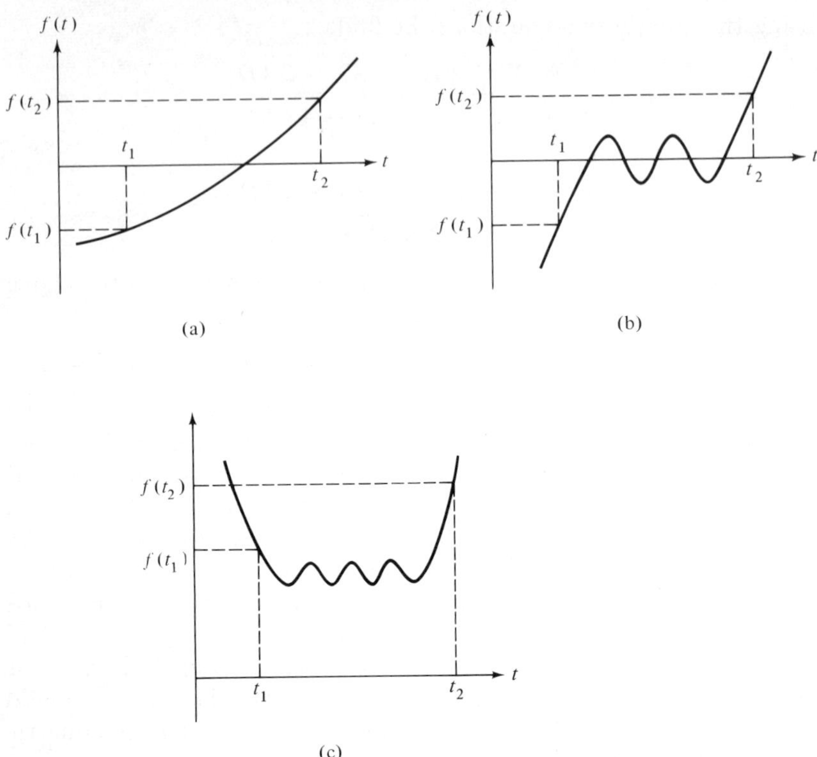

Figure 4.10 The graphical location of a real root

done by writing a Fortran program that evaluates $f(t)$ for successive values of t and tests for sign changes in $f(t)$. Whenever two successive evaluations produce a sign change in $f(t)$, there must be an odd number of real roots that lie between these two values of t. Care must be taken in the selection of the incremental value of t so that the situation shown in Figure 4.10b does not result. This may be avoided by choosing a very coarse increment to find the general location of successive sign changes and then using a finer increment to establish an upper and lower limit for t as shown in Figure 4.10a. It thus follows that if a continuous function $f(t)$ changes sign in the interval $t_1 < t < t_2$, and if the derivatives of $f(t)$ do not change sign in the interval, then $f(t) = 0$ has a single real root that lies between t_1 and t_2. The special case shown in Figure 4.10c in which there are no sign changes in $f(t)$ corresponds to the case of complex roots. The problem of finding complex roots will be covered at a later point in this chapter.

Since the method described above gives only a rough approximation of the real roots, the second part of the problem is to improve on the initial approximations until the desired degree of accuracy has been obtained. One method of doing this is simply to continue the above procedure in the vicinity of the initial approximations to the real roots, using finer and finer increments, until the increment $t_2 - t_1$ is equal to or less than the required accuracy. One systematic procedure for doing this is known as the Method of Half-Interval Search or the Bisection Method.

Figure 4.11 shows that once we have found the approximate location of a real root, the bisection method proceeds by calculating the value of the function at the midpoint of the interval ($T = (t_1 + t_2)/2$). We then test to see whether $|t_2 - t_1| < \text{ACC}$. If it is less than the accuracy specified, T is the root. If it is not less than the specified accuracy, we test to see whether $f(T)*f(t_1)$ is positive or negative. If $f(T)*f(t_1)$ is positive, the root lies between T and t_2 and we repeat the process with the location of t_1 changed to T. If $f(T)*f(t_1)$ is negative, the root lies between t_1 and T and we repeat the process with t_2 set equal to T. Thus in each iteration we either find the root to the desired degree of accuracy, or we narrow the range to half of the previous increment.

Figure 4.12 is a Fortran program that uses the bisection method to improve on the initial approximation of a real root when $f(t) = 0.9 - \epsilon - t/T$. First, computer statements S.0001 and S.0002 are used to print the heading and subheadings. Computer statements S.0003 and S.0004 then read in the initial values of t_1 and t_2 and the required root accuracy ACC. Computer statement S.0005 calculates the value of the function at t_1, and the iteration count variable I is then set equal to zero. Computer statement S.0007 begins the first iteration by setting $I = 1$. We next test to see whether $|t_2 - t_1| \leq \text{ACC}$. If $|t_2 - t_1|$ is not less than or equal to ACC, we go to statement 20 (computer statement S.0009) and compute the value of T. The function is then evaluated for this value of T, and the numerical values of t_1, t_2, and FT ($f(T)$) are printed. Computer statement S.0013 is used to determine whether FT has the same sign as FT 1 ($f(t_1)$). If it has the same sign, we transfer to statement 6 (S.0016)

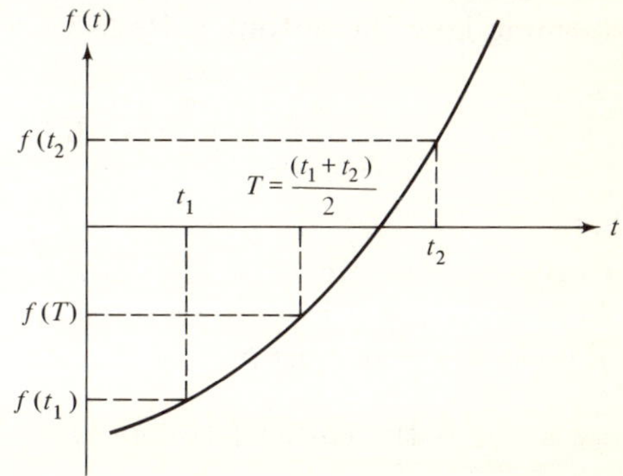

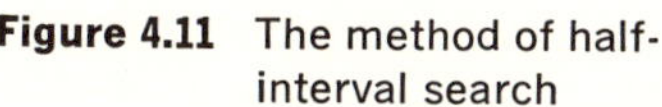
Figure 4.11 The method of half-interval search

```
       C      HALF INTERVAL METHOD
       C          USED
       C      TO FIND THE RISE TIME
       C
S.0001        WRITE(6,1)
S.0002      1 FORMAT('1THE HALF INTERVAL METHOD IS USED TO FIND THE RISE TIME',/
             1///,7X,'T1',13X,'T2',13X,'T',13X,'F(T)',//)
S.0003        READ(5,2)T1,T2,ACC
S.0004      2 FORMAT(3E15.7)
S.0005        FT1=0.9-EXP(-T1)
S.0006        I=0
S.0007      3 I=I+1
S.0008        IF(T2-T1-ACC) 10,10,20
S.0009     20 T=(T1+T2)/2.
S.0010        FT=0.9-EXP(-T)
S.0011        WRITE(6,4)T1,T2,T,FT
S.0012      4 FORMAT(4E15.7)
S.0013        IF(FT*FT1)5,10,6
S.0014      5 T2=T
S.0015        GO TO 3
S.0016      6 T1=T
S.0017        GO TO 3
S.0018     10 WRITE(6,7)T,I
S.0019      7 FORMAT(///,' THE ROOT =',E15.7,' SECONDS',//,' AND WAS FOUND IN ',
             1I5,' ITERATIONS')
S.0020        RETURN
S.0021        END
```

Figure 4.12 A Fortran program that uses the half-interval method to find the root

and set $T1 = T$ and then transfer to statement 3 to start the next iteration. If $FT*FT1$ is negative, we transfer to statement 5 setting $T2 = T$ and, in turn, transfer to computer statement S.0007 for the next iteration. When either $|t_2 - t_1| \leq$ ACC or $FT*FT1 = 0$, control transfers to statement 10 (S.0018), and the value of the root, T, and the number of iterations are printed.

The Calculation of the Rise Time of a Single-Stage Amplifier

The reader should recognize that although the above program has been written to solve for the root of a specific function, only computer statements S.0005 and S.0010 have to be modified to solve for the roots of any other function. The particular problem solved here is the calculation of the rise time of the single-stage amplifier shown in Figure 4.13.

In order to find the rise time, we use a step-voltage excitation and determine how the output voltage varies as a function of time. Thus

$$G(s) = F(s)*H(s) = \frac{1}{s}\frac{\omega_1}{s+\omega_1}$$

where

$G(s)$ is the transform of the output response.
$F(s)$ is the input excitation transform.
$H(s)$ is the transmittance or transfer function.

$\omega_1 = \dfrac{1}{RC}$ is the corner frequency.

and

R = the parallel combination of r_p, R_L, and R_g.

C = the sum of the shunt capacitors.

Taking the inverse transform, we find the output voltage

$$v_{g2}(t) = g(t) = \mathcal{L}^{-1}[(G(s)] = 1 - \epsilon^{-t/RC}$$

Figure 4.14a shows a plot of the output response and defines the rise time as the time it takes the output voltage to go from 10 to 90 percent of its final value.

In this example, it is analytically easy to find $t_{0.1}$ and $t_{0.9}$. For $t = t_{0.1}$ we have

$$0.1 = 1 - \epsilon^{-t_{0.1}/RC}$$

or

$$\begin{aligned} t_{0.1} &= RC\, Ln(10/9) \\ &= 0.105\ RC \text{ seconds} \end{aligned}$$

and for $t = t_{0.9}$ we have

$$0.9 = 1 - \epsilon^{-t_{0.9}/RC}$$

or

$$t_{0.9} = 2.3\ RC \text{ seconds}$$

The rise time is therefore equal to

$$\begin{aligned} T_{RT} &= t_{0.9} - t_{0.1} \\ &= 2.2\ RC \text{ seconds} \end{aligned}$$

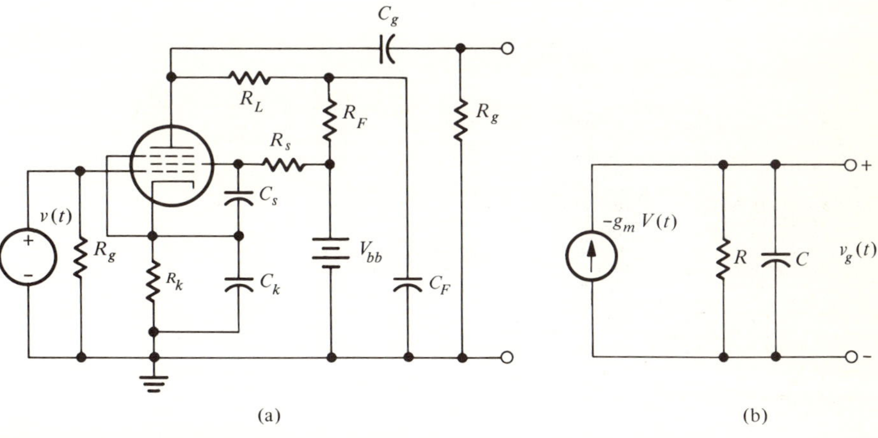

Figure 4.13 (a) Single-stage R-C coupled amplifier; (b) the high-frequency equivalent circuit

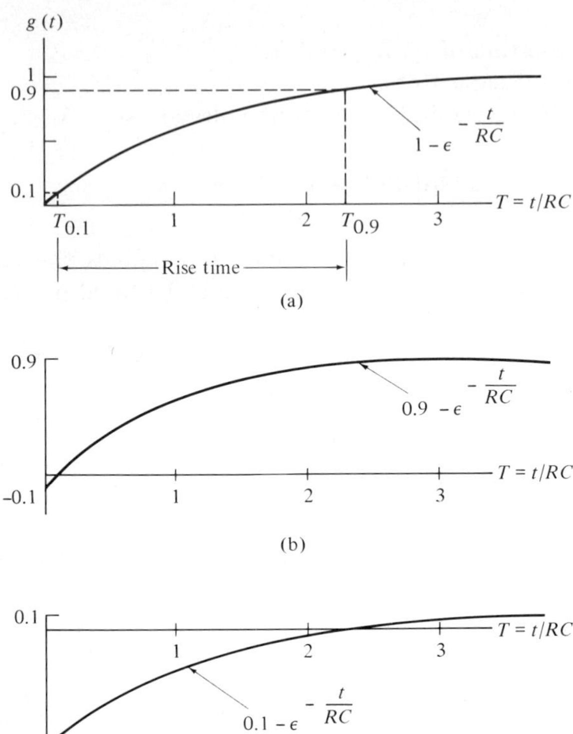

(c)

Figure 4.14 The calculation of the rise time

THE HALF INTERVAL METHOD IS USED TO FIND THE RISE TIME

T1	T2	T	F (T)
0.0	0.5000000E 01	0.2500000E 01	0.8179150E 00
0.0	0.2500000E 01	0.1250000E 01	0.6134952E 00
0.0	0.1250000E 01	0.6250000E 00	0.3647385E 00
0.0	0.6250000E 00	0.3125000E 00	0.1683843E 00
0.0	0.3125000E 00	0.1562500E 00	0.4465467E-01
0.0	0.1562500E 00	0.7812500E-01	-0.2484882E-01
0.7812500E-01	0.1562500E 00	0.1171875E 00	0.1058155E-01
0.7812500E-01	0.1171875E 00	0.9765625E-01	-0.6960630E-02
0.9765625E-01	0.1171875E 00	0.1074219E 00	0.1853287E-02
0.9765625E-01	0.1074219E 00	0.1025391E 00	-0.2542913E-02
0.1025391E 00	0.1074219E 00	0.1049805E 00	-0.3421307E-03
0.1049805E 00	0.1074219E 00	0.1062012E 00	0.7562637E-03
0.1049805E 00	0.1062012E 00	0.1055908E 00	0.2072453E-03
0.1049805E 00	0.1055908E 00	0.1052856E 00	-0.6741285E-04
0.1052856E 00	0.1055908E 00	0.1054382E 00	0.6991625E-04
0.1052856E 00	0.1054382E 00	0.1053619E 00	0.1251698E-05
0.1052856E 00	0.1053619E 00	0.1053238E 00	-0.3308058E-04
0.1053238E 00	0.1053619E 00	0.1053429E 00	-0.1591444E-04
0.1053429E 00	0.1053619E 00	0.1053524E 00	-0.7331371E-05

THE ROOT = 0.1053524E 00 SECONDS

AND WAS FOUND IN 20 ITERATIONS

Figure 4.15 The output results of the program of Figure 4.12 for the calculation of $T_{0.1}$

```
C         HALF INTERVAL METHOD
C             USED
C         TO FIND THE RISE TIME
C
          WRITE(6,1)
        1 FORMAT('1THE HALF INTERVAL METHOD IS USED TO FIND THE RISE TIME',/
        1///,7X,'T1',13X,'T2',13X,'T',13X,'F(T)',//)
          READ(5,2)T1,T2,ACC
        2 FORMAT(3E15.7)
          FT1=0.1-EXP(-T1)
          I=0
        3 I=I+1
          IF(T2-T1-ACC) 10,10,20
       20 T=(T1+T2)/2.
          FT=0.1-EXP(-T)
          WRITE(6,4)T1,T2,T,FT
        4 FORMAT(4E15.7)
          IF(FT*FT1)5,10,6
        5 T2=T
          GO TO 3
        6 T1=T
          GO TO 3
       10 WRITE(6,7)T,I
        7 FORMAT(///,' THE ROOT =',E15.7,' SECONDS',//,' AND WAS FOUND IN ',
        1I5,' ITERATIONS')
          RETURN
          END
```

Figure 4.16 The program of Figure 4.12 is modified to calculate $T_{0.9}$

Using the bisection method, $T_{0.1}$ may be found by finding the real root of the function

$$0.9 - \epsilon^{-T} = 0; T = t/RC$$

which is shown in Figure 4.14b.

Figure 4.15 shows the output results of the program of Figure 4.12 for starting values of $T1 = 0.0$ and $T2 = 5.0$ and an accuracy of $T_2 - T_1 = 1.0\text{E}-5$. The reader is encouraged to hand calculate and graphically follow the first five iterations.

THE HALF INTERVAL METHOD IS USED TO FIND THE RISE TIME

T1	T2	T	F(T)
0.0	0.5000000E 01	0.2500000E 01	0.1791495E-01
0.0	0.2500000E 01	0.1250000E 01	-0.1865048E 00
0.1250000E 01	0.2500000E 01	0.1875000E 01	-0.5335498E-01
0.1875000E 01	0.2500000E 01	0.2187500E 01	-0.1219690E-01
0.2187500E 01	0.2500000E 01	0.2343750E 01	0.4032850E-02
0.2187500E 01	0.2343750E 01	0.2265625E 01	-0.3765225E-02
0.2265625E 01	0.2343750E 01	0.2304687E 01	0.2099872E-03
0.2265625E 01	0.2304687E 01	0.2285156E 01	-0.1758218E-02
0.2285156E 01	0.2304687E 01	0.2294922E 01	-0.7693172E-03
0.2294922E 01	0.2304687E 01	0.2299805E 01	-0.2784729E-03
0.2299805E 01	0.2304687E 01	0.2302246E 01	-0.3391504E-04
0.2302246E 01	0.2304687E 01	0.2303467E 01	0.8809566E-04
0.2302246E 01	0.2303467E 01	0.2302856E 01	0.2712011E-04
0.2302246E 01	0.2302856E 01	0.2302551E 01	-0.3397465E-05
0.2302551E 01	0.2302856E 01	0.2302704E 01	0.1186132E-04
0.2302551E 01	0.2302704E 01	0.2302628E 01	0.4231930E-05
0.2302551E 01	0.2302628E 01	0.2302589E 01	0.4172325E-06
0.2302551E 01	0.2302589E 01	0.2302570E 01	-0.1490116E-05
0.2302570E 01	0.2302589E 01	0.2302580E 01	-0.5364418E-06

THE ROOT = 0.2302580E 01 SECONDS

AND WAS FOUND IN 20 ITERATIONS

Figure 4.17 The output results of the program of Figure 4.16

The value of $T_{0.9}$ may be similarly found by finding the root of

$$0.1 - \epsilon^{-T} = 0; \; T = t/RC$$

which is shown in Figure 4.14c. Figure 4.16 shows the modified Fortran program and Figure 4.17 shows the corresponding results. The reader may observe that the difference between these two results yields the correct time and that the evaluation of each root required 20 iterations.

The Calculation of the Rise Time of a Three-Stage Amplifier

In order to show the bisection method to better advantage, consider the calculation of the rise time of three identical cascaded stages similar to the one shown in Figure 4.13. For this network we would have

$$G(s) = F(s)*H(s) = \frac{1}{s}\frac{\omega_1^3}{(s + \omega_1)^3}$$

and the output time response would be given by

$$v_g(t) = g(t) = \mathcal{L}^{-1}[G(s)]$$

or

$$g(\tau) = 1 - \left(1 + \tau + \frac{1}{2}\tau^2\right)\epsilon^{-\tau}; \; \tau = t/RC$$

This would require the solution of

$$0.1 = \left[1 - \left(1 + \tau + \frac{1}{2}\tau^2\right)\epsilon^{-\tau}\right]_{\tau=T_{1.0}}$$

```
C        HALF INTERVAL METHOD
C            USED
C        TO FIND THE RISE TIME
C
         WRITE(6,1)
       1 FORMAT('1THE HALF INTERVAL METHCD IS USED TC FINC THE RISE TIME',/
        1///,7X,'T1',13X,'T2',13X,'T',13X,'F(T)',//)
         READ(5,2)T1,T2,ACC
       2 FORMAT(3E15.7)
         FT1=C.9-(1.+T1+0.5*T1**2)*EXP(-T1)
         I=0
       3 I=I+1
         IF(T2-T1-ACC)10,10,20
      2C T=(T1+T2)/2.
         FT=0.9-(1.+T+0.5*T**2)*EXP(-T)
         WRITE(6,4)T1,T2,T,FT
       4 FORMAT(4E15.7)
         IF(FT*FT1)5,10,6
       5 T2=T
         GO TO 3
       6 T1=T
         GO TO 3
      1C WRITE(6,7)T,I
       7 FORMAT(///,' THE ROOT =',E15.7,' SECCNDS',//,' AND WAS FOUND IN ',
        1I5,' ITERATIONS')
         RETURN
         END
```

Figure 4.18 A Fortran program for calculating $T_{0.1}$ of the three-stage amplifier

or

$$\left[0.9 - \left(1 + \tau + \frac{1}{2}\tau^2\right)\epsilon^{-\tau}\right]_{\tau=T_{0.1}} = 0$$

and

$$0.9 = \left[1 - \left(1 + \tau + \frac{1}{2}\tau^2\right)\epsilon^{-\tau}\right]_{\tau=T_{0.9}}$$

or

$$\left[0.1 - \left(1 + \tau + \frac{1}{2}\tau^2\right)\epsilon^{-\tau}\right]_{\tau=T_{0.9}} = 0$$

Note that the analytic solution of these two equations is much more difficult than the previous equations.

Figure 4.18 is the Fortran program for finding $T_{0.1}$, and Figure 4.19 is the program for finding $T_{0.9}$. Figures 4.20 and 4.21 show the respective results. Observe that both programs require 21 iterations to find the root and that this is one more than the previous programs. This may be explained by realizing the fact that the initial location of the root was given as $T1 = 0.0$ and $T2 = 10.0$ for this problem and $T1 = 0.0$ and $T2 = 5.0$ for the previous problem and, thus, for the same specified root accuracy, one more bisection is required. The difference between these results shows that the rise time of a three-stage amplifier is 4.22*RC* seconds.[8]

[8]For futher discussion of this problem see Joseph Pettit and Malcolm McWhorter, *Electronic Amplifier Circuits* (New York: McGraw-Hill Book Company, 1961), Sec. 4–9; and G. E. Valley, Jr. and H. Wallman, eds., *Vacuum Tube Amplifiers* (New York: McGraw-Hill Book Company, 1948), Sec. 2.3.

```
C        HALF INTERVAL METHOD
C             USED
C        TO FIND THE RISE TIME
C
         WRITE(6,1)
       1 FORMAT('1THE HALF INTERVAL METHOD IS USED TO FIND THE RISE TIME',/
        1///,7X,'T1',13X,'T2',13X,'T',13X,'F(T)',//)
         READ(5,2)T1,T2,ACC
       2 FORMAT(3E15.7)
         FT1=C.1-(1.+T1+0.5*T1**2)*EXP(-T1)
         I=0
       3 I=I+1
         IF(T2-T1-ACC)10,10,20
      2C T=(T1+T2)/2.
         FT=0.1-(1.+T+0.5*T**2)*EXP(-T)
         WRITE(6,4)T1,T2,T,FT
       4 FORMAT(4E15.7)
         IF(FT*FT1)5,10,6
       5 T2=T
         GO TO 3
       6 T1=T
         GO TO 3
      1C WRITE(6,7)T,I
       7 FORMAT(///,' THE ROOT =',E15.7,' SECONDS',//,' AND WAS FOUND IN ',
        1I5,' ITERATIONS')
         RETURN
         END
```

Figure 4.19 A Fortran program for calculating $T_{0.9}$ of the three-stage amplifier

```
THE HALF INTERVAL METHOD IS USED TO FIND THE RISE TIME


        T1                 T2                  T                  F(T)


 0.0               0.1000000E 02   0.5000000E 01   0.7753479E 00
 0.0               0.5000000E 01   0.2500000E 01   0.3561868E 00
 0.0               0.2500000E 01   0.1250000E 01   0.3153229E-01
 0.0               0.1250000E 01   0.6250000E 00  -0.7434309E-01
 0.6250000E 00     0.1250000E 01   0.9375000E 00  -0.3082818E-01
 0.9375000E 00     0.1250000E 01   0.1093750E 01  -0.1671672E-02
 0.1093750E 01     0.1250000E 01   0.1171875E 01   0.1447165E-01
 0.1093750E 01     0.1171875E 01   0.1132812E 01   0.6279349E-02
 0.1093750E 01     0.1132812E 01   0.1113281E 01   0.2272904E-02
 0.1093750E 01     0.1113281E 01   0.1103516E 01   0.2927184E-03
 0.1093750E 01     0.1103516E 01   0.1098633E 01  -0.6913543E-03
 0.1098633E 01     0.1103516E 01   0.1101074E 01  -0.1997948E-03
 0.1101074E 01     0.1103516E 01   0.1102295E 01   0.4649162E-04
 0.1101074E 01     0.1102295E 01   0.1101685E 01  -0.7677078E-04
 0.1101685E 01     0.1102295E 01   0.1101990E 01  -0.1502037E-04
 0.1101990E 01     0.1102295E 01   0.1102142E 01   0.1579523E-04
 0.1101990E 01     0.1102142E 01   0.1102066E 01   0.3576279E-06
 0.1101990E 01     0.1102066E 01   0.1102028E 01  -0.7271767E-05
 0.1102028E 01     0.1102066E 01   0.1102047E 01  -0.3457069E-05
 0.1102047E 01     0.1102066E 01   0.1102057E 01  -0.1549721E-05


THE ROOT =  0.1102057E 01 SECONDS

AND WAS FOUND IN     21 ITERATIONS
```

Figure 4.20 The output results of the program of Figure 4.18

```
THE HALF INTERVAL METHOD IS USED TO FIND THE RISE TIME


        T1                 T2                  T                  F(T)


 0.0               0.1000000E 02   0.5000000E 01  -0.2465206E-01
 0.5000000E 01     0.1000000E 02   0.7500000E 01   0.7974321E-01
 0.5000000E 01     0.7500000E 01   0.6250000E 01   0.4829999E-01
 0.5000000E 01     0.6250000E 01   0.5625000E 01   0.1904958E-01
 0.5000000E 01     0.5625000E 01   0.5312500E 01  -0.6810427E-03
 0.5312500E 01     0.5625000E 01   0.5468750E 01   0.9672582E-02
 0.5312500E 01     0.5468750E 01   0.5390625E 01   0.4622936E-02
 0.5312500E 01     0.5390625E 01   0.5351562E 01   0.2003372E-02
 0.5312500E 01     0.5351562E 01   0.5332031E 01   0.6693602E-03
 0.5312500E 01     0.5332031E 01   0.5322266E 01  -0.3755093E-05
 0.5322266E 01     0.5332031E 01   0.5327148E 01   0.3333092E-03
 0.5322266E 01     0.5327148E 01   0.5324707E 01   0.1649261E-03
 0.5322266E 01     0.5324707E 01   0.5323486E 01   0.8064508E-04
 0.5322266E 01     0.5323486E 01   0.5322876E 01   0.3838539E-04
 0.5322266E 01     0.5322876E 01   0.5322571E 01   0.1740456E-04
 0.5322266E 01     0.5322571E 01   0.5322418E 01   0.6794930E-05
 0.5322266E 01     0.5322418E 01   0.5322342E 01   0.1549721E-05
 0.5322266E 01     0.5322342E 01   0.5322304E 01  -0.1072884E-05
 0.5322304E 01     0.5322342E 01   0.5322323E 01   0.2384186E-06
 0.5322304E 01     0.5322323E 01   0.5322313E 01  -0.4172325E-06


THE ROOT =  0.5322313E 01 SECONDS

AND WAS FOUND IN     21 ITERATIONS
```

Figure 4.21 The output results of the program of Figure 4.19

4.4 THE METHOD OF FALSE POSITION

Another method for improving on the initial approximation of a root is the method of false position (*Regula Falsi*). This method starts in much the same way that the bisection method started in that two values of time, t_1 and t_2, which bracket the root, are initially found. The points $(t_1,f(t_1))$ and $(t_2,f(t_2))$ are then used to determine the equation of the straight line that passes through these points, and the root of this straight line equation is used as a better approximation to the actual root.

To determine the value of the unknown coefficients a and b shown in Figure 4.22, we note that at

$$t = t_1,\ f(t_1) = at_1 + b$$

and at

$$t = t_2,\ f(t_2) = at_2 + b$$

Since t_1, t_2, $f(t_1)$, and $f(t_2)$ are known, we find by solving the above equations that

$$a = \frac{f(t_1) - f(t_2)}{t_1 - t_2}$$

and

$$b = \frac{t_1 f(t_2) - t_2 f(t_1)}{t_1 - t_2}$$

Thus, the equation of the straight line approximating the function is given by

$$f(t) = \frac{f(t_1) - f(t_2)}{t_1 - t_2}\,t + \frac{t_1 f(t_2) - t_2 f(t_1)}{t_1 - t_2}$$

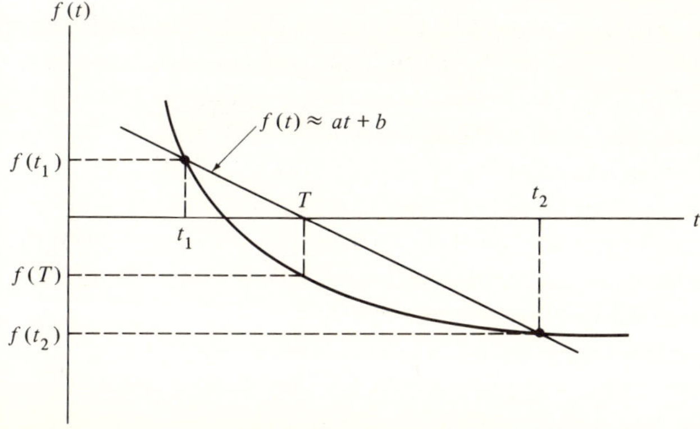

Figure 4.22 The method of false position

and the root of this equation would correspond to $f(T) = 0$, or

$$0 = \frac{f(t_1) - f(t_2)}{t_1 - t_2} T + \frac{t_1 f(t_2) - t_2 f(t_1)}{t_1 - t_2}$$

Solving for T, we find

$$T = \frac{t_1 f(t_2) - t_2 f(t_1)}{f(t_2) - f(t_1)}$$

Alternately, by referring to Figure 4.22 and using the law of similar triangles we find

$$\frac{T - t_1}{t_2 - t_1} = \frac{f(t_1)}{f(t_1) - f(t_2)}$$

or

$$T = t_1 + \frac{f(t_1)}{f(t_1) - f(t_2)} (t_2 - t_1)$$

which is identical to the above expression.

At this point the given function is evaluated at $t = T$, and the absolute value of $f(T)$ is tested to see whether it is equal to or less than some specified accuracy. If it is, T is the root. If it is not, we test to see whether $f(T)*f(t_1)$ is positive or negative. If $f(T)*f(t_1)$ is positive, the root lies between T and t_2, so we set $t_1 = T$ and repeat the process. If $f(T)*f(t_1)$ is negative, the root lies between t_1 and T, and we repeat the process with t_2 set equal to T. Thus, in each iteration, we either find $f(T)$ equal to or less than some specified accuracy and declare T as the root, or we narrow the brackets on the range of the root and repeat the process.

The Calculation of the Rise Time of a Single-Stage Amplifier

Figure 4.23 shows a Fortran program that uses the method of false position to find the value of $T_{0.1}$ in the calculation of the rise time of a single-stage amplifier. Figure 4.24 shows the corresponding results. Note that the convergence to the root is from one side and that only five iterations are required to produce an accuracy of $|f(T)| \leq 1.E-5$.

Figure 4.25 shows a modification of the program of Figure 4.23 so that the value of $T_{0.9}$ can be computed. Figure 4.26 gives the corresponding output results. Note that convergence again takes place from one side and that in this case 37 iterations are required to produce the same accuracy as specified in the calculation of $T_{0.1}$. The reader is urged to sketch the process for several iterations so that he may better understand why the calculation of $T_{0.9}$ converges at a slower rate than the calculation of $T_{0.1}$. In general, the steeper the slope of $f(t)$, the more rapid the convergence.

```
C       FALSE POSITION METHOD
C             USED
C       TO FIND THE RISE TIME
C
        WRITE(6,1)
      1 FORMAT('1THE FALSE POSITION METHOD IS USED TO FIND THE RISE TIME',
       1////,7X,'T1',13X,'T2',13X,'T',13X,'F(T)',//)
        READ(5,2)T1,T2,ACC
      2 FORMAT(3E15.7)
        I=0
      3 I=I+1
        FT1=0.9-EXP(-T1)
        FT2=0.9-EXP(-T2)
        T=T1+FT1*(T2-T1)/(FT1-FT2)
        FT=0.9-EXP(-T)
        WRITE(6,4)T1,T2,T,FT
      4 FORMAT(4E15.7)
        IF(ABS(FT)-ACC) 10,10,5
      5 IF(FT*FT1) 6,10,7
      6 T2=T
        GO TO 3
      7 T1=T
        GO TO 3
     10 WRITE(6,8)T,I
      8 FORMAT(///,' THE ROOT =',E15.7,' SECONDS',//,' AND WAS FOUND IN ',
       1I5,' ITERATIONS')
        RETURN
        END
```

Figure 4.23 A Fortran program for calculating $T_{0.1}$ of the single-stage amplifier

```
THE FALSE POSITION METHOD IS USED TO FIND THE RISE TIME



      T1              T2              T               F(T)


 0.0             0.5000000E 01  0.5033919E 00  0.2955231E 00
 0.0             0.5033919E 00  0.1272724E 00  0.1950622E-01
 0.0             0.1272724E 00  0.1064985E 00  0.1023591E-02
 0.0             0.1064985E 00  0.1054195E 00  0.5304813E-04
 0.0             0.1054195E 00  0.1053635E 00  0.2622604E-05



THE ROOT =   0.1053635E 00 SECONDS

AND WAS FOUND IN      5 ITERATIONS
```

Figure 4.24 The output results of the program of Figure 4.23

```
C       FALSE POSITION METHOD
C             USED
C       TO FIND THE RISE TIME
C
        WRITE(6,1)
      1 FORMAT('1THE FALSE POSITION METHOD IS USED TO FIND THE RISE TIME',
       1////,7X,'T1',13X,'T2',13X,'T',13X,'F(T)',//)
        READ(5,2)T1,T2,ACC
      2 FORMAT(3E15.7)
        I=0
      3 I=I+1
        FT1=0.1-EXP(-T1)
        FT2=0.1-EXP(-T2)
        T=T1+FT1*(T2-T1)/(FT1-FT2)
        FT=0.1-EXP(-T)
        WRITE(6,4)T1,T2,T,FT
      4 FORMAT(4E15.7)
        IF(ABS(FT)-ACC) 10,10,5
      5 IF(FT*FT1)6,10,7
      6 T2=T
        GO TO 3
      7 T1=T
        GO TO 3
     10 WRITE(6,8)T,I
      8 FORMAT(///,' THE ROOT =',E15.7,' SECONDS',//,' AND WAS FOUND IN ',
       1I5,' ITERATIONS')
        RETURN
        END
```

Figure 4.25 A Fortran program for calculating $T_{0.9}$ of the single-stage amplifier

THE FALSE POSITION METHOD IS USED TO FIND THE RISE TIME

T1	T2	T	F (T)
0.0	0.5000000E 01	0.4530526E 01	0.8922493E-01
0.0	0.4530526E 01	0.4121886E 01	0.8378601E-01
0.0	0.4121886E 01	0.3770838E 01	0.7696718E-01
0.0	0.3770838E 01	0.3473764E 01	0.6899983E-01
0.0	0.3473764E 01	0.3226406E 01	0.6030004E-01
0.0	0.3226406E 01	0.3023809E 01	0.5138430E-01
0.0	0.3023809E 01	0.2860493E 01	0.4275941E-01
0.0	0.2860493E 01	0.2730752E 01	0.3482968E-01
0.0	0.2730752E 01	0.2629010E 01	0.2785015E-01
0.0	0.2629010E 01	0.2550097E 01	0.2192593E-01
0.0	0.2550097E 01	0.2489448E 01	0.1704419E-01
0.0	0.2489448E 01	0.2443177E 01	0.1311558E-01
0.0	0.2443177E 01	0.2408084E 01	0.1001239E-01
0.0	0.2408084E 01	0.2381588E 01	0.7596254E-02
0.0	0.2381588E 01	0.2361654E 01	0.5735815E-02
0.0	0.2361654E 01	0.2346697E 01	0.4315257E-02
0.0	0.2346697E 01	0.2335498E 01	0.3237665E-02
0.0	0.2335498E 01	0.2327126E 01	0.2424121E-02
0.0	0.2327126E 01	0.2320873E 01	0.1812160E-02
0.0	0.2320873E 01	0.2316208E 01	0.1353025E-02
0.0	0.2316208E 01	0.2312731E 01	0.1009405E-02
0.0	0.2312731E 01	0.2310139E 01	0.7525086E-03
0.0	0.2310139E 01	0.2308208E 01	0.5606413E-03
0.0	0.2308208E 01	0.2306769E 01	0.4175305E-03
0.0	0.2306769E 01	0.2305698E 01	0.3108382E-03
0.0	0.2305698E 01	0.2304902E 01	0.2313852E-03
0.0	0.2304902E 01	0.2304309E 01	0.1721978E-03
0.0	0.2304309E 01	0.2303866E 01	0.1280308E-03
0.0	0.2303866E 01	0.2303538E 01	0.9524822E-04
0.0	0.2303538E 01	0.2303293E 01	0.7075071E-04
0.0	0.2303293E 01	0.2303111E 01	0.5257130E-04
0.0	0.2303111E 01	0.2302976E 01	0.3904104E-04
0.0	0.2302976E 01	0.2302875E 01	0.2890825E-04
0.0	0.2302875E 01	0.2302800E 01	0.2145767E-04
0.0	0.2302800E 01	0.2302744E 01	0.1585484E-04
0.0	0.2302744E 01	0.2302702E 01	0.1162291E-04
0.0	0.2302702E 01	0.2302671E 01	0.8583069E-05

THE ROOT = 0.2302671E 01 SECONDS

AND WAS FOUND IN 37 ITERATIONS

Figure 4.26 The output results of the program of Figure 4.25

The False Position Method with Ordinate Halving

There are several variations of the false position method that attempt to eliminate convergence to the root from one side and, in general, tend to converge more rapidly. One such method uses ordinate halving as shown in Figure 4.27. This method proceeds as before with the calculation of $f(t_1)$ and $f(t_2)$ from the bracketed root values of t_1 and t_2. The value of T is then calculated from the linear approximation based on these starting points. At this point the value of $f(T)$ is computed and compared in sign to the value of $f(t_1)$. If they are opposite in sign, as in the case shown in Figure 4.27, t_2 is replaced by T and the new value of $f(t_2)$ is found. The new point $(t_2,f(t_2))$ and the point $(t_1,f(t_1)/2)$, which is computed using half of the ordinate value of the old point, are now used to find the next approximation to the root (called $T1$ in Figure 4.27). The function is then evaluated, using this new approximation to the root, and is compared in sign to $f(t_1)$. If it is of the opposite sign, as in

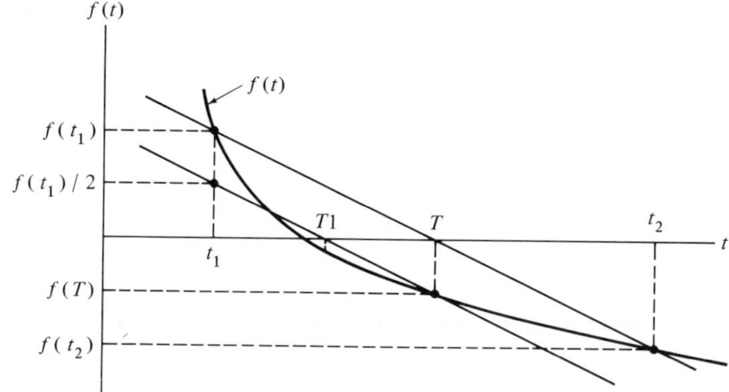

Figure 4.27 The method of false position with ordinate halving

```
C
C          FALSE POSITION METHOD
C          WITH ORDINATE HALVING
C                USED
C          TO FIND THE RISE TIME
C
           WRITE(6,1)
         1 FORMAT('1THE FALSE POSITION METHOD WITH ORDINATE HALVING IS USED T
          1O FIND THE RISE TIME',////,7X,'T1',13X,'T2',13X,'T',13X,'F(T)',//)
           READ(5,2)T1,T2,ACC
         2 FORMAT(3E15.7)
           I=0
         3 I=I+1
           FT1=0.9-EXP(-T1)
           FT2=0.9-EXP(-T2)
           T=T1+FT1*(T2-T1)/(FT1-FT2)
           FT=0.9-EXP(-T)
           WRITE(6,4)T1,T2,T,FT
         4 FORMAT(4E15.7)
           IF(ABS(FT)-ACC) 10,10,5
         5 IF(FT*FT1)6,10,7
         6 T2=T
           FT2=0.9-EXP(-T2)
           T=T1+FT1/2.*(T2-T1)/(FT1/2.-FT2)
           FT=0.9-EXP(-T)
           IF(FT*FT1) 15,10,16
        15 T2=T
           GO TO 3
        16 T1=T
           GO TO 3
         7 T1=T
           FT1=0.9-EXP(-T1)
           T=T1+FT1*(T2-T1)/(FT1-FT2/2.)
           FT=0.9-EXP(-T)
           IF(FT*FT1)15,10,16
        10 WRITE(6,8)T,I
         8 FORMAT(///,' THE ROOT =',E15.7,' SECONDS',//,' AND WAS FOUND IN',I
          15,' ITERATIONS')
           RETURN
           END
```

Figure 4.28 A Fortran program that uses ordinate halving to calculate $T_{0.1}$ of the single-stage amplifier

Figure 4.27, t_2 is set equal to the new T and the iteration is repeated. When either $|f(T)| \leq \text{ACC}$ or $f(T) = 0$, the value of T, the root, and the number of iterations are printed.

Figure 4.28 shows the Fortran program that uses the ordinate halving technique to find the value of $T_{0.1}$ in the calculation of the rise time of a single-stage amplifier, and Figure 4.29 gives the program results. Note

```
THE FALSE POSITION METHOD WITH ORDINATE HALVING IS USED TO FIND THE RISE TIME


       T1                T2                T               F(T)


 0.0             0.5000000E 01  0.5033919E 00  0.2955231E 00
 0.7284486E-01   0.5033919E 00  0.1122175E 00  0.6150126E-02
 0.1006991E 00   0.1122175E 00  0.1053765E 00  0.1436472E-04
 0.1053447E 00   0.1053765E 00  0.1053605E 00  0.0


THE ROOT =  0.1053605E 00 SECONDS

AND WAS FOUND IN     4 ITERATIONS
```

Figure 4.29 The output results of the program of Figure 4.28

```
C          FALSE POSITION METHOD
C          WITH ORDINATE HALVING
C                 USED
C          TO FIND THE RISE TIME
C
           WRITE(6,1)
         1 FORMAT('1THE FALSE POSITION METHOD WITH ORDINATE HALVING IS USED T
          1O FIND THE RISE TIME',////,7X,'T1',13X,'T2',13X,'T',13X,'F(T)',//)
           READ(5,2)T1,T2,ACC
         2 FORMAT(3E15.7)
           I=0
         3 I=I+1
           FT1=0.1-EXP(-T1)
           FT2=0.1-EXP(-T2)
           T=T1+FT1*(T2-T1)/(FT1-FT2)
           FT=0.1-EXP(-T)
           WRITE(6,4)T1,T2,T,FT
         4 FORMAT(4E15.7)
           IF(ABS(FT)-ACC) 10,10,5
         5 IF(FT*FT1)6,10,7
         6 T2=T
           FT2=0.1-EXP(-T2)
           T=T1+FT1/2.*(T2-T1)/(FT1/2.-FT2)
           FT=0.1-EXP(-T)
           IF(FT*FT1) 15,10,16
        15 T2=T
           GO TO 3
        16 T1=T
           GO TO 3
         7 T1=T
           FT1=0.1-EXP(-T1)
           T=T1+FT1*(T2-T1)/(FT1-FT2/2.)
           FT=0.1-EXP(-T)
           IF(FT*FT1)15,10,16
        10 WRITE(6,8)T,I
         8 FORMAT(///,' THE ROOT =',E15.7,' SECONDS',//,' AND WAS FOUND IN',I
          15,' ITERATIONS')
           RETURN
           END
```

Figure 4.30 A Fortran program that uses ordinate halving to calculate $T_{0.9}$ of the single-stage amplifier

that convergence to the root takes place from both sides. Note also that four iterations are needed to find the root in contrast with five iterations that were required by the program of Figure 4.23. Since each iteration of the ordinate halving program is equivalent to two iterations of the false position program, we see that the ordinate halving program has not increased the rate of convergence for this particular root. The reader is encouraged to examine these two results in greater detail and see that the same amount of computations are required for roughly the same accuracy.

```
THE FALSE POSITION METHOD WITH ORDINATE HALVING IS USED TO FIND THE RISE TIME


       T1               T2               T               F(T)


0.0              0.5000000E 01   0.4530526E 01   0.8922493E-01
0.0              0.3780864E 01   0.3482181E 01   0.6925964E-01
0.0              0.3017720E 01   0.2855624E 01   0.4248006E-01
0.0              0.2609304E 01   0.2534905E 01   0.2073073E-01
0.0              0.2423268E 01   0.2393039E 01   0.8648276E-02
0.0              0.2347915E 01   0.2336409E 01   0.3325760E-02
0.0              0.2319265E 01   0.2315009E 01   0.1234710E-02
0.0              0.2308674E 01   0.2307117E 01   0.4521608E-03
0.0              0.2304800E 01   0.2304233E 01   0.1645684E-03
0.0              0.2303390E 01   0.2303184E 01   0.5978346E-04
0.0              0.2302876E 01   0.2302800E 01   0.2145767E-04
0.0              0.2302689E 01   0.2302661E 01   0.7569790E-05


THE ROOT =   0.2302661E 01 SECONDS

AND WAS FOUND IN    12 ITERATIONS
```

Figure 4.31 The output results of the program of Figure 4.30

```
C         FALSE POSITION METHOD
C         WITH ORDINATE HALVING
C                USED
C         TO FIND THE RISE TIME
C
          WRITE(6,1)
        1 FORMAT('1THE FALSE POSITION METHOD WITH ORDINATE HALVING IS USED T
         1O FIND THE RISE TIME',////,7X,'T1',13X,'T2',13X,'T',13X,'F(T)',//)
          READ(5,2) T1,T2,ACC
        2 FORMAT(3E15.7)
          I=0
        3 I=I+1
          FT1=0.9-(1.+T1+0.5*T1**2)*EXP(-T1)
          FT2=0.9-(1.+T2+0.5*T2**2)*EXP(-T2)
          T=T1+FT1*(T2-T1)/(FT1-FT2)
          FT=0.9-(1.+T+0.5*T**2)*EXP(-T)
          WRITE(6,4) T1,T2,T,FT
        4 FORMAT(4E15.7)
          IF(ABS(FT)-ACC) 10,10,5
        5 IF(FT*FT1)6,10,7
        6 T2=T
          FT2=0.9-(1.+T2+0.5*T2**2)*EXP(-T2)
          T=T1+FT1/2.*(T2-T1)/(FT1/2.-FT2)
          FT=0.9-(1.+T+0.5*T**2)*EXP(-T)
          IF(FT*FT1) 15,10,16
       15 T2=T
          GO TO 3
       16 T1=T
          GO TO 3
        7 T1=T
          FT1=0.9-(1.+T1+0.5*T1**2)*EXP(-T1)
          T=T1+FT1*(T2-T1)/(FT1-FT2/2.)
          FT=0.9-(1.+T+0.5*T**2)*EXP(-T)
          IF(FT*FT1) 15,10,16
       10 WRITE(6,8) T,I
        8 FORMAT(///,' THE ROOT =',E15.7,' SECONDS',//,' AND WAS FOUND IN',I
         15    ITERATIONS )
          RETURN
          END
```

Figure 4.32 A Fortran program that uses ordinate halving to calculate $T_{0.1}$ of the three-stage amplifier

Figure 4.30 gives the Fortran program that was used to calculate $T_{0.9}$, and Figure 4.31 shows the corresponding results. Note that even though ordinate halving was used, convergence to the root is from one side. This can best be explained by graphically following each iteration and seeing that the reason for this is due to the small value of the slope of $f(t)$ in

```
C     FALSE POSITION METHOD
C     WITH ORDINATE HALVING
C           USED
C     TO FIND THE RISE TIME
C
      WRITE(6,1)
    1 FORMAT('1THE FALSE POSITION METHOD WITH ORDINATE HALVING IS USED T
     1O FIND THE RISE TIME',////,7X,'T1',13X,'T2',13X,'T',13X,'F(T)',//)
      READ(5,2) T1,T2,ACC
    2 FORMAT(3E15.7)
      I=0
    3 I=I+1
      FT1=0.1-(1.+T1+0.5*T1**2)*EXP(-T1)
      FT2=0.1-(1.+T2+0.5*T2**2)*EXP(-T2)
      T=T1+FT1*(T2-T1)/(FT1-FT2)
      FT=0.1-(1.+T+0.5*T**2)*EXP(-T)
      WRITE(6,4) T1,T2,T,FT
    4 FORMAT(4E15.7)
      IF(ABS(FT)-ACC) 10,10,5
    5 IF(FT*FT1)6,10,7
    6 T2=T
      FT2=0.1-(1.+T2+0.5*T2**2)*EXP(-T2)
      T=T1+FT1/2.*(T2-T1)/(FT1/2.-FT2)
      FT=0.1-(1.+T+0.5*T**2)*EXP(-T)
      IF(FT*FT1) 15,10,16
   15 T2=T
      GO TO 3
   16 T1=T
      GO TO 3
    7 T1=T
      FT1=0.1-(1.+T1+0.5*T1**2)*EXP(-T1)
      T=T1+FT1*(T2-T1)/(FT1-FT2/2.)
      FT=0.1-(1.+T+0.5*T**2)*EXP(-T)
      IF(FT*FT1) 15,10,16
   10 WRITE(6,8) T,I
    8 FORMAT(///,' THE ROOT =',E15.7,' SECONDS',//,' AND WAS FOUND IN',I
     15,' ITERATIONS')
      RETURN
      END
```

Figure 4.33 A Fortran program that uses ordinate halving to calculate $T_{0.9}$ of the three-stage amplifier

```
THE FALSE POSITION METHOD WITH ORDINATE HALVING IS USED TO FIND THE RISE TIME


      T1              T2              T               F(T)


0.0             0.1000000E 02  0.1002777E 01 -0.1918697E-01
0.1002777E 01   0.1371799E 01  0.1092498E 01 -0.1922309E-02
0.1092498E 01   0.1109388E 01  0.1102036E 01 -0.5841255E-05


 THE ROOT =  0.1102036E 01 SECONDS

 AND WAS FOUND IN    3 ITERATIONS
```

Figure 4.34 The output results of the program of Figure 4.32

the vicinity of t_2. Note, however, that the number of iterations has been materially lowered by the use of ordinate halving.

Figures 4.32 and 4.33 show the Fortran programs for finding $T_{0.1}$ and $T_{0.9}$ in the case of the three-stage amplifier. Figures 4.34 and 4.35 show that there has been a considerable reduction (over the method used in Figures 4.20 and 4.21) in the number of iterations required to find either root.

THE FALSE POSITION METHOD WITH ORDINATE HALVING IS USED TO FIND THE RISE TIME

T1	T2	T	F(T)
0.0	0.1000000E 02	0.9024994E 01	0.9389144E-01
0.0	0.7467017E 01	0.6862898E 01	0.6714618E-01
0.0	0.5971819E 01	0.5737462E 01	0.2523851E-01
0.0	0.5432761E 01	0.5388589E 01	0.4487991E-02
0.0	0.5335377E 01	0.5330050E 01	0.5331635E-03
0.0	0.5323742E 01	0.5323159E 01	0.5799532E-04
0.0	0.5322472E 01	0.5322409E 01	0.6139278E-05

THE ROOT = 0.5322409E 01 SECONDS

AND WAS FOUND IN 7 ITERATIONS

Figure 4.35 The output results of the program of Figure 4.33

4.5 THE NEWTON-RAPHSON METHOD

The Newton-Raphson method is one of the best methods that can be used to improve the accuracy of an initial approximation to a real root. The iteration formula may be derived by writing a Taylor's expansion that is based on the initial approximation t_1 to the root, as follows:

$$f(t_1 + h) = f(t_1) + hf'(t_1) + \frac{h^2}{2!}f''(t_1) + \cdots$$

We may now regard h as the exact correction that is needed to make $t_1 + h$ equal to the root and thus we may write

$$0 \equiv f(t_1) + hf'(t_1) + \frac{h^2}{2!}f''(t_1) + \cdots$$

If the correction h is small, we may discard the higher-order terms in h and use the first two terms to give us an approximation of the required correction. Solving for h we find

$$h = -\frac{f(t_1)}{f'(t_1)}$$

and therefore

$$t_2 = t_1 + h = t_1 - \frac{f(t_1)}{f'(t_1)}$$

should be a better approximation to the root if the process is converging.[9] This procedure may be applied repeatedly until the required degree of

[9] Note that this value of h, when added to t_1, will not be the exact value of the root since only the first two terms were used in the calculation.

accuracy has been obtained and the repetitive process may be generalized by the following iteration formula:

$$t_{n+1} = t_n - \frac{f(t_n)}{f'(t_n)}$$

with n corresponding to the number of the iteration.

The process may be interpreted geometrically by referring to Figure 4.36. Starting with the initial approximation t_1 to the root, we construct a tangent line at $(t_1, f(t_1))$. The intersection of the tangent line with the abscissa gives the new approximation to the root. We may see that this gives the same result mathematically as the Taylor's approach by expressing the derivative of $f(t)$ at $t = t_1$ in terms of $f(t_1)$ and h. From Figure 4.36 we have

$$f'(t_1) = -\frac{f(t_1)}{h}$$

or

$$h = -\frac{f(t_1)}{f'(t_1)}$$

and

$$t_2 = t_1 + h$$

$$= t_1 - \frac{f(t_1)}{f'(t_1)}$$

Using the new value t_2, we construct a tangent line at $(t_2, f(t_2))$ and the intersection of this line with the abscissa gives us t_3, and so on. The iteration may be terminated whenever the $|f(t_n)|$ is less than or equal to some specified value, the $|t_n - t_{n-1}|$ is less than or equal to some specified

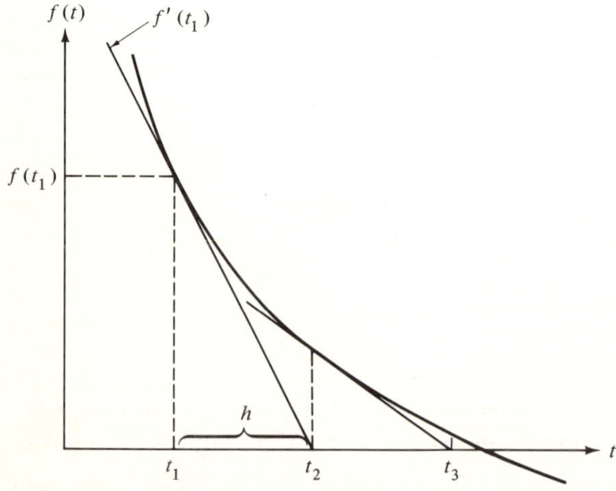

Figure 4.36 The Newton-Raphson method of finding a real root

value, or, since the process may fail to converge, whenever the number of iterations reaches a specified number.

The Calculation of the Rise Time

Figures 4.37 and 4.38 give the Fortran programs for finding $T_{0.1}$ and $T_{0.9}$ in the calculation of the rise time of a single-stage amplifier. Figures 4.39 and 4.40 give the respective results. Note that in both programs the process converges very rapidly from a starting value of $t = 0$ and that, in general, the next correction is less than the square of the preceding correction. Note also that the program was made to terminate whenever ABS(COR) $\leq$ ACC.

Figures 4.41 and 4.42 show the Fortran programs for finding $T_{0.1}$ and

```
C       NEWTON-RAPHSON METHOD
C               USED
C       TO FIND THE RISE TIME
C
        WRITE(6,1)
      1 FORMAT('1 THE NEWTON-RAPHSON METHOD IS USED TO FIND THE RISE TIME'
       1,////,4X,'TIME',13X,'F(T)',8X,'CORRECTION',/)
        READ(5,2)T,ACC
      2 FORMAT(2E15.7)
        I=0
      3 I=I+1
        FT=0.1-EXP(-T)
        FPT=EXP(-T)
        COR=-FT/FPT
        WRITE(6,4)T,FT,COR
      4 FORMAT(3E15.7)
        T=T+COR
        IF(ABS(COR)-ACC)10,10,3
     10 WRITE(6,5)I,ACC
      5 FORMAT(///,' THE NUMBER OF ITERATIONS REQUIRED =',I5,//,' BASED ON
       1 AN ACCURACY =',E15.5)
        RETURN
        END
```

Figure 4.37 A Fortran program that uses the Newton-Raphson method to calculate $T_{0.1}$ of the single-stage amplifier

```
C       NEWTON-RAPHSON METHOD
C               USED
C       TO FIND THE RISE TIME
C
        WRITE(6,1)
      1 FORMAT('1 THE NEWTON-RAPHSON METHOD IS USED TO FIND THE RISE TIME'
       1,////,4X,'TIME',13X,'F(T)',8X,'CORRECTION',/)
        READ(5,2)T,ACC
      2 FORMAT(2E15.7)
        I=0
      3 I=I+1
        FT=0.9-EXP(-T)
        FPT=EXP(-T)
        COR=-FT/FPT
        WRITE(6,4)T,FT,COR
      4 FORMAT(3E15.7)
        T=T+COR
        IF(ABS(COR)-ACC)10,10,3
     10 WRITE(6,5)I,ACC
      5 FORMAT(///,' THE NUMBER OF ITERATIONS REQUIRED =',I5,//,' BASED ON
       1 AN ACCURACY =',E15.5)
        RETURN
        END
```

Figure 4.38 A Fortran program that uses the Newton-Raphson method to calculate $T_{0.9}$ of the single-stage amplifier

```
THE NEWTON-RAPHSON METHOD IS USED TO FIND THE RISE TIME


  TIME                    F(T)              CORRECTION

0.0                -0.1000000E 00   0.1000000E 00
0.1000000E 00 -0.4837453E-02   0.5346213E-02
0.1053462E 00 -0.1293421E-04   0.1437114E-04
0.1053606E 00  0.0                 0.0


THE NUMBER OF ITERATIONS REQUIRED =      4

BASED ON AN ACCURACY =       0.10000E-04
```

Figure 4.39 The output results of the program of Figure 4.37

```
THE NEWTON-RAPHSON METHOD IS USED TO FIND THE RISE TIME


  TIME                    F(T)              CORRECTION

0.0                -0.9000000E 00   0.9000000E 00
0.9000000E 00 -0.3065697E 00   0.7540398E 00
0.1654039E 01 -0.9127575E-01   0.4771947E 00
0.2131233E 01 -0.1869088E-01   0.1574754E 00
0.2288708E 01 -0.1397431E-02   0.1378172E-01
0.2302489E 01 -0.9596348E-05   0.9595430E-04
0.2302585E 01 -0.5960464E-07   0.5960463E-06


THE NUMBER OF ITERATIONS REQUIRED =      7

BASED ON AN ACCURACY =       0.10000E-04
```

Figure 4.40 The output results of the program of Figure 4.38

```
C        NEWTON-RALPHSON METHOD
C               USED
C        TO FIND THE RISE TTME
C
         WRITE(6,1)
       1 FORMAT('1THE NEWTON-RALPHSON METHOD IS USED TO FIND THE RISE TIME'
        1.////,4X,'TIME',13X,'F(T)',8X,'F PRIME(T)',2X,'CORRECTION',/)
         READ(5,2)T,ACC
       2 FORMAT(2E15.7)
         I=0
       3 I=I+1
         FT=0.9-(1.+T+0.5*T**2)*EXP(-T)
         FPT=0.5*T**2*EXP(-T)
         COR=-FT/FPT
         WRITE(6,4)T,FT,FPT,COR
       4 FORMAT(4E15.7)
         T=T+COR
         IF(ABS(COR)-ACC) 10,10,3
      10 WRITE(6,5)I,ACC
       5 FORMAT(///,' THE NUMBER OF ITERATIONS REQUIRED =',I5,//,' BASED ON
        1 AN ACCURACY =',E15.5)
         RETURN
         END
```

Figure 4.41 A Fortran program that uses the Newton-Raphson method to calculate $T_{0.1}$ of the three-stage amplifier

$T_{0.9}$ for the case of the three-stage amplifier. Figures 4.43 and 4.44 show that this method is more efficient than the ordinate halving method of Figures 4.32 and 4.33. Note also, as in the last example, that each correction is less than the square of the preceding correction and we have second-order convergence. Let us now consider the convergence problem in more detail.

```
C       NEWTON-RALPHSON METHOD
C              USED
C       TO FIND THE RISE TIME
C
        WRITE(6,1)
      1 FORMAT('1THE NEWTON-RALPHSON METHOD IS USED TO FIND THE RISE TIME'
       1,////,4X,'TIME',13X,'F(T)',8X,'F PRIME(T)',2X,'CORRECTION',/)
        READ(5,2)T,ACC
      2 FORMAT(2E15.7)
        I=0
      3 I=I+1
        FT=0.1-(1.+T+0.5*T**2)*EXP(-T)
        FPT=0.5*T**2*EXP(-T)
        COR=-FT/FPT
        WRITE(6,4)T,FT,FPT,COR
      4 FORMAT(4E15.7)
        T=T+COR
        IF(ABS(COR)-ACC)10,10,3
     10 WRITE(6,5)I,ACC
      5 FORMAT(///,' THE NUMBER OF ITERATIONS REQUIRED =',I5,//,' BASED ON
       1 AN ACCURACY =',E15.5)
        RETURN
        END
```

Figure 4.42 A Fortran program that uses the Newton-Raphson method to calculate $T_{0.9}$ of the three-stage amplifier

THE NEWTON-RALPHSON METHOD IS USED TO FIND THE RISE TIME

TIME	F(T)	F PRIME(T)	CORRECTION
0.1000000E 01	-0.1969862E-01	0.1839397E 00	0.1070928E 00
0.1107092E 01	0.1016438E-02	0.2025504E 00	-0.5018197E-02
0.1102074E 01	0.1907349E-05	0.2017282E 00	-0.9455043E-05

THE NUMBER OF ITERATIONS REQUIRED = 3

BASED ON AN ACCURACY = 0.10000E-04

Figure 4.43 The output results of the program of Figure 4.41

THE NEWTON-RALPHSON METHOD IS USED TO FIND THE RISE TIME

TIME	F(T)	F PRIME(T)	CORRECTION
0.1000000E 01	-0.8196986E 00	0.1839397E 00	0.4456344E 01
0.5456344E 01	0.8887291E-02	0.6354970E-01	-0.1398479E 00
0.5316495E 01	-0.4034042E-03	0.6938988E-01	0.5813587E-02
0.5322308E 01	-0.8344650E-06	0.6913865E-01	0.1206944E-04
0.5322319E 01	-0.5960464E-07	0.6913811E-01	0.8621098E-06

THE NUMBER OF ITERATIONS REQUIRED = 5

BASED ON AN ACCURACY = 0.10000E-04

Figure 4.44 The output results of the program of Figure 4.42

The Convergence of the Newton-Raphson Method

The rate of convergence of the Newton-Raphson method can be found by letting ϵ_n correspond to the error at the n-th iteration or

$$\epsilon_n = T - t_{n+1}$$

where T is the exact root and t_{n+1} is the $(n + 1)$-th approximation to the root that has been found by using the Newton-Raphson formula

$$t_{n+1} = t_n - \frac{f(t_n)}{f'(t_n)}$$

Similarly, we let ϵ_{n-1} correspond to the error at the $(n - 1)$-th iteration or

$$\epsilon_{n-1} = T - t_n$$

If we now subtract both sides of the general Newton-Raphson iteration formula from the exact root, we get

$$T - t_{n+1} = T - t_n + \frac{f(t_n)}{f'(t_n)}$$

or

$$\begin{aligned} \epsilon_n &= \epsilon_{n-1} + \frac{f(t_n)}{f'(t_n)} \\ &= \epsilon_{n-1} + \frac{f(T - \epsilon_{n-1})}{f'(T - \epsilon_{n-1})} \end{aligned}$$

Expanding the numerator and the denominator of the second term in a Taylor series, we get

$$f(T - \epsilon_{n-1}) = f(T) - \epsilon_{n-1}f'(T) + \frac{(\epsilon_{n-1})^2}{2!}f''(T) + \cdots$$

$$f'(T - \epsilon_{n-1}) = f'(T) - \epsilon_{n-1}f''(T) + \frac{(\epsilon_{n-1})^2}{2!}f'''(T) + \cdots$$

Since T is the exact root, $f(T) \equiv 0$ and thus we may write

$$\epsilon_n = \epsilon_{n-1} + \frac{-\epsilon_{n-1}f'(T) + ((\epsilon_{n-1})^2/2!)f''(T) + \cdots}{f'(T) - \epsilon_{n-1}f''(T) + ((\epsilon_{n-1})^2/2!)f'''(T) + \cdots}$$

If we now divide out the second term in the right member we find

$$\epsilon_n = \epsilon_{n-1} - \epsilon_{n-1} - \frac{(\epsilon_{n-1})^2}{2!}\frac{f''(T)}{f'(T)} + \text{higher order terms of } \epsilon_{n-1}.$$

or

$$\epsilon_n \approx -\frac{(\epsilon_{n-1})^2}{2!}\frac{f''(T)}{f'(T)}$$

which says that if the error is small, the error at the n-th iteration is proportional to the square of the error at the previous iteration. The rate of convergence is thus said to be second order or quadratic. Note that this was the condition that we found in the solutions of Figures 4.39 and 4.40 in contrast with the first-order convergence that was found in the solutions of Figures 4.24 and 4.26.

From the above discussion, one may see why it is possible to terminate the iteration procedure based on $|f(t_{n+1}) - f(t_n)| \leq \text{ACC}$. For example, if the error at the n-th iteration is 0.1, the $(n + 1)$-th correction will be less than the square of 0.1, and the $(n + 2)$-th correction will be less than the square of 0.01, and so on. The effect of cumulative corrections can therefore be studied from the following table:

ITERATION	CORRECTION (LESS THAN)
n	0.1
$n+1$	0.01
$n+2$	0.0001
$n+3$	0.00000001
.	.
.	.
.	.

We thus see that if we terminate at the n-th correction, the sum of all the following corrections cannot be greater than 0.1. Neglecting round-off error, each iteration will therefore effectively double the number of significant figures of accuracy.

Unfortunately, the single criterion we used for the termination of the iteration procedure is usually not sufficient. Figure 4.45 shows four possible cases in which the Newton-Raphson method fails to converge to the desired root. Figures 4.45a and 4.45b are examples that show why the method may oscillate and Figures 4.45c and 4.45d show examples in which the method diverges or converges on a root that we were not seeking.

Thus, in addition to the termination criterion $|f(t_{n+1}) - f(t_n)| \leq \text{ACC}$, we should also specify the maximum number of iterations permissible. Whenever the program reaches the maximum number of iterations, it can be terminated, or a new starting value can be automatically introduced. In addition, the program may also include a test on the magnitude

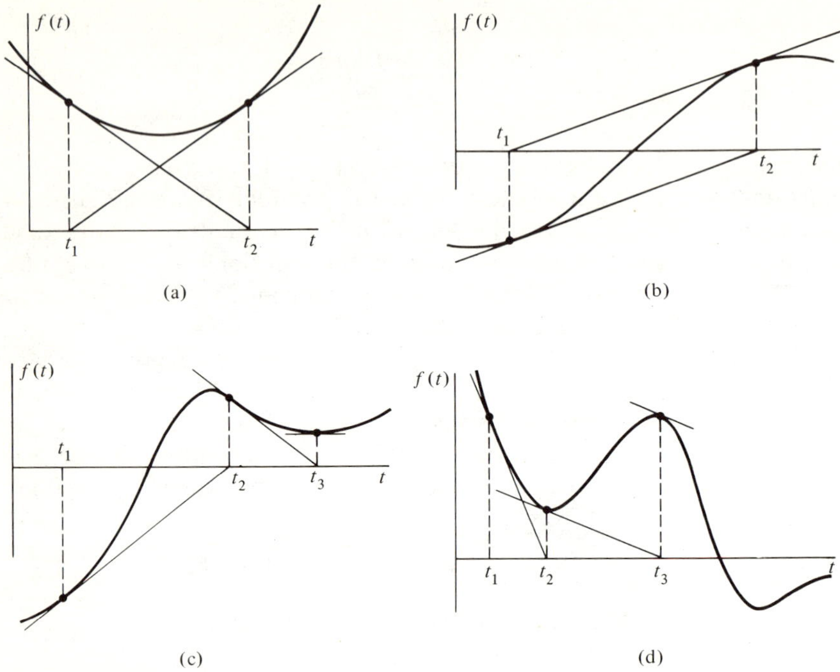

Figure 4.45 Newton-Raphson convergence

of the derivative and whenever the value of the derivative becomes too small, a shift in the root approximation can be made.

Finding the Roots of Polynomials

We have already seen how the Newton-Raphson method may be used to find the roots of a transcendental equation. We now show that this method is equally suited to finding the roots of a polynomial of the form

$$P_n(s) = a_n s^n + a_{n-1}s^{n-1} + \cdots + a_1 s + a_0 = 0$$

In order to show this, let us first recall the *Remainder Theorem* which states that if a polynomial $P_n(s)$ is divided by a linear divisor $s - s_1$, the remainder r_1 equals the value of the polynomial evaluated for $s = s_1$. The proof of this follows from the ability to express this division in terms of the quotient $Q_{n-1}(s)$ and the remainder r_1. Thus

$$\frac{P_n(s)}{s - s_1} = Q_{n-1}(s) + \frac{r_1}{s - s_1}$$

or

$$P_n(s) = Q_{n-1}(s)(s - s_1) + r_1$$

If we now evaluate $P_n(s)$ at $s = s_1$, we find

$$\begin{aligned} P_n(s_1) &= Q_{n-1}(s_1)(s_1 - s_1) + r_1 \\ &= r_1 \end{aligned}$$

The Newton-Raphson method also requires the value of the derivative of the polynomial $P_n'(s_1)$. This may be found from a second division by dividing the first quotient $Q_{n-1}(s)$ by $s - s_1$, thereby obtaining a new quotient $Q_{n-2}(s)$ and a new remainder r_2

$$\frac{Q_{n-1}(s)}{s - s_1} = Q_{n-2}(s) + \frac{r_2}{s - s_1}$$

or

$$Q_{n-1}(s) = Q_{n-2}(s)(s - s_1) + r_2$$

If we now differentiate

$$P_n(s) = Q_{n-1}(s)(s - s_1) + r_1$$

we obtain

$$P_n'(s) = Q_{n-1}'(s)(s - s_1) + Q_{n-1}(s)$$

Substituting the value of $Q_{n-1}(s)$, we found above, yields

$$P_n'(s) = Q_{n-1}'(s)(s - s_1) + Q_{n-2}(s)(s - s_1) + r_2$$

and evaluating $P_n'(s)$ at $s = s_1$ we obtain the value of the derivative

$$P_n'(s_1) = r_2$$

For a simple numerical example let us find the zeros of the characteristic equation

$$(s + 1)(s + 2)(s + 3) = s^3 + 6s^2 + 11s + 6 = 0$$

Starting with an initial approximation to the root of $s_1 = 0$, we divide $s - s_1$ into $P_3(s)$

$$\begin{array}{r|l} & s^2 + 6s + 11 \\ \hline s & s^3 + 6s^2 + 11s + 6 \\ & s^3 \\ \hline & \quad 6s^2 + 11s + 6 \\ & \quad 6s^2 \\ \hline & \qquad 11s + 6 \\ & \qquad 11s \\ \hline & \qquad\quad 6 = r_1 = P_3(0) \end{array}$$

We now divide the quotient $Q_2(s) = s^2 + 6s + 11$ by $s - s_1$ where $s_1 = 0$.

$$
\begin{array}{r|l}
 & s + 6 \\
\hline
s & s^2 + 6s + 11 \\
 & s^2 \\
\hline
 & \quad 6s + 11 \\
 & \quad 6s \\
\hline
 & \qquad 11 = r_2 = P_3'(0)
\end{array}
$$

Using the Newton-Raphson formula we find

$$
\begin{aligned}
s_2 &= s_1 - \frac{P_3(s_1)}{P_3'(s_1)} \\
&= 0 - \frac{6}{11} = -0.545
\end{aligned}
$$

Repeating the process with $s_2 = -0.545$ we have

$$
\begin{array}{r|l}
 & s^2 + 5.455s + 8.027 \\
\hline
s + 0.545 & s^3 + 6.000s^2 + 11s + 6 \\
 & s^3 + 0.545s^2 \\
\hline
 & \quad 5.455s^2 + 11.000s + 6 \\
 & \quad 5.455s^2 + 2.973s \\
\hline
 & \qquad 8.027s + 6.000 \\
 & \qquad 8.027s + 4.375 \\
\hline
 & \qquad\qquad 1.625 = P_3(-0.545)
\end{array}
$$

and dividing the quotient by $s + 0.545$ we find

$$
\begin{array}{r|l}
 & s + 4.91 \\
\hline
s + 0.545 & s^2 + 5.455s + 8.027 \\
 & s^2 + 0.545s \\
\hline
 & \quad 4.91s + 8.027 \\
 & \quad 4.91s + 2.675 \\
\hline
 & \qquad 5.352 = P_3'(-0.545)
\end{array}
$$

Using the Newton-Raphson formula for the second time, we find

$$
\begin{aligned}
s_3 &= s_2 - \frac{P_3(s_2)}{P_3'(s_2)} \\
&= -0.545 - \frac{1.625}{5.352} \\
&= -0.545 - 0.303 \\
&= -0.848
\end{aligned}
$$

Repeating the process once more with $s_3 = -0.848$, we have

$$
\begin{array}{r|l}
 & s^2 + 5.152s + 6.63 \\ \hline
s + 0.848 & s^3 + 6.000s^2 + 11s + 6 \\
 & s^3 + 0.848s^2 \\ \hline
 & \quad 5.152s^2 + 11.00s + 6 \\
 & \quad 5.152s^2 + 4.37s \\ \hline
 & \qquad\qquad 6.63s + 6.00 \\
 & \qquad\qquad 6.63s + 5.62 \\ \hline
 & \qquad\qquad\qquad 0.38 = P_3(-0.848)
\end{array}
$$

and dividing the quotient by $s + 0.848$, we find

$$
\begin{array}{r|l}
 & s + 4.304 \\ \hline
s + 0.848 & s^2 + 5.152s + 6.63 \\
 & s^2 + 0.848s \\ \hline
 & \quad 4.304s + 6.63 \\
 & \quad 4.304s + 3.65 \\ \hline
 & \qquad\qquad 2.98 = P_3'(-0.848)
\end{array}
$$

Using the Newton-Raphson formula for the third time, we find

$$
\begin{aligned}
s_4 &= s_3 - \frac{P_3(s_3)}{P_3'(s_3)} \\
&= -0.848 - \frac{0.38}{2.98} \\
&= -0.848 - 0.128 \\
&= -0.976
\end{aligned}
$$

At this point we can see that we are rapidly approaching the root, and the process may be repeated until the -1.0 root has been found to the desired accuracy. In turn, each of the remaining roots may be obtained by successively applying the Newton-Raphson formula to evaluate the roots of the successive quotients.

In developing our algorithm, we may recognize that the long division calculations above may be carried out more efficiently by using synthetic division. For a general n-th degree polynomial

$$P_n(s) = a_n s^n + a_{n-1}s^{n-1} + \cdots + a_1 s + a_0$$

the synthetic division process would be as follows

$$
\begin{array}{llllllll}
 & a_n & a_{n-1} & a_{n-2} & \cdots & a_2 & a_1 & a_0 \quad \underline{|s_1} = \text{the initial} \\
 & & b_n s_1 & b_{n-1}s_1 & \cdots & b_3 s_1 & b_2 s_1 & b_1 s_1 \qquad \text{approximation} \\ \hline
 & & & & & & & \qquad\qquad \text{of the root} \\
a_n = b_n & & b_{n-1} & b_{n-2} & \cdots & b_2 & b_1 & | \; a_0 + b_1 s_1 = P_n(s_1) \\
 & & c_{n-1}s_1 & c_{n-2}s_1 & \cdots & c_2 s_1 & c_1 s_1 & \\ \hline
b_n = c_{n-1} & & c_{n-2} & c_{n-3} & \cdots & c_1 & | \; b_1 + c_1 s_1 = P_n'(s_1) &
\end{array}
$$

where

$$
\begin{aligned}
b_n &= a_n \\
b_{n-1} &= a_{n-1} + a_n s_1 \\
b_{n-2} &= a_{n-2} + (a_{n-1} + a_n s_1)s_1 = a_{n-2} + b_{n-1}s_1 \\
&\cdot \\
&\cdot \\
&\cdot \\
b_i &= a_i + b_{i+1}s_1 \\
&\cdot \\
&\cdot \\
&\cdot \\
b_1 &= a_1 + b_2 s_1 \\
b_0 &= a_0 + b_1 s_1 = P_n(s_1)
\end{aligned}
$$

and

$$
\begin{aligned}
c_{n-1} &= b_n \\
c_{n-2} &= b_{n-1} + b_n s_1 \\
c_{n-3} &= b_{n-2} + c_{n-2}s_1 \\
&\cdot \\
&\cdot \\
&\cdot \\
c_i &= b_{i+1} + c_{i+1}s_1 \\
&\cdot \\
&\cdot \\
&\cdot \\
c_1 &= b_2 + c_2 s_1 \\
c_0 &= b_1 + c_1 s_1 = P'_n(s_1)
\end{aligned}
$$

In terms of the second iteration of the previous numerical example, we would have

$$
\begin{array}{cccc|l}
1 & 6 & 11 & 6 & -0.545 \\
 & -0.545 & -2.973 & -4.375 & \\
\hline
1 & 5.455 & 8.027 & 1.625 & = P_3(-0.545) \\
 & -0.545 & -2.675 & & \\
\hline
1 & 4.91 & 5.352 & & = P'_3(-0.545)
\end{array}
$$

and

$$
\begin{aligned}
s_3 &= s_2 - \frac{P_3(s_2)}{P'_3(s_2)} \\
&= -0.545 - \frac{1.625}{5.352} \\
&= -0.848 \text{ as before}
\end{aligned}
$$

We now calculate the number of operations saved by using synthetic division. In the long division process

$$
\begin{array}{r|l}
 & a_n s_{n-1} + \cdots \\
\cline{2-2}
s - s_1 & a_n s^n + a_{n-1}s^{n-1} + \cdots + a_1 s + a_0 \\
 & a_n s^n - a_n s_1 \\
\cline{2-2}
 & \qquad (a_{n-1} + a_n s_1)s^{n-1} + \cdots \\
 & \qquad\qquad \cdot \\
 & \qquad\qquad\quad \cdot \\
 & \qquad\qquad\qquad \cdot
\end{array}
$$

there would be $2n$ multiplications and $2n$ subtractions. Since dividing $P_n(s)$ by $s - s_1$ is equivalent to evaluating $P_n(s_1)$ by evaluating each of the following right-member terms separately and summing the individual results

$$P_n(s_1) = a_n(s_1)^n + a_{n-1}(s_1)^{n-1} + \cdots + a_1 s_1 + a_0$$

This would require $n \times (n + 1)/2$ multiplications and n additions. The number of these operations could be reduced if we first calculated and stored successive values of s_1; $s_1, s_1^2, s_1^3, \ldots, s_1^n$ and then multiplied each s_1^i by its corresponding coefficient a_i and, in turn, summed the individual results. This would require $2n - 1$ multiplications and n additions.

Synthetic division, on the other hand, would require only n multiplications and n additions. We thus see that if n is large and the time required for the multiplication operation is long compared with the time required for addition (or subtraction), synthetic division would require approximately one half the computing time as compared to the other methods discussed. We should also recognize that the synthetic division evaluation of the polynomial is equivalent to the evaluation of the nested form of the polynomial

$$P_n(s) = (\cdots((a_n s + a_{n-1})s + a_{n-2})s + \cdots)s + a_0$$

and thus problems where a polynomial must be evaluated for a given value of s should always be carried out using the nested form.

Figure 4.46 is a Fortran program that uses the Newton-Raphson method and the synthetic division scheme discussed above in order to evaluate all of the real roots of a polynomial. Initially we read in the value of the accuracy required in finding the roots, the number of coefficients N, and the N coefficients. Note that since we are not able to use zero as a subscript, the coefficient of X to the zero power is called $A(1)$, the coefficient of X to the first power is called $A(2)$, . . . , and the

```
                C       NEWTON'S METHOD IS USED TO FIND ALL OF THE
                C       REAL ROOTS OF A POLYNOMINAL OF ANY ORDER.
                C
                C       INPUT INFORMATION
                C          N = THE NUMBER OF POLYNOMINAL COEFFICIENTS ( N MAX = 20).
                C
                C          A(1)  = COEFFICIENT OF S TO THE ZEROTH POWER
                C          A(2)  = COEFFICIENT OF S TO THE FIRST POWER
                C          A(3)  = COEFFICIENT OF S TO THE SECOND POWER
                C                .
                C                .
                C                .
                C          A(N)  = COEFFICIENT OF S TO THE N-1 POWER (HIGHEST POWER)
                C
S.0001                  DIMENSION A(20),B(20),C(19),S(101)
S.0002                  READ(5,1) ACC,N,(A(I),I=1,N)
S.0003                1 FORMAT(E15.7,I2,/,(5E15.7))
S.0004                  WRITE(6,2) (A(I),I=1,N)
S.0005                2 FORMAT('1NEWTONS METHOD IS USED TO FIND ALL OF THE REAL ROOTS OF A
                       1 POLYNOMINAL OF ANY ORDER',///,' THE COEFFICIENTS  A(1),A(2),...,A
                       2(N)  ARE',//,(5E15.7))
S.0006                5 S(1)=0.0
S.0007                  S(2)=-A(1)/A(2)
S.0008                  B(N)=A(N)
S.0009                  DO 10 I=2,100
S.0010                  IF(ABS(S(I)-S(I-1))-ACC) 30,30,20
S.0011               20 DO 40 J=2,N
S.0012                  K=N+1-J
S.0013               40 B(K)=A(K)+S(I)*B(K+1)
S.0014                  C(N-1)=B(N)
S.0015                  DO 60 L=3,N
S.0016                  M=N+1-L
S.0017               60 C(M)=B(M+1)+S(I)*C(M+1)
S.0018               10 S(I+1)=S(I)-B(1)/C(1)
S.0019               30 ITER=I-2
S.0020                  WRITE(6,3)S(I),ITER
S.0021                3 FORMAT(//,' VALUE OF ROOT =',E15.7,//,' THE NUMBER OF ITERATIONS
                       1REQUIRED =',I4,//)
S.0022                  IF(N-2) 70,70,80
S.0023               80 N=N-1
S.0024                  DO 90 I=1,N
S.0025                  B(I)=B(I+1)
S.0026               90 WRITE(6,4) I,B(I)
S.0027                4 FORMAT(' B(',I2,') =',E15.7)
S.0028                  DO 100 I=1,N
S.0029              100 A(I)=B(I)
S.0030                  GO TO 5
S.0031               70 STOP
S.0032                  END
```

Figure 4.46 A program that uses the Newton-Raphson method to find all the roots of a polynomial known to have real roots only

coefficient of X to the $N - 1$ power is called $A(N)$. We next print out the heading and the numerical values of the N coefficients. Computer statement S.0006 then sets S(1), the initial approximation of the first root we seek, equal to zero, and the next statement approximates the second approximation as $-A(1)/A(2)$.

We may better understand this second approximation by considering the following three numerical examples:

(1) $(X + 1)(X + 10) = X^2 + 11X + 10$
(2) $(X + 1)(X + 10)(X + 100) = X^3 + 111X^2 + 1110X + 1000$
(3) $(X + 1)(X + 2)(X + 3) = X^3 + 6X^2 + 11X + 6$

In the first example we may approximate the smallest root by using the last two terms of the quadratic, $11X + 10$, to solve for the approximate root $X = -10/11 = -0.909$ $(-A(1)/A(2))$. If we now form the linear polynomial, $X + 11$, from the first two terms, we see that we can solve for an approximation to the largest root. Repeating this technique with the second numerical example, we approximate the smallest root from $1110X + 1000$ as $-1000/1100 = -0.909$, the middle-sized root from $111X + 1110$ as -10, and the largest root from $X + 111$ as -111. We may thus see from these two examples that it is possible to approximate the smallest root from $-A(1)/A(2)$. We should note, however, that this is only an approximation to the smallest root and that the accuracy of the approximation is a function of the magnitude spacing of the respective roots. If all the roots are relatively close together, as in the third example, the approximation is poor.

Following the general synthetic division scheme described above, computer statement S.0008 sets $B(N) = A(N)$ and the DO 10 loop allows 99 iterations to find each root. Initially, for each iteration, we test to see whether the difference between the previous two approximations to the root is less than or equal to the desired accuracy. If the difference is greater than the desired accuracy, the DO 40 loop calculates the numerical values of the B's $(B(N-1), \ldots, B(1))$ and the DO 60 loop then calculates the numerical values of the C's $(C(N-2), \ldots, C(1))$.[10] Computer statement S.0018 then uses the Newton-Raphson formula to calculate the next approximation to the root. This iteration loop is executed either for 99 iterations or until the root has been calculated to the specified accuracy. At either point computer statement S.0020 then prints out the root and the number of iterations required. Computer statement S.0022 now tests to see whether all the roots have been found. If not, the value of N is reduced by 1 and the DO 90 loop is used to print the numerical values of the quotient coefficients. The DO 100 loop then changes the numerical values of the A's of the original polynomial to the values of the coefficients of the quotient polynomial and the GO TO 5 statement transfers control to computer statement S.0006 so that the second root may be extracted from the first quotient. This process is repeated, extracting one root at a time from each of the successive quotients until all the roots have been found.

Figure 4.47 gives the results for the characteristic equation

$$(X+1)(X+2)(X+3)(X+4)(X+5) = X^5 + 15X^4 + 85X^3 + 225X^2 + 274X + 120 = 0$$

Note that the roots obtained are not exact. Note also that each successive root is obtained from a quotient that depends on the degree of accuracy

[10] Note $C(1)$ corresponds to c_0 and $B(1)$ corresponds to b_0 in the general synthetic division process since Fortran does not permit the use of zero subscripts.

```
NEWTONS METHOD IS USED TO FIND ALL OF THE REAL ROOTS OF A POLYNOMINAL OF ANY ORDER

THE COEFFICIENTS  A(1),A(2),....,A(N) ARE

 0.1200000E 03  0.2740000E 03  0.2250000E 03  0.8500000E 02  0.1500000E 02
 0.1000000E 01

VALUE OF ROOT = -0.1000000E 01

THE NUMBER OF ITERATIONS  REQUIRED =    6

B( 1) =  0.1200000E 03
B( 2) =  0.1540000E 03
B( 3) =  0.7100000E 02
B( 4) =  0.1400000E 02
B( 5) =  0.1000000E 01

VALUE OF ROOT = -0.2000005E 01

THE NUMBER OF ITERATIONS  REQUIRED =    7

B( 1) =  0.5999991E 02
B( 2) =  0.4699998E 02
B( 3) =  0.1200000E 02
B( 4) =  0.1000000E 01

VALUE OF ROOT = -0.2999968E 01

THE NUMBER OF ITERATIONS  REQUIRED =    7

B( 1) =  0.2000023E 02
B( 2) =  0.9000036E 01
B( 3) =  0.1000000E 01

VALUE OF ROOT = -0.4000089E 01

THE NUMBER OF ITERATIONS  REQUIRED =    6

B( 1) =  0.4999948E 01
B( 2) =  0.1000000E 01

VALUE OF ROOT = -0.4999948E 01

THE NUMBER OF ITERATIONS  REQUIRED =    1
```

Figure 4.47 The output results of the program of Figure 4.46

in finding the previous roots. To eliminate this error, a better scheme would use these root values as initial approximations to the respective roots and then repeat the process, using the original polynomial in each case to improve the accuracy of the initial approximations. In this way none of the roots obtained would be a function of the accuracy of any other root.

Special Polynomial Cases

Let us now investigate some special cases using the Newton-Raphson method that will prove useful, especially in hand computations. Consider

$$f(X) = X^2 - N = 0$$

where N is a positive quantity. The positive root of this quadratic is the $\sqrt{N}$. Using the Newton-Raphson formula to find the root, we derive the following iteration formula to find the square root of N:

$$\begin{aligned} X_{n+1} &= X_n - \frac{f(X_n)}{f'(X_n)} \\ &= X_n - \frac{X_n^2 - N}{2X_n} \\ &= \frac{X_n^2 + N}{2X_n} \\ &= \frac{1}{2}\left(X_n + \frac{N}{X_n}\right) \end{aligned}$$

As a simple numerical example, let us find the square root of $N = 4$, starting with an initial approximation of $X_1 = 1.0$.

ITERATION	APPROXIMATE ROOT
1	$X_2 = \frac{1}{2}\left(1.0 + \frac{4}{1.0}\right) = \frac{1}{2}(1.0 + 4.0) = 2.5000$
2	$X_3 = \frac{1}{2}\left(2.5 + \frac{4}{2.5}\right) = \frac{1}{2}(2.5 + 1.6) = 2.0500$
3	$X_4 = \frac{1}{2}\left(2.050 + \frac{4}{2.050}\right) = \frac{1}{2}(2.050 + 1.951) = 2.0005$

At this point we see that we have the root accurate to four significant figures, and one more iteration would give us an accuracy of eight significant figures.

As a second example, consider the calculation of $\sqrt{37}$. If we start with an initial approximation equal to one, we will find the process will converge slowly until the correction becomes small. For hand calculation, then, since 36 is the largest integer square less than 37, let us choose $X_1 = 6$, thus

ITERATION	APPROXIMATE ROOT
1	$X_2 = \frac{1}{2}\left(6.000 + \frac{37}{6.000}\right) = \frac{1}{2}(6.000 + 6.166) = 6.083$
2	$X_3 = \frac{1}{2}\left(6.083 + \frac{37}{6.083}\right) = \frac{1}{2}(6.083 + 6.083) = 6.083$

and we see that we have the root accurate to four significant figures in one iteration.

It is instructive to examine how this process works. If we choose as an initial approximation a value less than the square root ($X_1 = 6$) and divide that into N (37), we must obtain a number that is greater than the square root (6.166). The approximation is then taken as the arithmetic mean of these two values. Since the Newton-Raphson method was used to derive the iteration formula, once the correction (error) becomes small we have second-order convergence and successive corrections will be less than the square of the previous correction.

Another second-order formula that avoids division in finding the square root is

$$X_{n+1} = \frac{X_n}{2}\left(3 - \frac{X_n^2}{N}\right)$$

Here a more accurate first approximation to the root is needed, and we require

$$X_n < \sqrt{5N}$$

The following two formulas may also be used to find the square root:

$$X_{n+1} = \frac{1}{8}\left(3X_n + \frac{6N}{X_n} - \frac{N^2}{X_n^3}\right)$$

$$X_{n+1} = \frac{X_n}{8}\left(15 - \frac{10X_n^2}{N} + \frac{3X_n^4}{N^2}\right)$$

Both of these formulas possess third-order convergence. In other words, once the root has been found to a moderate accuracy, the number of significant figures triples for each additional iteration.

Let us next consider the following special case

$$f(X) = X^3 - N = 0$$

where N is a positive number. Using the Newton-Raphson method, we may derive an iteration formula for finding the cube root of N:

$$\begin{aligned} X_{n+1} &= X_n - \frac{f(X_n)}{f'(X_n)} \\ &= X_n - \frac{X_n^3 - N}{3X_n^2} \\ &= \frac{1}{3}\left(2X_n + \frac{N}{X_n^2}\right) \end{aligned}$$

or in the case of an n-th order polynomial

$$f(X) = X^n - N = 0$$

we may derive an iteration formula for finding the n-th root of N:

$$\begin{aligned} X_{i+1} &= X_i - \frac{f(X_i)}{f'(X_i)} \\ &= X_i - \frac{X_i^n - N}{n\, X_i^{n-1}} \\ &= \frac{1}{n}\left((n-1)X_i + \frac{N}{X_i^{n-1}}\right) \end{aligned}$$

where i is the number of the iteration. The reader is urged to test these formulas with simple numerical examples.

The Newton-Raphson Method Using the Second Derivative

At the beginning of this section we saw how the Newton-Raphson formula

$$t_{n+1} = t_n - \frac{f(t_n)}{f'(t_n)}$$

was derived from the Taylor's expansion of $f(t_1 + h)$. We also showed in Figures 4.45c and 4.45d that the method might diverge if $f'(t_n)$ was very small or equal to zero. In an attempt to avoid this problem, we may include one more term of the Taylor's expansion in the calculation of the correction h. Since one more term is to be included, this should yield a more accurate correction, and the process should converge more rapidly. Starting from

$$f(t_1 + h) = f(t_1) + hf'(t_1) + \frac{h^2}{2!}f''(t_1) + \cdots$$

we may again assume $t_1 + h$ is an exact root or

$$0 \equiv f(t_1) + hf'(t_1) + \frac{h^2}{2!}f''(t_1) + \cdots$$

Neglecting the higher-order terms for small h yields

$$0 = \frac{f''(t_1)}{2!}h^2 + f'(t_1)h + f(t_1)$$

The value of the correction therefore requires the solution of a quadratic and the investigation of both roots. This difficulty may be avoided if we rewrite the above equation as

$$0 = h\left(\frac{f''(t_1)}{2!}h + f'(t_1)\right) + f(t_1)$$

and approximate the h associated with the second derivative as[11]

$$-\frac{f(t_1)}{f'(t_1)}$$

thus

$$0 = h\left(-\frac{f''(t_1)}{2!}\frac{f(t_1)}{f'(t_1)} + f'(t_1)\right) + f(t_1)$$

and

$$h = -\frac{f(t_1)}{f'(t_1) - \dfrac{1}{2}\dfrac{f(t_1)}{f'(t_1)}f''(t_1)}$$

We may thus write the general iteration formula as

$$t_{n+1} = t_n - \frac{f(t_n)}{f'(t_n) - \dfrac{1}{2}\dfrac{f(t_n)}{f'(t_n)}f''(t_n)}$$

where n is the number of the iteration.

Figures 4.48 and 4.49 give the Fortran programs for finding $T_{0.1}$ and $T_{0.9}$ in the calculation of the rise time of a single-stage amplifier. Figures 4.50 and 4.51 give the respective results. Note that in both cases the process converges more rapidly than the usual Newton-Raphson formula, which was used in obtaining the results of Figures 4.39 and 4.40.

[11] For additional discussion see M. L. James, G. M. Smith, and J. C. Wolford, *Applied Numerical Methods for Digital Computation with Fortran*, Scranton, Pa.: International Textbook Co., 1967), Sec. 3–5.

```
C         NEWTON-RAPHSON METHOD
C                 USING
C         THE SECOND DERIVATIVE
C
          WRITE(6,1)
        1 FORMAT('1THE SECOND DERIVATIVE NEWTON-RAPHSON METHOD IS USED TO FI
         1ND THE RISE TIME',////,4X,'TIME',13X,'F(T)', 9X,'F PRIME(T)', 1X,'
         2F PRIME PRIME (T)',2X,'CORRECTION',//)
          READ(5,2) T,ACC
        2 FORMAT(2E15.7)
          DO 10 I=1,100
          FT=0.9-EXP(-T)
          FPT=EXP(-T)
          FPPT=-EXP(-T)
          COR=-FT/(FPT-(FPPT*FT/(2.*FPT)))
          WRITE(6,3)T,FT,FPT,FPPT,COR
        3 FORMAT(5E15.7)
          T=T+COR
          IF(ABS(COR)-ACC)20,20,10
       10 CONTINUE
       20 WRITE(6,5)I,ACC
        5 FORMAT(///,' THE NUMBER OF ITERATIONS REQUIRED =',I5,//,' BASED ON
         1 AN ACCURACY =',E15.7)
          RETURN
          END
```

Figure 4.48 A Fortran program that uses the second derivative Newton-Raphson method to calculate $T_{0.1}$ of the single-stage amplifier

```
C         NEWTON-RAPHSON METHOD
C                USING
C         THE SECOND DERIVATIVE
C
          WRITE(6,1)
        1 FORMAT('1THE SECOND DERIVATIVE NEWTON-RAPHSON METHOD IS USED TO FI
         1ND THE RISE TIME',////,4X,'TIME',13X,'F(T)', 9X,'F PRIME(T)', 1X,'
         2F PRIME PRIME (T)',2X,'CORRECTION',///)
          READ(5,2) T,ACC
        2 FORMAT(2E15.7)
          DO 10 I=1,100
          FT=0.1-EXP(-T)
          FPT=EXP(-T)
          FPPT=-EXP(-T)
          COR=-FT/(FPT-(FPPT*FT/(2.*FPT)))
          WRITE(6,3) T,FT,FPT,FPPT,COR
        3 FORMAT(5E15.7)
          T=T+COR
          IF(ABS(COR)-ACC)20,20,10
       10 CONTINUE
       20 WRITE(6,5)I,ACC
        5 FORMAT(///,' THE NUMBER OF ITERATIONS REQUIRED =',I5,//,' BASED ON
         1 AN ACCURACY =',E15.7)
          RETURN
          END
```

Figure 4.49 A Fortran program that uses the second derivative Newton-Raphson method to calculate $T_{0.9}$ of the single-stage amplifier

```
THE SECOND DERIVATIVE NEWTON-RAPHSON METHOD IS USED TO FIND THE RISE TIME


   TIME              F(T)            F PRIME(T) F PRIME PRIME (T)   CORRECTION


 0.0             -0.1000000E 00  0.1000000E 01 -0.1000000E 01  0.1052632E 00
 0.1052632E 00 -0.8761883E-04  0.9000876E 00 -0.9000876E 00  0.9734950E-04
 0.1053605E 00  0.0            0.9000000E 00 -0.9000000E 00  0.0


THE NUMBER OF ITERATIONS REQUIRED =     3

BASED ON AN ACCURACY =  0.1000000E-04
```

Figure 4.50 The output results of the program of Figure 4.48

```
THE SECOND DERIVATIVE NEWTON-RAPHSON METHOD IS USED TO FIND THE RISE TIME


   TIME              F(T)            F PRIME(T) F PRIME PRIME (T)   CORRECTION


 0.0             -0.9000000E 00  0.1000000E 01 -0.1000000E 01  0.1636363E 01
 0.1636363E 01 -0.9468687E-01  0.1946868E 00 -0.1946868E 00  0.6426271E 00
 0.2278990E 01 -0.2387643E-02  0.1023876E 00 -0.1023876E 00  0.2359476E-01
 0.2302584E 01 -0.1788139E-06  0.1000001E 00 -0.1000001E 00  0.1788138E-05


THE NUMBER OF ITERATIONS REQUIRED =     4

BASED ON AN ACCURACY =  0.1000000E-04
```

Figure 4.51 The output results of the program of Figure 4.49

THE NEWTON-RALPHSON METHOD IS USED TO FIND THE ROOTS

TTME	F(T)	F PRTME(T)	CORRECTION
0.0	0.2000000E 01	0.5000000E 01	-0.4000000E 00
-0.4000000E 00	0.5760002E 00	0.2280001E 01	-0.2526315E 00
-0.6526315E 00	0.1625805E 00	0.1056732E 01	-0.1538521E 00
-0.8064836E 00	0.4469490E-01	0.4993792E 00	-0.8950090E-01
-0.8959845E 00	0.1194382E-01	0.2404890E 00	-0.4966471E-01
-0.9456492E 00	0.3113747E-02	0.1175642E 00	-0.2648550E-01
-0.9721346E 00	0.7972717E-03	0.5806065E-01	-0.1373170E-01
-0.9858663E 00	0.2021790E-03	0.2886677E-01	-0.7003862E-02
-0.9928702E 00	0.5149841E-04	0.1441193E-01	-0.3573319E-02
-0.9964435E 00	0.1239777E-04	0.7151604E-02	-0.1733564E-02
-0.9981770E 00	0.2861023E-05	0.3656387E-02	-0.7824725E-03
-0.9989594E 00	0.9536743E-06	0.2084732E-02	-0.4574563E-03
-0.9994168E 00	0.0	0.1167297E-02	0.0

THE NUMBER OF ITERATIONS REQUIRED = 13

BASED ON AN ACCURACY = 0.10000E-04

Figure 4.52 The Newton-Raphson results of finding the two equals roots of $T^3 + 4T^2 + 5T + 2 = 0$

THE SECOND DERIVATIVE NEWTON-RAPHSON METHOD IS USED TO FIND THE ROOTS

TIME	F(T)	F PRIME(T)	F PRIME PRIME (T)	CORRECTION
0.0	0.2000000E 01	0.5000000E 01	0.8000000E 01	-0.5882351E 00
-0.5882351E 00	0.2393646E 00	0.1332181E 01	0.4470590E 01	-0.2572302E 00
-0.8454654E 00	0.2757168E-01	0.3807125E 00	0.2927208E 01	-0.1003641E 00
-0.9458295E 00	0.3092766E-02	0.1171446E 00	0.2325024E 01	-0.3577404E-01
-0.9816034E 00	0.3442764E-03	0.3780842E-01	0.2110380E 01	-0.1220836E-01
-0.9938118E 00	0.3814697E-04	0.1249123E-01	0.2037129E 01	-0.4066568E-02
-0.9978783E 00	0.3814697E-05	0.4256248E-02	0.2012731E 01	-0.1137261E-02
-0.9990156E 00	0.9536743E-06	0.1972198E-02	0.2005907E 01	-0.6412498E-03
-0.9996568E 00	-0.9536743E-06	0.6866455E-03	0.2002060E 01	0.4591672E-03
-0.9991976E 00	0.9536743E-06	0.1605988E-02	0.2004815E 01	-0.9435462E-03
-0.1000141E 01	0.0	-0.2813339E-03	0.1999153E 01	0.0

THE NUMBER OF ITERATIONS REQUIRED = 11

BASED ON AN ACCURACY = 0.1000000E-04

Figure 4.53 The second derivative Newton-Raphson method is used to find the roots of $T^3 + 4T^2 + 5T + 2 = 0$

Figures 4.52 and 4.53 show the respective results of the two formulas when a double root is encountered

$$(T + 1)^2(T + 2) = T^3 + 4T^2 + 5T + 2 = 0$$

Note how much faster the second derivative scheme of Figure 4.53 converges.

Let us now turn to the case of finding the roots of a general polynomial where the values of both the first and second derivative are required.

Continuing the discussion of the remainder theorem above by dividing the second quotient $Q_{n-2}(s)$ by $s - s_1$ to obtain a new quotient $Q_{n-3}(s)$ and a new remainder r_3, we have

$$\frac{Q_{n-2}(s)}{s - s_1} = Q_{n-3}(s) + \frac{r_3}{s - s_1}$$

or

$$Q_{n-2}(s) = Q_{n-3}(s)(s - s_1) + r_3$$

If we now take the derivative of

$$P_n'(s) = Q_{n-1}'(s)(s - s_1) + Q_{n-2}(s)(s - s_1) + r_2$$

and

$$Q_{n-1}(s) = Q_{n-2}(s)(s - s_1) + r_2$$

we obtain

$$P_n''(s) = Q_{n-1}''(s)(s - s_1) + Q_{n-1}'(s) + Q_{n-2}'(s)(s - s_1) + Q_{n-2}(s)$$

and

$$Q_{n-1}'(s) = Q_{n-2}'(s)(s - s_1) + Q_{n-2}(s)$$

or

$$P_n''(s) = Q_{n-1}''(s)(s - s_1) + 2Q_{n-2}'(s)(s - s_1) + 2Q_{n-2}(s)$$

Substituting the value of $Q_{n-2}(s)$, we found above, yields

$$P_n''(s) = Q_{n-1}''(s)(s - s_1) + 2Q_{n-2}'(s)(s - s_1) + 2Q_{n-3}(s)(s - s_1) + 2r_3$$

Evaluating $P_n''(s)$ at $s = s_1$ we find that

$$P_n''(s_1) = 2r_3$$

which is two times the remainder obtained in dividing the second quotient $Q_{n-2}(s)$ by $s - s_1$. It is also interesting to note that successive divisions of successive quotients will evaluate higher order derivatives and, in general, the $(i + 1)$ remainder will be given by

$$r_{i+1} = \frac{1}{i!} P_n{}^i(s_1); \; i = 1, 2, 3, \ldots$$

where $P_n{}^i(s_1)$ is the value of the i derivative evaluated at $s = s_1$.

For a general n degree polynomial

$$P_n(s) = a_n s^n + a_{n-1} s^{n-1} + \cdots + a_1 s + a_0$$

the synthetic division process would be as follows:

$$\begin{array}{lllllllll}
 & a_n & a_{n-1} & a_{n-2} & \cdots & a_3 & a_2 & a_1 & a_0 \;\underline{|s_1} = \text{the initial approximation of the root.} \\
 & & b_n s_1 & b_{n-1} s_1 & \cdots & b_4 s_1 & b_3 s_1 & b_2 s_1 & b_1 s_1 \\
\hline
a_n = & b_n & b_{n-1} & b_{n-2} & \cdots & b_3 & b_2 & b_1 & \big| \; a_0 + b_1 s_1 = P_n(s_1) \\
 & & c_{n-1} s_1 & c_{n-2} s_1 & \cdots & c_3 s_1 & c_2 s_1 & c_1 s_1 & \\
\hline
b_n = & c_{n-1} & c_{n-2} & c_{n-3} & \cdots & c_2 & c_1 & \big| \; b_1 + c_1 s_1 = P'_n(s_1) & \\
 & & d_{n-2} s_1 & d_{n-3} s_1 & \cdots & d_2 s_1 & d_1 s_1 & & \\
\hline
c_{n-1} = & d_{n-2} & d_{n-3} & d_{n-4} & \cdots & d_1 & \big| \; c_1 + d_1 s_1 = \frac{1}{2!} P''_n(s_1) & &
\end{array}$$

where the b's and c's are as before and

$$\begin{aligned}
d_{n-2} &= c_{n-1} \\
d_{n-3} &= c_{n-2} + d_{n-2} s_1 \\
&\;\;\vdots \\
d_i &= c_{i+1} + d_{i+1} s_1 \\
&\;\;\vdots \\
d_1 &= c_2 + d_2 s_1 \\
d_0 &= c_1 + d_1 s_1 = \frac{1}{2!} P''_n(s_1)
\end{aligned}$$

The reader is encouraged to modify the program of Figure 4.46 and include the second derivative term as discussed above.

4.6 THE EVALUATION OF COMPLEX ROOTS

The Newton-Raphson Method

The Newton-Raphson method may be used to find complex roots provided the initial approximation is taken as complex. For example, consider the polynomial

$$P_2(X) = X^2 - 2X + 2 = 0$$

Using $X_1 = 0 + j1$ as an initial approximation to the root, we have

$$\begin{array}{lll|l}
1.0 & -2.0 & +2.0 & \underline{0.0 + j1.0} \\
 & 0.0 + j1.0 & -1.0 - j2.0 & \\
\hline
1.0 & -2.0 + j1.0 & \multicolumn{2}{|l}{+1.0 - j2.0 = P_2(0.0 + j1.0)} \\
 & 0.0 + j1.0 & & \\
\hline
1.0 & \multicolumn{3}{|l}{-2.0 + j2.0 = P_2'(0.0 + j1.0)}
\end{array}$$

so that

$$\begin{aligned}
X_2 &= 0.0 + j1.0 - \frac{1.0 - j2.0}{-2.0 + j2.0} \\
&= 0.0 + j1.0 + 0.75 - j0.25 \\
&= 0.75 + j0.75
\end{aligned}$$

Repeating the process again, we obtain

$$\begin{array}{lll|l}
1.0 & -2.00 & +2.000 & \underline{0.75 + j0.75} \\
 & 0.75 + j0.75 & -1.500 - j0.375 & \\
\hline
1.0 & -1.25 + j0.75 & \multicolumn{2}{|l}{0.5 - j0.375 = P_2(0.75 + j0.75)} \\
 & 0.75 + j0.75 & & \\
\hline
1.0 & \multicolumn{3}{|l}{-0.50 + j1.50 = P_2'(0.75 + j0.75)}
\end{array}$$

so that

$$\begin{aligned}
X_3 &= 0.75 + j0.75 - \frac{0.5 - j0.375}{-0.5 + j1.50} \\
&= 0.75 + j0.75 + 0.325 + j0.225 \\
&= 1.075 + j0.975
\end{aligned}$$

Repeating the process once more, we find

$$\begin{array}{lll|l}
1.0 & -2.000 & +2.0000 & \underline{1.075 + j0.975} \\
 & 1.075 + j0.975 & -1.9449 + j0.14625 & \\
\hline
 & -0.925 + j0.975 & \multicolumn{2}{|l}{0.0551 + j0.14625 = P_2(1.075 + j0.975)} \\
 & 1.075 + j0.975 & & \\
\hline
1.0 & \multicolumn{3}{|l}{0.150 + j1.950 = P_2'(1.075 + j0.975)}
\end{array}$$

and

$$\begin{aligned}
X_4 &= 1.075 + j0.975 - \frac{0.0551 + j0.14625}{0.150 + j1.950} \\
&= 1.075 + j0.975 - 0.074 + j0.022 \\
&= 1.001 + j0.997
\end{aligned}$$

At this point we can see that the process is rapidly converging on the root $1.000 + j1.000$ and that the Newton-Raphson formula can be used to find complex roots, provided that the initial approximation of the root is complex. We note, however, that approximately four times the work is involved since each complex multiplication involves four real multiplications and two additions.

Lin's Method

The main difficulty in using the Newton-Raphson method to calculate complex roots is that complex numbers are involved in the multiplications, divisions, and additions. There are several alternate methods for finding complex roots that are based on using a quadratic divisor (rather than a linear divisor) and successively correcting the coefficients until it divides the polynomial with zero remainder. The quadratic divisor thus becomes a factor of the polynomial and the quadratic formula can be used to find two of the roots. Since complex roots will occur in conjugate pairs

$$\begin{aligned}(X - a + jb)(X - a - jb) &= X^2 - 2aX + a^2 + b^2 \\ &= X^2 + AX + B\end{aligned}$$

the quadratic divisor will have real coefficients and all multiplications, additions, and divisions will be real. Once the quadratic divisor and its roots have been found, the resulting quotient can be factored so that a new quadratic divisor is involved and the process can be repeated until all of the roots of the polynomial are found. Lin's method is one method that is based on the use of a quadratic divisor.[12] Before we develop this method, let us first review the remainder theorem and synthetic division for a quadratic divisor.

To explore this possibility, we first construct the quadratic divisor

$$D_2(s) = s^2 - As - B$$

where A and B are real coefficients even when the roots of $D_2(s)$ are complex

$$\begin{aligned}(s - a - jb)(s - a + jb) &= s^2 - 2as + a^2 + b^2 \\ &= s^2 - As - B\end{aligned}$$

since complex roots occur in conjugate pairs. The division of $P_n(s)$ by $D_2(s)$ yields

$$\frac{P_n(s)}{D_2(s)} = Q_{n-2}(s) + \frac{r_1 s + r_0}{D_2(s)}$$

or

$$P_n(s) = Q_{n-2}(s)D_2(s) + r_1 s + r_0$$

where $Q_{n-2}(s)$ is the resulting quotient, and $r_1 s + r_0$ is the linear re-

[12] Shih-Nge Lin, "A Method of Successive Approximations of Evaluating the Real and Complex Roots of Cubic and Higher-order Equations," *Journal of Mathematics and Physics*, Vol. 20 (1941), pp. 231–242.

mainder. Evaluating $P_n(s)$ at $s = a + jb$ and $a - jb$, yields

$$P_n(a + jb) = r_1(a + jb) + r_0$$

and

$$P_n(a - jb) = r_1(a - jb) + r_0 = \text{the conjugate of } P_n(a + jb)$$

since

$$D_2(a + jb) = D_2(a - jb) = 0.$$

The derivative $P'_n(s)$ may now be evaluated from a second division by dividing the first quotient $Q_{n-2}(s)$ by $D_2(s)$, obtaining a new quotient $Q_{n-4}(s)$ and a new linear remainder $r_3s + r_2$

$$\frac{Q_{n-2}(s)}{D_2(s)} = Q_{n-4}(s) + \frac{r_3s + r_2}{D_2(s)}$$

or

$$Q_{n-2}(s) = Q_{n-4}(s)D_2(s) + r_3s + r_2$$

If we now differentiate

$$P_n(s) = Q_{n-2}(s)D_2(s) + r_1s + r_0$$

we obtain

$$P'_n(s) = Q'_{n-2}(s)D_2(s) + Q_{n-2}(s)D'_2(s) + r_1$$

Substituting the value of $Q_{n-2}(s)$ and realizing $D'_2(s) = 2s - A = 2s - 2a$ yields

$$P'_n(s) = Q'_{n-2}(s)D_2(s) + (Q_{n-4}(s)D_2(s) + r_3s + r_2)(2s - 2a) + r_1$$

Evaluating $P'_n(s)$ at $s = a + jb$ and $s = a - jb$ we obtain the values of the derivatives

$$P'_n(a + jb) = 2jb(r_3(a + jb) + r_2) + r_1$$

and

$$P'_n(a - jb) = -2jb(r_3(a - jb) + r_2) + r_1$$

The reader may also verify that, as in the case of linear division, subsequent quotient divisions by $D_2(s)$ will yield remainders that can be used to evaluate higher-order derivatives.

Division by a quadratic divisor can also be carried out by the process of synthetic division as outlined by the general scheme that follows.

$$\begin{array}{rllclllll|l}
a_n & a_{n-1} & a_{n-2} & \cdots & a_4 & a_3 & a_2 & a_1 & a_0 & A,\ B \\
 & b_nA & b_{n-1}A & \cdots & b_3A & b_2A & b_1A & b_0A & & \\
 & & b_nB & \cdots & b_4B & b_3B & b_2B & b_1B & b_0B & \\
\hline
b_n = a_n & b_{n-1} & b_{n-2} & \cdots & b_2 & b_1 & b_0 & r_1 & r_0 & \\
 & c_{n-2}A & c_{n-3}A & \cdots & c_1A & c_0A & & & & \\
 & & c_{n-2}B & \cdots & c_2B & c_1B & c_0B & & & \\
\hline
c_{n-2} = b_n & c_{n-3} & c_{n-4} & \cdots & c_0 & r_3 & r_2 & & &
\end{array}$$

For a simple numerical example, let us evaluate the characteristic equation

$$\begin{aligned} P_4(s) &= (s + 1 + j1)(s + 1 - j1)(s + 2)^2 \\ &= s^4 + 6s^3 + 14s^2 + 16s + 8 \end{aligned}$$

and its derivative at $s = 1 + j1$. We first form the quadratic divisor

$$\begin{aligned} D_2(s) &= (s - (1 + j1))(s - (1 - j1)) \\ &= s^2 - 2s + 2 \\ &= s^2 - As - B \end{aligned}$$

finding $A = 2$ and $B = -2$. Using the synthetic division scheme which we described above, we find

$$\begin{array}{rrr|rrl} 1 & 6 & 14 & 16 & 8 & \underline{|2, \ -2} \\ & 2 & 16 & 56 & & \\ & & -2 & -16 & -56 & \\ \hline 1 & 8 & 28 & 56 & -48 & \\ & 2 & & & & \\ & & -2 & & & \\ \hline 1 \,| & 10 & 26 & & & \end{array}$$

The value of $P_4(1 + j1)$ is then found from

$$P_n(a + jb) = r_1(a + jb) + r_0$$

or

$$\begin{aligned} P_4(1 + j1) &= 56(1 + j1) - 48 \\ &= 8 + j56 \end{aligned}$$

and the value of $P_4'(1 + j1)$ is found from

$$\begin{aligned} P_n'(a + jb) &= 2jb(r_3(a + jb) + r_2) + r_1 \\ P_4'(1 + j1) &= 2j(10(1 + j1) + 26) + 56 \\ &= 36 + j72 \end{aligned}$$

Let us now turn to Lin's method of finding the roots of a polynomial. Starting with the polynomial

$$P_n(s) = a_n s^n + a_{n-1}s^{n-1} + \cdots + a_1 s + a_0$$

we choose a trial divisor

$$D_2(s) = s^2 - A_1 s - B_1$$

that is based on the last three terms of $P_n(s)$

$$a_2 s^2 + a_1 s + a_0$$

or

$$s^2 + \frac{a_1}{a_2}s + \frac{a_0}{a_2}$$

where

$$A_2 = A_1 - \frac{r_1}{b_0}$$

$$B_2 = B_1 - \frac{r_0}{b_0}$$

Repeating the process with this new divisor will not result in a zero remainder. Note that the values of A_2 and B_2 satisfy the last two columns of the synthetic division process (with the divisor $s^2 - A_1s - B_1$) and result in a zero remainder but, in general, they yield a finite remainder for a division by the new divisor $(s^2 - A_2s - B_2)$, since the last two columns yield equations for this division that are different from the first set. This iteration process may be terminated whenever both remainders, r_1 and r_0, become negligibly small.

Figure 4.54 is a Fortran program for finding the roots of a quartic. Using Lin's method and synthetic division, the quartic is factored into two quadratics and the quadratic formula is used to find the four roots. Initially the five coefficients of the quartic are read into memory. The heading and the quartic are then printed. Computer statements S.0006 through S.0010 now normalize the coefficients of the quartic with respect to A_4. Computer statements S.0011 and S.0012 choose the coefficients of the initial trial divisor and statements S.0013 and S.0014 initially set the last value of r_1 (RLAST1) and r_0 (RLAST0) to a large value so that they will not terminate the process during the first iteration.

The DO 10 loop is the main iteration loop in which Lin's method is used to factor the quartic. The notation used is as follows:

$$\begin{array}{lllll|l}
A4 & A3 & A2 & A1 & A0 & \underline{A,\ B} \\
 & A4*A & B1*A & B0*A & & \\
 & & A4*B & B1*B & B0*B & \\
\hline
A4 & B1 & B0 & |\ R1 & R0 &
\end{array}$$

$$\left.\begin{array}{l} A = A - R1/B0 \\ B = B - R0/B0 \end{array}\right\} \text{New divisor coefficients}$$

Whenever the corrections for both $R1$ and $R0$ become negligibly small or 100 iterations are performed, the value of $B1$ and $B0$ (the normalized coefficients of one of the quadratics), the value of A and B (the normalized coefficients of the other quadratic), and I, the number of iterations, are printed. Computer statements S.0031 through S.0049 then use the quadratic formula to calculate the roots and these values are printed by statements S.0047 and S.0049.

Figure 4.55 shows the corresponding output. Note that a total of nine iterations was required to find these roots, but this number will

```
C       LIN'S METHOD AND THE QUADRATIC FORMULA
C       ARE USED TO FIND THE ROOTS OF A QUARTIC
C
C       INPUT COEFFICIENTS:
C          A4 = COEFFICIENT OF S TO THE FOURTH POWER
C          A3 = COEFFICIENT OF S TO THE THIRD POWER
C          A2 = COEFFICIENT OF S TO THE SECOND POWER
C          A1 = COEFFICIENT OF S TO THE FIRST POWER
C          A0 = COEFFICIENT OF S TO THE ZEROTH POWER
C
        DIMENSION BQ(2),CQ(2)
        READ(5,1) A4,A3,A2,A1,A0
      1 FORMAT(5E15.7)
        WRITE(6,2) A4,A3,A2,A1,A0
      2 FORMAT('1     LINS METHOD AND THE QUADRATIC FORMULA ARE USED TO FIND
       1 THE ROOTS OF A QUARTIC',///, E15.7,'    S**4    +',E15.7,'    S**3    +',
       2E15.7,'   S**2    +',E15.7,'   S    +',E15.7,//)
        A0=A0/A4
        A1=A1/A4
        A2=A2/A4
        A3=A3/A4
        A4=A4/A4
C
C       THE TRIAL DIVISOR IS COMPUTED FROM THE COEFFICIENTS
C       OF THE LAST THREE TERMS
C
        A =-A1/A2
        B =-A0/A2
C
C       LIN'S METHOD
C
        RLAST1=1.E10
        RLAST0=1.E10
        DO 10 I=1,100
        B1=A4*A+A3
        B0=B1*A+A4*B+A2
        R1=B1*B+B0*A+A1
        R0=B0*B+A0
        IF(ABS(RLAST1-R1)-1.E-5) 20,20,15
     20 IF(ABS(RLAST0-R0)-1.E-5) 30,30,15
     15 A=A-R1/B0
        B=B-(A0+B0*B)/B0
        RLAST1=R1
     10 RLAST0=R0
        WRITE(6,3)
      3 FORMAT('1100 ITERATIONS HAVE BEEN PERFORMED',///,' DOES NOT CONVER
       1GE')
        GO TO 40
     30 WRITE(6,4) B1,B0,A,B,I
      4 FORMAT(///,' THE TWO QUADRATICS ARE',///,'     (S + ',E15.7,' )S + ',
       1E15.7,/,' AND',/,'     (S + ',E15.7,' )S + ',E15.7.//,' AND WERE FOUN
       1D IN',I4,' ITERATIONS',///,' THE ROOTS ARE',//)
C
C       THE CALCULATION OF THE ROOTS OF EACH EQUATION
C
        BQ(1)=B1
        BQ(2)=-A
        CQ(1)=B0
        CQ(2)=-B
        DO 50 I=1,2
        D=BQ(I)**2-4.*CQ(I)
        IF(D)60,70,70
     70 S1RE=(-BQ(I)+D**0.5)*0.5
        S2RE=(-BQ(I)-D**0.5)*0.5
        S1IM=0.0
        S2IM=0.0
        GO TO 80
     60 S1RE=-BQ(I)*0.5
        S2RE=S1RE
        S1IM=(-D)**0.5*0.5
        S2IM=-S1IM
     80 WRITE(6,5) S1RE,S1IM,S2RE,S2IM
     50 CONTINUE
      5 FORMAT(7X,'S1RE',12X,'S1IM',10X,'S2RE', 7X,'S2IM',//,4(E15.7),//)
     40 STOP
        END
```

Figure 4.54 A program that uses Lin's method and the quadratic formula to find the roots of a quartic

```
LINS METHOD AND THE QUADRATIC FORMULA ARE USED TO FIND THE ROOTS OF A QUARTIC

0.1000000E 01  S**4  +  0.2200000E 01  S**3  +  0.5750000E 01  S**2  +  0.1480000E 01  S  +  0.1320


THE TWO QUADRATICS ARE

 (S +   0.2012155E 01 )S +    0.5113906E 01
AND
 (S +  -0.1878452E 00 )S +   -0.2581198E 00

AND WERE FOUND IN    9 ITERATIONS

THE ROOTS ARE

      S1RE            S1IM            S2RE         S2IM

 0.1006077E 01  0.2025267E 01 -0.1006077E 01 -0.2025267E 01

      S1RE            S1IM            S2RE         S2IM

-0.9392256E-01  0.4992976E 00 -0.9392256E-01 -0.4992976E 00
```

Figure 4.55 The output results of the program of Figure 4.54

vary as a function of the spacing of the roots. The reader is encouraged to test this program by using quartics that are formed from only real roots such as those in the following equation:

$$(S + 1)(S + 2)(S + 3)(S + 4) = S^4 + 10S^3 + 35S^2 + 50S + 24 = 0$$

and quartics that are formed from roots similar to the following

$$(S + 0.1)(S + 1 - j1)(S + 1 + j1)(S + 10) = S^4 + 12.1S^3 + 23.2S^2 + 22.2S + 2 = 0$$

The latter case will reveal that Lin's method does not always converge.[14] The reader is also encouraged to modify this program so that it is capable of factoring higher order polynomials.

Bairstow's Method

Bairstow's method is similar to Lin's method in that the coefficients of a quadratic divisor are successively corrected until the quadratic divisor becomes an exact factor of the polynomial. It is also similar to the Newton-Raphson method in that first derivatives are used to obtain the corrections, and thus it generally converges faster than Lin's method.

[14] For an explanation of why Lin's method fails to converge and how it may be corrected, see B. J. Ley, S. Lutz, and C. Rehberg, *Linear Circuit Analysis* (New York: McGraw-Hill Book Company, 1959), pp. 219–26.

Starting with the polynomial

$$P_n(s) = a_n s^n + a_{n-1} s^{n-1} + \cdots + a_1 s + a_0$$

we divide by our trial divisor

$$D_2(s) = s^2 - A_1 s - B_1$$

obtaining the quotient $Q_{n-2}(s)$ and the linear remainder $r_1 s + r_0$. The quotient $Q_{n-2}(s)$ is then divided by the quadratic divisor, obtaining a new quotient $Q_{n-4}(s)$ and a new remainder $r_3 s + r_2$. These two divisions may be carried out by using the synthetic division scheme we discussed in Lin's method (see page 205). Since in general a division by $s^2 - A_1 s - B_1$ will not yield a zero remainder, we must, as in Lin's method, change A_1 and B_1 so that r_1 and r_0 become equal to zero. To force this behavior we change A_1 to A_2 and B_1 to B_2 where

$$\begin{aligned} A_2 &= A_1 + \Delta A_1 \\ B_2 &= B_1 + \Delta B_1 \end{aligned}$$

so that

$$\begin{aligned} r_1(A_1 + \Delta A_1, B_1 + \Delta B_1) &= 0 \\ r_0(A_1 + \Delta A_1, B_1 + \Delta B_1) &= 0 \end{aligned}$$

Now, similar to the Newton-Raphson method, we write a Taylor's expansion of these two equations through the linear terms, obtaining

$$r_1 + \frac{\partial r_1}{\partial A_1} \Delta A_1 + \frac{\partial r_1}{\partial B_1} \Delta B_1 = 0$$

$$r_0 + \frac{\partial r_0}{\partial A_1} \Delta A_1 + \frac{\partial r_0}{\partial B_1} \Delta B_1 = 0$$

Solving for ΔA_1 and ΔB_1 we find

$$\Delta A_1 = \frac{-r_1 \dfrac{\partial r_0}{\partial B_1} + r_0 \dfrac{\partial r_1}{\partial B_1}}{\text{DENOM}}$$

$$\Delta B_1 = \frac{-r_0 \dfrac{\partial r_1}{\partial A_1} + r_1 \dfrac{\partial r_0}{\partial A_1}}{\text{DENOM}}$$

where

$$\text{DENOM} = \begin{vmatrix} \dfrac{\partial r_1}{\partial A_1} & \dfrac{\partial r_1}{\partial B_1} \\[2ex] \dfrac{\partial r_0}{\partial A_1} & \dfrac{\partial r_0}{\partial B_1} \end{vmatrix}$$

Householder shows that the partial derivatives may be evaluated from the remainders r_3 and r_2 by using the following expressions[15]

$$\frac{\partial r_1}{\partial A_1} = A_1 r_3 + r_2 \qquad \frac{\partial r_0}{\partial A_1} = B_1 r_3$$

$$\frac{\partial r_1}{\partial B_1} = r_3 \qquad \frac{\partial r_0}{\partial B_1} = r_2$$

We may thus compute the new values of A_2 and B_2 and repeat the process until both remainders, r_1 and r_0, become negligibly small.

Figure 4.56 is a Fortran program that uses Bairstow's method and the quadratic formula to find the roots of a quartic. This program is very similar to the program of Figure 4.54, except for computer statements S.0015 through S.0034 where Bairstow's method is used instead of Lin's method. The DO 10 loop is again the main iteration loop and follows the synthetic division scheme and notation used below.

$$\begin{array}{llll|ll}
A4 & A3 & A2 & A1 & A0 & \underline{|A, B} \\
 & A4*A & B1*A & B0*A & & \\
 & & A4*B & B1*A & B0*B & \\
\hline
A4 & B1 & B0 & R1 & R0 & \\
 & A4*A & & & & \\
 & & A4*B & & & \\
\hline
A4 \,| & R3 & R2 & & &
\end{array}$$

where

$$\left.\begin{aligned} A &= A + \frac{-\text{R1*DR0DB} + \text{R0*DR1DB}}{\text{DENOM}} \\ B &= B + \frac{-\text{R0*DR1DA} + \text{R1*DR0DA}}{\text{DENOM}} \end{aligned}\right\} \text{New divisor coefficients.}$$

$$\text{DENOM} = \text{DR1DA*DR0DB} - \text{DR0DA*DR1DB}$$

and

$$\text{DR1DA} = \frac{\partial r_1}{\partial A} \qquad \text{DR1DB} = \frac{\partial r_1}{\partial B}$$

$$\text{DR0DA} = \frac{\partial r_0}{\partial A} \qquad \text{DR0DB} = \frac{\partial r_0}{\partial B}$$

In order to see how the process converges, the values of the derivatives and the new values of A and B are printed each iteration of the DO 10 loop. Whenever the correction for both R1 and R0 become negligibly small or 100 iterations are performed, the process is terminated, and the remainder of the program is identical to that of Figure 4.54.

[15] Alston S. Householder, *Principles of Numerical Analysis* (New York: McGraw-Hill Book Company, 1953), pp. 140–41.

```
C       BARISTOW'S METHOD AND THE QUADRATIC FORMULA
C       ARE USED TO FIND THE ROOTS OF A QUARTIC
C
C       INPUT COEFFICIENTS:
C          A4 = COEFFICIENT OF S TO THE FOURTH POWER
C          A3 = COEFFICIENT OF S TO THE THIRD POWER
C          A2 = COEFFICIENT OF S TO THE SECOND POWER
C          A1 = COEFFICIENT OF S TO THE FIRST POWER
C          A0 = COEFFICIENT OF S TO THE ZEROTH POWER
C
        DIMENSION BQ(2),CQ(2)
        READ(5,1) A4,A3,A2,A1,A0
      1 FORMAT(5E15.7)
        WRITE(6,2) A4,A3,A2,A1,A0
      2 FORMAT('1BAIRSTOW S METHOD AND QUADRATIC  FORMULA ARE USED TO FIND
       1 THE ROOTS OF A QUARTIC',///, E15.7,'   S**4  +',E15.7,'   S**3  +',
       2E15.7,'   S**2  +',E15.7,'   S    +',E15.7,//)
        A0=A0/A4
        A1=A1/A4
        A2=A2/A4
        A3=A3/A4
        A4=A4/A4
C
C       THE TRIAL DIVISOR IS COMPUTED FROM THE COEFFICIENTS
C       OF THE LAST THREE TERMS
C
        A =-A1/A2
        B =-A0/A2
C
C       BAIRSTOW'S METHOD
C
        RLAST1=1.E10
        RLAST0=1.E10
        DO 10 I=1,100
        B1=A4*A+A3
        B0=B1*A+A4*B+A2
        R1=B1*B+B0*A+A1
        R0=B0*B+A0
        R3=B1+A4*A
        R2=B0+A4*B
        IF(ABS(RLAST1-R1)-1.E-5) 20,20,15
     20 IF(ABS(RLAST0-R0)-1.E-5) 30,30,15
     15 DR1DA=A*R3+R2
        DR0DA=B*R3
        DR1DB=R3
        DR0DB=R2
        DENOM=DR1DA*DR0DB-DR0DA*DR1DB
        A=A+(-R1*DR0DB+R0*DR1DB)/DENOM
        B=B+(-R0*DR1DA+R1*DR0DA)/DENOM
        WRITE(6,88) DR1DA,DR0DA,DR1DB,DENOM,A,B
     88 FORMAT(5X,'DR1DA',10X,'DR0DA',10X,'DR1DB',10X,'DENOM',12X,'A',14X,
       1'B',/,6E15.7,/)
        RLAST1=R1
     10 RLAST0=R0
        WRITE(6,3)
      3 FORMAT('1100 ITERATIONS HAVE BEEN PERFORMED',///,' DOES NOT CONVER
       1GE')
        GO TO 40
     30 WRITE(6,4) B1,B0,A,B,I
      4 FORMAT(///,' THE TWO QUADRATICS ARE',///,'   (S + ',E15.7,' )S + ',
       1E15.7,/,' AND',/,'   (S + ',E15.7,' )S + ',E15.7,//,' AND WERE FOUN
       1D IN',I4,' ITERATIONS',///,' THE ROOTS ARE',//)
C
C       THE CALCULATION OF THE ROOTS OF EACH EQUATION
C
        BQ(1)=B1
        BQ(2)=-A
        CQ(1)=B0
        CQ(2)=-B
        DO 50 I=1,2
        D=BQ(I)**2-4.*CQ(I)
        IF(D) 60,70,70
     70 S1RE=(-BQ(I)+D**0.5)*0.5
        S2RE=(-BQ(I)-D**0.5)*0.5
        S1IM=0.0
        S2IM=0.0
        GO TO 80
     60 S1RE=-BQ(I)*0.5
        S2RE=S1RE
        S1IM=(-D)**0.5*0.5
        S2IM=-S1IM
     80 WRITE(6,5) S1RE,S1IM,S2RE,S2IM
     50 CONTINUE
      5 FORMAT(7X,'S1RE',12X,'S1IM',10X,'S2RE', 7X,'S2IM',//,4(E15.7),//)
     40 STOP
        END
```

Figure 4.56 A program that uses Bairstow's method and the quadratic formula to find the roots of a quartic

```
BAIRSTOW S METHOD AND QUADRATIC  FORMULA ARE USED TO FIND THE ROOTS OF A QUARTIC

0.1000000E 01  S**4  +  0.2200000E 01  S**3  +  0.5750000E 01  S**2  +  0.1480000E 01  S   +  0.1320000E 01

    DR1DA          DRODA           DR1DB          DENOM            A              B
0.4357098E 01 -0.3868668E 00  0.1685216E 01  0.2152618E 02 -0.1868218E 00 -0.2588261E 00

    DR1DA          DRODA           DR1DB          DENOM            A              B
0.4515038E 01 -0.4727085E 00  0.1826356E 01  0.2278944E 02 -0.1878452E 00 -0.2581199E 00

    DR1DA          DRODA           DR1DB          DENOM            A              B
0.4513097E 01 -0.4708906E 00  0.1824309E 01  0.2277367E 02 -0.1878452E 00 -0.2581197E 00


THE TWO QUADRATICS ARE

 (S +   0.2012155E 01 )S +   0.5113906E 01
AND
 (S +  -0.1878452E 00 )S +  -0.2581197E 00

AND WERE FOUND IN   4 ITERATIONS

THE ROOTS ARE

      S1RE           S1IM           S2RE          S2IM

-0.1006077E 01  0.2025267E 01 -0.1006077E 01 -0.2025267E 01

      S1RE           S1IM           S2RE          S2IM

-0.9392262E-01  0.4992976E 00 -0.9392262E-01 -0.4992976E 00
```

Figure 4.57 The output results of the program of Figure 4.56

Figure 4.57 shows the computer results for the same quartic that was used in Figure 4.54. Note that the results are identical to those of Figure 4.55 except that only four iterations were needed instead of nine iterations. The reader is urged to test this program for quartics that have all real roots and for quartics that are similar to

$$s^4 + 12.1s^3 + 23.2s^2 + 22.2s + 2 = 0$$

The reader is also encouraged to modify this program so that it is capable of finding the roots of higher-order polynomials.

PROBLEMS

4-1. In Section 4.2 we pointed out a number of shortcomings in the programs used to find the roots of a quadratic equation. List additional shortcomings and tell how the program may be modified to avoid possible error.

4-2. In Section 4.2 it was pointed out that error occurred in using the program of Figure 4.3 to evaluate the roots of a quadratic if B^2 was large compared with $4AC$. The program of Figure 4.5 included an alternative method of solution that minimized the roundoff error.

Another method of solution for minimizing this type of error is based on computing $X1$ from

$$\begin{aligned} X1 &= \frac{-B + \sqrt{B^2 - 4AC}}{2A} \\ &= \frac{-B + \sqrt{B^2 - 4AC}}{2A} * \frac{B + \sqrt{B^2 - 4AC}}{B + \sqrt{B^2 - 4AC}} \\ &= \frac{-2C}{B + \sqrt{B^2 - 4AC}} \end{aligned}$$

If B is positive and $X2$ from

$$\begin{aligned} X2 &= \frac{-B - \sqrt{B^2 - 4AC}}{2A} \\ &= \frac{-B - \sqrt{B^2 - 4AC}}{2A} * \frac{B - \sqrt{B^2 - 4AC}}{B - \sqrt{B^2 - 4AC}} \\ &= \frac{-2C}{B - \sqrt{B^2 - 4AC}} \end{aligned}$$

if B is negative.

Modify the program of Figure 4.5 to include this scheme rather than the one we previously discussed. Test your new program by using the same data sets that were used in Figure 4.6. Explain any differences in the results.

4-3. Generate appropriate tables that may be used in the solution of cubic equations.

4-4. In the solution of the cubic equation we used the identity

$$\sinh^{-1} X = \ln (X + (X^2 + 1)^{0.5})$$

Write a Fortran program based on this identity for computing the inverse hyperbolic sine function.

4-5. Repeat Problem 4-4 for the following identities:

a. $\cosh^{-1} X = \ln (X + (X^2 - 1)^{0.5});\ X \geq 1$

b. $\tanh^{-1} X = 0.5 \ln ((1 + X)/(1 - X));\ 0 \leq X^2 < 1$

c. $\operatorname{csch}^{-1} X = \ln \left(\frac{1}{X} + \left(\frac{1}{X^2} + 1\right)^{0.5}\right);\ X \neq 0$

d. $\operatorname{sech}^{-1} X = \ln \left(\frac{1}{X} + \left(\frac{1}{X^2} - 1\right)^{0.5}\right);\ 0 < X \leq 1$

e. $\coth^{-1} X = 0.5 \ln ((X + 1)/(X + 1));\ X^2 > 1$

4-6. Write a Fortran program based on the approximation that was used for computing the $\sin^{-1} X$ in determining the roots of a cubic equation.

4-7. Rewrite the program of Figure 4.8 to make it a more efficient program.

4-8. Rewrite the program of Figure 4.8 to improve the accuracy of the results.

4-9. Using the program of Figure 4.8, write a Fortran program for computing the roots of a quartic equation.

4-10. Discuss any shortcomings of the program of Figure 4.8.

4-11. The bisection method discussed in Section 4.3 was based on evaluating the function at the midpoint of the interval of the two points bracketing the real root ($T = (t_1 + t_2)/2$) and then testing to see whether $f(T)*f(t_1)$ was positive or negative. If $f(T)*f(t_1)$ was positive, the root lies between T and t_2 and we repeat the process with the location of t_1 changed to T. On the other hand, if $f(T)*f(t_1)$ is negative, the root lies between t_1 and T and we repeat the process with t_2 set equal to T.

In an attempt to improve the convergence of the bisection method, investigate the following modifications:

a. Evaluate the function at

$$T = \frac{t_1 + t_2}{4}$$

b. Test to see whether $f(T)*f(t_1)$ is positive or negative.

c. If $f(T)*f(t_1)$ is positive, the root lies between T and t_2. Change t_1 to T and repeat the procedure, but this time evaluate the function at a new T equal to

$$\text{new } T = 3\,\frac{t_1 + t_2}{4}, \text{ etc.}$$

d. If $f(T)*f(t_1)$ is negative, the root lies between t_1 and T. Change t_2 to T and repeat the process by evaluating the function at a new T equal to

$$\text{new } T = \frac{t_1 + t_2}{4}, \text{ etc.}$$

Using the program of Figure 4.12, incorporate the above procedure and investigate the rate of convergence in terms of the single stage and three stage amplifier problems.

4-12. Would there be any advantage in using the factors 1/10 and 9/10 instead of 1/4 and 3/4 in improving the rate of convergence of the program of Problem 4-11?

4-13. A simplification of the false position method used in the program of Figure 4.23 results when the step that determines whether $f(T)*f(t_1)$ is positive or negative is omitted and, instead, the location of t_1 is made equal to T. The process is then repeated with the location of t_2 equal to the new T, and so on. Modify the program of Figure 4.23 accordingly and investigate the subsequent behavior. Hint: Under certain conditions this latter procedure will have a convergence problem.

4-14. The false position method is sometimes used in conjunction with the bisection method. Write a Fortran program and investigate the behavior of a rootfinding method that alternately uses false position and bisection.

4-15. Note that the accuracy in the false position method has been specified in a way different from the accuracy in the bisection method. Modify the bisection program of Figure 4.12 so that the accuracy is determined in the same way as in the program of Figure 4.23. Compare the corresponding results.

4-16. Rerun the Newton-Raphson programs of Figure 4.37 and 4.38 using starting values of $T = 1.0$, 2.0 5.0, and 10.0. Compare the results and the number of iterations.

4-17. The normalized output response of n identical single-pole stages due to a unit step input excitation is given by

$$G(s) = \frac{1}{s(s+1)^n}$$

The corresponding time response is given by

$$g(t) = 1 - \epsilon^{-t} \sum_{I=0}^{n-1} \frac{t^I}{I!}$$

Use each of the following methods to compute the rise time of a ten stage cascade:

a. The bisection method of Figure 4.12.
b. The false position method of Figure 4.23.
c. The false position method of Figure 4.28 (ordinate halving).
d. The Newton-Raphson method of Figure 4.37.

4-18. The normalized time response of a simple-shunt peaked interstage is given by

$$g(t) = 1 - \left(\cos \frac{\sqrt{4K-1}}{2K} \frac{t}{RC} + \frac{1-2K}{\sqrt{4K-1}} \sin \frac{\sqrt{4K-1}}{2K} \frac{t}{RC}\right) \epsilon^{-\frac{t}{2KRC}}$$

Using each of the methods of Problem 4-17, compute the rise time for values of $K = 0.25$, 0.5, and 1.0. Note that the case of $K = 0.25$ corresponds to the critically damped case. Hint: Compute the rise time in terms of normalized time $t/(R*C)$.

4-19. Modify the program of Figure 4.46 so that after all the roots have been found, the process is repeated for each root, using the original polynomial and the value of the root previously found as the first approximation.

4-20. Modify the Newton-Raphson program of Figure 4.46 so that the accuracy is based on $\text{ABS}(P_n(s)) \leqq \text{ACC}$. Note that this method of termination would be a much more desirable criterion for roots located in a region where $\text{ABS}(P_n(s))$ is very small.

4-21. Descartes' rule of sign may be used to determine the number of positive and negative real roots of higher-order polynomials. The number of positive real roots is equal to or less (by an even number) than the number of sign changes of the coefficients of the polynomial. For example, if

$$P_4(s) = 3s^4 + 2s^2 - 6s - 8 = 0$$

there is a maximum of one positive real root since the successive signs of the coefficients are +, + −, and − (hence one sign change). The maximum number of negative real roots can be found from the number of sign changes of the coefficients of $P_m(-s)$.

For example, if

$$P_4(s) = 3s^4 + 2s^2 - 6s - 8$$

then

$$P_4(-s) = 3s^4 + 2s^2 + 6s - 8$$

and there is thus a maximum of one negative root since the successive signs of the coefficients $P_4(-s)$ are $+$, $+$, $+$, and $-$.

Write a Fortran program that reads in the coefficients of a n-th order polynomial and determines the maximum number of positive and negative real roots.

4-22. Using synthetic division and the Newton-Raphson method, hand compute the real root (s) of

a. $P_3(s) = 2s^3 + 3s - 5$
b. $P_3(s) = s^3 - 3s^2 - 9s - 4.5$
c. $P_5(s) = 2s^5 + 9s^4 + 18s^3 + 25s^2 + 20s + 5$

4-23. Using the method discussed in Section 4.5 (page 192), estimate each of the roots of the polynomials of Problem 4-22.

4-24. Using the method of synthetic division, write a Fortran program that may be used to obtain the data needed to plot a rough graph of a general n-th order polynomial.

4-25. Using the Newton-Raphson method for the special polynomial case, compute the cube root of 28 longhand to five significant figures. Examine the corrections at each step. Do the corrections show second-order convergence?

4-26. Repeat Problem 4-25 using each of the third-order formulas on page 196.

4-27. Hand compute the square root of 5 using the following relation

$$X_n = \frac{X_{n-1}}{2}\left(3 - \frac{{X_{n-1}}^2}{N}\right); N = 5$$

Hint: Use several different starting values.

4-28. It is possible to find the square root of a complex number using DeMoivre's theorem

$$\sqrt{N} = \sqrt{a + jb} = \sqrt{R}\left(\cos\frac{\theta}{2} + j\sin\frac{\theta}{2}\right)$$

where

$$R = \sqrt{a^2 + b^2}$$

$$\theta = \tan^{-1}\frac{b}{a}$$

Write a Fortran program for computing the square root of a complex number. Test the accuracy of your program with several sets of input data.

4-29. Modify the program of Figure 4.48 so that a quadratic equation involving the variable h (see page 197) must be solved in order to obtain the correction. Examine the corresponding results.

4-30. Using the remainder theorem and synthetic division, evaluate by hand calculation each of the polynomials and their derivatives of Problem 4-22 for

a. $s = -2$

b. $s = 1 + j1$

4-31. Show that the iterative relation

$$X_n = X_{n-1}(2 - AX_{n-1})$$

may be used to evaluate $1/A$. Note that this algorithm avoids division and would therefore be very useful when a computer that does not permit the arithmetic operation of division is used.

4-32. The secant method for finding a real root is a variation of the Newton-Raphson method. It avoids computing the derivative of the function by using a secant line (instead of a tangent line) that is determined by the last two function point values. Write a Fortran program that uses the secant line method to determine the next approximation to the root. Compare the results obtained by this program with the results of the program of Figure 4.37. For a fair basis of comparison, use as the first two function point values in the secant program

$$t = 0.0 \qquad F(t) = -0.9$$

and

$$t = 0.9 \qquad F(t) = -0.3065697$$

4-33. Another variation of the Newton-Raphson formula is the constant-slope or von Mises formula.

$$X_{n+1} = X_n - \frac{f(X_n)}{f'(X_0)}$$

where $f'(X_0)$ is the initial value of the derivative. Write a Fortran program that uses this iterative relation and compare the results obtained by this program with those of Figure 4.37.

4-34. Write a Fortran program that uses the Newton-Raphson method to evaluate complex as well as real roots of a general n-th order polynomial.

4-35. Modify the program of Problem 4-34 to include the features of the program of Problem 4-19.

4-36. Modify the program of Figure 4.46 and include the second derivative term as discussed in Section 4.5.

Note that this is similar to the way we choose the initial approximation to the root in the Newton-Raphson method. In an analogous way, the roots of this initial divisor will approximate the two smallest roots of $P_n(s)$ and the accuracy of this approximation will depend upon the magnitude spacing of the respective roots. If all the roots are relatively far apart, this will be a good approximation.

We next divide $D_2(s)$ into $P_n(s)$, obtaining the quotient $Q_{n-2}(s)$ and the linear remainder $r_1s + r_0$, as shown in the synthetic division process below.

$$
\begin{array}{rcccccccl}
 & a_n & a_{n-1} & a_{n-2} & \cdots & a_2 & a_1 & a_0 & \underline{|A_1, B_1} \\
 & & b_{n-2}A_1 & b_{n-3}A_1 & \cdots & b_1A_1 & b_0A_1 & & \\
 & & & b_{n-2}B_1 & \cdots & b_2B_1 & b_1B_1 & b_0B_1 & \\
\hline
b_{n-2} = & a_n & b_{n-3} & b_{n-4} & \cdots & b_0 & |\ r_1 & r_0 &
\end{array}
$$

In general, $s^2 - A_1s - B_1$ will not be an exact factor of $P_n(s)$ and we will have a remainder

$$
\begin{aligned}
r_1 &= a_1 + b_0A_1 + b_1B_1 \\
r_0 &= a_0 + b_0B_1
\end{aligned}
$$

Since we would like r_1 and r_0 to be equal to zero, we may force this behavior in these two equations by changing A_1 to A_2 in the equation for r_1, and B_1 to B_2 in the equation for r_0, where[13]

$$
\begin{aligned}
A_2 &= A_1 + \Delta A_1 \\
B_2 &= B_1 + \Delta B_1
\end{aligned}
$$

Thus

$$
\begin{aligned}
0 &= a_1 + b_0(A_1 + \Delta A_1) + b_1B_1 \\
0 &= a_0 + b_0(B_1 + \Delta B_1)
\end{aligned}
$$

If we now subtract these latter two equations from the former two equations, we may solve for the respective corrections, or

$$\Delta A_1 = -\frac{r_1}{b_0}$$

$$\Delta B_1 = -\frac{r_0}{b_0}$$

If the process is convergent, we will therefore obtain a better approximation to the divisor

$$D_2(s) = s^2 - A_2s - B_2$$

[13] Note that as the process converges, the corrections ΔA and ΔB become small and thus it will not matter that we have not changed B_1 to B_2 in the equation for r_1. This will, however, greatly simplify our calculations.

4-37. Horner's method for finding the roots of a polynomial is based on successive polynomial transformations. Each transformation shifts the previous polynomial equation closer to the origin, or forces one of the roots of the transformed polynomial to approach zero. The amount of shifting required to make the root of the transformed polynomial exactly equal to zero is the value of the root of the original polynomial. The coefficients of the polynomial

$$P_n(s) = a_0 + a_1 s + a_2 s^2 + \cdots + a_n s^n$$

are the successive remainders that are obtained from the synthetic division process described in Section 4.5 (successive divisions by $s - s_1$ where $s_1 = 0$). The proof of this follows from successive differentiation of the above polynomial

$$\begin{aligned} P'_n(s) &= a_1 + 2a_2 s + 3a_3 s^2 + \cdots + n a_n s^{n-1} \\ P''_n(s) &= 2a_2 + 3{*}2a_3 s + \cdots + n(n-1)a_n s^{n-2} \\ &\vdots \\ P_n{}^n(s) &= n! a_n \end{aligned}$$

and the evaluation of each of these at $s = 0$

$$\begin{aligned} a_0 &= P_n(0) \\ a_1 &= P'_n(0) \\ a_2 &= \frac{1}{2!} P''_n(0) \\ &\vdots \\ a_n &= \frac{1}{n!} P_n{}^n(0) \end{aligned}$$

If we now divide $F_n(S)$ by $s - s_1$, the remainders obtained will be equal to the coefficients of the transformed polynomial for $s = s_1$

$$\begin{aligned} F(S) &= a_0 + a_1 S + \cdots + a_n S^n \\ \text{or } F(s - s_1) &= a_0 + a_1(s - s_1) + \cdots + a_n(s - s_1)^n \end{aligned}$$

where the transformation is given by

$$S = s - s_1$$

Write a Fortran program for computing the successive derivatives of a general n-th order polynomial.

4-38. Modify the program of Problem 4-37 so that Horner's method may be used to find the real roots of a polynomial.

4-39. Use the Lin's method program of Figure 4.54 to evaluate the following polynomials

a. $s^4 + 10s^3 + 35s^2 + 50s + 24 = 0$

b. $s^4 + 12.1s^3 + 23.2s^2 + 22.2s + 2 = 0$

4-40. Modify the program of Figure 4.54 so that it is capable of finding the roots of a general n-th degree polynomial (n is even).

4-41. Modify the program of Problem 4-40 to include the features of the program of Problem 4-19.

4-42. Modify the program of Figure 4.54 so that if n is odd, a real root is first determined by the Newton-Raphson method.

4-43. Modify the program of Figure 4.56 so that it is capable of finding the roots of a general n-th degree polynomial. Hint: See Problems 4-41 and 4-42.

4-44. Use the Bairstow method program of Figure 4.56 to find the roots of the polynomials of Problem 4-39.

CHAPTER 5

Numerical Integration

5.1 INTRODUCTION

The solution of many different electrical engineering problems requires the evaluation of definite integrals. For example, integration is performed in the evaluation of

1. The RMS value of a current or voltage waveform.
2. The Fourier coefficients for a periodic waveform.
3. The Fourier integrals for a nonperiodic waveform.
4. The convolution integral, and so on.

If the integrand is expressible in terms of simple algebraic and transcendental functions, then calculus can be employed in the evaluation. On the other hand, if it is not possible to express the integrand in closed form or in terms of elementary functions, or if a digital computer is to be used in obtaining the solution, numerical methods of integration must be used instead.

Numerical integration or numerical *quadrature* is a way of approximating the area under a curve. Consider the curve $f(x)$ shown in Figure 5.1 and let it be required to evaluate

$$\int_a^b f(x)\, dx$$

which is the area bounded by the curve $f(x)$, the x-axis, and the ordinates erected at $x = a$ and $x = b$. The reader may recall from his study of elementary calculus that the integral calculus was invented to solve the problem of finding the area bounded by curves. This was accomplished by dividing the given area into an infinite number of infinitesimal ele-

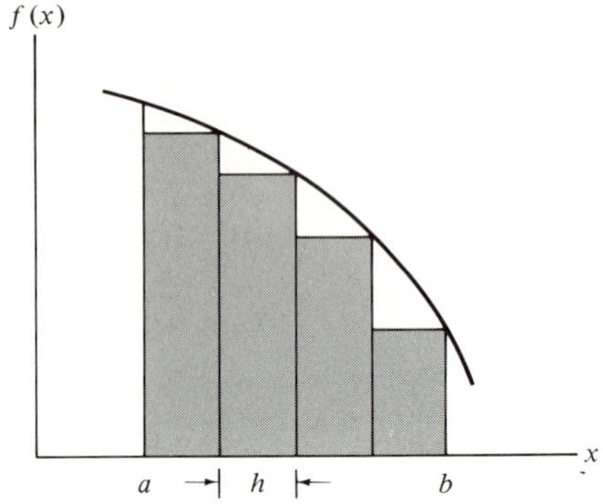

Figure 5.1 The calculation of the area under a curve

ments and summing the area of these elements to obtain the overall area. The integral sign is the long S used by early mathematicians to indicate sum. Thus, in order to find this area, let us divide the interval $x = a$ to $x = b$ into n equal subintervals so that each has a width $h = (b - a)/n$. In addition, let us erect ordinates at the interval endpoints and complete the rectangles as shown by the shaded areas. It should now be clear that the sum of these n element areas is an approximation of the total area and that in the limit as n goes to infinity, the sum of these element areas will equal the area under the curve or mathematically

$$\int_a^b f(x)\,dx \approx \sum_{I=1}^{n} hf(a + I * h)$$

This method is thus equivalent to evaluating the definite integral from its definition rather than from the theorems of integral calculus.

We will see later that although it is possible to write a Fortran program that uses this method to calculate definite integrals, the result can only be an approximation to the true area since only a finite number of intervals can be used. At this point, the reader may jump to the conclusion that this error can be made as small as required by simply reducing the width of the intervals (increasing the number of intervals). We will show later that this is not true because of round-off error.

This chapter will also consider the limitation of other methods such as the trapezoid rule and Simpson's rule which are special cases of the n-point Newton-Cotes formulas.

5.2 THE EULER FORMULA

The numerical quadrature method discussed in the previous section is called the Euler formula or the rectangle rule. It is clear from an examination of Figure 5.1 why an error will exist. The error that occurs in a given

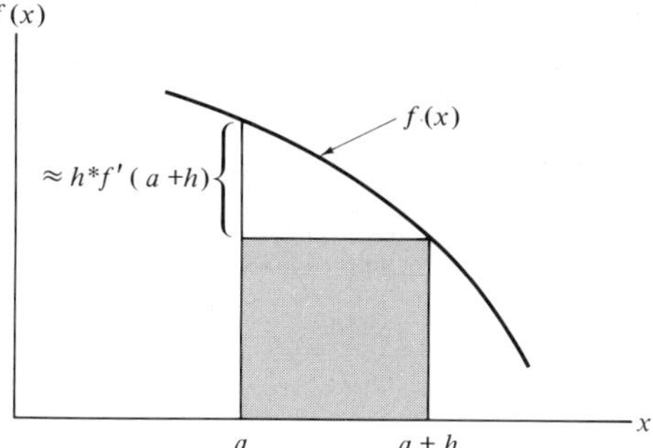

Figure 5.2 The rectangle rule error

interval can be approximated by referring to Figure 5.2. Here we have

$$\int_a^{a+h} f(x)\,dx \approx h * f(a+h)$$

This is less than the true area by the curvilinear triangle that is located between $f(x)$ and the shaded rectangle. In order to calculate the area of this triangle, we may recognize that the derivative of $f(x)$ at $x = a + h$ is approximately coincident with the hypotenuse of this triangle; therefore, the altitude of the triangle may be approximated by $h * f'(a + h)$. Since the base of the triangle equals h, the area, and hence the approximate error, will be given by

$$\frac{1}{2} h^2 f'(a+h)$$

It can be proved mathematically that the exact error is equal to

$$\frac{1}{2} h^2 f'(\epsilon)$$

where $a \leq \epsilon \leq a + h$.

Silicon Controlled Half-Wave Rectifier

In order to show how this integration scheme can be applied to a practical problem, let us consider the silicon-controlled rectifier shown in Figure 5.3. The operation of this circuit is very similar to that of a thyratron rectifier. If no signal is applied to the gate, the SCR (silicon controlled rectifier) will essentially be an open circuit and no current will flow to the load. During the half cycle in which the anode voltage is positive with respect to the cathode, the SCR can be fired by the application of a proper gate signal. Once fired, the device will act as a virtual short circuit until the polarity of the anode to cathode voltage is reversed. It is thus possible, by means of the 180° phase shift circuit shown, to

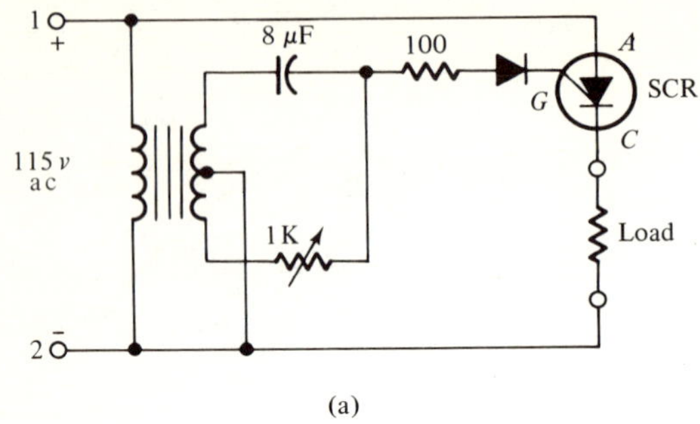

(a)

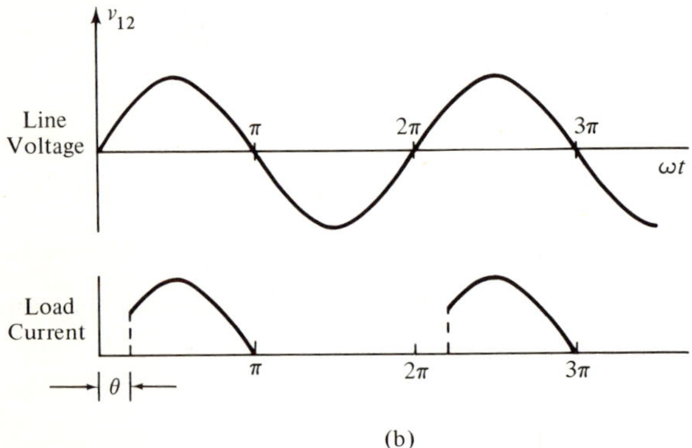

(b)

Figure 5.3 (a) Silicon-controlled half-wave rectifier; (b) the line voltage and load current

control the point at which the SCR fires and thus control the average load current. For example, if $\theta = 0°$ in Figure 5.3b, the average current will be equal to

$$
\begin{aligned}
I_{\text{ave}} &= \frac{1}{2\pi} \int_{\omega t = \theta}^{\pi} I_{\max} \sin \omega t \, d\omega t \\
&= \frac{I_{\max}}{2\pi} [-\cos \omega t]_{\theta}^{\pi} \\
&= \frac{I_{\max}}{2\pi} (1 + \cos \theta) \\
&= 0.318 \, I_{\max}
\end{aligned}
$$

or

$$I_{\text{ave}}/I_{\max} = 0.318$$

while if $\theta = 90°$, $I_{\text{ave}}/I_{\max} = 0.159$.

Figure 5.4 is a Fortran program that calculates the normalized value of the current for 10° increment changes in θ. First the program heading

```
       C      SILICCN CONTROLLED RECTIFIER
       C
       C      THE CALCULATION OF THE AVERAGE CURRENT
       C                USING
       C      EULER'S INTEGRATION FORMULA
       C
S.0001        WRITE (6,1)
S.0002      1 FORMAT('1THE CALCULATION OF THE AVERAGE CURRENT',//,' PRODUCED BY
             110 DEGREE INCREMENT FIRING',//,' OF A SILICON CONTROLLED RECTIFIER
             2',///,5X,'THETA',5X,'AVE I',3X,'EXACT AVE I',3X,'ERROR',//)
S.0003        H=5.0
S.0004        THETA=-10.0
S.0005        DO 10 I=1,19
S.0006        THETA=THETA+10.0
S.0007        K=THETA/5.+1.
S.0008        ARG=THETA
S.0009        SUM=0.0
S.0010        DO 20 J=K,36
S.0011        ARG=ARG+H
S.0012     20 SUM=SUM+SIN(ARG/57.29578)
S.0013        AVEI=H*SUM/360.
S.0014        EAVEI=(1.+COS(THETA/57.29578))/(2.*3.141592)
S.0015        ERROR=EAVEI-AVEI
S.0016     10 WRITE(6,2) THETA,AVEI,EAVEI,ERROR
S.0017      2 FORMAT(F10.2,2F10.4,E14.3)
S.0018        RETURN
S.0019        END
```

Figure 5.4 The Euler method of calculating the average current using a step size of $H = 5°$

and subheadings are printed. Computer statement S.0003 then declares h equal to 5° and computer statement S.0004 sets THETA equal to $-10°$. Entering the DO 10 loop, computer statement S.0006 sets THETA equal to 0° for the first iteration. The dummy variable K is then introduced to determine the initial value of the iteration variable J, which is used in the DO 20 loop. ARG is then made equal to THETA or 0°, and the variable SUM is set equal to zero. The DO 20 loop now calculates the area under the curve for this value of θ by summing the area of the 36 rectangles that approximate the actual area. Computer statements S.0013 and S.0014 compute the Euler average and the exact average, respectively. Computer statements S.0016 and S.0017 print out the value of θ,

THE CALCULATION OF THE AVERAGE CURRENT

PRODUCED BY 10 DEGREE INCREMENT FIRING

'F A SILICON CCNTRCLLED RECTIFIER

THETA	AVE I	EXACT AVE I	ERROR
0.0	0.3181	0.3183	0.204E-03
10.00	0.3145	0.3159	0.141E-02
20.00	0.3061	0.3087	0.257E-02
30.00	0.2933	0.2970	0.366E-02
40.00	0.2764	0.2811	0.464E-02
50.00	0.2560	0.2615	0.549E-02
60.00	0.2326	0.2387	0.617E-02
70.00	0.2069	0.2136	0.666E-02
80.00	0.1798	0.1868	0.696E-02
90.00	0.1521	0.1592	0.705E-02
100.00	0.1246	0.1315	0.692E-02
110.00	0.0981	0.1047	0.659E-02
120.00	0.0735	0.0796	0.606E-02
130.00	0.0515	0.0569	0.536E-02
140.00	0.0327	0.0372	0.449E-02
150.00	0.0178	0.0213	0.349E-02
160.00	0.0072	0.0096	0.238E-02
170.00	0.0012	0.0024	0.121E-02
180.00	-0.0012	0.0	0.121E-02

Figure 5.5 The output results of the program of Figure 5.4

```
C       SILICCN CONTROLLED RECTIFIER
C
C       THE CALCULATION OF THE AVERAGE CURRENT
C                       USING
C       EULER'S INTEGRATION FORMULA
C
        WRITE (6,1)
      1 FORMAT('1THE CALCULATION OF THE AVERAGE CURRENT',//,' PRODUCED BY
     110 DEGREE INCREMENT FIRING',//,' OF A SILICON CONTROLLED RECTIFIER
     2',///,5X,'THETA',5X,'AVE I',3X,'EXACT AVE I',3X,'ERROR',//)
        H=1.0
        THETA=-10.0
        DO 10 I=1,19
        THETA=THETA+10.0
        K=THETA/1.+1.
        ARG=THETA
        SUM=0.0
        DO 20 J=K 180
        ARG=ARG+H
     20 SUM=SUM+SIN(ARG/57.19578)
        AVEI=H*SUM/360.
        EAVEI=(1.+COS(THETA/57.29578))/(2.*3.141592)
        ERROR=EAVEI-AVEI
     10 WRITE(6,2) THETA,AVEI,EAVEI,ERROR
      2 FORMAT(F10.2,2F12.4,E12.3)
        RETURN
        END
```

Figure 5.6 The Euler method of calculating the average current using a step size of $H = 1°$

the Euler average current, the exact average current, and the error of the Euler average value. The DO 10 loop is now executed for a second time, but now with a value of THETA = 10°. This process is repeated, each time with THETA incremented by 10°, until THETA = 180°.

Figure 5.5 shows the output results of this integration. Note that the error is maximum for THETA = 90° and minimum for THETA = 0°. This can be explained by referring to Figure 5.1 and noting that when the slope is negative, the area calculated is less than the true area, but when the slope is positive, the calculated area is larger. Thus for $\theta = 0°$, we find that due to the symmetry of the sine function, positive and negative slopes occur for equal lengths of time and thus maximum canceling of the two errors takes place and a minimum error results. In a similar

THE CALCULATION OF THE AVERAGE CURRENT

PRODUCED BY 10 DEGREE INCREMENT FIRING

OF A SILICON CONTROLLED RECTIFIER

THETA	AVE I	EXACT AVE I	ERROR
0.0	0.3177	0.3183	0.577E-03
10.00	0.3151	0.3159	0.822E-03
20.00	0.3076	0.3087	0.107E-02
30.00	0.2957	0.2970	0.131E-02
40.00	0.2795	0.2811	0.153E-02
50.00	0.2597	0.2615	0.173E-02
60.00	0.2368	0.2387	0.189E-02
70.00	0.2116	0.2136	0.202E-02
80.00	0.1847	0.1868	0.209E-02
90.00	0.1570	0.1592	0.212E-02
100.00	0.1294	0.1315	0.209E-02
110.00	0.1027	0.1047	0.200E-02
120.00	0.0777	0.0796	0.186E-02
130.00	0.0552	0.0569	0.165E-02
140.00	0.0358	0.0372	0.140E-02
150.00	0.0202	0.0213	0.110E-02
160.00	0.0088	0.0096	0.759E-03
170.00	0.0020	0.0024	0.389E-03
180.00	0.0001	0.0	0.638E-04

Figure 5.7 The output results of the program of Figure 5.6

```
C        SILICON CONTROLLED RECTIFIER
C
C        THE CALCULATION OF THE AVERAGE CURRENT
C                          USING
C        EULER'S INTEGRATION FORMULA
C
         WRITE(6,1)
       1 FORMAT('1THE CALCULATION OF THE AVERAGE CURRENT',//,' PRODUCED BY
        110 DEGREE INCREMENT FIRING',//,' OF A SILICON CONTROLLED RECTIFIER
        2',///,5X,'THETA',5X,'AVE I',3X,'EXACT AVE I',3X,'ERROR',//)
         H=0.1
         THETA=-10.0
         DO 10 I=1,19
         THETA=THETA+10.0
         K=THETA/.1+1.
         ARG=THETA
         SUM=0.0
         DO 20 J=K,1800
         ARG=ARG+H
      20 SUM=SUM+SIN(ARG/57.19578)
         AVEI=H*SUM/360.
         EAVEI=(1.+COS(THETA/57.29578))/(2.*3.141592)
         ERROR=EAVEI-AVEI
      10 WRITE(6,2)THETA,AVEI,EAVEI,ERROR
       2 FORMAT(F10.2,2F12.4,E12.3)
         RETURN
         END
```

Figure 5.8 The Euler method of calculating the average current using a step size of $H = 0.1°$

way, we can see that from THETA = 90° to THETA = 180° the slope is always negative, and the areas calculated are all less than the true areas and since the curvilinear triangles decrease in size as THETA approaches 180°, the maximum error occurs for THETA = 90°.

Since the accuracy of the above results is good to about two significant figures, we will attempt to reduce the error by rerunning the problem with a smaller value of h. Figure 5.6 shows the previous Fortran program modified for a step size of $h = 1°$ and Figure 5.7 shows the corresponding output. Note that for this value of h, the results are accurate to almost three significant figures. In an attempt to improve the accuracy further, a step size of 0.1° was used in the program of Figure 5.8. The results shown in Figure 5.9 reveal that the accuracy has not materially improved,

THE CALCULATION OF THE AVERAGE CURRENT

PRODUCED BY 10 DEGREE INCREMENT FIRING

OF A SILICON CONTROLLED RECTIFIER

THETA	AVE I	EXACT AVE I	ERROR
0.0	0.3177	0.3183	0.573E-03
10.00	0.3153	0.3159	0.601E-03
20.00	0.3081	0.3087	0.636E-03
30.00	0.2963	0.2970	0.677E-03
40.00	0.2804	0.2811	0.721E-03
50.00	0.2607	0.2615	0.763E-03
60.00	0.2379	0.2387	0.804E-03
70.00	0.2128	0.2136	0.836E-03
80.00	0.1859	0.1868	0.857E-03
90.00	0.1583	0.1592	0.863E-03
100.00	0.1307	0.1315	0.853E-03
110.00	0.1039	0.1047	0.821E-03
120.00	0.0788	0.0796	0.765E-03
130.00	0.0562	0.0569	0.686E-03
140.00	0.0366	0.0372	0.590E-03
150.00	0.0209	0.0213	0.470E-03
160.00	0.0093	0.0096	0.330E-03
170.00	0.0022	0.0024	0.172E-03
180.00	-0.0000	0.0	0.201E-05

Figure 5.9 The output results of the program of Figure 5.8

but the time for computing the integral has increased by 50 times. Round-off error is the main reason for the small reduction in the error. If the step size is reduced further, the reader will find that the error will eventually increase.

5.3 THE TRAPEZOIDAL RULE

The Trapezoidal Rule is the simplest scheme of a more general method of integration that is based on using a polynomial approximation of the actual function. The particular method described here uses a linear polynomial $F(t) = a + bt$ to approximate the actual function $f(t)$, as shown in Figure 5.10. Using the calculus, the crosshatched area A_{12} can be found from

$$A_{12} = \int_{t=t_1}^{t_1+h} (a + bt)\, dt = [at + bt^2/2]_{t_1}^{t_1+h}$$
$$= ah + b(h^2 + 2ht_1)/2$$

To determine the value of a and b, we note that at

$$t = t_1, \qquad F(t_1) = f(1) = a + bt_1$$

and at

$$t = t_2, \qquad F(t_2) = f(2) = a + bt_2 = a + b(t_1 + h)$$

Since $f(1)$, $f(2)$, and h are known, we find by solving the above equations that

$$a = f(1) - \frac{f(2) - f(1)}{h}\, t_1$$

and

$$b = \frac{f(2) - f(1)}{h}$$

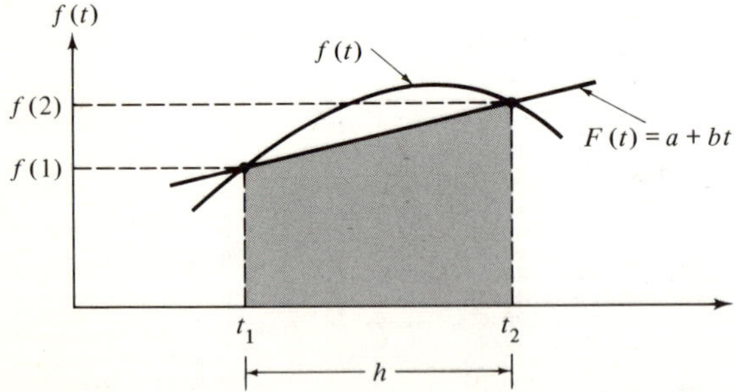

Figure 5.10 The linear polynomial approximation

When these coefficients are substituted in A_{12} above, we find

$$A_{12} = \frac{h}{2}(f(1) + f(2))$$

Similarly the next area A_{23} would be found from

$$A_{23} = \frac{h}{2}(f(2) + f(3))$$

and the total area between t_1 and t_n would be approximated by

$$\int_{t_1}^{t_n} f(t)dt \approx \frac{h}{2}(f(1) + 2f(2) + 2f(3) + \cdots + 2f(n-1) + f(n))$$

Silicon Controlled Half-Wave Rectifier

In order to see how much more effective the trapezoidal method is over the Rectangle Rule, let us apply this technique in the solution of the silicon controlled rectifier problem that we discussed in the last section. Figure 5.11 shows the corresponding Fortran program which is identical to the program of Figure 5.4 except for computer statements S.0012 and S.0013 where the trapezoidal rule is used instead of the rectangle rule. Figure 5.12 shows that the results are accurate to almost four significant figures.

Figure 5.13 is the same program as Figure 5.11, except that the step size has been changed to $h = 1°$. Figure 5.14 shows that the corresponding

```
               C       SILICON CONTROLLED RECTIFIER
               C
               C       THE CALCULATION OF THE AVERAGE CURRENT
               C                   USING
               C         THE TRAPEZOIDAL RULE
               C
S.0001                 WRITE(6,1)
S.0002               1 FORMAT('1THE CALCULATION OF THE AVERAGE CURRENT',//,' PRODUCED BY
                      110 DEGREE INCREMENT FIRING',//,' OF A SILICON CONTROLLED RECTIFIER
                      2',///,5X,'THETA',5X,'AVE I',3X,'EXACT AVE I',3X,'ERROR',//)
S.0003                 H=5.0
S.0004                 THETA=-10.0
S.0005                 DO 10 I=1,19
S.0006                 THETA=THETA+10.0
S.0007                 K=THETA/5.+1.
S.0008                 ARG=THETA
S.0009                 SUM=0.0
S.0010                 DO 20 J=K,36
S.0011                 ARG=ARG+H
S.0012              20 SUM=SUM+SIN((ARG-H)/57.29578)+SIN(ARG/57.29578)
S.0013                 AVEI=5.*SUM/(2.*360.)
S.0014                 EAVEI=(1.+COS(THETA/57.29578))/(2.*3.141592)
S.0015                 ERROR=EAVEI-AVEI
S.0016              10 WRITE(6,2)THETA,AVEI,EAVEI,ERROR
S.0017               2 FORMAT(F10.2,2F12.4,E12.3)
S.0018                 RETURN
S.0019                 END
```

Figure 5.11 The trapezoidal rule used to calculate the average current with a step size of $H = 5°$

```
THE CALCULATION OF THE AVERAGE CURRENT

PRODUCED BY 10 DEGREE INCREMENT FIRING

OF A SILICON CONTROLLED RECTIFIER

   THETA       AVE I     EXACT AVE I     ERROR

    0.0       0.3181        0.3183    0.205E-03
   10.00      0.3157        0.3159    0.203E-03
   20.00      0.3085        0.3087    0.199E-03
   30.00      0.2968        0.2970    0.191E-03
   40.00      0.2809        0.2811    0.181E-03
   50.00      0.2613        0.2615    0.168E-03
   60.00      0.2386        0.2387    0.154E-03
   70.00      0.2135        0.2136    0.138E-03
   80.00      0.1867        0.1868    0.120E-03
   90.00      0.1591        0.1592    0.102E-03
  100.00      0.1314        0.1315    0.844E-04
  110.00      0.1047        0.1047    0.666E-04
  120.00      0.0795        0.0796    0.507E-04
  130.00      0.0568        0.0569    0.362E-04
  140.00      0.0372        0.0372    0.237E-04
  150.00      0.0213        0.0213    0.136E-04
  160.00      0.0096        0.0096    0.612E-05
  170.00      0.0024        0.0024    0.153E-05
  180.00     -0.0006        0.0       0.605E-03
```

Figure 5.12 The output results of the program of Figure 5.11

output results are accurate to almost five significant figures. When the accuracy of these two results are compared respectively to those of Figures 5.4 and 5.6, we find that for the same step size the trapezoidal method requires additional computer time, but its accuracy is far superior. Putting it another way, if the same amount of computer time is to be used by each method in the calculation of a given area (this would mean different h sizes to make the computation times the same), the trapezoidal method will give more accurate results. It thus follows that there is no advantage in using Euler's method over the trapezoidal method. The reader should also be aware that, as in the case of Euler's

```
C       SILICON CONTROLLED RECTIFIER
C
C       THE CALCULATION OF THE AVERAGE CURRENT
C                USING
C          THE TRAPEZOIDAL RULE
C
        WRITE (6,1)
      1 FORMAT('1THE CALCULATION OF THE AVERAGE CURRENT',//,' PRODUCED BY
     110 DEGREE INCREMENT FIRING',//,' OF A SILICON CONTROLLED RECTIFIER
     2',///,5X,'THETA',5X,'AVE I',3X,'EXACT AVE I',3X,'ERROR',//)
        H=1.0
        THETA=-10.0
        DO 10 I=1,19
        THETA=THETA+10.0
        K=THETA/1.+1.
        ARG=THETA
        SUM=0.0
        DO 20 J=K,180
        ARG=ARG+H
     20 SUM=SUM+SIN((ARG-H)/57.29578)+SIN(ARG/57.29578)
        AVEI=H*SUM/(2.*360)
        EAVEI=(1.+COS(THETA/57.29578))/(2.*3.141592)
        ERROR=EAVEI-AVEI
     10 WRITE(6,2) THETA,AVEI,EAVEI,ERROR
      2 FORMAT(F10.2,2F12.4,E12.3)
        RETURN
        END
```

Figure 5.13 The trapezoidal rule used to calculate the average current with a step size of $H = 1°$

THE CALCULATION OF THE AVERAGE CURRENT

PRODUCED BY 10 DEGREE INCREMENT FIRING

OF A SILICON CONTROLLED RECTIFIER

THETA	AVE I	EXACT AVE I	ERROR
0.0	0.3183	0.3183	0.113E-04
10.00	0.3159	0.3159	0.111E-04
20.00	0.3087	0.3087	0.108E-04
30.00	0.2970	0.2970	0.104E-04
40.00	0.2811	0.2811	0.978E-05
50.00	0.2614	0.2615	0.906E-05
60.00	0.2387	0.2387	0.840E-05
70.00	0.2136	0.2136	0.751E-05
80.00	0.1868	0.1868	0.662E-05
90.00	0.1591	0.1592	0.584E-05
100.00	0.1315	0.1315	0.495E-05
110.00	0.1047	0.1047	0.405E-05
120.00	0.0796	0.0796	0.322E-05
130.00	0.0568	0.0569	0.238E-05
140.00	0.0372	0.0372	0.164E-05
150.00	0.0213	0.0213	0.607E-06
160.00	0.0096	0.0096	0.275E-06
170.00	0.0024	0.0024	0.664E-07
180.00	-0.0000	0.0	0.242E-04

Figure 5.14 The output results of the program of Figure 5.13

method, because of roundoff error the accuracy can not be increased without limits by further reducing the step size.

The Calculation of the RMS Value

For another example that requires the use of integration, consider the nonlinear RL circuit shown in Figure 5.15. In order to find the power supplied to this circuit, it is necessary to find the RMS or effective value of the current waveform shown. This follows from the fact that the instantaneous power dissipated by the resistor is given by

$$p = vi = i^2R$$

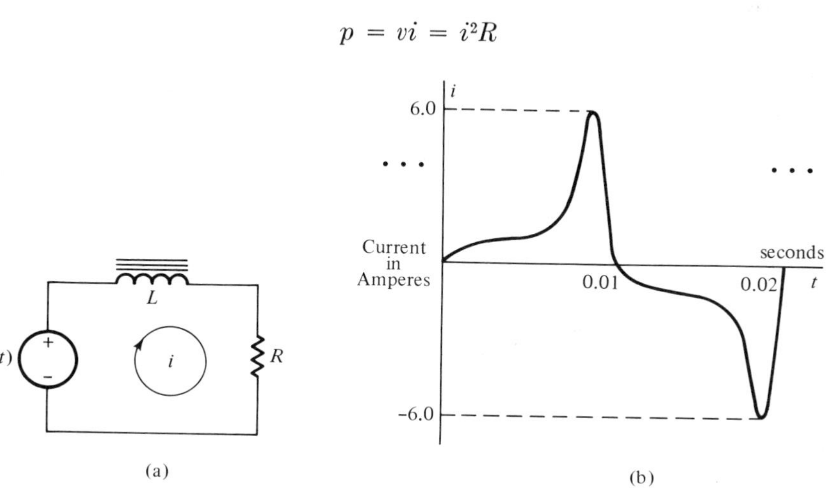

Figure 5.15 (a) A nonlinear RL circuit; (b) one-cycle plot of the current waveform

where i is the periodic time varying current flowing in the resistor and iR is the time varying voltage across the resistor. The average power dissipated by the resistor is therefore

$$P_{\text{ave}} = \frac{1}{T} \int_0^T i^2 R \, dt$$

where T is the period of the waveform. Since the RMS or effective value of the current is defined as that magnitude of constant current that produces the same amount of heating in the resistor, it thus follows that

$$I_{\text{RMS}}^2 R = \frac{1}{T} \int_0^T i^2 R \, dt$$

or

$$I_{\text{RMS}} = \sqrt{\frac{1}{T} \int_0^T i^2 \, dt}$$

Figure 5.16 is a Fortran program for computing the RMS value of the current waveform shown in Figure 5.15b. Computer statements S.0002 and S.0003 print the heading and subheading. Computer statements S.0004 and S.0005 read in the values of the current waveform. Since 41 points are read in and the period T of the wave equals 0.02 seconds, the value of h equals $T/(N - 1) = 0.0005$ seconds. The 41 data values are now printed by computer statements S.0006 and S.0007 and the dummy variable SUM is set equal to zero. The DO 10 loop is then executed 40 times. Note that the current values are squared and when the final SUM value is multiplied by h (0.0005) and divided by 2, the result is equivalent to the trapezoidal method of finding the area under the squared current waveform. Computer statement S.0011 is used to find the root of the mean squared value or the RMS value of current, and computer statements S.0012 and S.0013 print the result.

```
               C      THE CALCULATION OF THE RMS VALUE OF CURRENT
               C                      USING
               C         THE TRAPEZOIDAL RULE
               C
S.0001                DIMENSION C(41)
S.0002                WRITE(6,1)
S.0003              1 FORMAT('1THE CALCULATION OF THE RMS VALUE OF CURRENT',///,' THE IN
                     1PUT DATA    I(1), I(2),..., I(41)',//)
S.0004                READ(5,2) (C(I),I=1,41)
S.0005              2 FORMAT(10F5.2)
S.0006                WRITE(6,4) (C(I),I=1,41)
S.0007              4 FORMAT(10F7.2)
S.0008                SUM=0.0
S.0009                DO 10 I=2,41
S.0010             10 SUM=SUM+C(I-1)**2+C(I)**2
S.0011                CRMS=SQRT(50.*0.0005*SUM/2.)
S.0012                WRITE(6,3)CRMS
S.0013              3 FORMAT(///,' I RMS =',F6.3)
S.0014                RETURN
S.0015                END
```

Figure 5.16 A Fortran program for calculating the RMS value of current using the trapezoidal rule

```
THE CALCULATION OF THE RMS VALUE OF CURRENT

THE INPUT DATA    I(1), I(2),...., I(41)

  0.0     0.50    0.70    0.80    0.92    0.98    1.00    1.00    1.00    1.00
  1.01    1.07    1.20    1.50    2.02    3.20    4.80    6.00    4.90    2.70
  0.0    -0.50   -0.70   -0.80   -0.92   -0.98   -1.00   -1.00   -1.00   -1.00
 -1.01   -1.07   -1.20   -1.50   -2.02   -3.20   -4.80   -6.00   -4.90   -2.70
  0.0


I RMS = 2.426
```

Figure 5.17 The output results of the program of Figure 5.16

Figure 5.17 shows the computed result. Note that the RMS value of current is given as 2.426 amperes. At this point we have no way of telling how accurate this result is. We do know, however, that because the shape of the squared curve is concave downward, the trapezoidal method will yield a result that is less than the true value. To insure accurate results we can rerun the problem, each time halving the value of h, until round-off error exceeds the truncation error or the desired accuracy is obtained.

5.4 SIMPSON'S RULE

Simpson's Rule uses a second-degree polynomial

$$F(t) = a + bt + ct^2$$

to approximate the actual function $f(t)$ as shown in Figure 5.18. The crosshatched area shown, A_{123}, may be approximated by evaluating the following integral

$$A_{123} \approx \int_{t_1}^{t_3} (a + bt + ct^2)\, dt$$

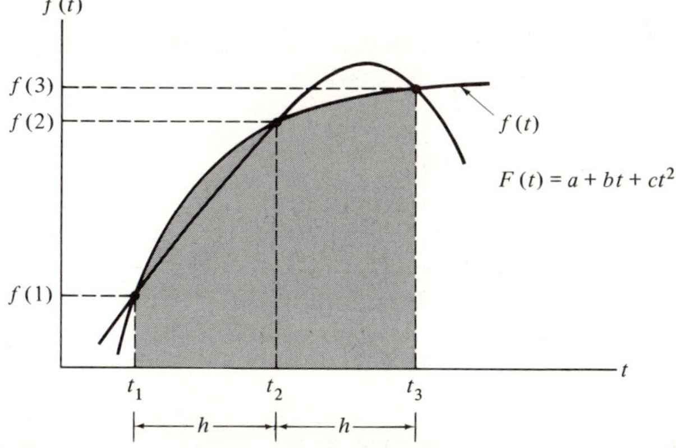

Figure 5.18 The parabolic approximation

To simplify the integration, we first make a change in the variable letting

$$s = t - t_1$$

or

$$t = s + t_1$$

so that

$$dt = ds$$

thus

$$A_{123} \approx \int_0^{2h} F(s)\, ds$$

where

$$F(s) = A + Bs + Cs^2$$

with

$$\begin{aligned} A &= a + bt_1 + ct_1^2 \\ B &= b + 2t_1c \end{aligned}$$

and

$$C = c$$

Evaluating the integral we get

$$\begin{aligned} A_{123} &\approx [As + Bs^2/2 + Cs^3/3]_0^{2h} \\ &= 2Ah + 2Bh^2 + \frac{8}{3}Ch^3 \end{aligned}$$

To determine the value of A, B, and C we note that at

$$\begin{aligned} t = t_1 \text{ or } s &= 0, & F(0) &= f(1) = A + 0 + 0 \\ t = t_2 \text{ or } s &= h, & F(h) &= f(2) = A + Bh + Ch^2 \\ t = t_3 \text{ or } s &= 2h, & F(2h) &= f(3) = A + 2Bh + 4Ch^2 \end{aligned}$$

Since $f(1)$, $f(2)$, $f(3)$, and h are known, we find by solving the above equations that

$$\begin{aligned} A &= f(1) \\ B &= \frac{-3f(1) + 4f(2) - f(3)}{2h} \\ C &= \frac{f(1) - 2f(2) + f(3)}{2h^2} \end{aligned}$$

When these coefficients are substituted in A_{123} above, we find

$$A_{123} \approx \frac{h}{3}(f(1) + 4f(2) + f(3))$$

Similarly the next area A_{345} would be found from

$$A_{345} \approx \frac{h}{3} (f(3) + 4f(4) + f(5))$$

and if n is an odd number, the total area between t_1 and t_n would be approximated by

$$\int_{t_1}^{t_n} f(t)\, dt \approx \frac{h}{3} (f(1) + 4f(2) + 2f(3) + \cdots + 4f(n-1) + f(n))$$

The Calculation of the RMS Value

In order to see how much more effective this method is over the trapezoidal rule, let us recalculate the RMS value of the current waveform shown in Figure 5.15b. Figure 5.19 shows the corresponding Fortran program which is identical to the program of Figure 5.16, except for computer statements S.0009 through S.0011 where Simpson's rule is used instead of the trapezoidal rule. Figure 5.20 shows that the RMS value of current equals 2.439 amperes. The reader should verify that this is a more accurate result than the value found by the trapezoidal rule.

```
        C      THE CALCULATION OF THE RMS VALUE OF CURRENT
        C                     USING
        C                 SIMPSON'S RULE
        C
S.0001         DIMENSION C(41)
S.0002         WRITE(6,1)
S.0003       1 FORMAT('1THE CALCULATION OF THE RMS VALUE OF CURRENT',///,' THE IN
              1PUT DATA     I(1), I(2),..., I(41)',//)
S.0004         READ(5,2) (C(I),I=1,41)
S.0005       2 FORMAT(10F5.2)
S.0006         WRITE(6,4) (C(I),I=1,41)
S.0007       4 FORMAT(10F6.2)
S.0008         SUM=0.0
S.0009         DO 10 I=1,39,2
S.0010      10 SUM=SUM+C(I)**2+4.*C(I+1)**2+C(I+2)**2
S.0011         CRMS=SQRT(50.*0.0005*SUM/3.)
S.0012         WRITE(6,3)CRMS
S.0013       3 FORMAT(///,' I RMS =',F6.3)
S.0014         RETURN
S.0015         END
```

Figure 5.19 A Fortran program for calculating the RMS value of current using Simpson's rule

```
THE CALCULATION OF THE RMS VALUE OF CURRENT


THE INPUT DATA    I(1), I(2),..., I(41)


 0.0   0.50  0.70  0.80  0.92  0.98  1.00  1.00  1.00  1.00
 1.01  1.07  1.20  1.50  2.02  3.20  4.80  6.00  4.90  2.70
 0.0  -0.50 -0.70 -0.80 -0.92 -0.98 -1.00 -1.00 -1.00 -1.00
-1.01 -1.07 -1.20 -1.50 -2.02 -3.20 -4.80 -6.00 -4.90 -2.70
 0.0



I RMS = 2.439
```

Figure 5.20 The output results of the program of Figure 5.19

The Calculation of the Fourier Coefficients

The Fourier series representation of a periodic function is one of the more important topics studied by electrical engineers. This series permits functions that satisfy certain conditions to be represented in terms of a d-c term and an infinite series of harmonically related sinusoids. Stated mathematically, we have

$$\begin{aligned} f(t) &= C_0 + 2C_1 \cos (\omega_1 t + \phi_1) + 2C_2 \cos (2\omega_1 t + \phi_2) + \cdots \\ &\qquad + 2C_n \cos (n\omega_1 t + \phi_n) + \cdots \\ &= C_0 + \sum_{n=1}^{\infty} 2C_n \cos (n\omega_1 t + \phi_n) \end{aligned}$$

where

$$\begin{aligned} \omega_1 &= 2\pi f_1 \\ &= 2\pi/T \end{aligned}$$

and f_1 is called the fundamental frequency of the waveform and T is the period.

The above form of the Fourier series is often referred to as the magnitude and phase form. It is also possible to write the series in its sine-cosine form:[1]

$$f(t) = A_0 + \sum_{n=1}^{\infty} 2(A_n \cos n\omega_1 t - B_n \sin n\omega_1 t)$$

where

$$\begin{aligned} A_n &= C_n \cos \phi_n \\ B_n &= C_n \sin \phi_n \\ C_n &= \sqrt{A_n{}^2 + B_n{}^2} \\ \phi_n &= \tan^{-1} (B_n/A_n) \end{aligned}$$

and the coefficients A_n and B_n are evaluated from

$$\begin{aligned} A_n &= \frac{1}{T} \int_0^T f(t) \cos n\omega_1 t \, dt; \; n = 0, 1, 2, \ldots \\ B_n &= -\frac{1}{T} \int_0^T f(t) \sin n\omega_1 t \, dt; \; n = 1, 2, 3, \ldots \end{aligned}$$

Figure 5.21 is a general program for finding the Fourier coefficients. Computer statements S.0002 and S.0003 print out the output heading. Computer statements S.0004 and S.0005 read in NUMB, the number of equally spaced data points; T, the period of the waveform; and NCOEF,

[1] See B. James Ley, Samuel G. Lutz, and Charles F. Rehberg, *Linear Circuit Analysis* (New York: McGraw-Hill Book Company, 1959), Ch. 6.

```
          C      THE COEFFICIENTS OF A FOURIER SERIES THAT REPRESENT AN ARBITRARY PERIODIC
          C      WAVEFORM ARE DETERMINED BY THIS PROGRAM. THE EQUATION OF THE WAVEFORM DOES
          C      NOT HAVE TO BE KNOWN.
          C      THE REQUIRED INPUT DATA MUST SPECIFY THE VALUES OF THE WAVEFORM AT NUMB
          C      EQUALLY SPACED POINTS, STARTING AT TIME = 0 AND ENDING AT TIME = T, WHERE
          C      T IS THE PERIOD OF THE WAVEFORM. THE NUMBER OF POINTS NUMB, THE NUMBER OF
          C      COEFFICIENTS DESIRED NCOEF, AND THE PERIOD T MUST ALSO BE FED IN AS DATA.
          C      NUMB MUST BE AN ODD NUMBER SINCE SIMPSON'S RULE IS USED TO EVALUATE THE
          C      COEFFICIENTS.
          C
S.0001           DIMENSION F(100)
S.0002           WRITE(6,1)
S.0003         1 FORMAT('1THE FOURIER COEFFICIENTS',///)
S.0004           READ(5,2)NUMB,T,NCOEF
S.0005         2 FORMAT(I5,E15.7,I5)
S.0006           READ(5,3)(F(I),I=1,NUMB)
S.0007         3 FORMAT(8F10.4)
S.0008           NINT=NUMB-1
S.0009           ARG=2.*3.141592/T
S.0010           TINC=T/NINT
S.0011           A0=0.0
S.0012           DO 10 I=1,NINT,2
S.0013        10 A0=A0+F(I)+4.*F(I+1)+F(I+2)
S.0014           A0=A0*TINC/(3.*T)
S.0015           WRITE(6,5)A0
S.0016         5 FORMAT('0A( 0) =',E10.3)
S.0017           DO 20 N=1,NCOEF
S.0018           A=0.0
S.0019           B=0.0
S.0020           TIME=-TINC
S.0021           DO 30 J=1,NINT,2
S.0022           TIME=TIME+2.*TINC
S.0023           A=A+F(J)*COS(N*ARG*(TIME-TINC))+4.*F(J+1)*COS(N*ARG*TIME)
                1+F(J+2)*COS(N*ARG*(TIME+TINC))
S.0024        30 B=B+F(J)*SIN(N*ARG*(TIME-TINC))+4.*F(J+1)*SIN(N*ARG*TIME)+
                1F(J+2)*SIN(N*ARG*(TIME+TINC))
S.0025           A=A*TINC/(3.*T)
S.0026           B=-B*TINC/(3.*T)
S.0027           C=SQRT(A*A+B*B)
S.0028           IF(A)21,22,23
S.0029        22 IF(B)24,25,25
S.0030        24 THETA =-90.
S.0031           GO TO 28
S.0032        25 THETA =+90.
S.0033           GO TO 28
S.0034        23 THETA =(ATAN(B/A))*180./3.
S.0035           GO TO 28
S.0036        21 IF(B)26,27,27
S.0037        26 THETA =-180.+(ATAN(B/A))*180./3.1415927
S.0038           GO TO 28
S.0039        27 THETA =180.+(ATAN(B/A))*180./3.1415927
S.0040        28 CONTINUE
S.0041        50 WRITE(6,6)N,A,N,B,N,C,N,THETA
S.0042         6 FORMAT('0A('I2,') =',E10.3,3X,'B(',I2,') =',E10.3, 3X,'C(', I2,') ='
                1,E10.3,3X,'THETA(',I2,') =',F7.2).
S.0043        20 CONTINUE
S.0044           RETURN
S.0045           END
```

Figure 5.21 A general program for finding the Fourier coefficients of a periodic waveform

the number of coefficients required. Computer statements S.0006 and S.0007 then read in NUMB equally spaced data points. The number of intervals, NINT, is then computed as one less than the number of data points and the argument, ARG, is set equal to 2π divided by the period T. The time of one increment, TINC, is then determined and $A0$ is set equal to zero. The DO 10 loop and computer statement S.0014 now calculates the dc coefficient $A0$, and computer statements S.0015 and S.0016 print out the numerical result. The DO 20 loop is then executed NCOEF times and during each iteration the corresponding cosine-sine coefficients, A_n and B_n, and the magnitude coefficient C_n, and its phase angle ϕ_n are computed and printed.

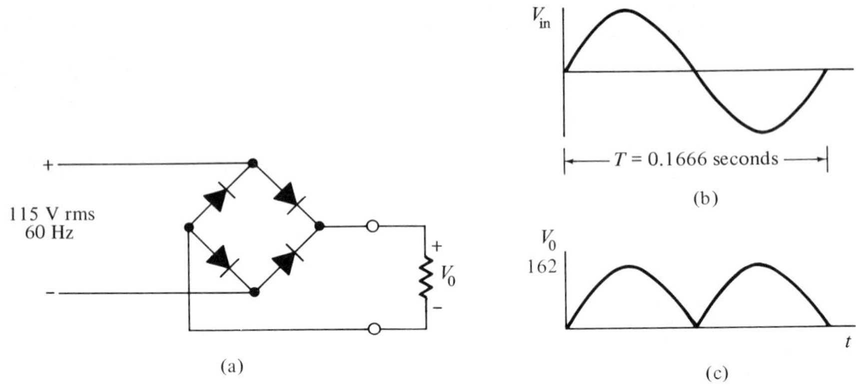

Figure 5.22 (a) A full-wave bridge rectifier; (b) the input waveform; (c) the output waveform

For a simple test of this program, consider the full-wave rectifier circuit shown in Figure 5.22. The input voltage is sinusoidal and has a frequency of 60 cycles per second and a magnitude of 115 volts RMS. The output voltage is as shown with a peak swing of 162 volts. In order to calculate the Fourier coefficients of this waveform, the wave was sampled every 9° and the numerical values of these 41 points were read in as data.

Figure 5.23 shows the output results. Note that since this function is an even function, only A_n coefficients should be present and thus the phase angles should be either 0° or 180°. In this respect, the computer results indicate a small error and an increase of data points would obviously lead to better results. Note also that in this example it is possible to calculate all the coefficients analytically. It may be shown that the A

```
THE FOURIER COEFFICIENTS


A( 0) = 0.103E 03
A( 1) =-0.174E-03   B( 1) = 0.195E-03   C( 1) = 0.261E-03   THETA( 1) = 131.79
A( 2) =-0.344E 02   B( 2) =-0.280E-03   C( 2) = 0.344E 02   THETA( 2) =-180.00
A( 3) =-0.610E-04   B( 3) =-0.296E-04   C( 3) = 0.673E-04   THETA( 3) =-154.11
A( 4) =-0.685E 01   B( 4) =-0.103E-03   C( 4) = 0.685E 01   THETA( 4) =-180.00
A( 5) = 0.261E-04   B( 5) =-0.356E-05   C( 5) = 0.263E-04   THETA( 5) =  -8.14
A( 6) =-0.287E 01   B( 6) = 0.271E-04   C( 6) = 0.287E 01   THETA( 6) = 180.00
A( 7) =-0.346E-04   B( 7) = 0.331E-05   C( 7) = 0.347E-04   THETA( 7) = 174.54
A( 8) =-0.156E 01   B( 8) = 0.220E-04   C( 8) = 0.156E 01   THETA( 8) = 180.00
A( 9) = 0.710E-04   B( 9) = 0.107E-03   C( 9) = 0.125E-03   THETA( 9) =  59.08
A(10) =-0.867E 00   B(10) =-0.122E-04   C(10) = 0.857E 00   THETA(10) =-180.00
```

Figure 5.23 The Fourier coefficients of the full-wave rectifier of Figure 5.22

coefficients have the following values:

$$A_0 = \frac{2}{\pi} V_{\max}$$

and

$$A_n = \frac{2}{\pi} V_{\max} \left(\frac{1}{n^2 - 1} \right), \quad n = 2, 4, 6, \ldots$$

The verification of these results is left as an exercise for the reader.

The Frequency Spectrum of a Snap Diode Circuit

One type of a frequency spectrum plot is based on the trigonometric form of the Fourier series discussed in the last problem.

$$f(t) = C_0 + \sum_{n=1}^{\infty} 2C_n \cos (n\omega_1 t + \phi_n)$$

The spectrum plot consists of two graphs, one for the magnitude C_n versus frequency and one for the phase ϕ_n versus frequency. Each graph is a histogram in which the numerical value of the magnitude and phase is indicated for each harmonic.

Figure 5.24 shows a shunt-series snap diode circuit that generates a pulse that is less than one nanosecond in duration.[2] Such a circuit is useful in producing a frequency spectrum that is rich in high-order har-

[2] See John Giorgis, "Understanding Snap Diodes," EEE (Electronic Equipment Engineering), Vol. 11, (November 1963) pp. 60–64; or *Transistor Manual*, Seventh Edition, The General Electric Company, (1964), pp. 451–455.

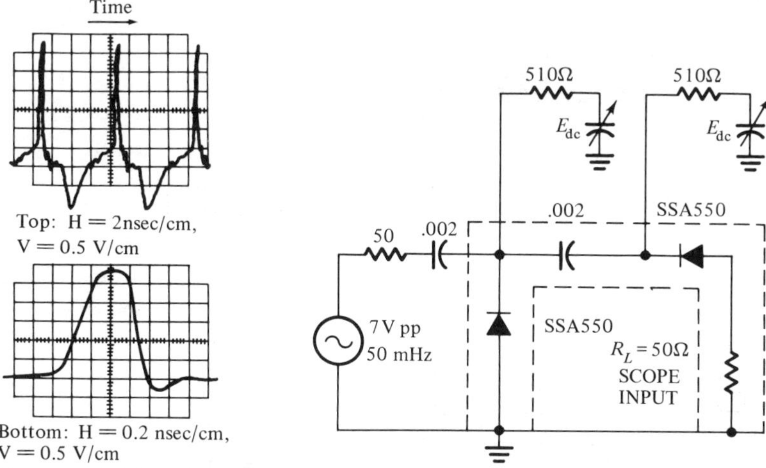

Figure 5.24 The shunt-series snap diode circuit (Courtesy, General Electric Co.)

monics that would be needed in the measuring of very high frequencies. For example, if we have an electronic counter that is capable of measuring to only 100 mHz and we would like to accurately measure a frequency in the vicinity of 1750 mHz, we could use the snap diode circuit shown to generate the required heterodyne frequency. This would be done by feeding the 100 mHz clock signal of the counter to the snap diode circuit and filtering out the 17th harmonic (1700 mHz). The 17th harmonic would then be mixed with the frequency we were trying to measure and the resulting signal would be fed through a low-pass 100 mHz filter to the electronic counter. This heterodyned signal could then be measured to eight significant figures, and the true frequency reading of the unknown signal would be equal to the counter reading plus 1700 mHz.

In order to show the richness of the harmonics generated by the circuit of Figure 5.24, the waveform was sampled every 0.4 nanoseconds, and the 21 points were read in as data to the Fortran program of Figure 5.21. Figure 5.25 shows the output results and Figure 5.26 is the frequency

```
THE FOURIER COEFFICIENTS


A( 0) =-0.123E 01
A( 1) = 0.455E 00   B( 1) = 0.231E 00   C( 1) = 0.510E 00   THETA( 1) =  26.96
A( 2) = 0.248E 00   B( 2) =-0.192E-01   C( 2) = 0.249E 00   THETA( 2) =  -4.41
A( 3) = 0.143E 00   B( 3) = 0.157E 00   C( 3) = 0.213E 00   THETA( 3) =  47.68
A( 4) = 0.195E 00   B( 4) = 0.157E 00   C( 4) = 0.250E 00   THETA( 4) =  38.81
A( 5) = 0.105E 00   B( 5) = 0.933E-01   C( 5) = 0.140E 00   THETA( 5) =  41.64
A( 6) = 0.658E-03   B( 6) = 0.121E 00   C( 6) = 0.121E 00   THETA( 6) =  89.69
A( 7) = 0.546E-02   B( 7) = 0.160E 00   C( 7) = 0.160E 00   THETA( 7) =  88.04
A( 8) =-0.192E-01   B( 8) = 0.510E-01   C( 8) = 0.545E-01   THETA( 8) = 110.68
A( 9) =-0.108E 00   B( 9) = 0.926E-01   C( 9) = 0.143E 00   THETA( 9) = 139.53
A(10) = 0.418E 00   B(10) = 0.165E-04   C(10) = 0.418E 00   THETA(10) =   0.00
A(11) =-0.108E 00   B(11) =-0.926E-01   C(11) = 0.143E 00   THETA(11) = 220.48
A(12) =-0.192E-01   B(12) =-0.510E-01   C(12) = 0.545E-01   THETA(12) = 249.34
A(13) = 0.549E-02   B(13) =-0.160E 00   C(13) = 0.160E 00   THETA(13) = -88.03
A(14) = 0.695E-03   B(14) =-0.121E 00   C(14) = 0.121E 00   THETA(14) = -89.67
A(15) = 0.105E 00   B(15) =-0.933E-01   C(15) = 0.140E 00   THETA(15) = -41.62
A(16) = 0.195E 00   B(16) =-0.157E 00   C(16) = 0.250E 00   THETA(16) = -38.80
A(17) = 0.143E 00   B(17) =-0.157E 00   C(17) = 0.213E 00   THETA(17) = -47.67
A(18) = 0.248E 00   B(18) = 0.192E-01   C(18) = 0.249E 00   THETA(18) =   4.42
A(19) = 0.455E 00   B(19) =-0.231E 00   C(19) = 0.510E 00   THETA(19) = -26.95
A(20) =-0.123E 01   B(20) =-0.142E-03   C(20) = 0.123E 01   THETA(20) = 180.01
```

Figure 5.25 The Fourier coefficients of the snap diode circuit of Figure 5.24

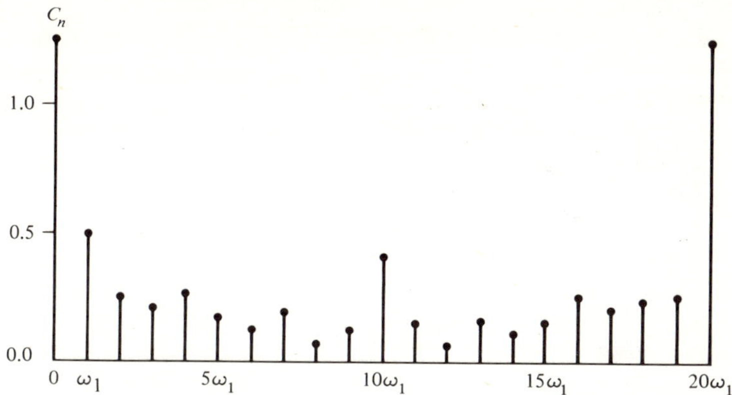

Figure 5.26 The frequency spectrum of the snap diode circuit of Figure 5.24

spectrum plot of the magnitude function. An examination of this histogram reveals the richness of the harmonic content of the waveform.

At this point the author would like to indicate to those readers who may be involved in more complicated evaluations of the Fourier Series, Fourier Integral, Convolution integral, Power Spectrum estimation, or other signal processing operations, that there exists a more efficient algorithm for computing the Fourier coefficients.[3] This more efficient method is commonly known as the "fast Fourier transform" (FFT) and requires approximately $N \log_2 N$ operations for computing the finite Fourier transform from a series of N (complex) data points, in contrast to the approximate N^2 operations that would normally be needed. For a further discussion of the Cooley-Tukey FFT algorithm, including the derivation and its application in signal processing (including Fortran computer programs), the reader is directed to the two special issues devoted to the FFT of the IEEE Transactions on Audio and Electroacoustics, June 1967 and June 1969.

5.5 TRUNCATION ERRORS IN THE NEWTON-COTES FORMULAS

We have seen that the trapezoidal rule is a method of integration that is based on passing a first-degree polynomial through two consecutive points of the function $f(t)$ and finding the area under this straight line as an approximation to the true area. Similarly, Simpson's rule approximates the true area by finding the area under a second-degree poly-

[3] J. W. Cooley and J. W. Tukey, "An Algorithm for the Machine Calculation of Complex Fourier Series," *Mathematics of Computation*, 19 (April, 1965), pp. 297–301.

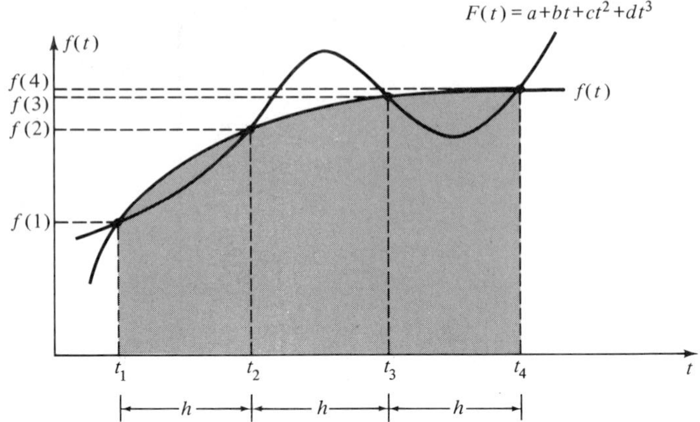

Figure 5.27 The third-degree polynomial approximation

nomial that passes through three consecutive points of $f(t)$. These two methods are special cases of the n-point Newton-Cotes closed formulas.

The four-point (three equal interval) formula can be derived by using

$$F(t) = a + bt + ct^2 + dt^3$$

to approximate the function $f(t)$ as shown in Figure 5.27. The crosshatched area A_{1234} shown may be approximated by evaluating the following integral

$$A_{1234} \approx \int_{t_1}^{t_4} (a + bt + ct^2 + dt^3)\, dt$$

To simplify the integration, we again make a change in the variable by letting

$$s = t - t_1$$

or

$$t = s + t_1$$

so that

$$dt = ds$$

Thus

$$A_{1234} \approx \int_0^{3h} F(s)\, ds$$

where

$$F(s) = A + Bs + Cs^2 + Ds^3$$

with

$$A = a + bt_1 + ct_1^2 + dt_1^3$$
$$B = b + 2ct_1 + 3dt_1^2$$
$$C = c + 3dt_1$$

and

$$D = d$$

Evaluating the integral, we find

$$A_{1234} \approx \left[As + \frac{Bs^2}{2} + \frac{Cs^3}{3} + \frac{Ds^4}{4}\right]_0^{3h}$$
$$= 3Ah + \frac{9Bh^2}{2} + 9Ch^3 + \frac{81Dh^4}{4}$$

To determine the value of A, B, C, and D we note that at

$$\begin{aligned}
t &= t_1 \text{ or } s = 0, & F(0) &= f(1) = A + 0 + 0 + 0\\
t &= t_2 \text{ or } s = h, & F(h) &= f(2) = A + Bh + Ch^2 + Dh^3\\
t &= t_3 \text{ or } s = 2h, & F(2h) &= f(3) = A + 2Bh + 4Ch^2 + 8Dh^3\\
t &= t_4 \text{ or } s = 3h, & F(3h) &= f(4) = A + 3Bh + 9Ch^2 + 27Dh^3
\end{aligned}$$

Since $f(1)$, $f(2)$, $f(3)$, $f(4)$, and h are known, we find by solving the above equations that

$$\begin{aligned}
A &= f(1)\\
B &= \frac{1}{h}\left(-\frac{11}{6}f(1) + 3f(2) - \frac{3}{2}f(3) + \frac{1}{3}f(4)\right)\\
C &= \frac{1}{h^2}\left(f(1) - \frac{5}{2}f(2) + 2f(3) - \frac{1}{2}f(4)\right)\\
D &= \frac{1}{h^3}\left(-\frac{1}{6}f(1) + \frac{1}{2}f(2) - \frac{1}{2}f(3) + \frac{1}{6}f(4)\right)
\end{aligned}$$

When these coefficients are substituted in A_{1234} above, we find

$$A_{1234} \approx \frac{3h}{8}(f(1) + 3f(2) + 3f(3) + f(4))$$

which is commonly called Simpson's ⅜th rule.

In a similar way we can derive the next seven formulas.

$$\int_{t_1}^{t_5} f(t)\,dt \approx \frac{2h}{45}(7f(1) + 32f(2) + 12f(3) + 32f(4) + 7f(5))$$

$$\int_{t_1}^{t_6} f(t)\,dt \approx \frac{5h}{288}(19f(1) + 75f(2) + 50f(3) + 50f(4) + 75f(5) + 19f(6))$$

$$\int_{t_1}^{t_7} f(t)\,dt \approx \frac{h}{140}(41f(1) + 216f(2) + 27f(3) + 272f(4) + 27f(5) + 216f(6) + 41f(7))$$

$$\int_{t_1}^{t_8} f(t)\,dt \approx \frac{7h}{17280}(751f(1) + 3577f(2) + 1323f(3) + 2989f(4) + 2989f(5) + 1323f(6) + 3577f(7) + 751f(8))$$

$$\int_{t_1}^{t_9} f(t)\,dt \approx \frac{4h}{14175}(989f(1) + 5888f(2) - 928f(3) + 10496f(4) - 4540f(5) + 10496f(6) - 928f(7) + 5888f(8) + 989f(9))$$

$$\int_{t_1}^{t_{10}} f(t)\,dt \approx \frac{9h}{89600}(2857(f(1) + f(10)) + 15741(f(2) + f(9)) + 1080(f(3) + f(8)) + 19344(f(4) + f(7)) + 5778(f(5) + f(8)))$$

$$\int_{t_1}^{t_{11}} f(t)\,dt \approx \frac{5h}{299376}(16067(f(1) + f(11)) + 106300(f(2) + f(10)) - 48525(f(3) + f(9)) + 272400(f(4) + f(8)) - 260550(f(5) + f(7)) + 427368f(6))$$

A simplification and a corresponding loss of accuracy of the seven-point (six equal interval) formula leads to the well known Weddle's formula

$$\int_{t_1}^{t_7} f(t)\,dt \approx \frac{3h}{10}(f(1) + 5f(2) + f(3) + 6f(4) + f(5) + 5f(6) + f(7))$$

which is useful in hand calculations because of the simpler coefficients.

Let us now consider the local truncation error involved in the application of these formulas. We have approximated $f(t)$ by the general n-th degree polynomial

$$F(t) = a_1 + a_2t + a_3t^2 + \cdots + a_{n+1}t^n$$

It follows that if $f(t)$ was a first-degree polynomial, the use of the trapezoidal rule would produce zero error because $F(t)$ would be a first-degree polynomial that would be coincident with $f(t)$ and therefore finding the area under $F(t)$ would be identical to finding the area under $f(t)$. On the other hand, if $f(t)$ had been a higher-order polynomial, the use of the trapezoidal formula, or any other higher-order Newton-Cotes formula whose order was less than the order of $f(t)$, would produce a truncation error. This may be expressed mathematically as follows:

$$f(t) = F(t) + (t - t_1)(t - t_2) \cdots (t - t_{n+1}) \frac{f^{n+1}(\epsilon)}{(n + 1)!}$$

where the second term in the right member corresponds to the truncation error that occurs for using an n-th degree polynomial approximation and f^{n+1} is the $(n + 1)$th derivative of $f(t)$ evaluated for $t_1 \leq \epsilon \leq t_2$.[4]

Let us now determine the local truncation error that occurs when a first-degree polynomial is used in approximating the area under $f(t)$

[4] See Bruce Arden, *Digital Computing* (Reading, Pa.: Addison-Wesley Publishing Company, 1963), p. 127.

between t_1 and t_2

$$\int_{t_1}^{t_2} f(t)\,dt = \int_{t_1}^{t_2} (a + bt)\,dt + \int_{t_1}^{t_2} (t - t_1)(t - t_2)\frac{f^2(\epsilon)}{2!}\,dt$$

In order to simplify the evaluation of the truncation error integral, let us make a change in the variable by letting

$$t = t_1 + hs$$

so that

$$dt = h\,ds$$

Thus, we have

$$\begin{aligned} TE_1 &= h\int_{s=0}^{1} h^2 s(s-1)\frac{f^2(\epsilon)}{2!}\,ds \\ &= h^3\frac{f^2(\epsilon)}{2!}\left[\frac{s^3}{3} - \frac{s^2}{2}\right]_0^1 \\ &= -\frac{h^3}{12}f^2(\epsilon),\ t_1 \le \epsilon \le t_2 \end{aligned}$$

As a numerical example, let us find the area under ϵ^{-t} between $t = 0$ and $t = 1$. Using the calculus, we find

$$\begin{aligned} \int_0^1 \epsilon^{-t}\,dt &= [-\epsilon^{-t}]_0^1 = -\epsilon^{-1} + 1 = -0.36787 + 1 \\ &= 0.63213 \end{aligned}$$

Using the trapezoidal rule, with $h = 1$, we find

$$\begin{aligned} \int_0^1 \epsilon^{-t}\,dt &\approx \frac{h}{2}(f(0) + f(1)) = \frac{1}{2}(1 + 0.36787) \\ &= 0.68394 \end{aligned}$$

or the error in using the trapezoidal rule equals

$$0.63213 - 0.68394 = -0.05181$$

It is also clear, from an examination of Figure 5.28, why the trapezoidal method gives us an answer that is larger than the true area.

Using the error formula derived above, we find

$$\left|\frac{h^3}{12}f^2(\epsilon)\right| = \frac{1}{12}f^2(\epsilon),\ 0 \le \epsilon \le 1$$

Since the second derivative of ϵ^{-t} is also ϵ^{-t}, the maximum value that

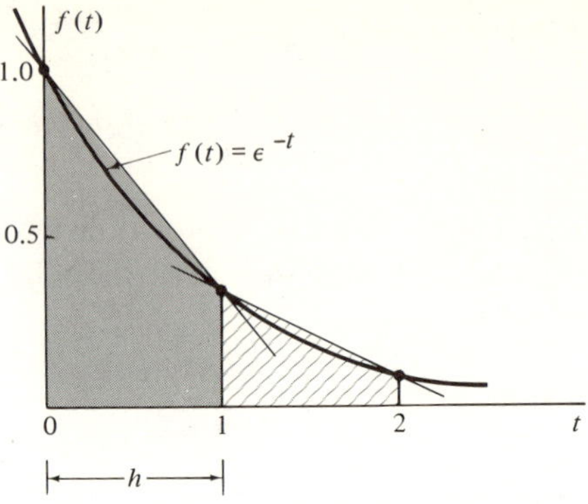

Figure 5.28 The trapezoidal method of finding the area under $f(t) = \epsilon^{-t}$

$f^2(\epsilon)$ can have is 1, and thus the maximum truncation error is

$$TE_{1\,\max} = \frac{1}{12} = 0.0825$$

which is greater than 0.05181.

When a second-degree polynomial is used in approximating the area under $f(t)$ between t_1 and t_3, we have

$$\int_{t_1}^{t_3} f(t)\,dt = \int_{t_1}^{t_3} (a + bt + ct^2)\,dt + \int_{t_1}^{t_3} (t - t_1)(t - t_2)(t - t_3)\frac{f^3(\epsilon)}{3!}\,dt$$

If we again let

$$t = t_1 + hs$$

so that

$$dt = h\,ds$$

we find

$$\begin{aligned} TE_2 &= h\int_{s=0}^{2} h^3 s(s-1)(s-2)\frac{f^3(\epsilon)}{3!}\,ds \\ &= \frac{h^4}{3!}f^3(\epsilon)\left[\frac{s^4}{4} - s^3 + s^2\right]_0^2 \\ &= 0 \end{aligned}$$

This does not mean that the truncation error is equal to zero, but rather that Simpson's rule yields correct integration for an $f(t)$ corresponding to a cubic. We will see below that all odd-point Newton-Cotes formulas have a similar jump in the error term. To calculate the truncation error then, we must consider the next term as the dominant term (true for h small). Thus

$$TE_2 = \int_{t_1}^{t_3} (t - t_1)(t - t_2)(t - t_3)(t - t_4)\frac{f^4(\epsilon)}{4!}\,dt$$

or

$$\begin{aligned} TE_2 &= h \int_{s=0}^{2} h^4 s(s-1)(s-2)(s-3) \frac{f^4(\epsilon)}{4!}\, ds \\ &= \frac{h^5}{4!} f^4(\epsilon) \left[\frac{s^5}{5} - \frac{6s^4}{4} + \frac{11s^3}{3} - \frac{6s^2}{2} \right]_0^2 \\ &= -\frac{h^5}{90} f^4(\epsilon),\ t_1 \le \epsilon \le t_3 \end{aligned}$$

As a simple numerical example, let us again consider the area under ϵ^{-t}, but this time between $t = 0$ and $t = 2$. Using the calculus, we find

$$\begin{aligned} \int_0^2 \epsilon^{-t}\, dt &= [-\epsilon^{-t}]_0^2 = -\epsilon^{-2} + 1 = -0.13533 + 1 \\ &= 0.86467 \end{aligned}$$

Using Simpson's rule, with $h = 1$, we find

$$\begin{aligned} \int_0^2 \epsilon^{-t}\, dt &\approx \frac{h}{3}(f(0) + 4f(1) + f(2)) \\ &= \frac{1}{3}(1 + 4(0.36787) + 0.13533) \\ &= 0.86894 \end{aligned}$$

or the error in using Simpson's rule equals

$$0.86467 - 0.86894 = -0.00427$$

Using the error formula derived above, we find

$$\left| \frac{h^5}{90} f^4(\epsilon) \right| = \frac{1}{90} f^4(\epsilon),\ 0 \le \epsilon \le 2$$

Since the fourth derivative of ϵ^{-t} is also ϵ^{-t}, the maximum value that $f^4(\epsilon)$ can have is 1, and thus the maximum truncation error is

$$TE_{2\,\max} = \frac{1}{90} = 0.0111$$

which is greater than 0.00427.

In a similar way we can show that the four-point (three equal interval) approximation has an error equal to

$$-\frac{3h^5}{80} f^4(\epsilon),\ t_1 \le \epsilon \le t_4$$

the five-point approximation has an error equal to

$$-\frac{8h^7}{945}f^6(\epsilon),\ t_1 \leq \epsilon \leq t_5$$

the six-point approximation has an error equal to

$$-\frac{275h^7}{12096}f^6(\epsilon),\ t_1 \leq \epsilon \leq t_6$$

the seven-point approximation has an error equal to

$$-\frac{9h^9}{1400}f^8(\epsilon),\ t_1 \leq \epsilon \leq t_7$$

the eight-point approximation has an error equal to

$$-\frac{8183h^9}{518400}f^8(\epsilon),\ t_1 \leq \epsilon \leq t_8$$

the nine-point approximation has an error equal to

$$-\frac{2368h^{11}}{467775}f^{10}(\epsilon),\ t_1 \leq \epsilon \leq t_9$$

the ten-point approximation has an error equal to

$$-\frac{173h^{11}}{14620}f^{10}(\epsilon),\ t_1 \leq \epsilon \leq t_{10}$$

and the eleven-point approximation has an error equal to

$$-\frac{1346350h^{13}}{326918592}f^{12}(\epsilon),\ t_1 \leq \epsilon \leq t_{11}$$

Since all the error terms are a function of the point spacing h and the smoothness of $f(t)$ (the derivatives are a measure of the smoothness), and since we get a jump in the error term for all odd-point formulas, it thus follows that one should use only odd-point polynomials in approximating the area under a curve.

To further emphasize this point, let us find the area under ϵ^{-t} between $t = 0$ and $t = 2$ by using the trapezoidal rule, and let us compare this result to the result we previously found with Simpson's rule. Since a point spacing of $h = 1$ was previously used, let us use this same spacing with the trapezoidal rule. Figure 5.28 shows that the trapezoidal rule will have to be applied twice (in contrast to the single application of

Simpson's rule) to determine the area. Thus we have

$$\begin{aligned}\int_0^2 \epsilon^{-t}\, dt &\approx \frac{h}{2}(f(0) + f(1)) + \frac{h}{2}(f(1) + f(2)) \\ &= \frac{h}{2}(f(0) + 2f(1) + f(2)) \\ &= \frac{1}{2}(1 + 2(0.36787) + 0.13533) \\ &= 0.93554\end{aligned}$$

or the total error in applying the trapezoidal rule is

$$0.86467 - 0.93554 = -0.07087$$

which is considerably larger than the error produced in using Simpson's rule (0.00427). Note also that in the double application of the trapezoidal rule, two multiplications and two additions were required. This is identical to the number of mathematical operations required in the single application of Simpson's rule.

In order to estimate the error that occurs in the double application of the trapzeoidal rule, it is necessary to find the maximum error that may occur in each interval and sum the two for the overall maximum error. Thus, for $0 \leq t \leq 1$

$$\left| \frac{h^3}{12} f^2(\epsilon) \right|, 0 \leq \epsilon \leq 1 = \frac{1}{12}(1) = 0.0825$$

and for $1 \leq t \leq 2$,

$$\left| \frac{h^3}{12} f^2(\epsilon) \right|, 1 \leq \epsilon \leq 2 = \frac{1}{12}(0.36787) = 0.0307$$

and the overall maximum truncation error is

$$TE_{1\,\max} = 0.0825 + 0.0307 = 0.1132$$

which is greater than 0.07087.

The last example thus points out that the trapezoidal rule may be applied $(n - 1)$ times to approximate the area under $f(t)$ in the range of t_1 to t_n. The estimate of the total truncation error, however, is the sum of the local truncation errors that occur in each interval

$$\begin{aligned}\mathrm{TE}_1 &= -\frac{h^3}{12} f^2(\epsilon_1) - \frac{h^3}{12} f^2(\epsilon_2) - \cdots - \frac{h^3}{12} f^2(\epsilon_{n-1}) \\ &= -\frac{h^3}{12}(f^2(\epsilon_1) + f^2(\epsilon_2) + \cdots + f^2(\epsilon_{n-1}))\end{aligned}$$

If $f^2(t)$ is continuous in the overall interval, there is for one value of ϵ a value of $f^2(\epsilon)$ that represents the mean value of the derivatives and thus we may write

$$TE_{1\,\max} = -\frac{h^3}{12}(n-1)f^2(\epsilon),\ t_1 \leq \epsilon \leq t_n$$

In a similar way the total truncation error that occurs for the repeated application of Simpson's rule is

$$TE_2 = -\frac{h^5}{90}f^4(\epsilon_1) - \frac{h^5}{90}f^4(\epsilon_2) - \cdots - \frac{h^5}{90}f^4(\epsilon_{(n-1)/2})$$

and

$$TE_{2\,\max} = -\frac{h^5}{90}\frac{(n-1)}{2}f^4(\epsilon),\ t_1 \leq \epsilon \leq t_n$$

where $f^4(\epsilon)$ is the mean value of the derivatives in the overall interval. We will see later how we may estimate the value of the derivatives when the function is not so well behaved as the ϵ^{-t} function.

5.6 THE EVALUATION OF THE FOURIER INTEGRAL

In Section 5.4 we developed a general Fortran program for evaluating the Fourier coefficients of a periodic waveform. We now extend this to cover the case of a nonrepetitive wave. The Fourier transform of a nonperiodic time function $f(t)$ may be found from the evaluation of the following integral

$$f(t) = F(j\omega) = \int_{t=-\infty}^{+\infty} f(t)\epsilon^{-j\omega t}\,dt$$

where, in general, the frequency spectrum $F(j\omega)$ is complex and may be expressed in terms of its real and imaginary part by

$$F(j\omega) = A(\omega) + jB(\omega)$$

It thus follows that

$$A(\omega) = \int_{-\infty}^{+\infty} f(t)\cos\omega t\,dt$$

$$B(\omega) = -\int_{-\infty}^{+\infty} f(t)\sin\omega t\,dt$$

and

$$|F(j\omega)| = \sqrt{A(\omega)^2 + B(\omega)^2}$$

$$\phi(\omega) = \tan^{-1}\frac{B(\omega)}{A(\omega)}$$

Figure 5.29 is a general program for evaluating the Fourier integral. First the heading is printed, and then computer statements S.0004 and S.0005 read in the lower time limit of the waveform (TINIT), the upper time limit of the waveform (TMAX), the increment of time between data points (TINCR), and the number of data points N. These N data points are then read into memory by computer statements S.0006 and S.0007 and printed out by computer statements S.0008 and S.0009. The values of the starting frequency WLOW, the maximum frequency WMAX, and the frequency increment WINCR are now read in and the subheadings "FREQUENCY," "A(W)," "B(W)," "F(JW)," and

```
              C       THE EVALUATION OF THE FOURIER INTEGRAL FOR AN ARBITRARY FUNCTION
              C
S.0001                DIMENSION F(200),FJW(200),PHI(200)
S.0002                WRITE(6,1)
S.0003              1 FORMAT('1THE FREQUENCY SPECTRUM IS COMPUTED FOR THE FOLLOWING INPU
                     1T DATA',////)
S.0004                READ(5,2)TINIT,TMAX,TINCR,N
S.0005              2 FORMAT(3E15.7,I5)
S.0006                READ(5,3)(F(I),I=1,N)
S.0007              3 FORMAT(5E15.7)
S.0008                WRITE(6,4)N,(F(I),I=1,N)
S.0009              4 FORMAT(///,' THE NUMBER OF DATA POINTS =',I5,//,' THE INPUT DATA
                     1F(1), F(2),..., F(N)  IS',//,5(E15.7))
S.0010                READ(5,5)WLOW,WMAX,WINCR
S.0011              5 FORMAT(3E15.7)
S.0012                WRITE(6,6)
S.0013              6 FORMAT(///,4X,'FREQUENCY',7X,'A(W)',10X,'B(W)',9X,'F(JW)',9X,'PHAS
                     1E',//)
S.0014                LIMIT1=(WMAX-WLOW)/WINCR+1.
S.0015                N=N-1
S.0016                W=WLOW-WINCR
S.0017                DO 10 I=1,LIMIT1
S.0018                W=W+WINCR
S.0019                A=0.0
S.0020                B=0.0
S.0021                T=-2.*TINCR+TINIT
S.0022                DO 20 J=1,N,2
S.0023                T=T+2.*TINCR
              C
              C       INTEGRATION BY SIMPSON'S RULE
              C
S.0024                C=F(J)*COS(W*T)
S.0025                D=4.*F(J+1)*COS(W*(T+TINCR))
S.0026                E=F(J+2)*COS(W*(T+2.*TINCR))
S.0027                A=A+C+D+E
S.0028                FF=F(J)*SIN(W*T)
S.0029                G=4.*F(J+1)*SIN(W*(T+TINCR))
S.0030                H=F(J+2)*SIN(W*(T+2.*TINCR))
S.0031             20 B=B+FF+G+H
S.0032                A=TINCR*A/3.
S.0033                B=-TINCR*B/3.
S.0034                FJW(I)=SQRT(A*A+B*B)
S.0035                IF(A)21,22,23
S.0036             22 IF(B)24,25,25
S.0037             24 PHI(I)=-90.
S.0038                GO TC 28
S.0039             25 PHI(I)=+90.
S.0040                GO TC 28
S.0041             23 PHI(I)=(ATAN(B/A))*180./3.1415927
S.0042                GO TC 28
S.0043             21 IF(B)26,27,27
S.0044             26 PHI(I)=-180.+(ATAN(B/A))*180./3.1415927
S.0045                GO TC 28
S.0046             27 PHI(I)=180.+(ATAN(B/A))*180./3.1415927
S.0047             28 CONTINUE
S.0048             10 WRITE(6,7)W,A,B,FJW(I),PHI(I)
S.0049              7 FORMAT(5(2X,E12.4))
S.0050                RETURN
S.0051                END
```

Figure 5.29 A general program for finding the frequency spectrum of a nonperiodic waveform

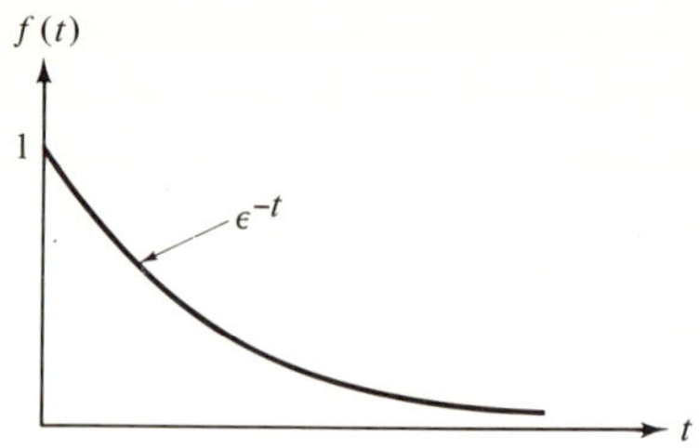

Figure 5.30 The time function $f(t) = \epsilon^{-t}$

"PHASE" are printed so that their numerical values can be printed each iteration of the DO 10 loop.

The upper limit of the DO 10 loop and DO 20 loop is next computed, and the frequency variable ω is initially set equal to WLOW $-$ WINCR. The DO 10 loop is now entered, and each iteration of this loop computes the Fourier coefficients for a specified ω. In the execution of this loop, computer statements S.0019 through S.0033 use Simpson's rule to evaluate the two integrals. After $A(\omega)$ and $B(\omega)$ have been computed, computer statement S.0034 determines the magnitude, $|F(j\omega)|$ and computer statements S.0035 through S.0046 evaluate the phase angle $\phi(\omega)$. Computer statements S.0048 and S.0049 are then used to print the results. The DO 10 loop is now executed a second time with

$$\text{W} = \text{WLOW} + \text{WINCR}$$

and this is continued until W is equal to WMAX.

In order to illustrate the operation of this program, the frequency spectrum of the waveform shown in Figure 5.30 was computed. Note that in this example $f(t)$ is a function that is nonzero over an infinite time interval. Care must therefore be exercised in choosing the value of TMAX, since we can not numerically integrate over an infinite time. We may thus reason that since $f(t)$ will be equal to ϵ^{-10} for $t = 10$ seconds, the value of $f(t) \cos \omega t$ and $f(t) \sin \omega t$ will be essentially equal to zero, and thus a value of TMAX $=$ 10 seconds may be initially used in attempting to evaluate the two integrals. We may also choose TINCR $=$ 0.1 seconds to insure accurate integration by Simpson's rule over this time range.

```
          C        COMPUTER GENERATED DATA FOR FREQUENCY SPECTRUM PROBLEM
          C
S.0001             DIMENSION F(101)
S.0002             T=-0.1
S.0003             DO 10 I=1,101
S.0004             T=T+0.1
S.0005          10 F(I)=EXP(-T)
S.0006             WRITE(6,1)(F(I),I=1,101)
S.0007           1 FORMAT(5E15.7)
S.0008             WRITE(7,1)(F(I),I=1,101)
S.0009             RETURN
S.0010             END
```

Figure 5.31 Computer generated data for the program of Figure 5.29

In order to discuss the accuracy of our results and since 101 data points are needed, let us ensure the accuracy of the input data by using the program of Figure 5.31. Note that this program prints as well as punches the required data. In addition, in order to obtain sufficient data for a plot of the frequency spectrum, let us choose WLOW = −10.0 rps, WMAX = +10.0 rps, and WINCR = 0.25 radians per second.

Figure 5.32 shows the computer output and Figure 5.33 is the corresponding frequency spectrum plot. Note that in this example it is analytically possible to use the calculus to evaluate our two integrals. Thus we would be able to find that our exact solutions are

$$A(\omega) = \frac{1}{1 + \omega^2}$$

$$B(\omega) = \frac{-\omega}{1 + \omega^2}$$

and

$$|F(j\omega)| = \frac{1}{1 + \omega^2}$$

$$\phi(\omega) = -\tan^{-1} \omega$$

If we now evaluate these exact solutions for the frequency range

```
THE FREQUENCY SPECTRUM IS COMPUTED FOR THE FOLLOWING INPUT DATA



THE NUMBER OF DATA POINTS =  101

THE INPUT DATA  F(1), F(2),...., F(N)  IS

 0.1C00000E 01  0.9048374E 00  0.8187308E 00  0.7408183E 00  0.6703202E 00
 0.6065308E 00  0.5488117E 00  0.4965854E 00  0.4493291E 00  0.4065698E 00
 0.3678796E 00  0.3328713E 00  0.3011945E 00  0.2725323E 00  0.2465975E 0C
 0.2231308E 00  0.2018971E 00  0.1826842E 00  0.1652997E 00  0.1495694E 0C
 0.1353360E 00  0.1224571E 00  0.1108038E 00  0.1002595E 0C  C.9071869E-01
 0.8208573E-01  0.7427418E-01  0.6720608E-01  0.6081069E-01  0.5502382E-01
 0.4978764E-01  0.4504974E-01  0.4076272E-01  0.3688365E-01  0.3337373E-01
 0.3019781E-01  0.2732413E-01  0.2472391E-01  0.2237113E-01  0.2024225E-01
 0.1831595E-01  0.1657297E-01  0.1499585E-01  0.1356881E-01  C.1227758E-01
 0.1110922E-01  0.1005204E-01  0.9095468E-02  0.8229923E-02  0.7446751E-02
 0.6738096E-02  0.6096885E-02  0.5516697E-02  0.4991718E-02  0.4516695E-02
 0.4086874E-02  0.3697961E-02  0.3346056E-02  0.3027638E-02  0.2739522E-02
 0.2478823E-02  0.2242933E-02  0.2029491E-02  0.1836360E-02  0.1661609E-02
 0.1503486E-02  0.1360412E-02  0.1230952E-02  0.1113812E-02  C.1007820E-02
 0.9119131E-03  0.8251334E-03  0.7466120E-03  0.6755630E-03  0.6112750E-03
 0.5531048E-03  0.5004704E-03  0.4528442E-03  0.4097510E-03  0.3707579E-03
 0.3354759E-03  0.3035516E-03  0.2746650E-03  0.2485272E-03  0.2248768E-03
 0.2034771E-03  0.1841138E-03  0.1665931E-03  0.1507398E-03  C.1363951E-03
 0.1234154E-03  0.1116710E-03  0.1010441E-03  0.9142858E-04  0.8272805E-04
 0.7485546E-04  0.6773205E-04  0.6128653E-04  0.5545438E-04  C.5017722E-04
 0.4540227E-04
```

Figure 5.32 The output results of the program of Figure 5.29 for the waveform of Figure 5.30

FREQUENCY	A(W)	B(W)	F(JW)	PHASE
-0.1000E 02	0.9694E-02	0.9961E-01	0.1001E 00	0.8444E 02
-0.9750E 01	0.1022E-01	0.1021E 00	0.1026E 00	0.8428E 02
-0.9500E 01	0.1078E-01	0.1046E 00	0.1052E 00	0.8412E 02
-0.9250E 01	0.1138E-01	0.1073E 00	0.1079E 00	0.8395E 02
-0.9000E 01	0.1204E-01	0.1102E 00	0.1108E 00	0.8376E 02
-0.8750E 01	0.1274E-01	0.1132E 00	0.1139E 00	0.8358E 02
-0.8500E 01	0.1351E-01	0.1164E 00	0.1172E 00	0.8338E 02
-0.8250E 01	0.1435E-01	0.1198E 00	0.1206E 00	0.8317E 02
-0.8000E 01	0.1526E-01	0.1234E 00	0.1243E 00	0.8295E 02
-0.7750E 01	0.1627E-01	0.1272E 00	0.1282E 00	0.8271E 02
-0.7500E 01	0.1736E-01	0.1312E 00	0.1324E 00	0.8246E 02
-0.7250E 01	0.1857E-01	0.1356E 00	0.1368E 00	0.8220E 02
-0.7000E 01	0.1991E-01	0.1402E 00	0.1416E 00	0.8191E 02
-0.6750E 01	0.2139E-01	0.1451E 00	0.1467E 00	0.8162E 02
-0.6500E 01	0.2305E-01	0.1504E 00	0.1522E 00	0.8129E 02
-0.6250E 01	0.2489E-01	0.1561E 00	0.1581E 00	0.8094E 02
-0.6000E 01	0.2696E-01	0.1623E 00	0.1645E 00	0.8057E 02
-0.5750E 01	0.2930E-01	0.1689E 00	0.1714E 00	0.8016E 02
-0.5500E 01	0.3194E-01	0.1761E 00	0.1790E 00	0.7972E 02
-0.5250E 01	0.3497E-01	0.1839E 00	0.1872E 00	0.7923E 02
-0.5000E 01	0.3841E-01	0.1924E 00	0.1962E 00	0.7871E 02
-0.4750E 01	0.4240E-01	0.2017E 00	0.2061E 00	0.7813E 02
-0.4500E 01	0.4703E-01	0.2118E 00	0.2170E 00	0.7748E 02
-0.4250E 01	0.5242E-01	0.2230E 00	0.2291E 00	0.7677E 02
-0.4000E 01	0.5880E-01	0.2353E 00	0.2426E 00	0.7597E 02
-0.3750E 01	0.6636E-01	0.2490E 00	0.2577E 00	0.7508E 02
-0.3500E 01	0.7545E-01	0.2642E 00	0.2747E 00	0.7406E 02
-0.3250E 01	0.8648E-01	0.2811E 00	0.2941E 00	0.7290E 02
-0.3000E 01	0.9997E-01	0.3000E 00	0.3162E 00	0.7157E 02
-0.2750E 01	0.1168E 00	0.3212E 00	0.3418E 00	0.7002E 02
-0.2500E 01	0.1379E 00	0.3448E 00	0.3714E 00	0.6820E 02
-0.2250E 01	0.1649E 00	0.3712E 00	0.4062E 00	0.6604E 02
-0.2000E 01	0.2000E 00	0.4000E 00	0.4472E 00	0.6343E 02
-0.1750E 01	0.2461E 00	0.4308E 00	0.4961E 00	0.6026E 02
-0.1500E 01	0.3077E 00	0.4615E 00	0.5547E 00	0.5631E 02
-0.1250E 01	0.3902E 00	0.4878E 00	0.6247E 00	0.5134E 02
-0.1000E 01	0.5000E 00	0.5000E 00	0.7071E 00	0.4500E 02
-0.7500E 00	0.6400E 00	0.4800E 00	0.8000E 00	0.3687E 02
-0.5000E 00	0.7999E 00	0.4000E 00	0.8944E 00	0.2657E 02
-0.2500E 00	0.9412E 00	0.2353E 00	0.9701E 00	0.1404E 02
0.0	0.9999E 00	0.0	0.9999E 00	0.0
0.2500E 00	0.9412E 00	-0.2353E 00	0.9701E 00	-0.1404E 02
0.5000E 00	0.7999E 00	-0.4000E 00	0.8944E 00	-0.2657E 02
0.7500E 00	0.6400E 00	-0.4800E 00	0.8000E 00	-0.3687E 02
0.1000E 01	0.5000E 00	-0.5000E 00	0.7071E 00	-0.4500E 02
0.1250E 01	0.3902E 00	-0.4878E 00	0.6247E 00	-0.5134E 02
0.1500E 01	0.3077E 00	-0.4615E 00	0.5547E 00	-0.5631E 02
0.1750E 01	0.2461E 00	-0.4308E 00	0.4961E 00	-0.6026E 02
0.2000E 01	0.2000E 00	-0.4000E 00	0.4472E 00	-0.6343E 02
0.2250E 01	0.1649E 00	-0.3712E 00	0.4062E 00	-0.6604E 02
0.2500E 01	0.1379E 00	-0.3448E 00	0.3714E 00	-0.6820E 02
0.2750E 01	0.1168E 00	-0.3212E 00	0.3418E 00	-0.7002E 02
0.3000E 01	0.9997E-01	-0.3000E 00	0.3162E 00	-0.7157E 02
0.3250E 01	0.8648E-01	-0.2811E 00	0.2941E 00	-0.7290E 02
0.3500E 01	0.7545E-01	-0.2642E 00	0.2747E 00	-0.7406E 02
0.3750E 01	0.6636E-01	-0.2490E 00	0.2577E 00	-0.7508E 02
0.4000E 01	0.5880E-01	-0.2353E 00	0.2426E 00	-0.7597E 02
0.4250E 01	0.5242E-01	-0.2230E 00	0.2291E 00	-0.7677E 02
0.4500E 01	0.4703E-01	-0.2118E 00	0.2170E 00	-0.7748E 02
0.4750E 01	0.4240E-01	-0.2017E 00	0.2061E 00	-0.7813E 02
0.5000E 01	0.3841E-01	-0.1924E 00	0.1962E 00	-0.7871E 02
0.5250E 01	0.3497E-01	-0.1839E 00	0.1872E 00	-0.7923E 02
0.5500E 01	0.3194E-01	-0.1761E 00	0.1790E 00	-0.7972E 02
0.5750E 01	0.2930E-01	-0.1689E 00	0.1714E 00	-0.8016E 02
0.6000E 01	0.2696E-01	-0.1623E 00	0.1645E 00	-0.8057E 02
0.6250E 01	0.2489E-01	-0.1561E 00	0.1581E 00	-0.8094E 02
0.6500E 01	0.2305E-01	-0.1504E 00	0.1522E 00	-0.8129E 02
0.6750E 01	0.2139E-01	-0.1451E 00	0.1467E 00	-0.8162E 02
0.7000E 01	0.1991E-01	-0.1402E 00	0.1416E 00	-0.8191E 02
0.7250E 01	0.1857E-01	-0.1356E 00	0.1368E 00	-0.8220E 02
0.7500E 01	0.1736E-01	-0.1312E 00	0.1324E 00	-0.8246E 02
0.7750E 01	0.1627E-01	-0.1272E 00	0.1282E 00	-0.8271E 02
0.8000E 01	0.1526E-01	-0.1234E 00	0.1243E 00	-0.8295E 02
0.8250E 01	0.1435E-01	-0.1198E 00	0.1206E 00	-0.8317E 02
0.8500E 01	0.1351E-01	-0.1164E 00	0.1172E 00	-0.8338E 02
0.8750E 01	0.1274E-01	-0.1132E 00	0.1139E 00	-0.8358E 02
0.9000E 01	0.1204E-01	-0.1102E 00	0.1108E 00	-0.8376E 02
0.9250E 01	0.1138E-01	-0.1073E 00	0.1079E 00	-0.8395E 02
0.9500E 01	0.1078E-01	-0.1046E 00	0.1052E 00	-0.8412E 02
0.9750E 01	0.1022E-01	-0.1021E 00	0.1026E 00	-0.8428E 02
0.1000E 02	0.9694E-02	-0.9961E-01	0.1001E 00	-0.8444E 02

Figure 5.32 Continued

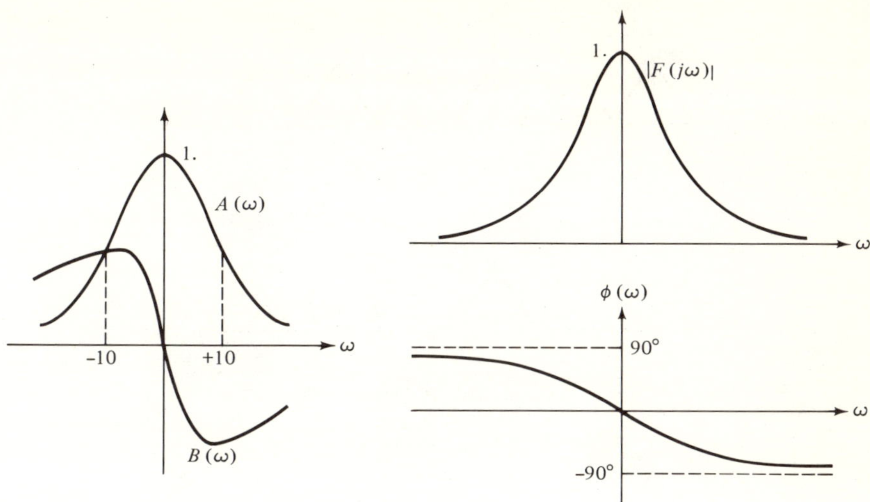

Figure 5.33 Frequency spectrum plots of the time function of Figure 5.30

$-10 \leq \omega \leq +10$, we find that our computer results are less than 1 percent in error. The reader may verify that the accuracy of these results would not be materially improved by rerunning the problem with TMAX = 20 seconds. The reader would find that using a value of TMAX = 5 seconds and/or TINCR = 0.5 seconds, however, would result in considerable loss in accuracy. In other words, the two main causes of inaccurate results are the choice of TINCR and TMAX. The programmer may therefore obtain accurate results by varying these two quantities and rerunning the program until consistent results are obtained.

The reader should also note that since $f(t)$ was neither an even function nor an odd function of time, both $A(\omega)$ and $B(\omega)$ terms resulted. The reader should realize, however, that the calculation of the frequency spectrum for negative values of ω was unnecessary, since the magnitude function will always be an even function of frequency and the phase function will always be an odd function of frequency.

Quasi-Recurrent Waves

In order to see how the frequency spectrum of a quasi-recurrent waveform approaches that of a periodic waveform, let us consider the three waveforms shown in Figure 5.34. Let us first find the Fourier series of

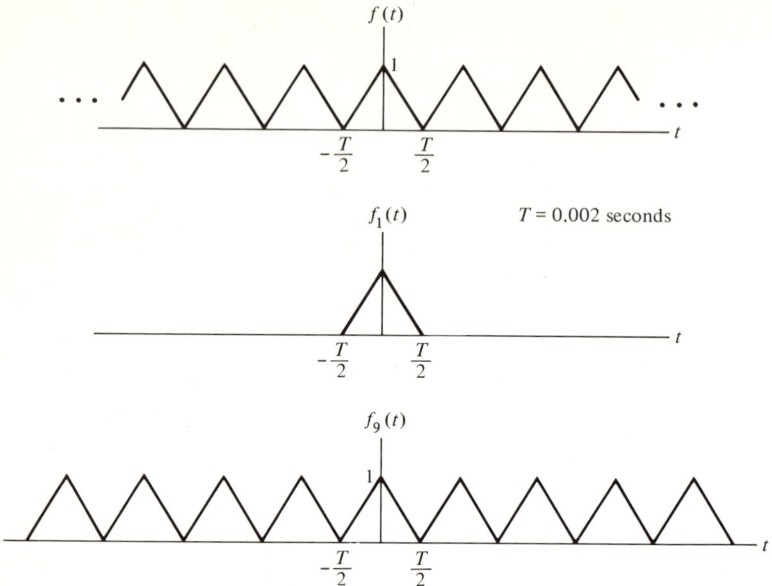

Figure 5.34 A periodic train of triangular pulses; a singular triangular pulse; and a train of nine triangular pulses

the periodic wave. Using the exponential form of the Fourier series, we may write

$$f(t) = \sum_{n=-\infty}^{+\infty} \boldsymbol{F}_n \epsilon^{jn\omega_1 t}$$

where

$$\omega_1 = \frac{2\pi}{T} = \frac{2\pi}{2\ 10^{-3}} = 3141.59 \text{ rps}$$

and

$$\boldsymbol{F}_n = \frac{1}{T} \int_{-T/2}^{T/2} f(t) \epsilon^{-jn\omega_1 t}\, dt$$

Evaluating the integral for $\boldsymbol{F}_n$ we find

$$\begin{aligned} \boldsymbol{F}_n &= 0 & n &= \pm 2, \pm 4, \pm 6, \ldots \\ &= -\frac{4}{n^2\omega_1{}^2 T} = -\frac{T}{n^2\pi^2} & n &= \pm 1, \pm 3, \pm 5, \ldots \end{aligned}$$

and

$$F_0 = \frac{T}{4} \text{ for } n = 0.$$

Thus the Fourier series of a periodic train of these pulses would be

```
THE FREQUENCY SPECTRUM IS COMPUTED FOR THE FOLLOWING INPUT DATA


THE NUMBER OF DATA POINTS =    21

THE INPUT DATA  F(1), F(2),..., F(N)   IS

 0.0              0.9999996E-01  0.2000000E 00  0.3000000E 00  0.4000000E 00
 0.5000000E 00    0.6000000E 00  0.7000000E 00  0.8000000E 00  0.9000000E 00
 0.1000000E 01    0.9000000E 00  0.8000000E 00  0.7000000E 00  0.6000000E 00
 0.5000000E 00    0.4000000E 00  0.3000000E 00  0.2000000E 00  0.9999996E-01
 0.0
```

FREQUENCY	A(W)	B(W)	F(JW)	PHASE
0.0	0.1000E-02	0.0	0.1000E-02	0.0
0.1000E 03	0.9992E-03	-0.7475E-10	0.9992E-03	-0.4287E-05
0.2000E 03	0.9967E-03	-0.1272E-09	0.9967E-03	-0.7310 I-05
0.3000E 03	0.9925E-03	-0.3239E-09	0.9925E-03	-0.1870E-04
0.4000E 03	0.9867E-03	-0.4828E-09	0.9867E-03	-0.2803E-04
0.5000E 03	0.9793E-03	-0.5404E-09	0.9793E-03	-0.3162E-04
0.6000E 03	0.9704E-03	-0.6119E-09	0.9704E-03	-0.3613E-04
0.7000E 03	0.9598E-03	-0.8007E-09	0.9598E-03	-0.4780E-04
0.8000E 03	0.9478E-03	-0.7808E-09	0.9478E-03	-0.4720E-04
0.9000E 03	0.9343E-03	-0.7490E-09	0.9343E-03	-0.4593E-04
0.1000E 04	0.9194E-03	-0.8404E-09	0.9194E-03	-0.5237E-04
0.1100E 04	0.9031E-03	-0.8523E-09	0.9031E-03	-0.5407E-04
0.1200E 04	0.8856E-03	-0.9100E-09	0.8856E-03	-0.5887E-04
0.1300E 04	0.8669E-03	-0.9159E-09	0.8669E-03	-0.6054E-04
0.1400E 04	0.8470E-03	-0.9318E-09	0.8470E-03	-0.6304 -04
0.1500E 04	0.8260E-03	-0.9656E-09	0.8260E-03	-0.6698E-04
0.1600E 04	0.8040E-03	-0.1065E-08	0.8040E-03	-0.7589E-04
0.1700E 04	0.7812E-03	-0.1093E-08	0.7812E-03	-0.8015E-04
0.1800E 04	0.7575E-03	-0.9656E-09	0.7575E-03	-0.7303E-04
0.1900E 04	0.7331E-03	-0.1093E-08	0.7331E-03	-0.8540E-04
0.2000E 04	0.7081E-03	-0.1037E-08	0.7081E-03	-0.8392E-04
0.2100E 04	0.6824E-03	-0.1079E-08	0.6824E-03	-0.9058E-04
0.2200E 04	0.6564E-03	-0.1103E-08	0.6564E-03	-0.9625E-04
0.2300E 04	0.6299E-03	-0.1192E-08	0.6299E-03	-0.1084E-03
0.2400E 04	0.6032E-03	-0.1121E-08	0.6032E-03	-0.1064E-03
0.2500E 04	0.5763E-03	-0.1190E-08	0.5763E-03	-0.1183E-03
0.2600E 04	0.5493E-03	-0.1218E-08	0.5493E-03	-0.1270E-03
0.2700E 04	0.5223E-03	-0.1200E-08	0.5223E-03	-0.1316E-03
0.2800E 04	0.4954E-03	-0.1113E-08	0.4954E-03	-0.1287E-03
0.2900E 04	0.4687E-03	-0.1049E-08	0.4687E-03	-0.1282E-03
0.3000E 04	0.4422E-03	-0.1166E-08	0.4422E-03	-0.1511E-03
0.3100E 04	0.4160E-03	-0.1041E-08	0.4160E-03	-0.1434E-03
0.3200E 04	0.3902E-03	-0.1041E-08	0.3902E-03	-0.1529E-03
0.3300E 04	0.3649E-03	-0.1021E-08	0.3649E-03	-0.1603E-03
0.3400E 04	0.3402E-03	-0.1154E-08	0.3402E-03	-0.1944E-03
0.3500E 04	0.3161E-03	-0.1044E-08	0.3161E-03	-0.1893E-03
0.3600E 04	0.2926E-03	-0.1037E-08	0.2926E-03	-0.2030E-03
0.3700E 04	0.2699E-03	-0.1021E-08	0.2699E-03	-0.2168E-03
0.3800E 04	0.2480E-03	-0.1049E-08	0.2480E-03	-0.2424E-03
0.3900E 04	0.2269E-03	-0.9358E-09	0.2269E-03	-0.2363E-03
0.4000E 04	0.2066E-03	-0.8841E-09	0.2066E-03	-0.2452E-03
0.4100E 04	0.1873E-03	-0.8484E-09	0.1873E-03	-0.2596E-03
0.4200E 04	0.1689E-03	-0.7987E-09	0.1689E-03	-0.2710E-03
0.4300E 04	0.1514E-03	-0.8484E-09	0.1514E-03	-0.3210E-03
0.4400E 04	0.1350E-03	-0.7689E-09	0.1350E-03	-0.3264E-03
0.4500E 04	0.1195E-03	-0.7351E-09	0.1195E-03	-0.3525E-03
0.4600E 04	0.1050E-03	-0.7749E-09	0.1050E-03	-0.4227 -03
0.4700E 04	0.9158E-04	-0.6417E-09	0.9158E-04	-0.4015E-03
0.4800E 04	0.7914E-04	-0.6954E-09	0.7914E-04	-0.5035E-03
0.4900E 04	0.6769E-04	-0.6000E-09	0.6769E-04	-0.5078E-03
0.5000E 04	0.5724E-04	-0.4709E-09	0.5724E-04	-0.4713E-03
0.5100E 04	0.4777E-04	-0.3775E-09	0.4777E-04	-0.4527E-03
0.5200E 04	0.3926E-04	-0.5106E-09	0.3926E-04	-0.7452E-03
0.5300E 04	0.3168E-04	-0.3676E-09	0.3168E-04	-0.6647E-03
0.5400E 04	0.2502E-04	-0.3517E-09	0.2502E-04	-0.8054E-03
0.5500E 04	0.1923E-04	-0.3954E-09	0.1923E-04	-0.1178E-02
0.5600E 04	0.1429E-04	-0.2722E-09	0.1429E-04	-0.1091E-02
0.5700E 04	0.1016E-04	-0.2702E-09	0.1016E-04	-0.1524 -02
0.5800E 04	0.6793E-05	-0.3258E-09	0.6793E-05	-0.2748E-02
0.5900E 04	0.4158E-05	-0.4570E-10	0.4158E-05	-0.6297E-03
0.6000E 04	0.2208E-05	-0.4053E-09	0.2208E-05	-0.1052E-01

Figure 5.35 The frequency spectrum of Figure 5.34b

0.6100E 04	0.8973E-06	-0.2245E-09	0.8973E-06	-0.1434E-01
0.6200E 04	0.1796E-06	-0.2623E-09	0.1796E-06	-0.8369E-01
0.6300E 04	0.7359E-08	-0.2186E-09	0.7362E-08	-0.1701E 01
0.6400E 04	0.3321E-06	-0.2126E-09	0.3321E-06	-0.3668E-01
0.6500E 04	0.1105E-05	-0.3378E-10	0.1105E-05	-0.1752E-02
0.6600E 04	0.2277E-05	-0.8742E-10	0.2277E-05	-0.2199 -02
0.6700E 04	0.3801E-05	-0.4768E-1C	0.3801E-05	-0.7188E-03
0.6800E 04	0.5627E-05	0.3974E-11	0.5627E-05	0.4046E-04
0.6900E 04	0.7709E-05	-0.9189E-11	0.7709E-05	-0.6830E-04
0.7000E 04	0.1000E-04	-0.1465E-10	0.1000E-04	-0.8395E-04
0.7100E 04	0.1246E-04	-0.4967E-11	0.1246E-04	-0.2285E-04
0.7200E 04	0.1503E-04	0.7947E-11	0.1503E-04	0.3029E-04
0.7300E 04	0.1769E-04	-0.3179E-10	0.1769E-04	-0.1029E-03
0.7400E 04	0.2039E-04	-0.7749E-10	0.2039E-04	-0.2177E-03
0.7500E 04	0.2309E-04	-0.8146E-10	0.2309E-04	-0.2021E-03
0.7600E 04	0.2576E-04	-0.3378E-1C	0.2576E-04	-0.7511E-04
0.7700E 04	0.2837E-04	-0.1450E-09	0.2837E-04	-0.2929E-03
0.7800E 04	0.3088E-04	-0.1073E-09	0.3088E-04	-0.1991E-03
0.7900E 04	0.3327E-04	-0.1152E-09	0.3327E-04	-0.1984E-03
0.8000E 04	0.3552E-04	-0.1748E-09	0.3552E-04	-0.2820E-03
0.8100E 04	0.3760E-04	-0.9139E-1C	0.376CE-C4	-0.1393E-03
0.8200E 04	0.3949E-04	-0.1113E-09	0.3949E-04	-0.1614E-03
0.83C0E 04	0.4118E-04	-0.1093E-09	0.4118E-04	-0.1521E-03
0.8400E 04	0.4265E-04	-0.3775E-10	0.4265E-04	-0.5071E-04
0.8500E 04	0.4390E-04	-0.1748E-09	0.439CE-04	-0.2282E-03
0.86C0E 04	0.4491E-04	-0.1669E-09	0.4491E-04	-0.2129E-03
0.87C0E 04	0.4569E-04	-0.6954E-10	0.4569E-04	-0.8720E-04
0.88C0E 04	0.4623E-04	-0.1252E-09	0.4623E-04	-0.1551E-03
0.8900E 04	0.4652E-04	-0.9139E-1C	0.4652E-04	-0.1126E-03
0.9000E 04	0.4658E-04	-0.1987E-1C	0.4658E-04	-0.2444E-04
0.9100E 04	0.4640E-04	-0.2245E-09	0.4640E-04	-0.2772E-03
0.9200E 04	0.4600E-04	-0.1907E-09	0.4600E-04	-0.2376E-C3
0.9300E 04	0.4538E-04	-0.2583E-10	0.4538E-04	-0.3261E-04
0.9400E 04	0.4455E-04	-0.2424E-09	0.4455E-04	-0.3117E-03
0.95C0E 04	0.4353E-04	-0.2027E-09	0.4353E-04	-0.2667E-03
0.9600E 04	0.4233E-04	-0.1093E-09	0.4233E-04	-0.1479E-03
0.9700E 04	0.4096E-04	-0.1152E-09	0.4096E-04	-0.1612E-03
0.9800E 04	0.3945E-04	-0.1490E-09	0.3945E-04	-0.2164E-03
0.9900E 04	0.3779E-04	-0.1748E-09	0.3779E-04	-0.2651E-03
0.1000E 05	0.3603E-04	-0.1391E-09	0.3603E-04	-0.2212E-03

Figure 5.35 Continued

given by

$$f(t) = \cdots - \frac{T}{9\pi^2}\epsilon^{-j3\omega_1 t} - \frac{T}{\pi^2}\epsilon^{-j\omega_1 t} + \frac{T}{4} - \frac{T}{\pi^2}\epsilon^{j\omega_1 t} - \frac{T}{9\pi^2}\epsilon^{j3\omega_1 t} + \cdots$$

$$= \frac{T}{4} - \frac{T}{\pi^2} \sum_{\substack{n=-\infty \\ (\text{odd})}}^{+\infty} \frac{1}{n^2}\epsilon^{jn\omega_1 t}$$

An examination of this series shows that there are no even-order harmonics and the amplitude of the odd-order harmonics goes down as $1/n^2$.

Let us now turn to the program of Figure 5.29 and use this to obtain the data to plot the frequency spectrums of the nonperiodic waves. Figures 5.35 and 5.36 are the respective computer outputs, and Figure 5.37 gives the corresponding plots. Note that in this plot the magnitude of the frequency spectrum of the single pulse has been multiplied by a factor of nine.

These results may be readily verified by first computing the frequency spectrum of the single pulse.

$$F_1(j\omega) = \int_{-\infty}^{+\infty} f_1(t)\epsilon^{-j\omega t}\,dt$$

```
THE FREQUENCY SPECTRUM IS COMPUTED FOR THE FOLLOWING INPUT DATA



THE NUMBER OF DATA POINTS =  181

THE INPUT DATA  F(1), F(2),..., F(N)  IS

0.0             0.9999996E-01  0.2000000E 00  0.3000000E 00  0.4000000E 00
0.5000000E 00   0.6000000E 00  0.7000000E 00  0.8000000E 00  0.9000000E 00
0.1000000E 01   0.9000000E 00  0.8000000E 00  0.7000000E 00  0.6000000E 00
0.5000000E 00   0.4000000E 00  0.3000000E 00  0.2000000E 00  0.9999996E-01
0.0             0.9999996E-01  0.2000000E 00  0.3000000E 00  0.4000000E 00
0.5000000E 00   0.6000000E 00  0.7000000E 00  0.8000000E 00  0.9000000E 00
0.1000000E 01   0.9000000E 00  0.8000000E 00  0.7000000E 00  0.6000000E 00
0.5000000E 00   0.4000000E 00  0.3000000E 00  0.2000000E 00  0.9999996E-01
0.0             0.9999996E-01  0.2000000E 00  0.3000000E 00  0.4000000E 00
0.5000000E 00   0.6000000E 00  0.7000000E 00  0.8000000E 00  0.9000000E 00
0.1000000E 01   0.9000000E 00  0.8000000E 00  0.7000000E 00  0.6000000E 00
0.5000000E 00   0.4000000E 00  0.3000000E 00  0.2000000E 00  0.9999996E-01
0.0             0.9999996E-01  0.2000000E 00  0.3000000E 00  0.4000000E 00
0.5000000E 00   0.6000000E 00  0.7000000E 00  0.8000000E 00  0.9000000E 00
0.1000000E 01   0.9000000E 00  0.8000000E 00  0.7000000E 00  0.6000000E 00
0.5000000E 00   0.4000000E 00  0.3000000E 00  0.2000000E 00  0.9999996E-01
0.0             0.9999996E-01  0.2000000E 00  0.3000000E 00  0.4000000E 00
0.5000000E 00   0.6000000E 00  0.7000000E 00  0.8000000E 00  0.9000000E 00
0.1000000E 01   0.9000000E 00  0.8000000E 00  0.7000000E 00  0.6000000E 00
0.5000000E 00   0.4000000E 00  0.3000000E 00  0.2000000E 00  0.9999996E-01
0.0             0.9999996E-01  0.2000000E 00  0.3000000E 00  0.4000000E 00
0.5000000E 00   0.6000000E 00  0.7000000E 00  0.8000000E 00  0.9000000E 00
0.1000000E 01   0.9000000E 00  0.8000000E 00  0.7000000E 00  0.6000000E 00
0.5000000E 00   0.4000000E 00  0.3000000E 00  0.2000000E 00  0.9999996E-01
0.0             0.9999996E-01  0.2000000E 00  0.3000000E 00  0.4000000E 00
0.5000000E 00   0.6000000E 00  0.7000000E 00  0.8000000E 00  0.9000000E 00
0.1000000E 01   0.9000000E 00  0.8000000E 00  0.7000000E 00  0.6000000E 00
0.5000000E 00   0.4000000E 00  0.3000000E 00  0.2000000E 00  0.9999996E-01
0.0             0.9999996E-01  0.2000000E 00  0.3000000E 00  0.4000000E 00
0.5000000E 00   0.6000000E 00  0.7000000E 00  0.8000000E 00  0.9000000E 00
0.1000000E 01   0.9000000E 00  0.8000000E 00  0.7000000E 00  0.6000000E 00
0.5000000E 00   0.4000000E 00  0.3000000E 00  0.2000000E 00  0.9999996E-01
0.0             0.9999996E-01  0.2000000E 00  0.3000000E 00  0.4000000E 00
0.5000000E 00   0.6000000E 00  0.7000000E 00  0.8000000E 00  0.9000000E 00
0.1000000E 01   0.9000000E 00  0.8000000E 00  0.7000000E 00  0.6000000E 00
0.5000000E 00   0.4000000E 00  0.3000000E 00  0.2000000E 00  0.9999996E-01
0.0
```

FREQUENCY	A(W)	B(W)	F(JW)	PHASE
0.0	0.9000E-02	0.0	0.9000E-02	0.0
0.1000E 03	0.7840E-02	-0.1153E-06	0.7840E-02	-0.8429E-03
0.2000E 03	0.4885E-02	-0.1368E-06	0.4885E-02	-0.1605E-02
0.3000E 03	0.1435E-02	-0.1078E-06	0.1435E-02	-0.4305E-02
0.4000E 03	-0.1121E-02	-0.3908E-07	0.1121E-02	-0.1800E 03
0.5000E 03	-0.1997E-02	0.5020E-07	0.1997E-02	0.1800E 03
0.6000E 03	-0.1328E-02	0.8995E-07	0.1328E-02	0.1800E 03
0.7000E 03	0.2509E-04	0.5757E-07	0.2509E-04	0.1315E 00
0.8000E 03	0.1049E-02	0.6924E-08	0.1049E-02	0.3783E-03
0.9000E 03	0.1157E-02	-0.5147E-07	0.1157E-02	-0.2549E-02
0.1000E 04	0.4502E-03	-0.5585E-07	0.4502E-03	-0.7108E-02
0.1100E 04	-0.4637E-03	-0.8760E-08	0.4637E-03	-0.1800E 03
0.1200E 04	-0.9321E-03	0.4071E-07	0.9321E-03	0.1800E 03
0.1300E 04	-0.6855E-03	0.5603E-07	0.6855E-03	0.1800E 03
0.1400E 04	0.2894E-04	0.2416E-07	0.2894E-04	0.4784E-01
0.1500E 04	0.6656E-03	-0.4371E-07	0.6656E-03	-0.3762E-02
0.1600E 04	0.7767E-03	-0.6098E-07	0.7767E-03	-0.4498E-02
0.1700E 04	0.3125E-03	-0.4656E-07	0.3125E-03	-0.8536E-02
0.1800E 04	-0.3675E-03	0.3167E-08	0.3675E-03	0.1800E 03
0.1900E 04	-0.7623E-03	0.5513E-07	0.7623E-03	0.1800E 03
0.2000E 04	-0.5847E-03	0.5381E-07	0.5847E-03	0.1800E 03
0.2100E 04	0.3992E-04	0.2011E-07	0.3992E-04	0.2886E-01
0.2200E 04	0.6606E-03	-0.4448E-07	0.6606E-03	-0.3858E-02
0.2300E 04	0.8119E-03	-0.5984E-07	0.8119E-03	-0.4223E-02
0.2400E 04	0.3404E-03	-0.2439E-07	0.3404E-03	-0.4105E-02

Figure 5.36 The frequency spectrum of Figure 5.34c

0.2500E 04	-0.4692E-03	0.6095E-07	0.4692E-03	0.1800E 03
0.2600E 04	-0.1052E-02	0.1258E-06	0.1052E-02	0.1800E 03
0.2700E 04	-0.9041E-03	0.9253E-07	0.9041E-03	0.1800E 03
0.2800E 04	0.9945E-04	-0.8491E-07	0.9945E-04	-0.4892E-01
0.2900E 04	0.1613E-02	-0.3705E-06	0.1613E-02	-0.1316E-01
0.3000E 04	0.2997E-02	-0.6558E-06	0.2997E-02	-0.1254E-01
0.3100E 04	0.3658E-02	-0.8491E-06	0.3658E-02	-0.1330E-01
0.3200E 04	0.3354E-02	-0.8492E-06	0.3354E-02	-0.1451E-01
0.3300E 04	0.2289E-02	-0.6897E-06	0.2289E-02	-0.1726E-01
0.3400E 04	0.9696E-03	-0.4205E-06	0.9696E-03	-0.2485E-01
0.3500E 04	-0.7568E-04	-0.1451E-06	0.7568E-04	-0.1799E 03
0.3600E 04	-0.5506E-03	0.4199E-07	0.5506E-03	0.1800E 03
0.3700E 04	-0.4845E-03	0.1090E-06	0.4845E-03	0.1800E 03
0.3800E 04	-0.1418E-03	0.8462E-07	0.1418E-03	0.1800E 03
0.3900E 04	0.1703E-03	0.1910E-07	0.1703E-03	0.6427E-02
0.4000E 04	0.2707E-03	-0.2298E-07	0.2707E-03	-0.4863E-02
0.4100E 04	0.1640E-03	-0.2241E-07	0.1640E-03	-0.7830E-02
0.4200E 04	-0.1956E-04	0.1516E-08	0.1956E-04	0.1800E 03
0.4300E 04	-0.1392E-03	0.1720E-07	0.1392E-03	0.1800E 03
0.4400E 04	-0.1342E-03	0.2674E-07	0.1342E-03	0.1800E 03
0.4500E 04	-0.4083E-04	0.9491E-08	0.4083E-04	0.1800E 03
0.4600E 04	0.5610E-04	-0.1483E-07	0.5610E-04	-0.1515E-01
0.4700E 04	0.9102E-04	-0.2518E-07	0.9102E-04	-0.1585E-01
0.4800E 04	0.5599E-04	-0.2469E-07	0.5599E-04	-0.2527E-01
0.4900E 04	-0.8092E-05	-0.1506E-07	0.8092E-05	-0.1799E 03
0.5000E 04	-0.5079E-04	-0.3052E-08	0.5079E-04	-0.1800E 03
0.5100E 04	-0.4851E-04	0.1253E-07	0.4851E-04	0.1800E 03
0.5200E 04	-0.1413E-04	0.4909E-08	0.1413E-04	0.1800E 03
0.5300E 04	0.2075E-04	0.8737E-09	0.2075E-04	0.2413E-02
0.5400E 04	0.3223E-04	-0.8382E-08	0.3223E-04	-0.1490E-01
0.5500E 04	0.1888E-04	-0.3246E-08	0.1888E-04	-0.9853E-02
0.5600E 04	-0.3048E-05	0.3844E-08	0.3048E-05	0.1799E 03
0.5700E 04	-0.1586E-04	0.8851E-08	0.1586E-04	0.1800E 03
0.5800E 04	-0.1367E-04	0.1216E-07	0.1367E-04	0.1799E 03
0.5900E 04	-0.3364E-05	0.3246E-08	0.3364E-05	0.1799E 03
0.6000E 04	0.4414E-05	0.2775E-08	0.4414E-05	0.3602E-01
0.6100E 04	0.4906E-05	-0.1550E-08	0.4906E-05	-0.1810E-01
0.6200E 04	0.1471E-05	-0.4901E-08	0.1471E-05	-0.1909E 00
0.6300E 04	0.6385E-07	-0.4776E-08	0.6403E-07	-0.4278E 01
0.6400E 04	0.2475E-05	-0.6068E-08	0.2475E-05	-0.1405E 00
0.6500E 04	0.4775E-05	-0.6549E-08	0.4775E-05	-0.7857E-01
0.6600E 04	0.2095E-05	-0.2126E-08	0.2095E-05	-0.5815E-01
0.6700E 04	-0.5373E-05	-0.1626E-08	0.5373E-05	-0.1800E 03
0.6800E 04	-0.1136E-04	0.2726E-08	0.1136E-04	0.1800E 03
0.6900E 04	-0.8903E-05	0.1977E-08	0.8903E-05	0.1800E 03
0.7000E 04	0.2549E-05	0.8464E-09	0.2549E-05	0.1903E-01
0.7100E 04	0.1497E-04	-0.4013E-09	0.1497E-04	-0.1536E-02
0.7200E 04	0.1746E-04	-0.3213E-08	0.1746E-04	-0.1054E-01
0.7300E 04	0.5605E-05	0.5305E-08	0.5605E-05	0.5423E-01
0.7400E 04	-0.1331E-04	0.4307E-08	0.1331E-04	0.1800E 03
0.7500E 04	-0.2460E-04	0.1565E-07	0.2460E-04	0.1800E 03
0.7600E 04	-0.1745E-04	0.5291E-08	0.1745E-04	0.1800E 03
0.7700E 04	0.5288E-05	-0.5094E-08	0.5288E-05	-0.5519E-01
0.7800E 04	0.2735E-04	-0.1054E-07	0.2735E-04	-0.2208E-01
0.7900E 04	0.3049E-04	-0.1404E-07	0.3049E-04	-0.2638E-01
0.8000E 04	0.9103E-05	-0.7377E-08	0.9103E-05	-0.4643E-01
0.8100E 04	-0.2326E-04	0.5654E-08	0.2326E-04	0.1800E 03
0.8200E 04	-0.4196E-04	0.1008E-07	0.4196E-04	0.1800E 03
0.8300E 04	-0.2933E-04	0.5788E-08	0.2933E-04	0.1800E 03
0.8400E 04	0.1001E-04	-0.5647E-08	0.1001E-04	-0.3230E-01
0.8500E 04	0.4905E-04	-0.1770E-07	0.4905E-04	-0.2068E-01
0.8600E 04	0.5555E-04	-0.1318E-07	0.5555E-04	-0.1360E-01
0.8700E 04	0.1635E-04	0.1049E-07	0.1635E-04	0.3676E-01
0.8800E 04	-0.4850E-04	0.3082E-07	0.4850E-04	0.1800E 03
0.8900E 04	-0.9287E-04	0.4277E-07	0.9287E-04	0.1800E 03
0.9000E 04	-0.7119E-04	0.1402E-07	0.7119E-04	0.1800E 03
0.9100E 04	0.3155E-04	-0.5367E-07	0.3155E-04	-0.9745E-01
0.9200E 04	0.1857E-03	-0.1406E-06	0.1857E-03	-0.4339E-01
0.9300E 04	0.3287E-03	-0.2349E-06	0.3287E-03	-0.4094E-01
0.9400E 04	0.3977E-03	-0.2825E-06	0.3977E-03	-0.4070E-01
0.9500E 04	0.3629E-03	-0.2676E-06	0.3629E-03	-0.4226E-01
0.9600E 04	0.2428E-03	-0.2013E-06	0.2428E-03	-0.4751E-01
0.9700E 04	0.9294E-04	-0.1123E-06	0.9294E-04	-0.6923E-01
0.9800E 04	-0.2511E-04	-0.2357E-07	0.2511E-04	-0.1799E 03
0.9900E 04	-0.7488E-04	0.2127E-07	0.7488E-04	0.1800E 03
0.1000E 05	-0.5917E-04	0.3772E-07	0.5917E-04	0.1800E 03

Figure 5.36 Continued

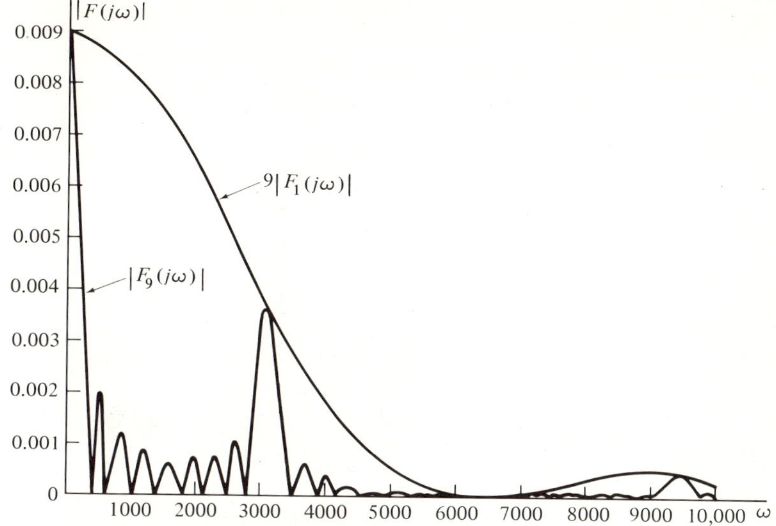

Figure 5.37 The frequency spectrums of Figure 5.34 (b) and (c)

Since $f_1(t)$ is an even function of time, it therefore follows that

$$F_1(j\omega) = \int_{-\infty}^{+\infty} f_1(t) \cos \omega t \, dt$$

$$= 2 \int_0^{T/2} \left(1 - \frac{2}{T} t\right) \cos \omega t \, dt$$

$$= \frac{4}{T} \frac{1 - \cos \frac{\omega T}{2}}{\omega^2}$$

Using the trigonometric identity

$$2 \sin^2 X = 1 - \cos 2X$$

we may write

$$F_1(j\omega) = \frac{4}{T} \frac{2 \sin^2 \frac{\omega T}{4}}{\omega^2}$$

$$= \frac{T}{2} \left(\frac{\sin \frac{\omega T}{4}}{\frac{\omega T}{4}} \right)^2$$

We may thus see that at $\omega = 0$

$$F_1(j0) = \frac{T}{2} = 0.1\ 10^{-2}$$

and $F_1(j\omega)$ will be equal to zero at

$$\begin{aligned}\omega &= \pm n4\pi/T \\ &= \pm n6283.18 \text{ rps} \qquad n = \pm 1, \pm 2, \ldots\end{aligned}$$

The frequency spectrum of the train of nine pulses may now be found by multiplying the frequency spectrum of the single pulse by the repetition factor $R_9(j\omega)$[5]

$$F_9(j\omega) = F_1(j\omega)R_9(j\omega)$$

where

$$|R_9(j\omega)| = \left| \frac{\sin 9 \dfrac{\omega T}{2}}{\sin \dfrac{\omega T}{2}} \right|$$

Thus

$$|F_9(j\omega)| = \left| \frac{T}{2} \left(\frac{\sin \dfrac{\omega T}{4}}{\dfrac{\omega T}{4}} \right)^2 \right| \left| \frac{\sin 9 \dfrac{\omega T}{2}}{\sin \dfrac{\omega T}{2}} \right|$$

In light of the above discussion, we may reexamine Figure 5.37 and note that $F_1(j\omega)$ and $F_9(j\omega)$ equal zero at $\omega = n6283.18$ rps. We may also observe that at $\omega = 3141.59$ rps (ω_1) and $\omega = 9424.68$ rps ($3\omega_1$) the amplitudes of $F_9(j\omega)$ are nine times the corresponding amplitudes of $F_1(j\omega)$. Thus we may see that as the number of pulses increases, the maximum peaks become larger and narrower. When we realize that the frequency spectrum term of a nonperiodic wave is a coefficient density term, we see that in the limit, the frequency spectrum of the nonperiodic wave will approach the frequency spectrum of the periodic wave. The reader is encouraged to extend this work by rerunning the program for waves that contain a larger number of pulses.

5.7 OBTAINING THE TIME RESPONSE FROM THE FREQUENCY RESPONSE

This section concerns itself with finding the impulse response of a system from its frequency response. Once the impulse response has been found, the response to any arbitrary forcing function can then be found by using the convolution integral.[6]

[5] See Ley, Lutz, and Rehberg, *Linear Circuit Analysis*, pp. 406–11.

[6] For an early reference see Gordon S. Brown and Donald P. Campbell, *Principles of Servomechanisms* (New York: John Wiley & Sons, Inc., 1948), Ch. 11.

Let us first consider how we can obtain the impulse response of a system from its frequency response. Writing the system transfer function as the ratio of two rational polynomials

$$H(s) = \frac{a_m s^m + a_{m-1}s^{m-1} + \cdots + a_1 s + a_0}{b_n s^n + b_{n-1}s^{n-1} + \cdots + b_1 s + b_0}$$

where the a's and b's are real constants, it follows that if $H(s)$ has no poles on the $j\omega$ axis or in the right half plane,

$$\operatorname*{Lim}_{s \to \infty} H(s) = 0$$

The inverse transform of $H(s)$ is called the impulse response $h(t)$ and, in general, would be given by

$$h(t) = \frac{1}{2\pi j} \int_{C-j\infty}^{C+j\infty} H(s)\epsilon^{st}\, ds$$

If we now use the imaginary axis as the path of integration and let $s = j\omega$

$$H(s) = Re\{H(j\omega)\} + jIm\{H(j\omega)\}$$

and

$$\epsilon^{st} = \epsilon^{j\omega t} = \cos \omega t + j \sin \omega t$$

The substitution of these latter two equations in the equation for the impulse response yields

$$h(t) = \frac{1}{2\pi} \int_{-\infty}^{\infty} [Re\,\{H(j\omega)\} \cos \omega t - Im\{H(j\omega)\} \sin \omega t]\, d\omega$$
$$+ \frac{1}{2\pi} \int_{-\infty}^{\infty} [Re\,\{H(j\omega)\} \sin \omega t + Im\{H(j\omega)\} \cos \omega t]\, d\omega$$

Because the $Re\,\{H(j\omega)\}$ is an even function in ω and $Im\{H(j\omega)\}$ is an odd function in ω, it follows that the second integral reduces to zero and the first integral may be expressed as the sum of two integrals

$$h(t) = \frac{1}{\pi} \int_0^{\infty} Re\{H(j\omega)\} \cos \omega t\, d\omega - \frac{1}{\pi} \int_0^{\infty} Im\{H(j\omega)\} \sin \omega t\, d\omega$$

Since $h(t)$ must be zero for negative time, the two integrals are numerically equal for all times, so we may write

$$h(t) = \frac{2}{\pi} \int_0^{\infty} Re\,\{H(j\omega)\} \cos \omega t\, d\omega$$

The above expression therefore allows us to compute the impulse response of a system from its frequency response.

Once we have found the impulse response of the system, the response to any arbitrary forcing function may be found by the use of the convolution integral; thus,

$$F_0(t) = \int_0^t h(t - \tau)F_i(\tau)\, d\tau$$

where

$F_i(t)$ is the input forcing function.
$h(t)$ is the system impulse response.
$F_0(t)$ is the output time response.

and

τ is the dummy variable of integration.

Note that this equation implies that we can obtain the output time response, $F_0(t)$, to any input forcing function, $F_i(t)$, if we know the system impulse response, $h(t - \tau)$.

For a simple example to illustrate the use of the convolution integral, let us assume that

$$H(s) = \frac{1}{s + 1}$$

so that the impulse response $h(t)$ will be given by ϵ^{-t}. Let us further assume that the input forcing function is a unit step and thus

$$F_0(s) = H(s)F_i(s)$$
$$= \frac{1}{s(s + 1)}$$

Taking the inverse transform of this latter expression, we find that the corresponding output time response would be given by

$$F_0(t) = 1 - \epsilon^{-t}$$

Let us now recompute the output time response by using the convolution integral

$$F_0(t) = \int_0^t h(t - \tau)F_i(\tau)\, d\tau$$

Figure 5.38 graphically pictures the steps involved in using Simpson's rule to evaluate the integral for $t = 2\Delta\tau$. Figure 5.38a shows the graph of the impulse response and its values at $\tau = 0$, $\Delta\tau$, and $2\Delta\tau$ units; Figure 5.38b shows the impulse response "folded" about the line $\tau = 0$; Figure 5.38c shows the translation of Figure 5.38b for the value of $t = 2\Delta\tau$; Figure 5.38d shows the input forcing function; and Figure 5.38e shows

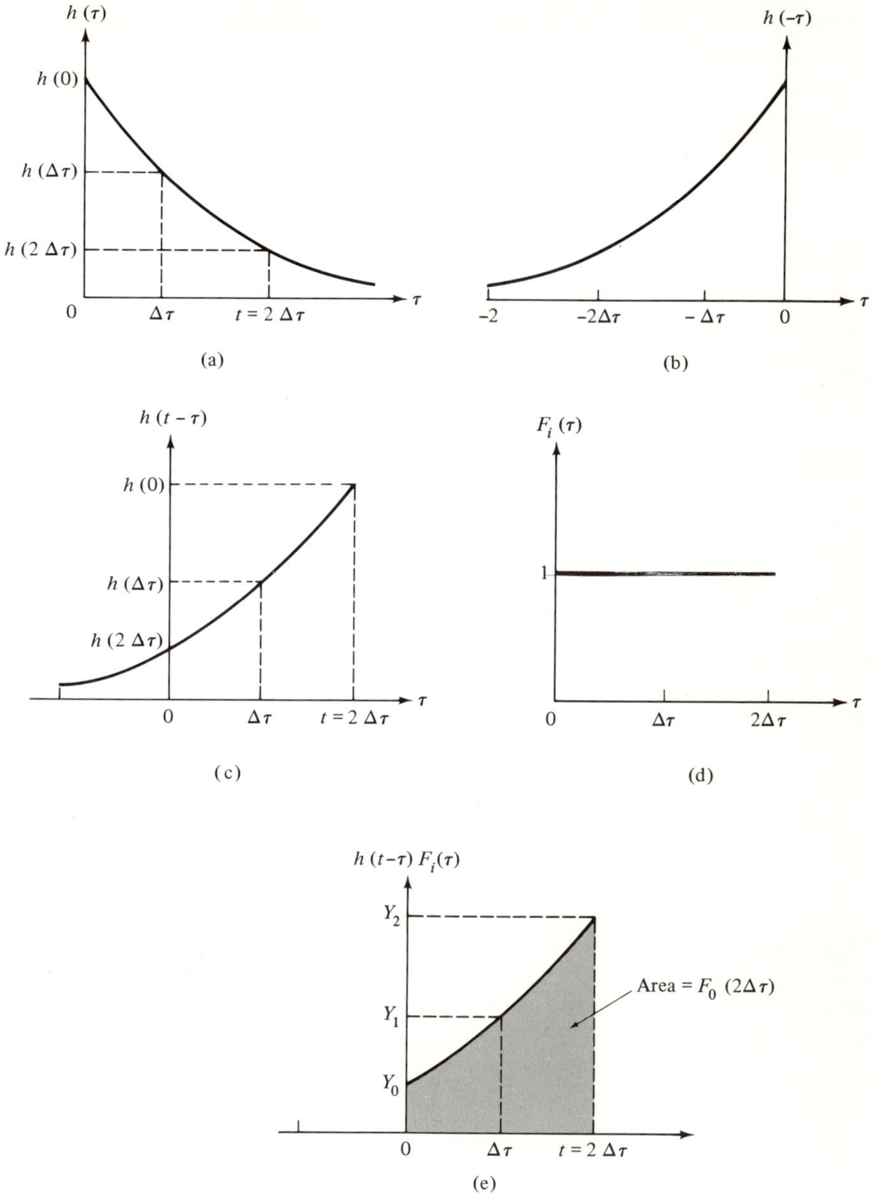

Figure 5.38 A graphical interpretation of the convolution integral

the area that is equivalent to $F_0(2\Delta\tau)$. Thus we have

$$\begin{aligned} Y_0 &= h(2\Delta\tau - 0\Delta\tau)F_i(0\Delta\tau) = h(2\Delta\tau)F_i(0) \\ Y_1 &= h(\Delta\tau)F_i(\Delta\tau) \\ Y_2 &= h(0)F_i(2\Delta\tau) \end{aligned}$$

and the value of $F_0(2\Delta\tau)$ would be given by

$$F_0(2\Delta\tau) = \frac{\Delta\tau}{3}(Y_0 + 4Y_1 + Y_2)$$

Extending the above to the general case, the area found by Simpson's rule would be given by

$$F_0(K\Delta\tau) = \frac{\Delta\tau}{3}(Y_0 + 4Y_1 + 2Y_2 + \cdots + 2Y_{N-2} + 4Y_{N-1} + Y_N)$$

where

$$Y_N = h(t - \tau)F_i(\tau)$$

with

$$t = K\Delta\tau;\ K = 2, 4, 6, \ldots$$

and

$$\tau = N\Delta\tau;\ N = 0, 1, 2, \ldots, K$$

The computation would therefore be carried out by first choosing a value of K and then letting N vary from 0 to K to obtain the output time response $F_0(K\Delta\tau)$.

Figure 5.39 gives the Fortran program for finding the impulse response of a system from its frequency response and then obtaining the output time response to a given input forcing function. Initially the program prints the output heading and then computer statements S.0004 and S.0005 read in the number of input frequency values NUMBPT, the time increment DELTAT ($\Delta\tau$), and the maximum value of time TMAX. This is followed by computer statements S.0006 through S.0009 which read in and print the real part of the system transfer function ($Re\ \{H(j\omega)\}$).

Computer statements S.0011 through S.0024 are then used to compute the impulse response. Note that the DO 20 loop uses Simpson's rule to evaluate the integral

$$h(t) = \frac{2}{\pi}\int_0^\infty Re\{H(j\omega)\}\cos\omega t\,d\omega$$

Computer statements S.0026 and S.0027 then read in the values of the input forcing function and the calculation of the forced response is then evaluated by computer statements S.0028 through S.0044. The reader is encouraged to hand compute several iterations of the DO 30 loop in order to better understand how the convolution integral has been evaluated.

```
LEVEL  15JUL68                         IBM OS/360 BASIC FORTRAN IV (E) COMPILATION
      C     THIS PROGRAM FINDS THE IMPULSE RESPONSE OF A SYSTEM FROM THE REAL
      C     PART OF THE SYSTEM TRANSFER FUNCTION. IT THEN USES THE IMPULSE
      C     RESPONSE TO OBTAIN THE TIME RESPONSE CAUSED BY AN ARBITRARY FORCING
      C     FUNCTION.
      C
      C     NUMBPT = NUMBER OF DATA POINTS SUPPLIED FOR  RE(W(JW)). NUMBPT
      C     MUST BE AN ODD NUMBER.
      C     DELTAT = THE TIME INCREMENT USED IN THE CALCULATION OF THE IMPULSE
      C     RESPONSE.
      C     TMAX = THE LENGTH OF TIME OVER WHICH THE FORCED RESPONSE IS FOUND.
      C
            DIMENSION REH(100),W(100),H(100),FI(100),HFI(100)
            WRITE(6,1)
          1 FORMAT('1THE CALCULATION OF THE IMPULSE RESPONSE',////,' THE FOLLO
           1WING INPUT DATA REPRESENTS THE REAL PART OF THE SYSTEM TRANSFER FU
           2NCTION  H(JW)',//)
            READ(5,2)NUMBPT,DELTAT,TMAX
          2 FORMAT(I5,2E10.3)
            READ(5,3)(W(I),REH(I),I=1,NUMBPT)
          3 FORMAT(4E15.7)
            WRITE(6,4)(REH(I),W(I),I=1,NUMBPT)
          4 FORMAT(' RE H(JW)',7X,'FREQUENCY',//,(2E15.7))
      C
      C     THE CALCULATION OF THE IMPULSE RESPONSE
      C
            TIME=-DELTAT
            WINCR=W(NUMBPT)/(NUMBPT-1.)
            LIMIT1=TMAX/DELTAT+1.
            WRITE(6,5)
          5 FORMAT('0',5X,'TIME',5X,'IMPULSE RESPONSE',//)
            DO 10 I=1,LIMIT1
            TIME=TIME+DELTAT
            SUM=0.0
            LIMIT2=NUMBPT-2
            DO 20 J=1,LIMIT2,2
            XX=W(J)*TIME
         20 SUM=SUM+REH(J)*COS(W(J)*TIME)+4.*REH(J+1)*COS(W(J+1)*TIME)+REH(J+2
           1)*COS(W(J+2)*TIME)
            H(I)=(2./3.14159)*WINCR*SUM/3.
         10 WRITE(6,6)TIME,H(I)
          6 FORMAT(F10.2,F15.4)
      C
      C     THE INPUT FREQUENCY FUNCTION FI(T)
      C
            READ(5,7)(FI(I),I=1,LIMIT1)
          7 FORMAT(8F10.4)
      C
      C     THE CALCULATION OF THE FORCED RESPONSE
      C
            WRITE(6,8)
          8 FORMAT(////,' THE OUTPUT TIME RESPONSE',//,5X,'TIME',5X,'OUTPUT RE
           1SPONSE',//)
            TIME=0.0
            LIMIT3=TMAX/DELTAT+2.
            DO 30 I=4,LIMIT3,2
            LIMIT4=I-1
            DO 40 J=1,LIMIT4
            K=I-J
         40 HFI(J)=H(K)*FI(J)
            LIMIT5=LIMIT4-2
            SUM=0.0
            DO 50 M=1,LIMIT5,2
         50 SUM=SUM+HFI(M)+4.*HFI(M+1)+HFI(M+2)
            OUTPUT=DELTAT*SUM/3.
            TIME=TIME+2.*DELTAT
         30 WRITE(6,9)TIME,OUTPUT
          9 FORMAT(F10.2,F15.4)
            RETURN
            END
```

Figure 5.39 A computer program for finding the impulse response of a system from its frequency response and then obtaining the output time response to an input forcing function

```
      DIMENSION REH(30),FREQ(30)
      W=-0.5
      DO 10 I=1,30
      W=W+0.5
      FREQ(I)=W
      REH(I)=1./(1.+W*W)
   10 REH(I)=1./(1.+W*W)
      WRITE(6,20)(FREQ(I),REH(I),I=1,30)
      WRITE(7,20)(FREQ(I),REH(I),I=1,30)
   20 FORMAT(4E15.7)
      STOP
      END
```

Figure 5.40 The computer is used to generate the input data for the program of Figure 5.39

THE CALCULATION OF THE IMPULSE RESPONSE

THE FOLLOWING INPUT DATA REPRESENTS THE REAL PART OF THE SYSTEM TRANSFER FUNCTION H(JW)

RE H(JW)	FREQUENCY
0.1000000E 01	0.0
0.8000000E 00	0.5000000E 00
0.5000000E 00	0.1000000E 01
0.3080000E 00	0.1500000E 01
0.2000000E 00	0.2000000E 01
0.1379300E 00	0.2500000E 01
0.9999996E-01	0.3000000E 01
0.7547098E-01	0.3500000E 01
0.5882350E-01	0.4000000E 01
0.4705800E-01	0.4500000E 01
0.3846100E-01	0.5000000E 01
0.3200000E-01	0.5500000E 01
0.2702702E-01	0.6000000E 01
0.2312100E-01	0.6500000E 01
0.2000000E-01	0.7000000E 01
0.1746720E-01	0.7500000E 01
0.1538460E-01	0.8000000E 01
0.1365180E-01	0.8500000E 01
0.1219500E-01	0.9000000E 01
0.1095880E-01	0.9500000E 01
0.9900987E-02	0.1000000E 02
0.8988757E-02	0.1050000E 02
0.8196719E-02	0.1100000E 02
0.7504687E-02	0.1150000E 02
0.6896548E-02	0.1200000E 02

THE OUTPUT TIME RESPONSE

TIME	OUTPUT RESPONSE
0.20	0.1812
0.40	0.3295
0.60	0.4500
0.80	0.5501
1.00	0.6305
1.20	0.6970
1.40	0.7513
1.60	0.7950
1.80	0.8313
2.00	0.8601
2.20	0.8836
2.40	0.9026
2.60	0.9172
2.80	0.9290
3.00	0.9377
3.20	0.9439
3.40	0.9481
3.60	0.9499
3.80	0.9499
4.00	0.9479
4.20	0.9436
4.40	0.9374
4.60	0.9284
4.80	0.9166
5.00	0.9016

TIME	IMPULSE RESPONSE
0.0	0.9460
0.10	0.9139
0.20	0.8335
0.30	0.7392
0.40	0.6596
0.50	0.6008
0.60	0.5517
0.70	0.5008
0.80	0.4478
0.90	0.4004
1.00	0.3631
1.10	0.3326
1.20	0.3025
1.30	0.2708
1.40	0.2413
1.50	0.2178
1.60	0.1993
1.70	0.1816
1.80	0.1625
1.90	0.1437
2.00	0.1285
2.10	0.1172
2.20	0.1069
2.30	0.0952
2.40	0.0829
2.50	0.0726
2.60	0.0653
2.70	0.0592
2.80	0.0519
2.90	0.0434
3.00	0.0358
3.10	0.0305
3.20	0.0264
3.30	0.0215
3.40	0.0150
3.50	0.0085
3.60	0.0038
3.70	0.0003
3.80	-0.0039
3.90	-0.0098
4.00	-0.0163
4.10	-0.0218
4.20	-0.0260
4.30	-0.0307
4.40	-0.0373
4.50	-0.0452
4.60	-0.0527
4.70	-0.0590
4.80	-0.0657
4.90	-0.0746
5.00	-0.0856

Figure 5.41 The output results of the program of Figure 5.39

Figure 5.40 gives the computer program that was used to supply the real part of the frequency response data. Note that the system transfer function used is the same as the one that was used in the illustrative example above

$$H(s) = \frac{1}{s+1}$$

or

$$H(j\omega) = \frac{1}{1+j\omega} = \frac{1-j\omega}{1+\omega^2}$$

so that

$$Re\,\{H(j\omega)\} = \frac{1}{1+\omega^2}$$

Figure 5.41 gives the computer results for a unit step input excitation. Note that the value of the impulse response at $t = 1.0$ units is computed as 0.3631 instead of 0.3678 and the impulse response for larger values of time are actually negative. These errors are primarily due to round-off errors and the use of too large a frequency increment in the evaluation of the integral. Note also that the errors in the impulse response are reflected in the output time response.

PROBLEMS

5-1. Using a small value of h in a numerical integration scheme will involve a large number of mathematical operations and, as a consequence, numerous rounding errors. The truncation error will therefore be negligible compared to the round-off error. In order to investigate this effect, write a program that uses the Euler formula to compute

$$\int_0^{\pi/2} \sin x\, dx$$

for $h = \pi/4, \pi/8, \ldots$, and $\pi/8192$. Have your program print out the value of h, the computed value of the integral, and the error (exact value $= 1.0000000$). Plot the error as a function of h and comment on the resulting plot.

5-2. Rerun the program of Figure 5.8 using a step size of $h = 0.01°$ and $h = 0.001°$. Comment on your results.

5-3. Repeat Problem 5-1 using the trapezoidal rule.

5-4. Repeat Problem 5-2 for Figure 5.13.

5-5. Rerun the program of Figure 5.16 using half the step size. Use linear interpolation to obtain the additional data points. Comment on the accuracy of your result.

5-6. What simple argument can you give to prove that the result of Figure 5.20 is more accurate than the result of Figure 5.16?

5-7. Using Simpson's rule (with $h = 0.001$), determine the average and RMS voltage values of the following periodic waveforms:

$$V_{\text{ave}} = \frac{1}{T}\int_0^T v\,dt \qquad\qquad V_{\text{RMS}} = \sqrt{\frac{1}{T}\int_0^T v^2\,dt}$$

a.	$v = 100 \sin 100\pi t$	$0.00 \leq t \leq 0.02$
b.	$v = 100 \sin 100\pi t$	$0.00 \leq t \leq 0.01$
	$= -100 \sin 100\pi t$	$0.01 \leq t \leq 0.02$
c.	$v = 100 \sin 100\pi t$	$0.00 \leq t \leq 0.01$
	$= 0$	$0.01 \leq t \leq 0.02$
d.	$v = 100$	$0.00 < t < 0.01$
	$= -100$	$0.01 < t < 0.02$
e.	$v = 100$	$0.00 < t < 0.01$
	$= 0$	$0.01 < t < 0.02$
f.	$v = 100$	$0.00 < t < 0.005$
	$= 0$	$0.005 < t < 0.02$
g.	$v = 20000t$	$0.00 \leq t \leq 0.005$
	$= -20000t + 200$	$0.005 \leq t \leq 0.015$
	$= 20000t - 400$	$0.015 \leq t \leq 0.02$
h.	$v = 20000t$	$0.00 \leq t \leq 0.005$
	$= -20000t + 200$	$0.005 \leq t \leq 0.01$
	$= 20000t - 200$	$0.01 \leq t \leq 0.015$
	$= -20000t + 400$	$0.015 \leq t \leq 0.02$
i.	$v = 50000t$	$0.00 \leq t \leq 0.002$
	$= 100$	$0.002 \leq t \leq 0.008$
	$= -50000t + 500$	$0.008 \leq t \leq 0.012$
	$= -100$	$0.012 \leq t \leq 0.018$
	$= 50000t - 10000$	$0.018 \leq t \leq 0.02$

5-8. Using the program of Figure 5.21, calculate the frequency spectrum of each of the waveforms of Problem 5-7.

5-9. The twelve-point method of computing the Fourier series coefficients of an arbitrary waveform has long been used in hand computation because it involves only 65 mathematical operations. The method is based on the twelve ordinate values of the waveform

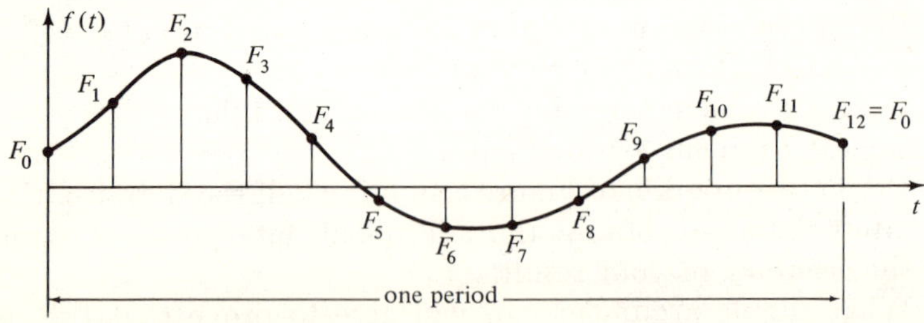

and the evaluation of the following sums and differences:

	F_0	F_1	F_2	F_3	F_4	F_5	F_6
		F_{11}	F_{10}	F_9	F_8	F_7	
Sum	s_0	s_1	s_2	s_3	s_4	s_5	s_6
Difference		d_1	d_2	d_3	d_4	d_5	

Using the sum terms we now calculate

	s_0	s_1	s_2	s_3	S_0	S_1
	s_6	s_5	s_4		S_2	S_3
Sum	S_0	S_1	S_2	S_3	S_7	S_8
Difference	D_0	D_1	D_2			

Using the difference terms we now calculate

	d_1	d_2	d_3		
	d_5	d_4		S_4	D_0
Sum	S_4	S_5	S_6	S_6	D_2
Difference	D_3	D_4		D_5	D_6

The coefficients of the Fourier series

$$f(t) = A_0 + \sum_{n=1}^{6} 2(A_n \cos n\omega_1 t - B_n \sin n\omega_1 t)$$

may now be computed from the following relations:

$$A_0 = \frac{S_7 + S_8}{12} \qquad -B_1 = \frac{0.5S_4 + 0.866S_5 + S_6}{12}$$

$$A_1 = \frac{D_0 + 0.866D_1 + 0.5D_2}{12} \qquad -B_2 = \frac{0.866(D_3 + D_4)}{12}$$

$$A_2 = \frac{S_0 + 0.5S_1 - 0.5S_2 - S_3}{12} \qquad -B_3 = \frac{D_5}{12}$$

$$A_3 = \frac{D_6}{12} \qquad -B_4 = \frac{0.866(D_3 - D_4)}{12}$$

$$A_4 = \frac{S_0 - 0.5S_1 - 0.5S_2 + S_3}{12} \qquad -B_5 = \frac{0.5S_4 - 0.866S_5 + S_6}{12}$$

$$A_5 = \frac{D_0 - 0.866D_1 + 0.5D_2}{12}$$

$$A_6 = \frac{S_7 - S_8}{12}$$

Write a Fortran program that uses the twelve-point method of computing the Fourier series coefficients.

5-10. Using the program of Problem 5-9, determine the Fourier coefficients of the waveform of Figure 5.22c. Compare these results with those of Figure 5.23.

5-11. If a periodic function possesses a finite discontinuity, the magnitude of the Fourier coefficients will decrease as $1/n$. If a periodic function is continuous but its first derivative has a finite discontinuity, the magnitude of the Fourier coefficients will decrease as $1/n^2$, and so on. In general, the smoother the variations of the function, the faster the magnitude of the Fourier coefficients approach zero with increasing n.

The above behavior may be used to reduce random noise that is superimposed on a smooth periodic signal. Since the desired function is smooth, its Fourier coefficients will decrease rapidly to zero. On the other hand, the Fourier coefficients of the unsmooth random noise will decrease very slowly. The combined signal will therefore consist almost entirely of noise beyond a given Fourier harmonic.

Use the above discussion and the program of Figure 5.21 to extract the desired signal from the sum of the following two waveforms. In addition, have the program print out the required data so that you can plot the original (noisy) and extracted waveforms. The desired signal

$$v(t) = 100 \sin 100\pi t$$
$$0.0 \leq t \leq 0.02$$

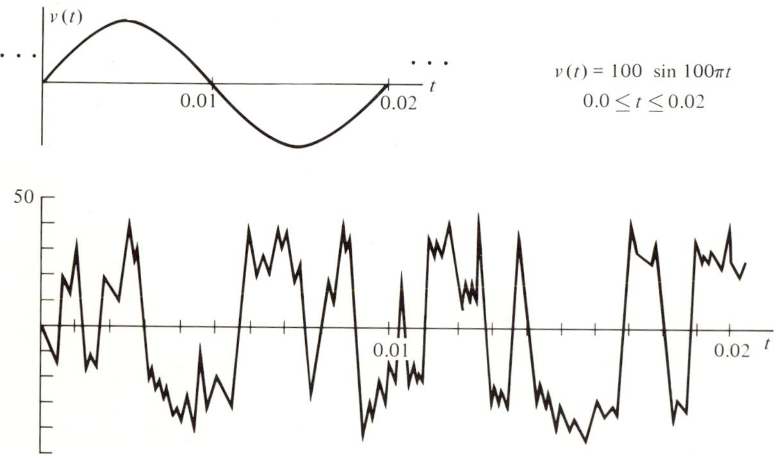

Random noise Signal

5-12. Using the method discussed in Problem 5-11, modify the program of Figure 5.21 so that it may be used to extract the desired signal from a noisy repetitive waveform.

5-13. Evaluate the definite integral

$$\int_0^2 (x^3 - 3x + 3)\, dx$$

a. Using the calculus.
b. Using Euler's rule with $h = 0.5$.
c. Using the trapezoidal rule with $h = 0.5$.
d. Using Simpson's rule with $h = 0.5$.
e. Compare and explain any differences in your results.

5-14. Repeat Problem 5-13 for

$$\int_0^2 (4x^3 - 3x^2 + x - 5)\, dx$$

5-15. Evaluate the definite integral of Problem 5-13 using Simpson's rule with $h = 0.5$ and 1.0. Are these two results consistent with the error term of Simpson's rule? Explain.

5-16. Repeat Problem 5-15 for the definite integral of Problem 5-14.

5-17. Using the four-point Newton-Cotes integration formula, find the area under the sin x curve over the interval 0 to π. Use a mesh size of $h = \pi/3$ and calculate the maximum error bound.

5-18. Repeat Problem 5-17 using the five-point Newton-Cotes formula with $h = \pi/4$.

5-19. Derive the truncation error term in the five-point Newton-Cotes formula.

5-20. The trapezoidal rule is to be applied once in the numerical integration of

$$\int_{-a}^{0} \epsilon^x\, dx;\ a > 0$$

Determine the largest value of a for which the absolute value of the error is less than ½. Compare this result to that obtained if Simpson's rule is used instead.

5-21. The function $f(t) = \epsilon^{-t}$ is to be integrated over the range of $t = 1.0$ to $t = 2.0$. Using the trapezoidal rule, determine the value of h that should be used to limit the maximum truncation error to 0.00001.

5-22. Repeat Problem 5-21 using Simpson's rule.

5-23. Repeat Problem 5-15 for the definite integral

$$\int_0^2 \frac{dx}{x^2 - 3}$$

Hint: How may you avoid the singularity?

5-24. The trapezoidal rule may be used in multiple integration problems by shifting the single integration formula.[7] For example,

[7] See John M. McCormick and Mario G. Salvadori, *Numerical Methods in Fortran* (Englewood Cliffs, N.J.: Prentice-Hall, Inc., 1964), Sec. 8.2.

$$\int_{X_I}^{X_{I+1}} \int_{Y_J}^{Y_{J+1}} F(X,Y)\, dY\, dX$$
$$= \frac{k}{2}\frac{h}{2}(F_{I+1,J+1} + F_{I,J+1}) + \frac{h}{2}(F_{I+1,J} + F_{I,J})$$
$$= \frac{kh}{4}(F_{I,J} + F_{I+1,J} + F_{I,J+1} + F_{I+1,J+1})$$

where h is the X step size and k is the Y step size. Using this double integration formula, evaluate

a. $\int_0^2 \int_0^1 (X + Y)\, dY\, dX$ using $h = 2$ and $k = 1$.

b. $\int_0^2 \int_0^3 XY(X - Y)\, dY\, dX$ using $h = k = 0.5$

5-25. Repeat Problem 5-24 using Simpson's double integration formula

$$\int_{X_{I-1}}^{X_{I+1}} \int_{Y_{J-1}}^{Y_{J+1}} F(X,Y)\, dY\, dX = \frac{hk}{9}(F_{I+1,J+1} + F_{I+1,J-1} + F_{I-1,J+1} + F_{I-1,J-1} + 4(F_{I,J+1} + F_{I,J-1} + F_{I-1,J+1} + F_{I-1,J-1}) + 16F_{I,J})$$

5-26. Write a general Fortran program that uses Simpson's rule to evaluate definite double integrals. Test your program using the examples of Problem 5-24.

5-27. Rerun the program of Figure 5.29 (see Figure 5.30) using
a. TINCR = 0.1 and TMAX = 20 seconds.
b. TINCR = 0.5 and TMAX = 5 seconds.
Discuss the accuracy of these results.

5-28. Use the program of Figure 5.29 to obtain the frequency spectrum of a train of one, three, and nine pulses defined as follows

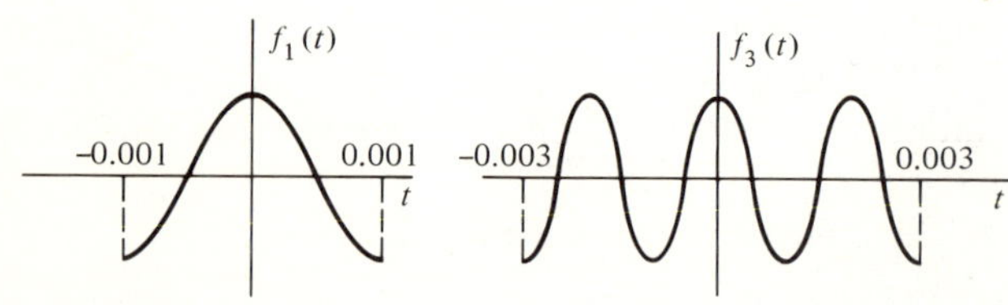

$f_1(t) = \cos 1000\pi t$; $-0.001 \leq t \leq 0.001$

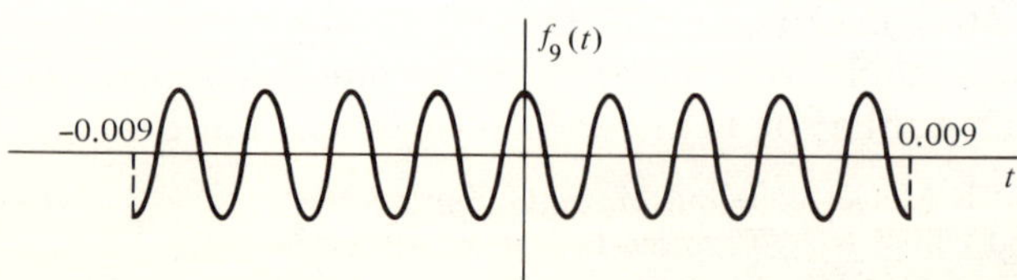

5-29. In order to better understand how the convolution integral is evaluated in the program of Figure 5.39, hand compute several iterations of the DO 30 loop and justify your results.

5-30. Use the program of Figure 5.39 to obtain the impulse response of the following transform functions:

a. $H(s) = \dfrac{1}{s^2 - 1}$ b. $H(s) = \dfrac{s + 5}{s^2 + 2s + 4}$

Now determine the corresponding output for the following input functions:

(a) A unit step input. (b) $f(t) = t \quad ; 0 \leq t \leq 5$
$\qquad = -t + 10; 5 \leq t \leq 10$

5-31. The autocorrelation function is defined by

$$R_{AA}(\tau) = \lim_{T \to 0} \frac{1}{2T} \int_{-T}^{+T} F_A(t) F_A(t - \tau)\, dt$$

in which the waveform $F_A(t)$ is multiplied by itself, delayed by τ seconds, and integrated over the period T. The autocorrelation function thus allows us to extract desired information that may be buried deep in noise by computing the above integration for many different values of τ.

Write a Fortran program for computing the autocorrelation function. Test your program by using a modification of the example shown in Problem 5-11.

5-32. The crosscorrelation function

$$R_{AB}(\tau) = \lim_{T \to 0} \frac{1}{2T} \int_{-T}^{+T} F_B(t) F_A(t - \tau)\, dt$$

permits the comparison between two signals from different random processes. Note that this function is very similar to the autocorrelation function (see Problem 5-31), except that in this function the first input signal is multiplied by a delayed version of another input signal. The cross-correlation function thus may be used to describe the amount of conformity between two different signals.

Write a Fortran program for computing the cross-correlation between two different functions. Test your program by supplying suitable input data.

CHAPTER 6

Simultaneous Linear Algebraic Equations

6.1 INTRODUCTION

A large number of electrical engineering problems, either directly or indirectly, require the solution of simultaneous linear algebraic equations. For example, if we write Kirchhoff's voltage law for the two-mesh network shown in Figure 6.1, we obtain two simultaneous linear equations

$$
\begin{aligned}
6I_1 - 3I_2 &= 10 \\
-3I_1 + 9I_2 &= -5
\end{aligned}
$$

These two equations have the general form

$$
\begin{aligned}
a_{11}I_1 + a_{12}I_2 &= V_1 \\
a_{21}I_1 + a_{22}I_2 &= V_2
\end{aligned}
$$

where

I_1 = the current in mesh one.

I_2 = the current in mesh two.

V_1 = the algebraic summation of all voltage sources in mesh one.

V_2 = the algebraic summation of all voltage sources in mesh two.

a_{11} = the self resistance of mesh one.

a_{22} = the self resistance of mesh two.

$a_{12} = a_{21}$ = the resistance mutual to mesh one and two.

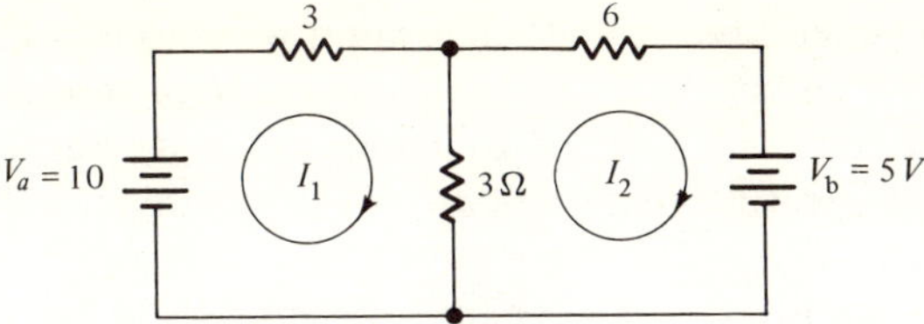

Figure 6.1 A two-mesh network

The solution of these equations may be found by applying Cramer's rule, thus

$$I_1 = \frac{\begin{vmatrix} V_1 & a_{12} \\ V_2 & a_{22} \end{vmatrix}}{\Delta} \qquad I_2 = \frac{\begin{vmatrix} a_{11} & V_1 \\ a_{21} & V_2 \end{vmatrix}}{\Delta}$$

where

$$\Delta = \begin{vmatrix} a_{11} & a_{12} \\ a_{21} & a_{22} \end{vmatrix}$$

and a unique solution exists provided both V's do not equal zero and the determinant $\Delta \neq 0$.

Figure 6.2 is a Fortran program for solving two-mesh networks of this type. Initially the heading is printed, and the six coefficients are read in and printed out. Computer statements S.0008 and S.0009 then test to see whether there is an infinite number of solutions or if there is no unique solution. If either of these two conditions is detected, the corresponding condition is printed as part of the output and control is returned to computer statement S.0003 so that the next set of coefficients can be read in. On the other hand, if there is a unique solution to the problem,

```
         C      A PROGRAM THAT SOLVES THE SIMULTANECUS LINFAB ECUATIONS
         C      CF A TWO-MESH NETWORK.
         C
S.0001          WRITE(6,1)
S.0002        1 FORMAT('1TWO-MESH NETWORK SCLUTION',///)
S.0003        2 READ(5,3) A11,A12,V1
S.0004        3 FORMAT(3E15.7)
S.0005          READ(5,3) A21,A22,V2
S.0006          WRITE(6,4) A11,A12,V1,A21,A22,V2
S.0007        4 FORMAT(' A11 =',E15.7,'   A12 =',E15.7,'   V1 =',E15.7,//,' A21 =',
               1E15.7,'   A22 =',E15.7,'   V2 =',E15.7,///)
S.0008          IF(A21/A22-A11/A12) 10,20,10
S.0009       20 IF(V1/A12-V2/A22) 30,40,30
S.0010       40 WRITE(6,5)
S.0011        5 FORMAT(' THIS SYSTEM HAS AN INFINITE NUMBER CF SCLUTICNS',////)
S.0012          GO TC 2
S.0013       30 WRITE(6,6)
S.0014        6 FORMAT(' THIS SYSTEM HAS NO UNIQUE SOLUTION',////)
S.0015          GO TO 2
S.0016       10 DELTA=A11*A22-A21*A12
S.0017          C1=(V1*A22-V2*A12)/DELTA
S.0018          C2=(A11*V2-A21*V1)/DELTA
S.0019          WRITE(6,7) C1,C2
S.0020        7 FORMAT(' I1 =',E15.7,'   I2 =',E15.7,////)
S.0021          GO TO 2
S.0022          RETURN
S.0023          END
```

Figure 6.2 A Fortran program for solving two-mesh resistive networks

the value of DELTA is first calculated and then computer statements S.0017 and S.0018 compute the values of I_1 (called C1 because of mode) and I_2 (C2). These results, in turn, are printed by computer statements S.0019 and S.0020 and control is returned to S.0003 so that the next set of equations can be solved.

Figure 6.3 shows the corresponding results for four different data sets. Notice that although the answers to the first data set do not appear to be correct, they can be easily verified by computing the individual element drops and summing the voltages about the two meshes. Since both voltages were equal to zero in the second set, the currents should also be zero. The network for the third and fourth set was modified by changing the middle branch resistance to -2 ohms. Since this causes the value of DELTA to be equal to zero, the third set has no unique solution. Since the fourth set has zero excitation, this results in an infinite number of solutions. The reader may verify this fact by arbitrarily choosing any value for I_1 and then, using the first equation, he may solve for the value of I_2. These results may then be substituted to show that they satisfy both equations.

```
TWO-MESH NETWORK SOLUTION


A11 =  0.6000000E 01   A12 = -0.3000000E 01   V1 =  0.1000000E 02
A21 = -0.3000000E 01   A22 =  0.9000000E 01   V2 = -0.5000000E 01


I1 =  0.1666666E 01   I2 =  0.0


A11 =  0.6000000E 01   A12 = -0.3000000E 01   V1 =  0.0
A21 = -0.3000000E 01   A22 =  0.9000000E 01   V2 =  0.0


I1 =  0.0              I2 =  0.0


A11 =  0.1000000E 01   A12 =  0.2000000E 01   V1 =  0.1000000E 02
A21 =  0.2000000E 01   A22 =  0.4000000E 01   V2 = -0.5000000E 01


THIS SYSTEM HAS NO UNIQUE SOLUTION


A11 =  0.1000000E 01   A12 =  0.2000000E 01   V1 =  0.0
A21 =  0.2000000E 01   A22 =  0.4000000E 01   V2 =  0.0


THIS SYSTEM HAS AN INFINITE NUMBER OF SOLUTIONS
```

Figure 6.3 The output results of the program of Figure 6.2

6.2 THE DETERMINANT SOLUTION

In the last section we used Cramer's rule in the solution of a two-mesh network. We may also use a determinant solution in solving higher-order systems. For example, a three-mesh network would be described by the following set of equations

$$\begin{aligned} a_{11}I_1 + a_{12}I_2 + a_{13}I_3 &= V_1 \\ a_{21}I_1 + a_{22}I_2 + a_{23}I_3 &= V_2 \\ a_{31}I_1 + a_{32}I_2 + a_{33}I_3 &= V_3 \end{aligned}$$

and by Cramer's rule if $\Delta \neq 0$, the solution would be given by

$$I_1 = \frac{\begin{vmatrix} V_1 & a_{12} & a_{13} \\ V_2 & a_{22} & a_{23} \\ V_3 & a_{32} & a_{33} \end{vmatrix}}{\Delta}, \quad I_2 = \frac{\begin{vmatrix} a_{11} & V_1 & a_{13} \\ a_{21} & V_2 & a_{23} \\ a_{31} & V_3 & a_{33} \end{vmatrix}}{\Delta}, \quad \text{and } I_3 = \frac{\begin{vmatrix} a_{11} & a_{12} & V_1 \\ a_{21} & a_{22} & V_2 \\ a_{31} & a_{32} & V_3 \end{vmatrix}}{\Delta}$$

where

$$\Delta = \begin{vmatrix} a_{11} & a_{12} & a_{13} \\ a_{21} & a_{22} & a_{23} \\ a_{31} & a_{32} & a_{33} \end{vmatrix}$$

Each determinant in the above solution may be evaluated by expanding the determinant about any column (or row) into three terms where each term is the product of an element and its cofactor. For example,

$$\begin{aligned} \Delta = \begin{vmatrix} a_{11} & a_{12} & a_{13} \\ a_{21} & a_{22} & a_{23} \\ a_{31} & a_{32} & a_{33} \end{vmatrix} &= a_{11}\Delta_{11} + a_{21}\Delta_{21} + a_{31}\Delta_{31} \\ &\text{or} \\ &= a_{12}\Delta_{12} + a_{22}\Delta_{22} + a_{32}\Delta_{32} \\ &\text{or} \\ &= a_{13}\Delta_{13} + a_{23}\Delta_{23} + a_{33}\Delta_{33} \\ &\text{or} \\ &= a_{11}\Delta_{11} + a_{12}\Delta_{12} + a_{13}\Delta_{13} \\ &\text{or} \\ &= a_{21}\Delta_{21} + a_{22}\Delta_{22} + a_{23}\Delta_{23} \\ &\text{or} \\ &= a_{31}\Delta_{31} + a_{32}\Delta_{32} + a_{33}\Delta_{33} \end{aligned}$$

where the cofactors are given by

$$\Delta_{IJ} = (-1)^{I+J}M_{IJ}$$

and the minor M_{IJ} is formed by deleting the I row and the J column of Δ.

If we choose to evaluate Δ by expanding about the first column, we obtain

$$\Delta = \begin{vmatrix} a_{11} & a_{12} & a_{13} \\ a_{21} & a_{22} & a_{23} \\ a_{31} & a_{32} & a_{33} \end{vmatrix} = a_{11}\Delta_{11} + a_{21}\Delta_{21} + a_{31}\Delta_{31}$$

$$= a_{11}\begin{vmatrix} a_{22} & a_{23} \\ a_{32} & a_{33} \end{vmatrix} - a_{21}\begin{vmatrix} a_{12} & a_{13} \\ a_{32} & a_{33} \end{vmatrix} + a_{31}\begin{vmatrix} a_{12} & a_{13} \\ a_{22} & a_{23} \end{vmatrix}$$

and using the difference of the diagonal products rule to evaluate the second-order determinants, we find

$$\Delta = a_{11}(a_{22}a_{33} - a_{32}a_{23}) - a_{21}(a_{12}a_{33} - a_{32}a_{13}) + a_{31}(a_{12}a_{23} - a_{22}a_{13})$$

The evaluation of Δ would thus require nine multiplications and five additions. In a similar way, each of the numerator determinants of I_1, I_2, and I_3 could be evaluated, and thus the total number of arithmetic operations required to solve for the currents would be equal to 36 multiplications, three divisions, and 20 additions.

Since a computer solution of a third-order determinant is relatively simple, let us consider the solution of the four-mesh network shown in Figure 6.4. Writing Kirchhoff's voltage law for each of the four meshes, we obtain

$$\begin{aligned} 4I_1 - 2I_2 - I_3 \quad\quad &= 10 \\ -2I_1 + 9I_2 \quad\quad - 5I_4 &= -5 \\ - I_1 \quad + 6I_3 - 3I_4 &= 0 \\ - 5I_2 - 3I_3 + 10I_4 &= 0 \end{aligned}$$

and writing these four equations in general form, we have

$$\begin{aligned} a_{11}I_1 + a_{12}I_2 + a_{13}I_3 + a_{14}I_4 &= V_1 \\ a_{21}I_1 + a_{22}I_2 + a_{23}I_3 + a_{24}I_4 &= V_2 \\ a_{31}I_1 + a_{32}I_2 + a_{33}I_3 + a_{34}I_4 &= V_3 \\ a_{41}I_1 + a_{42}I_2 + a_{43}I_3 + a_{44}I_4 &= V_4 \end{aligned}$$

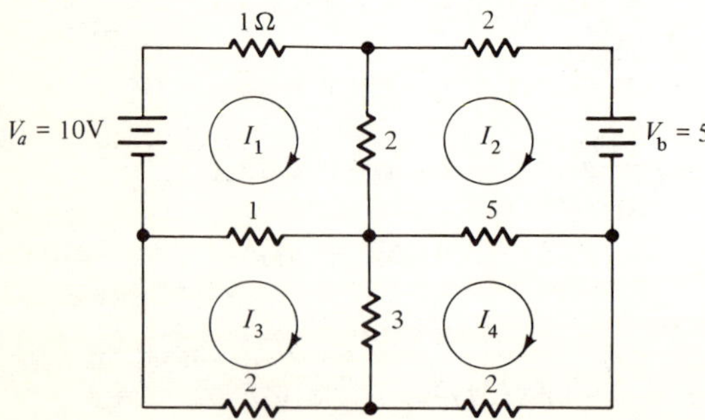

Figure 6.4 A four-mesh network

and by Cramer's rule, if $\Delta \neq 0$, the solution would be given by

$$I_1 = \frac{\begin{vmatrix} V_1 & a_{12} & a_{13} & a_{14} \\ V_2 & a_{22} & a_{23} & a_{24} \\ V_3 & a_{32} & a_{33} & a_{34} \\ V_4 & a_{42} & a_{43} & a_{44} \end{vmatrix}}{\Delta} \qquad I_2 = \frac{\begin{vmatrix} a_{11} & V_1 & a_{13} & a_{14} \\ a_{21} & V_2 & a_{23} & a_{24} \\ a_{31} & V_3 & a_{33} & a_{34} \\ a_{41} & V_4 & a_{43} & a_{44} \end{vmatrix}}{\Delta}$$

$$I_3 = \frac{\begin{vmatrix} a_{11} & a_{12} & V_1 & a_{14} \\ a_{21} & a_{22} & V_2 & a_{24} \\ a_{31} & a_{32} & V_3 & a_{34} \\ a_{41} & a_{42} & V_4 & a_{44} \end{vmatrix}}{\Delta} \qquad I_4 = \frac{\begin{vmatrix} a_{11} & a_{12} & a_{13} & V_1 \\ a_{21} & a_{22} & a_{23} & V_2 \\ a_{31} & a_{32} & a_{33} & V_3 \\ a_{41} & a_{42} & a_{43} & V_4 \end{vmatrix}}{\Delta}$$

where

$$\Delta = \begin{vmatrix} a_{11} & a_{12} & a_{13} & a_{14} \\ a_{21} & a_{22} & a_{23} & a_{24} \\ a_{31} & a_{32} & a_{33} & a_{34} \\ a_{41} & a_{42} & a_{43} & a_{44} \end{vmatrix}$$

We will now write a computer program that will first evaluate Δ by expanding the determinant about the first column

$$\begin{aligned}
\begin{vmatrix} a_{11} & a_{12} & a_{13} & a_{14} \\ a_{21} & a_{22} & a_{23} & a_{24} \\ a_{31} & a_{32} & a_{33} & a_{34} \\ a_{41} & a_{42} & a_{43} & a_{44} \end{vmatrix}
&= a_{11} \begin{vmatrix} a_{22} & a_{23} & a_{24} \\ a_{32} & a_{33} & a_{34} \\ a_{42} & a_{43} & a_{44} \end{vmatrix} - a_{21} \begin{vmatrix} a_{12} & a_{13} & a_{14} \\ a_{32} & a_{33} & a_{34} \\ a_{42} & a_{43} & a_{44} \end{vmatrix} \\
&\quad + a_{31} \begin{vmatrix} a_{12} & a_{13} & a_{14} \\ a_{22} & a_{23} & a_{24} \\ a_{42} & a_{43} & a_{44} \end{vmatrix} - a_{41} \begin{vmatrix} a_{12} & a_{13} & a_{14} \\ a_{22} & a_{23} & a_{24} \\ a_{32} & a_{33} & a_{34} \end{vmatrix} \\
&= a_{11} \; a_{22} \begin{vmatrix} a_{33} & a_{34} \\ a_{43} & a_{44} \end{vmatrix} - a_{11} \; a_{32} \begin{vmatrix} a_{23} & a_{24} \\ a_{43} & a_{44} \end{vmatrix} \\
&\quad + a_{11} \; a_{42} \begin{vmatrix} a_{23} & a_{24} \\ a_{33} & a_{34} \end{vmatrix} - a_{21} \; a_{12} \begin{vmatrix} a_{33} & a_{34} \\ a_{43} & a_{44} \end{vmatrix} \\
&\quad + a_{21} \; a_{32} \begin{vmatrix} a_{13} & a_{14} \\ a_{43} & a_{44} \end{vmatrix} - a_{21} \; a_{42} \begin{vmatrix} a_{13} & a_{14} \\ a_{33} & a_{34} \end{vmatrix} \\
&\quad + a_{31} \; a_{12} \begin{vmatrix} a_{23} & a_{24} \\ a_{43} & a_{44} \end{vmatrix} - a_{31} \; a_{22} \begin{vmatrix} a_{13} & a_{14} \\ a_{43} & a_{44} \end{vmatrix} \\
&\quad + a_{31} \; a_{42} \begin{vmatrix} a_{13} & a_{14} \\ a_{23} & a_{24} \end{vmatrix} - a_{41} \; a_{12} \begin{vmatrix} a_{23} & a_{24} \\ a_{33} & a_{34} \end{vmatrix} \\
&\quad + a_{41} \; a_{22} \begin{vmatrix} a_{13} & a_{14} \\ a_{33} & a_{34} \end{vmatrix} - a_{41} \; a_{32} \begin{vmatrix} a_{13} & a_{14} \\ a_{23} & a_{24} \end{vmatrix}
\end{aligned}$$

$$
\begin{aligned}
= \; & a_{11} \; a_{22}(a_{33} \; a_{44} - a_{43} \; a_{34}) - a_{11} \; a_{32}(a_{23} \; a_{44} - a_{43} \; a_{24}) \\
+ & a_{11} \; a_{42}(a_{23} \; a_{34} - a_{33} \; a_{24}) - a_{21} \; a_{12}(a_{33} \; a_{44} - a_{43} \; a_{34}) \\
+ & a_{21} \; a_{32}(a_{13} \; a_{44} - a_{43} \; a_{14}) - a_{21} \; a_{42}(a_{13} \; a_{34} - a_{33} \; a_{14}) \\
+ & a_{31} \; a_{12}(a_{23} \; a_{44} - a_{43} \; a_{24}) - a_{31} \; a_{22}(a_{13} \; a_{44} - a_{43} \; a_{14}) \\
+ & a_{31} \; a_{42}(a_{13} \; a_{24} - a_{23} \; a_{14}) - a_{41} \; a_{12}(a_{23} \; a_{34} - a_{33} \; a_{24}) \\
+ & a_{41} \; a_{22}(a_{13} \; a_{34} - a_{33} \; a_{14}) - a_{41} \; a_{32}(a_{13} \; a_{24} - a_{23} \; a_{14})
\end{aligned}
$$

and then evaluating each of the four remaining determinants by repeating the above process, each time replacing each of the respective columns of Δ by the right-member voltage column matrix.

Figure 6.5 is the corresponding Fortran program that solves a general four-mesh network using the determinant expansion scheme discussed above. Initially the heading is printed and the coefficients are read into the computer. Computer statements S.0008 and S.0009 are then used to print the coefficients. The outermost DO 12 loop is then entered and each iteration of this loop will evaluate one of the five determinants. Initially IGO is set equal to one so that computer statement S.0011 transfers control to statement 13 (S.0030). Computer statements S.0030 through S.0055 are then used to evaluate Δ. Initially SN is set equal to -1 and we enter the middle DO 12 loop. Here we set I equal to one and SCALE1 equal to a_{11}. We next enter the lower DO 12 loop, setting J equal to one, and then test to see if I equals J. Since I does not equal J for the first iteration of this loop, control is returned to computer statement S.0034, and J becomes equal to two. This time I does not equal J, and, as a result, SN becomes equal to plus one, and SCALE is set equal to $a_{11}a_{22}$. The DO 22 loop then sets each of the four elements of $N(K)$ to one. Entering the DO 26 loop, we initially set M equal to one and test to see whether M is equal to I. Since M will be equal to I for the first iteration of this loop, control is transferred to statement 25 and $N(1)$ is set equal to zero. During the second iteration of the DO 26 loop, M will not be equal to I, and we therefore test to see whether M is equal to J. Since M will be equal to J for the second iteration of this loop, $N(2)$ is set equal to zero. During the third and fourth iteration, we find that M is not equal to either I or J and therefore $N(3)$ and $N(4)$ remain equal to one. We next set $N1$ equal to zero and enter the DO 30 loop. In the first iteration of this loop, since $N1$ is equal to zero, control is transferred to statement 29, and we test to see whether $N(1)$ is greater than zero. Since $N(1)$ is not greater than zero for this iteration, L becomes equal to two, and we execute the DO 30 loop a second time. We now find $N(2)$ is equal to zero, and thus L becomes equal to three. This time we find $N(3)$ is greater than zero and we therefore set $N1$ equal to 3. In the fourth iteration of the DO 30 loop, since $N1$ equals 3, control is transferred to statement 28 and since $N(4)$ is greater than zero, we set $N2$ equal to 4.

```
LEVEL: 15JUL68                              IBM OS/360 BASIC FORTRAN IV (E) COMPILATION

             C        DETERMINANT SOLUTION
             C              OF
             C        A FOUR-MESH NETWORK
             C
S.0001                DIMENSION A(4,4),V(4),SUM(5),B(4),N(4),C(4)
S.0002                WRITE(6,1)
S.0003              1 FORMAT('1FOUR-MESH NETWORK SOLUTION',///)
S.0004                DO 10 I=1,4
S.0005             10 READ(5,2) A(I,1),A(I,2),A(I,3),A(I,4),V(I)
S.0006              2 FORMAT(5E15.7)
S.0007                DO 11 I=1,4
S.0008             11 WRITE(6,3) I,A(I,1),I,A(I,2),I,A(I,3),I,A(I,4),I,V(I)
S.0009              3 FORMAT(' A(',I1,',1) =',E15.7,' A(',I1,',2) =',E15.7,' A(',I1,',
                  13) =',E15.7,' A(',I1,',4) =',E15.7,' V(',I1,') =',E15.7,/)
S.0010                DO 12 IGO=1,5
S.0011                GO TO (13,14,15,16,17),IGO
S.0012             14 DO 18 II=1,4
S.0013                B(II)=A(II,1)
S.0014             18 A(II,1)=V(II)
S.0015                GO TO 13
S.0016             15 DO 19 JJ=1,4
S.0017                A(JJ,1)=B(JJ)
S.0018                B(JJ)=A(JJ,2)
S.0019             19 A(JJ,2)=V(JJ)
S.0020                GO TO 13
S.0021             16 DO 20 KK=1,4
S.0022                A(KK,2)=B(KK)
S.0023                B(KK)=A(KK,3)
S.0024             20 A(KK,3)=V(KK)
S.0025                GO TO 13
S.0026             17 DO 21 LL=1,4
S.0027                A(LL,3)=B(LL)
S.0028                B(LL)=A(LL,4)
S.0029             21 A(LL,4)=V(LL)
S.0030             13 SUM(IGO)=0.0
S.0031                SN=-1.
S.0032                DO 12 I=1,4
S.0033                SCALE1=A(I,1)
S.0034                DO 12 J=1,4
S.0035                IF(I-J) 5,12,5
S.0036              5 SN=-SN
S.0037                SCALE= SN*SCALE1*A(J,2)
S.0038                DO 22 K=1,4
S.0039             22 N(K)=1
S.0040                DO 26 M=1,4
S.0041                IF(M-I) 24,25,24
S.0042             24 IF(M-J)26,25,26
S.0043             25 N(M)=0
S.0044             26 CONTINUE
S.0045                N1=0
S.0046                DO 30 L=1,4
S.0047                IF(N1)28,29,28
S.0048             29 IF(N(L)) 30,30,31
S.0049             31 N1=L
S.0050                GO TO 30
S.0051             28 IF(N(L))30,30,32
S.0052             32 N2=L
S.0053                SUM(IGO)=SUM(IGO)+(A(N1,3)*A(N2,4)-A(N2,3)*A(N1,4))*SCALE
S.0054                GO TO 12
S.0055             30 CONTINUE
S.0056             12 CONTINUE
S.0057                DELTA=SUM(1)
S.0058                DO 35 I=1,4
S.0059                C(I)=SUM(I+1)/DELTA
S.0060             35 WRITE(6,4) I,C(I)
S.0061              4 FORMAT(//,' 1(',I1,') =',E15.7,/)
S.0062                RETURN
S.0063                END
```

Figure 6.5 The determinant solution of a four-mesh network

Computer statement S.0053 now evaluates SUM(1), and we see that this is equal to

$$a_{11}a_{22}(a_{33}a_{44} - a_{43}a_{34})$$

which is the first term of the fully expanded form of Δ.

Having completed the second iteration of the inner DO 12 loop, we transfer to computer statement S.0034 and J becomes equal to three. This iteration of the lower DO 12 loop will make

$$\text{SUM}(1) = a_{11}a_{22}(a_{33}a_{44} - a_{43}a_{34}) - a_{11}a_{32}(a_{23}a_{44} - a_{43}a_{24})$$

where the second term is the contribution to SUM(1) due to this third iteration. In a similar way, the fourth iteration of the lower DO 12 loop adds

$$+a_{11}a_{42}(a_{23}a_{34} - a_{33}a_{24})$$

to SUM(1). At this point control is transferred to the middle DO 12 loop, and the next four iterations of the lower DO 12 loop add the next three terms of the expanded form of Δ to SUM(1). This process is continued until all 12 terms have been generated and summed. Control is then transferred to the outer DO 12 loop (computer statement S.0010) and IGO is set equal to two. The computed GO TO statement now transfers control to statement 14 and the DO 18 loop is used to form the numerator determinant of I_1. Control is then transferred to computer statement S.0030 and, as before, the middle and lower DO 12 loops are used to evaluate this determinant. This process is continued until all five determinants have been evaluated and then the DO 35 loop is used to evaluate and print the four current values.

Figure 6.6 shows the computer results for the network shown in Figure 6.4. The reader may readily verify the accuracy of these results by summing the element drops around each of the meshes of Figure 6.4. The reader will also find it worthwhile to modify this program so that the program will be capable of solving networks with a greater number of meshes.

We now point out that although it is possible to extend this method so that it can be used to solve for the currents in a network that contains a greater number of meshes, it is generally not advisable to do so when the number of meshes becomes greater than five or six. For example, recall that in the evaluation of the three-mesh network we required 36 multiplications, three divisions, and 20 additions. In the evaluation of the four-mesh network we require 240 multiplications, four divisions, and 165 additions. This means that $(n/2)(n + 1)!$ multiplications are required to obtain the solution by using expansion by minors to evaluate the determinants. When we contrast this with the approximate n^3 multiplications that are required in a Gauss-Jordan method of solution, we see that as far as multiplications are concerned, it does not pay to use a determinant solution for large n. Putting it another way, there would be approximately 25 million multiplications required for a determinant solution of a ten-mesh network in contrast to approximately 1000 multi-

```
FOUR-MESH NETWORK SOLUTION

A(1,1) =  0.4000000E 01  A(1,2) = -0.2000000E 01  A(1,3) = -0.1000000E 01  A(1,4) =  0.0            V(1) =  0.1000000E 02

A(2,1) = -0.2000000E 01  A(2,2) =  0.9000000E 01  A(2,3) =  0.0            A(2,4) = -0.5000000E 01  V(2) = -0.5000000E 01

A(3,1) = -0.1000000E 01  A(3,2) =  0.0            A(3,3) =  0.6000000E 01  A(3,4) = -0.3000000E 01  V(3) =  0.0

A(4,1) =  0.0            A(4,2) = -0.5000000E 01  A(4,3) = -0.3000000E 01  A(4,4) =  0.1000000E 02  V(4) =  0.0


I(1) =  0.2761851E 01


I(2) =  0.2205071E 00


I(3) =  0.6063946E 00


I(4) =  0.2921720E 00
```

Figure 6.6 The output results of the program of Figure 6.5

plications that would be required by a Gauss-Jordan solution. If 100 microseconds are required for each multiplication operation, this would mean that the determinant multiplications would take approximately 40 hours in contrast to the 0.1 seconds required for the multiplications needed in the Gauss-Jordan solution.

6.3 THE GAUSS ELIMINATION METHOD

In order to understand how the Gauss elimination method works, consider the following set of linear equations:

$$\begin{aligned} a_{11}X_1 + a_{12}X_2 + a_{13}X_3 &= a_{14} \\ a_{21}X_1 + a_{22}X_2 + a_{23}X_3 &= a_{24} \\ a_{31}X_1 + a_{32}X_2 + a_{33}X_3 &= a_{34} \end{aligned}$$

Using the arithmetic operations of multiplication and addition, we force these equations into the following triangular form:

$$\begin{aligned} X_1 + b_{12}X_2 + b_{13}X_3 &= b_{14} \\ X_2 + b_{23}X_3 &= b_{24} \\ X_3 &= b_{34} \end{aligned}$$

The solution may now be found by starting with the last equation and solving back to obtain

$$\begin{aligned} X_3 &= b_{34} \\ X_2 &= b_{24} - b_{23}X_3 \\ X_1 &= b_{14} - b_{12}X_2 - b_{13}X_3 \end{aligned}$$

It should be clear at this point that this method may also be applied to find the solution of higher-order simultaneous equations and all that remains to be explained is how we force the equations into the triangular form.

Using the three original equations to illustrate the general procedure for obtaining the triangular form, we start by

(1) Dividing the first equation by a_{11}. This normalizing step will lessen the possibility of overflow or underflow in the subsequent calculations. In the event that $a_{11} = 0$ (or the magnitude of a_{11} is very small), we must interchange the first equation with one of the other two to avoid dividing by zero. At this time we will ignore this possibility and assume $a_{11} \neq 0$; thus we have

$$\begin{aligned} X_1 + \frac{a_{12}}{a_{11}}X_2 + \frac{a_{13}}{a_{11}}X_3 &= \frac{a_{14}}{a_{11}} \\ a_{21}X_1 + a_{22}X_2 + a_{23}X_3 &= a_{24} \\ a_{31}X_1 + a_{32}X_2 + a_{33}X_3 &= a_{34} \end{aligned}$$

and (2) manipulating the equations so that the coefficient of the X_1 term in the second and third equation equals zero. This may be accomplished by multiplying the first equation obtained by step (1) by a_{21} and then a_{31} and, in turn, subtracting each of these new equations from the original second and third equations; thus

$$\begin{aligned} X_1 \quad + \frac{a_{12}}{a_{11}} X_2 + \quad \frac{a_{13}}{a_{11}} X_3 &= \frac{a_{14}}{a_{11}} \\ \left(a_{22} - a_{21} \frac{a_{12}}{a_{11}}\right) X_2 + \left(a_{23} - a_{21} \frac{a_{13}}{a_{11}}\right) X_3 &= a_{24} - a_{21} \frac{a_{14}}{a_{11}} \\ \left(a_{32} - a_{31} \frac{a_{12}}{a_{11}}\right) X_2 + \left(a_{33} - a_{31} \frac{a_{13}}{a_{11}}\right) X_3 &= a_{34} - a_{31} \frac{a_{14}}{a_{11}} \end{aligned}$$

Steps (1) and (2) are now repeated on the resulting second and third equations, obtaining

$$\begin{aligned} X_1 + \frac{a_{12}}{a_{11}} X_2 + \frac{a_{13}}{a_{11}} X_3 &= \frac{a_{14}}{a_{11}} \\ X_2 + C_{23} X_3 &= C_{24} \\ C_{33} X_3 &= C_{34} \end{aligned}$$

where

$$C_{23} = \frac{a_{23} - a_{21} \dfrac{a_{13}}{a_{11}}}{a_{22} - a_{21} \dfrac{a_{12}}{a_{11}}}$$

$$C_{24} = \frac{a_{24} - a_{21} \dfrac{a_{14}}{a_{11}}}{a_{22} - a_{21} \dfrac{a_{12}}{a_{11}}}$$

$$C_{33} = \left(a_{33} - a_{31} \frac{a_{13}}{a_{11}}\right) - \left(a_{32} - a_{31} \frac{a_{12}}{a_{11}}\right) C_{23}$$

and

$$C_{34} = \left(a_{34} - a_{31} \frac{a_{14}}{a_{11}}\right) - \left(a_{32} - a_{31} \frac{a_{12}}{a_{11}}\right) C_{24}$$

Step (1) is now repeated on the resulting third equation, obtaining the required triangular form

$$\begin{aligned} X_1 + \frac{a_{12}}{a_{11}} X_2 + \frac{a_{13}}{a_{11}} X_3 &= \frac{a_{14}}{a_{11}} \\ X_2 + C_{23} X_3 &= C_{24} \\ X_3 &= D_{34} \end{aligned}$$

where

$$D_{34} = \frac{C_{34}}{C_{33}}$$

For a simple numerical example that illustrates this procedure consider the solution of the following set of equations:

$$\begin{aligned} -2X_1 - X_2 + X_3 &= -1 \\ X_1 + X_2 + X_3 &= 6 \\ 3X_1 + X_2 - X_3 &= 2 \end{aligned}$$

Applying step (1) we find

$$\begin{aligned} X_1 + 0.5X_2 - 0.5X_3 &= 0.5 \\ X_1 + X_2 + X_3 &= 6.0 \\ 3X_1 + X_2 - X_3 &= 2.0 \end{aligned}$$

Applying step (2) yields

$$\begin{aligned} X_1 + 0.5X_2 - 0.5X_3 &= 0.5 \\ 0.5X_2 + 1.5X_3 &= 5.5 \\ -0.5X_2 + 0.5X_3 &= 0.5 \end{aligned}$$

Repeating the procedure on the resulting second and third equations yields

$$\begin{aligned} X_1 + 0.5X_2 - 0.5X_3 &= 0.5 \\ X_2 + 3.0X_3 &= 11.0 \\ -0.5X_2 + 0.5X_3 &= 0.5 \end{aligned}$$

and

$$\begin{aligned} X_1 + 0.5X_2 - 0.5X_3 &= 0.5 \\ X_2 + 3.0X_3 &= 11.0 \\ 2.0X_3 &= 6.0 \end{aligned}$$

Finally, repeating step (1) on the resulting third equation we obtain the required triangular form:

$$\begin{aligned} X_1 + 0.5X_2 - 0.5X_3 &= 0.5 \\ X_2 + 3.0X_3 &= 11.0 \\ X_3 &= 3.0 \end{aligned}$$

These equations may now be solved back to obtain the solution, thus

$$\begin{aligned} X_3 &= 3.0 \\ X_2 &= 11.0 - 3.0(3.0) = 2.0 \\ X_1 &= 0.5 - 0.5(2.0) + 0.5(3.0) = 1.0 \end{aligned}$$

Figure 6.7 is a general Fortran program that uses the Gauss elimination method to solve simultaneous linear equations that may be tenth order or less. Initially the heading is printed and the number of equations N and the coefficient minimum magnitude EPS are read into memory. Computer statement S.0006 then computes the value of M and computer statements S.0007 and S.0008 are used to read in the elements of the augmented matrix. The dummy variables KK and JJ are then set equal to zero before entering the outer DO 10 loop. Computer statements S.0012 through S.0022 are used to interchange rows if the magnitude of the pivot coefficient is less than or equal to EPS. For example, in the first iteration of the outer DO 10 loop, I equals one, and initially JJ, LL, and KK are set equal to one. Computer statement S.0015 then tests to

```
          C      SOLUTION OF SIMULTANEOUS LINEAR ALGEBRAIC EQUATIONS
          C                           BY
          C                 THE GAUSS ELIMINATION METHOD
          C
S.0001           DIMENSION A(10,11),X(10)
S.0002           WRITE(6,1)
S.0003         1 FORMAT('1THE SOLUTION OF SIMULTANEOUS LINEAR EQUATIONS',/,16X,'BY'
                1,/,5X,'THE GAUSS ELIMINATION METHOD',////)
S.0004           READ(5,2)N,EPS
S.0005         2 FORMAT(I5,E15.7)
S.0006           M=N+1
S.0007           READ(5,3)((A(I,J),J=1,M),I=1,N)
S.0008         3 FORMAT(5E15.7)
S.0009           KK=0
S.0010           JJ=0
S.0011           DO 10 I=1,N
S.0012           JJ=KK+1
S.0013           LL=JJ
S.0014           KK=KK+1
S.0015        20 IF(ABS(A(JJ,KK))-EPS)21,21,22
S.0016        21 JJ=JJ+1
S.0017           GO TO 20
S.0018        22 IF(LL-JJ)23,24,23
S.0019        23 DO 25 MM=1,M
S.0020           ATEMP=A(LL,MM)
S.0021           A(LL,MM)=A(JJ,MM)
S.0022        25 A(JJ,MM)=ATEMP
          C
          C      THE EQUATIONS ARE FORCED INTO THE TRIANGLE FORM
          C
S.0023        24 DIV=A(I,I)
S.0024           DO 11 J=I,M
S.0025        11 A(I,J)=A(I,J)/DIV
S.0026           K=I+1
S.0027           IF(K-M)12,13,13
S.0028        12 DO 10 L=K,N
S.0029           AMULT=A(L,I)
S.0030           DO 10 J=I,M
S.0031        10 A(L,J)=A(L,J)-A(I,J)*AMULT
          C
          C      BACK SUBSTITUTION
          C
S.0032        13 X(N)=A(N,M)
S.0033           L=N
S.0034           DO 30 J=2,N
S.0035           SUM=0.0
S.0036           I=M+1-J
S.0037           DO 31 K=I,N
S.0038        31 SUM=SUM+A(I-1,K)*X(K)
S.0039           L=L-1
S.0040        30 X(L)=A(I-1,M)-SUM
S.0041           WRITE(6,40)(I,X(I),I=1,N)
S.0042        40 FORMAT(//,' X(',I2,') =',E15.7,/)
S.0043           STOP
S.0044           END
```

Figure 6.7 A program that uses the Gauss elimination method to solve simultaneous linear algebraic equations

see whether $|A_{11}| > \text{EPS}$ (0.1E-5). If $|A_{11}| \leq \text{EPS}$, control is transferred to statement 21, and JJ becomes equal to two. We then test to see if $|A_{21}| > \text{EPS}$. Assuming that it is, control is transferred to statement 22 where we test to see if LL equals JJ. Since LL will not equal JJ for our assumption, control is transferred to statement 23 (S.0019) and the DO 25 loop is used to interchange row one and two.

Computer statements S.0023 through S.0025 are then used to carry out step (1) of the elimination procedure. Computer statement S.0026 now sets K equal to $I + 1$ $(=2)$ and we next test to see whether K equals M. Since K will not be equal to M in the first iteration of the outer DO 10 loop, we enter the middle DO 10 loop and set L equal to K (or 2) and begin to carry out step (2) of the elimination procedure. This is done by making the dummy variable AMULT equal to A_{21} before entering the lower DO 10 loop. The lower DO 10 loop then forms a new row two with $a_{21} = 0$. Returning to the middle DO 10 loop, L becomes equal to three, and AMULT becomes equal to A_{31}. The execution of the lower DO 10 loop forms a new row three with $a_{31} = 0$. This process is repeated until the middle DO loop forms new rows two through N; with a_{21}, a_{31}, . . . , a_{N1} equal to zero. Control is then returned to the outer DO 10 loop, and we repeat the procedure, using the last $N - 1$ equations that resulted from the first iteration of the outer DO 10 loop. This outer DO 10 iteration continues (N times) until the original equations are forced into the required triangular form. Computer statements S.0032 through S.0040 are then used to solve the equations back and computer statements S.0041 and S.0042 are used to print the output results.

Let us now use this program to solve for the branch currents of the network shown in Figure 6.8. Since there are ten unknown branch currents, Kirchhoff's current and voltage laws must be used to provide a suitable set of ten equations. These may be formulated by first choosing the arbitrary positive direction of current for each of the branches as indicated by the arrows in the network. The current law may then be used at each of the independent nodes to obtain a set of $N - 1$ $(6 - 1 = 5)$

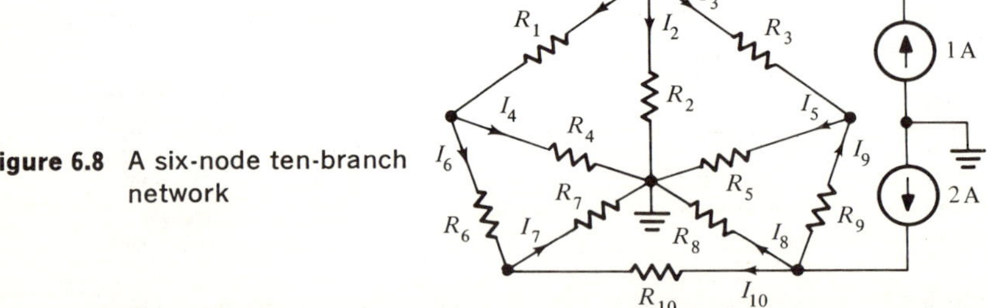

Figure 6.8 A six-node ten-branch network

independent equations

$$
\begin{array}{lllllllll}
I_1 + I_2 + I_3 & & & & & & & = 1.0 \\
 & & & & & +I_8 + I_9 + I_{10} & & = 2.0 \\
-I_1 & +I_4 & & -I_6 & & & & = 0.0 \\
-I_3 & & +I_5 & & & -I_9 & & = 0.0 \\
 & & & I_6 + I_7 & & & - I_{10} & = 0.0
\end{array}
$$

The remaining $B - N + 1$ $(10 - 6 + 1 = 5)$ independent equations can then be obtained by writing Kirchhoff's voltage law for each of the elementary meshes

$$
\begin{array}{llllll}
 & & & -R_7I_7 + R_8I_8 & -R_{10}I_{10} & = 0 \\
 & & -R_5I_5 & + R_8I_8 - R_9I_9 & & = 0 \\
R_2I_2 - R_3I_3 & & -R_5I_5 & & & = 0 \\
-R_1I_1 + R_2I_2 & -R_4I_4 & & & & = 0 \\
 & -R_4I_4 & & -R_6I_6 + R_7I_7 & & = 0
\end{array}
$$

Since $R_1 = 1$ ohm, $R_2 = 2$ ohms, . . . , and $R_{10} = 10$ ohms; the augmented matrix of this network thus becomes

$$
\left[\begin{array}{cccccccccc|c}
1.0 & 1.0 & 1.0 & 0.0 & 0.0 & 0.0 & 0.0 & 0.0 & 0.0 & 0.0 & 1.0 \\
0.0 & 0.0 & 0.0 & 0.0 & 0.0 & 0.0 & 0.0 & 1.0 & 1.0 & 1.0 & 2.0 \\
-1.0 & 0.0 & 0.0 & 1.0 & 0.0 & -1.0 & 0.0 & 0.0 & 0.0 & 0.0 & 0.0 \\
0.0 & 0.0 & -1.0 & 0.0 & 1.0 & 0.0 & 0.0 & 0.0 & -1.0 & 0.0 & 0.0 \\
0.0 & 0.0 & 0.0 & 0.0 & 0.0 & 1.0 & 1.0 & 0.0 & 0.0 & -1.0 & 0.0 \\
0.0 & 0.0 & 0.0 & 0.0 & 0.0 & 0.0 & -7.0 & 8.0 & 0.0 & -10.0 & 0.0 \\
0.0 & 0.0 & 0.0 & 0.0 & -5.0 & 0.0 & 0.0 & 8.0 & 9.0 & 0.0 & 0.0 \\
0.0 & 2.0 & -3.0 & 0.0 & -5.0 & 0.0 & 0.0 & 0.0 & 0.0 & 0.0 & 0.0 \\
-1.0 & 2.0 & 0.0 & -4.0 & 0.0 & 0.0 & 0.0 & 0.0 & 0.0 & 0.0 & 0.0 \\
0.0 & 0.0 & 0.0 & -4.0 & 0.0 & -6.0 & 7.0 & 0.0 & 0.0 & 0.0 & 0.0
\end{array}\right]
$$

Coefficient Matrix (first ten columns); Column Matrix (last column)

Augmented Matrix

Figure 6.9 shows the output results for this network. The accuracy of these answers may be checked by substituting these values into the original ten equations. For example, substituting the results for I_8, I_9, and I_{10} in equation two yields

$$0.9256420 + 0.5778688 + 0.496892 = 2.0000000$$

which satisfies the equation exactly. It will be left as an exercise for the reader to check the remaining equations.

To further aid the reader in understanding the Gauss elimination process, the program of Figure 6.7 was modified by inserting the following

```
THE SOLUTION OF SIMULTANEOUS LINEAR EQUATIONS
                      BY
         THE GAUSS ELIMINATION METHOD


X( 1) =  0.2403544E 00

X( 2) =  0.8966507E 00

X( 3) = -0.1370052E 00

X( 4) =  0.3882380E 00

X( 5) =  0.4408636E 00

X( 6) =  0.1478835E 00

X( 7) =  0.3486057E 00

X( 8) =  0.9256420E 00

X( 9) =  0.5778688E 00

X(10) =  0.4964892E 00
```

Figure 6.9 The output results of the program of Figure 6.7 for the network of Figure 6.8

WRITE statement after computer statements S.0008 and S.0026

```
      WRITE(6,60)((A(II,JJ),JJ=1,5),II=1,4)
   60 FORMAT(5E15.7)
```

The augmented matrix data of the network of Figure 6.4 was then used as the input data to this modified program. Figure 6.10 shows the corresponding output results. The reader is encouraged to check each of these printed results by hand-calculating each of the statements of the modified program.

The program of Figure 6.7 may also be used in the solution of nodal equations. For example, consider once again the network shown in Figure 6.8. Choosing the node voltages indicated in Figure 6.11, we write the nodal equations by applying Kirchhoff's current law at each of the independent nodes:

$$
\begin{aligned}
G_2V_{10} + G_3(V_{10} - V_{20}) + G_1(V_{10} - V_{50}) &= 1.0\\
G_5V_{20} + G_9(V_{20} - V_{30}) + G_3(V_{20} - V_{10}) &= 0.0\\
G_8V_{30} + G_{10}(V_{30} - V_{40}) + G_9(V_{30} - V_{20}) &= 2.0\\
G_7V_{40} + G_6(V_{40} - V_{50}) + G_{10}(V_{40} - V_{30}) &= 0.0\\
G_4V_{50} + G_1(V_{50} - V_{10}) + G_6(V_{50} - V_{40}) &= 0.0
\end{aligned}
$$

```
THE SOLUTION OF SIMULTANEOUS LINEAR EQUATIONS
                       BY
    THE GAUSS ELIMINATION METHOD


 0.4000000E 01 -0.2000000E 01 -0.1000000E 01  0.0            0.1000000E 02
-0.2000000E 01  0.9000000E 01  0.0           -0.5000000E 01 -0.5000000E 01
-0.1000000E 01  0.0            0.6000000E 01 -0.3000000E 01  0.0
 0.0           -0.5000000E 01 -0.3000000E 01  0.1000000E 02  0.0
 0.1000000E 01 -0.5000000E 00 -0.2500000E 00  0.0            0.2500000E 01
-0.2000000E 01  0.9000000E 01  0.0           -0.5000000E 01 -0.5000000E 01
-0.1000000E 01  0.0            0.6000000E 01 -0.3000000E 01  0.0
 0.0           -0.5000000E 01 -0.3000000E 01  0.1000000E 02  0.0
 0.1000000E 01 -0.5000000E 00 -0.2500000E 00  0.0            0.2500000E 01
 0.0            0.1000000E 01 -0.6250000E-01 -0.6250000E 00  0.0
 0.0           -0.5000000E 00  0.5750000E 01 -0.3000000E 01  0.2500000E 01
 0.0           -0.5000000E 01 -0.3000000E 01  0.1000000E 02  0.0
 0.1000000E 01 -0.5000000E 00 -0.2500000E 00  0.0            0.2500000E 01
 0.0            0.1000000E 01 -0.6250000E-01 -0.6250000E 00  0.0
 0.0            0.0            0.1000000E 01 -0.5792350E 00  0.4371585E 00
 0.0            0.0           -0.3312500E 01  0.6875000E 01  0.0
 0.1000000E 01 -0.5000000E 00 -0.2500000E 00  0.0            0.2500000E 01
 0.0            0.1000000E 01 -0.6250000E-01 -0.6250000E 00  0.0
 0.0            0.0            0.1000000E 01 -0.5792350E 00  0.4371585E 00
 0.0            0.0            0.0            0.1000000E 01  0.2921718E 00

X( 1) =  0.2761851E 01


X( 2) =  0.2205070E 00


X( 3) =  0.6063945E 00


X( 4) =  0.2921718E 00
```

Figure 6.10 The output results for the program of Figure 6.7 for the solution of the network of Figure 6.4

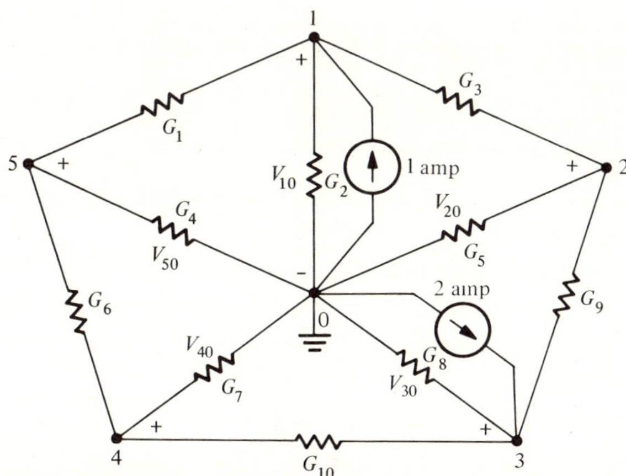

Figure 6.11 A five-independent-node network

```
THE SOLUTION OF SIMULTANEOUS LINEAR EQUATIONS
                BY
     THE GAUSS ELIMINATION METHOD



 0.1333332E 01 -0.3333333E 00  0.0            0.0           -0.1000000E 01  0.1000000E 01
-0.3333333E 00  0.6444443E 00 -0.1111110E 00  0.0            0.0            0.0
 0.0           -0.1111110E 00  0.3351111E 00 -0.9999996E-01  0.0            0.2000000E 01
 0.0            0.0           -0.9999995E-01  0.4095238E 00 -0.1666666E 00  0.0
-0.1000000E 01  0.0            0.0           -0.1666666E 00  0.1416667E 01  0.0
 0.1000000E 01 -0.1818182E 00  0.0            0.0           -0.5454549E 00  0.5454549E 00
-0.3333333E 00  0.6444443E 00 -0.1111110E 00  0.0            0.0            0.0
 0.0           -0.1111110E 00  0.3351111E 00 -0.9999996E-01  0.0            0.2000000E 01
 0.0            0.0           -0.9999996E-01  0.4095238E 00 -0.1666666E 00  0.0
-0.1000000E 01  0.0            0.0           -0.1666666E 00  0.1416667E 01  0.0
 0.1000000E 01 -0.1818182E 00  0.0            0.0           -0.5454549E 00  0.5454549E 00
 0.0            0.1000000E 01 -0.1903113E 00  0.0           -0.3114188E 00  0.3114188E 00
 0.0           -0.1111110E 00  0.3351111E 00 -0.9999996E-01  0.0            0.2000000E 01
 0.0            0.0           -0.9999995E-01  0.4095238E 00 -0.1666666E 00  0.0
 0.0           -0.1818182E 00  0.0           -0.1666666E 00  0.8712121E 00  0.5454549E 00
 0.1000000E 01 -0.1818182E 00  0.0            0.0           -0.5454549E 00  0.5454549E 00
 0.0            0.1000000E 01 -0.1903113E 00  0.0           -0.3114188E 00  0.3114188E 00
 0.0            0.0            0.1000000E 01 -0.3174951E 00 -0.1098599E 00  0.6459762E 01
 0.0            0.0           -0.9999995E-01  0.4095238E 00 -0.1666666E 00  0.0
 0.0            0.0           -0.3450207E-01 -0.1666666E 00  0.8145904E 00  0.6020765E 00
 0.1000000E 01 -0.1818182E 00  0.0            0.0           -0.5454549E 00  0.5454549E 00
 0.0            0.1000000E 01 -0.1903113E 00  0.0           -0.3114188E 00  0.3114188E 00
 0.0            0.0            0.1000000E 01 -0.3174951E 00 -0.1098599E 00  0.6459762E 01
 0.0            0.0            0.0            0.1000000E 01 -0.4702611E 00  0.1709951E 01
 0.0            0.0            0.0           -0.1776525E 00  0.8107890E 00  0.8255976E 00
 0.1000000E 01 -0.1818182E 00  0.0            0.0           -0.5454549E 00  0.5454549E 00
 0.0            0.1000000E 01 -0.1903113E 00  0.0           -0.3114188E 00  0.3114188E 00
 0.0            0.0            0.1000000E 01 -0.3174951E 00 -0.1098599E 00  0.6459762E 01
 0.0            0.0            0.0            0.1000000E 01 -0.4702611E 00  0.1709951E 01
 0.0            0.0            0.0            0.0            0.1000000E 01  0.1552947E 01


X( 1) =  0.1793301E 01


X( 2) =  0.2204315E 01


X( 3) =  0.7405132E 01


X( 4) =  0.2440242E 01


X( 5) =  0.1552947E 01
```

Figure 6.12 The output results of the program of Figure 6.7 for the nodal formulation of Figure 6.11

Rearranging the above equations we find

$$
\begin{aligned}
(G_1 + G_2 + G_3)V_{10} - G_3V_{20} \qquad\qquad\qquad\qquad\qquad\qquad\qquad\qquad -G_1V_{50} &= 1.0 \\
-G_3V_{10} + (G_3 + G_5 + G_9)V_{20} - G_9V_{30} \qquad\qquad\qquad\qquad\qquad\qquad &= 0.0 \\
-G_9V_{20} + (G_8 + G_9 + G_{10})V_{30} - G_{10}V_{40} \qquad\qquad &= 2.0 \\
-G_{10}V_{30} + (G_6 + G_7 + G_{10})V_{40} - G_6V_{50} &= 0.0 \\
-G_1V_{10} \qquad\qquad\qquad\qquad\qquad\qquad -G_6V_{40} + (G_1 + G_4 + G_6)V_{50} &= 0.0
\end{aligned}
$$

Figure 6.12 gives the computer results for the same resistor values that were used previously. The reader may readily verify the accuracy of the nodal solution by using the solution printed for the branch currents shown in Figure 6.9. Note that the augmented matrix has been printed after each iteration step so that the reader may be able to check his hand computed solution.

6.4 THE GAUSS-JORDAN ELIMINATION METHOD

The Gauss-Jordan method forces the set of simultaneous linear equations to have zero coefficients above and below and nonzero coefficients on the main diagonal. For example, in the case of the set of three equations

$$\begin{aligned} a_{11}X_1 + a_{12}X_2 + a_{13}X_3 &= a_{14} \\ a_{21}X_1 + a_{22}X_2 + a_{23}X_3 &= a_{24} \\ a_{31}X_1 + a_{32}X_2 + a_{33}X_3 &= a_{34} \end{aligned}$$

the Gauss-Jordan method would force these into the following diagonal form:

$$\begin{aligned} X_1 \qquad\qquad &= b_{14} \\ X_2 \qquad &= b_{24} \\ X_3 &= b_{34} \end{aligned}$$

As a consequence, the values of X_1, X_2, and X_3 can be found directly without recourse to back substitution.

In order to understand how the equations can be forced into the diagonal form, let us illustrate the procedure in terms of the following numerical example:

$$\begin{aligned} -2X_1 - X_2 + X_3 &= -1 \\ X_1 + X_2 + X_3 &= 6 \\ 3X_1 + X_2 - X_3 &= 2 \end{aligned}$$

Using only the coefficients, we first form the augmented matrix

$$\begin{bmatrix} -2 & -1 & 1 & -1 \\ 1 & 1 & 1 & 6 \\ 3 & 1 & -1 & 2 \end{bmatrix}$$

In order to make the a_{21} coefficient zero, we form a new row two by multiplying the second equation by a_{11} (-2) and subtract from this a_{21} (1) times the first equation; thus,

$$(-2(1) - 1(-2)) = 0 \quad (-2(1) - 1(-1)) = -1 \quad (-2(1) - 1(1)) = -3 \quad (-2(6) - 1(-1)) = -11$$

We now make the a_{31} coefficient zero by forming a new row three. This is done by multiplying the third equation by a_{11} (-2) and subtracting from it a_{31} (3) times the first equation; thus,

$$(-2(3) - 3(-2)) = 0 \quad (-2(1) - 3(-1)) = 1 \quad (-2(-1) - 3(1)) = -1 \quad (-2(2) - 3(-1)) = -1$$

At this point our augmented matrix has been changed to

$$\begin{bmatrix} -2 & -1 & 1 & -1 \\ 0 & -1 & -3 & -11 \\ 0 & 1 & -1 & -1 \end{bmatrix}$$

In order to make this new a_{12} coefficient equal to zero, we form a new row one by multiplying the new row two by the a_{12} (-1) coefficient and subtract this from the new a_{22} (-1) coefficient times row one; thus,

$$(-1(-2) - (-1)(0)) = 2 \qquad (-1(-1) - (-1)(-1)) = 0$$

$$(-1(1) - (-1)(-3)) = -4 \qquad (-1(-1) - (-1)(-11) = -10$$

We now make the new a_{32} coefficient zero by forming a new row three. We do this by multiplying the new row three by the new a_{22} (-1) and subtracting from it the new a_{32} (1) times the new second equation; thus

$$(-1(0) - 1(0)) = 0 \qquad (-1(1) - 1(-1)) = 0$$

$$(-1(-1) - 1(-3)) = 4 \qquad (-1(-1) - 1(-11) = 12$$

Our augmented matrix is now equal to

$$\begin{bmatrix} 2 & 0 & -4 & -10 \\ 0 & -1 & -3 & -11 \\ 0 & 0 & 4 & 12 \end{bmatrix}$$

We now make the a_{13} element of this matrix zero by forming a new row one. This is done by multiplying the first equation by a_{33} (4) and subtracting a_{13} (-4) times the third equation; thus

$$(4(2) - (-4)(0)) = 8 \qquad (4(0) - (-4)(0)) = 0$$

$$(4(-4) - (-4)(4)) = 0 \qquad (4(-10) - (-4)(12) = 8$$

Finally we make a_{23} equal to zero by multiplying the second equation by a_{33} (4) and subtracting from this a_{23} (-3) times the third equation; thus

$$(4(0) - (-3)(0)) = 0 \qquad (4(-1) - (-3)(0)) = -4$$

$$(4(-3) - (-3)(4)) = 0 \qquad (4(-11) - (-3)(12)) = -8$$

Our equations are thus forced into the diagonal form

$$\begin{bmatrix} 8 & 0 & 0 & 8 \\ 0 & -4 & 0 & -8 \\ 0 & 0 & 4 & 12 \end{bmatrix}$$

and we may solve each of these directly, finding

$$X_1 = 1$$
$$X_2 = 2$$
$$X_3 = 3$$

We may now develop the computer algorithm for this process by recognizing that each new element was formed by carrying out the following mathematical operations

$$A * B - C * D$$

where A, B, C, and D correspond to specific matrix elements. In order to find exactly what elements are involved, let us consider the process in terms of the following symbolic matrix:

$$\begin{bmatrix} a_{11} & a_{12} & a_{13} & a_{14} \\ a_{21} & a_{22} & a_{23} & a_{24} \\ a_{31} & a_{32} & a_{33} & a_{34} \end{bmatrix}$$

Using row one and coefficient a_{11}, we first formed a new row two

$$(a_{11}a_{21} - a_{21}a_{11}) \quad (a_{11}a_{22} - a_{21}a_{12}) \quad (a_{11}a_{23} - a_{21}a_{13}) \quad (a_{11}a_{24} - a_{21}a_{14})$$
$$= 0$$

and then a new row three

$$(a_{11}a_{31} - a_{31}a_{11}) \quad (a_{11}a_{32} - a_{31}a_{12}) \quad (a_{11}a_{33} - a_{31}a_{13}) \quad (a_{11}a_{34} - a_{31}a_{14})$$
$$= 0$$

In these two steps, row one is called the pivot row and a_{11} is called the pivot coefficient. Let us therefore define the following integer mode variables

K = The pivot row.
I = The new row being formed.
J = The element being formed.

From an examination of either of these two new rows, we see that the following elements are involved in the generation of any new element

$$A_{KK}A_{IJ} - A_{IK}A_{KJ}$$

Thus, in the case of N simultaneous equations, we first let

$$K = 1, 2, \ldots, N$$

and for each $K \neq I$ we form a new row

$$I = 1, 2, \ldots, N; I \neq K$$

and for each I we form the new element

$$J = 1, 2, \ldots, N + 1$$

We thus see that our program will require a nested set of three DO loops to force the equations into the diagonal form.

We should recognize, however, that only multiplication and subtraction are involved in the generation of a new element and, as a consequence, if our coefficients are large to begin with, we can produce overflow while if our coefficients are small initially, the multiplication process can lead to underflow. To avoid either of these two possibilities, we therefore first normalize the pivot row coefficients with respect to the pivot coefficient before carrying out the process.

In the case of the numerical example above

$$\begin{bmatrix} -2 & -1 & 1 & -1 \\ 1 & 1 & 1 & 6 \\ 3 & 1 & -1 & 2 \end{bmatrix}$$

we would therefore first normalize the pivot row $(K = 1)$; thus, we would start the process with the augmented matrix

$$\begin{bmatrix} 1 & 0.5 & -0.5 & 0.5 \\ 1 & 1 & 1 & 6 \\ 3 & 1 & -1 & 2 \end{bmatrix}$$

forming a new row two and three obtaining

$$\begin{bmatrix} 1 & 0.5 & -0.5 & 0.5 \\ 0 & 0.5 & 1.5 & 5.5 \\ 0 & -0.5 & 0.5 & 0.5 \end{bmatrix}$$

At this point we would normalize the coefficients of row two with respect to the pivot coefficient a_{22} (0.5); thus

$$\begin{bmatrix} 1 & 0.5 & -0.5 & 0.5 \\ 0 & 1 & 3.0 & 11.0 \\ 0 & -0.5 & 0.5 & 0.5 \end{bmatrix}$$

We would now form a new row one and three obtaining

$$\begin{bmatrix} 1 & 0 & -2.0 & -5.0 \\ 0 & 1 & 3.0 & 11.0 \\ 0 & 0 & 2.0 & 6.0 \end{bmatrix}$$

At this point we would normalize the coefficients of row three with respect to the pivot coefficient a_{33} (2.0); thus,

$$\begin{bmatrix} 1 & 0 & -2.0 & -5.0 \\ 0 & 1 & 3.0 & 11.0 \\ 0 & 0 & 1.0 & 3.0 \end{bmatrix}$$

Finally we would form a new row one and two obtaining

$$\begin{bmatrix} 1 & 0 & 0 & 1.0 \\ 0 & 1 & 0 & 2.0 \\ 0 & 0 & 1.0 & 3.0 \end{bmatrix}$$

from which we see that our solution is given by the respective terms in column four.

In addition to minimizing the possibility of overflow or underflow, we should also recognize that the normalization process above reduces the number of mathematical operations that must be performed since the generation of a new element after normalization is given by the simpler expression

$$A_{IJ} - A_{IK}A_{KJ}$$

Thus, in the case of the old algorithm, for each pivot row (a total of N), $N - 1$ new rows had to be formed and since each element required two multiplications, a total of

$$N(N - 1)(2(N + 1)) = 2N(N^2 - 1) \approx 2N^3$$

multiplications were required. In the case of the normalized algorithm, for each pivot row we required $N + 1$ divisions, but only $(N - 1)(N + 1)$ multiplications. Assuming a division operation is equivalent in computer time to a multiplication operation, we thus require a total of

$$\underbrace{N(N + 1)}_{\text{Divisions}} + \underbrace{N(N - 1)(N + 1)}_{\text{Multiplications}} = N^2 + 1 + N(N^2 - 1) \approx N^3$$

equivalent multiplications and therefore for large N we cut the number of multiplication operations in half.

Figure 6.13 is a general Fortran program that uses the Gauss-Jordan elimination scheme to solve simultaneous linear algebraic equations. Note that computer statements S.0001 through S.0022 are essentially identical to the same number statements in the Gauss elimination program of Figure 6.7. These statements serve to interchange rows if the pivot coefficient magnitude is too small. The DO 30 loop is used to normalize the pivot row coefficients and the inner and lower DO 10 loops are used to generate the new rows. Note that the notation corresponds to the notation used in our earlier discussion. After the limits of the DO 10 loops are satisfied, the DO 40 loop is used to print the results.

Figure 6.14 gives the output results for the six-node, ten-branch network of Figure 6.8. A comparison of these results with those of Figure 6.9 will show that they are in close agreement.

In order for the reader to follow through the individual calculations that are required in using the Gauss-Jordan elimination process, the

```
          C        SOLUTION OF SIMULTANEOUS LINEAR ALGEBRAIC EQUATIONS
          C                                   BY
          C                         THE GAUSS-JORDAN METHOD
          C
S.0001             DIMENSION A(10,11),X(10)
S.0002             WRITE(6,1)
S.0003           1 FORMAT('1THE SOLUTION OF SIMULTANEOUS LINEAR EQUATIONS',/,16X,'BY'
                  1,/,5X,'THE GAUSS-JORDAN METHOD',////)
S.0004             READ(5,2)N,EPS
S.0005           2 FORMAT(I5,E15.7)
S.0006             M=N+1
S.0007             READ(5,3)((A(I,J),J=1,M),I=1,N)
S.0008           3 FORMAT(5E15.7)
S.0009             WRITE(6,60)((A(I,J),J=1,5),I=1,4)
S.0010          60 FORMAT(5E15.7)
S.0011             KK=0
S.0012             JJ=0
S.0013             DO 10 K=1,N
S.0014             JJ=KK+1
S.0015             LL=JJ
S.0016             KK=KK+1
S.0017          20 IF(ABS(A(JJ,KK))-EPS)21,21,22
S.0018          21 JJ=JJ+1
S.0019             GO TO 20
S.0020          22 IF(LL-JJ)23,24,23
S.0021          23 DO 25 MM=1,M
S.0022             ATEMP=A(LL,MM)
S.0023             A(LL,MM)=A(JJ,MM)
S.0024          25 A(JJ,MM)=ATEMP
          C
          C        THE EQUATIONS ARE FORCED INTO THE DIAGONAL FORM
          C
S.0025          24 DO 30 LJ=1,M
S.0026             J=M+1-LJ
S.0027          30 A(K,J)=A(K,J)/A(K,K)
S.0028             DO 10 I=1,N
S.0029             WRITE(6,60)((A(II,JJ),JJ=1,5),II=1,4)
S.0030             DO 10 LJ=1,M
S.0031             J=M+1-LJ
S.0032             IF(I-K)31,10,31
S.0033          31 A(I,J)=A(I,J)-A(I,K)*A(K,J)
S.0034          10 CONTINUE
S.0035             DO 40 I=1,N
S.0036             X(I)=A(I,M)
S.0037          40 WRITE(6,4)I,X(I)
S.0038           4 FORMAT(//,' X(',I2,') =',E15.7,/)
S.0039             STOP
S.0040             END
```

Figure 6.13 A program that uses the Gauss-Jordan elimination method to solve simultaneous linear algebraic equations

```
THE SOLUTION OF SIMULTANEOUS LINEAR EQUATIONS
                     BY
    THE GAUSS-JORDAN METHOD




X( 1) =  0.2403439E 00

X( 2) =  0.8966520E 00

X( 3) = -0.1370050E 00

X( 4) =  0.3882219E 00

X( 5) =  0.4408640E 00

X( 6) =  0.1478834E 00

X( 7) =  0.3486056E 00

X( 8) =  0.9256418E 00

X( 9) =  0.5778688E 00

X(10) =  0.4964892E 00
```

Figure 6.14 The output results of the program of Figure 6.13 for the network of Figure 6.8

program of Figure 6.13 was modified by adding WRITE statements that would print out the augmented matrix after statements S.0008 and S.0026. The program was then used to solve for the currents of the network of Figure 6.4. Figure 6.15 gives the corresponding results. The reader is encouraged to check each of these results by hand calculating the statements of the modified program.

The Equivalent T-Network of a Two-Port

We now turn to another example that utilizes the Gauss-Jordan elimination method to solve a set of simultaneous equations. Let us consider the two-port special case of the K-port equivalent circuit theorem.[1] This theorem states that any linear, passive, unilateral or bilateral network that has K independent ports is completely represented by K^2 admittance parameters as far as measurements made at the K ports are concerned. In the case of a reciprocal network, only $K(K + 1)/2$ parameters are needed. Thus in the special case of a two-port reciprocal network, only

[1] See B. James Ley, Samuel G. Lutz, and Charles F. Rehberg, *Linear Circuit Analysis* (New York: McGraw-Hill Book Company, 1959), Sec. 9-9.

```
THE SOLUTION OF SIMULTANEOUS LINEAR EQUATIONS
                      BY
     THE GAUSS-JORDAN METHOD


 0.4000000E 01 -0.2000000E 01 -0.1000000E 01  0.0            0.1000000E 02
-0.2000000E 01  0.9000000E 01  0.0           -0.5000000E 01 -0.5000000E 01
-0.1000000E 01  0.0            0.6000000E 01 -0.3000000E 01  0.0
 0.0           -0.5000000E 01 -0.3000000E 01  0.1000000E 02  0.0
 0.1000000E 01 -0.5000000E 00 -0.2500000E 00  0.0            0.2500000E 01
-0.2000000E 01  0.9000000E 01  0.0           -0.5000000E 01 -0.5000000E 01
-0.1000000E 01  0.0            0.6000000E 01 -0.3000000E 01  0.0
 0.0           -0.5000000E 01 -0.3000000E 01  0.1000000E 02  0.0
 0.1000000E 01 -0.5000000E 00 -0.2500000E 00  0.0            0.2500000E 01
-0.2000000E 01  0.9000000E 01  0.0           -0.5000000E 01 -0.5000000E 01
-0.1000000E 01  0.0            0.6000000E 01 -0.3000000E 01  0.0
 0.0           -0.5000000E 01 -0.3000000E 01  0.1000000E 02  0.0
 0.1000000E 01 -0.5000000E 00 -0.2500000E 00  0.0            0.2500000E 01
 0.0            0.8000000E 01 -0.5000000E 00 -0.5000000E 01  0.0
-0.1000000E 01  0.0            0.6000000E 01 -0.3000000E 01  0.0
 0.0           -0.5000000E 01 -0.3000000E 01  0.1000000E 02  0.0
 0.1000000E 01 -0.5000000E 00 -0.2500000E 00  0.0            0.2500000E 01
 0.0            0.8000000E 01 -0.5000000E 00 -0.5000000E 01  0.0
 0.0           -0.5000000E 00  0.5750000E 01 -0.3000000E 01  0.2500000E 01
 0.0           -0.5000000E 01 -0.3000000E 01  0.1000000E 02  0.0
 0.1000000E 01 -0.5000000E 00 -0.2500000E 00  0.0            0.2500000E 01
 0.0            0.1000000E 01 -0.6250000E-01 -0.6250000E 00  0.0
 0.0           -0.5000000E 00  0.5750000E 01 -0.3000000E 01  0.2500000E 01
 0.0           -0.5000000E 01 -0.3000000E 01  0.1000000E 02  0.0
 0.1000000E 01  0.0           -0.2812500E 00 -0.3125000E 00  0.2500000E 01
 0.0            0.1000000E 01 -0.6250000E-01 -0.6250000E 00  0.0
 0.0           -0.5000000E 00  0.5750000E 01 -0.3000000E 01  0.2500000E 01
 0.0           -0.5000000E 01 -0.3000000E 01  0.1000000E 02  0.0
 0.1000000E 01  0.0           -0.2812500E 00 -0.3125000E 00  0.2500000E 01
 0.0            0.1000000E 01 -0.6250000E-01 -0.6250000E 00  0.0
 0.0           -0.5000000E 00  0.5750000E 01 -0.3000000E 01  0.2500000E 01
 0.0           -0.5000000E 01 -0.3000000E 01  0.1000000E 02  0.0
 0.1000000E 01  0.0           -0.2812500E 00 -0.3125000E 00  0.2500000E 01
 0.0            0.1000000E 01 -0.6250000E-01 -0.6250000E 00  0.0
 0.0            0.0            0.5718750E 01 -0.3312500E 01  0.2500000E 01
 0.0           -0.5000000E 01 -0.3000000E 01  0.1000000E 02  0.0
 0.1000000E 01  0.0           -0.2812500E 00 -0.3125000E 00  0.2500000E 01
 0.0            0.1000000E 01 -0.6250000E-01 -0.6250000E 00  0.0
 0.0            0.0            0.1000000E 01 -0.5792350E 00  0.4371585E 00
 0.0            0.0           -0.3312500E 01  0.6875000E 01  0.0
 0.1000000E 01  0.0            0.0           -0.4754098E 00  0.2622951E 01
 0.0            0.1000000E 01 -0.6250000E-01 -0.6250000E 00  0.0
 0.0            0.0            0.1000000E 01 -0.5792350E 00  0.4371585E 00
 0.0            0.0           -0.3312500E 01  0.6875000E 01  0.0
 0.1000000E 01  0.0            0.0           -0.4754098E 00  0.2622951E 01
 0.0            0.1000000E 01  0.0           -0.6612021E 00  0.2732240E-01
 0.0            0.0            0.1000000E 01 -0.5792350E 00  0.4371585E 00
 0.0            0.0           -0.3312500E 01  0.6875000E 01  0.0
 0.1000000E 01  0.0            0.0           -0.4754098E 00  0.2622951E 01
 0.0            0.1000000E 01  0.0           -0.6612021E 00  0.2732240E-01
 0.0            0.0            0.1000000E 01 -0.5792350E 00  0.4371585E 00
 0.0            0.0           -0.3312500E 01  0.6875000E 01  0.0
 0.1000000E 01  0.0            0.0           -0.4754098E 00  0.2622951E 01
 0.0            0.1000000E 01  0.0           -0.6612021E 00  0.2732240E-01
 0.0            0.0            0.1000000E 01 -0.5792350E 00  0.4371585E 00
 0.0            0.0            0.0            0.1000000E 01  0.2921718E 00
 0.1000000E 01  0.0            0.0            0.0            0.2761851E 01
 0.0            0.1000000E 01  0.0           -0.6612021E 00  0.2732240E-01
 0.0            0.0            0.1000000E 01 -0.5792350E 00  0.4371585E 00
 0.0            0.0            0.0            0.1000000E 01  0.2921718E 00
 0.1000000E 01  0.0            0.0            0.0            0.2761851E 01
 0.0            0.1000000E 01  0.0            0.0            0.2205069E 00
 0.0            0.0            0.1000000E 01 -0.5792350E 00  0.4371585E 00
 0.0            0.0            0.0            0.1000000E 01  0.2921718E 00
 0.1000000E 01  0.0            0.0            0.0            0.2761851E 01
 0.0            0.1000000E 01  0.0            0.0            0.2205069E 00
 0.0            0.0            0.1000000E 01  0.0            0.6063945E 00
 0.0            0.0            0.0            0.1000000E 01  0.2921718E 00
```

Figure 6.15 The output results of the program of Figure 6.13 for the solution of the network of Figure 6.4

```
X( 1) =  0.2761851E 01

X( 2) =  0.2205069E 00

X( 3) =  0.6063945E 00

X( 4) =  0.2921718E 00
```

Figure 6.15 Continued

$2(2 + 1)/2 = 3$ parameters are required to fully represent the network as far as measurements at either port are concerned.

The algorithm that is needed to determine the equivalent network is based on writing the mesh (or cut-set) equations for an N-mesh network and including the two ports as part of the first and second mesh respectively. Thus, we have

$$\begin{array}{l}
a_{11} \quad I_1 + a_{12} \quad I_2 + \cdots + a_{1N} \quad I_N = V_1 \\
a_{21} \quad I_1 + a_{22} \quad I_2 + \cdots + a_{2N} \quad I_N = V_2 \\
a_{31} \quad I_1 + a_{32} \quad I_2 + \cdots + a_{3N} \quad I_N = 0 \\
\quad \vdots \\
a_{N1} \quad I_1 + a_{N2} \quad I_2 + \cdots + a_{NN} \quad I_N = 0
\end{array}$$

We may now use the Gauss-Jordan elimination scheme to force the last $N - 2$ equations into the following form

$$\begin{array}{l}
b_{31} \quad I_1 + b_{32} \quad I_2 + I_3 + 0 + \cdots = 0 \\
b_{41} \quad I_1 + b_{42} \quad I_2 \quad + I_4 + 0 + \cdots = 0 \\
\quad \vdots \\
b_{N1} \quad I_1 + b_{N2} \quad I_2 + 0 + \cdots \quad + I_N = 0
\end{array}$$

It is now possible to solve these equations for $I_3, I_4, \ldots$, and I_N in terms of I_1 and I_2. When these values are substituted in the first two equations, we obtain

$$\begin{aligned}
b_{11}I_1 + b_{12}I_2 &= V_1 \\
b_{21}I_1 + b_{22}I_2 &= V_2
\end{aligned}$$

Writing the mesh equations for the two-port equivalent T shown in Figure 6.16 yields

$$\begin{aligned}
(R_1 + R_3)I_1 \qquad - R_3 I_2 &= V_1 \\
-R_3 I_1 + (R_2 + R_3) I_2 &= V_2
\end{aligned}$$

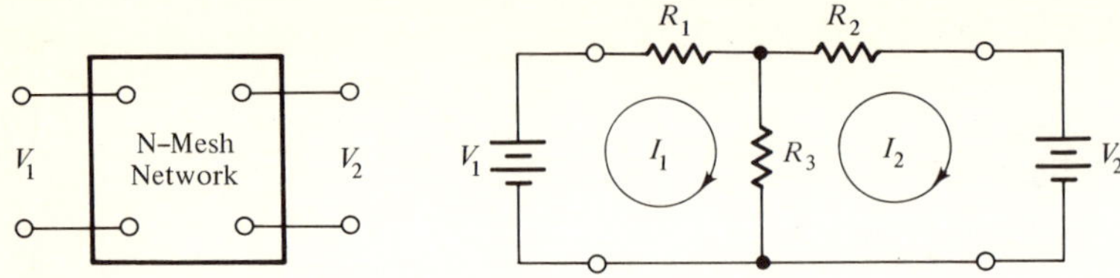

Figure 6.16 A two-port network and its equivalent T

```
C         THIS PROGRAM CONVERTS AN N-MESH, TWO INDEPENDENT PORT,
C         RECIPROCAL NETWORK INTO AN EQUIVALENT T-NETWORK.
C
          DIMENSION A(20,21)
          READ(5, 1) N
        1 FORMAT(I5)
          M=N+1
C
C         READ IN AUGMENTED MATRIX
C
          READ(5, 2) ((A(I,J),J=1,M),I=1,N)
        2 FORMAT(5E15.7)
          WRITE(6, 3)
        3 FORMAT('1EQUIVALENT T-NETWORK',///)
C
C         THE GAUSS-JORDAN METHOD IS USED TO SOLVE FOR
C         I(3), I(4), ..., AND I(N)  IN TERMS OF I(1) AND I(2).
C
          DO 10 I=3,N
          DIV=A(I,I)
          DO 20 J=1,M
       20 A(I,J)=A(I,J)/DIV
          DO 10 K=3,N
          AMULT=A(K,I)
          IF(I-K) 30, 10, 30
       30 DO 10 L=1,M
          A(K,L)=A(K,L)-A(I,L)*AMULT
       10 CONTINUE
C
C         THE CURRENTS  I(3), I(4), ..., I(N)   ARE SUBSTITUTED
C         INTO EQUATIONS ONE AND TWO.
C
          DO 40 I=3,N
       40 A(1,1)=A(1,1)-A(1,I)*A(I,1)
          DO 50 I=3,N
       50 A(1,2)=A(1,2)-A(1,I)*A(I,2)
          DO 60 I=3,N
       60 A(2,1)=A(2,1)-A(2,I)*A(I,1)
          DO 70 I=3,N
       70 A(2,2)=A(2,2)-A(2,I)*A(I,2)
C
C         THE EQUATIONS OF THE EQUIVALENT T ARE PRINTED
C
          WRITE(6,4)A(1,1),A(1,2),A(1,M),A(2,1),A(2,2),A(2,M)
        4 FORMAT(E15.7,' I(1) + ',E15.7,' I(2) =',E15.7,//,E15.7,' I(1) + ',
         1E15.7,' I(2) =',E15.7,///)
C
C         THE ELEMENTS OF THE EQUIVALENT T ARE COMPUTED.
C
          R3=ABS(A(1,2))
          R1=A(1,1)-R3
          R2=A(2,2)-R3
          WRITE(6,5) R1,R2,R3
        5 FORMAT(' R1 =',E15.7,//,' R2 =',E15.7,//,' R3 =',E15.7)
          STOP
          END
```

Figure 6.17 A Fortran program that uses the Gauss-Jordan method to find the equivalent T-network of a two-port N-mesh network

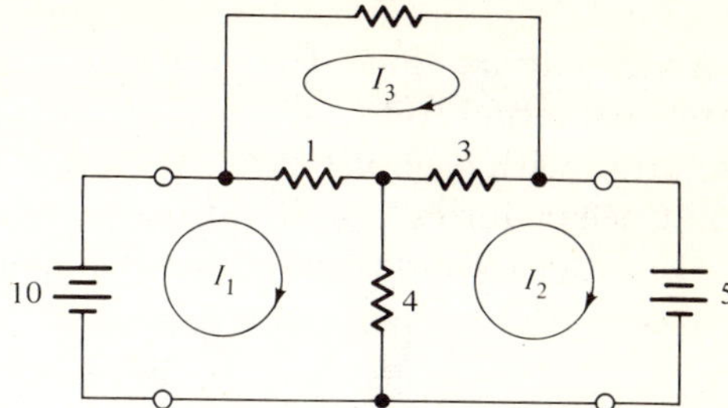

Figure 6.18 A three-mesh resistive network

It thus follows that

$$\begin{aligned} R_3 &= -b_{12} = -\ b_{21} \\ R_1 &= b_{11} - R_3 \\ R_2 &= b_{22} - R_3 \end{aligned}$$

Figure 6.17 is a Fortran program for finding the equivalent T-network of an N-mesh resistive network. Using the augmented matrix data of the network of Figure 6.18 as input for this program, we obtain the equivalent T-network from the output results of Figure 6.19. The reader may readily verify this result by first writing the three mesh equations and then using the third equation to solve for I_3 in terms of I_1 and I_2 and, in turn, substituting this value in equations one and two to find the equivalent T-network.

6.5 THE GAUSS-SEIDEL METHOD

The last two elimination methods for solving simultaneous linear algebraic equations are generally limited to the solution of up to 15 equations. The reader should recognize that this limitation is due mainly to the propagation of round-off error. Since many engineering problems may require the solution of thousands of simultaneous equations, an

```
EQUIVALENT T-NETWORK


 0.4833333E 01 I(1) +  -0.4500000E 01 I(2) =  0.1000000E 02

-0.4499999E 01 I(1) +   0.5500000E 01 I(2) = -0.5000000E 01


R1 =  0.3333330E 00

R2 =  0.1000000E 01

R3 =  0.4500000E 01
```

Figure 6.19 The output results of the program of Figure 6.17 for the network of Figure 6.18

alternate method must be considered. The Gauss-Seidel method is such an alternative, since it is an iterative method that starts with an initial approximation to the solution of the equations and then improves on the solution with each iteration. It is thus obvious that computational errors that occur during any iteration are not propagated.

In order to understand how this method works, consider the following two equations

$$\begin{aligned} 5I_1 - 2I_2 &= 8 \\ -2I_1 + 10I_2 &= 6 \end{aligned}$$

which have the solution $I_1 = 2$ and $I_2 = 1$. We now assume that I_2 is equal to zero and solve for I_1 from the first equation. We thus find that $I_1 = 1.6$. Using this value of I_1 in the second equation, we solve for I_2 obtaining

$$\begin{aligned} I_2 &= \frac{6 + 3.2}{10} \\ &= 0.92 \end{aligned}$$

so that at the end of the first iteration we have

$$I_1 = 1.6 \text{ and } I_2 = 0.92.$$

Repeating the process a second time, we use $I_2 = 0.92$ in the first equation and solve for I_1 obtaining

$$\begin{aligned} I_1 &= \frac{8 + 1.84}{5} \\ &= 1.968 \end{aligned}$$

and using this value in the second equation, we find

$$\begin{aligned} I_2 &= \frac{6 + 3.936}{10} \\ &= 0.9936 \end{aligned}$$

Repeating the process a third time, we find

$$I_1 = 1.9974 \text{ and } I_2 = 0.99948$$

and thus, in three iterations, we have obtained a solution accurate to almost four significant figures.

This method may be extended to a larger number of equations. For example, in the case of n simultaneous equations, we would initially assume $I_2 = I_3 = , \ldots , I_n = 0$ and solve for I_1 using the first equation. Using this value of I_1 and $I_3 = I_4 = , \ldots , I_n = 0$, we would next solve for I_2 using the second equation. This would be repeated until the values of all n currents were found. Then, using the values obtained from the first iteration, a second iteration should yield more accurate results if the process is converging. Additional iterations would continue to improve the accuracy of the results, and the procedure would be terminated whenever the latest values of the currents agreed with the previous iteration values to within the specified accuracy.

As we have already suggested above, unfortunately this procedure does not always converge. For example, if we reversed the order of the two equations above, we would find that at the end of the first iteration

$$I_1 = -3 \text{ and } I_2 = -12$$

and at the end of the second iteration

$$I_1 = -63 \text{ and } I_2 = -161.5$$

so that clearly the method is not converging. The exact criterion for convergence is difficult to define; however, a sufficient but not necessary condition for convergence is that

$$\sum_{I=1}^{n} |a_{II}| \geq \sum_{\substack{J=1 \\ J \neq I}}^{n} |a_{IJ}|$$

The number of iterations required to achieve a specified accuracy is also difficult to predict. In general, the fewer the equations, the more accurate the initial approximations, and the smaller the summation terms with respect to the diagonal term in the above inequality, the more rapidly the process will tend to converge.

Figure 6.20 is a Fortran program that uses the Gauss-Seidel method to solve a set of up to ten linear algebraic equations. First the heading is printed and the number of equations N and the accurary EPS are read into memory. Computer statements S.0007 and S.0008 then read in the elements of the augmented matrix and the DO 10 loop initially sets all elements of $X(I)$ to a value of zero. The DO 20 loop, the main iteration loop, is then entered and is capable of being executed 100 times. During each iteration of this loop, the dummy variable SWITCH is set equal to zero. The DO 30 loop is then used to find the values of the n unknowns. In the first iteration of this loop, $J = 1$ and so SUM is set equal to A_{1M}. The DO 40 loop then sets $K = 1$ and tests to see if $J = K$. Since it does in the first iteration, control is transferred to statement 40 and the DO 40 loop is executed a second time with $K = 2$. Since $K \neq J$ this iteration, SUM is made equal to

$$\text{SUM} = \text{A}_{1M} - \text{A}_{12} * \text{X}(2)$$

```
                  C         SOLUTION OF SIMULTANEOUS LINEAR ALGEBRAIC EQUATIONS
                  C                                  BY
                  C                       THE GAUSS-SEIDEL METHOD
                  C
S.0001                      DIMENSION A(10,11),X(10)
S.0002                      WRITE(6,1)
S.0003                    1 FORMAT('1THE SOLUTION OF SIMULTANEOUS LINEAR EQUATIONS',/,16X,'BY'
                           1,/,5X,'THE GAUSS-SEIDEL METHOD',////)
S.0004                      READ(5,2) N,EPS
S.0005                    2 FORMAT(I5,E15.7)
S.0006                      M=N+1
S.0007                      READ(5,3) ((A(I,J),J=1,M),I=1,N)
S.0008                    3 FORMAT(5E15.7)
S.0009                      DO 10 I=1,N
S.0010                   10 X(I)=0.0
                  C
                  C         THE MAIN ITERATION LOOP
                  C
S.0011                      DO 20 I=1,100
S.0012                      SWITCH=0.0
S.0013                      DO 30 J=1,N
S.0014                      SUM=A(J,M)
S.0015                      DO 40 K=1,N
S.0016                      IF (J-K)50,40,50
S.0017                   50 SUM=SUM-A(J,K)*X(K)
S.0018                   40 CONTINUE
S.0019                      XNEXT=SUM/A(J,J)
S.0020                      IF (ABS(XNEXT-X(J))-EPS) 30,30,60
S.0021                   60 SWITCH=1.0
S.0022                   30 X(J)=XNEXT
S.0023                      IF(SWITCH) 70,70,20
S.0024                   20 CONTINUE
S.0025                   70 WRITE(6,4) I,(J,X(J),J=1,N)
S.0026                    4 FORMAT(//,' THE NUMBER OF ITERATIONS REQUIRED =',I4,//,(' X(',I2,'
                           1) =',E15.7))
S.0027                      STOP
S.0028                      END
```

Figure 6.20 A program that uses the Gauss-Seidel method to solve a set of simultaneous linear algebraic equations

The execution of the DO 40 loop is continued until

$$\text{SUM} = \text{A}_{1M} - \text{A}_{12} * \text{X}(2) - \text{A}_{13} * \text{X}(3) - , \; . \; . \; . \; , -\text{A}_{1N} * \text{X}(\text{N})$$

At this point XNEXT is computed from

$$\text{XNEXT} = \text{SUM}/\text{A}_{11}$$

This is equivalent to solving the first equation for $X(1)$. Computer statement S.0020 then tests to see if the difference between the value of $X(1)$ in this iteration compared to the previous value of $X(1)$ is less than or greater than EPS. If it is greater than EPS, the value of SWITCH is made equal to one. Computer statement S.0022 then sets $X(1) =$ XNEXT. The DO 30 loop is then executed a second time to find the value of $X(2)$ and set the variable SWITCH to one if the correction of $X(2)$ is greater than EPS. The execution of the DO 30 loop is repeated until all unknowns are found. At this point computer statement S.0023 tests to see whether any of the N corrections exceed EPS. If any correction is greater than EPS, the DO 20 loop is again executed. This is repeated until a maximum of 100 iterations occur or until all N values of the unknowns compared with their previous iteration values are less than EPS. When either of these two conditions occur, computer statements S.0025 and S.0026 print the number of iterations and the values of the unknowns.

```
THE SOLUTION OF SIMULTANEOUS LINEAR EQUATIONS
                    BY
    THE GAUSS-SEIDEL METHOD



THE NUMBER OF ITERATIONS REQUIRED =   15

X( 1) =   0.2761720E 01
X( 2) =   0.2203997E 00
X( 3) =   0.6063024E 00
X( 4) =   0.2920905E 00
```

Figure 6.21 The output results of the program of Figure 6.20 for the network of Figure 6.4

Figure 6.21 shows the computer results for the network shown in Figure 6.4. The reader should note that 15 iterations were required to achieve an accuracy of approximately four significant figures. In order to see that the process may not converge or the number of iterations required to obtain a specific accuracy may differ, the reader is urged to rearrange the order of these equations and rerun the problem. The reader should also note that the DIMENSION statement may easily be changed to allow this program to solve a set of more than ten equations.

6.6 SIMULTANEOUS LINEAR EQUATIONS WITH COMPLEX COEFFICIENTS

The solution of simultaneous linear algebraic equations with complex coefficients is often encountered in the field of electrical engineering. For example, simultaneous complex coefficient equations result when the phasor method is used to solve for the steady-state solution of a set of simultaneous integrodifferential equations. As a numerical example, consider the six-mesh network shown in Figure 6.22. Writing the phasor

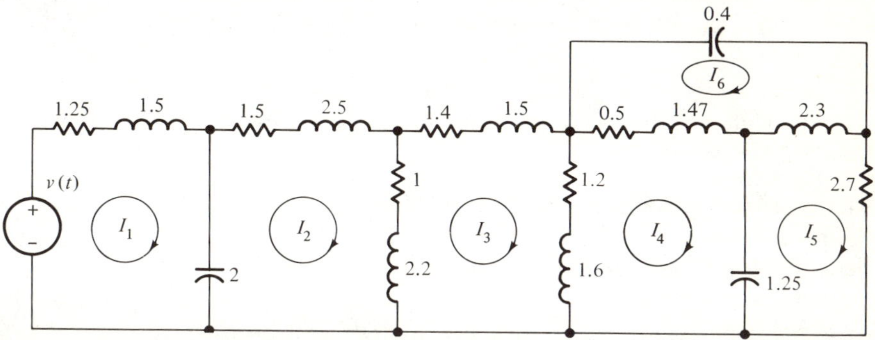

Figure 6.22 A six-mesh network, where $v(t) = 100 \cos(\omega t + 60°)$; $\omega = 1$ radian per second

mesh equations for the sinusoidal steady-state positive mode, we have

$$
\begin{aligned}
(1.25 + j1.0)I_1 + j0.5I_2 &= 50 + j86.6 \\
j0.5I_1 + (2.5 + j4.2)I_2 - (1.0 + j2.2)I_3 &= 0 \\
-(1.0 + j2.2)I_2 + (3.6 + j5.3)I_3 - (1.2 + j1.6)I_4 &= 0 \\
-(1.2 + j1.6)I_3 + (1.7 + j2.27)I_4 + j0.8I_5 - (0.5 + j1.47)I_6 &= 0 \\
j0.8I_4 + (2.7 + j1.5)I_5 - j2.3I_6 &= 0 \\
-(0.5 + j1.47)I_4 - j2.3I_5 + (0.5 + j1.27)I_6 &= 0
\end{aligned}
$$

The corresponding augmented matrix is therefore given by

$$
\begin{bmatrix}
(1.25 + j1.0) & j0.5 & 0 & 0 & 0 & 0 & (50 + j86.6) \\
j0.5 & (2.5 + j4.2) & -(1.0 + j2.2) & 0 & 0 & 0 & 0 \\
0 & -(1.0 + j2.2) & (3.6 + j5.3) & -(1.2 + j1.6) & 0 & 0 & 0 \\
0 & 0 & -(1.2 + j1.6) & (1.7 + j2.27) & j0.8 & -(0.5 + j1.47) & 0 \\
0 & 0 & 0 & j0.8 & (2.7 + j1.5) & -j2.3 & 0 \\
0 & 0 & 0 & -(0.5 + j1.47) & -j2.3 & (0.5 + j1.27) & 0
\end{bmatrix}
$$

The problem we face then is to solve this set of equations so that we can find the solutions $I_1(j\omega)$, $I_2(j\omega)$, . . . , $I_6(j\omega)$. If real coefficients were involved, we could use the Gauss elimination or the Gauss-Jordan method which we have just discussed. Since complex numbers are involved, our previous programs will have to be modified (unless our computer is capable of complex arithmetic), or we will have to transform the augmented matrix so that only real terms are involved. The latter alternative is more efficient. Let us therefore transform our N by $N + 1$ (6 by 7) complex number matrix to a new real matrix of size $2N$ by $2N + 1$ (12 by 13) and solve this matrix by using the Gauss-Jordan method.

In order to develop the algorithm to accomplish this transformation, let us consider this process in terms of the simple case of the following three phasor equations:

$$
\begin{aligned}
&(Z_{11R} + jZ_{11I})(I_{1R} + jI_{1I}) + (Z_{12R} + jZ_{12I})(I_{2R} + jI_{2I}) \\
&\qquad + (Z_{13R} + jZ_{13I})(I_{3R} + jI_{3I}) = V_{1R} + jV_{1I} \\
&(Z_{21R} + jZ_{21I})(I_{1R} + jI_{1I}) + (Z_{22R} + jZ_{22I})(I_{2R} + jI_{2I}) \\
&\qquad + (Z_{23R} + jZ_{23I})(I_{3R} + jI_{3I}) = V_{2R} + jV_{2I} \\
&(Z_{31R} + jZ_{31I})(I_{1R} + jI_{1I}) + (Z_{32R} + jZ_{32I})(I_{2R} + jI_{2I}) \\
&\qquad + (Z_{33R} + jZ_{33I})(I_{3R} + jI_{3I}) = V_{3R} + jV_{3I}
\end{aligned}
$$

Multiplying out each of these terms we get

$$
\begin{aligned}
&((Z_{11R}I_{1R} - Z_{11I}I_{1I}) + j(Z_{11I}I_{1R} + Z_{11R}I_{1I}))\\
&\qquad + ((Z_{12R}I_{2R} - Z_{12I}I_{2I}) + j(Z_{12I}I_{2R} + Z_{12R}I_{2I}))\\
&\qquad + ((Z_{13R}I_{3R} - Z_{13I}I_{3I}) + j(Z_{13I}I_{3R} + Z_{13R}I_{3I})) = V_{1R} + jV_{1I}\\
&((Z_{21R}I_{1R} - Z_{21I}I_{1I}) + j(Z_{21I}I_{1R} + Z_{21R}I_{1I}))\\
&\qquad + ((Z_{22R}I_{2R} - Z_{22I}I_{2I}) + j(Z_{22I}I_{2R} + Z_{22R}I_{2I}))\\
&\qquad + ((Z_{23R}I_{3R} - Z_{23I}I_{3I}) + j(Z_{23I}I_{3R} + Z_{23R}I_{3I})) = V_{2R} + jV_{2I}\\
&((Z_{31R}I_{1R} - Z_{31I}I_{1I}) + j(Z_{31I}I_{1R} + Z_{31R}I_{1I}))\\
&\qquad + ((Z_{32R}I_{2R} - Z_{32I}I_{2I}) + j(Z_{32I}I_{2R} + Z_{32R}I_{2I}))\\
&\qquad + ((Z_{33R}I_{3R} - Z_{33I}I_{3I}) + j(Z_{33I}I_{3R} + Z_{33R}I_{3I})) = V_{3R} + jV_{3I}
\end{aligned}
$$

Since the equality of two complex numbers requires that both the real and imaginary parts must be equal to one another, we therefore equate the real and imaginary parts in each of these three equations, obtaining the following set of six real equations:

$$
\begin{aligned}
Z_{11R}I_{1R} - Z_{11I}I_{1I} + Z_{12R}I_{2R} - Z_{12I}I_{2I} + Z_{13R}I_{3R} - Z_{13I}I_{3I} &= V_{1R}\\
Z_{11I}I_{1R} + Z_{11R}I_{1I} + Z_{12I}I_{2R} + Z_{12R}I_{2I} + Z_{13I}I_{3R} + Z_{13R}I_{3I} &= V_{1I}\\
Z_{21R}I_{1R} - Z_{21I}I_{1I} + Z_{22R}I_{2R} - Z_{22I}I_{2I} + Z_{23R}I_{3R} - Z_{23I}I_{3I} &= V_{2R}\\
Z_{21I}I_{1R} + Z_{21R}I_{1I} + Z_{22I}I_{2R} + Z_{22R}I_{2I} + Z_{23I}I_{3R} + Z_{23R}I_{3I} &= V_{2I}\\
Z_{31R}I_{1R} - Z_{31I}I_{1I} + Z_{32R}I_{2R} - Z_{32I}I_{2I} + Z_{33R}I_{3R} - Z_{33I}I_{3I} &= V_{3R}\\
Z_{31I}I_{1R} + Z_{31R}I_{1I} + Z_{32I}I_{2R} + Z_{32R}I_{2I} + Z_{33I}I_{3R} + Z_{33R}I_{3I} &= V_{3I}
\end{aligned}
$$

From the pattern of the above equations, we may deduce a general algorithm for transforming the complex augmented matrix. First we see all elements in the odd rows of the new matrix alternate in sign, beginning with a positive sign and ending with a negative sign. Next we observe that all imaginary coefficients in the even rows of the real matrix are the same as the elements diagonally above, except for the sign change we have already noted. We may thus see that in the case of these three equations, if the initial complex matrix is read in as $[A(I,J)]$

$$
\begin{bmatrix}
Z_{11R} & Z_{11I} & Z_{12R} & Z_{12I} & Z_{13R} & Z_{13I} & V_{1R} & V_{1I}\\
Z_{21R} & Z_{21I} & Z_{22R} & Z_{22I} & Z_{23R} & Z_{23I} & V_{2R} & V_{2I}\\
Z_{31R} & Z_{31I} & Z_{32R} & Z_{32I} & Z_{33R} & Z_{33I} & V_{3R} & V_{31}
\end{bmatrix}
$$

this will correspond to a general N by $2N + 2$ complex matrix or in this case a 3 by 8 matrix

$$
\begin{bmatrix}
a_{11} & a_{12} & a_{13} & a_{14} & a_{15} & a_{16} & a_{17} & a_{18}\\
a_{21} & a_{22} & a_{23} & a_{24} & a_{25} & a_{26} & a_{27} & a_{28}\\
a_{31} & a_{32} & a_{33} & a_{34} & a_{35} & a_{36} & a_{37} & a_{38}
\end{bmatrix}
$$

```
C        SOLUTION OF SIMULTANEOUS LINEAR ALGEBRAIC EQUATIONS
C                               WITH
C                    COMPLEX COEFFICIENTS
C                              USING
C             THE GAUSS-JORDAN METHOD
C
C         THE MAXIMUM NUMBER OF EQUATIONS IS N = 10.
C
C
C        THE FOLLOWING SECTION CONVERTS THE ORIGINAL A MATRIX (WITH COMPLEX
C        COEFFICIENTS) INTO THE B MATRIX (WITH REAL COEFFICIENTS).
C
       DIMENSION A(10,22),B(20,22),X(10),PHI(10)
       WRITE(6,1)
     1 FORMAT('1THE SOLUTION OF SIMULTANEOUS LINEAR EQUATIONS',/,16X,'USI
     1NG',/,5X,'THE GAUSS-JORDAN METHOD',////)
       READ(5,2) N,EPS
     2 FORMAT(I5,E15.7)
       MA=2*N-1
       MB=2*N+1
       MC=2*N
       MD=2*N+2
       READ(5,3) ((A(I,J),J=1,MD),I=1,N)
     3 FORMAT(5E15.7)
C
C        SET UP ODD ROWS
C
       IK=-1
       DO 10 I=1,MA,2
       IK=IK+1
       DO 10 J=1,MB,2
       LJ=I-IK
       B(I,J)=A(LJ,J)
    10 B(I,J+1)=-A(LJ,J+1)
C
C        SET UP EVEN ROWS
C
       DO 20 I=2,MC,2
       DO 20 J=1,MB,2
       LJ=I/2
       B(I,J)=A(LJ,J+1)
    20 B(I,J+1)=A(LJ,J)
C
C        THE FOLLOWING SECTION USES THE GAUSS-JORDAN METHOD
C        TO SOLVE THE B MATRIX SYSTEM OF EQUATIONS
C
       JJ=1
       DO 30 I=1,MC
       KK=I-1
       IF(ABS(B(I,I))-EPS)35,40,35
    40 KK=KK+1
       LL=0
    45 LL=LL+1
       DO 50 J=1,MB
       BTEMP=B(KK,J)
       KL=KK+LL
       B(KK,J)=B(KL,J)
    50 B(KL,J)=BTEMP
       IF(ABS(B(I,I))-EPS)35,45,35
    35 BTEMP=B(I,I)
C
C        NORMALIZE THE PIVOT ROW
C
       DO 60 K=1,MB
    60 B(I,K)=B(I,K)/BTEMP
C
C        CREATE ZEROS BELOW THE DIAGONAL
C
       JJ=JJ+1
       DO 30 J=JJ,MC
       BTEMP=B(J,I)
       DO 30 K=1,MB
    30 B(J,K)=B(J,K)-BTEMP*B(I,K)
```

Figure 6.23 A Fortran program that uses the Gauss-Jordan method to sc've simultaneous linear algebraic equations with complex coefficients

```
              C
              C     CREATE ZEROS ABOVE THE DIAGONAL
              C
S.0045              B(MC,MB)=B(MC,MB)/B(MC,MC)
S.0046              B(MC,MC)=1.
S.0047              JJ=1
S.0048              KK=2
S.0049           56 DO 55 I=1,JJ
S.0050              BTEMP=B(I,KK)
S.0051              DO 55 J=KK,MB
S.0052           55 B(I,J)=B(I,J)-BTEMP*B(KK,J)
S.0053              JJ=JJ+1
S.0054              KK=KK+1
S.0055              IF(JJ-MC)56,57,56
S.0056           57 CONTINUE
S.0057              K=0
S.0058              DO 65 I=1,MC,2
S.0059              K=K+1
S.0060              X(K)=SQRT(B(I+1,MB)**2+B(I,MB)**2)
S.0061              PHASE=ATAN(B(I+1,MB)/B(I,MB))*57.29578
S.0062              IF(B(I,MB))66,67,67
S.0063           66 PHASE=PHASE+180.0
S.0064           67 PHI(K)=PHASE
S.0065           65 WRITE(6,4)K,B(I,MB),K,B(I+1,MB),X(K),PHI(K)
S.0066            4 FORMAT(//,' I(',I2,') RE =',E15.7,'  I(',I2,') IM =',E15.7,'    MAG
                   1 =',E15.7,'    PHASE =',E15.7)
S.0067              STOP
S.0068              END
```

Figure 6.23 Continued

The transformed real matrix $[B(I,J)]$ will, in general, correspond to a $2N$ by $2N + 1$ matrix which, in the case of three equations, will be a 6 by 7 matrix

$$\begin{bmatrix} a_{11} & -a_{12} & a_{13} & -a_{14} & a_{15} & -a_{16} & a_{17} \\ a_{12} & a_{11} & a_{14} & a_{13} & a_{16} & a_{15} & a_{18} \\ a_{21} & -a_{22} & a_{23} & -a_{24} & a_{25} & -a_{26} & a_{27} \\ a_{22} & a_{21} & a_{24} & a_{23} & a_{26} & a_{25} & a_{28} \\ a_{31} & -a_{32} & a_{33} & -a_{34} & a_{35} & -a_{36} & a_{37} \\ a_{32} & a_{31} & a_{34} & a_{33} & a_{36} & a_{35} & a_{38} \end{bmatrix}$$

Figure 6.23 is a general Fortran program for solving up to tenth order simultaneous linear equations with complex coefficients. First the heading is printed, and the number of equations N and the coefficient minimum magnitude EPS are read into memory. Based on the value of N, the constants MA, MB, MC, and MD are computed and this is followed by reading in the elements of the complex matrix $[A(I,J)]$. Computer statements S.0012 through to S.0023 are then used to form the real matrix $[B(I,J]$.

The Gauss-Jordan method then solves the real matrix equations and this solution is followed by the DO 65 loop which computes the magnitude and phase of each of the mesh currents and prints the results.

Figure 6.24 is the corresponding output for the network of Figure 6.22. The reader may verify these results by summing the voltages around the outside loop of the network. Since this summation will involve all of the mesh currents except I_4, it will serve as a quick check on the solution.

```
THE SOLUTION OF SIMULTANEOUS LINEAR EQUATIONS
                    USING
     THE GAUSS-JORDAN METHOD


I( 1) RE =  0.5714381E 02  I( 1) IM =  0.2508484E 02   MAG =  0.6240724E 02   PHASE =  C.237CC38E C2

I( 2) RE = -0.38CC103E C1  I( 2) IM = -0.7310123E 01   MAG =  0.8238852E 01   PHASE =  C.2425327E C3

I( 3) RE = -0.6507902E 00  I( 3) IM = -0.4232084E 01   MAG =  0.4281827E 01   PHASE =  C.2612576E C3

I( 4) RE =  0.1135756E 01  I( 4) IM = -0.4026322E 01   MAG =  0.4183442E 01   PHASE = -C.742471CE C2

I( 5) RE =  0.1126590E C1  I( 5) IM =  0.1196630F 01   MAG =  0.1643511E 01   PHASE =  C.4672681E C2

I( 6) RE =  0.2534521E 01  I( 6) IM = -0.1942570E 01   MAG =  0.3193332E 01   PHASE = -C.3746815E C2
```

Figure 6.24 The output results of the program of Figure 6.23 for the network of Figure 6.22

PROBLEMS

6-1. Modify the program of Figure 6.5 so that it is capable of evaluating a second- to a sixth-order determinant.

6-2. Using the appropriate parts of the program of Problem 6-1, modify the program of Figure 6.5 so that it is capable of solving second- to sixth-order simultaneous linear algebraic equations.

6-3. Assuming 100 microseconds are required for each computer multiplication operation, estimate the time that it will take to solve a set of six simultaneous equations using the program of Problem 6-2.

6-4. In Section 6.3 we forced a set of three simultaneous equations to have the following form

$$\begin{aligned} X_1 + b_{12}X_2 + b_{13}X_3 &= b_{14} \\ X_2 + b_{23}X_3 &= b_{24} \\ X_3 &= b_{34} \end{aligned}$$

The solution of the original set was then obtained by solving back from the last of these three equations.

Alternately we could have forced the original equations into the following triangular form

$$\begin{aligned} X_1 \qquad\qquad\quad &= C_{14} \\ C_{21}X_1 + \quad X_2 \qquad &= C_{24} \\ C_{31}X_1 + C_{32}X_2 + X_3 &= C_{34} \end{aligned}$$

and using the first equation in this set, solve successively to obtain the solution of the original set of equations.

Discuss any advantages or disadvantages in using this alternative procedure.

6-5. Using the method discussed in Problem 6-4, hand compute the solution for the following set of equations

$$\begin{aligned} -2X_1 - X_2 + X_3 &= -1 \\ X_1 + X_2 + X_3 &= 6 \\ 3X_1 + X_2 - X_3 &= 2 \end{aligned}$$

6-6. The Gauss elimination method may also be used in the evaluation of a determinant. Since the value of the resulting triangular determinant

$$\begin{vmatrix} 1 & b_{12} & b_{13} & b_{14} & \cdots & b_{1N} \\ & 1 & b_{23} & b_{24} & \cdots & b_{2N} \\ & & 1 & b_{34} & \cdots & b_{3N} \\ & & & \cdot & & \\ & & & & \cdot & \\ & & & & & \cdot \\ & & & & 1 & b_{N-1.N} \\ & & & & & 1 \end{vmatrix}$$

will always be equal to one, the value of the original determinant is equal to the product of the normalizing pivot element values, modified by a sign change if an odd number of row interchanges have been made.

Using the Gauss elimination method of Section 6.3, hand compute the value of the following determinant

$$\begin{bmatrix} 2 & 2 & 1 & 1 \\ 3 & 0 & 1 & 2 \\ 1 & 3 & 4 & 1 \\ 1 & 2 & 1 & 0 \end{bmatrix}$$

6-7. Using the appropriate parts of the program of Figure 6.7, write a Fortran program that is capable of evaluating up to a tenth-order determinant.

6-8. Compute the number of arithmetic operations required in using the Gauss elimination method to evaluate a set of N linear algebraic equations.

6-9. Round-off error accumulation depends approximately on the square root of the number of arithmetic operations performed. For example, in the case of 100 arithmetic operations a loss of one significant figure may result (factor of 10).

Compute the number of operations involved in evaluating the determinant of Problem 6-6 and compare the possible loss of significant digits to that loss that would result in using the method of Section 6.2.

6-10. Using the one-tree branch cut sets shown below for the network of Figure 6.8, define the positive sense of each of the cut set voltages and write the corresponding cut set equations. Now use the program of Figure 6.7 to obtain the solution to your equations. Compare your results with those of Figures 6.9 and 6.12. Note that tree branches are indicated by heavy lines.

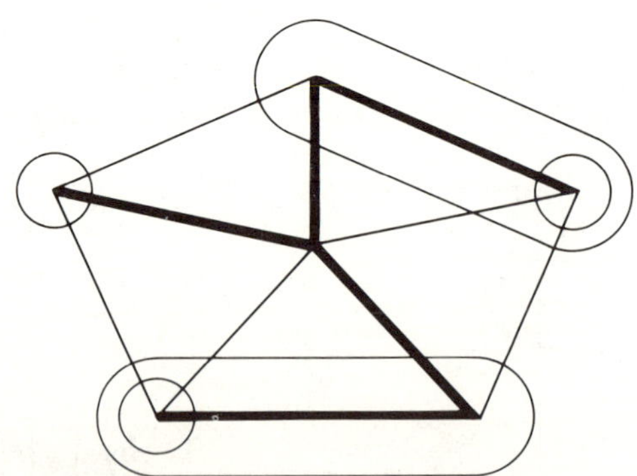

6-11. Repeat Problem 6-6 using the Gauss-Jordan method. Are there any disadvantages in using this method rather than the Gauss elimination method?

6-12. Repeat Problem 6-11 without normalizing the pivot row (no division). Compare the results obtained at each step with those obtained in Problem 6-11. What are your conclusions?

6-13. Repeat Problem 6-10 using the Gauss-Jordan method.

6-14. Using the Gauss-Jordan method, hand compute the solution for the network of Figure 6.4 (see Figure 6.15).

6-15. In Section 6.4 we introduced pivot row normalization to minimize the possibility of overflow or underflow. It also can be shown that error propagation will be reduced if we always use the largest magnitude divisor. The largest magnitude divisor can be found at each step of the elimination process by examining the elements of column J of the augmented matrix for each of the remaining rows $(I \geq J)$. The row in which the largest magnitude element is located is then exchanged with row J.

Modify the program of Figure 6.13 to include the largest magnitude divisor.

6-16. Using the program of Problem 6-15, determine the solution for the network of Figure 6.8 and compare your results with those of Figure 6.14.

6-17. The Gauss-Jordan method may also be used to find the inverse of a matrix A. This may be done by forming a double-square augmented matrix that contains the elements of A on the left and the unit matrix on the right. The Gauss-Jordan reduction procedure is then applied to the entire matrix to force the A matrix part to have element values that correspond to those of the unit matrix. As a consequence of the reduction process, the right-hand side will then correspond to the inverse matrix.

Using the above procedure, hand compute the inverse of matrix A

$$[A] = \begin{bmatrix} 3 & -2 & 4 \\ 2 & -1 & 3 \\ 1 & 1 & -2 \end{bmatrix}$$

Hint: First form the double-square matrix

$$\begin{bmatrix} 3 & -2 & 4 & 1 & 0 & 0 \\ 2 & -1 & 3 & 0 & 1 & 0 \\ 1 & 1 & -2 & 0 & 0 & 1 \end{bmatrix}$$

6-18. Write a Fortran program that uses the method of Problem 6-17 to obtain the inverse of matrix A. Limit the program to handle a maximum array size of 10 by 10.

6-19. Modify the program of Figure 6.17 so that it is capable of finding the equivalent K-port parameters. Assume that a symmetrical network exists.

6-20. In the solution of simultaneous linear algebraic equations, the results obtained may often be incorrect due to the round-off errors. One method of improving the accuracy of the results is based on the following procedure:

a. Given

$$\begin{array}{l} a_{11}\ X_1 + a_{12}\ X_2 + \cdots + a_{1N}\ X_N = a_{1,N+1} \\ a_{21}\ X_1 + a_{22}\ X_2 + \cdots + a_{2N}\ X_N = a_{2,N+1} \\ \vdots \\ a_{N1}\ X_1 + a_{N2}\ X_2 + \cdots + a_{NN}\ X_N = a_{N,N+1} \end{array}$$

compute the approximate solutions $A_1, A_2, \ldots, A_N$

b. Substitute the values obtained in part a. to evaluate the sum of the left member terms

$$\begin{array}{l} a_{11}\ A_1 + a_{12}\ A_2 + \cdots + a_{1N}\ A_N = \mathrm{SUM}_1 \\ a_{21}\ A_1 + a_{22}\ A_2 + \cdots + a_{2N}\ A_N = \mathrm{SUM}_2 \\ \vdots \\ a_{N1}\ A_1 + a_{N2}\ A_2 + \cdots + a_{NN}\ A_N = \mathrm{SUM}_N \end{array}$$

c. Substitute the exact solution into the original equations

$$\begin{array}{l} X_1 = A_1 + C_1 \\ X_2 = A_2 + C_2 \\ \vdots \\ X_N = A_N + C_N \end{array}$$

and subtract from these equations, the equations obtained in part b., thus yielding the following correction equations

$$\begin{array}{l} a_{11}\ C_1 + a_{12}\ C_2 + \cdots + a_{1N}\ C_N = a_{1,N+1} - \mathrm{SUM}_1 \\ a_{21}\ C_1 + a_{22}\ C_2 + \cdots + a_{2N}\ C_N = a_{2,N+1} - \mathrm{SUM}_2 \\ \vdots \\ a_{N1}\ C_1 + a_{N2}\ C_2 + \cdots + a_{NN}\ C_N = a_{N,N+1} - \mathrm{SUM}_N \end{array}$$

d. Solve the resulting set of equations in part c. to obtain the corrections for the approximate solutions.

Using the Gauss elimination method and the procedure described above, hand compute the solution for the following set of equations.

$$
\begin{aligned}
-2X_1 - X_2 + X_3 &= -1 \\
X_1 + 1.5X_2 + X_3 &= 7 \\
3X_1 + X_2 - X_3 &= 2
\end{aligned}
$$

f. Round each arithmetic operation to two significant figures.
g. Round each arithmetic operation to three significant figures.

6-21. Using the correction procedure of Problem 6-20, modify the program of Figure 6.7 accordingly.

6-22. Using the modified program of Problem 6-21, compute the solution for the network of Figure 6.8 and compare your results with those of Figure 6.9.

6-23. Using the equivalent K-port theorem of Section 6.4, hand compute the equivalent T-network for the four-mesh network shown below.

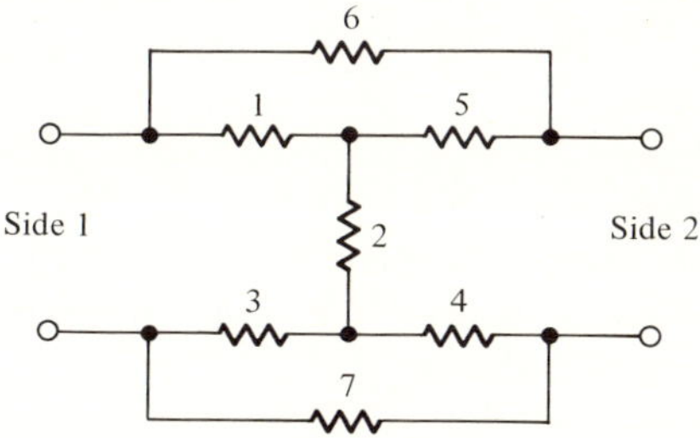

6-24. Hand compute the equivalent four-port network for the eight-mesh network shown below.

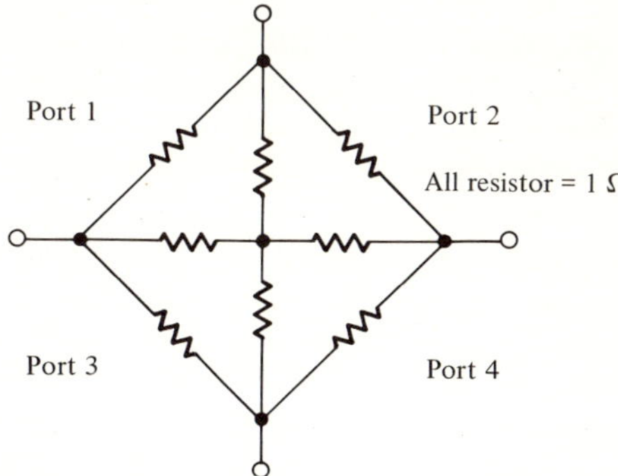

6-25. Use the Gauss-Seidel method to hand compute the solution for the network of Figure 6.4.

6-26. Determine the number of iterations required by the program of Figure 6.20 to obtain
a. five significant figures of accuracy
b. six significant figures of accuracy
c. seven significant figures of accuracy
for the solution to the network of Figure 6.4.

6-27. Use the Gauss-Seidel program of Figure 6.20 to obtain the solution to the network of Figure 6.8.

6-28. Jacobi's method is a modification of the Gauss-Seidel Method and uses the same approximations for the solutions during any given iteration. The results obtained from the previous iteration are therefore used in the computation of the next iteration. Modify the program of Figure 6.20 accordingly.

6-29. Repeat Problem 6-25 using the method of Problem 6-28.

6-30. Investigate the convergence problem of the method of Problem 6-28.

6-31. Use the program of Figure 6.23 to compute the insertion loss of the network of Figure 3.39 at
a. $f = 270$ kHz
b. $f = 280$ kHz
c. $f = 296$ kHz.
Compare these results to those of Figure 3.45.

6-32. Using the method of Section 6.6, hand-compute the solution to the following sets of equations:
a. $(2 + j3)I_1 + (2 - j2)I_2 = 2 + j1$
$(4 + j6)I_1 + (3 - j6)I_2 = -2 - j5$
b. $(2 - j1)I_1 - (1 - j2)I_2 = 0 - j100$
$-(1 - j2)I_1 + (3 + j3)I_2 = 0$

6-32. Using the method of Section 6.6, write a Fortran program that is capable of evaluating determinants that contain complex elements.

6-33. Use the program of Figure 6.23 to modify the program of Figure 6.17 so that it is capable of finding the equivalent T-network when complex impedances are involved.

6-34. Modify the program of Figure 6.23 so that it also computes and prints the sum of the voltages around each mesh.

6-35. Modify the program of Problem 6-18 so that it is capable of finding the inverse of a matrix that contains complex elements.

CHAPTER 7

Active Networks

7.1 INTRODUCTION

In Chapter 3 the reader was introduced to a number of passive network examples. In this chapter we will be concerned with active networks. Our goal is to show not only how the computer may be used to analyze existing active networks, but also to show the important role the computer may play in the actual design.

Although the author believes, at this writing, that vacuum-tube active networks are equally as important as transistor networks, we will nevertheless confine our discussion to transistor circuits alone. This should not prove to be a serious limitation since most vacuum-tube circuits are very similar to their transistor counterparts and, in addition, they usually are much simpler to analyze and design.

The reader should realize that it is impossible to do justice in one chapter to such an important branch of electrical engineering as active networks. The following discussion should serve as only an introduction to how the computer may be used in the analysis and design of small and large signal devices. The material, however, should whet the appetite of the more interested readers so that they will be able to go further alone.

7.2 TRANSISTOR BIASING

One of the first problems encountered in the design of a transistor amplifier is the establishment and maintenance of the operating point Q. A good biasing circuit must therefore maintain the operating point values under changing temperature conditions as well as in the replacement of the transistor itself.

The Operating Point

Before we discuss the various biasing circuits that may be used, let us consider how we may locate the operating point. Figure 7.1a shows one of the simplest biasing schemes. Summing the voltages around the supply-collector loop yields

$$V_{CC} - R_C I_C - V_{CE} = 0$$

or

$$I_C = -\frac{1}{R_C} V_{CE} + \frac{V_{CC}}{R_C}$$

This is the equation of a straight line and we may therefore construct the load line shown in Figure 7.1b by realizing that when

$$I_C = 0,\ V_{CE} = V_{CC}, \quad \text{and when} \quad V_{CE} = 0,\ I_C = \frac{V_{CC}}{R_C}$$

Since the base current is given by

$$I_B = \frac{V_{CC} - V_{BE}}{R_B} \approx \frac{V_{CC}}{R_B}$$

the operating point Q is therefore located at the intersection of the load line and the I_B base current I_C-V_{CE} characteristic curve. This follows from the fact that the collector current of the transistor must be the same as the current flowing through R_C.

The graphical procedure described above may also be performed by the computer program of Figure 7.2. This program initially reads in the number of I_C-V_{CE} data points N, the supply voltage V_{CC}, and the collector resistor R_C. This is followed by reading in the V_{CE}-I_C point values that correspond to the I_B base-current characteristic curve. For example, for a base current of 0.4 millamps in the curve of Figure 7.3, the following

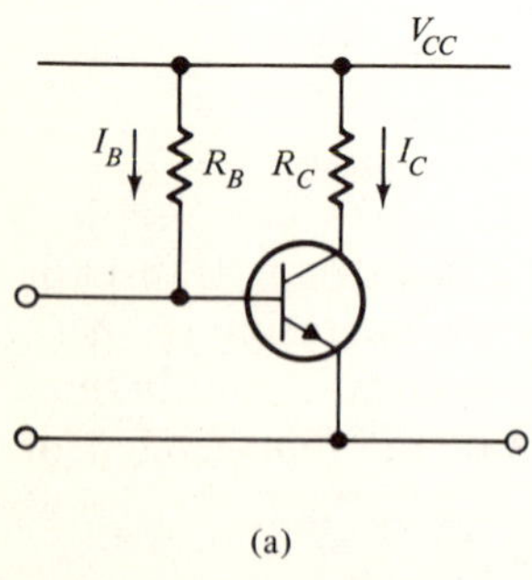

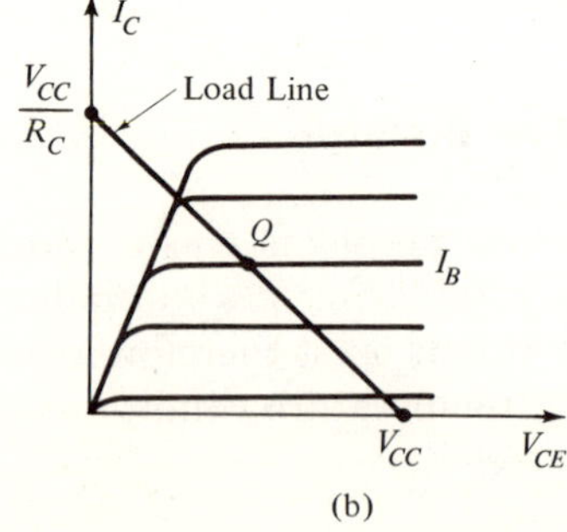

Figure 7.1 A simple bias circuit

```
C        THE CALCULATION OF THE TRANSISTOR OPERATING POINT
C        AS A FUNCTION OF VCC AND RC.
C
         DIMENSION X(100),Y(100)
         READ(5,1)N,VCC,RC
       1 FORMAT(I5,2E15.7)
         READ(5,2)(X(I),Y(I),I=1,N)
       2 FORMAT(4E15.7)
         DO 10 I=2,N
         YIC=-X(I)/RC+VCC/RC
         IF(YIC-Y(I))20,20,10
      10 CONTINUE
      20 A11=1.
         A12=1./RC
         V1=VCC/RC
         A21=1.
         A22=(Y(I)-Y(I-1))/(X(I-1)-X(I))
         V2=(X(I)*Y(I-1)-X(I-1)*Y(I))/(X(I)-X(I-1))
         DELTA=A11*A22-A21*A12
         CI=(V1*A22-V2*A12)/DELTA
         VCE=(A11*V2-A21*V1)/DELTA
         WRITE(6,7)VCC,RC,VCE,CI
       7 FORMAT('1THE LOAD LINE IS DETERMINED BY',//,' VCC =',E15.7,'  AND
        1RC =',E15.7,///,' THE TRANSISTOR OPERATING POINT IS LOCATED AT',//
        2,' VCE =',E15.7,/,' IC =',E15.7)
         STOP
         END
```

Figure 7.2 A program for locating the transistor's operating point

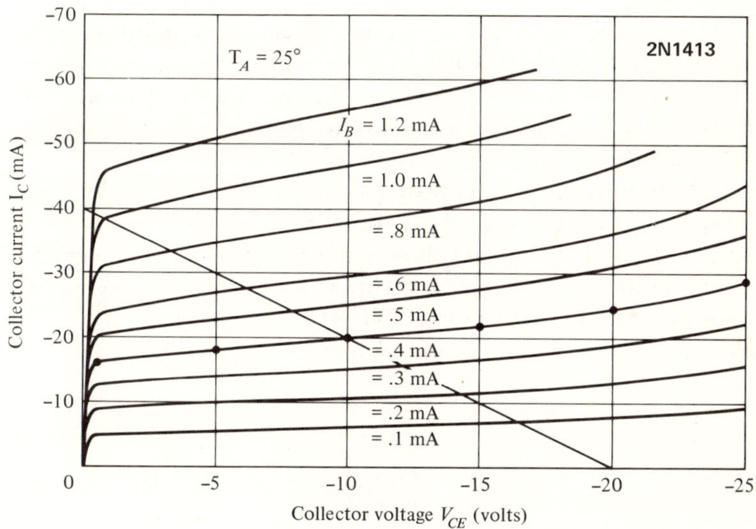

Figure 7.3 The collector characteristics for the 2N1413 transistor

seven points were used to represent this characteristic curve (note that in the case of a PNP transistor, only magnitude values are used):

I	*X(I)* OR V_{CE}	*Y(I)* OR I_C
1	0.0	0.0
2	0.150	0.016
3	5.0	0.018
4	10.0	0.020
5	15.0	0.022
6	20.0	0.0245
7	25.0	0.0285

The DO 10 loop is now used to find the value of I that corresponds to the nearest value of collector current that is equal to or less than the operating point value. This value of I is found by noting in Figure 7.3 that for V_{CE} values less than the operating point value, the value of I_C of the load line must be greater than the value of I_C of the characteristic curve. Computer statement S.0007 thus first computes YIC (I_C) from

$$I_C = -\frac{1}{R_C} V_{CE} + \frac{V_{CC}}{R_C}$$

or

$$YIC = \frac{-X(I)}{RC} + \frac{VCC}{RC}$$

We then test to see whether the value of YIC is greater than $Y(I)$. If the value of YIC is greater than $Y(I)$, this means that the load line collector current is greater than the collector current of the characteristic curve and we should therefore repeat the iteration for the next point values. Whenever the value of the load line collector current is less than or equal to the point value, control is transferred to statement 20 and computer statements S.0010 through S.0018 are used to compute the operating point.

The method of computing the operating point is based on taking the two successive point values of the characteristic curve that are closest to the operating point and then using these values to determine the equation of the straight line passing through these two points. Referring to Figure 7.4, we thus see that for the points $X(I-1), Y(I-1)$ and $X(I), Y(I)$ to determine a straight line equation of the form

$$y = mx + b$$

it requires that

$$\begin{aligned} Y(I) &= mX(I) + b \\ Y(I-1) &= mX(I-1) + b \end{aligned}$$

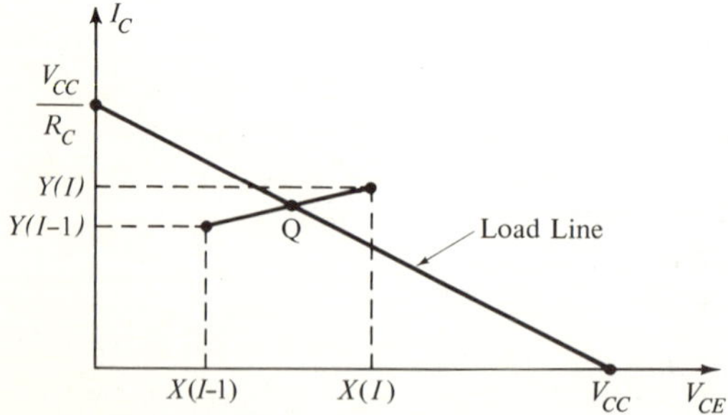

Figure 7.4 The graphical solution of two simultaneous linear equations

or

$$m = \frac{Y(I) - Y(I-1)}{X(I) - X(I-1)}$$

and

$$b = \frac{X(I) * Y(I-1) - X(I-1) * Y(I)}{X(I) - X(I-1)}$$

The two equations that must be solved therefore correspond to the equation of the load line

$$I_C = -\frac{1}{R_C} V_{CE} + \frac{V_{CC}}{R_C}$$

or

$$I_C + \frac{1}{R_C} V_{CE} = \frac{V_{CC}}{R_C}$$

and the equation of the straight line passing through the two points on the characteristic curve

$$y - mx = b$$

or

$$I_C + \frac{Y(I) - Y(I-1)}{X(I-1) - X(I)} V_{CE} = \frac{X(I) * Y(I-1) - X(I-1) * Y(I)}{X(I) - X(I-1)}$$

These two equations are of the form

$$a_{11}I_1 + a_{12}I_2 = V_1$$
$$a_{21}I_1 + a_{22}I_2 = V_2$$

Since we have already written a program to solve equations of this type, we may identify the corresponding constants and take advantage of the statements in the program of Figure 6.2. After the solution has been found, we may use computer statements S.0019 and S.0020 to print the results.

Figure 7.5 gives the computer results. The reader may readily verify the results by referring to the graphical solution shown in Figure 7.3. The reader is urged to modify the previous program so that it is capable of finding the operating point for more than one base-current characteristic curve. Such a program could then be extended to permit interpo-

```
THE LOAD LINE IS DETERMINED BY

VCC =   0.20CCCCCE 02   AND RC =   0.5000000E 03

THE TRANSISTOR OPERATING POINT IS LOCATED AT

VCE =   0.10CCCCCE 02
IC =  0.1999999E-C1
```

Figure 7.5 The output results of the program of Figure 7.2

lation between defined base-current curves. This type of program would thus allow the reader to simulate laboratory environmental measurements. For example, it would be possible without building any circuits to learn the effect on the operating point for such things as transistor replacement, temperature variation changes in the characteristic curves, and resistor and supply tolerance variations.

The General Biasing Network

One of the main problems with the bias circuit of Figure 7.1 is that its thermal stability is very poor. In fact, when this method of biasing is employed with certain types of transistors, thermal runaway results. We therefore turn to the investigation of other types of bias circuits, bearing in mind for a particular type of transistor the variation of the current gain among transistors of the same type; the variation of the current gain, the base-to emitter voltage drop, and the collector leakage current with a variation in temperature; and the effect of supply voltage and resistor tolerance.

In order to investigate several different one-supply biasing circuits, consider the biasing network shown in Figure 7.6. By short-circuiting and open-circuiting appropriate resistors, we may observe that this network may be reduced to any one of the biasing networks shown in Figure 7.7. Let us therefore derive a general expression for I_C in the network of Figure 7.6 so that we may investigate any one of the biasing networks shown in Figure 7.7.

Using Thévenin's theorem, we first redraw Figure 7.6 as shown in Figure 7.8. Writing Kirchhoff's voltage law for the collector-supply loop and the base-emitter loop, we have

$$V_{CC} - R_C(I_C + I_F) = V_{C0}$$

and

$$V_{BE} + R_E I_E = V_{B0}$$

Writing Kirchhoff's current law for the cut set containing the collector,

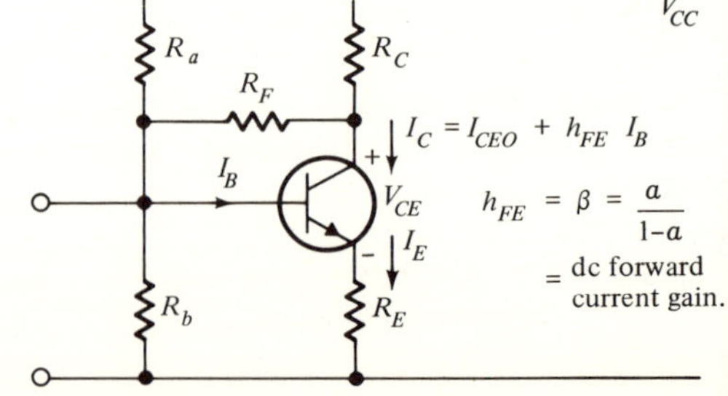

Figure 7.6 A general one-supply biasing circuit

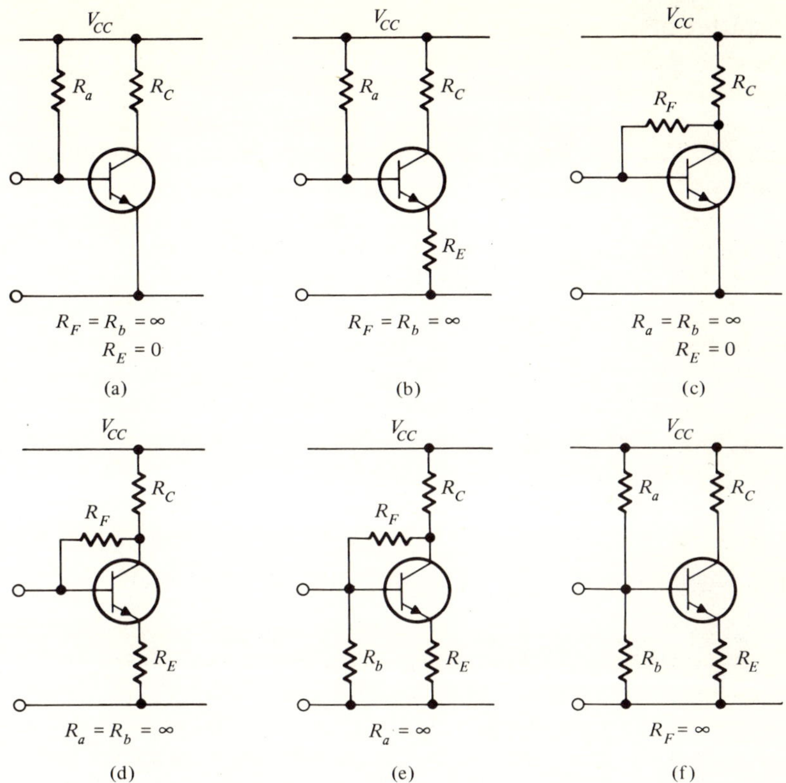

Figure 7.7 Variations of the bias circuit shown in Figure 7.6

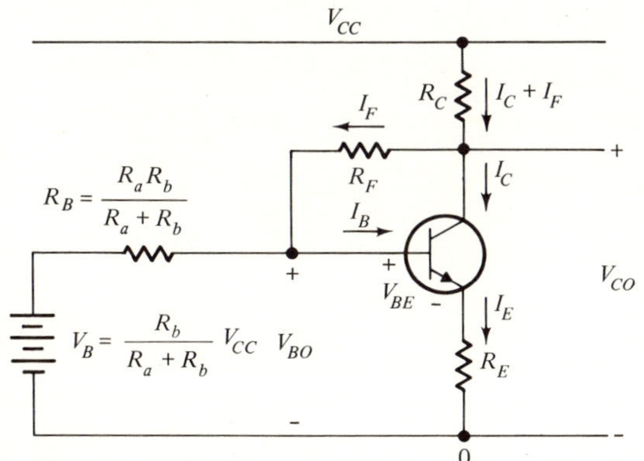

Figure 7.8 Thévenin's theorem is used in redrawing the circuit of Figure 7.6

base, and emitter branches yields

$$I_E = I_B + I_C$$

Substituting for V_{C0} and V_{B0} in the collector-supply and base-emitter equations gives

$$\begin{aligned} V_{CC} - R_C(I_C + I_F) - R_F I_F - V_{BE} - R_E I_E &= 0 \\ V_B - R_B(I_B - I_F) - V_{BE} - R_E I_E &= 0 \end{aligned}$$

Solving for I_B from

$$I_C = I_{CE0} + \beta I_B$$

yields

$$I_B = \frac{I_C - I_{CE0}}{\beta}$$

Substituting the latter and rearranging the equations, we obtain

$$\begin{cases} -R_B I_F + \dfrac{R_B}{\beta} I_C + R_E I_E = V_B - V_{BE} + \dfrac{I_{CE0}}{\beta} R_B \\ -\left(1 + \dfrac{1}{\beta}\right) I_C + I_E = -\dfrac{I_{CE0}}{\beta} \\ (R_C + R_F) I_F + R_C I_C + R_E I_E = V_{CC} - V_{BE} \end{cases}$$

Using determinants to solve for I_C we find

$$I_C = \frac{\begin{vmatrix} -R_B & V_B - V_{BE} + \dfrac{I_{CE0}}{\beta} R_B & R_E \\ 0 & -\dfrac{I_{CE0}}{\beta} & 1 \\ (R_C + R_F) & V_{CC} - V_{BE} & R_E \end{vmatrix}}{\begin{vmatrix} -R_B & \dfrac{R_B}{\beta} & R_E \\ 0 & -\left(1 + \dfrac{1}{\beta}\right) & 1 \\ (R_C + R_F) & R_C & R_E \end{vmatrix}}$$

$$= \frac{A + B}{C + D}$$

where

$$A = \frac{R_B R_E I_{CE0}}{\beta} + (R_C + R_F)\left(V_B - V_{BE} + \frac{I_{CE0}}{\beta} R_B\right)$$

$$B = R_E(R_C + R_F)\,\frac{I_{CE0}}{\beta} + R_B(V_{CC} - V_{BE})$$

$$C = R_B R_E\left(1 + \frac{1}{\beta}\right) + \frac{R_B}{\beta}\,(R_C + R_F) + R_C R_B$$

$$D = R_E(R_C + R_F)\left(1 + \frac{1}{\beta}\right)$$

Figure 7.9 gives the Fortran program for computing the operating point. Initially the heading and subheadings are printed and then the parameter values are read into memory by two READ statements. Thévenin's equivalent values of R_B and V_B are then computed and this is followed by computing the numerator and denominator terms of I_C. Computer statements S.0013 and S.0014 are then used to compute the values of I_C and V_{C0}. The element values and the computed values of I_C and V_{C0} are then printed, and control is returned to statement 2 so that the operating point can be computed for the next set of data.

In order to test the above program, let us now investigate the commonly used biasing scheme of Figure 7.7f. This may be done by making the value of R_F ($= 10^7$ ohms) large compared to the other circuit element values. Figure 7.10 shows the results of five different sets of input data. The third data set may be regarded as the reference and considered as the typical nominal room temperature values. The first and fourth data sets therefore show the effect of transistor beta variations. The second data set shows the effect of a higher leakage current and the fifth data set shows the effect of a higher beta value and a lower V_{BE} drop which could be caused by a higher temperature operation.

```
          C          A GENERAL
          C       TRANSISTOR BIASING PROGRAM.
S.0001            WRITE(6,1)
S.0002          1 FORMAT('1TRANSISTOR BIASING PROGRAM',//,5X,'RC',9X,'RF',9X,'RA',9X
               1,'RBB',8X,'RE',8X,'BETA',8X,'VBE',8X,'CEO',8X,'VCC',8X,'IC',8X,'VC
               20',//)
S.0003          2 READ(5,3)RC,RF,RA,RBB,RE
S.0004          3 FORMAT(5E15.7)
S.0005            READ(5,4)B,VBE,CEO,VCC
S.0006          4 FORMAT(4E15.7)
S.0007            RB=RA*RBB/(RA+RBB)
S.0008            VB=RBB*VCC/(RA+RBB)
S.0009            A=RB*RE*CEO/B+(RC+RF)*(VB-VBE+CEO*RB/B)
S.0010            BB=RE*(RC+RF)*CEO/B+RB*(VCC-VBE)
S.0011            C=RB*RE*(1.+1./B)+RB*(RC+RF)/B+RC*RB
S.0012            D=RE*(RC+RF)*(1.+1./B)
S.0013            CI=(A+BB)/(C+D)
S.0014            VCO=VCC-RC*CI
S.0015            WRITE(6,5)RC,RF,RA,RBB,RE,B,VBE,CEO,VCC,CI,VCO
S.0016          5 FORMAT(11E11.3)
S.0017            GO TO 2
S.0018            STOP
S.0019            END
```

Figure 7.9 A general program for designing transistor biasing circuits

RANSISTOR BIASING PROGRAM

RC	RF	RA	RBB	RE	BETA	VBE	CEO	VCC	IC	VCC
0.100E 05	0.100E 08	0.200E 05	0.500E 04	0.400E 04	0.200E 03	0.650E 00	0.500E-05	0.250E 02	0.108E-02	0.142E 02
0.100E 05	0.100E 08	0.200E 05	0.500E 04	0.400E 04	0.100E 03	0.650E 00	0.200E-04	0.250E 02	0.107E-02	0.143E 02
0.100E 05	0.100E 08	0.200E 05	0.500E 04	0.400E 04	0.100E 03	0.650E 00	0.500E-05	0.250E 02	0.107E-02	0.143E 02
0.100E 05	0.100E 08	0.200E 05	0.500E 04	0.400E 04	0.500E 02	0.650E 00	0.500E-05	0.250E 02	0.105E-02	0.145E 02
0.100E 05	0.100E 08	0.200E 05	0.500E 04	0.400E 04	0.200E 03	0.450E 00	0.500E-05	0.250E 02	0.113E-02	0.137E 02

IHC217I

Figure 7.10 The output results for the program of Figure 7.9

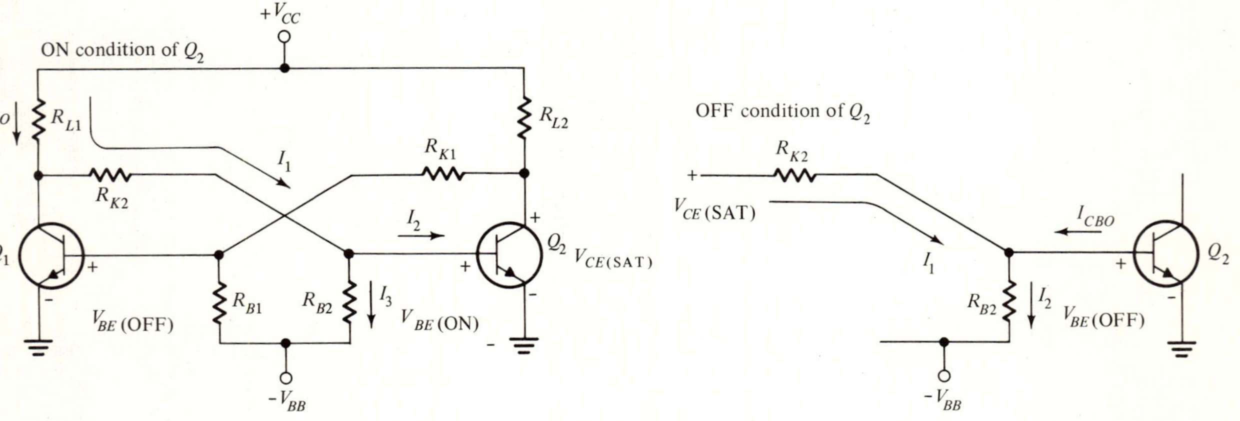

Figure 7.11 A two-supply flip-flop showing the ON condition of Q_2

Figure 7.12 The OFF condition of Q_2

At this point the reader should realize that there exist many different formal procedures for determining the stability of the operating point. Many of these methods are very lengthy and restrictive, and others lead to very inaccurate results.[1] As a consequence, most designers usually arrive at the final biasing circuit by trial-and-error environmental tests, realizing that an increase in the value of R_E and/or a reduction in the R_a-R_b combination will improve the temperature stability of the circuit. The reader should now realize that it would be a simpler matter to use the above computer program to simulate these time-consuming laboratory tests. Such "computer laboratory testing" would not only provide much better laboratory notebook data, but it would also be better able to pinpoint the very sensitive elements. Furthermore, the ease of this type of laboratory investigation would allow many more different circuit variations to be considered, so that the reduction in the total component costs could produce sizeable production run savings as well as reducing quality control inspection costs.

7.3 WORST-CASE FLIP-FLOP DESIGN

A basic flip-flop or bistable multivibrator is shown in Figure 7.11. This circuit has two stable states; Q_1 on and Q_2 off, or Q_1 off and Q_2 on. It may be made to change from one state to the other state by the application of a trigger pulse (the trigger circuitry is not shown). Two successive trigger pulses will therefore return the flip-flop to its original state. For this reason the circuit is often called a binary circuit and is widely used in electronic counters as a frequency divider and pulse counter. Counting or dividing by other than a power of 2 can be achieved by cascading an appropriate number of flip-flops and using the necessary feedback. The flip-flop can also be used as a memory device and perhaps its greatest use can be found in digital computers where it is used as a storage unit.

Worst-Case Design of Two-Supply Grounded Emitter Flip-Flop[2]

One of the first steps in the design of a transistor flip-flop is to determine the element values that will insure turn-on and turn-off conditions. The following computer program investigates the "worst-case" d-c design for

[1] For one of the best biasing procedures see J. Mulligan, Jr., and S. Shamis, "Transistor Amplifier Stages with Prescribed Gain and Sensitivity," *AIEE Commun. Electronics*, Vol. 80, (July,.1961), p. 335.

[2] This discussion follows the general design scheme outlined in Joseph A. Walston and John R. Miller, *Transistor Circuit Design* (New York: McGraw-Hill Book Company, 1963), pp. 369–77.

two relatively inexpensive (approximately 25 cents) silicon epoxy transistors, the 2N3646 (NPN) and the 2N3639 (PNP).

Worst-case d-c design is accomplished by deriving the ON and OFF equations. These equations contain all the transistor parameters, resistor parameters, and supply voltages specified at their worst-case extremes. In each of these equations the base-return resistor R_B is taken as the dependent variable and the coupling resistor R_K is taken as the independent variable. The solution of these equations thus gives the range in the value of R_B for a given value of R_K that will insure turn-on and turn-off conditions.

Although the magnitudes of the supply voltages are usually determined by other circuit requirements, they should be made much larger than the saturation voltages of the transistors, $V_{CE(\mathrm{SAT})}$ and $V_{BE(\mathrm{ON})}$, to minimize the effects of variations of these voltages. In the following discussion both V_{CC} and V_{BB} will be assigned a nominal value of ten volts.

The collector saturation current

$$I_{C(\mathrm{SAT})} \approx \frac{V_{CC}}{R_L}$$

will influence the value of R_L. For minimum power drain it would be desirable to make R_L as large as possible; however, the value of f_t and, therefore, the switching speed could be reduced. For $V_{CC} = 10$ volts, R_L was chosen as 1 K ohm to give $I_{C(\mathrm{SAT})} = 10$ ma and maximum f_t (see manufacturer's data sheet). This choice of V_{CC} and R_L also insures that the load line does not cross any breakdown region on the collector characteristics and that the total circuit as well as transistor dissipations are well within limits.

To analyze the flip-flop operation, assume that Q_2 is on and Q_1 is off as shown in Figure 7.11. With Q_1 off, enough current must flow from V_{CC} through R_{L1} and R_{K2} to forward-bias the base of Q_2 and saturate Q_2. Under this condition the collector voltage of Q_2 will be $V_{CE(SAT)}$. The level of this voltage and the base-supply voltage V_{BB} along with the $R_{K1} - R_{B1}$ divider network must hold Q_1 off. The network must therefore remain in this stable state until some disturbance is introduced (trigger pulse) to start the turn-on of Q_1 or the turn-off of Q_2. At this point regeneration can occur and the opposite state will follow (Q_1 on, Q_2 off).

In order to derive the design equation for holding Q_2 on, let us write Kirchhoff's current law at the base of Q_2, thus

$$I_1 = I_2 + I_3$$

$$\frac{V_{CC} - R_{L1}I_{CB0} - V_{BE(\mathrm{ON})}}{R_{L1} + R_{K2}} = \frac{1}{h_{FE}}\left[\frac{V_{CC} - V_{CE(\mathrm{SAT})}}{R_{L2}} - \frac{V_{CE(\mathrm{SAT})} - V_{BE(\mathrm{OFF})}}{R_{K1}}\right] + \frac{V_{BB} + V_{BE(\mathrm{ON})}}{R_{B2}}$$

Solving for R_{B2}, we have

$$R_{B2} = \frac{V_{BB} + V_{BE(\text{ON})}}{\left\{ \frac{V_{CC} - R_{L1}I_{CBO} - V_{BE(\text{ON})}}{R_{L1} + R_{K2}} - \frac{1}{h_{FE}} \left[\frac{V_{CC} - V_{CE(\text{SAT})}}{R_{L2}} - \frac{V_{CE(\text{SAT})} - V_{BE(\text{OFF})}}{R_{K1}} \right] \right\}}$$

The smallest value of R_{B2} which can be used must therefore be that value which still causes the transistor to be in saturation when all voltage and resistor values are at their worst-case extremes. Thus, the worst-case extreme for the on condition of Q_2 is

$$\underline{R}_{B2} = \frac{\bar{V}_{BB} + \bar{V}_{BE(\text{ON})}}{\left\{ \frac{\underline{V}_{CC} - \bar{R}_{L1}\bar{I}_{CBO} - \bar{V}_{BE(\text{ON})}}{\bar{R}_{K2} + \bar{R}_{L1}} - \frac{1}{\underline{h}_{FE}} \left[\frac{\bar{V}_{CC} - \underline{V}_{CE(\text{SAT})}}{\underline{R}_{L2}} - \frac{\underline{V}_{CE(\text{SAT})} - V_{BE(\text{OFF})}}{\bar{R}_{K1}} \right] \right\}}$$

where the underlines indicate the minimum, and the overlines the maximum.

Figure 7.12 represents the appropriate portion of the flip-flop that must be considered in determining the parameters that influence the off condition of Q_2. The application of Kirchhoff's current law at the base of Q_2 yields

$$I_2 = I_1 + I_{CB0}$$

$$\frac{V_{BB} + V_{BE(\text{OFF})}}{R_{B2}} = \frac{V_{CE(\text{SAT})} - V_{BE(\text{OFF})}}{R_{K2}} + I_{CB0}$$

Solving for R_{B2}, we have

$$R_{B2} = \frac{V_{BB} + V_{BE(\text{OFF})}}{\frac{V_{CE(\text{SAT})} - V_{BE(\text{OFF})}}{R_{K2}} + I_{CB0}}$$

The largest value of R_{B2} which can be used is that value which still causes the transistor to be cut off when all voltage and resistor values are at their worst-case extremes. Thus the worst-case extreme for the off condition of Q_2 is:

$$\bar{R}_{B2} = \frac{\underline{V}_{BB} + V_{BE(\text{OFF})}}{\frac{\bar{V}_{CE(\text{SAT})} - V_{BE(\text{OFF})}}{\underline{R}_{K2}} + \bar{I}_{CB0}}$$

Let us now assume for simplicity that all similar resistors are equal and that the two transistors are identical. Thus

$$\begin{aligned} R_{L1} &= R_{L2} = R_L \\ R_{B1} &= R_{B2} = R_B \\ R_{K1} &= R_{K2} = R_K \\ Q_1 &= Q_2 \end{aligned}$$

Ignoring the temperature effects on the resistors, but not the transistors, we note that the following conditions exist at temperature extremes:

LOW TEMPERATURE	HIGH TEMPERATURE
$V_{BE(\mathrm{ON})}$ is a maximum	I_{CB0} is a maximum
h_{FE} is a minimum	$V_{CE(\mathrm{SAT})}$ is a maximum
$V_{CE(\mathrm{SAT})}$ is a minimum	

It follows that worst-case turn-on occurs at low temperature, and worse-case turn-off occurs at high temperature.

We now introduce the following relations that may be substituted for the resistor and power supply maximum and minimum values:

$$\bar{R} = R(1 + RT) \qquad \text{and} \qquad \bar{V} = V(1 + VT)$$
$$\underline{R} = R(1 - RT) \qquad\qquad\qquad \underline{V} = V(1 - VT)$$

where R and V are the nominal resistance and voltage values and RT and VT are the resistance and voltage tolerance expressed as a decimal fraction. Substituting the above, the off and on equations may thus be written as

Off Equation

$$R_B \text{ nominal} = \frac{V_{BB}(1 - VT) + V_{BE(\mathrm{OFF})}}{(1 + RT)\left[\dfrac{\bar{V}_{CE(\mathrm{SAT})} - V_{BE(\mathrm{OFF})}}{R_K(1 - RT)} + \bar{I}_{CB0}\right]}$$

On Equation

$$R_B \text{ nominal} = \frac{V_{BB}(1 + VT) + \bar{V}_{BE(\mathrm{ON})}}{(1 - RT)\left\{\dfrac{V_{CC}(1 - VT) - R_L(1 + RT)\bar{I}_{CBO} - \bar{V}_{BE(\mathrm{ON})}}{R_K(1 + RT) + R_L(1 + RT)} - \dfrac{1}{\underline{h}_{FE}}\left[\dfrac{V_{CC}(1 + VT) - \underline{V}_{CE(\mathrm{SAT})}}{R_L(1 - RT)} - \dfrac{\underline{V}_{CE(\mathrm{SAT})} - V_{BE(\mathrm{OFF})}}{R_K(1 + RT)}\right]\right\}}$$

Figure 7.13 shows the Fortran design program. Initially, computer statement S.0004 reads into memory the run number RUN(I), the collector resistance RL(I), the base-supply voltage VBB(I), the collector supply voltage VCC(I), and the resistor and supply voltage tolerances. Note that these values correspond to the parameters that will be used in 12 different computer runs. Computer statement S.0006 next reads in the base-emitter off voltage VBEOFF(I), the base-emitter on voltage VBEON(I), the minimum collector-emitter saturation voltage VCESN(I), and the maximum collector-emitter saturation voltage VCESF(I). This is followed by computer statement S.0008, which reads in the collector to base-leakage current XICBO(I), the d-c current gain HFE(I), the low value of temperature TLO(I), and the high value of temperature THI(I). Computer statements S.0010 and S.0011 are then used to print the output heading and subheadings.

This is followed by the DO 90 loop that computes the values of RBON and RBOFF for R_K values ranging from 1 to 35 K ohms in 2 K ohm increment steps. Note that whenever the value of RBON is greater than RBOFF, computation is halted and control is returned to the DO 90

```
          C       WORSE CASE DESIGN OF TWO SUPPLY
          C       GROUNDED EMITTER FLIP FLOP
          C
S.0001            DIMENSION RUN(12),RL(12),VBB(12),VCC(12),TV(12),TR(12)
S.0002            DIMENSION VBEOFF(12),VBEON(12),VCESN(12),VCESF(12)
S.0003            DIMENSION XICBO(12),HFE(12),TLO(12),THI(12)
S.0004            READ(5,10)(RUN(I),RL(I),VBB(I),VCC(I),TV(I),TR(I),I=1,12)
S.0005         10 FORMAT(6F6.2)
S.0006            READ(5,20)(VBEOFF(I),VBEON(I),VCESN(I),VCESF(I),I=1,12)
S.0007         20 FORMAT(4F6.2)
S.0008            READ(5,30)(XICBO(I),HFE(I),TLO(I),THI(I),I=1,12)
S.0009         30 FORMAT(E10.2,3F10.2)
S.0010            WRITE(6,40)
S.0011         40 FORMAT(1H1,10X,'WORSE CASE D-C DESIGN OF TWO SUPPLY',/,11X,'GROUND
                 1ED EMITTER FLIP FLOP',/,11X,'FOR 2N3646(NPN) AND 2N3639(PNP) TRANS
                 2ISTORS')
S.0012            WRITE(6,50)
S.0013         50 FORMAT(//,10X,'RUNS 1 THROUGH 6 ARE FOR THE 2N3646 ',/,10X,'RUNS
                 17 THROUGH 12 ARE FOR THE 2N3639 TRANSISTOR')
S.0014            DO 90 I=1,12
S.0015            WRITE(6,65)RUN(I),TV(I),TR(I),TLO(I),THI(I)
S.0016         65 FORMAT(/,3X,4H RUN,F5.1,2X,3HTV=,F4.2,2X,3HTR=,F4.2,2X,4HTLO=,F4.1
                 1,2X,4HTHI=,F4.1)
S.0017            WRITE(6,70)
S.0018         70 FORMAT(/15X,3H RK,6X,4HRBON,4X,5HRBOFF/)
S.0019            RK=1000.
S.0020         12 A=VBB(I)*(1.+TV(I))+VBEON(I)
S.0021            B=VCC(I)*(1.-TV(I))-RL(I)*(1.+TR(I))*XICBO(I)-VBEON(I)
S.0022            C=(VCC(I)*(1.+TV(I))-VCESN(I))/(RL(I)*(1.-TR(I)))
S.0023            D=(-VBEOFF(I)+VCESN(I))/(RK*(1.+TR(I)))
S.0024            G=RK*(1.+TR(I))+RL(I)*(1.+TR(I))
S.0025            RBON=A/((1.-TR(I))*((B/G)-((1./HFE(I))*(C-D))))
S.0026            E=VBB(I)*(1.-TV(I))+VBEOFF(I)
S.0027            F=(-VBEOFF(I)+VCESF(I))/(RK*(1.-TR(I)))
S.0028            RBOFF=E/((F+XICBO(I))*(1.+TR(I)))
S.0029            WRITE(6,80)RK,RBON,RBOFF
S.0030         80 FORMAT(10X,F10.1,F10.1,F10.1)
S.0031            RK=RK+2000.
S.0032            IF(36000.-RK)90,15,15
S.0033         15 IF(RBON-RBOFF)12,12,90
S.0034         90 CONTINUE
S.0035            STOP
S.0036            END
```

Figure 7.13 A Fortran program for investigating the worst-case design of a two-supply flip-flop

Table 7.1 Input Data

	RUN(*I*)	$R_L(I)$	$V_{BB}(I)$	$V_{CC}(I)$	*TV*(*I*)	*TR*(*I*)	$V_{BE(OFF)}$	$V_{BE(ON)}$	V_{CESN}	V_{CESF}	XICBO*	HFE	TLO	TH1
2N3646 NPN	1	1kΩ	−10V	10V	.05	.05	0.0	.72	.13	.20	.0005	50	25°	25°
	2	↓	↓	↓	↓	.10	↓	↓	↓	↓	↓	↓	↓	↓
	3	↓	↓	↓	↓	.20	↓	↓	↓	↓	↓	↓	↓	↓
	4	↓	↓	↓	.02	.10	↓	↓	↓	↓	↓	↓	↓	↓
	5	↓	↓	↓	.05	↓	↓	.80	.11	.15	.002	30	0°	65°
	6	↓	↓	↓	.02	↓	↓	.80	.11	.15	.002	30	0°	65°
2N3639 PNP	7	1kΩ	−10V	10V	.05	.05	0.0	.90	.08	.08	.00008	75	25°	25°
	8	↓	↓	↓	↓	.10	↓	↓	↓	↓	↓	↓	↓	↓
	9	↓	↓	↓	↓	.20	↓	↓	↓	↓	↓	↓	↓	↓
	10	↓	↓	↓	.02	.10	↓	↓	↓	↓	↓	↓	↓	↓
	11	↓	↓	↓	.05	.10	↓	.95	.07	.08	.001	30	0°	65°
	12	↓	↓	↓	.02	.10	↓	.95	.07	.08	.001	30	0°	65°

*Current in mA.

loop so that the next run may be computed. Note also that the numerical values of RK, RBON, and RBOFF are printed for each computation of the RBON and RBOFF equations. Table 7.1 shows the input data that were supplied for these 12 runs, and Figure 7.14 shows the corresponding computer results.

```
        WORSE CASE D-C DESIGN OF TWO SUPPLY
        GROUNDED EMITTER FLIP FLOP
        FOR 2N3646(NPN) AND 2N3639(PNP) TRANSISTORS


        RUNS 1 THROUGH 6 ARE FOR THE 2N3646
        RUNS 7 THROUGH 12 ARE FOR THE 2N3639 TRANSISTOR

RUN   1.0   TV=0.05   TR=0.05   TLO=25.0   THI=25.0

             RK         RBON       RBOFF

           1000.0      2978.8     42874.4
           3000.0      6306.1    128016.6
           5000.0     10045.1    212359.4
           7000.0     14277.4    295914.2
           9000.0     19107.8    378691.7
          11000.0     24672.4    460702.9
          13000.0     31152.6    541958.9
          15000.0     38794.6    622468.5
          17000.0     47941.5    702242.8
          19000.0     59086.5    781292.6
          21000.0     72964.5    859626.9
          23000.0     90721.7    937255.7
          25000.0    114247.9   1014188.4
          27000.0    146900.6   1090434.0
          29000.0    195268.5   1166002.0
          31000.0    274291.1   1240901.0
          33000.0    426631.5   1315141.0
          35000.0    842607.9   1388729.0

RUN   2.0   TV=0.05   TR=0.10   TLO=25.0   THI=25.0

             RK         RBON       RBOFF

           1000.0      3313.3     38776.4
           3000.0      7060.6    115809.2
           5000.0     11330.8    192156.6
           7000.0     16241.9    267827.5
           9000.0     21950.2    342830.9
          11000.0     28666.8    417175.5
          13000.0     36684.8    490870.4
          15000.0     46422.8    563923.4
          17000.0     58501.0    636342.2
          19000.0     73878.1    708136.7
          21000.0     94119.4    779314.4
          23000.0    121966.6    849882.7
          25000.0    162698.5    919850.0
          27000.0    227949.5    989223.7
          29000.0    349389.7   1058011.0
          31000.0    654472.4   1126219.0
          33000.0   2851321.0   1193857.0

RUN   3.0   TV=0.05   TR=0.20   TLO=25.0   THI=25.0

             RK         RBON       RBOFF

           1000.0      4123.8     31603.5
           3000.0      8930.2     94433.4
           5000.0     14601.1    156765.7
           7000.0     21393.2    218606.4
           9000.0     29675.7    279960.9
          11000.0     39999.6    340835.3
          13000.0     53225.7    401235.0
          15000.0     70777.7    461165.7
          17000.0     95193.4    520632.4
          19000.0    131476.7    579640.9
          21000.0    191059.6    638196.0
          23000.0    306994.2    696303.6
          25000.0    630957.4    753968.5
          27000.0   6608543.0    811195.7
```

Figure 7.14 The output results of the program of Figure 7.13

RUN 4.0 TV=0.02 TR=0.10 TLO=25.0 THI=25.0

RK	RBON	RBOFF
1000.0	3106.6	40000.9
3000.0	6592.3	119466.4
5000.0	10529.0	198224.7
7000.0	15010.6	276285.2
9000.0	20158.7	353657.1
11000.0	26133.9	430349.4
13000.0	33153.1	506371.6
15000.0	41515.8	581731.6
17000.0	51648.8	656437.2
19000.0	64180.8	730498.9
21000.0	80077.9	803924.3
23000.0	100905.8	876721.2
25000.0	129380.2	948897.9
27000.0	170658.0	1020462.3
29000.0	235879.2	1091422.0
31000.0	354386.2	1161784.0
33000.0	636577.2	1231557.0
35000.0	2178656.0	1300749.0

RUN 5.0 TV=0.05 TR=0.10 TLO= 0.0 THI=65.0

RK	RBON	RBOFF
1000.0	3514.9	51203.8
3000.0	7881.3	150052.7
5000.0	13447.1	244425.7
7000.0	20785.6	334619.9
9000.0	30904.7	420906.6
11000.0	45754.1	503534.5
13000.0	69662.2	582731.2
15000.0	114556.2	658706.2
17000.0	229682.9	731652.6
19000.0	1171685.0	801747.6

RUN 6.0 TV=0.02 TR=0.10 TLO= 0.0 THI=65.0

RK	RBON	RBOFF
1000.0	3285.9	52820.7
3000.0	7308.3	154791.3
5000.0	12342.3	252144.4
7000.0	18825.1	345186.8
9000.0	27487.3	434198.4
11000.0	39650.1	519435.6
13000.0	57972.9	601133.3
15000.0	88722.4	679507.4
17000.0	151027.1	754757.4
19000.0	344645.7	827066.0
21000.0	-7046176.0	896602.4
23000.0	-373397.0	963523.1
25000.0	-207291.2	1027972.8
27000.0	-150069.9	1090086.0
29000.0	-121098.7	1149987.0
31000.0	-103598.7	1207793.0
33000.0	-91882.9	1263611.0
35000.0	-83490.2	1317542.0

Figure 7.14 Continued

A preliminary examination of these results shows that for some larger values of R_K, the $R_{B(\mathrm{ON})}$ values become negative. This is so because for some values of R_K the base current is insufficient to switch the transistor to saturation. Another observation, which is especially true for the 2N3646 transistor, is that for high values of R_K, the $R_{B(\mathrm{OFF})}$ values become relatively constant. This is caused by a relatively high value of I_{CBO} and the constancy can be seen from an examination of the $R_{B(\mathrm{OFF})}$ equation. Note that in this equation, the first term in the denominator $((V_{CE(\mathrm{SAT})} - V_{BE(\mathrm{OFF})})/R_K)$ is much smaller than I_{CBO} for $R_K > 11\ K$.

RUN 7.0 TV=0.05 TR=0.05 TLO=25.0 THI=25.0

RK	RBON	RBOFF
1000.0	3038.0	107338.6
3000.0	6310.2	321405.9
5000.0	9843.9	534664.1
7000.0	13672.1	747115.8
9000.0	17833.0	958767.9
11000.0	22372.2	1169623.0
13000.0	27343.5	1379688.0
15000.0	32811.8	1588966.0
17000.0	38855.5	1797461.0
19000.0	45570.5	2005178.0
21000.0	53075.3	2212121.0
23000.0	61517.8	2418294.0
25000.0	71085.4	2623703.0
27000.0	82019.6	2828349.0
29000.0	94634.9	3032240.0
31000.0	109351.9	3235377.0
33000.0	126743.2	3437767.0
35000.0	147610.9	3639413.0

RUN 8.0 TV=0.05 TR=0.10 TLO=25.0 THI=25.0

RK	RBON	RBOFF
1000.0	3372.7	97071.8
3000.0	7035.2	290692.8
5000.0	11025.6	483620.2
7000.0	15390.2	675856.3
9000.0	20184.3	867406.6
11000.0	25474.6	1058274.0
13000.0	31342.2	1248462.0
15000.0	37887.2	1437975.0
17000.0	45233.9	1626815.0
19000.0	53539.3	1814988.0
21000.0	63004.3	2002496.0
23000.0	73889.8	2189342.0
25000.0	86541.6	2375530.0
27000.0	101427.6	2561065.0
29000.0	119196.9	2745948.0
31000.0	140776.9	2930183.0
33000.0	167540.6	3113775.0
35000.0	201611.2	3296726.0

RUN 9.0 TV=0.05 TR=0.20 TLO=25.0 THI=25.0

RK	RBON	RBOFF
1000.0	4178.2	79103.4
3000.0	8805.6	236931.6
5000.0	13957.7	394256.9
7000.0	19729.1	551081.2
9000.0	26238.7	707407.0
11000.0	33637.8	863237.2
13000.0	42122.2	1018574.1
15000.0	51949.6	1173419.0
17000.0	63466.0	1327776.0
19000.0	77148.0	1481646.0
21000.0	93669.9	1635032.0
23000.0	114018.1	1787936.0
25000.0	139696.1	1940360.0
27000.0	173113.1	2092307.0
29000.0	218388.9	2243779.0
31000.0	283197.9	2394778.0
33000.0	383657.1	2545306.0
35000.0	560342.5	2695365.0

Figure 7.14 Continued

Additional insight can be gained by examining the plotted results. The curves plotted in Figure 7.15 correspond to runs 1, 2, and 3 and give the effect of resistance tolerance variation. As shown for run 3, any point within the cross-hatched area would be a valid solution to the design and as we would expect, the best range of design values of R_B and R_K are obtained for the 5 percent tolerance case.

```
RUN  10.0   TV=0.02   TR=0.10   TLO=25.0   THI=25.0

          RK        RBON      RBOFF

        1000.0     3165.3   100137.2
        3000.0     6584.4   299872.6
        5000.0    10288.6   498892.4
        7000.0    14315.0   697199.2
        9000.0    18707.7   894798.4
       11000.0    23519.0  1091693.0
       13000.0    28811.8  1287887.0
       15000.0    34662.1  1483384.0
       17000.0    41163.0  1678189.0
       19000.0    48429.2  1872303.0
       21000.0    56604.6  2065732.0
       23000.0    65871.0  2258479.0
       25000.0    76462.6  2450547.0
       27000.0    88685.3  2641940.0
       29000.0   102947.6  2832662.0
       31000.0   119806.1  3022715.0
       33000.0   140041.1  3212105.0
       35000.0   164779.8  3400833.0

RUN  11.0   TV=0.05   TR=0.10   TLO= 0.0   THI=65.0

          RK        RBON      RBOFF

        1000.0     3633.2    96078.2
        3000.0     8169.2   281961.4
        5000.0    13989.4   459925.3
        7000.0    21729.8   630465.1
        9000.0    32528.2   794036.4
       11000.0    48643.2   951057.5
       13000.0    75283.3  1101914.0
       15000.0   127760.1  1246962.0
       17000.0   279046.5  1386531.0
       19000.0  5295744.0  1520926.0

RUN  12.0   TV=0.02   TR=0.10   TLO= 0.0   THI=65.0

          RK        RBON      RBOFF

        1000.0     3395.0    99112.3
        3000.0     7569.7   290865.5
        5000.0    12825.2   474449.2
        7000.0    19643.8   650374.5
        9000.0    28845.1   819111.2
       11000.0    41942.2   981090.9
       13000.0    62074.0  1136711.0
       15000.0    96988.7  1286340.0
       17000.0   172416.6  1430316.0
       19000.0   456313.9  1568955.0
       21000.0-1314277.0  1702547.0
       23000.0  -310446.1  1831364.0
       25000.0  -188573.7  1955656.0
       27000.0  -141096.7  2075658.0
       29000.0  -115823.9  2191588.0
       31000.0  -100130.7  2303650.0
       33000.0   -89438.3  2412035.0
       35000.0   -81684.7  2516921.0
```

Figure 7.14 Continued

The curves plotted in Figure 7.16 correspond to runs 5 and 6 and give the effect of power supply tolerance variation. It is quite apparent that power supply tolerance variation from 5 to 2 percent is negligible for R_B value calculations. This figure also points out that in an instrument such as an electronic counter where most of the supply power is used by flip-flop decades, it is not necessary to have a well regulated supply voltage. This observation therefore suggests that a much cheaper power supply can be used. The flip-flop decades can be supplied from a pre-regulator and only the low-drain amplifier and d-c level circuits require post-supply regulation.

Figure 7.17 corresponds to the curves plotted for the results of runs

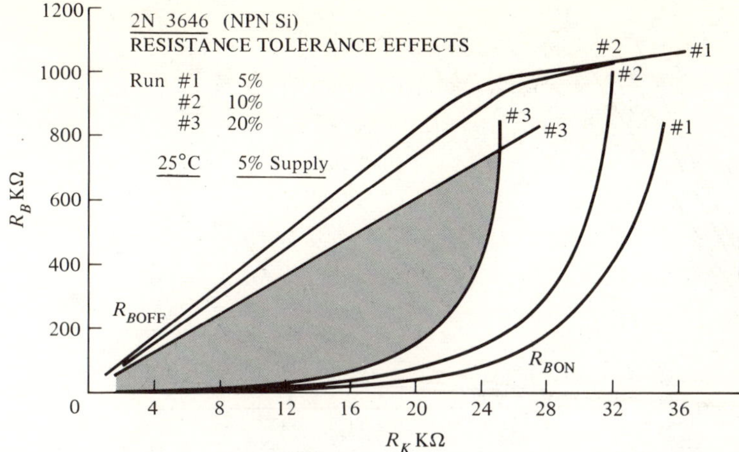

Figure 7.15 $R_{B(\mathrm{OFF})}$ and $R_{B(\mathrm{ON})}$ values are plotted for the first three computer runs

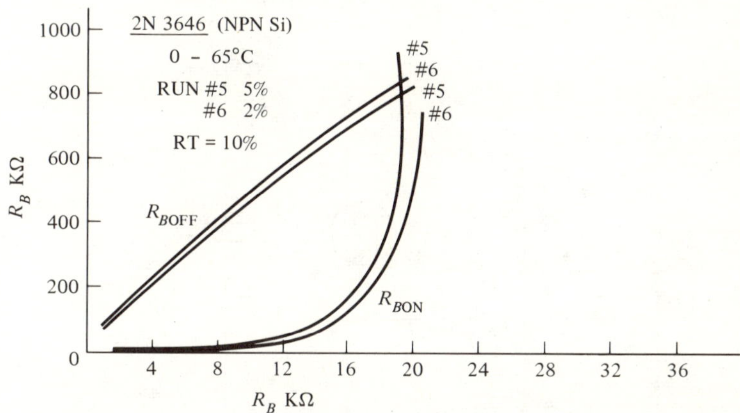

Figure 7.16 $R_{B(\mathrm{OFF})}$ and $R_{B(\mathrm{ON})}$ values are plotted for computer runs 5 and 6

7, 8, and 9 and give the effect of resistance tolerance variation for the 2N3639 (PNP) transistor. It is quite obvious that resistance tolerance variations have a much smaller effect for this transistor than for the 2N3646 transistor (see Figure 7.15). It is left as an exercise for the reader to explain why.

Worst-Case Design of Single-Supply Common-Emitter Flip-Flop

A basic common emitter single supply flip flop is shown in Figure 7.18. The worst-case design of this circuit closely follows that of the two-supply circuit. It is therefore left as an exercise for the reader to show that the

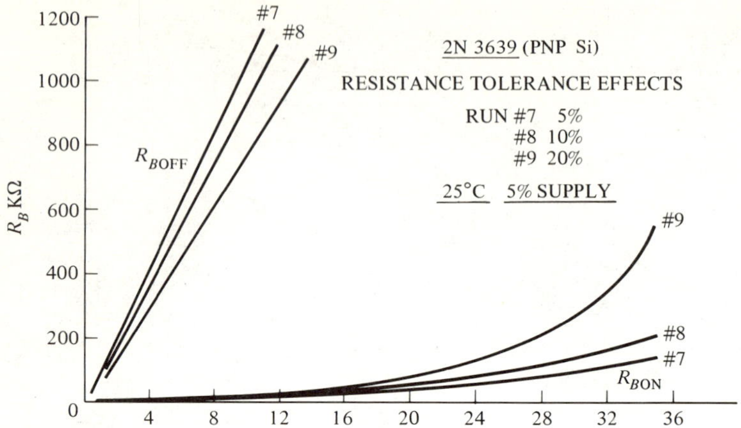

Figure 7.17 $R_{B(\text{OFF})}$ and $R_{B(\text{ON})}$ values are plotted for computer runs 7, 8, and 9

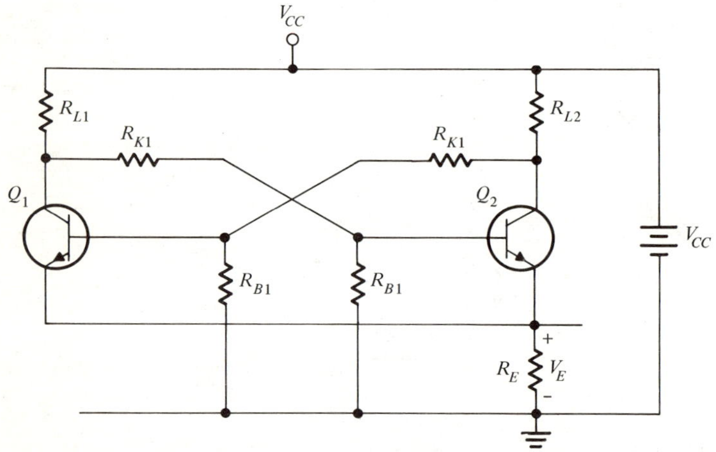

Figure 7.18 A common emitter one-supply flip-flop

on and off equations for Q_2 are:

On Equation

$$\underline{R}_B = \frac{V_E + \bar{V}_{BE(\text{ON})}}{\left\{\dfrac{\underline{V}_{CC} - V_E - \bar{R}_{L1}\bar{I}_{CB01} - \bar{V}_{BE(\text{ON})}}{\bar{R}_{K2} + \bar{R}_{L1}} - \dfrac{1}{\underline{h}_{FE}}\left[\dfrac{\bar{V}_{CC} - V_E - \underline{V}_{CE(\text{SAT})}}{\underline{R}_{L2}} - \dfrac{\underline{V}_{CE(\text{SAT})} - V_{BE(\text{OFF})}}{\bar{R}_{K1}}\right]\right\}}$$

Off Equation

$$\bar{R}_B = \frac{V_E + V_{BE(\text{OFF})}}{\dfrac{\bar{V}_{CE(\text{SAT})} - V_{BE(\text{OFF})}}{\underline{R}_{K2}} + \bar{I}_{CBO}}$$

where

$$V_E = \frac{(\underline{V}_{CC} - \underline{V}_{CE(\text{SAT})})(\underline{h}_{FE2} + 1)\bar{R}_E}{\bar{h}_{FE2}R_{L2} + (\underline{h}_{FE2} + 1)\bar{R}_E}$$

Figure 7.19 shows the corresponding computer program. Although this program is very similar to the previous program, it differs in that the program computes one run at a time. It is therefore necessary to resubmit the source deck with each data change. Figure 7.20 shows the effect of varying the emitter-resistor for the same data used in run 1 in Table 7.1. Figure 7.21 shows a plot of these results. It may be seen from the cross-

```
C     WORSE CASE D-C DESIGN OF SINGLE SUPPLY
C     COMMON EMITTER FLIP FLOP
C
      DIMENSION TRANS(80)
      READ(5,10)(TRANS(I),I=1,8)
   10 FORMAT(8A1)
      READ(5,15)RL,VCC,RE,TV,TR
   15 FORMAT(5F6.2)
      READ(5,20)VBEOFF,VBFCN,VCESN,VCESF
   20 FORMAT(4F6.2)
      READ(5,25)XICBO,HFE,TLO,THI
   25 FORMAT(E10.2,3F6.2)
      WRITE(6,30)(TRANS(I),I=1,8)
   30 FORMAT(1H1,10X,'WORSE CASE D-C DESIGN OF SINGLE SUPPLY',/,11X,'COM
     1MON EMITTER FLIP FLOP',/,11X,'FOR ',8A1,' TRANSISTOR',//)
      WRITE(6,35)RL,VCC,RE,TV,TR
   35 FORMAT(' RL =',F6.0,2X,'VCC =',F6.2,2X,'RE =',F6.2,2X,'TV =',F6.2,
     12X,'TR =',F6.2)
      WRITE(6,40)VBEOFF,VBEON,VCESN,VCESF
   40 FORMAT(' VBEOFF=',F6.2,2X,'VBEON=',F6.2,2X,'VCESN=',F6.2,2X,'VCESF
     1=',F6.2)
      WRITE(6,45)XICBO,HFE,TLO,THI
   45 FORMAT(' XICBO=',E10.2,2X,' HFE =',F6.2,2X,' TLO =',F6.2,2X,' THI
     1=',F6.2,//)
      WRITE(6,50)
   50 FORMAT(/,15X,3H RK,6X,4HRBON,4X,5HRBOFF,/)
      RK=1000.
   55 X=(VCC*(1.-TV)-VCESN)*(HFE+1.)*RE*(1.+TR)
      Y=HFE*RL*(1.-TR)+(HFE+1.)*RE*(1.+TR)
      VE=X/Y
      A=VE+VBEON
      B=VCC*(1.-TV)-VE-RL*(1.+TR)*XICBO-VBEON
      C=RK*(1.+TR)+RL*(1.+TR)
      D=(VCC*(1.+TV)-VE-VCESN)/(RL*(1.-TR))
      E=(-VBEOFF+VCESN)/(RK*(1.+TR))
      RBON=A/((B/C)-((1./HFE)*(D-E))*(1.-TR))
      F=VE+VBEOFF
      G=(-VBEOFF+VCESF)/(RK*(1.-TR))
      RBOFF=F/((G+XICBO)*(1.+TR))
      WRITE(6,60)RK,RBON,RBOFF
   60 FORMAT(10X,3F10.1)
      RK=RK+2000.
      IF(36000.-RK)70,65,65
   65 IF(RBON-RBOFF)55,55,70
   70 CONTINUE
      STOP
      END
```

Figure 7.19 A Fortran program for investigating the worst-case design of a one-supply flip-flop

```
          WORSE CASE D-C DESIGN OF SINGLE SUPPLY
          COMMON EMITTER FLIP FLOP
          FOR 2N3646    TRANSISTOR

RL = 1000.  VCC = 10.00  RE = 56.00  TV =  0.05  TR =  0.05
VBEOFF=  0.0   VBEON=  0.72  VCESN=  0.13  VCESF=  0.20
XICBO=  0.50E-06   HFE = 50.00   TLO = 25.00   THI = 25.00

              RK         RBON      RBOFF

            1000.0       343.0     2511.2
            3000.0       724.3     7498.0
            5000.0      1150.5    12438.1
            7000.0      1630.2    17332.0
            9000.0      2174.0    22180.3
           11000.0      2795.7    26983.8
           13000.0      3513.4    31743.0
           15000.0      4351.2    36458.5
           17000.0      5341.8    41131.0
           19000.0      6531.5    45761.0
           21000.0      7986.8    50349.1
           23000.0      9807.9    54695.9
           25000.0     12152.5    59401.9
           27000.0     15284.4    63867.7
           29000.0     19680.0    68293.8
           31000.0     26297.3    72680.7
           33000.0     37390.4    77029.0
           35000.0     59821.8    81339.1

          WORSE CASE D-C DESIGN OF SINGLE SUPPLY
          COMMON EMITTER FLIP FLOP
          FOR   2N36 TRANSISTOR

RL = 1000.  VCC = 10.00  RE =100.00  TV =  0.05  TR =  0.05
VBEOFF=  0.0   VBEON=  0.72  VCESN=  0.13  VCESF=  0.20
XICBO=  0.50E-06   HFE = 50.00   TLO = 25.00   THI = 25.00

              RK         RBON      RBOFF

            1000.0       471.2     4284.4
            3000.0       995.6    12792.5
            5000.0      1582.4    21220.7
            7000.0      2243.5    29570.2
            9000.0      2993.9    37842.1
           11000.0      3853.1    46037.3
           13000.0      4846.6    54157.1
           15000.0      6008.5    62202.3
           17000.0      7385.7    70174.0
           19000.0      9044.1    78073.3
           21000.0     11079.6    85901.1
           23000.0     13637.2    93658.4
           25000.0     16947.5   101346.2
           27000.0     21400.0   108965.3
           29000.0     27709.2   116516.7
           31000.0     37342.4   124001.3
           33000.0     53866.1   131419.9
           35000.0     88789.1   138773.6
```

Figure 7.20 The output results of the program of Figure 7.19

hatched area that the range in R_B-R_K is increased for higher values of R_E. This is to be expected since higher R_E values provide greater back bias, V_E. Referring to the design equations for $R_{B(\mathrm{ON})}$ and $R_{B(\mathrm{OFF})}$, we see that an increase in V_E will increase the numerators of both equations. The value of R_E cannot be increased without bound, however, since an increase in R_E will reduce the collector swings and also adversely affect the back bias breakdown of the base-emitter diode.

```
         WORSE CASE D-C DESIGN OF SINGLE SUPPLY
         COMMON EMITTER FLIP FLOP
         FOR 2N3646   TRANSISTOR


RL = 1000.  VCC = 10.00  RE =180.00  TV =  0.05  TR =  0.05
VBEOFF=  0.0   VBEON=  0.72  VCESN=  0.13  VCESF=  0.20
XICBO=  0.50E-06   HFE = 50.00   TLO = 25.00   THI = 25.00


            RK          RBON       RBOFF

          1000.0        706.9      7133.7
          3000.0       1494.9     21300.1
          5000.0       2378.3     35333.5
          7000.0       3375.6     49235.8
          9000.0       4510.4     63008.7
         11000.0       5813.2     76654.1
         13000.0       7324.4     90173.9
         15000.0       9098.1    103569.6
         17000.0      11209.5    116842.8
         19000.0      13765.0    129995.5
         21000.0      16921.2    143029.2
         23000.0      20918.3    155945.5
         25000.0      26143.7    168745.9
         27000.0      33266.6    181432.1
         29000.0      43549.6    194005.6
         31000.0      59695.4    206467.7
         33000.0      88717.6    218820.1
         35000.0     156235.1    231064.1
```

Figure 7.20 Continued

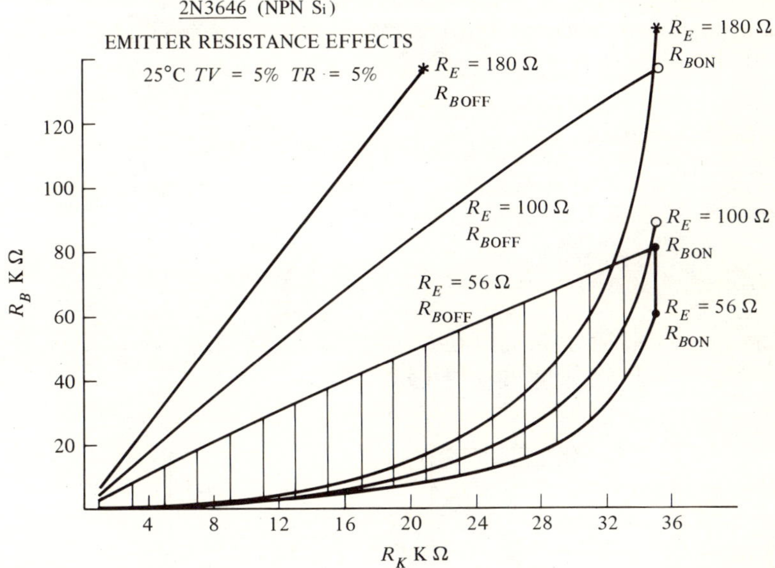

Figure 7.21 $R_{B(\mathrm{OFF})}$ and $R_{B(\mathrm{ON})}$ values are plotted for three different emitter resistances

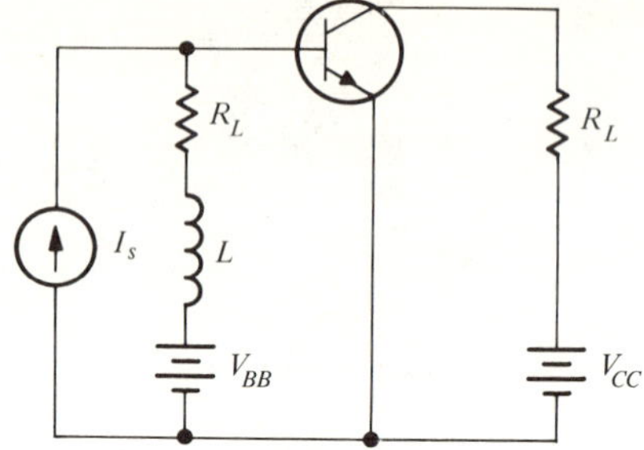

Figure 7.22 As implified shunt-peaked transistor amplifier stage

7.4 SHUNT-PEAKED TRANSISTOR AMPLIFIER DESIGN

Figure 7.22 shows a simplified shunt-peaked transistor amplifier stage whose equivalent circuit contains a two-reactance two-terminal coupling network. Such a circuit affords us one degree of correction and we may therefore adjust the value of L, relative to other circuit element values, so that we improve the amplitude response, the phase response, or the time response. This section will be concerned only with the design of a maximally flat amplitude response. The design problem will therefore fall into two separate parts. The simpler case will be that where the nominal gain has been specified and the more complex case will be where the three dB bandwidth has been specified.

Nominal Gain Specified

Figure 7.23 shows the unilateral high-frequency equivalent[3] of the circuit shown in Figure 7.22. The value of C_i is given by

$$C_i = \frac{1}{\omega_t r_e'} + C_c\left(1 + \frac{\alpha_0 R_L}{r_e'}\right)$$

[3] See Mohammed Ghausi, *Principles and Design of Linear Active Circuits* (New York: McGraw-Hill Book Company, 1965), Sec. 9-5.

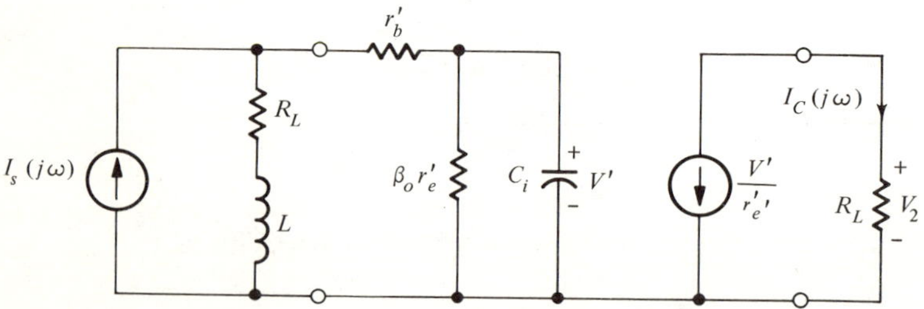

Figure 7.23 The unilateral high-frequency equivalent of the circuit shown in Figure 7.22

where

ω_t is the gain-bandwidth product frequency where the short-circuit forward current transfer ratio is equal to unity.
r_e' is the incremental resistance of the emitter-base diode.
C_c is the collector depletion-layer capacitance.
α_0 is the d-c or nominal common base short-circuit current gain.

The value of the resistance R_i is given by

$$R_i = \beta_0 r_e'$$

where β_0 is the d-c or nominal common emitter short-circuit current gain.

The forward current transfer ratio of the amplifier is thus given by

$$A(j\omega) = \frac{I_0(j\omega)}{I_s(j\omega)} = K \frac{j\omega + R_L/L}{(j\omega)^2 + \left(\dfrac{R_L + r_b'}{L} + \dfrac{1}{R_iC_i}\right) j\omega + \dfrac{R_L + r_b' + R_i}{R_iC_iL}}$$

where

$$K = -\frac{1}{r_e'C_i}$$

The d-c gain may be obtained from the above expression by setting ω equal to zero; thus

$$A(0) = \frac{-\beta_0 R_L}{R_i + r_b' + R_L}$$

It is therefore possible to write the normalized forward current transfer ratio as

$$\frac{A(j\omega)}{A_0} = \frac{1 + j\dfrac{\omega L}{R_L}}{1 - \dfrac{R_iC_iL}{(R_L + r_b' + R_i)}\,\omega^2 + j\omega\,\dfrac{L + R_iC_iR_L + R_iC_ir_b'}{(R_L + r_b' + R_i)}}$$

so that the normalized amplitude function squared would be given by

$$|H|^2 = \left|\frac{A(j\omega)}{A_0}\right|^2 = \frac{1 + (L/R_L)^2\omega^2}{\left\{1 + \left(\dfrac{(L + R_iC_iR_L + R_iC_ir_b')^2}{(R_L + r_b' + R_i)^2} - 2\,\dfrac{R_iC_iL}{(R_L + r_b' + R_i)}\right)\omega^2 + \dfrac{(R_iC_iL)^2}{(R_L + r_b' + R_i)^2}\,\omega^4\right\}}$$

Using the derivative adjustment technique,[4] we equate the ω^2 coefficients

[4] See W. Lynch, *The Role Played by Derivative Adjustment in Broadband Amplifiers Design*, Proc. Symposium on Modern Network Synthesis, Brooklyn Polytechnic Institute, 1952.

to find the value of L that will give a maximally flat amplitude response

$$\left(\frac{L}{R_L}\right)^2 = \left(\frac{L + R_iC_iR_L + R_iC_ir_b'}{R_L + r_b' + R_i}\right)^2 - 2\frac{R_iC_iL}{(R_L + r_b' + R_i)}$$

Letting

$$A = R_L + r_b' + R_i$$
$$C = R_iC_i$$
$$B = C(R_L + r_b')$$

we have

$$\left(\frac{L}{R_L}\right)^2 = \left(\frac{L+B}{A}\right)^2 - 2\frac{CL}{A}$$

Solving for L we get

$$L = \frac{\left(\frac{B}{A^2} - \frac{C}{A}\right) + \sqrt{\left(\frac{B}{A^2} - \frac{C}{A}\right)^2 + \frac{B^2}{A^2}\left(\frac{1}{R_L{}^2} - \frac{1}{A^2}\right)}}{\left(\frac{1}{R_L{}^2} - \frac{1}{A^2}\right)}$$

or

$$L = \frac{AA + BB}{CC}$$

where

$$AA = \frac{B}{A^2} - \frac{C}{A}$$
$$BB = \sqrt{\left(\frac{B}{A^2} - \frac{C}{A}\right)^2 + \frac{B^2}{A^2}\left(\frac{1}{R_L{}^2} - \frac{1}{A^2}\right)}$$
$$CC = \frac{1}{R_L{}^2} - \frac{1}{A^2}$$

To find the expression for $\omega_{3\text{dB}}$, we let $\omega = \omega_{3\text{dB}}$ in the normalized amplitude function squared expression and then, in turn, set this result equal to ½.

$$\left|\frac{A(j\omega_{3\text{dB}})}{A_0}\right|^2 = \frac{1}{2}$$

This yields

$$\omega_{3\text{dB}}^4\left(\frac{CL}{A}\right)^2 - \left(2\left(\frac{L}{R_L}\right)^2 + 2\left(\frac{CL}{A}\right) - \left(\frac{B+L}{A}\right)^2\right)\omega_{3\text{dB}}^2 - 1 = 0$$

$$\omega_{3\text{dB}}^4 E - (CD + BD - AD)\omega_{3\text{dB}}^2 - 1 = 0$$

$$\omega_{3\text{dB}}^4 E - D\omega_{3\text{dB}}^2 - 1 = 0$$

$$\omega_{3\text{dB}} = \sqrt{\frac{D + \sqrt{D^2 + 4E}}{2E}}$$

Design Procedure. Assuming that the nominal d-c gain A_0 has been specified and that all other transistor parameters are known, we first calculate R_L.

1. We thus find from the d-c gain expression

$$R_L = \frac{A_0(R_i + r_b')}{\beta_0 - A_0}$$

2. All the parameters are now known and we therefore calculate the numerical value of L from the appropriate expression above.
3. Using the value of L found in step 2, we now compute the numerical value of ω_{3dB}.

Figure 7.24 shows the computer design program. Initially the specified d-c gain is read into memory. This is followed by reading in the numerical values of β_0, r_b', C_c, f_t, and the emitter current CUR. Computer statements S.0005 and S.0006 then print the output heading and the numerical values of the above parameters. The values of ω_t, r_e', and R_i are then computed by computer statements S.0007 through S.0009. The values of R_L and C_i are computed next and this is followed by computer statements

```
                C      SHUNT PEAKED TRANSISTOR AMPLIFIER
                C      FROM HYBRID-PI EQUIVALENT CIRCUIT
                C      NOMINAL GAIN SPECIFIED
                C
S.0001                 READ(5,10) AO
S.0002              10 FORMAT(F15.4)
S.0003                 READ (5,20) BO,RBP,CC,FT,CUR
S.0004              20 FORMAT(2F15.4,3E15.7)
S.0005                 WRITE (6,30) BO,FT,CC,CUR,RBP,AO
S.0006              30 FORMAT('1SHUNT PEAKED TRANSISTOR AMPLIFIER',/,' FRCM HYBRID-PI EQU
                      1IVALENT CIRCUIT',//,' BO =',F6.2,3X,'FT =',E15.7,3X,'CC =',E15.7,/
                      2/,'  I COLECTOR =',E15.7,3X,'RBP =',F6.2,3X,'AO =',F6.2)
S.0007                 WT=6.28*FT
S.0008                 RE=.025/CUR
S.0009                 RI=BO*RE
S.0010                 RL=AO*(RI+RBP)/(BO-AO)
S.0011                 CI=(1.+(RL+RE)*(CC*WT))/(RE*WT)
S.0012                 A=RL+RBP+RI
S.0013                 C=RI*CI
S.0014                 B=C*(RL+RBP)
S.0015                 AA=B/A**2-C/A
S.0016                 BB=SQRT(AA**2+((B**2)/(A**2))*((1./RL**2)-1./A**2))
S.0017                 CC=(1./(RL**2))-(1./(A**2))
S.0018                 AL=(AA+BB)/CC
S.0019                 WRITE (6,40) AL,RL,CI
S.0020              40 FORMAT('OL =',E15.7,3X,'RL =',E15.7,3X,'CI =',E15.7)
S.0021                 E=((C*AL)/A)**2
S.0022                 AD=((AL+B)/A)**2
S.0023                 BD=2.*(C*AL)/A
S.0024                 CD=2.*((AL/RL)**2)
S.0025                 D=CD+BD-AD
S.0026                 W3DB=SQRT((D+SQRT(D**2+4.*E))/(2.*E))
S.0027                 F3DB=W3DB/6.28
S.0028                 GBWP=W3DB*AO
S.0029                 GBWPN=GBWP/WT
S.0030                 WRITE(6,65) F3DB,GBWPN
S.0031              65 FORMAT('OF3DB =',E15.7,/,'OGBWPN =',E15.7)
S.0032                 STOP
S.0033                 END
```

Figure 7.24 A program for designing a maximally flat shunt-peaked amplifier response for a given nominal gain

S.0012 through S.0017, which are used to compute the constants A, C, B, AA, BB, and CC which were introduced in the development of the equation for computing the value of L. Computer statement S.0018 is now able to carry out step 2 of the design procedure and computer statements S.0019 and S.0020 print the results: L (called AL for mode purposes), R_L, and C_i.

The constants E, AD, BD, CD, and D, used in the calculation of $\omega_{3\mathrm{dB}}$, are next computed by computer statements S.0021 through S.0025. The value of $\omega_{3\mathrm{dB}}$ and $f_{3\mathrm{dB}}$ are then determined. This is followed by the calculation of the gain-bandwidth product GBWP and the normalized gain-bandwidth product GBWPN and, in turn, by the printing of the numerical values of $f_{3\mathrm{dB}}$ and GBWPN.

Figure 7.25 shows the computer output results for three different sets of input data. Note that, except for f_t, in each case the transistor parameters are the same. Note also that the value computed for R_L is different in case 1 and case 2 because the nominal gains specified for these two cases were different, but the value of R_L is the same in cases 2 and 3 since

```
SHUNT PEAKEC TRANSISTOR AMPLIFIER
FROM HYBRIC-PI EQUIVALENT CIRCUIT

BO = 27.00    FT =  0.2CCCCCCE 09    CC =  0.3000000E-11

 I COLECTOR =  C.4999999E-C2   RBP = 5C.CC    AO =  5.00

L =  0.3341C29E-C6    RL =  C.4204544E 02    CI =  0.1874630E-09

F3DB =  0.3833840E 08

GBWPN =  0.958460IE CC


SHUNT PEAKED TRANSISTOR AMPLIFIER
FROM HYBRID-PI EQUIVALENT CIRCUIT

BO = 27.00    FT =  0.2000000E 09    CC =  0.3000000E-11

 I COLECTOR =  0.4999999E-02   RBP = 50.00    AO = 10.00

L =  0.1394611E-05    RL =  0.1088235E 03    CI =  0.2275298E-09

F3DB =  0.1724998E 08

GBWPN =  0.8624998E 00


SHUNT PEAKED TRANSISTOR AMPLIFIER
FROM HYBRID-PI EQUIVALENT CIRCUIT

BO = 27.00    FT =  0.5000000E 09    CC =  0.3000000E-11

 I COLECTOR =  0.4999999E-02   RBP = 50.00    AO = 10.00

L =  0.8090032E-06    RL =  0.1088235E 03    CI =  0.1319884E-09

F3DB =  0.2973666E 08

GBWPN =  0.5947334E 0C
```

Figure 7.25 The output results of the program of Figure 7.24

$A_0 = 10$ for both cases. In addition, note also the improvement in the bandwidth in cases 2 and 3 when a higher value f_t transistor is used.

Since f_t is a crude approximation of the cyclic gain-bandwidth product, an interesting extension of this problem is to examine the normalized gain-bandwidth product ratio

$$\text{GBWPN} = \frac{A_0 f_{3\text{dB}}}{f_t} = \frac{A_0 \omega_{3\text{dB}}}{\omega_t}$$

for various values of A_0. The reader will find that one does not get an equal exchange of bandwidth for gain and that this property is typical of most transistor amplifiers.

Bandwidth Specified

In the case where the bandwidth of the amplifier has been specified, we find that because of the form of the design equations it is not possible to obtain a direct solution. This type of design therefore requires a converging iterative procedure. The reader should therefore recognize the natural ability of the computer to solve a problem of this type.

One such method of solution proceeds by assuming that the gain-bandwidth product is equal to ω_t

$$\frac{\text{G} * \text{BW}}{\omega_t} = \frac{A_0 \omega_{3\text{dB}}}{\omega_t} \equiv 1$$

This gives

$$A_0 = \frac{\omega_t}{W}$$

as an initial value for A_0, where initially W is equal to $\omega_{3\text{dB}}$. Using the design procedure for the case where the nominal gain is specified, we now compute the value of $\omega_{3\text{dB}}$ (this is called BW in the computer program of Figure 7.26 and is computed by computer statement S.0021) and compare this with the specified bandwidth $\omega_{3\text{dB}}$ (see computer statements S.0022 and S.0025). The difference, DW, is then multiplied by 0.75 and added to the previous value of W (initially the previous value of W was equal to $\omega_{3\text{dB}}$). The equation

$$A_0 = \frac{\omega_t}{W}$$

is now used to compute a new value of A_0. This process is then repeated

```
C     SHUNT PEAKED TRANSISTOR AMPLIFIER
C     FROM HYBRID-PI EQUIVALENT CIRCUIT
C     BANDWIDTH SPECIFIED
C
      READ(5,10)BO,RBP,CC,WT,CUR,W3DB
   10 FORMAT(2F9.4,4E12.3)
      WRITE(6,25)BO,WT,CC,CUR,RBP,W3DB
   25 FORMAT('1SHUNT PEAKED TRANSISTOR AMPLIFIER',/,' FROM HYBRID-PI EQU
     1IVALENT CIRCUIT',//,' BO =',F6.2,3X,'WT =',E15.7,3X,'CC =',E15.7,/
     2,' I COLECTOR =',E15.7,3X,'RBP =',F6.2,3X,'W3DB =',E15.7)
      AO=WT/W3DB
      REP=.025/CUR
      RI=BO*REP
      W=W3DB
      DO 20 I=1,10
      RL=AO*(RBP+RI)/(BO-AO)
      CI=(1.+(RL+REP)*WT*CC)/(REP*WT)
      Z=RI*RL*CI
      Y=2.*RL+RBP+RI
      X=RL+RBP+RI
      A=(Z/((RBP+RI)*Y))
      B=-RI*RL
      C=RBP+RL
      D=SQRT((RI*RL)**2+((C**2)*(RBP+RI)*Y))
      AL=A*(B+D)
      F=SQRT((.25*(X/Z)**2)+(RL/AL)**2)
      BW=SQRT((X/Z)*((.5*X/Z)+F))
      DW=W3DB-BW
      WRITE(6,30)AO,RL,AL,BW,DW
   30 FORMAT(5E15.7)
      IF(DW)50,60,50
   50 W=W+.75*DW
   20 AO=WT/W
   60 CONTINUE
      WRITE(6,45)AL,RL,CI
   45 FORMAT('OL =',E15.7,3X,'RL =',E15.7,3X,'CI =',E15.7)
      WRITE(6,55)AO
   55 FORMAT('OAO =',E15.7)
      GBWP=W3DB*AO
      GBWPN=GBWP/WT
      WRITE(6,65)W3DB,GBWPN
   65 FORMAT('OW3DB =',E15.7,/,'OGBWPN =',E15.7)
      STOP
      END
```

Figure 7.26 A program for designing a maximally flat shunt-peaked amplifier for a given bandwidth

until the correction becomes very small or until ten iterations have been performed.

Figure 7.26 shows the computer program. Note that it is very similar to the previous program, but it differs in a minor way in the calculation of L and ω_{3dB}. The value of L is found from the equation

$$L = \frac{R_i R_L C_i}{(r_b' + R_L)(2R_L + r_b' + R_i)} \Big(-R_i R_L + \sqrt{(R_i R_L)^2 + (r_b' + R_L)^2 (r_b' + R_i)(2R_L + r_b' + R_i)}\Big)$$

which is written in terms of new computer program variables as

$$L = \frac{Z}{(r_b' + R_L)Y} [B + \sqrt{(R_i R_L)^2 + C^2 (r_b' + R_i) Y}]$$

and finally as

$$AL = A(B + D)$$

The value of $\omega_{3\text{dB}}$ is found from the equation

$$\omega_{3\text{dB}} = \sqrt{\left(\frac{R_L + r_b' + R_i}{R_i R_L C_i}\right)\left[\frac{1}{2}\left(\frac{R_L + r_b' + R_i}{R_i R_L C_i}\right) \sqrt{\frac{1}{4}\left(\frac{R_L + r_b' + R_i}{R_i R_L C_i}\right)^2 \left(\frac{R_L}{L}\right)^2}\right]}$$

or

$$= \sqrt{\frac{X}{Z}\left[\frac{1}{2}\left(\frac{X}{Z}\right) + \sqrt{\frac{1}{4}\left(\frac{X}{Z}\right)^2 + \left(\frac{R_L}{AL}\right)^2}\right]}$$

and finally, in terms of a new variable F, as

$$\text{BW} = \sqrt{\frac{X}{Z}\left(\frac{1}{2}\left(\frac{X}{Z}\right) + F\right)}$$

Figure 7.27 gives the computer results. The case considered here corresponds to case 1 in the previous problem, except that in this problem the bandwidth has been specified instead of the nominal gain. An examination of the numerical values of A_0, R_L, BW, and DW (these values are printed each iteration) show that the program is very effective and that convergence is quite rapid. Investigation of other cases reveals that the speed of convergence varies with the initial data, but an accuracy better than 0.05 percent is usually obtained in less than ten iterations.

In order to show the effectiveness of the original 0.75 choice for the speed-up constant, the program of Figure 7.26 was modified so that computer statement S.0026 would use a larger and a smaller speed-up value. Figure 7.28 shows the respective results for a 0.9 and a 0.6 speed-up value. The results obtained for these two values clearly indicate the

```
SHUNT PEAKEC TRANSISTOR AMPLIFIER
FROM HYBRIC-PI EQUIVALENT CIRCLIT

BO = 27.CC    WT =  0.1256CCCE 10    CC =  C.3000000E-11
I COLECTOR =  0.4999999E-C2   RBP = 5C.CC    W3DB =  0.2405000E 09
 0.5222453E C1  C.4436469E C2  C.359C371E-C6  0.2301579E 09  0.1034214E 08
 0.5059280E 01  0.4265889E 02  C.34C6225E-06  0.2378538E 09  0.2646192E 07
 0.5019156E C1  C.4224332E C2  C.3361995E-06  0.2398176E 09  0.6823840E 06
 0.5008912E 01  0.4213748E C2  C.335C770E-C6  0.2403236E 09  0.1764480E 06
 0.500627CE C1  C.4211C17E C2  C.3347878E-06  0.2404546E 09  0.4539200E 05
 0.5005591E 01  C.421C316E C2  C.3347135E-C6  0.2404881E 09  0.1190400E 05
 0.5005413E 01  0.421C132E C2  0.3346937E-06  0.2404972E 09  0.2848000E 04
 0.5005370E C1  C.421C088E C2  C.3346893E-C6  0.2404990E 09  0.9600000E 03
 0.5005356E 01  0.421CC72E C2  C.3346873E-C6  0.2405000E 09 -0.3200000E 02
 0.5005357E C1  C.421CC77E C2  0.3346880E-06  0.2404997E 09  0.2560000E 03

L =  0.3346880E-C6    RL =  C.421CC77E C2    CI =  0.1874960E-09

AO =  0.5CC5353E C1

W3DB =  0.2405CCCE C9

GBWPN =  0.9584293E CC
```

Figure 7.27 The output results of the program of Figure 7.26

```
SHUNT PEAKED TRANSISTOR AMPLIFIER
FROM HYBRID-PI EQUIVALENT CIRCUIT

BO = 27.00    WT =  0.1256000E 10    CC =  0.3000000E-11
I COLECTOR =  0.4999999E-02   RBP = 50.00   W3DB =  0.2405000E 09
 0.5222453E 01  0.4436469E 02  0.3590371E-06  0.2301579E 09  0.1034214E 08
 0.5027863E 01  0.4233336E 02  0.3371555E-06  0.2393892E 09  0.1110832E 07
 0.5007821E 01  0.4212621E 02  0.3349577E-06  0.2403775E 09  0.1224640E 06
 0.5005622E 01  0.4210350E 02  0.3347169E-06  0.2404867E 09  0.1332800E 05
 0.5005383E 01  0.4210101E 02  0.3346906E-06  0.2404985E 09  0.1536000E 04
 0.5005355E 01  0.4210072E 02  0.3346873E-06  0.2405000E 09 -0.1600000E 02
 0.5005355E 01  0.4210072E 02  0.3346873E-06  0.2405000E 09 -0.1600000E 02
 0.5005356E 01  0.4210072E 02  0.3346873E-06  0.2405000E 09 -0.1600000E 02
 0.5005356E 01  0.4210072E 02  0.3346873E-06  0.2405000E 09 -0.1600000E 02
 0.5005356E 01  0.4210072E 02  0.3346873E-06  0.2405000E 09 -0.1600000E 02

L =  0.3346873E-06   RL =  0.4210072E 02   CI =  0.1874960E-09

AO =  0.5005357E 01

W3DB =  0.2405000E 09

GBWPN =  0.9584301E 00
```

```
SHUNT PEAKED TRANSISTOR AMPLIFIER
FROM HYBRID-PI EQUIVALENT CIRCUIT

BO = 27.00    WT =  0.1256000E 10    CC =  0.3000000E-11
I COLECTOR =  0.4999999E-02   RBP = 50.00   W3DB =  0.2405000E 09
 0.5222453E 01  0.4436469E 02  0.3590371E-06  0.2301579E 09  0.1034214E 08
 0.5091094E 01  0.4298947E 02  0.3441585E-06  0.2363173E 09  0.4182736E 07
 0.5039826E 01  0.4245723E 02  0.3384732E-06  0.2388022E 09  0.1697760E 07
 0.5019310E 01  0.4224492E 02  0.3362167E-06  0.2398098E 09  0.6901600E 06
 0.5011019E 01  0.4215924E 02  0.3353076E-06  0.2402195E 09  0.2805440E 06
 0.5007655E 01  0.4212448E 02  0.3349394E-06  0.2403858E 09  0.1142080E 06
 0.5006288E 01  0.4211038E 02  0.3347896E-06  0.2404538E 09  0.4624000E 05
 0.5005734E 01  0.4210464E 02  0.3347292E-06  0.2404810E 09  0.1900800E 05
 0.5005507E 01  0.4210229E 02  0.3347041E-06  0.2404924E 09  0.7568000E 04
 0.5005417E 01  0.4210135E 02  0.3346943E-06  0.2404968E 09  0.3152000E 04

L =  0.3346943E-06   RL =  0.4210135E 02   CI =  0.1874965E-09

AO =  0.5005380E 01

W3DB =  0.2405000E 09

GBWPN =  0.9584344E 00
```

Figure 7.28 The output results of the program of Figure 7.26 with speed-up factor changes of (a) 0.9; and (b) 0.6

advantage of using a 0.9 speed-up factor. The reader is urged to experiment with other choices of speed-up constants since this particular concept is an important one that often is required in other unrelated programs.

7.5 SCHMITT TRIGGER DESIGN

A Schmitt trigger circuit is a bistable network whose state depends on the amplitude of the input signal. For this reason, Schmitt trigger circuits are widely used as amplitude comparators and squaring circuits.[5] A current-mode Schmitt is generally much faster than a voltage-mode Schmitt and may possess switching times in the order of 1 nanosecond.

[5] See Jacob Millman and Herbert Taub, *Pulse, Digital, and Switching Waveforms* (New York: McGraw-Hill Book Company, 1965), Sec. 10-11.

The main reasons for the current-mode Schmitt to be faster in switching are that this circuit does not operate in the saturation mode, and a lower impedance circuit produces a smaller voltage swing.

Current-Mode Schmitt

In order to understand the operation of the current-mode Schmitt, let us consider the network shown in Figure 7.29. In normal operation one transistor will be off and the other transistor will be on, and the particular state will depend on whichever base is at the higher potential. For a crude understanding of the operation of this circuit, let us assume that the base voltages will be very small (approximately zero volts). Neglecting the base-emitter diode drop will therefore cause the emitters to be essentially at ground potential; hence,

$$I_E \approx \frac{V_{EE}}{R_E}$$

If we now assume that $-V_{EE} = -10$ volts and $R_E = 1000$ ohms, we are talking about a switching current (I_E) of 10 milliamperes.

Assume now that Q_1 is off and Q_2 is on. This means that the current through R_{L1} also flows through R_{B2} (except for the small base drive current required to keep Q_2 on). Thus

$$V_{B2\max} = \left(\frac{V_{CC} - V_{D1}}{R_{L1} + R_{B2}}\right) R_{B2} = \frac{12 - 5.6}{390 + 50} 50 = 0.73 \text{ volts}$$

In order to hold Q_1 off, the base voltage $V_{B1} < V_{B2\max}$. When the value of V_{B1} is made more positive than $V_{B2\max}$, Q_1 will turn on, switching the

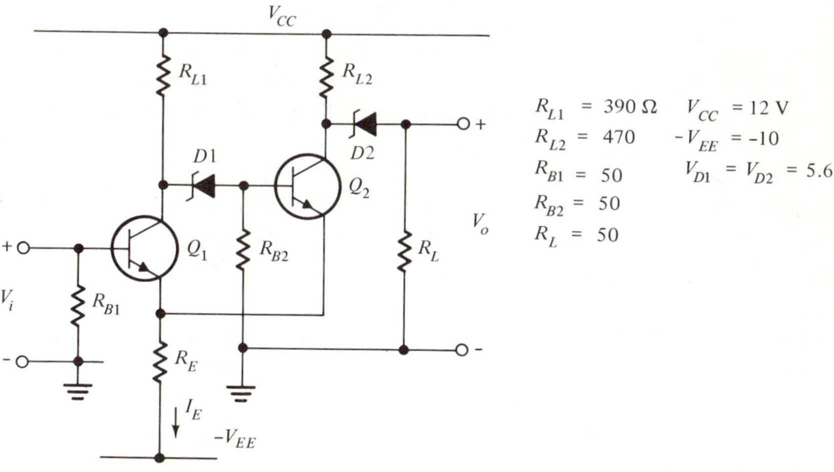

Figure 7.29 A current-mode Schmitt trigger

current I_E from Q_2 to Q_1. The base voltage of Q_2 therefore becomes equal to

$$V_{B2\min} = \frac{V_{CC} - I_E R_{L1} - V_{D1}}{R_{L1} + R_{B2}} R_{B2}$$
$$= \frac{12 - 10^{-2}\,390 - 5.6}{390 + 50}\,50 = 0.29 \text{ volt}$$

We may therefore see that in order to switch the Schmitt trigger, the input base voltage must swing greater than $V_{B2\max}$ and less than $V_{B2\min}$. The difference of these two levels is the minimum sensitivity to operate the switch.

The output voltage swing may similarly be calculated. With Q_2 off, the output voltage will be given by

$$V_{0\max} = \frac{V_{CC} - V_{D2}}{R_{L2} - R_L} R_L = \frac{12 - 5.6}{470 + 50}\,50 = 0.62 \text{ volts}$$

and with Q_2 on, the load current will be reduced by approximately I_E or the output voltage swing will be approximately $I_E R_L = 0.50$ volts. In practice, a small trimming resistor potentiometer is added to R_{L2} so that the d-c output levels can be adjusted to produce the correct driving levels. For example, at very low sinusoidal driving frequencies, it is often necessary to add a second Schmitt trigger to achieve very fast rise and fall times. In order to have the second Schmitt trigger properly, the small trimming resistor can be adjusted to provide the proper d-c levels.

Voltage-Mode Schmitt

Let us now consider the operation of the voltage-mode Schmitt trigger shown in Figure 7.30.[6] Assuming that the input voltage V_i is zero and Q_1 is off and Q_2 is on, the drop across $R_E(V_E)$ will be due to the current flowing in Q_2 and this potential will therefore be sufficient to keep Q_1 cut off. If we now increase the input voltage to a value of $V_E + V_{BE1(\mathrm{ON})}$, Q_1 will begin to conduct and thus increase the value of V_E. At the same time, the collector voltage of Q_1 will decrease and both of these changes will have the effect of reducing the base drive of Q_2. The reduction of the base current of Q_2 will cause the collector current of Q_2 to decrease and this, in turn, will cause V_E to decrease. This resulting reduction of V_E will thus increase the base drive of Q_1. The above behavior will therefore be regenerative while the two transistors are in their active region, and this will continue until Q_1 snaps on and Q_2 snaps off.

[6] The following discussion is based on the article "A Simple Change in the Basic Schmitt Trigger and a Simple Design Procedure Enable Improvements of Tenfold and More in Hysteresis Levels" by William E. Zrubeck, EEE (Electronic Equipment Engineering), December 1963, pp. 41–44.

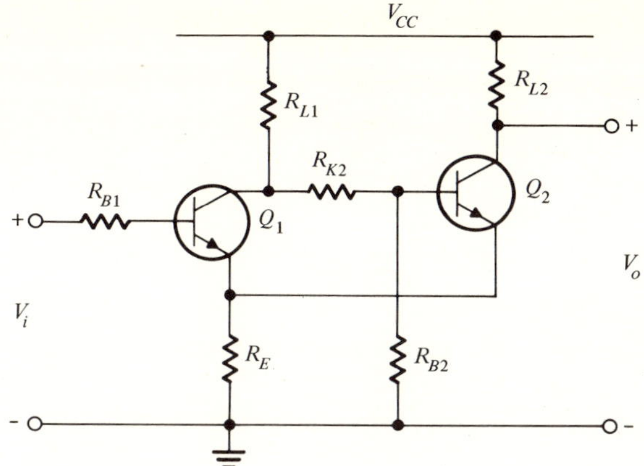

Figure 7.30 A voltage-mode Schmitt trigger

In the above discussion we assumed that the loop gain of the circuit was greater than one. This is what produced the snap action in switching from one state to the other. Figure 7.31a shows a plot of the output voltage versus the input voltage for loop gains ≤ 1. Note that for a loop gain less than one, the circuit behaves as an amplifier and the smaller the loop gain, the larger the input voltage required to drive Q_2 to cutoff (high voltage output state).

Figure 7.31b shows the effect of the input signal for loop gains greater than one. Note that as the input voltage is raised from zero volts, Q_1 will remain off and Q_2 on until the input voltage reaches V_T. At this point regeneration will take place and the circuit will switch to the opposite state. Increasing the input voltage beyond this point causes no further change in the state. However, if the input voltage is now reduced to V_L, we see that we will switch back to the initial state (Q_1 off and Q_2 on).

An examination of Figure 7.31 points out that a Schmitt trigger circuit exhibits hysteresis for loop gains greater than unity, and minimum hysteresis occurs for a loop gain equal to unity. The basic problem,

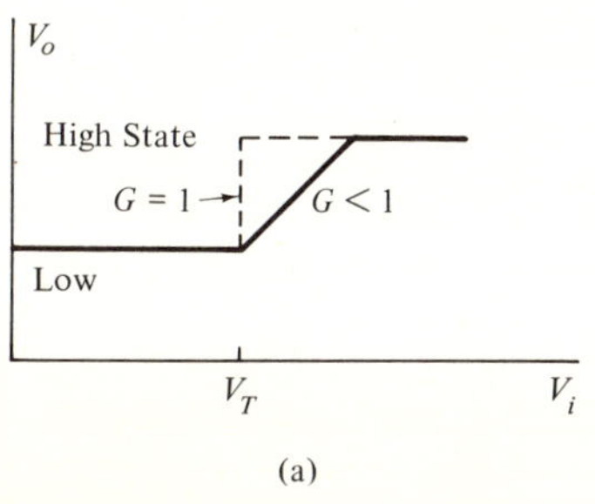

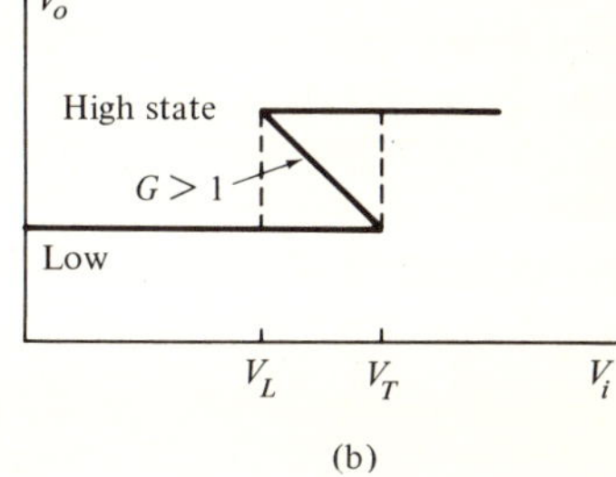

Figure 7.31 Schmitt trigger response

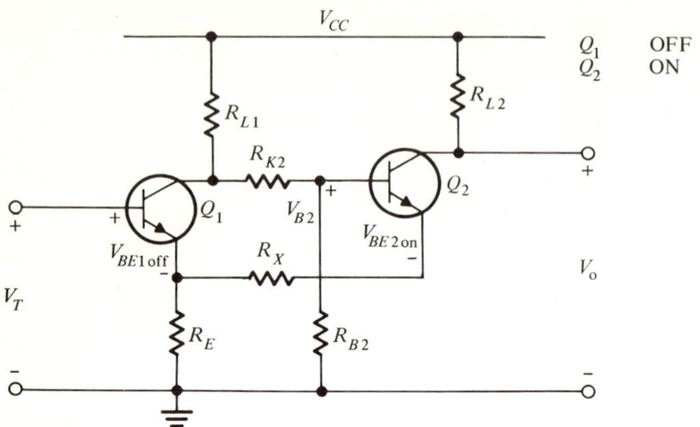

Figure 7.32 The minimum hysteresis Schmitt

therefore, in the design of a minimum hysteresis Schmitt is to obtain a unity loop gain. In many other applications, however, the design of the Schmitt requires a certain amount of hysteresis (noisy signals, and so on). The general design of a Schmitt trigger therefore falls into two classes: the minimum hysteresis design and the predictable hysteresis design.

Let us now derive the necessary design equations. Figure 7.32 shows a modification of the circuit of Figure 7.30 to which we have added the resistor R_X so that we may be able to adjust the loop gain to unity. Assuming Q_1 is off and Q_2 is on, we may write Kirchhoff's current law at the emitter of Q_1 and the base of Q_2

Q_1 *off*

$$\frac{V_T - V_{BE1(\text{OFF})}}{R_E} + \frac{V_T - V_{BE1(\text{OFF})} + V_{BE2(\text{ON})} - V_{B2}}{R_X} = 0$$

$$\frac{V_{B2} - V_{CC}}{R_{L1} + R_{K2}} + \frac{V_{B2}}{R_{B2}} + \frac{V_{B2} - V_{BE2(\text{ON})} + V_{BE1(\text{OFF})} - V_T}{\beta R_X} = 0$$

If we now assume that Q_2 is off and Q_1 is on, we may write Kirchhoff's current law at the collector of Q_1 and the base of Q_2

Q_2 *off*

$$\frac{V_{C1} - V_{CC}}{R_{L1}} + \frac{V_{C1} - V_{BE2(\text{OFF})} + V_{BE1(\text{ON})} - V_T}{R_{K2}} + \alpha \frac{V_T - V_{BE1(\text{ON})}}{R_E} = 0$$

$$\frac{V_T - V_{BE1(\text{ON})} + V_{BE2(\text{OFF})}}{R_{B2}} + \frac{V_T - V_{BE1(\text{ON})} + V_{BE2(\text{OFF})} - V_{C1}}{R_{K2}} = 0$$

Solving the above equations for R_{L1}, R_{K2}, and R_X, we find

$$R_{L1} = \frac{R_{B2}R_E(V_{CC} - V_{C1})}{R_E(V_T - V_{BE1(ON)} + V_{BE2(OFF)}) + \alpha R_{B2}(V_T - V_{BE1(ON)})}$$

$$R_{K2} = \frac{R_{B2}(V_{C1} - V_T + V_{BE1(OFF)} - V_{BE2(OFF)})}{V_T - V_{BE1(ON)} + V_{BE2(OFF)}}$$

and

$$R_X = \frac{R_E}{V_T - V_{BE2(OFF)}}\left(\frac{V_{CC}}{W+1} - (V_T + V_{BE1(ON)} - V_{BE2(OFF)})\right) - R_{B2}(1-\alpha)\frac{W}{W+1}$$

where

$$W = \frac{R_{L1} + R_{K2}}{R_{B2}}$$

and for minimum hysteresis

$$V_{C1} = V_T \text{ in } R_{L1} \text{ and } R_{K2} \text{ equations.}$$

for predictable hysteresis

$$\left.\begin{aligned} V_T &= V_L \\ V_{C1} &= V_T \end{aligned}\right\} \text{in } R_{L1} \text{ and } R_{K2} \text{ equations and } V_T = V_T \text{ in } R_X \text{ equations.}$$

Referring once again to Figure 7.32, we may now write the equation for the loop gain

$$G = \alpha \frac{R_{L1}(R_{B2} + R_{K2})/(R_{L1} + R_{K2} + R_{B2})}{R_E R_X/(R_E + R_X)} \frac{R_{B2}}{R_{B2} + R_{K2}} \frac{R_E}{R_E + R_X}$$

$$= \alpha \frac{R_{L1}R_{B2}}{R_X(R_{L1} + R_{K2} + R_{B2})}$$

In the minimum hysteresis design, we may therefore use the above equation to determine the value of R_X since for this case $G \equiv 1$; thus,

$$R_X = \alpha \frac{R_{L1}R_{B2}}{R_{L1} + R_{K2} + R_{B2}}$$

Note that the value of R_{L2} does not affect this calculation and it may therefore be chosen arbitrarily as long as saturation of Q_2 is still possible. Similarly, the values of R_E and R_{B2} may also be arbitrarily chosen. In general, the value of R_E would be chosen in the same way we would choose the value of R_E in the design of the single supply flip-flop

$$R_E = \frac{V_{E2} R_{L2}}{V_{CC} - V_{E2}}$$

where V_{E2} corresponds to the emitter voltage with Q_2 on. Usually V_{E2} has a numerical value between 0.5 and 1.5 volts. Finally, since the value of R_{B2} determines the temperature stability, we would generally choose $R_{B2} = 5 * R_E$ to insure stable temperature operation.

Figure 7.33 gives the Schmitt trigger Fortran design program. Initially the values of R_E, V_{CC}, V_T, V_L, β, $V_{BE(ON)}$, $V_{BE(OFF)}$, and R_{L2} are read into the computer memory. Computer statement S.0003 then computes the value of ALPHA and computer statement S.0004 sets R_{B2} equal to five times the value of R_E. We then test to see whether V_L is equal to zero. If the value of V_L is equal to zero, control is transferred to statement 20 and we compute the minimum hysteresis design. If the value of V_L is not equal to zero, we transfer to statement 10 and compute the predictable hysteresis design. In either case, after the design values have been ob-

```
                C       VOLTAGE-MODE SCHMITT TRIGGER DESIGN
                C       MINIMUM HYSTERESIS CASE
                C           AND
                C       PREDICTABLE HYSTERESIS CASE.
                C
S.0001                5 READ(5,1)RE,VCC,VT,VL,B,VBEON,VBEOFF,RL2
S.0002                1 FORMAT(8E10.3)
S.0003                  ALPHA=B/(B+1.)
S.0004                  RB2=5.*RE
S.0005                  IF(VL)10,20,10
                C
                C       MINIMUM HYSTERESIS CASE.
                C
S.0006               20 RL1=RB2*RE*(VCC-VT)/(RE*(VT-VBEON+VBEOFF) +ALPHA*RB2*(VT-VBEON))
S.0007                  RK2=RB2*(VBEON-VBEOFF)/(VT-VBEON+VBEOFF)
S.0008                  W=(RL1+RK2)/RB2
S.0009                  RX=RE/(VT-VBEOFF)*(VCC/(W+1.)-VT-VBEON+VBEOFF)-RB2*(1.-ALPHA)*
                       1W/(W+1.)
                C
                C       COMPUTE LOOP GAIN AND CHECK TO SEE IF IT IS UNITY.
                C       IF LOOP GAIN IS NOT UNITY, COMPUTE NEW VALUE OF RX.
                C
S.0010                  G=ALPHA*(RL1*RB2)/((RL1+RK2+RB2)*RX)
S.0011                  IF(G-1.)30,50,30
S.0012               30 RXN=ALPHA*(RL1*RB2)/(RL1+RK2+RB2)
S.0013                  WRITE(6,2)VCC,VT,RL2,RE,B,VBEOFF,VBEON,RL1,RK2,RB2,G,RX,RXN
S.0014                2 FORMAT('1MINIMUM HYSTERESIS SCHMITT',////,' VCC =',E11.3,'  VT =',
                       1E11.3,'  RL2 =',E11.3,'  RE =',E11.3,//,' BETA =',E11.3,'  VBEOFF
                       2=',E11.3,'  VBEON =',E11.3,///,'  RL1 =',E11.3,//,'  RK2 =',E11.3,
                       3//,'  RB2 =',E11.3,//,'  GAIN =',E11.3,//,'  RX =',E11.3,//,'  RXN
                       4 =',E11.3,' CORRECTED RX VALUE')
S.0015                  GO TO 50
                C
                C       PREDICTABLE HYSTERESIS CASE.
                C
S.0016               10 RL1=RB2*RE*(VCC-VL)/(RE*(VL-VBEON+VBEOFF) +ALPHA*RB2*(VL-VBEON))
S.0017                  RK2=RB2*(VBEON-VBEOFF)/(VL-VBEON+VBEOFF)
S.0018                  W=(RL1+RK2)/RB2
S.0019                  RX=RE/(VT-VBEOFF)*(VCC/(W+1.)-VT-VBEON+VBEOFF)-
                       1RB2*(1.-ALPHA)*W/(W+1.)
S.0020                  WRITE(6,3)VCC,VT,VL,RL2,RE,RB2,B,VBEOFF,VBEON,RL1,RK2,RX
S.0021                3 FORMAT('1PREDICTABLE HYSTERESIS SCHMITT',////,' VCC =',E11.3,'  VT
                       1 =',E11.3,'  VL =',E11.3,//,'  RL2 =',E11.3,'  RE =',E11.3,'  RB2
                       2=',E11.3,//,'  BETA =',E11.3,'  VBEOFF =',E11.3,'  VBEON =',E11.3,
                       3///,'  RL1 =',E11.3,//,'  RK2 =',E11.3,//,'  RX =',E11.3)
S.0022               50 GO TO 5
S.0023                  END
```

Figure 7.33 A program for designing a voltage-mode Schmitt trigger

```
PREDICTABLE HYSTERESIS SCHMITT


VCC =  0.200E C2  VT =  0.120E 02  VL =  C.100E 02

 RL2 =  0.100E 04  RE =  0.100E 04  RB2 =  0.500E 04

 BETA =  0.100E 03  VBEOFF =  0.0        VBEON =  0.60CE OC

 RL1 =  0.894E C3

 RK2 =  0.319E 03

 RX =  0.282E 03


MINIMUM HYSTERESIS SCHMITT


VCC =  0.200E C2  VT =  0.120E 02  RL2 =  0.100E 04  RE =  0.10CE 04

BETA =  0.10CE 03  VBECFF =  0.0        VBEON =  0.60CE CC

 RL1 =  0.590E C3

 RK2 =  C.263E C3

 RB2 =  0.5CCE 04

 GAIN =  C.136E 01

 RX =  0.367E C3

 RXN =  0.499E C3 CORRECTED RX VALUE

IHC217I
```

Figure 7.34 The output results of the program of Figure 7.33

tained, we transfer to statement 5 so that the next set of data values may be processed.

In order to test the program of Figure 7.33 the following two data sets were used:

	R_E	V_{CC}	V_T	V_L	β	$V_{BE(\text{ON})}$	$V_{BE(\text{OFF})}$	R_{L2}
1)	1000.0	20.0	12.0	10.0	100.0	0.60	0.0	1000.0
2)	1000.0	20.0	12.0	0.0	100.0	0.60	0.0	1000.0

Figure 7.34 gives the computer results and Figures 7.35a and 7.35b show the corresponding two networks. The reader is encouraged to verify these results by hand calculating each of the program design statements. It is also suggested that the reader make changes in the input data so that he may determine the influence of the individual parameter tolerances.

In addition, the reader is urged to explore the temperature stability of this circuit. This may be done by realizing that the temperature stability mainly depends on the changes in I_{CE0} and $V_{BE(\text{ON})}$. Referring to Figure 7.36, we may therefore compute the change in the trigger voltage

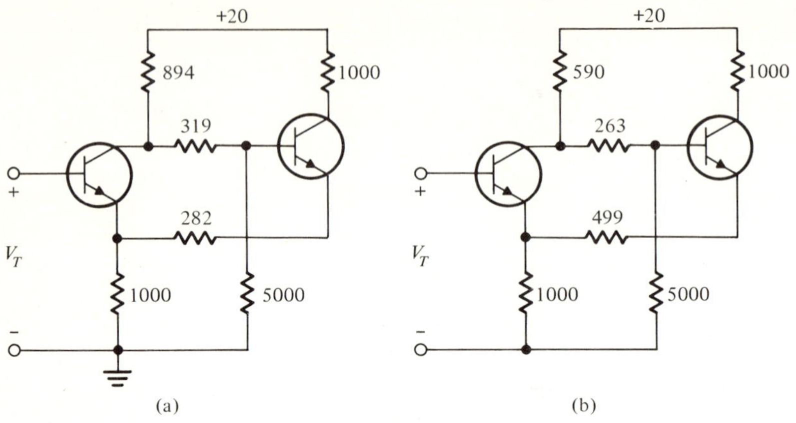

Figure 7.35 Schmitt triggers designed by the program of Figure 7.33

by assuming Q_1 is off and Q_2 is on. Thus from Figure 7.36 we have

$$\Delta I_{C2} = \frac{(A + B)\Delta I_{CEO2} - \beta \Delta V_{BE(\text{ON})2}}{A + (1 + \beta)B}$$

where

$$A = \frac{R_{B2}(R_{L1} + R_{K2})}{R_{B2} + R_{L1} + R_{K2}}$$

$$B = R_E + R_X$$

The change in the trigger voltage of Q_1 will therefore be equal to

$$\Delta V_{T1} = R_E * \Delta I_{C2} + \Delta V_{BE(\text{ON})1}$$

For computational purposes, the reader may assume that I_{CEO} ap-

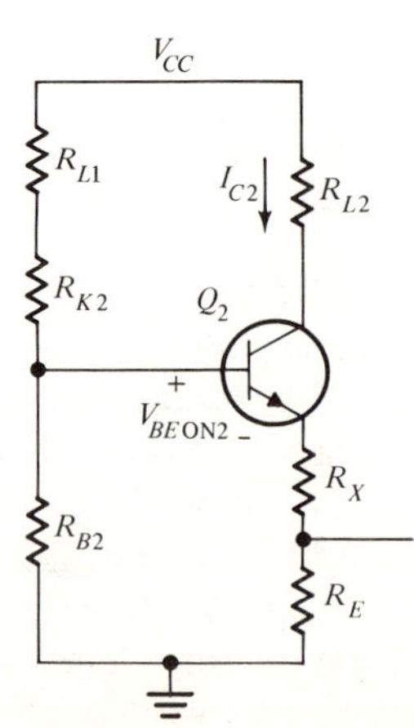

Figure 7.36 A circuit for investigating temperature stability

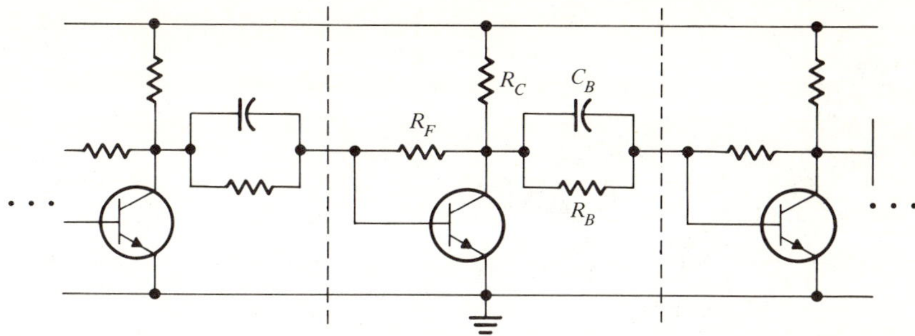

Figure 7.37 A cascade of R_B-C_B feedback stages

proximately doubles for every 10°C increase in temperature and $V_{BE(\text{ON})}$ decreases approximately 25 millivolts for every 10°C increase in temperature.

7.6 THE R-C FEEDBACK TRANSISTOR AMPLIFIER

In addition to reducing the forward transmission A circuit variations, nonlinear distortion, noise, and impedance variations, negative nominal feedback is also used to improve the frequency response of an amplifier. One very useful, but not widely known, type of feedback circuit is the R-C configuration shown in Figure 7.37.[7] Two major advantages of this type of circuit are that no inductances are required and the gain bandwidth product is essentially constant, independent of the nominal gain.

In order to analyze the behavior of one stage in a cascade of this type of feedback amplifier, we may assume that the stage is driven by a current source and loaded by the parallel R_B-C_B network since the input impedance of the following stage will be considerably smaller (due to shunt feedback) than the impedance of the R_B-C_B coupling network. The circuit shown in Figure 7.38 may therefore be used as the unilateral

[7] See M. S. Ghausi, *Principles and Design of Linear Active Circuits* (New York: McGraw-Hill Book Company, 1965), Sec. 12-10.

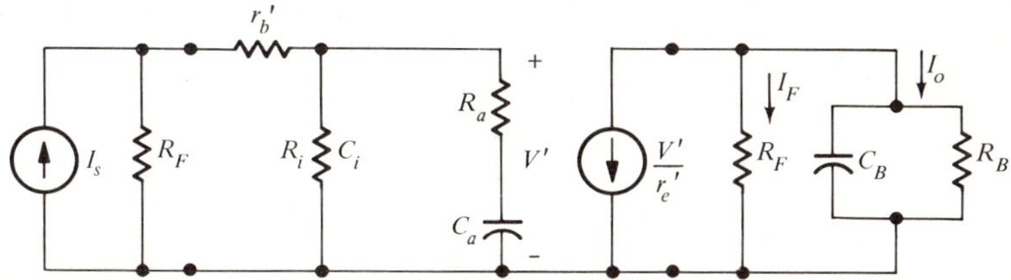

Figure 7.38 The unilateral forward circuit

forward equivalent network. The forward current transfer ratio is thus given by

$$A(s) = \frac{I_0(s)}{I_s(s)}$$

$$= \frac{-1}{r'_e C_i\left(1 + \dfrac{r'_b}{R_F}\right)} \frac{s + 1/R_B C_B}{s^2 + \left(p_i + p_a + \dfrac{C_c}{r'_e C_i C_B} + \dfrac{1}{R_F C_i(1 + r'_b/R_F)}\right)s + p_a\left(p_i + \dfrac{1}{R_F C_i(1 + r'_b/R_F)}\right)}$$

where

$$R_a = \frac{C_B}{C_c} r'_e \qquad C_a = C_c \frac{R_B}{r'_e} \qquad C_i = \frac{1}{r'_e \omega_t} + C_c$$

$$p_a = \frac{1}{R_a C_a} = \frac{R_F + R_B}{R_F R_B C_B} \qquad p_i = \frac{1}{R_i C_i} = \frac{1}{\beta_0 r'_e C_i}$$

This equation may further be simplified as follows

$$A(s) = \frac{-A_0 p_{x1} p_{x2}(s + p_{01})}{p_{01}(s + p_{x1})(s + p_{x2})}$$

where p_{x1} and p_{x2} are the poles and $p_{01} = \dfrac{1}{R_B C_B}$ is the zero of the forward current transfer ratio and

$$A_0 = \beta_0 \left(\frac{R_F}{R_F + r'_b + R_i}\right)\left(\frac{R_F}{R_F + R_B}\right)$$

is the nominal gain.

An examination of Figure 7.38 will show that the feedback function β for the closed-loop amplifier will be given by

$$\beta = \frac{I_F}{I_0} = \frac{\dfrac{Z_B}{R_F + Z_B} \dfrac{V'}{r'_e}}{\dfrac{R_F}{R_F + Z_B} \dfrac{V'}{r'_e}} = \frac{Z_B}{R_F} = \frac{1}{R_F} \frac{R_B}{1 + sR_B C_B}$$

$$= \frac{R_B}{R_F} \frac{p_{01}}{s + p_{01}}$$

The closed-loop current transfer ratio is therefore given by

$$G(s) = \frac{A(s)}{1 - \beta A(s)} = \frac{A_0 p_{x1} p_{x2}}{p_{01}} \frac{s + p_{01}}{s^2 + (p_{x1} + p_{x2})s + p_{x1} p_{x2}\left(1 + A_0 \dfrac{R_B}{R_F}\right)}$$

and the low-frequency closed-loop current gain is thus

$$G_0 = \frac{\beta_0(R_F/(R_F + R_B))(R_F/(R_F + r_b' + R_i))}{1 + \beta_0(R_F/(R_F + R_B))(R_F/(R_F + r_b' + R_i))R_B/R_F}$$

If we now rewrite $G(s)$ as

$$G = K\frac{s + c}{s^2 + as + b}$$

where

$$K = a \text{ constant}$$

and

$$a = p_{x1} + p_{x2} \qquad b = p_{x1}p_{x2}\left(1 + A_0\frac{R_B}{R_F}\right) \qquad c = p_{01}$$

the conditions for a maximally flat amplitude response are given by

$$|G^2| = K'\frac{\omega^2 + c^2}{(b - \omega^2)^2 + a^2\omega^2} = \frac{K'c^2}{b^2}\frac{1 + (1/c^2)\omega^2}{1 + ((a^2 - 2b)/b^2)\omega^2 + \omega^4/b^2}$$

Equating coefficients, we thus find

$$\frac{1}{c^2} = \frac{a^2 - 2b}{b^2}$$

or

$$\left(\frac{b}{c}\right)^2 = a^2 - 2b$$

The 3-dB bandwidth is thus given by

$$\omega_{3\text{dB}} = \sqrt{\frac{1}{2}\left(\frac{b}{c}\right)^2 + \sqrt{\left(\frac{1}{2}\left(\frac{b}{c}\right)^2\right)^2 + b^2}}$$

Design Procedure. Assuming that the nominal gain G_0 has been specified and that all other transistor parameters are known, we first calculate R_F.

1. Experimentally it has been found that the gain bandwidth product is not critically dependent on the value of R_B and a choice of R_B ranging from 100 to 400 ohms is a reasonable one. Specifying an appropriate value for R_B, the value of R_F may thus be found from the solution of the equation for G_0. Thus

$$R_F = A + \sqrt{A^2 + \frac{R_B(R_i + r_b')}{\beta_0/G_0 - 1}}$$

where

$$A = \frac{1}{2}\frac{(\beta_0 + 1)R_B + R_i + r_b'}{\beta_0/G_0 - 1}$$

2. The value of C_B is found from the solution of the closed-loop gain expression and the maximally flat coefficient equation and is given by

$$\left(\left(\frac{\left(1 + A_0\frac{R_B}{R_F}\right)(R_F + R_B)}{R_F}\right)^2 - 1\right)C_B{}^2 + 2\left(\frac{A_0\frac{R_B}{R_F}}{AK1}\left(\frac{R_F + R_B}{R_F R_B}\right) - \frac{C_c R_i R_F(1 + r_b'/R_F)}{r_e'(R_i + R_F(1 + r_b'/R_F))}\right)C_B - \frac{AK2^2}{AK1^2} = 0$$

where

$$AK2 = \frac{C_c}{C_i r_e'} + \frac{R_F + R_B}{R_F R_B}$$

$$AK1 = \frac{1}{R_i C_i} + \frac{1}{C_i R_F(1 + r_b'/R_F)}$$

or

$$BC_B{}^2 + HC_B + D = 0$$

so that

$$C_B = -\frac{H}{2B} + \sqrt{\left(\frac{H}{2B}\right)^2 - \frac{D}{B}}$$

3. The 3-dB bandwidth is then computed by substituting the values of b and c in the above expression for $\omega_{3\text{dB}}$.

Figure 7.39 shows the corresponding Fortran design program. Initially the program reads in values of FT, RBP, CC, and BO. Assuming a collector current of 5 milliamperes, REP is then set equal to 5.0 ohms. For other values of collector current, the value of the emitter resistance is found from

$$r_e' = \frac{0.025}{I_e} \text{ (room temperature)}$$

The outer DO 10 loop is then used to compute the design for values of R_B = 100, 200, 300, and 400 ohms. The inner DO 10 loop determines the design values for each of the above R_B values, for G_0 values of 5.0, 10.0, 15.0, and 20.0.

```
              C      R-C FEEDBACK DESIGN
              C
S.0001               READ(5,1)FT,RBP,CC,BO
S.0002             1 FORMAT(4E15.7)
S.0003               REP=5.0
S.0004               RI=BO*REP
S.0005               CI=1./(REP*6.28*FT)+CC
S.0006               WRITE(6,2)FT,RBP,CC,BO
S.0007             2 FORMAT('1R-C FEEDBACK DESIGN',//,' FT =',E15.7,4X,'RBP =',E15.7,//
                    1,' CC =',E15.7,4X,'BO =',E15.7,///)
S.0008               RB=0.0
S.0009               DO 10 L=1,4
S.0010               RB=RB+100.
S.0011               GO=0.0
S.0012               DO 10 M=1,4
S.0013               GO=GO+5.0
S.0014               A=0.5*(((BO+1.)*RB+RI+RBP)/((BO/GO)-1.))
S.0015               RF=A+(A*A+(RB*(RI+RBP))/((BO/GO)-1.))**0.5
S.0016                AC=BO*((RF*RF)/((RF+RBP+RI)*(RF+RB)))
S.0017               FO=RB/RF
S.0018               AK1=1./(RI*CI)+1./(CI*RF*(1.+RBP/RF))
S.0019               AK2=CC/(CI*REP)+(RF+RB)/(RF*RB)
S.0020               B=(((1.+ AC*FO)*(RF+RB))/RF)**2-1.
S.0021               F=2.*(( AC*FO*(RF+RB))/(AK1*RF*RB))
S.0022               E=2.*(CC*RI*RF*(1.+(RBP/RF)))/(REP*(RI+RF*(1.+(RBP/RF))))
S.0023               H=F-E
S.0024               D=-(AK2*AK2)/(AK1*AK1)
S.0025               CB=-H/(2.*B)+((H/(2.*B))**2-D/B)**0.5
S.0026               B2=((RF+RB)/(RF*RB*CB*CI))*(1./RI+1./(RF*(1.+RBP/RF)))
S.0027               B3=1.+((BO*RF*RB)/((RF+RI+RBP)*(RF+RB)))
S.0028               B1=B2*B3
S.0029               P01=1./(RB*CB)
S.0030               B5=(((0.5*B1*B1)/(P01*P01))**2+B1*B1)**0.5
S.0031               W3=((0.5*B1*B1)/(P01*P01)+B5)**0.5
S.0032               W3N=(BO*W3)/(6.28*FT)
S.0033               GBW=(W3*GO)/(6.28*FT)
S.0034            10 WRITE(6,3)RF,RB,CB,GO,W3,W3N,GBW
S.0035             3 FORMAT(' RF =',E15.7,2X,'RB =',E15.7,2X,'CB =',E15.7,/ ,' GO =',E1
                    15.7,2X,'W3 =',E15.7, /,' W3N =',E15.7,2X,'GBW =',E15.7,/)
S.0036               STOP
S.0037               END
```

Figure 7.39 Transistor R-C feedback amplifier design program

Computer statements S.0014 and S.0015 are used to compute design step 1. Computer statements S.0016 through S.0025 are used in computing C_B (design step 2). Computer statements S.0026 through S.0031 are used in the computation of the 3-dB bandwidth. The normalized value of the 3-dB bandwidth and the gain bandwidth product are then computed by computer statements S.0032 and S.0033. The 16 results (one each iteration) are then printed by computer statements S.0034 and S.0035.

Figure 7.40 gives the computer results for a set of typical transistor values. Figure 7.41 shows a graph of these corresponding results. An examination of this graph readily shows that this particular type of feedback network essentially allows for a direct exchange of gain for bandwidth. Since this is not a typical property of most transistor amplifiers, the reader is encouraged to reinvestigate the shunt-peaked amplifier of Section 7.4 (or other transistor configurations) in order to reinforce this important property of the R-C feedback configuration. The reader is encouraged to extend this program so that it is capable of obtaining the circuit design when the bandwidth, instead of the nominal gain, is specified. The reader is referred to Section 7.4 (Bandwidth Specified) for one possible method of attack.

```
R-C FEEDBACK DESIGN

FT =   0.5CCCCCCE C9      RBP =   0.500C0C0E 02

CC =   0.2CCCCCCE-11      BO =   0.6C0C00CE 02


RF =   0.59174C2E C3   RB =   0.1C0C0C0E 03   CB =   0.2096848E-1C
GC =   0.5CCCCCCE C1   W3 =   C.6843180E 09
W3N =   C.13C7614E 02   GBW =   0.1C89678E C1

RF =   0.12954C4E C4   RB =   0.10CC0C0E 03   CB =   C.4020989E-1C
GC =   0.1CCCCCCE C2   W3 =   C.3571766E 09
W3N =   C.6825C29E 01   GBW =   0.1137505E C1

RF =   0.2155412E C4   RB =   0.10CC0CCE 03   CB =   0.6319809E-1C
GC =   0.15CCCCCE C2   W3 =   C.2381488E 09
W3N =   C.455C614E 01   GBW =   0.1137653E C1

RF =   0.323C417E C4   RB =   0.1C000CCE 03   CB =   0.9114751E-1C
GC =   0.2CCCCCCE C2   W3 =   C.1764972E 09
W3N =   0.3372556E 01   GBW =   0.1124186E 01

RF =   C.1146459E C4   RB =   0.2C0C0C0E 03   CB =   0.1518269E-1C
GC =   0.5CCCCCCE C1   W3 =   C.6533594E 09
W3N =   C.1248457E 02   GBW =   0.104038CE C1

RF =   0.2515565E C4   RB =   0.20CC0C0E 03   CB =   0.3C60054E-1C
GC =   0.1CCCCCCE C2   W3 =   C.3325850E 09
W3N =   C.6355126E 01   GBW =   0.1059187E C1

RF =   C.4188898F C4   RB =   C.2CCC000E 03   CB =   0.4919259E-1C
GC =   0.15CCCCCE C2   W3 =   C.22C7968E 09
W3N =   0.4219C46E 01   GBW =   0.1054761E C1

RF =   0.628C57CF C4   RB =   C.2C000C0E 03   CB =   0.7208693E-1C
GC =   0.2CCCCCCE C2   W3 =   C.1639721E 09
W3N =   C.3133224E 01   GBW =   0.1044408F 01

RF =   C.1701C65E C4   RB =   C.3C0000CE 03   CB =   C.1349270E-1C
GC =   0.5CCCCCCE C1   W3 =   C.6347185E 09
W3N =   C.1212838E 02   GBW =   C.101C697E 01

RF =   0.3735622E C4   RB =   C.3C000C0E 03   CB =   C.2784938E-1C
GC =   C.1CCCCCCE C2   W3 =   C.32066C0E 09
W3N =   C.612726CE 01   GBW =   0.1021210F C1

RF =   0.6222281E C4   RB =   C.3C000C0E 03   CB =   C.4524599E-1C
GC =   C.15CCCCCE C2   W3 =   C.213C392E C9
W3N =   0.4C7C812E 01   GBW =   C.1C17703E C1

RF =   0.933C6C9E C4   RB =   C.3C0C0C0E 03   CB =   0.6679444F-1C
GC =   0.2CCCCCCE C2   W3 =   C.1587164E C9
W3N =   C.3C32797E 01   GBW =   0.1C1C932F 01

RF =   0.2255643E C4   RB =   0.4CC00C0E 03   CB =   0.1271937F-1C
GC =   0.5CCCCCCE C1   W3 =   C.6344777E 09
W3N =   C.1193269F 02   GBW =   0.9943913F CC

RF =   0.4955648F C4   RB =   C.4CCC0C0E 03   CB =   C.2662126F-1C
GC =   C.1CCCCCCF C2   W3 =   C.3145441F 09
W3N =   C.6C1C397F 01   GBW =   C.1001733F 01

RF =   C.8255637E C4   RB =   C.4CC0CCCE 03   CB =   C.4351229E-1C
GC =   0.15CCCCCE C2   W3 =   C.2C92057E 09
W3N =   C.3957561E 01   GBW =   0.9993905F CC

RF =   C.1238062F C5   RB =   C.4CCC0CCF 03   CB =   C.6449771E-1C
GC =   0.2CCCCCCE C2   W3 =   C.1562056F 09
W3N =   C.298482CF C1   GBW =   0.9949405F CC
```

Figure 7.40 The output results of the program of Figure 7.39

7.7 THE SHUNT-SERIES TRANSISTOR FEEDBACK PAIR

Most of our present-day broadband integrated circuit amplifiers make use of the shunt-series feedback configuration. This section concerns itself with the design of the maximally flat shunt-series feedback pair. Our subsequent discussion should not only enable the reader to design

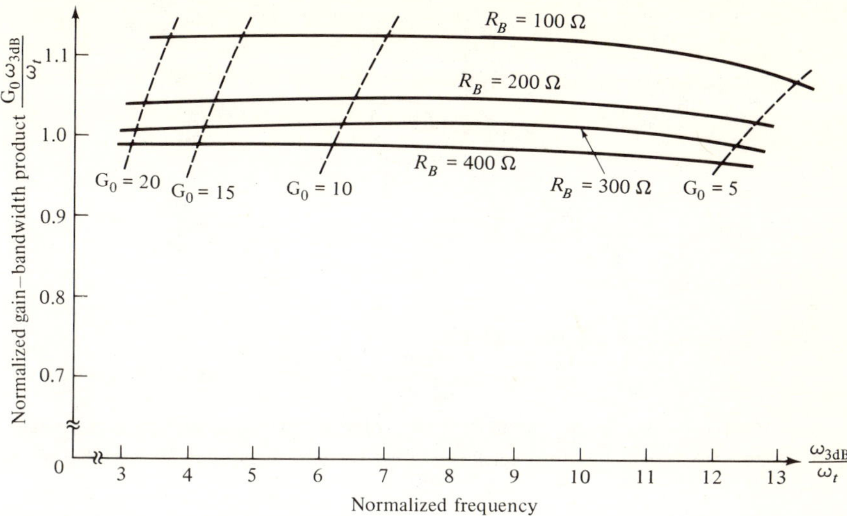

Figure 7.41 The plotted results of Figure 7.40

discrete component shunt-series feedback amplifiers, but it should also allow him to make the necessary modifications on existing integrated circuit feedback pairs. For example, by shunting (resistor and/or capacitor) existing elements (via terminals), the reader will be able to change the frequency or time behavior of the pair. In addition, it will also be possible to employ shunting so that the cascade of two or more integrated feedback pairs will have an overall maximally flat amplitude or maximally flat time delay response.

Since the mathematical development of the shunt-series pair is quite lengthy, the following treatment will be in the form of an outline that will serve to explain the computer program.[8] The outline is divided into five parts, and the numbers assigned to the various equations correspond to the computer statement numbers.

1. The shunt-series feedback pair and its equivalent circuit.
2. The forward current gain.
3. The feedback factor β.
4. The closed-loop gain.
5. Determination of the optimum value of R_F.

1. *The Shunt-Series Feedback Pair.* Figure 7.42 shows a simplified schematic of the shunt-series feedback pair. The reader should note that

[8] For a more detailed discussion see M. S. Ghausi, "Optimum Design of the Shunt-Series Feedback Pair with a Maximally Flat Magnitude Response," *IRE Trans. Circuit Theory*, Vol. CT-8 (December, 1961), pp. 448–453.

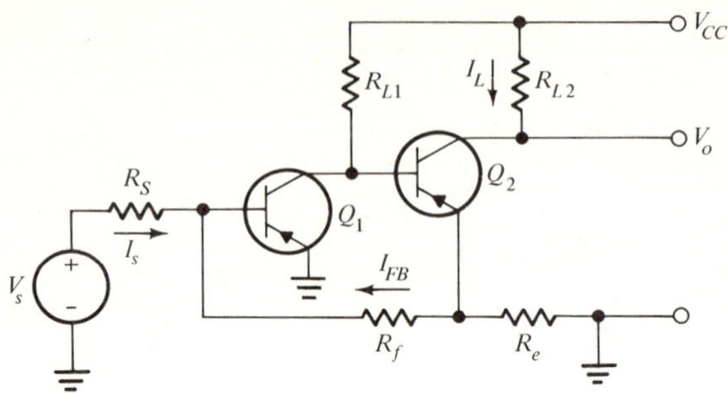

Figure 7.42 The shunt-series feedback pair

nominal negative feedback exists for this circuit since a positive increase in the source voltage, V_s, will cause a negative increase of the voltage at the emitter of Q_2. This, in turn, will cause a feedback current to flow through R_F, which is opposite in sense to the input or source current, hence nominal negative feedback. The reader should also note that the use of feedback reduces the input impedance of the amplifier.

Using the appropriate unilateral high-frequency equivalents for Q_1 and Q_2, we include the loading effects of R_F and construct the overall forward equivalent circuit of Figure 7.43.

2. *The Forward Equivalent Circuit.* The forward current gain of Figure 7.43 is thus given by

$$A(s) = \frac{I_L(s)}{I_s(s)}$$

$$= K \frac{1}{s^2 + \left\{\left(s_{p1} + s_{p2} + \dfrac{C_{ob1}}{C_{i1}C_{i2}r'_{e1}} + \dfrac{1}{R_pC_{i1}(1 + r'_b/R_p)}\right) s + s_{p2}\left(s_{p1} + \dfrac{1}{R_pC_{i1}(1 + r'_b/R_p)}\right)\right\}}$$

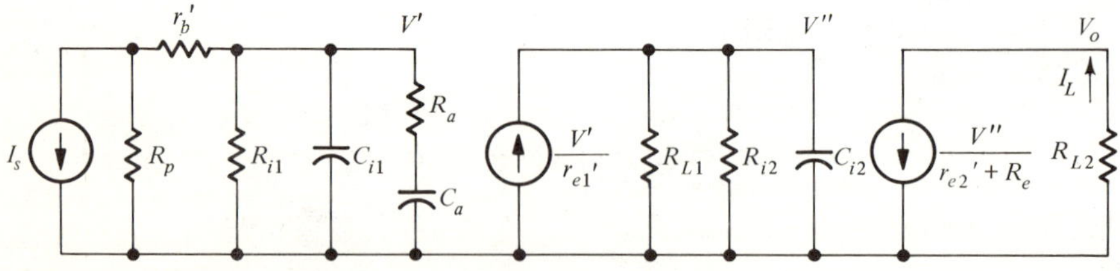

Figure 7.43 The forward high-frequency equivalent of the circuit of Figure 7.42

where

$$K = \frac{1}{(R_e + r'_{e2})C_{i2}r'_{e1}C_{i1}(1 + r'_b/R_p)}$$

$$s_{p1} = \frac{1}{R_{i1}C_{i1}} \qquad \text{(S.0022)}^{9}$$

$$s_{p2} = \frac{R_{L1} + R_{i2}}{R_{L1}R_{i2}C_{i2}} \qquad \text{(S.0023)}$$

$$R_{i1} = \beta_{01}r'_{e1} \qquad \text{(S.0019)}$$

$$R_{i2} = \beta_{02}(r'_{e2} + R_e) \qquad \text{(S.0020)}$$

$$C_{i1} = \frac{1 + r'_{e1}C_{0b1}\omega_{t1}}{r'_{e1}\omega_{t1}} \qquad \text{(S.0015)}$$

$$C_{i2} = \frac{1 + (R_{L2} + r'_{e2} + R_e)C_{0b2}\omega_{t2}}{(R_e + r'_{e2})\omega_{t2}} \qquad \text{(S.0017)}$$

$$R_p = \frac{(R_e + R_f)R_s}{R_e + R_s + R_f} \qquad \text{(S.0018)}$$

We may now compute the poles of the forward circuit by finding the roots of the denominator.

$$s^2 + \left(s_{p1} + s_{p2} + \frac{C_{0b1}}{C_{i1}C_{i2}r'_{e1}} + \frac{1}{R_pC_{i1}(1 + r'_b/R_p)}\right)s + s_{p2}\left(s_{p1} + \frac{1}{R_pC_{i1}(1 + r'_b/R_p)}\right) = 0$$

We may thus let

$$C = \frac{C_{0b1}}{C_{i1}C_{i2}r'_{e1}} + s_{p2} \qquad \text{(S.0024)}$$

$$D = s_{p1} + \frac{1}{R_pC_{i1}(1 + r'_b/R_p)} \qquad \text{(S.0025)}$$

$$E = C + D \qquad \text{(S.0026)}$$

$$F = s_{p2}D \qquad \text{(S.0027)}$$

The poles of the forward current transfer ratio are therefore found from

$$s^2 + Es + F = 0$$

or

$$s_1 = \frac{-E + \sqrt{E^2 - 4F}}{2} \qquad \text{(S.0028)}$$

$$s_2 = \frac{-E - \sqrt{E^2 - 4F}}{2} \qquad \text{(S.0029)}$$

[9] (S.00xx) numbers refer to computer statements in program.

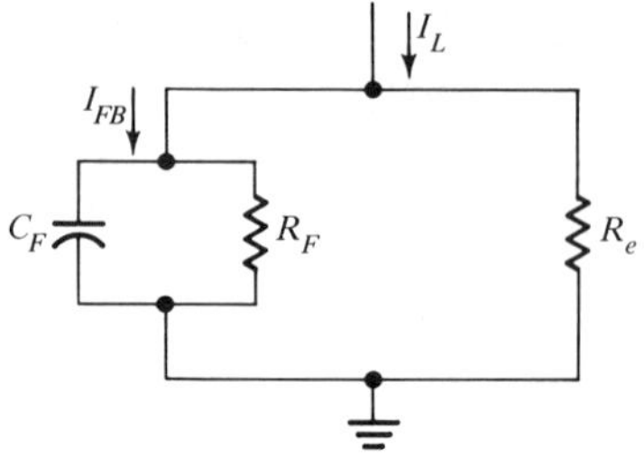

Figure 7.44 Calculation of the feedback factor β

To obtain the midband gain we thus let

$$R_{L1} = R_{i2} \tag{S.0021}$$

and set $\omega = 0$, yielding

$$A_0 = \frac{\beta_{01}\beta_{02}}{2}\left(\frac{R_p}{R_p + r_b' + R_{i1}}\right) \tag{S.0030}$$

3. *The Feedback Factor* β. Since the forward current transfer ratio contains two poles and no zeros, the condition for a maximally flat amplitude response will generally lead to a small value of nominal loop gain (βA_0). If we introduce the capacitor C_F across the feedback resistor, we may obtain a larger value of loop gain and still have a maximally flat response.

Referring to Figure 7.44, we may thus see that since the feedback factor β is defined by

$$\beta = \frac{I_{FB}}{I_L}$$

we may compute

$$I_{FB} = I_L \frac{R_e}{R_e + \dfrac{R_F(1/j\omega C_F)}{R_F + 1/j\omega C_F}}$$

or

$$\beta = \frac{s + 1/R_F C_F}{s + (R_e + R_F)/(R_e R_F C_F)}$$

and for most practical cases

$$\beta_0 = \frac{R_e}{R_e + R_F} \approx \frac{R_e}{R_F} \tag{S.0031}$$

$$\approx \frac{1}{G_0} \tag{S.0016}$$

4. *The Closed Loop Gain.* The general single-loop feedback transmission equation is given by

$$G(s) = \frac{A(s)}{1 - \beta A(s)}$$

with

$$G_0 = \frac{A_0}{1 + \beta_0 A_0}$$

$$F_0 = 1 + \beta_0 A_0 = \frac{A_0}{G_0} \qquad \text{(S.0032)}$$

and

$$A(s) = \frac{A_0 s_1 s_2}{(s + s_1)(s + s_2)}$$

Thus we may write for the closed loop transmission

$$G(s) = \frac{A_0 s_1 s_2}{s^2 + \left(s_1 + s_2 + \dfrac{\beta_0 A_0 s_1 s_2}{1/R_F C_F}\right) s + (1 + \beta_0 A_0) s_1 s_2}$$

Note that because the zero was in the feedback path, it does not appear as a zero in the closed loop transmission expression.

Solving for the poles of the closed loop expression, we have

$$s^2 + \underbrace{\left(s_1 + s_2 + \frac{\beta_0 A_0 s_1 s_2}{1/R_F C_F}\right)}_{B} s + \underbrace{(1 + \beta_0 A_0) s_1 s_2}_{C} = 0$$

or

$$s_{x1},\ s_{x2} = \frac{-B \pm \sqrt{B^2 - 4C}}{2} = -a \pm jb$$

Thus for the condition of maximal flatness the poles would lie on a semicircle of radius $\omega_{3\text{dB}}$ as shown in Figure 7.45.

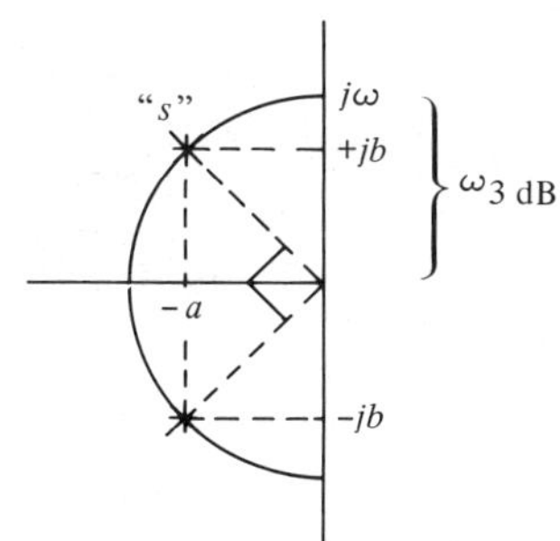

Figure 7.45 The location of the closed-loop poles for maximally flat amplitude response

The 3-dB bandwidth is therefore equal to

$$\omega_{3\text{dB}} = \sqrt{a^2 + b^2}$$

or

$$= \sqrt{s_1 s_2(1 + \beta_0 A_0)} \tag{S.0033}$$

Since $\tan^{-1} 45° = 1 = b/a$, this gives $a = b$ or

$$s_1 + s_2 + \frac{\beta_0 A_0 s_1 s_2}{1/R_F C_F} = \sqrt{2(1 + \beta_0 A_0) s_1 s_2} = \omega_{3\text{dB}} \sqrt{2}$$

from which we may calculate the value of C_F

$$C_F = \frac{\sqrt{2}\,\omega_{3\text{dB}} - (s_1 + s_2)}{R_F(s_1 + s_2 + \beta_0 A_0 s_1 s_2)} \tag{S.0034}$$

5. *Determination of the Optimum Value of R_F.* Starting with

$$\omega_{3\text{dB}} = s_1 s_2 (1 + \beta_0 A_0)$$

we find

$$(\omega_{3\text{dB}})^2 = s_1 s_2 \frac{A_0}{G_0}$$

If we thus specify G_0, this gives R_e in terms of R_F. Substituting the values of s_1, s_2, and $\beta_0 A_0$, and then differentiating to maximize, we find

$$R_{F\text{opt}} = \sqrt{G_0 \left(\frac{r_b' R_s}{R_s + r_b'}\right)\left(\frac{1 + (R_{L2} + r_e')(C_{c2}\omega_{t2})}{C_{c2}\omega_{t2}}\right)}$$

If we now let

$$A = \frac{r_b' R_s}{r_b' + R_s} \tag{S.0012}$$

and

$$B = \frac{1 + (R_{L2} + r_{e2}')(C_{c2}\omega_{t2})}{C_{c2}\omega_{t2}} \tag{S.0013}$$

we have

$$R_F = \sqrt{G_0 A B} \tag{S.0014}$$

Design Procedure. Assuming then that the overall nominal gain G_0 has been specified, the design procedure may be carried out in terms of the following four steps.

1. Determine $R_{F\text{opt}}$ from Eq. (S.0014).
2. Determine R_e from Eq. (S.0016).
3. Make $R_{L1} = R_{i2} = \beta_{02}(R_e + r_{e2}')$, Eqs. (S.0019) and (S.0020).
4. Calculate C_F from Eq. (S.0034).

The computer program for this design is shown in Figure 7.46. The reader will find that the statements of this program will very closely follow the above discussion. We will therefore avoid a detailed description of how this program carries out the design procedure.

Figure 7.47 shows the computer output for three different sets of input data. The reader should note that the main difference between the first and second data sets is the specified value of G_0. As a consequence, it is not surprising to see that for the higher nominal gain we obtain a higher value of A_0 and R_F and a lower value of F_0 and f_{3dB}. The main difference

```
C      MAXIMALLY FLAT RESPONSE
C                OF A
C      SHUNT SERIES TRANSISTOR
C            FEEDBACK PAIR
C
       WRITE(6,10)
    10 FORMAT('1MAXIMALLY FLAT RESPONSE'/10X,'OF A '/' SHUNT SERIES TRANS
      1ISTOR'/5X,'FEEDBACK PAIR'//)
       READ(5,20)FT1,B01,COB1,RB1
       READ(5,20)FT2,B02,COB2,RB2
    20 FORMAT(4E10.2)
       READ(5,30)RS,RL2,G0,CUR
    30 FORMAT(4F10.2)
       RE1=0.025/CUR
       RE2=RE1
       WT1=2.*3.14*FT1
       WT2=2.*3.14*FT2
       A=RB1*RS/(RS+RB1)
       B=(1.+(RL2+RE2)*(COB2*WT2))/(COB1*WT2)
       RF=SQRT(G0*A*B)
       CI1=(1.+RE1*(COB1*WT1))/(RE1*WT1)
       REE=RF/G0
       CI2=(1.+(RL2+RE2+REE)*(COB2*WT2))/((REE+RE2)*WT2)
       RP=((REE+RF)*RS)/(RS+REE+RF)
       RI1=B01*RE1
       RI2=B02*(RE2+REE)
       RL1=RI2
       SP1=1./(RI1*CI1)
       SP2=(RL1+RI2)/(RL1*RI2*CI2)
       C=((COB1)/(CI1*CI2*RE1))+SP2
       D=1./((RP*CI1)*(1.+(RB1/RF)))+SP1
       E=C+D
       F=SP2*D
       S1=(-E+SQRT(E**2-4.*F))/2.
       S2=(-E-SQRT(E**2-4.*F))/2.
       A0=(B01*B02*RP)/(2.*(RP+RB1+RI1))
       BF=1./G0
       F0=A0/G0
       W3DB=SQRT(S1*S2*(1.+BF*A0))
       CF=((SQRT(2.)*W3DB)-(S1+S2))/(RF*(BF*A0*S1*S2))
       F3DB=W3DB/6.28
       WRITE(6,35)FT1,B01,COB1
    35 FORMAT(' FT1 =',E15.7,3X,'B01 =',E15.7,3X,'COB1=',E15.7)
       WRITE(6,40)FT2,B02,COB2
    40 FORMAT(' FT2 =',E15.7,3X,'B02 =',E15.7,3X,'COB2=',E15.7)
       WRITE(6,45)RB1,RB2,RL2
    45 FORMAT(' RB1 =',E15.7,3X,'RB2 =',E15.7,3X,'RL2 =',E15.7)
       WRITE(6,50)CUR,RS,G0
    50 FORMAT(' CUR =',E15.7,3X,'RS  =',E15.7,3X,' G0 =',E15.7,//)
       WRITE(6,70) CI1,RI1,SP1
    70 FORMAT(' CI1 =',E15.7,3X,'RI1 =',E15.7,3X,'SP1 =',E15.7)
       WRITE(6,75)S1,S2
    75 FORMAT('  S1 =',E15.7,3X,' S2 =',E15.7)
       WRITE(6,80)CI2,RI2,SP2
    80 FORMAT(' CI2 =',E15.7,3X,'RI2 =',E15.7,3X,'SP2 =',E15.7)
       WRITE(6,85)RF,CF,REE
    85 FORMAT('  RF =',E15.7,3X,' CF =',E15.7,3X,'REE =',E15.7)
       WRITE(6,90)A0,F0,F3DB
    90 FORMAT('  A0 =',E15.7,3X,' F0 =',E15.7,3X,'F3DB =',E15.7)
       STOP
       END
```

Figure 7.46 A Fortran program for designing a maximally flat shunt-series feedback pair

```
MAXIMALLY FLAT RESPONSE
          OF A
SHUNT SERIES TRANSISTOR
     FEEDBACK PAIR

FT1 =  0.4200000E 09     BO1 =  0.5700000E 02     COB1=  0.2000000E-11
FT2 =  0.4499999E 09     BO2 =  0.5000000E 02     COB2=  0.2000000E-11
RB1 =  0.7500000E 02     RB2 =  0.7500000E 02     RL2 =  0.5000000E 03
CUR =  0.4999999E-02     RS  =  0.1000000E 04      GO =  0.2000000E 02

CI1 =  0.7782652E-10     RI1 =  0.2850000E 03     SP1 =  0.4508453E 08
 S1 = -0.6750976E 07      S2 = -0.2784013E 09
CI2 =  0.2717719E-10     RI2 =  0.2688656E 04     SP2 =  0.2737096E 08
 RF =  0.9754629E 03      CF =  0.8980933E-11     REE =  0.4877313E 02
 AO =  0.8326118E 03      FO =  0.4163058E 02     F3DB =  0.4507331E 08
```

```
MAXIMALLY FLAT RESPONSE
          OF A
SHUNT SERIES TRANSISTOR
     FEEDBACK PAIR

FT1 =  0.4200000E 09     BO1 =  0.5700000E 02     COB1=  0.2000000E-11
FT2 =  0.4499999E 09     BO2 =  0.5000000E 02     COB2=  0.2000000E-11
RB1 =  0.7500000E 02     RB2 =  0.7500000E 02     RL2 =  0.5000000E 03
CUR =  0.4999999E-02     RS  =  0.1000000E 04      GO =  0.1000000E 02

CI1 =  0.7782652E-10     RI1 =  0.2350000E 03     SP1 =  0.4508453E 08
 S1 = -0.5535104E 07      S2 = -0.3462136E 09
CI2 =  0.2030139E-10     RI2 =  0.3593781E 04     SP2 =  0.2663458E 08
 RF =  0.6397563E 03      CF =  0.8774051E-11     REE =  0.6897563E 02
 AO =  0.7757883E 03      FO =  0.7757883E 02     F3DB =  0.6183067E 08
```

```
MAXIMALLY FLAT RESPONSE
          OF A
SHUNT SERIES TRANSISTOR
     FEEDBACK PAIR

FT1 =  0.4699999E 09     BO1 =  0.5700000E 02     COB1=  0.2000000E-11
FT2 =  0.4499999E 09     BO2 =  0.5000000E 02     COB2=  0.2000000E-11
RB1 =  0.7500000E 02     RB2 =  0.7500000E 02     RL2 =  0.5000000E 03
CUR =  0.4999999E-02     RS  =  0.1000000E 04      GO =  0.2000000E 02

CI1 =  0.6975985E-10     RI1 =  0.2850000E 03     SP1 =  0.5029787E 08
 S1 = -0.6804352E 07      S2 = -0.3081562E 09
CI2 =  0.2717719E-10     RI2 =  0.2688656E 04     SP2 =  0.2737096E 08
 RF =  0.9754629E 03      CF =  0.8664539E-11     REE =  0.4877313E 02
 AO =  0.8326118E 03      FO =  0.4163058E 02     F3DB =  0.4760797E 08
```

Figure 7.47 The output results of the program of Figure 7.46

between the first and third data sets is in the increase in the value of f_{t1} and, as a consequence, the obvious increase in f_{3dB}. The reader is urged to study these results further and draw additional conclusions in regard to parameter and specification changes.

7.8 ACTIVE FILTER DESIGN

The availability of inexpensive microcircuit unity-gain operational amplifiers has led to their widespread use in the design of active filters. Figure 7.48 shows the basic two-pole lowpass or highpass section. Note that each section contains only two resistors, two capacitors, and one

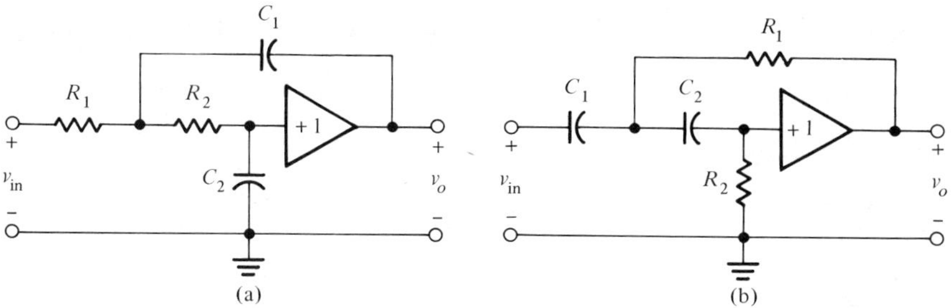

Figure 7.48 Two-pole lowpass and highpass R-C active networks

unity-gain active microcircuit. By connecting $N/2$ such sections in cascade an N-pole filter results. It is thus possible, by suitable selection of the R's and C's, to design N-pole Butterworth or Chebyshev lowpass or highpass active filters.[10]

Butterworth Design

A maximally flat or Butterworth lowpass response may be obtained by forcing the poles of the overall transfer function to be equally spaced on a unit circle in the normalized complex frequency plane, as shown in Figure 7.49. The resulting phase angles are thus given by

$$P_I = \frac{\pi}{2} + \frac{\pi}{2N}(2I - 1),\ I = 1 \text{ to } N$$

and the pole locations by

$$\sigma_I + j\omega_I = \cos P_I + j \sin P_I$$

[10] The following discussion is based on the article "Get Something Extra in Filter Design," written by Russell Kincaid and Frederick Shirley, *Electronic Design*, Vol. 17, No. 13 (June 21, 1969), pp. 114–121.

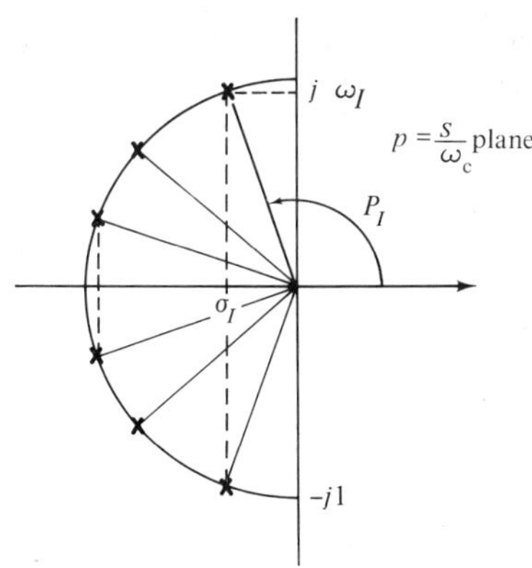

Figure 7.49 Maximally flat or Butterworth pole location ($N = 6$)

The normalized transfer function of a lowpass pole pair may therefore be written as

$$\frac{1}{p^2 - 2\sigma_I p + 1}$$

This, in turn, may be equated to the transfer function of the two-pole lowpass filter section of Figure 7.48.

$$\frac{1}{R_1C_1R_2C_2p^2 + (R_1 + R_2)C_2p + 1}$$

It thus follows that

$$R_1C_1R_2C_2 = 1$$

and

$$(R_1 + R_2)C_2 = -2\sigma_I$$

Choosing C_1 and C_2 as standard values with

$$C_1 = C \text{ (maximum allowable } C)$$

and

$$C_2 = \frac{C}{M} \text{ } (M > 1 \text{ for realizability})$$

we thus find that the normalized resistance values are given by

$$\frac{-\sigma_I M}{C}\left(1 + \left(1 \pm \frac{1}{\sigma_I{}^2 M}\right)^{1/2}\right)$$

or

$$R_1 = \frac{-\sigma_I M}{2\pi f_c C}\left(1 + \left(1 - \frac{1}{\sigma_I{}^2 M}\right)^{1/2}\right)$$

$$R_2 = \frac{-\sigma_I M}{2\pi f_c C}\left(1 - \left(1 - \frac{1}{\sigma_I{}^2 M}\right)^{1/2}\right)$$

where f_c is the cutoff frequency and $M \geq 1/\sigma_I{}^2$

Similarly, the normalized transfer function of the highpass pole pair is obtained by replacing p by $1/p$ or

$$\frac{1}{\dfrac{1}{p^2} - 2\sigma_I \dfrac{1}{p} + 1}$$

and this is equated to the transfer function of the two-pole highpass filter section of Figure 7.48

$$\frac{1}{\dfrac{1}{R_1C_1R_2C_2}\dfrac{1}{p^2} - 2\sigma_I \dfrac{1}{p} + 1}$$

to obtain

$$C_1 = C_2 = C$$

$$R_1 = \frac{-\sigma_I}{2\pi f_c C} \qquad \text{and} \qquad R_2 = \frac{1}{2\pi f_c \sigma_I C}$$

After the element values of each section of the N-pole lowpass or highpass Butterworth filter have been found, the frequency response of the composite filter can be obtained by adding the individual dB losses of each of the sections. The dB loss for the lowpass N-pole would be found from

$$\text{DBL} = \sum_{J=1}^{N/2} -10 \text{ Log}_{10}((1 - R1(J) * C1(J) * R2(J) * C2(J) * \omega^2)^2 + ((R1(J) + R2(J)) * C_2(J) * \omega)^2)$$

and the dB loss from the highpass N-pole would be found from

$$\text{DBL} = \sum_{J=1}^{N/2} -10 \text{ Log}_{10}((1 - 1/(R1(J) * C1(J) * R2(J) * C2(J) * \omega^2))^2 + ((C1(J) + C2(J))/(C1(J) * R2(J) * C2(J) * \omega))^2)$$

Figure 7.50 is the Fortran design program. Initially the number of poles N, the cutoff frequency F, and the maximum allowable capacitance C are read into the computer memory. Computer statements S.0004 and S.0005 are then used to print the heading, the number of poles, and the cutoff frequency. Computer statement S.0006 next specifies the value of the constant PI (π) and S.0007 computes the required number of filter sections NN. The DO 10 loop is then used to determine the element values of each section.

Each of the section element values is computed by first calculating the angle of the pole location (S.0009) and then the real (RREAL $= \sigma_I$) and imaginary (AIMAG $= \omega_I$) value of the pole. The constant

$$\text{CST} = \frac{-\sigma_I}{2\pi f_c C}$$

is then found by computer statement S.0013 and the value of C_1 (C1(I)) is set equal to C.

Computer statements S.0015 through S.0020 are used to define the standard capacitance scale values. The value of $N2$ is then computed by computer statement S.0021. For example, if $C = 68$, ALOG10(68) would be equal to 1.832 and $N2$ would therefore be equal to 1. The two DO 20 loops are now used to find the value of C_2 ($C2(I)$). In the first iteration

```
LEVEL   15JUL68                               IBM OS/360 BASIC FORTRAN IV (E) COMPILATION

            C      RC ACTIVE LOW-PASS BUTTERWORTH FILTER DESIGN.
            C
            C      N = NUMBER OF POLES (MUST BE EVEN).
            C      F = CUTOFF FREQUENCY IN KHZ.
            C      C = MAXIMUM CAPACITANCE IN NANOFARADS.
            C
S.0001             DIMENSION CC(25),R1(5),R2(5),C1(5),C2(5),W(20),DBMAG(20),
                  1PLOT(51,101)
S.0002             READ(5,1)N,F,C
S.0003           1 FORMAT(I5,2E15.7)
S.0004             WRITE(6,2)N,F
S.0005           2 FORMAT('1LOWPASS BUTTERWORTH RC ACTIVE FILTER',///,' NUMBER OF POL
                  1ES =',I4,//,' 3-DB CUTOFF FREQUENCY =',E15.7,' KHZ',//,' C IN NANO
                  2FARADS AND R IN OHMS',//)
S.0006             PI=3.1415926
S.0007             NN=N/2
S.0008             DO 10 I=1,NN
S.0009             P=PI/2.+PI/(2.*N)*(2.*I-1.)
S.0010             RREAL=COS(P)
S.0011             AIMAG=SIN(P)
S.0012             AMAG=RREAL*RREAL+AIMAG*AIMAG
S.0013             CST=-RREAL/(2.*PI*F*1.E3*C*1.E-9)
S.0014             C1(I)=C
S.0015             CC(1)=1.0
S.0016             CC(2)=1.5
S.0017             CC(3)=2.2
S.0018             CC(4)=3.3
S.0019             CC(5)=4.7
S.0020             CC(6)=6.8
S.0021             N2=ALOG10(C)
S.0022             DO 20 J=1,100
S.0023             JJ=J-1
S.0024             DO 20 K=1,6
S.0025             KK=7-K
S.0026             C2(I)=CC(KK)*10.**(N2-JJ)
S.0027             AM=C/C2(I)
S.0028             IF(AM-AMAG/(RREAL*RREAL))20,20,30
S.0029          20 CONTINUE
S.0030          30 R1(I)=CST*AM/AMAG*(1.+SQRT(1.-AMAG/(RREAL*RREAL*AM)))
S.0031             R2(I)=CST*AM/AMAG*(1.-SQRT(1.-AMAG/(RREAL*RREAL*AM)))
S.0032          10 WRITE(6,3)I,C1(I),R1(I),C2(I),R2(I)
S.0033           3 FORMAT(' SECTION',I3,///,' C1 =',E15.7,4X,'R1 =',E15.7,//,' C2 =',
                  1E15.7,4X,'R2 =',E15.7,////)
            C
            C      CALCULATION OF THE AMPLITUDE FUNCTION.
            C
S.0034             DO 40 I=1,20
S.0035          40 W(I)=0.1*I
S.0036             WRITE(6,50)
S.0037          50 FORMAT(////,' FREQUENCY RESPONSE',///,3X,'DB LOSS',7X,'FREQUENCY')
S.0038             DO 60 I=1,20
S.0039             DBL=0.0
S.0040             DO 70 J=1,NN
S.0041             AA=(1.-R1(J)*C1(J)*R2(J)*C2(J)*1.E-12*(2.*PI*W(I)*F)**2)**2
S.0042             BB=((R1(J)+R2(J))*C2(J)*1.E-6*(2.*PI*W(I)*F))**2
S.0043          70 DBL=-10.*ALOG10(AA+BB)+DBL
S.0044             DBMAG(I)=DBL
S.0045             FREQ=W(I)*F*1.E3
S.0046          60 WRITE(6,200)DBL,FREQ
S.0047         200 FORMAT(2E15.7)
            C
            C      THE PLOT ROUTINE.
            C
S.0048             READ(5,8)PLUS,STAR,DOT,BLANK
S.0049           8 FORMAT(4A1)
            C
            C      THE GRID LINES ARE FORMED.
            C
S.0050             DO 55 I=1,51
S.0051             DO 55 J=1,101
S.0052          55 PLOT(I,J)=BLANK
S.0053             AN=0.0
S.0054             DO 65 I=1,10
S.0055             AN=AN+0.1
S.0056             K=50.*(ALOG10(AN)+1.)+1.5
S.0057             DO 65 J=1,51
S.0058          65 PLOT(J,K)=DOT
S.0059             AN=0.0
```

Figure 7.50 A program for designing Butterworth lowpass R-C active filters

```
S.0060          DO 72 I=1,10
S.0061          AN=AN+1.0
S.0062          K=50.*(ALOG10(AN)+1.)+1.5
S.0063          DO 72 J=1,51
S.0064       72 PLOT(J,K)=DOT
S.0065          DO 75 I=1,51,5
S.0066          DO 75 J=1,101
S.0067       75 PLOT(I,J)=DOT
          C
          C     THE ELEMENTS OF PLOT CORRESPONDING TO THE PLOTTED POINTS ARE DETERMINED.
          C
S.0068          DO 80 I=1,20
S.0069          J=50.*(ALOG10(W(I))+1.)+1.5
S.0070          KP=ABS(DBMAG(I))+1.5
S.0071       80 PLOT(KP,J)=STAR
          C
          C     THE GRAPH IS PRINTED.
          C
S.0072          WRITE(6,9)
S.0073        9 FORMAT('1 RC ACTIVE LOW-PASS BUTTERWORTH FILTER RESPONSE',//,9X,'
               1.1',47X,'1.0',46X,'10.0',2X,'W/WC',//)
S.0074          L=0
S.0075          F=5.0
S.0076          DF=-5.0
S.0077          DO 90 I=1,51
S.0078          IF(L-L/5*5)100,100,110
S.0079      100 F=F+DF
S.0080          WRITE(6,120)F,(PLOT(I,J),J=1,101)
S.0081      120 FORMAT(F8.1,2X,101A1)
S.0082          GO TO 90
S.0083      110 WRITE(6,130)(PLOT(I,J),J=1,101)
S.0084      130 FORMAT(10X,101A1)
S.0085       90 L=L+1
S.0086          STOP
S.0087          END
```

Figure 7.50 Continued

of the outer DO 20 loop, $J = 1$ and $JJ = 0$. The inner DO 20 loop then initially sets $K = 1$ and $KK = 6$. The value of $C2(1)$ is next computed from

$$
\begin{aligned}
\mathrm{C2(I)} &= \mathrm{CC(KK)} * 10. ** (\mathrm{N2} - \mathrm{JJ}) \\
&= 6.8 * 10. ** (1 - 0) \\
&= 68
\end{aligned}
$$

The value of M (AM for mode purposes) is found from computer statement S.0027 and this value is compared to the value of $1/\sigma_I{}^2$. If $M < (1/\sigma_I{}^2)$, the execution of the DO 20 loops is continued. Whenever a value of M is found that is greater than the value of $(1/\sigma_I{}^2)$, control is transferred to statement 30 and the value of R_1(R1(I)) and R_2 (R2(I)) is computed. Computer statements S.0032 and S.0033 are then used to print the element values for this section. The execution of the DO 10 loop is then repeated until all section element values have been found.

After all the section element values have been determined, computer statements S.0034 through S.0047 are used to compute the amplitude function. The amplitude response of the composite filter is calculated for frequency values of $W(I) = \omega/\omega_c = 0.1, 0.2, \ldots, 1.0, 1.1, \ldots,$ and 2.0. Computer statements S.0048 through S.0085 (see Section 11.4) are then used to produce the graph of the overall amplitude response.

```
LOWPASS BUTTERWORTH RC ACTIVE FILTER

NUMBER OF POLES =    8

3-DB CUTOFF FREQUENCY =   0.10000000E 01  KHZ

C IN NANOFARADS AND R IN OHMS

SECTION   1

C1 =   0.68000000E 02      R1 =   0.1957853E 05

C2 =   0.22000000E 01      R2 =   0.8648266E 04


SECTION   2

C1 =   0.68000000E 02      R1 =   0.9043543E 04

C2 =   0.15000000E 02      R2 =   0.2746005E 04


SECTION   3

C1 =   0.68000000E 02      R1 =   0.2858896E 04

C2 =   0.46999998E 02      R2 =   0.2772273E 04


SECTION   4

C1 =   0.68000000E 02      R1 =   0.5083254E 04

C2 =   0.46999998E 02      R2 =   0.1559166E 04



FREQUENCY RESPONSE

   DB LOSS            FREQUENCY
  0.5304813E-05    0.9999995E 02
  0.7152557E-05    0.1999999E 03
  0.1370907E-04    0.2999998E 03
  0.2002716E-04    0.3999998E 03
 -0.4100800E-04    0.4999998E 03
 -0.1187325E-02    0.5999998E 03
 -0.1436806E-01    0.6999995E 03
 -0.1204977E 00    0.7999995E 03
 -0.7381983E 00    0.8999995E 03
 -0.3010176E 01    0.9999995E 03
 -0.7477825E 01    0.1099999E 04
 -0.1289758E 02    0.1199999E 04
 -0.1829556E 02    0.1299999E 04
 -0.2340021E 02    0.1399999E 04
 -0.2818105E 02    0.1499999E 04
 -0.3266142E 02    0.1599999E 04
 -0.3687256E 02    0.1699999E 04
 -0.4084383E 02    0.1799999E 04
 -0.4460057E 02    0.1899999E 04
 -0.4816473E 02    0.1999999E 04
```

Figure 7.51 The output results of the program of Figure 7.50

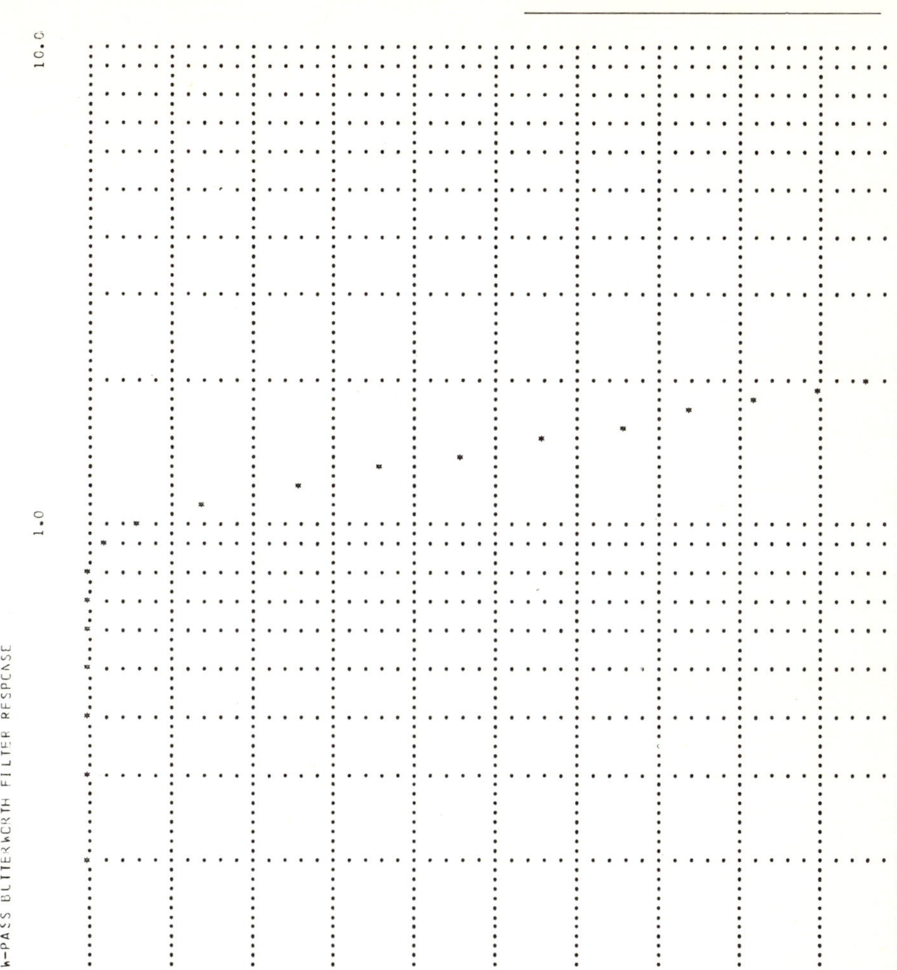

Figure 7.51 Continued

Figure 7.51 shows the computer results for the case of $N = 8$, $F =$ 1 kHz, and $C = 68\ 10^{-9}$ farads. Figure 7.52 gives the corresponding schematic diagram of the composite filter.

Chebyshev Design

The two-pole lowpass or highpass filter sections of Figure 7.48 may also be used in the design of Chebyshev filters. The element values of

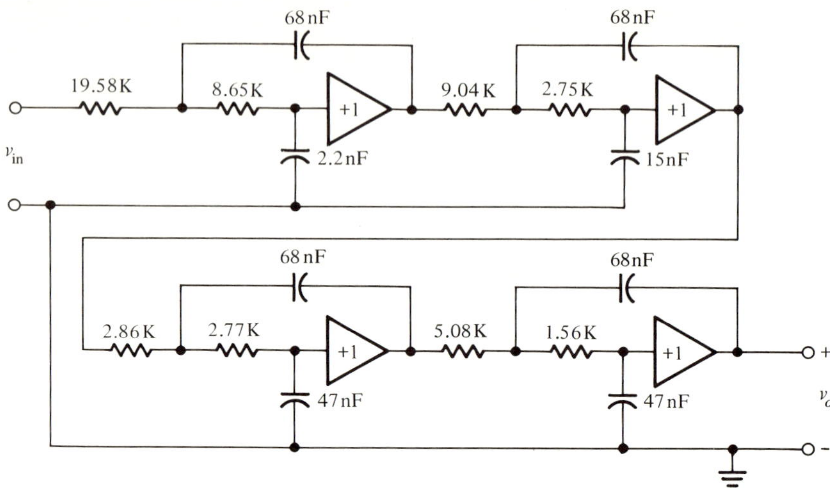

Figure 7.52 Eight-pole Butterworth lowpass active filter

each section are again based on the required locations of the poles of the overall transfer function.

For the case of a Chebyshev response having R dB ripple in the passband, the poles are required to be located on an ellipse in the normalized complex frequency plane, as shown in Figure 7.53. The semiminor and the semimajor axis of the ellipse are given by

$$X1 = \frac{D - 1/D}{2}$$

$$X2 = \frac{D + 1/D}{2}$$

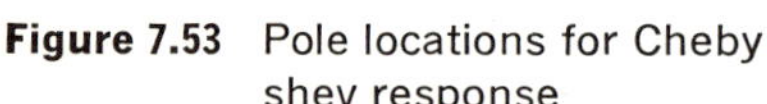

Figure 7.53 Pole locations for Chebyshev response

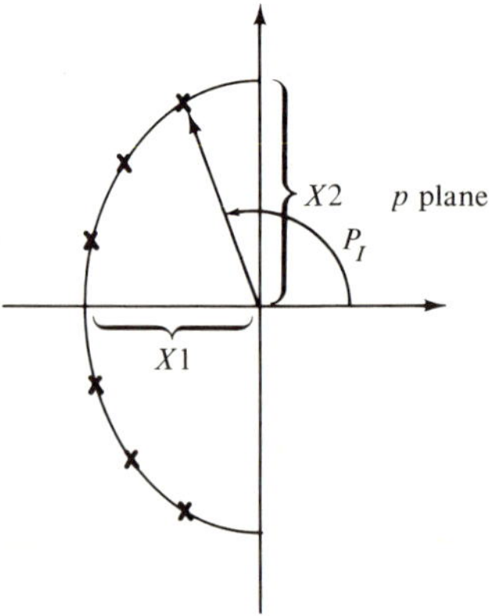

where

$$D = \left(\frac{1}{\epsilon} + \left(\frac{1}{\epsilon^2} + 1\right)^{1/2}\right)^{1/N}$$

and

$$\epsilon = (10^{R/10} - 1)^{1/2}$$

The locations of the poles are therefore given by

$$\sigma_I + j\omega_I = X1 \cos P_I + jX_2 \sin P_I$$

where

$$P_I = \frac{\pi}{2} + \frac{\pi}{2N}(2I - 1), \qquad I = 1 \text{ to } N$$

Once the location of the poles has been established, the synthesis of the individual sections is identical to the procedure used in the Butterworth case that was described above. The modification of the previous program is therefore left as an exercise for the reader.

PROBLEMS

7-1. Modify the program of Figure 7.2 so that the values of R_B and the other collector characteristic curve data points are read into the computer memory. Based on the values of V_{CC}, R_B, and V_{BE} (assume 0.6 volts for silicon) have the computer determine the base current I_B. Using this computed value, now have the program determine the two adjacent collector characteristic curves that will be used in the calculation of the operating point. For example, if I_B were computed to be equal to 0.46 mA. (see Figure 7.3), the 0.4 and the 0.5 mA. characteristic curves would be used separately to determine the respective operating points for a value of I_B equal to 0.4 and 0.5 mA. The actual operating point should then be found by linearly interpolating between these two points.

Test the above program by using the same data that was used in Figures 7.3 and 7.5, but with an R_B value equal to

a. 20 kilohms b. 50 kilohms c. 100 kilohms

7-2. Modify the program of Problem 7-1 so that you can determine the change in the operating point due to a change in the values of V_{CC}, R_B, and R_C. Determine the worst case variation for a $\pm 10\%$ change in

$$V_{CC} = 20.0,\ R_B = 50{,}000,\ \text{and } R_C = 500$$

7-3. Using the collector characteristics shown in Figure 7.3, write a Fortran program that computes

a. The incremental current gain $\left.\frac{\Delta I_C}{\Delta I_B}\right|_{V_{CE}=\text{cst}}$

b. The incremental output impedance $\left.\frac{\Delta V_{CE}}{\Delta I_C}\right|_{I_B=\text{cst}}$

Test your program for the following operating points

	V_{CE} (volts)	I_B (mA)
a.	−10.0	−0.40
b.	−17.0	−0.15
c.	−5.0	−0.80

7-4. Modify the program of Figure 7.2 so that it is capable of determining the operating point of the circuit of Figure 7.7c. Determine the worst case variation for a ±10% change in

$$V_{CC} = 20.0,\ R_F = 20{,}000, \text{ and } R_C = 470$$

7-5. Repeat Problem 7-4 for the circuit of Figure 7.7b. Use a value of $R_E = 330$ ohms.

7-6. Choose suitable resistor values and use the following collector characteristics to determine the stability of the operating point as a function of temperature for the circuits shown in
a. Figure 7.1
b. Figure 7.7c
c. Figure 7.7b
Assume that the resistor values and the base-to-emitter voltage remain constant with temperature changes. (See Problems 7-4 and 7-5.) What conclusions can you draw?

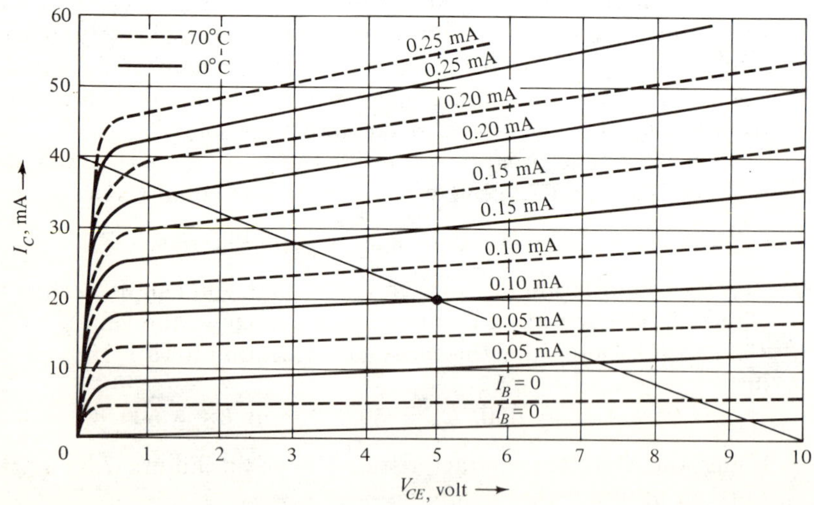

7-7. Rerun the program of Figure 7.9 for each of the biasing circuits shown in Figures 7.6 and 7.7b,d,e. In each case choose suitable values of resistors so that the operating point is essentially the same. Use the same variations in β, V_{BE}, and I_{CEO} that were used in Figure 7.10 and in addition investigate the change in the operating point due to a $\pm 5\%$ change in the supply voltage. What conclusions can you draw?

7-8. The nominal input resistance, output resistance, voltage gain, and current gain of a single-stage common-base transistor amplifier can be computed from the following relations:[10]

$$R_i = r_e + r_b \frac{r_c(1 - \alpha_0) + R_L}{r_b + r_c + R_L} \qquad R_o = r_c - r_b \frac{\alpha_0 r_c - r_e - R_s}{r_e + r_b + R_s}$$

$$A_i = \frac{\alpha_0 r_c + r_b}{r_c + r_b + R_L} \qquad A_v = \frac{(\alpha_0 r_c + r_b)R_L}{r_b((1 - \alpha_0)r_c + r_e + R_L) + r_e(r_c + R_L)}$$

Write an appropriate Fortran program to compute the above quantities. Test your program using the following input data:

$$r_e = 50 \text{ ohms} \qquad r_b = 150 \text{ ohms} \qquad r_c = 10^6 \text{ ohms}$$
$$\alpha_o = 0.98 \qquad R_L = 5 \text{ kilohms} \qquad R_s = 1 \text{ kilohm}$$

7-9. Repeat Problem 7-8 for the common-collector transistor stage where

$$R_i = r_b + r_c \frac{R_L + r_e}{(1 - \alpha_0)r_c + R_L + r_e}$$

$$R_o = r_e + \frac{(1 - \alpha_0)r_c(R_s + r_b)}{R_s + r_b + r_c}$$

$$A_v = \frac{r_c R_L}{r_b((1 - \alpha_0)r_c + R_L + r_e) + r_c(R_L + r_e)}$$

$$A_i = \frac{r_c}{r_c(1 - \alpha_0) + R_L + r_e}$$

7-10. Repeat Problem 7-8 for the common-emitter transistor stage where

$$R_i = r_b + r_e \frac{r_c + R_L}{r_c(1 - \alpha_0) + R_L + r_e}$$

$$R_o = r_c(1 - \alpha_0) + r_e \frac{R_s + r_b + \alpha_0 r_c}{R_s + r_b + r_e}$$

$$A_v = \frac{-R_L(\alpha_0 r_c - r_e)}{r_b(r_c(1 - \alpha_0) + r_e + R_L) + r_e(r_c + R_L)}$$

$$A_i = \frac{r_e - \alpha_0 r_c}{r_c(1 - \alpha_0) + r_e + R_L}$$

[10] See M. S. Ghausi, *Principles and Design of Linear Active Circuits*, pp. 281–83.

7-11. Rerun the program of Figure 7.13 to see whether it is possible to operate the 2N3646 transistor flip-flop over the 0° to 65° C temperature range using 20 percent tolerance resistors, 10 percent tolerance supply voltage, and an R_B value of 560 kilohms and an R_K value of 22 kilohms.

7-12. Repeat run #5 in Table 7.1 to determine the largest power supply variation that can be used with $R_B = 500$ kilohms and $R_K = 15$ kilohms.

7-13. Explain why the resistance tolerance variation (see Figure 7.17) has a much smaller effect when 2N3639 transistors are used instead of 2N3646 transistors (see Figure 7.15).

7-14. Rerun the program of Figure 7.13 to obtain additional information that is needed in specifying an optimum R_B-R_K resistance combination for both the 2N3646 and 2N3639 transistors. Assume that the operation will be over the 0° to 65° C temperature range. Give the reasons for your final choice.

7-15. Derive the on and off equations for the worst case single supply common-emitter flip-flop.

7-16. Rerun the program of Figure 7.19 to determine the effects of resistance tolerance, supply voltage tolerance, and temperature changes. Determine an optimum circuit and state the reasons for your choice of element values and tolerances.

7-17. Modify the program of Figure 7.19 so that it also computes the collector voltage swing.

7-18. Using hand computation, verify the results shown in Figure 7.25a.

7-19. Using the program of Figure 7.24, obtain data and plot the normalized gain-bandwidth product ratio versus the normalized bandwidth $\omega_{3dB}/\omega_\beta$ ($\omega_\beta = \omega_t/\beta$) for A_o values ranging from 5 to 25.

7-20. Rerun the program of Figure 7.24 and investigate the effects of other parameter variations. State your conclusions.

7-21. Explain why it is not possible to obtain a direct solution in the design of a shunt-peaked amplifier when the bandwidth is specified. Is this true for the corresponding vacuum-tube case?

7-22. Mathematically justify the assumption that the gain-bandwidth product is equal to ω_t.

7-23. Derive the inductance equation used in the design of the shunt-peaked amplifier.

7-24. Write an appropriate program for obtaining the data for plotting the frequency response of the shunt-peaked amplifier of Figure 7.22.

7-25. Make the necessary changes in the design equations and modify the program of Figure 7.24 so that the value of the load resistance may be specified independently of the resistor in series with the inductor L.

7-26. Determine the number of iterations for convergence if no speed-up factor is used in the program of Figure 7.26. In addition, investigate other speed-up factor choices.

7-27. Repeat Problem 7-26 using the results of Figure 7.25b as input data.

7-28. Devise other means of speeding up the convergence of the program of Figure 7.26.

7-29. Write a program for determining the minimum sensitivity and the output voltage swing of the current mode Schmitt of Figure 7.29.

7-30. Derive the voltage mode Schmitt equations for R_{L1}, R_{K2}, and R_X.

7-31. Rerun the program of Figure 7.33 and determine the effects of a 10 percent resistance tolerance and a 5 percent supply voltage tolerance on the resulting design.

7-32. Write a Fortran program that determines the switching voltages of a voltage mode Schmitt.

7-33. Investigate the temperature stability of a voltage mode Schmitt trigger circuit (see Figure 7.36).

7-34. Rerun the program of Figure 7.39 using an f_t value of 0.80E+09. Compare these results with those of Figure 7.40.

7-35. Modify the program of Figure 7.39 so that the circuit can be designed based on specifying the value of $\omega_{3\text{dB}}$ rather than the nominal gain.

7-36. Write an appropriate program for obtaining the frequency response of the shunt-series feedback pair of Figure 7.42.

7-37. Using the program of Problem 7-36, plot the effects of varying C_F for the input data of Figure 7.47a.

7-38. Rerun the program of Figure 7.46 and investigate the effects of other parameter variations. State your conclusions.

7-39. Modify the program of Figure 7.46 so that the circuit can be designed in terms of the value of $\omega_{3\text{dB}}$ rather than the nominal gain.

7-40. Using the program of Figure 7.50 design a 6-pole Butterworth lowpass filter that has a cutoff frequency of 10 kHz and uses a maximum capacitor size of 33 nF.

7-41. Modify the program of Figure 7.50 so that it will be able to compute the section element values for lowpass or highpass active filters. Hint: Use a value of $K = 1$ to represent that a lowpass design is required and $K = 2$ to correspond to a highpass design. Read the appropriate value of K as part of your data.

7-42. Test the program of Problem 7-41 by designing a highpass filter that has a cutoff frequency of 10 kHz and uses a maximum capacitor size of 33 nF.

7-43. Repeat Problem 7-41 for the Chebyshev case.

7-44. Test your program of Problem 7-43 by using it to design:
   a. Lowpass filter, $f_c = 10$ kHz, $C = 33$ nF, and $R = 3$ dB.
   b. Highpass filter, $f_c = 10$ kHz, $C = 33$ nF, and $R = 3$ dB.

CHAPTER 8

Subprograms

8.1 INTRODUCTION

In previous chapters we have concentrated on writing Fortran programs that were designed to solve specific problems. In the course of our work, we have found that in writing new programs it was often necessary to repeat the type of computation that was required in some earlier program. For example, computer statements S.0009 through S.0022 in the Gauss elimination program of Figure 6.7 were also used in the Gauss-Jordan program of Figure 6.13. It would thus greatly simplify the writing of new programs if we could make free use of parts of previous programs.

This simplification may be accomplished by incorporating the required statements as part of a subprogram. Therefore, instead of writing out the required statements each time they are to be used, the routine can be used any number of times within the same program by simply supplying the appropriate arguments and calling on the subprogram by name. It is thus possible to extend this technique to a very long and involved program by breaking it into several smaller ones, each of which is given a name and written as a subprogram, and then using a short calling program to call in the appropriate subprograms. In addition to the advantage of permitting several people to be working at the same time, each person writing one or more of the subprograms, this technique also allows each of the subprograms to be checked out individually before it is combined, and thus greatly simplifies the debugging problem.

At this point the reader should realize that the use of subprograms is not a new idea, for we already have been making free use of Fortran-supplied subprograms such as SIN(X), COS(X), ATAN(X), and so on (see Section 2.3). What is a new idea is the ability of the programmer to incorporate subprograms of his own design and use these as freely as the Fortran-supplied subprograms.

The remainder of this chapter is used to discuss the three different kinds of subprograms that may be used with Fortran: arithmetic statement functions, function subprograms, and subroutine subprograms. The first two of these classes are function subprograms and return only a single result to the calling program. The subroutine subprogram, however, is a multivalued function and is capable of returning several different results to the calling program. In addition to our discussion of the three types of subprograms, we will also include a discussion of the use of some of the subroutines of the System 360 Scientific Subroutine Package.

8.2 ARITHMETIC STATEMENT FUNCTIONS

The form of the arithmetic statement function is

$$\text{NAME}\ (A_1, A_2, \ldots, A_N) = \text{Arithmetic Expression}$$

where NAME is the calling name of the function and $A_1, A_2, \ldots, A_N$ are the dummy arguments of the function. The NAME may contain from one through six alphabetic or numeric characters, but the first one must be alphabetic. In the predefined specification, the Fortran translator recognizes that the mode of the value returned should be integer if the NAME begins with the letters I, J, K, L, M, or N; otherwise the mode of the value returned is real. There must also be at least one argument and each argument must be a nonsubscripted variable name. The arithmetic expression must not include any subscripted variables and must contain all of the dummy variables enclosed by the parentheses.

Note that the arithmetic statement function is a one-statement subprogram and may be used only in the program where it is defined. Since the arithmetic statement function serves only to define the arithmetic operations involved in evaluating the function and does not actually involve any computation, it must be thought of as a declaration type of statement and therefore must not be labeled.

The arithmetic statement function is called in the same way that we call Fortran library functions. The name of the function is followed by parentheses that enclose the actual (not dummy) arguments and this is included in one or more of the arithmetic expressions of the program. The actual arguments of the calling statements may be constants, subscripted or nonsubscripted variables, arithmetic expressions, or the name of some other subprogram. The dummy arguments are replaced by the actual arguments during execution and must agree in number, order, and mode with the actual arguments of each calling statement.

```
          C        THE ARITHMETIC FUNCTION IS USED IN A GENERAL SIMPSON'S RULE
          C        PROGRAM FOR INTEGRATING A SPECIFIED FUNCTION F(X).
          C
          C           XLOWER = LOWER LIMIT OF INTEGRATION.
          C           XUPPER = UPPER LIMIT OF INTEGRATION.
          C           NUMB = THE NUMBER OF INTEGRATION INTERVALS. MUST BE EVEN.
          C
S.0001             F(X)=EXP(-X)
S.0002             XLOWER=0.0
S.0003             XUPPER=1.0
S.0004             NUMB=100
S.0005             DX=(XUPPER-XLOWER)/NUMB
S.0006             SUM=0.0
S.0007             X=XLOWER
S.0008             DO 10 I=1,NUMB,2
S.0009             SUM=SUM+F(X)+4.*F(X+DX)+F(X+2.*DX)
S.0010          10 X=X+2.*DX
S.0011             AREA=DX*SUM/3.
S.0012             WRITE(6,1)XLOWER,XUPPER,AREA
S.0013           1 FORMAT('1AREA UNDER F(X), BETWEEN  X =',E15.7,'  AND  X =',E15.7,/
                  1/,' AREA =',E15.7)
S.0014             STOP
S.0015             END
```

Figure 8.1 An arithmetic statement function is used in a general integration routine

Figure 8.1 is a simple numerical example that illustrates the use of an arithmetic statement function. This program is a general integration routine that employs Simpson's rule to integrate the function F(X). Initially we specify the function F(X) by means of the arithmetic statement function shown in computer statement S.0001. We next specify the lower limit of integration, the upper limit of integration, and the number of integration intervals, NUMB. Since Simpson's rule is used, the value of NUMB must be an even number. The integration interval, DX, is then computed by computer statement S.0005, and this, in turn, is followed by setting SUM equal to zero and X equal to XLOWER. The DO 10 loop and computer statement S.0011 are then used to compute the value of the integral. Note that the function F(X) as specified in computer statement S.0001 is evaluated for three different arguments in computer statement S.0009. Note also that because we have used the arithmetic statement function (S.0001), it is not necessary to write the function out completely each time it is evaluated. After the value of the integral has been found, computer statements S.0012 and S.0013 are used to print the lower and upper limits of integration and the numerical value obtained for the integral.

Figure 8.2 shows the computer result. An examination of this result will show that it is quite accurate, since the numerical value of this integral to ten significant figures is equal to 0.6321205588. Note also that

```
AREA UNDER F(X), BETWEEN  X =   0.0              AND  X =   0.1000000E 01

AREA =   0.6321169E 00
```

Figure 8.2 The output results of the program of Figure 8.1

other functions may be similarly integrated by simply changing the right member terms of computer statements S.0001 through S.0004.

Frequency Response of Modified Shunt-Peaked Network

For a second example that illustrates the use of arithmetic statement functions, consider the modified shunt-peaked network shown in Figure 8.3. This three-reactance two-terminal network permits two degrees of correction in the frequency response. The values of L and C_1 may thus be chosen so that both corrections are made to reduce either the frequency distortion (amplitude function) or the phase distortion (phase function), or one correction can be used to reduce the frequency distortion and the other correction to reduce the phase distortion.

One method of correction is the derivative adjustment technique,[1] which is based on the properties that all amplitude functions (magnitude response) are even functions and all phase functions (phase response) are odd functions. Thus, as far as the magnitude and phase responses are concerned, all odd derivatives of the magnitude response and all even derivatives of the phase response are identically equal to zero at zero frequency (midband). Setting successive even derivatives of the magnitude response equal to zero will thus flatten out the magnitude response and reduce the frequency distortion in the vicinity of the origin. Similarly, since the phase function is an odd function, its first derivative will be an even function and, therefore, setting successive even derivatives of $d\theta/d\omega$ equal to zero will produce a linear phase response (minimum phase distortion).

A practical scheme for correcting the magnitude response, which avoids the need for finding successive derivatives, is based on finding

[1] W. Lynch, *The Role Played by Derivative Adjustment in Broadband Amplifier Design*, Proceedings of Symposium on Modern Network Synthesis, Brooklyn Polytechnic Institute, 1952.

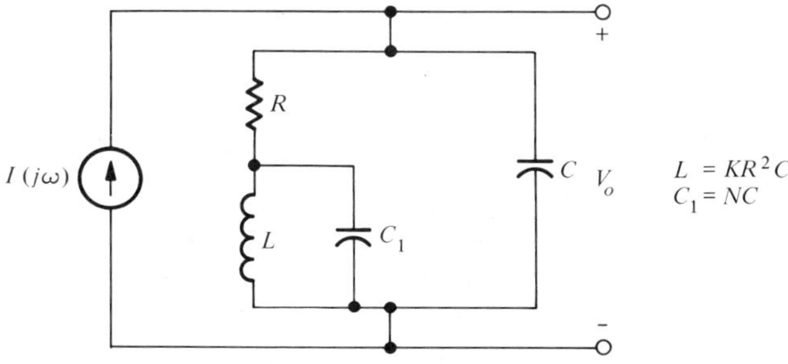

Figure 8.3 Modified shunt-peaked network

the square of the normalized magnitude function response. In the case of the network of Figure 8.3, we would first write the normalized output response

$$H(\phi) = \frac{V_o(\phi)}{V_o(0)} = \frac{Z_{dp}}{R}$$

$$= \frac{1 - KN\phi^2 + jK\phi}{1 - (1 + N)K\phi^2 + j\phi(1 - KN\phi^2)}$$

where

$$\phi = RC\omega = \frac{\omega}{\omega_1}$$

$$K = \frac{L}{R^2C}$$

and

$$N = \frac{C_1}{C}$$

The square of the magnitude of the amplitude function would then be given by

$$|H(\phi)|^2 = \frac{1 + (K^2 - 2NK)\phi^2 + N^2K^2\phi^4}{\left(\begin{array}{r}1 + (1 - 2K(1 + N))\phi^2 + (K^2 + 2NK^2 + N^2K^2 - 2NK)\phi^4 \\ + N^2K^2\phi^6\end{array}\right)}$$

The scheme now requires that we equate n pairs of coefficients of like order, beginning with the lowest, to force the first $2n + 1$ derivatives of the magnitude function ($|H(\phi)|$) equal to zero, thus

$$K^2 - 2NK = 1 - 2K(1 + N)$$

and

$$N^2K^2 = K^2 + 2NK^2 + N^2K^2 - 2NK$$

requires

$$K_a = \sqrt{2} - 1 = 0.414$$

and

$$N_a = \frac{1}{2\sqrt{2}} = 0.3535$$

The values of L and C_1 would therefore be made equal respectively to 0.414 R^2C and 0.3535 C. This adjustment would produce the flattest monotonic magnitude response (maximally flat response).

The phase response may similarly be corrected without the calculation of successive derivatives. This is done by first writing the phase function in the following general form

$$\theta(\phi) = \tan^{-1}(a_1\phi + a_3\phi^3 + a_5\phi^5 + a_7\phi^7 + \cdots)$$

First-degree correction is then found by forcing a_3 to be equal to

$$a_3 = \frac{1}{3} a_1^3,$$

second-degree correction results by forcing a_5 to be equal to

$$a_5 = \frac{2}{15} a_1^5,$$

third-degree correction results by forcing a_7 to be equal to

$$a_7 = \frac{17}{315} a_1^7, \text{ and so on}$$

Applying this to the network of Figure 8.3, we find

$$\theta(\phi) = -\tan^{-1}\left[\frac{(1-K)\phi - (2NK - K^2(N+1))\phi^3 + N^2K^2\phi^5}{(1-NK\phi^2)^2}\right]$$

or

$$\begin{aligned}\theta(\phi) &= -\tan^{-1}((1-K)\phi + K^2(1-N)\phi^3 + NK^3(2-N)\phi^5 + \cdots)\\ &= -\tan^{-1}(a_1\phi + a_3\phi^3 + a_5\phi^5 + \cdots)\end{aligned}$$

Two degrees of correction of the phase function are thus found by setting a_3 equal to $a_1^3/3$ and a_5 equal to $2a_1^5/15$. The solution of these two equations yields

$$\begin{aligned}K_p &= 0.35\\ N_p &= 0.22\end{aligned}$$

and thus the network requires $L = 0.35\ R^2C$ and $C_1 = 0.22\ C$ to produce a maximally flat delay.

Since two degrees of correction are possible with the modified shunt-peaked network, it is possible to produce partial correction in both the magnitude and phase function. The values of L and C_1 for partial correction are found from solving the equations that result from one degree of correction in both the magnitude and phase functions. Thus, from the magnitude function square we have

$$K^2 - 2NK = 1 - 2K(1 + N)$$

and from the phase response we have

$$K^2(1 - N) = \frac{1}{3}(1 - K)^3$$

Solving these two equations yields

$$K_{ap} = 0.414$$
$$N_{ap} = 0.608$$

Figure 8.4 is a Fortran program for finding the magnitude and phase response of the modified shunt-peaked network for the three adjustments described above. Computer statements S.0001 through S.0004 are arithmetic function statements that respectively define the real and imaginary terms of the numerator and denominator of $H(\phi)$. Computer statements S.0005 and S.0006 then print the output heading. Computer statements S.0007 and S.0008 now read in the values of AN and AK (N and K) and these values are then printed by computer statements S.0009 and S.0010. The dummy frequency variable W is then set equal to zero and the DO 10 loop is entered so that the magnitude and phase response can be computed for 20 normalized frequency values. During each iteration of the DO 10 loop, W is set equal to $W + 1.0$ and ϕ is made equal to $W/10.0$. This corresponds to normalized frequency values running from 0.1 to 2.0. The amplitude function is next calculated by computer statement S.0015, and the phase function is computed by computer statements S.0016 through S.0022. The values of PHI, AMP, and PHASE are then printed

```
               C        FREQUENCY RESPONSE OF MODIFIED
               C        SHUNT PEAKED NETWORK.
               C
S.0001                  A(PHI)=1.-AK*AN*PHI**2
S.0002                  B(PHI)=AK*PHI
S.0003                  C(PHI)=1.-(1.+AN)*AK*PHI**2
S.0004                  D(PHI)=PHI*(1.-AK*AN*PHI**2)
S.0005                  WRITE(6,1)
S.0006                1 FORMAT('1MODIFIED SHUNT PEAKED NETWORK FREQUENCY RESPONSE',//)
S.0007                2 READ(5,3)AK,AN
S.0008                3 FORMAT(2E15.7)
S.0009                  WRITE(6,4)AN,AK
S.0010                4 FORMAT(//,' AN =',E15.7,' AK =',E15.7,//,4X,'FREQUENCY',2X,'MAGNIT
                       1UDE',8X,'PHASE',//)
S.0011                  W=0.0
S.0012                  DO 10 I=1,20
S.0013                  W=W+1.0
S.0014                  PHI=W/10.0
S.0015                  AMP=SQRT((A(PHI)**2+B(PHI)**2)/(C(PHI)**2+D(PHI)**2))
S.0016                  PHASEN=ATAN(B(PHI)/A(PHI))*57.2957795
S.0017                  IF(A(PHI))5,6,6
S.0018                5 PHASEN=PHASEN+180.0
S.0019                6 PHASED=ATAN(D(PHI)/C(PHI))*57.2957795
S.0020                  IF(C(PHI))7,8,8
S.0021                7 PHASED=PHASED+180.0
S.0022                8 PHASE=PHASEN-PHASED
S.0023               10 WRITE(6,9)PHI,AMP,PHASE
S.0024                9 FORMAT(3E15.7)
S.0025                  GO TO 2
S.0026                  END
```

Figure 8.4 Arithmetic statement functions are used to find the frequency response of the modified shunt-peaked network

and the calculations are continued until all 20 normalized frequency values have been found. Control is then transferred to statement 2 so that the frequency response for the next set of K and N values can be computed.

Figure 8.5 is the corresponding computer output. Figure 8.6 is the corresponding plot of the three magnitude responses. Note that the flattest response results (minimum frequency distortion) when both corrections are used to improve the magnitude function. Note also that for the case of one correction in the magnitude response and one correction in the phase response, we produce a flatter magnitude response (but not monotonic) than for the case where both corrections are made in the phase function.

```
MODIFIED SHUNT PEAKED NETWORK FREQUENCY RESPONSE


AN =   0.3534999E CO AK =   C.4140000F CO

   FREQUENCY  MAGNITUDE          PHASE


 0.9999996E-01   0.9999968E CO -0.3360043E 01
 0.2C00C0CF CC   0.9999871E CO -0.6735420E 01
 C.3C00CCCE CC   0.9999650E CO -0.1014259E 02
 0.4C00CC0E CC   0.9999C76E CO -C.1360C30E 02
 0.5CC00CCE CC   0.9997539E CO -0.1713C40E 02
 0.6C00CCCE CC   0.9993746E CO -C.2075865E 02
 0.7C00C0CE CC   C.9985302E CO -0.2451468E 02
 0.8C000CCE CC   0.9968201F CO -0.2843120E 02
 0.9C00CCCE OC   C.9936245F CO -0.3254214E 02
 0.1CC0CC0E C1   0.988C539E CO -C.3687811E 02
 0.1099999F C1   C.9789283E CO -C.4145972E 02
 0.1200CCCE C1   0.9648290E CO -0.4628787E 02
 0.1299999E C1   0.9442695E CO -0.5133202E 02
 0.1400CC0E C1   C.916C1C4E CO -0.5652037E 02
 0.1500CCCE C1   C.8794748E CC -C.6173598E 02
 0.1599999E C1   C.8351063E CC -C.6682487E 02
 0.1700CC0E 01   0.7844712E 00 -0.7161751E 02
 0.1799999F C1   C.7299988F CC -0.7595692E 02
 0.190UCCCE 01   C.6744351E CO -C.7972511E 02
 0.2C00C0OE 01   C.6202856F CO -0.8285698E 02

AN =   0.22CCCCCE CO AK =   C.35CCC0CE CO

   FREQUENCY  MAGNITUDE          PHASE


 0.9999996E-01   C.9991127E CO -0.3724449E 01
 0.2C00C0CF OC   0.9964467F CC -C.7450285E 01
 0.3C00CCCE CC   0.9919944E 00 -0.1117881E 02
 0.4C00CCCE CC   C.9857410E 00 -C.1491129E 02
 0.5CC0C0OE CC   C.9776643E CO -C.1864873E 02
 0.6C00C0OE CC   C.9677334E CO -C.2239180E 02
 0.7C00CC0F OC   0.9559106E CO -0.2614044E 02
 0.8C00CC0E OC   0.9421502E CO -C.2989360E 02
 0.9C00CCCF OC   C.9264047E CC -0.3364870E 02
 0.1C000CCF C1   C.9086291F CC -C.3740121E 02
 0.1099999E 01   C.8887922E CO -0.4114418E 02
 0.12C0CCCF 01   C.8668868E CO -0.4486758E 02
 0.1299999E 01   0.8429433E CO -0.4855811E 02
 0.14C0CC0E 01   0.8170428E CC -0.5219919E 02
 0.1500000F C1   C.7893298F CC -C.5577092E 02
 0.1599999E C1   C.76CC169E CO -C.5925069E 02
 0.1700CCCF 01   C.7293845E 00 -0.6261420E 02
 0.1799999E 01   C.6977726E CO -C.6583655E 02
 0.1900CC0E C1   C.6655647E CO -0.6889355E 02
 0.2CC0CCCE C1   0.6331675E CO -0.7176326E 02

AN =   0.608CCCCE CO AK =   0.4140C00E CO

   FREQUENCY  MAGNITUDE          PHASE


 0.9999996E-01   C.1CCCCC3E 01 -C.3357565E 01
 0.2C00CCCE CC   C.1CCCC86E C1 -C.6716053E 01
 0.3C00CCCE CC   C.1CCC463E C1 -C.1CC7994E 02
 0.4C00CCCE CO   C.1CC1483E 01 -0.1346106E 02
 0.5C00CCCE CC   C.1CC3593E C1 -C.1688290E 02
 0.6C00C00E CC   C.10073C6E C1 -C.2038582E 02
 0.7C00C00E CC   C.1C13114E C1 -C.2403316E 02
 0.8000CC0E CC   C.1021338E C1 -C.2791879E 02
 0.9C00CCCE CC   C.1031835E C1 -C.3217471E 02
 0.1C00C00E 01   C.1043456E 01 -0.3697375E 02
 C.1099999E 01   0.1053163E C1 -C.4251639E 02
 0.1200CCCE 01   C.105493CE 01 -C.4897667E 02
 0.1299999E C1   C.1C39491E C1 -C.5637749E 02
 0.1400CC0F C1   C.9971567E CC -0.6441133E 02
 0.1500CCCE 01   0.9246718E CC -C.7235620E 0?
 0.1599999E C1   C.6304147E CO -0.7929257F 02
 0.1700CCCE 01   C.730212CE CO -C.8453558E 02
 0.1799999E 01   C.6378250E CO -0.8788983F 02
 0.1900C00E 01   0.560C685E CO -0.8957326F 02
 0.2C00CCCE C1   C.4979414E CC -0.8999797E 02

IHC217I
```

Figure 8.5 The output response of the program of Figure 8.4

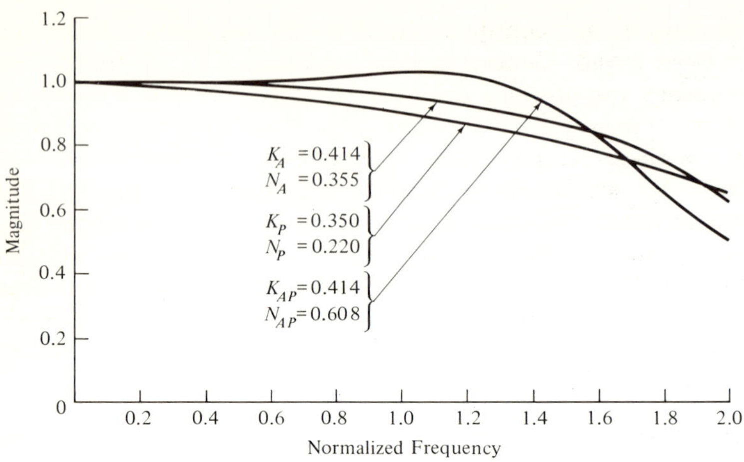

Figure 8.6 Modified shunt-peaked network magnitude responses

Figure 8.7 is the corresponding plot of the phase response. Note how linear the delay is when both corrections are made to reduce phase distortion.

At this point the reader should realize that because of our use of arithmetic statement functions, the program in Figure 8.4 may be used to obtain the frequency response of a general transfer function. This follows from the fact that a general transfer function may be written in the following form

$$T(j\omega) = \frac{A(\omega) + jB(\omega)}{C(\omega) + jD(\omega)}$$

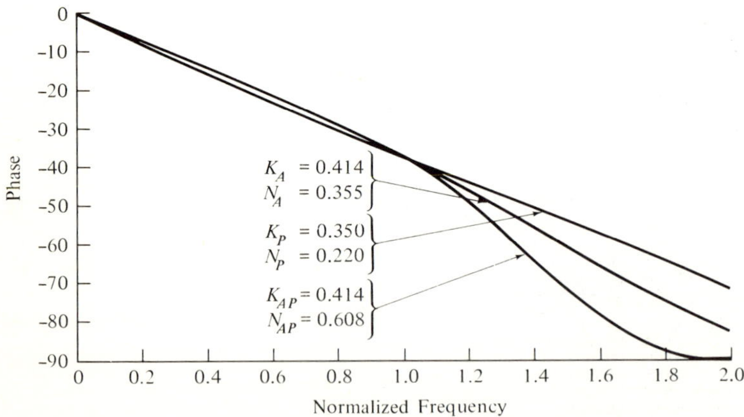

Figure 8.7 Modified shunt-peaked network phase response

where

$$\begin{aligned} A(\omega) &= a_o + a_2\omega^2 + a_4\omega^4 + \cdots \\ B(\omega) &= b_1\omega + b_3\omega^3 + b_5\omega^5 + \cdots \\ C(\omega) &= c_o + c_2\omega^2 + c_4\omega^4 + \cdots \\ D(\omega) &= d_1\omega + d_3\omega^3 + d_5\omega^5 + \cdots \end{aligned}$$

It is therefore a simple matter to use arithmetic statement functions (computer statements S.0001 through S.0004) to define these terms so that the magnitude and phase response may be computed over the desired frequency range.

8.3 FUNCTION SUBPROGRAMS

In the last section we pointed out that the arithmetic statement function was a one-statement subprogram that could be used only in the program where it was defined. A function subprogram must therefore be used when the particular function cannot be defined by a single arithmetic statement. The function subprogram is thus a separate program that defines a group of statements that can be called on by the main program, or some other subprogram, to evaluate the particular function that it defines.

Since the function subprogram is a separate or independent program, it must be combined with other programs before it can be used. The function subprogram may be combined with other programs by placing its punch cards after the END statement of the main (or calling) program or the END statement of some other function subprogram or subroutine subprogram, and the overall deck is processed as a single source program.

The general form of the function subprogram is

```
FUNCTION NAME (A1, A2, . . . , AN)
                  .
                  .
                  .
RETURN
END
```

where NAME is the calling name of the function and $A_1, A_2, \ldots, A_N$ are the arguments of the function. The actual NAME of the function is arbitrary, but it must be chosen according to the Fortran rules that are used in naming other functions or variables. There must also be at least one argument and if there is more than one, the arguments must be separated by commas. The arguments may be considered to be dummy variables and may be either nonsubscripted variable names, array names (without subscripts), or names of subroutine subprograms or other function subprograms. During execution, these dummy arguments are replaced by the actual arguments of the calling statement and the actual arguments must agree in number, order, and mode with the dummy arguments. The actual arguments of the calling statement may be any type of subscripted or nonsubscripted variable, arithmetic expression, or the name of some other subprogram. When the dummy argument is an array name, an appropriate DIMENSION statement must be used in the subprogram as well as in the calling program, and the size of these arrays must be the same except where adjustable dimensions are used.

There must also be at least one RETURN statement in each function subprogram. The RETURN statement is the last logical statement of a subprogram and signifies that the computation of the function has been completed and control should be returned to the calling program.

Physically the END statement is the last statement of the subprogram. The END statement is used to inform the Fortran complier that the end of the subprogram has been reached.

Function subprograms are called in the same way that arithmetic statement functions are called. The function is called by writing the name of the subprogram, followed by parentheses that enclose the actual arguments, in some arithmetic expression.

Inverse Sine and Inverse Cosine Functions

For a simple numerical example that illustrates the use of the function subprogram, consider the calculation of the ARC SIN (X) or the ARC COS (X). Figure 8.8 shows how these two functions may be obtained from the ATAN function. An examination of Figure 8.8a shows that

$$\text{SIN}\ \alpha = X$$

or

$$\alpha = \text{SIN}^{-1}\ (X)$$

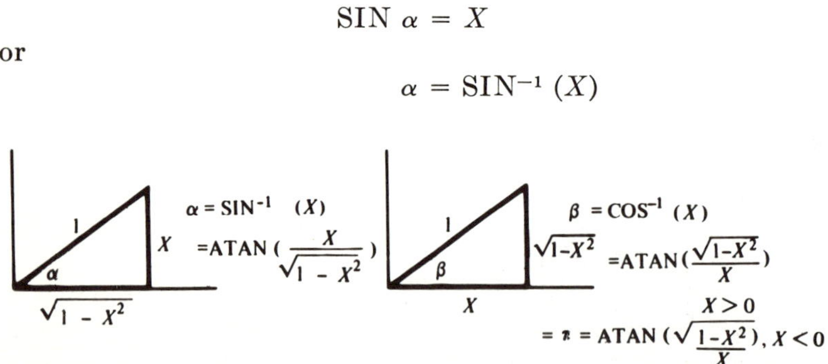

Figure 8.8 The inverse sine and the inverse cosine are defined in terms of the inverse tangent

Since α is also given by

$$\alpha = \text{ATAN}\left(\frac{X}{\sqrt{1 - X^2}}\right)$$

we see that we may write

$$\text{ARC SIN } (X) = \text{ATAN}\left(\frac{X}{\sqrt{1 - X^2}}\right)$$

Similarly,

$$\text{ARC COS } (X) = \text{ATAN}\left(\frac{\sqrt{1 - X^2}}{X}\right)$$

where X^2 is less than or equal to 1.

Figure 8.9 shows a Fortran program that includes a main or calling program for testing the ARCSIN function subprogram. Initially the main program prints the heading and subheading. This is followed by computer statement S.0003 which reads in the value of the argument X. The dummy variable W is then used to obtain the ARC SIN (X). Note that the right member of computer statement S.0005 contains the name of the function subprogram, ARCSIN, followed by the argument X in parenthesis. This statement thus calls on the function ARCSIN to evaluate the inverse sine of X. Control is therefore transferred to the function ARCSIN subprogram where the dummy argument X in the subprogram is, in this case, also the name of the actual argument X in the main program.

```
              C        TEST PROGRAM FOR FUNCTION ARC SIN (X) SUBPROGRAM.
              C        X**2 LESS THAN OR EQUAL TO 1.
              C
S.0001                 WRITE(6,1)
S.0002                1 FORMAT('1THE CALCULATION OF THE ARC SIN (X)',///,7X,'X',9X,'ARC SI
                       1N (X)',6X,'DEG SIN',//)
S.0003                2 READ(5,3)X
S.0004                3 FORMAT(E15.7)
S.0005                 W=ARCSIN(X)
S.0006                 DEGSIN=W*57.2957795
S.0007                 WRITE(6,4)X,W,DEGSIN
S.0008                4 FORMAT(3E15.7)
S.0009                 GO TO 2
S.0010                 END

S.0001                 FUNCTION ARCSIN(X)
S.0002                 IF(X)1,2,1
S.0003                2 ARCSIN=0.0
S.0004                 RETURN
S.0005                1 Z=ABS(X)
S.0006                 W=1-Z**2
S.0007                 IF(W)4,5,4
S.0008                5 ARCSIN=1.5707963
S.0009                 GO TO 6
S.0010                4 ARCSIN=ATAN(Z/SQRT(1.-Z**2))
S.0011                6 IF(Z*X)7,8,8
S.0012                7 ARCSIN=-ARCSIN
S.0013                8 RETURN
S.0014                 END
```

Figure 8.9 The ARCSIN function subprogram

The subprogram initially tests to see whether X is equal to zero. If X is equal to zero, ARCSIN is set equal to zero and this function value is returned to the calling program. If X is not equal to zero, control is transferred to statement 1, and the dummy variable Z is set equal to the absolute value of X. The variable W is then computed as

$$W = 1. - Z ** 2$$

Note that this variable W in the subprogram has nothing to do with the variable W in the main calling program. Likewise, the statement labels in the subprogram are independent of the statement numbers in the main program.

The value of W is now tested to see whether it is equal to zero. If W is equal to zero, ARCSIN is set equal to 1.5707963 (90°) and control is transferred to statement 6. If the value of W is not equal to zero, control is transferred to statement 4 and the ARCSIN is computed from the ATAN function. Statement 6 then tests to see if X is positive or negative. If $Z*X$ is negative, ARCSIN is changed in sign before the numerical value of the function is returned to the calling statement.

After the inverse sine of X has been computed, control is returned from the subprogram and the inverse sine of X is computed in degrees (DEGSIN). Computer statements S.0007 and S.0008 are then used to print the value of the argument, the ARC SIN (X), and DEGSIN. Computer statement S.0009 then transfers control to statement 2 so that the next data card can be processed.

Figure 8.10 gives the computer results for several different values of X. The reader may readily verify the accuracy of these results.

Figures 8.11 and 8.12 are the corresponding Fortran program and computer results for the function ARCCOS. The reader is encouraged to write similar programs for obtaining the ARC SINH (X) and the ARC COSH (X).

```
THE CALCULATION OF THE ARC SIN (X)

       X               ARC SIN (X)          DEG SIN

 0.1C0000CCE 01   C.1570796E 01   0.8999997E 02
 0.8715574E CC    C.1058369E 01   0.6064C05E 02
 0.8660254E 0C    C.1047196E 01   0.5999992E 02
 0.7071068E 0C    C.7853981E C0   C.4499998E 02
 0.5C00C00E 0C    C.52359987E C0  0.2999998E 02
 0.342020C1E CC   C.349C658E C0   0.1999998E 02
 0.1736482E 0C    C.1745323E C0   0.9999994E 01
 0.0              C.0             0.0
-0.1736482E 0C   -C.1745328E C0  -0.9999994E 01
-0.3420201E CC   -C.349C658E C0  -0.1999998E 02
-0.5C00C00E CC   -C.5235987E C0  -C.2999998E 02
-0.7071068E CC   -C.7853981E C0  -0.4499998E 02
-0.8660254E 0C   -0.1047196E 01  -0.5999992E 02
-0.8715574E CC   -C.1058369E C1  -0.6064C05E 02
-0.1C00C00E 01   -C.1570796E 01  -0.8999997E 02

HC217I
```

Figure 8.10 The output results of the program of Figure 8.9

```
C        TEST PROGRAM FOR FUNCTICN ARC CCS (X) SUBPRCGRAM.
C        X**2 LESS THAN OR EQUAL TO 1.
C
         WRITE(6,1)
       1 FORMAT('1THE CALCULATION OF THE ARC COS (X)',///,7X,'X',9X,'ARC CO
        1S (X)',6X,'DEG COS',//)
       2 READ(5,3)X
       3 FORMAT(E15.7)
         W=ARCCOS(X)
         DEGCOS=W*57.2957795
         WRITE(6,4)X,W,DEGCOS
       4 FORMAT(3E15.7)
         GO TO 2
         END

         FUNCTION ARCCOS(X)
         IF(X)1,2,1
       2 ARCCOS=0.15707963
         RETURN
       1 Z=ABS(X)
         W=1-Z**2
         IF(W)4,5,4
       5 ARCCOS=0.0
         GO TO 6
       4 ARCCOS=ATAN(SQRT(1.-Z**2)/Z)
       6 IF(Z*X)7,8,8
       7 ARCCOS=3.1415926-ARCCOS
       8 RETURN
         END
```

Figure 8.11 The ARCCOS function subprogram

```
THE CALCULATION OF THE ARC COS (X)


      X             ARC COS (X)         DEG COS


 0.1000000E 01    0.0                0.0
 0.8715574E 00    0.5124265E 00      0.2935986E 02
 0.8660254E 00    0.5235988E 00      0.3000000E 02
 0.7071068E 00    0.7853981E 00      0.4499998E 02
 0.5000000E 00    0.1047196E 01      0.5999992E 02
 0.3420201E 00    0.1221730E 01      0.6999998E 02
 0.1736482E 00    0.1396263E 01      0.7999997E 02
 0.0              0.1570796E 00      0.8999996E 01
-0.1736482E 00    0.1745329E 01      0.9999997E 02
-0.3420201E 00    0.1919862E 01      0.1100000E 03
 0.5000000E 00    0.2094396E 01      0.1200000E 03
-0.7071068E 00    0.2356194E 01      0.1349999E 03
-0.8660254E 00    0.2617993E 01      0.1500000E 03
-0.8715574E 00    0.2629166E 01      0.1506401E 03
-0.1000000E 01    0.3141592E 01      0.1800000E 03

HC217I
```

Figure 8.12 The output results of the program of Figure 8.11

The Four-Quadrant Arctangent Function

The Fortran library function ATAN (X) is a single argument two-quadrant subprogram that is useful only in the $\pm\pi/2$ radian range. For example, in the case of the argument $X = B/A$, ATAN (B/A) will yield correct results as long as A remains positive. If the value of A should be negative, however, ATAN (B/A) will still return a result between $+\pi/2$ and $-\pi/2$ radians and this result will therefore be in error.

Many Fortran libraries also supply a four-quadrant arctangent function, ATAN2 (X,Y). This function requires two arguments that correspond to the opposite and adjacent sides of a right triangle and thus permit the computation of angles that range in value between $-\pi$ and $+\pi$ radians.

The reader should recall, from some of our previous examples, that only the single argument arctangent function was available, and from time to time it was necessary to add additional Fortran statements to extend the range of this library function. Let us therefore write a function subprogram for ATAN2, thus making it available for all subsequent programs.

Figure 8.13 shows a test calling program and the ATAN2 function subprogram. Initially the calling program prints the heading and subheadings. The argument B is then set equal to -20.0 and we enter the outer DO 10 loop setting $B = B + 10.0$ and $A = -20.0$. Thus, during the first iteration of the outer DO 10 loop, $B = -10.0$ as we enter the inner DO 10 loop. The inner DO 10 loop then causes A to vary from -15.0 to $+15.0$ in 5.0 unit steps. For each of these A argument values, the ATAN2 function subprogram is called on to compute the value of the arctangent.

The subprogram determines the value of the function by examining the dummy arguments X and Y. First we test to see if Y is less than zero, equal to zero, or greater than zero. If Y is greater than zero, a two-quadrant situation results, and control is transferred to statement 3 where the value of ATAN2 is determined from the single argument function ATAN. If the value of Y is equal to zero, however, control is therefore transferred to statement 2 and X is tested to see whether it is positive,

```
C     TEST PROGRAM FOR FUNCTION ATAN2(X,Y)
C
      WRITE(6,1)
    1 FORMAT('1FOUR QUADRANT CALCULATION OF THE ARC TAN (B/A)',///,' INV
     1ERSE TAN (B/A)',4X,'DEGREES',11X,'A',14X,'B',///)
      B=-20.0
      DO 10 J=1,3
      B=B+10.0
      A=-20.0
      DO 10 I=1,7
      A=A+5.0
      PHI=ATAN2(B,A)
      DEG=PHI*57.2957795
   10 WRITE(6,5)PHI,DEG,A,B
    5 FORMAT(4F15.7)
      STOP
      END

      FUNCTION ATAN2(X,Y)
      IF(Y)1,2,3
    2 IF(X)4,8,5
    4 ATAN2=-1.5707963
      RETURN
    5 ATAN2=1.5707963
      RETURN
    3 ATAN2=ATAN(X/Y)
      RETURN
    1 IF(X)6,7,7
    6 ATAN2=-3.1415927+ATAN(X/Y)
      RETURN
    7 ATAN2= 3.1415927+ATAN(X/Y)
      RETURN
    8 ATAN2=0.0
      RETURN
      END
```

Figure 8.13 A four-quadrant arctangent function subprogram

```
FOUR QUADRANT CALCULATION OF THE ARC TAN (B/A)

INVERSE TAN (B/A)    DEGREES              A                  B


-0.2553590E 01 -0.1463099E 03 -0.1500000E 02 -0.1000000E 02
-0.2356194E 01 -0.1350000E 03 -0.1000000E 02 -0.1000000E 02
-0.2034444E 01 -0.1165650E 03 -0.5000000E 01 -0.1000000E 02
-0.1570796E 01 -0.8999997E 02  0.0           -0.1000000E 02
-0.1107148E 01 -0.6343491E 02  0.5000000E 01 -0.1000000E 02
-0.7853981E 00 -0.4499998E 02  0.1000000E 02 -0.1000000E 02
-0.5880026E 00 -0.3369006E 02  0.1500000E 02 -0.1000000E 02
 0.3141592E 01  0.1800000E 03 -0.1500000E 02  0.0
 0.3141592E 01  0.1800000E 03 -0.1000000E 02  0.0
 0.3141592E 01  0.1800000E 03 -0.5000000E 01  0.0
 0.0            0.0            0.0            0.0
 0.0            0.0            0.5000000E 01  0.0
 0.0            0.0            0.1000000E 02  0.0
 0.0            0.0            0.1500000E 02  0.0
 0.2553590E 01  0.1463099E 03 -0.1500000E 02  0.1000000E 02
 0.2356194E 01  0.1350000E 03 -0.1000000E 02  0.1000000E 02
 0.2034444E 01  0.1165650E 03 -0.5000000E 01  0.1000000E 02
 0.1570796E 01  0.8999997E 02  0.0            0.1000000E 02
 0.1107148E 01  0.6343491E 02  0.5000000E 01  0.1000000E 02
 0.7853981E 00  0.4499998E 02  0.1000000E 02  0.1000000E 02
 0.5880026E 00  0.3369006E 02  0.1500000E 02  0.1000000E 02
```

Figure 8.14 The output results of the program of Figure 8.13

negative, or equal to zero. Control is thus transferred to statement 5, 4, or 8, and ATAN2 is set equal to $+1.5707963$ $(+90°)$, -1.5707963, or 0.0, respectively.

If the value of Y is negative, control is transferred to statement 1 where we test to see if X is positive or negative. If X is negative, we add -3.1415927 $(-180.0°)$ to the single argument function result while if X is positive, we add $+3.1415927$ to ATAN (X/Y).

After the ARC TAN (B/A) has been computed, control is returned to the main program and the inverse tangent of B/A is computed in degrees. Computer statements S.0011 and S.0012 then print the value of the ARC TAN (B/A), DEG, and the two arguments A and B. The execution of the two DO 10 loops thus computes the inverse tangent function for 21 different arguments, in which A varies from -15.0 to $+15.0$ in 5.0 unit steps and B is equal to -10.0, 0.0, and $+10.0$ for each of the A values.

Figure 8.14 gives the corresponding computer results. The reader may readily verify that the range of argument used checks the computation of the arctangent function in each of the four quadrants as well as the special cases when $Y = 0.0$.

The Gamma Function Subprogram

As an example of another type of function subprogram that the reader may find useful, consider the computation of the gamma function which is partially graphed in Figure 8.15.

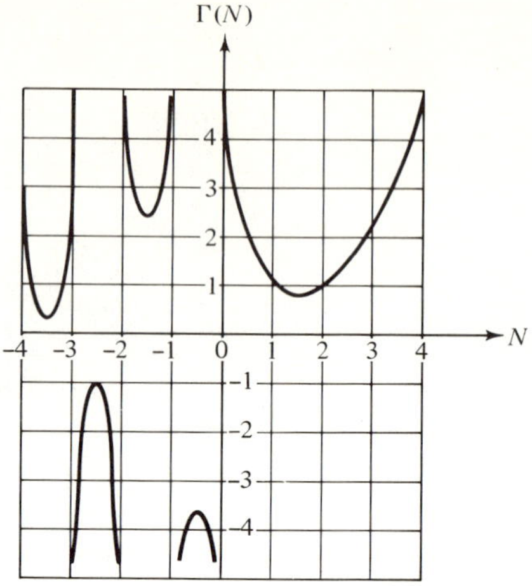

Figure 8.15 The Gamma function

The reader may recall that the gamma function is a generalization of the factorial and is defined by

$$\Gamma(A) = \int_0^\infty X^{A-1}\epsilon^{-X}\,dX$$

If a limited table of gamma function values is available, however, it is not necessary to evaluate the above integral to determine the value of the gamma function for an argument value that is not within the range of the table. Instead, the simple recurrence relation

$$\Gamma(A + 1) = A * \Gamma(A)$$

may be used to find the value of the gamma function for other argument values. For example, if we have table values of the gamma function for $1 \leq A \leq 2$, we may find the value of the gamma function for $A = 2.6$ from

$$\Gamma(2.6) = 1.6 * \Gamma(1.6)$$

Similarly,

$$\Gamma(6.25) = 5.25 * 4.25 * 3.25 * 2.25 * \Gamma(1.25)$$

We may also find the value of the gamma function for arguments less than 1 by realizing that the above recurrence relation may also be written as

$$\Gamma(A) = \frac{\Gamma(A + 1)}{A}$$

Thus,

$$\Gamma(0.5) = \frac{\Gamma(1.5)}{0.5}$$

and

$$\Gamma(-0.5) = \frac{\Gamma(0.5)}{-0.5}$$
$$= \frac{\Gamma(1.5)}{-0.5 * 0.5}$$

and so on.

Figure 8.16 shows a test calling program and the GAMMA function subprogram. Initially the program reads into memory and prints as output the 201 gamma function values. These values represent the value of the gamma function for arguments ranging in value from $1.000 \leq A \leq 2.000$ in 0.005 unit steps. Computer statements S.0007 and S.0008 are then used to read in the value of the argument (A) and computer statement S.0009 calls on the function GAMMA subprogram to compute the function value.

Information is usually supplied to the subprogram by means of the argument values. In this example, however, we have used the COMMON statement to transfer values between the main program and its subprogram. It is therefore not necessary to include VALUES as one of the arguments of the subprogram since its element values occupy common storage and are available to both the main and the subprogram. Note also that the DIMENSION statement must precede the COMMON statement.

```
              C       TEST PROGRAM FOR FUNCTION GAMMA.
              C
S.0001                DIMENSION VALUES(201)
S.0002                COMMON VALUES
S.0003                READ(5,1)(VALUES(I),I=1,201)
S.0004              1 FORMAT(5E15.7)
S.0005                WRITE(6,2)(VALUES(I),I=1,201)
S.0006              2 FORMAT('1A TABLE OF NUMERICAL VALUES OF THE GAMMA FUNCTION',//,' F
                     1OR THE INTERVAL ONE TO TWO, IN 0.005 INCREMENT STEPS',//,(5E15.7))
S.0007              3 READ(5,4)ARG
S.0008              4 FORMAT(E15.7)
S.0009                ANS=GAMMA(ARG)
S.0010                WRITE(6,5)ARG,ANS
S.0011              5 FORMAT(//,' GAMMA (',E15.7,')  =  ',E15.7)
S.0012                GO TO 3
S.0013                END

S.0001                FUNCTION GAMMA(ARG)
S.0002                DIMENSION VALUES(201)
S.0003                COMMON VALUES
S.0004                PROD=1.0
S.0005                A=ARG
S.0006              1 IF(A-2)2,2,3
S.0007              3 A=A-1.0
S.0008                PROD=PROD*A
S.0009                GO TO 1
S.0010              2 IF(A-1.0)4,5,5
S.0011              4 PROD=PROD/A
S.0012                A=A+1.0
S.0013                GO TO 2
S.0014              5 I=(A-1.0)/0.005+1.0
S.0015                GAMMA=PROD*(VALUES(I)+(VALUES(I+1)-VALUES(I))/0.005*
                     1(A-(1.+0.005*(I-1))))
S.0016                RETURN
S.0017                END
```

Figure 8.16 The GAMMA function subprogram

The value of the function is computed by first setting the dummy variable PROD equal to 1. The dummy argument A is then introduced to replace ARG. We next test to see whether the argument is less than or greater than 2. If the argument is greater than 2, control is transferred to statement 3 and A is made equal to $A - 1$. The value of PROD then becomes equal to PROD*A and control is transferred to statement 1. If the new value of A is still greater than 2, the previous steps are repeated until A becomes equal to or less than 2. Note that PROD is equivalent to the product of successive A values in the repeated application of the recurrence relation

$$\Gamma(A + 1) = A * \Gamma(A)$$

Whenever A becomes less than or equal to 2, control is transferred to statement 2 and we test to see whether A is less than 1. If A is not less than 1, control is transferred to statement 5 and we compute the value of the integer variable I

$$I = \frac{A - 1.0}{0.005} + 1.0$$

For example, if the current value of A is 1.600, I would be equal to

$$\begin{aligned} I &= \frac{1.600 - 1.0}{0.005} + 1.0 \\ &= 121 \end{aligned}$$

Thus VALUES(I) would correspond to

$$\text{VALUES(I)} = \text{VALUES(121)} = \Gamma(1.600)$$

since the 121st gamma function value of the 201 values read in is $\Gamma(1.600)$.

At this point we may therefore see that if the original value of A or ARG was equal to 2.6, PROD would now be equal to 1.600 and A would be equal to 1.600; thus,

$$\begin{aligned} \text{GAMMA} &= 1.600 * (\Gamma(1.600) \\ &\quad + \frac{\Gamma(1.605) - \Gamma(1.600)}{0.005} (1.600 - (1.0 + 0.005 * 120)) \\ &= 1.600 * \Gamma(1.600) \end{aligned}$$

Note that the expression

$$\begin{aligned} &\text{VALUES(I)} \\ &\quad + \frac{\text{VALUES(I + 1)} - \text{VALUES(I)}}{0.005} (\text{A} - (1.0 + 0.005 * (\text{I} - 1))) \end{aligned}$$

```
A TABLE OF NUMERICAL VALUES OF THE GAMMA FUNCTION

FOR THE INTERVAL ONE TO TWO, IN 0.005 INCREMENT STEPS

 0.1000000E 01  0.9971384E 00  0.9943259E 00  0.9915613E 00  0.9888442E 00
 0.9861740E 00  0.9835499E 00  0.9809716E 00  0.9784382E 00  0.9759493E 00
 0.9735042E 00  0.9711026E 00  0.9687436E 00  0.9664270E 00  0.9641520E 00
 0.9619183E 00  0.9597253E 00  0.9575725E 00  0.9554595E 00  0.9533857E 00
 0.9513507E 00  0.9493542E 00  0.9473954E 00  0.9454743E 00  0.9435902E 00
 0.9417427E 00  0.9399314E 00  0.9381559E 00  0.9364161E 00  0.9347112E 00
 0.9330409E 00  0.9314049E 00  0.9298031E 00  0.9282347E 00  0.9266996E 00
 0.9251974E 00  0.9237278E 00  0.9222904E 00  0.9208850E 00  0.9195112E 00
 0.9181687E 00  0.9168572E 00  0.9155765E 00  0.9143262E 00  0.9131058E 00
 0.9119155E 00  0.9107549E 00  0.9096234E 00  0.9085211E 00  0.9074475E 00
 0.9064025E 00  0.9053858E 00  0.9043971E 00  0.9034362E 00  0.9025031E 00
 0.9015971E 00  0.9007185E 00  0.8998666E 00  0.8990416E 00  0.8982430E 00
 0.9974707E 00  0.8967245E 00  0.8960042E 00  0.8953096E 00  0.8946404E 00
 0.8939967E 00  0.8933781E 00  0.8927844E 00  0.8922155E 00  0.8916712E 00
 0.8911514E 00  0.8906559E 00  0.8901845E 00  0.8897371E 00  0.3893135E 00
 0.3889136E 00  0.8885370E 00  0.8881841E 00  0.8878543E 00  0.8875476E 00
 0.8872638E 00  0.8870029E 00  0.8867646E 00  0.8865490E 00  0.8863558E 00
 0.8861849E 00  0.8860362E 00  0.8859097E 00  0.8858051E 00  0.8857223E 00
 0.8856614E 00  0.8856221E 00  0.8856043E 00  0.8856080E 00  0.8856331E 00
 0.8856794E 00  0.8857470E 00  0.8858356E 00  0.8859451E 00  0.8860756E 00
 0.8862269E 00  0.8863990E 00  0.8865917E 00  0.8868050E 00  0.8870388E 00
 0.8872930E 00  0.8875676E 00  0.8878624E 00  0.8881777E 00  0.8885130E 00
 0.8888683E 00  0.8892438E 00  0.8896392E 00  0.8900545E 00  0.8904897E 00
 0.8909448E 00  0.8914196E 00  0.8919141E 00  0.8924282E 00  0.8929620E 00
 0.8935153E 00  0.8940881E 00  0.8946806E 00  0.8952923E 00  0.8959237E 00
 0.8965743E 00  0.8972442E 00  0.8979335E 00  0.8986419E 00  0.3993698E 00
 0.9001167E 00  0.9008830E 00  0.9016683E 00  0.9024729E 00  0.9032965E 00
 0.9041392E 00  0.9050010E 00  0.9058819E 00  0.9067818E 00  0.9077008E 00
 0.9086387E 00  0.9095957E 00  0.9105717E 00  0.9115666E 00  0.9125805E 00
 0.9136135E 00  0.9146653E 00  0.9157362E 00  0.9168260E 00  0.9179348E 00
 0.9190625E 00  0.9202092E 00  0.9213749E 00  0.9225595E 00  0.9237631E 00
 0.9249856E 00  0.9262273E 00  0.9274879E 00  0.9287674E 00  0.9300660E 00
 0.9313838E 00  0.9327205E 00  0.9340762E 00  0.9354511E 00  0.9368449E 00
 0.9382581E 00  0.9396904E 00  0.9411418E 00  0.9426123E 00  0.9441022E 00
 0.9456112E 00  0.9471394E 00  0.9486870E 00  0.9502538E 00  0.9518402E 00
 0.9534458E 00  0.9550709E 00  0.9567153E 00  0.9583793E 00  0.9600627E 00
 0.9617658E 00  0.9634884E 00  0.9652306E 00  0.9669927E 00  0.9687743E 00
 0.9705757E 00  0.9723969E 00  0.9742380E 00  0.9760989E 00  0.9779798E 00
 0.9798807E 00  0.9818016E 00  0.9337425E 00  0.9857036E 00  0.9876850E 00
 0.9896865E 00  0.9917084E 00  0.9937506E 00  0.9958133E 00  0.9978964E 00
 0.1000000E 01

GAMMA (  0.1000000E 01)  =    0.1000000E 01

GAMMA (  0.1099999E 01)  =    0.9513507E 00

GAMMA (  0.1150000E 01)  =    0.9330408E 00

GAMMA (  0.1152499E 01)  =    0.9322230E 00

GAMMA (  0.3000000E 01)  =    0.1999619E 01

GAMMA (  0.5000000E 00)  =    0.1772453E 01

GAMMA ( -0.1500000E 01)  =    0.2363271E 01

IHC217I
```

Figure 8.17 The output results of the program of Figure 8.16

in computer statement S.0015 uses linear interpolation to evaluate the function value for arguments that are not exact table values.

If the value of A is less than 1 (S.0010), control is transferred to statement 4 and PROD is made equal to PROD/A. The value of A is then incremented by 1 and control is transferred to statement 2 to determine if the new value of A is still less than 1. These steps are repeated until A

is greater than 1. Note that this is equivalent to finding the product of successive $1/A$ values in the repeated application of the recurrence relation

$$\Gamma(A) = \frac{\Gamma(A+1)}{A}$$

Whenever A becomes greater than 1, control is transferred to statement 5 and the function value is determined as described above.

Figure 8.17 shows the computer results for several different argument values. It is suggested that the reader hand check one or more of these results to verify the accuracy of the computer program. The reader may also verify that computer results will agree with gamma function values obtained by direct integration (calculus) to approximately four or five significant figures.

Simpson's Integration Function Subprogram

For another example of a function subprogram, consider the general integration function, SIMP (NUMB,XINCR), shown in Figure 8.18. Note that the subscripted variable Y has been DIMENSIONED in both the main as well as the function subprogram.

```
LEVEL  15JUL68                                IBM OS/360 BASIC FORTRAN IV (E) COMPILATION

                C        CALCULATION OF AREA UNDER  EXP(-T)  BETWEEN
                C        T = 0  AND  T=1,  USING FUNCTION SIMP.
                C
S.0001                   DIMENSION Y(257)
S.0002                   COMMON Y
                C
                C        GENERATION OF INPUT DATA.
                C
S.0003                   T=0.0
S.0004                   TINCR=1./256.
S.0005                   DO 10 I=1,257
S.0006                   Y(I)=EXP(-T)
S.0007               10  T=T+TINCR
                C
                C        CALCULATION OF AREA.
                C
S.0008                   AREA=SIMP(257,TINCR)
S.0009                   WRITE(6,20)AREA
S.0010               20  FORMAT('1AREA UNDER   EXP(-T) , BETWEEN  T = 0   AND T = 1 , EQUALS'
                        1,//,' AREA =',E15.7)
S.0011                   STOP
S.0012                   END


S.0001                   FUNCTION SIMP(NUMB,XINCR)
S.0002                   DIMENSION Y(257)
S.0003                   COMMON Y
S.0004                   LIMIT=NUMB-1
S.0005                   C4=0.0
S.0006                   C2=0.0
S.0007                   DO 10 I=2,LIMIT,2
S.0008                   C4=C4+Y(I)
S.0009               10  C2=C2+Y(I-1)+Y(I+1)
S.0010                   SIMP=(4.*C4+C2)*XINCR/3.
S.0011                   RETURN
S.0012                   END
```

Figure 8.18 The SIMP function subprogram

Referring first to the calling program, we see that the dimension of Y is given as 257. Computer statements S.0003 through S.0007 are therefore used to provide the 257 ordinate values of the function ϵ^{-t} for the time range of 0.0 to 1.0 seconds. In other words, for a simple test problem we are integrating the function ϵ^{-t} between $t = 0.0$ and $t = 1.0$ seconds, using the function subprogram SIMP. It is therefore necessary for us to provide this subprogram with the appropriate 257 ordinate values. The reader should realize that the integration of some other function would also require the calling program to supply NUMB ordinate values, spaced TINCR (XINCR) seconds apart. The reader should also realize that the dimension size specified for Y can be easily increased if the number of ordinate points, NUMB, exceeds 257.

Computer statement S.0008 is used to calculate the area under the curve. The area is thus found by calling in the function subprogram SIMP (NUMB,XINCR). Note that in this example, NUMB is specified as a numerical constant (257) and TINCR is the actual argument that represents the time spacing between points and corresponds to the dummy argument, XINCR, of the subprogram.

Control is thus transferred to the subprogram with NUMB = 257, XINCR = 1./256., and the 257 elements of Y common to both the main and subprogram. Computer statement S.0004 in the subprogram is then used to compute the limit value of the DO 10 integration loop. Note that this value should be one less than the number of ordinate points. The values of $C4$ and $C2$ are now set equal to zero and the DO 10 loop is executed (for this calling program) 128 times. During the first iteration of the DO 10 loop, $C4$ is set equal to $Y(2)$ and $C(2)$ is made equal to $Y(1) + Y(3)$. During the second iteration of the DO 10 loop, $C4$ is equal to $Y(2) + Y(4)$ and $C2$ is equal to $Y(1) + 2Y(3) + Y(5)$ and therefore after 126 iterations

$$4. * C4 + C2 = Y(1) + 4Y(2) + 2Y(3) + 4Y(4) + \cdots + 2Y(255) + 4Y(256) + Y(257)$$

When the expression above is multiplied by XINCR and divided by 3., the value of SIMP is equivalent to the area that would be found by using Simpson's rule. After the value of SIMP has been computed, control is returned to the calling program, and computer statements S.0009 and S.0010 are used to print the results.

Figure 8.19 shows that the computed result is accurate to almost five significant figures since the exact area to ten significant figures is equal

```
AREA UNDER  EXP(-T) , BETWEEN  T = 0  AND T = 1 , EQUALS

AREA =  0.63211138E 00
```

Figure 8.19 The output results of the program of Figure 8.18

to 0.6321205588. The reader should realize that the computer error may be due to round-off as much as truncation since single precision computation has been used throughout. He should also recognize that in many cases an accurate answer may not be necessary. In addition, in integrating other types of functions, there may be no way of recognizing the accuracy of a given answer.

In the interest of saving computer time and also controlling the accuracy, it is fruitful to modify the above subprogram so that it will perform the integration with the largest possible step size. One way to do this would be to compute the area initially with a very coarse step size and then repeat the process using half the previous step size. If the difference of these two results were greater than some specified difference, the procedure could be repeated as often as necessary, each time halving the step size and each time comparing the new result with the previous result to determine whether the process should be repeated again. Note that such a technique would require each new step to perform twice the amount of computation that was performed in the preceding step. The second step size would therefore require twice the amount of computation needed for the first step size, or the total computation at the end of the second step size would be three times the amount required for the first step. Similarly, the third step would require four times the amount of computation needed for the first step or the total computation at the end of the third step would be seven times the amount required in the first step, and so on.

In order to avoid this binary buildup of computation time, we should not treat each new step size as a new integration because values from previous step sizes are always included. For example, consider Figure 8.20 in which we have defined some arbitrary function in terms of 17 ordinate values. Let us now use Simpson's rule to integrate this function

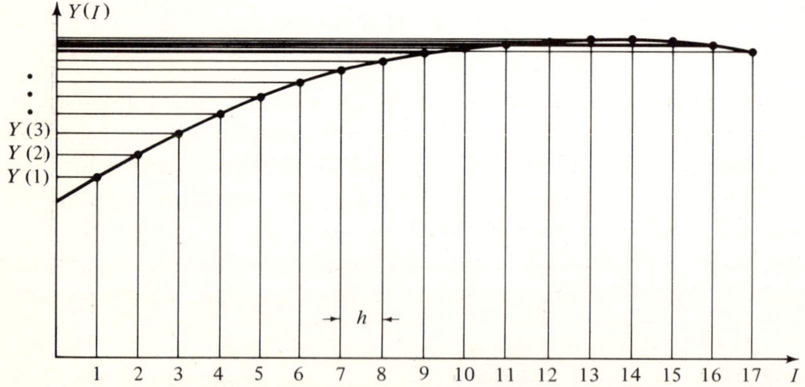

Figure 8.20 An arbitrary function defined in terms of 17 ordinate values

over the interval shown, starting initially with a step size equal to 8h and each time repeating the process until the step size is equal to h. Thus the successive SUM values would be given by:

STEP SIZE	SUM VALUE
8h	Y(1) + 4Y(9) + Y(17)
4h	Y(1) + 4Y(5) + 2Y(9) + 4Y(13) + Y(17)
2h	Y(1) + 4Y(3) + 2Y(5) + 4Y(7) + 2Y(9) + 4Y(11) + 2Y(13) + 4Y(15) + Y(17)
h	Y(1) + 4Y(2) + 2Y(3) + 4Y(4) + 2Y(5) + 4Y(6) + 2Y(7) + 4Y(8) + 2Y(9) + 4Y(10) + 2Y(11) + 4Y(12) + 2Y(13) + 4Y(14) + 2Y(15) + 4Y(16) + Y(17)

Note that for any step size, the terms involving the factor of four appear in each of the subsequent steps with the four-factor changed to two. It follows that once the function is evaluated for any given time, it is not necessary to repeat this evaluation in subsequent steps.

```
C        CALCULATION OF AREA UNDER  EXP(-T)  BETWEEN
C        X = 0  AND  X=1,  USING FUNCTION SIMPI.
C
         DIMENSION Y(257)
         COMMON Y
C
C        GENERATION OF INPUT DATA.
C
         T=0.0
         TINCR=1./256.
         DO 10 I=1,257
         Y(I)=EXP(-T)
      10 T=T+TINCR
C
C        CALCULATION OF AREA.
C
         ACC=1.E-4
         AREA=SIMPI(257,TINCR,ACC)
         WRITE(6,20) AREA,ACC
      20 FORMAT('1AREA UNDER  EXP(-T) , BETWEEN  X = 0  AND X = 1 , EQUALS'
        1,//,' AREA =',E15.7,//,' ACCURACY SPECIFIED =',E15.7)
         STOP
         END
```

```
         FUNCTION SIMPI(NUMB,TINCR,ACC)
         DIMENSION Y(257)
         COMMON Y
         OLDA4=0.0
         SUMOLD=Y(1)+Y(NUMB)
         LIMIT=NUMB-1
         INCR=2*LIMIT
         INIT=LIMIT
         AREA=0.0
      40 INCR=INCR/2
         INIT=INIT/2+1
         C4=0.0
         DO 20 I=INIT,LIMIT,INCR
      20 C4=C4+Y(I)
         SUMOLD=SUMOLD+OLDA4
         SUM=SUMOLD+4.*C4
         OLDA4=2.*C4
         H=(INIT-1)*TINCR
         AREAN=SUM*H/3.
         IF(ABS(AREAN-AREA)-ACC)30,30,50
      50 AREA=AREAN
         GO TO 40
      30 SIMPI=AREAN
         RETURN
         END
```

Figure 8.21 The SIMPI function subprogram

```
AREA UNDER  EXP(-T) , BETWEEN  X = C  AND X = 1 , EQUALS

AREA =  0.6321211E C0

ACCURACY SPECIFIED =  0.9999999E-04



AREA UNDER  EXP(-T) , BETWEEN  X = 0  AND X = 1 , EQUALS

AREA =  0.6321201E 00

ACCURACY SPECIFIED =  0.9999994E-06
```

Figure 8.22 The output results of the program of Figure 8.21

Figure 8.21 is a Fortran program that includes this function subprogram modification. Figure 8.22 shows the computer results for the integration of ϵ^{-t} for two different values of ACC. Note that for a specified ACC = 0.1E-03, the computed result is accurate to five significant figures while a value of ACC = 0.1E-05 produced six significant figures of accuracy. The reader is referred to Problems 5-1 through 5-4 for further discussion of integration accuracy.

8.4 SUBROUTINE SUBPROGRAMS

In the last section we pointed out that the function subprogram returned a single value to the calling program. It is therefore not possible to write a function subprogram to perform such operations as matrix inversion, since this operation requires returning more than one value to the main or calling program.

The subroutine subprogram is similar to the function subprogram, but differs in that it is not limited to returning a single value to the calling program. In fact a subroutine subprogram need not return any values to the calling program. For example, a subroutine subprogram may be used to print specific messages or instructions that are based on the values of its various arguments. Since such a subprogram performs no computations, it obviously can not return any values to the calling program.

The subroutine subprogram, like the function subprogram, is a separate subprogram and may contain any Fortran statements except a function statement or another subroutine statement. The variable names and statement numbers (labels) are also independent of the names and labels of the main program or other subprograms. The general form of the subroutine subprogram is

$$\text{SUBROUTINE NAME } (A_1, A_2, \ldots, A_N)$$

where NAME is the name of the subprogram and must be chosen according to the Fortran rules for naming variables. Note, however, that since a subroutine name does not correspond to the value(s) that is (are) returned to the calling program, the first letter of the name has no significance as far as mode is concerned.

The arguments A_1, A_2, . . . , A_N are dummy arguments and may be either nonsubscripted variable names, array names (without subscripts), or names of subroutine subprograms, or other function subprograms. When the dummy argument is an array name, an appropriate DIMENSION statement must be used in the subprogram as well as in the calling program, and the size of these arrays must be the same except when adjustable dimensions are used. When the argument is a subprogram name, it must be written without subscripts (arguments), and an EXTERNAL statement is required to show that it is the name of some other subprogram and not the name of a variable. For example, if GAMMA and FACT are subprogram names and FACT is an argument of GAMMA, then an EXTERNAL statement is required as follows:

Calling Program

```
      .
      .
      .
      EXTERNAL FACT
      .
      .
      .
      CALL GAMMA(FACT,ARG)
      .
      .
      .
```

On the other hand, if the calling statement was

CALL GAMMA(FACT(X),ARG)

an EXTERNAL statement would not be required because FACT(X) would first be computed and its value would correspond to the first argument.

The RETURN statement must again be the last logical statement encountered in the subprogram, and the END statement is required to indicate that it is physically the last statement of the subprogram.

Subroutine subprograms are not called in the same way that function subprograms are called because certain arguments of the subroutine subprogram, rather than its name, are used to return the values to the calling program. The general form of the calling statement is

CALL NAME $(A_1, A_2, \ldots, A_N)$

where NAME is the name of the subroutine subprogram and A_1, A_2,

. . . , A_N are the actual arguments that are substituted for the dummy arguments of the subroutine. These arguments may be any arithmetic expression, but must correspond in number, mode, and order to those of the subroutine. Subroutine arguments are therefore used to supply the necessary data to the subprogram as well as to return the required values to the calling program.

Continued Fraction Expansion Subroutine

The inverse tangent function may be evaluated from its power-series expansion:

$$\text{ARC TAN }(X) = X - \frac{X^3}{3} + \frac{X^5}{5} - \frac{X^7}{7} + \cdots$$

One of the important properties of an alternating converging series is that the sum of the first N terms of the series differs from the value of the infinite series by less than the absolute value of the $(N + 1)$th term. It is therefore a simple matter to write a Fortran program that generates and tests successive terms of the series. Whenever the absolute value of a given term is less than or equal to some specified value, the sum of the preceding terms may be used to determine the value of the function.

Although the above method will work quite well for the

$$\text{SIN }(X) = X - \frac{X^3}{3!} + \frac{X^5}{5!} - \frac{X^7}{7!} + \cdots$$

because we are dividing by the factorial of a number, the reader will find that this scheme will be quite unsuitable for the evaluation of ARC TAN (X) because of the slow convergence. A much faster convergence scheme for the evaluation of the arc tangent (as well as other functions) function makes use of its continued fraction expansion:

$$\text{ARC TAN }(X) = \cfrac{X}{1 + \cfrac{X^2}{3 + \cfrac{4X^2}{5 + \cfrac{9X^2}{7 + \cfrac{16X^2}{9 + \cdots}}}}}$$

Before we continue with the evaluation of the continued fraction expansion for the ARC TAN (X), let us first consider some of the general properties of a continued fraction expansion. If the number of terms in a continued fraction expansion is finite and if the fraction is truncated after the N-th term,

$$f_N = \frac{X_N}{Y_N} = a_1 + \cfrac{b_2}{a_2 + \cfrac{b_3}{a_3 + \cfrac{b_4}{a_4 + \cfrac{b_5}{a_5 + \cdots \cdots + \cfrac{b_N}{a_N}}}}}$$

the continued fraction expansion is called the N-th convergent of f. In general, the larger the value of N, the more accurate the approximation.

It may be shown that the N-th convergent may be evaluated by the following matrix equation:[2]

$$\begin{bmatrix} X_N & X_{N-1} \\ Y_N & Y_{N-1} \end{bmatrix} = \begin{bmatrix} a_1 & 1 \\ b_1 & 0 \end{bmatrix} \begin{bmatrix} a_2 & 1 \\ b_2 & 0 \end{bmatrix} \begin{bmatrix} a_3 & 1 \\ b_3 & 0 \end{bmatrix} \cdots \begin{bmatrix} a_N & 1 \\ b_N & 0 \end{bmatrix}$$

It is thus possible to evaluate the N-th convergent of the continued fraction expansion of the inverse tangent function by identifying the terms in the continued fraction expansion of the arc tangent function with the corresponding a_i and b_i terms in the above matrix equation. We thus find:

a_1	a_2	a_3	a_4	a_5	$\cdots$	a_N
0	1	3	5	7	$\cdots$	$(2N - 3)$

and

b_1	b_2	b_3	b_4	b_5	$\cdots$	b_N
1	X	X^2	$4X^2$	$9X^2$	$\cdots$	$(N - 2)^2X^2$ for $N > 2$

In order to develop the computer algorithm for finding the N-th convergent, we recognize that the product of the first $N - 1$ matrices is also

[2] See L. M. Milne-Thomson, *The Calculus of Finite Differences* (New York: The Macmillan Company, 1951), p. 108.

a 2 by 2 matrix so we write for $N > 2$

$$\begin{bmatrix} X_N & X_{N-1} \\ Y_N & Y_{N-1} \end{bmatrix} = \begin{bmatrix} X_{N-1} & X_{N-2} \\ Y_{N-1} & Y_{N-2} \end{bmatrix} \begin{bmatrix} a_N & 1 \\ b_N & 0 \end{bmatrix}$$

Evaluating the above equation yields

$$X_N = a_N X_{N-1} + b_N X_{N-2}$$

and

$$Y_N = a_N Y_{N-1} + b_N Y_{N-2}$$

It therefore follows that since

$$\begin{bmatrix} X_2 & X_1 \\ Y_2 & Y_1 \end{bmatrix} = \begin{bmatrix} 0 & 1 \\ 1 & 0 \end{bmatrix} \begin{bmatrix} 1 & 1 \\ X & 0 \end{bmatrix} = \begin{bmatrix} X & 0 \\ 1 & 1 \end{bmatrix}$$

all subsequent values of X_N and Y_N can be found by repeated application of the above equations for X_N and Y_N.

Figure 8.23 is a Fortran test program and ARCTAN subroutine that employs the matrix technique to evaluate the N-th convergent of the continued fraction expansion of the arc tangent function. Initially the main program reads in the accuracy factor, EPS, and the dividing increment, DINCR. Computer statements S.0003 and S.0004 are then used to print the heading and subheadings. The variable X is then set equal to zero, and we enter the DO 10 loop to obtain 50 values of the arc tangent function. During each iteration of the DO 10 loop, X is set equal to $X + 1.0$ and the argument, ARG, is made equal to X/DINCR. We thus see that for a value of DINCR = 10.0, the arc tangent function will be computed for arguments equal to 0.1, 0.2, . . . , 5.0.

During each iteration of the DO 10 loop, the value of K is set equal to zero before computer statement S.0010 calls the ARCTAN (ARG,EPS, K,TAN2) subroutine. The variable K is used to keep track of the number of matrix multiplications needed to evaluate the arc tangent function to the specified accuracy. The argument TAN2 is used to return the function value to the main program.

Turning now to the subroutine, we see that computer statements S.0002 and S.0003 introduce two factors, FACT1 and FACT2. These factors are used later to prevent numerical overflow that may occur when high accuracies are specified. To start the matrix multiplication scheme, computer statements S.0004 through S.0007 are used to specify the ele-

```
C         TEST PROGRAM FOR SUBROUTINE ARCTAN.
C
          READ (5,1) EPS,DINCR
        1 FORMAT (3F15.7)
          WRITE (6,2)
        2 FORMAT ('1ARC TANGENT TABLE',//,2X,'ARGUMENT X ',3X,'ARC TANGENT (X
         1)',3X,'ITERATIONS',//)
          X=0.0
          DO 10 I=1,50
          X=X+1.0
          ARG=X/DINCR
          K=0
          CALL ARCTAN(ARG,EPS,K,TAN2)
       10 WRITE(6,3) ARG,TAN2,K
        3 FORMAT (2E15.7,I10)
          WRITE(6,4) EPS
        4 FORMAT(///,' SPECIFIED ACCURACY EPS =',E15.7)
          STOP
          END
```

```
          SUBROUTINE ARCTAN (X,EPS,K,TAN2)
          FAC1=1.E60
          FAC2=1.E40
          X1=0.0
          Y1=1.0
          X2=X
          Y2=1.0
          DO 10 N=3,200
          AN=2*N-3
          BN=(N-2)**2*X**2
          XN=X2*AN+X1*BN
          YN=Y2*AN+Y1*BN
          X1=X2
          X2=XN
          Y1=Y2
          Y2=YN
          IF (X2-FAC1) 20, 30, 30
       20 IF (Y2-FAC1)40,30,30
       30 X1=X1/FAC2
          X2=X2/FAC2
          Y1=Y1/FAC2
          Y2=Y2/FAC2
       40 TAN2=X2/Y2
          TAN1=X1/Y1
          K=K+1
          IF (ABS(TAN2-TAN1)-EPS)50,50,10
       10 CONTINUE
       50 RETURN
          END
```

Figure 8.23 A continued fraction subroutine for the evaluation of the arctangent function

ments of the matrix that result from the first matrix multiplication:

$$\begin{bmatrix} X_2 & X_1 \\ Y_2 & Y_1 \end{bmatrix} = \begin{bmatrix} 0 & 1 \\ 1 & 0 \end{bmatrix} \begin{bmatrix} 1 & 1 \\ X & 0 \end{bmatrix} = \begin{bmatrix} X & 0 \\ 1 & 1 \end{bmatrix}$$

The DO 10 loop is then used to perform any additional matrix multiplications that may be required to obtain the desired accuracy:

$$\begin{bmatrix} X_N & X_{N-1} \\ Y_N & Y_{N-1} \end{bmatrix} = \begin{bmatrix} X_{N-1} & X_{N-2} \\ Y_{N-1} & Y_{N-2} \end{bmatrix} \begin{bmatrix} a_N & 1 \\ b_N & 0 \end{bmatrix}$$

For the first iteration of the DO 10 loop, N is set equal to 3 and thus

$$\begin{bmatrix} X_{N-1} & X_{N-2} \\ Y_{N-1} & Y_{N-2} \end{bmatrix} = \begin{bmatrix} X_2 & X_1 \\ Y_2 & Y_1 \end{bmatrix}$$

Note that these values have already been specified by computer statements S.0004 through S.0007. Computer statements S.0009 and S.0010

```
AN = 2 * N - 3
BN = (N - 2) ** 2 * X ** 2
```

are now used to determine the values of a_3 and b_3. This is followed by computer statements S.0011 and S.0012

```
XN = X2 * AN + X1 * BN
YN = Y2 * AN + Y1 * BN
```

which are used to compute the values of X_3 and Y_3.

To prevent memory overflow (which occurs for high-accuracy specifications), we avoid the use of subscripted variables for X and Y and use computer statements S.0013 through S.0016 to set the element values of matrix

$$\begin{bmatrix} X_{N-1} & X_{N-2} \\ Y_{N-1} & Y_{N-2} \end{bmatrix}$$

to their new values that are needed for the next iteration of the DO 10 loop ($N = 4$). We thus see that the variables X_2, X_1, Y_2, and Y_1 are used successively to store the respective values of X_{N-1}, X_{N-2}, Y_{N-1}, and Y_{N-2}.

Computer statements S.0017 through S.0022 are now used to prevent numerical overflow. This is done by first testing to see whether X_2 (which equals $X_N = X_3$ for the first iteration of the DO 10 loop) is equal to or greater than FACT1 (1.E60). If X_2 is less than FACT1, we test to see whether Y_2 is equal to or greater than FACT1. If both X_2 and Y_2 are less than FACT1, we transfer control to statement 40 and compute the N-th convergent of the arc tangent function. Computer statement S.0024 then computes TAN1 which is the $N - 1$th convergent of the function. The variable K is then set equal to $K + 1$ and corresponds to 1 for the first iteration of this loop. The variable K thus keeps track of the number of matrix multiplications that are required and the value of $K + 2$ corresponds to N, which is the value of the convergent approximation.

We then test to see whether the ABS(TAN2 − TAN1) is greater than the specified accuracy. Note that since we have an alternating series and since TAN2 corresponds to the N-th convergent of the continued fraction expansion of the arc tangent function and TAN1 corresponds to the $(N - 1)$th convergent, their difference will be less than the error produced by truncating the approximation at the N-th term. If the difference is less than EPS, the function has been computed to the required accuracy and the function value is thus returned to the calling program. If the ABS(TAN2 − TAN1) is greater than EPS, control is transferred to statement 10 and we continue the iteration of the DO 10 loop.

```
ARC TANGENT TABLE

 ARGUMENT X       ARC TANGENT (X)     ITERATIONS

 0.9999996E-01   0.9966856E-01             2
 0.2000000E 00   0.1973953E 00             3
 0.3000000E 00   0.2914564E 00             4
 0.4000000E 00   0.3805063E 00             4
 0.5000000E 00   0.4636475E 00             5
 0.6000000E 00   0.5404196E 00             6
 0.7000000E 00   0.6107261E 00             7
 0.8000000E 00   0.6747406E 00             7
 0.9000000E 00   0.7328150E 00             8
 0.1000000E 01   0.7853981E 00             9
 0.1099999E 01   0.8329806E 00             9
 0.1200000E 01   0.8760576E 00            10
 0.1299999E 01   0.9151009E 00            10
 0.1400000E 01   0.9505476E 00            12
 0.1500000E 01   0.9827939E 00            12
 0.1599999E 01   0.1012196E 01            12
 0.1700000E 01   0.1039073E 01            14
 0.1799999E 01   0.1063698E 01            14
 0.1900000E 01   0.1086318E 01            14
 0.2000000E 01   0.1107148E 01            16
 0.2099999E 01   0.1126377E 01            16
 0.2200000E 01   0.1144168E 01            17
 0.2299999E 01   0.1160668E 01            18
 0.2400000E 01   0.1176005E 01            19
 0.2500000E 01   0.1190289E 01            19
 0.2599999E 01   0.1203621E 01            20
 0.2700000E 01   0.1216089E 01            21
 0.2799999E 01   0.1227773E 01            22
 0.2900000E 01   0.1238735E 01            24
 0.3000000E 01   0.1249045E 01            23
 0.3099999E 01   0.1258752E 01            25
 0.3200000E 01   0.1267912E 01            25
 0.3299999E 01   0.1276560E 01            25
 0.3400000E 01   0.1284743E 01            26
 0.3500000E 01   0.1292498E 01            27
 0.3599999E 01   0.1299849E 01            27
 0.3700000E 01   0.1306833E 01            29
 0.3799999E 01   0.1313471E 01            29
 0.3900000E 01   0.1319795E 01            29
 0.4000000E 01   0.1325817E 01            31
 0.4099999E 01   0.1331562E 01            31
 0.4200000E 01   0.1337051E 01            33
 0.4299999E 01   0.1342300E 01            32
 0.4400000E 01   0.1347319E 01            34
 0.4500000E 01   0.1352127E 01            34
 0.4599999E 01   0.1356736E 01            35
 0.4700000E 01   0.1361156E 01            37
 0.4799999E 01   0.1365401E 01            37
 0.4900000E 01   0.1369478E 01            37
 0.5000000E 01   0.1373401E 01            39


SPECIFIED ACCURACY EPS =  0.9999994E-06
```

Figure 8.24 The output results of the program of Figure 8.23

If, in computer statements S.0017 and S.0018, we found that either X_2 or Y_2 (X_N or Y_N) were equal to or greater than 1.E60, control would be transferred to statement 30, and computer statements S.0019 through S.0022 would be used to reduce the value of all the elements of the matrix before the next iteration of the DO 10 loop could take place.

Figure 8.24 shows the corresponding computer output. The reader may readily verify that all of these results are accurate to six significant figures. Note, however, that larger argument values require a greater number of multiplications to obtain the same accuracy. By using double precision, the reader may readily extend the accuracy to 15 significant figures.

Subroutine COMPLX

Subroutine COMPLX is a subprogram that has been designed to perform required arithmetic operations on the complex numbers A and B. The call statement for the subroutine is

CALL COMPLX(AR,AI,AM,AA,BR,BI,BM,BA,CR,CI,CM,CA,I,J,K)

where the second letter (R, I, M, or A) in the name of the arguments A, B, or C represents respectively the real part, the imaginary part, the magnitude, and the argument of the complex number.

Since some of the operations are more easily carried out in rectangular form and others in polar form, the subprogram takes advantage of this fact, but it does not require any special input form for A or B. The value of the two operands may therefore be transferred to the subprogram in any one of the following four forms:

A	B	I
Rectangular	Rectangular	1
Polar	Polar	2
Rectangular	Polar	3
Polar	Rectangular	4

For example, if the variable I is equal to 3, this means that A will be supplied in rectangular form (AR and AI), and B will be supplied in polar form (BM and BA).

The mathematical operation is specified in terms of the numerical value of J as follows:

J	OPERATION
1	Conversion of rectangular A to polar A.
2	Conversion of polar A to rectangular A.
3	Addition $C = A + B$
4	Subtraction $C = A - B$
5	Multiplication $C = A * B$
6	Division $C = A/B$

Similarly, the numerical value of the variable K is used to specify the form of the output:

K	FORM
1	Rectangular
2	Polar
3	Both

Figure 8.25 shows a Fortran test program that includes the subroutine COMPLX. Initially the main program reads in $A1$, $A2$, $B1$, $B2$, I, J, and K. Depending on the value of I, we then transfer to statements

```
          C     A TEST PROGRAM FOR SUBROUTINE COMPLEX.
          C
S.0001       70 READ(5,6) A1,A2,B1,B2,I,J,K
S.0002        6 FORMAT(4E15.7,3I5)
S.0003          GO TO(1, 2, 3, 4), I
S.0004        1 AR=A1
S.0005          AI=A2
S.0006          BR=B1
S.0007          BI=B2
S.0008          GO TO 5
S.0009        2 AM=A1
S.0010          AA=A2
S.0011          BM=B1
S.0012          BA=B2
S.0013          GO TO 5
S.0014        3 AR=A1
S.0015          AI=A2
S.0016          BM=B1
S.0017          BA=B2
S.0018          GO TO 5
S.0019        4 AM=A1
S.0020          AA=A2
S.0021          BR=B1
S.0022          BI=B2
S.0023        5 CALL COMPLX(AR,AI,AM,AA,BR,BI,BM,BA,CR,CI,CM,CA,I,J,K)
S.0024          GO TO(10,20,30,40,50,60),J
S.0025       10 WRITE(6, 11) AR,AI,AM,AA
S.0026       11 FORMAT(' RECTANGULAR TO POLAR CONVERSION',//,E15.7,'  +  J ',E15.7
               1,'  =  ',E15.7,'  EXP (',E15.7,')',////)
S.0027          GO TO 70
S.0028       20 WRITE(6, 21) AM,AA,AR,AI
S.0029       21 FORMAT(' POLAR TO RECTANGULAR CONVERSION',//,E15.7,'  EXP (',E15.7
               1,')  =  ',E15.7,'  +  J ',E15.7,////)
S.0030          GO TO 70
S.0031       30 WRITE(6, 31) AR,AI,BR,BI
S.0032       31 FORMAT(' THE SUM OF TWO COMPLEX NUMBERS',//,10X,'C  =  A  +  B',//
               2,' A  =',E15.7,'  +  J ',E15.7,//,' B  =',E15.7,'  +  J ',E15.7,//
               3,' EQUALS',//)
S.0033          IF(K-2) 32, 33, 34
S.0034       32 WRITE(6,35)CR,CI
S.0035       35 FORMAT(' C  =',E15.7,'  +  J ',E15.7,////)
S.0036          GO TO 70
S.0037       33 WRITE(6, 36) CM,CA
S.0038       36 FORMAT(' C  =',E15.7,'  EXP (',E15.7,')',////)
S.0039          GO TO 70
S.0040       34 WRITE(6,37)CR,CI,CM,CA
S.0041       37 FORMAT(' C  =',E15.7,'  +  J ',E15.7,//,4X,'=',E15.7,'  EXP (',E15
               1.7,')',////)
S.0042          GO TO 70
S.0043       40 WRITE(6,41)AR,AI,BR,BI
S.0044       41 FORMAT(' THE DIFFERENCE OF TWO COMPLEX NUMBERS',//,10X,'C  =  A  -
               1  B',///,' A  =',E15.7,'  +  J ',E15.7,//,' B  =',E15.7,'  +  J ',
               2E15.7,//,' EQUALS',//)
S.0045          IF(K-2) 32, 33, 34
S.0046       50 WRITE(6,51)AM,AA,BM,BA
S.0047       51 FORMAT(' THE PRODUCT OF TWO COMPLEX NUMBERS',//,10X,'C  =  A  *  B
               1',///,' A  =',E15.7,'  EXP (',E15.7,')',//,' B  =',E15.7,'  EXP ('
               3E15.7,')',////)
S.0048          IF (K-2)32,33,34
S.0049       60 WRITE(6,61) AM,AA,BM,BA
S.0050       61 FORMAT(' THE RATIO OF TWO COMPLEX NUMBERS',//,10X,'C  =  A  /  B',
               1///,' A  =',E15.7,'  EXP (',E15.7,')',//,' B  =',E15.7,'  EXP (',E
               215.7,')',////)
S.0051          IF(K-2) 32, 33, 34
S.0052          STOP
S.0053          END
```

Figure 8.25 A test program for subroutine COMPLX

1, 2, 3, or 4. For example, if I is equal to 3, this would mean that A would be supplied in rectangular form (or $A1 = AR$ and $A2 = AI$), and B would be supplied in polar form (or $B1 = BM$ and $B2 = BA$). Computer statements S.0014 through S.0017 would therefore equate these actual arguments to the corresponding dummy arguments of the subroutine. Computer statement S.0023 would then call in the subroutine.

Turning now to the subroutine, we see that for the case of $I = 3$,

```
              C        A SUBROUTINE FOR DOING COMPLEX ARITHMETIC.
              C
S.0001                 SUBROUTINE COMPLX(AR,AI,AM,AA,BR,BI,BM,BA,CR,CI,CM,CA,I,J,K)
              C
              C        GENERATION OF MISSING INPUT DATA.
              C
S.0002                 GO TO(1,2,3,4),I
S.0003               1 BM=SQRT(BR**2+BI**2)
S.0004                 BA=ATAN2(BI,BR)
S.0005               3 AM=SQRT(AR**2+AI**2)
S.0006                 AA=ATAN2(AI,AR)
S.0007                 GO TO 5
S.0008               2 BR=BM*COS(BA)
S.0009                 BI=BM*SIN(BA)
S.0010               4 AR=AM*COS(AA)
S.0011                 AI=AM*SIN(AA)
              C
              C        ARITHMETIC OPERATIONS.
              C
S.0012               5 GO TO(6,6,7,8,9,10),J
S.0013               7 CR=AR+BR
S.0014                 CI=AI+BI
S.0015                 GO TO 6
S.0016               8 CR=AR-BR
S.0017                 CI=AI-BI
S.0018                 GO TO 6
S.0019               9 CM=AM*BM
S.0020                 CA=AA+BA
S.0021                 GO TO 6
S.0022              10 CM=AM/BM
S.0023                 CA=AA-BA
S.0024               6 GO TO(11,11,12,12,13,13),J
S.0025              11 RETURN
S.0026              12 GO TO(14,15,15),K
S.0027              15 CM=SQRT(CR**2+CI**2)
S.0028                 CA=ATAN2(CI,CR)
S.0029              14 RETURN
S.0030              13 GO TO(16,17,16),K
S.0031              16 CR=CM*COS(CA)
S.0032                 CI=CM*SIN(CA)
S.0033              17 RETURN
S.0034                 END
```

```
S.0001                 FUNCTION ATAN2(X,Y)
S.0002                 IF (Y)1,2,3
S.0003               2 IF(X)4,8,5
S.0004               4 ATAN2=-1.5707963
S.0005                 RETURN
S.0006               5 ATAN2=1.5707963
S.0007                 RETURN
S.0008               3 ATAN2=ATAN(X/Y)
S.0009                 RETURN
S.0010               1 IF (X)6,7,7
S.0011               6 ATAN2=-3.1415927+ATAN(X/Y)
S.0012                 RETURN
S.0013               7 ATAN2= 3.1415927+ATAN(X/Y)
S.0014                 RETURN
S.0015               8 ATAN2=0.0
S.0016                 RETURN
S.0017                 END
```

Figure 8.25 Continued

only the values of *AR*, *AI*, *BM*, and *BA* are supplied. The reader should realize that it does not matter that the values of *AM*, *AA*, *BR*, and *BI* are not supplied since these may be easily computed by computer statements S.0002 through S.0011.

The arithmetic operation is next determined by computer statement S.0012. For example, if J is equal to 5 (multiplication), control would be transferred to statement 9 where *CM* and *CA* would be computed. We would then transfer to statements 6 and 13. Since the required operation is multiplication (for $J = 5$), the results would automatically be available in polar form. Computer statement S.0030 would therefore determine if

```
RECTANGULAR TO POLAR CONVERSION

 0.3000000E 01  +  J   0.4000000E 01  =    0.5000000E 01  EXP (  0.9272949E 00)


POLAR TO RECTANGULAR CONVERSION

 0.5000000E 01  EXP (  0.9272953E 00)  =    0.2999999E 01  +  J   0.4000000E 01


THE SUM OF TWO COMPLEX NUMBERS

          C  =  A  +  B

A  =  0.3000000E 01  +  J   0.4272952E 01

B  = -0.6000000E 01  +  J  -0.8000000E 01

EQUALS

C  =  0.4784441E 01  EXP ( -0.2248534E 01)


THE DIFFERENCE OF TWO COMPLEX NUMBERS

          C  =  A  -  B

A  =  0.2999999E 01  +  J   0.4000000E 01

B  =  0.6000000E 01  +  J   0.1000000E 01

EQUALS

C  = -0.3000001E 01  +  J   0.3000000E 01


THE PRODUCT OF TWO COMPLEX NUMBERS

          C  =  A  *  B

A  =  0.5000000E 01  EXP (  0.9272949E 00)

B  =  0.5000000E 01  EXP (  0.9272953E 00)


C  =  0.2500000E 02  EXP (  0.1854589E 01)


THE RATIO OF TWO COMPLEX NUMBERS

          C  =  A  /  B

A  =  0.5000000E 01  EXP (  0.9272949E 00)

B  =  0.5000000E 01  EXP (  0.9272953E 00)


C  =  0.1000000E 01  +  J  -0.3576278E-06

   =  0.1000000E 01  EXP ( -0.3576279E-06)

IHC217I
```

Figure 8.26 The output results of the program of Figure 8.25

rectangular form is required ($K = 1$ or 3) before the results were returned to the calling statement.

Returning to the main program, we next see that computer statement S.0024 transfers control to statement 50 (for multiplication or $J = 5$) and computer statements S.0046 and S.0047 print the arithmetic operation and the numerical values of A and B (in polar form). Computer statement S.0048 would then test to see what output form had been specified and control would be transferred to statement 32, 33, or 34 to print the results.

In order to test this program the following six data cards were used:

A1	A2	B1	B2	I	J	K
0.3000000E+01	0.4000000E+01			1	1	3
0.5000000E+01	0.9272953E+00			2	2	2
0.3000000E+01	0.4272953E+01	−0.6000000E+01	−0.8000000E+01	1	3	2
0.5000000E+01	0.9272953E+00	0.6000000E+01	0.1000000E+01	4	4	1
0.3000000E+01	0.4000000E+01	0.5000000E+01	0.9272953E+00	3	5	2
0.3000000E+01	0.4000000E+01	0.5000000E+01	0.9272953E+00	3	6	3

It is suggested that the reader verify the accuracy of the computer results shown in Figure 8.26.

In order to show how subroutine COMPLX can be used by a general program, consider the calculation of the frequency response of the RC coupling network shown in Figure 3.52. Figure 8.27 shows the main calling program and subroutine COMPLX. Initially the main program reads in the values of R and C. The heading and subheadings are then printed by computer statements S.0003 and S.0004. The variables AR and AI are next set equal to 1.0 and 0.0 and correspond to the real and imaginary part of the numerator. The real part of the denominator, BR, is then set equal to 1.0 and the corner frequency $WP1$ is computed. Com-

```
             C      DECIBEL-MAGNITUDE AND PHASE RESPONSE OF R-C COUPLING NETWORK.
             C      USING SUBROUTINE COMPLX.
             C
S.0001              READ(5,1)R,C
S.0002            1 FORMAT(2E15.7)
S.0003              WRITE(6,2)
S.0004            2 FORMAT('1MAGNITUDE-PHASE RESPONSE',///,' FREQUENCY',8X,'MAGNITUDE'
                   1,6X,'PHASE',//)
S.0005              AR=1.0
S.0006              AI=0.0
S.0007              BR=1.0
S.0008              WP1=1./(R*C)
S.0009              W=WP1/32.
S.0010              DO 10 I=1,9
S.0011              W=2.*W
S.0012              BI=W/WP1
S.0013              CALL COMPLX(AR,AI,AM,AA,BR,BI,BM,BA,GR,GI,GM,GA,1,6,2)
S.0014              GDB=20.*ALOG10(GM)
S.0015              ANG=GA*57.2957795
S.0016           10 WRITE(6,3)W,GDB,ANG
S.0017            3 FORMAT(3E15.7)
S.0018              STOP
S.0019              END
```

Figure 8.27 Subroutine COMPLX is used to find the frequency response of an R-C coupling network

```
         C     A SUBROUTINE FOR DOING COMPLEX ARITHMETIC.
         C
S.0001         SUBROUTINE COMPLX(AR,AI,AM,AA,BR,BI,BM,BA,CR,CI,CM,CA,I,J,K)
         C
         C     GENERATION OF MISSING INPUT DATA.
         C
S.0002         GO TO(1,2,3,4),I
S.0003       1 BM=SQRT(BR**2+BI**2)
S.0004         BA=ATAN2(BR,BI)
S.0005       3 AM=SQRT(AR**2+AI**2)
S.0006         AA=ATAN2(AR,AI)
S.0007         GO TO 5
S.0008       2 BR=BM*COS(BA)
S.0009         BI=BM*SIN(BA)
S.0010       4 AR=AM*COS(AA)
S.0011         AI=AM*SIN(AA)
         C
         C     ARITHMETIC OPERATIONS.
         C
S.0012       5 GO TO(6,6,7,8,9,10),J
S.0013       7 CR=AR+BR
S.0014         CI=AI+BI
S.0015         GO TO 6
S.0016       8 CR=AR-BR
S.0017         CI=AI-BI
S.0018         GO TO 6
S.0019       9 CM=AM*BM
S.0020         CA=AA+BA
S.0021         GO TO 6
S.0022      10 CM=AM/BM
S.0023         CA=AA-BA
S.0024       6 GO TO(11,11,12,12,13,13),J
S.0025      11 RETURN
S.0026      12 GO TO(14,15,15),K
S.0027      15 CM=SQRT(CR**2+CI**2)
S.0028         CA=ATAN2(CR,CI)
S.0029      14 RETURN
S.0030      13 GO TO(16,17,16),K
S.0031      16 CR=CM*COS(CA)
S.0032         CI=CM*SIN(CA)
S.0033      17 RETURN
S.0034         END

S.0001         FUNCTION ATAN2(X,Y)
S.0002         IF(Y)1,2,3
S.0003       2 IF(X)4,8,5
S.0004       4 ATAN2=-1.5707963
S.0005         RETURN
S.0006       5 ATAN2=1.5707963
S.0007         RETURN
S.0008       3 ATAN2=ATAN(X/Y)
S.0009         RETURN
S.0010       1 IF(X)6,7,7
S.0011       6 ATAN2=-3.1415927+ATAN(X/Y)
S.0012         RETURN
S.0013       7 ATAN2= 3.1415927+ATAN(X/Y)
S.0014         RETURN
S.0015       8 ATAN2=0.0
S.0016         RETURN
S.0017         END
```

Figure 8.27 Continued

puter statement S.0009 now sets W equal to a frequency value that is five octaves below the corner frequency.

The DO 10 loop is then used to compute

$$G = \frac{AR + jAI}{BR + jBI} = \frac{1.0 + j0.0}{1.0 + jW/WP1}$$
$$= GM\underline{|GA}$$

at one octave spacing for four octaves below to four octaves above the corner frequency. During each iteration of the DO 10 loop, subroutine COMPLX is called. Note that a value of $I = 1$ is used to tell the sub-

```
MAGNITUDE-PHASE RESPONSE

FREQUENCY              MAGNITUDE          PHASE

0.6250001E 00 -0.1692364E-01  0.3576342E 01
0.1250000E 01 -0.6732404E-01  0.7125034E 01
0.2500000E 01 -0.2632783E 00  0.1403626E 02
0.5000000E 01 -0.9690961E 00  0.2656506E 02
0.1000000E 02 -0.3010297E 01  0.4499997E 02
0.2000000E 02 -0.6989698E 01  0.6343491E 02
0.4000000E 02 -0.1230448E 02  0.7596371E 02
0.8000000E 02 -0.1812912E 02  0.8287494E 02
0.1600000E 03 -0.2409930E 02  0.8642363E 02
```

Figure 8.28 The output results of the program sf Figure 8.27

routine that the values of A and B are supplied in rectangular form. A value of $J = 6$ indicates that the required mathematical operation is division, and $K = 2$ specifies that the output is required in polar form.

Figure 8.28 shows the corresponding computer results. It may readily be seen that these results are identical to those of Figure 3.54.

8.5 SYSTEM/360 SCIENTIFIC SUBROUTINE PACKAGE VERSION III[3]

This section has been written to acquaint the reader, especially one who has access to other than IBM 360 computers, with the System/360 Scientific Subroutine Package. The System/360 Scientific Subroutine Package is a collection of over 250 Fortran subroutines that may be called on by the programmer's main source program to perform various statistical and mathematical computations. This package thus provides the programmer, from the beginner to the expert, with a powerful tool to solve engineering problems that might otherwise prove to be too complicated or too lengthy. The non-IBM/360 user should also be aware that the program manual gives a copy of the complete subroutine and lists its specifications. Since all subroutines are written in Fortran, it is a simple matter to copy these or modify them so that they may meet general program needs.

All subroutines are used in essentially the way that was described in Section 8.4:

1. All arrays used by the subroutine must be defined by a DIMENSION statement.
2. The main program must provide all input data in appropriate form.
3. The CALL statement in the main program transfers control to the subroutine where the dummy arguments in the subroutine are replaced by the numerical values of the actual arguments in the main program.

[3] *System/360 Scientific Subroutine Package (360A-CM-03X) Version III Programmer's Manual,* IBM Application Program, H20-0205-3, IBM Data Processing Division, White Plains, New York.

4. The subroutine may return more than one result to the calling statement via appropriate dummy arguments.

Special precaution is required for matrix operations because all arrays are represented as *single* subscripted variables that are stored column-wise. The subroutines can therefore operate on any size array (limited only by available memory) since the routines do not contain maximum dimensions for data arrays.

The following examples represent a variety of problems that utilize the subroutine package in obtaining their solution. No attempt is made to use all the subroutines that may prove useful to electrical engineers; however, a sufficient number of examples have been selected to show the reader how to use these subroutines and at the same time encourage him to make further use of the remaining ones, including the writing of modifications if necessary.

Finding the Roots of Real Polynomials

In the solution of many electrical engineering problems it is often necessary to find the roots of real polynomials. For example, consider the series-shunt-peaking network shown in Figure 8.29 in which we are required to find $v_2(t)$ for a step input current excitation.

The Laplace transform method of solution first requires that we write the complex frequency domain equations. Assuming zero initial signal conditions, we have

$$\left(sC + \frac{1}{sL_2} + G\right) V_1(s) - \frac{1}{sL_2} V_2(s) - GV_3(s) = I(s)$$

$$-\frac{1}{sL_2} V_1(s) + \left(sC_2 + \frac{1}{sL_2}\right) V_2(s) = 0$$

$$-GV_1(s) + \left(G + \frac{1}{sL_1}\right) V_3(s) = 0$$

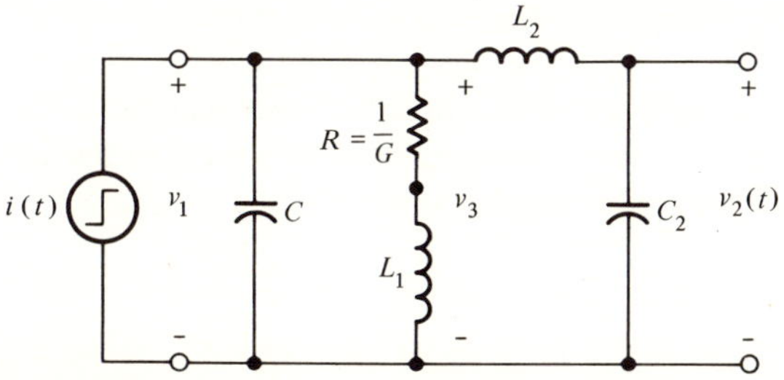

Figure 8.29 The series-shunt-peaking network

Solving for $V_2(s)$ we find

$$V_2(s) = \frac{L_1L_2\left(s + \dfrac{R}{L_1}\right)I(s)}{\left(\begin{array}{l}L_1L_2^2C_2Cs^4 + L_2^2C_2RCs^3 + (L_1L_2C + L_1L_2C_2 + L_2^2C_2)s^2 \\ \qquad + (L_2RC + L_2C_2R)s + L_2\end{array}\right)}$$

Since this network has three degrees of correction, let us make all corrections in the phase function so that we will obtain a maximally flat time delay; thus, we let

$$C_2 = C_0 \qquad L_1 = 0.2\ C_0R^2$$
$$C = 0.5\ C_0 \qquad L_2 = 0.7\ C_0R^2$$

If we now let $\omega_0 = \dfrac{1}{RC_0}$ and $p = \dfrac{s}{\omega_0}$, we may write for a unit step excitation

$$V_2(p) = \frac{R(p + 5.0)}{p(0.35p^4 + 1.75p^3 + 5.0p^2 + 7.5p + 5.0)}$$

and before we can find the $\mathcal{L}^{-1}[V_2(p)] = v_2\left(\dfrac{t}{RC_0}\right)$, we must first find the roots of the quartic. These roots may be found by using subroutine POLRT.[4]

Figure 8.30 is a Fortran program that may be used as a general calling program. The comment statements at the beginning of this program point out that subroutine POLRT may be called by the following statement:[5]

CALL POLRT(XCOF,COF,M,ROOTR,ROOTI,IER)

where XCOF is a vector of $M + 1$ elements which correspond to the coefficients of the polynomial ($P_M(X) = XCOF(1) + XCOF(2)\ X + \cdots + XCOF(M + 1)\ X^M$), COF is a working vector of length $M + 1$, M is the order of the polynomial, ROOTR and ROOTI are resulting vectors of length M that contain respectively the real and imaginary roots of the polynomial, and IER is the error code.

Initially computer statement S.0001 dimensions the arrays A, COF, ROOTR, and ROOTI. Note that A is the actual argument and corresponds to the dummy argument XCOF. Computer statements S.0002 through S.0006 are then used to read in the order of the polynomial and the polynomial coefficients. Computer statements S.0007 through S.0009 then print the heading and the numerical values of the coefficients.

Control is then transferred to the subroutine to determine the roots of the polynomial. After the roots have been found, computer statement

[4] *System/360 Scientific Subroutine Package*, p. 181.
[5] *Ibid.*, p. 447.

```
        C      THIS PROGRAM COMPUTES THE REAL AND COMPLEX ROOTS OF A REAL POLYNOMIAL
        C      USING THE SUBROUTINE POLRT.
        C
        C      THE FORM OF THE SUBROUTINE IS
        C      CALL POLRT(XCOF,COF,M,ROOTR,ROOTI,IER)
        C      WHERE
        C      XCOF - A VECTOR OF M + 1 COEFFICIENTS OF THE POLYNOMIAL ORDERED FROM
        C      THE SMALLEST TO THE LARGEST POWER. XCOF(1) IS THE SMALLEST.
        C      COF - A WORKING VECTOR OF LENGTH M + 1.
        C      M - THE ORDER OF THE POLYNOMIAL.
        C      ROOTR - A RESULTING VECTOR OF LENGTH M CONTAINING THE REAL ROOTS OF
        C      THE POLYNOMIAL.
        C      ROOTI - A RESULTING VECTOR OF LENGTH M CONTAINING THE COMPLEX ROOTS
        C      OF THE POLYNOMIAL.
        C      IER - REPRESENTS THE ERROR CODE WHERE
        C            IER = 0 NO ERROR
        C            IER = 1 M IS LESS THAN ONE
        C            IER = 2 M IS GREATER THAN 36
        C            IER = 3 INDICATES IT IS NOT POSSIBLE TO DETERMINE THE ROOTS
        C                  AFTER 500 ITERATIONS ON 8 STARTING VALUES
        C            IER = 4 THE HIGH ORDER COEFFICIENT IS ZERO
        C
        C      ................................................................
        C
S.0001         DIMENSION A(37),COF(37),ROOTR(36),ROOTI(36)
S.0002       1 READ(5,2)M
S.0003       2 FORMAT(I2)
S.0004         J=M+1
S.0005         READ(5,3)(A(I),I=1,J)
S.0006       3 FORMAT(5E15.7)
S.0007         WRITE(6,4)M
S.0008       4 FORMAT('1THE REAL AND COMPLEX ROOTS OF A REAL POLYNOMIAL ARE FOUND
              1 BY USING SUBROUTINE POLRT',//,' THE POLYNOMIAL IS OF ORDER ',I2,/
              2/' THE INPUT COEFFICIENTS ARE A(1),A(2),....,A(M+1)',//)
S.0009         WRITE(6,3)(A(I),I=1,J)
S.0010         CALL POLRT(A,COF,M,ROOTR,ROOTI,IER)
S.0011         IF(IER-1)10,20,30
S.0012      20 WRITE(6,40)
S.0013      40 FORMAT('0ERROR - THE ORDER OF THE POLYNOMIAL IS LESS THAN ONE')
S.0014         GO TO 1
S.0015      30 IF(IER-3)50,60,70
S.0016      50 WRITE(6,80)
S.0017      80 FORMAT('0ERROR - THE ORDER OF THE POLYNOMIAL IS GREATER THAN 36')
S.0018         GO TO 1
S.0019      70 WRITE(6,90)
S.0020      90 FORMAT('0ERROR - THE HIGH ORDER COEFFICIENT IS ZERO')
S.0021         GO TO 1
S.0022      60 WRITE(6,100)
S.0023     100 FORMAT('0ERROR - UNABLE TO DETERMINE ROOTS WITH 500 ITERATIONS,   T
              1HOSE ALREADY FOUND ARE')
S.0024      10 WRITE(6,110)
S.0025     110 FORMAT(///,6X,'REAL ROOTS',3X,'COMPLEX ROOTS',//)
S.0026         DO 120 I=1,M
S.0027     120 WRITE(6,130)ROOTR(I),ROOTI(I)
S.0028     130 FORMAT(1X,2E15.7)
S.0029         GO TO 1
S.0030         RETURN
S.0031         END
```

Figure 8.30 A program that uses subroutine POLRT to find the roots of a real polynomial

S.0011 is used to determine the error code value. If the error code is equal to zero, control is transferred to statement 10 and the subheadings REAL ROOTS and COMPLEX ROOTS are printed. This is followed by the DO 120 loop which is used to print the M roots. If the value of IER is equal to 1, control is transferred to statement 20 and the error message that indicates the order of the polynomial has been specified as being less than 1 is printed. Similarly, if IER is greater than 1, control is transferred to statement 30.

Computer statement S.0015 then tests to see whether IER is equal to 2, 3, or 4. If IER is equal to 2, control is transferred to statement 50 and the error message printed indicates that the polynomial is of order

```
THE REAL AND COMPLEX ROOTS OF A REAL POLYNOMIAL ARE FOUND BY USING SUBROUTINE POLRT

THE POLYNOMIAL IS OF ORDER  4

THE INPUT COEFFICIENTS ARE A(1),A(2),...,A(M+1)

 0.5000000E 01  0.7500000E 01  0.5000000E 01  0.1750000E 01  0.3500000E 00


     REAL ROOTS   COMPLEX ROOTS

 -0.1078757E 01 -0.2107114E 01
 -0.1078757E 01  0.2107114E 01
 -0.1421242E 01 -0.7276187E 00
 -0.1421242E 01  0.7276187E 00

IHC217I
```

Figure 8.31 The output results of the program of Figure 8.30

greater than 36. If IER is equal to 3, control is transferred to statement 60 where the error message printed indicates that it was not possible to determine the roots after 500 iterations and 8 starting values. Finally, if IER is equal to 4, control is transferred to statement 70 and the error message indicates that the high order coefficient has been specified as being equal to zero.

Figure 8.31 shows the corresponding roots of the quartic obtained from the network of Figure 8.29. The verification of these results is left as an exercise for the reader. The reader should also note that this program is capable of finding the roots of several sets of polynomials. All that is needed is the required input data. The reader is urged to experiment with this program by finding the roots of several of the polynomials that gave us trouble in Chapter 4.

Finding the Roots of Simultaneous Linear Equations Using Subroutine SIMQ

Figure 8.32 shows a Fortran calling program for finding the roots of the four-mesh network shown in Figure 6.4. Initially the number of equations N and the coefficients of the left members are read into memory. This is followed by reading in the right-member terms. The dummy variable K is then set equal to zero and the two DO 10 loops are used to store the double subscripted variable C, columnwise, as a single-subscripted variable A. Computer statements S.0011 through S.0014 are then used to print the original matrices.

The dummy variable KS is now set equal to zero so that a normal solution will be obtained from the subroutine. Computer statement S.0016 calls in the subroutine SIMQ using the statement

CALL SIMQ(A,B,N,KS)[6]

[6] *System/360 Scientific Subroutine Package,* p. 120.

```
          C      THIS PROGRAM FINDS THE ROOTS OF SIMULTANEOUS LINEAR EQUATIONS
          C      USING THE SUBROUTINE SIMQ(A,B,N,KS).
          C
S.0001           DIMENSION A(100),B(10),C(10,10)
S.0002         1 READ(5,2)N,((C(I,J),J=1,N),I=1,N)
S.0003         2 FORMAT(I5,/,(5E15.7))
S.0004           READ(5,3)(B(I),I=1,N)
S.0005         3 FORMAT(5E15.7)
S.0006           K=0
S.0007           DO 10 J=1,N
S.0008           DO 10 I=1,N
S.0009           K=K+1
S.0010        10 A(K)=C(I,J)
S.0011           WRITE(6,4)N,((I,J,C(I,J),J=1,N),I=1,N)
S.0012         4 FORMAT(///,' SOLUTION OF ',I2,' SIMULTANEOUS LINEAR ALGEBRAIC EQUA
                1TIONS',//,' COEFFICIENT MATRIX',//,4(2X,'C(',I2,',',I2,') =',E14.7
                2))
S.0013           WRITE(6,5)(I,B(I),I=1,N)
S.0014         5 FORMAT(///,' RIGHT MEMBER MATRIX',//,(2X,'B(',I2,') =',E15.7))
S.0015           KS=0
S.0016           CALL SIMQ(A,B,N,KS)
S.0017           WRITE(6,6)(I,B(I),I=1,N)
S.0018         6 FORMAT(////,' SOLUTION MATRIX',//,(2X,'X(',I2,') =',E15.7))
S.0019           STOP
S.0020           END
```

Figure 8.32 A program for finding the roots of simultaneous linear equations using subroutine SIMQ(A,B,N,KS)

where A is a single subscripted array that represents the coefficient matrix of the network stored columnwise, B is a vector of length N corresponding to the right member terms, N is the number of simultaneous equations, and KS is an output digit that is 0 for a normal solution and 1 for a singular set of equations. The subroutine method of solution uses the Gauss elimination technique with largest pivotal divisor.

After the subroutine has returned the solution (vector B), computer statements S.0017 and S.0018 are used to print the results. Figure 8.33 shows the computer output. The reader may readily verify that these results agree with those of Figure 6.6.

Subroutine DLGAM Is Used to Compute ln $\Gamma(X)$

In Section 8.3 we wrote a function subprogram for calculating the gamma function. In this example we use the IBM 360 subroutine DLGAM to compute the natural logarithm of the gamma function. This subroutine is based on using the Euler-McLaurin seventh-order expansion[7]

$$\ln \Gamma(X) \approx \left(X - \frac{1}{2}\right) \ln X - X + \frac{1}{2} \ln (2\pi) + \frac{1}{12X} - \frac{1}{360X^3} + \frac{1}{1260X^5} - \frac{1}{1680X^7}$$

[7] See Milton Abramowitz and Irene A. Stegun, eds., *Handbook of Mathematical Functions,* National Bureau of Standards Applied Mathematics Series 55 (Washington, D.C.: U.S. Government Printing Office, 1964), Equation 6.1.41.

```
SOLUTION OF   4 SIMULTANEOUS LINEAR ALGEBRAIC EQUATIONS

COEFFICIENT MATRIX

 C( 1, 1) = 0.4000000E 01  C( 1, 2) =-0.2000000E 01  C( 1, 3) =-0.1000000E 01  C( 1, 4) = 0.0
 C( 2, 1) =-0.2000000E 01  C( 2, 2) = 0.9000000E 01  C( 2, 3) = 0.0            C( 2, 4) =-0.5000000E 01
 C( 3, 1) =-0.1000000E 01  C( 3, 2) = 0.0            C( 3, 3) = 0.6000000E 01  C( 3, 4) =-0.3000000E 01
 C( 4, 1) = 0.0            C( 4, 2) =-0.5000000E 01  C( 4, 3) =-0.3000000E 01  C( 4, 4) = 0.1000000E 02


RIGHT MEMBER MATRIX

 B( 1) =  0.1000000E 02
 B( 2) = -0.5000000E 01
 B( 3) =  0.0
 B( 4) =  0.0


SOLUTION MATRIX

 X( 1) =  0.2761851E 01
 X( 2) =  0.2205070E 00
 X( 3) =  0.6063945E 00
 X( 4) =  0.2921718E 00
```

Figure 8.33 The output results of the program of Figure 8.32

Since the above expression is very accurate for values of X greater than 18, the variable X is replaced by the variable $Z = K + X$, where K is an integer greater than 18 for values of X less than 18. The $\ln \Gamma(X)$ is then evaluated using the above equation and $\ln X + \ln (X + 1), \ldots, + \ln (X + K - 1)$ is subtracted to obtain the required result.

For very large values of X, that is for $10^{10} \leq X \leq 10^{70}$ the lower order terms in the above equation are neglected and the $\ln \Gamma(X)$ is computed from

$$\ln \Gamma(X) = X(\ln X - 1)$$

This subroutine may therefore be used to compute the double-precision natural logarithm of the gamma function for $10^{-9} < X < 10^{70}$.

Figure 8.34 is the Fortran calling program that is used to generate a limited table of $\ln \Gamma(X)$ values, for arguments between 1.0 and 2.0. Initially the double-precision variable XX is set equal to $1.0 - 0.005$. The DO 10 loop is then used to compute the $\ln \Gamma(X)$ in 0.005 unit steps from $XX = 1.0$ to $XX = 2.0$.

During the first iteration of the DO 10 loop the value of XX is made equal to 1.0. Computer statement S.0005 then calls in subroutine DLGAM(XX,DLNG,IER),[8] where the variable XX is the argument of the gamma function, DLNG is the resulting natural logarithm of the gamma function for this argument, and IER is the variable name of the error code. Computer statement S.0006 then tests to see whether the error code IER is less than, equal to, or greater than zero. If the error code is negative (-1), control is transferred to statement 20 and the

[8] *System/360 Scientific Subroutine Package*, p. 362.

```
         C       SUBROUTINE DLGAM IS USED TO COMPUTE THE DOUBLE PRECISION NATURAL
         C       LOGARITHM OF THE GAMMA FUNCTION, FOR ARGUMENTS BETWEEN ONE AND TWO,
         C       IN C.C05 UNIT STEPS.
         C
S.0001           DOUBLE PRECISION XX,DLNG,R(201)
S.0002           XX=1.C-C.005
S.0003           DO 1C I=1,201
S.0004           XX=XX+C.CC5
S.0005           CALL DLGAM(XX,DLNG,IER)
S.0006           IF(IER)20,1C,3C
S.0007        1C R(I)=DLNG
S.0008           WRITE(6,1)(R(I),I=1,201)
S.0009         1 FORMAT('1NATURAL LOGARITHM OF THE GAMMA FUNCTION FOR ARGUMENT VALU
                1ES BETWEEN ONE AND TWO',/,' IN 0.005 INCREMENT STEPS',//,(5D20.1C)
                2)
S.0010           GO TO 40
S.0011        2C WRITE(6,2)
S.0012         2 FORMAT('1XX IS WITHIN 10**(-9) OF BEING ZERO OR XX IS NEGATIVE.',/
                1' DLNG IS SET EQUAL TO -1.0D75.')
S.0013           GO TO 4C
S.0014        3C WRITE(6,3)
S.0015         3 FORMAT('1XX IS GREATER THAN 10**70. DLNG IS SET EQUAL TO + 1.0D75'
                1)
S.0016        4C STOP
S.0017           END
```

Figure 8.34 Subroutine DLGAM is used to find the natural logarithm of the Gamma function for arguments ranging from 10^{-9} to 10^{70}

```
NATURAL LOGARITHM OF THE GAMMA FUNCTION FOR ARGUMENT VALUES BETWEEN ONE AND TWO
IN 0.005 INCREMENT STEPS

   -0.2988397817D-08   -0.2865568929D-02   -0.5690309701D-02   -0.8474519827D-02   -0.1121848592D-01
   -0.1392250676D-01   -0.1658685337D-01   -0.1921180907D-01   -0.2179764956D-01   -0.2434464698D-01
   -0.2685306959D-01   -0.2932318380D-01   -0.3175525026D-01   -0.3414952792D-01   -0.3650627206D-01
   -0.3882573481D-01   -0.4110816512D-01   -0.4335380889D-01   -0.4556290898D-01   -0.4773570530D-01
   -0.4987243481D-01   -0.5197333163D-01   -0.5403862702D-01   -0.5606354952D-01   -0.5806332489D-01
   -0.6002317625D-01   -0.6194832408D-01   -0.6383898625D-01   -0.6569537810D-01   -0.6751771246D-01
   -0.6930619971D-01   -0.7106104779D-01   -0.7278246226D-01   -0.7447064635D-01   -0.7612580097D-01
   -0.7774812477D-01   -0.7933781416D-01   -0.8089506336D-01   -0.8242006443D-01   -0.8391300731D-01
   -0.8537407981D-01   -0.8680346773D-01   -0.8820135481D-01   -0.8956792281D-01   -0.9090335150D-01
   -0.9220781874D-01   -0.9348150048D-01   -0.9472457078D-01   -0.9593720186D-01   -0.9711956413D-01
   -0.9827182619D-01   -0.9939415489D-01   -0.1004867153D 00   -0.1015496709D 00   -0.1025831833D 00
   -0.1035874126D 00   -0.1045625172D 00   -0.1055086538D 00   -0.1064259777D 00   -0.1073146425D 00
   -0.1081748003D 00   -0.1090066017D 00   -0.1098101956D 00   -0.1105857298D 00   -0.1111333502D 00
   -0.1120532017D 00   -0.1127454275D 00   -0.1134101694D 00   -0.1140475680D 00   -0.1146577623D 00
   -0.1152408902D 00   -0.1157970082D 00   -0.1163264914D 00   -0.1168292337D 00   -0.1173054478D 00
   -0.1177552650D 00   -0.1181733154D 00   -0.1185762280D 00   -0.1189476305D 00   -0.1192931494D 00
   -0.1196129100D 00   -0.1199070367D 00   -0.1201756525D 00   -0.1204188792D 00   -0.1206368379D 00
   -0.1208296482D 00   -0.1209974287D 00   -0.1211402972D 00   -0.1212583701D 00   -0.1213517631D 00
   -0.1214205905D 00   -0.1214649658D 00   -0.1214859016D 00   -0.1214808093D 00   -0.1214524995D 00
   -0.1214001817D 00   -0.1213239646D 00   -0.1212239557D 00   -0.1211002619D 00   -0.1209529891D 00
   -0.1207822420D 00   -0.1205881249D 00   -0.1203707408D 00   -0.1201301920D 00   -0.1198665800D 00
   -0.1195800053D 00   -0.1192705677D 00   -0.1189383662D 00   -0.1185834987D 00   -0.1182060627D 00
   -0.1178061545D 00   -0.1173838699D 00   -0.1169393038D 00   -0.1164725504D 00   -0.1159837031D 00
   -0.1154728544D 00   -0.1149400963D 00   -0.1143855199D 00   -0.1138092156D 00   -0.1132112732D 00
   -0.1125917817D 00   -0.1119508293D 00   -0.1112885037D 00   -0.1106048918D 00   -0.1099000797D 00
   -0.1091741531D 00   -0.1084271969D 00   -0.1076592953D 00   -0.1068705319D 00   -0.1060609857D 00
   -0.1052307510D 00   -0.1043798975D 00   -0.1035085103D 00   -0.1026166698D 00   -0.1017044559D 00
   -0.1007719478D 00   -0.9981922420D-01   -0.9884636320D-01   -0.9785344228D-01   -0.9684053835D-01
   -0.9580772778D-01   -0.9475508634D-01   -0.9368268927D-01   -0.9259061127D-01   -0.9147892646D-01
   -0.9034770845D-01   -0.8919703031D-01   -0.8802696459D-01   -0.8683758329D-01   -0.8562895793D-01
   -0.8440115950D-01   -0.8315425849D-01   -0.8188832486D-01   -0.8060342817D-01   -0.7929963736D-01
   -0.7797702097D-01   -0.7663564704D-01   -0.7527553312D-01   -0.7389689632D-01   -0.7249965325D-01
   -0.7108392009D-01   -0.6964976254D-01   -0.6819724587D-01   -0.6672643488D-01   -0.6523739394D-01
   -0.6373018699D-01   -0.6220487752D-01   -0.6066152860D-01   -0.5910020287D-01   -0.5752096255D-01
   -0.5592386945D-01   -0.5430898495D-01   -0.5267637005D-01   -0.5102608531D-01   -0.4935819092D-01
   -0.4767274665D-01   -0.4596981189D-01   -0.4424944564D-01   -0.4251170651D-01   -0.4075665272D-01
   -0.3898434213D-01   -0.3719483221D-01   -0.3538818007D-01   -0.3356444244D-01   -0.3172367570D-01
   -0.2986593586D-01   -0.2799127856D-01   -0.2609975912D-01   -0.2419143248D-01   -0.2226635323D-01
   -0.2032457564D-01   -0.1836615362D-01   -0.1639114076D-01   -0.1439959028D-01   -0.1239155510D-01
   -0.1036708781D-01   -0.8326240652D-02   -0.6269065565D-02   -0.4195614161D-02   -0.2105937734D-02
   -0.8726440406D-07
```

Figure 8.35 The output results of the program of Figure 8.34

following error message is printed:

XX IS WITHIN 10**(−9) OF BEING ZERO OR
XX IS NEGATIVE, DLNG IS SET TO −1.0D75.

If the value of IER is greater than zero (+1), control is transferred to statement 30 and the following error message is printed:

XX IS GREATER THAN 10**70. DLNG IS SET TO +1.0D75.

If the value of IER is equal to zero, no errors have been detected and control is transferred to statement 10 where the value of $R(I)$ is made equal to $\ln \Gamma(XX)$.

If no errors are detected, the DO 10 loop is executed 201 times and the value of the $\ln \Gamma(XX)$ is thus computed in 0.005 unit steps for $XX = 1.0$ to $XX = 2.0$. After all 201 values have been found, computer statements S.0008 and S.0009 are used to print the results.

Figure 8.35 gives the corresponding table of values. By referring to the *Handbook of Mathematical Functions*, the reader may verify that the accuracy of these results, for this range of the argument, is approximately five significant figures.

Subrountine BESJ Is Used to Compute N-th Order Bessel Functions of the First Kind

Bessel functions occur frequently in field and communication theory. For example, the sideband frequencies of angular modulation are given by

$$\begin{aligned} Y = A\{J_0(\Delta\theta) \cos \omega_0 t &- J_1(\Delta\theta)[\sin (\omega_0 + p)t + \sin (\omega_0 - p)t] \\ &- J_2(\Delta\theta)[\cos (\omega_0 + 2p)t + \cos (\omega_0 - 2p)t] \\ &+ J_3(\Delta\theta)[\sin (\omega_0 + 3p)t + \sin (\omega_0 - 3p)t]\} \end{aligned}$$

where

ω_0—is the frequency of the unmodulated carrier.
p—is the modulation frequency.
$J_n(\Delta\theta)$—is the Bessel function of the first kind and n-th order with argument $\Delta\theta$.

This example deals with the IBM 360 subroutine BESJ that may be used to evaluate Bessel functions of the first kind. The evaluation is based on using the recurrence formula

$$F_{n+1}(X) + F_{n-1}(X) = \frac{2n}{X} F_n(X)$$

to find the Bessel function

$$J_n(X) = \frac{F_n(X)}{\alpha}$$

where

$$\alpha = F_0(X) + 2 \sum_{m=1}^{M-2} F_{2m}(X)$$

with M initialized as M_0 where M_0 is the greater of

$$M_A = \begin{cases} X + 6; \ X < 5 \\ 1.4X + \dfrac{60}{X}; \ X \geq 5 \end{cases}$$

or

$$M_B = n + \frac{X}{4} + 2$$

Values of α and $J_n(X)$ are thus computed and compared until

$$|J_n(X)_M - J_n(X)_{M+3}| \leq \delta |J_n(X)_{M+3}|$$

If the value of M should reach

$$M_{\max} = \begin{cases} 20 + 10X - \dfrac{X^2}{3}; \ X \leq 15 \\ 90 + \dfrac{X}{2}; \ X > 15 \end{cases}$$

before the desired accuracy has been obtained, execution is halted.

Figure 8.36 shows the Fortran program for generating a partial table of zero-order Bessel functions of the first kind. Initially the values of the

```
C         SUBROUTINE BESJ IS USED TO OBTAIN BESSEL FUNCTIONS FOR ARGUMENTS
C         BETWEEN 0.1 AND 20.0, IN 0.1 UNCREMENT STEPS.
C
          DIMENSION R(200)
          D=1.E-4
          X=0.0
          N=0
          DO 10 I=1,200
          X=X+0.1
          CALL BESJ(X,N,BJ,D,IER)
          IF(IER-1)10,20,30
       10 R(I)=BJ
          WRITE(6,1)N,(R(I),I=1,200)
        1 FORMAT('1A TABLE OF J BESSEL FUNCTIONS OF ORDER ',I2,//,' FOR ARGU
         1MENTS BETWEEN 0.1 AND 20.0, IN 0.1 INCREMENT STEPS',//,(5E15.7))
          GO TO 70
       20 WRITE(6,2)
        2 FORMAT('0ERROR - N IS NEGATIVE')
          GO TO 70
       30 IF(IER-3)40,50,60
       40 WRITE(6,3)
        3 FORMAT('0ERROR - X IS NEGATIVE OR ZERO')
          GO TO 70
       50 WRITE(6,4)
        4 FORMAT('0ERROR - REQUIRED ACCURACY NOT OBTAINED')
          GO TO 70
       60 WRITE(6,5)
        5 FORMAT('0ERROR - N MUST BE GREATER THAN OR EQUAL TO ZERO, BUT IT M
         1UST BE LESS THAN ',//,5X,'20 + 10*X - X**2/3 FOR X LESS THAN OR EQ
         1UAL TO 15',//,5X,'90 + X/2  FOR X GREATER THAN 15')
       70 STOP
          END
```

Figure 8.36 Subroutine BESJ is used to compute the values of the N-th order Bessel functions of the first kind

required accuracy D, the argument X, and the order of the Bessel function are declared by computer statements S.0002 through S.0004. The DO 10 loop is then used to compute the value of the Bessel function for argument values ranging from $X = 0.1$ to $X = 20.0$. Each iteration of the DO 10 loop therefore calls on the subroutine

BESJ(X,N,BJ,D,IER)[9]

where

X—is the argument of the Bessel function.
N—is the order of the Bessel function.
BJ—is the value computed by the subroutine.
D—is the specified accuracy.
IER—is the error code.
IER = 0—no error.
= 1—N is negative.
= 2—X is negative or zero.
= 3—required accuracy is not obtained.
= 4—range of N compared to X is not correct.

[9] *Ibid.*, p. 363.

```
A TABLE OF J BESSEL FUNCTIONS OF ORDER  0

FOR ARGUMENTS BETWEEN 0.1 AND 20.0, IN 0.1 INCREMENT STEPS

 0.9975019E 00  0.9900253E 00  0.9776266E 00  0.9603983E 00  0.9384699E 00
 0.9120052E 00  0.8812010E 00  0.8462873E 00  0.8075240E 00  0.7651978E 00
 0.7196221E 00  0.6711331E 00  0.6200868E 00  0.5668562E 00  0.5118290E 00
 0.4554040E 00  0.3979870E 00  0.3399886E 00  0.2818207E 00  0.2238938E 00
 0.1666102E 00  0.1103660E 00  0.5554320E-01  0.2511482E-02 -0.4837992E-01
-0.9680086E-01 -0.1424455E 00 -0.1850325E 00 -0.2243081E 00 -0.2600486E 00
-0.2920610E 00 -0.3201848E 00 -0.3442936E 00 -0.3642938E 00 -0.3801264E 00
-0.3917679E 00 -0.3992295E 00 -0.4025561E 00 -0.4018264E 00 -0.3971506E 00
-0.3886717E 00 -0.3765597E 00 -0.3610141E 00 -0.3422603E 00 -0.3205467E 00
-0.2961427E 00 -0.2693363E 00 -0.2404317E 00 -0.2097447E 00 -0.1776042E 00
-0.1443420E 00 -0.1102981E 00 -0.7581151E-01 -0.4121848E-01 -0.6852143E-02
 0.2696298E-01  0.5991244E-01  0.9169441E-01  0.1220258E 00  0.1506382E 00
 0.1772343E 00  0.2017405E 00  0.2238063E 00  0.2433053E 00  0.2600900E 00
 0.2740397E 00  0.2850614E 00  0.2930936E 00  0.2981010E 00  0.3000794E 00
 0.2990525E 00  0.2950729E 00  0.2882196E 00  0.2785999E 00  0.2663445E 00
 0.2516073E 00  0.2345651E 00  0.2154149E 00  0.1943702E 00  0.1716598E 00
 0.1475270E 00  0.1222259E 00  0.9601694E-01  0.6916839E-01  0.4195066E-01
 0.1463419E-01 -0.1251167E-01 -0.3922256E-01 -0.6524199E-01 -0.9032279E-01
-0.1142289E 00 -0.1367387E 00 -0.1576466E 00 -0.1767631E 00 -0.1939216E 00
-0.2089722E 00 -0.2217901E 00 -0.2322717E 00 -0.2403380E 00 -0.2459338E 00
-0.2490291E 00 -0.2496178E 00 -0.2477186E 00 -0.2433749E 00 -0.2366526E 00
-0.2276412E 00 -0.2164506E 00 -0.2032111E 00 -0.1880704E 00 -0.1711994E 00
-0.1527782E 00 -0.1330037E 00 -0.1120810E 00 -0.9022707E-01 -0.6766701E-01
-0.4462885E-01 -0.2134465E-01  0.1953706E-02  0.2503628E-01  0.4767672E-01
 0.6965405E-01  0.9075850E-01  0.1107869E 00  0.1295512E 00  0.1468740E 00
 0.1625980E 00  0.1765796E 00  0.1886941E 00  0.1988370E 00  0.2069221E 00
 0.2128853E 00  0.2166839E 00  0.2182976E 00  0.2177261E 00  0.2149913E 00
 0.2101371E 00  0.2032264E 00  0.1943426E 00  0.1835882E 00  0.1710835E 00
 0.1569639E 00  0.1413804E 00  0.1244998E 00  0.1064971E 00  0.8755839E-01
 0.6787813E-01  0.4765720E-01  0.2709834E-01  0.6407596E-02 -0.1420851E-01
-0.3454731E-01 -0.5440621E-01 -0.7359338E-01 -0.9192348E-01 -0.1092176E 00
-0.1253135E 00 -0.1400587E 00 -0.1533160E 00 -0.1649621E 00 -0.1748923E 00
-0.1830170E 00 -0.1892698E 00 -0.1935986E 00 -0.1959734E 00 -0.1963820E 00
-0.1948320E 00 -0.1913503E 00 -0.1859838E 00 -0.1787971E 00 -0.1698720E 00
-0.1593072E 00 -0.1472169E 00 -0.1337304E 00 -0.1189894E 00 -0.1031475E 00
-0.8636779E-01 -0.6862364E-01 -0.5069253E-01 -0.3215862E-01 -0.1340700E-01
 0.5374365E-02  0.2399868E-01  0.4228199E-01  0.6004505E-01  0.7711256E-01
 0.9332097E-01  0.1085113E 00  0.1225393E 00  0.1352730E 00  0.1465914E 00
 0.1563900E 00  0.1645792E 00  0.1710857E 00  0.1758540E 00  0.1788458E 00
 0.1800399E 00  0.1794341E 00  0.1770433E 00  0.1729004E 00  0.1670551E 00
```

Figure 8.37 The output results of the program of Figure 8.36

Computer statements S.0010 and S.0011 are used to print the 200 values of the Bessel function. Computer statements S.0013 through S.0024 are used to print the appropriate error messages. The results shown in Figure 8.37 are accurate to approximately five significant figures.

Subroutine GAUSS Is Used to Generate Normally Distributed Random Numbers

Later in Chapter 10 we will discuss Monte Carlo methods and the use of random number generators. This example calls on the IBM 360 subroutine GAUSS to generate normally distributed random numbers having a specified standard deviation and mean. The procedure is based on using uniformly distributed random numbers provided by IBM subroutine RANDU to generate a normally distributed number

$$Y = \frac{\sum_{I=1}^{K} X_I - K/2}{\sqrt{K/12}}$$

where X_I is a uniformly distributed random number that has a numerical value between 0 and 1.0 and K is the number of uniformly distributed numbers used in the process. Once the above number has been obtained, the required mean and standard deviation is obtained from the following relation

$$Y' = Y{*}S + AM$$

where

Y'—is the required normally distributed random number.
S—is the required standard deviation.
AM—is the required mean value.

Figure 8.38 shows the program for generating 100 normally distributed random numbers that have a standard deviation of 1.0 and a mean value of zero. Each iteration of the DO 10 loop calls on subroutine GAUSS (IX,S,AM,V) to generate the required number.[10] The variable IX must be used initially to supply an odd number containing one through nine digits. After the first normally distributed random number has been computed, the variable IX is used to store this result so that it can be used in the generation of the next number. The variable S is the name of the argument that corresponds to the value of the standard deviation,

[10] *Ibid.*, p. 77.

```
C       GENERATING NORMALLY DISTRIBUTED RANDOM NUMBERS
C       USING SUBROUTINE GAUSS.
C
        DIMENSION RN(100)
        IX=123456789
        S=1.0
        AM=0.0
        V=0.5
        DO 10 I=1,100
        CALL GAUSS(IX,S,AM,V)
     10 RN(I)=V
        WRITE(6,20)(RN(I),I=1,100)
     20 FORMAT('1ONE HUNDRED NORMALLY DISTRIBUTED RANDOM NUMBERS',///,(5E1
       15.7))
        ST1=0.0
        DO 30 I=1,100
        IF(ABS(RN(I))-1.0)40,40,30
     40 ST1=ST1+1.0
     30 CONTINUE
        WRITE(6,50)ST1
     50 FORMAT(///,' NUMBER OF RANDOM NUMBERS WITHIN ONE STANDARD DEVIATIO
       1N EQUALS',F4.0)
        STOP
        END
```

Figure 8.38 Subroutine GAUSS is used to generate normally distributed random numbers

the variable AM is the name of the value of the required mean, and V corresponds to the value of the computed normal random number.

After the 100 normally distributed random numbers produced by the subroutine have been printed, the DO 30 loop is used to determine how closely the numbers produced correspond to a true normal distribution. This is done by determining the number of random numbers that have values between -1.0 and 1.0.

Figure 8.39 shows the numerical values of the 100 numbers that were generated and the fact that 65 of these had values between -1.0 and $+1.0$. The reader is referred to Section 10.4 for a more detailed discussion of how these numbers may be used in engineering designs.

```
ONE HUNDRED NORMALLY DISTRIBUTED RANDOM NUMBERS


 0.5935078E 00 -0.3675137E 00 -0.3237658E 00 -0.4696112E 00 -0.9991541E 00
 0.6398382E 00 -0.4454718E 00 -0.7218990E 00 -0.2966337E 00 -0.9211559E 00
 0.1067357E 01 -0.1328621E 00 -0.1604595E 01 -0.4655352E 00  0.4403105E 00
-0.8628941E-01 -0.1166204E 01  0.1654166E 01 -0.3852844E-01  0.9652901E-01
 0.8378735E 00  0.1479015E 00  0.9814291E 00  0.1568031E 00  0.2894802E 00
-0.1211661E 01  0.9132671E 00  0.8982420E 00  0.1329451E 00  0.4098368E 00
-0.1767429E 01  0.1870623E 00 -0.6248951E 00 -0.1089543E 01 -0.5225735E 00
-0.1048027E 01  0.8530331E-01 -0.1750043E 01  0.2484169E 00 -0.8158169E 00
 0.3953657E 00  0.4508028E 00 -0.1791363E 01 -0.1062598E 01 -0.5003681E 00
-0.4020462E 00  0.1083667E 01  0.3278446E 00  0.6549454E 00 -0.1611280E 00
 0.1661589E 01 -0.1465775E 01 -0.8193684E 00 -0.2616489E 01 -0.2070045E 00
 0.7977438E 00  0.4585276E 00  0.5043383E 00 -0.1609027E 01 -0.5778790E 00
-0.1067024E 01 -0.1463683E 01 -0.5688992E 00  0.7735624E 00 -0.1389193E 01
 0.1876902E 01 -0.1048530E 01  0.1280782E 01  0.5613585E 00 -0.1394272E-01
 0.5266190E-01  0.4349899E 00 -0.8353329E-01 -0.6137552E 00 -0.1102231E 01
-0.1210107E 01 -0.1295319E 00  0.1662423E 01 -0.1287627E 01  0.1961709E 01
 0.1180193E 01  0.4620504E 00 -0.1215440E 01  0.6291447E 00 -0.3350134E 00
-0.1504904E 01 -0.5350876E 00  0.5334129E 00  0.1206679E 01 -0.4659843E 00
-0.8334723E 00 -0.1521774E 01  0.7496109E 00  0.1413770E 01 -0.6350174E 00
 0.3297701E 00 -0.1699389E 01  0.3182316E-01  0.5983620E 00  0.1016816E 01


NUMBER OF RANDOM NUMBERS WITHIN ONE STANDARD DEVIATION EQUALS 65.
```

Figure 8.39 The output results of the program of Figure 8.38

Subroutine PLOT and Synchronous Timing

Figure 8.40 shows a simplified diagram of a resistance welding circuit. Since most resistance welds require a very large welding current at low voltage, a welding transformer must be employed which generally has a one- or two-turn secondary winding. Note that the back-to-back connection of the two silicon controlled rectifiers permits an a-c current to flow in the transformer primary winding. Note also that it is important that the average transformer current be equal to zero, otherwise core saturation could result. The actual amount of heat produced in making a weld is therefore determined by the number of cycles of welding current flow.

In the welding of such metals as aluminum, a welding current as short as two or three cycles is used. This requires special control circuit considerations to insure that the same amount of heat is used in each spot weld. Not only must the control circuit allow only an even number of half-cycles of current flow (so that the average transformer current is zero), but it must also insure that the circuit is closed at its power factor angle.

In order to better understand this latter point, let us use the equivalent circuit of Figure 8.41 to determine the transformer primary current. Writing Kirchhoff's voltage law for the weld cycle, we have

$$Ri + L\frac{di}{dt} = V_{\max} \sin(\omega t + \phi);\ t \geq 0$$

The actual primary current will therefore be equal to

$$i = i_{ss} + i_t$$

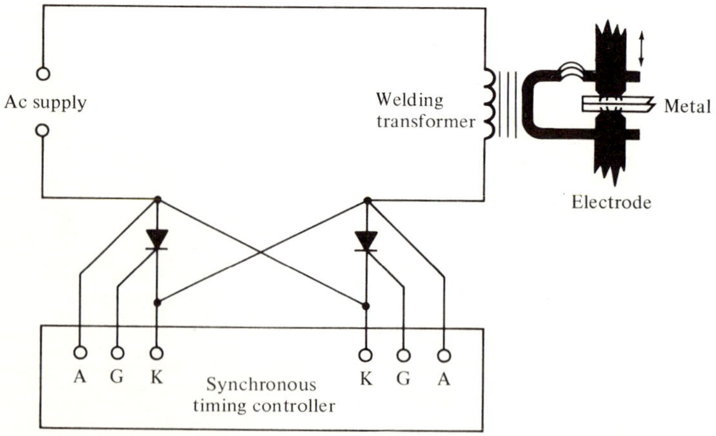

Figure 8.40 Resistance welding circuit

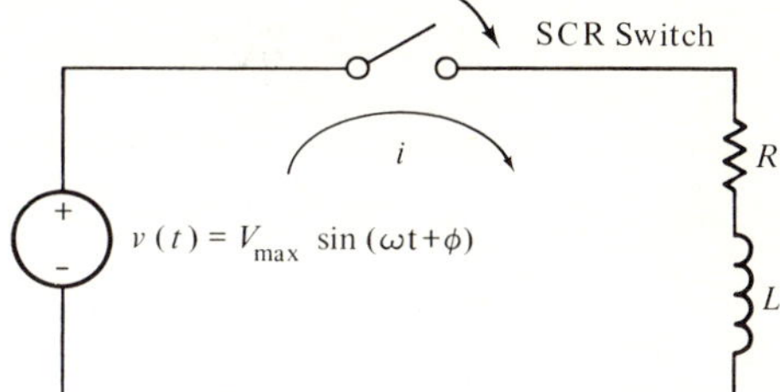

Figure 8.41 Welding-cycle equivalent circuit of Figure 8.40

where the steady-state current is given by

$$i_{ss} = Im\{\boldsymbol{I}\epsilon^{j\omega t}\} = I_{max}\sin(\omega t + \phi - \theta)$$

with

$$\boldsymbol{I} = \frac{V_{max}}{\sqrt{R^2 + (\omega L)^2}}\epsilon^{j(\phi-\theta)} = I_{max}\epsilon^{j(\phi-\theta)}$$

and

$$\theta = \tan^{-1}\frac{\omega L}{R}$$

and the transient current is given by

$$i_t = -I_{max}\sin(\phi - \theta)\epsilon^{-\frac{Rt}{L}}$$

The complete solution for the current is therefore equal to

$$i = i_{ss} + i_t = I_{max}\left[\sin(\omega t + \phi - \theta) - \sin(\phi - \theta)\epsilon^{-\frac{Rt}{L}}\right]$$

An examination of this latter equation will show that no transient current will flow if $\phi = \theta$ (the power factor angle) and a maximum transient current will flow if $\phi = \theta + \pi/2$. We may therefore see that if the circuit is closed at different points in the cycle for different spot welds, the amount of heat produced will vary from weld to weld. Note that if the circuit is closed at the same point in the cycle for each weld, the amount of heat produced would be the same (for the same number of cycles) if it were not for the transformer core saturation. Since our equivalent circuit did not take core saturation into consideration we therefore require a synchronous timing controller so that each weld cycle will start at the power factor angle ($\phi = \theta$).

Figure 8.42 is a Fortran program that uses subroutine PLOT to graph the normalized supply voltage and normalized welding current for four different values of ϕ (different cycle closure points). Initially the program defines $R = 7.54$ ohms and $L = 0.02$ henries, so that $\theta = \pi/4$ radians for $\omega = 377$ rps. We then set PHI (ϕ) equal to $-3.14159/4.$ ($-\pi/4$) and

```
                C       SYNCHRONOUS SWITCH CLOSING
                C       USING SUBROUTINE PLOT.
                C
S.0001                  DIMENSION A(150)
S.0002                  R=7.54
S.0003                  AL=0.02
S.0004                  THETA=ATAN(377.*AL/R)
S.0005                  PHI=-3.14159/4.
S.0006                  NO=0
S.0007                  DO 5 K=1,4
S.0008                  PHI=PHI+3.14159/4.
S.0009                  T=-0.0005
S.0010                  DO 10 I=1,50
S.0011                  T=T+0.0005
S.0012               10 A(I)=T
S.0013                  T=-0.0005
S.0014                  DO 20 I=51,100
S.0015                  T=T+0.0005
S.0016               20 A(I)=SIN(377.*T+PHI)
S.0017                  T=-0.0005
S.0018                  DO 30 I=101,150
S.0019                  T=T+0.0005
S.0020               30 A(I)=SIN(377.*T+PHI-THETA)-SIN(PHI-THETA)*EXP(-R*T/AL)
S.0021                  NO=NO+1
S.0022                  N=50
S.0023                  M=3
S.0024                  NL=0
S.0025                  NS=0
S.0026                5 CALL   PLOT(NO,A,N,M,NL,NS)
S.0027                  STOP
S.0028                  END
```

Figure 8.42 Subroutine PLOT is used in the graphing of the voltage and current waveforms in a synchronous switch circuit closing

NO equal to zero. The DO 5 loop is then executed four times to produce the four different graphs. During the first execution of the DO 5 loop, PHI is made equal to zero. The DO 10 loop is then used to store the graph time values (at 0.0005 second intervals) in the first 50 elements of the variable A. The DO 20 loop then uses the next 50 elements of A to store the corresponding normalized values of the supply voltage. The DO 30 loop, in turn, uses the last 50 elements of A to store the normalized transformer current.

Computer statement S.0021 then sets NO equal to 1 so that the first chart will be labelled CHART 1. The program next calls in subroutine

PLOT(NO,A,N,M,NL,NS)[11]

where

NO—is the number of the chart.
A—is the matrix of data to be plotted. The first column of the matrix contains the base or independent variable values and the successive columns contain the cross-variables (the maximum number is 9).
N—is the number of rows in A.
M—is the number of columns in A.
NL—is the number of lines in the plot.
NS—is the code for sorting the base variable data.
  0—Sorting is not necessary, data is in ascending order.
  1—Sorting is necessary.

[11] *Ibid.*, p. 452.

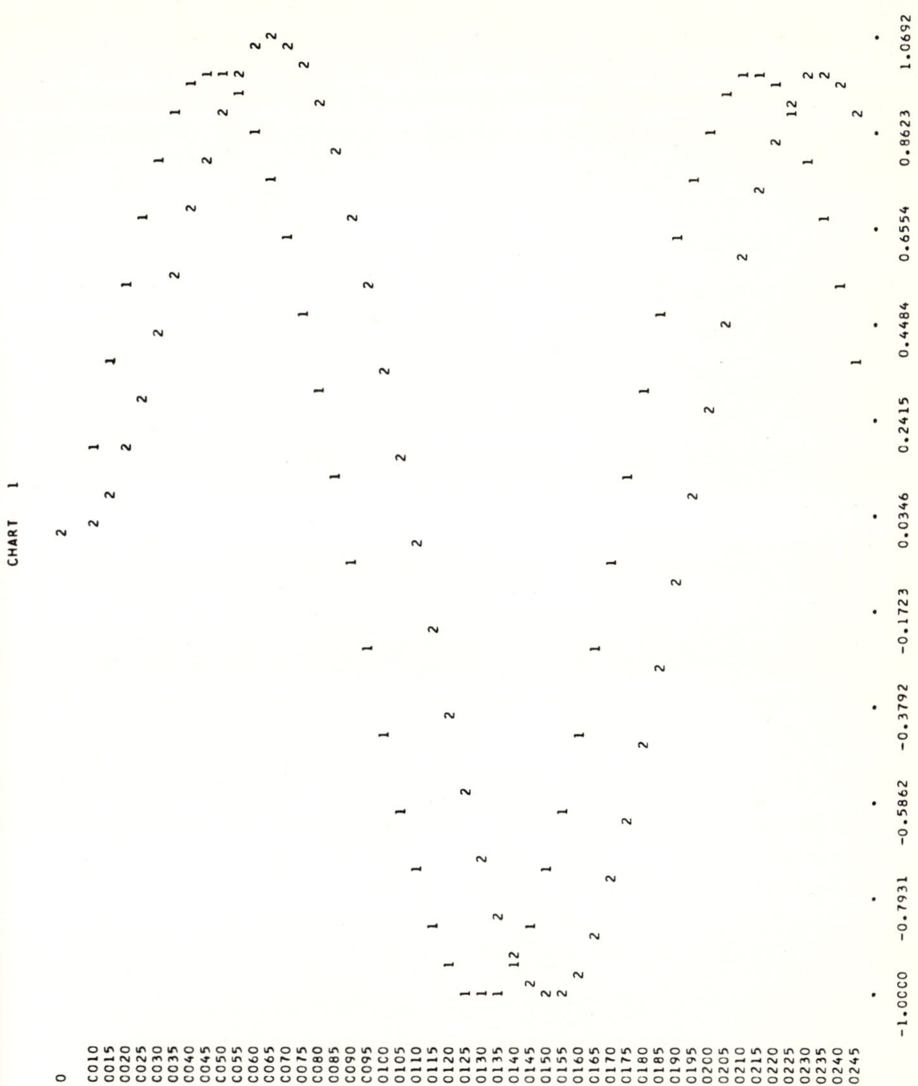

Figure 8.43 The output results of the program of Figure 8.42

Figure 8.43 shows the corresponding plots. Note that the numbers 1 and 2 are used as the respective plotting symbols for the voltage and current curves and that

CHART 1 is for a value of $\phi = 0°$
CHART 2 is for a value of $\phi = 45°$ (the power factor angle)

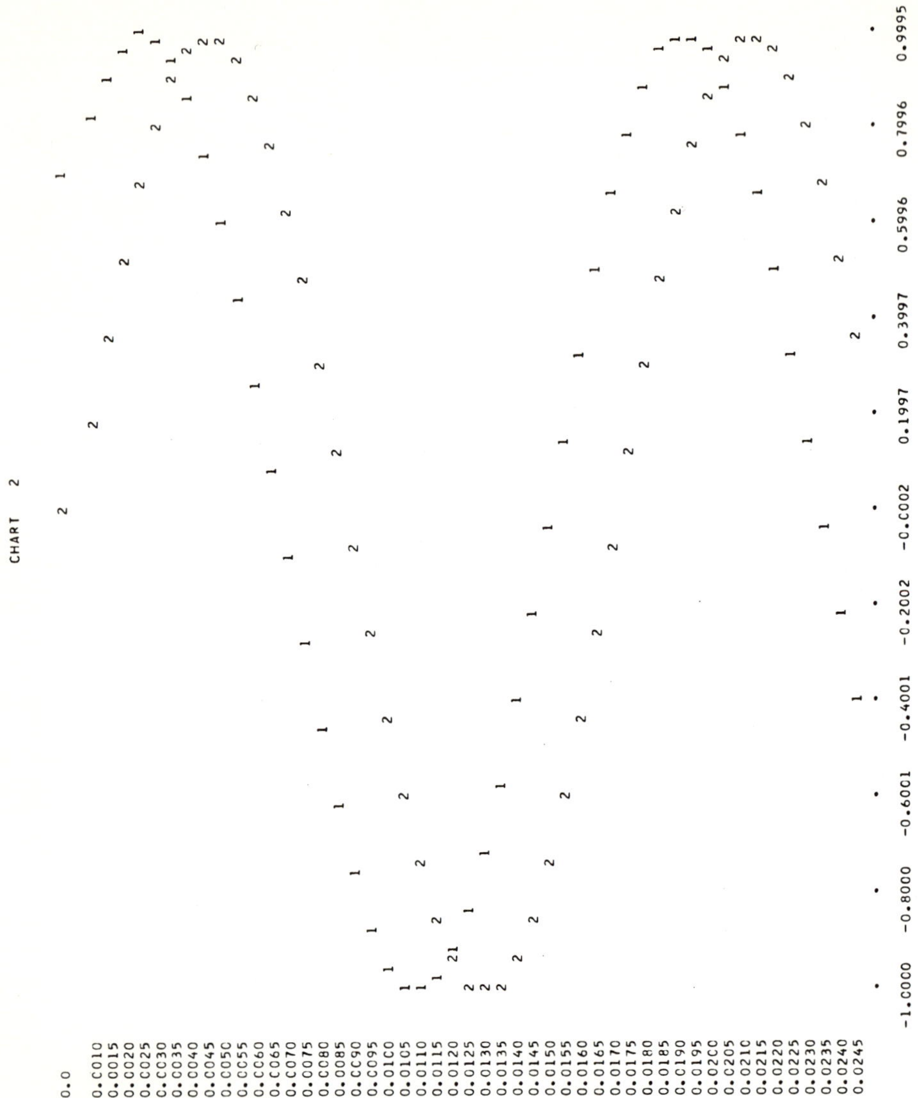

Figure 8.43 Continued

CHART 3 is for a value of $\phi = 90°$ (not shown)
CHART 4 is for a value of $\phi = 135°$ (maximum transient current).

An examination of these plots will show that when the circuit is closed at its power factor angle (CHART 2) the average current will be equal to zero, while if the circuit is closed at any other point in the cycle the

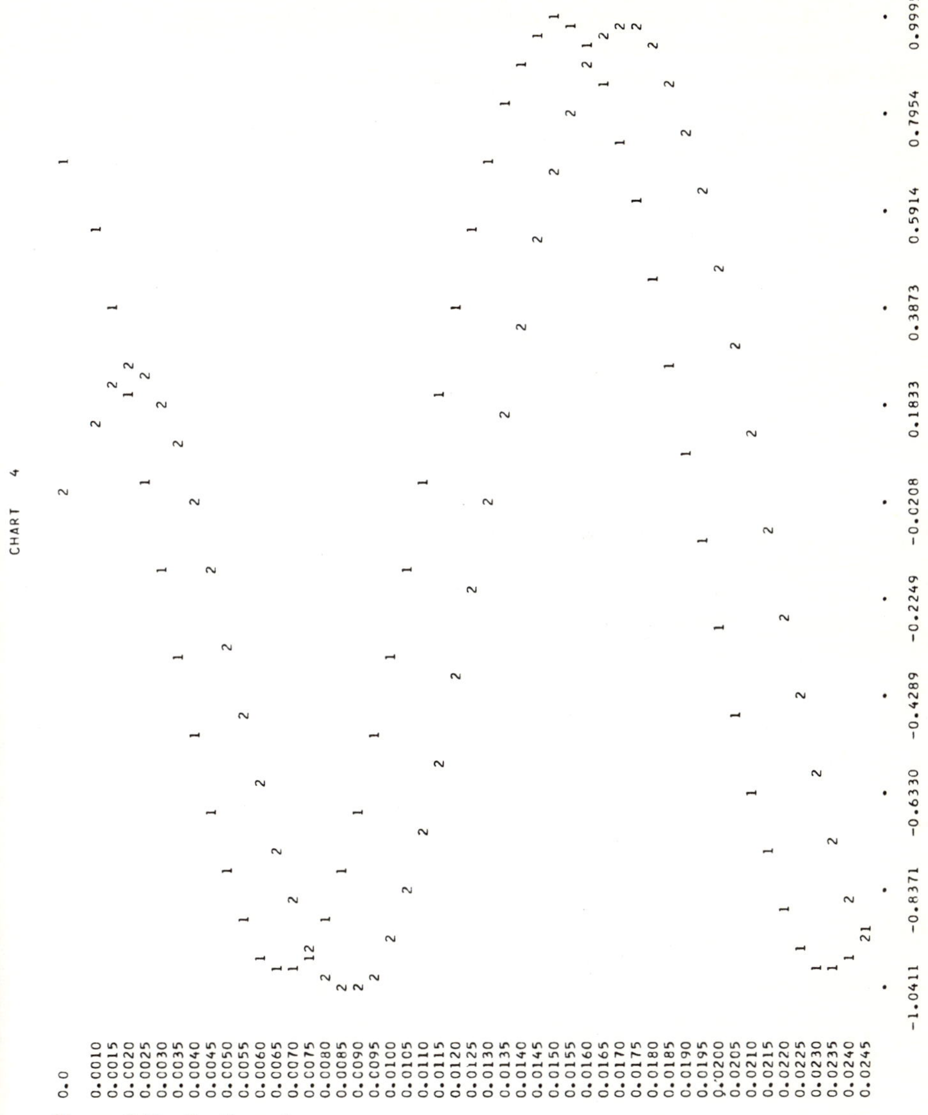

CHART 4

-1.0411 -0.8371 -0.6330 -0.4289 -0.2249 -0.0208 0.1833 0.3873 0.5914 0.7954 0.9995

0.0
0.0010
0.0015
0.0020
0.0025
0.0030
0.0035
0.0040
0.0045
0.0050
0.0055
0.0060
0.0065
0.0070
0.0075
0.0080
0.0085
0.0090
0.0095
0.0100
0.0105
0.0110
0.0115
0.0120
0.0125
0.0130
0.0135
0.0140
0.0145
0.0150
0.0155
0.0160
0.0165
0.0170
0.0175
0.0180
0.0185
0.0190
0.0195
0.0200
0.0205
0.0210
0.0215
0.0220
0.0225
0.0230
0.0235
0.0240
0.0245

Figure 8.43 Continued

transient term will be present and maximum current unbalance will occur at $\phi = \theta + \pi/2$ (CHART 4).

PROBLEMS

8-1. Using the arithmetic statement function, write a general program for evaluating a polynomial and its first derivative. Use the nested form in the evaluation process and arbitrarily limit the polynomial to the tenth order.

8-2. Modify the programs of Figures 4.12, 4.16, 4.18 and 4.19 so that a single program may be used to find the roots of a general function that has been defined in terms of an arithmetic statement function.

8-3. Repeat Problem 8-2 for the programs of Figures 4.23 and 4.25.

8-4. Repeat Problem 8-2 for the programs of Figures 4.28, 4.30, 4.32, and 4.33.

8-5. Repeat Problem 8-2 for the programs of Figures 4.37, 4.38, 4.41, and 4.42.

8-6. Use the program of Figure 8.1 to find the average and RMS values of the following waveforms:

a. $v(t) = 10 \sin 100\pi t \qquad 0.0 \le t \le 0.02$

b. $v(t) = 10 \sin 100\pi t \qquad 0.0 \le t \le 0.01$
$\quad = 0 \qquad 0.01 \le t \le 0.02$

c. $v(t) = 10 \sin 100\pi t \qquad 0.00 \le t \le 0.01$
$\quad = -10 \sin 100\pi t \qquad 0.01 \le t \le 0.02$

d. $v(t) = 10 \qquad 0.00 < t < 0.01$
$\quad = -10 \qquad 0.01 < t < 0.02$

e. $v(t) = 2000t \qquad 0.00 \le t \le 0.005$
$\quad = -2000t + 20 \qquad 0.005 \le t \le 0.015$
$\quad = 2000t - 40 \qquad 0.015 \le t \le 0.020$

f. $v(t) = 2000t \qquad 0.00 \le t \le 0.005$
$\quad = -2000t + 20 \qquad 0.005 \le t \le 0.010$
$\quad = 2000t - 20 \qquad 0.010 \le t \le 0.015$
$\quad = -2000t + 40 \qquad 0.015 \le t \le 0.020$

8-7. Use the program of Figure 8.4 to obtain data for plotting the frequency response of the series-peaking network

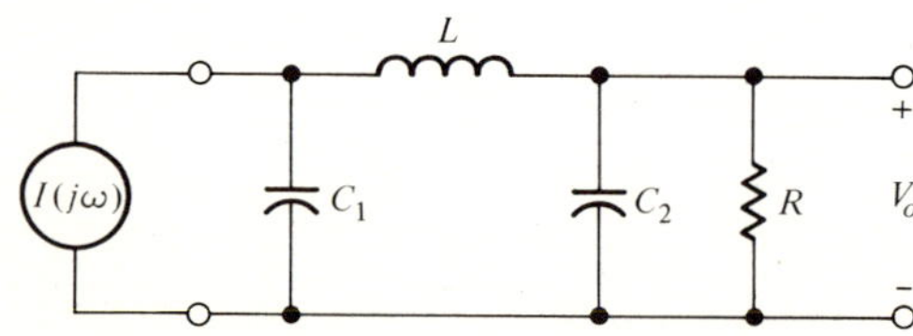

$$H(\phi) = \frac{1}{1 - \dfrac{KN}{N+1}\phi^2 + j\left(\phi - \dfrac{KN}{(N+1)^2}\phi^3\right)}$$

= the normalized output response

where

$$C = C_1 + C_2 \quad N = \frac{C_1}{C_2} \quad K = \frac{L}{R^2C} \quad \text{and} \quad \phi = \omega RC$$

Plot the response for

a. The flattest monotonic magnitude response, $N_a = 3.0$ and $K_a = 2/3$.
b. Maximally flat delay, $N_p = 5.0$ and $K_p = 0.48$.
c. Correction of both magnitude and phase function, $N_{ap} = 2.0$ and $K_{ap} = 0.75$.

8-8. Repeat Problem 8-7 for the linear phase adjustment of the Dietzold network shown below

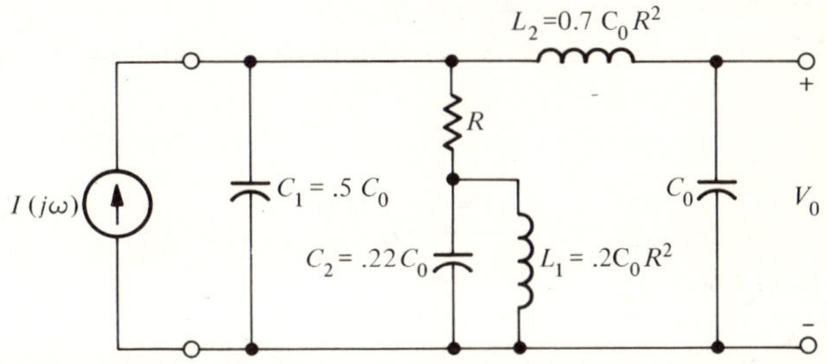

8-9. Repeat Problem 8-8 for the linear phase adjustment of the series-shunt peaking network

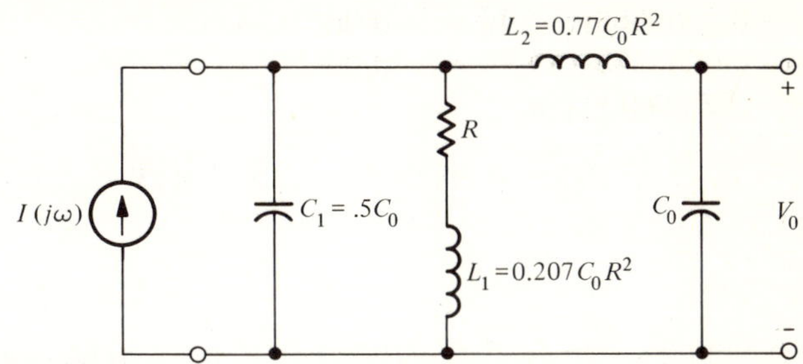

8-10. Use the program of Figure 8.4 to obtain data for plotting the frequency response of the following transfer functions:

a. $G(j\omega) = \dfrac{K}{1 + j\omega T_1}$

b. $G(j\omega) = K\dfrac{1 + j\omega T_1}{1 + j\omega T_2}$

c. $G(j\omega) = K\dfrac{1}{(1 + j\omega T_1)(1 + j\omega T_2)}$

d. $G(j\omega) = K\dfrac{1 + j\omega T_1}{(j\omega)^2(1 + j\omega T_2)}$

e. $G(j\omega) = K\dfrac{(1 + j\omega T_2)(1 + j\omega T_3)}{(1 + j\omega T_1)(1 + j\omega T_4)}$

Use values of $K = 10.0$, $T_1 = 10.0$, $T_2 = 5.0$, $T_3 = 2.5$, and $T_4 = 1.0$.

8-11. Using the arithmetic statement function, write a program for evaluating each of the following functions:[12]

a. $\text{Log}_{10} X = A_1Y + A_3Y^3 + A_5Y^5 + A_7Y^7 + A_9Y^9$

where

$A_1 = 0.8685917 \quad A_5 = 0.1775221 \quad A_9 = 0.1913377$
$A_3 = 0.2893355 \quad A_7 = 0.0943765$

b. $\epsilon^{-X} = 1 + A_1X + A_2X^2 + A_3X^3 + A_4X^4 + A_5X^5 + A_6X^6 + A_7X^7$

where

$A_1 = -1.0000000 \quad A_3 = -0.1666653 \quad A_5 = -0.0083014$
$A_2 = 0.5000000 \quad A_4 = 0.0416573 \quad A_6 = 0.0013299$
$A_7 = -0.0001413$

c. $\dfrac{\sin X}{X} = 1 + A_2X^2 + A_4X^4 + A_6X^6 + A_8X^8 + A_{10}X^{10}$

where

$A_2 = -0.1666666 \quad A_6 = -0.0001984$
$A_4 = 0.0083333 \quad A_8 = 0.00000275$
$A_{10} = -0.0000000239$

d. $\dfrac{\tan X}{X} = 1 + A_2X^2 + A_4X^4 + A_6X^6 + A_8X^8 + A_{10}X^{10} + A_{12}X^{12}$

where

$A_2 = 0.3333314 \quad A_6 = 0.0533741 \quad A_{10} = 0.0029005$
$A_4 = 0.1333924 \quad A_8 = 0.0245651 \quad A_{12} = 0.0095168$

e. $\arcsin X = \pi/2 - (1 - X)^{0.5}(A_o + A_1X + A_2X^2 + A_3X^3 + A_4X^4 + A_5X^5 + A_6X^6 + A_7X^7)$

where

$\pi = 3.14159265$
$A_0 = 1.5707963 \quad A_2 = 0.0889790 \quad A_4 = 0.0308919$
$A_1 = -0.2145988 \quad A_3 = -0.0501743 \quad A_5 = -0.0170881$
$A_6 = 0.0066701 \quad A_7 = -0.0012625$

Test each of the above programs for appropriate values of X.

8-12. Repeat Problem 8-11 by writing each of the programs as a function subprogram.

[12] See Abramowitz and Stegun, eds., *Handbook of Mathmetical Functions*, pp. 68, 71, 76, 81.

8-13. Run each of the following programs and comment on the results:

```
   B = -20.0
   DO 10 J=1,3
   B = B + 10.0
   A = -20.0
   DO 10 I = 1,7
   A = A + 5.0
   PHI = ATAN(B/A)
   DEG = PHI*57.2957795
10 WRITE(6,5)PHI, DEG, A, B
 5 FORMAT(4E15.7)
   STOP
   END
```

```
   B = -20.0
   DO 10 J=1,3
   B = B + 10.0
   A = -20.0
   DO 10 I = 1,7
   A = A + 5.0
   PHI = ATAN(B/A)
   IF(A) 20, 30, 30
20 PHI = PHI + 3.14159265
30 DEG = PHI*57.2957795
10 WRITE(6,5)PHI, DEG, A, B
 5 FORMAT(4E15.7)
   STOP
   END
```

8-14. Figure 8.13 contains a function subprogram for a four-quadrant arc tangent function. Is it necessary to have a six- or eight-quadrant arc tangent function?

8-15. Explain how computer statement S.0015 in the subprogram of Figure 8.16 permits the evaluation of the gamma function for argument values that are not exact table values.

8-16. Hand compute the value of

a. $\Gamma(1.1)$ b. $\Gamma(3.0)$ c. $\Gamma(-1.5)$ d. $\Gamma(1.1525)$ e. $\Gamma(0.5)$

by using the recurrence relation

$$\Gamma(A + 1) = A*\Gamma(A);\ A > 1 \text{ or } \Gamma(A) = \frac{\Gamma(A + 1)}{A};\ A < 1$$

Compare your results with those shown in Figure 8.17.

8-17. Use the function subprogram SIMP (Figure 8.18) to compute the RMS value of the waveform of Figure 5.15.

8-18. Repeat Problem 8-17 using the function subprogram of Figure 8.21.

8-19. Using the trapezoidal rule, write a program similar to that of Figure 8.21.

8-20. Using double-precision arithmetic, investigate the result of Figure 8.19 and see whether the major source of error in the program of Figure 8.18 is due to round-off error.

8-21. Modify the program of Figure 8.21 so that it prints out the value of H. Rerun this program for several different values of ACC. Verify your computer results by hand computation.

8-22. Write a function subprogram of your own choosing that will prove useful in subsequent work.

8-23. Write a subroutine subprogram that uses a power series expansion to compute the value of the sin X to a specified accuracy.

8-24. Modify the continued fraction expansion program of Figure 8.23 so that it may be used to compute the following functions:[13]

a. $$\ln(1+Z) = \cfrac{Z}{1+\cfrac{Z}{2+\cfrac{Z}{3+\cfrac{4Z}{4+\cfrac{4Z}{5+\cfrac{9Z}{6+\cdots}}}}}}$$

b. $$\epsilon^{Z} = \cfrac{1}{1-\cfrac{Z}{1+\cfrac{Z}{2-\cfrac{Z}{3+\cfrac{Z}{2-\cfrac{Z}{5+\cfrac{Z}{2-\cdots}}}}}}}$$

c. $$\tanh Z = \cfrac{Z}{1+\cfrac{Z^2}{3+\cfrac{Z^2}{5+\cfrac{Z^2}{7+\cdots}}}}$$

d. $$\text{arc tanh } Z = \cfrac{Z}{1-\cfrac{Z^2}{3-\cfrac{4Z^2}{5-\cfrac{9Z^2}{7-\cdots}}}}$$

8-25. Can the continued fraction expansion program of Figure 8.23 be modified to compute the input resistance of the ladder network (see Figure 3.28)?

[13] *Handbook of Mathematical Functions.* See Formulas 4.1.39, 4.2.40, 4.5.70, and 4.6.35.

8-26. Rerun the program of Figure 8.23 without using the FACT1 and FACT2 variables. Explain why they are needed.

8-27. Modify the program of Figure 8.23 to include double-precision arithmetic operations. Test the resulting program to see whether it is capable of computing the arc tangent function to fifteen significant figures of accuracy.

8-28. Use subroutine COMPLX (Figure 8.25) to determine the frequency response of the modified shunt-peaked network of Figure 8.3 (see Figures 8.6 and 8.7).

8-29. Repeat Problem 8-28 for the network of Problem 8-7.

8-30. Use the program of Figure 8.30 to find the roots of the following polynomials:

a. $P_2(s) = s^2 + 2.0\ 10^6 s + 100$

b. $P_2(s) = s^2 + 5.0\ 10^9 s + 100$

c. $P_4(s) = s^4 + 6s^3 + 14s^2 + 16s + 8$

d. $P_4(s) = s^4 + 10s^3 + 35s^2 + 50s + 24$

e. $P_4(s) = s^4 + 12.1s^3 + 23.2s^2 + 22.2s + 2$

f. $P_7(s) = (s + 1 + j1)(s + 1 - j1)(s + 1 + j2)(s + 1 - j2)(s + 1)(s + 2)(s + 3)$

8-31. Use subroutine SIMQ (see Figure 8.32) to find the solution to the equations of

a. the network of Figure 6.8.

b. the network of Figure 6.11.

8-32. Repeat Problem 8-31 using subroutine MINV and GMPRD.

8-33. Investigate the accuracy of the results of Figure 8.34 for argument values ranging from 10^{-9} to 10^{70}.

8-34. Rerun the program of Figure 8.36 for values of N equal to 1, 2, and 3.

8-35. Rerun the program of Figure 8.36 for values of D equal to 1.E-5 and 1.E-6.

8-36. Rerun the program of Figure 8.38 for at least three other starting values and compare the number of numbers generated within one standard deviation for each of these starting values.

8-37. Modify the program of Figure 8.42 so that it is capable of plotting

a. The frequency response of Figures 8.6 and 8.7.

b. The frequency response of Problem 8-7.

8-38. Write at least three different calling programs that make use of System/360 Scientific Subroutines not discussed in Section 8.5.

CHAPTER 9

Matrices

9.1 INTRODUCTION

This chapter concerns itself with some of the more common matrix operations and their topological applications. Before we begin a formal discussion of this material, let us first recall some of the basic properties of matrices.

A matrix is a rectangular array of elements, consisting of M rows and N columns (referred to as an M by N matrix). A matrix by itself has no numerical value nor can it be evaluated since its sole purpose is to store an ordered collection of facts. For example, if we consider the M mesh equations of an M-mesh network:

$$\begin{array}{ccccccccc} Z_{11}i_1 & + & Z_{12}i_2 & + & \cdots & + & Z_{1M}i_M & = & v_1 \\ Z_{21}i_1 & + & Z_{22}i_2 & + & \cdots & + & Z_{2M}i_M & = & v_2 \\ \cdot & & \cdot & & & & \cdot & & \cdot \\ \cdot & & \cdot & & & & \cdot & & \cdot \\ \cdot & & \cdot & & & & \cdot & & \cdot \\ Z_{M1}i_1 & + & Z_{M2}i_2 & + & \cdots & + & Z_{MM}i_M & = & v_M \end{array}$$

we may use the mesh-impedance matrix to describe the passive elements of the network and how they are connected.

$$[Z_{\text{mesh}}] = \begin{bmatrix} Z_{11} & Z_{12} & \cdots & Z_{1M} \\ Z_{21} & Z_{22} & \cdots & Z_{2M} \\ \cdot & \cdot & & \cdot \\ \cdot & \cdot & & \cdot \\ \cdot & \cdot & & \cdot \\ Z_{M1} & Z_{M2} & \cdots & Z_{MM} \end{bmatrix}$$

Note that the mesh-impedance matrix is a square matrix since the number of rows equals the number of columns. Note also that we must not confuse the mesh-impedance determinant with the mesh-impedance matrix since the former may be evaluated, but the latter is only an ordered collection of network information.

In a similar way we can describe the mesh currents and the excitation voltages in terms of column matrices ($N = 1$)

$$[i_{\text{mesh}}] = \begin{bmatrix} i_1 \\ i_2 \\ \cdot \\ \cdot \\ \cdot \\ i_M \end{bmatrix} \qquad [v] = \begin{bmatrix} v_1 \\ v_2 \\ \cdot \\ \cdot \\ \cdot \\ v_M \end{bmatrix}$$

Using these three matrices it is thus possible to write the original M-mesh equations as a single matrix equation

$$\begin{bmatrix} Z_{11} & Z_{12} & \cdots & Z_{1M} \\ Z_{21} & Z_{22} & \cdots & Z_{2M} \\ \cdot & \cdot & & \cdot \\ \cdot & \cdot & & \cdot \\ \cdot & \cdot & & \cdot \\ Z_{M1} & Z_{M2} & \cdots & Z_{MM} \end{bmatrix} \begin{bmatrix} i_1 \\ i_2 \\ \cdot \\ \cdot \\ \cdot \\ i_M \end{bmatrix} = \begin{bmatrix} v_1 \\ v_2 \\ \cdot \\ \cdot \\ \cdot \\ v_M \end{bmatrix}$$

or

$$[Z_{\text{mesh}}][i_{\text{mesh}}] = [v]$$

The above equation indicates that the product of $[Z_{\text{mesh}}]$ and $[i_{\text{mesh}}]$ is equal to the voltage excitation matrix. This implies that the matrix resulting from the product of $[Z_{\text{mesh}}]$ and $[i_{\text{mesh}}]$ has the same number of rows and columns as $[v]$ and therefore the corresponding elements of each matrix must be equal to each other.

It is also possible to solve the above matrix equation for $[i_{\text{mesh}}]$. This may be done by first premultiplying both sides of the equation by the inverse of the mesh-impedance matrix

$$[Z_{\text{mesh}}]^{-1}[Z_{\text{mesh}}][i_{\text{mesh}}] = [Z_{\text{mesh}}]^{-1}[v]$$

and then recognizing that

$$[Z_{\text{mesh}}]^{-1}[Z_{\text{mesh}}] = [I]$$

where $[I]$, the identity matrix, is a diagonal matrix in which $Z_{IJ} = 0$ whenever $I \neq J$ and $Z_{IJ} = 1$ whenever $I = J$. Since the product of $[I]$ times $[i_{\text{mesh}}]$ is equal to $[i_{\text{mesh}}]$, it follows that

$$[i_{\text{mesh}}] = [Z_{\text{mesh}}]^{-1}[v]$$

The above two mathematical operations (matrix multiplication and matrix inversion), along with matrix addition, subtraction, and transposition, play an important role in our ability to program a digital computer to formulate and solve the algebraic or integrodifferential equations that describe a given linear or nonlinear network. Let us therefore consider these basic matrix operations in more detail.

9.2 BASIC MATRIX OPERATIONS

Addition and Subtraction

Two matrices, $[A]$ and $[B]$, may be added or subtracted provided they have the same number of rows and columns. The sum of two M by N matrices is found by adding the corresponding elements of $[A]$ and $[B]$. Thus if

$$[A] = \begin{bmatrix} a_{11} & a_{12} & a_{13} \\ a_{21} & a_{22} & a_{23} \end{bmatrix} \qquad \text{and} \qquad [B] = \begin{bmatrix} b_{11} & b_{12} & b_{13} \\ b_{21} & b_{22} & b_{23} \end{bmatrix}$$

their sum, $[C]$, would be equal to

$$[C] = [A] + [B] = \begin{bmatrix} a_{11} + b_{11} & a_{12} + b_{12} & a_{13} + b_{13} \\ a_{21} + b_{21} & a_{22} + b_{22} & a_{23} + b_{23} \end{bmatrix}$$

In general, provided the matrices have the same dimensions, the following Fortran statements may be used to find the sum:

```
   DO 10 I=1,M
   DO 10 J=1,N
10 C(I,J)=A(I,J) + B(I,J)
```

The difference of two M by N matrices is found by subtracting the corresponding elements of $[A]$ and $[B]$, thus

$$[C] = [A] - [B]$$

or

```
   DO 10 I=1,M
   DO 10 J=1,N
10 C(I,J)=A(I,J) - B(I,J)
```

Multiplication

Matrix multiplication requires that the premultiplying matrix $[A]$ have as many columns as there are rows in the postmultiplying matrix $[B]$. In other words, if matrix $[A]$ is an MA by NA matrix and matrix $[B]$ is an

MB by *NB* matrix and *NA* equals *MB*, the resulting matrix, $[C]$, will be an *MA* by *NB* matrix with each element equal to

$$C_{IJ} = \sum_{K=1}^{NA} a_{IK}b_{KJ} \text{ where } I = 1, 2, \ldots, MA \text{ and } J = 1, 2, \ldots, NB.$$

or

$$C_{IJ} = \sum_{I=1}^{MA} \sum_{J=1}^{NB} \sum_{K=1}^{NA} a_{IK}b_{KJ}$$

where initially C_{IJ} is set equal to zero for each I and J value before computing the summation with respect to K. Thus if

$$[A] = \begin{bmatrix} a_{11} & a_{12} & a_{13} \\ a_{21} & a_{22} & a_{23} \end{bmatrix} \quad \text{and} \quad [B] = \begin{bmatrix} b_{11} & b_{12} \\ b_{21} & b_{22} \\ b_{31} & b_{32} \end{bmatrix}$$

their product, $[C]$, would be equal to

$$[C] = [A][B] = \begin{bmatrix} a_{11}b_{11} + a_{12}b_{21} + a_{13}b_{31} & a_{11}b_{12} + a_{12}b_{22} + a_{13}b_{32} \\ a_{21}b_{11} + a_{22}b_{21} + a_{23}b_{31} & a_{21}b_{12} + a_{22}b_{22} + a_{23}b_{32} \end{bmatrix}$$

In general, providing $[A]$ and $[B]$ are conformable,[1] the following Fortran statements may be used to find the product:

```
      DO 10 I = 1, MA
      DO 10 J=1,NB
      C(I,J)=0.0
      DO 10 K=1,NA
   10 C(I,J)=C(I,J) + A(I,K)*B(K,J)
```

Transposition

There are no restrictions on transposing a matrix. If matrix $[A]$ is an *M* by *N* matrix, its transpose, $[C]$ ($[A]_T$), will be an *N* by *M* matrix whose rows are the columns of $[A]$, and vice versa. Thus if

$$[A] = \begin{bmatrix} a_{11} & a_{12} & a_{13} \\ a_{21} & a_{22} & a_{23} \end{bmatrix}$$

its transpose will be

$$[C] = [A]_T = \begin{bmatrix} a_{11} & a_{21} \\ a_{12} & a_{22} \\ a_{13} & a_{23} \end{bmatrix}$$

[1] If the number of columns of $[A]$ equal the number of rows of $[B]$, then $[A]$ and $[B]$ are said to be conformable in the order $[A][B]$ and their product $[C] = [A][B]$ exists. Note that $[A][B] \neq [B][A]$ in general, since if $[A]$ and $[B]$ are conformable in the order $[A][B]$, they will in general not be conformable in the order $[B][A]$.

The following Fortran statements may therefore be used to find the transpose of a general M by N matrix:

```
      DO 10 I=1,M
      DO 10 J=1,N
   10 C(J,I)=A(I,J)
```

Inversion

In Section 9.1 we pointed out that the identity or unit matrix, $[I]$, was a square matrix which had all off-diagonal elements equal to zero and all principal diagonal elements equal to one. The inverse of the matrix $[A]$ (denoted as $[A]^{-1}$) has the property that

$$[A][A]^{-1} = [A]^{-1}[A] = [I]$$

and the only restriction in determining the inverse of a matrix is that $[A]$ be nonsingular. That is, $[A]$ must be a square matrix and its determinant must be nonzero (Det $[A] \neq 0$).

One method of determining the inverse of a matrix is based on using the adjoint matrix which is the transpose of the cofactor matrix, thus

$$[A]_A \equiv \text{cof } [A]_T$$

and

$$[A]^{-1} = \frac{[A]_A}{|A|} = \begin{bmatrix} \Delta_{11}/\Delta & \Delta_{21}/\Delta & \cdots & \Delta_{N1}/\Delta \\ \Delta_{12}/\Delta & \Delta_{22}/\Delta & \cdots & \Delta_{N2}/\Delta \\ \cdot & \cdot & & \cdot \\ \cdot & \cdot & & \cdot \\ \cdot & \cdot & & \cdot \\ \Delta_{1N}/\Delta & \Delta_{2N}/\Delta & \cdots & \Delta_{NN}/\Delta \end{bmatrix}$$

After the rows and columns have been interchanged, each element is replaced by its cofactor divided by the value of the determinant. Although it is possible to write a computer program for this procedure, it is to our advantage to look at an alternate method that utilizes one of the numerical techniques that we have already programmed in Chapter 6.

This alternate method for finding the inverse of a matrix makes use of the Gauss-Jordan reduction method. The method is based on the property that the product of a matrix and its inverse is equal to the identity matrix

$$[A][A]^{-1} = [I]$$

Therefore, if

$$[A] = \begin{bmatrix} a_{11} & a_{12} \\ a_{21} & a_{22} \end{bmatrix} \quad \text{and} \quad [I] = \begin{bmatrix} 1 & 0 \\ 0 & 1 \end{bmatrix}$$

we must find the element values of $[A]^{-1}$

$$[A]^{-1} = \begin{bmatrix} b_{11} & b_{12} \\ b_{21} & b_{22} \end{bmatrix}$$

in order to find the inverse of $[A]$.

These element values may be found by using the rules of matrix multiplication; thus,

$$[A][A]^{-1} = [I]$$

corresponds to

$$\begin{bmatrix} a_{11} & a_{12} \\ a_{21} & a_{22} \end{bmatrix} \begin{bmatrix} b_{11} & b_{12} \\ b_{21} & b_{22} \end{bmatrix} = \begin{bmatrix} 1 & 0 \\ 0 & 1 \end{bmatrix}$$

or

$$\begin{bmatrix} a_{11}b_{11} + a_{12}b_{21} & a_{11}b_{12} + a_{12}b_{22} \\ a_{21}b_{11} + a_{22}b_{21} & a_{21}b_{12} + a_{22}b_{22} \end{bmatrix} = \begin{bmatrix} 1 & 0 \\ 0 & 1 \end{bmatrix}$$

It follows that there are two different sets of equations involving the unknowns b_{11}, b_{21}, b_{12}, and b_{22}

$$\begin{cases} a_{11}b_{11} + a_{12}b_{21} = 1 \\ a_{21}b_{11} + a_{22}b_{21} = 0 \end{cases}$$

$$\begin{cases} a_{11}b_{12} + a_{12}b_{22} = 0 \\ a_{21}b_{12} + a_{22}b_{22} = 1 \end{cases}$$

These equations may now be solved as a set of equations using the Gauss-Jordan reduction method. This may be done by first realizing that the solution to the first set of equations would start with the augmented matrix

$$\begin{bmatrix} a_{11} & a_{12} & 1 \\ a_{21} & a_{22} & 0 \end{bmatrix}$$

Using the Gauss-Jordan method of solution, this matrix would then be forced into the following form

$$\begin{bmatrix} 1 & 0 & b_{11} \\ 0 & 1 & b_{21} \end{bmatrix}$$

where the right column terms would contain the solution of the set.

In a similar way, the second set of equations would start with the augmented matrix

$$\begin{bmatrix} a_{11} & a_{12} & 0 \\ a_{21} & a_{22} & 1 \end{bmatrix}$$

and be forced into the following form

$$\begin{bmatrix} 1 & 0 & b_{12} \\ 0 & 1 & b_{22} \end{bmatrix}$$

where again the right column terms would contain the solution of this set of equations.

From the above procedure we see that it is not necessary to solve these sets of equations individually. Since the same reduction steps are required for each separate solution, it is therefore possible to solve both sets of equations at the same time. This is done by including the right member constant terms of each set in the augmented matrix. The solution of the set of equations thus requires us to form a double square augmented matrix that contains the elements of $[A]$ and the unit matrix

$$\begin{bmatrix} a_{11} & a_{12} & 1 & 0 \\ a_{21} & a_{22} & 0 & 1 \end{bmatrix}$$

We now use the Gauss-Jordan reduction procedure to force the $[A]$ matrix to have element values that correspond to that of the unit matrix. As a consequence of this operation, the right-hand members will contain the solution to both sets of equations or they will correspond to the elements of $[A]^{-1}$.

As a simple numerical example that illustrates this procedure, let us assume that

$$[A] = \begin{bmatrix} 3 & -2 \\ 2 & -1 \end{bmatrix}$$

In order to find $[A]^{-1}$, we must first form the augmented matrix

$$\begin{bmatrix} 3 & -2 & 1 & 0 \\ 2 & -1 & 0 & 1 \end{bmatrix}$$

Using the Gauss-Jordan procedure, we now normalize row 1 with respect to its pivot element value and then, in turn, eliminate a_{21} by forming a new row two; thus,

$$\begin{bmatrix} 1 & -\frac{2}{3} & \frac{1}{3} & 0 \\ 0 & \frac{1}{3} & -\frac{2}{3} & 1 \end{bmatrix}$$

Next we normalize row two with respect to its pivot element value and eliminate a'_{12} $(-2/3)$ by forming a new row one; thus,

$$\begin{bmatrix} 1 & 0 & -1 & 2 \\ 0 & 1 & -2 & 3 \end{bmatrix}$$

We note that

$$[A]^{-1} = \begin{bmatrix} -1 & 2 \\ -2 & 3 \end{bmatrix}$$

and it further follows that

$$[A][A]^{-1} = [I]$$

since

$$\begin{bmatrix} 3 & -2 \\ 2 & -1 \end{bmatrix} \begin{bmatrix} -1 & 2 \\ -2 & 3 \end{bmatrix} = \begin{bmatrix} -3+4 & 6-6 \\ -2+2 & 4-3 \end{bmatrix}$$
$$= \begin{bmatrix} 1 & 0 \\ 0 & 1 \end{bmatrix}$$

Before we consider the Fortran program for this operation, it is advantageous to introduce a slight modification in the procedure and thereby save considerable computer storage.

The simplest first approach in writing a computer program for this procedure would be to set up an N by $2N$ augmented matrix whose left half corresponds to the matrix we wish to invert and whose right half corresponds to the unit matrix. The Gauss-Jordan elimination process would then force the left half to be equal to the unit matrix and the right half to be equal to the inverse matrix. Note, however, that during the elimination procedure after the Kth elimination, the first K columns of the left half would correspond to the first K columns of the unit matrix, and the last $N - K$ columns of the right half would be unchanged.

Obviously it is possible to save considerable storage by using only an N by $N + 1$ augmented matrix. Our program would therefore require that at each stage of the elimination procedure, a column of the unit matrix would be stored in the $N + 1$ column of the augmented matrix. For example, at the beginning of the Ith stage of elimination, the Ith column of the unit matrix would be stored in the $N + 1$ column of the augmented matrix. At the end of the Kth stage of the elimination procedure, the contents of the $N + 1$ column would be transferred to the Kth column before we started the next stage of elimination. At each iteration of the elimination procedure, the contents of the $N + 1$ column would be transferred to successive columns of the augmented matrix so that at the end of the process, the first N columns of the augmented matrix would contain the inverse matrix rather than the unit matrix.

Using the above example to illustrate the modified procedure, we would have at the beginning of the first stage of the elimination

$$\begin{bmatrix} 3 & -2 & 1 \\ 2 & -1 & 0 \end{bmatrix}$$

The elimination process would then yield the following matrix

$$\begin{bmatrix} 1 & -\frac{2}{3} & \frac{1}{3} \\ 0 & \frac{1}{3} & -\frac{2}{3} \end{bmatrix}$$

and before going on to the second stage of the elimination, we would transfer the contents of the 3rd column ($N + 1$ column) to the 1st (K column)

$$\begin{bmatrix} \frac{1}{3} & -\frac{2}{3} & \frac{1}{3} \\ -\frac{2}{3} & \frac{1}{3} & -\frac{2}{3} \end{bmatrix}$$

At the beginning of the second stage (Kth stage) of elimination, we would make the 3rd column ($N + 1$ column) equal to the 2nd column of the unit matrix

$$\begin{bmatrix} \frac{1}{3} & -\frac{2}{3} & 0 \\ -\frac{2}{3} & \frac{1}{3} & 1 \end{bmatrix}$$

The elimination procedure at this stage would then produce the following matrix

$$\begin{bmatrix} -1 & 0 & 2 \\ -2 & 1 & 3 \end{bmatrix}$$

and, in turn, in transferring the contents of the 3rd column to the 2nd column, we would be able to obtain the inverse matrix from the first N columns

$$\begin{bmatrix} -1 & 2 & 2 \\ -2 & 3 & 3 \end{bmatrix}$$

Figure 9.1 is a Fortran program in which each of the above matrix operations has been written as a subroutine. The program initially reads in the value of I which is used to indicate the matrix operation desired. Computer statement S.0005 then transfers control to statement 10, 20, 30, 40, or 50, depending on whether the user has requested matrix addition, subtraction, multiplication, transposition, or inversion.

Computer statements S.0006 through S.0023 are used in the matrix addition operation. Computer statements S.0006 through S.0012 read into memory and then print the element values of both matrices. Computer statements S.0013 and S.0014 then determine whether the matrices have the same dimensions. If they do not have the same dimensions, computer statements S.0015 and S.0016 print that the matrices are not compatible before control is transferred to statement 100 so that the next set of data can be processed. If the matrices are compatible, control is transferred to statement 7 and computer statement S.0020, in turn, is used to call in subroutine ADDM. Note that subroutine ADDM contains the Fortran statements for addition that were discussed above. After the sum of the two matrices has been found, control is returned to computer statement S.0021 in the main program, and the results are printed before going on to the next set of data.

Computer statements S.0024 through S.0035, S.0036 through S.0044, and S.0045 through S.0051 are used respectively in matrix subtraction, multiplication, and transposition. Since these operations are carried out in a way similar to the operation of matrix addition, it is left as an exercise for the reader to determine how each operation is performed.

Computer statements S.0052 through S.0062 are used in matrix inversion. The matrix that is to be inverted is initially read into memory and the numerical values of the elements are printed. Computer statement S.0055 then tests to see whether the matrix is square. If $[A]$ is a square matrix, computer statement S.0059 calls in subroutine DINVM.

Turning now to the subroutine, we see that the DO 5 loops are used to store the elements of the A matrix in the first N columns of the D matrix. The number of columns, M, of the augmented matrix is now computed, and the dummy variables KK and JJ are set equal to zero. We then enter the DO 10 loop, setting K initially equal to one. The DO 11 loop, in turn, sets each of the elements of the Mth $(N + 1)$ column of the augmented matrix equal to zero, and computer statement S.0012 then changes the value of $D(K,M)$ to one. Note that this is equivalent, in the modified procedure, to storing the Kth column of the unit matrix in the $N + 1$ (M) column of the augmented matrix. The reader may now observe that computer statements S.0013 through S.0032 are essentially identical to computer statements S.0012 through S.0031 in the Gauss-Jordan elimination program of Figure 6.13. These statements will therefore carry out the Kth stage of elimination. At the end of the Kth stage of the elimination process, computer statements S.0035 and S.0036 are used to transfer the contents of the Mth column to the Kth column before the next stage of the elimination procedure is carried out.

Figure 9.2 shows the numerical results obtained in testing each of these subroutines. The reader is urged to hand check a portion of each one of these results in order to verify the accuracy of the individual subroutines.

In order to show how these matrix operation subroutines may be used by a general calling program, consider once again the solution of the four-mesh network shown in Figure 6.4. The matrix equation for this network is

$$[A][C] = [V]$$

where $[A]$ is the mesh-impedance matrix, $[C]$ is the mesh-current matrix, and $[V]$ is the voltage excitation matrix. The matrix solution for the

```
C        MATRIX OPERATIONS
C
         DIMENSION A(10,10),B(10,10),C(10,10),D(10,11)
         COMMON A,B,C
   100   READ(5,1)I
     1   FORMAT(I5)
         GO TO (10,20,30,40,50),I
C
C        MATRIX ADDITION
C
    10   READ(5,2) MA,NA,((A(I,J),J=1,NA),I=1,MA)
     2   FORMAT(2I5,/,(5E15.7))
         READ(5,2) MB,NB,((B(I,J),J=1,NB),I=1,MB)
         WRITE(6,3) ((I,J,A(I,J),J=1,NA),I=1,MA)
     3   FORMAT('1MATRIX ADDITION:  C = A + B',///,' A MATRIX',//,
        15(2X,'A('I2,','I2,') =',E14.7))
         WRITE(6,4) ((I,J,B(I,J),J=1,NB),I=1,MB)
     4   FORMAT(///,' B MATRIX',//,5(2X,'B(',I2,',',I2,') =',E14.7))
         IF(MA-MB)5,6,5
     6   IF(NA-NB)5,7,5
     5   WRITE(6,8)
     8   FORMAT(///,' MATRICES ARE NOT COMPATIBLE',////)
         GO TO 100
     7   M=MA
         N=NA
         CALL ADDM(A,B,C,M,N)
         WRITE(6,9) ((I,J,C(I,J),J=1,N),I=1,M)
     9   FORMAT(///,' C MATRIX',//,5(2X,'C(',I2,',',I2,') =',E14.7))
         GO TO 100
C
C        MATRIX SUBTRACTION.
C
    20   READ(5,2) MA,NA,((A(I,J),J=1,NA),I=1,MA)
         READ(5,2) MB,NB,((B(I,J),J=1,NB),I=1,MB)
         WRITE(6,21) ((I,J,A(I,J),J=1,NA),I=1,MA)
    21   FORMAT('1MATRIX SUBTRACTION:  C = A - B',///,' A MATRIX',//,
        15(2X,'A(',I2,',',I2,') =',E14.7))
         WRITE(6,4) ((I,J,B(I,J),J=1,NB),I=1,MB)
         IF(MA-MB)5,22,5
    22   IF(NA-NB)5,23,5
    23   M=MA
         N=NA
         CALL SUBM(A,B,C,M,N)
         WRITE(6,9) ((I,J,C(I,J),J=1,N),I=1,M)
         GO TO 100
C
C        MATRIX MULTIPLICATION
C
    30   READ(5,2) MA,NA,((A(I,J),J=1,NA),I=1,MA)
         READ(5,2) MB,NB,((B(I,J),J=1,NB),I=1,MB)
         WRITE(6,31) ((I,J,A(I,J),J=1,NA),I=1,MA)
    31   FORMAT('1MATRIX MULTIPLICATION:  C = A*B',///,' A MATRIX',//,
        15(2X,'A(',I2,',',I2,') =',E14.7))
         WRITE(6,4) ((I,J,B(I,J),J=1,NB),I=1,MB)
         IF(NA-MB)5,32,5
    32   CALL PRODM(A,B,C,MA,NB,NA)
         WRITE(6,9) ((I,J,C(I,J),J=1,NB),I=1,MA)
         GO TO 100
C
C        MATRIX TRANSPOSITION.
C
    40   READ(5,2) M,N,((A(I,J),J=1,N),I=1,M)
         WRITE(6,41) ((I,J,A(I,J),J=1,N),I=1,M)
    41   FORMAT('1MATRIX TRANSPOSITION;  C = A TRANSPOSE',///,' A MATRIX',/
        1/,5(2X,'A(',I2,',',I2,') =',E14.7))
         CALL TRAM(A,C,M,N)
         WRITE(6,42) ((I,J,C(I,J),J=1,N),I=1,M)
    42   FORMAT(///,' A TRANSPOSE',//,5(2X,'C(',I2,',',I2,') =',E14.7))
         GO TO 100
C
C        MATRIX INVERSION.
C
    50   READ(5,2) M,N,((A(I,J),J=1,N),I=1,M)
         WRITE(6,51) ((I,J,A(I,J),J=1,N),I=1,M)
    51   FORMAT('1MATRIX INVERSION:  D = A INVERSE',///,' A MATRIX',//,
        15(2X,'A(',I2,',',I2,') =',E14.7))
         IF(M-N)52,53,52
    52   WRITE(6,54)
    54   FORMAT(///,' MATRIX IS NOT SQUARE',////)
         GO TO 100
    53   CALL DINVM(A,D,N)
         WRITE(6,55) ((I,J,D(I,J),J=1,M),I=1,M)
    55   FORMAT(///,' A INVERSE',//,5(2X,'D(',I2,',',I2,') =',E14.7))
         GO TO 100
         END
```

Figure 9.1 A test program for matrix subroutines ADDM, SUBM, PRODM, TRAM, and DINVM

```
      SUBROUTINE ADDM(A,B,C,M,N)
      DIMENSION A(10,10),B(10,10),C(10,10)
      DO 10 I=1,M
      DO 10 J=1,N
   10 C(I,J)=A(I,J)+B(I,J)
      RETURN
      END
```

```
      SUBROUTINE SUBM(A,B,C,M,N)
      DIMENSION A(10,10),B(10,10),C(10,10)
      DO 10 I=1,M
      DO 10 J=1,N
   10 C(I,J)=A(I,J)-B(I,J)
      RETURN
      END
```

```
      SUBROUTINE PRODM(A,B,C,MA,NB,NA)
      DIMENSION A(10,10),B(10,10),C(10,10)
      DO 10 I=1,MA
      DO 10 J=1,NB
      C(I,J)=0.0
      DO 10 K=1,NA
   10 C(I,J)=C(I,J)+A(I,K)*B(K,J)
      RETURN
      END
```

```
      SUBROUTINE TRAM(A,C,M,N)
      DIMENSION A(10,10),C(10,10)
      DO 10 I=1,M
      DO 10 J=1,N
   10 C(J,I)=A(I,J)
      RETURN
      END
```

```
      SUBROUTINE DINVM(A,D,N).
      DIMENSION A(10,10),D(10,11)
      DO 5 I=1,N
      DO 5 J=1,N
    5 D(I,J)=A(I,J)
      M=N+1
      KK=0
      JJ=0
      DO 10 K=1,N
      DO 11 J=1,N
   11 D(J,M)=0.0
      D(K,M)=1.0
      JJ=KK+1
      LL=JJ
      KK=KK+1
   20 IF(ABS(D(JJ,KK))-1.E-4) 21,21,22
   21 JJ=JJ+1
      GO TO 20
   22 IF(LL-JJ)23,24,23
   23 DO 25 MM=1,M
      DTEMP=D(LL,MM)
      D(LL,MM)=D(JJ,MM)
   25 D(JJ,MM)=DTEMP
C
C     THE EQUATIONS ARE FORCED INTO THE DIAGONAL FORM.
C
   24 DIV=D(K,K)
      DO 30 LJ=1,M
      J=M+1-LJ
   30 D(K,J)=D(K,J)/DIV
      DO 12 I=1,N
      FAC=D(I,K)
      DO 12 LJ=1,M
      J=M+1-LJ
      IF (I-K)31,12,31
   31 D(I,J)=D(I,J)-FAC*D(K,J)
   12 CONTINUE
      DO 40 J=1,N
   40 D(J,K)=D(J,M)
   10 CONTINUE
      RETURN
      END
```

Figure 9.1 Continued

```
MATRIX ADDITION     C = A + B

A MATRIX

A( 1, 1) = 0.5000000E 01  A( 1, 2) = 0.1000000E 02  A( 1, 3) = 0.1500000E 02  A( 1, 4) = 0.2000000E 02  A( 1, 5) = 0.2500000E 02
A( 2, 1) = 0.3000000E 02  A( 2, 2) = 0.3500000E 02  A( 2, 3) = 0.4000000E 02  A( 2, 4) = 0.4500000E 02  A( 2, 5) = 0.5000000E 02
A( 3, 1) = 0.5500000E 02  A( 3, 2) = 0.6000000E 02  A( 3, 3) = 0.6500000E 02  A( 3, 4) = 0.7000000E 02  A( 3, 5) = 0.7500000E 02
A( 4, 1) = 0.8000000E 02  A( 4, 2) = 0.8500000E 02  A( 4, 3) = 0.9000000E 02  A( 4, 4) = 0.9500000E 02  A( 4, 5) = 0.1000000E 03
A( 5, 1) = 0.1050000E 03  A( 5, 2) = 0.1100000E 03  A( 5, 3) = 0.1150000E 03  A( 5, 4) = 0.1200000E 03  A( 5, 5) = 0.1250000E 03

B MATRIX

B( 1, 1) = 0.1250000E 03  B( 1, 2) = 0.1000000E 03  B( 1, 3) = 0.7500000E 02  B( 1, 4) = 0.5000000E 02  B( 1, 5) = 0.2500000E 02
B( 2, 1) = 0.1200000E 03  B( 2, 2) = 0.9500000E 02  B( 2, 3) = 0.7000000E 02  B( 2, 4) = 0.4500000E 02  B( 2, 5) = 0.2000000E 02
B( 3, 1) = 0.1150000E 03  B( 3, 2) = 0.9000000E 02  B( 3, 3) = 0.6500000E 02  B( 3, 4) = 0.4000000E 02  B( 3, 5) = 0.1500000E 02
B( 4, 1) = 0.1100000E 03  B( 4, 2) = 0.8500000E 02  B( 4, 3) = 0.6000000E 02  B( 4, 4) = 0.3500000E 02  B( 4, 5) = 0.1000000E 02
B( 5, 1) = 0.1050000E 03  B( 5, 2) = 0.8000000E 02  B( 5, 3) = 0.5500000E 02  B( 5, 4) = 0.3000000E 02  B( 5, 5) = 0.5000000E 01

C MATRIX

C( 1, 1) = 0.1300000E 03  C( 1, 2) = 0.1100000E 03  C( 1, 3) = 0.9000000E 02  C( 1, 4) = 0.7000000E 02  C( 1, 5) = 0.5000000E 02
C( 2, 1) = 0.1500000E 03  C( 2, 2) = 0.1300000E 03  C( 2, 3) = 0.1100000E 03  C( 2, 4) = 0.9000000E 02  C( 2, 5) = 0.7000000E 02
C( 3, 1) = 0.1700000E 03  C( 3, 2) = 0.1500000E 03  C( 3, 3) = 0.1300000E 03  C( 3, 4) = 0.1100000E 03  C( 3, 5) = 0.9000000E 02
C( 4, 1) = 0.1900000E 03  C( 4, 2) = 0.1700000E 03  C( 4, 3) = 0.1500000E 03  C( 4, 4) = 0.1300000E 03  C( 4, 5) = 0.1100000E 03
C( 5, 1) = 0.2100000E 03  C( 5, 2) = 0.1900000E 03  C( 5, 3) = 0.1700000E 03  C( 5, 4) = 0.1500000E 03  C( 5, 5) = 0.1300000E 03

MATRIX SUBTRACTION     C = A - B

A MATRIX

A( 1, 1) = 0.5000000E 01  A( 1, 2) = 0.1000000E 02  A( 1, 3) = 0.1500000E 02  A( 1, 4) = 0.2000000E 02  A( 1, 5) = 0.2500000E 02
A( 2, 1) = 0.3000000E 02  A( 2, 2) = 0.3500000E 02  A( 2, 3) = 0.4000000E 02  A( 2, 4) = 0.4500000E 02  A( 2, 5) = 0.5000000E 02
A( 3, 1) = 0.5500000E 02  A( 3, 2) = 0.6000000E 02  A( 3, 3) = 0.6500000E 02  A( 3, 4) = 0.7000000E 02  A( 3, 5) = 0.7500000E 02
A( 4, 1) = 0.8000000E 02  A( 4, 2) = 0.8500000E 02  A( 4, 3) = 0.9000000E 02  A( 4, 4) = 0.9500000E 02  A( 4, 5) = 0.1000000E 03
A( 5, 1) = 0.1050000E 03  A( 5, 2) = 0.1100000E 03  A( 5, 3) = 0.1150000E 03  A( 5, 4) = 0.1200000E 03  A( 5, 5) = 0.1250000E 03

B MATRIX

B( 1, 1) = 0.1250000E 03  B( 1, 2) = 0.1000000E 03  B( 1, 3) = 0.7500000E 02  B( 1, 4) = 0.5000000E 02  B( 1, 5) = 0.2500000E 02
B( 2, 1) = 0.1200000E 03  B( 2, 2) = 0.9500000E 02  B( 2, 3) = 0.7000000E 02  B( 2, 4) = 0.4500000E 02  B( 2, 5) = 0.2000000E 02
B( 3, 1) = 0.1150000E 03  B( 3, 2) = 0.9000000E 02  B( 3, 3) = 0.6500000E 02  B( 3, 4) = 0.4000000E 02  B( 3, 5) = 0.1500000E 02
B( 4, 1) = 0.1100000E 03  B( 4, 2) = 0.8500000E 02  B( 4, 3) = 0.6000000E 02  B( 4, 4) = 0.3500000E 02  B( 4, 5) = 0.1000000E 02
B( 5, 1) = 0.1050000E 03  B( 5, 2) = 0.8000000E 02  B( 5, 3) = 0.5500000E 02  B( 5, 4) = 0.3000000E 02  B( 5, 5) = 0.5000000E 01
```

Figure 9.2 The output results of the program of Figure 9.1

```
C MATRIX

C( 1, 1) =-0.1200000E 03  C( 1, 2) =-0.9000000E 02  C( 1, 3) =-0.6000000E 02  C( 1, 4) =-0.3000000E 02  C( 1, 5) = 0.0
C( 2, 1) =-0.9000000E 02  C( 2, 2) =-0.6000000E 02  C( 2, 3) =-0.3000000E 02  C( 2, 4) = 0.0             C( 2, 5) = 0.3000000E 02
C( 3, 1) =-0.6000000E 02  C( 3, 2) =-0.3000000E 02  C( 3, 3) = 0.0             C( 3, 4) = 0.3000000E 02  C( 3, 5) = 0.6000000E 02
C( 4, 1) =-0.3000000E 02  C( 4, 2) = 0.0             C( 4, 3) = 0.3000000E 02  C( 4, 4) = 0.6000000E 02  C( 4, 5) = 0.9000000E 02
C( 5, 1) = 0.0             C( 5, 2) = 0.3000000E 02  C( 5, 3) = 0.6000000E 02  C( 5, 4) = 0.9000000E 02  C( 5, 5) = 0.1200000E 03


MATRIX MULTIPLICATION    C = A*B


A MATRIX

A( 1, 1) = 0.5000000E 01  A( 1, 2) = 0.1000000E 02  A( 1, 3) = 0.1500000E 02  A( 1, 4) = 0.2000000E 02  A( 1, 5) = 0.2500000E 02
A( 2, 1) = 0.3000000E 02  A( 2, 2) = 0.3500000E 02  A( 2, 3) = 0.4000000E 02  A( 2, 4) = 0.4500000E 02  A( 2, 5) = 0.5000000E 02
A( 3, 1) = 0.5500000E 02  A( 3, 2) = 0.6000000E 02  A( 3, 3) = 0.6500000E 02  A( 3, 4) = 0.7000000E 02  A( 3, 5) = 0.7500000E 02
A( 4, 1) = 0.8000000E 02  A( 4, 2) = 0.8500000E 02  A( 4, 3) = 0.9000000E 02  A( 4, 4) = 0.9500000E 02  A( 4, 5) = 0.1000000E 03
A( 5, 1) = 0.1050000E 03  A( 5, 2) = 0.1100000E 03  A( 5, 3) = 0.1150000E 03  A( 5, 4) = 0.1200000E 03  A( 5, 5) = 0.1250000E 03


B MATRIX

B( 1, 1) = 0.1250000E 03  B( 1, 2) = 0.1000000E 03  B( 1, 3) = 0.7500000E 02  B( 1, 4) = 0.5000000E 02  B( 1, 5) = 0.2500000E 02
B( 2, 1) = 0.1200000E 03  B( 2, 2) = 0.9500000E 02  B( 2, 3) = 0.7000000E 02  B( 2, 4) = 0.4500000E 02  B( 2, 5) = 0.2000000E 02
B( 3, 1) = 0.1150000E 03  B( 3, 2) = 0.9000000E 02  B( 3, 3) = 0.6500000E 02  B( 3, 4) = 0.4000000E 02  B( 3, 5) = 0.1500000E 02
B( 4, 1) = 0.1100000E 03  B( 4, 2) =-0.8500000E 02  B( 4, 3) = 0.6000000E 02  B( 4, 4) = 0.3500000E 02  B( 4, 5) = 0.1000000E 02
B( 5, 1) = 0.1050000E 03  B( 5, 2) = 0.8000000E 02  B( 5, 3) = 0.5500000E 02  B( 5, 4) = 0.3000000E 02  B( 5, 5) = 0.5000000E 01


C MATRIX

C( 1, 1) = 0.8375000E 04  C( 1, 2) = 0.6500000E 04  C( 1, 3) = 0.4625000E 04  C( 1, 4) = 0.2750000E 04  C( 1, 5) = 0.8750000E 03
C( 2, 1) = 0.2275000E 05  C( 2, 2) = 0.1775000E 05  C( 2, 3) = 0.1275000E 05  C( 2, 4) = 0.7750000E 04  C( 2, 5) = 0.2750000E 04
C( 3, 1) = 0.3712500E 05  C( 3, 2) = 0.2900000E 05  C( 3, 3) = 0.2087500E 05  C( 3, 4) = 0.1275000E 05  C( 3, 5) = 0.4625000E 04
C( 4, 1) = 0.5150000E 05  C( 4, 2) = 0.4025000E 05  C( 4, 3) = 0.2900000E 05  C( 4, 4) = 0.1775000E 05  C( 4, 5) = 0.6500000E 04
C( 5, 1) = 0.6587500E 05  C( 5, 2) = 0.5150000E 05  C( 5, 3) = 0.3712500E 05  C( 5, 4) = 0.2275000E 05  C( 5, 5) = 0.8375000E 04


MATRIX TRANSPOSITION    C = A TRANSPOSE


A MATRIX

A( 1, 1) = 0.5000000E 01  A( 1, 2) = 0.1000000E 02  A( 1, 3) = 0.1500000E 02  A( 1, 4) = 0.2000000E 02  A( 1, 5) = 0.2500000E 02
A( 2, 1) = 0.3000000E 02  A( 2, 2) = 0.3500000E 02  A( 2, 3) = 0.4000000E 02  A( 2, 4) = 0.4500000E 02  A( 2, 5) = 0.5000000E 02
A( 3, 1) = 0.5500000E 02  A( 3, 2) = 0.6000000E 02  A( 3, 3) = 0.6500000E 02  A( 3, 4) = 0.7000000E 02  A( 3, 5) = 0.7500000E 02
A( 4, 1) = 0.8000000E 02  A( 4, 2) = 0.8500000E 02  A( 4, 3) = 0.9000000E 02  A( 4, 4) = 0.9500000E 02  A( 4, 5) = 0.1000000E 03
A( 5, 1) = 0.1050000E 03  A( 5, 2) = 0.1100000E 03  A( 5, 3) = 0.1150000E 03  A( 5, 4) = 0.1200000E 03  A( 5, 5) = 0.1250000E 03
```

Figure 9.2 Continued

```
A TRANSPOSE

C( 1, 1) = 0.5000000E 01  C( 1, 2) = 0.3000000E 02  C( 1, 3) = 0.5500000E 02  C( 1, 4) = 0.8000000E 02  C( 1, 5) = 0.1050000E 03
C( 2, 1) = 0.1000000E 02  C( 2, 2) = 0.3500000E 02  C( 2, 3) = 0.6000000E 02  C( 2, 4) = 0.8500000E 02  C( 2, 5) = 0.1100000E 03
C( 3, 1) = 0.1500000E 02  C( 3, 2) = 0.4000000E 02  C( 3, 3) = 0.6500000E 02  C( 3, 4) = 0.9000000E 02  C( 3, 5) = 0.1150000E 03
C( 4, 1) = 0.2000000E 02  C( 4, 2) = 0.4500000E 02  C( 4, 3) = 0.7000000E 02  C( 4, 4) = 0.9500000E 02  C( 4, 5) = 0.1200000E 03
C( 5, 1) = 0.2500000E 02  C( 5, 2) = 0.5000000E 02  C( 5, 3) = 0.7500000E 02  C( 5, 4) = 0.1000000E 03  C( 5, 5) = 0.1250000E 03


MATRIX INVERSION    D = A INVERSE


A MATRIX

A( 1, 1) = 0.1000000E 01  A( 1, 2) = 0.1000000E 01  A( 1, 3) = 0.1000000E 01  A( 1, 4) = 0.1000000E 01  A( 1, 5) = 0.1000000E 01
A( 2, 1) = 0.1000000E 01  A( 2, 2) = 0.2000000E 01  A( 2, 3) = 0.3000000E 01  A( 2, 4) = 0.4000000E 01  A( 2, 5) = 0.5000000E 01
A( 3, 1) = 0.1000000E 01  A( 3, 2) = 0.3000000E 01  A( 3, 3) = 0.6000000E 01  A( 3, 4) = 0.1000000E 02  A( 3, 5) = 0.1500000E 02
A( 4, 1) = 0.1000000E 01  A( 4, 2) = 0.4000000E 01  A( 4, 3) = 0.1000000E 02  A( 4, 4) = 0.2000000E 02  A( 4, 5) = 0.3500000E 02
A( 5, 1) = 0.1000000E 01  A( 5, 2) = 0.5000000E 01  A( 5, 3) = 0.1500000E 02  A( 5, 4) = 0.3500000E 02  A( 5, 5) = 0.7000000E 02


A INVERSE

D( 1, 1) = 0.5000000E 01  D( 1, 2) =-0.1000000E 02  D( 1, 3) = 0.1000000E 02  D( 1, 4) =-0.5000000E 01  D( 1, 5) = 0.1000000E 01
D( 2, 1) =-0.1000000E 02  D( 2, 2) = 0.3000000E 02  D( 2, 3) =-0.3500000E 02  D( 2, 4) = 0.1900000E 02  D( 2, 5) =-0.4000000E 01
D( 3, 1) = 0.1000000E 02  D( 3, 2) =-0.3500000E 02  D( 3, 3) = 0.4600000E 02  D( 3, 4) =-0.2700000E 02  D( 3, 5) = 0.6000000E 01
D( 4, 1) =-0.5000000E 01  D( 4, 2) = 0.1900000E 02  D( 4, 3) =-0.2700000E 02  D( 4, 4) = 0.1700000E 02  D( 4, 5) =-0.4000000E 01
D( 5, 1) = 0.1000000E 01  D( 5, 2) =-0.4000000E 01  D( 5, 3) = 0.6000000E 01  D( 5, 4) =-0.4000000E 01  D( 5, 5) = 0.1000000E 01

IHC217I
```

Figure 9.2 Continued

```
C     MATRIX SOLUTION OF FOUR MESH NETWORK.
C
      DIMENSION A(4,4),D(4,5),C(4,1),V(4,1)
      DO 10 I=1,4
   10 READ(5,1)A(I,1),A(I,2),A(I,3),A(I,4),V(I,1)
    1 FORMAT(5E15.7)
      DO 11 I=1,4
   11 WRITE(6,3)I,A(I,1),I,A(I,2),I,A(I,3),I,A(I,4),I,V(I,1)
    3 FORMAT(' A(',I1,',','1) =',E15.7,'  A(',I1,',','2) =',E15.7,'  A('
     1I1,',3) =',E15.7,'  A(',I1,',4) =',E15.7,'  V(',I1,') =',E15.7)
      CALL DINVM(A,D,4)
      WRITE(6,60)((D(I,J),J=1,4),I=1,4)
   60 FORMAT(//,' INVERSE MATRIX',//,(4E15.7))
      DO 20 I=1,4
      DO 20 J=1,4
   20 A(I,J)=D(I,J)
      CALL PRODM(A,V,C,4,1,4)
      DO 30 I=1,4
   30 WRITE(6,4)I,C(I,1)
    4 FORMAT(//,' I(',I1,') =',E15.7,/)
      RETURN
      END

      SUBROUTINE PRODM(A,B,C,MA,NB,NA)
      DIMENSION A(4,4),B(4,1),C(4,1)
      DO 10 I=1,MA
      DO 10 J=1,NB
      C(I,J)=0.0
      DO 10 K=1,NA
   10 C(I,J)=C(I,J)+A(I,K)*B(K,J)
      RETURN
      END

      SUBROUTINE DINVM(A,D,N)
      DIMENSION A(4,4),D(4,5)
      DO 5 I=1,N
      DO 5 J=1,N
    5 D(I,J)=A(I,J)
      M=N+1
      KK=0
      JJ=0
      DO 10 K=1,N
      DO 11 J=1,N
   11 D(J,M)=0.0
      D(K,M)=1.0
      JJ=KK+1
      LL=JJ
      KK=KK+1
   20 IF(ABS(D(JJ,KK))-1.E-4)21,21,22
   21 JJ=JJ+1
      GO TO 20
   22 IF(LL-JJ)23,24,23
   23 DO 25 MM=1,M
      DTEMP=D(LL,MM)
      D(LL,MM)=D(JJ,MM)
   25 D(JJ,MM)=DTEMP
C
C     THE EQUATIONS ARE FORCED INTO THE DIAGONAL FORM.
C
   24 DIV=D(K,K)
      DO 30 LJ=1,M
      J=M+1-LJ
   30 D(K,J)=D(K,J)/DIV
      DO 12 I=1,N
      FAC=D(I,K)
      DO 12 LJ=1,M
      J=M+1-LJ
      IF(I-K)31,12,31
   31 D(I,J)=D(I,J)-FAC*D(K,J)
   12 CONTINUE
      DO 40 J=1,N
   40 D(J,K)=D(J,M)
   10 CONTINUE
      RETURN
      END
```

Figure 9.3 A Fortran program that uses subroutines DINVM and PRODM to obtain the solution of the network equations of Figure 6.4

```
A(1,1) =  0.4000000E 01  A(1,2) = -0.2000000E 01  A(1,3) = -0.1000000E 01  A(1,4) =  0.0            V(1) =  0.1000000E 02
A(2,1) = -0.2000000E 01  A(2,2) =  0.9000000E 01  A(2,3) =  0.0            A(2,4) = -0.5000000E 01  V(2) = -0.5000000E 01
A(3,1) = -0.1000000E 01  A(3,2) =  0.0            A(3,3) =  0.6000000E 01  A(3,4) = -0.3000000E 01  V(3) =  0.0
A(4,1) =  0.0            A(4,2) = -0.5000000E 01  A(4,3) = -0.3000000E 01  A(4,4) =  0.1000000E 02  V(4) =  0.0

INVERSE MATRIX

 0.3406835E 00  0.1289966E 00  0.1047408E 00  0.9592056E-01
 0.1289966E 00  0.2138918E 00  0.8820277E-01  0.1334068E 00
 0.1047409E 00  0.8820277E-01  0.2425578E 00  0.1168687E 00
 0.9592056E-01  0.1334068E 00  0.1168687E 00  0.2017640E 00

I(1) =  0.2761851E 01

I(2) =  0.2205067E 00

I(3) =  0.6063942E 00

I(4) =  0.2921718E 00
```

Figure 9.4 The output results of the program of Figure 9.3

mesh currents is found by multiplying both sides of this equation by the inverse of the mesh-impedance matrix, thus

$$[A]^{-1}[A][C] = [A]^{-1}[V]$$

or

$$[C] = [A]^{-1}[V]$$

The solution of the mesh currents requires that we first find the inverse of the mesh-impedance matrix and then multiply this result by the voltage excitation matrix.

Figure 9.3 is a Fortran program that calls on subroutines DINV and PRODM to obtain the solution for the mesh currents. Initially the DO 10 loop reads in the elements of the A and V matrices, and the DO 11 loop prints out the corresponding numerical element values. Computer statement S.0008 then calls subroutine DINVM to compute $[A]^{-1}$. The numerical element values of the inverse matrix are next printed by computer statements S.0009 and S.0010. We then use the DO 20 loops to transfer the contents of the first four columns of matrix D into the four columns of matrix A. Note that matrix A will now be equivalent to $[A]^{-1}$. Computer statement S.0014 now calls on subroutine PRODM to multiply $[A]^{-1}$ by $[V]$. After the product of the two matrices has been found, the DO 30 loop is used to print the result. Figure 9.4 shows the corresponding computer results, and the reader may readily verify that these results are identical to those of Figure 6.6.

9.3 MATRIX FORMULATION OF SINUSOIDAL STEADY-STATE NETWORK EQUATIONS

In this section we will discuss how matrix methods may be used to formulate the phasor equations of an arbitrary network. After we have obtained these equations by using a suitable computer program, we may use a program similar to Figure 6.23 to obtain the solutions of the unknowns.

Mesh Formulation

The matrix method of formulating mesh equations is particularly suited for networks that contain an involved combination of mutual inductances. In order to better understand this method of formulation, let us illustrate the procedure in terms of the simple network of Figure 9.5. Initially the connection matrix must be read into the computer memory. Since the connection matrix is a rectangular array that describes

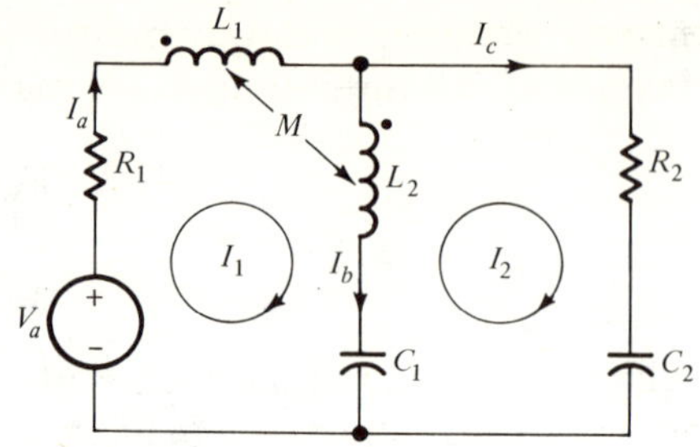

Figure 9.5 A simple two-mesh current network

the branches associated with the individual meshes, it contains the required topological information.

For the network of Figure 9.5, the connection matrix would be given by

$$\text{branches}\begin{cases} a \\ b \\ c \end{cases}\overset{\overbrace{\quad 1 \qquad 2 \quad}^{\text{meshes}}}{\begin{bmatrix} 1 & 0 \\ 1 & -1 \\ 0 & 1 \end{bmatrix}} = [C]$$

Note that the connection matrix has as many rows as there are branches and as many columns as there are meshes. If a branch is not a member of a mesh, the corresponding element is 0. If a mesh contains a particular branch, the corresponding element of the matrix is either +1 or −1. If the positive sense of the branch current and mesh current are the same, the element is +1, otherwise it is −1.

After the connection matrix has been read into the computer memory, we read in the branch impedance matrix. This matrix represents the self-impedance of each branch and any mutual impedance that may exist between branches. For the network of Figure 9.5, the branch impedance matrix would be given by

$$\begin{array}{c} \\ a \\ b \\ c \end{array}\begin{array}{c} \begin{array}{ccc} a & b & c \end{array} \\ \begin{bmatrix} R_1 + sL_1 & sM & 0 \\ sM & sL_2 + \dfrac{1}{sC_1} & 0 \\ 0 & 0 & R_2 + \dfrac{1}{sC_2} \end{bmatrix} \end{array} = [Z_{\text{branch}}]$$

Note that branches a and b have a mutual impedance sM, and branch c has zero mutual impedance with respect to branches a and b.

Although the above representation of the branch impedance matrix is suitable for analytical manipulation, it is not suitable for computer formulation by computers that are not able to do complex arithmetic. In order to get around this problem, we represent the element information in terms of three matrices—the Z-branch resistive matrix, the Z-branch inductive matrix, and the Z-branch capacitive matrix. For the network of Figure 9.5, these three matrices would be given by

$$[Z_{\text{branch } R}] = \begin{bmatrix} R_1 & 0 & 0 \\ 0 & 0 & 0 \\ 0 & 0 & R_2 \end{bmatrix}$$

$$[Z_{\text{branch } L}] = \begin{bmatrix} L_1 & M & 0 \\ M & L_2 & 0 \\ 0 & 0 & 0 \end{bmatrix}$$

$$[Z_{\text{branch } C}] = \begin{bmatrix} 0 & 0 & 0 \\ 0 & \dfrac{1}{C_1} & 0 \\ 0 & 0 & \dfrac{1}{C_2} \end{bmatrix}$$

Note that two separate matrices are used to store the inductive and capacitive element information. This is done mainly to increase the efficiency of operation of the program when we require the mesh equations, and ultimately the network response, for several different frequencies. Once the frequency has been specified, it is only a matter of multiplying each element of the appropriate matrix by ω or $1/(-\omega)$ to obtain the required complex elements. This approach greatly reduces the computer execution time since most of the matrix operations are performed prior to introducing any frequency values.

After the connection matrix and the branch impedance element matrices are read into the computer memory, the computer program must perform the following matrix operations:[2]

$$[C]_T[Z_{\text{branch}}][C][I_{\text{mesh}}] = [C]_T[V_{g,\,\text{branch}}]$$
$$[Z_{\text{mesh}}][I_{\text{mesh}}] = [V_{\text{mesh}}]$$

For example, in the case of the network of Figure 9.5, we would have

$$[C]_T[Z_{\text{branch}}] = \begin{bmatrix} 1 & 1 & 0 \\ 0 & -1 & 1 \end{bmatrix} \begin{bmatrix} R_1 + sL_1 & sM & 0 \\ sM & sL_2 + \dfrac{1}{sC_1} & 0 \\ 0 & 0 & R_2 + \dfrac{1}{sC_2} \end{bmatrix}$$

$$= \begin{bmatrix} R_1 + s(L_1 + M) & s(L_2 + M) + \dfrac{1}{sC_1} & 0 \\ -sM & -\left(sL_2 + \dfrac{1}{sC_1}\right) & R_2 + \dfrac{1}{sC_2} \end{bmatrix}$$

[2] See B. James Ley, Samuel G. Lutz, and Charles F. Rehberg, *Linear Circuit Analysis* (New York: McGraw-Hill Book Company, 1959) Sec. 2-12.

so that

$$[Z_{\text{mesh}}] = [C]_T[Z_{\text{branch}}][C]$$

$$= \begin{bmatrix} R_1 + s(L_1 + M) & s(L_2 + M) + \dfrac{1}{sC_1} & 0 \\ -sM & -\left(sL_2 + \dfrac{1}{sC_1}\right) & R_2 + \dfrac{1}{sC_2} \end{bmatrix} \begin{bmatrix} 1 & 0 \\ 1 & -1 \\ 0 & 1 \end{bmatrix}$$

$$= \begin{bmatrix} R_1 + s(L_1 + L_2 + 2M) + \dfrac{1}{sC_1} & -s(L_2 + M) + \dfrac{1}{sC_1} \\ -s(L_2 + M) + \dfrac{1}{sC_1} & R_2 + sL_1 \dfrac{1}{sC_1} + \dfrac{1}{sC_2} \end{bmatrix}$$

and

$$[V_{\text{mesh}}] = [C]_T[V_{g,\text{ branch}}]$$

$$= \begin{bmatrix} 1 & 1 & 0 \\ 0 & -1 & 1 \end{bmatrix} \begin{bmatrix} V_a \\ 0 \\ 0 \end{bmatrix} = \begin{bmatrix} V_a \\ 0 \\ 0 \end{bmatrix}$$

or

$$[Z_{\text{mesh}}][I_{\text{mesh}}] = [V_{\text{mesh}}]$$

Alternately, using the computer, the matrix operation

$$[C]_T[Z_{\text{branch}}]$$

may be carried out in terms of $[Z_{\text{branch }R}]$, $[Z_{\text{branch }L}]$, and $[Z_{\text{branch }C}]$. Thus

$$[C]_T[Z_{\text{branch }R}] = \begin{bmatrix} 1 & 1 & 0 \\ 0 & -1 & 1 \end{bmatrix} \begin{bmatrix} R_1 & 0 & 0 \\ 0 & 0 & 0 \\ 0 & 0 & R_2 \end{bmatrix} = \begin{bmatrix} R_1 & 0 & 0 \\ 0 & 0 & R_2 \end{bmatrix}$$

$$[C]_T[Z_{\text{branch }L}] = \begin{bmatrix} 1 & 1 & 0 \\ 0 & -1 & 1 \end{bmatrix} \begin{bmatrix} L_1 & M & 0 \\ M & L_2 & 0 \\ 0 & 0 & 0 \end{bmatrix} = \begin{bmatrix} L_1 + M & L_2 - M & 0 \\ -M & -L_2 & 0 \end{bmatrix}$$

and

$$[C]_T[Z_{\text{branch }C}] = \begin{bmatrix} 1 & 1 & 0 \\ 0 & -1 & 1 \end{bmatrix} \begin{bmatrix} 0 & 0 & 0 \\ 0 & \dfrac{1}{C_1} & 0 \\ 0 & 0 & \dfrac{1}{C_2} \end{bmatrix} = \begin{bmatrix} 0 & \dfrac{1}{C_1} & 0 \\ 0 & -\dfrac{1}{C_1} & \dfrac{1}{C_2} \end{bmatrix}$$

Solving now for

$$[C]_T[Z_{\text{branch}}][C] = [Z_{\text{mesh}}]$$

we find

$$[C]_T[Z_{\text{branch }R}][C] = \begin{bmatrix} R_1 & 0 & 0 \\ 0 & 0 & R_2 \end{bmatrix} \begin{bmatrix} 1 & 0 \\ 1 & -1 \\ 0 & 1 \end{bmatrix} = \begin{bmatrix} R_1 & 0 \\ 0 & R_2 \end{bmatrix}$$

$$[C]_T[Z_{\text{branch } L}][C]$$
$$= \begin{bmatrix} L_1 + M & L_2 + M & 0 \\ -M & -L_2 & 0 \end{bmatrix} \begin{bmatrix} 1 & 0 \\ 1 & -1 \\ 0 & 1 \end{bmatrix} = \begin{bmatrix} L_1 + L_2 + 2M & -(L_2 + M) \\ -(L_2 + M) & L_2 \end{bmatrix}$$

and

$$[C]_T[Z_{\text{branch } C}][C] = \begin{bmatrix} 0 & \frac{1}{C_1} & 0 \\ 0 & -\frac{1}{C_1} & \frac{1}{C_2} \end{bmatrix} \begin{bmatrix} 1 & 0 \\ 1 & -1 \\ 0 & 1 \end{bmatrix} = \begin{bmatrix} \frac{1}{C_1} & -\frac{1}{C_1} \\ -\frac{1}{C_1} & \frac{1}{C_1} + \frac{1}{C_2} \end{bmatrix}$$

Multiplying each element in these latter two matrices by ω and $1/(-\omega)$ respectively yields

$$\omega * [C]_T[Z_{\text{branch } L}][C] = \begin{bmatrix} \omega(L_1 + L_2 + 2M) & -\omega(L_2 + M) \\ -\omega(L_2 + M) & \omega L_2 \end{bmatrix}$$

and

$$-\frac{1}{\omega} * [C]_T[Z_{\text{branch } C}][C] = \begin{bmatrix} -\frac{1}{\omega C_1} & \frac{1}{\omega C_1} \\ \frac{1}{\omega C_1} & -\frac{1}{\omega C_1} - \frac{1}{\omega C_2} \end{bmatrix}$$

Finally realizing that these two latter matrices contain purely imaginary elements, we now add these to $[C]_T[Z_{\text{branch } R}][C]$, thus obtaining

$$[Z_{\text{mesh}}] =$$
$$\begin{bmatrix} R_1 + j\omega(L_1 + L_2 + 2M) - j\frac{1}{j\omega C_1} & -j\omega(L_2 + M) + j\frac{1}{\omega C_1} \\ -j\omega(L_2 + M) + j\frac{1}{\omega C_1} & R_2 + j\omega L_2 - j\left(\frac{1}{\omega C_1} + \frac{1}{\omega C_2}\right) \end{bmatrix}$$

Figure 9.6 is the corresponding Fortran program. Initially the transpose of the connection matrix is read into memory rather than the connection matrix. The DO 10 loops are then used to produce the connection matrix. The output heading and subheading and the element values of the connection matrix are then printed. This is followed by the DO 20 loop that is used to read in $[Z_{\text{branch } R}]$, $[Z_{\text{branch } L}]$, and $[Z_{\text{branch } C}]$ and compute $[C]_T[Z_{\text{branch } R}][C]$, $[C]_T[Z_{\text{branch } L}][C]$, and $[C]_T[Z_{\text{branch } C}][C]$.

Computer statements S.0038 and S.0039 are then used to read in the number of frequency calculations N and the corresponding frequency values $W(I)$. This is followed by the execution of the DO 40 loop. During each iteration of the outer DO 40 loop, the DO 50 and DO 60 loops are used to obtain the values of the imarinary part of each of the elements of $[Z_{\text{mesh}}]$ for one of the specified frequencies. The inner DO 40 loop is

```
C       MATRIX FORMULATION OF SINUSOIDAL STEADY STATE MESH EQUATIONS
C
C          NB = NUMBER OF BRANCHES (MAXIMUM = 20.)
C          NM = NUMBER OF MESHES (MAXIMUM = 10.)
C
C
        DIMENSION CT (10,20),C(20,10),ZB(20,20),T(20,20),ZMESH(3,10,10),
       1ZMESHW(2,10, 10),W(50)
        READ (5,1)NB,NM,1(CT(I,J),J=1,NF),I=1,NM)
      1 FORMAT (2I5,/,(16F5.0))
C
C       TRANSPOSE CT TO C, THE CONNECTION MATRIX.
C
        DO 10 I=1,NB
        DO 10 J=1,NM
     10 C(I,J)=CT(J,I)
        WRITE (6,3)
      3 FORMAT ('1INPUT MATRICES',///' CONNECTION MATRIX',//)
        DO 15 I=1,NB
     15 WRITE(6,4)(C(I,J),J=1,NM)
      4 FORMAT(20F5.0)
C
C       COMPUTE MESH IMPEDANCES,
C
        DO 20 I=1,3
        READ(5,5)((ZB(I,J),J=1,NB),I=1,NB)
      5 FORMAT (5E15.7)
        GO TO (21,22,23),L
     21 WRITE (6,6)
      6 FORMAT(//,' Z-BRANCH RESISTIVE MATRIX',//)
        GO TO 24
     22 WRITE (6,7)
      7 FORMAT (//' Z-BRANCH INDUCTIVE MATRIX',//)
        GO TO 24
     23 WRITE (6,8)
      8 FORMAT (//,' Z-BRANCH CAPACITIVE MATRIX',//)
     24 DO 25 I=1,NB
     25 WRITE (6,9)(ZB(I,J),J=1,NB)
      9 FORMAT (8E15.7)
        DO 30 I=1,NM
        DO 30 J=1,NB
        T(I,J)=0.0
        DO 30 K=1,NB
     30 T(I,J)=CT(I,K)*ZB(K,J)+T(I,J)
        DO 35 I=1,NM
        DO 35 J=1,NM
        ZMESH(L,I,J)=0.0
        DO 35 K=1,NB
     35 ZMESH(L,I,J)=T(I,K)*C(K,J)+ZMESH(L,I,J)
     20 CONTINUE
C
C       GENERATION OF THE PHASOR EQUATIONS
C
        READ(5,11)N,(W(I),I=1,N)
     11 FORMAT(I10,/,(5E15.7))
        DO 40 I=1.N
        WRITE(6,12)W(I)
     12 FORMAT('1PHASOR EQUATIONS',//,' FREQUENCY =',E15.7,///)
        DO 50 J=1,NM
        DO 50 K=1,NM
        ZMESHW(1,J,K)=ZMESH(2,J,K)*W(I)
     50 ZMESHW(2,J,K)=ZMESH(3,J,K)/(-W(I))
        DO 60 J=1,NM
        DO 60 K=1,NM
     60 ZMESHW(1,J,K)=ZMESHW(1,J,K) +ZMESHW(2,J,K)
        DO 40 I=1,NM
     40 WRITE(6,13)L,(ZMESH(1,L,K),ZMESHW(1,L,K),K,K=1,NM)
     13 FORMAT(' V-MESH(',I2,') = ',('(',E15.7,' +J ',E15.7,') I(',I2,')
       1 +  '))
        STOP
        END
```

Figure 9.6 A program that uses matrix methods to formulate steady-state mesh equations

then used to print these results. The number of iterations of the outer DO 40 loop is therefore a function of the number of frequencies that have been specified.

Figure 9.7 shows the computer results for the network of Figure 6.22. Note that only one frequency calculation has been specified by this data

```
INPUT MATRICES

CONNECTION MATRIX

 1.   0.   0.   0.   0.   0.
 1.  -1.   0.   0.   0.   0.
 0.   1.   0.   0.   0.   0.
 0.   1.  -1.   0.   0.   0.
 0.   0.   1.   0.   0.   0.
 0.   0.   1.  -1.   0.   0.
 0.   0.   0.   1.   0.  -1.
 0.   0.   0.   1.  -1.   0.
 0.   0.   0.   0.   1.  -1.
 0.   0.   0.   0.   1.   0.
 0.   0.   0.   0.   0.   1.

Z-BRANCH RESISTIVE MATRIX

 0.1250000E 01  0.0            0.0            0.0            0.0            0.0            0.0            0.0
 0.0            0.0            0.0
 0.0            0.0            0.0            0.0            0.0            0.0            0.0            0.0
 0.0            0.0            0.0
 0.0            0.0            0.1500000E 01  0.0            0.0            0.0            0.0            0.0
 0.0            0.0            0.0
 0.0            0.0            0.0            0.1000000E 01  0.0            0.0            0.0            0.0
 0.0            0.0            0.0
 0.0            0.0            0.0            0.0            0.1400000E 01  0.0            0.0            0.0
 0.0            0.0            0.0
 0.0            0.0            0.0            0.0            0.0            0.1200000E 01  0.0            0.0
 0.0            0.0            0.0
 0.0            0.0            0.0            0.0            0.0            0.0            0.5000000E 00  0.0
 0.0            0.0            0.0
 0.0            0.0            0.0            0.0            0.0            0.0            0.0            0.0
 0.0            0.0            0.0
 0.0            0.0            0.0            0.0            0.0            0.0            0.0            0.0
 0.0            0.0            0.0
 0.0            0.0            0.0            0.0            0.0            0.0            0.0            0.0
 0.0            0.2700000E 01  0.0
 0.0            0.0            0.0            0.0            0.0            0.0            0.0            0.0
 0.0            0.0            0.0
```

Figure 9.7 The output results of the program of Figure 9.6 for the network of Figure 6.22

```
Z-BRANCH INDUCTIVE MATRIX

0.1500000E 01  0.0            0.0            0.0            0.0            0.0            0.0            0.0
0.0            0.0            0.0
0.0            0.0            0.0            0.0            0.0            0.0            0.0            0.0
0.0            0.0            0.0
0.0            0.0            0.2500000E 01  0.0            0.0            0.0            0.0            0.0
0.0            0.0            0.0
0.0            0.0            0.0            0.2200000E 01  0.0            0.0            0.0            0.0
0.0            0.0            0.0
0.0            0.0            0.0            0.0            0.1500000E 01  0.0            0.0            0.0
0.0            0.0            0.0
0.0            0.0            0.0            0.0            0.0            0.1599999E 01  0.0            0.0
0.0            0.0            0.0
0.0            0.0            0.0            0.0            0.0            0.0            0.1469999E 01  0.0
0.0            0.0            0.0
0.0            0.0            0.0            0.0            0.0            0.0            0.0            0.0
0.0            0.0            0.0
0.0            0.0            0.0            0.0            0.0            0.0            0.0            0.0
0.2299999E 01  0.0            0.0
0.0            0.0            0.0            0.0            0.0            0.0            0.0            0.0
0.0            0.0            0.0
0.0            0.0            0.0            0.0            0.0            0.0            0.0            0.0
0.0            0.0            0.0


Z-BRANCH CAPACITIVE MATRIX

0.0            0.0            0.0            0.0            0.0            0.0            0.0            0.0
0.0            0.0            0.0
0.0            0.5000000E 00  0.0            0.0            0.0            0.0            0.0            0.0
0.0            0.0            0.0
0.0            0.0            0.0            0.0            0.0            0.0            0.0            0.0
0.0            0.0            0.0
0.0            0.0            0.0            0.0            0.0            0.0            0.0            0.0
0.0            0.0            0.0
0.0            0.0            0.0            0.0            0.0            0.0            0.0            0.0
0.0            0.0            0.0
0.0            0.0            0.0            0.0            0.0            0.0            0.0            0.0
0.0            0.0            0.0
0.0            0.0            0.0            0.0            0.0            0.0            0.0            0.0
0.0            0.0            0.0
0.0            0.0            0.0            0.0            0.0            0.0            0.0            0.8000000E 00
0.0            0.0            0.0
0.0            0.0            0.0            0.0            0.0            0.0            0.0            0.0
0.0            0.0            0.0
0.0            0.0            0.0            0.0            0.0            0.0            0.0            0.0
0.0            0.0            0.0
0.0            0.0            0.0            0.0            0.0            0.0            0.0            0.0
0.0            0.0            0.2500000E 01
```

Figure 9.7 Continued

```
PHASOR EQUATIONS

FREQUENCY =   0.1000000E 01


V-MESH( 1) =   (  0.1250000E 01 +J   0.1000000E 01)  I( 1)  +
(  0.0              +J   0.5000000E 00)  I( 2)  +
(  0.0              +J   0.0           )  I( 3)  +
(  0.0              +J   0.0           )  I( 4)  +
(  0.0              +J   0.0           )  I( 5)  +
(  0.0              +J   0.0           )  I( 6)  +
V-MESH( 2) =   (  0.0            +J   0.5000000E 00)  I( 1)  +
(  0.2500000E 01 +J   0.4200000E 01)  I( 2)  +
( -0.1000000E 01 +J  -0.2200000E 01)  I( 3)  +
(  0.0              +J   0.0           )  I( 4)  +
(  0.0              +J   0.0           )  I( 5)  +
(  0.0              +J   0.0           )  I( 6)  +
V-MESH( 3) =   (  0.0            +J   0.0           )  I( 1)  +
( -0.1000000E 01 +J  -0.2200000E 01)  I( 2)  +
(  0.3599999E 01 +J   0.5299999E 01)  I( 3)  +
( -0.1200000E 01 +J  -0.1599999E 01)  I( 4)  +
(  0.0              +J   0.0           )  I( 5)  +
(  0.0              +J   0.0           )  I( 6)  +
V_MESH( 4) =   (  0.0            +J   0.0           )  I( 1)  +
(  0.0              +J   0.0           )  I( 2)  +
( -0.1200000E 01 +J  -0.1599999E 01)  I( 3)  +
(  0.1700000E 01 +J . 0.2269999E 01)  I( 4)  +
(  0.0              +J   0.8000000E 00)  I( 5)  +
( -0.5000000E 00 +J  -0.1469999E 01)  I( 6)  +
V-MESH( 5) =   (  0.0            +J   0.0           )  I( 1)  +
(  0.0              +J   0.0           )  I( 2)  +
(  0.0              +J   0.0           )  I( 3)  +
(  0.0              +J   0.8000000E 00)  I( 4)  +
(  0.2700000E 01 +J   0.1499999E 01)  I( 5)  +
(  0.0              +J  -0.2299999E 01)  I( 6)  +
V-MESH( 6) =   (  0.0            +J   0.0           )  I( 1)  +
(  0.0              +J   0.0           )  I( 2)  +
(  0.0              +J   0.0           )  I( 3)  +
( -0.5000000E 00 +J  -0.1469999E 01)  I( 4)  +
(  0.0              +J  -0.2299999E 01)  I( 5)  +
(  0.5000000E 00 +J   0.1269999E 01)  I( 6)  +
```

Figure 9.7 Continued

set. The accuracy of these results may be readily verified by referring to Section 6.6. The reader is encouraged to use this program in formulating the phasor equations of a more complicated network where involved combinations of mutual inductance exist. The reader may further wish to extend this program by combining it with the program of Figure 6.23 so that the resulting program will be capable of formulating as well as solving an arbitrary network for various specified frequencies.

Nodal Formulation[3]

The matrix formulation of the nodal equations of a given network is quite similar to that of the matrix mesh equation formulation. The method is somewhat more complicated, however, since it involves the inversion of $[Z_{\text{branch}}]$ to obtain $[Y_{\text{branch}}]$.

In order to outline the complete formulation procedure, let us use the simplified network of Figure 9.8. The branch impedance matrix of this

[3] Ley, Lutz, Rehberg, *Linear Circuit Analysis*, Sec. 2-13.

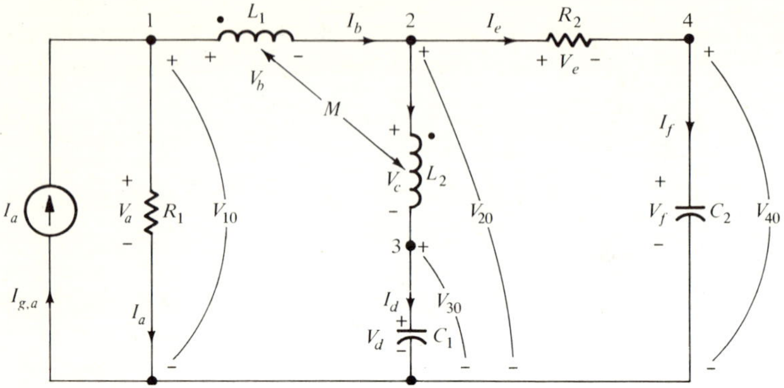

Figure 9.8 A simple four-node voltage network

network would thus be given by

$$
\begin{array}{c} \\ a \\ b \\ c \\ d \\ e \\ f \end{array}
\begin{array}{c}
\begin{array}{cccccc} a & b & c & d & e & f \end{array} \\
\begin{bmatrix}
R_1 & 0 & 0 & 0 & 0 & 0 \\
0 & sL_1 & sM & 0 & 0 & 0 \\
0 & sM & sL_2 & 0 & 0 & 0 \\
0 & 0 & 0 & \dfrac{1}{sC_1} & 0 & 0 \\
0 & 0 & 0 & 0 & R_2 & 0 \\
0 & 0 & 0 & 0 & 0 & \dfrac{1}{sC_2}
\end{bmatrix}
\end{array} = [Z_{\text{branch}}]
$$

The branch admittance matrix may now be found from

$$[Y_{\text{branch}}] = [Z_{\text{branch}}]^{-1}$$

and thus, for the network of Figure 9.8, we would have

$$
Y_{\text{branch}} = \begin{bmatrix}
\dfrac{1}{R_1} & 0 & 0 & 0 & 0 & 0 \\
0 & \dfrac{\Gamma_1}{s} & -\dfrac{\Gamma_{12}}{s} & 0 & 0 & 0 \\
0 & -\dfrac{\Gamma_{12}}{s} & \dfrac{\Gamma_2}{s} & 0 & 0 & 0 \\
0 & 0 & 0 & sC_1 & 0 & 0 \\
0 & 0 & 0 & 0 & \dfrac{1}{R_2} & 0 \\
0 & 0 & 0 & 0 & 0 & sC_2
\end{bmatrix}
$$

where

$$\Gamma_1 = \frac{L_2}{L_1L_2 - M^2}, \qquad \Gamma_2 = \frac{L_1}{L_1L_2 - M^2}, \qquad \text{and} \qquad \Gamma_{12} = \frac{M}{L_1L_2 - M^2}$$

The junction matrix may now be introduced to supply the required topological information. For the network of Figure 9.8, the junction matrix would be

$$\text{Branches}\left\{\begin{matrix} a \\ b \\ c \\ d \\ e \\ f \end{matrix}\right.\overbrace{\begin{bmatrix} 1 & 0 & 0 & 0 \\ 1 & -1 & 0 & 0 \\ 0 & 1 & -1 & 0 \\ 0 & 0 & 1 & 0 \\ 0 & 1 & 0 & -1 \\ 0 & 0 & 0 & 1 \end{bmatrix}}^{\substack{\text{Nodes} \\ 1 \quad\;\; 2 \quad\;\; 3 \quad\;\; 4}} = [J]$$

Note that the junction matrix has as many rows as there are branches and as many columns as there are independent nodes. If all the independent nodes are defined as being positive with respect to the reference node, the plus end of a branch incident on a given independent node will be noted by a +1 in the corresponding element of the junction matrix. If the minus end of a branch is incident on a given node, the corresponding element value will be −1. If a branch is not incident on a given node, the element value will be 0.

After the junction matrix and the branch impedance element matrices are read into the computer memory, the computer program must perform the following matrix operations:

$$\begin{aligned} [J]_T[Z_{\text{branch}}]^{-1}[J][V_{\text{node}}] &= [J]_T[I_{g,\text{branch}}] \\ [Y_{\text{node}}][V_{\text{node}}] &= [I_{\text{node}}] \end{aligned}$$

In the case of the network of Figure 9.8 we would thus find

$$\begin{bmatrix} \dfrac{1}{R_1} + \dfrac{\Gamma_1}{s} & -\dfrac{\Gamma_1}{s} - \dfrac{\Gamma_{12}}{s} & \dfrac{\Gamma_{12}}{s} & 0 \\ -\dfrac{\Gamma_1}{s} - \dfrac{\Gamma_{12}}{s} & \dfrac{\Gamma_1}{s} + \dfrac{\Gamma_2}{s} + 2\dfrac{\Gamma_{12}}{s} + \dfrac{1}{R_2} & -\dfrac{\Gamma_{12}}{s} - \dfrac{\Gamma_2}{s} & -\dfrac{1}{R_2} \\ \dfrac{\Gamma_{12}}{s} & -\dfrac{\Gamma_{12}}{s} - \dfrac{\Gamma_2}{s} & \dfrac{\Gamma_2}{s} + sC_1 & 0 \\ 0 & -\dfrac{1}{R_2} & 0 & \dfrac{1}{R_2} + sC_2 \end{bmatrix} \begin{bmatrix} V_{10} \\ V_{20} \\ V_{30} \\ V_{40} \end{bmatrix} = \begin{bmatrix} I_a \\ 0 \\ 0 \\ 0 \end{bmatrix}$$

The computer program for finding the nodal phasor equations is left as an exercise for the reader. (Hint: use the mesh equation formulation program of Figure 9.6 as a guide and refer to Sections 6.6 and 9.2 and SUBROUTINE MINC (Figure 9.13) in the programming of the inverse of a matrix.)

Cut-Set Formulation[4]

The matrix formulation of cut-set equations is very similar to the matrix formulation of nodal equations. The procedure begins by supplying the required topological information in terms of the cut-set matrix $[K]$. To illustrate the complete formulation procedure let us once again use the network of Figure 9.8, but this time let us introduce cut-set voltages instead of node voltages. Referring to the arbitrary defined cut-sets shown in Figure 9.9 we write the following cut-set matrix:

$$\text{Branches}\left\{\begin{array}{c} a \\ b \\ c \\ d \\ e \\ f \end{array}\right.\overset{\overbrace{\quad 1 \quad\; 2 \quad\; 3 \quad\; 4 \quad}^{\text{Cut-sets}}}{\begin{bmatrix} 1 & 0 & 0 & 0 \\ 0 & 0 & -1 & 0 \\ 1 & -1 & 1 & 0 \\ 0 & 1 & 0 & 0 \\ 1 & 0 & 1 & -1 \\ 0 & 0 & 0 & 1 \end{bmatrix}} = [K]$$

[4] Ley, Lutz, Rehberg, *Linear Circuit Analysis*, Sec. 2-14.

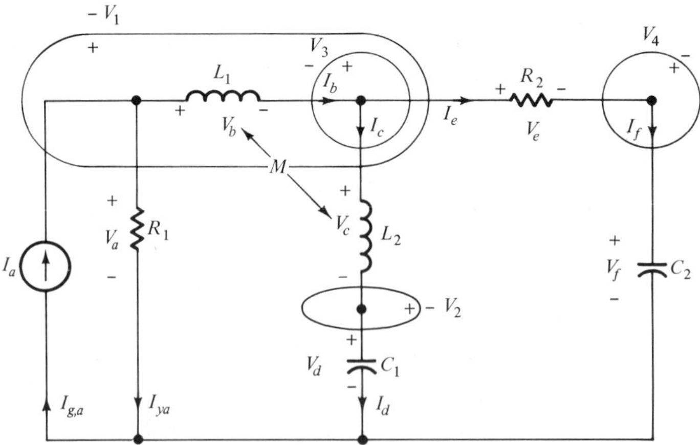

Figure 9.9 A simple four-cut-set voltage network

Note that the cut-set matrix has as many rows as there are branches and as many columns as there are cut-sets. Note also that the elements of the matrix are either $+1$, -1, or 0. If a branch is not a member of a given cut-set, the corresponding element of the matrix is 0. If a branch is a member of a given cut-set, the corresponding element of the cut-set matrix is $+1$ if both the branch voltage and cut-set voltage act in the same sense and -1 if they are of the opposite sense.

Since the network of Figure 9.9 has been defined in the same manner as Figure 9.8, the branch impedance matrix and the branch admittance matrix are also the same. It follows that the required matrix operations are

$$[K]_T[Z_{\text{branch}}]^{-1}[K][V_{\text{cut-set}}] = [K]_T[I_{g,\text{branch}}]$$
$$[Y_{\text{cut-set}}][V_{\text{cut-set}}] = [I_{\text{cut-set}}]$$

and in the case of the network of Figure 9.9 we would find

$$\begin{bmatrix} \frac{1}{R_1}+\frac{\Gamma_2}{s}+\frac{1}{R_2} & -\frac{\Gamma_2}{s} & \frac{\Gamma_{12}}{s}+\frac{\Gamma_2}{s}+\frac{1}{R_2} & -\frac{1}{R_2} \\ -\frac{\Gamma_2}{s} & \frac{\Gamma_2}{s}+sC_1 & -\frac{\Gamma_{12}}{s}-\frac{\Gamma_2}{s} & 0 \\ \frac{\Gamma_{12}}{s}+\frac{\Gamma_2}{s}+\frac{1}{R_2} & -\frac{\Gamma_{12}}{s}-\frac{\Gamma_2}{s} & \frac{\Gamma_1}{s}+2\frac{\Gamma_{12}}{s}+\frac{\Gamma_2}{s}+\frac{1}{R_2} & -\frac{1}{R_2} \\ -\frac{1}{R_2} & 0 & -\frac{1}{R_2} & \frac{1}{R_2}+sC_2 \end{bmatrix} \begin{bmatrix} V_1 \\ V_2 \\ V_3 \\ V_4 \end{bmatrix} = \begin{bmatrix} I_a \\ 0 \\ 0 \\ 0 \end{bmatrix}$$

The computer program for finding the cut-set phasor equations is left as an exercise for the reader.

9.4 MATRIX FORMULATED STEADY-STATE A-C ANALYSIS PROGRAM

For many years the author's wife, who is also a teacher, has preached that the writing of term papers or term projects is a valuable learning technique. For many years this advice was largely ignored since the author thought that such devices had no place in the teaching of electrical engineering. About five years ago, however, he decided to require each student in his computer courses to submit a term project. For the computer course from which this text was developed, the students were required to use some of the numerical techniques discussed in class in the investigation of an electrical engineering problem of their own choosing. The a-c circuit analysis program discussed in this section was submitted by Robert Guarino and is one example of the type of work a student is capable of doing on his own.

This author has also had success in using this term project technique with the sophomore version of this course. Most students agree that the term project plays a major role in helping to achieve the goal outlined in the preface of this book. A side benefit has also been the pleasure experienced by the author when grading this assignment. This pleasure is one of the rewards that only a teacher can receive.

Let us now turn to the problem of developing the necessary background so that we may be able to make use of Robert Guarino's a-c circuit analysis program. This program is based on our supplying simple input matrices that are used to generate more complicated matrices that are needed in the formulation of the network nodal equations. The network is arbitrarily limited to ten nodes and twenty branches and may contain resistors, capacitors, inductors, voltage and current generators, and voltage-controlled current sources.

Figure 9.10 shows a standard network branch and the conventions adopted by this program to indicate the positive senses of voltages and currents. The branch current is designated as i_r and the total voltage across the branch as e_r. The generated voltage is called E_r and the passive voltage drop is designated as $V_r = J_r Y_{rr}$, where Y_{rr} is an element of the branch admittance matrix and J_r is the current flowing through this element. Each branch may also contain a voltage-controlled current source that is designated as I_r.

Referring to Figure 9.10, we may thus write

$$J_r = I_r + i_r = Y_{rr} V_r$$

and

$$V_r = -E_r + e_r$$

Figure 9.10 The standard branch

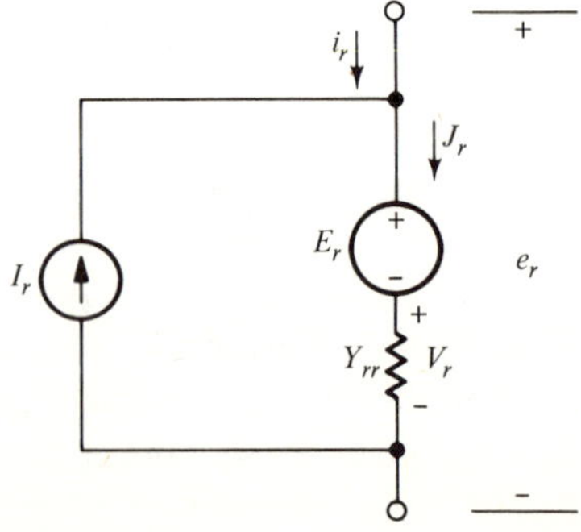

When a network of M such branches is connected, the following matrix equations may be used to describe the circuit

$$[J] = [I_Y] + [i_{\text{branch}}]$$
$$[V_{\text{branch}}] = [-E_{\text{gen}}] + [e_{\text{branch}}]$$
$$[J] = [Y_{\text{branch}}][V_{\text{branch}}]$$

where $[J]$, $[I_Y]$, $[i_{\text{branch}}]$, $[V_{Y\ \text{branch}}]$, $[E_{\text{gen}}]$, and $[e_{\text{branch}}]$ are M by 1 matrices and $[Y_{\text{branch}}]$ is an M by M matrix called the branch-admittance matrix.

Substituting the first two matrix equations into the third matrix equation yields

$$[I_Y] + [i_{\text{branch}}] = [Y_{\text{branch}}]([e_{\text{branch}}] - [E_{\text{gen}}])$$

or

$$[I_Y] + [Y_{\text{branch}}][E_{\text{gen}}] + [i_{\text{branch}}] = [Y_{\text{branch}}][e_{\text{branch}}]$$

If we now write Kirchhoff's current law for each of the independent nodes we have

$$[K]_T[i_{\text{branch}}] = [0]$$

where $[K]_T$ is the transpose of the junction matrix.[5] If we now multiply both sides of the previous equation by $[K]_T$ we find

$$[K]_T([I_Y] + Y_{\text{branch}}][E_{\text{gen}}]) + [K]_T[i_{\text{branch}}] = [K]_T[Y_{\text{branch}}][e_{\text{branch}}]$$

or

$$[K]_T([I_Y] + [Y_{\text{branch}}[]E_{\text{gen}}]) = [K]_T[Y_{\text{branch}}][e_{\text{branch}}]$$

since $[K]_T[i_{\text{branch}}] \equiv 0$

The junction matrix may also be used to express the branch voltages in terms of the node-to-datum voltages (node voltages)

$$[e_{\text{branch}}] = [K][v_{\text{node}}]$$

Using this latter relation, the former equation may be written as

$$[K]_T([I_Y] + [Y_{\text{branch}}][E_{\text{gen}}]) + [K]_T[i_{\text{branch}}] = [K]_T[Y_{\text{branch}}][K][v_{\text{node}}]$$

Solving this equation for the node voltages yields

$$[v_{\text{node}}] = ([K]_T[Y_{\text{branch}}][K])^{-1}[K]_T([I_Y] + [Y_{\text{branch}}][E_{\text{gen}}])$$

Assuming that we are first supplied with the numerical values of $[Y_{\text{branch}}]$, $[K]$, $[I_Y]$, and $[E_{\text{gen}}]$, we may use this last equation to solve for the unknown node voltages.

For a simple example that shows how we may formulate the nodal equations and solve for the node voltages, consider the network shown in Figure 9.11a. The first step in the formulation procedure requires us to

[5] For a more detailed discussion of this matrix formulation see Ley, Lutz, and Rehberg, *Linear Circuit Analysis*, Sec. 2-13.

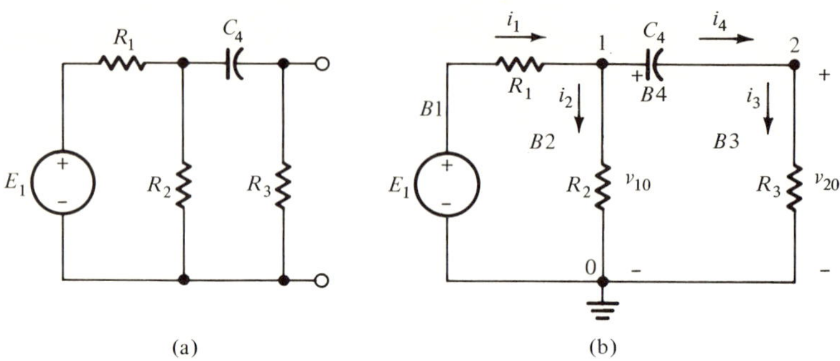

Figure 9.11 Numbering the branches and selecting the positive sense of branch currents and node voltages

number the individual branches and specify the positive sense of the branch currents and node voltages. Let us therefore define these quantities as shown in Figure 9.11b, noting that branch $B1$ contains both the generator E_1 and the resistor R_1.

The input matrices that must be supplied for this network are therefore

$$[E_{\text{gen}}] = \begin{bmatrix} E_1 \\ 0 \\ 0 \\ 0 \end{bmatrix}; \; [I_Y] = \begin{bmatrix} 0 \\ 0 \\ 0 \\ 0 \end{bmatrix}; \; [Y_{\text{branch}}] = \begin{bmatrix} G_1 & 0 & 0 & 0 \\ 0 & G_2 & 0 & 0 \\ 0 & 0 & G_3 & 0 \\ 0 & 0 & 0 & j\omega C_4 \end{bmatrix},$$

and

$$[K] = \begin{bmatrix} 1 & 0 \\ 1 & 0 \\ 0 & 1 \\ 1 & -1 \end{bmatrix}$$

Note that in the junction matrix $[K]$, each row corresponds to one of the branches and each column corresponds to one of the independent nodes. If the plus end of branch X is incident on a positive node Y, then $+1$ appears in the X row and the Y column. If the minus end of the branch X is incident on the positive node Y, then a -1 appears in the X row and the Y column. If branch X is not incident on node Y, then a 0 appears in the X row and the Y column.

Substituting each of these input matrices into the nodal equation

$$[K]_T([I_Y] + [Y_{\text{branch}}][E_{\text{gen}}]) = [K]_T[Y_{\text{branch}}][K][v_{\text{node}}]$$

and carrying out the matrix arithmetic operations, yields the nodal network equations

$$\begin{bmatrix} G_1 + G_2 + j\omega C_4 & -j\omega C_4 \\ -j\omega C_4 & G_3 + j\omega C_4 \end{bmatrix} \begin{bmatrix} v_{10} \\ v_{20} \end{bmatrix} = \begin{bmatrix} G_1 E_1 \\ 0 \end{bmatrix}$$

or

$$[Y_{\text{node}}][v_{\text{node}}] = [i]$$

where $[i]$ is the matrix that contains the applied node currents. Finally, multiplying each side of this equation by $[Y_{\text{node}}]^{-1}$ yields

$$[v_{\text{node}}] = [Y_{\text{node}}]^{-1}[i]$$

Let us now consider another example that involves a dependent current source. Referring to Figure 9.12, we see that the required input matrices would be

$$[I_y] = \begin{bmatrix} 0 \\ 0 \\ 0 \\ 0 \\ 0 \end{bmatrix}, \quad [E_{\text{gen}}] = \begin{bmatrix} E_1 \\ 0 \\ 0 \\ 0 \\ 0 \end{bmatrix}, \quad [K] = \begin{bmatrix} 1 & 0 \\ 1 & 0 \\ 1 & 0 \\ 0 & 1 \\ 0 & 1 \end{bmatrix},$$

$$[Y_{\text{branch}}] = \begin{bmatrix} g_b' & 0 & 0 & 0 & 0 \\ 0 & G_i & 0 & 0 & 0 \\ 0 & 0 & j\omega C_i & 0 & 0 \\ 0 & 0 & g_e' & 0 & 0 \\ 0 & 0 & 0 & 0 & G_L \end{bmatrix}$$

Note that we have made the dependent current generator a separate branch (B4). Note also the off diagonal term in the branch admittance matrix. This term is required since $[J] = [Y_{\text{branch}}][V_{\text{branch}}]$ or $J_4 = g_e'V_3$. As a consequence, the element in the fourth row and third column of the branch admittance matrix must be g_e'.[6] Substituting these input matrices into the nodal equation and carrying out the matrix arithmetic operations yields the nodal network equations

$$\begin{bmatrix} g_b' + G_i + j\omega C_i & 0 \\ g_e' & G_L \end{bmatrix} \begin{bmatrix} v_{10} \\ v_{20} \end{bmatrix} = \begin{bmatrix} E_1 g_b' \\ 0 \end{bmatrix}$$

[6] Note also that $J_4 = g_e'V_2$ and therefore the off diagonal term g_e' could alternately be located in the fourth row and second column.

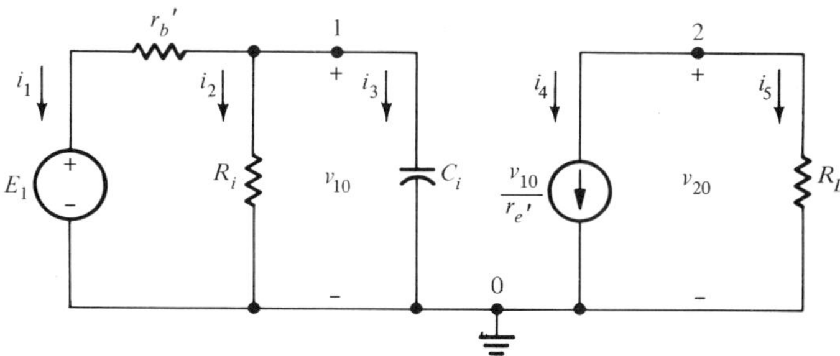

Figure 9.12 Small signal transistor equivalent circuit

Figure 9.13 gives the corresponding Fortran a-c circuit analysis program. An examination of this program will show that eight subroutines have been used; YMATP, YMATW, MINC, CMPRD, RITEG, GTPRD, GMPRD, and GMADD. The first five of these subroutines have been written by Robert Guarino and the last three have been obtained from the IBM/360 Scientific Subroutine Package. A short description of each subroutine follows:

SUBROUTINE YMATP(YR,YI,TY,PO,EL,B,N,W). This subroutine is used to form matrices that correspond to the real and imaginary parts of $[Y_{\text{branch}}]$ evaluated at $\omega = 1.0$ radians per second. The real part matrix is called *YR* and the imaginary part matrix is called *YI*.

SUBROUTINE YMATW(YR,YI,PO,EL,N,W). This subroutine is used to form matrices that correspond to the real and imaginary parts of $[Y_{\text{branch}}]$ evaluated at the required frequency ω. The real part and the imaginary part matrix are respectively called *YR* and *YI*. Note also that this subroutine is used to evaluate only the frequency sensitive terms since the constant terms are evaluated by subroutine YMATP.

SUBROUTINE MINC(A,B,L,M,N). This subroutine uses the Gauss-Jordan elimination method to invert a complex matrix. The required inputs are:

A—the real part of the matrix to be inverted.
B—the imaginary part of the matrix to be inverted.
L—a working vector of length N.
M—a working vector of length N.
N—the order of the matrix to be inverted.

The resulting matrix has its real part stored in *A* and its imaginary part stored in *B*.

SUBROUTINE CMPRD(AR,AI,BR,BI,RR,RI,N,M,L). This subroutine performs the multiplication of two complex matrices. The required inputs are:

AR—the real part of the premultiplying matrix.
AI—the imaginary part of the premultiplying matrix.
BR—the real part of the postmultiplying matrix.
BI—the imaginary part of the postmultiplying matrix.
RR—the real part of the resulting matrix.
RI—the imaginary part of the resulting matrix.
N—the number of rows in the premultiplying matrix.
M—the number of columns in the premultiplying matrix, or the number of rows in the postmultiplying matrix.
L—the number of columns in the postmultiplying matrix.

```
C     A-C CIRCUIT ANALYSIS PROGRAM
C
      DIMENSION FI(20),GI(10),STORR(20,10),STORI(20,10)
      DIMENSION H(20),A(200),YR(400),YI(400),
     1TY(20),PO(20),EL(20),B(20,10),
     1XIJ(20),W(20),E(20),
     1 TEMR(200),XLR(100),TEMI(200),XLI(100),
     1L(10),MA(10),BER(10),BEI(10),
     1ROR(10),ROI(10)
C
C
C
    1 READ(5,100) (H(I),I=1,20)
  100 FORMAT(20A4)
      WRITE(6,200) (H(I),I=1,20)
  200 FORMAT('1',20A4)
C
C
C
      READ(5,110) N,M
  110 FORMAT(2I5)
      DO 10 I=1,M
      READ(5,120) (XIJ(J),J=1,N)
  120 FORMAT(16F5.0)
      DO 10 J=1,N
      IJ=J+(I-1)*N
   10 A(IJ)=XIJ(J)
      WRITE(6,210)
  210 FORMAT('0THE A MATRIX IS')
      CALL RITEG(A,XIJ,N,M)
C
C
C
      DO 20 I=1,N
   20 READ(5,130)TY(I),PO(I),EL(I),(B(I,J),J=1,9)
  130 FORMAT(2F10.0,E15.7,9F5.0)
      WA=1.
      CALL YMATP(YR,YI,TY,PO,EL,B,N,WA)
      WRITE(6,220)
  220 FORMAT('0THE REAL PART OF Y FOR W = 1, IS')
      CALL RITEG(YR,XIJ,N,N)
      WRITE(6,230)
  230 FORMAT('0THE IMAGINARY PART OF Y FOR W = 1, IS')
      CALL RITEG(YI,XIJ,N,N)
C
C
C
      READ(5,140) (E(J),J=1,N)
  140 FORMAT(8F10.4)
      WRITE(6,240)
  240 FORMAT('0THE E MATRIX TRANSPOSED IS',/)
      WRITE(6,250) (E(J),J=1,N)
  250 FORMAT(10F5.0)
C
C
C
      READ(5,145) (FI(J),J=1,N)
  145 FORMAT(8F10.4)
      WRITE(6,245)
  245 FORMAT('0THE I MATRIX TRANSPOSED IS',/)
      WRITE(6,250) (FI(J),J=1,N)
C
C
C
      READ(5,150) NA
  150 FORMAT(I10)
      READ(5,160) (W(I),I=1,NA)
  160 FORMAT(8E10.2)
CC
C
C
      DO 40 K=1,NA
      WA=W(K)
      CALL YMATW(YR,YI,TY,PO,EL,N,WA)
      CALL GTPRD(A,YR,TEMR,N,M,N)
      CALL GMPRD(TEMR,A,XLR,M,N,M)
      CALL GTPRD(A,YI,TEMI,N,M,N)
      CALL GMPRD(TEMI,A,XLI,M,N,M)
      CALL MINC(XLR,XLI,L,MA,M)
      LO=1
```

Figure 9.13 A-C circuit analysis program

```
      CALL GMPRD(TEMR,E,ROR,M,N,LO)
      CALL GMPRD(TEMI,E,ROI,M,N,LO)
      CALL GTPRD(A,FI,GI,N,M,LO)
      CALL GMADD(GI,ROR,ROR,M,LO)
      CALL CMPRD(XLR,XLI,ROR,ROI,RER,REI,M,M,LO)
      DO 30 I=1,M
      STORR(K,I)=RER(I)
   30 STORI(K,I)=REI(I)
   40 CONTINUE
C
C
C
      DO 50 K=1,M
      WRITE(6,260) K
  260 FORMAT('0NODE', I3)
      WRITE(6,270)
  270 FORMAT(23X,'NODE VOLTAGE',11X,'MAGNITUDE',6X,'MAGNITUDE(DB)',
     115X,'PHASE',11X,'FREQUENCY')
      DO 60 I=1,NA
      XR=ABS(STORR(I,K))
      XI=ABS(STORI(I,K))
      XAG=SQRT(XR**2+XI**2)
      DB=20.0*ALOG10(XAG)
      PHASE=57.3*ATAN(STORI(I,K)/STORR(I,K))
   60 WRITE(6,280) STORR(I,K),STORI(I,K),XAG,DB,PHASE,W(I)
  280 FORMAT(1P1E19.6,1P1E15.6,'J',1P4E20.6)
   50 CONTINUE
      GO TO 1
      END

      SUBROUTINE YMATW(YR,YI,TY,PO,EL,N,W)
      DIMENSION YR(1),YI(1),TY(1),PO(1),EL(1)
      DO 15 I=1,N
      IX=PO(I)
      J=TY(I)
      IJ=IX+(IX-1)*N
      GO TO (15,5,10,15),J
    5 YI(IJ)=W*EL(I)
      GO TO 15
   10 YI(IJ)=(-1.0)/(W*EL(I))
   15 CONTINUE
      RETURN
      END

      SUBROUTINE YMATP(YR,YI,TY,PO,EL,B,N,W)
      DIMENSION YR(1),YI(1),TY(1),PO(1),EL(1)
      DIMENSION B(20,10)
      MA=N*N
      DO 25 I=1,MA
      YR(I)=0.0
   25 YI(I)=0.0
      DO 65 I=1,N
      IX=PO(I)
      J=TY(I)
      IJ=IX+(IX-1)*N
      GO TO (30,35,40,45),J
   30 YR(IJ)=1./EL(I)
      GO TO 65
   35 YI(IJ)=W*EL(I)
      GO TO 65
   40 YI(IJ)=(-1.0)/(W*EL(I))
      GO TO 65
   45 DO 60 K=1,10
      IF(B(I,K))50,65,55
   50 IY=-B(I,K)
      IX=PO(I)
      IJ=IX+(IY-1)*N
      YR(IJ)=(1.0)/EL(I)
      GO TO 60
   55 IY=B(I,K)
      IX=PO(I)
      IJ=IX+(IY-1)*N
      YR(IJ)=(1.)/EL(I)
   60 CONTINUE
   65 CONTINUE
      RETURN
      END
```

Figure 9.13 Continued

```
      SUBROUTINE RITEG(A,XIJ,N,M)
      DIMENSION A(1),XIJ(1)
      DO 10 J=1,N
      DO 5 I=1,M
      IJ=J+(I-1)*N
    5 XIJ(I)=A(IJ)
      WRITE(6,50) (XIJ(IJ),IJ=1,M)
   50 FORMAT(10E11.3)
   10 CONTINUE
      RETURN
      END

      SUBROUTINE CMPRD(AR,AI,BR,BI,RR,RI,N,M,L)
      DIMENSION AR(1),BR(1),AI(1),BI(1),RR(1),RI(1)
      IR=0
      IK=-M
      DO 10 K=1,L
      IK=IK+M
      DO 10 J=1,N
      IR=IR+1
      JI=J-N
      IB=IK
      RR(IR)=0.0
      RI(IR)=0.0
      DO 10 I=1,M
      JI=JI+N
      IB=IB+1
      RR(IR)=RR(IR)+AR(JI)*BR(IB)-AI(JI)*BI(IB)
   10 RI(IR)=RI(IR)+AR(JI)*BI(IB)+AI(JI)*BR(IB)
      RETURN
      END

      SUBROUTINE MINC(A,B,L,M,N)
      DIMENSION A(1),B(1),L(1),M(1)
      NK=-N
      DO 80 K=1,N
      NK=NK+N
      L(K)=K
      M(K)=K
      KK=NK+K
      BIGA=A(KK)
      BIGB=B(KK)
      BIG1=ABS(A(KK))
      BIG2=ABS(B(KK))
      BIGS=BIG1**2+BIG2**2
      BIGA=A(KK)
      DO 20 J=K,N
      IZ=N*(J-1)
      DO 20 I=K,N
      IJ=IZ+I
      BAG1=ABS(A(IJ))
      BAG2=ABS(B(IJ))
      BIGSA=BAG1**2+BAG2**2
   10 IF(ABS(BIGS)-ABS(BIGSA)) 15,20,20
   15 BIGS=BIGSA
      BIGA=A(IJ)
      BIGB=B(IJ)
      L(K)=I
      M(K)=J
   20 CONTINUE
C
C
C
C     INTERCHANGE ROWS
      J=L(K)
      IF(J-K) 35,35,25
   25 KI=K-N
      DO 30 I=1,N
      KI=KI+N
      HOLDA=-A(KI)
      HOLDB=-B(KI)
      JI=KI-K+J
      A(KI)=A(JI)
      B(KI)=B(JI)
      A(JI)=HOLDA
   30 B(JI)=HOLDB
```

Figure 9.13 Continued

```
C
C      INTERCHANGE COLUMNS
C
    35 I=M(K)
       IF (I-K) 45,45,38
    38 JP=N*(I-1)
       DO 40 J=1,N
       JK=NK+J
       JI=JP+J
       HOLDA=-A(JK)
       HOLDB=-B(JK)
       A(JK)=A(JI)
       B(JK)=B(JI)
       A(JI)=HOLDA
    40 B(JI)=HOLDB
    45 IF(BIGS) 48,46,48
C
C      DIVIDE COLUMN BY MINUS PIVOT (VALUE OF PIVOT ELEMENT IS
C      CONTAINED IN BIGA)
C
    46 D=0.0
    48 DO 55 I=1,N
       IF (I-K) 50,55,50
    50 IK=NK+I
       HOLDA=A(IK)
       A(IK)=(A(IK)*BIGA+B(IK)*BIGB)/(-BIGS)
       B(IK)=(B(IK)*BIGA-HOLDA*BIGB)/(-BIGS)
    55 CONTINUE
C
C      REDUCE MATRIX
C
       DO 65 I=1,N
       IK=NK+I
       IJ=I-N
       DO 65 J=1,N
       IJ=IJ+N
       IF (I-K) 60,65,60
    60 IF(J-K) 62,65,62
    62 KJ=IJ-I+K
       A(IJ)=A(IK)*A(KJ)-B(IK)*B(KJ)+A(IJ)
       B(IJ)=A(IK)*B(KJ)+B(IK)*A(KJ)+B(IJ)
    65 CONTINUE
C
C      DIVIDE ROWS BY PIVOT
C
       KJ=K-N
       DO 75 J=1,N
       KJ=KJ+N
       IF (J-K) 70,75,70
    70 HOLDA=A(KJ)
       A(KJ)=(A(KJ)*BIGA+B(KJ)*BIGB)/BIGS
       B(KJ)=(B(KJ)*BIGA-HOLDA*BIGB)/BIGS
    75 CONTINUE
C
C      REPLACE PIVOT BY RECIPROCAL
C
       A(KK)=BIGA/BIGS
       B(KK)=-BIGB/BIGS
    80 CONTINUE
```

Figure 9.13 Continued

SUBROUTINE RITEG(A,XIJ,N,M). This subroutine is used to print out an N by M matrix that has been stored as a single subscripted vector. The required inputs are:

A—the matrix to be printed.
XIJ—a working vector of length N.
N—the number of rows of the matrix.
M—the number of columns of the matrix.

SUBROUTINE GTPRD(A,B,R,N,M,L). This subroutine premultiplies a matrix by the transpose of another matrix. The required inputs are:

A—the name of the first matrix.
B—the name of the second matrix.

```
C
C        FINAL ROW AND COLUMN INTERCHANGE
C
         K=N
     100 K= (K-1)
         IF (K) 150,150,105
     105 I=L (K)
         IF (I-K) 120,120,108
     108 JQ=N* (K-1)
         JR =N* (I-1)
         DO  110  J=1,N
         JK=JQ+ J
         HOLDA=A (JK)
         HOLDB=B (JK)
         JI=JR+J
         A (JK) =-A (JI)
         B (JK) =-B (JI)
         A (JI) =HOLDA
     110 B(JI) =HOLDB
     120 J=M (K)
         IF (J-K) 100, 100, 125
     125 KI =K- N
         DO  130  I=1,N
         KI =KI+ N
         HOLDA=A (KI)
         HOLDB=B (KI)
         JI=KI-K +J
         A (KI) =-A (JI)
         B(KI) =-B(JI)
         A (JI) =HOLDA
     130 B(JI) =HOLDB
         GO  TO  100
     150 RETURN
         END
```

Figure 9.13 Continued

R—the name of the resulting matrix.
N—the number of rows in A and B.
M—the number of columns in A and rows in R.
L—the number of columns in B and R.

SUBROUTINE GMPRD(A,B,R,N,M,L). This subroutine multiplies matrix *A* by matrix *B*. The required inputs are:

A—the name of the first matrix.
B—the name of the second matrix.
R—the name of the resulting matrix.
N—the number of rows in A.
M—the number of columns in A and rows in B.
L—the number of columns in B.

SUBROUTINE GMADD(A,B,R,N,M). The subroutine adds matrix *A* and matrix *B*. The required inputs are:

A—the name of the first matrix.
B—the name of the second matrix.
R—the name of the resulting matrix.
N—the number of rows in A, B, and R.
M—the number of columns in A, B, and R.

Following is a short description of some of the variables used in this program:

H(I)—this is an A-format variable that is used in printing the output heading.

M—corresponds to the number of independent nodes of the network.

N—corresponds to the number of branches of the network.

A—corresponds to the junction matrix $[K]$. The program reads in the individual columns of $[K]$ and then internally forms $[K]$. This procedure is used to save data cards since a network usually has many more branches than nodes.

TY—is a code number used to represent a particular element type.
- 1—corresponds to a resistor.
- 2—corresponds to a capacitor.
- 3—corresponds to an inductor.
- 4—corresponds to a controlled current source.

PO—corresponds to the number of the branch in which the element is located.

EL—corresponds to the numerical value of the element. The numerical value of resistors must be given in ohms, capacitors in farads, inductors in henries and controlled sources must have an EL value that is the inverse of the coefficient of the controlled variable. For example, if the controlled source is specified as $g_e'V$, the EL value would be r_e'.

B—corresponds to the column number of A in which the element appears. For resistors, capacitors, and inductors, the value of B is specified as 0 and the elements are placed on the main diagonal. The value of zero is used because in most computers a blank is read as zero; therefore, for these elements, the value of B need not be punched. For example, in the previous problem the value of B would be specified as 3 since the controlled current source required a term in the third column of A. It should also be realized that a controlled current source may depend on more than one branch voltage. For example, if the controlled source depended on branch voltages V_2, V_5, and $-V_7$; the value punched on the data card for B would be +2,+5,−7,0. Note that a sign must be included and a zero indicates that no more values follow for this branch. Note also that in practice a zero need not be punched since a blank is read as a zero. No spaces may therefore be skipped as these would be read as zeros.

E—a vector corresponding to the source voltages in volts.

FI—a vector corresponding to source currents in amperes.

NA—the number of frequency evaluations required.

W—the frequencies in radians per second at which the node voltages are to be evaluated.

Summary: the program will set up matrix $[Y_{\text{branch}}]$ and solve for the node voltages $[v_{\text{node}}]$, using the matrix nodal equation

$$[v_{\text{node}}] = ([K]_T[Y_{\text{branch}}][K])^{-1}[K]_T([I_Y] + [Y_{\text{branch}}][E_{\text{gen}}])$$

at each of the specified frequencies. The A matrix, the Y_{branch} matrix, evaluated at $\omega = 1.0$ rps, and the E_{gen} and I_Y matrices will also be printed with the output. Further details of the program will be left as an exercise for the reader.

Test Amplifier

In order to see how we may use this program, let us now consider several examples. Figure 9.14a shows a simplified equivalent circuit for a one-stage transistor amplifier whose voltage transfer ratio is given by

$$\frac{V_0}{V_s} = \frac{\beta_0 R_L}{(R_s + \beta_0 r'_e)}\left[\frac{1}{R_L C_L s + 1}\right]$$

This amplifier therefore has a d-c or nominal gain of

$$A_0 = \frac{\beta_0 R_L}{R_s + \beta_0 r'_e}$$

and a single high-frequency pole at

$$s = -\frac{1}{R_L C_L}$$

or

$$\omega_{3\text{dB}} = \frac{1}{R_L C_L}$$

Choosing the positive sense of the branch currents and node voltages as shown in Figure 9.14b, and assuming that

$$\begin{aligned} \beta_0 &= 20 & R_1 &= 1000 \text{ ohms} \\ r'_e &= 5 \text{ ohms} & R_2 &= \beta_0 r'_e = 100 \text{ ohms} \\ C_L &= 10^{-9} \text{ farads} & R_3 &= 1000 \text{ ohms} \end{aligned}$$

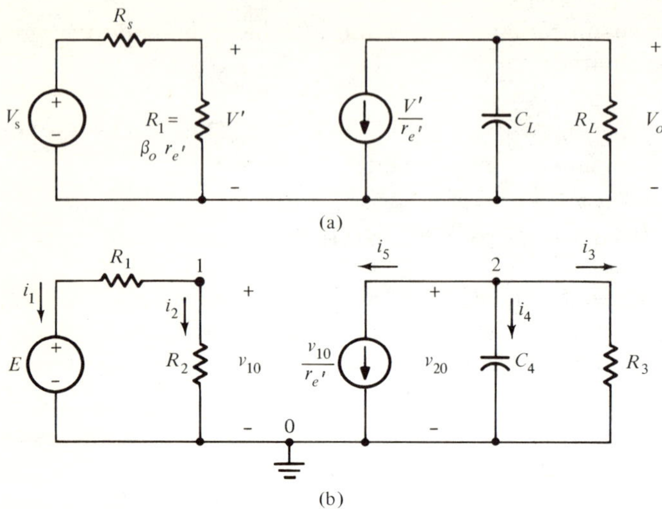

Figure 9.14 Simplified equivalent circuit of a one-stage transistor amplifier

it follows that

$$A_0 = \frac{V_0(0)}{V_s(0)} = \frac{20(1000)}{1000 + 100} = 18.18$$

and

$$\omega_{3\,\text{dB}} = \frac{1}{R_L C_L} = \frac{1}{10^3 10^{-9}} = 10^6 \text{ rps.}$$

We now turn to the a-c circuit analysis program in order to compute the frequency response. The following data cards were used to supply the necessary input information:

DATA CARD	CARD INFORMATION								CARD FORMAT
1	TEST AMPLIFIER								20A4
2	5	2							2I5
3	1.	1.	0.	0.	0.				16F5.0
4	0.	0.	1.	1.	1.				"
5		1.		1.		0.1E+04			2F10.0,E15.7,9F5.0
6		1.		2.		0.1E+03			"
7		1.		3.		0.1E+04			"
8		2.		4.		0.1E−08			"
9		4.		5.		0.5E+01	2.		"
10		1.0000		0.0000		0.0000	0.0000	0.0000	8F10.4
11		0.0000		0.0000		0.0000	0.0000	0.0000	"
12		12							I10
13	1.00E05	2.00E05	4.00E05	5.00E05	7.00E05	9.00E05	1.00E06	2.00E06	8E10.2
14	4.00E06	5.00E06	7.00E06	9.00E06					

Note that:

—The format of the first data card permits the printing of an 80-space free-form heading.
—The second data card reads in the number of branches and the number of independent nodes of the network.
—The third and fourth cards read in the two columns of the junction matrix.
—The fifth through ninth cards read in the network element values and their locations.
—The tenth and eleventh cards read in the elements of the E_{gen} and I_Y matrices.
—The twelfth card reads in the number of frequency calculations.
—The thirteenth and fourteenth cards read in the specific frequencies.

Figure 9.15 shows the computer output. Note that for each frequency the real and imaginary part and the magnitude and phase of each node voltage is tabulated. In addition, the A and Y_{branch} matrices, evaluated at $\omega = 1.0$ rps, and the E_{gen} and I_Y matrices are printed out. Figure 9.16 shows the corresponding graph of the output voltage. The reader may readily verify that these results agree with the transfer function calculations of $A_0 = 18.18$ and $\omega_{3\text{dB}} = 10^6$ rps.

Shunt-Peaked Amplifier

As a second example, let us consider the evaluation of the frequency response of the shunt-peaked amplifier we discussed in Section 7.4. Figure 9.17a shows the simplified equivalent circuit and Figure 9.17b gives the chosen positive sense of the branch currents and node voltages. Using the same transistor data values that were used in the calculation of Figure 7.25a,

$$\beta_0 = 27 \qquad f_t = 200 \text{ MHz} \qquad C_c = 3 \text{ pF}$$
$$r_e' = 5 \text{ ohms and } r_b' = 50 \text{ ohms}$$

the circuit element values for a maximally-flat magnitude response becomes

$$R_1 = R_4 = 42 \text{ ohms} \qquad R_2 = 50 \text{ ohms} \qquad R_3 = 135 \text{ ohms}$$
$$C_1 = 187 \text{ pF and } L_1 = 0.334\ \mu\text{H}.$$

The following data cards were therefore required to supply the necessary information to the a-c circuit analysis program of Figure 9.13:

DATA CARD	CARD INFORMATION								CARD FORMAT
1	SHUNT-PEAKED AMPLIFIER								20A4
2	8	4							2I5
3	1.	1.	1.	0.	0.	0.	0.	0.	16F5.0
4	0.	−1.	0.	0.	0.	0.	1.	0.	"
5	0.	0.	−1.	1.	0.	1.	0.	0.	"
6	0.	0.	0.	0.	1.	0.	0.	1.	"
7		2.		1.	0.				2F10.0,E15.7,9F5.0
8		1.		2.	0.4200000E+02				"
9		1.		3.	0.5000000E+02				"
10		1.		4.	0.1350000E+03				"
11		1.		5.	0.4200000E+02				"
12		2.		6.	0.1870000E−09				"
13		3.		7.	0.3340000E−06				"
14		4.		8.	0.5000000E+01			6.	"
15	0.	0.	0.	0.	0.	0.	0.	0.	8F10.4
16	1.	0.	0.	0.	0.	0.	0.	0.	"
17		12							I10
18	1.00E07	2.00E07	4.00E07	5.00E07	7.00E07	9.00E07	1.00E08	2.00E08	8E10.2
19	4.00E08	5.00E08	7.00E08	9.00E08					

Note that in this example we are using a current source rather than a voltage source and card number seven corresponds to a zero admittance, which is obtained by using a capacitor of zero farads, in branch one. Note also that card number fourteen corresponds to the controlled source

```
TEST AMPLIFIER

THE A MATRIX IS
 0.100E 01   0.0
 0.100E 01   0.0
 0.0         0.100E 01
 0.0         0.100E 01
 0.0         0.100E 01

THE REAL PART OF Y FOR W = 1, IS
 0.100E-02   0.0         0.0         0.0         0.0
 0.0         0.100E-01   0.0         0.0         0.0
 0.0         0.0         0.100E-02   0.0         0.0
 0.0         0.0         0.0         0.0         0.0
 0.0         0.200E 00   0.0         0.0         0.0

THE IMAGINARY PART OF Y FOR W = 1, IS
 0.0         0.0         0.0         0.0         0.0
 0.0         0.0         0.0         0.0         0.0
 0.0         0.0         0.0         0.0         0.0
 0.0         0.0         0.0         0.100E-08   0.0
 0.0         0.0         0.0         0.0         0.0

THE E MATRIX TRANSPOSED IS

  1.    0.    0.    0.    0.

THE I MATRIX TRANSPOSED IS

  0.    0.    0.    0.    0.
```

Figure 9.15 The output results of the program of Figure 9.13 for the test amplifier circuit of Figure 9.14

```
                  NODE VOLTAGE          MAGNITUDE      MAGNITUDE(DB)         PHASE        FREQUENCY
  9.090906E-02  -1.162823E-09J        9.090900E-02    -2.082784E 01     -7.329275E-07    1.000000E 05
  9.090906E-02  -3.387842E-09J        9.090900E-02    -2.082784E 01     -2.135356E-06    2.000000E 05
  9.090906E-02   0.0          J       9.090900E-02    -2.082784E 01      0.0             4.000000E 05
  9.090912E-02   0.0          J       9.090906E-02    -2.082784E 01      C.0             5.000000E 05
  9.090912E-02   0.0          J       9.090906E-02    -2.082784E 01      0.0             7.000000E 05
  9.090912E-02  -1.038190E-08J        9.090906E-02    -2.082784E 01     -6.543708E-06    9.000000E 05
  9.090912E-02   0.0          J       9.090906E-02    -2.082784E 01      0.0             1.000000E 06
  9.090900E-02  -1.127474E-08J        9.090894E-02    -2.082787E 01     -7.106471E-06    2.000000E 06
  9.090906E-02  -1.768587E-08J        9.090900E-02    -2.082784E 01     -1.114740E-05    4.000000E 06
  9.090918E-02   1.156385E-08J        9.090912E-02    -2.082784E 01      7.288681E-06    5.000000E 06
  9.090912E-02   6.013202E-09J        9.090906E-02    -2.082784E 01      3.790116E-06    7.000000E 06
  9.090900E-02   0.0          J       9.090894E-02    -2.082787E 01      C.0             9.000000E 06
NODE  2
                  NODE VOLTAGE          MAGNITUDE      MAGNITUDE(DB)         PHASE        FREQUENCY
 -1.800180E 01   1.800180E 00J        1.809157E 01     2.514951E 01     -5.711010E 00    1.000000E 05
 -1.748251E 01   3.496505E 00J        1.782872E 01     2.502238E 01     -1.131077E 01    2.CCC000E 05
 -1.567399E 01   6.269593E 00J        1.688139E 01     2.454814E 01     -2.180299E 01    4.000000E 05
 -1.454548E 01   7.272734E 00J        1.626233E 01     2.422365E 01     -2.656697E 01    5.000000E 05
 -1.220257E 01   8.541800E 00J        1.489514E 01     2.346088E 01     -3.499457E 01    7.000000E 05
 -1.004521E 01   9.040690E 00J        1.351445E 01     2.261597E 01     -4.199028E 01    9.000000E 05
 -9.090913E 00   9.090913E 00J        1.285649E 01     2.218243E 01     -4.500330E 01    1.000000E 06
 -3.636366E 00   7.272730E 00J        8.131158E 00     1.820303E 01     -6.343958E 01    2.000000E 06
 -1.069520E 00   4.278077E 00J        4.409739E 00     1.288826E 01     -7.596930E 01    4.000000E 06
 -6.993019E-01   3.496509E 00J        3.565752E 00     1.104302E 01     -7.869579E 01    5.000000E 06
 -3.636371E-01   2.545458E 00J        2.571301E 00     8.203058E 00     -8.187587E 01    7.000000E 06
 -2.217297E-01   1.995566E 00J        2.007846E 00     6.054608E 00     -8.366589E 01    9.000000E 06

IHC217I
```

Figure 9.15 Continued

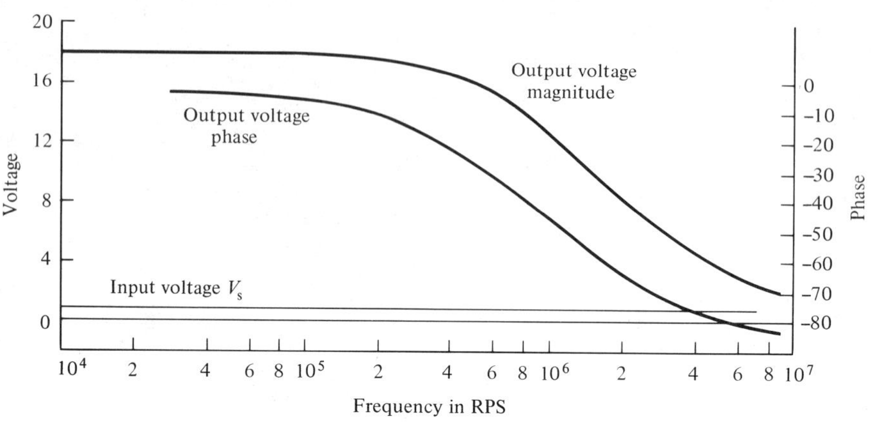

Figure 9.16 The output results of Figure 9.15

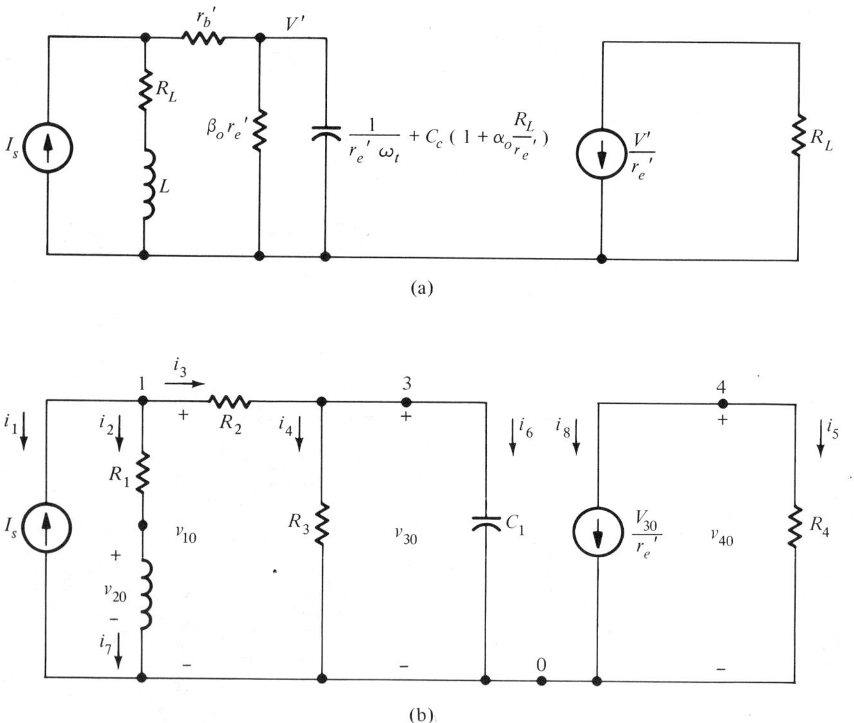

Figure 9.17 Simplified equivalent circuit of a shunt-peaked amplifier

```
SHUNT-PEAKED AMPLIFIER DESIGN.

THE A MATRIX IS
0.100E 01  0.0         0.0         0.0
0.100E 01 -0.100E 01   0.0         0.0
0.100E 01  0.0        -0.100E 01   0.0
0.0        0.0         0.100E 01   0.0
0.0        0.0         0.0         0.100E 01
0.0        0.0         0.100E 01   0.0
0.0        0.100E 01   0.0         0.0
0.0        0.0         0.0         0.100E 01

THE REAL PART OF Y FOR W = 1, IS
0.0        0.0         0.0         0.0         0.0         0.0         0.0         0.0
0.0        0.238E-01   0.0         0.0         0.0         0.0         0.0         0.0
0.0        0.0         0.200E-01   0.0         0.0         0.0         0.0         0.0
0.0        0.0         0.0         0.741E-02   0.0         0.0         0.0         0.0
0.0        0.0         0.0         0.0         0.238E-01   0.0         0.0         0.0
0.0        0.0         0.0         0.0         0.0         0.0         0.0         0.0
0.0        0.0         0.0         0.0         0.0         0.0         0.0         0.0
0.0        0.0         0.0         0.0         0.0         0.200E 00   0.0         0.0

THE IMAGINARY PART OF Y FOR W = 1, IS
0.0        0.0         0.0         0.0         0.0         0.0         0.0         0.0
0.0        0.0         0.0         0.0         0.0         0.0         0.0         0.0
0.0        0.0         0.0         0.0         0.0         0.0         0.0         0.0
0.0        0.0         0.0         0.0         0.0         0.0         0.0         0.0
0.0        0.0         0.0         0.0         0.0         0.0         0.0         0.0
0.0        0.0         0.0         0.0         0.0         0.187E-09   0.0         0.0
0.0        0.0         0.0         0.0         0.0         0.0        -0.299E 07   0.0
0.0        0.0         0.0         0.0         0.0         0.0         0.0         0.0

THE E MATRIX TRANSPOSED IS

 0.   0.   0.   0.   0.   0.   0.   0.

THE I MATRIX TRANSPOSED IS

 1.   0.   0.   0.   0.   0.   0.   0.

NODE  1
                   NODE VOLTAGE              MAGNITUDE        MAGNITUDE(DB)        PHASE            FREQUENCY
      3.429314E 01  1.044957E 00J       3.430904E 01      3.070816E 01      1.745466E 00      1.000000E 07
      3.449314E 01  2.068365E 00J       3.454510E 01      3.076772E 01      3.432851E 00      2.000000E 07
      3.521068E 01  3.970337E 00J       3.543381E 01      3.098834E 01      6.433933E 00      4.000000E 07
      3.572360E 01  4.814620E 00J       3.604657E 01      3.113727E 01      7.676299E 00      5.000000E 07
      3.696577E 01  6.229821E 00J       3.748705E 01      3.147762E 01      9.566836E 00      7.000000E 07
      3.837373E 01  7.248199E 00J       3.905226E 01      3.183292E 01      1.069705E 01      9.000000E 07
      3.909621E 01  7.611255E 00J       3.983018E 01      3.200424E 01      1.101736E 01      1.000000E 08
      4.479826E 01  7.808276E 00J       4.547365E 01      3.315518E 01      9.887975E 00      2.000000E 08
      4.845093E 01  4.998328E 00J       4.870805E 01      3.375200E 01      5.890381E 00      4.000000E 08
      4.899548E 01  4.112323E 00J       4.916776E 01      3.383359E 01      4.798092E 00      5.000000E 08
      4.948325E 01  3.006102E 00J       4.957446E 01      3.390514E 01      3.476695E 00      7.000000E 08
      4.968668E 01  2.359363E 00J       4.974265E 01      3.393457E 01      2.718837E 00      9.000000E 08
```

Figure 9.18 The output results of the program of Figure 9.13 for the network of Figure 9.17

```
NODE  2
                  NODE VOLTAGE           MAGNITUDE     MAGNITUDE(DB)        PHASE         FREQUENCY
1.329312E-01   2.716552E 00J         2.719802E 00     8.690743E 00     8.720491E 01    1.000000E 07
5.299174E-01   5.400174E 00J         5.426111E 00     1.468977E 01     8.440170E 01    2.000000E 07
2.068506E 00   1.053601E 01J         1.074101E 01     2.062090E 01     7.879359E 01    4.000000E 07
3.223852E 00   1.292252E 01J         1.331858E 01     2.248914E 01     7.599754E 01    5.000000E 07
6.097463E 00   1.716333E 01J         1.823309E 01     2.521721E 01     7.046802E 01    7.000000E 07
9.568012E 00   2.061668E 01J         2.272870E 01     2.713147E 01     6.510916E 01    9.000000E 07
1.143824E 01   2.199466E 01J         2.479108E 01     2.788589E 01     6.252803E 01    1.000000E 08
2.858768E 01   2.578255E 01J         3.849567E 01     3.170845E 01     4.204964E 01    2.000000E 08
4.266322E 01   1.841042E 01J         4.646505E 01     3.334270E 01     2.334331E 01    4.000000E 08
4.510312E 01   1.545688E 01J         4.768288E 01     3.356723E 01     1.891608E 01    5.000000E 08
4.741319E 01   1.152345E 01J         4.879344E 01     3.376721E 01     1.366148E 01    7.000000E 08
4.841194E 01   9.123518E 00J         4.926413E 01     3.385060E 01     1.067334E 01    9.000000E 08

NODE  3
                  NODE VOLTAGE           MAGNITUDE     MAGNITUDE(DB)        PHASE         FREQUENCY
2.496005E 01  -9.455378E-01J         2.497792E 01     2.795110E 01    -2.168456E 00    1.000000E 07
2.490366E 01  -1.898085E 00J         2.497588E 01     2.795039E 01    -4.358807E 00    2.000000E 07
2.464185E 01  -3.845930E 00J         2.494016E 01     2.793797E 01    -8.871414E 00    4.000000E 07
2.441379E 01  -4.837640E 00J         2.488846E 01     2.791995E 01    -1.120890E 01    5.000000E 07
2.371376E 01  -6.810090E 00J         2.467224E 01     2.784416E 01    -1.602411E 01    7.000000E 07
2.266632E 01  -8.666657E 00J         2.426669E 01     2.770020E 01    -2.092630E 01    9.000000E 07
2.202240E 01  -9.511853E 00J         2.398877E 01     2.760014E 01    -2.336206E 01    1.000000E 08
1.409663E 01  -1.358968E 01J         1.958044E 01     2.583644E 01    -4.395422E 01    2.000000E 08
5.341109E 00  -1.096844E 01J         1.219976E 01     2.172701E 01    -6.404077E 01    4.000000E 08
3.623329E 00  -9.393110E 00J         1.006772E 01     2.005861E 01    -6.891122E 01    5.000000E 08
1.947695E 00  -7.133594E 00J         7.394704E 00     1.737840E 01    -7.473409E 01    7.000000E 08
1.204284E 00  -5.693199E 00J         5.819175E 00     1.529722E 01    -7.806195E 01    9.000000E 08

NODE  4
                  NODE VOLTAGE           MAGNITUDE     MAGNITUDE(DB)        PHASE         FREQUENCY
-2.096648E 02   7.938319E 00J        2.098151E 02     4.643672E 01    -2.168452E 00    1.000000E 07
-2.091911E 02   1.594392E 01J        2.097978E 02     4.643602E 01    -4.358804E 00    2.000000E 07
-2.069917E 02   3.230588E 01J        2.094975E 02     4.642357E 01    -8.871424E 00    4.000000E 07
-2.050762E 02   4.063623E 01J        2.090635E 02     4.640556E 01    -1.120890E 01    5.000000E 07
-1.991959E 02   5.720482E 01J        2.072472E 02     4.632976E 01    -1.602411E 01    7.000000E 07
-1.903974E 02   7.279997E 01J        2.038406E 02     4.618579E 01    -2.092628E 01    9.000000E 07
-1.849884E 02   7.989961E 01J        2.015050E 02     4.608575E 01    -2.336203E 01    1.000000E 08
-1.184118E 02   1.141536E 02J        1.644761E 02     4.432205E 01    -4.395427E 01    2.000000E 08
-4.486536E 01   9.213496E 01J        1.024780E 02     4.021260E 01    -6.404077E 01    4.000000E 08
-3.043599E 01   7.890218E 01J        8.456339E 01     3.854420E 01    -6.891122E 01    5.000000E 08
-1.636066E 01   5.992224E 01J        6.211557E 01     3.536400E 01    -7.473409E 01    7.000000E 08
-1.011599E 01   4.782292E 01J        4.388112E 01     3.378281E 01    -7.806195E 01    9.000000E 08
```

IHC217I

Figure 9.18 Continued

and the element in the eighth row and sixth column of the branch admittance matrix must be g'_e.[7]

Figure 9.18 shows the computer output. Figure 9.19 is the corresponding graph of the output voltage. Note that these results agree with those of Figure 7.25a, since a nominal output voltage of 210 units across 42 ohms is equivalent to a nominal current gain of $A_0 = 5.$ and a 3 dB bandwidth of 240 megaradians per second corresponds to a normalized gain-bandwidth product of 0.954.

Shunt-Series Feedback Pair

Let us now use the a-c circuit analysis program to calculate the frequency response of the maximally-flat feedback pair of Section 7.7. Figure 9.20 gives the equivalent network for the shunt-series feedback pair with

$$\begin{aligned}
R_1 &= R_s = 1000 \text{ ohms} \\
R_2 &= r'_{b1} = 75 \\
R_3 &= R_F = 1000 \\
R_4 &= \beta_{01} r'_{e1} = 57(5) = 285 \\
R_5 &= R_{L1} = 2700 \\
R_6 &= r'_{b2} = 75 \\
R_7 &= \beta_{02} r'_{e2} = 50(5) = 250 \\
R_3 &= R_E = 50 \\
R_9 &= R_{L2} = 500
\end{aligned}$$

$$\begin{aligned}
C_1 &= \frac{1}{r'_{e1}\omega_{t1}} = 75.7 \text{ pF} \\
C_2 &= C_{c1} = 2 \text{ pF} \\
C_3 &= \frac{1}{r'_{e2}\omega_{t2}} = 70.6 \text{ pF} \\
C_4 &= C_{c2} = 2 \text{ pF} \\
C_5 &= C_F = 1 \text{ pF}
\end{aligned}$$

The required input data to the circuit analysis program of Figure 9.13 thus becomes:

[7] We could also write $J_8 = g'_e V_4$ and therefore the off diagonal term g'_e could alternately be located in the eighth row and fourth column.

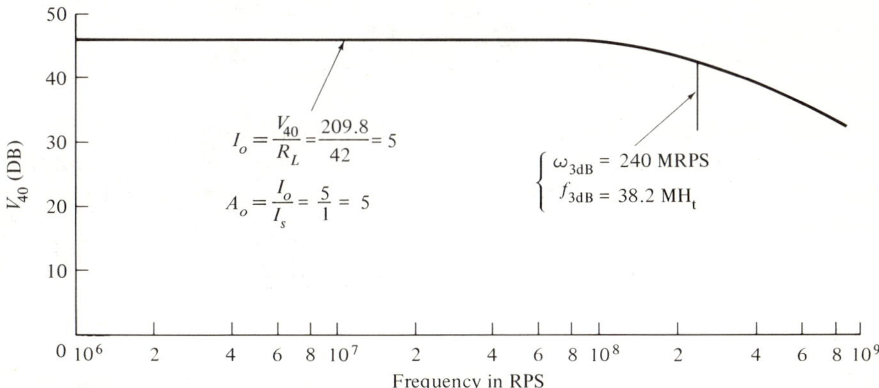

Figure 9.19 The output voltage of the network of Figure 9.17

DATA CARD	CARD INFORMATION	CARD FORMAT
1	MAXIMALLY FLAT SHUNT-SERIES FEEDBACK PAIR	20A4
2	16 6	2I5
3	1. 1. 1. 0. 0. 0. 0. 0. 0. 0. 0. 0. 0. 1. 0. 0.	16F5.0
4	0. −1. 0. 1. 0. 0. 0. 0. 0. 1. 1. 0. 0. 0. 0. 0.	"
5	0. 0. 0. 0. 1. 1. 0. 0. 0. 0. −1. 0. 0. 0. 1. 0.	"
6	0. 0. 0. 0. 0. −1. 1. 0. 0. 0. 0. 1. 1. 0. 0. 0.	"
7	0. 0. −1. 0. 0. 0. −1. 1. 0. 0. 0. −1. 0. −1. 0. −1.	"
8	0. 0. 0. 0. 0. 0. 0. 0. 1. 0. 0. 0. −1. 0. 0. 1.	"
9	1. 1. 0.1000000E+04	2F10.0,E15.7,9F5.0
10	1. 2. 0.7500000E+02	"
11	1. 3. 0.1000000E+04	"
12	1. 4. 0.2850000E+03	"
13	1. 5. 0.2700000E+04	"
14	1. 6. 0.7500000E+02	"
15	1. 7. 0.2500000E+03	"
16	1. 8. 0.5000000E+02	"
17	1. 9. 0.5000000E+02	"
18	2. 10. 0.7570000E−10	"
19	2. 11. 0.2000000E−11	"
20	2. 12. 0.7060000E−10	"
21	2. 13. 0.2000000E−11	"
22	2. 14. 0.1000000E−11	"
23	4. 15. 0.5000000E+01 4.	"
24	4. 16. 0.5000000E+01 7.	"
25	1. 0. 0. 0. 0. 0. 0. 0.	8F10.4
26	0. 0. 0. 0. 0. 0. 0. 0.	"
27	0. 0. 0. 0. 0. 0. 0. 0.	8F10.4
28	0. 0. 0. 0. 0. 0. 0. 0.	"
29	16	I10
30	1.00E07 2.00E07 3.00E07 4.00E07 5.00E07 6.00E07 7.00E07 8.00E07	8E10.2
31	9.00E07 1.00E08 2.00E08 3.00E08 4.00E08 5.00E08 7.00E08 9.00E08	"

Note that input cards 23 and 24 correspond respectively to controlled sources

$$J_{14} = g'_{e1} V_4$$
$$J_{15} = g'_{e2} V_7$$

Figure 9.21 gives the computer results, and Figure 9.22 shows the corresponding graph of the output voltage. The accuracy of the nominal gain ($G_0 = 20$) obtained by the a-c analysis program may be easily verified from

$$I_s = \frac{V_s - v_{10}}{R_s} = \frac{1.0 - 0.006}{1000}$$

and

$$I_0 = \frac{v_{60}}{R_{L2}} = \frac{10.05}{500}$$

so that

$$G_0 = \frac{I_0}{I_s} = 20.0$$

The reader will find, on the other hand, that the value of $\omega_{3\text{dB}}$ does not agree with the value computed in Section 7.7. One reason for this disagreement is that the value of C_F supplied is 1.0 instead of 8.9 pf. Note that since the value used here is considerably smaller than that value

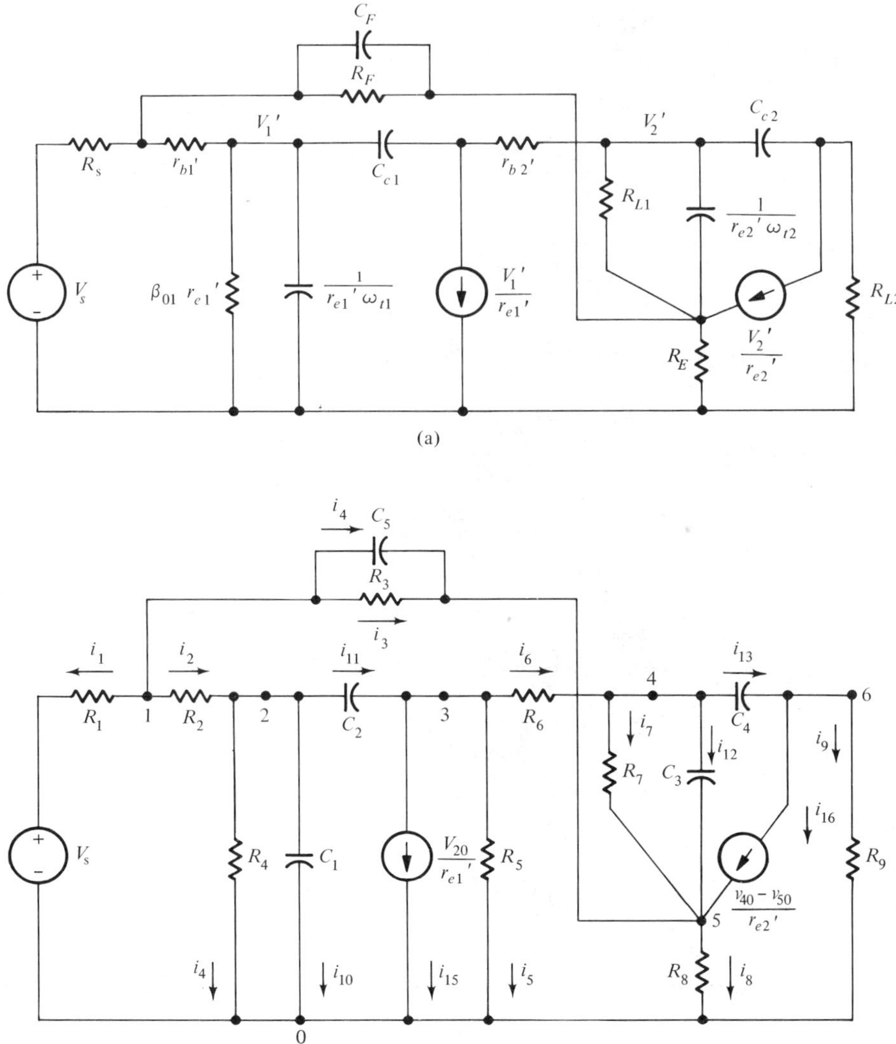

Figure 9.20 Equivalent circuit of a shunt-series feedback pair

```
MAXIMALLY FLAT SERIES-SHUNT FEEDBACK PAIR

THE A MATRIX IS
 0.1COE 01  0.C          0.0          0.0          0.0          0.C
 0.1COE 01 -0.100E 01    0.0          0.0          C.0          C.C
 0.1COE 01  0.C          0.0          0.0         -0.1COE C1    C.C
 0.0        0.1COE 01    0.0          0.0          0.0          0.C
 C.0        0.C          0.100E 01    C.0          0.0          0.C
 C.0        0.C          0.100E 01   -0.100E C1    0.0          0.C
 0.0        0.C          0.0          0.100E C1   -0.1COE C1    0.C
 0.0        0.C          0.0          0.0          C.1COE 01    0.C
 0.0        0.C          0.0          0.0          C.0          0.1CCE 01
 C.0        0.1COE 01    0.0          0.0          0.0          C.C
 C.0        0.1COE 01   -0.100E 01    0.0          0.0          C.C
 C.0        0.C          0.0          0.100E 01   -0.1COE 01    C.C
 C.0        0.C          0.0          0.1COE C1    0.0         -0.1CCE 01
 0.1COE 01  0.C          0.0          0.0         -0.100E 01    0.C
 0.0        0.C          0.100E 01    0.0          0.0          0.C
 0.0        0.C          0.0          0.0         -0.1COE C1    0.1CCE 01

THE REAL PART CF Y FOR W = 1, IS
 0.1COE-02  0.C          0.0          0.0          0.0          0.C          0.0          0.0          0.0          0.0
 0.0        0.C          0.0          C.0          0.0          C.C
 0.0        0.133E-01    0.0          0.0          C.0          0.C          0.0          0.0          0.0          0.0
 0.0        0.C          0.0          0.0          0.0          0.C
 C.0        0.C          0.100E-02    0.0          0.0          0.C          0.0          C.0          0.0          0.0
 0.0        0.C          0.0          0.0          C.0          C.C
 0.0        0.C          0.0          0.351E-C2    0.0          0.C          0.C          0.C          0.0          C.0
 C.0        0.C          0.0          0.0          0.0          0.C
 0.0        0.C          0.0          0.0          C.370E-C3    0.C          0.0          C.0          0.0          0.0
 C.0        0.C          0.0          0.0          0.0          0.C
 0.0        0.C          0.0          0.0          0.0          C.133E-01    C.C          0.C          0.0          0.0
 0.0        0.C          0.0          0.0          0.0          0.C
 0.0        0.C          0.0          0.0          0.0          C.C          0.4C0E-02    0.0          0.0          0.0
 0.0        0.C          0.0          0.0          C.0          C.C
 0.0        0.C          0.0          C.0          C.0          0.C          C.0          0.20CE-01    0.0          0.0
 0.0        0.C          0.0          0.0          0.0          0.C
 0.0        0.C          0.0          0.0          C.0          0.C          0.0          C.0          0.2CCE-02    0.0
 0.0        0.C          0.0          0.0          0.0          C.C
 0.0        0.C          0.0          0.0          0.0          C.C          0.0          0.0          0.0          C.0
 0.0        0.C          0.0          0.0          0.0          0.C
 0.0        0.C          0.0          0.0          C.0          0.C          0.0          0.0          0.0          0.0
 0.0        0.C          0.0          0.0          0.0          0.C
 0.0        0.C          0.0          0.0          0.0          0.C          0.C          C.0          0.0          0.0
 0.0        0.C          0.0          0.0          0.0          0.C
 0.0        0.C          0.0          C.0          0.0          0.C          0.0          C.0          0.0          0.C
 0.0        0.C          0.0          0.0          C.0          0.C
 0.0        0.C          0.0          0.0          C.0          0.C          C.0          C.0          0.0          0.0
 0.0        0.C          0.0          C.0          0.0          0.C
 0.0        0.C          0.0          0.2COE 00    C.0          0.C          0.0          C.0          0.0          0.0
 0.0        0.C          0.0          0.0          0.0          C.C
 0.0        0.C          0.0          0.0          0.0          0.C          0.2COE 00    0.C          0.0          0.0
 0.0        0.C          0.0          C.0          0.0          0.C
```

Figure 9.21 The output results of the program of Figure 9.13 for the network of Figure 9.20

```
THE IMAGINARY PART OF Y FOR W = 1, IS
 0.0        0.C        0.0        0.0        0.0        0.C        0.0        0.0        0.0        0.0
 0.0        0.C        0.0        0.0        0.0        0.C
 0.0        0.C        0.0        0.0        0.0        0.C        0.0        0.0        0.0        0.0
 0.0        0.C        0.0        0.0        0.0        C.C
 0.0        0.C        0.0        0.0        0.0        0.C        0.0        0.0        0.0        0.0
 0.0        0.C        0.0        0.0        0.0        0.C
 0.0        0.C        0.0        0.0        0.0        0.C        0.0        0.0        0.0        C.C
 0.0        0.C        0.0        0.0        0.0        C.C
 0.0        0.0        0.0        0.0        0.0        C.C        0.0        0.0        0.0        0.0
 C.0        0.C        0.0        0.0        0.0        0.C
 0.0        0.C        0.0        0.0        0.0        0.C        0.0        0.0        0.0        C.C
 0.0        0.C        0.0        0.0        0.0        C.C
 0.0        0.C        0.0        0.0        0.0        0.C        0.0        C.0        0.0        0.C
 0.0        0.C        0.0        0.0        0.0        0.C
 C.0        0.C        0.0        0.0        0.0        0.C        0.0        0.0        0.0        0.0
 0.0        0.C        0.0        0.0        0.0        0.C
 0.0        0.C        0.0        0.0        0.0        0.C        0.0        0.0        0.0        0.0
 0.0        0.C        0.0        0.0        0.0        0.C
 0.0        0.C        0.0        0.0        0.0        0.C        0.0        C.0        0.0        0.757E-10
 0.0        0.C        0.0        0.0        0.0        0.C
 0.0        0.C        0.0        0.0        0.0        0.C        0.0        C.0        0.0        0.0
 0.200E-11  0.C        0.0        C.0        0.0        0.C
 0.0        0.C        0.0        0.0        0.0        0.C        0.0        0.0        0.C        0.C
 0.0        0.706E-10  0.0        0.0        0.0        0.C
 0.0        0.0        0.0        0.0        0.0        0.C        0.0        0.0        0.0        0.0
 0.0        0.C        0.200E-11  0.0        0.0        C.C
 0.0        0.C        0.0        0.0        0.0        0.C        0.0        0.0        0.C        0.0
 0.0        0.C        0.0        0.100E-11  0.0        C.C
 0.0        0.C        0.0        0.0        0.0        0.0        0.0        0.0        0.0        C.0
 C.0        0.C        0.0        0.0        0.0        0.C
 0.0        0.C        0.0        0.0        0.0        0.C        0.0        C.0        0.C        0.0
 0.0        0.C        0.0        0.0        0.0        0.C

THE E MATRIX TRANSPOSED IS

 1.  0.  0.  C.  0.  0.  0.  0.  0.  0.
 0.  0.  0.  0.  0.  0.

THE I MATRIX TRANSPOSED IS

 0.  0.  0.  0.  0.  0.  0.  0.  0.  0.
 0.  0.  0.  0.  0.  0.
```

Figure 9.21 Continued

NODE 1

	NODE VOLTAGE	MAGNITUDE	MAGNITUDE(DB)	PHASE	FREQUENCY
5.203992E-03	3.686824E-03J	6.377630E-03	-4.390681E 01	3.531866E 01	1.000000E 07
5.446438E-03	7.397208E-03J	9.185985E-03	-4.073747E 01	5.364034E 01	2.000000E 07
5.866546E-03	1.115302E-02J	1.260183E-02	-3.799130E 01	6.225993E 01	3.000000E 07
6.489333E-03	1.497232E-02J	1.631814E-02	-3.574657E 01	6.657185E 01	4.000000E 07
7.349286E-03	1.886711E-02J	2.024795E-02	-3.387238E 01	6.872264E 01	5.000000E 07
8.490186E-03	2.284076E-02J	2.436767E-02	-3.226372E 01	6.961435E 01	6.000000E 07
9.964641E-03	2.688403E-02J	2.867132E-02	-3.085103E 01	6.966768E 01	7.000000E 07
1.183142E-02	3.097149E-02J	3.315442E-02	-2.958916E 01	6.909767E 01	8.000000E 07
1.415196E-02	3.505715E-02J	3.780584E-02	-2.844881E 01	6.802187E 01	9.000000E 07
1.698329E-02	3.907101E-02J	4.260253E-02	-2.741129E 01	6.651131E 01	1.000000E 08
6.646764E-02	5.067552E-02J	8.358198E-02	-2.155772E 01	3.732495E 01	2.000000E 08
8.366174E-02	2.050187E-02J	8.613712E-02	-2.129617E 01	1.377038E 01	3.000000E 08
7.893145E-02	5.995400E-03J	7.915878E-02	-2.203000E 01	4.343991E 00	4.000000E 08
7.422370E-02	6.430310E-04J	7.422644E-02	-2.258882E 01	4.964012E-01	5.000000E 08
6.910074E-02	-2.847524E-03J	6.915933E-02	-2.320297E 01	-2.359900E 00	7.000000E 08
6.673527E-02	-4.074208E-03J	6.685948E-02	-2.349672E 01	-3.493845E 00	9.000000E 08

NODE 2

	NODE VOLTAGE	MAGNITUDE	MAGNITUDE(DB)	PHASE	FREQUENCY
4.146714E-03	1.413045E-03J	4.380856E-03	-4.716882E 01	1.881856E 01	1.000000E 07
4.416574E-03	2.814605E-03J	5.237184E-03	-4.561803E 01	3.251103E 01	2.000000E 07
4.870024E-03	4.192032E-03J	6.425746E-03	-4.384151E 01	4.072427E 01	3.000000E 07
5.512230E-03	5.530261E-03J	7.808227E-03	-4.214894E 01	4.509682E 01	4.000000E 07
6.349441E-03	6.810609E-03J	9.311270E-03	-4.061981E 01	4.701042E 01	5.000000E 07
7.387605E-03	8.009866E-03J	1.089654E-02	-3.925423E 01	4.731770E 01	6.000000E 07
8.631475E-03	9.098820E-03J	1.254157E-02	-3.803294E 01	4.651328E 01	7.000000E 07
1.008106E-02	1.004229E-02J	1.422938E-02	-3.693626E 01	4.489291E 01	8.000000E 07
1.173037E-02	1.079842E-02J	1.594388E-02	-3.594812E 01	4.263431E 01	9.000000E 07
1.356248E-02	1.132066E-02J	1.766630E-02	-3.505708E 01	3.985481E 01	1.000000E 08
2.899807E-02	-1.217574E-03J	2.902362E-02	-3.074495E 01	-2.404505E 00	2.000000E 08
1.996154E-02	-1.449640E-02J	2.467000E-02	-3.215662E 01	-3.599042E 01	3.000000E 08
1.152459E-02	-1.493465E-02J	1.886425E-02	-3.448720E 01	-5.234760E 01	4.000000E 08
7.264663E-03	-1.310116E-02J	1.498051E-02	-3.648946E 01	-6.099582E 01	5.000000E 08
3.657323E-03	-9.895388E-03J	1.054963E-02	-3.953525E 01	-6.972086E 01	7.000000E 08
2.239920E-03	-7.832360E-03J	8.146353E-03	-4.178072E 01	-7.404578E 01	9.000000E 08

NODE 3

	NODE VOLTAGE	MAGNITUDE	MAGNITUDE(DB)	PHASE	FREQUENCY
-1.106850E 00	3.223875E-02J	1.107318E 00	8.854461E-01	-1.668481E 00	1.000000E 07
-1.109058E 00	6.517756E-02J	1.110971E 00	9.140579E-01	-3.363558E 00	2.000000E 07
-1.112538E 00	9.951371E-02J	1.116980E 00	9.609048E-01	-5.111733E 00	3.000000E 07
-1.116973E 00	1.359354E-01J	1.125214E 00	1.024699E 00	-6.939267E 00	4.000000E 07
-1.121896E 00	1.751117E-01J	1.135479E 00	1.103580E 00	-8.872108E 00	5.000000E 07
-1.126668E 00	2.176640E-01J	1.147500E 00	1.195053E 00	-1.093522E 01	6.000000E 07
-1.130461E 00	2.641345E-01J	1.160908E 00	1.295953E 00	-1.315230E 01	7.000000E 07
-1.132229E 00	3.149253E-01J	1.175210E 00	1.402308E 00	-1.554485E 01	8.000000E 07
-1.130724E 00	3.702265E-01J	1.189791E 00	1.509410E 00	-1.813103E 01	9.000000E 07
-1.124517E 00	4.299155E-01J	1.203896E 00	1.611774E 00	-2.092393E 01	1.000000E 08
-6.138727E-01	9.473124E-01J	1.128822E 00	1.052511E 00	-5.706032E 01	2.000000E 08
-8.701068E-02	7.356415E-01J	7.407694E-01	-2.606338E 00	-8.326054E 01	3.000000E 08
3.339323E-02	4.926923E-01J	4.938225E-01	-6.128584E 00	8.612889E 01	4.000000E 08
5.300694E-02	3.583164E-01J	3.622158E-01	-8.820645E 00	8.159100E 01	5.000000E 08
5.016419E-02	2.296685E-01J	2.350830E-01	-1.257557E 01	7.768462E 01	7.000000E 08
[illegible]	[illegible]	[illegible]	-1.515413E 01	7.544678E 01	9.000000E 08

Figure 9.21 Continued

specified by the analysis of Section 7.7, peaking in the output response should be expected. Additional computer runs will show, however, that a value of $C_F \approx 3$ pf is required to produce a maximal-flat response. Since a larger value of C_F will reduce the bandwidth further from that shown in Figure 9.22, this value of C_F makes the bandwidth discrepancy

NODE 4

NODE VOLTAGE		MAGNITUDE	MAGNITUDE(DB)	PHASE	FREQUENCY
-1.075441E 00	5.266357E-02J	1.076729E 00	6.421263E-01	-2.803698E 00	1.000000E 07
-1.073804E 00	1.058668E-01J	1.079009E 00	6.605017E-01	-5.631026E 00	2.000000E 07
-1.070821E 00	1.601302E-01J	1.082726E 00	6.903747E-01	-8.505593E 00	3.000000E 07
-1.066099E 00	2.159305E-01J	1.087746E 00	7.305471E-01	-1.145078E 01	4.000000E 07
-1.059081E 00	2.736728E-01J	1.093868E 00	7.793002E-01	-1.448969E 01	5.000000E 07
-1.049036E 00	3.336511E-01J	1.100817E 00	8.343001E-01	-1.764481E 01	6.000000E 07
-1.035067E 00	3.959936E-01J	1.108230E 00	8.925948E-01	-2.093727E 01	7.000000E 07
-1.016121E 00	4.606003E-01J	1.115640E 00	9.504790E-01	-2.438623E 01	8.000000E 07
-9.910292E-01	5.270634E-01J	1.122468E 00	1.003478E 00	-2.800763E 01	9.000000E 07
-9.585951E-01	5.945961E-01J	1.128028E 00	1.046396E 00	-3.181277E 01	1.000000E 08
-2.244084E-01	9.360766E-01J	9.625998E-01	-3.310845E-01	-7.652428E 01	2.000000E 08
1.762394E-01	5.338144E-01J	5.621548E-01	-5.002881E 00	7.173454E 01	3.000000E 08
1.767319E-01	2.836704E-01J	3.342199E-01	-9.519351E 00	5.808054E 01	4.000000E 08
1.355929E-01	1.751824E-01J	2.215272E-01	-1.309146E 01	5.226357E 01	5.000000E 08
8.126330E-02	9.250039E-02J	1.231261E-01	-1.819299E 01	4.870366E 01	7.000000E 08
5.484881E-02	6.192799E-02J	8.272523E-02	-2.164722E 01	4.847264E 01	9.000000E 08

NODE 5

NODE VOLTAGE		MAGNITUDE	MAGNITUDE(DB)	PHASE	FREQUENCY
-9.750575E-01	4.749323E-02J	9.762135E-01	-2.091042E-01	-2.788772E 00	1.000000E 07
-9.736154E-01	9.547693E-02J	9.782856E-01	-1.906867E-01	-5.601172E 00	2.000000E 07
-9.709818E-01	1.444247E-01J	9.816639E-01	-1.607431E-01	-8.460814E 00	3.000000E 07
-9.668007E-01	1.947701E-01J	9.862245E-01	-1.204839E-01	-1.139109E 01	4.000000E 07
-9.605695E-01	2.468837E-01J	9.917889E-01	-7.161540E-02	-1.441512E 01	5.000000E 07
-9.516279E-01	3.010334E-01J	9.981065E-01	-1.646250E-02	-1.755528E 01	6.000000E 07
-9.391630E-01	3.573422E-01J	1.004848E 00	4.200335E-02	-2.083286E 01	7.000000E 07
-9.222203E-01	4.157233E-01J	1.011590E 00	1.000905E-01	-2.426692E 01	8.000000E 07
-8.997411E-01	4.758142E-01J	1.017807E 00	1.533086E-01	-2.787347E 01	9.000000E 07
-8.706396E-01	5.369089E-01J	1.022880E 00	1.964903E-01	-3.166376E 01	1.000000E 08
-2.079775E-01	8.481474E-01J	8.732746E-01	-1.176983E 00	-7.622766E 01	2.000000E 08
1.562691E-01	4.858661E-01J	5.103781E-01	-5.842158E 00	7.217592E 01	3.000000E 08
1.579962E-01	2.594321E-01J	3.037561E-01	-1.034950E 01	5.866251E 01	4.000000E 08
1.213920E-01	1.609581E-01J	2.016023E-01	-1.391008E 01	5.298087E 01	5.000000E 08
7.276952E-02	8.570850E-02J	1.124337E-01	-1.898206E 01	4.967119E 01	7.000000E 08
4.911369E-02	5.782000E-02J	7.586372E-02	-2.239931E 01	4.965825E 01	9.000000E 08

NODE 6

NODE VOLTAGE		MAGNITUDE	MAGNITUDE(DB)	PHASE	FREQUENCY
1.003157E 01	-6.281017E-01J	1.005121E 01	2.004436E 01	-3.583014E 00	1.000000E 07
9.991546E 00	-1.260283E 00J	1.007071E 01	2.006119E 01	-7.189558E 00	2.000000E 07
9.922152E 00	-1.900345E 00J	1.010249E 01	2.008855E 01	-1.084310E 01	3.000000E 07
9.819136E 00	-2.551441E 00J	1.014521E 01	2.012521E 01	-1.456690E 01	4.000000E 07
9.676620E 00	-3.215734E 00J	1.019696E 01	2.016940E 01	-1.838402E 01	5.000000E 07
9.487190E 00	-3.893900E 00J	1.025520E 01	2.021889E 01	-2.231673E 01	6.000000E 07
9.241855E 00	-4.584518E 00J	1.031648E 01	2.027061E 01	-2.638612E 01	7.000000E 07
8.930596E 00	-5.283421E 00J	1.037642E 01	2.032094E 01	-3.061118E 01	8.000000E 07
8.542913E 00	-5.982987E 00J	1.042964E 01	2.036539E 01	-3.500784E 01	9.000000E 07
8.068972E 00	-6.671515E 00J	1.046983E 01	2.039879E 01	-3.958722E 01	1.000000E 08
-2.996833E-01	-8.777882E 00J	8.782994E 00	1.887285E 01	8.805104E 01	2.000000E 08
-3.284197E 00	-3.756736E 00J	4.989889E 00	1.396181E 01	4.884309E 01	3.000000E 08
-2.524405E 00	-1.343377E 00J	2.859594E 00	9.126083E 00	2.802199E 01	4.000000E 08
-1.748004E 00	-4.806440E-01J	1.812880E 00	5.167377E 00	1.537564E 01	5.000000E 08
-9.058651E-01	1.180327E-02J	9.059420E-01	-8.579920E-01	-7.465670E-01	7.000000E 08
-5.273703E-01	1.132000E-01J	5.393826E-01	-5.362062E 00	-1.211560E 01	9.000000E 08

IHC217I

Figure 9.21 Continued

even larger. In attempting to reconcile this difference, the reader will find, upon further investigation, that the results found by the a-c analysis program are more accurate than those found by the previous analysis (Section 7.7). This is mainly due to the fact that the a-c analysis program uses the exact nodal network in the analysis, while our previous analysis

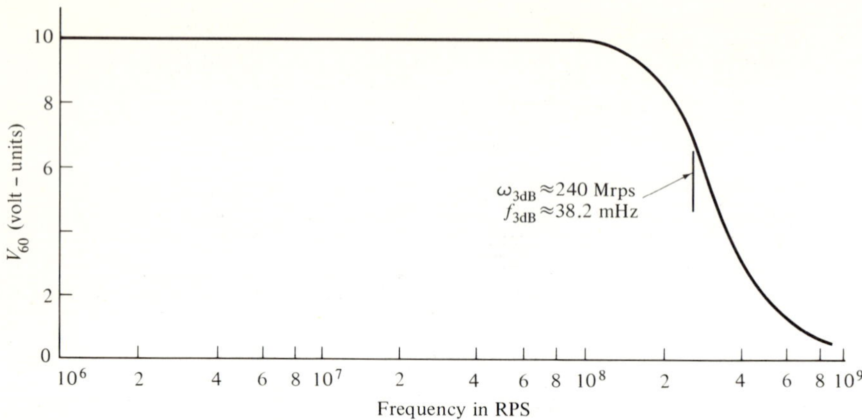

Figure 9.22 The output voltage of the network of Figure 9.21

made a number of simplifying assumptions so that the network could be treated in terms of the single-loop feedback expression

$$G(s) = \frac{A(s)}{1 - \beta A(s)}$$

Since our earlier approach was used to avoid the solution of a multi-node network, a detailed study of these two approaches to multi-node feedback networks will bring out the real power of this a-c circuit analysis program.

PROBLEMS

9-1. Indicate whether the following statements are true or false:

a. Adding a 3 by 2 matrix to a 2 by 3 matrix will produce the sum matrix, which will be a 3 by 3 matrix.

b. The inverse of

$$\begin{bmatrix} 1 & 0 & 1 \\ 0 & 3 & 2 \\ 1 & 2 & 1 \end{bmatrix} \text{ is } \begin{bmatrix} \frac{1}{4} & -\frac{1}{2} & \frac{3}{4} \\ -\frac{1}{2} & 0 & \frac{1}{2} \\ \frac{3}{4} & \frac{1}{2} & -\frac{3}{4} \end{bmatrix}$$

c. The adjoint of

$$\begin{bmatrix} 2 & 2 \\ -1 & 3 \end{bmatrix} \text{ is } \begin{bmatrix} 3 & -1 \\ 2 & 2 \end{bmatrix}$$

d. The branch impedance matrix of the two port is

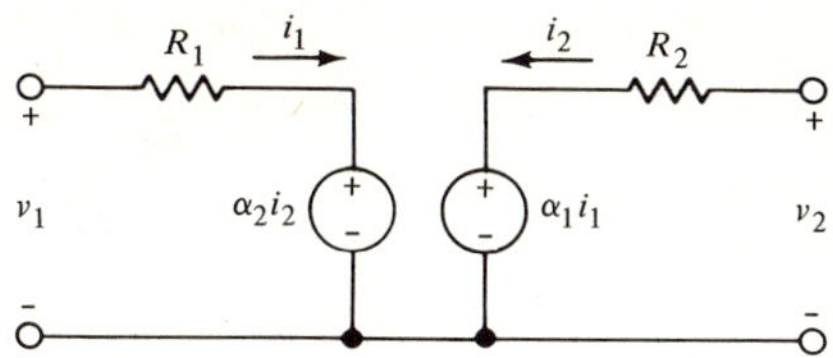

e. A unit matrix is a matrix whose determinant is equal to one.
f. Matrix A and matrix B are conformable with respect to the matrix multiplication operation $[A][B]$, provided that the number of rows of A equals the number of columns of B.
g. Each column of a general mesh impedance matrix must contain at least one nonzero element.
h. Matrix A will always possess an inverse if it is a square matrix whose determinant does not equal to zero.
i. The matrix $\begin{bmatrix} 1 & 0 & 1 \\ 0 & 1 & 0 \\ 1 & 0 & 1 \end{bmatrix}$ is equivalent to the unit matrix.
j. The adjoint of a symmetric matrix is symmetric.

9-2. Given

$$[A] = \begin{bmatrix} 2 & -2 & 4 & 1 \\ 2 & -1 & 3 & -1 \\ 1 & 1 & -2 & 1 \\ 1 & 1 & -1 & 2 \end{bmatrix} \quad \text{and} \quad [B] = \begin{bmatrix} 2 & 2 & 1 & 1 \\ 3 & 0 & 1 & 2 \\ 1 & 3 & 4 & 1 \\ 1 & 2 & 1 & 0 \end{bmatrix}$$

determine:

a. $[A] + [B]$
b. $[A] - [B]$
c. $[A][B]$
d. $[B][A]$
e. $[A]_T$
f. $[A]^{-1}$

by using the program of Figure 9.1. Verify each answer by hand computation of each of the matrix operations.

9-3. Use the matrix values of Problem 9-2 and the program of Figure 9.1 to show that $[[A][B]]_T = [B]_T[A]_T$.

9-4. Modify the program of Figure 9.3 so that it will determine the solution of

a. The network of Figure 6.8.
b. The network of Figure 6.11.

9-5. The phasor equation

$$(ZR + jZI)(IR + jII) = VR + jVI$$

may be written in matrix form as

$$\begin{bmatrix} ZR & -ZI \\ ZI & ZR \end{bmatrix} \begin{bmatrix} IR \\ II \end{bmatrix} = \begin{bmatrix} VR \\ VI \end{bmatrix}$$

The obvious advantage of using this matrix equation is that only real arithmetic operations are involved. Use a similar approach to formulate a system of N simultaneous phasor equations.

9-6. Using the technique described in Problem 9-5, devise a method for finding the inverse of a complex element matrix. Hint: If $Z = A + jB$ and $Z^{-1} = C + jD$, then

$$[Z][Z]^{-1} = [I]$$

or

$$\begin{bmatrix} A & -B \\ B & A \end{bmatrix} \begin{bmatrix} C \\ D \end{bmatrix} = \begin{bmatrix} 1 \\ 0 \end{bmatrix}$$

Write an appropriate program for finding the inverse of a complex element matrix.

9-7. Using the procedure of Problem 9-6, hand compute the inverse of

$$\begin{bmatrix} 1 + j1 & -j2 \\ -j2 & 2 + j2 \end{bmatrix}$$

Verify your solution by using the program of Problem 9-6.

9-8. The phasor equation

$$(ZR + jZI)(IR + jII) = VR + jVI$$

may also be written as

$$\begin{aligned} (ZI^{-1}ZR + ZR^{-1}ZI)IR &= ZI^{-1}VR + ZR^{-1}VI \\ (ZI^{-1}ZR + ZR^{-1}ZI)II &= ZI^{-1}VI - ZR^{-1}VR \end{aligned}$$

Use a similar approach to formulate the matrix equations of a system of N simultaneous phasor equations.

9-9. Using the procedure described in Problem 9-8, devise a method for finding the inverse of a complex element matrix. Hint: If $Z = A + jB$ and $Z^{-1} = C + jD$, then

$$[Z][Z]^{-1} = [I]$$

or

$$C = \frac{1}{A + BA^{-1}B} \quad \text{and} \quad D = \frac{-B}{A(A + BA^{-1}B)}$$

Write an appropriate program for finding the inverse of a complex matrix.

9-10. Using the procedure of Problem 9-9, hand compute the inverse of the matrix in Problem 9-7 and verify your solution by using the program of Problem 9-9.

9-11. A matrix may contain elements that are matrices themselves. It is therefore possible to subdivide or partition a given matrix. For example,

$$[A] = \left[\begin{array}{ccc|c} a_{11} & a_{12} & a_{13} & a_{14} \\ a_{21} & a_{22} & a_{23} & a_{24} \\ a_{31} & a_{32} & a_{33} & a_{34} \\ \hline a_{41} & a_{42} & a_{43} & a_{44} \end{array}\right] = \begin{bmatrix} A_{11} & A_{12} \\ A_{21} & A_{22} \end{bmatrix}$$

where

$$[A_{11}] = \begin{bmatrix} a_{11} & a_{12} & a_{13} \\ a_{21} & a_{22} & a_{23} \\ a_{31} & a_{32} & a_{33} \end{bmatrix} \quad [A_{12}] = \begin{bmatrix} a_{14} \\ a_{24} \\ a_{34} \end{bmatrix} \quad \begin{matrix} [A_{21}] = [a_{41} \quad a_{42} \quad a_{43}] \\ \\ [A_{22}] = [a_{44}] \end{matrix}$$

It is also possible to partition the two matrices involved in a matrix multiplication operation. The matrices involved must be conformably partitioned, however, with the partitioned columns of the first matrix equal to the partitioned rows in the second matrix. For example,

$$\left[\begin{array}{ccc|c} 1 & 1 & -1 & 0 \\ 1 & 1 & -1 & 2 \\ 2 & 1 & 1 & 2 \\ \hline 3 & -1 & 0 & 1 \end{array}\right] \left[\begin{array}{cccc} 2 & 1 & 1 & 1 \\ 4 & 1 & -1 & 1 \\ 2 & 1 & 1 & 0 \\ \hline 1 & -1 & -1 & 1 \end{array}\right] = \begin{bmatrix} A_{11} & A_{12} \\ A_{21} & A_{22} \end{bmatrix} \begin{bmatrix} B_{11} \\ B_{21} \end{bmatrix}$$

$$= \begin{bmatrix} A_{11}B_{11} + A_{12}B_{21} \\ A_{21}B_{11} + A_{22}B_{21} \end{bmatrix}$$

Using matrix partitioning, hand compute the product of the two matrices shown.

9-12. Write a program for carrying out the matrix multiplication operation described in Problem 9-11. The input data should specify how the two matrices are to be partitioned. Test your program by using the numerical values given in Problem 9-11.

9-13. A two-port network may be described in terms of the terminal values of I_1, V_1, I_2, and V_2.

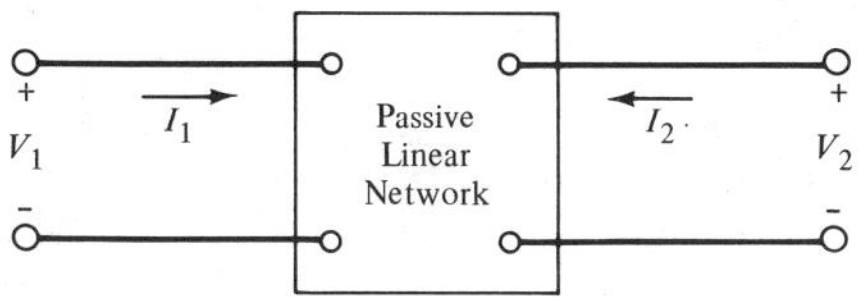

In matrix notation we thus have

a. The y-parameters

$$\begin{bmatrix} I_1 \\ I_2 \end{bmatrix} = \begin{bmatrix} y_{11} & y_{12} \\ y_{21} & y_{22} \end{bmatrix} \begin{bmatrix} V_1 \\ V_2 \end{bmatrix}$$

b. The z-parameters

$$\begin{bmatrix} V_1 \\ V_2 \end{bmatrix} = \begin{bmatrix} z_{11} & z_{12} \\ z_{21} & z_{22} \end{bmatrix} \begin{bmatrix} I_1 \\ I_2 \end{bmatrix}$$

c. The g-parameters

$$\begin{bmatrix} I_1 \\ V_2 \end{bmatrix} = \begin{bmatrix} g_{11} & g_{12} \\ g_{21} & g_{22} \end{bmatrix} \begin{bmatrix} V_1 \\ I_2 \end{bmatrix}$$

d. The h-parameters

$$\begin{bmatrix} V_1 \\ I_2 \end{bmatrix} = \begin{bmatrix} h_{11} & h_{12} \\ h_{21} & h_{22} \end{bmatrix} \begin{bmatrix} I_1 \\ V_2 \end{bmatrix}$$

e. The ABCD parameters

$$\begin{bmatrix} V_1 \\ I_1 \end{bmatrix} = \begin{bmatrix} A & B \\ C & D \end{bmatrix} \begin{bmatrix} V_2 \\ -I_2 \end{bmatrix}$$

The interconnection of two four-terminal networks therefore leads to five separate cases:

1. The cascade connection.

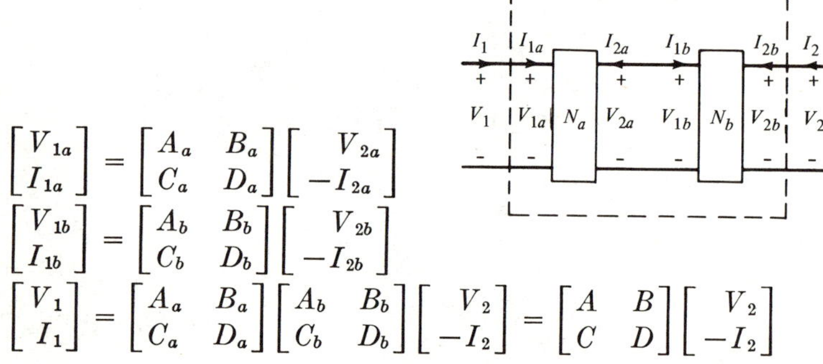

$$\begin{bmatrix} V_{1a} \\ I_{1a} \end{bmatrix} = \begin{bmatrix} A_a & B_a \\ C_a & D_a \end{bmatrix} \begin{bmatrix} V_{2a} \\ -I_{2a} \end{bmatrix}$$

$$\begin{bmatrix} V_{1b} \\ I_{1b} \end{bmatrix} = \begin{bmatrix} A_b & B_b \\ C_b & D_b \end{bmatrix} \begin{bmatrix} V_{2b} \\ -I_{2b} \end{bmatrix}$$

$$\begin{bmatrix} V_1 \\ I_1 \end{bmatrix} = \begin{bmatrix} A_a & B_a \\ C_a & D_a \end{bmatrix} \begin{bmatrix} A_b & B_b \\ C_b & D_b \end{bmatrix} \begin{bmatrix} V_2 \\ -I_2 \end{bmatrix} = \begin{bmatrix} A & B \\ C & D \end{bmatrix} \begin{bmatrix} V_2 \\ -I_2 \end{bmatrix}$$

2. The parallel or shunt-shunt connection.

$$V_1 = V_{1a} = V_{1b} \qquad I_1 = I_{1a} + I_{1b}$$

$$V_2 = V_{2a} = V_{2b} \qquad I_2 = I_{2a} + I_{2b}$$

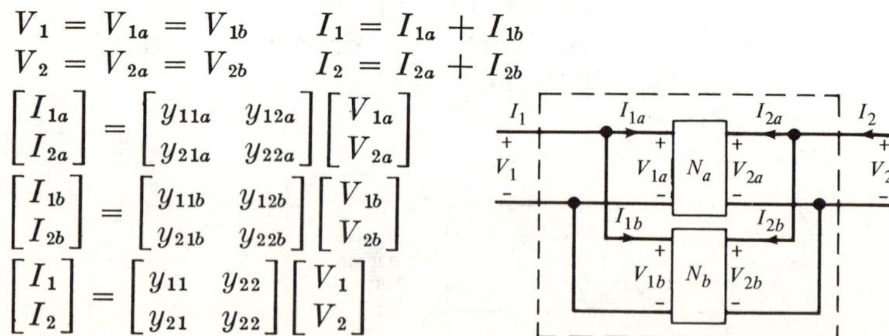

$$\begin{bmatrix} I_{1a} \\ I_{2a} \end{bmatrix} = \begin{bmatrix} y_{11a} & y_{12a} \\ y_{21a} & y_{22a} \end{bmatrix} \begin{bmatrix} V_{1a} \\ V_{2a} \end{bmatrix}$$

$$\begin{bmatrix} I_{1b} \\ I_{2b} \end{bmatrix} = \begin{bmatrix} y_{11b} & y_{12b} \\ y_{21b} & y_{22b} \end{bmatrix} \begin{bmatrix} V_{1b} \\ V_{2b} \end{bmatrix}$$

$$\begin{bmatrix} I_1 \\ I_2 \end{bmatrix} = \begin{bmatrix} y_{11} & y_{22} \\ y_{21} & y_{22} \end{bmatrix} \begin{bmatrix} V_1 \\ V_2 \end{bmatrix}$$

3. The series connection.

$$V_1 = V_{1a} + V_{1b} \qquad I_1 = I_{1a} = I_{1b}$$
$$V_2 = V_{2a} + V_{2b} \qquad I_2 = I_{2a} = I_{2b}$$
$$\begin{bmatrix} V_{1a} \\ V_{2a} \end{bmatrix} = \begin{bmatrix} z_{11a} & z_{12a} \\ z_{21a} & z_{22a} \end{bmatrix} \begin{bmatrix} I_{1a} \\ I_{2a} \end{bmatrix}$$
$$\begin{bmatrix} V_{1b} \\ V_{2b} \end{bmatrix} = \begin{bmatrix} z_{11b} & z_{12b} \\ z_{21b} & z_{22b} \end{bmatrix} \begin{bmatrix} I_{1b} \\ I_{2b} \end{bmatrix}$$
$$\begin{bmatrix} V_1 \\ V_2 \end{bmatrix} = \begin{bmatrix} z_{11} & z_{12} \\ z_{21} & z_{22} \end{bmatrix} \begin{bmatrix} I_1 \\ I_2 \end{bmatrix}$$

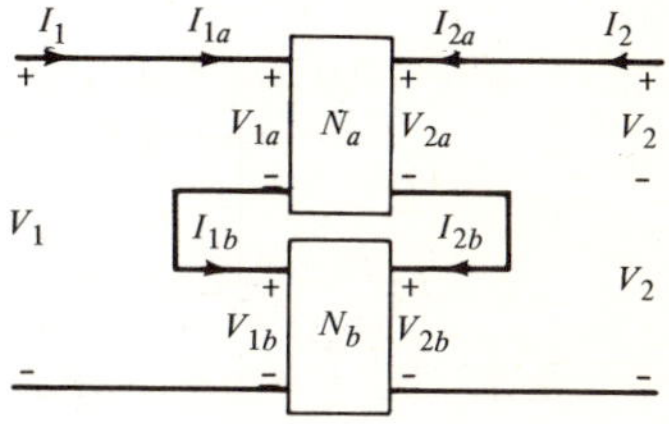

4. The series-parallel connection.

$$I_1 = I_{1a} = I_{1b} \qquad V_1 = V_{1a} + V_{1b}$$
$$V_2 = V_{2a} = V_{2b} \qquad I_2 = I_{2a} + I_{2b}$$
$$\begin{bmatrix} V_{1a} \\ I_{2a} \end{bmatrix} = \begin{bmatrix} h_{11a} & h_{12a} \\ h_{21a} & h_{22a} \end{bmatrix} \begin{bmatrix} I_{1a} \\ V_{2a} \end{bmatrix}$$
$$\begin{bmatrix} V_{1b} \\ I_{2b} \end{bmatrix} = \begin{bmatrix} h_{11b} & h_{12b} \\ h_{21b} & h_{22b} \end{bmatrix} \begin{bmatrix} I_{1b} \\ V_{2b} \end{bmatrix}$$
$$\begin{bmatrix} V_1 \\ I_2 \end{bmatrix} = \begin{bmatrix} h_{11} & h_{12} \\ h_{21} & h_{22} \end{bmatrix} \begin{bmatrix} I_1 \\ V_2 \end{bmatrix}$$

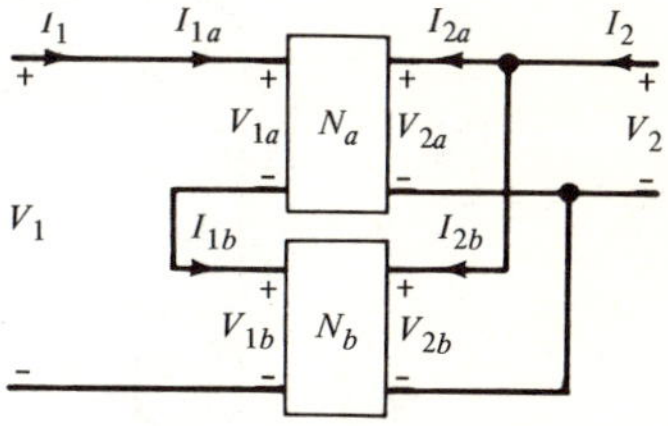

5. The parallel-series connection.

$$V_1 = V_{1a} = V_{1b} \qquad I_1 = I_{1a} + I_{1b}$$
$$I_2 = I_{2a} = I_{2b} \qquad V_2 = V_{2a} + V_{2b}$$
$$\begin{bmatrix} I_{1a} \\ V_{2a} \end{bmatrix} = \begin{bmatrix} g_{11a} & g_{12a} \\ g_{21a} & g_{22a} \end{bmatrix} \begin{bmatrix} V_{1a} \\ I_{2a} \end{bmatrix}$$
$$\begin{bmatrix} I_{1b} \\ V_{2b} \end{bmatrix} = \begin{bmatrix} g_{11b} & g_{12b} \\ g_{21b} & g_{22b} \end{bmatrix} \begin{bmatrix} V_{1b} \\ I_{2b} \end{bmatrix}$$
$$\begin{bmatrix} I_1 \\ V_2 \end{bmatrix} = \begin{bmatrix} g_{11} & g_{12} \\ g_{21} & g_{22} \end{bmatrix} \begin{bmatrix} V_1 \\ I_2 \end{bmatrix}$$

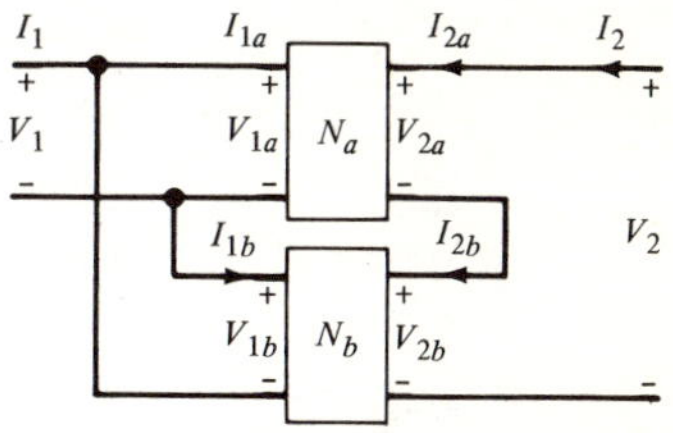

Write a general program for computing the equivalent network parameters. Use the code $I = 1, 2, 3, 4$, or 5 to describe the cascade, parallel, series, series-parallel, or parallel-series connection respectively. Test your program by using a suitable number of examples.

9-14. Use the program of Figure 9.6 to formulate the sinusoidal steady-state mesh equations of the following networks. Note that the polarity dots or the sense of inductive coupling between each pair of coils is indicated by a separate pair of symbols.

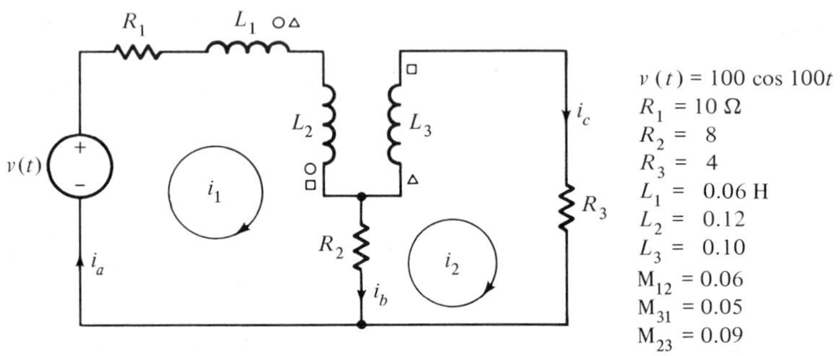

a.

$v(t) = 100 \cos 10t$, $L_1 = 0.05$, $L_2 = 0.10$, $L_3 = 0.06$, $L_4 = 1.0$, $L_5 = 0.05$

$M_{12} = M_{13} = M_{14} = M_{15} = M_{23} = M_{24} = M_{25} = M_{34} = M_{35} = M_{45} = 0.02$ H

b.

9-15. Combine the program of Figure 9.6 with the program of Figure 6.23 so that the resulting program will be capable of formulating, as well as solving, an arbitrary network for various specified frequencies.

9-16. Using the program of Problem 9-15, determine the phasor solution of the networks of Problem 9-14.

9-17. Write a program similar to Figure 9.6 for formulating nodal sinusoidal steady-state equations.

9-18. Using matrix techniques, formulate the nodal equations of the following network:

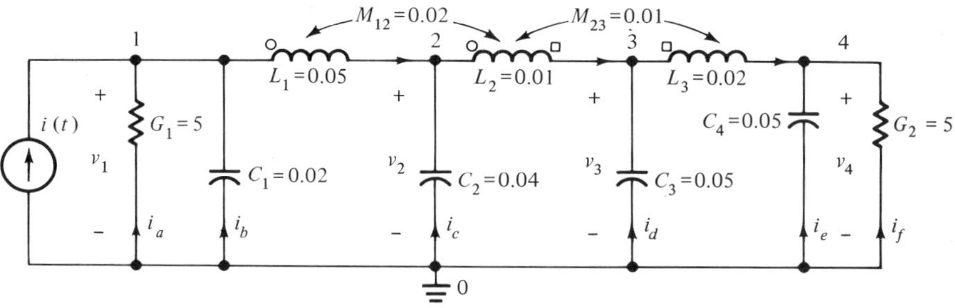

Verify your formulation by using the program of Problem 9-17.

9-19. Write a program similar to Figure 9.6 for formulating cut-set sinusoidal steady-state equations. Test your program by choosing cut-set voltages for the network of Problem 9-18.

9-20. Combine the program of Problem 9-17 with the program of Figure 6.23 so that the resulting program will be capable of formulating as well as solving nodal phasor equations.

9-21. Repeat Problem 9-20 using the program of Problem 9-19.

9-22. Can the program of Figure 9.13 be used to find the d-c behavior of a network? Note that such a program would be useful in the d-c coupling of transistor stages.

9-23. Use the ac circuit analysis program of Figure 9.13 to determine the value of $\boldsymbol{I}_5$ in the network of Figure 6.22.

9-24. Use the a-c circuit analysis program of Figure 9.13 to determine the ratio of $V_2(j\omega)/V_1(j\omega)$ over the frequency range $\omega = 1000$ to 100,000 rps for

a. $R = 0$ ohms. b. $R = 20$ ohms. c. $R = 40$ ohms.

Note that this four-pole Butterworth filter was designed for a three-dB frequency of $\omega_{3dB} = 10{,}000$ rps with $R = 0$ (lossless elements).

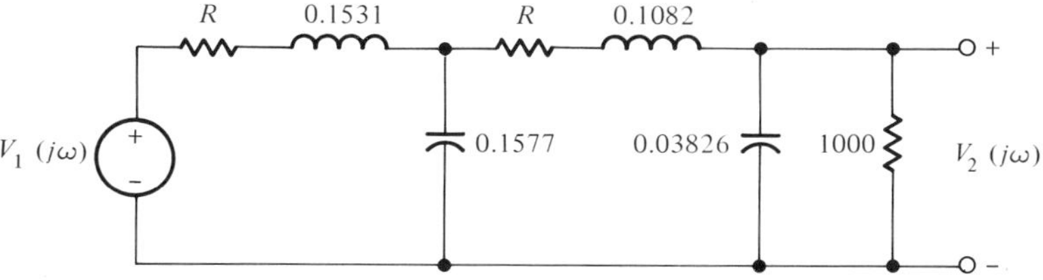

9-25. Use the a-c circuit analysis program of Figure 9.13 to determine the dissipationless insertion loss of the network of Figure 3.39. Consider the frequency range of f = 270 to 296 kHz.

9-26. Repeat Problem 9-25 using a minimum Q = 300, 200, 150, 100, and 50 for each of the four coils.

9-27. Use the a-c circuit analysis program of Figure 9.13 to determine the frequency response of the twin-tee network shown below. $\left(f_o = \dfrac{1}{4\pi RC}\right)$

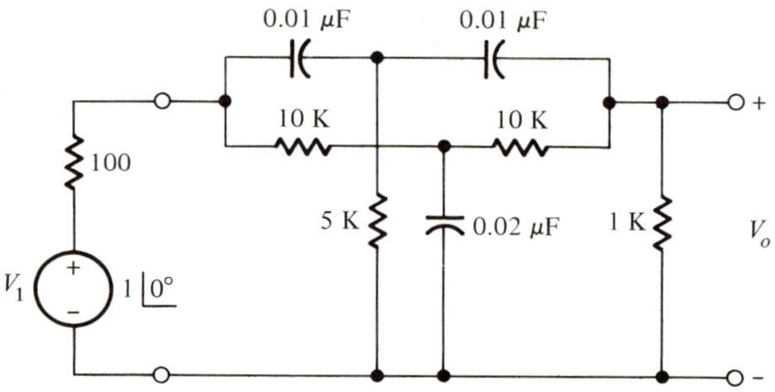

9-28. What effect do the generator and load resistance values have on the frequency response of the network of Problem 9-27?

9-29. Use the a-c circuit analysis program of Figure 9.13 to determine the frequency response of the vacuum-tube amplifier shown below

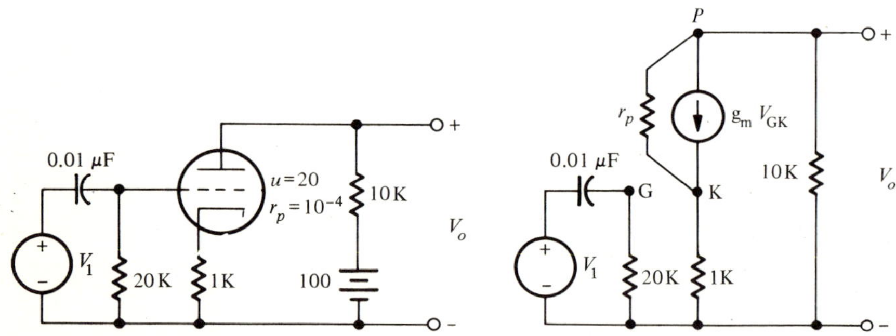

9-30. Rerun Problem 9-29 using a 0.1 μF coupling capacitor.

9-31. Rerun Problem 9-30 with a 0.1 μF capacitor bypass across the 1,000-ohm cathode resistor.

9-32. Determine the lower three-dB frequency of the transistor amplifier shown.

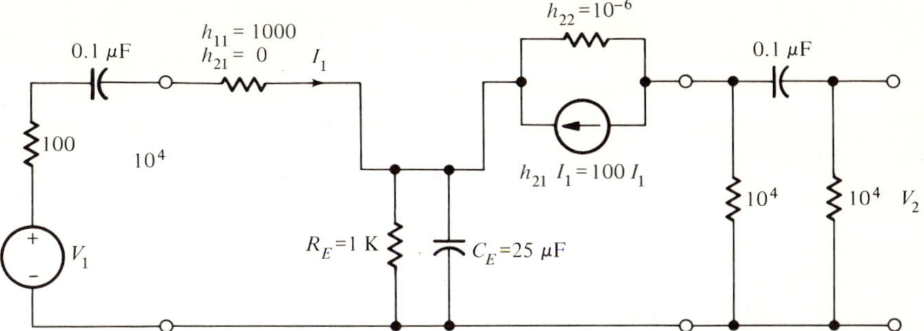

9-33. Use the a-c circuit analysis program of Figure 9.13 to determine the frequency response of the twin-tee active filter shown.

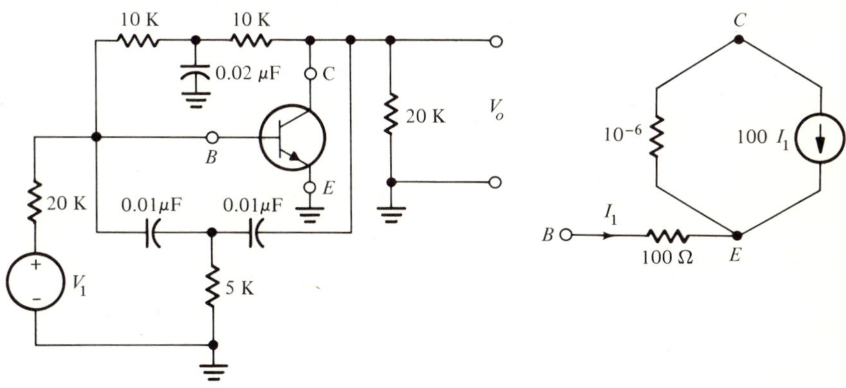

9-34. Use the a-c circuit analysis program of Figure 9.13 to determine the upper and lower three-dB frequency of the FET stage shown.

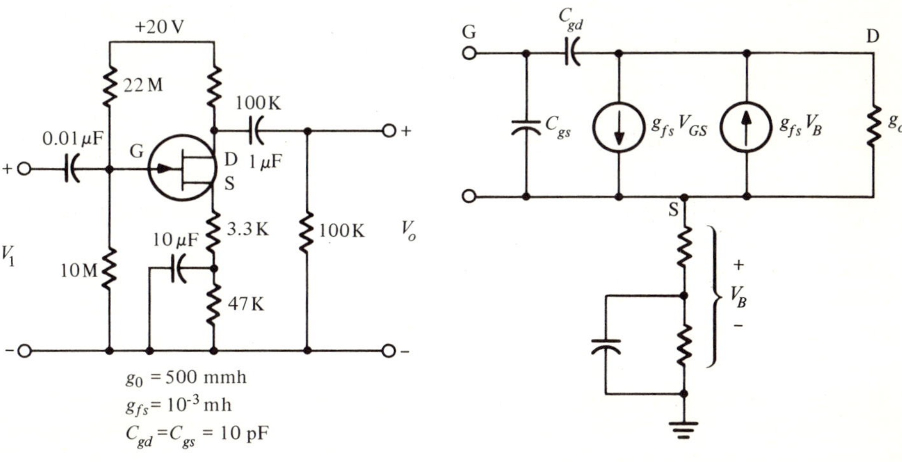

CHAPTER 10

Random Number Generators and Monte Carlo Analysis

10.1 INTRODUCTION

In Chapter 7 we considered the worst-case design of a flip-flop. The design was based on defining the tolerance, or range, of each circuit element and using the maximum or minimum value at its worst extreme. In a production fabrication of a design based on worst-case extremes, it is very unlikely that all elements would have their worst values at the same time. A Monte Carlo analysis is based on the use of random numbers and simulates a random selection of circuit element values based on specified probabilistic distributions. The analysis is repeated many times in order to provide statistical assurance of proper circuit operation.

It is interesting to realize that shot noise in a diode is a random phenomenon and noise impulses can be counted and translated into a table of random numbers. The use of a random number table is obviously undesirable and in most cases we would prefer to have the computer generate the required random numbers. Since some numerical algorithm must be used, the number generated by a computer will not be truly random. Numbers generated by the computer are called pseudo-random numbers and for all practical purposes they satisfy the tests of randomness. Since the sequence of computer generated random numbers is predictable, this is more of an advantage than a disadvantage in debugging a problem or in comparing the solutions to a problem that has been solved by using several different random number generators.

In general, a computer random number generator produces numbers between 0 and 1. When each decimal digit of the random number has the same probability of occurrence, the numbers generated are called uniformly distributed and would have a distribution as shown in Figure 10.1a. The programmer may change the range of the numbers generated by scaling. If MAX represents the maximum value required and MIN represents the minimum value, the random number generated should be multiplied by (MAX − MIN) and MIN should be added to this result to shift the origin from zero to MIN; thus

$$RN_{scaled} = (MAX - MIN) * RN + MIN$$

will produce numbers in the required range. For example, if it is required to generate random numbers between −10 and +20, the following scaling is required

$$RN_{scaled} = (20 - (-10)) * RN + (-10)$$

Thus, when RN = 0, the scaled number $RN_{scaled} = -10$ and when RN = +1, the scaled number $RN_{scaled} = +20$.

Two other types of distributions are shown in Figure 10.1b and c. Later in this chapter we will discuss how other types of distributions can be obtained from a uniform distribution.

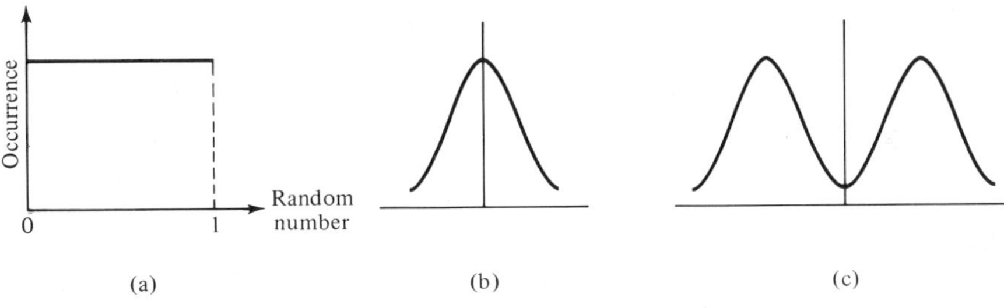

Figure 10.1 Some types of random number distributions: (a) uniform distribution; (b) normal distribution; (c) bimoded distribution

10.2 SOME FORTRAN ALGORITHMS FOR GENERATING PSEUDO-RANDOM NUMBERS

The Inner Product Method

One of the earliest methods for generating random numbers is based on the center-squaring technique. This method starts by first selecting a *suitable* large number, squaring it, and saving the middle part. The middle or inner product is then squared, and so on.

As a numerical example, consider the four digit number 1111. Since a four digit number multiplied by itself will in general produce an eight digit number, the two most significant and the two least significant digits must be discarded, and the four middle digits saved will be the next random number as shown below:

```
     1111          2343     4896
     1111          2343     4896
     ----          ----     ----
     1111          7029     etc.
    1111          9372
   1111          7029
  1111          4686
 --------      --------
 01234321      05489649
   \__/          \__/
 the middle
part is saved
```

The first Fortran random number generator we will discuss will generate a four digit random number based on the integer-mode properties of the IBM 1620 computer. In the IBM 1620 machine, the *four least significant digits* are preserved in integer-mode arithmetic. For computational purposes we may regard the four digit number as being made from two parts, I and J, as follows:

$$\text{NR} = \underbrace{11}_{I}\ \underbrace{11}_{J}$$

where NR stands for number random. We may now write NR as

$$\text{NR} = I * 100 + J$$

so that

$$(\text{NR})^2 = I^2 * 10^4 + 2 * I * J * 100 + J^2$$

Because of the integer mode property of the IBM 1620 machine, all we need do to retain the middle part of the above product is to divide the product by 100. This automatically will drop the least two significant digits since they will be the fractional part and are discarded in integer mode arithmetic; and the two most significant digits will also be lost since the IBM 1620 computer preserves only the four least significant digits in integer-mode arithmetic. Thus,

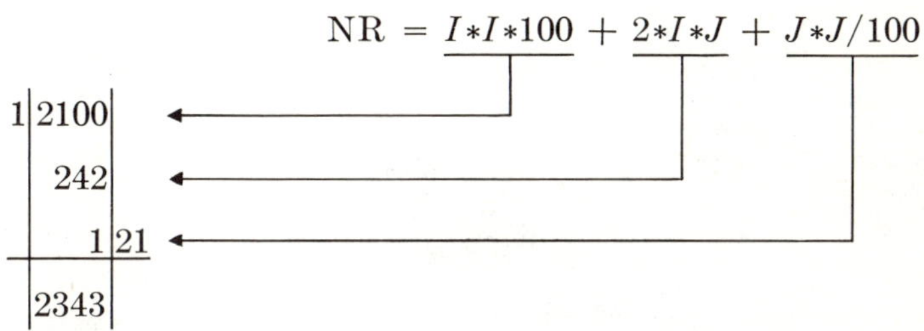

In order to fully understand the importance of the special integer-mode property of the IBM 1620 machine, the reader is urged to repeat the above process and hand calculate the next three terms.

An IBM 1620 Fortran II program is shown in Figure 10.2. Note that the initial value of I and J is first read in from a data card. The DO 10 loop then generates 20 random numbers based on the algorithm described above. Observe that during each iteration the random number NR is punched out in I4 format code. In order to repeat the algorithm it is necessary to obtain new values for I and J. This is accomplished by dividing the random number NR by 100. This automatically determines the new value of I because the fractional part is dropped in integer mode arithmetic. The variable I100 is now formed by multiplying I by 100. I100 is now subtracted from NR to obtain the value of J. This process is repeated for 20 iterations, and at that point control is passed to the statement following the scope label 10. The GO TO 5 statement reads in the next set of I and J values and the process is continued until the data cards are exhausted. Figure 10.3 shows the starting value and the corresponding 20 random numbers that were generated for four different data sets. An examination of the individual results show the importance of the starting value. The fourth set shows that not only are the output numbers nonrandom, but the process very quickly degenerates into a

Figure 10.2 Fortran II random number center-squaring program

```
C     RANDOM NUMBER GENERATOR
C     CENTER SQUARING METHOD
C
    5 READ 1, I, J
    1 FORMAT(2I2)
      DO 10 K=1,20
      NR=I*I*100+2*I*J+J*J/100
      PUNCH 2, NR
    2 FORMAT(I4)
      I=NR/100
      I100=I*100
   10 J=NR-I100
      GO TO 5
      END
```

Output Set	I	II	III	IV
Starting Value	1111	1963	1964	9999
	2343	8533	8572	9800
	4896	8120	4791	400
	9708	9344	9536	1600
	2452	3103	9352	5600
	123	6286	4599	3600
	151	5137	1508	9600
	228	3887	2740	1600
	519	1087	5076	5600
	2693	1815	7657	3600
	2522	2942	6296	9600
	3604	6553	6396	1600
	9888	9418	9088	5600
	7725	6987	5917	3600
	6756	8181	108	9600
	6435	9287	116	1600
	4092	2483	134	5600
	7444	1652	179	3600
	4131	7291	320	9600
	651	1587	1024	1600
	4238	5153	485	5600

Figure 10.3 Four sets of random numbers

four-number cycle. This particular effect is unpredictable, and it is impossible to tell when it might occur. For example, consider the first output set that appears to be fairly random. Changing the DO 10 statement in the previous program to

DO 10 K=1,50

and using the same starting value of 1111 produces the 50 random numbers shown in Figure 10.4. An examination of the occurrence of the most significant digit will give some idea of the distribution. The following table lists the ten decimal digits and the number of times they occurred in the most significant place:

DECIMAL DIGIT	0	1	2	3	4	5	6	7	8	9
DIGIT OCCURRENCE	7	4	7	3	9	3	5	5	3	4

Since there are ten decimal digits and since 50 numbers were generated, the random occurrence of each digit should be five. It is obvious that these numbers are not very uniform. It may also be observed that the 50th number generated was equal to 9600. This corresponds to one of the numbers in the four-number cycle of the previous fourth data set. Thus, with a starting value of 1111, 50 numbers were generated before a four-number cycle occurred while with a starting value of 9999, only two numbers were obtained before the four-number cycle occurred.

2343	3604	9606	4322	4195
4896	9888	2752	6796	5980
9708	7725	5735	1856	7604
2452	6756	8902	4447	8208
123	6435	2456	7758	3712
151	4092	319	1865	7789
228	7444	1017	4782	6685
519	4131	342	8675	6892
2693	651	1169	2556	4996
2522	4238	3665	5331	9600

Figure 10.4 Fifty random numbers from a starting value of 1111

Two other disadvantages of the center-squaring method are that the sequence can very easily degenerate to a zero number at any point or a long sequence of very small numbers will be produced. All three of these disadvantages can be delayed by using a larger starting number, but when and if they will occur is impossible to predict.

Even though the above disadvantages rule out the inner product method as a useful random number generator, it is instructive to modify the algorithm developed above (which was based on the integer mode properties of the IBM 1620 computer) so that it can be used on other machines. Figure 10.5 shows a similar program that was written for the IBM 360 computer. Note that the basic difference is that the IBM 360 computer preserves more than four digits in integer mode arithmetic operations. It is therefore necessary to modify the following statement so that only four figures will be preserved in any arithmetic operation:

$$\mathrm{NR} = \mathrm{I}*\mathrm{I}*100 + 2*\mathrm{I}*\mathrm{J} + \mathrm{J}*\mathrm{J}/100$$

Since I is a two-digit number, $I*I*100$ could be a six-digit number. To prevent this, a new variable INR1 is introduced (see statement S.0006). This variable is a two-digit number (since the fractional part is dropped in integer-mode arithmetic) that contains the two most significant digits of $I*I*100$. When INR1 is multiplied by 10000 and subtracted from $I*I*100$, only the four least significant figures will be preserved. Similarly the term $2*I*J$ could have as many as five digits. The variable INR2 is introduced (see statement S.0007) and it will contain the most significant digit of $2*I*J$. When INR2 is multiplied by 10000 and subtracted from $2*I*J$, only the four least significant figures will be preserved.

The third term $J*J/100$ requires no modification because it can not be more than two digits. The above expression is thus changed to

$$\mathrm{NR1} = \mathrm{I}*\mathrm{I}*100 - \mathrm{INR1}*10000 + 2*\mathrm{I}*\mathrm{J} - \mathrm{INR2}*10000 + \mathrm{J}*\mathrm{J}/100$$

```
             C       RANDOM NUMBER GENERATOR
             C       CENTER SQUARING METHOD
             C
S.0001              5 READ(5,1)I,J
S.0002              1 FORMAT(2I2)
S.0003                WRITE(6,4)I,J
S.0004              4 FORMAT(2I10)
S.0005                DO 10 K=1,20
S.0006                INR1=I*I*100/10000
S.0007                INR2=2*I*J/10000
S.0008                NR1=I*I*100-INR1*10000+2*I*J-INR2*10000+J*J/100
S.0009                NR=NR1-NR1/10000*10000
S.0010                WRITE(6,2)NR
S.0011              2 FORMAT(I10)
S.0012                I=NR/100
S.0013                I100=I*100
S.0014             10 J=NR-I100
S.0015                GO TO 5
S.0016                RETURN
S.0017                END
```

Figure 10.5 IBM 360 random number center-squaring program

11	11	19	68
2343		8730	
4896		2129	
9708		5326	
2452		3662	
123		4102	
151		8264	
228		2936	
519		6200	
2693		4400	
2522		3600	
3604		9600	
9888		1600	
7725		5600	
6756		3600	
6435		9600	
4092		1600	
7444		5600	
4131		3600	
651		9600	
4238		1600	
19	63	99	99
8533		9800	
8120		400	
9344		1600	
3103		5600	
6286		3600	
5137		9600	
3887		1600	
1087		5600	
1815		3600	
2942		9600	
6553		1600	
9418		5600	
6987		3600	
8181		9600	
9287		1600	
2483		5600	
1652		3600	
7291		9600	
1586		1600	
5153		5600	

HC2171

Figure 10.6 Four sets of random numbers obtained from the IBM 360 center-squaring program of Figure 10.5

which is computer statement S.0008. It is now necessary to recognize that the sum of these modified three terms may overflow into the fifth place. To prevent this, NR1 is first divided by 10000 so that it will contain the overflow digit, and then this result is multiplied by 10000 and, in turn, subtracted from NR1. This modified IBM 360 program will now yield the same results as the IBM 1620 program. Identical results for the previous starting values are shown in Figure 10.6. The reader is encouraged to use the modified algorithm and hand calculate three numbers based on a starting value of 9080. The reader is also urged to modify the above program so that six-digit random numbers can be produced.

Lehmer's Method

Lehmer's method of generating random numbers by the Eniac computer[1] starts with any eight-digit number that, in turn, is multiplied by any two-digit number to produce a ten-digit number. The left two digits

[1] See IBM Data Processing Techniques, *Random Number Generation and Testing*, C20-8011, p. 5.

```
C
C     LEHMER SYSTEM
C     EIGHT DIGIT
C     RANDOM NUMBER GENERATOR
C
    1 READ(5,2)ISEED,MULT
    2 FORMAT(I8,I2)
      WRITE(6,3)ISEED,MULT
    3 FORMAT('1LEHMER SYSTEM'/'0EIGHT DIGIT'/'0RANDOM NUMBER GENERATOR'/
     1///'0ISEED =',I9,5X,'MULT =',I3,////,9X,'N',9X,'NL2D',6X,'NR8D',10
     2X,'NR',//)
      DO 10 K=1,20
      N=ISEED*MULT
      NL2D=N/100000000
      NR8D=N-NL2D*100000000
      NR=NR8D-NL2D
      ISEED=NR
   10 WRITE(6,4)N,NL2D,NR8D,NR
    4 FORMAT(I15,I7,2I13)
      GO TO 1
      END
```

Figure 10.7 Program for generating eight-digit Lehmer random numbers

(the two most significant digits) are removed and subtracted from the remaining eight-digit number. This result is the first random number and it, in turn, is multiplied by the same two-digit multiplier to repeat the process and generate additional random numbers. A Fortran program based on this method is shown in Figure 10.7. Note that the eight-digit starting number ISEED and the two-digit multiplier MULT are first read in. The DO 10 loop then generates 20 random numbers by first forming the ten-digit number N which is the product of ISEED and MULT. The left two digits, NL2D, are then removed. These two digits are then multiplied by 100000000 and the result subtracted from N to yield NR8D, the remaining eight digits. Finally the random number is formed by subtracting NL2D from NR8D and this, in turn, becomes the new value of ISEED so that the algorithm can be executed again. The corresponding output for four sets of ISEED and MULT are shown in Figure 10.8.

Note that the first data set uses a two-digit multiplier of 93, so that in the second iteration 82800703 is multiplied by 93. This produces an overflow (a number larger than 2147483647) and a corresponding negative result. In order to avoid a negative product (for IBM 360 machines), the multiplier must be less than or equal to 21, as in the last three sets. A comparison of the last two results reveals the importance of the value of ISEED. This method will be considered further when we deal with the testing of the distribution of random number generators.

Fibonacci Series Method

Another method of generating random numbers is based on Fibonacci numbers.[2] This method is very fast since it basically requires only the

[2] *Ibid.*, p. 9.

```
LEHMER SYSTEM

EIGHT DIGIT

RANDOM NUMBER GENERATOR


ISEED = 12345678        MULT = 93


          N           NL2D        NR8D           NR

     1148148054        11       48148054       48148043
      182800703         1       82800703       82800702
     -889469306        -8      -89469306      -89469298
      269289878         2       69289878       69289876
    -2145976124       -21      -45976124      -45976103
       19189717         0       19189717       19189717
     1784643681        17       84643681       84643664
     -718073840        -7      -18073840      -18073833
    -1680866469       -16      -80866469      -80866453
     1069354463        10       69354463       69354453
    -2139970463       -21      -39970463      -39970442
      577716190         5       77716190       77716185
    -1362329387       -13      -62329387      -62329374
    -1501664486       -15       -1664486       -1664471
     -154795803        -1      -54795803      -54795802
     -801042290        -8       -1042290       -1042282
      -96932226         0      -96932226      -96932226
     -424762426        -4      -24762426      -24762422
     1992062050        19       92062050       92062031
      -28165709         0      -28165709      -28165709


LEHMER SYSTEM

EIGHT DIGIT

RANDOM NUMBER GENERATOR


ISEED = 12345678        MULT = 20


          N           NL2D        NR8D           NR

      246913560         2       46913560       46913558
      938271160         9       38271160       38271151
      765423020         7       65423020       65423013
     1308460260        13        8460260        8460247
      169204940         1       69204940       69204939
     1384098780        13       84098780       84098767
     1681975340        16       81975340       81975324
     1639506480        16       39506480       39506464
      790129280         7       90129280       90129273
     1802585460        18        2585460        2585442
       51708840         0       51708840       51708840
     1034176800        10       34176800       34176790
      683535800         6       83535800       83535794
     1670715880        16       70715880       70715864
     1414317280        14       14317280       14317266
      286345320         2       86345320       86345318
     1726906360        17       26906360       26906343
      538126860         5       38126860       38126855
      762537100         7       62537100       62537093
     1250741860        12       50741860       50741848
```

Figure 10.8 Output of the random number program of Figure 10.7

LEHMER SYSTEM

EIGHT DIGIT

RANDOM NUMBER GENERATOR

ISEED = 12345678 MULT = 21

N	NL2D	NR8D	NR
259259238	2	59259238	59259236
1244443956	12	44443956	44443944
933322824	9	33322824	33322815
699779115	6	99779115	99779109
2095361289	20	95361289	95361269
2002586649	20	2586649	2586629
54319209	0	54319209	54319209
1140703389	11	40703389	40703378
854770938	8	54770938	54770930
1150189530	11	50189530	50189519
1053979899	10	53979899	53979889
1133577669	11	33577669	33577658
705130818	7	5130818	5130811
107747031	1	7747031	7747030
162687630	1	62687630	62687629
1316440209	13	16440209	16440196
345244116	3	45244116	45244113
950126373	9	50126373	50126364
1052653644	10	52653644	52653634
1105726314	11	5726314	5726303

LEHMER SYSTEM

EIGHT DIGIT

RANDOM NUMBER GENERATOR

ISEED = 11111111 MULT = 21

N	NL2D	NR8D	NR
233333331	2	33333331	33333329
699999909	6	99999909	99999903
2099997963	20	99997963	99997943
2099956803	20	99956803	99956783
2099092443	20	99092443	99092423
2080940883	20	80940883	80940863
1699758123	16	99758123	99758107
2094920247	20	94920247	94920227
1993324767	19	93324767	93324748
1959819708	19	59819708	59819689
1256213469	12	56213469	56213457
1180482597	11	80482597	80482586
1690134306	16	90134306	90134290
1892820090	18	92820090	92820072
1949221512	19	49221512	49221493
1033651353	10	33651353	33651343
706678203	7	6678203	6678196
140242116	1	40242116	40242115
845084415	8	45084415	45084407
946772547	9	46772547	46772538

Figure 10.8 Continued

```
C
C     FIBONACCI NUMBERS
C
      DIMENSION NF(45)
      NF(1)=1
      NF(2)=1
      DO 10 I=3,45
   10 NF(I)=NF(I-1)+NF(I-2)
      WRITE(6,1) (NF(I),I=1,45)
    1 FORMAT('1A SEQUENCE OF FORTY FIVE FIBONACCI NUMBERS',///,(9I11))
      RETURN
      END
```

Figure 10.9 A program for generating Fibonacci numbers

```
A SEQUENCE OF FORTY FIVE FIBONACCI NUMBERS

          1          1          2          3          5          8         13         21         34
         55         89        144        233        377        610        987       1597       2584
       4181       6765      10946      17711      28657      46368      75025     121393     196418
     317811     514229     832040    1346269    2178309    3524578    5702887    9227465   14930352
   24157817   39088169   63245986  102334155  165580141  267914296  433494437  701408733 1134903170
```

Figure 10.10 The first 45 numbers of the Fibonacci series

```
C     FIBONACCI NUMBER
C     FOUR DIGIT
C     RANDOM NUMBER GENERATOR
C
    1 READ(5,2) LASTNF,NF
    2 FORMAT(2I8)
      WRITE(6,3)
    3 FORMAT('1FIBONACCI NUMBER'/'0FOUR DIGIT'/'0RANDOM NUMBER GENERATOR
     1',////,3X,'LASTNF',4X,'NF',4X,'NEXTNF',5X,'NR',///)
      LASTNF=LASTNF-LASTNF/10000*10000
      NF=NF-NF/10000*10000
      DO 10 K=1,20
      NEXTNF=LASTNF+NF
      NR=NEXTNF-NEXTNF/10000*10000
      WRITE(6,4) LASTNF,NF,NEXTNF,NR
      LASTNF=NF
   10 NF=NR
    4 FORMAT(I7,I8,I9,I8)
      GO TO 1
      END
```

Figure 10.11 A program for generating random numbers based on the Fibonacci number sequence

addition of the two previous numbers. The Fibonacci number sequence starts with the numbers 1, 1. The next number in the sequence is obtained by taking the sum of the two previous numbers, so that, in general,

$$\mathrm{NF}(I) = \mathrm{NF}(I-1) + \mathrm{NF}(I-2),\ I \geq 3, 4, \ldots$$

A program for generating the first 45 numbers in the sequence is shown in Figure 10.9 and the corresponding output is shown in Figure 10.10.

An examination of the output shown in Figure 10.10 will reveal a random behavior in the right-most digits of the larger numbers. A program for generating random numbers based on the Fibonacci number sequence is shown in Figure 10.11. Note that NF (number Fibonacci) and LASTNF (the previous number in the Fibonacci sequence) are initially read in as data. The right-most four digits (or some larger integer may be used) of LASTNF and NF are then obtained by computer statement numbers S.0005 and S.0006, respectively. The DO 10 loop then generates the next number in the sequence (NEXTNF) by taking the sum of LASTNF and NF. The last four digits of this result are used as the random number (S.0009). Observe that only two numbers have to be stored in order to generate the next random number. The numerical value of LASTNF must be changed to the numerical value of NF and NF to NR before the loop may be executed again. The output results for two different data sets are shown in Figure 10.12. It may be observed that the randomness of the results depends on the starting values. Because only addition is used, one may see why the numbers run up and down. The

FIBONACCI NUMBER
FOUR DIGIT
RANDOM NUMBER GENERATOR

LASTNF	NF	NEXTNF	NR
7465	352	7817	7817
352	7817	8169	8169
7817	8169	15986	5986
8169	5986	14155	4155
5986	4155	10141	141
4155	141	4296	4296
141	4296	4437	4437
4296	4437	8733	8733
4437	8733	13170	3170
8733	3170	11903	1903
3170	1903	5073	5073
1903	5073	6976	6976
5073	6976	12049	2049
6976	2049	9025	9025
2049	9025	11074	1074
9025	1074	10099	99
1074	99	1173	1173
99	1173	1272	1272
1173	1272	2445	2445
1272	2445	3717	3717

FIBONACCI NUMBER
FOUR DIGIT
RANDOM NUMBER GENERATOR

LASTNF	NF	NEXTNF	NR
7817	8169	15986	5986
8169	5986	14155	4155
5986	4155	10141	141
4155	141	4296	4296
141	4296	4437	4437
4296	4437	8733	8733
4437	8733	13170	3170
8733	3170	11903	1903
3170	1903	5073	5073
1903	5073	6976	6976
5073	6976	12049	2049
6976	2049	9025	9025
2049	9025	11074	1074
9025	1074	10099	99
1074	99	1173	1173
99	1173	1272	1272
1173	1272	2445	2445
1272	2445	3717	3717
2445	3717	6162	6162
3717	6162	9879	9879

Figure 10.12 The output results of the program of Figure 10.11

reader may greatly improve this random number generator by changing the program so that only every I number is used. The value of I should be sufficiently large enough to produce good statistical results.

Power Residue Method

The power residue method is by far the most popular present-day scheme for generating pseudo-random numbers. This method is based on modulus arithmetic and the concept of congruence. Two numbers, A and B, are said to be congruent modulo M when their difference, $A - B$, is exactly divisible by the integer modulus M, or

$$A \equiv B \pmod{M}$$

For example,

$$10 \equiv 2 \pmod{4}$$

and

$$10 \equiv 4 \pmod{3}$$

This is equivalent to saying

$$\frac{A - B}{M} = I$$

where I is an integer and either A or B is a residue.

There are M residue classes of integers that are mutually congruent modulo M. For example,

$$\begin{array}{llllll} \ldots, & 10, 6, & 2, & -2, & -6, & \ldots \\ \ldots, & 9, 5, & 1, & -3, & -7, & \ldots \\ \ldots, & 8, 4, & 0, & -4, & -8, & \ldots \\ \ldots, & 7, 3, & -1, & -5, & -9, & \ldots \end{array}$$

are the four residue classes that are mutually congruent modulo 4. The smallest positive number in any set is called the least positive residue (in this example, the least positive residues are 0, 1, 2, and 3) and any integer is congruent modulo M to one of them. A digital computer represents numbers in residue form since a register can store only a given number of bits. Thus an N-bit register would store integers in terms of their least positive residue modulo 2^N.

The power residue method of generating random numbers is based on the relation

$$NR_1 = X * NR_0 \pmod{R^N}$$

where NR_1 is the first (next) random number generated, X is an appropriate multiplier, NR_0 is the starting value (or the last random number generated), R is the base or radix of the number system (2 for binary, 10 for decimal, and so on), and N is the number of significant digits. This is equivalent to multiplying NR_0 by X and storing only the N right-most digits.

One such power residue method for choosing an appropriate multiplier and generating random numbers is based on the following procedure:[3]

1. Choose a starting value NR_0 equal to any integer not divisible by 2 or 5.
2. Choose a multiplier constant IX, of the form

$$IX = 200*IT \pm IR$$

   where IT is any integer and IR is equal to any one of the following values: 3, 11, 13, 19, 21, 27, 29, 37, 53, 59, 61, 67, 69, 77, 83, or 91. The value of IX should be made approximately equal to $10^{N/2}$, where N is the required number of digits.
3. The starting number NR_0 is now multiplied by IX to produce a number 1.5 N digits long. The higher order digits are discarded and the N low-order digits are used for the value of NR_1.
4. The next random number is found by multiplying NR_1 by IX, and so on. This procedure will produce 5 10^{N-2} terms before repeating.

As a simple example that illustrates the procedure, let $NR_0 = 1111$ and $N = 4$. Thus

1. $NR_0 = 1111$.
2. Since $IX = 200*IT \pm IR$ should be approximately equal to $10^{N/2} = 100$, let $IT = 1$ and $IR = -91$, so
   $IX = 200*1 - 91 = 109$
3. Thus
   $NR_1 = IX*NR_0 \pmod{10^4}$
   $= 1099$
4. $NR_2 = IX*NR_1 \pmod{10^4}$
   $= 9791$
5. $NR_3 = IX*NR_2 \pmod{10^4}$
   $= 7219$
6. And so on.

A program for generating 500 numbers is shown in Figure 10.13. First the starting value, IT, and IR are read into memory. The multiplier

[3] *Ibid.*, pp. 3–11.

```
C        RANDOM NUMBER GENERATOR
C        POWER RESIDUE METHOD
C
         READ(5,1)NR,IT,IR
       1 FORMAT(3I5)
         IX=200*IT+IR
         DO 10 K=1,500
         IRN=IX*NR/10000
         NR=IX*NR-IRN*10000
      10 WRITE(6,2)NR
       2 FORMAT(I10)
         RETURN
         END
```

Figure 10.13 Power residue four-digit random number generator

1099	3899	6699	9499	2299	5099	7899
9791	4991	191	5391	591	5791	991
7219	4019	819	7619	4419	1219	8019
6871	8071	9271	471	1671	2871	4071
8939	9739	539	1339	2139	2939	3739
4351	1551	8751	5951	3151	351	7551
4259	9059	3859	8659	3459	8259	3059
4231	7431	631	3831	7031	231	3431
1179	9979	8779	7579	6379	5179	3979
8511	7711	6911	6111	5311	4511	3711
7699	499	3299	6099	8899	1699	4499
9191	4391	9591	4791	9991	5191	391
1819	8619	5419	2219	9019	5819	2619
8271	9471	671	1871	3071	4271	5471
1539	2339	3139	3939	4739	5539	6339
7751	4951	2151	9351	6551	3751	951
4859	9659	4459	9259	4059	8859	3659
9631	2831	6031	9231	2431	5631	8831
9779	8579	7379	6179	4979	3779	2579
5911	5111	4311	3511	2711	1911	1111
4299	7099	9899	2699	5499	8299	
8591	3791	8991	4191	9391	4591	
6419	3219	19	6819	3619	419	
9671	871	2071	3271	4471	5671	
4139	4939	5739	6539	7339	8139	
1151	8351	5551	2751	9951	7151	
5459	259	5059	9859	4659	9459	
5031	8231	1431	4631	7831	1031	
8379	7179	5979	4779	3579	2379	
3311	2511	1711	911	111	9311	
899	3699	6499	9299	2099	4899	
7991	3191	8391	3591	8791	3991	
1019	7819	4619	1419	8219	5019	
1071	2271	3471	4671	5871	7071	
6739	7539	8339	9139	9939	739	
4551	1751	8951	6151	3351	551	
6059	859	5659	459	5259	59	
431	3631	6831	31	3231	6431	
6979	5779	4579	3379	2179	979	
711	9911	9111	8311	7511	6711	
7499	299	3099	5899	8699	1499	
7391	2591	7791	2991	8191	3391	
5619	2419	9219	6019	2819	9619	
2471	3671	4871	6071	7271	8471	
9339	139	939	1739	2539	3339	
7951	5151	2351	9551	6751	3951	
6659	1459	6259	1059	5859	659	
5831	9031	2231	5431	8631	1831	
5579	4379	3179	1979	779	9579	
8111	7311	6511	5711	4911	4111	
4099	6899	9699	2499	5299	8099	
6791	1991	7191	2391	7591	2791	
219	7019	3819	619	7419	4219	
3871	5071	6271	7471	8671	9871	
1939	2739	3539	4339	5139	5939	
1351	8551	5751	2951	151	7351	
7259	2059	6859	1659	6459	1259	
1231	4431	7631	831	4031	7231	
4179	2979	1779	579	9379	8179	
5511	4711	3911	3111	2311	1511	
699	3499	6299	9099	1899	4699	
6191	1391	6591	1791	6991	2191	
4819	1619	8419	5219	2019	8819	
5271	6471	7671	8871	71	1271	
4539	5339	6139	6939	7739	8539	
4751	1951	9151	6351	3551	751	
7859	2659	7459	2259	7059	1859	
6631	9831	3031	6231	9431	2631	
2779	1579	379	9179	7979	6779	
2911	2111	1311	511	9711	8911	
7299	99	2899	5699	8499	1299	
5591	791	5991	1191	6391	1591	
9419	6219	3019	9819	6619	3419	
6671	7871	9071	271	1471	2671	
7139	7939	8739	9539	339	1139	
8151	5351	2551	9751	6951	4151	
8459	3259	8059	2859	7659	2459	
2031	5231	8431	1631	4831	8031	
1379	179	8979	7779	6579	5379	
311	9511	8711	7911	7111	6311	

Figure 10.14 Five hundred numbers produced by the random number generator of Figure 10.13 ($NR_0 = 1111$, $IT = 1$, and $IR = -91$)

Figure 10.15 Power residue six-digit random number generator

```
C       RANDOM NUMBER GENERATOR
C       POWER RESIDUE METHOD
C
        READ(5,1) NR,IT,IR
      1 FORMAT(3I10)
        IX=200*IT+IR
        DO 10 K=1,50
        IRN=IX*NR/1000000
        NR=IX*NR-IRN*1000000
     10 WRITE(6,2) NR
      2 FORMAT(I10)
        RETURN
        END
```

value *IX* is then calculated. The DO 10 loop generates four-digit numbers by first finding *IRN*, which is the overflow of the first four digits of the product of *IX* and *NR*. The lower four digits of $IX*NR$ are then obtained by multiplying *IRN* by 10,000 and subtracting it from $IX*NR$. The process is repeated, each time printing out the value of *NR* (the lower-four digits), until 500 numbers have been found.

The resulting 500 numbers are shown in Figure 10.14. Note that the lower-order digits are far from random. A closer examination of the procedure will reveal that for a starting value of 1111 and a multiplier of 109, the least significant digit will be either a 1 or a 9. Other starting values and multipliers will show a similar behavior with the least significant digit being an even number or an odd number, with a maximum cycle of four numbers. In general, the period (cycle) of the digits of a given digit position will increase in order from right to left (least significant to most significant position).

Figure 10.15 is a modification of the program of Figure 10.13 and generates six-digit numbers rather than four-digit numbers. This program uses a starting value of 123456 and a multiplier of 1109 (with $IT = 6$ and $IR = -91$). The first 50 numbers are shown in Figure 10.16. Note again the lack of randomness of the lower-order digits and, in particular, the least significant digit. Note also that the least significant digit is either a 4 or a 6 because the least significant digits of the starting value and the multiplier were 6 and 9 respectively.

Figure 10.16 Fifty numbers produced by the random number generator of Figure 10.15 ($NR_0 = 123456$, $IT = 6$, and $IR = -91$)

912704	303296
188736	355264
308224	987776
820416	443584
841344	934656
50496	533504
64	655936
70976	433024
712384	223616
33856	990144
546304	69696
851136	292864
909824	786176
994816	869184
250944	925056
296896	887104
257664	798336
749376	354624
57984	278016
304256	319744
419904	596096
673536	70464
951424	144576
129216	334784
300544	275456

```
C       SYSTEM 360 SCIENTIFIC SUBROUTINE
C       POWER RESIDUE RANDOM NUMBER GENERATOR
C
C       1000 RANDOM NUMBERS
C
        DIMENSION RN(1000)
        ISEED=123456789
        DO 10 I=1,1000
        IRN=ISEED*65539
        IF(IRN) 2,3,3
      2 IRN=IRN+2147483647+1
      3 RN(I)=IRN
        ISEED=IRN
        RN(I)=RN(I)*.4656613E-9
     10 CONTINUE
        WRITE(6,4)
      4 FORMAT('1 ONE THOUSAND RANDOM NUMBERS',//)
        WRITE(6,5) (RN(I),I=1,1000)
      5 FORMAT(5F11.7)
        RETURN
        END
```

Figure 10.17 System 360 power residue seven-digit random number generator

The above program could easily be modified to generate a number that contained more digits. Similarly it could be changed by dividing the value of NR by 100 which would cause the two least significant digits to be dropped and thus would produce a four-digit number that would be more random than the four-digit generator of Figure 10.13.

The IBM 360 Power Residue Generator

A power residue random number generator especially designed for the IBM System 360 is shown in Figure 10.17.[4] The output of the first 1000 numbers (for a starting value, ISEED = 123456789) is shown in Figure 10.18. Note that unlike any of the previous generators we have discussed, statements have been added in this program to produce a floating point random number that has a numerical value between 0 and 1 (this is the usual range of random-number generators). An examination of the output will show a lack of randomness in the least significant digits, but a fairly uniform distribution for each of the more significant digit positions. The testing of the distribution of this and other random number generators will be considered in the next section.

10.3 TESTING THE DISTRIBUTION OF A RANDOM NUMBER GENERATOR

Many tests have been devised to determine the randomness of a random number generator. A number of these tests depend on the laws of probability to predict a certain frequency of occurrence. For example,

[4] IBM Application Program, *System 360 Scientific Subroutine Package* (360A-CM-03X) Version II Programmer's Manual, H20-0205-2, p. 54.

if a random number generator generates numbers between zero and one, the probability of generating a number between 0 and 0.5 (equivalent to false, no, tail, logical 0, etc.) should be the same as the probability of generating a number between 0.5 and 1 (equivalent to true, yes, head, logical 1, and so on). Similarly, by multiplying a random number by 5 and adding 1, a generator can be made to produce numbers that range from 1 to 6. These, in turn, can be used to represent the toss of a die. The first random number can be used to represent the value of the first die, and the second random number can correspond to the value of the second die. The randomness of a random number generator can then be tested, based on the laws of probability of the game of dice. For example, if the first two numbers correspond to the first toss of the dice and if the point is 5, the chance of making the point is 2/5, and the probability of throwing a 2 or a 12 is 1/36.

In much the same way, a random number generator can be used to simulate a poker hand and since the expected frequency of getting two pairs or filling an inside straight can be predicted from the laws of probability, the randomness of the generator can be checked.

It is important to realize that if a random number generator passes some of the tests for randomness, it may not pass others. Thus, the use of such a generator might be equivalent to playing with loaded dice in a Monte Carlo analysis, and it is therefore wise to know the characteristics of the random number generator being used or to repeat a Monte Carlo solution using several different generators. The following are some of the tests that can be used to help determine the characteristics of a given generator.

The Averaging Test

One of the simplest tests that can be applied to determine the distribution of a random-number generator is the averaging of several hundred (thousand) numbers. For example, the average of the first 100 numbers should have the same value as the average of any other 100 numbers, and in the case of a uniformly distributed number, the average value should be equal to 0.5 for a numerical range of 0 to 1.

This particular form of statistical analysis results in the *arithmetic mean* and may be calculated from the general relation

$$\text{MEAN} = \frac{\sum_{I=1}^{I=N} \text{NR}(I)}{N}$$

ONE THOUSAND RANDOM NUMBERS

0.7746705	0.1306216	0.8116952	0.6945766	0.8622029
0.9220276	0.7723391	0.3357657	0.0636622	0.3599014
0.5864489	0.2795600	0.3994397	0.8804175	0.6875478
0.2015291	0.0212440	0.3137015	0.6910132	0.3227652
0.7174720	0.3999442	0.9424168	0.0550031	0.8482673
0.5945762	0.9330512	0.2471213	0.0852666	0.2875075
0.9576459	0.1583077	0.3310328	0.5614274	0.3892686
0.2827659	0.1931767	0.6141678	0.9464158	0.1509846
0.3881654	0.9701296	0.3272896	0.2325706	0.4498177
0.6057703	0.5862626	0.0656421	0.1174890	0.1141553
0.6275307	0.7377863	0.7789413	0.0335707	0.1909525
0.8435792	0.3429021	0.4651994	0.7050769	0.0436674
0.9163109	0.1048594	0.3823578	0.3504121	0.6612523
0.8138044	0.9315562	0.2656971	0.2065769	0.8535873
0.2623307	0.8916979	0.9892111	0.9099857	0.5570141
0.1522131	0.9001515	0.0309905	0.0845791	0.2285602
0.6101496	0.6038561	0.1317894	0.3560314	0.9500830
0.4362149	0.4265413	0.0933142	0.7210132	0.4862514
0.4283891	0.1940724	0.3089322	0.1069418	0.8612612
0.2050912	0.4791960	0.0293545	0.8633624	0.9159847
0.7256459	0.1100128	0.1292632	0.7854643	0.5494168
0.2273214	0.4191774	0.4691716	0.0424332	0.0320543
0.8104270	0.5740726	0.1505933	0.7369057	0.0660944
0.7644149	0.9916396	0.0701032	0.4958626	0.3442469
0.6027176	0.5180836	0.6840421	0.4415003	0.4926224
0.9822313	0.4597861	0.9186343	0.3737313	0.9746789
0.4844922	0.1348425	0.4486256	0.4781706	0.8313932
0.6848232	0.6263994	0.5949881	0.9323339	0.2391101
0.0436551	0.1099386	0.2667359	0.6109685	0.2651869
0.0924055	0.1677500	0.1748506	0.5393530	0.6624624
0.1205963	0.7614166	0.4831328	0.0460478	0.9280909
0.1581154	0.5716739	0.0442045	0.1183618	0.3123301
0.3087251	0.0413795	0.9697499	0.4460547	0.9487584
0.6777877	0.5279003	0.0073134	0.6527764	0.3108376
0.9900380	0.1426928	0.9458091	0.3966198	0.8314358
0.4730366	0.3552979	0.8744571	0.0490612	0.4242527
0.1039656	0.8055187	0.8974218	0.1348621	0.7323757
0.1804954	0.4915900	0.3250816	0.5261787	0.2313378
0.6524181	0.6324680	0.1230450	0.2460575	0.3689406
0.9991252	0.6742862	0.0535903	0.2529644	0.0354741
0.9361646	0.2977213	0.3608459	0.4855835	0.6658873
0.6250744	0.7574561	0.9190664	0.6972933	0.9121621
0.1973324	0.9745352	0.0712195	0.6564991	0.2980192
0.8796227	0.5955638	0.6567781	0.5805943	0.5725631
0.2100292	0.1071080	0.7523848	0.5503368	0.5305578
0.2303149	0.6068686	0.5683783	0.9484516	0.5753050
0.9157648	0.3168442	0.6591816	0.1034908	0.6883107
0.1984466	0.9358831	0.1692792	0.1727266	0.3328466
0.4425404	0.6596221	0.9748685	0.9126123	0.7018569
0.9975299	0.6690677	0.0357374	0.1928139	0.8352473
0.2761582	0.1397237	0.3529174	0.8599914	0.9836916

0.1622258	0.1201309	0.2607528	0.4833390	0.5532578
0.9694965	0.8376534	0.3024816	0.2639631	0.8794442
0.9009976	0.4909875	0.8369459	0.6027874	0.0842106
0.0801770	0.7231666	0.6174062	0.1959385	0.6189741
0.9503978	0.1316198	0.2361377	0.2322475	0.2682451
0.5192435	0.7012547	0.5343367	0.8947277	0.5593357
0.3034649	0.7867680	0.9694239	0.8556314	0.2289732
0.6731560	0.9781780	0.8106632	0.0603773	0.0662941
0.8543695	0.5295700	0.4880935	0.1624312	0.5817456
0.0285927	0.9358051	0.3577369	0.7238151	0.1232582
0.2252131	0.2419540	0.4248061	0.3712507	0.4042491
0.0842379	0.8671850	0.4449695	0.8651518	0.1861842
0.3307390	0.3087760	0.8760047	0.4770439	0.9782202
0.5759265	0.6515772	0.7261248	0.4925528	0.4201938
0.0881879	0.7473823	0.6906028	0.4171758	0.2876297
0.9711963	0.2385097	0.6962912	0.9951594	0.7583361
0.5935814	0.7364633	0.0765472	0.8311126	0.2977509
0.3064916	0.1591907	0.1967137	0.7475963	0.7151095
0.5622902	0.9377556	0.5659219	0.9557308	0.6410877
0.2449487	0.6999037	0.9948834	0.6701669	0.0670497
0.3707963	0.6213306	0.3998169	0.7529256	0.0002018
0.2248790	0.3474581	0.0608376	0.2379023	0.8798751
0.1381301	0.9099042	0.2162545	0.1063889	0.7040426
0.2487550	0.1561466	0.6980848	0.7831888	0.4163696
0.4495177	0.9497802	0.6530213	0.3701054	0.3434402
0.7296920	0.2871901	0.1559125	0.3507637	0.7013694
0.0513434	0.9957355	0.5123225	0.1123157	0.0629910
0.3671041	0.6357056	0.5102967	0.3404300	0.4499097
0.6355873	0.7643367	0.8657335	0.3153709	0.1006235
0.7654020	0.6868006	0.2321855	0.2119070	0.1817724
0.1834704	0.4648703	0.1379887	0.6440992	0.6226970
0.9392885	0.0314581	0.7351520	0.1277888	0.1503646
0.7520889	0.1592515	0.1867065	0.6869869	0.4415457
0.4663913	0.8244365	0.7490970	0.0746533	0.7060462
0.5643972	0.0319674	0.1122291	0.3856680	0.3039465
0.3526660	0.3804773	0.1086699	0.2289231	0.3937091
0.3019464	0.2682966	0.8922616	0.9389005	0.6030480
0.1681834	0.5816685	0.9763594	0.6231396	0.9516028
0.1013601	0.0437350	0.3501687	0.7073970	0.0928636
0.1906087	0.3078797	0.1317996	0.0198803	0.9330844
0.4195840	0.1197446	0.9422110	0.5755652	0.9734914
0.6608615	0.2037454	0.2747190	0.8146052	0.4151597
0.1595113	0.2266302	0.8881766	0.3433995	0.0667894
0.3101399	0.2597349	0.7671503	0.2652876	0.6873729
0.7366489	0.2335368	0.7713898	0.5264531	0.2162919
0.5596736	0.4114105	0.4314054	0.8857387	0.4317828
0.6190486	0.8282456	0.3960367	0.9340091	0.0217233
0.7242581	0.1500386	0.3319090	0.9411069	0.2094598
0.7867961	0.6356388	0.9326671	0.0752537	0.0575179
0.6678236	0.4892800	0.9252681	0.1480882	0.5611167
0.0339058	0.1533036	0.6151499	0.3104460	0.3263272
0.1639489	0.0467478	0.8049468	0.4089504	0.2091819
0.5745367	0.5645931	0.2166681	0.2187607	0.3625506
0.2064568	0.9757853	0.9966004	0.1975344	0.2158030

Figure 10.18 One thousand numbers produced by the random number generator of Figure 10.17

0.5170075	0.1598178	0.3058392	0.3966740	0.6274918
0.1948842	0.5218781	0.3773106	0.5669602	0.0059649
0.9331473	0.5452001	0.8728746	0.3304464	0.1268067
0.7868228	0.5796762	0.3966517	0.1628252	0.4070854
0.9770854	0.1987426	0.3986878	0.6034427	0.0324656
0.7638091	0.2906646	0.8697051	0.6022488	0.7861471
0.2966433	0.7045355	0.5574238	0.0037231	0.0055234
0.9996330	0.9480872	0.6918265	0.6181736	0.4826032
0.3320564	0.6489098	0.9049510	0.5895175	0.3925456
0.0496157	0.7647834	0.1421591	0.9699039	0.5399916
0.5108135	0.2049563	0.6324165	0.9498925	0.0076061
0.4966036	0.9111673	0.9975706	0.7849178	0.7313709
0.3239652	0.3614527	0.2530289	0.2650988	0.3133323
0.4941049	0.1446378	0.4208825	0.2235546	0.5533850
0.3083181	0.8694432	0.4417953	0.8257828	0.9785385
0.4391862	0.8282704	0.0169464	0.6472441	0.7309474
0.5604874	0.7843980	0.6620014	0.9124264	0.5165457
0.8874360	0.6757056	0.0673083	0.3225000	0.3292248
0.0728493	0.4740717	0.1887863	0.8660721	0.4973556
0.1894850	0.6607097	0.2588933	0.6069715	0.3117900
0.4079958	0.6418648	0.1792260	0.2985730	0.1784034
0.3832632	0.6939487	0.7143233	0.0404006	0.8134929
0.5173527	0.7826796	0.0399029	0.1953007	0.8126782
0.1183630	0.3960742	0.3111781	0.3024002	0.0137988
0.3611901	0.0429516	0.0069982	0.6554245	0.8695633
0.3185596	0.0852873	0.6446863	0.1005325	0.8010176
0.9013137	0.1987225	0.0805122	0.6945702	0.4428111
0.4057347	0.4491073	0.0430311	0.2162210	0.9100461
0.5142876	0.8953103	0.7432735	0.4018481	0.7216270
0.7131293	0.7841328	0.2866333	0.6626039	0.3959231
0.4121044	0.9093173	0.7469043	0.2979304	0.0649034
0.7080466	0.6641482	0.6124704	0.6974881	0.6726949
0.7587771	0.4984080	0.1614531	0.4830474	0.4452055
0.3238062	0.9359877	0.7016705	0.7861334	0.4017658
0.3353940	0.3964716	0.3602830	0.5934529	0.3181703
0.5679449	0.5441362	0.1533136	0.0226545	0.7561045
0.3327360	0.1914761	0.1542318	0.2021059	0.8245491
0.1283411	0.3491042	0.9395549	0.4953922	0.5163587
0.6396220	0.1905031	0.3864202	0.6039937	0.1461792
0.4411316	0.3311770	0.0168773	0.1206697	0.5721226
0.3467085	0.9311473	0.4665072	0.4187173	0.3137386
0.1139758	0.3602074	0.1354624	0.0709075	0.2062832
0.5995321	0.7406436	0.0480725	0.6226414	0.3031966
0.2154070	0.5636711	0.4433632	0.5871394	0.5325670
0.9111471	0.6737796	0.8423532	0.9901029	0.3594385
0.2457047	0.2392807	0.2243426	0.1925286	0.1360881
0.0837711	0.2778332	0.9130590	0.9778551	0.6495993
0.0969000	0.7350056	0.5379332	0.6125487	0.8338926

0.4904169	0.4374679	0.2110549	0.3291181	0.0752146
0.4892241	0.2584132	0.1474626	0.5590568	0.0271767
0.1315463	0.5446998	0.0342643	0.6032875	0.8613456
0.7384866	0.6788083	0.4264704	0.4495469	0.8590473
0.1033617	0.9187435	0.5372061	0.9545444	0.8924116
0.7635090	0.5497097	0.4261372	0.6094359	0.8213863
0.4433584	0.2677279	0.6161413	0.2872972	0.1785117
0.4853950	0.3057640	0.4660285	0.0442953	0.0715151
0.0304329	0.5389608	0.9598697	0.9085706	0.8125955
0.6984372	0.8772635	0.9776464	0.9755064	0.0242209
0.4107667	0.2466131	0.7827778	0.4771482	0.8178887
0.0129986	0.3169926	0.3849685	0.4568771	0.2765459
0.5473808	0.7953713	0.8458004	0.9164602	0.8865571
0.0712011	0.4481919	0.0483415	0.2563221	0.1028595
0.3102570	0.935[illegible]64	0.8225256	0.5128957	0.6746430
0.4317966	0.5189921	0.2277633	0.6957707	0.1245744
0.4855092	0.7918554	0.3817294	0.1634071	0.5448785
0.7986066	0.8877330	0.1389377	0.8440289	0.8137336
0.2861469	0.3932434	0.7841921	0.1659613	0.9380389
0.1345819	0.3651410	0.9796089	0.5913842	0.7318248
0.0654909	0.8245217	0.3307124	0.5635790	0.4050620
0.3581607	0.5034076	0.7969949	0.2513058	0.3348807
0.7475315	0.4712628	0.0997933	0.3573940	0.2462237
0.2607965	0.3487655	0.7454238	0.3336531	0.2931042
0.7557468	0.8965436	0.5775399	0.3963468	0.1802211
0.5142046	0.4632379	0.1515651	0.7403694	0.0779506
0.8043779	0.1247121	0.5088714	0.9308199	0.0050756
0.6530747	0.8727682	0.3589358	0.2987009	0.5617827
0.6823881	0.0382939	0.0682098	0.1847039	0.3143352
0.2236761	0.5130390	0.0651485	0.7735401	0.0549042
0.3675629	0.7112399	0.9593731	0.3550786	0.4961134
0.7809734	0.2208196	0.2961563	0.7895615	0.0719621
0.3257191	0.3066556	0.9084619	0.6908705	0.9690657
0.5965586	0.8577607	0.7775362	0.9453710	0.6744004
0.5380630	0.1587774	0.1100906	0.2315476	0.3984697
0.3068898	0.2551117	0.7686616	0.3159641	0.9778303
0.0233043	0.3393525	0.8263767	0.9040874	0.9871341
0.7860179	0.8319002	0.9172403	0.0163390	0.8428707
0.9101737	0.8752049	0.0596665	0.4811540	0.3499255
0.7691674	0.4656749	0.8715427	0.0381816	0.3852051
0.9675969	0.3387345	0.3240345	0.8955958	0.4572648
0.6832265	0.9839752	0.7546127	0.6730990	0.2452800
0.4137883	0.2752090	0.9271591	0.0860735	0.1720063
0.2573986	0.9962566	0.6610415	0.9999397	0.0502648
0.3021309	0.3604028	0.4432386	0.4156059	0.5056881
0.2918750	0.2000575	0.5734698	0.6403009	0.6805767
0.3207525	0.7993242	0.9091725	0.2611175	0.3841522

Figure 10.18 Continued

```
C        THE ARITHMETIC MEAN TEST
C        FOR RANDOMNESS
C           OF THE
C        FOUR-DIGIT POWER RESIDUE
C        RANDOM NUMBER GENERATOR
C
         DIMENSION NR(500)
C
C        GENERATION OF 500 RANDOM NUMBERS
C
         N=1111
         DO 10 K=1,500
         NR(K)=109*N-109*N/10000*10000
      10 N=NR(K)
C
C        CALCULATION OF THE ARITHMETIC MEAN
C
         INIT=1
         LIMIT=100
         DO 20 J=1,5
         ISUM=0
         DO 30 I=INIT,LIMIT
      30 ISUM=ISUM+NR(I)
         MEAN=ISUM/100
         WRITE(6,1) INIT,LIMIT,MEAN
       1 FORMAT('OTHE ARITHMETIC MEAN OF THE RANDOM NUMBERS',I4,' THROUGH',
        1I5,' IS',I10)
         INIT=INIT+100
      20 LIMIT=LIMIT+100
         RETURN
         END
```

Figure 10.19 A program for applying the arithmetic mean test for randomness to the random number generator of Figure 10.13

```
C
C        THE ARITHMETIC MEAN TEST
C           FOR RANDOMNESS
C             OF THE
C        SYSTEM 360 SCIENTIFIC SUBROUTINE
C        POWER RESIDUE RANDOM NUMBER GENERATOR
         DIMENSION RN(1000)
C
C        GENERATION OF 1000 RANDOM NUMBERS
C
         ISEED=123456789
         DO 10 I=1,1000
         IRN=ISEED*65539
         IF(IRN) 2,3,3
       2 IRN=IRN+2147483647+1
       3 RN(I)=IRN
         ISEED=IRN
         RN(I)=RN(I)*.4656613E-9
      10 CONTINUE
C
C        THE CALCULATION OF THE ARITHMETIC MEAN
C
         INIT=1
         LIMIT=100
         DO 20 J=1,10
         SUM=0.
         DO 30 I=INIT,LIMIT
      30 SUM=SUM+RN(I)
         ARMEAN=SUM/100
         WRITE(6,1) INIT,LIMIT,ARMEAN
       1 FORMAT('OTHE ARITHMETIC MEAN OF THE RANDOM NUMBERS',I4,' THROUGH '
        1I5,' IS',F11.7)
         INIT=INIT+100
      20 LIMIT=LIMIT+100
         RETURN
         END
```

Figure 10.20 A program for applying the arithmetic mean test for randomness to the random number generator of Figure 10.17

```
THE ARITHMETIC MEAN OF THE RANDOM NUMBERS   1 THROUGH  100 IS        5445
THE ARITHMETIC MEAN OF THE RANDOM NUMBERS 101 THROUGH  200 IS        4645
THE ARITHMETIC MEAN OF THE RANDOM NUMBERS 201 THROUGH  300 IS        5045
THE ARITHMETIC MEAN OF THE RANDOM NUMBERS 301 THROUGH  400 IS        5445
THE ARITHMETIC MEAN OF THE RANDOM NUMBERS 401 THROUGH  500 IS        4445
```

(a)

```
THE ARITHMETIC MEAN OF THE RANDOM NUMBERS   1 THROUGH   100 IS  0.4819323
THE ARITHMETIC MEAN OF THE RANDOM NUMBERS 101 THROUGH   200 IS  0.4646562
THE ARITHMETIC MEAN OF THE RANDOM NUMBERS 201 THROUGH   300 IS  0.5420539
THE ARITHMETIC MEAN OF THE RANDOM NUMBERS 301 THROUGH   400 IS  0.4898274
THE ARITHMETIC MEAN OF THE RANDOM NUMBERS 401 THROUGH   500 IS  0.4567685
THE ARITHMETIC MEAN OF THE RANDOM NUMBERS 501 THROUGH   600 IS  0.4837883
THE ARITHMETIC MEAN OF THE RANDOM NUMBERS 601 THROUGH   700 IS  0.4798147
THE ARITHMETIC MEAN OF THE RANDOM NUMBERS 701 THROUGH   800 IS  0.4545593
THE ARITHMETIC MEAN OF THE RANDOM NUMBERS 801 THROUGH   900 IS  0.4981732
THE ARITHMETIC MEAN OF THE RANDOM NUMBERS 901 THROUGH  1000 IS  0.5177698
```

(b)

Figure 10.21 (a) The arithmetic mean of 100 number sets of the random number generator of Figure 10.13; (b) the arithmetic mean of 100 number sets of the random number generator of Figure 10.17

It is a measure of the *central tendency* and yields a middle or mean value of the distribution.

Figure 10.19 shows a program for finding the arithmetic mean of the first five sets of one hundred numbers generated by the program of Figure 10.13. Figure 10.20 shows a similar program for finding the arithmetic mean of the first ten sets of one hundred numbers generated by the program of Figure 10.17. The corresponding mean values are shown in Figure 10.21a and b, respectively. Although both of these results yield arithmetic mean values for each 100 number sets that are very close to the theoretical predicted value of a uniformly distributed random number, it should be clear that this is a necessary, but not sufficient, test for uniform distribution. For example, if all numbers generated had the same value of 0.5, this test would be passed, but the numbers obviously would not be uniformly distributed between 0 and 1.

The Frequency Test

The frequency test determines how often each of the ten decimal digits, zero through nine, occur in each digit position. Thus, in the case of the example cited in the last paragraph, where each of the numbers generated had the same value of 0.5, the frequency test would clearly indicate the lack of a uniform distribution.

In order to illustrate this test, the investigation will be limited to the most significant digit of the 1000 seven-digit random numbers generated by the program of Figure 10.17.

Figure 10.22 is a program for determining the frequency of occurrence of the ten decimal digits, zero through nine, in the most significant digit position. First, 1000 numbers are generated by computer statements S.0001 through S.0010. It may be noted that this will produce the same sequence of numbers that were generated by the program of Figure 10.17. The DO 20 loop initially sets each of the elements of the frequency count variable $F(I)$ to zero. The element $F(1)$ is used to store the frequency of occurrence of the digit zero, $F(2)$ the occurrence of digit one, and so on.

The DO 30 loop then individually examines the 1000 numbers. Each number is multiplied by 10, and 1 is added to that result to produce a

```
         C      THE FREQUENCY TEST
         C         FOR RANDOMNESS
         C               OF THE
         C      SYSTEM 360 SCIENTIFIC SUBROUTINE
         C      POWER RESIDUE RANDOM NUMBER GENERATOR
         C
S.0001          DIMENSION RN(1000),F(10)
         C
         C      GENERATION OF 1000 RANDOM NUMBERS
         C
S.0002          ISEED=123456789
S.0003          DO 10 I=1,1000
S.0004          IRN=ISEED*65539
S.0005          IF(IRN) 2,3,3
S.0006        2 IRN=IRN+2147483647+1
S.0007        3 RN(I)=IRN
S.0008          ISEED=IRN
S.0009          RN(I)=RN(I)*.4656613E-9
S.0010       10 CONTINUE
         C
         C      THE CALCULATION OF THE FREQUENCY DISTRIBUTION
         C
S.0011          DO 20 I=1,10
S.0012       20 F(I)=0.0
S.0013          DO 30 I=1,1000
S.0014          K=RN(I)*10.+1.
S.0015       30 F(K)=F(K)+1.
S.0016          WRITE(6,6)
S.0017        6 FORMAT('1THE FREQUENCY DISTRIBUTION '/' OF 1000 RANDOM NUMBERS'//)
S.0018          A=0.0
S.0019          B=.1
S.0020          DO 40 I=1,10
S.0021          WRITE(6,7) A,B,F(I)
S.0022        7 FORMAT('0TOTAL NUMBER OF NUMBERS BETWEEN',F4.1,' AND ',F4.1,' = ',
               1F6.0)
S.0023          A=A+.1
S.0024       40 B=B+.1
         C
         C      CHI-SQUARE CALCULATION
         C      9 DEGREES OF FREEDOM
         C
S.0025          CHISQU=0.0
S.0026          DO 50 I=1,10
S.0027       50 CHISQU=CHISQU+((F(I)-100.)**2)/100.
S.0028          WRITE(6,8)CHISQU
S.0029        8 FORMAT(////,'  CHI-SQUARE  =',F7.3)
S.0030          RETURN
S.0031          END
```

Figure 10.22 The frequency test for randomness of 1000 numbers generated by the random number generator of Figure 10.17

numerical value for K. Thus, if the value of the most significant digit equals zero, K will have a value of 1; if the value of the most significant digit equals one, K will have a value of 2; and so on. To store the actual count the statement

$$F(K) = F(K) + 1$$

is used. If the value of $K = 3$, for example, this statement will add 1 to the count element $F(3)$.

The DO 40 loop prints out the frequency of occurrence of the numbers generated between 0.0 – 0.1, 0.1 – 0.2, . . . , and 0.9 – 1.0. These results are shown in Figure 10.23. Since 1000 numbers were generated and tested, and since the probability of occurrence of each of the ten decimal digits is equal to 0.1 for a uniformly distributed random number, the theoretical occurrence of each possible event is 100. An examination of the output shows that only two of the ten events have a value equal to 100. Since the range of these events runs from a low of 82 to a high of 126, one may have doubts in using this generator to produce uniformly distributed random numbers.

In attempting to analyze the output further, one might (without thinking) find that the average of the ten events exactly equals 100. Further thought will show that this must always be true because there are ten events and 1000 numbers. It is thus clear that some standard statistical test must be applied in order to determine whether this generator produces uniformly distributed random numbers.

One such statistical test is the Chi-square (χ^2) calculation. This is a statistic based on the expected frequency of occurrence of numbers in

```
THE FREQUENCY DISTRIBUTION
OF 1000 RANDOM NUMBERS


TOTAL NUMBER OF NUMBERS BETWEEN 0.0 AND  0.1 =    99.

TOTAL NUMBER OF NUMBERS BETWEEN 0.1 AND  0.2 =   114.

TOTAL NUMBER OF NUMBERS BETWEEN 0.2 AND  0.3 =   100.

TOTAL NUMBER OF NUMBERS BETWEEN 0.3 AND  0.4 =   126.

TOTAL NUMBER OF NUMBERS BETWEEN 0.4 AND  0.5 =    95.

TOTAL NUMBER OF NUMBERS BETWEEN 0.5 AND  0.6 =    85.

TOTAL NUMBER OF NUMBERS BETWEEN 0.6 AND  0.7 =   100.

TOTAL NUMBER OF NUMBERS BETWEEN 0.7 AND  0.8 =    89.

TOTAL NUMBER OF NUMBERS BETWEEN 0.8 AND  0.9 =    82.

TOTAL NUMBER OF NUMBERS BETWEEN 0.9 AND  1.0 =   110.


 CHI-SQUARE   = 16.680
```

Figure 10.23 The frequency distribution of 1000 numbers generated by the program of Figure 10.17

each interval and the actual frequency of occurrence and is defined by

$$\chi^2 = \sum_{I=1}^{I=N} \frac{(\text{Actual Freq. of Occ.} - \text{Expected Freq. of Occ.})^2}{\text{Expected Freq. of Occ.}}$$

so that for our case

$$\chi^2 = \sum_{I=1}^{I=10} \frac{(F(I) - 100)^2}{100}$$

A zero value for Chi-square will correspond to the theoretical expected value and a large value of Chi-square will establish that the numbers are not uniformly distributed. In order to interpret the calculated value of χ^2, a χ^2 distribution table may be used.[5] A partial table is shown in Table 10.1 and includes the 5 percent significance levels that are usually used by statisticians and the 1 percent levels that are sometimes specified.

Returning to the program of Figure 10.22, the DO 50 loop is used to find the value of χ^2 and this result ($\chi^2 = 16.680$) is printed in the output shown in Figure 10.23. In order to interpret this value of χ^2, the number of degrees of freedom must first be defined as the number of independent frequency pairs being compared and in our example since $N - 1$ equals 9, we see that the number of degrees of freedom is nine.

Table 10.1 shows that $\chi^2 = 16.92$ for nine degrees of freedom at the 5 percent significance level. Since the calculated value is 16.68 and it is less than the table value, the hypothesis of a uniform distribution is accepted. Actually the accuracy of the test depends on the sample size (we have used only 1000 numbers) and the starting value ISEED. Since no test was made to determine the distribution of the six lesser significant digits, no accurate conclusion can be drawn for the test made on this random-number generator.

One should also be aware that the Chi-square test was applied to interpret the results of the frequency test. The Chi-square test results may be misleading, however, since we would expect that in the examination of the most significant digit of 1000 uniformly distributed random numbers, each decimal digit would occur 100 times. It is obvious, however, that if the first 100 numbers had 0 as the most significant digit, the second 100 numbers had 1 as the most significant digit, . . . , and the last 100 numbers had 9 as the most significant digit, the frequency test would be passed, but the numbers would be far from random.

[5] A Chi-square distribution table can usually be found in any book on statistics. See also, *Handbook of Mathematical Functions*, National Bureau of Standards, Applied Mathematics Series 55 (Washington, D.C.: U.S. Government Printing Office, June 1964), p. 985.

Table 10.1 Chi-square (χ^2) Values at the 1 Percent and 5 Percent Levels of Significance

DEGREES OF FREEDOM	1%	5%
1	6.63	3.84
2	9.21	5.99
3	11.34	7.81
4	13.28	9.49
5	15.09	11.07
6	16.81	12.59
7	18.48	14.07
8	20.09	15.51
9	21.66	16.92
10	23.21	18.31
11	24.73	19.68
12	26.22	21.03
13	27.69	22.36
14	29.14	23.68
15	30.58	25.00
16	32.00	26.30
17	33.41	27.59
18	34.81	28.87
19	36.19	30.14
20	37.57	31.41
21	38.93	32.67
22	40.29	33.92
23	41.64	35.17
24	42.98	36.42
25	44.31	37.65
26	45.64	38.89
27	46.96	40.11
28	48.28	41.34
29	49.59	42.56
30	50.89	43.77
40	63.69	55.76
50	76.15	67.50
60	88.38	79.08
70	100.43	90.53
80	112.33	101.88
90	124.12	113.15
100	135.81	124.34

The Gap Interval Test

Another test for determining the randomness of a uniformly distributed random-number generator that would overcome the difficulty cited in the preceding paragraph is the gap interval test.

Since the decimal number system is defined in terms of the ten decimal digits zero through nine, the generation of any one of the digits, in each

```
      C     THE GAP INTERVAL TEST
      C     FOR RANDOMNESS
      C     OF THE
      C     FOUR-DIGIT POWER RESIDUE
      C     RANDOM NUMBER GENERATOR
      C
            DIMENSION NR(500)
      C
      C     GENERATION OF 500 RANDOM NUMBERS
      C
            N=1111
            DO 5 K=1,500
            NR(K)=109*N-109*N/10000*10000
          5 N=NR(K)
      C
      C     THE GAP INTERVAL TEST
      C
            DO 10 I=1,10
            IDIGIT=I-1
            WRITE(6,1)IDIGIT
          1 FORMAT('1GAP INTERVAL TEST FOR DIGIT ',I3,//,' THE SUCCESSIVE GAP
           1LENGTHS EQUAL',///)
            L=0
            M=0
      C
      C     FINDING THE GAP INTERVALS OF EACH OF THE TEN DECIMAL DIGITS
      C     OF THE MOST SIGNIFICANT DIGIT OF 500 RANDOM NUMBERS
      C
            DO 20 K=1,500
            N=NR(K)/1000
            L=L+1
            IF(IDIGIT-N)20,30,20
      C
      C     THE VALUE OF L IS EQUAL TO THE LENGTH OF THE GAP
      C
         30 WRITE(6,2)L
          2 FORMAT(I8)
            L=0
      C
      C     THE FINAL VALUE OF M IS EQUAL TO THE TOTAL NUMBER OF GAPS
      C
            M=M+1
         20 CONTINUE
            IAVE=500/M
            WRITE(6,3)IAVE,M
          3 FORMAT(//,' THE AVERAGE GAP LENGTH =',I4,/,' THE NUMBER OF GAPS ='
           1,I4)
         10 CONTINUE
            RETURN
            END
```

Figure 10.24 A program that applies the gap interval test for randomness to the random number generator of Figure 10.13

digit position, should be equally probable. For example, if we examine the most significant digit position of K random numbers, we would expect that the frequency of occurrence of the number 0 (or any of the nine other decimal digits) would correspond to an average interval (gap) of 10. Figure 10.24 is a program for determining the length of each gap, the number of gaps, and the average gap length for each of the ten decimal digits of the most significant digit of the 500 numbers generated by the program of Figure 10.13. Figure 10.25 shows that all results are exactly equal to the expected average length of 10 and thus it may be assumed that as far as the most significant digit is concerned, this generator produces uniformly distributed random numbers. The reader is encouraged to check the successive gap length results by referring to the output shown (the 500 numbers that were generated) in Figure 10.14.

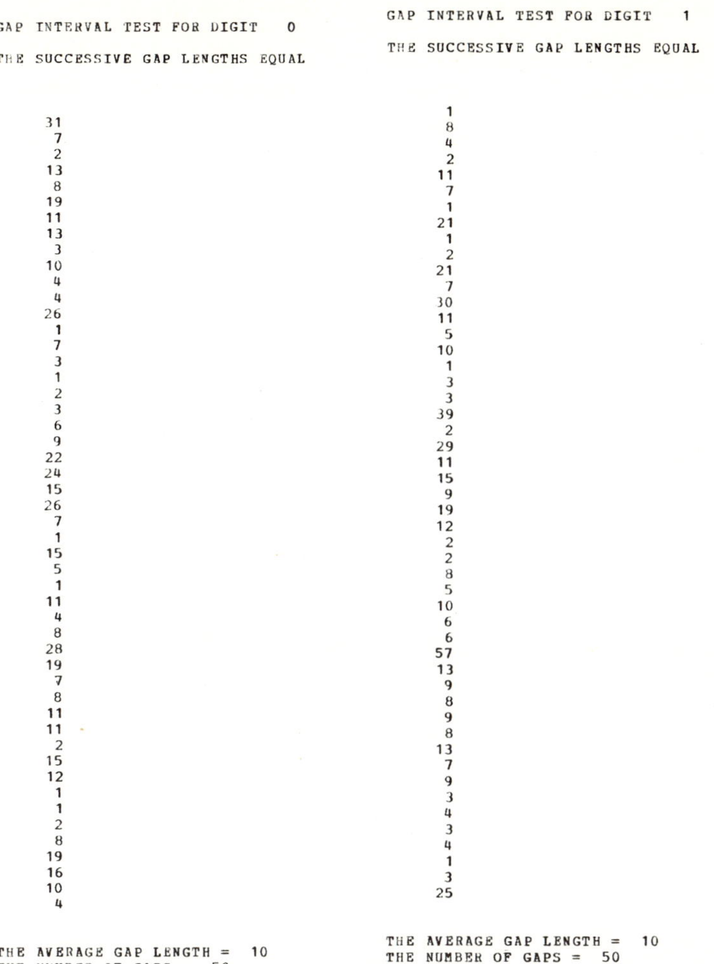

```
GAP INTERVAL TEST FOR DIGIT    0

THE SUCCESSIVE GAP LENGTHS EQUAL


   31
    7
    2
   13
    8
   19
   11
   13
    3
   10
    4
    4
   26
    1
    7
    3
    1
    2
    3
    6
    9
   22
   24
   15
   26
    7
    1
   15
    5
    1
   11
    4
    8
   28
   19
    7
    8
   11
   11
    2
   15
   12
    1
    1
    2
    8
   19
   16
   10
    4

THE AVERAGE GAP LENGTH =   10
THE NUMBER OF GAPS =   50
```

```
GAP INTERVAL TEST FOR DIGIT    1

THE SUCCESSIVE GAP LENGTHS EQUAL


    1
    8
    4
    2
   11
    7
    1
   21
    1
    2
   21
    7
   30
   11
    5
   10
    1
    3
    3
   39
    2
   29
   11
   15
    9
   19
   12
    2
    2
    8
    5
   10
    6
    6
   57
   13
    9
    8
    9
    8
   13
    7
    9
    3
    4
    3
    4
    1
    3
   25

THE AVERAGE GAP LENGTH =   10
THE NUMBER OF GAPS =   50
```

Figure 10.25 The ten gap intervals of 500 numbers of the random number generator of Figure 10.13

The Serial Test

It has previously been pointed out that if a random-number generator passes a certain test for randomness, the numbers generated are not necessarily random. For example, all the above tests for a uniformly distributed random number would be passed by a generator that produced 1000 numbers so that in each digit position the integers 0 through 9 repeated 100 times. It is obvious that such a generator would be considered a perfect one by each of the above tests, but the numbers could not be considered random.

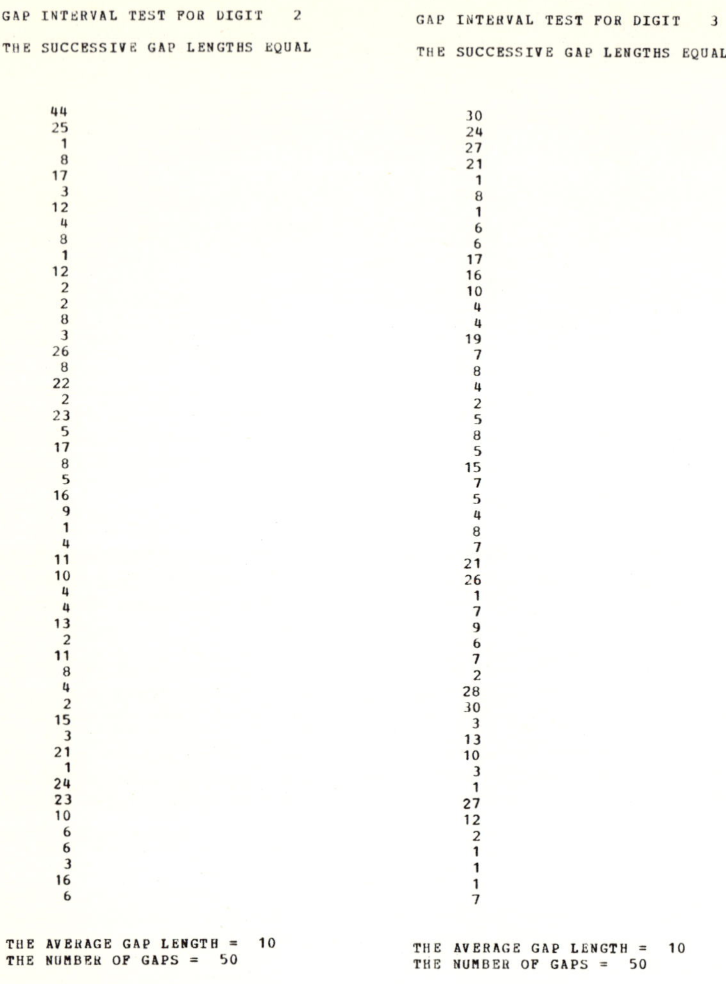

```
GAP INTERVAL TEST FOR DIGIT    2

THE SUCCESSIVE GAP LENGTHS EQUAL


     44
     25
      1
      8
     17
      3
     12
      4
      8
      1
     12
      2
      2
      8
      3
     26
      8
     22
      2
     23
      5
     17
      8
      5
     16
      9
      1
      4
     11
     10
      4
      4
     13
      2
     11
      8
      4
      2
     15
      3
     21
      1
     24
     23
     10
      6
      6
      3
     16
      6

THE AVERAGE GAP LENGTH =    10
THE NUMBER OF GAPS =    50
```

```
GAP INTERVAL TEST FOR DIGIT    3

THE SUCCESSIVE GAP LENGTHS EQUAL


     30
     24
     27
     21
      1
      8
      1
      6
      6
     17
     16
     10
      4
      4
     19
      7
      8
      4
      2
      5
      8
      5
     15
      7
      5
      4
      8
      7
     21
     26
      1
      7
      9
      6
      7
      2
     28
     30
      3
     13
     10
      3
      1
     27
     12
      2
      1
      1
      1
      7

THE AVERAGE GAP LENGTH =    10
THE NUMBER OF GAPS =    50
```

Figure 10.25 Continued

Some test is necessary to determine if some cyclic pattern exists (numbers increasing, numbers decreasing, odd number followed by even number, large number followed by small number, and so on). Such a test is the serial correlation test. In this test a count is made of the occurrence of each combination of two successive digits. For example, assume that 1001 numbers are generated by the program of Figure 10.17 and that for illustration purposes the test will be limited to the most significant digit. Since the first digit is a 7 (see Figure 10.18) and the second digit is a 1 (the most significant digit of the second number generated), a count of 1 is stored for the combination of the number 7 followed by the number 1.

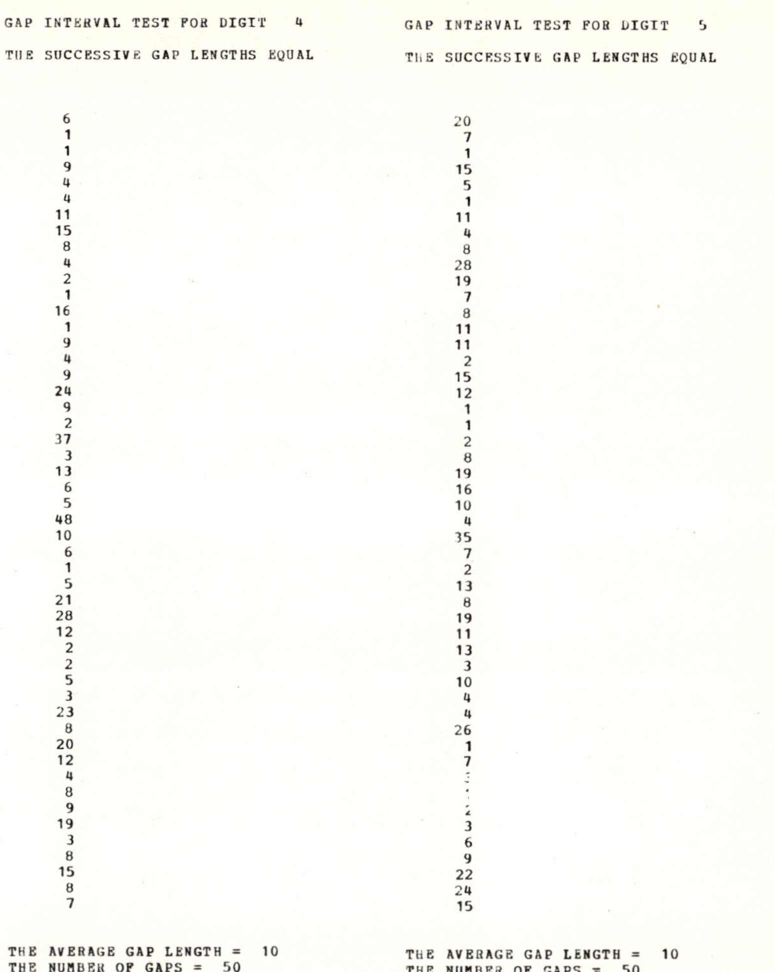

```
GAP INTERVAL TEST FOR DIGIT    4

THE SUCCESSIVE GAP LENGTHS EQUAL

     6
     1
     1
     9
     4
     4
    11
    15
     8
     4
     2
     1
    16
     1
     9
     4
     9
    24
     9
     2
    37
     3
    13
     6
     5
    48
    10
     6
     1
     5
    21
    28
    12
     2
     2
     5
     3
    23
     8
    20
    12
     4
     8
     9
    19
     3
     8
    15
     8
     7

THE AVERAGE GAP LENGTH =   10
THE NUMBER OF GAPS =   50
```

```
GAP INTERVAL TEST FOR DIGIT    5

THE SUCCESSIVE GAP LENGTHS EQUAL

    20
     7
     1
    15
     5
     1
    11
     4
     8
    28
    19
     7
     8
    11
    11
     2
    15
    12
     1
     1
     2
     8
    19
    16
    10
     4
    35
     7
     2
    13
     8
    19
    11
    13
     3
    10
     4
     4
    26
     1
     7
     [illegible]
     [illegible]
     [illegible]
     3
     6
     9
    22
    24
    15

THE AVERAGE GAP LENGTH =   10
THE NUMBER OF GAPS =   50
```

Figure 10.25 Continued

Similarly, since the second digit is a 1 and the third digit is an 8, a count of 1 is stored for the combination of the number 1 followed by the number 8, and so on.

The combination pair count can easily be stored in a 10 by 10 matrix S, where the ten rows correspond to the ten decimal digits 0 to 9, respectively, and the ten columns also correspond to the ten-decimal digits 0 to 9, respectively. The row is used to indicate the first decimal number of the pair and the column the second number of the pair. For example, the number 7 followed by the number 1 would cause a count of 1 to be added to the contents of element $S(8,2)$.

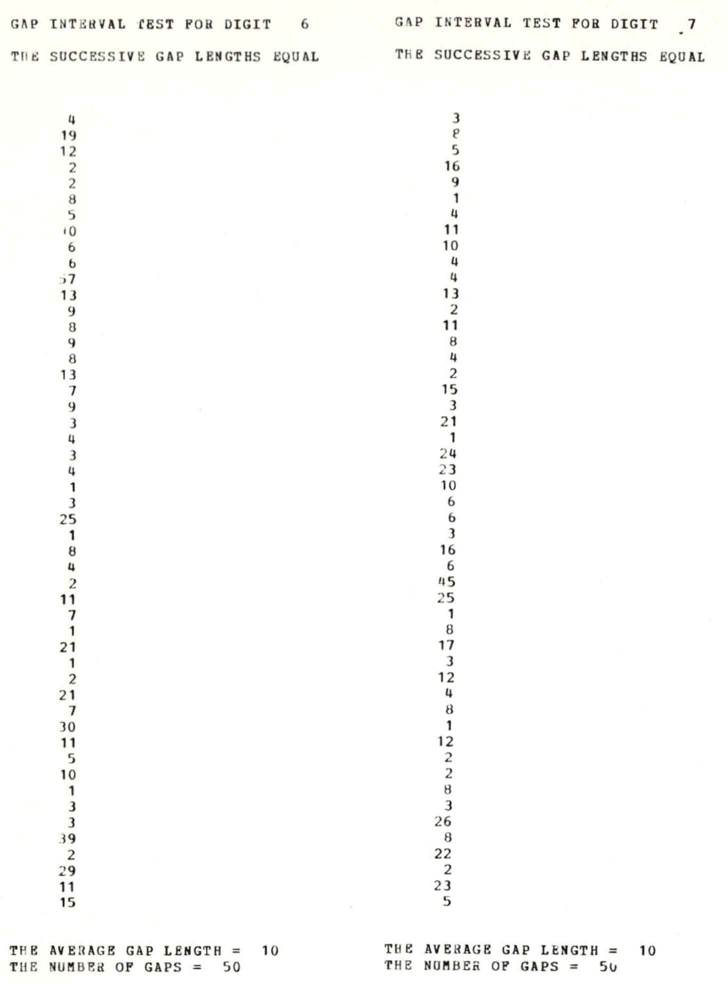

```
GAP INTERVAL TEST FOR DIGIT    6

THE SUCCESSIVE GAP LENGTHS EQUAL

     4
    19
    12
     2
     2
     8
     5
    10
     6
     6
    57
    13
     9
     8
     9
     8
    13
     7
     9
     3
     4
     3
     4
     1
     3
    25
     1
     8
     4
     2
    11
     7
     1
    21
     1
     2
    21
     7
    30
    11
     5
    10
     1
     3
     3
    39
     2
    29
    11
    15

THE AVERAGE GAP LENGTH =    10
THE NUMBER OF GAPS =    50
```

```
GAP INTERVAL TEST FOR DIGIT    7

THE SUCCESSIVE GAP LENGTHS EQUAL

     3
     8
     5
    16
     9
     1
     4
    11
    10
     4
     4
    13
     2
    11
     8
     4
     2
    15
     3
    21
     1
    24
    23
    10
     6
     6
     3
    16
     6
    45
    25
     1
     8
    17
     3
    12
     4
     8
     1
    12
     2
     2
     8
     3
    26
     8
    22
     2
    23
     5

THE AVERAGE GAP LENGTH =    10
THE NUMBER OF GAPS =    50
```

Figure 10.25 Continued

Figure 10.26 is a program for testing the serial correlation of 1000 number pairs that can be formed from 1001 numbers generated by the random-number generator of Figure 10.17. First, 1001 numbers are generated by computer statements S.0001 through S.0010. These numbers will be identical to the sequence shown in Figure 10.18, except for the additional number (1001). The DO 20 loop initially sets each of the elements of the serial pair association matrix S to zero. The element $S(1,1)$ is used to store the frequency of occurrence of a 0 digit followed by a 0 digit, $S(1,2)$ a 0 digit followed by a 1 digit, and so on.

The DO 30 loop then individually examines the most significant digit

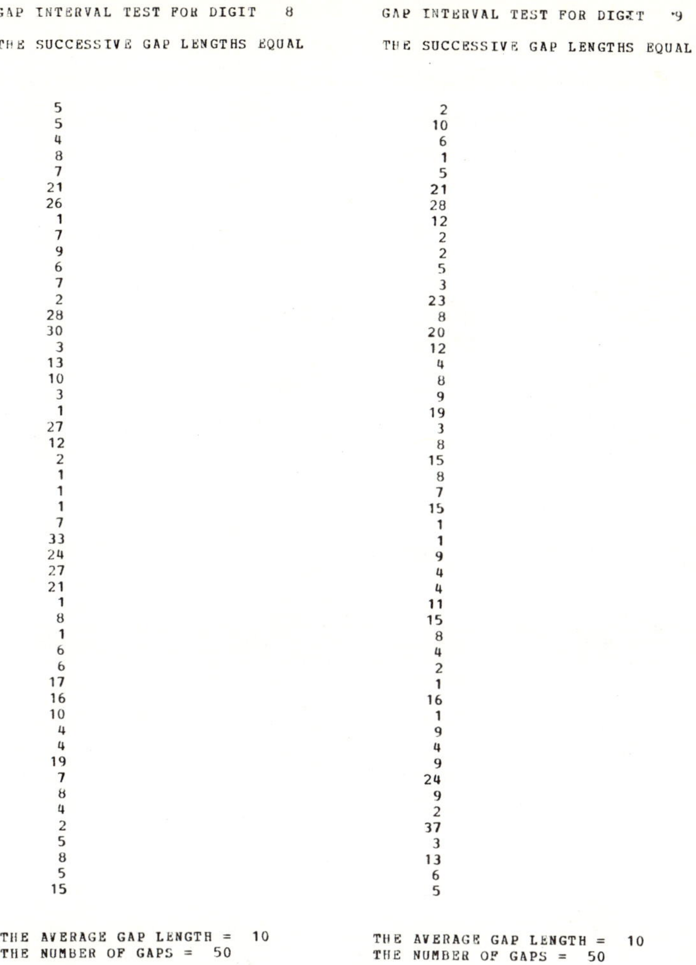

```
GAP INTERVAL TEST FOR DIGIT    8

THE SUCCESSIVE GAP LENGTHS EQUAL


   5
   5
   4
   8
   7
  21
  26
   1
   7
   9
   6
   7
   2
  28
  30
   3
  13
  10
   3
   1
  27
  12
   2
   1
   1
   1
   7
  33
  24
  27
  21
   1
   8
   1
   6
   6
  17
  16
  10
   4
   4
  19
   7
   8
   4
   2
   5
   8
   5
  15

THE AVERAGE GAP LENGTH =   10
THE NUMBER OF GAPS =   50
```

```
GAP INTERVAL TEST FOR DIGIT    9

THE SUCCESSIVE GAP LENGTHS EQUAL


   2
  10
   6
   1
   5
  21
  28
  12
   2
   2
   5
   3
  23
   8
  20
  12
   4
   8
   9
  19
   3
   8
  15
   8
   7
  15
   1
   1
   9
   4
   4
  11
  15
   8
   4
   2
   1
  16
   1
   9
   4
   9
  24
   9
   2
  37
   3
  13
   6
   5

THE AVERAGE GAP LENGTH =   10
THE NUMBER OF GAPS =   50
```

Figure 10.25 Continued

of 1000 successive pairs. First, a given number is multiplied by 10, and 1 is added to that result to produce a numerical value for K. Thus, if the value of the most significant digit of the first number equals 7, K will have a value of 8 and will correspond to the row subscript of the variable S. The next random number will then be multiplied by 10, and 1 will be added to that result to produce a numerical value for L, which in turn will correspond to the column subscript of the matrix S. The statement

$$S(K,L) = S(K,L) + 1.$$

is used to store the actual count of the occurrence of the number 7 fol-

```
C
C      THE SERIAL TEST
C        THE SERIAL TEST
C          FOR RANDOMNESS
C             OF THE
C      SYSTEM 360 SCIENTIFIC SUBROUTINE
C      POWER RESIDUE RANDOM NUMBER GENERATOR
C
       DIMENSION RN(1001),S(10,10)
C
C      GENERATION OF 1001 RANDOM NUMBERS
C
       ISEED=123456789
       DO 10 I=1,1001
       IRN=ISEED*65539
       IF(IRN) 2,3,3
     2 IRN=IRN+2147483647+1
     3 RN(I)=IRN
       ISEED=IRN
       RN(I)=RN(I)*.4656613E-9
    10 CONTINUE
C
C      INITIALIZE THE SERIAL PAIR ASSOCIATION MATRIX S
C
       DO 20 I=1,10
       DO 20 J=1,10
    20 S(I,J)=0.0
C
C      DETERMINE ELEMENT VALUES
C
       DO 30 I=1,1000
       K=RN(I)*10.+1.
       L=RN(I+1)*10.+1.
    30 S(K,L)=S(K,L)+1.
C
C      PRINT SERIAL PAIR ASSOCIATION MATRIX
C      THE ROW NUMBER IS THE FIRST NUMBER OF A PAIR
C      THE COLUMN NUMBER IS THE SECOND NUMBER OF THE PAIR
C
       WRITE(6,6)
     6 FORMAT('1SERIAL PAIR ASSOCIATION MATRIX'/'0OF 1001 RANDOM NUMBERS'
      1/'0THE ROW NUMBER IS THE FIRST DECIMAL DIGIT OF A PAIR'/'0THE CCLU
      2MN NUMBER IS THE SECOND DECIMAL DIGIT OF THE PAIR',///,35X,'COLUMN
      3 NUMBER'///17X,'0',4X,'1',4X,'2',4X,'3',4X,'4',4X,'5',4X,'6',4X,'7
      4',4X,'8',4X,'9',//)
       DO 40 I=1,10
       K=I-1
    40 WRITE(6,7)K,(S(I,J),J=1,10)
     7 FORMAT(' ROW NO. =',I2,3X,(10F5.0))
C
C      CHI-SQUARE CALCULATION
C      90 DEGREES OF FREEDOM
C
       CHISQU=0.0
       DO 50 I=1,10
       DO 50 J=1,10
    50 CHISQU=CHISQU+((S(I,J)-10.)**2)/10.
       WRITE(6,8)CHISQU
     8 FORMAT(////,' CHI-SQUARE  =',F7.3)
       RETURN
       END
```

Figure 10.26 A program that applies the serial test for randomness to the random number generator of Figure 10.17

lowed by the number 1. This process is continued until 1000 number pairs have been correlated.

The DO 40 loop is used to print the serial pair association matrix *S*. The results are shown in Figure 10.27. Since 1001 random numbers were generated and 1000 successive pairs were correlated, the 10 by 10 *S* matrix, which represents 100 possible events that are equally probable for a uniformly distributed random number, should have a theoretical element occurrence of 10. An examination of the output shows that only 14 of the 100 events have a value equal to 10. Since the range of these

```
SERIAL PAIR ASSOCIATION MATRIX

OF 1001 RANDOM NUMBERS

THE ROW NUMBER IS THE FIRST DECIMAL DIGIT OF A PAIR

THE COLUMN NUMBER IS THE SECOND DECIMAL DIGIT OF THE PAIR
```

	COLUMN NUMBER									
	0	1	2	3	4	5	6	7	8	9
ROW NO. = 0	10.	14.	9.	14	8.	1.	10.	13.	11.	9.
ROW NO. = 1	8.	12.	12.	22.	9.	12.	9.	11.	10.	9.
ROW NO. = 2	6.	12.	13.	14.	6.	11.	13.	8.	10.	7.
ROW NO. = 3	10.	12.	14.	20.	12.	8.	11.	13.	9.	17.
ROW NO. = 4	10.	12.	9.	12.	16.	5.	8.	4.	10.	9.
ROW NO. = 5	11.	6.	10.	10.	6.	6.	9.	9.	5.	13.
ROW NO. = 6	8.	12.	9.	10.	8.	8.	9.	10.	11.	15.
ROW NO. = 7	15.	9.	10.	7.	10.	12.	7.	6.	6.	7.
ROW NO. = 8	7.	10.	6.	9.	11.	8.	8.	5.	2.	16.
ROW NO. = 9	14.	15.	8.	8.	9.	14.	16.	9.	8.	9.

```
CHI-SQUARE   =116.199
```

Figure 10.27 The serial pair association matrix for the serial test of Figure 10.26

events runs from a low of 1 to a high of 22, one may again have doubts in using this generator to produce uniformly distributed random numbers.

A Chi-square calculation would correspond to the evaluation of

$$\chi^2 = \sum_{L=1}^{L=10} \sum_{K=1}^{K=10} \frac{(S,(K,L) - 10)^2}{10}$$

The two DO 50 loops are used to find the value of χ^2 and the result ($\chi^2 = 116.199$) is printed in the output shown in Figure 10.27. In order to interpret this result, the number of degrees of freedom are first calculated. For this type of array the number of degrees of freedom would be given by the relation Rows*(Columns − 1), which equals 90.

Table 10.1 shows that $\chi^2 = 113.15$ for 90 degrees of freedom at the 5 percent significance level. Since the calculated value is 116.199 and is more than the table value, the hypothesis of a uniform distribution must be rejected. It is quite possible, with another starting value or a larger sample, that this test would be passed.

Calculation of the Period

Using N digits, one can represent R^N different numbers, where R is the base or radix of the number system. For example, with four decimal digits, one can represent 10^4 or 10,000 different decimal numbers (0000, 0001, 0002, . . . , 9999). Since all the schemes for the computer gen-

```
              C
              C       THE CALCULATION OF THE PERIOD
              C       OF A RANDOM NUMBER GENERATOR
              C
S.0001                DIMENSION NR(1000)
              C
              C       GENERATION OF 1000 RANDOM NUMBERS
              C
S.0002                N=1111
S.0003                DO 5 K=1,1000
S.0004                NR(K)=109*N-109*N/10000*10000
S.0005              5 N=NR(K)
              C
              C       FINDING THE PERIOD
              C
S.0006                ICYCLE=0
S.0007                DO 10 I=2,1000
S.0008                L=I-1
S.0009                DO 10 K=1,L
S.0010                IF(NR(K)-NR(I)) 10,20,10
S.0011             20 NEXCYC=I-K
S.0012                IF(NEXCYC-ICYCLE) 30,40,30
S.0013             30 WRITE(6,1)I,K,NEXCYC
S.0014              1 FORMAT('1THE',I5,' NUMBER WAS FOUND TO BE THE SAME AS THE',I5,' NU
                     1MBER',//' THIS PERIOD =',I5)
S.0015                ICYCLE=NEXCYC
S.0016             10 CONTINUE
S.0017                WRITE(6,2)
S.0018              2 FORMAT('1NO CYCLE WAS FOUND')
S.0019             40 CONTINUE
S.0020                RETURN
S.0021                END
```

Figure 10.28 A program for determining the length of the period of repetition of the random number generator of Figure 10.13

eration of a random number are based on some numerical algorithm, it thus follows that if the iteration is repeated often enough the numbers must eventually repeat.[6] It is also obvious that once a number occurs for the second time, the following number must be exactly the same as the number that was produced after the original occurrence, and so on.

Figure 10.28 gives a program for determining the length of the period of repetition. This program determines the period of the random number generator of Figure 10.13. First, 1000 numbers are generated by computer statements S.0001 through S.0005. It will be noted that this will produce the same sequence of numbers as in Figure 10.14. The outside DO 10 loop then examines successive random numbers, each time comparing the number in question to all the previous numbers by means of the inner DO 10 loop. If, at any time during the execution of the inner DO 10 loop, a match is found, the period (NEXCYC) is calculated and compared to the previous period that was found. (Initially the previous period is given the value of zero.) Whenever the period calculated is different from the previous period, the numerical location of the two numbers in the sequence and the period is printed. Whenever the period calculated is equal to the previous period, the process is terminated because this means the number sequence will repeat in the same way from

[6] It is important to understand that although some of the early numbers may not occur again, there will be a sequence of numbers that will repeat indefinitely. See, for example, the first and fourth data sets in Figure 10.3.

```
THE  501 NUMBER WAS FOUND TO BE THE SAME AS THE     1 NUMBER

THIS PERIOD =  500
```

Figure 10.29 The period found for the random number generator of Figure 10.13

here on out. If no match is found in 1000 comparisons, the period is greater than 1000 and NO CYCLE WAS FOUND is printed. The LIMIT value of the iteration variable in the outer DO 10 loop may obviously be increased (from 1000) if the generator has a period greater than 1000.

Figure 10.29 shows that the period of this generator is exactly equal to 500 ($5*10^{D-2}$), and this can readily be verified by examining the output data of Figure 10.14. The reader is encouraged to modify this program and use it to determine the period of the center-squaring random number generator of Figure 10.5. The reader is also encouraged to investigate possible changes that would improve the efficiency of this program when it is applied to larger period generators such as the one described by Figure 10.17.

10.4 CHANGING THE DISTRIBUTION OF A RANDOM NUMBER GENERATOR

Many physical situations exhibit a randomness corresponding to a normal distribution. In other words, values close to the average or mean value will predominate. The central limit theorem states that the sum of a large number of independent random variables has a distribution that is approximately normal, regardless of the individual random variable distributions. Thus a normal, or Gaussian, distribution is often found in practice because it occurs whenever a large number of independent random variables are measured (or act) collectively. For example, if the value of f_t was measured at a given operating point for 1000 transistors of the same type, we would not expect to find a uniform distribution. In other words, if the range in the value of f_t ran from 400 to 1000 mHz, we would find that most of the transistors would have a value of about 800 mHz and very few would have a value of 400 or 1000 mHz.

From the above illustration, it is obvious that in the design of an electronic circuit we would not want to select randomly element values that were based on a uniform distribution. In general, all computer pseudorandom number generators produce uniformly distributed random numbers. Other distributions may be obtained from a uniformly distributed number by suitable computer programs.

The Normal Distribution

Figure 10.30 shows a normal distribution curve that has a mean value of zero. This is a plot of the normal probability density function versus the deviation from the mean. The normal frequency function for a mean value of zero and a standard deviation σ is given by

$$\phi_\sigma(x) = \frac{1}{\sqrt{2\pi\sigma}} \epsilon^{-\frac{1}{2}\left(\frac{x}{\sigma}\right)^2}$$

When the mean value is μ instead of zero, the probability density function becomes $\phi_\sigma(x - \mu)$. The standard deviation σ, is given by

$$\sigma = \sqrt{\frac{\sum_{I=1}^{I=N} x(I)^2}{N}}$$

where $x(I) = X(I) - \mu =$ the deviation from the mean value. Thus, if the standard deviation is calculated for a series of measurements and the parameter being measured is normally distributed, 68.26 percent of the measurements will fall within $\pm 1\sigma$ of the mean, 95.46 percent will fall within $\pm 2\sigma$ of the mean, and 99.74 percent will fall within $\pm 3\sigma$ of the mean value.

Normally distributed random numbers may be obtained by summing N uniformly distributed random numbers. In order to produce a normally distributed number with a mean value of zero and a variance equal to zero, $N\mu$ must be subtracted from the sum and the difference divided

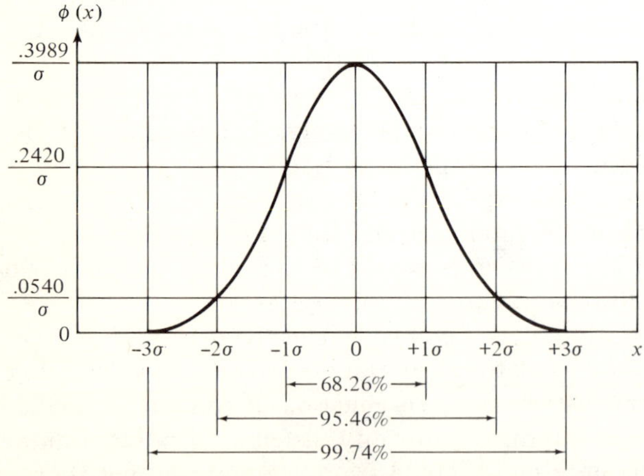

Figure 10.30 The normal distribution curve

by $\sqrt{N\sigma^2}$.[7] Since the uniformly distributed random numbers have a range of 0 to 1, the mean value of each probability density function of each random number is $\mu = 0.5$. The variance can be calculated from

$$\text{Var}(x) = \sigma^2 = E[x^2] - \mu^2$$

where the expected value E is given by

$$E[x^2] = \int_0^1 x^2 f(x)\, dx$$

For a uniformly distributed number between zero and one, $f(x) = 1$, thus

$$E[x^2] = \int_0^1 x^2\, dx = \frac{x^3}{3}\Big]_0^1 = \frac{1}{3}$$

and

$$\text{Var}(x) = \sigma^2 = \frac{1}{3} - \left(\frac{1}{2}\right)^2 = \frac{1}{12}$$

Therefore, normally distributed random numbers would be calculated from the general relation

$$\text{RNORM(I)} = \frac{\sum_{J=1}^{J=N} \text{RN}(J) - N\mu}{\sqrt{N\sigma^2}}$$

so that if 250 normally distributed numbers are to be produced (I = 1,250) and each one is calculated from a series of 20 (N = 20) uniformly distributed numbers having a mean value of $\mu = 0.5$, and $\sigma^2 = 1/12$, then

$$\text{RNORM(I)} = \frac{\sum_{J=1}^{J=20} \text{RN}(J) - 10.}{\sqrt{20./12.}}$$

Figure 10.31 is a program for generating 250 normally distributed random numbers from 5000 uniformly distributed random numbers. Computer statements S.0001 through S.0010 first generate 5000 uniformly distributed random numbers based on the generator of Figure 10.17. The DO 20 loop is used to produce the 250 numbers. For each normally distributed number, the DO 30 loop sums 20 uniformly distributed numbers. The statement

$$\text{RNORM(I)} = (\text{SUM} - 10.)/\text{SQRT}(20./12.)$$

[7] *Handbook of Mathematical Functions*, p. 952.

```
C       GENERATION OF 250 NORMALLY DISTRIBUTED
C             RANDOM NUMBERS
C               FROM 5,000
C       UNIFORMLY DISTRIBUTED RANDOM NUMBERS
C
        DIMENSION RN(5000),RNORM(250)
C
C       GENERATION OF 5,000 RANDOM NUMBERS
C
        ISEED=123456789
        DO 10 I=1,5000
        IRN=ISEED*65539
        IF(IRN)2,3,3
      2 IRN=IRN+2147483647+1
      3 RN(I)=IRN
        ISEED=IRN
        RN(I)=RN(I)*.4656613E-9
     10 CONTINUE
        ST1IN=0.0
        ST1OUT=0.0
        ST2IN=0.0
        ST2OUT=0.0
        N=1
        NN=20
        DENOM=SQRT(20./12.)
        DO 20 I=1,250
        SUM=0.0
        DO 30 J=N,NN
     30 SUM=SUM+RN(J)
        RNORM(I)=(SUM-10.)/DENOM
        IF(ABS(RNORM(I))-1.)40,40,50
     40 ST1IN=ST1IN+1.
        GO TO 60
     50 ST1OUT=ST1OUT+1.
     60 IF(ABS(RNORM(I))-2.)70,70,80
     70 ST2IN=ST2IN+1.
        GO TO 90
     80 ST2OUT=ST2OUT+1.
     90 N=N+20
     20 NN=NN+20
        WRITE(6,100)(RNORM(I),I=1,250)
    100 FORMAT(' TWO HUNDRED AND FIFTY NORMALLY DISTRIBUTED RANDOM NUMBERS
      1 ',///,5(F15.7))
C
C       CHI-SQUARE CALCULATION
C       ONE DEGREE OF FREEDOM
C
C       NUMBERS WITHIN ONE STANDARD DEVIATION
C
        CHISQ1=((ST1IN-170.)**2)/170.+((ST1OUT- 80.)**2)/80.
C
C       NUMBERS WITHIN TWO STANDARD DEVIATIONS
C
        CHISQ2=((ST2IN-240.)**2)/240.+((ST2OUT-10.)**2)/10.
        WRITE(6,105)ST1IN,ST1OUT,ST2IN,ST2OUT
    105 FORMAT('1THE NUMBER OF NUMBERS WITHIN ONE STANDARD DEVIATION =',F5
      1.0,/,'0THE NUMBER OF NUMBERS OUTSIDE ONE STANDARD DEVIATION =',F5.
      20,////,' THE NUMBER OF NUMBERS WITHIN TWO STANDARD DEVIATION =',F5
      3.0,/,'0THE NUMBER OF NUMBERS OUTSIDE TWO STANDARD DEVIATIONS ='F5.
      40,////)
        WRITE(6,110)CHISQ1,CHISQ2
    110 FORMAT(' NUMBERS WITHIN ONE STANDARD DEVIATION',/,'0CHI-SQUARE =',
      1F8.3,////,' NUMBERS WITHIN TWO STANDARD DEVIATIONS',/,'0CHI-SQUARE
      2 =',F8.3)
        RETURN
        END
```

Figure 10.31 A program for generating 250 normally distributed random numbers

is then used to calculate the normal random number. The statement

$$\text{IF(ABS(RNORM(I))} - 1.)40{,}40{,}50$$

is used to determine if the random number is within or outside one standard deviation of the mean value of zero. Statement number 40 is used to store the count of the number of numbers that are within one standard deviation and statement number 50 is used to store the count of the number of numbers that are outside one standard deviation. Similarly, statements 70 and 80 are used to store the count of the number of numbers that are within and outside two standard deviations of the mean. Statements 90 and 20 are used to increment both N and NN by 20 so that the next 20 uniformly distributed numbers can be summed. This process is repeated until 250 normally distributed numbers have been generated.

Figure 10.32 shows the printout of the 250 normally distributed random numbers. If the distribution of the numbers is truly normal, then 68.26 percent should lie within one standard deviation of the mean value of zero. Thus theoretically (.6826*250 = 170.6) 170 numbers should lie between −1 and +1. A Chi-square statistic can be evaluated from

$$\chi^2 = \frac{(\text{ST1IN} - 170.)^2}{170.} + \frac{(\text{ST1OUT} - 180.)^2}{80.}$$

TWO HUNDRED AND FIFTY NORMALLY DISTRIBUTED RANDOM NUMBERS

0.0861044	-0.2356251	-1.0639162	0.3815263	-0.5671420
-1.1662998	1.4742126	-2.7316408	0.7188294	-1.0323887
1.6576195	-0.2691687	0.4757395	0.1714000	1.2223396
-0.5768996	0.9963603	0.4654633	-0.8550962	-0.8174122
-1.2844210	-1.0163298	-0.6116872	-0.9127941	0.4769281
-1.7028313	-0.7169117	-0.4462449	1.0692787	0.5414015
0.3716156	-0.9858691	-2.0558195	0.4447113	0.6621760
-1.4370966	-1.8127604	0.0457212	-1.4980526	1.1827240
0.7023199	0.1727281	0.3229573	-0.6492279	-0.6899028
-1.1054659	0.5025703	1.3750620	0.5636110	0.0410141
-0.0461917	-1.1386538	0.5104598	0.0349456	-1.6206017
0.7954074	0.1846923	-0.4890127	-0.3060799	1.0099621
0.1346569	0.9918734	0.4048873	0.6480082	-0.1016137
0.3678519	-0.3175226	0.4168116	0.6020070	-0.7097669
0.1341473	-0.9612611	-0.7370623	0.5238194	0.0586605
-1.3265629	1.7080507	0.9755153	-1.1317377	1.0448370
-0.6142261	-0.5596862	0.1870658	1.2443857	-1.1992550
-0.3003387	-0.5424941	0.1245943	-0.9494638	0.4517269
-0.5677256	0.4260507	0.3461899	-0.8156378	-0.6700315
-1.7431412	0.1392526	-0.0789137	-0.5173305	0.9083641
0.0931444	2.3767157	0.5613350	0.5188678	-1.6628199
-1.9156017	-1.8463554	-0.3646873	-1.2382565	-0.1697238
0.5796979	-0.4056770	-1.3752670	-0.1747404	1.0065622
0.5222312	0.9451062	-1.6348801	-0.0884542	-0.2141397
-0.9890500	-0.0425026	-1.8891497	-1.8106909	-2.7063990
0.9283477	-0.9166451	0.7485390	0.6032688	-0.5082799
-0.3245471	-0.3497291	1.6955128	-0.3772889	0.9454069
-0.9609575	0.6361349	0.9877314	1.1919765	1.0745020
0.1536309	-1.0555077	-0.1725929	0.5092040	-0.5576148
1.5831470	0.4758097	-0.5940312	-0.4151325	-0.4829966
1.6780109	-0.6080955	-0.3213617	0.7393841	-1.5592623
0.8838114	-0.0652483	1.1889572	0.8269408	-0.0102068
0.1285427	0.2416479	-2.4502592	-0.6463838	2.6980209
-0.8385187	-1.4247313	-0.3658611	0.3503318	-0.4486487
-0.5088043	-0.4946831	1.6234341	0.2546589	2.4208546
0.3037922	0.6768350	0.4536948	-1.7924919	0.6066890
-1.7942848	0.4442223	0.4966266	-0.6699702	1.4409008
-0.6882939	-0.7399529	0.3762947	0.1363805	1.1280737
-2.0233459	0.2408944	-0.8976313	0.2343804	0.4642067
0.4056954	0.3895826	-0.9434662	-1.5141935	-1.1815329
0.1908503	-1.1898594	-1.1392193	-0.0373545	-0.3145404
-1.2741013	1.0013762	-0.6346124	-1.1575928	-0.0774127
0.7857369	-0.1358907	-0.4812555	-0.7897149	0.6276677
0.7621276	1.5866451	-0.8398573	0.1814775	2.4762402
-0.1650293	0.3584015	-0.5596440	-0.4346442	-1.3133593
2.1809988	0.2855792	1.6446123	2.1288662	0.4697663
-0.4848774	0.0777141	-0.2296047	-0.0740021	-0.3119653
0.5099892	0.9237921	-1.3815041	1.3784475	1.1796780
1.2977581	-1.9466658	0.0772199	-1.6598749	0.8128654
0.0754772	-1.4145498	0.2025123	-0.2714291	-0.1025984

Figure 10.32 Two hundred and fifty normally distributed random numbers produced by the program of Figure 10.31

```
THE NUMBER OF NUMBERS WITHIN ONE STANDARD DEVIATION = 173.

THE NUMBER OF NUMBERS OUTSIDE ONE STANDARD DEVIATION =  77.


THE NUMBER OF NUMBERS WITHIN TWO STANDARD DEVIATION = 239.

THE NUMBER OF NUMBERS OUTSIDE TWO STANDARD DEVIATIONS =  11.



NUMBERS WITHIN ONE STANDARD DEVIATION

CHI-SQUARE =    0.165


NUMBERS WITHIN TWO STANDARD DEVIATIONS

CHI-SQUARE =    0.104
```

Figure 10.33 Chi-square results of the normally distributed random number generator of Figure 10.31

Similarly, 95.46 percent or 240 numbers should lie within two standard deviations of the mean value of zero. Figure 10.33 shows an excellent correlation in the ideal normal distribution of Figure 10.30 with the one degree of freedom-5 percent significance level of the χ^2 table value of Table 10.1.

Since the central limit theorem states that the sum of a large number of independent random numbers has a distribution that is approximately normal, it is interesting to note the excellent results that have been obtained with a value of $N = 20$. The reader is encouraged to investigate the results of using a smaller as well as a larger value of N.

10.5 THE MONTE CARLO METHOD

In Chapter 7 we discussed the design of a flip-flop that was based on using the maximum or minimum value for each circuit element so that they would produce the worst-case extreme. Since it is extremely unlikely that all circuit elements would have their worst value at the same time, this type of design is equivalent to guarding against the one-in-a-million possibility. By statistically defining the nominal value and the range (tolerance) of each element and using a random number generator that would randomly select a numerical value for each element, the flip-flop could be analyzed a sufficient number of times to assure proper operation for the specified tolerances. This type of design would lead to a much lower cost (a larger element tolerance is usually a less expensive element) that still would not result in a large number of rejections from quality control inspections.

It should be pointed out, however, that there may be cases where a Monte Carlo approach would not be suitable. For example, in the design of an electronic firing device (or a fail-safe device) for a missile, we should realize that although the missile will operate only one time, it must nevertheless work the first time (or all the time). Since quality control inspections can not include element-aging effects, for example, these must be considered, and the design should be based on worst-case extremes. In other words, there are certain situations where we would want to design for that one-in-a-million possibility.

In order to illustrate the Monte Carlo method and the use of a random number generator, consider the simple integration problem below.

Monte Carlo Integration

Assume we wish to calculate the area under the curve

$$Y = 0.25X + 0.5$$

between $X = 0$ and $X = 2$, as shown in Figure 10.34.[8]

The following method is based on generating uniformly distributed random numbers that have a range of 0 to 1. The first number generated is scaled so that it corresponds to a random selection of a value of X between 0 and 2. This value of X is then used to calculate Y from the equation $Y = 0.25X + 0.5$. This value of Y is then compared to a second random number that corresponds to a random selection of a value of Y between 0 and 1. If the calculated value of Y is equal to or greater than the random selected Y, one count is added to the count variable N. This process is repeated for a sufficient number of times, and the area is found by taking the ratio of the count of the number of times a random point was selected on or below the curve to the total number of iterations, times

[8] Note that we could also use this method to calculate the value of π from the equation $X^2 + Y^2 = 1$ and the knowledge that the area of a circle equals πR^2. It should be recognized, however, that this method would not lead to a very accurate value of π.

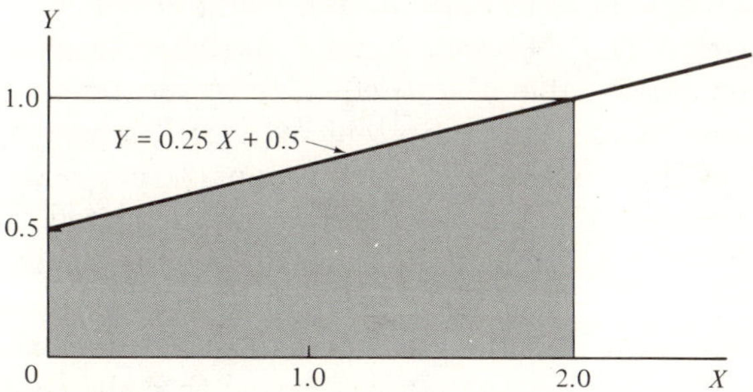

Figure 10.34 Finding the area under the curve $Y = 0.25X + 0.5$ between $X = 0$ and $X = 2$

Figure 10.35 A random number program for calculating the area under $Y = 0.25\,X + 0.5$ between $X = 0$ and $X = 2$

```
              C      RANDOM NUMBER CALCULATION
              C                OF
              C      AREA UNDER   Y = .25 X & .5
              C      BETWEEN   X = 0 AND 2
S.0001               READ(5,1)NR,IT,IR
S.0002             1 FORMAT(3I5)
S.0003               IX=200*IT&IR
S.0004               N=0
S.0005               DO 10 L=1,50
S.0006               IRN=IX*NR/10000
S.0007               NR=IX*NR-IRN*10000
S.0008               X=NR
S.0009               X=.0002*X
S.0010               Y=.25*X&.5
S.0011               IRN=IX*NR/10000
S.0012               NR=IX*NR-IRN*10000
S.0013               Y1=NR
S.0014               Y1=.0001*Y1
S.0015               IF(Y-Y1) 10,20,20
S.0016            20 N=N&1
S.0017            10 CONTINUE
S.0018               FLTN=N
S.0019               AREA=2.*FLTN/50.
S.0020               WRITE(6,30)AREA
S.0021            30 FORMAT(F8.4)
S.0022               RETURN
S.0023               END
```

the rectangle area value of 2. In other words, the method is based on the assumption that it is equally probable to select randomly any point within the rectangle area defined by $Y = 1$, $X = 2$, and the X and Y axis; thus, the probability of selecting points that are located on or below the curve in contrast to any point within the rectangle is in direct proportion to the area below the curve to that of the entire rectangle.

Figure 10.35 is the program that was used to calculate the area under this curve. Computer statements S.0001 through S.0003 are used to read in starting values and calculate the multiplying constant IX. Statement S.0004 sets the count variable N to zero. The DO 10 loop then randomly selects 50 points within the rectangle and determines the number that are located on or below the curve. Statements S.0006 through S.0008 use the power residue method of Figure 10.13 to find the first random number (four digits). Statement S.0009 scales the number so that we have a value of X between 0 and 2. Statement S.0010 uses this value of X to calculate the curve value of Y. Statements S.0011 through S.0014 are used to calculate a second random number that corresponds to a random selection of a value of Y between 0 and 1. Statement S.0015 determines whether the calculated value of Y is equal to or greater than the random selected Y value of the point. Statement S.0016 adds one count to the count variable N if the curve value of Y is equal to or greater than the random selected value of Y. This procedure is then repeated for an additional 49 random selected points and statement S.0019 is used to calculate the area.

The results of this program show that the area equals 1.4800. Since the exact answer is equal to 1.5000, an examination will reveal that the computer answer is a very accurate one in light of the fact that only 50 points were involved. For example, if only two points were used, the possibility of both points being below the curve, both points above the curve, or one point above and one point below, would lead to an area of 2, 0, or 1, respectively. The fact that 50 points were involved would correspond to $\frac{3}{4}$ of 50; in other words, 37.5 points should lie below the curve and 12.5 points above. Since the number of points must be an integer, this would correspond to 37 (or 38) points below the curve and 13 (or 12) points above the curve. Thus, assuming a perfect random selection, the area should have a value of $37*2/50 = 1.480$ (or $38*2/50 = 1.520$). The above analysis supplies additional evidence that this particular random number generator produces very uniformly distributed numbers.

Note that this simple Monte Carlo area example may be extended, using interpolation, to cover the case of more complex waves that are defined at only discrete points, and this, in turn, may be extended to calculate complex volumes that are defined in terms of discrete three-dimensional grid values.

Voltage Divider Design

For a simple electrical illustration of a Monte Carlo design, let us consider the voltage divider circuit shown in Figure 10.36. The problem is to design a voltage divider that will provide voltage divider ratios from 0.0 to 1.0 in 0.1 steps, maintaining a specified accuracy and a one megohm input resistance. To further simplify the problem, let us assume that the FET preamplifier has a negligible loading effect on the attenuator. It readily follows, then, that we require the following nominal resistor values:

NOMINAL VOLTAGE DIVIDER RATIO	R_1 NOMINAL (k-OHMS)	R_2 NOMINAL (k-OHMS)
0.0	0.0	1000.0
0.1	100.0	900.0
0.2	200.0	800.0
0.3	300.0	700.0
0.4	400.0	600.0
0.5	500.0	500.0
0.6	600.0	400.0
0.7	700.0	300.0
0.8	800.0	200.0
0.9	900.0	100.0
1.0	1000.0	0.0

Let us next consider what accuracy we want the voltage divider ratios to have. Since the nominal voltage divider ratio is given by

$$\mathrm{VRN} = \frac{R_1}{R_1 + R_2}$$

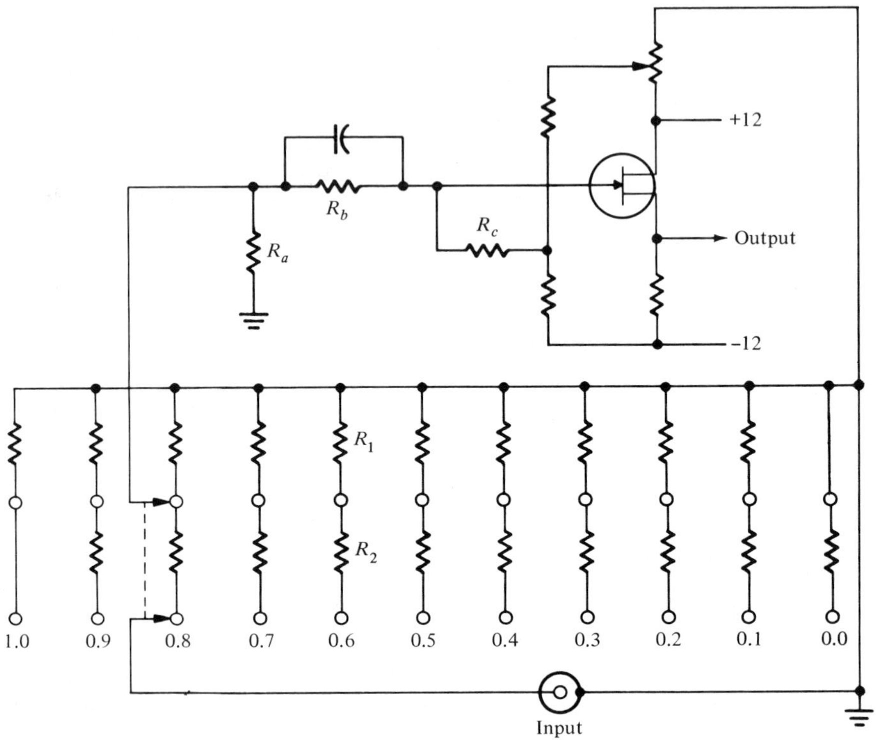

Figure 10.36 Attenuator-preamplifier

we may introduce the resistance tolerance RT and write

$$\text{VR} = \frac{R_1(1 \pm RT)}{R_1(1 \pm RT) + R_2(1 \pm RT)}$$

It thus follows that the voltage divider ratio will have its minimum value when R_1 has its minimum value and R_2 has its maximum value, or

$$\text{VRMIN} = \frac{R_1(1 - RT)}{R_1(1 - RT) + R_2(1 + RT)}$$

Similarly the voltage divider ratio will have its maximum value when R_1 has its maximum value and R_2 has its minimum value, or

$$\text{VRMAX} = \frac{R_1(1 + RT)}{R_1(1 + RT) + R_2(1 - RT)}$$

For a numerical example, consider the case of VRN $= 0.500$ where $R_1 = R_2 = 500$ kilohms. Since R_1 equals R_2 in this case, it thus follows that

$$VR = 0.500(1 \pm RT)$$

or the accuracy of the voltage divider ratio is equal to the accuracy of the resistors. Putting it another way, if we assume each resistor has a tolerance of $\pm 10\%$, the resistor values will range from 450 kilohms to 550 kilohms and therefore the worst-case extremes of the voltage divider ratio will be

$$\frac{450}{450 + 550} \leq \text{VR} \leq \frac{550}{550 + 450}$$

$$0.450 \leq \text{VR} \leq 0.550$$

which is $\equiv 10\%$ of the nominal value of 0.500. At this point we might conclude that if we want the voltage divider ratio to have an accuracy of $0.500 \pm 10\%$, the resistors should have a maximum tolerance of $\pm 10\%$. The reader should recognize that such a worst-case design expects that a value of $R_1 = 450$ kilohms will occur at the same time that $R_2 = 550$ kilohms or a value of $R_1 = 550$ kilohms will occur at the same time that $R_2 = 450$ kilohms. Since it is highly unlikely that we would obtain either extreme from a random selection of 500 kilohms $\pm$ 10% resistors, it follows from a production point of view that we are using resistors that are more accurate than they need be, and a Monte Carlo design should be used instead of a worst-case design.

Before we discuss the Monte Carlo design, we should realize that the overall problem we pose is really much more complicated than we might first suspect. For example, it does not follow that $\pm 10\%$ resistors will be accurate enough to provide a voltage divider ratio of $0.100 \pm 10\%$. In fact the accuracy of the resistors required will differ for each of the

```
C       VOLTAGE DIVIDER RATIO
C              VERSUS
C          RESISTANCE TOLERANCE
C
        WRITE(6,1)
      1 FORMAT('1WORST CASE VOLTAGE DIVIDER RATIO CALCULATIONS FOR A GIVEN
       1 RESISTANCE TOLERANCE',////,'  NOMINAL VR',3X,'MINIMUM VR',3X,'MAX
       2IMUM VR',3X,'PERCENT MIN',2X,'PERCENT MAX',3X,'R1 NOMINAL',3X,'R2
       3NOMINAL',//)
        RT=0.1
        R1=-1.E5
        R2=1.1E6
        DO 10 I=1,11
        R1=R1+1.E5
        R2=R2-1.E5
        VRN=R1/(R1+R2)
        VRMIN=R1*(1.-RT)/(R1*(1.-RT)+R2*(1.+RT))
        VRMAX=R1*(1.+RT)/(R1*(1.+RT)+R2*(1.-RT))
        PERMIN=(VRMIN-VRN)/VRN*100.
        PERMAX=(VRMAX-VRN)/VRN*100.
     10 WRITE(6,2) VRN,VRMIN,VRMAX,PERMIN,PERMAX,R1,R2
      2 FORMAT(F9.2,4F13.3,5X,3P2E13.4)
        RETURN
        END
```

Figure 10.37 A program that investigates the worst-case design of a voltage divider

```
WORST CASE VOLTAGE DIVIDER RATIO CALCULATIONS FOR A GIVEN RESISTANCE TOLERANCE


NOMINAL VR     MINIMUM VR     MAXIMUM VR     PERCENT MIN     PERCENT MAX     R1 NOMINAL     R2 NOMINAL


IHC210I PROGRAM INTERRUPT(P) OLD PSW IS      FF05000F8200C3E8

IHC210I PROGRAM INTERRUPT(P) OLD PSW IS      FF05000F8200C3FC
   0.0          0.0          0.0          0.0          0.0          0.0          100.0000E 04
   0.10         0.083        0.120      -16.667       19.565      100.0000E 03 900.0000E 03
   0.20         0.170        0.234      -15.094       17.021      200.0000E 03 800.0000E 03
   0.30         0.260        0.344      -13.461       14.583      300.0000E 03 700.0000E 03
   0.40         0.353        0.449      -11.765       12.245      400.0000E 03 600.0000E 03
   0.50         0.450        0.550      -10.000       10.000      500.0000E 03 500.0000E 03
   0.60         0.551        0.647       -8.163        7.843      600.0000E 03 400.0000E 03
   0.70         0.656        0.740       -6.250        5.769      700.0000E 03 300.0000E 03
   0.80         0.766        0.830       -4.255        3.774      800.0000E 03 200.0000E 03
   0.90         0.880        0.917       -2.174        1.852      900.0000E 03 100.0000E 03
   1.00         1.000        1.000        0.0          0.0       100.0000E 04   0.0
```

Figure 10.38 The output results of the program of Figure 10.37

voltage divider ratios. Before we go ahead then, let us discuss this part of the problem for completeness.

Figure 10.37 is a Fortran program that calculates the minimum and maximum voltage divider ratios which are based on a resistor tolerance of 10 percent. Figure 10.38 gives the corresponding results. Note that for $\pm 10\%$ resistors, we have an accuracy of approximately $\pm 18\%$ for a nominal voltage divider ratio of 0.100 and approximately $\pm 2\%$ for a nominal value of 0.900. We thus see that the accuracy required of the individual resistors is a function of the voltage divider ratio and the specified accuracy of the ratio, and each case must be treated individually. Note also that two PROGRAM INTERRUPTS occur in the calculation of the 0.000 voltage divider ratio. This is due to the division by zero (VRN = 0.0) in computer statements S.0012 and S.0013.

Let us now turn to the Monte Carlo design of the voltage divider and restrict our discussion to the 0.500 case. Figure 10.39 is a Fortran program that is based on the random selection of uniformly distributed resistors that have a $\pm 10\%$ tolerance. Initially the heading and subheadings are printed. Computer statements S.0004 through S.0012 are then used to generate 200 random numbers (see Figure 10.17). The DO 20 loop now computes and prints 100 random selections of R_1 and R_2 and the corresponding voltage divider ratio. This is repeated 100 times to establish statistically the effective range of the voltage divider ratio values that we might expect in a production run. In order to find the distribution, the DO 30 loop first sets each of the ten elements of the variable DIST(I) to zero. These elements will be used to store respectively the number of times a voltage ratio between 0.45 and 0.46, 0.46 and 0.47, . . . , or 0.54 and 0.55 occurs. The distribution is found by the DO 40 loop in which computer statement S.0022 determines the value of J from

$$J = VR(I)*100. - 44.$$

```
C      MONTE CARLO SOLUTION
C              OF
C      VOLTAGE DIVIDER DESIGN
C
C      RESISTANCE TOLERANCE 10 PERCENT
C             UNIFORMLY DISTRIBUTED
C
       DIMENSION RN(200),VR(100),DIST(10)
       WRITE(6,1)
     1 FORMAT('1MONTE CARLO DESIGN OF A VOLTAGE DIVIDER',///,' VOLTAGE DI
      1VIDER RATIO = 0.500',//,' NOMINAL VALUE OF R1 = 500K',/,' NOMINAL
      2VALUE OF R2 = 500K',//,' RESISTOR TOLERANCE = 10 PERCENT (UNIFORM
      4DISTRIBUTION)',///,5X,'R1',10X,'R2',10X,'VR',//)
C
C      GENERATION OF 200 RANDOM NUMBERS
C
       ISEED=123456789
       DO 10 I=1,200
       IRN=ISEED*65539
       IF(IRN) 2,3,3
     2 IRN=IRN+2147483647+1
     3 RN(I)=IRN
       ISEED=IRN
       RN(I)=RN(I)*.4656613E-9
    10 CONTINUE
C
C      THE RANDOM SELECTION OF R1 AND R2
C                  AND
C      THE CORRESPONDING VOLTAGE DIVIDER RATIO
C
       DO 20 I=1,100
       R1=1.0E5*RN(I)+4.5E5
       R2=1.0E5*RN(I+100)+4.5E5
       VR(I)=R1/(R1+R2)
    20 WRITE(6,4)R1,R2,VR(I)
     4 FORMAT(3E12.4)
C
C      THE CALCULATION OF THE VOLTAGE RATIO DISTRIBUTION
C
       DO 30 I=1,10
    30 DIST(I)=0.0
       DO 40 I=1,100
       J=VR(I)*100.-44.
    40 DIST(J)=DIST(J)+1.
       A=0.45
       B=0.46
       DO 50 I=1,10
       WRITE(6,5) A,B,DIST(I)
     5 FORMAT('0TOTAL NUMBER OF VOLTAGE DIVIDER RATIOS BETWEEN',F5.2,' AN
      1D',F5.2,' = ',F4.0)
       A=A+0.01
    50 B=B+0.01
       RETURN
       END
```

Figure 10.39 A Monte Carlo design of the voltage divider of Figure 10.37 in which uniformly distributed 10% resistors are used

For example, since the first random number is equal to 0.774605 (see Figure 10.18) and the 101st is equal to 0.7256459, R_1 will therefore be equal to

$$\begin{aligned} R_1 &= 1.0\text{E}5*\text{RN}(1)+4.5\text{E}5 \\ &= 1.0\text{E}5*0.774605+4.5\text{E}5 \\ &= 0.5275\text{E}6 \end{aligned}$$

and R_2 will be equal to

$$\begin{aligned} R_2 &= 1.0\text{E}5*\text{RN}(101)+4.5\text{E}5 \\ &= 1.0\text{E}5*0.7256459+4.5\text{E}5 \\ &= 0.5226\text{E}6 \end{aligned}$$

MONTE CARLO DESIGN OF A VOLTAGE DIVIDER

VOLTAGE DIVIDER RATIO = 0.500

NOMINAL VALUE OF R1 = 500K
NOMINAL VALUE OF R2 = 500K

RESISTOR TOLERANCE = 10 PERCENT (UNIFORM DISTRIBUTION)

R1	R2	VR
0.5275E 06	0.5226E 06	0.5023E 00
0.4631E 06	0.4610E 06	0.5011E 00
0.5312E 06	0.4629E 06	0.5343E 00
0.5195E 06	0.5285E 06	0.4957E 00
0.5362E 06	0.5049E 06	0.5150E 00
0.5422E 06	0.4727E 06	0.5342E 00
0.5272E 06	0.4919E 06	0.5173E 00
0.4836E 06	0.4969E 06	0.4932E 00
0.4564E 06	0.4542E 06	0.5012E 00
0.4860E 06	0.4532E 06	0.5175E 00
0.5086E 06	0.5310E 06	0.4892E 00
0.4780E 06	0.5074E 06	0.4851E 00
0.4899E 06	0.4651E 06	0.5130E 00
0.5380E 06	0.5237E 06	0.5068E 00
0.5188E 06	0.4566E 06	0.5319E 00
0.4702E 06	0.5264E 06	0.4718E 00
0.4521E 06	0.5492E 06	0.4515E 00
0.4814E 06	0.4570E 06	0.5130E 00
0.5191E 06	0.4996E 06	0.5096E 00
0.4923E 06	0.4844E 06	0.4989E 00
0.5217E 06	0.5103E 06	0.5056E 00
0.4900E 06	0.5018E 06	0.4940E 00
0.5442E 06	0.5184E 06	0.5122E 00
0.4555E 06	0.4941E 06	0.4797E 00
0.5348E 06	0.4993E 06	0.5172E 00
0.5095E 06	0.5482E 06	0.4817E 00
0.5433E 06	0.4960E 06	0.5228E 00
0.4747E 06	0.5419E 06	0.4670E 00
0.4585E 06	0.4874E 06	0.4848E 00
0.4788E 06	0.5475E 06	0.4665E 00
0.5458E 06	0.4984E 06	0.5227E 00
0.4658E 06	0.4635E 06	0.5013E 00
0.4831E 06	0.4949E 06	0.4940E 00
0.5061E 06	0.4978E 06	0.5041E 00
0.4889E 06	0.5331E 06	0.4784E 00
0.4783E 06	0.5165E 06	0.4798E 00
0.4693E 06	0.5126E 06	0.4779E 00
0.5114E 06	0.5095E 06	0.5009E 00
0.5446E 06	0.5432E 06	0.5006E 00
0.4651E 06	0.4739E 06	0.4953E 00
0.4888E 06	0.4544E 06	0.5183E 00
0.5470E 06	0.4610E 06	0.5427E 00
0.4827E 06	0.4767E 06	0.5032E 00
0.4733E 06	0.5111E 06	0.4808E 00
0.4950E 06	0.4765E 06	0.5095E 00
0.5106E 06	0.4592E 06	0.5265E 00
0.5086E 06	0.4668E 06	0.5215E 00
0.4566E 06	0.4675E 06	0.4941E 00
0.4617E 06	0.5039E 06	0.4782E 00
0.4614E 06	0.5162E 06	0.4720E 00
0.5128E 06	0.4621E 06	0.5260E 00
0.5238E 06	0.5261E 06	0.4989E 00
0.5279E 06	0.4983E 06	0.5144E 00
0.4534E 06	0.4546E 06	0.4993E 00
0.4691E 06	0.5428E 06	0.4636E 00
0.5344E 06	0.4654E 06	0.5345E 00
0.4843E 06	0.5072E 06	0.4885E 00
0.4965E 06	0.4544E 06	0.5221E 00
0.5205E 06	0.4618E 06	0.5299E 00
0.4544E 06	0.4812E 06	0.4856E 00
0.5416E 06	0.5309E 06	0.5050E 00
0.4605E 06	0.4541E 06	0.5035E 00
0.4882E 06	0.5470E 06	0.4716E 00
0.4850E 06	0.4946E 06	0.4951E 00
0.5161E 06	0.5449E 06	0.4865E 00
0.5314E 06	0.5178E 06	0.5065E 00
0.5432E 06	0.5028E 06	0.5193E 00
0.4765E 06	0.4567E 06	0.5106E 00
0.4707E 06	0.5153E 06	0.4774E 00
0.5354E 06	0.4811E 06	0.5267E 00
0.4762E 06	0.5490E 06	0.4645E 00
0.5392E 06	0.4643E 06	0.5373E 00
0.5489E 06	0.5446E 06	0.5020E 00
0.5410E 06	0.4891E 06	0.5252E 00
0.5057E 06	0.5331E 06	0.4868E 00
0.4652E 06	0.4973E 06	0.4833E 00
0.5400E 06	0.4855E 06	0.5266E 00
0.4531E 06	0.5374E 06	0.4574E 00
0.4585E 06	0.4549E 06	0.5019E 00
0.4729E 06	0.4924E 06	0.4899E 00
0.5110E 06	0.4604E 06	0.5261E 00
0.5104E 06	0.5306E 06	0.4903E 00
0.4632E 06	0.5397E 06	0.4618E 00
0.4856E 06	0.4635E 06	0.5117E 00
0.5450E 06	0.5232E 06	0.5102E 00
0.4996E 06	0.4680E 06	0.5163E 00
0.4927E 06	0.4992E 06	0.4967E 00
0.4593E 06	0.4825E 06	0.4877E 00
0.5221E 06	0.5026E 06	0.5095E 00
0.4986E 06	0.4731E 06	0.5131E 00
0.4928E 06	0.5152E 06	0.4889E 00
0.4694E 06	0.5332E 06	0.4682E 00
0.4809E 06	0.4623E 06	0.5099E 00
0.4607E 06	0.4746E 06	0.4926E 00
0.5361E 06	0.4869E 06	0.5241E 00
0.4705E 06	0.5499E 06	0.4611E 00
0.4979E 06	0.5174E 06	0.4904E 00
0.4529E 06	0.4554E 06	0.4987E 00
0.5363E 06	0.4753E 06	0.5302E 00
0.5416E 06	0.4535E 06	0.5442E 00

TOTAL NUMBER OF VOLTAGE DIVIDER RATIOS BETWEEN 0.45 AND 0.46 = 2.

TOTAL NUMBER OF VOLTAGE DIVIDER RATIOS BETWEEN 0.46 AND 0.47 = 7.

TOTAL NUMBER OF VOLTAGE DIVIDER RATIOS BETWEEN 0.47 AND 0.48 = 9.

TOTAL NUMBER OF VOLTAGE DIVIDER RATIOS BETWEEN 0.48 AND 0.49 = 13.

TOTAL NUMBER OF VOLTAGE DIVIDER RATIOS BETWEEN 0.49 AND 0.50 = 15.

TOTAL NUMBER OF VOLTAGE DIVIDER RATIOS BETWEEN 0.50 AND 0.51 = 19.

TOTAL NUMBER OF VOLTAGE DIVIDER RATIOS BETWEEN 0.51 AND 0.52 = 15.

TOTAL NUMBER OF VOLTAGE DIVIDER RATIOS BETWEEN 0.52 AND 0.53 = 12.

TOTAL NUMBER OF VOLTAGE DIVIDER RATIOS BETWEEN 0.53 AND 0.54 = 6.

TOTAL NUMBER OF VOLTAGE DIVIDER RATIOS BETWEEN 0.54 AND 0.55 = 2.

Figure 10.40 The output results of the program of Figure 10.39

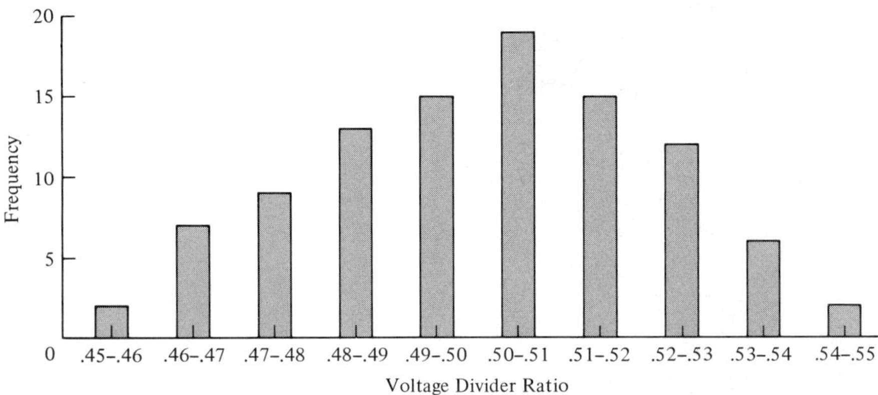

Figure 10.41 The distribution histogram for the program of Figure 10.39

This will produce a $VR(1)$ value of

$$\begin{aligned} VR(1) &= R1/(R1+R2) \\ &= 0.5275\text{E}6/(0.5275\text{E}6+0.5226\text{E}6) \\ &= 0.5023 \end{aligned}$$

and thus

$$\begin{aligned} \text{J} &= 0.5023*100.-44 \\ &= 6 \end{aligned}$$

In other words, since $VR(1) = 0.5023$, this corresponds to the sixth element of DIST(J); and computer statement S.0023 adds 1 to the count of this particular element of DIST. Finally, computer statements S.0024 through S.0030 are used to print the distribution.

Figure 10.40 gives the computer results and Figure 10.41 is the plot of the distribution. An examination of these two figures shows that there would be an 83 percent production yield if the accuracy of this voltage divider ratio was specified as $0.500 \pm 6\%$, even though 10-percent resistors are used.

Since the manufacture of large quantities of resistors will more nearly yield a normal distribution rather than a uniform distribution, the program of Figure 10.39 was modified and rerun. Figure 10.42 is the modified program. It is essentially identical to the previous program except for the following:

1. Computer statements S.0005 through S.0018 are used to produce 200 normally distributed random numbers. Note that the program of Figure 10.31 has been modified to conserve computer memory. Each iteration of the DO 10 loop is used to generate a single normally distributed number and in the execution of this loop, the DO 15 loop is used to supply the required 20 uniformly distributed num-

```
               C      MONTE CARLO SOLUTION
               C                 OF
               C      VOLTAGE DIVIDER DESIGN
               C
               C      RESISTANCE TOLERANCE 10 PERCENT
               C              NORMAL DISTRIBUTION
               C
S.0001                DIMENSION    RN(20),RNORM(200),VR(100),DIST(10)
S.0002                WRITE(6,1)
S.0003              1 FORMAT('1MONTE CARLO DESIGN OF A VOLTAGE DIVIDER',///,' VOLTAGE DI
                     1VIDER RATIO = 0.500',//,' NOMINAL VALUE OF R1 = 500K',/,' NOMINAL
                     2VALUE OF R2 = 500K',//,' RESISTOR TOLERANCE = 10 PERCENT (NORMALLY
                     4 DISTRIBUTED)',///,5X,'R1',10X,'R2',10X,'VR',//)
S.0004                ISEED=123456789
S.0005                DENOM=SQRT(20./12.)
               C
               C      GENERATION OF 200 NORMALLY DISTRIBUTED NUMBERS
               C
S.0006                DO 10 I=1,200
               C
               C      GENERATION OF 20 UNIFORMLY DISTRIBUTED RANDOM NUMBERS
               C
S.0007                DO 15 K=1,20
S.0008                IRN=ISEED*65539
S.0009                IF(IRN)2,3,3
S.0010              2 IRN=IRN+2147483647+1
S.0011              3 RN(K)=IRN
S.0012                ISEED=IRN
S.0013                RN(K)=RN(K)*.4656613E-9
S.0014             15 CONTINUE
               C
               C      GENERATION OF A NORMALLY DISTRIBUTED RANDOM NUMBER
               C
S.0015                SUM=0.0
S.0016                DO 16 J=1,20
S.0017             16 SUM=SUM+RN(J)
S.0018             10 RNORM(I)=(SUM-10.)/DENOM
               C
               C      THE RANDOM SELECTION OF R1 AND R2
               C                       AND
               C      THE CORRESPONDING VOLTAGE DIVIDER RATIO
               C
S.0019                DO 20 I=1,100
S.0020                R1=5.E5+RNORM(I)*5.E4/3.
S.0021                R2=5.E5+RNORM(I+100)*5.E4/3.
S.0022                VR(I)=R1/(R1+R2)
S.0023             20 WRITE(6,4)R1,R2,VR(I)
S.0024              4 FORMAT(3E12.4)
               C
               C      THE CALCULATION OF THE VOLTAGE RATIO DISTRIBUTION
               C
S.0025                DO 30 I=1,10
S.0026             30 DIST(I)=0.0
S.0027                DO 40 I=1,100
S.0028                J=VR(I)*100.-44.
S.0029             40 DIST(J)=DIST(J)+1.
S.0030                A=0.45
S.0031                B=0.46
S.0032                DO 50 I=1,10
S.0033                WRITE(6,5)A,B,DIST(I)
S.0034              5 FORMAT('0TOTAL NUMBER OF VOLTAGE DIVIDER RATIOS BETWEEN',F5.2,' AN
                     1D',F5.2,' = ',F6.0)
S.0035                A=A+0.01
S.0036             50 B=B+0.01
S.0037                RETURN
S.0038                END
```

Figure 10.42 A Monte Carlo design of the voltage divider of Figure 10.36 in which normally distributed 10% resistors are used

bers. This changes the dimension requirements for *RN* from 4000 to 20 in the generation of 200 normally distributed numbers.

2. Computer statements S.0019 through S.0024 calculate and print 100 random selections of R_1 and R_2 and the corresponding voltage divider ratio. The resistors are randomly computed on the basis that $\pm 3\sigma$ variation will fall within $\pm 10\%$ tolerance (see Figure 10.30 and note that $\pm 3\sigma$ variation will mean that 99.74 percent of all

MONTE CARLO DESIGN OF A VOLTAGE DIVIDER

VOLTAGE DIVIDER RATIO = 0.500

NOMINAL VALUE OF R1 = 500K
NOMINAL VALUE OF R2 = 500K

RESISTOR TOLERANCE = 10 PERCENT (NORMALLY DISTRIBUTED)

R1	R2	VR
0.5014E 06	0.5016E 06	0.4999E 00
0.4961E 06	0.5396E 06	0.4790E 00
0.4823E 06	0.5094E 06	0.4863E 00
0.5064E 06	0.5086E 06	0.4989E 00
0.4905E 06	0.4723E 06	0.5095E 00
0.4806E 06	0.4681E 06	0.5066E 00
0.5246E 06	0.4692E 06	0.5278E 00
0.4545E 06	0.4939E 06	0.4792E 00
0.5120E 06	0.4794E 06	0.5165E 00
0.4828E 06	0.4972E 06	0.4927E 00
0.5276E 06	0.5097E 06	0.5087E 00
0.4955E 06	0.4932E 06	0.5012E 00
0.5079E 06	0.4771E 06	0.5157E 00
0.5029E 06	0.4971E 06	0.5029E 00
0.5204E 06	0.5168E 06	0.5017E 00
0.4904E 06	0.5087E 06	0.4908E 00
0.5166E 06	0.5158E 06	0.5004E 00
0.5078E 06	0.4728E 06	0.5179E 00
0.4857E 06	0.4985E 06	0.4935E 00
0.4864E 06	0.4964E 06	0.4949E 00
0.4786E 06	0.4835E 06	0.4974E 00
0.4831E 06	0.4993E 06	0.4917E 00
0.4898E 06	0.4685E 06	0.5111E 00
0.4848E 06	0.4698E 06	0.5078E 00
0.5079E 06	0.4549E 06	0.5276E 00
0.4716E 06	0.5155E 06	0.4778E 00
0.4881E 06	0.4847E 06	0.5017E 00
0.4926E 06	0.5125E 06	0.4901E 00
0.5178E 06	0.5101E 06	0.5038E 00
0.5090E 06	0.4915E 06	0.5087E 00
0.5062E 06	0.4946E 06	0.5058E 00
0.4836E 06	0.4942E 06	0.4946E 00
0.4657E 06	0.5283E 06	0.4686E 00
0.5074E 06	0.4937E 06	0.5068E 00
0.5110E 06	0.5158E 06	0.4977E 00
0.4760E 06	0.4840E 06	0.4959E 00
0.4698E 06	0.5106E 06	0.4792E 00
0.5008E 06	0.5165E 06	0.4923E 00
0.4750E 06	0.5199E 06	0.4775E 00
0.5197E 06	0.5179E 06	0.5009E 00
0.5117E 06	0.5026E 06	0.5045E 00
0.5029E 06	0.4824E 06	0.5104E 00
0.5054E 06	0.4971E 06	0.5041E 00
0.4892E 06	0.5085E 06	0.4903E 00
0.4885E 06	0.4907E 06	0.4989E 00
0.4816E 06	0.5264E 06	0.4778E 00
0.5084E 06	0.5079E 06	0.5002E 00
0.5229E 06	0.4901E 06	0.5162E 00
0.5094E 06	0.4931E 06	0.5081E 00
0.5007E 06	0.4919E 06	0.5044E 00
0.4992E 06	0.5280E 06	0.4860E 00
0.4810E 06	0.4899E 06	0.4954E 00
0.5085E 06	0.4946E 06	0.5069E 00
0.5006E 06	0.5123E 06	0.4942E 00
0.4730E 06	0.4740E 06	0.4995E 00
0.5133E 06	0.5147E 06	0.4993E 00
0.5031E 06	0.4989E 06	0.5021E 00
0.4918E 06	0.5198E 06	0.4862E 00
0.4949E 06	0.5138E 06	0.4906E 00
0.5168E 06	0.4998E 06	0.5084E 00
0.5031E 06	0.5021E 06	0.5005E 00
0.5165E 06	0.5040E 06	0.5061E 00
0.5067E 06	0.4592E 06	0.5246E 00
0.5108E 06	0.4892E 06	0.5108E 00
0.4983E 06	0.5450E 06	0.4776E 00
0.5061E 06	0.4860E 06	0.5101E 00
0.4947E 06	0.4763E 06	0.5095E 00
0.5069E 06	0.4939E 06	0.5065E 00
0.5100E 06	0.5058E 06	0.5021E 00
0.4882E 06	0.4925E 06	0.4978E 00
0.5022E 06	0.4915E 06	0.5054E 00
0.4840E 06	0.4918E 06	0.4960E 00
0.4877E 06	0.5271E 06	0.4806E 00
0.5087E 06	0.5042E 06	0.5022E 00
0.5010E 06	0.5403E 06	0.4811E 00
0.4779E 06	0.5051E 06	0.4862E 00
0.5285E 06	0.5113E 06	0.5083E 00
0.5163E 06	0.5076E 06	0.5042E 00
0.4811E 06	0.4701E 06	0.5058E 00
0.5174E 06	0.5101E 06	0.5036E 00
0.4898E 06	0.4701E 06	0.5102E 00
0.4907E 06	0.5074E 06	0.4916E 00
0.5031E 06	0.5083E 06	0.4974E 00
0.5207E 06	0.4888E 06	0.5158E 00
0.4800E 06	0.5240E 06	0.4781E 00
0.4950E 06	0.4885E 06	0.5033E 00
0.4910E 06	0.4877E 06	0.5017E 00
0.5021E 06	0.5063E 06	0.4979E 00
0.4842E 06	0.5023E 06	0.4908E 00
0.5075E 06	0.5188E 06	0.4945E 00
0.4905E 06	0.4663E 06	0.5127E 00
0.5071E 06	0.5040E 06	0.5015E 00
0.5058E 06	0.4850E 06	0.5105E 00
0.4864E 06	0.5039E 06	0.4912E 00
0.4888E 06	0.5077E 06	0.4905E 00
0.4709E 06	0.5068E 06	0.4817E 00
0.5023E 06	0.5065E 06	0.4979E 00
0.4987E 06	0.4843E 06	0.5073E 00
0.4914E 06	0.4748E 06	0.5086E 00
0.5151E 06	0.4803E 06	0.5175E 00

TOTAL NUMBER OF VOLTAGE DIVIDER RATIOS BETWEEN 0.45 AND 0.46 = 0.

TOTAL NUMBER OF VOLTAGE DIVIDER RATIOS BETWEEN 0.46 AND 0.47 = 1.

TOTAL NUMBER OF VOLTAGE DIVIDER RATIOS BETWEEN 0.47 AND 0.48 = 8.

TOTAL NUMBER OF VOLTAGE DIVIDER RATIOS BETWEEN 0.48 AND 0.49 = 7.

TOTAL NUMBER OF VOLTAGE DIVIDER RATIOS BETWEEN 0.49 AND 0.50 = 30.

TOTAL NUMBER OF VOLTAGE DIVIDER RATIOS BETWEEN 0.50 AND 0.51 = 38.

TOTAL NUMBER OF VOLTAGE DIVIDER RATIOS BETWEEN 0.51 AND 0.52 = 13.

TOTAL NUMBER OF VOLTAGE DIVIDER RATIOS BETWEEN 0.52 AND 0.53 = 3.

TOTAL NUMBER OF VOLTAGE DIVIDER RATIOS BETWEEN 0.53 AND 0.54 = 0.

TOTAL NUMBER OF VOLTAGE DIVIDER RATIOS BETWEEN 0.54 AND 0.55 = 0.

Figure 10.43 The output results of the program of Figure 10.42

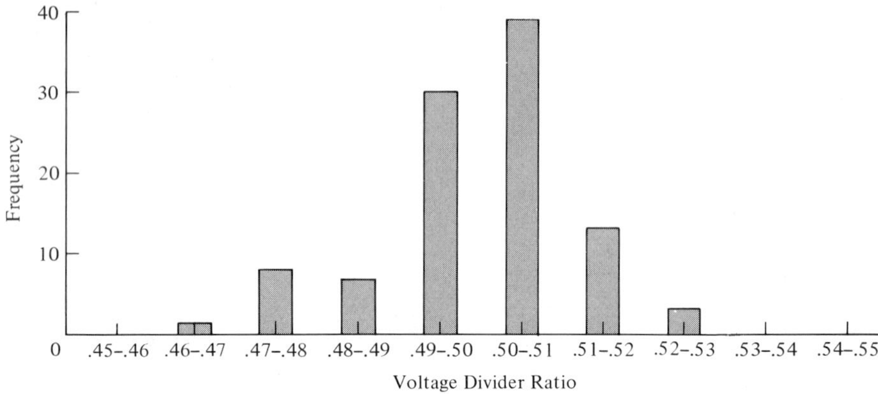

Figure 10.44 The distribution histogram for the program of Figure 10.43

resistors computed will be normally distributed and fall within $\pm 10\%$ of 500 kilohms).

Figure 10.43 gives the computer results and Figure 10.44 is the plot of the distribution. Note that in this case, for $\pm 10\%$ normally distributed resistors, there would be a 99% production yield if the accuracy of this divider ratio was specified as $0.500 \pm 6\%$. Note also that an 88 percent production yield would result if the specification was tightened to $0.500 \pm 4\%$. We thus see that in the production of this type of voltage divider it would be to our advantage, economically speaking, to use a Monte Carlo type of design rather than a worst-case type of design.

Digital-to-Analog Resistive Ladder

Let us again consider the binary $R-2R$ resistive ladder of Figure 3.31 and investigate the accuracy requirements of the ladder resistors. Generally the accuracy of the digital-to-analog conversion is specified in terms of the least significant bit (LSB). If we again assume that the logical "1" and "0" levels will be 10 volts and 0 volts, respectively, the least significant bit contribution to the output voltage will be

$$\frac{1}{1024}10 = 0.009765 \text{ volts}$$

If we thus specify a $\pm\frac{1}{3}$ LSB accuracy, this means that any one of the 1024 input combinations must not deviate more than

$$\pm\frac{1}{3}0.009765 = \pm 0.00325 \text{ volts}$$

from its nominal value.

```
              C        DIGITAL TO ANALOG
              C        RESISTIVE LADDER
              C          DESIGN
              C
S.0001                 DIMENSION R(20),V(10),RIN(10),ROUT(10),AN(10),AJ(10),CODE(10),
                      1VO(6),ERROR(6)
S.0002                 WRITE(6,1)
S.0003               1 FORMAT('1DIGITAL TO ANALOG RESISTIVE LADDER',///)
S.0004                 DO 40 I=1,11
S.0005                 READ(5,2) (CODE(M),M=1,10),VNOM
S.0006               2 FORMAT(10A1,F10.6)
S.0007                 READ(5,3) (V(M),M=1,10)
S.0008               3 FORMAT(5F10.2)
S.0009                 RT=0.0
S.0010                 DO 50 J=1,6
S.0011                 RT=RT+0.001
S.0012                 R(20)=2.0E4*(1.+RT)
S.0013                 DO 60 K=1,10
S.0014              60 R(K)=2.0E4*(1.+RT)
S.0015                 DO 70 L=11,19
S.0016              70 R(L)=1.0E4*(1.-RT)
S.0017                 RIN(10)=R(20)
S.0018                 DO 10 M=1,9
S.0019                 K=11-M
S.0020                 AN(K)=R(K)*RIN(K)/(R(K)+RIN(K))
S.0021              10 RIN(K-1)=R(K+9)+AN(K)
S.0022                 AN(1)=R(1)*RIN(1)/(R(1)+RIN(1))
S.0023                 AJ(1)=R(1)
S.0024                 ROUT(1)=R(1)
S.0025                 DO 20 M=2,10
S.0026                 ROUT(M)=R(M+9)+AJ(M-1)
S.0027              20 AJ(M)=R(M)*ROUT(M)/(R(M)+ROUT(M))
S.0028                 RATIO=1.
S.0029                 VOUT=RIN(1)/(RIN(1)+R(1))*V(1)
S.0030                 DO 30 N=2,10
S.0031                 A=ROUT(N)*RIN(N)/(ROUT(N)+RIN(N))
S.0032                 RATIO=RATIO*AJ(N-1)/(R(N+9)+AJ(N-1))
S.0033              30 VOUT=VOUT+A/(A+R(N))*V(N)*RATIO
S.0034                 ERROR(J)=VNOM-VOUT
S.0035              50 VO(J)=VOUT
S.0036              40 WRITE(6,4) (CODE(JJ),JJ=1,10),VNOM,(VO(JJ),JJ=1,6),(ERROR(JJ),JJ=1,
                      16)
S.0037               4 FORMAT(' DIGITAL CODE =',10A1,'   NOMINAL VOLTAGE =',F10.6,//,25X,'
                      1PERCENT TOLERANCE ',/,12X,'0.10',6X,'0.20',6X,'0.30',6X,'0.40',6X,
                      2'0.50',6X,'0.60',//,' VOLTAGE ',6F10.6,/,'    ERROR ',6F10.6,///)
S.0038                 RETURN
S.0039                 END
```

Figure 10.45 A Fortran program that computes the worst-case error of the digital-to-analog resistive ladder, as a function of resistor tolerance

In order to determine the ladder resistor tolerance required to obtain this type of accuracy, let us first consider the error produced for a worst-case design. A little thought will show that worst-case operation will result when all the $2R$ resistors are above their nominal value, and all R resistors are below their nominal value or vice versa. The program of Figure 3.33 was therefore modified to determine the errors produced by using various resistor tolerances. Figure 10.45 is a Fortran program that allows us to compute the errors for 11 different digital inputs using worst-case resistor tolerances of 0.05, 0.06, 0.07, 0.08, 0.09, and 0.10 percent. Computer statements S.0017 through S.0033 are essentially identical to computer statements S.0008 through S.0024 in the program of Figure 3.33. The three main additions are as follows:

1. The DO 40 loop that permits the investigation of 11 different digital inputs.

```
DIGITAL TO ANALOG RESISTIVE LADDER


DIGITAL CODE =C000000001  NOMINAL VOLTAGE =   0.009766

                              PERCENT TOLERANCE
              0.05      0.06      0.07      0.08      0.09      0.10

VOLTAGE   0.009794  0.009799  0.009805  0.009811  0.009816  0.009822
  ERROR  -0.00C028 -0.000033 -0.000039 -0.000045 -0.000050 -0.000056


DIGITAL CODE =C000000010  NOMINAL VOLTAGE =   0.019531

                              PERCENT TOLERANCE
              0.05      0.06      0.07      0.08      0.09      0.10

VOLTAGE   0.019578  0.019587  0.019597  0.019606  0.019615  0.019625
  ERROR  -0.00C047 -0.000056 -0.000066 -0.000075 -0.000084 -0.000094


DIGITAL CODE =C000000100  NOMINAL VOLTAGE =   0.039062

                              PERCENT TOLERANCE
              0.05      0.06      0.07      0.08      0.09      0.10

VOLTAGE   0.039141  0.039157  0.039173  0.039188  0.039204  0.039220
  ERROR  -0.00C079 -0.000095 -0.000111 -0.000126 -0.000142 -0.000158


DIGITAL CODE =CC00001000  NOMINAL VOLTAGE =   0.078125

                              PERCENT TOLERANCE
              0.05      0.06      0.07      0.08      0.09      0.10

VOLTAGE   0.078255  0.078281  0.078308  0.078334  0.078360  0.078386
  ERROR  -0.000130 -0.000156 -0.000183 -0.000209 -0.000235 -0.000261


DIGITAL CODE =C000010000  NOMINAL VOLTAGE =   0.156250

                              PERCENT TOLERANCE
              0.05      0.06      0.07      0.08      0.09      0.10

VOLTAGE   0.156458  0.156500  0.156542  0.156583  0.156625  0.156667
  ERROR  -0.000208 -0.000250 -0.000292 -0.000333 -0.000375 -0.000417


DIGITAL CODE =C000100000  NOMINAL VOLTAGE =   0.312500

                              PERCENT TOLERANCE
              0.05      0.06      0.07      0.08      0.09      0.10

VOLTAGE   0.312P12  0.312874  0.312937  0.312999  0.313062  0.313124
  ERROR  -0.000312 -0.000374 -0.000437 -0.000499 -0.000562 -0.000624


DIGITAL CODE =C001000000  NOMINAL VOLTAGE =   0.625000

                              PERCENT TOLERANCE
              0.05      0.06      0.07      0.08      0.09      0.10

VOLTAGE   0.625415  0.625499  0.625582  0.625665  0.625749  0.625832
  ERROR  -0.000415 -0.C00499 -0.000582 -0.000665 -0.000749 -0.000832
```

Figure 10.46 The output results of the program of Figure 10.45 for 0.05, 0.06, 0.07, 0.08, 0.09, and 0.10% resistor tolerance

```
DIGITAL CODE =0010000000  NOMINAL VOLTAGE =  1.250000

                              PERCENT TOLERANCE
           0.05       0.06       0.07       0.08       0.09       0.10

VOLTAGE   1.250415   1.250498   1.250581   1.250664   1.250748   1.250831
  ERROR  -0.000415  -0.000498  -0.000581  -0.000664  -0.000748  -0.000831


DIGITAL CODE =0100000000  NOMINAL VOLTAGE =  2.500000

                              PERCENT TOLERANCE
           0.05       0.06       0.07       0.08       0.09       0.10

VOLTAGE   2.499998   2.499998   2.499998   2.499997   2.499997   2.499997
  ERROR   0.000002   0.000002   0.000002   0.000003   0.000003   0.000003


DIGITAL CODE =1000000000  NOMINAL VOLTAGE =  5.000000

                              PERCENT TOLERANCE
           0.05       0.06       0.07       0.08       0.09       0.10

VOLTAGE   4.998332   4.997998   4.997665   4.997334   4.997001   4.996667
  ERROR   0.001668   0.002002   0.002335   0.002666   0.002999   0.003333


DIGITAL CODE =1111111111  NOMINAL VOLTAGE =  9.990233

                              PERCENT TOLERANCE
           0.05       0.06       0.07       0.08       0.09       0.10

VOLTAGE   9.990194   9.990190   9.[illegible]90184   9.990178   9.990175   9.990168
  ERROR   0.000039   0.000044   0.0[illegible]0050   0.000055   0.000058   0.000066
```

Figure 10.46 Continued

2. The DO 50 loop that permits the incrementing of the resistor tolerance from 0.05 to 0.10 percent in 0.01 percent steps.
3. The DO 60 and the DO 70 loops which compute the worst-case extremes of the ladder resistors.

Figure 10.46 shows the output results. Note that for the eleven digital inputs considered, only one result has an error greater than 0.00325 volts. This one exception corresponds to the case of a nominal input voltage of 5.00 volts and a resistor tolerance of 0.10 percent.

Figure 10.47 shows the corresponding results for higher resistor tolerances. Note that although there are more cases where the LSB accuracy is not obtained, it is difficult to pinpoint which inputs are producing the error. The results of this figure seem to indicate that the resistor tolerance of the more significant bits should be more tightly controlled.

In order to investigate this conclusion and in light of the fact that only 11 out of 1024 inputs were considered, the program of Figure 10.48 was developed. This program is a binary voltage generator that will allow us to investigate all 1024 digital inputs. Since the program is a brute-force approach and in addition, is quite lengthy, we will avoid a detailed discussion. The reader should experience no difficulty in following this program. Figure 10.49 shows part of the output. Note the systematic generation of each of the voltage inputs.

```
DIGITAL TO ANALOG RESISTIVE LADDER


DIGITAL CODE =0000000001   NOMINAL VOLTAGE =   0.009766

                          PERCENT TOLERANCE
            0.10       0.20       0.30       0.40       0.50       0.60

VOLTAGE    0.009822   0.009879   0.009936   0.009993   0.010051   0.010109
  ERROR   -0.000056  -0.000113  -0.000170  -0.000227  -0.000285  -0.000343


DIGITAL CODE =0000000010   NOMINAL VOLTAGE =   0.019531

                          PERCENT TOLERANCE
            0.10       0.20       0.30       0.40       0.50       0.60

VOLTAGE    0.019625   0.019718   0.019813   0.019907   0.020002   0.020097
  ERROR   -0.000094  -0.000187  -0.000282  -0.000376  -0.000471  -0.000566


DIGITAL CODE =0000000100   NOMINAL VOLTAGE =   0.039062

                          PERCENT TOLERANCE
            0.10       0.20       0.30       0.40       0.50       0.60

VOLTAGE    0.039220   0.039378   0.039536   0.039695   0.039855   0.040015
  ERROR   -0.000158  -0.000316  -0.000474  -0.000633  -0.000793  -0.000953


DIGITAL CODE =0000001000   NOMINAL VOLTAGE =   0.078125

                          PERCENT TOLERANCE
            0.10       0.20       0.30       0.40       0.50       0.60

VOLTAGE    0.078386   0.078648   0.078910   0.079173   0.079437   0.079701
  ERROR   -0.000261  -0.000523  -0.000785  -0.001048  -0.001312  -0.001576


DIGITAL CODE =0000010000   NOMINAL VOLTAGE =   0.156250

                          PERCENT TOLERANCE
            0.10       0.20       0.30       0.40       0.50       0.60

VOLTAGE    0.156667   0.157085   0.157503   0.157922   0.158341   0.158762
  ERROR   -0.000417  -0.000835  -0.001253  -0.001672  -0.002091  -0.002512


DIGITAL CODE =0000100000   NOMINAL VOLTAGE =   0.312500

                          PERCENT TOLERANCE
            0.10       0.20       0.30       0.40       0.50       0.60

VOLTAGE    0.313124   0.313750   0.314376   0.315003   0.315630   0.316257
  ERROR   -0.000624  -0.001250  -0.001876  -0.002503  -0.003130  -0.003757


DIGITAL CODE =0001000000   NOMINAL VOLTAGE =   0.625000

                          PERCENT TOLERANCE
            0.10       0.20       0.30       0.40       0.50       0.60

VOLTAGE    0.625832   0.626665   0.627498   0.628331   0.629163   0.629996
  ERROR   -0.000832  -0.001665  -0.002498  -0.003331  -0.004163  -0.004996
```

Figure 10.47 The output results of the program of Figure 10.45 for 0.10, 0.20, 0.30, 0.40, 0.50, and 0.60% resistor tolerance

```
DIGITAL CODE =0010000000  NOMINAL VOLTAGE =  1.250000

                          PERCENT TOLERANCE
            0.10      0.20      0.30      0.40      0.50      0.60

VOLTAGE   1.250831  1.251662  1.252492  1.253321  1.254148  1.254975
  ERROR  -0.000831 -0.001662 -0.002492 -0.003321 -0.004148 -0.004975


DIGITAL CODE =0100000000  NOMINAL VOLTAGE =  2.500000

                          PERCENT TOLERANCE
            0.10      0.20      0.30      0.40      0.50      0.60

VOLTAGE   2.499997  2.499993  2.499989  2.499980  2.499970  2.499958
  ERROR   0.000003  0.000007  0.000011  0.000020  0.000030  0.000042


DIGITAL CODE =1000000000  NOMINAL VOLTAGE =  5.000000

                          PERCENT TOLERANCE
            0.10      0.20      0.30      0.40      0.50      0.60

VOLTAGE   4.996667  4.993334  4.990004  4.986672  4.983343  4.980012
  ERROR   0.003333  0.006666  0.009996  0.013328  0.016657  0.019988


DIGITAL CODE =1111111111  NOMINAL VOLTAGE =  9.990233

                          PERCENT TOLERANCE
            0.10      0.20      0.30      0.40      0.50      0.60

VOLTAGE   9.990168  9.990109  9.990053  9.989995  9.989938  9.989880
  ERROR   0.000066  0.000124  0.000180  0.000238  0.000296  0.000354
```

Figure 10.47 Continued

Figure 10.50 is a Fortran program that incorporates the binary voltage generator of Figure 10.48. This program modifies the program of Figure 10.45 so that all binary inputs are considered. Whenever the LSB accuracy is exceeded, the digital input voltages are printed. Figure 10.51 is the corresponding output. These results show that the LSB accuracy is achieved if all the resistors are accurate to $\pm 0.09\%$. For a tolerance of 0.10%, six of the 1024 possible inputs produce an error greater than 0.00325 volts.

If we now modify the program of Figure 10.50 so that we can investigate the $\pm 0.20\%$ tolerance, we find that 523 of the 1024 possible inputs produce an error greater than 0.00325 volts. Since the first digital input that produces this excessive error is

$$0001011011$$

it now readily follows that the more significant bits should be more tightly controlled. To explore this possibility still further, the program of Figure 10.50 was modified to produce the program of Figure 10.52. Note that computer statements S.0157 through S.0165 hold the first four bits to a $\pm 0.05\%$ resistor tolerance and the remaining six bits to a $\pm 1.0\%$ resistor tolerance. Figure 10.53 shows that there are only six inputs that produce an error in excess of 0.00325 volts and the first input to do so is 1000100000.

```
C     BINARY VOLTAGE GENERATOR
C
      DIMENSION V(10)
      I2=0
      I3=0
      I4=0
      I5=0
      I6=0
      I7=0
      I8=0
      I9=0
      I10=0
      II2=0
      II3=0
      II4=0
      II5=0
      II6=0
      II7=0
      II8=0
      II9=0
      II10=0
      DO 100 I=1,1024
      I1=I-I/2*2
      IF(I1)5,5,6
    5 V(10)=0.0
      GO TO 7
    6 V(10)=10.0
    7 IF(I-2)8,9,9
    8 V(9)=0.0
      GO TO 10
    9 I2=I2+1
      IF(I2-2)11,11,12
   11 V(9)=10.0
      GO TO 10
   12 II2=II2+1
      IF(II2-2)13,14,14
   13 V(9)=0.0
      GO TO 10
   14 V(9)=0.0
      I2=0
      II2=0
   10 IF(I-4)15,16,16
   15 V(8)=0.0
      GO TO 17
   16 I3=I3+1
      IF(I3-4)18,18,19
   18 V(8)=10.0
      GO TO 17
   19 II3=II3+1
      IF(II3-4)20,21,21
   20 V(8)=0.0
      GO TO 17
   21 V(8)=0.0
      I3=0
      II3=0
   17 IF(I-8)22,23,23
   22 V(7)=0.0
      GO TO 24
   23 I4=I4+1
      IF(I4-8)25,25,26
   25 V(7)=10.0
      GO TO 24
   26 II4=II4+1
      IF(II4-8)27,28,28
   27 V(7)=0.0
      GO TO 24
   28 V(7)=0.0
      I4=0
      II4=0
   24 IF(I-16)29,30,30
   29 V(6)=0.0
      GO TO 31
   30 I5=I5+1
      IF(I5-16)32,32,33
   32 V(6)=10.0
      GO TO 31
   33 II5=II5+1
      IF(II5-16)34,35,35
   34 V(6)=0.0
      GO TO 31
   35 V(6)=0.0
      I5=0
      II5=0
   31 IF(I-32)36,37,37
   36 V(5)=0.0
      GO TO 38
   37 I6=I6+1
      IF(I6-32)39,39,40
   39 V(5)=10.0
      GO TO 38
   40 II6=II6+1
      IF(II6-32)41,42,42
   41 V(5)=0.0
      GO TO 38
   42 V(5)=0.0
      I6=0
      II6=0
   38 IF(I-64)43,44,44
   43 V(4)=0.0
      GO TO 45
   44 I7=I7+1
      IF(I7-64)46,46,47
   46 V(4)=10.0
      GO TO 45
   47 II7=II7+1
      IF(II7-64)48,49,49
   48 V(4)=0.0
      GO TO 45
   49 V(4)=0.0
      I7=0
      II7=0
   45 IF(I-128)50,51,51
   50 V(3)=0.0
      GO TO 52
   51 I8=I8+1
      IF(I8-128)53,53,54
   53 V(3)=10.0
      GO TO 52
   54 II8=II8+1
      IF(II8-128)55,56,56
   55 V(3)=0.0
      GO TO 52
   56 V(3)=0.0
      I8=0
      II8=0
   52 IF(I-256)57,58,58
   57 V(2)=0.0
      GO TO 59
   58 I9=I9+1
      IF(I9-256)60,60,61
   60 V(2)=10.0
      GO TO 59
   61 II9=II9+1
      IF(II9-256)62,63,63
   62 V(2)=0.0
      GO TO 59
   63 V(2)=0.0
      I9=0
      II9=0
   59 IF(I-512)64,65,65
   64 V(1)=0.0
      GO TO 66
   65 I10=I10+1
      IF(I10-512)67,67,68
   67 V(1)=10.0
      GO TO 66
   68 II10=II10+1
      IF(II10-512)69,70,70
   69 V(1)=0.0
      GO TO 66
   70 V(1)=0.0
      I10=0
      II10=0
   66 WRITE(6,1)(V(J),J=1,10)
    1 FORMAT(10F5.1)
  100 CONTINUE
      RETURN
      END
```

Figure 10.48 A ten-bit binary voltage generator

```
 0.0  0.0  0.0  0.0  0.0  0.0  0.0  0.0  0.0 10.0
 0.0  0.0  0.0  0.0  0.0  0.0  0.0  0.0 10.0  0.0
 0.0  0.0  0.0  0.0  0.0  0.0  0.0  0.0 10.0 10.0
 0.0  0.0  0.0  0.0  0.0  0.0  0.0 10.0  0.0  0.0
 0.0  0.0  0.0  0.0  0.0  0.0  0.0 10.0  0.0 10.0
 0.0  0.0  0.0  0.0  0.0  0.0  0.0 10.0 10.0  0.0
 0.0  0.0  0.0  0.0  0.0  0.0  0.0 10.0 10.0 10.0
 0.0  0.0  0.0  0.0  0.0  0.0 10.0  0.0  0.0  0.0
 0.0  0.0  0.0  0.0  0.0  0.0 10.0  0.0  0.0 10.0
 0.0  0.0  0.0  0.0  0.0  0.0 10.0  0.0 10.0  0.0
 0.0  0.0  0.0  0.0  0.0  0.0 10.0  0.0 10.0 10.0
 0.0  0.0  0.0  0.0  0.0  0.0 10.0 10.0  0.0  0.0
 0.0  0.0  0.0  0.0  0.0  0.0 10.0 10.0  0.0 10.0
 0.0  0.0  0.0  0.0  0.0  0.0 10.0 10.0 10.0  0.0
 0.0  0.0  0.0  0.0  0.0  0.0 10.0 10.0 10.0 10.0
 0.0  0.0  0.0  0.0  0.0 10.0  0.0  0.0  0.0  0.0
 0.0  0.0  0.0  0.0  0.0 10.0  0.0  0.0  0.0 10.0
 0.0  0.0  0.0  0.0  0.0 10.0  0.0  0.0 10.0  0.0
 0.0  0.0  0.0  0.0  0.0 10.0  0.0  0.0 10.0 10.0
 0.0  0.0  0.0  0.0  0.0 10.0  0.0 10.0  0.0  0.0
 0.0  0.0  0.0  0.0  0.0 10.0  0.0 10.0  0.0 10.0
 0.0  0.0  0.0  0.0  0.0 10.0  0.0 10.0 10.0  0.0
 0.0  0.0  0.0  0.0  0.0 10.0  0.0 10.0 10.0 10.0
 0.0  0.0  0.0  0.0  0.0 10.0 10.0  0.0  0.0  0.0
 0.0  0.0  0.0  0.0  0.0 10.0 10.0  0.0  0.0 10.0
 0.0  0.0  0.0  0.0  0.0 10.0 10.0  0.0 10.0  0.0
 0.0  0.0  0.0  0.0  0.0 10.0 10.0  0.0 10.0 10.0
 0.0  0.0  0.0  0.0  0.0 10.0 10.0 10.0  0.0  0.0
 0.0  0.0  0.0  0.0  0.0 10.0 10.0 10.0  0.0 10.0
 0.0  0.0  0.0  0.0  0.0 10.0 10.0 10.0 10.0  0.0
 0.0  0.0  0.0  0.0  0.0 10.0 10.0 10.0 10.0 10.0
 0.0  0.0  0.0  0.0 10.0  0.0  0.0  0.0  0.0  0.0
 0.0  0.0  0.0  0.0 10.0  0.0  0.0  0.0  0.0 10.0
 0.0  0.0  0.0  0.0 10.0  0.0  0.0  0.0 10.0  0.0
 0.0  0.0  0.0  0.0 10.0  0.0  0.0  0.0 10.0 10.0
                        •
                        •
                        •
10.0 10.0 10.0 10.0 10.0  0.0 10.0  0.0 10.0 10.0
10.0 10.0 10.0 10.0 10.0  0.0 10.0 10.0  0.0  0.0
10.0 10.0 10.0 10.0 10.0  0.0 10.0 10.0  0.0 10.0
10.0 10.0 10.0 10.0 10.0  0.0 10.0 10.0 10.0  0.0
10.0 10.0 10.0 10.0 10.0  0.0 10.0 10.0 10.0 10.0
10.0 10.0 10.0 10.0 10.0 10.0  0.0  0.0  0.0  0.0
10.0 10.0 10.0 10.0 10.0 10.0  0.0  0.0  0.0 10.0
10.0 10.0 10.0 10.0 10.0 10.0  0.0  0.0 10.0  0.0
10.0 10.0 10.0 10.0 10.0 10.0  0.0  0.0 10.0 10.0
10.0 10.0 10.0 10.0 10.0 10.0  0.0 10.0  0.0  0.0
10.0 10.0 10.0 10.0 10.0 10.0  0.0 10.0  0.0 10.0
10.0 10.0 10.0 10.0 10.0 10.0  0.0 10.0 10.0  0.0
10.0 10.0 10.0 10.0 10.0 10.0  0.0 10.0 10.0 10.0
10.0 10.0 10.0 10.0 10.0 10.0 10.0  0.0  0.0  0.0
10.0 10.0 10.0 10.0 10.0 10.0 10.0  0.0  0.0 10.0
10.0 10.0 10.0 10.0 10.0 10.0 10.0  0.0 10.0  0.0
10.0 10.0 10.0 10.0 10.0 10.0 10.0  0.0 10.0 10.0
10.0 10.0 10.0 10.0 10.0 10.0 10.0 10.0  0.0  0.0
10.0 10.0 10.0 10.0 10.0 10.0 10.0 10.0  0.0 10.0
10.0 10.0 10.0 10.0 10.0 10.0 10.0 10.0 10.0  0.0
10.0 10.0 10.0 10.0 10.0 10.0 10.0 10.0 10.0 10.0
 0.0  0.0  0.0  0.0  0.0  0.0  0.0  0.0  0.0  0.0
```

Figure 10.49 The beginning and end of the output of the program of Figure 10.48

Turning now to the Monte Carlo approach, we use the program of Figure 10.54 to investigate the random selection of $\pm 0.1\%$ normally distributed ladder resistors. Note that ten separate runs are made and for each run we print the numerical values of the ladder resistors and the number of times the error exceeds the LSB accuracy. Figure 10.55 shows the corresponding output and reveals that for each run all 1024 digital inputs produce an error less than 0.00325 volts. It thus follows that while a worst-case design (see Figure 10.51) produced six sets of digital inputs that exceeded the LSB accuracy for a $\pm 0.10\%$ tolerance and therefore was considered an unacceptable design, a Monte Carlo analysis reveals that we would have virtually a 100 percent production yield.

```
C        DIGITAL TC ANALCG
C        RESISTIVE LADDER
C          DESIGN
C
         DIMENSION R(2C),V(10),RIN(10),RCUT(10),AN(10),AJ(10),CODE(1C),
        1VO(6),ERROR(6)
         WRITE(6,1)
       1 FORMAT('1DIGITAL TC ANALCG RESISTIVE LADDER',///)
         RT=C.CCC4
         DO 8C J=1,6
         RT=RT+C.CCC1
         WRITE(6,11C)RT
     11C FORMAT(//,' TCLERANCE =',F8.4,//)
         I2=C
         I3=C
         I4=C
         I5=C
         I6=C
         I7=C
         I8=C
         I9=C
         I1C=C
         II2=C
         II3=C
         II4=C
         II5=C
         II6=C
         II7=C
         II8=C
         II9=C
         II1C=C
         DO 1CC I=1,1C24
         I1=I-I/2*2
         IF(I1)5,5,6
       5 V(1C)=C.C
         GO TO 7
       6 V(1C)=1C.C
       7 IF(I-2)8,9,9
       8 V(9)=C.C
         GO TC 1C
       9 I2=I2+1
         IF(I2-2)11,11,12
      11 V(9)=1C.C
         GO TC 1C
      12 II2=II2+1
         IF(II2-2)13,14,14
      13 V(9)=C.C
         GO TO 1C
      14 V(9)=C.C
         I2=C
         II2=C
      1C IF(I-4)15,16,16
      15 V(8)=C.C
         GO TC 17
      16 I3=I3+1
         IF(I3-4)18,18,19
      18 V(8)=1C.C
         GO TO 17
      19 II3=II3+1
         IF(II3-4)2C,21,21
      2C V(8)=C.C
         GO TC 17
      21 V(8)=C.C
         I3=C
         II3=C
      17 IF(I-8)22,23,23
      22 V(7)=C.C
         GO TC 24
      23 I4=I4+1
         IF(I4-8)25,25,26
      25 V(7)=1C.C
         GO TO 24
      26 II4=II4+1
         IF(II4-8)27,28,28
      27 V(7)=C.C
         GO TC 24
      28 V(7)=C.C
         I4=C
         II4=C
      24 IF(I-16)29,30,30
```

Figure 10.50 A Fortran program that incorporates the binary voltage generator of Figure 10.48

```
   25 V(6)=0.0
      GO TO 31
   30 I5=I5+1
      IF(I5-16)32,32,33
   32 V(6)=10.0
      GO TO 31
   33 II5=II5+1
      IF(II5-16)34,35,35
   34 V(6)=0.0
      GO TO 31
   35 V(6)=0.0
      I5=0
      II5=0
   31 IF(I-32)36,37,37
   36 V(5)=0.0
      GO TO 38
   37 I6=I6+1
      IF(I6-32)39,39,40
   39 V(5)=10.0
      GO TO 38
   40 II6=II6+1
      IF(II6-32)41,42,42
   41 V(5)=0.0
      GO TO 38
   42 V(5)=0.0
      I6=0
      II6=0
   38 IF(I-64)43,44,44
   43 V(4)=0.0
      GO TO 45
   44 I7=I7+1
      IF(I7-64)46,46,47
   46 V(4)=10.0
      GO TO 45
   47 II7=II7+1
      IF(II7-64)48,49,49
   48 V(4)=0.0
      GO TO 45
   49 V(4)=0.0
      I7=0
      II7=0
   45 IF(I-128)50,51,51
   50 V(3)=0.0
      GO TO 52
   51 I8=I8+1
      IF(I8-128)53,53,54
   53 V(3)=10.0
      GO TO 52
   54 II8=II8+1
      IF(II8-128)55,56,56
   55 V(3)=0.0
      GO TO 52
   56 V(3)=0.0
      I8=0
      II8=0
   52 IF(I-256)57,58,58
   57 V(2)=0.0
      GO TO 59
   58 I9=I9+1
      IF(I9-256)60,60,61
   60 V(2)=10.0
      GO TO 59
   61 II9=II9+1
      IF(II9-256)62,63,63
   62 V(2)=0.0
      GO TO 59
   63 V(2)=0.0
      I9=0
      II9=0
   59 IF(I-512)64,65,65
   64 V(1)=0.0
      GO TO 66
   65 I10=I10+1
      IF(I10-512)67,67,68
   67 V(1)=10.0
      GO TO 66
   68 II10=II10+1
      IF(II10-512)69,70,70
   69 V(1)=0.0
      GO TO 66
```

Figure 10.50 Continued

```
70 V(1)=0.0
   II0=0
   III0=0
66 B=0.5
   VNOM=((((((((((V(10)*B+V(9))*B+V(8))*B+V(7))*B+V(6))*B+V(5))*B+
  1V(4))*B+V(3))*B+V(2))*B+V(1))*B
   R(20)=2.0E4*(1.+RT)
   DO 81 K=1,10
81 R(K)=2.0E4*(1.+RT)
   DO 82 L=11,19
82 R(L)=1.0E4*(1.-RT)
   RIN(10)=R(20)
   DO 83 M=1,9
   K=11-M
   AN(K)=R(K)*RIN(K)/(R(K)+RIN(K))
83 RIN(K-1)=R(K+9)+AN(K)
   AN(1)=R(1)*RIN(1)/(R(1)+RIN(1))
   AJ(1)=R(1)
   ROUT(1)=R(1)
   DO 84 M=2,10
   ROUT(M)=R(M+9)+AJ(M-1)
84 AJ(M)=R(M)*ROUT(M)/(R(M)+ROUT(M))
   RATIO=1.
   VOUT=RIN(1)/(RIN(1)+R(1))*V(1)
   DO 85 N=2,10
   A=ROUT(N)*RIN(N)/(ROUT(N)+RIN(N))
   RATIO=RATIO*AJ(N-1)/(R(N+9)+AJ(N-1))
85 VOUT=VOUT+A/(A+R(N))*V(N)*RATIO
   ERROR(J)=VNOM-VOUT
   IF(ABS(ERROR(J))-0.00325)100,100,86
86 WRITE(6,87)(V(M),M=1,10)
87 FORMAT(10F6.1)
100 CONTINUE
80 CONTINUE
   STOP
   END
```

Figure 10.50 Continued

```
DIGITAL TO ANALOG RESISTIVE LADDER


TOLERANCE =   0.0005


TOLERANCE =   0.0006


TOLERANCE =   0.0007


TOLERANCE =   0.0008


TOLERANCE =   0.0009


TOLERANCE =   0.0010

  0.0   0.0  10.0  10.0  10.0  10.0  10.0  10.0  10.0  10.0
  0.0  10.0  10.0  10.0  10.0  10.0  10.0  10.0  10.0  10.0
 10.0   0.0   0.0   0.0   0.0   0.0   0.0   0.0   0.0   0.0
 10.0   0.0   0.0   0.0   0.0   0.0   0.0   0.0   0.0  10.0
 10.0  10.0   0.0   0.0   0.0   0.0   0.0   0.0   0.0   0.0
 10.0  10.0   0.0   0.0   0.0   0.0   0.0   0.0   0.0  10.0
```

Figure 10.51 The output results of the program of Figure 10.50

```
C       DIGITAL TO ANALOG
C       RESISTIVE LADDER
C         DESIGN
C
        DIMENSION R(20),V(10),RIN(10),ROUT(10),AN(10),AJ(10)
        WRITE(6,1)
      1 FORMAT('1DIGITAL TO ANALOG RESISTIVE LADDER',///)
        I2=0
        I3=0
        I4=0
        I5=0
        I6=0
        I7=0
        I8=0
        I9=0
        I10=0
        II2=0
        II3=0
        II4=0
        II5=0
        II6=0
        II7=0
        II8=0
        II9=0
        II10=0
        KOUNT=0
        DO 100 I=1,1024
        I1=I-I/2*2
        IF(I1)5,5,6
      5 V(10)=0.0
        GO TO 7
      6 V(10)=10.0
      7 IF(I-2)8,9,9
      8 V(9)=0.0
        GO TO 10
      9 I2=I2+1
        IF(I2-2)11,11,12
     11 V(9)=10.0
        GO TO 10
     12 II2=II2+1
        IF(II2-2)13,14,14
     13 V(9)=0.0
        GO TO 10
     14 V(9)=0.0
        I2=0
        II2=0
     10 IF(I-4)15,16,16
     15 V(8)=0.0
        GO TO 17
     16 I3=I3+1
        IF(I3-4)18,18,19
     18 V(8)=10.0
        GO TO 17
     19 II3=II3+1
        IF(II3-4)20,21,21
     20 V(8)=0.0
        GO TO 17
     21 V(8)=0.0
        I3=0
        II3=0
     17 IF(I-8)22,23,23
     22 V(7)=0.0
        GO TO 24
     23 I4=I4+1
        IF(I4-8)25,25,26
     25 V(7)=10.0
        GO TO 24
     26 II4=II4+1
        IF(II4-8)27,28,28
     27 V(7)=0.0
        GO TO 24
     28 V(7)=0.0
        I4=0
        II4=0
     24 IF(I-16)29,30,30
     29 V(6)=0.0
        GO TO 31
     30 I5=I5+1
        IF(I5-16)32,32,33
```

Figure 10.52 A modification of the program of Figure 10.50 in which the first four bits are held to $\pm 0.05\%$ tolerance and the last six bits to $\pm 1.0\%$ tolerance

```
32 V(6)=10.0
   GO TO 31
33 II5=II5+1
   IF(II5-16)34,35,35
34 V(6)=0.0
   GO TO 31
35 V(6)=0.0
   I5=0
   II5=0
31 IF(I-32)36,37,37
36 V(5)=0.0
   GO TO 38
37 I6=I6+1
   IF(I6-32)39,39,40
39 V(5)=10.0
   GO TO 38
40 II6=II6+1
   IF(II6-32)41,42,42
41 V(5)=0.0
   GO TO 38
42 V(5)=0.0
   I6=0
   II6=0
38 IF(I-64)43,44,44
43 V(4)=0.0
   GO TO 45
44 I7=I7+1
   IF(I7-64)46,46,47
46 V(4)=10.0
   GO TO 45
47 II7=II7+1
   IF(II7-64)48,49,49
48 V(4)=0.0
   GO TO 45
49 V(4)=0.0
   I7=0
   II7=0
45 IF(I-128)50,51,51
50 V(3)=0.0
   GO TO 52
51 I8=I8+1
   IF(I8-128)53,53,54
53 V(3)=10.0
   GO TO 52
54 II8=II8+1
   IF(II8-128)55,56,56
55 V(3)=0.0
   GO TO 52
56 V(3)=0.0
   I8=0
   II8=0
52 IF(I-256)57,58,58
57 V(2)=0.0
   GO TO 59
58 I9=I9+1
   IF(I9-256)60,60,61
60 V(2)=10.0
   GO TO 59
61 II9=II9+1
   IF(II9-256)62,63,63
62 V(2)=0.0
   GO TO 59
63 V(2)=0.0
   I9=0
   II9=0
59 IF(I-512)64,65,65
64 V(1)=0.0
   GO TO 66
65 I10=I10+1
   IF(I10-512)67,67,68
67 V(1)=10.0
   GO TO 66
68 II10=II10+1
   IF(II10-512)69,70,70
69 V(1)=0.0
   GO TO 66
70 V(1)=0.0
   I10=0
   II10=0
66 B=0.5
```

Figure 10.52 Continued

```
S.0156          VNOM=((((((((((V(10)*B+V(9))*B+V(8))*B+V(7))*B+V(6))*B+V(5))*B+
               1V(4))*B+V(3))*B+V(2))*B+V(1))*B
S.0157          R(20)=2.0E4*(1.+0.001)
S.0158          DO 81 K=1,4
S.0159       81 R(K)=2.0E4*(1.+0.0005)
S.0160          DO 88 K=5,10
S.0161       88 R(K)=2.0E4*(1.+0.01)
S.0162          DO 82 L=11,14
S.0163       82 R(L)=1.0E4*(1.-0.0005)
S.0164          DO 89 L=15,19
S.0165       89 R(L)=1.0E4*(1.-0.01)
S.0166          RIN(10)=R(20)
S.0167          DO 83 M=1,9
S.0168          K=11-M
S.0169          AN(K)=R(K)*RIN(K)/(R(K)+RIN(K))
S.0170       83 RIN(K-1)=R(K+9)+AN(K)
S.0171          AN(1)=R(1)*RIN(1)/(R(1)+RIN(1))
S.0172          AJ(1)=R(1)
S.0173          ROUT(1)=R(1)
S.0174          DO 84 M=2,10
S.0175          ROUT(M)=R(M+9)+AJ(M-1)
S.0176       84 AJ(M)=R(M)*ROUT(M)/(R(M)+ROUT(M))
S.0177          RATIO=1.
S.0178          VOUT=RIN(1)/(RIN(1)+R(1))*V(1)
S.0179          DO 85 N=2,10
S.0180          A=ROUT(N)*RIN(N)/(ROUT(N)+RIN(N))
S.0181          RATIO=RATIO*AJ(N-1)/(R(N+9)+AJ(N-1))
S.0182       85 VOUT=VOUT+A/(A+R(N))*V(N)*RATIO
S.0183          ERROR=VNOM-VOUT
S.0184          IF(ABS(ERROR)-0.00325)100,100,86
S.0185       86 WRITE(6,87)(V(M),M=1,10)
S.0186          KOUNT=KOUNT+1
S.0187       87 FORMAT(10F6.1)
S.0188      100 CONTINUE
S.0189          WRITE(6,90)KOUNT
S.0190       90 FORMAT(//,' KOUNT =',I5)
S.0191          STOP
S.0192          END
```

Figure 10.52 Continued

```
DIGITAL TO ANALOG RESISTIVE LADDER


 10.0   0.0   0.0   0.0  10.0   0.0   0.0   0.0   0.0   0.0
 10.0   0.0   0.0   0.0  10.0   0.0   0.0   0.0   0.0  10.0
 10.0   0.0   0.0   0.0  10.0  10.0   0.0   0.0   0.0   0.0
 10.0   0.0   0.0   0.0  10.0  10.0   0.0   0.0   0.0  10.0
 10.0  10.0   0.0   0.0  10.0   0.0   0.0   0.0   0.0   0.0
 10.0  10.0   0.0   0.0  10.0  10.0   0.0   0.0   0.0   0.0

KOUNT =     6
```

Figure 10.53 The output results of the program of Figure 10.52

Figure 10.56 shows a modification of computer statements S.0023 through S.0030 in the program of Figure 10.54. This modification permits us to investigate the random selection of $\pm 0.20\%$ normally distributed ladder resistors. Figure 10.57 gives the corresponding results. These results show that we could expect approximately a 90 percent production yield. This should be compared to the worst-case design that indicated 523 out of 1024 digital inputs would produce an error greater than 0.00325 volts. Figure 10.58 also points out that a Monte Carlo design, based on the first four bit resistors having a tolerance of $\pm 0.05\%$ and the last six a tolerance of $\pm 1.00\%$ would also be an acceptable design.

At this point the reader should realize that there is no unique or simple solution to this problem. For example, further investigation will show

```
          C      MONTE CARLO DIGITAL TO ANALOG
          C              RESISTIVE LADDER
          C                 DESIGN
          C
S.0001          DIMENSION RN(20),RNORM(200),V(10),R(20),RIN(10),ROUT(10),AN(10),
               1AJ(10)
S.0002          WRITE(6,1)
S.0003        1 FORMAT('1MONTE CARLO DESIGN OF A DIGITAL-TO-INALOG RESISTIVE LADDE
               1R',////)
S.0004          ISEED=123456789
S.0005          DENOM=SQRT(20./12.)
          C
          C     GENERATION OF 200 NORMALLY DISTRIBUTED NUMBERS
          C
S.0006          DO 80 I=1,200
          C
          C     GENERATION OF 20 UNIFORMLY DISTRIBUTED RANDOM NUMBERS
          C
S.0007          DO 81 K=1,20
S.0008          IRN=ISEED*65539
S.0009          IF(IRN)2,3,3
S.0010        2 IRN=IRN+2147483647+1
S.0011        3 RN(K)=IRN
S.0012          ISEED=IRN
S.0013          RN(K)=RN(K)*.4656613E-9
S.0014       81 CONTINUE
          C
          C     GENERATION OF A NORMALLY DISTRIBUTED RANDOM NUMBER
          C
S.0015          SUM=0.0
S.0016          DO 82 J=1,20
S.0017       82 SUM=SUM+RN(J)
S.0018       80 RNORM(I)=(SUM-10.)/DENOM
S.0019          N=0
S.0020          NN=10
S.0021          DO 200 KK=1,10
S.0022          KOUNT=0
          C
          C     RANDOM SELECTION OF LADDER RESISTORS
          C
S.0023          DO 101 J=1,10
S.0024          JN=J+N
S.0025      101 R(J)=2.0E4+RNORM(JN)*2.E1/3.
S.0026          R(20)=2.0E4+RNORM(NN+1)*2.E1/3.
S.0027          M=N+1
S.0028          DO 102 J=11,19
S.0029          JM=J+M
S.0030      102 R(J)=1.0E4+RNORM(JM)*1.0E1/3.
S.0031          WRITE(6,4)(R(I),I=1,20)
S.0032        4 FORMAT(//,'  R1 =',F7.1,'  R2 =',F7.1,'  R3 =',F7.1,'  R4 =',F7.1,
               1'  R5 =',F7.1,/,'  R6 =',F7.1,'  R7 =',F7.1,'  R8 =',F7.1,'  R9 ='
               2,F7.1,' R10 =',F7.1,/,' R11 =',F7.1,' R12 =',F7.1,' R13 =',F7.1,'
               3R14 =',F7.1,' R15 =',F7.1,/,' R16 =',F7.1,' R17 =',F7.1,' R18 =',F
               47.1,' R19 =',F7.1,' R20 =',F7.1,/)
          C
          C     BINARY VOLTAGE GENERATOR
          C
S.0033          I2=0
S.0034          I3=0
S.0035          I4=0
S.0036          I5=0
S.0037          I6=0
S.0038          I7=0
S.0039          I8=0
S.0040          I9=0
S.0041          I10=0
S.0042          II2=0
S.0043          II3=0
S.0044          II4=0
S.0045          II5=0
S.0046          II6=0
S.0047          II7=0
S.0048          II8=0
S.0049          II9=0
S.0050          II10=0
S.0051          DO 100 I=1,1024
S.0052          I1=I-I/2*2
S.0053          IF(I1)5,5,6
S.0054        5 V(10)=0.0
S.0055          GO TO 7
```

Figure 10.54 A Monte Carlo investigation using $\pm 0.1\%$ resistors

```
    6 V(10)=10.0
    7 IF(I-2)8,9,9
    8 V(9)=0.0
      GO TO 10
    9 I2=I2+1
      IF(I2-2)11,11,12
   11 V(9)=10.0
      GO TO 10
   12 II2=II2+1
      IF(II2-2)13,14,14
   13 V(9)=0.0
      GO TO 10
   14 V(9)=0.0
      I2=0
      II2=0
   10 IF(I-4)15,16,16
   15 V(8)=0.0
      GO TO 17
   16 I3=I3+1
      IF(I3-4)18,18,19
   18 V(8)=10.0
      GO TO 17
   19 II3=II3+1
      IF(II3-4)20,21,21
   20 V(8)=0.0
      GO TO 17
   21 V(8)=0.0
      I3=0
      II3=0
   17 IF(I-8)22,23,23
   22 V(7)=0.0
      GO TO 24
   23 I4=I4+1
      IF(I4-8)25,25,26
   25 V(7)=10.0
      GO TO 24
   26 II4=II4+1
      IF(II4-8)27,28,28
   27 V(7)=0.0
      GO TO 24
   28 V(7)=0.0
      I4=0
      II4=0
   24 IF(I-16)29,30,30
   29 V(6)=0.0
      GO TO 31
   30 I5=I5+1
      IF(I5-16)32,32,33
   32 V(6)=10.0
      GO TO 31
   33 II5=II5+1
      IF(II5-16)34,35,35
   34 V(6)=0.0
      GO TO 31
   35 V(6)=0.0
      I5=0
      II5=0
   31 IF(I-32)36,37,37
   36 V(5)=0.0
      GO TO 38
   37 I6=I6+1
      IF(I6-32)39,39,40
   39 V(5)=10.0
      GO TO 38
   40 II6=II6+1
      IF(II6-32)41,42,42
   41 V(5)=0.0
      GO TO 38
   42 V(5)=0.0
      I6=0
      II6=0
   38 IF(I-64)43,44,44
   43 V(4)=0.0
      GO TO 45
   44 I7=I7+1
      IF(I7-64)46,46,47
   46 V(4)=10.0
      GO TO 45
   47 II7=II7+1
      IF(II7-64)48,49,49
```

Figure 10.54 Continued

```
   48 V(4)=0.0
      GO TO 45
   49 V(4)=0.0
      I7=0
      II7=0
   45 IF(I-128)50,51,51
   50 V(3)=0.0
      GO TO 52
   51 I8=I8+1
      IF(I8-128)53,53,54
   53 V(3)=10.0
      GO TO 52
   54 II8=II8+1
      IF(II8-128)55,56,56
   55 V(3)=0.0
      GO TO 52
   56 V(3)=0.0
      I8=0
      II8=0
   52 IF(I-256)57,58,58
   57 V(2)=0.0
      GO TO 59
   58 I9=I9+1
      IF(I9-256)60,60,61
   60 V(2)=10.0
      GO TO 59
   61 II9=II9+1
      IF(II9-256)62,63,63
   62 V(2)=0.0
      GO TO 59
   63 V(2)=0.0
      I9=0
      II9=0
   59 IF(I-512)64,65,65
   64 V(1)=0.0
      GO TO 66
   65 I10=I10+1
      IF(I10-512)67,67,68
   67 V(1)=10.0
      GO TO 66
   68 II10=II10+1
      IF(II10-512)69,70,70
   69 V(1)=0.0
      GO TO 66
   70 V(1)=0.0
      I10=0
      II10=0
   66 B=0.5
      VNOM=(((((((((V(10)*B+V(9))*B+V(8))*B+V(7))*B+V(6))*B+V(5))*B+
     1V(4))*B+V(3))*B+V(2))*B+V(1))*B
      RIN(10)=R(20)
      DO 83 II=1,9
      K=11-II
      AN(K)=R(K)*RIN(K)/(R(K)+RIN(K))
   83 RIN(K-1)=R(K+9)+AN(K)
      AN(1)=R(1)*RIN(1)/(R(1)+RIN(1))
      AJ(1)=R(1)
      ROUT(1)=R(1)
      DO 84 L=2,10
      ROUT(L)=R(L+9)+AJ(L-1)
   84 AJ(L)=R(L)*ROUT(L)/(R(L)+ROUT(L))
      RATIO=1.
      VOUT=RIN(1)/(RIN(1)+R(1))*V(1)
      DO 85 L=2,10
      A=ROUT(L)*RIN(L)/(ROUT(L)+RIN(L))
      RATIO=RATIO*AJ(L-1)/(R(L+9)+AJ(L-1))
   85 VOUT=VOUT+A/(A+R(L))*V(L)*RATIO
      ERROR=VNOM-VOUT
      IF(ERROR-0.00325)100,86,86
   86 KNOUNT=KNOUNT+1
  100 CONTINUE
      N=N+20
      NN=NN+20
  200 WRITE(6,87)KNOUNT
   87 FORMAT(' KNOUNT =',I4)
      STOP
      END
```

Figure 10.54 Continued

```
MONTE CARLO DESIGN OF A DIGITAL-TO-ANALOG RESISTIVE LADDER



 R1  =20000.6  R2  =19998.4  R3  =19992.9  R4  =20002.5  R5  =19996.2
 R6  =19992.2  R7  =20009.8  R8  =19981.8  R9  =20004.8 R10  =19993.1
R11  = 9999.1 R12  =10001.6 R13  =10000.6 R14  =10004.1 R15  = 9998.1
R16  =10003.3 R17  =10001.6 R18  = 9997.1 R19  = 9997.3 R20  =20011.1

KNOUNT =    0


 R1  =19991.4  R2  =19993.2  R3  =19995.9  R4  =19993.9  R5  =20003.2
 R6  =19988.6  R7  =19995.2  R8  =19997.0  R9  =20007.1 R10  =20003.6
R11  = 9996.7 R12  = 9993.1 R13  =10001.5 R14  =10002.2 R15  = 9995.2
R16  = 9994.0 R17  =10000.2 R18  = 9995.0 R19  =10003.9 R20  =20002.5

KNOUNT =    0


 R1  =20004.7  R2  =20001.1  R3  =20002.2  R4  =19995.7  R5  =19995.4
 R6  =19992.6  R7  =20003.3  R8  =20009.2  R9  =20003.8 R10  =20000.3
R11  = 9996.2 R12  =10001.7 R13  =10000.1 R14  = 9994.6 R15  =10002.6
R16  =10000.6 R17  = 9998.4 R18  = 9999.0 R19  =10003.4 R20  =19999.7

KNOUNT =    0


 R1  =20001.2  R2  =20006.6  R3  =20002.7  R4  =20004.3  R5  =19999.3
 R6  =20002.4  R7  =19997.9  R8  =20002.8  R9  =20004.0 R10  =19995.3
R11  = 9996.8 R12  = 9997.5 R13  =10001.7 R14  =10000.2 R15  = 9995.6
R16  =10005.7 R17  =10003.2 R18  = 9996.2 R19  =10003.5 R20  =20000.9

KNOUNT =    0


 R1  =19995.9  R2  =19996.3  R3  =20001.2  R4  =20008.3  R5  =19992.0
 R6  =19998.0  R7  =19996.4  R8  =20000.8  R9  =19993.7 R10  =20003.0
R11  =10001.4 R12  =10001.2 R13  = 9997.3 R14  = 9997.8 R15  = 9994.2
R16  =10000.5 R17  = 9999.7 R18  = 9998.3 R19  =10003.0 R20  =19996.2

KNOUNT =    0


 R1  =20000.6  R2  =20015.8  R3  =20003.7  R4  =20003.5  R5  =19988.9
 R6  =19987.2  R7  =19987.7  R8  =19997.6  R9  =19991.7 R10  =19998.9
R11  = 9998.6 R12  = 9995.4 R13  = 9999.4 R14  =10003.4 R15  =10001.7
R16  =10003.1 R17  = 9994.5 R18  = 9999.7 R19  = 9999.3 R20  =20003.9

KNOUNT =    0


 R1  =19993.4  R2  =19999.7  R3  =19987.4  R4  =19987.9  R5  =19982.0
 R6  =20006.2  R7  =19993.9  R8  =20005.0  R9  =20004.0 R10  =19996.6
R11  = 9998.8 R12  =10005.6 R13  = 9998.7 R14  =10003.1 R15  = 9996.8
R16  =10002.1 R17  =10003.3 R18  =10004.0 R19  =10003.6 R20  =19997.8

KNOUNT =    0


 R1  =20001.0  R2  =19993.0  R3  =19998.8  R4  =20003.4  R5  =19996.3
 R6  =20010.6  R7  =20003.2  R8  =19996.0  R9  =19997.2 R10  =19996.8
R11  = 9998.0 R12  = 9998.9 R13  =10002.5 R14  = 9994.8 R15  =10002.9
R16  = 9999.8 R17  =10004.0 R18  =10002.8 R19  =10000.0 R20  =20011.2

KNOUNT =    0


 R1  =20000.9  R2  =20001.6  R3  =19983.7  R4  =19995.7  R5  =20018.0
 R6  =19994.4  R7  =19990.5  R8  =19997.6  R9  =20002.3 R10  =19997.0
R11  = 9998.3 R12  =10005.4 R13  =10000.8 R14  =10008.1 R15  =10001.0
R16  =10002.3 R17  =10001.5 R18  = 9994.0 R19  =10002.0 R20  =19996.6

KNOUNT =    0


 R1  =19988.0  R2  =20003.0  R3  =20003.3  R4  =19995.5  R5  =20009.6
 R6  =19995.4  R7  =19995.1  R8  =20002.5  R9  =20000.9 R10  =20007.5
R11  =10000.8 R12  = 9997.0 R13  =10000.8 R14  =10001.5 R15  =10001.4
R16  =10001.3 R17  = 9996.9 R18  = 9994.9 R19  = 9996.1 R20  =19986.5
```

Figure 10.55 The output results of the program of Figure 10.54 for a tolerance of $\pm 0.1\%$

Figure 10.56 A modification of the computer statements S.0023 to S.0030 in the program of Figure 10.54

```
S.0023          DO 101  J=1,10
S.0024          JN=J+N
S.0025      1C1 R(J)=2.0E4+RNCRM(JN)*4.E1/3.
S.0026          R(20)=2.0E4+RNORM(NN+1)*4.E1/3.
S.0027          M=N+1
S.0028          DO 102  J=11,19
S.0029          JM=J+M
S.0030      1C2 R(J)=1.0E4+RNCRM(JM)*2.0E1/3.
```

that the shunt $2R$ resistors are twice as sensitive as the series R resistors in producing error and it is thus possible to fabricate an acceptable worst-case design based on using $\pm 0.05\%$-20 kilohm nominal resistors and $\pm 0.10\%$-10 kilohm nominal resistors. What has been shown is that a Monte Carlo analysis makes it possible to specify a $\pm 1.0\%$ tolerance for the least six significant bit resistors and a $\pm 0.05\%$ tolerance for the four most significant bit resistors and still achieve the desired LSB accuracy. Such a specification economically lends itself to the production of large quantities of these ladder networks.

Before we leave this problem it is important to point out that while a Monte Carlo analysis was useful in determining the specification of a discrete element ladder, a worst-case analysis is needed in the fabrication of a thin film resistor ladder. In order to understand this point, consider the thin film ten-bit ladder network shown in Figure 10.59. This ladder is built by depositing microscopically thin films on an alumina (Al_2O_3) ceramic substrate approximately 0.8 by 0.8 inches. Initially, one side of the substrate has a 200-Angstrom (Å) thick vacuum deposited chromium resistive film applied. A two-layer conducting film of nickel and gold is then added by successively electroplating nickel and gold. The 50 microinch thick nickel film is required because gold will not adhere to the resistive chromium layer and the 50 microinch gold layer is needed to cover the nickel conducting layer because nickel oxydizes in air, making it difficult to solder the connecting leads.

The actual fabrication of the ladder requires photographic exposing and etching of the surfaces to remove the unwanted portions of resistor and conductor materials so that all that remains on the substrate is the resistor and conductor paths shown in Figure 10.59 (note that the conductors are darker in appearance than the resistors). Since the resistivity of the resistance material will vary across the substrate, it is obvious that no two boards will be exactly alike and it is not possible to fabricate resistors to a $\pm 0.09\%$ tolerance. It is therefore necessary to provide trimming resistors (see Figure 10.59) and trim each resistor individually. Since hand trimming each resistor is very time consuming and expensive, it is therefore important to know the maximum and minimum value that any resistor might have and still obtain the LSB accuracy. Clearly then, a worst-case and not a Monte Carlo type of analysis is needed to specify how close each of the resistors is to be trimmed.

```
MONTE CARLO DESIGN OF A DIGITAL-TO-INALOG RESISTIVE LADDER

 R1 =20001.1  R2 =19996.9  R3 =19985.8  R4 =20005.1  R5 =19992.4
 R6 =19984.4  R7 =20019.7  R8 =19963.6  R9 =20009.6 R10 =19986.2
R11 = 9998.2 R12 =10003.2 R13 =10001.1 R14 =10008.1 R15 = 9996.2
R16 =10006.6 R17 =10003.1 R18 = 9994.3 R19 = 9994.6 R20 =20022.1

KNOUNT =    0

 R1 =19982.9  R2 =19986.4  R3 =19991.8  R4 =19987.8  R5 =20006.4
 R6 =19977.3  R7 =19990.4  R8 =19994.0  R9 =20014.3 R10 =20007.2
R11 = 9993.4 R12 = 9986.3 R13 =10003.0 R14 =10004.4 R15 = 9990.4
R16 = 9987.9 R17 =10000.3 R18 = 9990.0 R19 =10007.9 R20 =20005.0

KNOUNT =    0

 R1 =20009.4  R2 =20002.3  R3 =20004.3  R4 =19991.3  R5 =19990.8
 R6 =19985.3  R7 =20006.7  R8 =20018.3  R9 =20007.5 R10 =20000.5
R11 = 9992.4 R12 =10003.4 R13 =10000.2 R14 = 9989.2 R15 =10005.3
R16 =10001.2 R17 = 9996.7 R18 = 9998.0 R19 =10006.7 R20 =19999.4

KNOUNT =    0

 R1 =20002.5  R2 =20013.2  R3 =20005.4  R4 =20008.6  R5 =19998.6
 R6 =20004.9  R7 =19995.8  R8 =20005.6  R9 =20008.0 R10 =19990.5
R11 = 9993.6 R12 = 9995.1 R13 =10003.5 R14 =10000.4 R15 = 9991.2
R16 =10011.4 R17 =10006.5 R18 = 9992.5 R19 =10007.0 R20 =20001.8

KNOUNT =    0

 R1 =19991.8  R2 =19992.5  R3 =20002.5  R4 =20016.6  R5 =19984.0
 R6 =19996.0  R7 =19992.8  R8 =20001.7  R9 =19987.3 R10 =20006.0
R11 =10002.8 R12 =10002.3 R13 = 9994.6 R14 = 9995.5 R15 = 9988.4
R16 =10000.9 R17 = 9999.5 R18 = 9996.6 R19 =10006.1 R20 =19992.4

KNOUNT =    0

 R1 =20001.2  R2 =20031.7  R3 =20007.5  R4 =20006.9  R5 =19977.8
 R6 =19974.5  R7 =19975.4  R8 =19995.1  R9 =19983.5 R10 =19997.7
R11 = 9997.3 R12 = 9990.8 R13 = 9998.8 R14 =10006.7 R15 =10003.5
R16 =10006.3 R17 = 9989.1 R18 = 9999.4 R19 = 9998.6 R20 =20007.7

KNOUNT =    0

 R1 =19986.8  R2 =19999.4  R3 =19974.8  R4 =19975.9  R5 =19963.9
 R6 =20012.4  R7 =19987.8  R8 =20010.0  R9 =20008.0 R10 =19993.2
R11 = 9997.7 R12 =10011.3 R13 = 9997.5 R14 =10006.3 R15 = 9993.6
R16 =10004.2 R17 =10006.6 R18 =10007.9 R19 =10007.2 R20 =19995.7

KNOUNT =    0

 R1 =20002.0  R2 =19985.9  R3 =19997.7  R4 =20006.8  R5 =19992.6
 R6 =20021.1  R7 =20006.3  R8 =19992.1  R9 =19994.5 R10 =19993.6
R11 = 9995.9 R12 = 9997.9 R13 =10004.9 R14 = 9989.6 R15 =10005.9
R16 = 9999.6 R17 =10007.9 R18 =10005.5 R19 = 9999.9 R20 =20022.4

KNOUNT =    0

 R1 =20001.7  R2 =20003.2  R3 =19967.3  R4 =19991.4  R5 =20036.0
 R6 =19988.8  R7 =19981.0  R8 =19995.1  R9 =20004.7 R10 =19994.0
R11 = 9996.7 R12 =10010.8 R13 =10001.7 R14 =10016.1 R15 =10002.0
R16 =10004.5 R17 =10003.0 R18 = 9988.0 R19 =10004.0 R20 =19993.2

KNOUNT =    0

 R1 =19976.1  R2 =20005.9  R3 =20006.6  R4 =19991.1  R5 =20019.2
 R6 =19990.8  R7 =19990.1  R8 =20005.0  R9 =20001.8 R10 =20015.0
R11 =10001.6 R12 = 9994.0 R13 =10001.6 R14 =10003.1 R15 =10002.7
R16 =10002.6 R17 = 9993.7 R18 = 9989.9 R19 = 9992.1 R20 =19973.0

KNOUNT =   40
```

Figure 10.57 The output results of the program of Figure 10.54 for a tolerance of $\pm 0.2\%$

```
MONTE CARLO DESIGN OF A DIGITAL-TO-ANALOG RESISTIVE LADDER

 R1 =20000.3  R2 =19999.2  R3 =19996.5  R4 =20001.3  R5 =19962.2
 R6 =19922.2  R7 =20098.3  R8 =19817.9  R9 =20047.9 R10 =19931.2
R11 = 9995.5 R12 =10007.9 R13 =10002.9 R14 =10020.4 R15 = 9980.8
R16 =10033.2 R17 =10015.5 R18 = 9971.5 R19 = 9972.7 R20 =20110.5

KNOUNT =    0


 R1 =19995.7  R2 =19996.6  R3 =19998.0  R4 =19997.0  R5 =20031.8
 R6 =19886.5  R7 =19952.2  R8 =19970.2  R9 =20071.3 R10 =20036.1
R11 = 9983.6 R12 = 9965.7 R13 =10007.4 R14 =10011.0 R15 = 9952.1
R16 = 9939.6 R17 =10001.5 R18 = 9950.1 R19 =10039.4 R20 =20024.8

KNOUNT =   20


 R1 =20002.3  R2 =20000.6  R3 =20001.1  R4 =19997.8  R5 =19954.0
 R6 =19926.3  R7 =20033.5  R8 =20091.7  R9 =20037.6 R10 =20002.7
R11 = 9981.0 R12 =10008.5 R13 =10000.6 R14 = 9973.0 R15 =10026.5
R16 =10006.2 R17 = 9983.7 R18 = 9989.8 R19 =10033.7 R20 =19996.9

KNOUNT =    0


 R1 =20000.6  R2 =20003.3  R3 =20001.3  R4 =20002.2  R5 =19993.2
 R6 =20024.5  R7 =19978.8  R8 =20027.8  R9 =20040.1 R10 =19952.7
R11 = 9984.0 R12 = 9987.7 R13 =10008.7 R14 =10001.0 R15 = 9955.8
R16 =10056.9 R17 =10032.5 R18 = 9962.3 R19 =10034.8 R20 =20008.9

KNOUNT =    0


 R1 =19997.9  R2 =19998.1  R3 =20000.6  R4 =20004.1  R5 =19920.0
 R6 =19980.0  R7 =19963.8  R8 =20008.3  R9 =19936.7 R10 =20030.1
R11 =10007.1 R12 =10005.8 R13 = 9986.4 R14 = 9988.8 R15 = 9941.9
R16 =10004.6 R17 = 9997.4 R18 = 9982.8 R19 =10030.3 R20 =19962.1

KNOUNT =    0


 R1 =20000.3  R2 =20007.9  R3 =20001.9  R4 =20001.7  R5 =19889.1
 R6 =19872.3  R7 =19876.9  R8 =19975.7  R9 =19917.4 R10 =19988.7
R11 = 9993.2 R12 = 9977.1 R13 = 9997.1 R14 =10016.8 R15 =10017.4
R16 =10031.5 R17 = 9945.5 R18 = 9997.1 R19 = 9992.9 R20 =20038.6

KNOUNT =    0


 R1 =19996.7  R2 =19999.9  R3 =19993.7  R4 =19994.0  R5 =19819.6
 R6 =20061.9  R7 -19938.9  R8 =20049.9  R9 =20040.2 R10 =19966.1
R11 = 9994.2 R12 =10028.3 R13 = 9993.7 R14 =10015.8 R15 = 9968.0
R16 =10021.2 R17 =10032.9 R18 =10039.7 R19 =10035.8 R20 =19978.4

KNOUNT =    0


 R1 =20000.5  R2 =19996.5  R3 =19999.4  R4 =20001.7  R5 =19962.8
 R6 =20105.5  R7 =20031.7  R8 =19960.4  R9 =19972.3 R10 =19967.8
R11 = 9989.9 R12 = 9994.6 R13 =10012.3 R14 = 9974.0 R15 =10029.5
R16 = 9997.8 R17 =10039.6 R18 =10027.6 R19 = 9999.7 R20 =20111.9

KNOUNT =    0


 R1 =20000.4  R2 =20000.8  R3 =19991.8  R4 =19997.8  R5 =20179.9
 R6 =19944.1  R7 =19905.0  R8 =19975.6  R9 =20023.4 R10 =19970.1
R11 = 9991.8 R12 =10027.1 R13 =10004.2 R14 =10040.3 R15 =10010.1
R16 =10022.6 R17 =10015.1 R18 = 9940.2 R19 =10020.2 R20 =19966.1

KNOUNT =    0


 R1 =19994.0  R2 =20001.5  R3 =20001.7  R4 =19997.8  R5 =20096.1
 R6 =19954.1  R7 =19950.7  R8 =20025.1  R9 =20009.1 R10 =20075.2
R11 =10004.0 R12 = 9985.0 R13 =10003.9 R14 =10007.7 R15 =10013.5
R16 =10013.0 R17 = 9968.6 R18 = 9949.5 R19 = 9960.6 R20 =19865.1

KNOUNT =    0
```

Figure 10.58 A modification of the program of Figure 10.54 that permits the investigation of holding the first four bit resistors to a tolerance of $\pm 0.05\%$ and the last six bit resistors to a tolerance of $\pm 1.0\%$

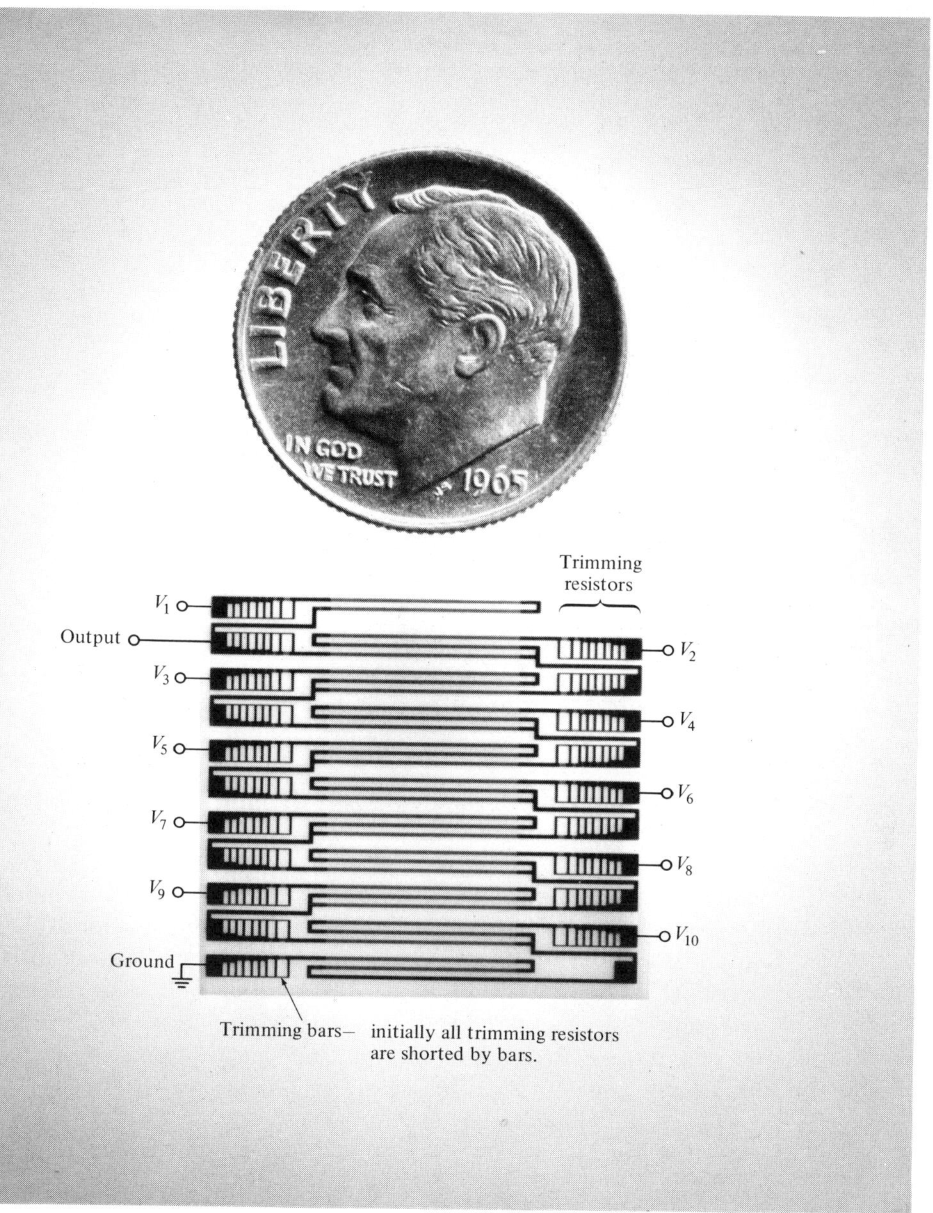

Figure 10.59 A ten-bit thin-film ladder network (Courtesy, Lockheed Electronics Co.)

PROBLEMS

10-1. Using the scaling technique discussed in Section 10.1, write appropriate scaling formulas for obtaining uniformly distributed random numbers that have the following maximum and minimum values:

	MAX	*MIN*
a.	−10	−20
b.	25	22
c.	−10.6	−13.4

10-2. Using the inner product algorithm equation

$$\text{NR} = \text{I}*\text{I}*100 + 2*\text{I}*\text{J} + \text{J}*\text{J}/100$$

hand compute and verify the first five numbers generated from a starting value of 1234.

10-3. Modify the inner product equation of Problem 10-2 so that it will produce six-digit random numbers.

10-4. Modify the program of Figure 10.5 so that it will produce numbers between 0 and 1.

10-5. Hand compute the appropriate computer statements in the program of Figure 10.5 and verify the first five numbers produced in Figure 10.6a.

10-6. Modify the program of Figure 10.5 so that it produces six-digit random numbers.

10-7. Modify the program of Figure 10.7 so that the three left digits are removed and subtracted from the remaining seven-digit number. Investigate the behavior of this generator and compare the randomness of the numbers it generates with those of Figure 10.8.

10-8. Modify the program of Figure 10.11 so that only every tenth number produced is used. Compare the randomness of the numbers generated with those produced by the program of Figure 10.11. Base your comparison on the frequency of occurrence of the most significant digit.

10-9. Determine the M residue classes of integers that are mutually congruent modulo M for the following values of M

a. 2 b. 5 c. 10

10-10. The smallest positive number in any of the sets of Problem 10-9 is called the least positive residue. Since we are usually interested in the integer that is equal to or greater than zero and less than the modulus, show that the least positive residue can be obtained by using the property of integer mode arithmetic and the following relation

$$N - N/M*M$$

10-11. Using the relation of Problem 10-10, compute the least positive residue for the following cases:

	N	M
a.	4567	5
b.	4567	9
c.	4567	15

10-12. Modify the relation of Problem 10-10 so that it is capable of finding the least positive residue when N is specified as a negative number.

10-13. Using the relation

$$NR_{n+1} = X*NR_n \pmod{R^N}$$

compute the first five numbers for each of the following cases:

	X	NR_0	R	N
a.	109	1234	10	4
b.	1109	123456	10	6

10-14. Using the generator of Figure 10.15, investigate the use of other starting values by examining the frequency of occurrence of the most significant digit.

10-15. Modify the program of Figure 10.15 so that it is capable of generating eight-digit random numbers.

10-16. Modify the program of Problem 10-15 so that the least two significant digits are disregarded.

10-17. Modify the program of Problem 10-15 so that the first two digits are subtracted from the number generated.

10-18. In Section 10.2 we stated that the power residue generator of Figure 10.15 would produce a least significant digit that was either an odd or an even number with a maximum cycle of four numbers. Explain the reason for this.

10-19. Modify the program of Figure 10.15 so that it produces random numbers between 0 and 1.

10-20. The throwing of two dice will produce numbers that range from 2 to 12. The probability of occurrence of these numbers will range from $\frac{1}{36}$ to $\frac{1}{6}$. Determine the probability of each number.

10-21. Using the probabilities determined in Problem 10-20, determine an appropriate method for testing the randomness of
a. The generator of Figure 10.15
b. The generator of Figure 10.17

10-22. Use the following number code to represent a card value

Number Code	02,	03,	04,	05,	06,	07,	08,	09,	10,	11,	12,	13,	14
Card Value	2,	3,	4,	5,	6,	7,	8,	9,	10,	J,	Q,	K,	A

and the following number code to represent the card suit

Clubs = 1xx Diamonds = 2xx Hearts = 3xx Spades = 4xx

For example, the number 212 would correspond to the queen of diamonds.

Now, using the generator of Figure 10.17 (or Figure 10.15) devise an appropriate program to determine the frequency of occurrence of dealing a three-of-a-kind poker hand. Use a minimum of 100 hands to establish the corresponding probability. This poker test is another test for determining the randomness of a given generator.

10-23. Modify the program of Figure 10.22 so that you can determine the frequency of occurrence of each of the digits of the random numbers produced by the generator used in the program of Figure 10.15.

10-24. Repeat Problem 10-23 for the generator of Figure 10.17.

10-25. Repeat the serial test of Figure 10.26 using other starting values. State your conclusion based on Chi-square values obtained.

10-26. Apply the serial test of Figure 10.26 to the generator of Figure 10.15.

10-27. Devise a test for determining how often the generator of Figure 10.17 produces an odd number followed by an odd number. (Use the most significant digit.)

10-28. Use the program of Figure 10.28 to find the period of the generator of Figure 10.15. Use the same starting value used in Figure 10.16.

10-29. Modify the generator of Figure 10.17 to produce a four-digit number and use the program of Figure 10.28 to determine the corresponding period.

10-30. Devise other tests that may be used to determine the randomness of a uniformly distributed number generator.

10-31. Rerun the program of Figure 10.31 using a series of five uniformly distributed random numbers to calculate each normally distributed number. Repeat using a series of ten uniformly distributed numbers. Compare your results with those of Figure 10.33.

10-32. An alternate method of computing normally distributed random numbers is based on using a triangular approximation to the normal probability distribution.[9]

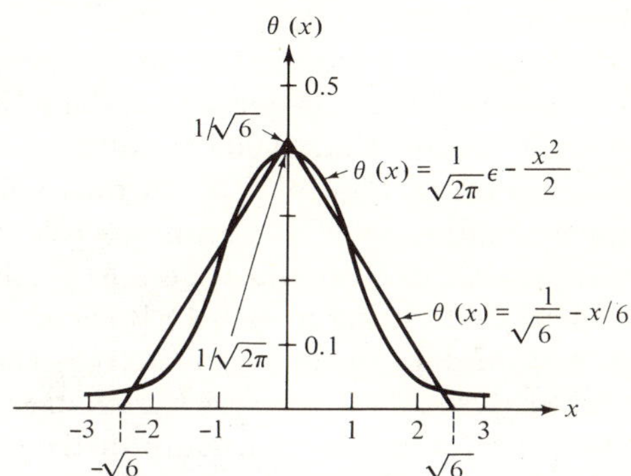

[9] V. A. Kinitsky, "A Triangular Approximation to the Normal Probability Distribution," *Systems Design*, April, 1964.

The advantage of using this alternate method is that it requires only one uniformly distributed number to generate one Gaussian number and thus considerable computer time is saved. The general relations for generating the numbers are as follows

$$\text{RNORM(I)} = \sqrt{6}\,(\sqrt{2*\text{RN(I)}} - 1);\; 0 \le \text{RN(I)} \le 0.5$$
$$\text{RNORM(I)} = \sqrt{6}\,(1 - \sqrt{2*(1 - \text{RN(I)})});\; 0.5 \le \text{RN(I)} \le 1.0$$

Write an appropriate program for generating normally distributed numbers that are based on this triangular approximation. Compare your results with those of Figure 10.33.

10-33. Use the Monte Carlo integration method described in Section 10.5 to compute the RMS value of the waveform of Figure 5.15.

10-34. Use the Monte Carlo integration method of Section 10.5 to compute the volume of a one-foot diameter sphere.

10-35. Modify the program of Figure 10.35 so that it uses the random number generator of Figure 10.17. Calculate the area using 500 random points and compare your results with those of Figure 10.35.

10-36. Determine the worst-case resistance tolerance required for the simple voltage divider of Section 10.5 to ensure a $\pm 10\%$ tolerance for each of the voltage divider ratios.

10-37. Repeat the calculations of Figure 10.39 for each of the other voltage divider ratios. What conclusions can you draw?

10-38. Repeat the calculations of Figure 10.42 using the generator of Problem 10-32.

10-39. Give a word description of the operation of the binary voltage generator program of Figure 10.48.

10-40. Verify that the six inputs shown in Figure 10.53 will produce an error in excess of 0.00325 volts.

10-41. Modify the program of Figure 10.50 so that you can investigate the $\pm 0.15\%$ tolerance case. Determine the number of inputs that produce an error greater than 0.00325 volts.

10-42. Modify the program of Figure 10.52 so that you can investigate other tolerance combinations. Determine a tolerance combination that will lead to a more economical design. Explain.

10-43. Repeat Problem 10-42 using a $\pm\frac{1}{2}$ LSB accuracy.

10-44. Assuming the third data set of Figure 7.10 as the reference, use a Monte Carlo analysis to determine the change in the operating point due to $\pm 10\%$ changes (normally distributed) in the input parameters.

10-45. Repeat the design of Figure 7.13 using a Monte Carlo worst-case approach. Compare your results.

CHAPTER 11

Computer Printer Plotting

11.1 INTRODUCTION

At this point in the text the reader should be willing to accept the fact that the high-speed digital computer offers the electrical engineer a means of investigating many different solutions to the same problem. It should also be apparent that it is neither difficult nor time-consuming to write a relatively simple program that will generate vast amounts of output. As a computer user, however, the reader must realize that it is both difficult and time-consuming to evaluate these large amounts of output, but such evaluation must be made in order to determine the best solution.

Automatic graphing or plotting of the output is a natural extension of the use of the computer and greatly simplifies the task of evaluating and understanding these solutions. Borrowing from an old cliché, we might say that one plot is worth a thousand pages of computer output.

The reader should be aware that there may be several different types of peripheral plotting equipment available in his computing center. The Calcomp digital plotter is one example of an electromechanical plotter. This particular machine produces an ink plot by moving a pen relative to the recording paper. In the Calcomp 700 series plotter, the pen may be moved one incremental step in any one of 16 possible directions (at 22.5° angles). By moving a half or a full step, a total of 24 different directions can be produced at a rate of 450 steps per second. Calcomp plotters may be used in either on-line or off-line operation. In on-line use, the plotter is directly connected to the computer and appropriate subroutines are used to control and generate the plot. In off-line use, magnetic tape generated previously by a computer program is used to operate the Calcomp to produce the desired plot.

Electronic plotters are similar in principle to the electro-mechanical plotters but differ by electronically producing the plot on the screen of a cathode ray tube. Since electronic speeds are involved, plotting rates as high as 100,000 increments per second are possible. It is also possible to get hard copy from a CRT display by recording on microfilm. In addition, a CRT display permits the use of a light pen to make necessary modifications in any given plot without resorting to other input devices. The light pen thus aids considerably in improving man–machine interaction.

It is also possible to produce plots by using the computer printer. Although most large computer installations provide a library printer routine, its format is often too restrictive to be of universal use (see Section 8.5, Subroutine PLOT). It is therefore desirable to be able to write one's own routine (or be able to modify existing routines) so that it may be incorporated into the source program. Since this latter method is based on using minimal computing equipment, we will concentrate on this type of plotting and use the remainder of this chapter to develop several different types of Fortran printer plotting routines and illustrate each of these in terms of electrical engineering examples.

11.2 HISTOGRAM PLOTS

The histogram or bar graph is one of the simplest plots that can be produced by the computer. It is very useful in displaying the results of any statistical problem in which some frequency distribution has been found as a function of some other system parameter.

Figure 11.1 is a Fortran program that plots the frequency distribution histograms of Figures 10.41 and 10.44. Initially the heading and plot code are printed and this step is followed by the printing of the top edge of the histogram. The values of AMARK(1), AMARK(2), BLANK, and N are then read into the computer memory. The variable AMARK(1) (+++) represents the symbol that is used in the plotting of the histogram of Figure 10.41 and AMARK(2) (***) is the symbol that is used in the plotting of Figure 10.44. The variable BLANK is used to store the computer representation of a blank, and the variable N stands for the largest value of the ordinate. For this plot the value of $N = 40$ was used, since the highest frequency value we will encounter is 38 (see Figure 10.44). Computer statements S.0008 and S.0009 now read in the frequency of occurrence (IRATIO(I)) of each of the ten voltage divider ratios. Note that the 0.45 to 0.46 occurrence of Figure 10.41 is read first,

```
C
C     THE DISTRIBUTION HISTOGRAM FOR THE VOLTAGE DIVIDER RATIO
C     OF PROGRAM FIGURE 10.5.6.
C
      DIMENSION IRATIO(20),COL(20),AMARK(2)
      WRITE(6,1)
    1 FORMAT('1',20X,'DISTRIBUTION HISTOGRAM OF VOLTAGE DIVIDER RATIOS',
     1//,' BASED ON 10 PERCENT RESISTORS',40X,'CODE- UNIFORM DISTRIBUTIO
     2N = +++',/,76X,'NORMAL DISTRIBUTION = ***',//,' FREQUENCY',/)
      WRITE(6,2)
    2 FORMAT(10X,111(1H.))
      READ(5,3)AMARK(1),AMARK(2),BLANK,N
    3 FORMAT(3A3,I5)
      READ(5,4)(IRATIO(I),I=1,20)
    4 FORMAT(20I2)
      DO 10 I=1,N
      K=N+1-I
      J=0
      DO 20 L=1,20
      J=J+1
      IF(IRATIO(L)-K) 30,40,40
   30 COL(L)=BLANK
      GO TO 50
   40 COL(L)=AMARK(J)
   50 IF(J-2) 20,60,60
   60 J=0
   20 CONTINUE
   10 WRITE(6,5)K,(COL(M),M=1,20)
    5 FORMAT(3X,I5,2X,10('.',1X,A3,2X,A3,1X),'.')
      WRITE(6,2)
      WRITE(6,6)
    6 FORMAT(/,10X,'.45 TO .46 .46 TO .47 .47 TO .48 .48 TO .49 .49 TO .
     150 .50 TO .51 .51 TO .52 .52 TO .53 .53 TO .54 .54 TO .55',//,25X,
     2'VOLTAGE DIVIDER RATIO')
      STOP
      END
```

Figure 11.1 A Program for plotting the frequency distribution histograms of Figures 10.41 and 10.44

the 0.45 to 0.46 occurrence of Figure 10.44 is second, the 0.46 to 0.47 of Figure 10.41 is third, and so on.

The plotting is now done by the DO 10 loop. Initially I is set equal to 1, and K is set equal to 40 ($K = 40 + 1 - 1 = 40$). The dummy variable J is now set equal to 0 and its value will be used to determine which symbol is to be plotted. We next enter the DO 20 loop, which is used to determine whether the symbol AMARK(1), AMARK(2), or a BLANK is to be printed. This is done by first setting $J = 1$ and then testing to see if IRATIO(1) (this corresponds to the frequency distribution of the 0.45 to 0.46 voltage divider ratio of Figure 10.41) is equal to or greater than K ($K = 40$). Since it will not be greater than 40 in the first iteration of the DO 10 loop, the variable COL(1) stores the computer representation of a blank. Control is then transferred to statement 50 where we determine whether J is equal to 2. Since $J \neq 2$, we return from the scope number to the DO 20 loop and increment L to 2. The variable J is now made equal to 2 and we test to see whether IRATIO(2) (this corresponds to the frequency distribution of the 0.45 to 0.46 voltage divider ratio of Figure 10.44) is equal to or greater than K ($K = 40$). Since it is not, COL(2) will store BLANK and statement 50 will test to see if J is equal to 2. Since J is equal to 2 this iteration, control is transferred to statement 60 and J is set equal to 0 again. The iteration of the DO 20 loop is re-

```
                    DISTRIBUTION HISTOGRAM OF VOLTAGE DIVIDER RATIOS

BASED ON 10 PERCENT RESISTORS                                 CODE- UNIFORM DISTRIBUTION = +++
                                                                    NORMAL DISTRIBUTION = ***

FREQUENCY

   ...............................................................................................................
40 .          .          .          .          .          .          .          .          .          .          .
39 .          .          .          .          .          .          .          .          .          .          .
38 .          .          .          .          .          .      *** .          .          .          .          .
37 .          .          .          .          .          .      *** .          .          .          .          .
36 .          .          .          .          .          .      *** .          .          .          .          .
35 .          .          .          .          .          .      *** .          .          .          .          .
34 .          .          .          .          .          .      *** .          .          .          .          .
33 .          .          .          .          .          .      *** .          .          .          .          .
32 .          .          .          .          .          .      *** .          .          .          .          .
31 .          .          .          .          .          .      *** .          .          .          .          .
30 .          .          .          .          .      *** .      *** .          .          .          .          .
29 .          .          .          .          .      *** .      *** .          .          .          .          .
28 .          .          .          .          .      *** .      *** .          .          .          .          .
27 .          .          .          .          .      *** .      *** .          .          .          .          .
26 .          .          .          .          .      *** .      *** .          .          .          .          .
25 .          .          .          .          .      *** .      *** .          .          .          .          .
24 .          .          .          .          .      *** .      *** .          .          .          .          .
23 .          .          .          .          .      *** .      *** .          .          .          .          .
22 .          .          .          .          .      *** .      *** .          .          .          .          .
21 .          .          .          .          .      *** .      *** .          .          .          .          .
20 .          .          .          .          .      *** .      *** .          .          .          .          .
19 .          .          .          .          .      *** . +++  *** .          .          .          .          .
18 .          .          .          .          .      *** . +++  *** .          .          .          .          .
17 .          .          .          .          .      *** . +++  *** .          .          .          .          .
16 .          .          .          .          .      *** . +++  *** .          .          .          .          .
15 .          .          .          .          . +++  *** . +++  *** . +++      .          .          .          .
14 .          .          .          .          . +++  *** . +++  *** . +++      .          .          .          .
13 .          .          .          . +++      . +++  *** . +++  *** . +++  *** .          .          .          .
12 .          .          .          . +++      . +++  *** . +++  *** . +++  *** . +++      .          .          .
11 .          .          .          . +++      . +++  *** . +++  *** . +++  *** . +++      .          .          .
10 .          .          .          . +++      . +++  *** . +++  *** . +++  *** . +++      .          .          .
 9 .          .          . +++      . +++      . +++  *** . +++  *** . +++  *** . +++      .          .          .
 8 .          .          . +++  *** . +++      . +++  *** . +++  *** . +++  *** . +++      .          .          .
 7 .          . +++      . +++  *** . +++  *** . +++  *** . +++  *** . +++  *** . +++      .          .          .
 6 .          . +++      . +++  *** . +++  *** . +++  *** . +++  *** . +++  *** . +++      . +++      .          .
 5 .          . +++      . +++  *** . +++  *** . +++  *** . +++  *** . +++  *** . +++      . +++      .          .
 4 .          . +++      . +++  *** . +++  *** . +++  *** . +++  *** . +++  *** . +++      . +++      .          .
 3 .          . +++      . +++  *** . +++  *** . +++  *** . +++  *** . +++  *** . +++  *** . +++      .          .
 2 . +++      . +++      . +++  *** . +++  *** . +++  *** . +++  *** . +++  *** . +++  *** . +++      . +++      .
 1 . +++      . +++  *** . +++  *** . +++  *** . +++  *** . +++  *** . +++  *** . +++  *** . +++      . +++      .
   ...............................................................................................................

   .45 TO .46 .46 TO .47 .47 TO .48 .48 TO .49 .49 TO .50 .50 TO .51 .51 TO .52 .52 TO .53 .53 TO .54 .54 TO .55

                  VOLTAGE DIVIDER RATIO
```

Figure 11.2 The output results of the program of Figure 11.1

peated until all 20 voltage divider ratios (IRATIO(I)) are compared to K ($K = 40$). The WRITE statement of computer statement S.0022 is now executed and since all IRATIO's were less than 40, a "BLANK" is printed for each voltage divider ratio for this ordinate value of K.

The DO 10 loop is then executed a second time and since all IRATIO's will be less than $K = 39$, a BLANK is again printed for each of the 20 ratios. In the third iteration of the DO 10 loop, however, K and IRATIO(12) both equal 38. Thus with $L = 12$ and $J = 2$, computer statement S.0015 tests and finds IRATIO(12) is equal to K. Control is therefore transferred to statement 40 and COL(12) stores the symbol AMARK(2) (***). We then test to see whether $J = 2$. Since it does, J is set equal to zero and we continue the execution of the DO 20 loop until all 20 IRATIO's are compared to K ($K = 38$). We now execute statement 10 and this time we print a "BLANK" for each of the voltage divider ratios except the 0.50 to 0.51 ratio, in which the symbol "***" is printed.

This process continues until the DO 10 loop has been executed 40 times. Whenever the value of IRATIO(I) is less than or equal to K, the symbol AMARK(1) or AMARK(2) is stored in the respective element of COL(L), otherwise a BLANK is stored. The symbol AMARK(1) is stored for odd values of L and the symbol AMARK(2) is stored for even values of L. The symbols stored by the variable COL(L) are printed each iteration of the DO 10 loop and 40 iterations produce the histogram shown in Figure 11.2.

The reader should now realize that if we had added a few WRITE statements to the programs that were used to obtain the information to plot Figures 10.41 and 10.44, we could have printed each of the uniformly distributed and normally distributed random numbers that were generated, each of the values of the resistors that were computed, each of the corresponding voltage divider ratios that were determined, and the frequency of occurrence of each of the voltage divider ratios. This would have resulted in many pages of computer output. It should be evident that an examination of Figure 11.2, rather than an examination of all the above possible output, would be much more desirable in helping us draw an intelligent conclusion about these two resistor distributions.

11.3 RECTANGULAR COORDINATE PLOTTING

We will first describe the general method of producing a rectangular coordinate plot and then use several examples to illustrate various modifications of the basic scheme. The method described below prints one line for each iteration of the main DO loop. Before entering the DO loop,

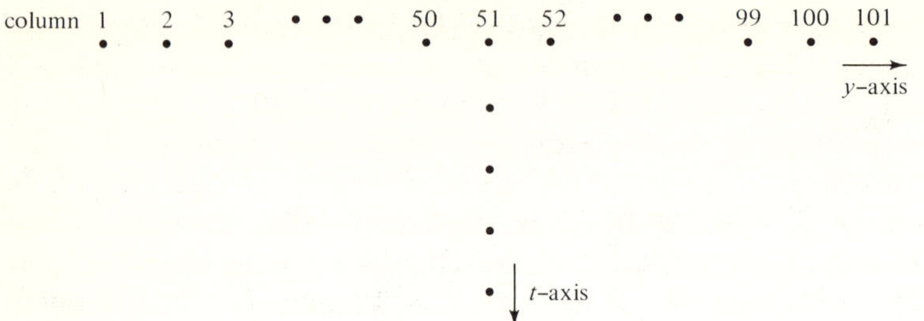

Figure 11.3 The *y*- and *t*-axes of a rectangular coordinate plot

the horizontal axis that represents the y variable is printed and this is formed by printing 101 dots. These DOTS are stored in the 101 elements of the variable COL(I). The printed line is therefore determined by what is stored in the respective elements of the subscripted variable COL.

The vertical axis represents the independent variable, say the time axis, and is formed each iteration by printing a DOT in column 51, as shown in Figure 11.3. In addition to the DOT that is printed in column 51, each line will also contain the symbol "*." This symbol is printed in the appropriate column (1 to 101) to correspond to the numerical value of the variable at a particular time.

To limit the value of y to the 101 columns provided, we first determine the maximum magnitude of the function $|y_{\max}|$ and then normalize the function with respect to its maximum value, $y(t)/|y_{\max}|$. This will result in a magnitude of $y(t)/|y_{\max}|$ equal to or less than 1. We next define the column J by

$$J = 50\left(\frac{y(t)}{|y_{\max}|} + 1\right) + 1.5$$

so that if $y(t)$ ranges from a value of $+|y_{\max}|$ to $-|y_{\max}|$, J would range from

$$101.5 \geq J \geq 1.5$$

Since J is an integer mode variable, the fractional part is dropped in rounding, yielding

$$101 \geq J \geq 1$$

and the value of J corresponds to one of the 101 columns we will be printing in.

Note that the horizontal or y axis may now be interpreted as $y(t)/|y_{\max}|$ with

$$-1 \leq \frac{y(t)}{|y_{\max}|} \leq +1$$

Negative values of $y(t)$ are therefore plotted in columns corresponding to

$$51 > J \geq 1$$

positive values of $y(t)$ are plotted in columns corresponding to

$$51 < J \leq 101$$

and $y(t) = 0$ corresponds to

$$J = 51$$

Let us now show why the 1.5 factor is needed in the arithmetic statement

$$J = 50\left(\frac{y(t)}{|y_{\max}|} + 1\right) + 1.5$$

To understand this, first recognize that the resolution of our plot is 0.02. In other words, the normalized difference between any two adjacent columns is $1/50$ or 0.02. Thus, for example, when $y(t)/|y_{\max}|$ exactly equals 0.98

$$\begin{aligned} J &= 50\left(\frac{y(t)}{|y_{\max}|} + 1\right) + 1.5 \\ &= 50(1.98) + 1.5 = 100.5 \end{aligned}$$

so that in rounding

$$J = 100$$

This results in the "*" being printed in column 100 to correspond to a value of $y(t)/|y_{\max}| = 0.98$. On the other hand, if $y(t)/|y_{\max}|$ is equal to 0.99,

$$\begin{aligned} J &= 50\left(\frac{y(t)}{|y_{\max}|} + 1\right) + 1.5 \\ &= 50(1.99) + 1.5 = 101 \end{aligned}$$

and a "*" would be printed in column 101 for a value of $y(t)/|y_{\max}|$ equal to 0.99. We thus see that if

$$0.97 \leq \frac{y(t)}{|y_{\max}|} < 0.99$$

this would result in a "*" being printed in column 100, but if

$$0.99 \leq \frac{y(t)}{|y_{\max}|} \leq 1.0$$

this would result in a "*" being printed in column 101. The 1.5 factor therefore produces the required rounding in the printing of $y(t)/|y_{\max}|$.

The use of A format code thus allows us to print a "." or a "*". For example, for $J = 51$, COL(J) = DOT cause a "." to be stored in memory for the element COL(51); while COL(J) = STAR will cause a "*" to be stored in memory for the appropriate element of COL(J) corresponding to $J = 50(y(t)/|y_{\max}| + 1) + 1.5$, and all other values of J will store a blank for the corresponding elements.

Using the above technique, a typical graph would be plotted by the following pseudo program:

```
      DIMENSION COL(101)
      READ(5,1)STAR,DOT,BLANK
    1 FORMAT(3A1)
      WRITE(6,2)
    2 FORMAT(' ....................................
     1.....................................')
      DO 3 I=1,101
    3 COL(I)=BLANK
      COL(51)=DOT
      T=0.0
      DELTA=
      YMAX=
      DO 4 I=1,50
      Y=
      J=50.*(Y/YMAX+1.)+1.5
      COL(J)=STAR
      WRITE(6,5)(COL(IJ),IJ=1,101)
    5 FORMAT(1H ,101A1)
      COL(J)=BLANK
      COL(51)=DOT
    4 T=T+DELTA
      END
```

The first statement reserves the necessary space in memory for the 101 elements of COL. The next two statements read into memory the symbols that will be used for STAR, DOT, and BLANK. Note that the first data card would contain a "*" in column one, a "." in column two, and a blank in column three. The WRITE statement, along with statement 2, is then used to print the 101 dots that correspond to the $y(t)|y_{\max}|$ axis. The DO 3 loop next sets each of the elements of COL to BLANK. COL(51) is then made to store a DOT. The variable T is then set equal to 0. The value of DELTA, the increment in time, is then computed by some expression and this is followed by the computation of YMAX. We next enter the DO 4 loop that will be executed 50 times and thus produce a graph that contains 50 lines. Each iteration of this loop will first find the value of Y for the corresponding value of time T and then the value of J from "J=50.*(Y/YMAX+1.)+1.5." The corresponding element of COL will then be set equal to STAR. The WRITE statement will then print one line of the graph with a DOT in column 51 and a "*" in column J. At this point we must now set COL(J) to BLANK so that the next "*" can be plotted and COL(51) must be set equal to DOT to take care of the case where J = 51. The scope statement 4 then increments the variable T by DELTA and we return to the DO 4 statement so that the next line can be plotted. This iteration is continued until 50 lines have been printed.

The Transient Response of the RLC Network

As a simple electrical example, consider the plotting of $i(t)$ for the source free RLC network shown in Figure 11.4, with $q(0) = Q$ and $i(0) = 0$. In order to plot $i(t)$, we must first find the solution of the differential equation of this circuit. Starting with the differential equation

$$\frac{d^2i}{dt^2} + \frac{R}{L}\frac{di}{dt} + \frac{1}{LC}i = 0,\ t \geq 0$$

we obtain the characteristic equation

$$s^2 + \frac{R}{L}s + \frac{1}{LC} = 0$$

and then compute the natural modes

$$s_1 = -\frac{R}{2L} + \sqrt{\left(\frac{R}{2L}\right)^2 - \frac{1}{LC}}$$

$$s_2 = -\frac{R}{2L} - \sqrt{\left(\frac{R}{2L}\right)^2 - \frac{1}{LC}}$$

Depending on the value of the discriminant $\left(\left(\frac{R}{2L}\right)^2 - \frac{1}{LC}\right)$, the solution can then be found from one of the following three forms:

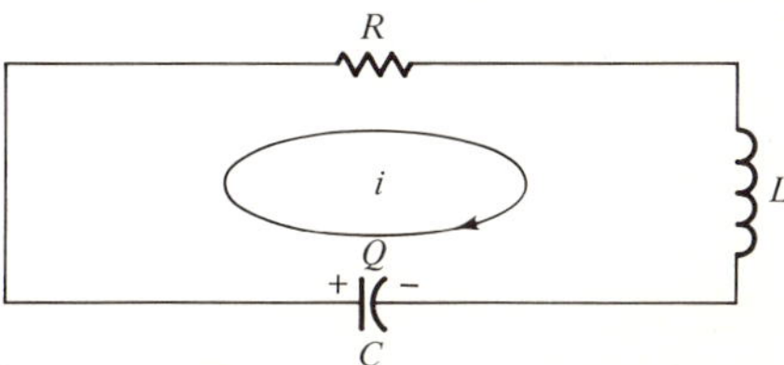

Figure 11.4 A source free R-L-C circuit

The Underdamped Case. For the underdamped case, the value of the discriminant will be less than zero and the solution will be given by

$$i(t) = I\epsilon^{-\alpha t} \sin \beta t$$

where

$$I = 2\pi F_o{}^2 \frac{Q}{F}$$

$$F_o = \frac{1}{2\pi \sqrt{LC}}$$

$$F = \sqrt{\frac{1}{LC} - \left(\frac{R}{2L}\right)^2} \Big/ (2\pi)$$

$$\alpha = \frac{R}{2L}$$

$$\beta = 2\pi F$$

A sketch of the general form of this solution is shown in Figure 11.5a.

In order to find the time it takes to reach the maximum value and the value of $i_{\max}$, we may differentiate $i(t)$ and set the derivative equal to zero; therefore,

$$\frac{di}{dt} = [I(-\alpha\epsilon^{-\alpha t} \sin \beta t + \beta \cos \beta t \; \epsilon^{-\alpha t})]_{t=T_{\max}} = 0$$

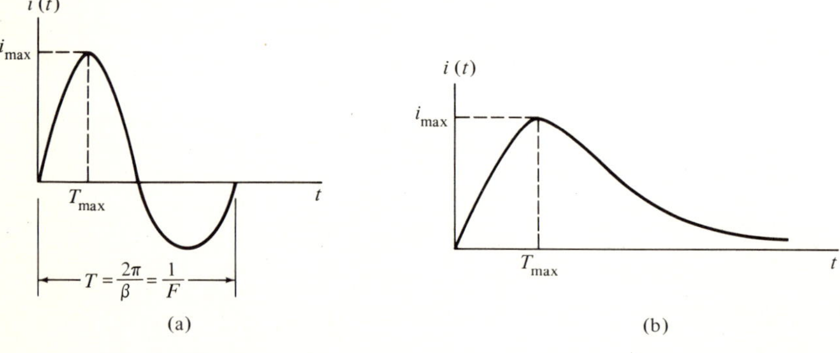

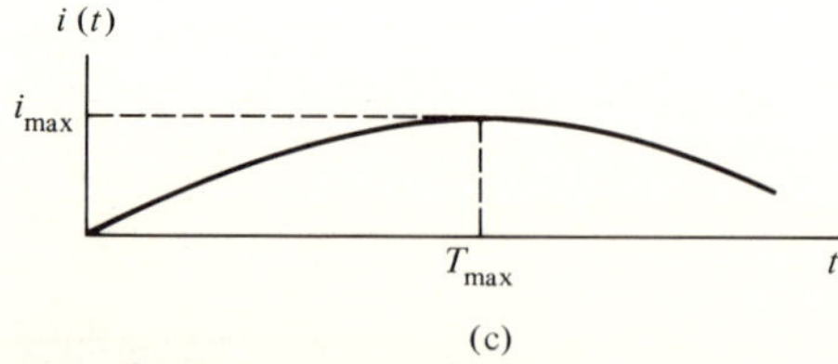

Figure 11.5 (a) The underdamped case; (b) the critical case; (c) the overdamped case

We thus find that

$$\tan \beta T_{\max} = \frac{\beta}{\alpha}$$

or

$$T_{\max} = \frac{1}{\beta} \tan^{-1}\left(\frac{\beta}{\alpha}\right)$$

and

$$i_{\max} = I\epsilon^{-\alpha T_{\max}} \sin \beta T_{\max}$$

The Critical Case. The critical case will occur when the value of the discriminant equals 0. The solution will be given by

$$i(t) = \frac{Q}{LC} t\epsilon^{-\alpha t}$$

where

$$\alpha = \frac{R}{2L}$$

A sketch of the general shape of the critical solution is shown in Figure 11.5b.

The maximum value of $i(t)$ and the time it takes to reach this value may again be found by differentiating. Thus we find that

$$\frac{di}{dt} = \frac{Q}{LC}(-\alpha t\epsilon^{-\alpha t} + \epsilon^{-\alpha t})$$

Setting the derivative equal to zero, we find

$$T_{\max} = \frac{1}{\alpha}$$

and

$$i_{\max} = \frac{Q}{LC}\frac{1}{\alpha}\epsilon^{-1}$$

The Overdamped Case. The overdamped case occurs when the discriminant is greater than zero and the solution is given by

$$i(t) = \frac{Qs_1s_2}{s_1 - s_2}(\epsilon^{s_1t} - \epsilon^{s_2t})$$

A sketch of the general shape of the overdamped solution is shown in Figure 11.5c.

Differentiating $i(t)$ to find $T_{\max}$, we have

$$\frac{di}{dt} = \frac{Qs_1s_2}{s_1 - s_2}(s_1\epsilon^{s_1t} - s_2\epsilon^{s_2t})$$

```
C        A PLOT ROUTINE
C        ILLUSTRATED BY THE RLC CIRCUIT
C        TIME RESPONSE
C
         DIMENSION COL(101)
         READ(5,1)STAR,DOT,BLANK
       1 FORMAT(3A1)
       2 READ(5,3)Q,R,AL,C
       3 FORMAT(4F10.6)
         WRITE(6,4)
       4 FORMAT(1H1,30X,'PLOT OF RLC CIRCUIT TIME RESPONSE',//,96X,'I(T)/IM
        1AX',/,' -1',48X,'0',48X,'+1')
         WRITE(6,5)
       5 FORMAT(' ..............................................................
        1...........................................')
         DO 6 I=1,101
       6 COL(I)=BLANK
         COL(51)=DOT
         ALPHA=R/(2.*AL)
         DISC=ALPHA**2-1./(AL*C)
         T=0.0
         IF(DISC)10,20,30
C
C        THE UNDERDAMPED CASE.
C
      10 F0=0.159155/SQRT(AL*C)
         BETA=SQRT(-DISC)
         F=0.159155*BETA
         TMAX=ATAN(BETA/ALPHA)/BETA
         DELTA=8.*TMAX/50.
         AI=6.2832*F0**2*Q/F
         AMAXI=AI*EXP(-ALPHA*TMAX)*SIN(BETA*TMAX)
         DO 11 I=1,50
         TI=AI*EXP(-ALPHA*T)*SIN(BETA*T)
         J=50.*(TI/AMAXI+1.)+1.5
         COL(J)=STAR
         WRITE(6,12)(COL(IJ),IJ=1,101)
      12 FORMAT(1H ,101A1)
         COL(J)=BLANK
         COL(51)=DOT
      11 T=T+DELTA
         WRITE(6,13)DELTA,Q,R,AL,C,AMAXI
      13 FORMAT('0DELTA =',F6.3,3X,'Q =',F6.2,3X,'R =',F6.2,3X,'L =',F6.2,3
        1X,'C =',F6.2,3X,'AMAXI =',F10.4)
         GO TO 2
C
C        THE CRITICALLY DAMPED CASE.
C
      20 DELTA=2./(50.*ALPHA)
         AMAXI=Q/(AL*C*2.71828*ALPHA)
         DO 21 I=1,50
         TI=Q/(AL*C)*T*EXP(-ALPHA*T)
         J=50.*(TI/AMAXI+1.)+1.5
         COL(J)=STAR
         WRITE(6,22)(COL(IJ),IJ=1,101)
      22 FORMAT(1H ,101A1)
         COL(J)=BLANK
         COL(51)=DOT
      21 T=T+DELTA
         WRITE(6,13)DELTA,Q,R,AL,C,AMAXI
         GO TO 2
C
C        THE OVERDAMPED CASE.
C
      30 BETA=SQRT(DISC)
         S1=-ALPHA+BETA
         S2=-ALPHA-BETA
         TMAX=ALOG(S2/S1)/(S1-S2)
         DELTA=2.*TMAX/50.
         AMAXI=((Q*S1*S2)/(S1-S2))*(EXP(S1*TMAX)-EXP(S2*TMAX))
         DO 31 I=1,50
         TI=((Q*S1*S2)/(S1-S2))*(EXP(S1*T)-EXP(S2*T))
         J=50.*(TI/AMAXI+1.)+1.5
         COL(J)=STAR
         WRITE(6,32)(COL(IJ),IJ=1,101)
      32 FORMAT(1H ,101A1)
         COL(J)=BLANK
         COL(51)=DOT
      31 T=T+DELTA
         WRITE(6,13)DELTA,Q,R,AL,C,AMAXI
         GO TO 2
         STOP
         END
```

Figure 11.6 A Fortran program that plots the time response of the R-L-C circuit

Setting the derivative equal to zero, we find

$$\frac{s_2}{s_1} = \frac{\epsilon^{s_1 T_{\max}}}{\epsilon^{s_2 T_{\max}}} = \epsilon^{T_{\max}(s_1 - s_2)}$$

and

$$T_{\max} = \frac{\ln s_2/s_1}{s_1 - s_2}$$

$$i_{\max} = \frac{Q s_1 s_2}{s_1 - s_2} \left(\epsilon^{s_1 T_{\max}} - \epsilon^{s_2 T_{\max}}\right)$$

Figure 11.6 shows a Fortran program that permits an investigation of the time response for various sets of R, L, C, and Q values. Computer statement S.0002 first reads in the symbols that will be used for the variables STAR, DOT, and BLANK. The second data card is then read, storing the numerical values of Q, R, L (called AL because of mode), and C. The graph heading, the labeling of the y-axis, and the 101 dots representing the y-axis are then printed. The DO 6 loop then sets all the elements of COL to BLANK. This is followed by the calculation of α and the discriminant and the setting of $t = 0$. We then test to see if the discriminant is less than 0, equal to 0, or greater than 0. If the discriminant is less than 0, control is transferred to statement 10 and we consider the underdamped case. The values of F_o, β, F, $T_{\max}$, DELTA, I (called AI because of mode), and $i_{\max}$ are then computed. Note that the value of DELTA is computed to allow approximately two complete oscillations of $i(t)$, independent of the actual values of R, L, or C.

The DO 11 loop is then used to print the 50 lines of the graph and this loop is essentially identical to the general printing loop we have previously discussed. After the 50 lines have been printed, the actual values of DELTA, Q, R, L, C, and IMAX are printed so that one may be able to determine actual current values from the plot. Control is then transferred to statement 2 so that another set of data may be read and plotted.

If the discriminant had been found equal to 0 (computer statement S.0016), control would have been transferred to statement 20 where the critically damped case would have been computed. Here the value of DELTA is made equal to 2/(50*ALPHA) so that 50 lines of plot will correspond to an overall time of 2*TMAX. Similarly, if the discriminant had been found to be greater than 0, control would have been transferred to statement 30 and the overdamped case would be plotted.

Figure 11.7 shows the corresponding output for three different sets of input data. The first plot is for the case of $R = 1.0$, $C = 1.0$, and $L = 1.0$. This corresponds to the underdamped case since $R < 2\sqrt{L/C}$. The second plot is for the values of $R = 2.0$, $C = 1.0$, and $L = 1.0$ and thus corresponds to the critically damped case. The last plot is for $R = 3.0$,

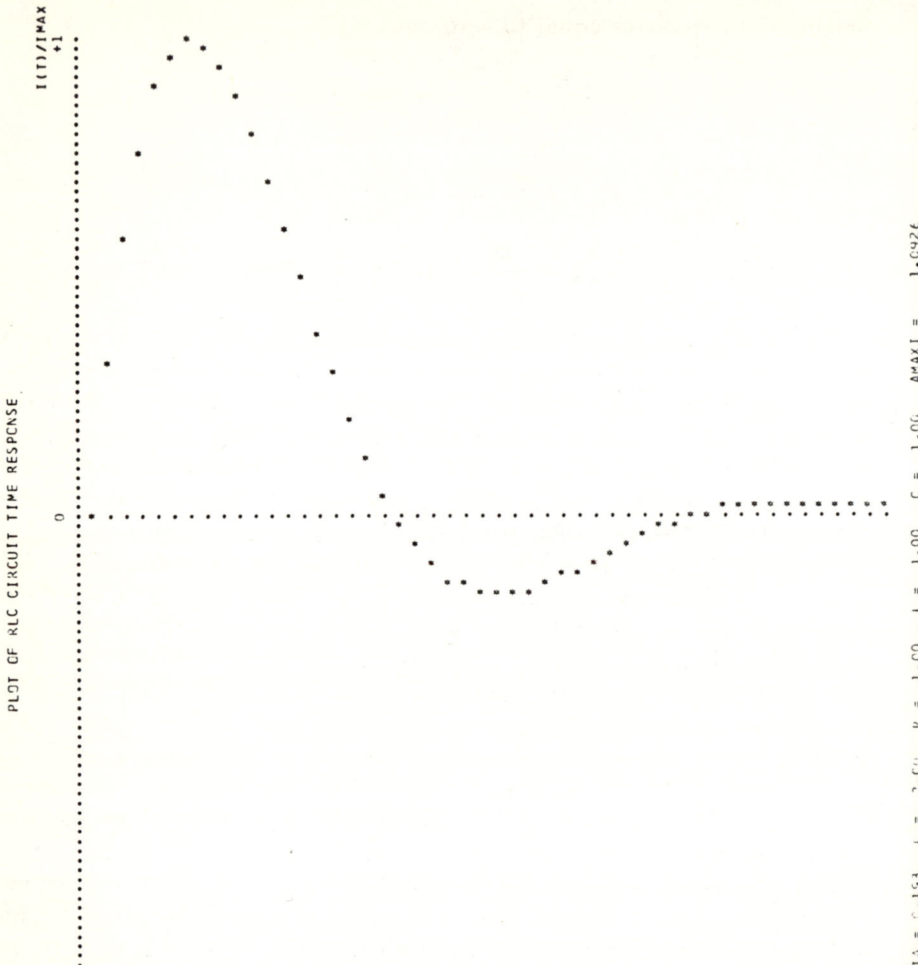

Figure 11.7 The output results of the program of Figure 11.6 for three different sets of R-L-C values

$C = 1.0$, and $L = 1.0$ and corresponds to the overdamped case. The reader may readily verify the accuracy of these plots.

Frequency Spectrum Plot

In Section 5.6 we discussed the evaluation of the Fourier integral and the plotting of the frequency spectrum of a quasi-recurrent waveform. We now modify the general plot scheme discussed above so that it may be used to plot the normalized frequency spectrum.

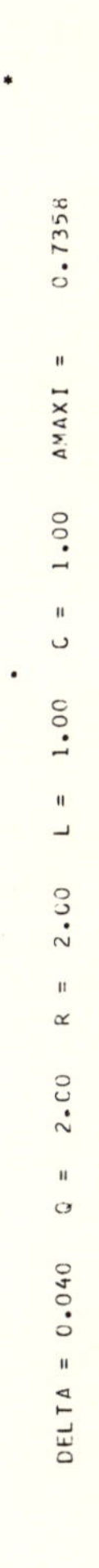

Figure 11.7 Continued

Figure 11.8 is a general Fortran program for plotting the normalized frequency spectrum. Initially the symbols that are used for the variables GRID, STAR, and BLANK are read into memory. This is followed by reading in the plot starting frequency FLOW, the maximum frequency FMAX, the frequency increment FINCR, the time increment between data points TINCR, the lower limit or initial starting time of the waveform TINIT, and the number of data points N. The N data points are then read into memory by computer statements S.0006 and S.0007.

Computer statements S.0010 through S.0013 are used to compute the

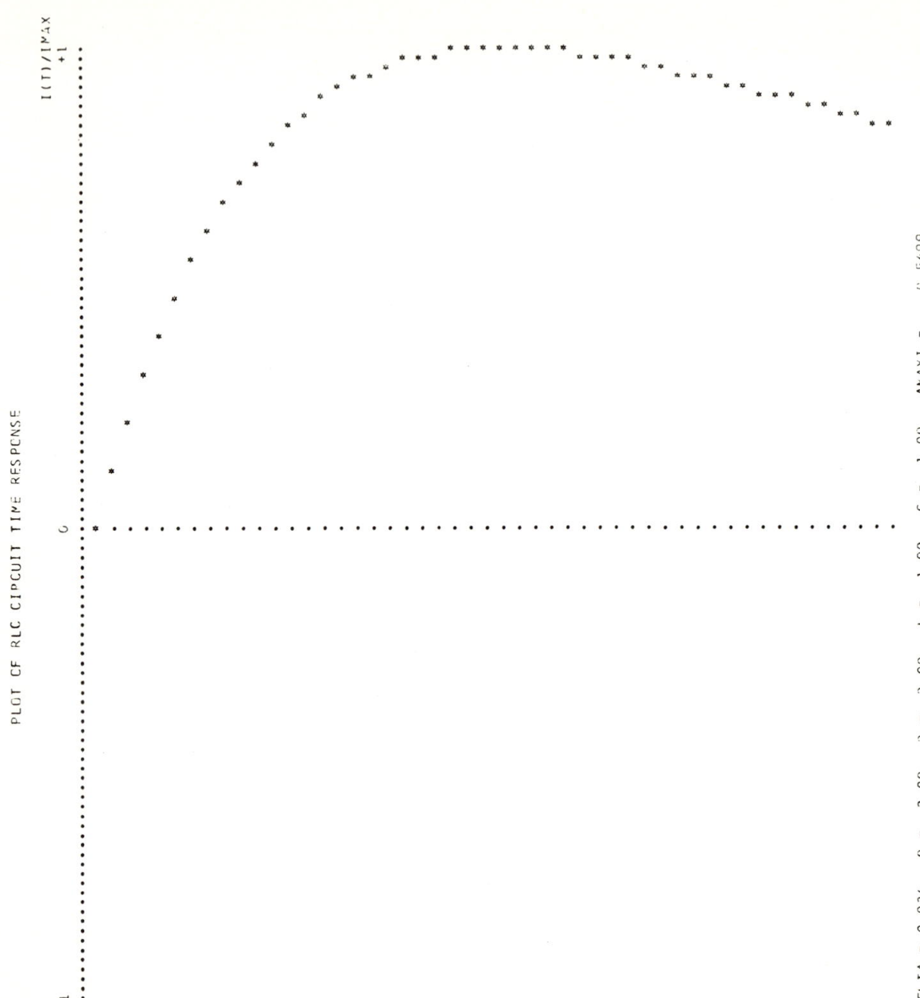

Figure 11.7 Continued

numerical values of each of the frequencies we have specified and the DO 40 loop is used to convert these values to angular frequencies. Computer statements S.0017 through S.0033 now employ Simpson's rule to evaluate the real and imaginary part of the frequency spectrum for each of the specified frequencies. This part of the program is essentially identical to computer statements S.0019 through S.0034 in the program of Figure 5.29.

After the magnitude of the frequency spectrum has been computed, computer statements S.0035 through S.0038 are used to determine the

```
C      THIS PROGRAM COMPUTES AND PLOTS THE FREQUENCY SPECTRUM.
C
       DIMENSION F(100),FREQ(100),Y(100),COL(101)
       READ (5,1)GRID,STAR,BLANK
     1 FORMAT(3A1)
       READ (5,2)FLOW,FMAX,FINCR,TINCR,TINIT,N
     2 FORMAT(5E15.7,I5)
       READ (5,3) (F(I),I=1,N)
       WRITE(6,3) (F(I),I=1,N)
     3 FORMAT(5E15.7)
       I=1
       FREQ(1)=FLOW
    20 I=I+1
       FREQ(I)=FREQ(I-1)+FINCR
       IF(FREQ(I)-FMAX) 20,30,30
    30 DO 40 J=1,I
    40 FREQ(J)=FREQ(J)*6.2831853
       N=N-1
       DO 50 K=1,I
       W=FREQ(K)
C
C      SIMPSON'S RULE IS USED TO COMPUTE THE SPECTRUM.
C
       A=0.0
       B=0.0
       T=-2.*TINCR+TINIT
       DO 60 J=1,N,2
       T=T+2.*TINCR
       C=F(J)*COS(W*T)
       D=4.*F(J+1)*COS(W*(T+TINCR))
       E=F(J+2)*COS(W*(T+2.*TINCR))
       A=A+C+D+E
       FF=F(J)*SIN(W*T)
       G=4.*F(J+1)*SIN(W*(T+TINCR))
       H=F(J+2)*SIN(W*(T+2.*TINCR))
    60 B=B+FF+G+H
       A=TINCR*A/3.
       B=-TINCR*B/3.
    50 Y(K)=SQRT(A*A+B*B)
C
C      THE MAXIMUM VALUE OF Y(K) IS FOUND.
C
       YMAX=Y(1)
       DO 70 J=2,I
       IF(YMAX-Y(J)) 75,70,70
    75 YMAX=Y(J)
    70 CONTINUE
       WRITE(6,80) YMAX
    80 FORMAT('1',60X,'FREQUENCY SPECTRUM PLOT OF Y/YMAX',10X,'YMAX =',E1
      15.7,////,9X,'FREQUENCY',5X,'0',23X,'0.25',21X,'0.50',21X,'0.75',21
      2X,'1.0')
       LINE=0
       DO 90 J=1,I
       IF(LINE-(LINE/10)*10) 95,95,100
    95 DO 110 K=1,101
   110 COL(K)=GRID
       GO TO 130
   100 DO 120 K=2,100
   120 COL(K)=BLANK
       COL(1)=GRID
       COL(26)=GRID
       COL(51)=GRID
       COL(76)=GRID
       COL(101)=GRID
   130 K=100.*Y(J)/YMAX+1.5
       COL(K)=STAR
       WRITE(6,140)FREQ(J),(COL(K),K=1,101)
   140 FORMAT(E20.6,3X,101A1)
    90 LINE=LINE+1
       STOP
       END
```

Figure 11.8 A general program that computes and plots the frequency spectrum of an arbitrary waveform

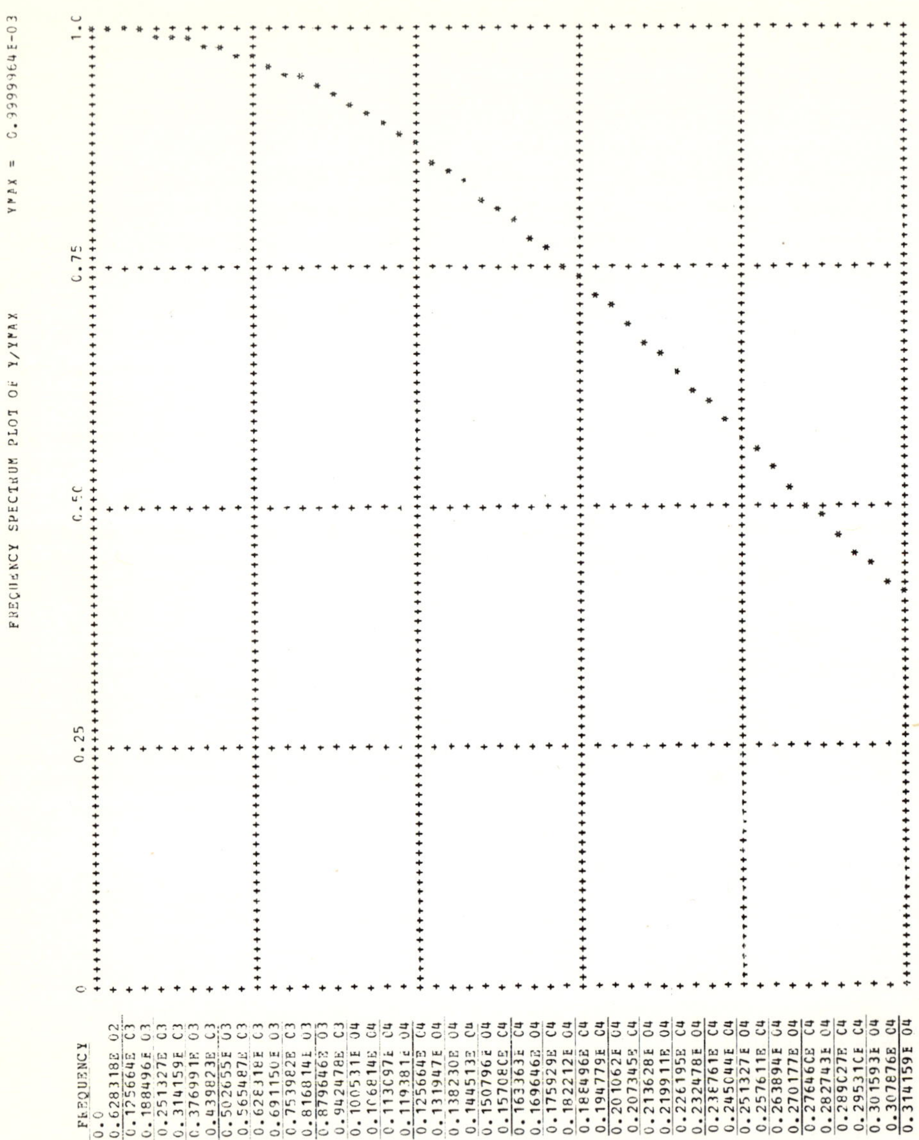

Figure 11.9 The output results of the program of Figure 11.8 for the single pulse waveform of Figure 5.34b

maximum value YMAX. The plot heading and Y-axis labeling are then printed. The dummy variable LINE, which is used to determine when the horizontal grid lines should be printed, is next set equal to 0. We then enter the DO 90 loop, which is the main printing iteration loop, printing one line for each iteration. During each iteration we test to see if LINE − LINE/10*10 is greater than 0, or less than or equal to 0. The reader may observe that starting with the first iteration and every tenth iteration thereafter, this expression will be equal to 0 and thus control will be transferred to statement 95 so that the horizontal grid line symbol can be stored in each of the elements of the subscripted variable COL. If the expression "LINE − LINE/10*10" is greater than 0 (this will occur for lines 2, 3, 4, 5, 6, 7, 8, 9, 10, 12, and so on), control is transferred to statement 100 where the DO 120 loop sets each of the elements of COL to BLANK. Computer statements S.0050 through S.0054 are then used to store the grid symbol that is used in printing the vertical grid lines, and statement 130 computes the value of K which specifies the column in which the symbol "*" is printed to represent the plotted point Y/YMAX. Computer statement S.0057 now prints the line and computer statement S.0059 increments the variable LINE before the iteration is repeated.

Figure 11.9 is the resulting plot for the single pulse waveform of Figure 5.34b. The reader may readily verify the accuracy of this plot by comparing it to the corresponding plot of Figure 5.37. Figure 11.10 is the corresponding spectrum plot for the nine pulses of Figure 5.34c. The reader may verify this plot by again referring to Figure 5.37.

Normalized Time Response Plot

The last two plots were printed so that the horizontal axis represented the normalized dependent variable I(T)/IMAX or Y/YMAX and the vertical axis corresponded to the independent variable T or W. Since the value of I(T)/IMAX (or Y/YMAX) was computed in a DO loop for a specific value of T (or W) that was changed by the increment DELTA (or FINCR) each iteration, it was possible to print one line of the graph each iteration.

If we desire to print the graph with the variables that represent the horizontal and vertical axis interchanged, we will find that it is not possible to print the graph one line at a time, since the value of I(T)/IMAX (or Y/YMAX) will not necessarily change in magnitude by a fixed increment. We will therefore have to store the plotted-point symbol for I(T)/IMAX, for each T, in a double-subscripted array as T is incremented over the required time range. The graph may then be plotted by printing the entire plotted-point array with a single WRITE statement.

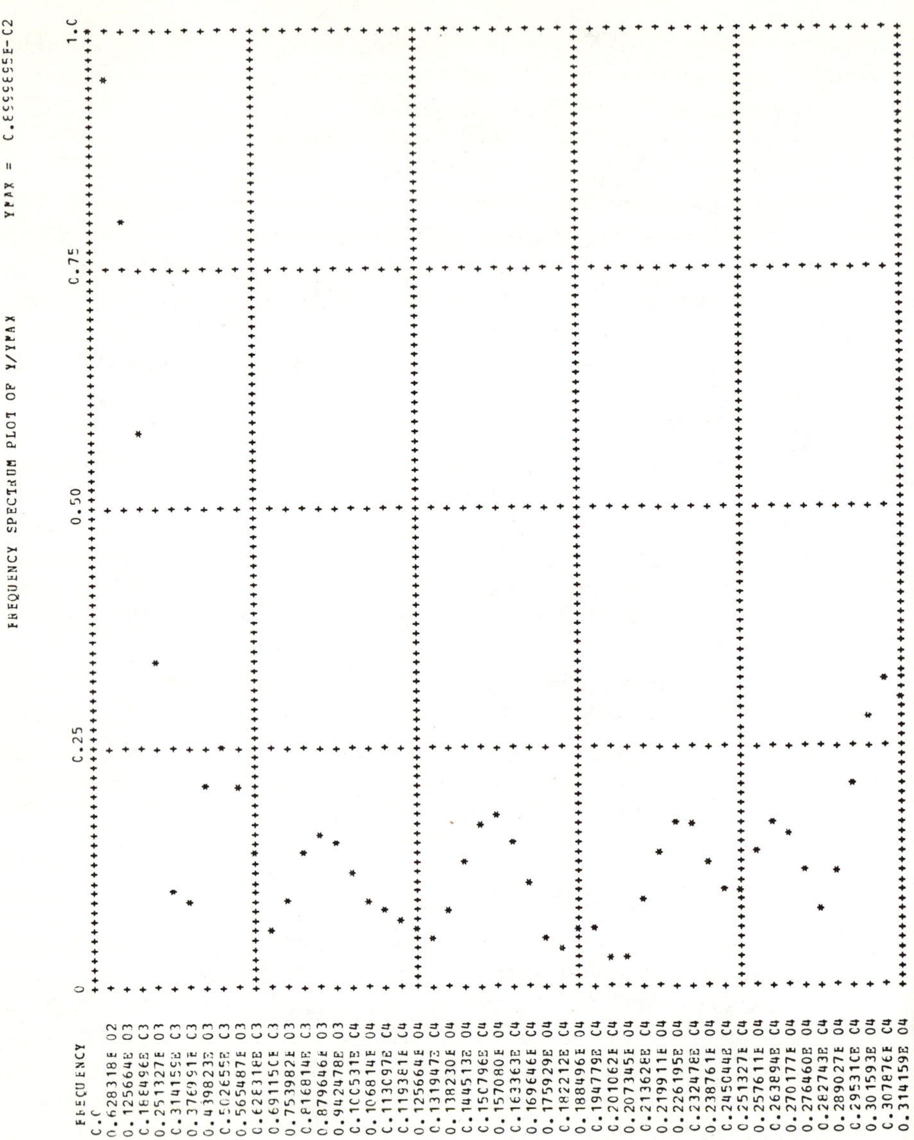

Figure 11.10 The output results of the program of Figure 11.8 for the nine pulse waveform of Figure 5.34c

As a consequence, additional memory is needed to produce a plot of this type.

Figure 11.11 is a Fortran program that plots the underdamped case of the normalized time response shown in Figure 11.7a. Note that the DIMENSION statement uses PLOT(41,101) to store the plotted-point symbol. This permits the printing of a rectangular grid plot that contains 41 by 101 grid points. The 101 columns correspond to the normalized values of T/TMAX, where TMAX is the maximum time and not the

```
C      NORMALIZED TIME RESPONSE PLOT
C
       DIMENSION PLOT(41,101),TI(50),T(50)
       READ (5,1) STAR,DOT,BLANK
     1 FORMAT (3A1)
       READ (5,2) Q,R,AL,C
     2 FORMAT (4F10.6)
       ALPHA=R/(2.*AL)
       DISC=ALPHA**2-1./(AL*C)
       TIME=0.0
       F0=.159155/SQRT(AL*C)
       BETA=SQRT(-DISC)
       F=.159155*BETA
       TMAX=ATAN(BETA/ALPHA)/BETA
       DELTA=8.*TMAX/50.
       AI=6.2832*F0**2*Q/F
       AMAXI=AI*EXP(-ALPHA*TMAX)*SIN(BETA*TMAX)
       DO 10 I=1,50
       T(I)=TIME
       TI(I)=AI*EXP(-ALPHA*TIME)*SIN(BETA*TIME)
    10 TIME=TIME+DELTA
C
C      THE GRID LINES ARE FORMED.
C
       DO 20 I=1,41
       DO 20 J=1,101
    20 PLOT(I,J)=BLANK
       DO 30 J=1,101,25
       DO 30 I=1,41
    30 PLOT(I,J)=DOT
       DO 40 I=1,41,10
       DO 40 J=1,101
    40 PLOT(I,J)=DOT
C
C      THE ELEMENTS OF PLOT CORRESPONDING TO THE PLOTTED POINTS ARE DETERMINED.
C
       FTIME=50.*DELTA
       DO 50 I=1,50
       J=100.*T(I)/FTIME+1.5
       K=20.*(1.-TI(I)/AMAXI)+1.5
    50 PLOT(K,J)=STAR
C
C      THE GRAPH IS PRINTED.
C
       WRITE(6,3) AMAXI,FTIME
     3 FORMAT('1',10X,'NORMALIZED TIME RESPONSE PLOT',//,15X,'YMAX =',E15
      1.7,4X,'TMAX =',E15.7,//,' Y(T)/YMAX',//)
       K=0
       F=1.5
       DF=0.5
       DO 60 I=1,41
       IF (K-K/10*10) 70,70,80
    70 F=F-DF
       WRITE(6,90) F,(PLOT(I,J),J=1,101)
    90 FORMAT(F8.1,2X,101A1)
       GO TO 60
    80 WRITE (6,100) (PLOT (I,J),J=1,101)
   100 FORMAT (10X,101A1)
    60 K=K+1
       WRITE(6,110)
   110 FORMAT(//10X,'0',23X,'0.25',22X,'0.50',22X,'0.75',22X,'1.00',/,10C
      1X,'T/TMAX')
       STOP
       END
```

Figure 11.11 A Fortran program that produces a normalized time response plot of an underdamped R-L-C circuit

time to reach the first maximum. For example, if a plotted point is located at a horizontal position that corresponds to column 51, this would be equivalent to a time of

$$\frac{51 - 1}{101 - 1} \text{TMAX} = 0.5 \text{ TMAX (see Figure 11.12)}$$

The 41 rows of this plot represent the normalized values of Y(T)/YMAX (I(T)/IMAX) with the first row or line corresponding to a normalized value of +1.0, the 11th line to a normalized value of +0.5, the 21st line to a value of 0, the 31st line to a value of −0.5, and the 41st line to a value of −1.0.

Computer statements S.0002 through S.0019 are identical to those found in the program of Figure 11.6 for the underdamped case. Computer statements S.0020 through S.0049 are the statements used to produce the plot. The DO 20 loop is used first to store a blank in each of the elements of PLOT. The DO 30 loop then stores a DOT in each line element for columns 1, 26, 51, 76, and 101. The DO 40 loop then completes the grid by storing a DOT in each column element for rows 1, 11, 21, 31, and 41.

The DO 50 loop is now used to locate and store the plotted-point symbol. This is done by first computing the maximum value of time from

FTIME = 50.*DELTA

Note that this is done here so that this plot will correspond exactly to that of Figure 11.7a. In general, any value of FTIME (final time) may be specified so that columns 1 through 101 will be equivalent to time values of 0 to FTIME with a time-axis resolution of FTIME/100 units. Computer statement S.0031 is used to determine the column location of the point from

J = 100.*T(I)/FTIME + 1.5

and computer statement S.0032 determines the row location of the point from

K = 20.*(1. − TI(I)/AMAXI) + 1.5

The scope statement 50 then stores the symbol STAR in the corresponding element of PLOT.

After the DO 50 loop has stored the 50 plotted points, computer statements S.0034 and S.0035 are used to print the plot heading and the normalized scale values of YMAX (AMAXI) and TMAX (FTIME). The dummy variable K is then set equal to 0 and F and DF are set equal to 1.5 and 0.5, respectively. We then enter the DO 60 loop, which is used to print one line of PLOT each iteration of the loop. During each iteration computer statement S.0040 is used to determine whether

K − K/10*10

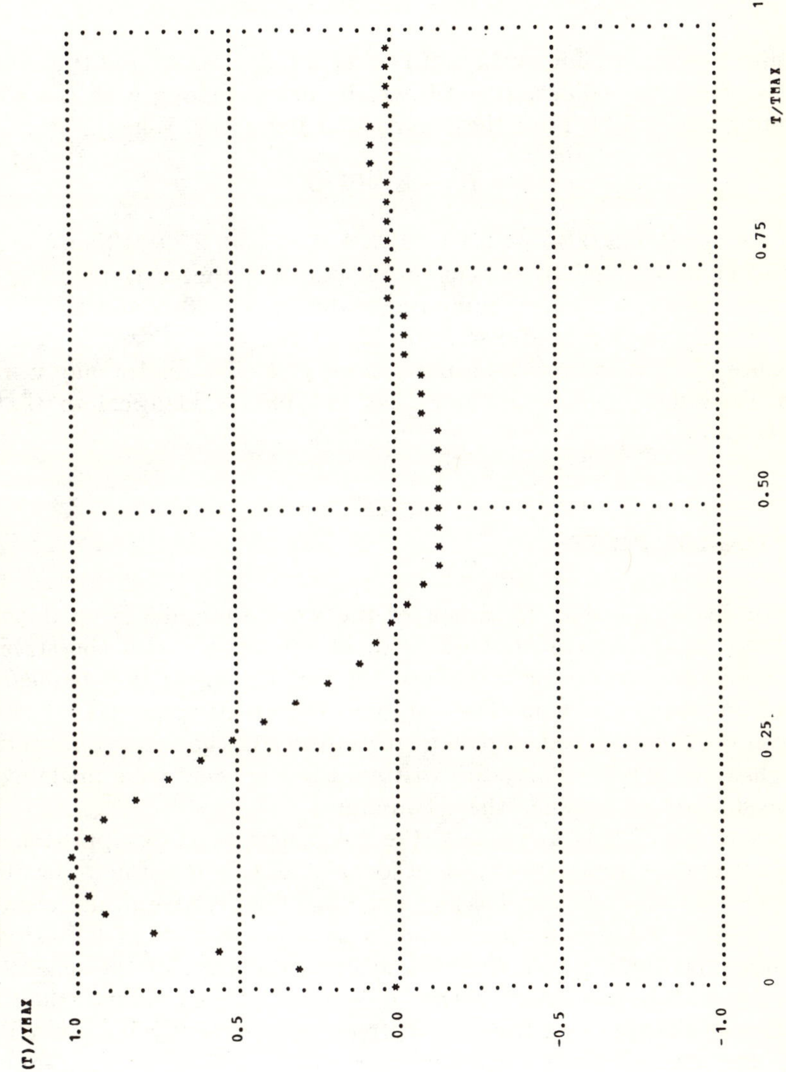

Figure 11.12 The normalized time response plot produced by the program of Figure 11.11

is equal to 0. If the above term is equal to 0, F is set equal to $F - DF$ so that in the first iteration of the DO 60 loop F will be equal to $+1.0$. Computer statements S.0042 and S.0043 then print the value of F and the first line of PLOT. The reader may observe that

$$\text{K} - \text{K}/10*10$$

will also be equal to 0 for values of $I = 11, 21, 31$, and 41 and thus values of $F = +0.5, 0, -0.5$, and -1.0 will be printed along with the corresponding line of PLOT for these particular iterations. When

$$\text{K} - \text{K}/10*10$$

does not equal 0 (S.0040 with I = 2, 3, 4, 5, 6, 7, 8, 9, 10, 12, and so on), control is transferred to statement 80 and only the corresponding line I will be printed. Finally, computer statements S.0048 and S.0049 are used to print the horizontal scale for T/TMAX.

Figure 11.12 is the resulting normalized plot. The reader may use the normalized scale values to show that this plot is identical to that of Figure 11.7a.

11.4 LOG-LOG PLOTS

Prior to World War II, much of the work that had been done by engineers in the active network field, in connection with the stability and response of feedback amplifiers, was not known to the engineers in the automatic control field. The war, however, greatly stimulated activity in both fields mentioned, and as a consequence of the many publications and classified reports that were written, the engineers were made aware of the similarities between these two areas.

As a result of this awareness, the frequency response approach was developed into a very powerful tool for the analysis of a general feedback system. This analysis was based on the fact that a magnitude response change of 20 dB/decade corresponded to a phase shift of 90°. It thus followed that for a stable feedback system, the slope of the magnitude response must be less than -40 dB/decade at unity gain. Since the effect of element changes as a function of frequency is readily seen in a Bode plot, the use of these log-magnitude and phase-versus-log-frequency plots became one of the primary tools in the analysis and design of feedback systems. The remainder of this section is used to discuss how the computer may be used to produce two different forms of a log-log plot.

Papoulis versus Butterworth Log-Log Response Plot

In Section 3.3 we developed the program of Figure 3.50 in order to compare the frequency response of one-to-eight-pole Papoulis and Butterworth filters. In this section we consider the computer plotting of the log-magnitude versus log-frequency response of the three-pole Papoulis and Butterworth filters.

Figure 11.13 is the Fortran program for producing this log-log plot. Computer statements S.0001 through S.0022 are essentially the same as computer statements S.0001 through S.0024 in the program of Figure

```
              C      PAPOULIS VERSUS BUTTERWORTH FILTERS
              C
S.0001               DIMENSION W(40),P(16),B(16),AP(40),AB(40),PLOT(51,101)
S.0002               WRITE(6,1)
S.0003             1 FORMAT('1A COMPARISON OF THE AMPLITUDE RESPONSE OF PAPOULIS VERSUS
                    1 BUTTERWORTH FILTERS',///)
S.0004               DO 20 I=1,40
S.0005            20 W(I)=0.1*I
S.0006               N=3
S.0007            41 WRITE(6,2)N
S.0008             2 FORMAT(//,10X,'POLYNOMINAL DEGREE =',I4)
S.0009               READ(5,3) (P(I),I=2,16,2)
S.0010             3 FORMAT(8F10.2)
S.0011               WRITE(6,4) (P(I),I=2,16,2)
S.0012             4 FORMAT(//,' THE COEFFICIENTS P(2), P(4), ..., P(16) ARE',//,8(3X,F
                    18.2))
S.0013               READ(5,3) (B(I),I=2,16,2)
S.0014               WRITE(6,5) (B(I),I=2,16,2)
S.0015             5 FORMAT(//,' THE COEFFICIENTS B(2), B(4), ..., B(16) ARE',//,8(3X,F
                    18.2))
              C
              C      CALCULATION OF THE AMPLITUDE FUNCTIONS
              C
S.0016               WRITE(6,6)
S.0017             6 FORMAT(//,2X,' FREQUENCY W',3X,'AMPLITUDE AP(W)',2X,'AMPLITUDE AB(
                    1W)',//,8X,'RADIANS',12X,'DB',15X,'DB',//)
S.0018               DO 50 I=1,40
S.0019               AP(I)=-10.0*ALOG10(1.+P(2)*W(I)**2+P(4)*W(I)**4+P(6)*W(I)**6+P(8)*
                    1W(I)**8+P(10)*W(I)**10+P(12)*W(I)**12+P(14)*W(I)**14+
                    2P(16)*W(I)**16)
S.0020               AB(I)=-10.0*ALOG10(1.+B(2)*W(I)**2+B(4)*W(I)**4+B(6)*W(I)**6+B(8)*
                    1W(I)**8+B(10)*W(I)**10+B(12)*W(I)**12+B(14)*W(I)**14+
                    2B(16)*W(I)**16)
S.0021            50 WRITE(6,7) W(I),AP(I),AB(I)
S.0022             7 FORMAT(3E17.5)
              C
              C      THE PLOT ROUTINE.
              C
S.0023               READ(5,8) PLUS,STAR,DOT,BLANK
S.0024             8 FORMAT(4A1)
              C
              C      THE GRID LINES ARE FORMED.
              C
S.0025               DO 55 I=1,46
S.0026               DO 55 J=1,101
S.0027            55 PLOT(I,J)=BLANK
S.0028               AN=0.0
S.0029               DO 60 I=1,10
S.0030               AN=AN+0.1
S.0031               K=50.*(ALOG10(AN)+1.)+1.5
S.0032               DO 60 J=1,46
S.0033            60 PLOT(J,K)=DOT
S.0034               AN=0.0
S.0035               DO 65 I=1,10
S.0036               AN=AN+1.0
S.0037               K=50.*(ALOG10(AN)+1.)+1.5
S.0038               DO 65 J=1,46
S.0039            65 PLOT(J,K)=DOT
S.0040               DO 70 I=1,46,5
S.0041               DO 70 J=1,101
S.0042            70 PLOT(I,J)=DOT
```

Figure 11.13 A program for producing a log-log plot of the three-pole Papoulis and Butterworth filter response

```
          C
          C      THE ELEMENTS OF PLOT CORRESPONDING TO THE PLOTTED POINTS ARE DETERMINED.
          C
S.0043           DO 80 I=1,40
S.0044           J=50.*(ALOG10(W(I))+1.)+1.5
S.0045           KP=ABS(AP(I))+1.5
S.0046           KB=ABS(AB(I))+1.5
S.0047           PLOT(KP,J)=PLUS
S.0048        80 PLOT(KB,J)=STAR
          C
          C      THE GRAPH IS PRINTED.
          C
S.0049           WRITE(6,9)
S.0050         9 FORMAT('1A COMPARISON OF THE AMPLITUDE RESPONSE OF PAPOULIS VERSUS
                1 BUTTERWORTH FILTERS',//,20X,'POLYNOMIAL DEGREE = 3',///,' MAGNITU
                2DE ',/,3X,'DB',/,9X,'0.1',47X,'1.0',46X,' 10.0',2X,'FREQUENCY',/)
S.0051           L=0
S.0052           F=5.0
S.0053           DF=-5.0
S.0054           DO 90 I=1,46
S.0055           IF(L-L/5*5) 100,100,110
S.0056       100 F=F+DF
S.0057           WRITE(6,120) F,(PLOT(I,J),J=1,101)
S.0058       120 FORMAT(F8.1,2X,101A1)
S.0059           GO TO 90
S.0060       110 WRITE(6,130) (PLOT(I,J),J=1,101)
S.0061       130 FORMAT(10X,101A1)
S.0062        90 L=L+1
S.0063           STOP
S.0064           END
```

Figure 11.13 Continued

3.50, except that in this program the response is computed for only the three-pole case.

Computer statements S.0023 through S.0062 constitute the plot routine. Initially the A-format variables PLUS, STAR, DOT, and BLANK are read into the computer memory. PLUS is the name of the variable that represents the symbol "+" that is used to indicate Papoulis filter points and STAR is the variable name for the symbol "*" that is used for the Butterworth response points. The variable DOT is the symbol "." that is used to form the grid lines and "BLANK" is the variable used to store the computer representation of a blank.

The grid lines are formed by first using the two DO 55 loops to blank each of the elements of the array PLOT. The double subscripted variable PLOT is used to store the grid lines and plotted point information. Since the dimension of PLOT is 46 by 101, the printing of this array will result in a plot that contains 46 lines and 101 columns. The vertical grid lines are formed by the DO 60 and DO 65 loops. The DO 60 loops are used to produce vertical grid lines that are spaced logarithmically on a 0.1 unit basis in the frequency range of 0.1 to 1.0 units. This is accomplished by first setting the variable AN equal to 0 so that in the first iteration of the outer DO 60 loop, AN will be equal to 0.1 (this will correspond to W(1) of a normalized frequency of 0.1 units). Computer statement S.0031 then computes K from

$$
\begin{aligned}
\mathrm{K} &= 50.*(\mathrm{ALOG10(AN)} + 1.) + 1.5 \\
&= 50.*(\mathrm{ALOG10}(0.1) + 1.) + 1.5 \\
&= 1.5
\end{aligned}
$$

so that in rounding, $K = 1$ and thus the execution of the inner DO 60 loops causes a DOT to be stored in the first column of each of the 46 rows of PLOT. In the second iteration of the outer DO 60 loop, AN will be equal to 0.2 so that

$$\begin{aligned} \text{K} &= 50.*(\text{ALOG10}(0.2) + 1.) + 1.5 \\ &= 50.*(-0.7 + 1.0) + 1.5 \\ &= 16.5 \end{aligned}$$

and thus a DOT will be stored in column 16 for each of the 46 rows of PLOT. The DO 60 iteration is continued until $I = 10$ (AN = 1.0 or W(10) = 1.0) and for this tenth iteration

$$\begin{aligned} \text{K} &= 50.*(\text{ALOG}(1.0) + 1.) + 1.5 \\ &= 51.5 \end{aligned}$$

and a DOT is thus stored in column 51 for each of the 46 rows.

In a similar way the two DO 65 loops are used to store the symbol for producing the vertical grid lines in the normalized frequency range of 1.0 to 10.0 units. The 101 columns of PLOT therefore correspond to the normalized frequency range of 0.1 to 10.0 units and the vertical grid lines are spaced logarithmically on a 0.1 unit basis in the 0.1 to 1.0 frequency range and on a 1.0 unit basis in the 1.0 to 10.0 range.

The DO 70 loops are used to store the symbol for producing the horizontal grid lines. Since the outer DO 70 loop is incremented by five each iteration, this corresponds to a horizontal grid spacing of 5 dB.

The DO 80 loop is used to store the plotted-point information. During each iteration of the DO 80 loop, the value of J is found from

$$\text{J} = 50.*(\text{ALOG10}(\text{W(I)}) + 1.) + 1.5$$

The value of J thus equals the column location of PLOT for the specific frequency $W(I)$. The value of KP is next found from

$$\text{KP} = \text{ABS(AP(I))} + 1.5$$

For example, during the first iteration, the value of I is equal to one and the value of $AP(1)$ corresponds to the magnitude of the three-pole Papoulis filter for the frequency $W(1)$ or 0.1 units. Since ABS(AP(1)) = 0.041931 dB, KP will be equal to 1. Thus, for a value of $J = 1$ and $KP = 1$, a PLUS is stored in PLOT(1,1). Similarly, KB will also be equal to 1 for $J = 1$, and a STAR is now stored in PLOT(1,1). Since the STAR is stored after the PLUS, only the STAR will appear in the printing of PLOT for this value of J. The execution of the DO 80 loop is continued until all 40 frequencies have been considered and the corresponding points stored.

Computer statements S.0049 through S.0062 are then used to print the plot. Initially the heading and frequency axis labeling are printed by computer statements S.0049 and S.0050. The dummy variable L is then set equal to 0 and this is followed by setting F and DF equal to 5.0 and -5.0, respectively. We then enter the DO 90 loop, which is executed 46 times to produce the plot. During each iteration of the DO 90 loop, computer statement S.0055 tests to see if

$$\mathrm{L - L/5{*}5}$$

is equal to 0. The value of L will be equal to zero for $I = 1, 6, 11, 16, 21, 26, 31, 36, 41$, and 46; and thus for each of these iterations, control is transferred to statement 100 so that the dB magnitude can be printed. In the first iteration, when control is transferred to statement 100, F is equal to 0, and the execution of computer statements S.0057 and S.0058 causes the value of 0.0 to be printed along with the first line of PLOT. Similarly, in the sixth iteration, F will be equal to -5.0 and thus the dB value of -5.0 will be printed along with the sixth line of PLOT, and so on. For all other values of I (2, 3, 4, 5, 7, etc.), control is transferred to statement 110 and only the corresponding line of PLOT is printed.

Figure 11.14 shows the computer output for this program. The reader should observe that it would be a simple matter to change either the horizontal or vertical scaling of this plot. For example, the dB scale could be changed from 0 to -45 dB to 0 to -50 dB by changing the DIMENSION statement of PLOT(46,101) to PLOT(51,101) and changing the upper limit of computer statements S.0025, S.0032, S.0038, S.0040, and S.0054 from 46 to 51. Similarly, the vertical resolution could be changed from 1.0 dB to 0.1 dB by changing computer statements S.0045 and S.0046 to

$$\mathrm{KP = 10.{*}ABS(AP(I)) + 1.5}$$
$$\mathrm{KB = 10.{*}ABS(AB(I)) + 1.5}$$

The frequency axis could also be changed by making similar changes for the column variable of PLOT. For example, for a desired frequency range of 0.1 to 100.0 (or three decades), computer statement S.0031 could be changed to

$$\mathrm{K = 30.{*}(ALOG10(AN) + 1.) + 1.5}$$

Thus when the frequency $AN = 0.1$, K would be equal to 1; when $AN = 1.0$, K would be equal to 31; when $AN = 10.0$, K would be equal to 61; and when $AN = 100.0$, K would be equal to 91. The 91 columns of PLOT would thus correspond to a frequency range of 0.1 to 100.0.

It is suggested that the reader determine what other statements would have to be modified for this particular frequency change.

Magnitude-Phase Transfer Function Plot

As a second example of a computer log-log plot, let us consider the lead-lag network shown in Figure 11.15. The open-circuit voltage transfer ratio of this network is given by

$$\begin{aligned} G(j\omega) &= \frac{V_2(j\omega)}{V_1(j\omega)} = \frac{R_2/(1 + j\omega R_2C_2)}{R_1/(1 + j\omega R_1C_1) + R_2/(1 + j\omega R_2C_2)} \\ &= \frac{R_2(1 + j\omega R_1C_1)}{R_1(1 + j\omega R_2C_2) + R_2(1 + j\omega R_1C_1)} \\ &= \frac{R_2(1 + j\omega R_1C_1)}{(R_1 + R_2)\left(1 + j\omega R_1C_1 \dfrac{R_2(C_1 + C_2)}{C_1(R_1 + R_2)}\right)} \\ &= R\,\frac{1 + j\omega/\omega_1}{1 + j\omega/\omega_2} \end{aligned}$$

where $R = \dfrac{R_2}{R_1 + R_2}$ $\qquad C = \dfrac{C_1}{C_1 + C_2}$

and $\omega_1 = \dfrac{1}{R_1C_1}$ $\qquad \omega_2 = \dfrac{C}{R}\,\omega_1 = \dfrac{C_1(R_1 + R_2)}{R_2(C_1 + C_2)}\,\dfrac{1}{R_1C_1}$

Writing this equation in polar form, we have

$$G(j\omega) = \frac{R\sqrt{1 + (\omega/\omega_1)^2}\;\underline{|\tan^{-1}(\omega/\omega_1)}}{\sqrt{1 + (\omega/\omega_2)^2}\;\underline{|\tan^{-1}(\omega/\omega_2)}}$$

The decibel magnitude and phase functions may now be written as separate expressions

$$\begin{aligned} G(j\omega)\Big|_{\text{dB}} &= 20\log_{10}\left[R\sqrt{\frac{1 + (\omega/\omega_1)^2}{1 + (\omega/\omega_2)^2}}\right] \\ \theta(j\omega) &= \tan^{-1}(\omega/\omega_1) - \tan^{-1}(\omega/\omega_2) \end{aligned}$$

An examination of these last two equations will show that this network will correspond to a leading doublet, a lagging doublet, or an all-pass network; depending on whether ω_1 is less than, greater than, or equal to ω_2. In order to examine each of these cases, let us plot the log-magnitude

```
A COMPARISON OF THE AMPLITUDE RESPONSE OF PAPOULIS VERSUS BUTTERWORTH FILTERS


          POLYNOMINAL DEGREE =    3

THE COEFFICIENTS P(2), P(4), ..., P(16) ARE

     1.00       -3.00        3.00        0.0        0.0        0.0        0.0        0.0

THE COEFFICIENTS B(2), B(4), ..., B(16) ARE

     0.0         0.0         1.00        0.0        0.0        0.0        0.0        0.0

  FREQUENCY W   AMPLITUDE AP(W)   AMPLITUDE AB(W)

     RADIANS          DB                DB

   0.10000E 00     -0.41931E-01     -0.41418E-05
   0.20000E 00     -0.15104E 00     -0.27749E-03
   0.30000E 00     -0.28525E 00     -0.31631E-02
   0.40000E 00     -0.39607E 00     -0.17748E-01
   0.50000E 00     -0.45078E 00     -0.67330E-01
   0.60000E 00     -0.45780E 00     -0.19804E 00
   0.70000E 00     -0.50243E 00     -0.48305E 00
   0.80000E 00     -0.78323E 00     -0.10111E 01
   0.90000E 00     -0.15716E 01     -0.18510E 01
   0.10000E 01     -0.30103E 01     -0.30103E 01
   0.11000E 01     -0.49587E 01     -0.44272E 01
   0.12000E 01     -0.71409E 01     -0.60053E 01
   0.13000E 01     -0.93460E 01     -0.76543E 01
   0.14000E 01     -0.11469E 02     -0.93092E 01
   0.15000E 01     -0.13470E 02     -0.10931E 02
   0.16000E 01     -0.15344E 02     -0.12499E 02
   0.17000E 01     -0.17097E 02     -0.14003E 02
   0.18000E 01     -0.18738E 02     -0.15442E 02
   0.19000E 01     -0.20280E 02     -0.16817E 02
   0.20000E 01     -0.21732E 02     -0.18129E 02
   0.21000E 01     -0.23104E 02     -0.19383E 02
   0.22000E 01     -0.24404E 02     -0.20583E 02
   0.23000E 01     -0.25640E 02     -0.21733E 02
   0.24000E 01     -0.26817E 02     -0.22835E 02
   0.25000E 01     -0.27941E 02     -0.23894E 02
   0.26000E 01     -0.29017E 02     -0.24912E 02
   0.27000E 01     -0.30048E 02     -0.25893E 02
   0.28000E 01     -0.31038E 02     -0.26838E 02
   0.29000E 01     -0.31991E 02     -0.27751E 02
   0.30000E 01     -0.32909E 02     -0.28633E 02
   0.31000E 01     -0.33795E 02     -0.29487E 02
   0.32000E 01     -0.34651E 02     -0.30313E 02
   0.33000E 01     -0.35478E 02     -0.31114E 02
   0.34000E 01     -0.36280E 02     -0.31892E 02
   0.35000E 01     -0.37057E 02     -0.32646E 02
   0.36000E 01     -0.37811E 02     -0.33380E 02
   0.37000E 01     -0.38543E 02     -0.34094E 02
   0.38000E 01     -0.39255E 02     -0.34788E 02
   0.39000E 01     -0.39947E 02     -0.35465E 02
   0.40000E 01     -0.40621E 02     -0.36125E 02

A COMPARISON OF THE AMPLITUDE RESPONSE OF PAPOULIS VERSUS BUTTERWORTH FILTERS

                    POLYNOMIAL DEGREE = 3
```

Figure 11.14 The output results of the program of Figure 11.13

and phase-versus-log-frequency response of this network for the following three sets of *R-C* values:

Case (a)	$R_1 = 1.0\text{E}6$	$C_1 = 1.0\text{E-}8$	$R_2 = 1.0\text{E}5$	$C_2 = 1.0\text{E-}9$
Case (b)	$R_1 = 1.0\text{E}5$	$C_1 = 1.0\text{E-}8$	$R_2 = 1.0\text{E}6$	$C_2 = 1.0\text{E-}7$
Case (c)	$R_1 = 1.0\text{E}5$	$C_1 = 1.0\text{E-}8$	$R_2 = 1.0\text{E}6$	$C_2 = 1.0\text{E-}9$

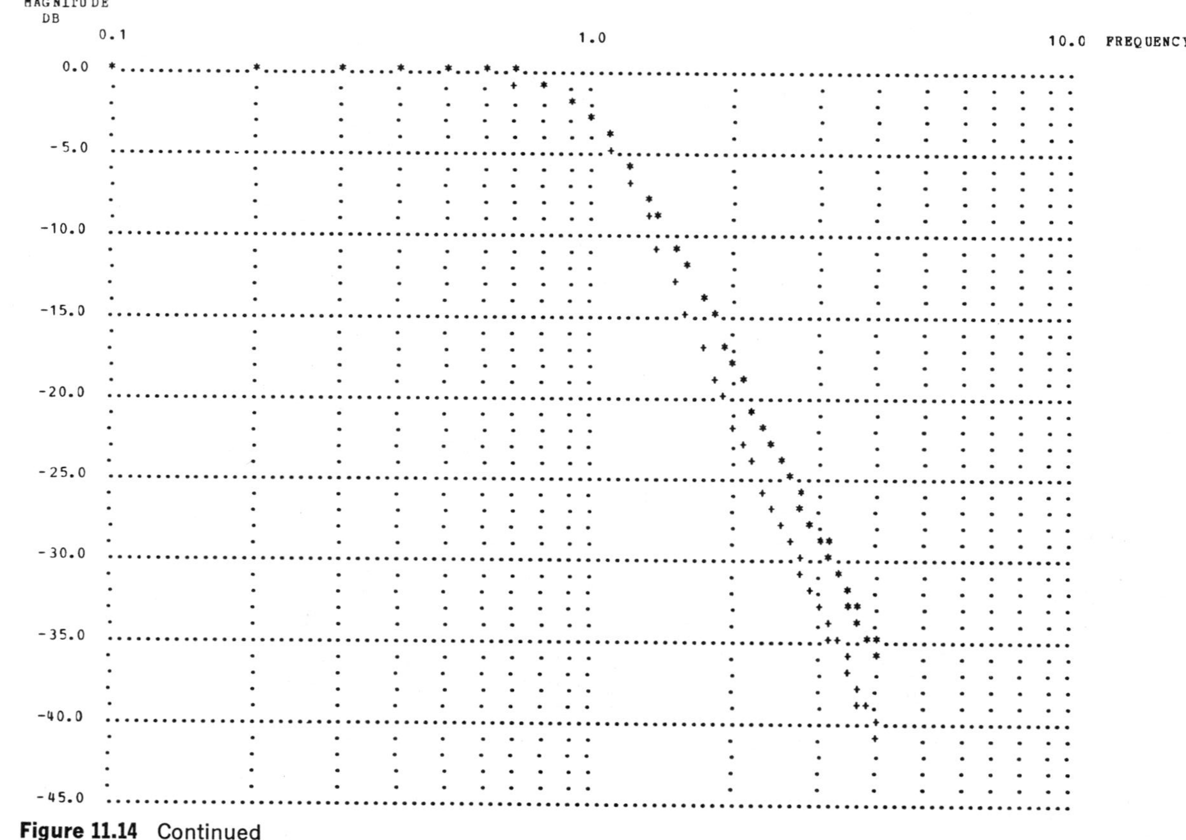

Figure 11.14 Continued

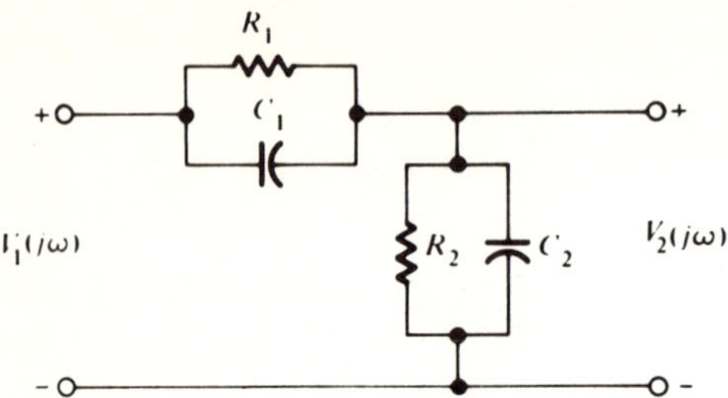

Figure 11.15 A lead-lag network

In case (a) we see that $\omega_1 = 100$ rps and $\omega_2 = 1000$ rps and, thus the network is a lead network; in case (b) we see that $\omega_1 = 1000$ rps and $\omega_2 = 100$ rps, and the network corresponds to a lag network; and in case (c), since $\omega_1 = \omega_2$, the network corresponds to an all-pass network. Let us now evaluate and plot these three cases for the following frequencies: $\omega = 10$, 25, 50, 100, 200, 500, 1000, 2000, and 10,000 radians per second.

Figure 11.16 is the Fortran program that computes and plots these three cases. Initially, we read in the A-format variables BLANK and

```
C       LOG MAGNITUDE AND PHASE
C       VERSUS LOG FREQUENCY PLOT.
C
        DIMENSION FREQ(9),GDB(3,9),ANG(3,9),IMAG(3,9),IPHASE(3,9),COL(101)
       1,SBL(3)
C
C       READ IN FORMAT VARIABLES
C
        READ(5,1)BLANK,DOT,N,M,(FREQ(I),I=1,M)
      1 FORMAT(2A1,2I4,/,(5E15.7))
C
C       THE DATA SETS ARE READ IN.
C
        GMIN=10.0
        DO 10 I=1,N
        READ(5,2)SBL(I),R1,R2,C1,C2
      2 FORMAT(1A1,4E15.7)
C
C       PRINT SYMBOLS USED IN GRAPHS AND DATA.
C
        WRITE(6,3)SBL(I),R1,C1,R2,C2
      3 FORMAT('1THE SYMBOL (',1A1,') REPRESENTS THE VALUES FOUND FROM THE
       1 FOLLOWING',//,'  R1 =',1PE11.4,'  C1 =',1PE11.4,'  R2 =',1PE11.4,
       2'  C2 =',1PE11.4,//,2X,'MAGNITUDE',8X,'PHASE',8X,'FREQUENCY',/)
C
C       THE CONSTANTS ARE EVALUATED.
C
        B=R2/(R1+R2)
        C=C1/(C1+C2)
        W1=1./(R1*C1)
        W2=1./(R1*B*(C1+C2))
C
C       THE MINIMUM MAGNITUDE IS FOUND.
C
        IF(C-B)20,30,30
     20 IF(GMIN-C)21,21,22
     30 IF(GMIN-B)21,21,23
     22 GMIN=C
        GO TO 21
     23 GMIN=B
     21 CONTINUE
```

Figure 11.16 A Fortran program for producing a log-log plot of the transfer function of the network of Figure 11.15

```
C
C      THE MAGNITUDE AND PHASE IS COMPUTED AND TABULATED FOR EACH FREQUENCY.
C
    40 DO 10 J=1,M
       GDB(I,J) = 20.*ALOG10(R*SQRT((1.+(FREQ(J)/W1)**2)/
      1(1.+(FREQ(J)/W2)**2)))
       ANG(I,J)=(ATAN(FREQ(J)/W1)-ATAN(FREQ(J)/W2))*57.29578
    10 WRITE(6,4)GDB(I,J),ANG(I,J),FREQ(J)
     4 FORMAT(3E15.7)
       GDBMIN=20.*ALOG10(GMIN)
       WRITE(6,602)GDBMIN
   602 FORMAT(' GDBMIN =',E15.7)
C
C      MAGNITUDE AND PHASE LOCATIONS ARE FOUND.
C
       DO 50 I=1,M
       DO 50 J=1,N
       IMAG(J,I)=100.*(1.-GDB(J,I)/GDBMIN)+1.5
    50 IPHASE(J,I)=46.5+ANG(J,I)/2.
C
C      PRINT DB MAGNITUDE VERSUS LOG FREQUENCY PLOT.
C
       WRITE(6,60)GDBMIN
    60 FORMAT('1',32X,'NORMALIZED MAGNITUDE IN DB VERSUS LOG FREQUENCY IN
      1 RADIANS',//,10X,'  GDB/GDBMIN',60X,'GDBMIN =',E15.7,//,9X,'1.0',7
      1X,'0.9',7X,'0.8',7X,'0.7',7X,'0.6',7X,'0.5',7X,'0.4',7X,'0.3',7X,'
      20.2',7X,'0.1',8X,'0',/)
       DO 80 I=1,101
    80 COL(I)=BLANK
       DO 90 I=1,61
       DO 100 K=1,M
       J=20.*(ALOG10(FREQ(K))-1.)+1.5
       IF(I-J)100,110,100
   110 DO 120 JJ=1,101
   120 COL(JJ)=DOT
       DO 130 MM=1,N
       NN=IMAG(MM,K)
   130 COL(NN)=SEL(MM)
       WRITE(6,6)(COL(LL),LL=1,101),FREQ(K)
     6 FORMAT(10X,101A1,F6.0)
       DO 135 MM=1,101
   135 COL(MM)=BLANK
       GO TO 90
   100 CONTINUE
       DO 140 II=1,101,10
   140 COL(II)=DOT
       WRITE(6,5)(COL(KK),KK=1,101)
     5 FORMAT(10X,101A1)
    90 CONTINUE
C
C      PRINT PHASE VERSUS LOG FREQUENCY PLOT.
C
       WRITE(6,7)
     7 FORMAT('1',32X,'ANGLE IN DEGREES VERSUS LOG FREQUENCY IN RADIANS',
      1//,8X,'-90',44X,'0',42X,'+90',2X,'FREQUENCY',/)
       DO 155 I=1,101
   155 COL(I)=BLANK
       DO 160 I=1,61
       DO 170 K=1,M
       J=20.*(ALOG10(FREQ(K))-1.)+1.5
       IF(I-J)170,180,170
   180 DO 190 JJ=1,91
   190 COL(JJ)=DOT
       DO 200 MM=1,N
       NN=IPHASE(MM,K)
   200 COL(NN)=SEL(MM)
       WRITE(6,9)(COL(II),II=1,91),FREQ(K)
     9 FORMAT(10X,91A1,F6.0)
       DO 205 IJ=1,101
   205 COL(IJ)=BLANK
       GO TO 160
   170 CONTINUE
       DO 210 II=1,91,5
   210 COL(II)=DOT
       WRITE(6,8)(COL(KK),KK=1,91)
     8 FORMAT(10X,91A1)
   160 CONTINUE
       STOP
       END
```

Figure 11.16 Continued

DOT, the number of cases N ($N = 3$), the number of frequencies M ($M = 9$), and the M frequencies (FREQ(I),I=1,M). The dummy variable GMIN is then set equal to some arbitrary value (10.0). We then enter the outer DO 10 loop. For each of our N cases, this outer loop reads in and prints the plotting symbol and the numerical values of the R's and C's of the network. This is followed by computer statements S.0010 through S.0013 which compute the transfer function constants R, C, ω_1, and ω_2. Computer statements S.0014 through S.0020 are then used to compute GMIN. We next enter the inner DO 10 loop and compute the dB magnitude and degree phase for each of the M frequencies.

After each of the N cases has been computed, the smallest value of GMIN is used to compute GDBMIN. This minimum dB value is needed in order to plot a normalized dB magnitude response (GDB/GDBMIN).

The two DO 50 loops are then used to find the magnitude and phase locations of the plotted points. This is done by first computing

$$\text{IMAG(J,I)} = 100.*(1. - \text{GDB(J,I)/GDBMIN}) + 1.5$$

and

$$\text{IPHASE(J,I)} = 46.5 + \text{ANG(J,I)}/2.$$

where the numerical value of I stands for one of the specific M frequencies and the value of J stands for one of the N cases. For example, since GDBMIN $= -20.827$ dB, we will find that for case (a) and frequency $\omega = 100.0$ rps,

$$\text{GDB(1,4)} = -17.8607 \text{ dB}$$

and

$$\text{ANG(1,4)} = 39.289°$$

Thus,

$$\begin{aligned}\text{IMAG(1,4)} &= 100.*(1. - (-17.8607/-20.827)) + 1.5\\ &= 100.*(1. - 0.857) + 1.5\\ &= 15.8\end{aligned}$$

and therefore in rounding IMAG(1,4) = 15. The symbol "+" will therefore be stored in column 15 for this particular frequency. Similarly,

$$\begin{aligned}\text{IPHASE(1,4)} &= 46.5 + 39.289/2.\\ &= 66.1445\end{aligned}$$

will cause the + symbol to be stored in column 66 in the phase plot (+40°).

Computer statements S.0033 through S.0056 are used to print the magnitude plot. Computer statements S.0033 and S.0034 first print the heading, the value of GDBMIN, and the horizontal axis scale. The

DO 80 loop then sets each of the elements of COL(I) to BLANK. The DO 90 loop is then entered and each iteration of this loop produces one line of a 61-line plot. During each iteration of the DO 90 loop, the DO 100 loop is used to determine whether the line should contain the plotted point symbols. This is determined by first computing

$$\mathrm{J} = 20.*(\mathrm{ALOG10(FREQ(K))} - 1.) + 1.5$$

For example, for the first iteration $K = 1$ and therefore FREQ(1) = 10 rps, the value of J is therefore equal to

$$\begin{aligned}\mathrm{J} &= 20.*(\mathrm{ALOG10(10)} - 1.) + 1.5\\ &= 1.5\end{aligned}$$

and thus, in rounding, $J = 1$. We then test to see if $I = J$. Since in the first iteration I will be equal to J, control is transferred to statement 110 where the DO 120 loop sets all of the elements of COL(JJ) equal to DOT. The DO 130 loop is then executed N times and each iteration stores the individual case symbols in the appropriate element of COL(NN). Computer statements S.0046 and S.0047 then print the graph line and the corresponding value of frequency. The DO 135 loop is then used to blank all the elements of COL(MM) and control is transferred to statement 90.

During the second iteration of the DO 90 loop, I is equal to 2. The first execution of the DO 100 loop therefore computes a value of $J = 1$ as before. Computer statement S.0040 thus transfers control to statement 100 since $I \neq J$ in this iteration. The value of K thus becomes equal to 2 and the value of J becomes equal to

$$\begin{aligned}\mathrm{J} &= 20.*(\mathrm{ALOG10(25)} - 1.) + 1.5\\ &= 20.*(1.398 - 1.) + 1.5\\ &= 9.46\end{aligned}$$

Since now the value of $J = 9$ for a value of $I = 2$, computer statement S.0040 again transfers control to statement 100. This will continue until $K = M$. During the last iteration of the DO 100 loop

$$\begin{aligned}\mathrm{J} &= 20.*\mathrm{ALOG10(FREQ(K))} - 1.) + 1.5\\ &= 20.*\mathrm{ALOG10}(4.0 - 1.) + 1.5\\ &= 61.5\end{aligned}$$

and therefore computer statement S.0040 once again transfers control to statement 100. Since the value of K would now be greater than M, control is transferred to computer statement S.0052. The DO 140 loop thus stores a DOT in COL(1), COL(11), . . . , and COL(101). Computer state-

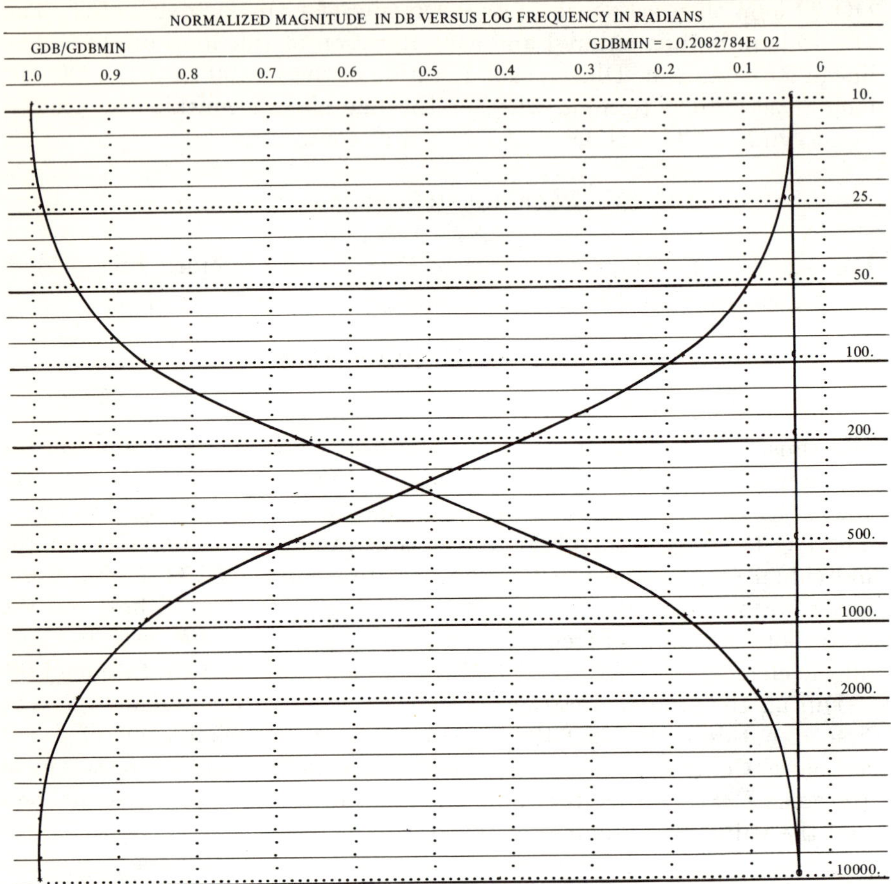

Figure 11.17 The magnitude plot produced by the program of Figure 11.16

ments S.0054 and S.0055 then print the second line of the graph which in this case contains only the grid line symbol.

It should therefore be evident that computer statements S.0039 and S.0040 determine whether the line will contain the plotted point symbols. The values of J in computer statement S.0039 will thus range in value from 1 through 61 ($\omega = 10$ through $\omega = 10{,}000$) and whenever the value of I corresponds to the value of J, the plotted point symbols will be printed. If the execution of the DO 100 loop fails to find a value of J equal to I, a line containing only the grid points will be printed. The graph lines are therefore printed on a log-frequency basis.

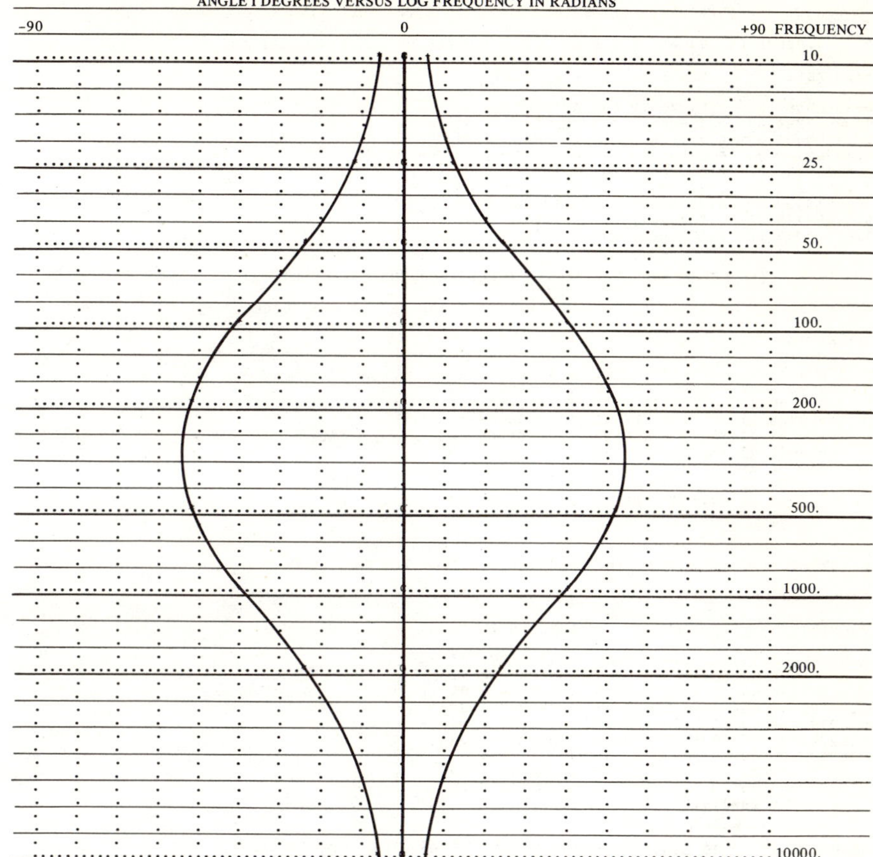

Figure 11.18 The phase plot produced by the program of Figure 11.16

Computer statements S.0057 through S.0080 are used to produce the phase plot. This graph is produced in much the same way that the magnitude graph was plotted, except that only 91 elements of COL are used for storage. The horizontal scale resolution is therefore equal to 2.0°.

Figures 11.17 and 11.18 are the respective magnitude and phase plots. The accuracy of these plots may be verified by the reader by comparing the plotted point values with the computed results shown in Figure 11.19. The reader should note that case (c) corresponds to the case where the pole and zero coincide and thus produces a voltage transfer ratio that is constant, independent of frequency. A practical application of case (c) would be the compensated voltage divider, where we pad the stray series and shunt capacitances associated with R_1 and R_2 so that $\omega_1 = \omega_2$.

```
THE SYMBOL (+) REPRESENTS THE VALUES FO NE FROM THE FOLLOWING

R1 = 1.0000E 06  C1 = 1.0000E-08  R2 = 1.0000E 05  C2 = 1.CCCOE-C9

MAGNITUDE          PHASE              FREQUENCY

-0.2078508E 02  0.5137647E 01   0.1000000E 02
-0.2056728E 02  0.1260414E 02   0.250000 E 02
-0.1986960E 02  0.2370262E 02   0.5CC000 E 02
-0.1786076E 02  0.3928938E 02   0.100000 E 03
-0.1400848E 02  0.5212498E 02   0.2CC000 E 03
-0.7647229E 01  0.5212497E 02   0.500000 E 03
-0.3794943E 01  0.3928937E 02   0.1CC0C0 E 04
-0.1786111E 01  0.2370267E 02   0.200000 E 04
-0.8706321E 00  0.5137668E 01   0.1C0000 E 05

THE SYMBOL (*) REPRESENTS THE VALUES FOUND FROM THE FOLLOWING

R1 = 1.0000E 05  C1 = 1.0000E-08  R2 = 1.0000E 06  C2 = 1.0CCOE-C7

MAGNITUDE          PHASE              FREQUENCY

-0.8706347E 00 -0.5137644E 01   0.1000000E 02
-0.1088427E 01 -0.1260413E 02   0.2500000E 02
-0.1786109E 01 -0.2370261E 02   0.5000000E 02
-0.3794938E 01 -0.3928937E 02   0.1000000E 03
-0.7647214E 01 -0.5212492E 02   0.2000000E 03
-0.1400848E 02 -0.5212495E 02   0.5000000E 03
-0.1786075E 02 -0.3928934E 02   0.1000000E 04
-0.1986958E 02 -0.2370261E 02   0.2000000E 04
-0.2078506E 02 -0.5137668E 01   0.1000000E 05


THE SYMBOL (O) REPRESENTS THE VALUES FOUND FROM THE FOLLOWING

R1 = 1.0000E 05  C1 = 1.0000E-08  R2 = 1.0000E 06  C2 = 1.0000E-C9

MAGNITUDE          PHASE              FREQUENCY

-0.8278540E 00  0.4268868E-06   0.1000000E 02
-0.8278540E 00  0.1067217E-05   0.250000 E 02
-0.8278540E 00  0.2134434E-05   0.500000 E 02
-0.8278540E 00  0.6830189E-05   0.100000 E 03
-0.8278540E 00  0.6830189E-05   0.200000 E 03
-0.8278540E 00  0.1707546E-04   0.500000 E 03
-0.8278540E 00  0.2049057E-04   0.100000 E 04
-0.8278540E 00  0.5464151E-04   0.200000 E 04
-0.8278540E 00  0.0             0.100000 E 05
GDBMIN = -0.2082784E 02
```

Figure 11.19 The output results of the program of Figure 11.16

11.5 POLAR-COORDINATE PLOTTING

Polar-coordinate or complex plane plotting is very common in many branches of electrical engineering. For example, field-intensity contours of antenna radiation patterns, conformal mapping problems, and graphical analysis of transfer functions are typical examples that require complex plane plots. The remainder of this section will consider how the computer may be used to produce two different graphical analyses of transfer function plots.

The Nyquist Plot

The Nyquist stability criterion is a graphical test that determines the stability of a feedback system from a polar plot of the open-loop transfer function, $\beta A(s)$, plotted for values of $s = j\omega$ for ω running from $-\infty$ to

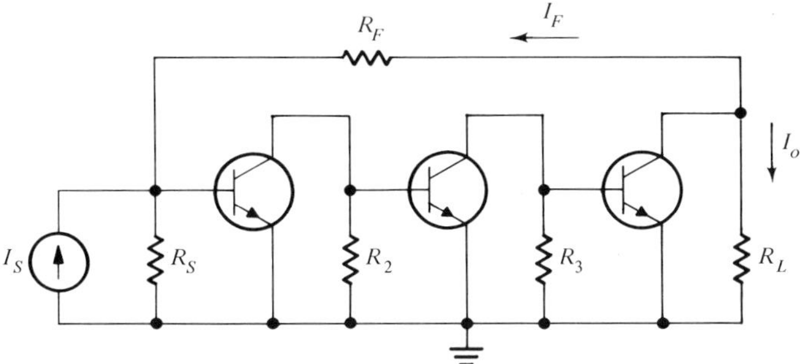

Figure 11.20 A three-stage feedback amplifier

$+\infty$. The system is unstable if the resulting plot encloses the $1 + j0$ point, and it is stable if it does not.

For a typical numerical illustration of the Nyquist criterion, let us consider the three-stage feedback amplifier shown in Figure 11.20. Using the unilateral hybrid-pi model, we may draw the forward equivalent circuit shown in Figure 11.21. The forward transmission function may then be written as

$$A(s) = -\frac{A_0 p_1 p_2 p_3}{(s + p_1)(s + p_2)(s + p_3)} = \frac{I_0(s)}{I_S(s)}$$

where

$$A_0 = \prod_{K=1}^{3} \beta_{0K} \frac{R_K}{R_K + R_{iK} + r'_{bK}}$$

and

$$p_K = \frac{\omega_{\beta K}}{D_K}\left(\frac{R_K + R_{iK} + r'_{bK}}{R_K + r'_{bK}}\right)$$

For the closed-loop amplifier, the feedback factor β may be approximated from Figure 11.20 as

$$\beta = \frac{I_F}{I_0} = \frac{R_L}{R_L + R_F}$$

or

$$\approx \frac{R_L}{R_F}$$

and thus the closed-loop transmission may be written as

$$G(s) = \frac{A(s)}{1 - \beta A(s)} = \frac{-A_0 p_1 p_2 p_3}{(s + p_1)(s + p_2)(s + p_3) + \beta A_0 p_1 p_2 p_3}$$

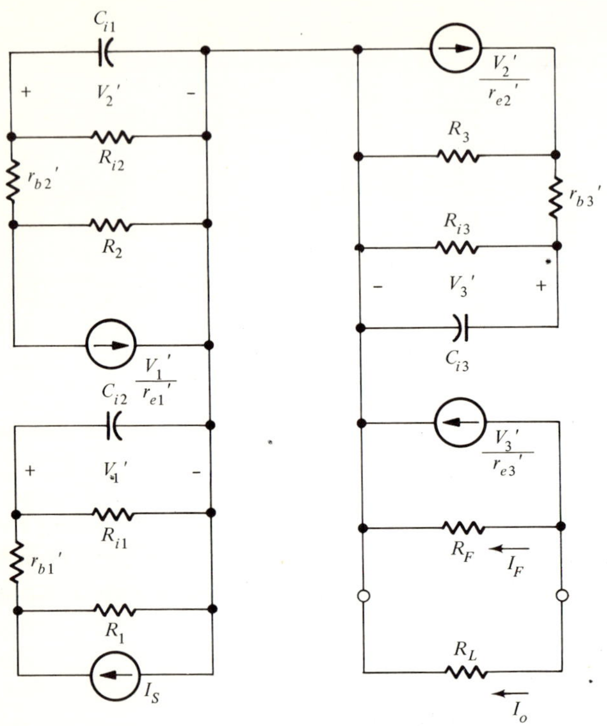

where: $R_{iK} = \dfrac{r_{eK}'}{1 - \alpha_{oK}} = \beta_{oK} r_{eK}' \quad ; \; K = 1, 2, 3$

$$C_{iK} = \frac{D_K}{r_{eK}' \, \omega_{tK}} \; : \; (D_K = 1 + (R_{LK} + r_{eK}') \, C_{cK} \, \omega_{tK})$$

$$R_1 = \frac{R_S R_F}{R_S + R_F}$$

Figure 11.21 The forward equivalent circuit of Figure 11.20

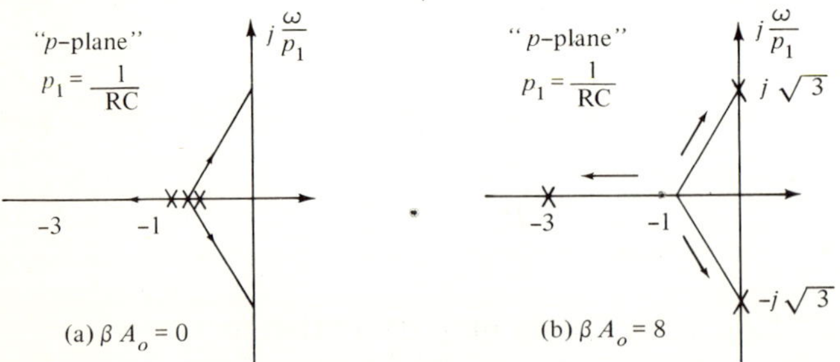

Figure 11.22 The effect of feedback on the poles of the network of Figure 11.20

If we now assume that the three stages are identical, or

$$p_1 = p_2 = p_3$$

we may write

$$A(p) = \frac{-A_0}{(p+1)^3}$$

or

$$G(p) = \frac{A(p)}{1 - \beta A(p)}$$

$$= \frac{-A_0}{p^3 + 3p^2 + 3p + \underbrace{1 + \beta A_0}_{F_0}}$$

$$= \frac{-A_0}{(p+1)^3 + \beta A_0}$$

where the normalized complex frequency variable p is given by

$$p = \frac{s}{p_1} = \frac{j\omega}{p_1}$$

An examination of this closed-loop transfer function shows that as the nominal feedback is increased from $F_0 = 1(\beta A_0 = 0$ or no feedback), the poles of $G(p)$ will move as shown in Figure 11.22a and reach threshold conditions for a value of $\beta A_0 = 8$ (see Figure 11.22b). It is therefore not possible to build a stable amplifier, using three identical stages, with a nominal loop gain, $\beta A_0 \geq 8$.

Consider now the three-stage vacuum tube amplifier shown in Figure 11.23. Assuming that we again have three identical stages and using

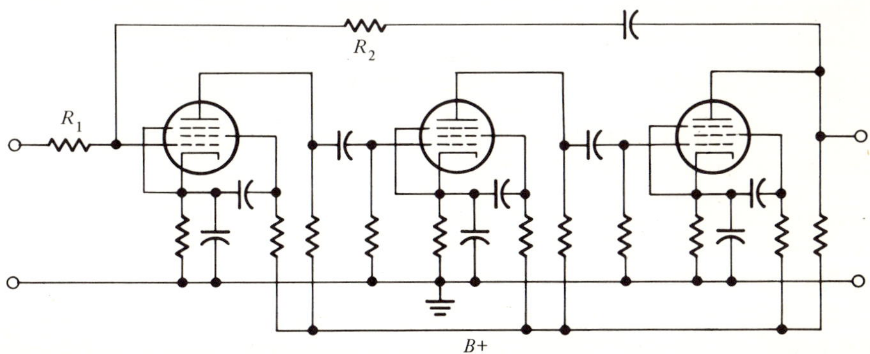

Figure 11.23 A three-stage vacuum tube feedback amplifier

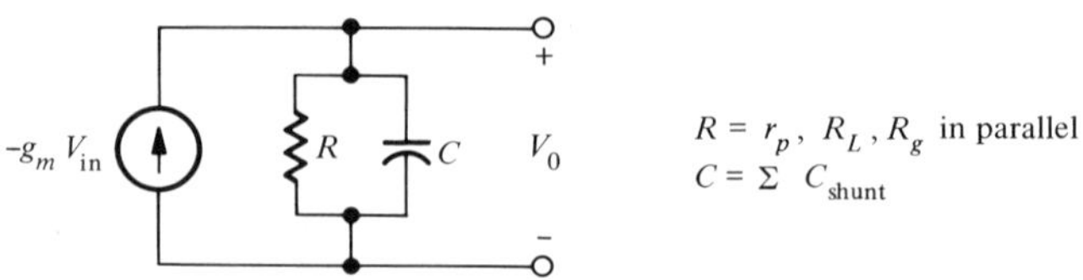

Figure 11.24 The high-frequency R-C amplifier equivalent circuit

Figure 11.24 as the high-frequency equivalent of a given stage, we may write the high-frequency stage transmission as

$$A_{st} = \frac{V_0}{V_{in}} = \frac{-g_m R}{1 + j\beta RC} = \frac{-A_{0st}}{1 + j\omega RC}$$

or

$$A_{st}(s) = \frac{-A_{0st}}{1 + sRC}$$

The magnitude and phase function may thus be written as

$$A_{st}(j\omega) = \frac{-A_{0st}}{\sqrt{1 + \left(\dfrac{\omega}{1/RC}\right)^2}} \quad \Bigg| \quad \theta_{st} = -\tan^{-1}\left(\frac{\omega}{1/RC}\right)$$

or

$$|A_{st}(j\omega)| = \frac{A_{0st}}{\sqrt{1 + \left(\dfrac{\omega}{p_1}\right)^2}} = A_{0st} \cos \theta_{st}$$

$$\theta_{st}(j\omega) = -\tan^{-1} \frac{\omega}{p_1}$$

The overall three-stage high-frequency transmission is thus given by

$$A(p) = \frac{-A_0}{(p + 1)^3}$$

where the overall nominal gain is called A_0 and the normalized complex frequency variable p is given by

$$p = \frac{s}{p_1} = \frac{j\omega}{p_1} = j\omega RC$$

Note that this overall transmission expression is identical to the forward transmission function of the three-stage transistor amplifier of Figure 11.20.

Since the transmission function of a single-stage vacuum-tube amplifier may be written as

$$A_{st} = -A_{0st} \cos \theta_{st} \left| \underline{\theta_{st} = -\tan^{-1}\left(\frac{\omega}{p_1}\right)} \right.$$

and the feedback factor β may be approximated from Figure 11.23 as

$$\beta = \frac{V_f}{V_0} = \frac{R_1}{R_1 + R_2}$$

the overall open-loop transmission, βA, is therefore given by

$$\beta A = -\beta A_0 \cos^3 \theta_{st} \left| \underline{\theta = -3\tan^{-1}\left(\frac{\omega}{p_1}\right)} \right.$$

Figure 11.25 shows the Nyquist plot for threshold conditions. Note the threshold conditions require the value of $|\beta A| = 1$ to occur at the same frequency that $\theta = -\pi$ radians. Thus at threshold $\Theta_{st} = -\pi/3$ and

$$|\beta A|_{th} = \frac{1}{\cos^3 \dfrac{\pi}{3}} = 8$$

Let us now plot a Nyquist diagram for a three-stage R-C cascaded amplifier with β = a scalar constant and βA_0 = 4 or 12 dB.

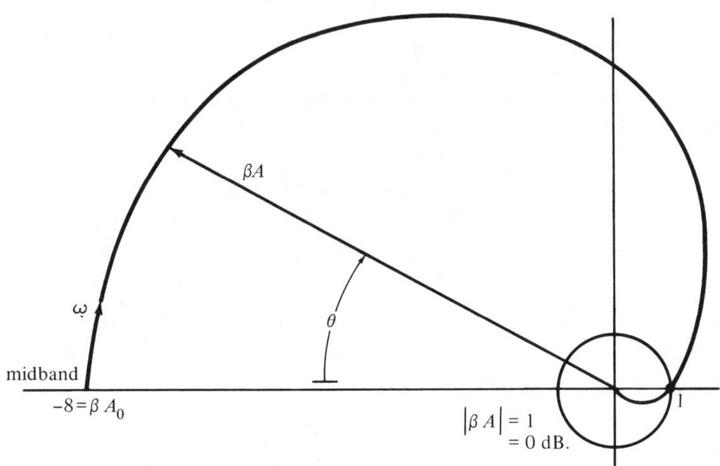

Figure 11.25 The three-stage amplifier Nyquist plot for threshold conditions

```
              C       NYQUIST PLOT OF THREE STAGE
              C       DIRECT COUPLED AMPLIFIER.
              C       NOMINAL LOOP GAIN = 4 OR 12 DB.
              C
S.0001                DIMENSION PLOT(81,81)
S.0002                READ (5,1) STAR,DOT,BLANK,PLUS,SCALE
S.0003              1 FORMAT(4A1,E15.7)
              C
              C       THE GRID LINES ARE FORMED.
              C
S.0004                DO 10 I=1,81
S.0005                DO 10 J=1,81
S.0006             10 PLOT(I,J)=BLANK
S.0007                DO 20 I=1,81
S.0008                DO 20 J=1,81,20
S.0009             20 PLOT(I,J)=DOT
S.0010                DO 30 I=1,81,20
S.0011                DO 30 J=1,81
S.0012             30 PLOT(I,J)=DOT
S.0013                READ (5,2) WINIT,WINCR,WMAX
S.0014              2 FORMAT(3E15.7)
S.0015                W=WINIT-WINCR
S.0016             40 W=W+WINCR
S.0017                THETAS=ATAN(W)
S.0018                THETA=-3.*THETAS
S.0019                BMAG=4.*(ABS(COS(THETAS)))**3
S.0020                X=-BMAG*COS(THETA)
S.0021                Y=-BMAG*SIN(THETA)
S.0022                J=40.*(X/SCALE+1.)+1.5
S.0023                I=40.*(1.-Y/SCALE)+1.5
S.0024                IF(I) 43,43,44
S.0025             44 IF (J) 43,43,45
S.0026             45 IF(J-81) 41,41,43
S.0027             41 IF (I-81) 42,42,43
S.0028             42 PLOT(I,J)=STAR
S.0029             43 IF (W-WMAX) 40,40,50
S.0030             50 WRITE(6,3) SCALE
S.0031              3 FORMAT('1NYQUIST PLOT OF THREE STAGE DIRECT COUPLED AMPLIFIER',//.
                     140X,'SCALE =',E15.7,//, 8X,' -1',17X,'-0.5',18X,'0',18X,'0.5',17X,
                     1'1.0',/)
S.0032                I=41.+40./SCALE
S.0033                PLOT(41,I)=PLUS
S.0034                K=0
S.0035                F=1.5
S.0036                DF=0.5
S.0037                DO 60 I=1,81
S.0038                IF(K-K/20*20) 70,70,80
S.0039             70 F=F-DF
S.0040                WRITE(6,90) F,(PLOT(I,J),J=1,81)
S.0041             90 FORMAT(F8.1,2X,81A1)
S.0042                GO TO 60
S.0043             80 WRITE(6,100) (PLOT(I,J),J=1,81)
S.0044            100 FORMAT(10X,81A1)
S.0045             60 K=K+1
S.0046                STOP
S.0047                END
```

Figure 11.26 A Fortran program that produces a Nyquist plot of a three-stage amplifier

Figure 11.26 is a Fortran program that produces the required Nyquist plot. Initially the A-format variables STAR, DOT, BLANK, and PLUS are read into memory along with the numerical value of the graph SCALE factor. The grid lines are then stored by the execution of the DO 10, DO 20, and DO 30 loops. Note that the DIMENSION PLOT (81,81) permits the printing of a graph that contains 81 lines and 81 columns. Each row or column will therefore have a normalized resolution of 0.025 units and the actual resolution will be determined by the value of SCALE.

Computer statements S.0013 and S.0014 next read in the initial value of the starting frequency WINIT, the frequency increment WINCR, and the maximum frequency WMAX. Computer statement S.0015 then sets W equal to WINIT—WINCR and statement 40 increments the value of W by WINCR so that for the first iteration W has the value of WINIT. Computer statements S.0017 and S.0018 now compute the value of θ_{st} and θ ($=3\theta_{st}$) and computer statement S.0019 computes the value of $|\beta A|$ for this corresponding value of W. Computer statements S.0020 and S.0021 then determine the real and imaginary parts of βA and computer statements S.0022 and S.0023 determine the plotted point location. For example, for $\omega = 0$ and SCALE = 5.0,

$$\theta = -3 \tan^{-1} \theta_{st} = -0.0$$
$$|\beta A| = 4 \cos^3 \theta_{st} = 4.0$$

and

$$X = -40.*0.0 = 0.0$$
$$Y = -40.*-1.0 = 4.0$$

thus,

$$J = 40.*\ (0.0/5.0 + 1.) + 1.5 = 41.5$$
$$I = (40.*\ (1. - (4.0)/5.0) + 1.5) = 9.5$$

This will cause computer statement S.0028 to store a STAR in PLOT (9,41). Note that computer statements S.0024 through S.0027 are required for a scale factor value that produces a value of $|X/\text{SCALE}|$ or $|Y/\text{SCALE}|$ that is greater than 1.0 since these will produce values of I and/or J that are negative or greater than 81.

Computer statement S.0029 next tests to see whether all the points have been plotted. If the value of W is less than WMAX, control is transferred to statement 40 so that the next point can be located and stored; otherwise, we go on to print the graph.

Computer statements S.0030 and S.0031 are used to print the Nyquist plot heading, the graph SCALE factor, and the normalized horizontal scale readings. Computer statements S.0032 and S.0033 then store a PLUS at point $1 + j0$. For example, with SCALE = 5.0, the $1 + j0$ point can be located by first computing I, thus

$$I = 41.0 + \frac{40.}{5.0}$$
$$= 49.0$$

This value of I thus causes a PLUS to be stored in PLOT(41,49) and thus the location PLOT(41,49) is equivalent to the point $1 + j0$.

Computer statements S.0034 through S.0045 are used to print the graph. Initially K is set equal to 0, F is set equal to 1.5, and DF is set equal to 0.5. The DO 60 loop is then executed 81 times to print the 81 lines of the graph. During each execution of the DO 60 loop, the expression

$$(\text{K} - \text{K}/20*20)$$

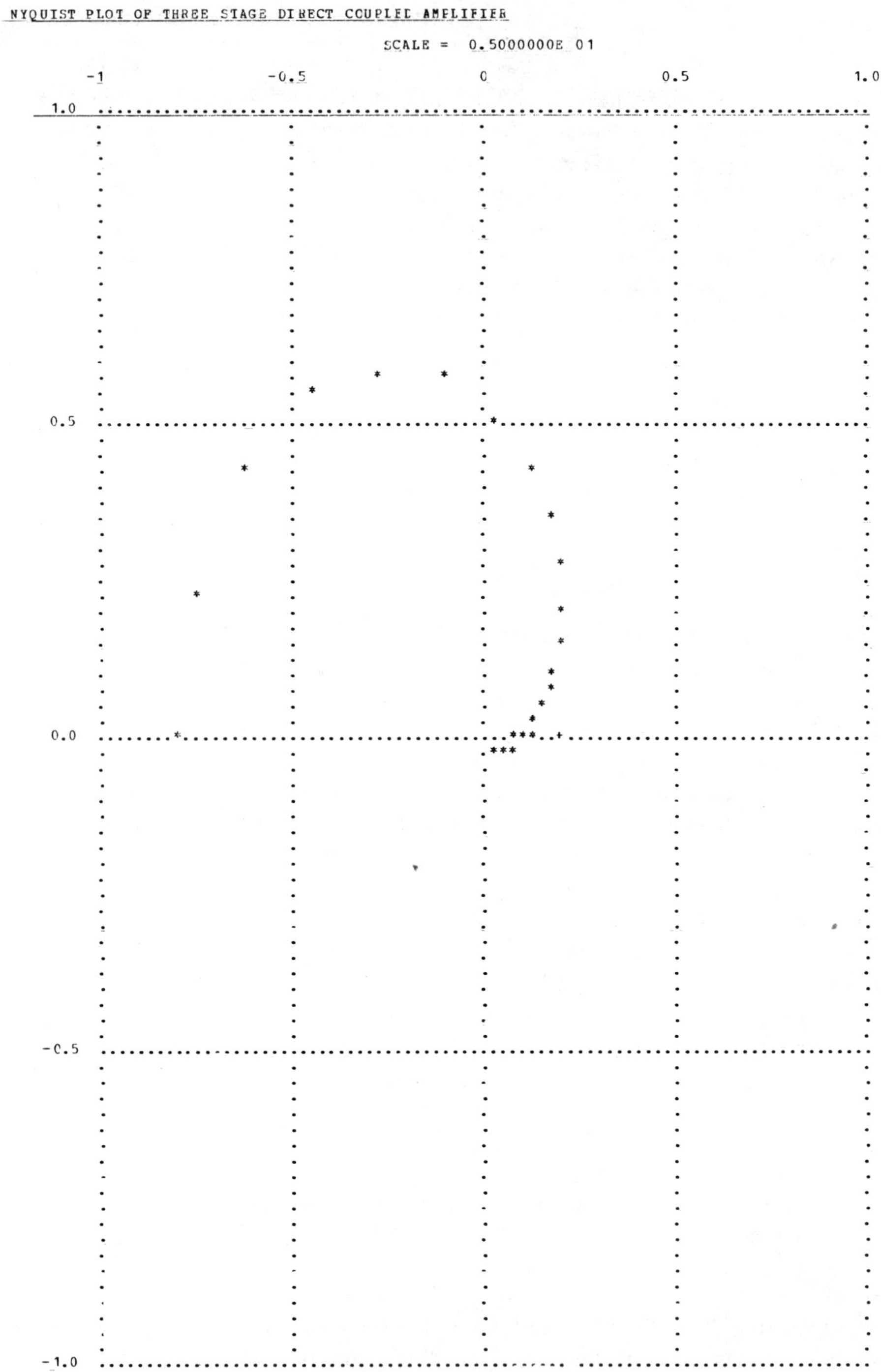

NYQUIST PLOT OF THREE STAGE DIRECT COUPLED AMPLIFIER

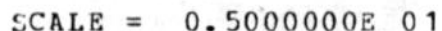

-1 -0.5 0 0.5 1.0

1.0

0.5

0.0

-0.5

-1.0

Figure 11.27 The output results of the program of Figure 11.26 for a scale factor SCALE = 5.0

is tested to see whether it is equal to 0. Since this expression will be equal to zero for $I = 1, 21, 41, 61$, and 81, control will be transferred to statement 70 for these values of I, and the normalized scale value F will be printed along with the corresponding line of PLOT. For all other values of I, only the corresponding line of PLOT will be printed.

Figure 11.27 shows the computer plot for a SCALE factor value of 5.0. Note that this graph has been plotted for a value of $|\beta A_0| = 4$ and since $|\beta A_0|_{th} = 8$, the $1 + j0$ point has not been enclosed. The closed-loop system would thus be a stable one.

In order to examine the gain and phase margins of this or more complicated Nyquist plots, the value of SCALE can be reduced. The reduction of SCALE will effectively produce a Zoom lens effect by blowing up the Nyquist plot in the vicinity of the origin. The gain and phase margin can then be determined by recalling from Figure 11.28 that the gain margin (in dB) is the negative dB value of βA at the frequency of phase crossover (the frequency where $\theta = -\pi$) and the phase margin (ψ) is the angle between the positive real axis and the βA vector at the frequency of loop-gain crossover (the frequency where $|\beta A| = 1$ or 0 dB).

Figure 11.29 shows the resulting plot for a SCALE factor of 1.0. Note that the value of βA is approximately equal to 0.5 at the frequency of phase crossover. The gain margin is thus equal to 6 dB ($|\beta A_0|_{th} - \beta A_0 = 18 - 12$ or 6 dB). Note also that the phase margin is more difficult to estimate because the spacing between printed lines is different from the spacing between printed columns. It is fairly simple, however, to approximate the phase margin by drawing a straight line between the point

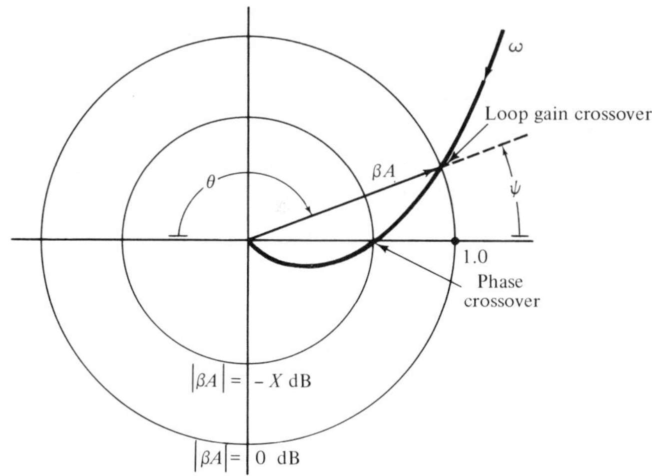

Figure 11.28 Gain and phase margins are defined

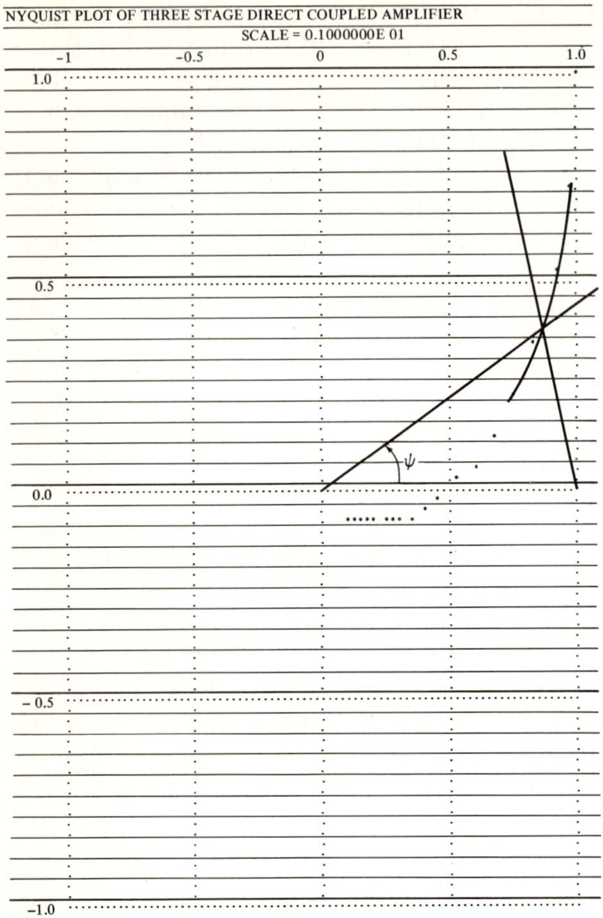

Figure 11.29 The output results of the program of Figure 11.26 for a scale factor SCALE = 1.0

$0.775 + j0.775$ and the point $1 + j0$ and then reading the X and Y intersection values of this straight line with the βA locus, as shown in Figure 11.29. For example, in this problem the intersection occurs at a value of $X = 0.875$ and $Y = 0.375$, thus

$$\psi = \tan^{-1} \frac{0.375}{0.875} = 23.1°$$

is the approximate value of the phase margin.

The reader will find that this Zoom lens effect is particularly useful in investigating the effect of element value changes on the shape of the βA locus in the vicinity of the $1 + j0$ point. For example, in the above problem one may readily see the effect produced by narrow banding one of the three stages or changing the elements of one or more plate decoupling networks so that they produce a lagging doublet effect and thus may be used as a high-frequency stabilizing device.

Root Locus Plot

A feedback control system, in which the input command signal $e(t)$ is used to control the output response $r(t)$, may be schematically represented by the block diagram shown in Figure 11.30. Operating at a very low power, the error detector senses the difference between the input command signal and the unity feedback output response signal. The error signal is then used as the input to the power amplifier which, in turn, is used to drive the load into correspondence.

In studying the operation of such a feedback system, the Laplace transform of the forward path open-loop transfer function is defined as

$$KG(s) = \frac{R(s)}{\epsilon(s)}$$

and the closed-loop transfer function as

$$H(s) = \frac{R(s)}{E(s)}$$

Since $\epsilon(s) = E(s) - R(s)$ we may write

$$H(s) = \frac{R(s)}{\epsilon(s) + R(s)} = \frac{1}{\epsilon(s)/R(s) + 1}$$
$$= \frac{KG(s)}{1 + KG(s)}$$

which is the fundamental relation between the closed-loop and the open-loop system.[1]

[1] The reader should recognize that although this is the unity negative feedback case, any other case can be reduced to this type by simple algebraic manipulation.

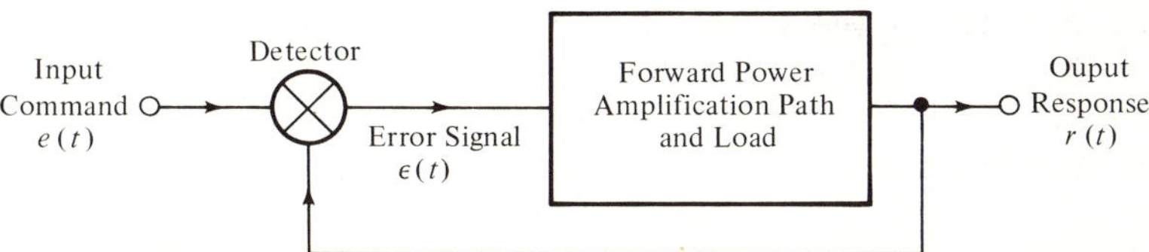

Figure 11.30 General block diagram of feedback control system

If the open-loop transfer function has no poles (no integration) at the origin, then we may let

$$KG(s) = K\frac{1 + a_1 s + \cdots + a_m s^m}{1 + b_1 s + \cdots + b_n s^n}$$

or

$$KG(0) = K$$

and the value K is thus the dc gain of the system when there is no feedback. Similarly, if the open-loop transfer function has a single pole at the origin, the open-loop transfer function may be written as

$$KG(s) = \frac{K}{s}\frac{1 + a_1 s + \cdots + a_m s^m}{1 + b_1 s + \cdots + b_n s^n}$$

or

$$\lim_{s \to 0} sKG(s) = K$$

and the value of K is thus the open-loop velocity gain of the system, and so on.

The root locus is defined as the locus of the poles of the closed-loop transfer function as the gain (or some other system element) is varied. Since

$$H(s) = \frac{KG(s)}{1 + KG(s)}$$

it follows that the zeros of the closed-loop system are identical to those of the open-loop system. The poles of the closed-loop system, however, are located at the values of s at which

$$KG(s) = -1$$

or

$$G(s) = \frac{-1}{K}$$

A root locus plot is therefore a very valuable graph because a knowledge of the locations of the poles of the closed-loop system gives information about the stability of the system as well as the actual behavior of the system.

One simple scheme for obtaining the root locus of the above feedback system is based on finding the values of s that makes

$$KG(s) \equiv -1 = \frac{a_1 + a_2 s + \cdots + a_{m-1}s^{m-2} + a_m s^{m-1}}{b_1 + b_2 s + \cdots + b_{n-1}s^{n-2} + b_n s^{n-1}}$$

where $K \geq 0$ and the a's and b's are real coefficients. This may be accomplished by evaluating $KG(s)$ for $s = \sigma + j\omega$, yielding

$$KG(\sigma + j\omega) = \frac{AR + jAI}{BR + jBI} = -1$$

where

$$AR = \sum_{KK=1}^{M} a_{KK}|\sigma + j\omega|^{KK} \cos(FK*\text{ARG})$$

$$AI = \sum_{KK=1}^{M} a_{KK}|\sigma + j\omega|^{KK} \text{SIN}(FK*\text{ARG})$$

$$BR = \sum_{LL=1}^{N} b_{LL}|\sigma + j\omega|^{LL} \cos(FL*\text{ARG})$$

$$BI = \sum_{LL=1}^{N} b_{LL}|\sigma + j\omega|^{LL} \sin(FL*\text{ARG})$$

and

$$FK = KK - 1$$
$$FL = LL - 1$$
$$\text{ARG} = \tan^{-1} \omega/\sigma$$

Since the real part of $KG(\sigma + j\omega)$ is given by

$$Re\{KG(\sigma + j\omega)\} = -1 = \frac{AR*BR + AI*BI}{BR*BR + BI*BI}$$

it follows that for a point to be on the locus,

$$AR*BR + AI*BI$$

must be negative (for various values of K). It also follows that the imaginary part of $KG(\sigma + j\omega)$ must be equal to zero at the same point, or

$$Im\{KG(\sigma + j\omega)\} = BR*AI - AR*BI = 0$$

It is thus possible to write a computer program that scans a defined portion of the complex plane (point-by-point) and determines whenever these two conditions are satisfied. An asterisk may therefore be stored in the appropriate elements of the PLOT storage matrix and the elements of matrix PLOT may later be printed to produce the root locus plot.

One disadvantage of the above method is that it is very time-consuming. The following discussion presents an alternative method developed by Julio Galvez for his computer course term project (also see the computer term project of Section 9.4). This method uses a more direct approach in which the roots of the characteristic equation

$$KG(s) + 1 = 0$$

are evaluated for various values of K.

Galvez first assumes that $G(s)$ may be written in the form $N(s)/D(s)$ and thus the characteristic equation can be reduced to

$$KN(s) + D(s) = 0$$

He then sets $K = 0$ and uses the Newton-Raphson method to compute the initial location of the poles. The value of K is then changed by a specified increment and the Newton-Raphson method is again used to determine the new location of the poles. Since the change in the values of the roots of the polynomial is very small each time K is changed, the old roots (previous) are used as initial approximations in obtaining the new

```
C       ROOT LOCUS PLOT
C
C
C       PURPOSE
C          FIND THE ROOT LOCUS OF A LINEAR CONTROL SYSTEM WITHIN
C          A SPECIFIED REGION IN THE S PLANE.
C
C       INPUT
C          CARD 1   COL 02-02   DEGREE OF THE POLYNOMIAL OF
C                               DENOMINATOR OF G(S)*H(S).
C          CARD 2   COL  01-10  COEFFICIENT OF HIGHEST DEGREE OF POL.
C                   COL  11-20  COEFFICIENT OF NEXT DEGREE
C                   COL  21-30  NEXT COEFFICIENTS
C          CARD 3               SAME AS CARD 1 BUT FOR NUMERATOR
C          CARD 4               SAME AS CARD 2 BUT FOR NUMERATOR
C          CARD 5   COL  01-10  INCREMENT OF K. K GOES FROM 0 TO 50*K.
C          CARD 6   COL  01-10  LOWEST VALUE OF INTEREST OF REAL AXIS.
C                   COL  11-20  HIGHEST VALUE OF INTEREST IN IMAGINARY AXIS.
C
C
C                               EXAMPLE
C
C
C
C                                         K
C          G(S)*H(*)  =  ---------------------------------
C                        1*S**3  + 20*S**2  + 64*S**1  + 0
C
C
C       COLUMN          000000000111111111122222222223333333333444444444
C                       123456789012345678901234567890123456789012345678
C
C
C       CARD 1           3
C       CARD 2          1.0            20.0          64.0          0.0
C       CARD 3           0
C       CARD 4          1.0
C       CARD 5          20.0
C       CARD 6          -27.0
C
C
C       THE FOLLOWING DIMENSIONS CAN BE INCREASED TO HANDLE LARGER DEGREE
C       POLYNOMIALS. FIVE ROOTS CAN BE HANDLED AT THE PRESENT TIME.
C
```

Figure 11.31 A program for plotting the root locus of the feedback control system of Figure 11.30

```
      DIMENSION CI(6),A(6),B(6),AR(6),AI(6),BR(6),BI(6),CR(6),RR(5,51),
     1RI(5,51),IDD(6),X(1),Y(1),PLOT(102),VAL(11)
      READ(5,10)BLANK,SIGNM,DOT,EYE
   10 FORMAT(4A1)
C
C     INITIALIZE THE OUTPUT MATRIX.
C
    1 DO 20 I=1,5
      DO 20 J=1,50
      RR(I,J)=0.0
   20 RI(I,J)=1.0
C
C
C     INPUT ALL PARAMETERS.
C
C
      READ(5,2000)NA
      IF(NA)190,190,30
   30 NDEG=NA
      NA=NA+1
      READ(5,1000)(A(I),I=1,NA)
      READ(5,2000)NB
      NB=NB+1
      READ(5,1000)(B(I),I=1,NB)
      READ(5,1000)XKINC
      READ(5,1000)XHIGH,YHIGH
      WRITE(6,6000)
      NN=NB-1
      DO 35 I=1,NN
   35 IDD(I)=NB-I
      WRITE(6,3100)(B(I),IDD(I),I=1,NB)
      WRITE(6,3200)
      NN=NA-1
      DO 36 I=1,NN
   36 IDD(I)=NA-I
      WRITE(6,3100)(A(I),IDD(I),I=1,NA)
      WRITE(6,4000)
      NDIFF=NA-NB
      NDIF1=NDIFF+1
C
C     COMPUTER ACCURACY REQUIRED. IT WILL BE ONE PART IN ONE THOUSANDTHS.
C
C
C
C     COMPUTATIONS START HERE FOR 50 POINTS ON ROOT LOCUS OF EACH ROOT.
C
C
      DELTA=ABS(XHIGH)
      IF(DELTA-ABS(YHIGH))50,50,40
   40 DELTA=ABS(YHIGH)
   50 DELTA=DELTA/100000.
      DO 160 IN=1,50
      XK=(IN-1)*XKINC
C
C     ROOT LOCUS PROGRAM FINDS ROOT LOCUS WITHIN A SPECIFIED REGION.
C
      N=NDEG
      N1=NDEG+1
C
C     COMPUTE THE POLYNOMIAL K*N(S)
C
      DO 60 I=1,NDIFF
      AR(I)=A(I)
   60 AI(I)=0.0
      DO 70 I=NDIF1,NA
      K=I-NDIF1+1
      AR(I)=A(I)+XK*B(K)
   70 AI(I)=0.0
C
C     USE SYNTHETIC DIVISION TO EVALUATE POLYNOMIAL
C     BY USING THIS WE CAN AUTOMATICALLY GET THE REDUCED POLYNOMIAL.
C
      BR(1)=AR(1)
      BI(1)=AI(1)
      CR(1)=AR(1)
      CI(1)=AI(1)
      DO 150 NROOT=1,NDEG
      C=RR(NROOT,IN)
      D=RI(NROOT,IN)
```

Figure 11.31 Continued

roots. Convergence to the new roots is therefore quite rapid (second-order), usually taking less than three or four iterations to obtain the desired degree of accuracy. After all roots are stored in memory as they are computed, for the specified range in K; the computer printer is then used to produce the root locus plot.

One of the difficulties that Galvez encountered in writing the above program was underflow. This underflow occurred whenever the loci lay on the real axis for any length of time. Since the imaginary part of the root is zero for this case, the real part of the answer becomes too accurate for the computer to handle and underflow occurs. A check for underflow

```
      C
      C     WE NOW USE NEWTON'S METHOD TO GET A BETTER APPROXIMATION
      C     WE USE IT FOR A MAXIMUM OF 200 TIMES.
      C
            DO 110 KK=1,200
            DO 80 I=2,N1
            BR(I)=AR(I)+BR(I-1)*C-BI(I-1)*D
         80 BI(I)=AI(I)+BR(I-1)*D+BI(I-1)*C
            DO 90 I=2,N
            CR(I)=BR(I)+CR(I-1)*C+CI(I-1)*D
         90 CI(I)=BI(I)+CR(I-1)*D+CI(I-1)*C
      C
      C     COMPUTE NEW ROOTS AND CHECK IF ACCURACY IS ACHIEVED.
      C
            DEN=CR(N)*CR(N)+CI(N)*CI(N)
            X1=-(BR(N1)*CR(N)+BI(N1)*CI(N))/DEN
            Y1=+(BR(N1)*CI(N)-BI(N1)*CR(N))/DEN
            C=C+X1
            D=D+Y1
            IF(ABS(X1)+ABS(Y1)-DELTA)130,130,110
        110 CONTINUE
      C
      C     IT GETS HERE IF NO CONVERGENCE AFTER 200 TRIES.
      C
            WRITE(6,120)XK
        120 FORMAT(F11.1,10X,' DOES NOT CONVERGE TRYING DIFFERENT K VALUE')
            GO TO 160
        130 N1=N1-1
      C
      C     USE THE REDUCED POLYNOMIAL PRODUCED BY SYNTHETIC DIVISION.
      C
      C
      C     SAVE THIS ROOT FOR NEXT APPROXIMATION.
      C
            N=N-1
            DO 140 I=2,N1
            AR(I)=BR(I)
        140 AI(I)=BI(I)
            RR(NROOT,IN)=C
            RI(NROOT,IN)=D
            RR(NROOT,IN+1)=C
            RI(NROOT,IN+1)=D
        150 CONTINUE
        160 WRITE(6,5000)XK,(RR(I,IN),RI(I,IN),I=1,NDEC)
            WRITE(6,6000)
       6000 FORMAT('1')
            CALL PLT(255,RR,ABS(XHIGH)/9.,XHIGH,RI,YHIGH,-YHIGH,BLANK,SIGNM,
           1DOT,EYE)
            GO TO 1
        190 CALL EXIT
       1000 FORMAT(8F10.0)
       2000 FORMAT(I2)
       3100 FORMAT(11X,5(F5.1,'*S**',I2,' + '))
       3200 FORMAT(' F(S) = ',79('-'))
       4000 FORMAT(8X,'K',14X,5(8X,'ROOT',8X),/,25X,5('  REAL      IMAGINARY
           1'),/)
       5000 FORMAT(F11.1,10X,10F10.3)
            STOP
            END
```

Figure 11.31 Continued

```
      SUBROUTINE PLT(N,X,XMAX,XMIN,Y,YMAX,YMIN,BLANK,SIGNM,DOT,EYE)
      DIMENSION X(1),Y(1),PLOT(102),VAL(11)
      YS=YMAX-YMIN
      XS=XMAX-XMIN
      DO 10 I=1,11
   10 VAL(I)=XMIN+(I-1)*XS/10.
      WRITE(6,100)(VAL(I),I=1,11)
      WRITE(6,101)
      LYZ=-50.*YMIN/YS+1.5
      LYZ=52-LYZ
      LXZ=-100.*XMIN/XS+1.5
      IF(LXZ)5,5,4
    4 IF(LXZ-101)15,15,5
    5 LXZ=102
   15 LIN=10
      DO 99 I=1,51
      IF(I-LYZ)25,45,25
   25 DO 32 II=1,101
   32 PLOT(II)=BLANK
      GO TO 65
   45 DO 55 II=1,101
   55 PLOT(II)=SIGNM
   65 PLOT(LXZ)=EYE
      DO 85 J=1,N
      K=50.*(Y(J)-YMIN)/YS+1.5
      IF(52-K-I)85,74,85
   74 K=100.*(X(J)-XMIN)/XS+1.5
      IF(K)85,85,75
   75 IF(K-101)76,76,85
   76 PLOT(K)=DOT
   85 CONTINUE
      IF(LIN-10)82,81,82
   81 YVAL=YMIN+YS*(51-I)/50.
      WRITE(6,103)YVAL,(PLOT(L),L=1,101),YVAL
      LIN=1
      GO TO 99
   82 WRITE(6,102)(PLOT(L),L=1,101)
      LIN=LIN+1
   99 CONTINUE
      WRITE(6,101)
      WRITE(6,100)(VAL(I),I=1,11)
  100 FORMAT(6X,11(E9.2,1X))
  101 FORMAT(10X,'*',10('+*********')'+*')
  102 FORMAT(10X,'*',101A1,'*')
  103 FORMAT(F10.2,'-',101A1,'-',E8.1)
      RETURN
      END
```

Figure 11.31 Continued

is therefore incorporated into the program to overcome this drawback. Another possible drawback occurs at the point of arrival or departure of the locus. Here the program may miss the point in question because the value of K at these points may not be calculated by the program. A judicious choice in the increment of K will therefore overcome this drawback.

Figure 11.31 is Galvez's root locus program. The reader should experience little difficulty in using this program because of the many comment statements that are used throughout. Figure 11.32 gives the computer plot for the example discussed in the initial comment statements. The reader may easily verify the accuracy of these results by direct calculation or by referring to Kuo[2] for his original discussion of this example.

[2] See Benjamin F. Kuo, *Automatic Control Systems*. Englewood Cliffs, New Jersey: Prentice-Hall, 1963, Chap. 8.

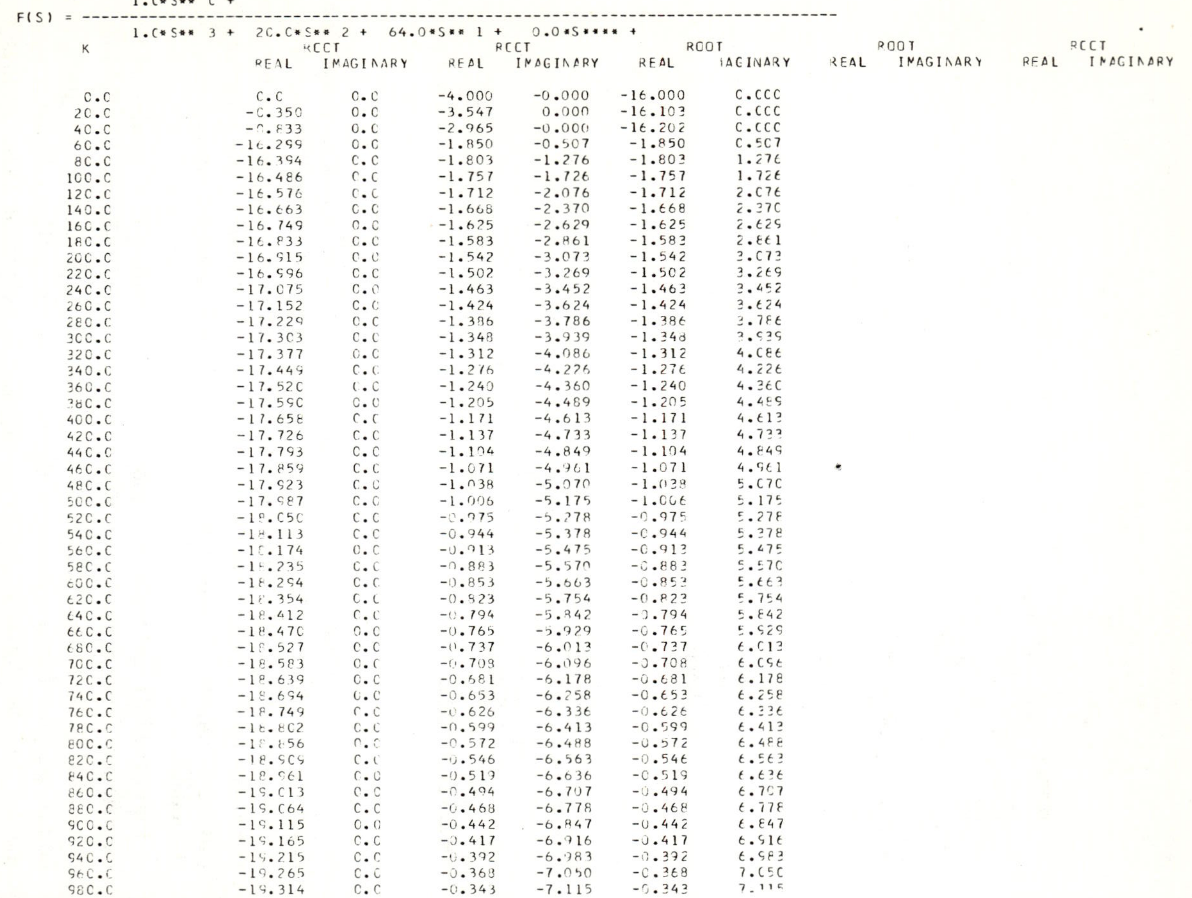

```
                 1.0*S** 0 +
F(S) = -------------------------------------------------------------
        1.0*S** 3 +  20.0*S** 2 +  64.0*S** 1 +   0.0*S**** +
```

K	ROOT REAL	ROOT IMAGINARY	ROOT REAL	ROOT IMAGINARY	ROOT REAL	ROOT IMAGINARY	ROOT REAL	ROOT IMAGINARY	ROOT REAL	ROOT IMAGINARY
0.0	0.0	0.0	-4.000	-0.000	-16.000	0.000				
20.0	-0.350	0.0	-3.547	0.000	-16.103	0.000				
40.0	-0.833	0.0	-2.965	-0.000	-16.202	0.000				
60.0	-16.299	0.0	-1.850	-0.507	-1.850	0.507				
80.0	-16.394	0.0	-1.803	-1.276	-1.803	1.276				
100.0	-16.486	0.0	-1.757	-1.726	-1.757	1.726				
120.0	-16.576	0.0	-1.712	-2.076	-1.712	2.076				
140.0	-16.663	0.0	-1.668	-2.370	-1.668	2.370				
160.0	-16.749	0.0	-1.625	-2.629	-1.625	2.629				
180.0	-16.833	0.0	-1.583	-2.861	-1.583	2.861				
200.0	-16.915	0.0	-1.542	-3.073	-1.542	3.073				
220.0	-16.996	0.0	-1.502	-3.269	-1.502	3.269				
240.0	-17.075	0.0	-1.463	-3.452	-1.463	3.452				
260.0	-17.152	0.0	-1.424	-3.624	-1.424	3.624				
280.0	-17.229	0.0	-1.386	-3.786	-1.386	3.786				
300.0	-17.303	0.0	-1.348	-3.939	-1.348	3.939				
320.0	-17.377	0.0	-1.312	-4.086	-1.312	4.086				
340.0	-17.449	0.0	-1.276	-4.226	-1.276	4.226				
360.0	-17.520	0.0	-1.240	-4.360	-1.240	4.360				
380.0	-17.590	0.0	-1.205	-4.489	-1.205	4.489				
400.0	-17.658	0.0	-1.171	-4.613	-1.171	4.613				
420.0	-17.726	0.0	-1.137	-4.733	-1.137	4.733				
440.0	-17.793	0.0	-1.104	-4.849	-1.104	4.849				
460.0	-17.859	0.0	-1.071	-4.961	-1.071	4.961				
480.0	-17.923	0.0	-1.038	-5.070	-1.038	5.070				
500.0	-17.987	0.0	-1.006	-5.175	-1.006	5.175				
520.0	-18.050	0.0	-0.975	-5.278	-0.975	5.278				
540.0	-18.113	0.0	-0.944	-5.378	-0.944	5.378				
560.0	-18.174	0.0	-0.913	-5.475	-0.913	5.475				
580.0	-18.235	0.0	-0.883	-5.570	-0.883	5.570				
600.0	-18.294	0.0	-0.853	-5.663	-0.853	5.663				
620.0	-18.354	0.0	-0.823	-5.754	-0.823	5.754				
640.0	-18.412	0.0	-0.794	-5.842	-0.794	5.842				
660.0	-18.470	0.0	-0.765	-5.929	-0.765	5.929				
680.0	-18.527	0.0	-0.737	-6.013	-0.737	6.013				
700.0	-18.583	0.0	-0.708	-6.096	-0.708	6.096				
720.0	-18.639	0.0	-0.681	-6.178	-0.681	6.178				
740.0	-18.694	0.0	-0.653	-6.258	-0.653	6.258				
760.0	-18.749	0.0	-0.626	-6.336	-0.626	6.336				
780.0	-18.802	0.0	-0.599	-6.413	-0.599	6.413				
800.0	-18.856	0.0	-0.572	-6.488	-0.572	6.488				
820.0	-18.909	0.0	-0.546	-6.563	-0.546	6.563				
840.0	-18.961	0.0	-0.519	-6.636	-0.519	6.636				
860.0	-19.013	0.0	-0.494	-6.707	-0.494	6.707				
880.0	-19.064	0.0	-0.468	-6.778	-0.468	6.778				
900.0	-19.115	0.0	-0.442	-6.847	-0.442	6.847				
920.0	-19.165	0.0	-0.417	-6.916	-0.417	6.916				
940.0	-19.215	0.0	-0.392	-6.983	-0.392	6.983				
960.0	-19.265	0.0	-0.368	-7.050	-0.368	7.050				
980.0	-19.314	0.0	-0.343	-7.115	-0.343	7.115				

Figure 11.32 The root locus plot produced by the program of Figure 11.31 for the function $G(s)*H(s) = \dfrac{K}{s^3 + 20s^2 + 64s}$

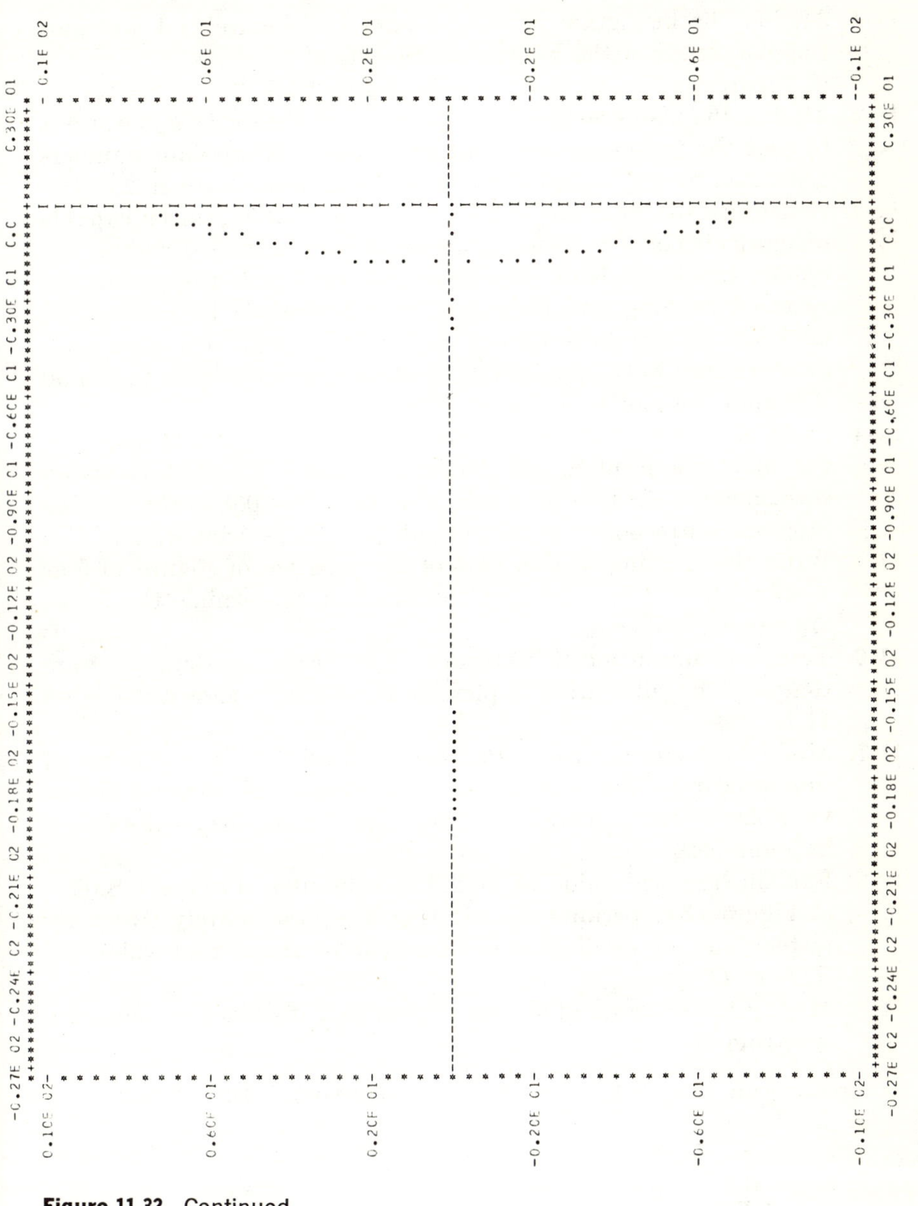

Figure 11.32 Continued

PROBLEMS

11-1. Modify the histogram plotting program of Figure 11.1 and plot the magnitude of the Fourier coefficients of
a. Figure 5.23 b. Figure 5.25

11-2. Modify the histogram plotting program of Figure 11.1 and use it to plot the frequency distribution of the 1000 random numbers generated by the program of Figure 10.22 (see Figure 10.23).

11-3. Generalize the program of Figure 11.1 so that it will be capable of producing a histogram plot comparison of two different frequency functions. Hint: The use of A-format code will permit the required heading and horizontal scale values to be printed. In addition, the normalization of the frequency scale will allow for a greater range in the magnitude of the frequency distributions and still limit the plot to a single printed page.

11-4. State how you would modify the program of Figure 11.6 so that the first line printed, the horizontal axis $I(T)/IMAX$, would correspond to the time $T = 0.0$. Note that $T = 0.0$, in the existing program, corresponds to the second line that is printed.

11-5. Write the plotting section part of the program of Figure 11.6 as a subroutine. Give detailed instructions concerning the input arguments, and so on.

11-6. Test your subroutine of Problem 11-5 by writing a calling program that uses the subroutine to plot the three graphs shown in Figure 11.7.

11-7. Modify the subroutine of Problem 11-5 so that it is capable of simultaneously plotting up to a maximum of three different variables. Test this modified program by plotting the results shown in Figure 11.7.

11-8. Explain how the value of DELTA, computer statement S.0021 in Figure 11.6, permits the plotting of approximately two complete cycles of oscillation independent of the actual values of R, L, or C.

11-9. Modify the program of Figure 11.8 so that it plots the frequency spectrum of

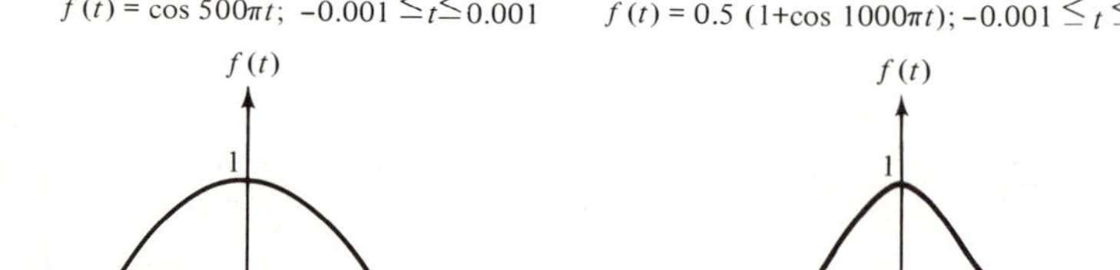

11-10. Repeat Problem 11-9 for a train of nine pulses.

11-11. Modify the normalized time response plot program of Figure 11.11 and use it to plot the normalized output voltage of the simple shunt-peaked network shown in Figure 13.29.

$$v(t) = \left(1 + 2Q^2 \frac{\epsilon^{-\frac{t}{2Q^2RC}}}{\sqrt{4Q^2 - 1}} \quad \sin\left(\frac{\sqrt{4Q^2 - 1}}{2Q^2RC} t + \psi\right)\right) * I * R$$

where $\psi = \tan^{-1} \sqrt{4Q^2 - 1} - \tan^{-1} \sqrt{4Q^2 - 1}/(-1)$. Use values of Q^2 equal to a. 0.25 b. 0.50 c. 0.707 d. 1.00.

11-12. Modify the program of Figure 11.11 so that the three plots shown in Figure 11.7 may be plotted as a single graph. Use +, −, and 0 as the respective plotting symbols.

11-13. Write the plotting section part of the program of Figure 11.11 as a subroutine. Give detailed instructions concerning the input arguments, and so on.

11-14. Test your subroutine of Problem 11-13 by writing a calling program that uses the subroutine to plot the normalized voltage response of Problem 11-11.

11-15. Modify the plotting program of Figure 11.13 so that the three-pole Papoulis and Butterworth filter responses are plotted with a magnitude resolution that is five times larger.

11-16. Modify the program of Figure 11.13 and use it to plot the magnitude response of the three cases of the lead-lag network plotted in Figure 11.17.

11-17. What program changes are necessary in the program of Figure 11.13 so that it can be used to produce a log frequency plot for the frequency range a. 0.1 to 100.0 rps b. 100.0 to 10,000 rps c. 0.001 to 0.1 rps

11-18. Repeat Problem 11-17 for the following magnitude ranges: a. 0.0 to 30.0 dB b. 0.0 to −40.0 dB c. −20.0 to +20.0 dB.

11-19. Use the program of Figure 11.16 to plot the magnitude and phase response of the following network transfer functions:

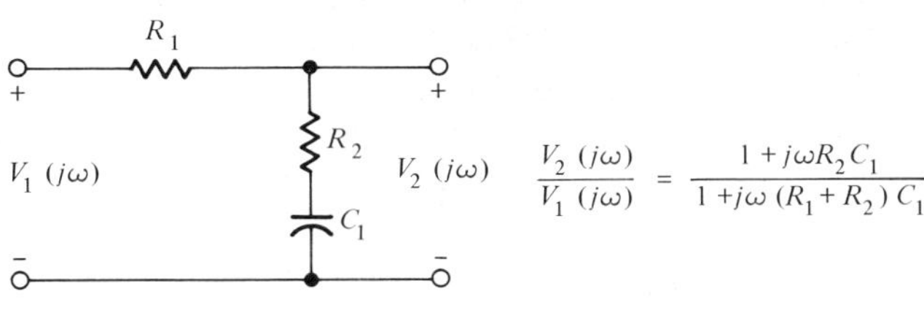

(a)

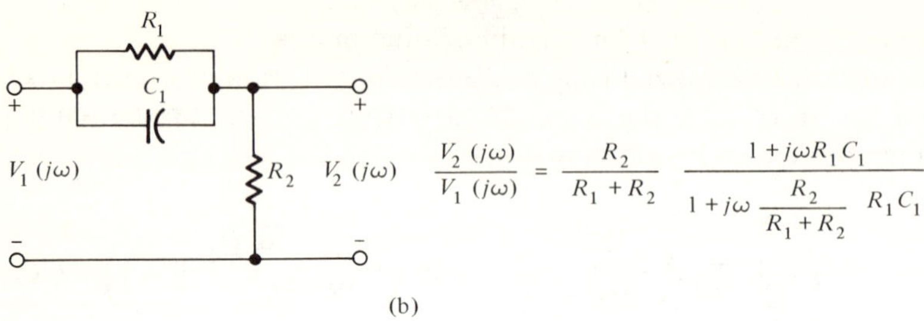

(b)

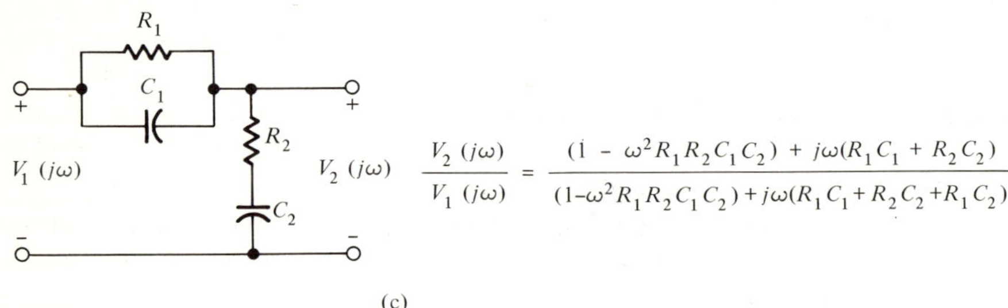

(c)

Use values of $R_1 = 10$ Kilohms, $R_2 = 1$ Kilohms, $C_1 = 0.1\ \mu$F, and $C_2 = 10.0\ \mu$F.

11-20. Write the plotting section part of the program of Figure 11.16 as a subroutine. Give detailed instructions concerning the input arguments, and so on.

11-21. Use the program of Figure 11.26 to plot the Nyquist diagram for the following loop transmissions:

a. $$\beta A(j\omega) = \frac{-20}{j\omega(1 + j\omega)(1 + j0.5\omega)}$$

b. $$\beta A(s) = \frac{-100}{(s + 4000)(s + 1000 - j1000)(s + 1000 + j1000)}$$

11-22. Write the plotting section of the program of Figure 11.26 as a subroutine. Give detailed instructions concerning the input arguments, and so on.

11-23. Using the Zoom lens feature in the program of Figure 11.26, determine whether the closed-loop response of the network transmission functions given in Problem 11-21 will be stable (for example, determine the gain and phase margins).

11-24. The Zoom lens feature in the program of Figure 11.26 may be used to blow up the Nyquist plot in the vicinity of the origin. Extend this feature so that any section of the plot may be enlarged.

11-25. Write a Fortran program for producing a root locus plot. Base the program on the point-by-point scanning method described in Section 11.5.

11-26. Test the program of Problem 11-25 by using the example data values of Figure 11.31. Compare the two output results and discuss the advantages or disadvantages of each program.

11-27. Use the root locus program of Figure 11.31 to plot the root locus of:

a. $G(s)*H(s) = \dfrac{K}{s(s+4)(s+16)}$; K increment $= 20$

b. $G(s)*H(s) = \dfrac{K(s+a)}{s(s^2+2s+2)}$; K increment $= 1.0$ and $a = 3, 2, 1,$ and 0.

c. $G(s)*H(s) = \dfrac{K}{s(s+a)(s^2+2s+2)}$; K increment $= 0.2$ and $a = 4, 3, 2,$ and 1.

Comment on the effect of the location of the zero in part b. and the location of the pole in part c.

CHAPTER 12

Polynomial Evaluation, Interpolation, and Approximation

12.1 INTRODUCTION

Polynomials are often encountered in electrical engineering research where the investigator develops an empirical formula based on actual data rather than theoretical considerations. This chapter concerns itself with a discussion of polynomials that have the following general form:

$$P_N(X) = a_1 + a_2X + a_3X^2 + \cdots + a_{N+1}X^N$$

We will first discuss some of the problems of evaluating polynomials of this type; we will then discuss several different methods that can be used to generate interpolating polynomials; and, lastly, we will discuss different ways of generating least-square polynomials.

12.2 POLYNOMIAL EVALUATION

We have already pointed out that the evaluation of polynomials having the form

$$P_N(X) = a_1 + a_2X + a_3X^2 + \cdots + a_{N+1}X^N$$

is common practice in the field of electrical engineering. For example, in the computation of the frequency response of a Butterworth, Chebyshev, or Papoulis type filter, we would be required to evaluate such a series. Similar type polynomials would also be encountered in the power series approximation of such functions as the sine, cosine, tangent, exponential, logarithm, and so on.

The accuracy of such polynomial evaluations will depend on several different factors. To begin with, we must know if the polynomial is based on a mathematical model that only approximates the physical system it represents. We must also be concerned with various computational errors:

1. Has there been an error produced in converting one of the polynomial coefficients from decimal to hexadecimal?
2. Have there been rounding errors caused by the limitation of the number of significant figures preserved by the computer?
3. Have there been significant truncation errors produced by truncating an infinite series at a point where we think the sum of the remainder terms will be negligible?
4. In iteration procedures, have we caused still further errors in the propagation of errors from one step to the next?

The complete error analysis of the evaluation of a polynomial is beyond the scope of this section; however, let us consider some of the errors that we may encounter and how they may be reduced or avoided. Referring first to the power series expansion for the inverse tangent function

$$\tan^{-1}(X) = X - \frac{X^3}{3} + \frac{X^5}{5} - \frac{X^7}{7} + \cdots ; |X| \leq 1$$

$$\tan^{-1}(X) = \frac{\pi}{2} - \frac{1}{X} + \frac{1}{3X^3} - \frac{1}{5X^5} + \cdots ; |X| > 1$$

we see that the series alternates in sign. An important property of a convergent alternating series is that the sum of the first N terms differs from the sum of the infinite series by less than the magnitude of the $N + 1$ term. It is thus possible to obtain a desired accuracy by truncating the series at the N-th term whenever the magnitude of the $N + 1$ term is less than the specified accuracy.

Term Summing

Figure 12.1 shows a program for computing the arctangent (X), for $|X| \leq 1$. Initially the heading and subheadings are printed and then computer statement S.0001 reads in the argument, X, and the specified accuracy ACC. The DO 10 loop is then used to compute the additional

```
          C         POWER SERIES COMPUTATION OF THE ARC TANGENT (X).
          C         ABSOLUTE VALUE OF X LESS THAN OR EQUAL TO ONE.
          C
S.0001            2 READ(5,1)X,ACC
S.0002            1 FORMAT(2E15.7)
S.0003              WRITE(6,3)X,ACC
S.0004            3 FORMAT('1POWER SERIES COMPUTATION OF THE ARC TANGENT (X)',//,' ARG
                   1UMENT =',E15.7,5X,'ACCURACY =',E15.7,///,' ITERATION',10X,'TERM',1
                   21X,'SUM',//)
S.0005              SUM=X
S.0006              TERM=X
S.0007              DO 10 I=2,100
S.0008              FN=2*(I-1)-1
S.0009              FD=2*I-1
S.0010              TERM=-TERM*X**2*FN/FD
S.0011              SUM=SUM+TERM
S.0012              IF(ABS(TERM)-ACC)20,20,10
S.0013           10 WRITE(6,4)I,TERM,SUM
S.0014            4 FORMAT(I5,5X,2E15.7,/)
S.0015           20 GO TO 2
S.0016              END
```

Figure 12.1 Power series computation of the arctangent function

terms of the series. During the first iteration of the DO 10 loop, $FN = 1.0$ and $FD = 3.0$. The value of TERM is therefore equal to $-X^3/3.$ and SUM equals $X - X^3/3$. Computer statement S.0012 then tests to see whether ABS(TERM)−ACC is greater than, less than, or equal to zero. If the value of ABS(TERM) is greater than ACC, control is transferred to statement 10 and the number of the iteration and the values of TERM and SUM are printed before the next term of the series is computed. If the value of ABS(TERM) is less than or equal to ACC, control is transferred to statement 20 and the series is computed for the next argument value.

```
POWER SERIES COMPUTATION OF THE ARC TANGENT (X)

ARGUMENT =  0.5000000E 00        ACCURACY =  0.9999999E-04


ITERATION           TERM            SUM


    2        -0.4166666E-01   0.4583333E 00

    3         0.6249998E-02   0.4645833E 00

    4        -0.1116071E-02   0.4634672E 00

    5         0.2170138E-03   0.4636841E 00



POWER SERIES COMPUTATION OF THE ARC TANGENT (X)

ARGUMENT =  0.5000000E 00        ACCURACY =  0.1000000E-04


ITERATION           TERM            SUM


    2        -0.4166666E-01   0.4583333E 00

    3         0.6249998E-02   0.4645833E 00

    4        -0.1116071E-02   0.4634672E 00

    5         0.2170138E-03   0.4636841E 00

    6        -0.4438916E-04   0.4636397E 00
```

Figure 12.2 The output results of the program of Figure 12.1

```
POWER SERIES COMPUTATION OF THE ARC TANGENT (X)

ARGUMENT =  0.5CCCOCOE CO       ACCURACY =  0.9999997E-07

ITERATION          TERM              SUM

    2        -C.4166666E-C1  0.4583333E 00
    3         C.6249998E-02  0.4645833E 00
    4        -C.1116C71E-C2  0.4634672E C0
    5         C.2170138E-C3  C.4636841E 00
    6        -0.4438916E-C4  0.4636397E 00
    7         C.9390C14E-C5  C.4636491E C0
    8        -C.2C34501E-C5  0.4636470E C0
    9         0.4487870E-C6  0.4636474E 00
   10        -C.1C03865E-C6  0.4636473E 00


POWER SERIES COMPUTATION OF THE ARC TANGENT (X)

ARGUMENT =  0.5CCC000E CO       ACCURACY =  0.9999997E-08

ITERATION          TERM              SUM

    2        -0.4166666E-01  0.4583333E 00
    3         C.6249998E-02  0.4645833E CC
    4        -0.1116C71E-02  0.4634672E C0
    5         0.2170138E-03  C.4636841E 00
    6        -C.4438916E-C4  0.4636397E C0
    7         C.939CC14E-C5  0.4636491E 00
    8        -C.2C34501E-05  C.4636470E C0
    9         0.4487870E-06  0.4636474E C0
   10        -C.1C03865E-06  0.4636473E 00
   11         0.227C647E-07  0.4636473E 00

IHC217I
```

Figure 12.2 Continued

Figure 12.2 gives the computer results for the argument $X = 0.5$ and ACC values of 1.0E-04, 1.0E-05, 1.0E-07, and 1.0E-08. Note that although we have truncated the series properly, we have not avoided additional round-off errors in obtaining the value of SUM. This follows from the fact that the SUM term is obtained by summing from the largest term of the series to the smallest. As a consequence, maximum round-off error results.

In order to illustrate this point more clearly, consider the program shown in Figure 12.3. Computer statements S.0002 through S.0004 are used to produce 101 numbers with the first number equal to 1.0 and each of the remaining numbers equal to 0.1E-07. Computer statements

```
                C         SUMMING A SERIES
                C
S.0001                    DIMENSION A(101)
S.0002                    DO 10 I=1,101
S.0003               10   A(I)=0.1E-7
S.0004                    A(1)=1.0
S.0005                    SUM=0.0
S.0006                    DO 20 I=1,101
S.0007               20   SUM=SUM+A(I)
S.0008                    WRITE(6,1)SUM
S.0009                  1 FORMAT('1SUMMING A SERIES',//' 1.0 + 0.1E-7 + ... + 0.1E-7',//,' Y
                         1IELDS',//' SUM =',E15.7,///)
S.0010                    SUM =0.0
S.0011                    A(1)=0.1E-7
S.0012                    A(101)=1.0
S.0013                    DO 30 I=1,101
S.0014               30   SUM=SUM+A(I)
S.0015                    WRITE(6,2)SUM
S.0016                  2 FORMAT(' SUMMING THE SERIES',//,' 0.1E-7 + 0.1E-7 +... +0.1E-7 + 1
                         11.0',//,' YIELDS',//' SUM =',E15.7)
S.0017                    STOP
S.0018                    END
```

Figure 12.3 The forward and reverse summing of a series of numbers

S.0005 through S.0009 are then used to sum these 101 numbers and print the result. Computer statements S.0010 through S.0016 then change the value of $A(1)$ to 0.1E-07 and $A(101)$ to 1.0 before repeating the previous summation.

Figure 12.4 shows the corresponding computer results. Note that in the first case when $A(1)$ is added to $A(2)$, the result is equal to

0.1000000E 01

because the computer preserves only seven significant figures. The total SUM for case one is therefore equal to 0.1000000E 01 and is obviously incorrect. In the second case, however, the sum of $A(1)$ and $A(2)$ is equal to 0.2000000E-07 and the sum of $A(1)$, $A(2)$, . . . , $A(100)$ is equal to 0.1000000E-05. The total SUM for case two is therefore equal to 0.1000001E 01, which is the correct result.

In order to avoid round-off errors in computing the SUM term, we should compute and store the individual terms. Whenever the value of ABS(TERM) − ACC is less than or equal to zero, the SUM should be obtained by summing from the last stored term to the first. The program

```
SUMMING A SERIES

1.0 + 0.1E-7 + ... + 0.1E-7

YIELDS

SUM =    0.1000000E 01


SUMMING THE SERIES

0.1E-7 + 0.1E-7 +... +0.1E-7 + 1.0

YIELDS

SUM =    0.1000001E 01
```

Figure 12.4 The output results of the program of Figure 12.3

```
C        MODIFIED POWER SERIES COMPUTATION OF ARC TANGENT (X)
C        ABSOLUTE VALUE OF X MUST BE LESS THAN OR EQUAL TO ONE.
C
         DIMENSION S(100)
       2 READ(5, 1) X,ACC
       1 FORMAT(2E15.7)
         WRITE(6, 3) X,ACC
       3 FORMAT('1POWER SERIES COMPUTATION OF THE ARC TANGENT (X)',//,' ARG
        1UMENT =',E15.7,5X,'ACCURACY =',E15.7,///,' ITERATION',10X,'TERM',1
        21X,'SUM',//)
         SUM=X
         S(1)=X
         DO 10 I=2,100
         FN=2*(I-1)-1
         FD=2*I-1
         S(I)=-S(I-1)*X**2*FN/FD
         SUM=SUM+S(I)
         IF(ABS(S(I))-ACC)20,20,15
      15 K=I
      10 WRITE(6,4)I,S(I),SUM
       4 FORMAT(I5,5X,2E15.7,/)
      20 SUM=0.0
         DO 30 I=1,K
         L=K+1-I
      30 SUM=SUM+S(L)
         WRITE(6,5)SUM
       5 FORMAT(//,' REVERSE SUM =',E15.7)
         GO TO 2
         END
```

Figure 12.5 A modified power series program for computing the arctangent function

of Figure 12.5 incorporates this last feature in the computation of the arctangent function and Figure 12.6 gives the corresponding results. Note that for this particular polynomial example, round-off error produced from summing from the largest to the smallest term is not significant. The reader is encouraged to investigate this problem further by evaluating the arctangent function for larger argument values and by computing the value of other polynomials such as the power series representation of the sine function. The reader should also be aware of positive and negative term cancellation in alternate sign series. In such cases where the value of $P_N(X)$ is much less than the individual term values, double precision should be used to minimize round-off errors.

Wolfe Summing

An alternate method for handling the previous problem has been suggested by Jack Wolfe.[1] Wolfe's method is also applicable to numerical integration schemes where we are summing a large number of individual addends that are each relatively small compared with the total sum. The method is based on using separate variables to store the partial sums of the individual addends at decade magnitude intervals.

Figure 12.7 shows a suitable Fortran program for this type of summing. Initially we read in the N numbers that we wish to sum. In this example

[1] Jack M. Wolfe, "Reducing Truncation Errors by Programming," *ACM* (Association for Computing Machinery), Vol. 7, No. 6 (June, 1964).

```
POWER SERIES COMPUTATION OF THE ARC TANGENT (X)

ARGUMENT =  0.5000000E 00        ACCURACY =  0.9999999E-04

ITERATION            TERM              SUM

   2         -0.4166666E-01   0.4583333E CC

   3          C.6249998E-02   0.4645833E 00

   4         -0.1116071E-02   0.4634672E CC

   5          C.2170138E-03   0.4636841E 00

REVERSE SUM =  C.4636843E 00



POWER SERIES COMPUTATION OF THE ARC TANGENT (X)

ARGUMENT =  0.5000000E 00        ACCURACY =  C.1CCCCCCE-C4

ITERATION            TERM              SUM

   2         -0.4166666E-01   0.4583333E CO

   3          C.6249998E-02   0.4645833E 00

   4         -0.1116071E-02   0.4634672E CC

   5          C.2170138E-03   0.4636841E 00

   6         -0.4438916E-04   0.4636397E CC

REVERSE SUM =  C.4636399E 00



POWER SERIES COMPUTATION OF THE ARC TANGENT (X)

ARGUMENT =  0.5000000E 00        ACCURACY =  C.9999997E-C7

ITERATION            TERM              SUM

   2         -0.4166666E-01   0.4583333E CC

   3          C.6249998E-02   0.4645833E 00

   4         -0.1116071E-02   0.4634672E CC

   5          C.2170138E-03   0.4636841E 00

   6         -0.4438916E-04   0.4636397E CC

   7          C.939CC14E-C5   0.4636491E 00

   8         -0.2034501E-05   0.4636470E CC

   9          C.4487670E-06   0.4636474E 00

  10         -0.1003865E-06   0.4636473E CC

REVERSE SUM =  0.4636475E 00

THC217I
```

Figure 12.6 The output results of the program of Figure 12.5

```
C         WOLFE SUMMING.
C
          DIMENSION A(100),SUM(50)
          READ(5,1)N,(A(I),I=1,N)
        1 FORMAT(I5,/,(5E15.7))
          BIG=A(1)
          SMALL=A(1)
          DO 10 I=2,N
          IF(BIG-A(I))2,3,3
        2 BIG=A(I)
        3 IF(SMALL-A(I))10,10,4
        4 SMALL=A(I)
       10 CONTINUE
          WRITE(6,60)BIG,SMALL
       60 FORMAT(//,' BIG =',E15.7,'    SMALL =',E15.7)
          M=0
          SMALL=ABS(SMALL)
          BIG=ABS(BIG)
          IF(SMALL-1)5,6,7
        5 SMALL=SMALL*10.
          M=M+1
          IF(SMALL-1)5,6,6
        7 SMALL=SMALL/10.
          M=M+1
          IF(SMALL-1)8,8,7
        6 M=-M
        8 K=0
          IF(BIG-1)9,11,12
        9 BIG=BIG*10.
          K=K+1
          IF(BIG-1)9,11,11
       12 BIG=BIG/10.
          K=K+1
          IF(BIG-1)13,13,12
       11 K=-K
       13 L=-M+K+1
          WRITE(6,61)L,M,K
       61 FORMAT(//,' L =',I10,'    M =',I10,'    K =',I10,//)
          DO 20 I=1,L
       20 SUM(I)=0.0
          DO 30 I=1,N
          DO 35 J=1,L
          IF(A(I)-10.**(M+J))21,21,35
       21 SUM(J)=A(I)+SUM(J)
          IF(SUM(J)-10.**(M+J))30,22,22
       22 SUM(J+1)=SUM(J)+SUM(J+1)
          SUM(J)=0.0
          GO TO 30
       35 CONTINUE
       30 CONTINUE
          TOTAL=0.0
          DO 40 I=1,L
       40 TOTAL=SUM(I)+TOTAL
          WRITE(6,14)TOTAL,(SUM(I),I=1,L)
       14 FORMAT(' TOTAL SUM =',E15.7,//,(5E15.7))
          STOP
          END
```

Figure 12.7 The use of partial sums in computing the sum of a sequence of numbers

the following 35 numbers are used:

0.2500000E-06 0.2500000E-06 0.2500000E-06 0.2500000E-06 0.2500000E-06
0.2500000E-04 0.2500000E-04 0.2500000E-04 0.2500000E-04 0.2500000E-04
0.2500000E-02 0.2500000E-02 0.2500000E-02 0.2500000E-02 0.2500000E-02
0.2500000E 00 0.2500000E 00 0.2500000E 00 0.2500000E 00 0.2500000E 00
0.2500000E 02 0.2500000E 02 0.2500000E 02 0.2500000E 02 0.2500000E 02

```
0.2500000E 04    0.2500000E 04    0.2500000E 04
                                            0.2500000E 04    0.2500000E 04
0.2500000E 06    0.2500000E 06    0.2500000E 06
                                            0.2500000E 06    0.2500000E 06
```

Computer statements S.0004 through S.0011 first determine the largest and smallest numbers. These two values are then printed by computer statements S.0012 and S.0013. Computer statements S.0014 through S.0024 are used to determine the power-of-ten exponent of the smallest number. For example, our data would reveal that SMALL = 0.2500000E-06. Computer statement S.0017 would therefore test to see whether SMALL − 1 is less than, equal to, or greater than zero. Since SMALL (0.2500000E-06) is less than a magnitude of one, computer statement S.0018 multiplies SMALL by ten and M is set equal to one. Computer statement S.0020 then tests to see if this new value of SMALL is still less than one. Since in our example, SMALL would still be less than one, computer statement S.0018 would again multiply SMALL by ten and M would be made equal to two. This procedure would continue until the new value of small became larger than one. At this time the value of M would be equal to seven and would correspond to the fact that the magnitude of the smallest number was between 1.0E-07 and 1.0E-06.

Using an approach similar to the above, computer statements S.0025 through S.0033 would determine the power-of-ten exponent of the largest number ($K = 6$ for this example). Computer statement S.0034 then computes the number of partial sum storage variables ($L = 14$ in this example). The DO 20 loop next sets each of these partial sum storage variables equal to an initial value of zero. The DO 30 loop then examines each of the N data numbers and adds the individual values to the appropriate partial sum variable.

After all the numbers have been stored, the DO 40 loop computes the total sum. Note that the summing is carried out from the lowest magnitude partial sum to the highest. Finally, computer statements S.0052 and S.0053 print the total SUM and the individual partial sum totals. Figure 12.8 shows the corresponding output. The reader's attention is

```
BIG =  0.25CCCCCE 06   SMALL =  0.2500C00E-06

L =          14  M =          -7  K =          6

TOTAL SUM =  C.1262626E 07

 0.2500CCCE-C6  C.9999994E-06  0.0            0.1249999E-03  0.0
 0.1249999E-01  C.25CCCC0E C0  C.1000000E 01  0.2500000E 02  0.1000000E 03
 0.2500CC0E C4  C.1CCCCCCE C5  0.2500000E 06  0.1000000E 07
```

Figure 12.8 The output results of the program of Figure 12.7

called to the added memory requirements and execution time of this particular method.

Nested Summing

At this point the reader should recall that in Section 4.5 we discussed the nested form for evaluating polynomials. At that time we pointed out that $N(N + 1)/2$ multiplications and N additions were required in the evaluation of an N-th order polynomial if we first evaluated each term separately and then summed these individual results. If the nested form is used, however, only N multiplications and N additions are required. The reader should therefore be aware that not only does nesting save computer time by doing fewer multiplications, but it also results in a smaller round-off summing error.

Arctangent Power Series Approximation. In the case of a seventh-order power series approximation of the arctangent function

$$F_7(X) = X - \frac{X^3}{3} + \frac{X^5}{5} - \frac{X^7}{7}$$

we would nest by writing

$$F_7(X) = \left(\left(\left(\left(\left(\left(-\frac{1}{7}X + 0\right)X + \frac{1}{5}\right)X + 0\right)X - \frac{1}{3}\right)X + 0\right)X + 1\right)X$$

$$= \left(\left(\left(-\frac{1}{7}X^2\right) + \frac{1}{5}\right)X^2 - \frac{1}{3}\right)X^2 + 1\Big)X$$

The reader is encouraged to compare the absolute round-off error that occurs in the evaluation of the arctangent function when the nested form is used in contrast with the methods used in Figures 12.5 and 12.7.

Converting Base R Numbers to their Decimal Equivalents. For another example that would show the advantage of nesting, consider the problem of converting base R numbers to their decimal equivalents. This problem is based on our ability to write any whole number in terms of the following general polynomial

$$N = a_{N+1}R^N + a_N R^{N-1} + \cdots + a_2 R + a_1 = \sum_{I=1}^{N+1} a_I R^{I-1}$$

where R is the base or radix of the number. In the decimal system the radix or base is equal to ten and ten different digits (a's) are defined

$$a = 0, 1, 2, 3, 4, 5, 6, 7, 8, \text{ or } 9$$

The decimal number 1384 is therefore actually a contraction and means

$$1*1000 + 3*100 + 8*10 + 4*1 = 1*10^3 + 3*10^2 + 8*10^1 + 4*10^0 = 1384$$

It is therefore not necessary to write the power of ten if we take advantage of the individual digit positions.

There are many reasons for using a number system that has a base other than ten. At one time certain Eskimo and North American Indian tribes used a number system that had a base equal to 5 (quinary system) and therefore they counted in terms of the digits on one hand. For example, the quinary number $(211)_5$ would be equal to the decimal number

$$\begin{aligned} N &= a_3 5^2 + a_2 5^1 + a_1 5^0 \\ &= 2*5^2 + 1*5 + 1*1 \\ &= (56)_{10} \end{aligned}$$

In the past, many computers have used the binary (2) number system because of the relative ease of driving a circuit into cutoff or saturation (two possible states). Other computers have used the octal (8) or biquinary (2-5) number system because of the relative ease of converting from these two systems to the binary system. At the present time the IBM system 360 computers use the hexadecimal (16) number system where 16 different digits (a's) must be defined. Since a single symbol must be used, the a's may be defined as

$$a = 0, 1, 2, \ldots, 9, A, B, C, D, E, \text{ or } F$$

It is thus possible to represent four binary digits in terms of a single hexadecimal digit.

Figure 12.9 shows a program for converting base R whole numbers to their decimal equivalents. Initially the heading is printed and the number of digits, the base, and the N digits of the number are read into the

```
C       CONVERTING BASE R NUMBERS TC THEIR DECIMAL EQUIVALENT.
C
        DIMENSION IA(20)
        WRITE(6,1)
      1 FORMAT('1CONVERTING BASE R NUMBERS TC THEIR DECIMAL EQUIVALENT',/)
      4 READ(5,2)N,IBASE,(IA(I),I=1,N)
      2 FORMAT(2I5,/,20I1)
        IDEC=0
        DO 10 I=1,N
     10 IDEC=IDEC*IBASE+IA(I)
        WRITE(6,3)IBASE,(IA(I),I=1,N)
      3 FORMAT('0BASE',I4,' NUMBER =',(20I1))
        WRITE(6,5)IDEC
      5 FORMAT(//,' DECIMAL EQUIVALENT =',I10,////)
        GO TO 4
        END
```

Figure 12.9 The nested form of a polynomial is used in the conversion of base R numbers to their decimal equivalents

```
CONVERTING BASE B NUMBERS TO THEIR DECIMAL EQUIVALENT

BASE    8 NUMBER =155

DECIMAL EQUIVALENT =        109


BASE    2 NUMBER =11001

DECIMAL EQUIVALENT =         25


BASE    6 NUMBER =12340

DECIMAL EQUIVALENT =       1860


IHC217I
```

Figure 12.10 The output results of the program of Figure 12.9

computer memory. The dummy variable IDEQ is used to store the decimal number equivalent and is initially set equal to zero. The execution of the DO 10 loop yields the corresponding decimal equivalent by the nested evaluation of the general N-th order polynomial. Computer statements S.0009 through S.0012 are then used to print the results.

Figure 12.10 gives the computer results for three different input data sets. The verification of these results are left as an exercise for the reader. He is encouraged to modify this program so that it is capable of handling the conversion of numbers whose base is greater than 10 (for example, hexadecimal and A format). The reader is also urged to extend this program so that it will be able to handle the fractional part of a number as well as to permit the conversion from any base to any other base, including negative bases.

Complementary Error Function. For another example that illustrates the use of the nested form in the evaluation of a polynomial, consider the ninth-degree least-squares polynomial that may be used in the evaluation of the complementary error function, ERFC(X). The error function, ERF(X), is defined as

$$\mathrm{ERF(X)} = \frac{2}{\sqrt{\pi}} \int_0^X \epsilon^{-X^2}\, dX$$

and the complementary error function as

$$\mathrm{ERFC}(X) = 1 - \mathrm{ERF}(X)$$

The evaluation of this function is needed in connection with semiconductor impurity diffusion problems, and so on.[2]

Using the method of least squares (see Section 12.4), an eight-degree polynomial

$$P_8(X) = A_1 + A_2X + A_3X^2 + A_4X^3 + A_5X^4 + A_6X^5 + A_7X^6 + A_8X^7 + A_9X^8$$

is computed from bivariate data of X and Ln ERFC(X) divided by X

$$P_8(X) = \frac{\text{Ln ERFC}(X)}{X}$$

for $X = 0.1$ to 4.9, in 0.1 unit steps. The coefficients of the least squares approximation are found to be

$$\begin{array}{llllll} A_1 = & -1.12835 & A_4 = & 2.13017\text{E-}02 & A_7 = & 9.38119\text{E-}05 \\ A_2 = & -0.636547 & A_5 = & -2.24511\text{E-}03 & A_8 = & -1.24439\text{E-}05 \\ A_3 = & -0.103679 & A_6 = & -1.67642\text{E-}04 & A_9 = & 5.98082\text{E-}07 \end{array}$$

and the approximation for ERFC(X) is therefore given by

$$\text{ERFC(X)} \approx \text{EXP}(A_1X + A_2X^2 + A_3X^3 + A_4X^4 + A_5X^5 + A_6X^6 + A_7X^7 + A_8X^8 + A_9X^9)$$

[2] G. W. McIver and J. L. Buie, "Approximation for the Complementary Error Function," *IEEE CADAR News*, Issue 8 (September, 1968), pp. 9–11.

```
          C      THE NESTED FORM IS USED TO EVALUATE A GENERAL POLYNOMIAL OF ORDER M.
          C
S.0001           DIMENSION A(100)
S.0002           READ(5,1)M
S.0003         1 FORMAT(I5)
S.0004           L=M+1
S.0005           READ(5,2)(A(I),I=1,L)
S.0006         2 FORMAT(5E15.7)
S.0007           WRITE(6,3)M
S.0008         3 FORMAT('1EVALUATION OF   P(X) = A(M+1)*X**M + A(M)*X**(M-1) + ... +
                1 A(I)*X + A(1)',//,' THE ORDER OF THE POLYNOMIAL IS =',I5,///,' TH
                2E POLYNOMIAL COEFFICIENTS ARE',//)
S.0009           DO 10 I=1,L
S.0010           K=I-1
S.0011        10 WRITE(6,4)K,A(I)
S.0012         4 FORMAT(' A(',I2,') =',E15.7,/)
S.0013           WRITE(6,5)
S.0014         5 FORMAT(//,7X,'X',14X,'F(X)',11X,'ERFC(X)',/)
S.0015           X=0.0
S.0016           DO 15 I=1,50
S.0017           X=X+0.1
S.0018           FX=0.0
S.0019        20 IF(L)30,30,40
S.0020        40 FX=FX*X+A(L)
S.0021           L=L-1
S.0022           GO TO 20
S.0023        30 ERFC=EXP(X*FX)
S.0024           WRITE(6,6)X,FX,ERFC
S.0025         6 FORMAT(3E15.7)
S.0026           L=M+1
S.0027        15 CONTINUE
S.0028           STOP
S.0029           END
```

Figure 12.11 The equivalent of the complementary error function from a least-squares polynomial approximation

```
EVALUATION OF  P(X) = A(M+1)*X**M + A(M)*X**(M-1) + ... + A(I)*X + A(1)

THE ORDER OF THE POLYNOMIAL IS =     8


THE POLYNOMIAL COEFFICIENTS ARE


A( 0) = -0.1128349E 01

A( 1) = -0.6365470E 00

A( 2) = -0.1036789E 00

A( 3) =  0.2130170E-01

A( 4) = -0.2245110E-02

A( 5) = -0.1676420E-03

A( 6) =  0.9381189E-04

A( 7) = -0.1244390E-04

A( 8) =  0.5980820E-06


      X                F(X)             ERFC(X)

 0.9999996E-01 -0.1193019E 01  0.8875399E 00
 0.1999999E 00 -0.1259639E 01  0.7773010E 00
 0.2999999E 00 -0.1328087E 01  0.6713760E 00
 0.3999999E 00 -0.1398252E 01  0.5716088E 00
 0.4999998E 00 -0.1470023E 01  0.4795001E 00
 0.5999998E 00 -0.1543300E 01  0.3961433E 00
 0.6999997E 00 -0.1617985E 01  0.3221982E 00
 0.7999997E 00 -0.1693987E 01  0.2578987E 00
 0.8999997E 00 -0.1771219E 01  0.2030920E 00
 0.9999996E 00 -0.1849604E 01  0.1572996E 00
 0.1099999E 01 -0.1929063E 01  0.1197957E 00
 0.1199999E 01 -0.2009527E 01  0.8968699E-01
 0.1299998E 01 -0.2090931E 01  0.6599301E-01
 0.1399998E 01 -0.2173214E 01  0.4771575E-01
 0.1499997E 01 -0.2256319E 01  0.3389554E-01
 0.1599997E 01 -0.2340193E 01  0.2365218E-01
 0.1699996E 01 -0.2424788E 01  0.1620995E-01
 0.1799995E 01 -0.2510058E 01  0.1090981E-01
 0.1899995E 01 -0.2595963E 01  0.7209793E-02
 0.1999994E 01 -0.2682462E 01  0.4677895E-02
 0.2099994E 01 -0.2769520E 01  0.2979579E-02
 0.2199993E 01 -0.2857103E 01  0.1862925E-02
 0.2299993E 01 -0.2945183E 01  0.1143229E-02
 0.2399992E 01 -0.3033728E 01  0.6885505E-03
 0.2499991E 01 -0.3122713E 01  0.4069766E-03
 0.2599991E 01 -0.3212113E 01  0.2360504E-03
 0.2699990E 01 -0.3301907E 01  0.1343429E-03
 0.2799990E 01 -0.3392072E 01  0.7501927E-04
 0.2899989E 01 -0.3482588E 01  0.4110148E-04
 0.2999989E 01 -0.3573437E 01  0.2209257E-04
 0.3099988E 01 -0.3664601E 01  0.1164982E-04
 0.3199987E 01 -0.3756065E 01  0.6026396E-05
 0.3299987E 01 -0.3847814E 01  0.3058048E-05
 0.3399986E 01 -0.3939832E 01  0.1522170E-05
 0.3499986E 01 -0.4032106E 01  0.7431899E-06
 0.3599985E 01 -0.4124625E 01  0.3559097E-06
 0.3699985E 01 -0.4217377E 01  0.1671744E-06
 0.3799984E 01 -0.4310349E 01  0.7701567E-07
 0.3899983E 01 -0.4403534E 01  0.3479800E-07
 0.3999983E 01 -0.4496920E 01  0.1542009E-07
 0.4099982E 01 -0.4590499E 01  0.6701374E-08
 0.4199982E 01 -0.4684262E 01  0.2856088E-08
 0.4299981E 01 -0.4778203E 01  0.1193742E-08
 0.4399981E 01 -0.4872313E 01  0.4892831E-09
 0.4499980E 01 -0.4966585E 01  0.1966623E-09
 0.4599979E 01 -0.5061012E 01  0.7751512E-10
 0.4699979E 01 -0.5155588E 01  0.2996053E-10
 0.4799978E 01 -0.5250307E 01  0.1135509E-10
 0.4899978E 01 -0.5345160E 01  0.4220101E-11
 0.4999977E 01 -0.5440141E 01  0.1537941E-11
```

Figure 12.12 The output results of the program of Figure 12.11

Figure 12.11 shows a Fortran program for computing the complementary error function based on this least squares approximation. Figure 12.12 gives the approximations obtained at the same X values that were used to determine the eight-degree least squares polynomial. The reader may verify from the previous reference that these results are accurate to approximately five significant figures.

12.3 INTERPOLATION

Polynomial Interpolation

The reader has undoubtedly used interpolation in connection with the evaluation of logarithms and trigonometric functions for argument values not listed in the table. In other cases, the reader may have referred to an instrument calibration listing to determine the error of a given reading. In still other cases the reader may have accumulated empirical data at certain discrete points and later used this data to estimate the values at other points. All these examples correspond to the general table-reference problem. For example, if we have the following table of values:

I	*X(I)*	*F(X)*
1	2	4
2	3	8
3	4	15
4	5	25
.	.	.
.	.	.
.	.	.

and we wish to estimate the value of $F(2.2)$, we might use a first-degree polynomial (linear interpolation) passing through the first two points to determine the function value for $X = 2.2$. This would be done by using the law of similar triangles in Figure 12.13 to write

$$\frac{2.2 - 2.0}{3.0 - 2.0} = \frac{F(2.2) - 4.0}{8.0 - 4.0}$$

or

$$\begin{aligned} F(2.2) &= 4.0 + 4.0(2.2 - 2.0) \\ &= 4.8 \end{aligned}$$

In a similar way we might use a first-degree polynomial passing through the points (3,8) and (4,15) to estimate the value of $F(3.6)$, and so on.

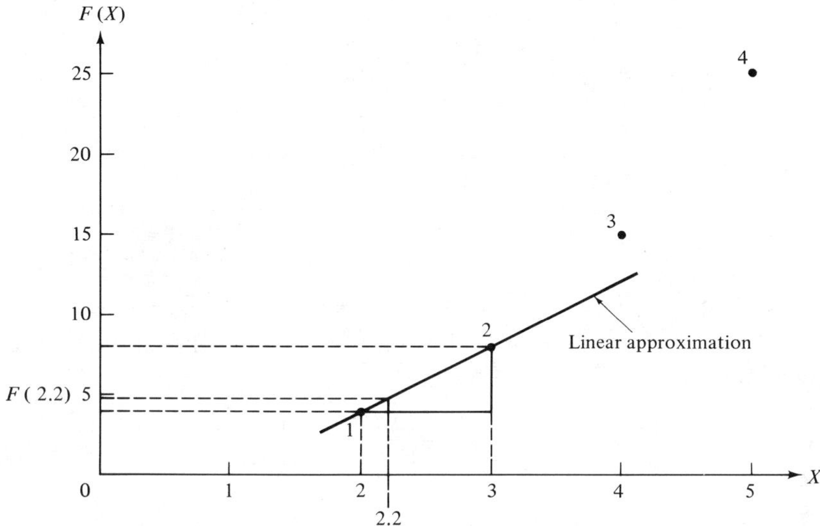

Figure 12.13 The concept of linear interpolation

The general problem of linear interpolation therefore requires us to find the equation of a straight line that passes through two successive points whose X values bracket the required XR, as shown in Figure 12.14. We must therefore find the value of the slope m and the intercept b in the general equation

$$F(X) = mX + b$$

These values can always be found from a knowledge of the locations of the bracketing points; thus,

$$\begin{cases} Y(I) = mX(I) + b \\ Y(I-1) = mX(I-1) + b \end{cases}$$

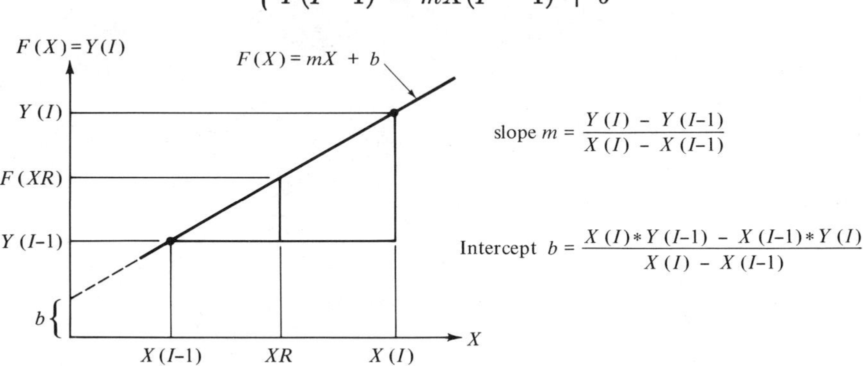

Figure 12.14 Determining the slope of an intercept of a first-degree interpolation polynomial

yields,

$$m = \frac{Y(I) - Y(I - 1)}{X(I) - X(I - 1)}$$

and

$$b = \frac{X(I)*Y(I-1) - X(I-1)*Y(I)}{X(I) - X(I-1)}$$

or

$$F(X) = \frac{Y(I) - Y(I-1)}{X(I) - X(I-1)}X + \frac{X(I)*Y(I-1) - X(I-1)*Y(I)}{X(I) - X(I-1)}$$

Note that this same equation will result from the application of the similar triangle law

$$\frac{XR - X(I-1)}{X(I) - X(I-1)} = \frac{F(XR) - Y(I-1)}{Y(I) - Y(I-1)}$$

or

$$\begin{aligned} F(XR) &= \frac{Y(I) - Y(I-1)}{X(I) - X(I-1)}(XR - X(I-1)) + Y(I-1) \\ &= \frac{Y(I) - Y(I-1)}{X(I) - X(I-1)}XR + \frac{X(I)*Y(I-1) - X(I-1)*Y(I)}{X(I) - X(I-1)} \end{aligned}$$

except that here we have evaluated the first-degree polynomial for the specific value $X = XR$.

In order to show how we may make general use of linear interpolation, consider the set of arbitrary points plotted in Figure 12.15. Assuming

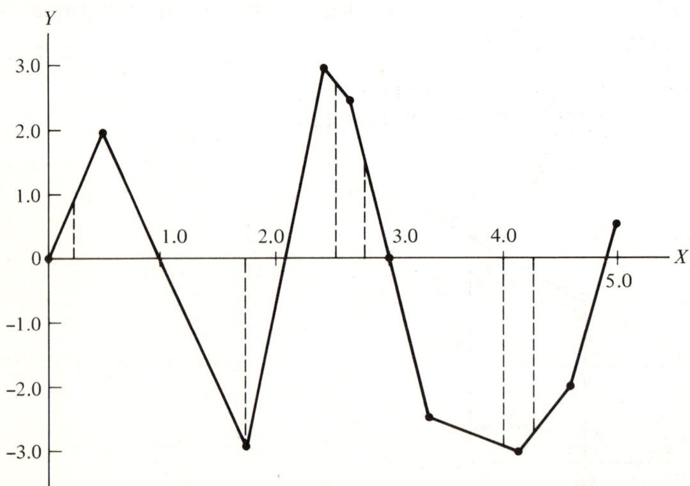

Figure 12.15 A set of arbitrary points is used to test the interpolation program of Figure 12.16

```
         C       LINEAR INTERPOLATION PROBLEM.
         C
S.0001           DIMENSION X(100),Y(100)
S.0002           READ(5,1)N,(X(I),Y(I),I=1,N)
S.0003         1 FORMAT(I5,/,(5E15.7))
S.0004           WRITE(6,2)
S.0005         2 FORMAT('1LINEAR INTERPOLATION',///,7X,'XR',9X,'F(XR)',/)
S.0006         3 READ(5,4)XR
S.0007         4 FORMAT(E15.7)
S.0008           DO 10 I=2,N
S.0009           IF(XR-X(I))20,20,10
S.0010        10 CONTINUE
S.0011        20 FXR=(XR-X(I-1))*(Y(I)-Y(I-1))/(X(I)-X(I-1))+Y(I-1)
S.0012           WRITE(6,30)XR,FXR
S.0013        30 FORMAT(2E15.7)
S.0014           GO TO 3
S.0015           END
```

Figure 12.16 A linear interpolation program

that the X and Y values of these points will be supplied as input data, we make use of the program of Figure 12.16 to estimate the Y values of various X values that may not correspond to the X values of any of the input data points.

Initially this program reads in the number of data points N and their corresponding X and Y values. The output heading and subheading is then printed. This is followed by reading in the numerical value of XR. The DO 10 loop is then used to locate the numbers of the bracketing points. For example, during the first iteration of the DO 10 loop, computer statement S.0009 compares the numerical value of XR with $X(1)$. If the value of XR is greater than $X(1)$, the iteration of the DO 10 loop is continued until a value of $X(I)$ is found which is less than or equal to XR. At this point, we will have found the value of I and the value of $X(I)$ that is equal to or greater than XR (see Figure 12.14). Control will therefore be transferred to statement 20 and the value of FXR will be estimated. Computer statements S.0012 and S.0013 will then print the values of XR and $F(XR)$ and control will be transferred to statement 3 so that the next interpolation can be made.

Figure 12.17 gives the computer results for several XR input values. The reader may readily verify the accuracy of these results by referring to Figure 12.15.

Strip Transmission Line. Let us now consider a more complicated example where it will be necessary to do a double linear interpolation in

```
LINEAR INTERPOLATION

        XR             F(XR)

   0.2000000E 00   0.8000000E 00
   0.1799999E 01  -0.3199999E 01
   0.2500000E 01   0.2699999E 01
   0.4200000E 01  -0.2979999E 01
   0.2799999E 01   0.1500001E 01
   0.4000000E 01  -0.3120000E 01

IHC217I
```

Figure 12.17 The output results for the program of Figure 12.16

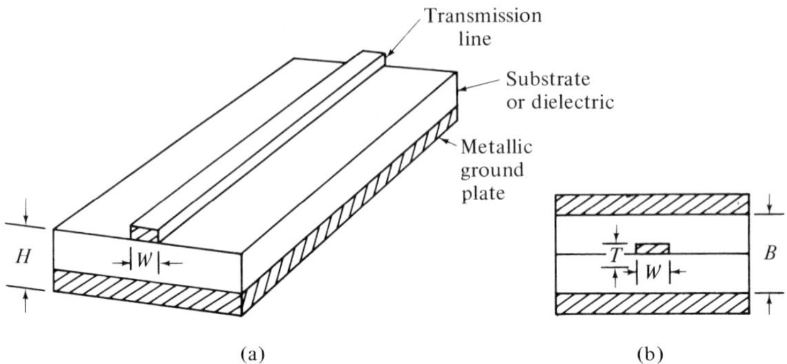

Figure 12.18 (a) Microstrip; (b) end view of strip transmission line

connection with the calculation of the characteristic impedance of a strip transmission line.

A microstrip may be made by taking a two-sided metallic clad fibreglass dielectric board and etching the metal from one side so that only a narrow transmission line remains, as shown in Figure 12.18a.[3] A strip transmission line may be made by bolting or cementing a one-sided metallic board above the microstrip, as shown in Figure 12.18b. Both of these configurations find wide use at microwave frequencies in connection with their use as filter elements and as transmission lines. In addition, microstrip transmission lines are often employed in high-frequency or high-speed active networks to preserve the signal waveshape from stage to stage.

The calculation of the characteristic impedance of narrow strip ($W < 2H$) and wide strip ($W > 2H$) microstrips can be computed from the following equations

$$Z_0 = \frac{377}{\pi\sqrt{(\epsilon_r + 1)/2}}\left(\ln\frac{8H}{W} + \frac{1}{8}\left(\frac{W}{2H}\right)^2 - \frac{1}{2}\frac{\epsilon_r - 1}{\epsilon_r + 1}\left(\ln\frac{\pi}{2} + \frac{1}{\epsilon_r}\ln\frac{4}{\pi}\right)\right); \quad W < 2H$$

$$Z_0 = \frac{377\sqrt{1/\epsilon_r/2}}{\dfrac{W}{2H} + 0.441 + \dfrac{\epsilon_r + 1}{2\pi\epsilon_r}\left(\begin{array}{l}(\ln(W/2H + 0.94) + 1.451) \\ \quad + 0.082(\epsilon_r - 1)/\epsilon_r^2\end{array}\right)}; \quad W > 2H$$

where

ϵ_r is the dielectric constant of the substrate or fibreglass board.
W is the width of the line.
H is the height of the substrate.

Similarly, the calculation of the characteristic impedance of tri-plate lines where the ratio of the width (W) of the centerstrip to the separation (B)

[3] See *Reference Data for Radio Engineers*, 4th Ed., (New York: International Telephone and Telegraph Corporation, 1962), pp. 595–600.

between the ground planes is $W/B > 0.35$ is found from

$$Z_0 = \frac{94.15}{\sqrt{\epsilon_r}\,(W/(B(1 - T/B)) + C'_f/0.0885\epsilon_r)}$$

where

$$C'_f = \frac{0.0885\epsilon_r}{\pi}\left(\frac{2}{1 - T/B}\ln\left(\frac{1}{1 - T/B} + 1\right) - \left(\frac{1}{1 - T/B}\ln\left(\frac{1}{(1 - T/B)^2} - 1\right)\right)\right) \text{ uuf/cm.}$$

and T is the thickness of the center strip.

When the ratio $W/B < 0.35$, the characteristic impedance can be computed from

$$Z_0 = \frac{60}{\sqrt{\epsilon_r}} \ln \frac{4B}{\pi W/B} \text{ ohms}$$

Although it is possible to hand calculate the characteristic impedance by using one of the appropriate formulas above, most engineers prefer to determine the value from an appropriate graph. Figure 12.19 shows the graph that should be used for the tri-plate case.

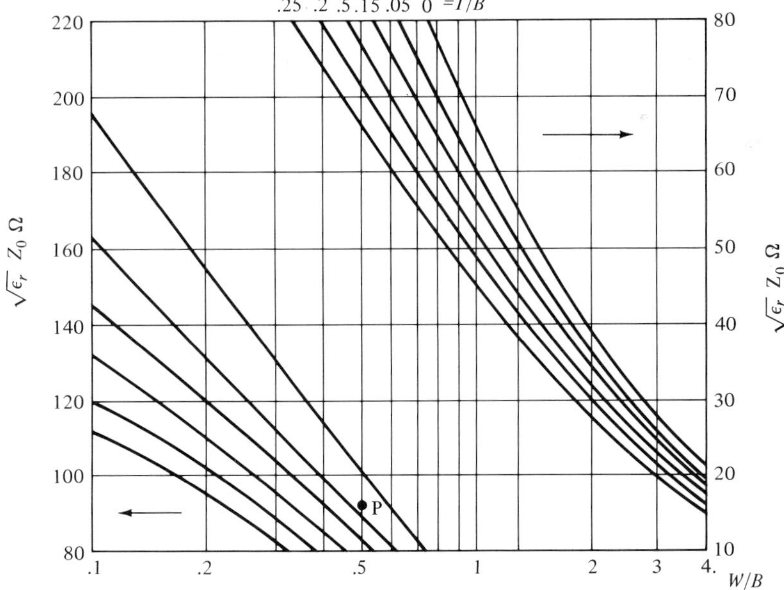

Figure 12.19 A plot of the characteristic impedance of a strip transmission line versus W/B ratios for T/B values of 0.00, 0.05, 0.10, 0.15, 0.20, and 0.25

```
C     INTERPOLATION OF STRIP
C     TRANSMISSION LINE CHART
C
      DIMENSION TBOX(11),TBOY(11),TB5X(11),TB5Y(11),TB10X(12),TB10Y(12),
     1TB15X(12),TB15Y(12),TB20X(12),TB20Y(12),TB25X(13),TB25Y(13)
      READ (5,1) (TBOX(I),TBOY(I),I=1,11)
    1 FORMAT(8F10.2)
      WRITE(6,1) (TBOX(I),TBOY(I),I=1,11)
      READ (5,1) (TB5X(I),TB5Y(I),I=1,11)
      WRITE(6,1) (TB5X(I),TB5Y(I),I=1,11)
      READ (5,1) (TB10X(I),TB10Y(I),I=1,12)
      WRITE(6,1) (TB10X(I),TB10Y(I),I=1,12)
      READ (5,1) (TB15X(I),TB15Y(I),I=1,12)
      WRITE(6,1) (TB15X(I),TB15Y(I),I=1,12)
      READ (5,1) (TB20X(I),TB20Y(I),I=1,12)
      WRITE(6,1) (TB20X(I),TB20Y(I),I=1,12)
      READ (5,1) (TB25X(I),TB25Y(I),I=1,13)
      WRITE(6,1) (TB25X(I),TB25Y(I),I=1,13)
    5 READ (5,2) W,B,T
    2 FORMAT(3F10.2)
      WRITE (6,8) W,B,T
    8 FORMAT('0W =',E15.7,' B =',E15.7,' T =',E15.7)
      WB=W/B
      TB=T/B
      WRITE (6,7) WB,TB
    7 FORMAT(' WB =',E15.7,' TB =',E15.7)
      V=0.0
      DO 10 I=1,5
      V=V+.05
      IF(TB-V) 3,3,10
    3 K=I
      GO TO 11
   10 CONTINUE
      K=5
   11 GO TO(40,50,60,70,80),K
   40 DO 41 I=2,11
      IF (WB-TBOX(I))42,42,41
   41 CONTINUE
   42 TBY1=(WB-TBOX(I-1))*(TBOY(I)-TBOY(I-1))/(TBOX(I)-TBOX(I-1))+
     1TBOY(I-1)
      DO 43 I=2,11
      IF(WB-TB5X(I)) 44,44,43
   43 CONTINUE
   44 TBY2=(WB-TB5X(I-1))*(TB5Y(I)-TB5Y(I-1))/(TB5X(I)-TB5X(I-1))+
     1TB5Y(I-1)
      Z=(TBY1-TBY2)*(.05-TB)/.05+TBY2
      WRITE (6,6) TBY1,TBY2,Z
    6 FORMAT(' TBY1 =',E15.7,' TBY2 =',E15.7,' Z =',E15.7)
      GO TO 90
   50 DO 51 I=2,11
      IF (WB-TB5X(I))52,52,51
   51 CONTINUE
   52 TBY1=(WB-TB5X(I-1))*(TB5Y(I)-TB5Y(I-1))/(TB5X(I)-TB5X(I-1))+
     1TB5Y(I-1)
      DO 53 I=2,12
      IF (WB-TB10X(I))54,54,53
   53 CONTINUE
   54 TBY2=(WB-TB10X(I-1))*(TB10Y(I)-TB10Y(I-1))/(TB10X(I)-TB10X(I-1))+
     1TB10Y(I-1)
      Z=(TBY1-TBY2)*(.1-TB)/.05+TBY2
      WRITE(6,6) TBY1,TBY2,Z
      GO TO 90
   60 DO 61 I=2,12
      IF (WB-TB10X(I))62,62,61
   61 CONTINUE
   62 TBY1=(WB-TB10X(I-1))*(TB10Y(I)-TB10Y(I-1))/(TB10X(I)-TB10X(I-1))+
     1TB10Y(I-1)
      DO 63 I=2,12
      IF(WB-TB15X(I)) 64,64,63
   63 CONTINUE
   64 TBY2=(WB-TB15X(I-1))*(TB15Y(I)-TB15Y(I-1))/(TB15X(I)-TB15X(I-1))+
     1TB15Y(I-1)
      Z=(TBY1-TBY2)*(.15-TB)/.05+TBY2
      WRITE (6,6) TBY1,TBY2,Z
      GO TO 90
   70 DO 71 I=2,12
      IF(WB-TB15X(I)) 72,72,71
   71 CONTINUE
   72 TBY1=(WB-TB15X(I-1))*(TB15Y(I)-TB15Y(I-1))/(TB15X(I)-TB15X(I-1))+
```

Figure 12.20 A program for computing the characteristic impedance of a strip transmission line

```
S.0070        DO 73 I=2,12
S.0071        IF(WB-TB20X(I))74,74,73
S.0072     73 CONTINUE
S.0073     74 TBY2=(WB-TB20X(I-1))*(TB20Y(I)-TB20Y(I-1))/(TB20X(I)-TB20X(I-1))+
             1TB20Y(I-1)
S.0074        Z=(TBY1-TBY2)*(.20-TB)/.05+TBY2
S.0075        WRITE(6,6)TBY1,TBY2,Z
S.0076        GO TO 90
S.0077     80 DO 81 I=2,12
S.0078        IF(WB-TB20X(I))82,82,81
S.0079     81 CONTINUE
S.0080     82 TBY1=(WB-TB20X(I-1))*(TB20Y(I)-TB20Y(I-1))/(TB20X(I)-TB20X(I-1))+
             1TB20Y(I-1)
S.0081        DO 83 I=2,13
S.0082        IF(WB-TB25X(I))84,84,83
S.0083     83 CONTINUE
S.0084     84 TBY2=(WB-TB25X(I-1))*(TB25Y(I)-TB25Y(I-1))/(TB25X(I)-TB25X(I-1))+
             1TB25Y(I-1)
S.0085        Z=(TBY1-TBY2)*(.25-TB)/.05+TBY2
S.0086        WRITE(6,6)TBY1,TBY2,Z
S.0087     90 CONTINUE
S.0088        GO TO 5
S.0089        RETURN
S.0090        END
```

Figure 12.20 Continued

For a numerical example that shows how we might use this graph, assume that we have two $\frac{1}{16}$ in. boards ($B = \frac{1}{8}$ in. $= 0.125$ in.) and we are using two-oz. copper clad board ($T = 0.0027$ in.). The ratio of T/B is therefore equal to 0.022. Etching a 0.062 in. line width ($W =$ 0.062 in.) will therefore result in a W/B ratio of 0.5. The value of T/B (0.022) and W/B (0.5) will thus correspond to the point P shown in Figure 12.19. This, in turn, will correspond to a $\sqrt{\epsilon_r}\, Z_0$ value of 94.0 ohms. If the dielectric constant is equal to 4.8, the resulting line will have a characteristic impedance of 43.4 ohms. Note that this value would be very close to what would be needed for very high-frequency transistor transmission line coupling ($r_b' = 50$ ohms).

Figure 12.20 is a Fortran program that computes the value of $\sqrt{\epsilon_r}\, Z_o$ from specified W, B, and T values. Initially computer statements S.0002 through S.0014 read in and print out the curve data. Note that TBOX(I) and TBOY(I) correspond to the X and Y values of the $T/B = 0.0$ curve and $I = 1$, 11 represents this particular curve in terms of eleven points, and so on. Computer statements S.0015 through S.0018 are then used to read in and print out the values of W, B, and T. The ratio of W to B and T to B are next computed and then printed by computer statements S.0019 through S.0021. The dummy variable V is then set equal to zero and we enter the DO 10 loop to determine what curve we should use in the interpolation. For example, if the ratio of T to B were equal to 0.14 and the ratio of W to B were equal to 1.0, the initial iteration of the DO 10 loop would find that T/B was greater than the value of V (0.05) and control would therefore return to the DO 10 loop. This time the value of I and V would be set equal to 2 and 0.10, respectively, and computer statement S.0026 would again find that the value of T/B was greater than V. In the third iteration of the DO 10 loop, however, the value of

T/B would be less than V and therefore control would be transferred to statement 3 where the value of K would be set to I or 3. Control would then be transferred to statement 11 and, in turn, to statement 60.

Computer statements S.0055 through S.0065 would now be used to compute the value of Z ($\sqrt{\epsilon_r}\, Z_o$). This would be done by first finding the W/B value of the TB10X(I) data point (on the $T/B = 0.10$ curve) that was equal to or greater than the value of $W/B = 1.0$ (the input data value). Computer statement S.0058 would then compute the value of TBY1 by linear interpolation. Note that this value of TBY1 would be the ordinate value ($\sqrt{\epsilon_r}\, Z_0$) of the $T/B = 0.10$ curve for a W/B ratio of 1.0. Computer statements S.0059 through S.0062 would then repeat the above to find the TBY2 ordinate value for $T/B = 0.15$ and $W/B = 1.0$. Computer statement S.0063 would then use interpolation once more to approximate the $\sqrt{\epsilon_r}\, Z_0$ value of the actual point ($T/B = 0.14$ and $W/B = 1.0$). Finally, computer statement S.0064 would be used to print the values of TBY1, TBY2, and Z ($\sqrt{\epsilon_r}\, Z_0$) before returning control to statement 5 so that the next set of data could be processed.

Figure 12.21 shows the computer input data for the six T/B curves.

T/B	Input Data							
0.0	0.10	193.00	0.30	130.00	0.50	100.00	0.75	80.00
	1.00	67.00	1.50	50.00	2.00	39.00	2.50	32.50
	3.00	28.00	3.50	24.00	4.00	22.00		
0.05	0.10	160.00	0.20	131.00	0.62	80.00	0.80	70.00
	1.00	60.00	1.50	45.00	2.00	37.00	2.50	30.00
	3.00	26.00	3.50	23.00	4.00	21.00		
0.10	0.10	142.00	0.20	119.00	0.30	104.00	0.53	80.00
	0.80	65.00	1.00	56.00	1.50	42.50	2.00	34.00
	2.50	28.00	3.00	24.50	3.50	22.00	4.00	19.00
0.15	0.10	130.00	0.15	118.00	0.25	102.00	0.45	80.00
	0.60	70.00	0.80	60.00	1.00	52.50	1.50	40.00
	2.00	32.00	2.50	27.00	3.00	23.00	4.00	18.00
0.20	0.10	119.00	0.15	110.00	0.25	95.00	0.38	80.00
	0.60	65.00	0.80	56.00	1.00	48.00	1.50	37.50
	2.00	30.00	2.50	25.00	3.00	22.00	4.00	17.00
0.25	0.10	110.00	0.15	102.00	0.20	92.50	0.25	88.00
	0.32	80.00	0.60	61.00	0.80	52.00	1.00	45.00
	1.50	34.00	2.00	28.00	2.50	23.00	3.00	20.00
	4.00	16.50						

Figure 12.21 The input data for the program of Figure 12.20

```
W =  0.1000000E 01 B =  0.1000000E 01 T =  0.2500000E 00
WB =  0.1000000E 01 TB =  0.2500000E 00
TBY1 =  0.4800000E 02 TBY2 =  0.4500000E 02 Z =  0.4500000E 02

W =  0.1000000E 01 B =  0.1000000E 01 T =  0.1400000E 00
WB =  0.1000000E 01 TB =  0.1400000E 00
TBY1 =  0.5600000E 02 TBY2 =  0.5250000E 02 Z =  0.5320000E 02

W =  0.1000000E 01 B =  0.1000000E 01 T =  0.5000000E-01
WB =  0.1000000E 01 TB =  0.5000000E-01
TBY1 =  0.6700000E 02 TBY2 =  0.6000000E 02 Z =  0.6000000E 02

W =  0.8000000E 00 B =  0.1000000E 01 T =  0.5000000E-01
WB =  0.8000000E 00 TB =  0.5000000E-01
TBY1 =  0.7739999E 02 TBY2 =  0.7000000E 02 Z =  0.7000000E 02

W =  0.8000000E 00 B =  0.1000000E 01 T =  0.2500000E 00
WB =  0.8000000E 00 TB =  0.2500000E 00
TBY1 =  0.5600000E 02 TBY2 =  0.5200000E 02 Z =  0.5200000E 02

W =  0.6000000E 00 B =  0.1000000E 01 T =  0.5000000E-01
WB =  0.6000000E 00 TB =  0.5000000E-01
TBY1 =  0.9200000E 02 TBY2 =  0.8242859E 02 Z =  0.8242859E 02

W =  0.6000000E 00 B =  0.1000000E 01 T =  0.2500000E 00
WB =  0.6000000E 00 TB =  0.2500000E 00
TBY1 =  0.6500000E 02 TBY2 =  0.6100002E 02 Z =  0.6100002E 02

W =  0.4000000E 00 B =  0.1000000E 01 T =  0.5000000E-01
WB =  0.4000000E 00 TB =  0.5000000E-01
TBY1 =  0.1150000E 03 TBY2 =  0.1067143E 03 Z =  0.1067143E 03

W =  0.4000000E 00 B =  0.1000000E 01 T =  0.1300000E 00
WB =  0.4000000E 00 TB =  0.1300000E 00
TBY1 =  0.9356522E 02 TBY2 =  0.8550002E 02 Z =  0.8872609E 02

W =  0.2000000E 01 B =  0.1000000E 01 T =  0.5000000E-01
WB =  0.2000000E 01 TB =  0.5000000E-01
TBY1 =  0.3900000E 02 TBY2 =  0.3700000E 02 Z =  0.3700000E 02

W =  0.2000000E 01 B =  0.1000000E 01 T =  0.1600000E 00
WB =  0.2000000E 01 TB =  0.1600000E 00
TBY1 =  0.3200000E 02 TBY2 =  0.3000000E 02 Z =  0.3159999E 02

W =  0.3500000E 01 B =  0.1000000E 01 T =  0.5000000E-01
WB =  0.3500000E 01 TB =  0.5000000E-01
TBY1 =  0.2400000E 02 TBY2 =  0.2300000E 02 Z =  0.2300000E 02

W =  0.3500000E 01 B =  0.1000000E 01 T =  0.1600000E 00
WB =  0.3500000E 01 TB =  0.1600000E 00
TBY1 =  0.2050000E 02 TBY2 =  0.1950000E 02 Z =  0.2029999E 02

W =  0.5000000E 00 B =  0.1000000E 01 T =  0.1600000E 00
WB =  0.5000000E 00 TB =  0.1600000E 00
TBY1 =  0.7666666E 02 TBY2 =  0.7181818E 02 Z =  0.7569696E 02

IHC217I
```

Figure 12.22 The output results of the program of Figure 12.20

Figure 12.22 gives the computer results for several input data sets. It is left as an exercise for the reader to use the information of Figure 12.21 to verify the results of Figure 12.22.

In the above discussion it should be quite obvious that our use of a straight line approximation based on points bracketing the point of interest will only lead to accurate results if the point of interest coincides with one of the bracketed points. The question of how accurate the $F(XR)$ value may be for an intermediate point is a difficult question to answer since we may have no way of knowing how the function may change between observed points. If the experimental data has been collected at sufficiently close intervals so that we may expect a smooth transition from one point to the next, then the use of a higher-order interpolating polynomial is advised. For example, an examination of the data plotted in Figure 12.13 would suggest that we may obtain more accurate results for $F(2.2)$ if we use a second-degree interpolating polynomial that is based on the first three points. In order to obtain this second-degree polynomial

$$F_2(X) = aX^2 + bX + c$$

we must find the values of the three coefficients. These values may be found from the knowledge of the locations of the three points, thus,

$$\begin{aligned}F_2(X(1)) &= Y(1) = aX(1)^2 + bX(1) + c\\ F_2(X(2)) &= Y(2) = aX(2)^2 + bX(2) + c\\ F_2(X(3)) &= Y(3) = aX(3)^2 + bX(3) + c\end{aligned}$$

or

$$\begin{aligned}4 &= 4a + 2b + c\\ 8 &= 9a + 3b + c\\ 15 &= 16a + 4b + c\end{aligned}$$

Solving these equations for a, b, and c yields

$$\begin{aligned}a &= 1.5\\ b &= -3.5\\ c &= 5.0\end{aligned}$$

thus

$$F_2(X) = 1.5X^2 - 3.5X + 5.0$$

and

$$\begin{aligned}F_2(2.2) &= 1.5(2.2)^2 - 3.5(2.2) + 5.0\\ &= 4.56\end{aligned}$$

In a similar way we may use $n+1$ points to determine a polynomial of order n that passes through all $n+1$ points

$$F_n(X) = a_{n+1}X^n + a_nX^{n-1} + \cdots + a_2X + a_1$$

and the $n+1$ coefficients would be found from a knowledge of the locations of the $n+1$ data points

$$\begin{aligned}F_n(X(1)) &= Y(1) = a_{n+1}X(1)^n + a_nX(1)^{n-1} + \cdots + a_2X(1) + a_1\\ F_n(X(2)) &= Y(2) = a_{n+1}X(2)^n + a_nX(2)^{n-1} + \cdots + a_2X(2) + a_2\\ &\vdots\\ F_n(X(n+1)) &= Y(n+1) = a_{n+1}X(n+1)^n + a_nX(n+1)^{n-1} + \cdots + a_2X(n+1) + a_1\end{aligned}$$

The $n+1$ linear equations above could now be solved for the $n+1$ unknowns $(a_{n+1}, a_n, \ldots, a_1)$ by using the Gauss-Jordan method or some other method discussed in Chapter 6. Once these coefficients have been found, the value of the function for any other XR not listed in the table could thus be determined.

The reader should be cautioned that the use of a higher degree polynomial does not necessarily result in more accurate interpolation. Putting it another way, the greater the number of points used in an interpolation, the more accurate the result. It should be noted that if most of the points are far removed from the point of interpolation, the additional points may, in some cases, actually reduce the accuracy of the interpolation. Putting it still another way, with a given number of points, the smaller the interval, the more accurate the polynomial interpolation in that interval.

Lagrange Interpolation

The previous discussion suggests that the coefficients of a higher-order interpolating polynomial may be obtained from the solution of simultaneous linear equations. The use of Lagrange polynomials avoids this procedure by expressing the previous interpolating polynomial

$$F_n(X) = a_{n+1}X^n + a_nX^{n-1} + \cdots + a_2X + a_1$$

in terms of the sum of the respective products of the $n+1$ ordinate values and appropriate Lagrange polynomials

$$\begin{aligned} F_n(X) &= Y(n+1)*L_n(X) + Y(n)*L_{n-1}(X) + \cdots + Y(1)*L_0(X) \\ &= \sum_{J=0}^{n} Y(J+1)*L_J(X) \end{aligned}$$

Such an equation therefore requires that each of the Lagrange polynomials, $L_J(X)$, be of degree n and satisfy the following conditions:

$$\left.\begin{aligned} L_J(X(I)) &= 1 \qquad \text{for } I = J+1 \\ L_J(X(I)) &= 0 \qquad \text{for } I \neq J+1 \end{aligned}\right\} I = 1, 2, \ldots, n+1.$$

These conditions may be satisfied by writing each of the Lagrange polynomials in terms of a constant K and n linear factors

$$\begin{aligned} L_J(X) = K(X - X(1))(X - X(2)) \\ \ldots (X - X(J))(X - X(J+2)) \ldots (X - X(n+1)) \end{aligned}$$

This resulting polynomial will therefore be of degree n and will be equal to zero for values of $X = X(I)$, with $I = 1, 2, \ldots, J, J+2, \ldots, n+1$. In order to satisfy the condition that

$$L_J(X(I)) = 1 \text{ for } I = J+1,$$

the constant K must be equal to

$$K = \frac{1}{\begin{pmatrix}(X(J+1) - X(1))(X(J+1) - X(2)) \cdots (X(J+1) - X(J)) \\ (X(J+1) - X(J+2)) \cdots (X(J+1) - X(n+1))\end{pmatrix}}$$

The Lagrange polynomials may therefore be written as

$$L_J(X) = \frac{\begin{pmatrix}(X-X(1))(X-X(2)) \cdots (X-X(J))(X-X(J+2)) \cdots \\ (X-X(n+1))\end{pmatrix}}{\begin{pmatrix}(X(J+1) - X(1))(X(J+1) - X(2)) \cdots (X(J+1) - X(J)) \\ (X(J+1) - X(J+2)) \cdots (X(J+1) - X(n+1))\end{pmatrix}}$$

$$= \prod_{\substack{I=1 \\ I \neq J+1}}^{n+1} \frac{X - X(I)}{X(J+1) - X(I)}$$

and the interpolating polynomial as

$$F_n(X) = \sum_{J=0}^{n} Y(J+1) * \prod_{\substack{I=1 \\ I \neq J+1}}^{n+1} \frac{X - X(I)}{X(J+1) - X(I)}$$

For an example that will show how we may use Lagrangian interpolation, consider the previous numerical example of determining a second-degree polynomial based on the first three data points of Figure 12.13.

I	1	2	3
$X(I)$	2	3	4
$F(X)$	4	8	15

First calculating the three Lagrange polynomials we find

$$L_0(X) = \frac{(X-3)(X-4)}{(2-3)(2-4)} = \frac{1}{2}(X-3)(X-4)$$

$$L_1(X) = \frac{(X-2)(X-4)}{(3-2)(3-4)} = -(X-2)(X-4)$$

$$L_2(X) = \frac{(X-2)(X-3)}{(4-2)(4-3)} = \frac{1}{2}(X-2)(X-3)$$

Using these to form a second-degree Lagrange interpolation polynomial, we obtain

$$F_2(X) = 4 * \frac{1}{2}(X-3)(X-4) + 8*(-(X-2)(X-4)) + 15 * \frac{1}{2}(X-2)(X-3)$$

This resulting polynomial may now be used to evaluate the function at $XR = 2.2$, yielding

$$\begin{aligned} F_2(2.2) &= 2(2.2 - 3)(2.2 - 4) - 8(2.2 - 2)(2.2 - 4) \\ &\qquad + 7.5(2.2 - 2)(2.2 - 3) \\ &= 4.56 \end{aligned}$$

The reader should note that this is the same numerical result that we obtained previously. This follows from the fact that the second-degree Lagrange interpolating polynomial is, in reality, identical with the previous polynomial we obtained by solving the simultaneous equations; thus,

$$\begin{aligned} F_2(X) &= 2(X - 3)(X - 4) - 8(X - 2)(X - 4) + 7.5(X - 2)(X - 3) \\ &= (2 - 8 + 7.5)X^2 - (14 - 48 + 37.5)X + (24 - 64 + 45) \\ &= 1.5X^2 - 3.5X + 5.0 \end{aligned}$$

Figure 12.23 shows a Fortran program that may be used to evaluate Lagrangian interpolation polynomials. Initially the number of points $M(= N + 1)$ and the corresponding X and Y values are read into memory. This is followed by printing the heading, subheadings, and input data. Computer statement S.0010 then reads in the value of XR at the point we wish to interpolate. The variable SUM is now set equal to zero and we enter the DO 10 loop to perform the interpolation.

During the first iteration of the DO 10 loop, K will be equal to 1, J will be equal to 0, and FAC will therefore be equal to $Y(1)$. During the first iteration of the DO 20 loop, computer statement S.0017 will find $(I - J - 1)$ is equal to 0 and control will be returned to computer statement S.0016. I will now be made equal to 2. During the second

```
          C      LAGRANGIAN INTERPOLATION.
          C
S.0001           DIMENSION X(20),Y(20)
S.0002           READ(5,1)M,(X(I),Y(I),I=1,M)
S.0003         1 FORMAT(I5,/,(6E13.6))
S.0004           WRITE(6,2)M
S.0005         2 FORMAT('1LAGRANGIAN INTERPOLATION',///,' NUMBER OF DATA POINTS =',
                1I4,//,' INPUT DATA',//,5X,'I',12X,'X(I)',10X,'Y(I)',/)
S.0006           WRITE(6,3)(I,X(I),Y(I),I=1,M)
S.0007         3 FORMAT(I6,6X,2E15.7)
S.0008           WRITE(6,4)
S.0009         4 FORMAT(///,' OUTPUT RESULTS',//,6X,'XR',12X,'F(XR)',/)
S.0010         5 READ(5,6)XR
S.0011         6 FORMAT(E15.7)
S.0012           SUM=0.0
S.0013           DO 10 K=1,M
S.0014           J=K-1
S.0015           FAC=Y(J+1)
S.0016           DO 20 I=1,M
S.0017           IF(I-J-1)30,20,30
S.0018        30 FAC=FAC*(XR-X(I))/(X(J+1)-X(I))
S.0019        20 CONTINUE
S.0020        10 SUM=SUM+FAC
S.0021           WRITE(6,7)XR,SUM
S.0022         7 FORMAT(2E15.7)
S.0023           GO TO 5
S.0024           END
```

Figure 12.23 Lagrangian interpolation program

iteration of the DO 20 loop, computer statement S.0017 will transfer control to statement 30 and FAC will be made equal to

$$\text{FAC} = \text{Y}(1)*(XR - X(2))/(X(1) - X(2))$$

The iteration of the DO 20 loop will continue for M iterations and we would thus find in the case of the first three points of the numerical example above, that FAC would be equal to

$$\text{FAC} = 4\,\frac{(2.2 - 3)(2.2 - 4)}{(2 - 3)(2 - 4)} = 2.88$$

at the end of the third iteration.

Computer statement S.0020 would now set SUM equal to SUM + FAC and the DO 10 loop would be executed a second time, yielding

$$\text{FAC} = Y(2) * \prod_{\substack{I=1\\ I\neq J+1}}^{M=N+1} \frac{XR - X(I)}{X(J+1) - X(I)}$$

After the DO 10 loop has been executed M $(= N + 1)$ times, the value of SUM would thus be equal to

$$\text{SUM} = F_N(XR) = \sum_{J=0}^{N} Y(J + 1) * \prod_{\substack{I=1\\ I\neq J+1}}^{M=N+1} \frac{XR - X(I)}{X(J+1) - X(I)}$$

Computer statements S.0021 and S.0022 would then be used to print the results, XR and SUM, before control would be transferred to statement 5 so that interpolation could be performed at the next data point.

Figure 12.24 shows the computer results for three different interpolations that were based on using the three data points of the previous

```
LAGRANGIAN INTERPOLATION

NUMBER OF DATA POINTS =     3

INPUT DATA

   I              X(I)              Y(I)

   1         C.2CCOCCCE 01    0.4CCOCCOE 01
   2         C.3CCCCCCE 01    0.8COOOCOE 01
   3         C.4CCOCCCE 01    0.1500UCOE 02


OUTPUT RESULTS

     XR                 F(XR)

 0.2200CCOE C1    C.4555599E C1
 0.3CCOCCCE 01    C.8CCCCCCE 01
 0.3900COOE 01    0.1416499E 02

IHC2171
```

Figure 12.24 The output results of the program of Figure 12.23

example. The reader is urged to hand calculate each of the statements of the computer program for the third data set in order to verify the accuracy of this result as well as to gain a better understanding of Lagrangian interpolation.

Newton Interpolation Formulas

Difference Table. The Newton, Gauss, Bessel, Stirling, and Everett formulas for obtaining an interpolating polynomial are examples of methods that are based on using difference tables that use either forward, backward, or central differences. In order to construct a forward difference table from $N + 1$ Y-values, let us first define the operator Δ as

$$\Delta Y(I) = Y(I + 1) - Y(I); I = 1, 2, \ldots, N$$

and refer to this as a first-order forward difference. Second-order differences may now be found by taking the first differences of ΔY, thus

$$\Delta^2 Y(I) = \Delta(\Delta Y(I)) = \Delta Y(I+1) - \Delta Y(I); I = 1, 2, \ldots, N-1$$

and, in general, a K-order difference would be given by

$$\Delta^K Y(I) = \Delta^{K-1} Y(I+1) - \Delta^{K-1} Y(I); I = 1, 2, \ldots, N-K+1$$

These successive differences may easily be obtained by forming a table that is developed by taking the difference (lower minus upper) of the left two adjacent terms, as shown below:

I	$Y(I)$	Δ	Δ^2	Δ^3
1	$Y(1)$			
		$\Delta Y(1)$		
2	$Y(2)$		$\Delta^2 Y(1)$	
		$\Delta Y(2)$		$\Delta^3 Y(1)$
3	$Y(3)$		$\Delta^2 Y(2)$	
		$\Delta Y(3)$		$\Delta^3 Y(2)$
4	$Y(4)$		$\Delta^2 Y(3)$	
.	.	.	.	.
.	.	.	.	.
.	.	.	.	.
$N-1$	$Y(N-1)$		$\Delta^2 Y(N-2)$	
		$\Delta Y(N-1)$		$\Delta^3 Y(N-2)$
N	$Y(N)$		$\Delta^2 Y(N-1)$	
		$\Delta Y(N)$		
$N+1$	$Y(N+1)$			

Figure 12.25 shows a computer program for producing a forward difference table of the first four differences that are based on as many as

```
        C       CONSTRUCTING A DIFFERENCE TABLE
        C
S.0001          DIMENSION Y(100),YD1(99),YD2(98),YD3(97),YD4(96)
S.0002          READ(5,1)M,(Y(I),I=1,M)
S.0003        1 FORMAT(I5,/,(5E15.7))
S.0004          N1=M-1
S.0005          DO 10 I=1,N1
S.0006       10 YD1(I)=Y(I+1)-Y(I)
S.0007          N2=M-2
S.0008          DO 20 I=1,N2
S.0009       20 YD2(I)=Y(I+2)-2.*Y(I+1)+Y(I)
S.0010          N3=M-3
S.0011          DO 30 I=1,N3
S.0012       30 YD3(I)=Y(I+3)-3.*Y(I+2)+3.*Y(I+1)-Y(I)
S.0013          N4=M-4
S.0014          DO 40 I=1,N4
S.0015       40 YD4(I)=Y(I+4)-4.*Y(I+3)+6.*Y(I+2)-4.*Y(I+1)+Y(I)
S.0016          WRITE(6,2)
S.0017        2 FORMAT('1DIFFERENCES OF THE FUNCTION Y',///,7X,'I',12X,'Y(I)',10X,
               1'YD1(I)',9X,'YD2(I)',9X,'YD3(I)',9X,'YD4(I)',/)
S.0018          WRITE(6,3)(I,Y(I),YD1(I),YD2(I),YD3(I),YD4(I),I=1,N4)
S.0019        3 FORMAT(/,I8,7X,5E15.7)
S.0020          WRITE(6,4)(I,Y(I),YD1(I),YD2(I),YD3(I),I=N3,N3)
S.0021        4 FORMAT(/,I8,7X,4E15.7)
S.0022          WRITE(6,5)(I,Y(I),YD1(I),YD2(I),I=N2,N2)
S.0023        5 FORMAT(/,I8,7X,3E15.7)
S.0024          WRITE(6,6)(I,Y(I),YD1(I),I=N1,N1)
S.0025        6 FORMAT(/,I8,7X,2E15.7)
S.0026          WRITE(6,7)(I,Y(I),I=M,M)
S.0027        7 FORMAT(/,I8,7X,E15.7)
S.0028          STOP
S.0029          END
```

Figure 12.25 A program for computing the first four forward differences

100 entries. Initially, the number of points M and their corresponding Y values are read into memory. The dummy variable $N1$ is then set equal to $M-1$ and the DO 10 loop forms the first-order differences. The dummy variable $N2$ is now made equal to $M-2$ and the DO 20 loop forms the second-order differences. This, in turn, is followed by the DO 30 and DO 40 loops that are used to form the third- and fourth-order differences. Computer statements S.0016 through S.0027 are then used to print the forward difference table.

DIFFERENCES OF THE FUNCTION Y

I	Y(I)	YD1(I)	YD2(I)	YD3(I)	YD4(I)
1	0.1000000E 01	0.2000000E 01	-0.3000000E 01	0.7000000E 01	-0.1300000E 02
2	0.3000000E 01	-0.1000000E 01	0.4000000E 01	-0.6000000E 01	0.5000000E 01
3	0.2000000E 01	0.3000000E 01	-0.2000000E 01	-0.1000000E 01	0.7000000E 01
4	0.5000000E 01	0.1000000E 01	-0.3000000E 01	0.6000000E 01	-0.1400000E 02
5	0.6000000E 01	-0.2000000E 01	0.3000000E 01	-0.8000000E 01	0.1900000E 02
6	0.4000000E 01	0.1000000E 01	-0.5000000E 01	0.1100000E 02	-0.2000000E 02
7	0.5000000E 01	-0.4000000E 01	0.6000000E 01	-0.9000000E 01	
8	0.1000000E 01	0.2000000E 01	-0.3000000E 01		
9	0.3000000E 01	-0.1000000E 01			
10	0.2000000E 01				

Figure 12.26 The output results of the program of Figure 12.25

Figure 12.26 shows the corresponding results for an arbitrary set of ten Y-values. The reader may readily verify the accuracy of any of these results by recalling that the successive differences are obtained by taking the difference (lower minus upper) of the left two adjacent terms.

Forward Difference Interpolation Formula. Let us now turn to a discussion of Newton's forward interpolation formula. The general interpolation polynomial is formed from the $M(=N+1)$ data points by using the following form for an N-th order polynomial

$$\begin{aligned} F_N(X) = {} & a_o + (X-X(1))a_1 + (X-X(1))(X-X(2))a_2 \\ & + \cdots + (X-X(1))(X-X(2)) \ldots (X-X(N))a_N \end{aligned}$$

where the coefficients a_o, a_1, . . . , and a_N are readily found from $F_N(X)$ by evaluating the polynomial at $X(1)$, $X(2)$, . . . , and $X(M)$. Thus

$$a_o = F_N(X(1)) = Y(1) = F[X(1)]$$

$$a_1 = \frac{F_N(X(2)) - F_N(X(1))}{X(2) - X(1)} = \frac{Y(2) - Y(1)}{X(2) - X(1)} = \frac{\Delta Y(1)}{X(2) - X(1)} = F[X(2),X(1)]$$

$$a_2 = \frac{\dfrac{F_N(X(3))-F_N(X(2))}{X(3) - X(2)} - \dfrac{F_N(X(2))-F_N(X(1))}{X(2) - X(1)}}{X(3) - X(1)} = \frac{F[X(3),X(2)] - F[X(2),X(1)]}{X(3) - X(1)} = F[X(3),X(2),X(1)]$$

and so on,

where the square brackets are used to denote divided differences. Newton's forward interpolation formula with divided differences may thus be written as

$$\begin{aligned} F_N(X) = {} & F[X(1)] + F[X(2),X(1)](X-X(1)) \\ & + F[X(3),X(2),X(1)](X-X(1))(X-X(2)) \\ & + F[X(4),X(3),X(2),X(1)](X-X(1))(X-X(2))(X-X(3)) \\ & + \cdots + F[X(M), \ldots ,X(2),X(1)](X-X(1))(X-X(2)) \\ & \cdots (X-X(N)) \end{aligned}$$

These divided differences may easily be found by finding the differences of the X and Y functions as shown in the following scheme

$$
\begin{array}{ccccccc}
 & & X(1) & Y(1) & & & \\
 & X(2)-X(1) & & & Y(2)-Y(1) & \dfrac{Y(2)-Y(1)}{X(2)-X(1)} & = \\
X(3)-X(1) & & X(2) & Y(2) & \longrightarrow & & \\
 & X(3)-X(2) & & & Y(3)-Y(2) & \dfrac{Y(3)-Y(2)}{X(3)-X(2)} & = \\
X(4)-X(2) & & X(3) & Y(3) & \longrightarrow & & \\
\cdot & X(4)-X(3) & & & Y(4)-Y(3) & \dfrac{Y(4)-Y(3)}{X(4)-X(3)} & = \\
\cdot & \cdot & X(4) & Y(4) & \longrightarrow & & \\
\cdot & \cdot & \cdot & \cdot & & & \\
 & \cdot & \cdot & \cdot & & &
\end{array}
$$

$$
\begin{array}{llll}
= F[X(2),X(1)] & & & \\
 & F[X(3),X(2)] - F[X(2),X(1)] \longrightarrow & \dfrac{F[X(3),X(2)] - F[X(2),X(1)]}{X(3) - X(1)} & = F[X(3),X(2),X(1)] \\
= F[X(3),X(2)] & & & \\
 & F[X(4),X(3)] - F[X(3),X(2)] \longrightarrow & \dfrac{F[X(4),X(3)] - F[X(3),X(2)]}{X(4) - X(2)} & = F[X(4),X(3),X(2)] \\
= F[X(4),X(3)] & & & \cdot \\
\quad\cdot & & & \cdot \\
\quad\cdot & & & \cdot \\
\quad\cdot & & &
\end{array}
$$

The divided difference table thus becomes

		FIRST DIV DIF	SECOND DIV DIF	THIRD DIV DIF
$X(1)$	$Y(1)$			
		$F[X(2),X(1)]$		
$X(2)$	$Y(2)$		$F[X(3),X(2),X(1)]$	
		$F[X(3),X(2)]$		$F[X(4),X(3),X(2),X(1)$
$X(3)$	$Y(3)$		$F[X(4),X(3),X(2)]$	.
		$F[X(4),X(3)]$	.	.
$X(4)$	$Y(4)$	.	.	.
.	.	.	.	
.	.	.		
.	.			

For a numerical example that will show how we may make use of Newton's forward interpolation formula, let us once again consider finding a second-degree polynomial that is based on the first three data points of Figure 12.13.

I	1	2	3
$X(I)$	2	3	4
$F(X(I))$	4	8	15

The first step requires that we compute the X and Y differences

		$X(I)$	$Y(I)$		
		2	4		
	1			$8-4=4 \longrightarrow \frac{4}{1}=4$	
2		3	8		$7-4=3 \longrightarrow \frac{3}{2}=1.5$
	1			$15-8=7 \longrightarrow \frac{7}{1}=7$	
		4	15		

in order to obtain the divided difference table

$X(I)$	$Y(I)$	FIRST DIV DIF	SECOND DIV DIF
2	$\boxed{4=a_0}$		
		$\boxed{4=a_1}$	
3	8		$\boxed{1.5=a_2}$
		7	
4	15		

The values of a_0, a_1, and a_2 are thus determined and Newton's forward interpolation polynomial becomes

$$F_2(X) = a_0 + (X - X(1))a_1 + (X - X(1))(X - X(2))a_2$$

or

$$F_2(XR) = 4 + (XR - 2)4 + (XR - 2)(XR - 3)1.5$$

This may now be used to estimate the value of the function at $XR = 2.2$,

$$\begin{aligned} F_2(2.2) &= 4 + (2.2 - 2)*4 + (2.2 - 2)(2.2 - 3)*1.5 \\ &= 4.56 \end{aligned}$$

which agrees with our two previous calculations.

Figure 12.27 is a general program for computing and evaluating Newton forward interpolation polynomials. Initially we read into memory the number of data points M and their corresponding X and Y values. The DO 10 loop then forms the zeroth divided differences. The variable K is next set equal to $M - 1$ and the DO 20 and DO 30 loops are used to generate the complete divided difference table. Computer statements S.0012 and S.0013 are then used to print these table values and computer statements S.0014 and S.0015 then read in the values of XR and N, the required degree of the interpolating polynomial. Computer statements S.0016 through S.0018 then set K equal to $N+1$, $FX = 0.0$, and A equal to 1.0. The DO 40 loop now evaluates the polynomial for the specified value of XR. For example, let us assume that N has been specified as two as it was in the previous numerical example. The DO 40 loop will thus be executed three times. The first execution of the DO 40 loop will

```
               C       NEWTON'S INTERPOLATION WITH DIVIDED DIFFERENCES.
               C
S.0001                 DIMENSION X(10),Y(10),DD(10,10)
S.0002                 READ(5,1)M,(X(I),Y(I),I=1,M)
S.0003               1 FORMAT(I5,/,(5E15.7))
S.0004                 DO 10 I=1,M
S.0005              10 DD(I,1)=Y(I)
S.0006                 K=M-1
S.0007                 DO 20 I=2,M
S.0008                 DO 30 J=1,K
S.0009                 L=J+I-1
S.0010              30 DD(J,I)=(DD(J+1,I-1)-DD(J,I-1))/(X(L)-X(J))
S.0011              20 K=K-1
S.0012                 WRITE(6,60)
S.0013              60 FORMAT(///,6X,'XR',12X,'F(XR)',10X,'N',/)
S.0014               3 READ(5,4)XR,N
S.0015               4 FORMAT(E15.7,I10)
S.0016                 K=N+1
S.0017                 FX=0.0
S.0018                 A=1.0
S.0019                 DO 40 I=1,K
S.0020                 FX=FX+DD(1,I)*A
S.0021              40 A=A*(XR-X(I))
S.0022                 WRITE(6,5)XR,FX,N
S.0023               5 FORMAT(2E15.7,I6)
S.0024                 GO TO 3
S.0025                 END
```

Figure 12.27 Newton forward interpolation program

```
     XR               F(XR)             N

0.22000000E 01  0.47999999E 01          1
0.22000000E 01  0.45599999E 01          2
0.22000000E 01  0.45599999E 01          3

IHC217I
```

Figure 12.28 The output results of the program of Figure 12.27

produce values of

$$FX = DD(1,1)$$

and

$$A = XR - X(1)$$

The second execution of the DO 40 loop will yield

$$FX = DD(1,1) + DD(1,2)*(XR-X(1))$$

and

$$A = (XR - X(1))*(XR - X(2))$$

and the third execution will produce a value of

$$FX = DD(1,1) + DD(1,2)*(XR-X(1)) + DD(1,3)*(XR-X(1))*(XR-X(2))$$

which is identical to the Newton second-degree interpolation polynomial

$$F_2(XR) = a_0 + a_1(XR-X(1)) + a_2(XR-X(1))(XR-X(2))$$

After the interpolation has been completed, computer statements S.0022 and S.0023 are used to print the results before control is transferred to statement 3 so that the next interpolation can be made.

Figure 12.28 shows the computer results of estimating the function value at $XR = 2.2$ for the four data points of Figure 12.13, when $N = 1$, 2, and 3. The reader is encouraged to hand calculate each of the computer statements for the case of $N = 3$ so that he will gain a better understanding of Newton's forward interpolation formula.

The reader will find, when he is carrying out the previous verification, that a considerable simplification in the computer program will result if we are willing to permit only equal-interval interpolation. For example, if

$$X(2) - X(1) = X(3) - X(2) = \cdots = X(M) - X(M-1) = H$$

we may rewrite Newton's forward interpolation formula with divided differences as

$$F_N(X) = Y(1) + (X-X(1))\frac{\Delta Y(1)}{H} + (X-X(1))(X-X(2))\frac{\Delta^2 Y(1)}{2!H^2} + (X - X(1))(X - X(2))(X - X(3))\frac{\Delta^3 Y(1)}{3!H^3} + \cdots + (X-X(1))(X-X(2)) \cdots (X-X(N))\frac{\Delta^N Y(1)}{N!H^N}$$

If we now make a change in the variable by letting

$$S = \frac{X - X(1)}{H}$$

The following additional simplification takes place

$$F_N(S) = Y(1) + S\Delta Y(1) + S(S-1)\frac{\Delta^2 Y(1)}{2!} + S(S-1)(S-2)\frac{\Delta^3 Y(1)}{3!}$$
$$+ \cdots + S(S-1) \cdots (S-N+1)\frac{\Delta^N Y(1)}{N!}$$

In order to see how much easier it is to use the above relation, let us again consider the previous numerical example. We first find the forward differences

X(I)	*Y(I)*	Δ	Δ^2
2	[4]		
		[4]	
3	8		[3]
		7	
4	15		

We now compute the value of S for $XR = 2.2$ and $H = 1$, obtaining

$$S = \frac{XR - X(1)}{H} = \frac{2.2 - 2.0}{1.0} = 0.2$$

Using the above form of the Newton interpolation formula for equal intervals, we find

$$F_2(2.2) = 4 + 0.2*4 + 0.2(0.2 - 1.0) * \frac{3}{2}$$
$$= 4.56 \text{ as before.}$$

It is left as an exercise for the reader to reprogram the previous computer program for the case of equal-interval interpolation.

Backward Difference Interpolation Formula. Note that the path through the difference table always starts at $Y(1)$ and then runs diagonally down and to the right. It thus follows that the results of a second-order interpolating polynomial, based on the first three points, would probably not yield results as accurate for interpolations in the vicinity of the fourth point, as a second-order polynomial formed from the second, third, and fourth points. Alternatively, we would probably obtain better results for interpolations in the vicinity of the fourth point if we used a backward second-order polynomial that was based on the fourth, third, and second points. In this latter case, the path through the difference table would start at the last point and move diagonally up and to the right.

The diagonal-difference table for backward equal-interval differences is analogous to the forward equal-interval difference table. Backward differences are defined in terms of the backward-difference operator ∇ (inverted delta) as follows:

First order backward difference $\nabla Y(I) = Y(I) - Y(I-1)$

Second order backward difference $\nabla^2 Y(I) = \nabla Y(I) - \nabla Y(I-1)$

$\vdots$

K-th order backward difference $\nabla^K Y(I) = \nabla^{K-1} Y(I) - \nabla^{K-1} Y(I-1)$

Using this backward-difference operator, the difference table may thus be constructed as below

$$
\begin{array}{ccccc}
 & & \nabla & \nabla^2 & \\
\vdots & \vdots & \vdots & \vdots & \\
X(M-3) & Y(M-3) & \cdot & \cdot & \\
 & & \nabla Y(M-2) & \cdot & \\
X(M-2) & Y(M-2) & & \nabla^2 Y(M-1) & \\
 & & \nabla Y(M-1) & & \cdots \\
X(M-1) & Y(M-1) & & \nabla^2 Y(M) & \\
 & & \nabla Y(M) & & \\
X(M) & Y(M) & & &
\end{array}
$$

and Newton's backward difference equal-interval interpolating polynomial may thus be written as

$$
\begin{aligned}
F_N(S) = Y(M) + S\nabla Y(M) + S(S+1)\frac{\nabla^2 Y(M)}{2!} \\
+ S(S+1)(S+2)\frac{\nabla^3 Y(M)}{3!} \\
+ \cdots + S(S+1)(S+2)\cdots(S+N-1)\frac{\nabla^N Y(M)}{N!}
\end{aligned}
$$

Using the last three points in the example of Figure 12.13 to estimate the value of $F_2(4.5)$ would thus require the construction of the backward difference table

		∇	∇^2
.	.		
.	.		
.	.		
3	8		
		7	
4	15		$3 = \nabla^2 Y(5)$
		$10 = \nabla Y(5)$	
5	$25 = Y(5)$		

Computing the value of S for $XR = 4.5$ and $H = 1.0$, we find

$$S = \frac{XR - X(M)}{H} = \frac{4.5 - 5.0}{1.0} = -0.5$$

and

$$F_2(-0.5) = 25 + (-0.5)*10 + (-0.5)(-0.5 + 1.0) * \frac{3}{2} = 19.625$$

Gauss Interpolation Formulas

In the last section we introduced Newton's backward difference interpolation polynomial to improve the accuracy of interpolation for points that had values close to those at the end of the table. In this section we introduce central differences to improve the accuracy of interpolation of points that have values close to those at the center of the table. In other words, a central difference interpolation polynomial will be based on points that are symmetrically located with respect to the point of interpolation.

If we now assume that $X(0)$ is the central point of our input data (equal interval spacing), the first central difference should be based on two points that are spaced H units apart and are symmetrically located with respect to $X(0)$; thus,

$$\delta Y(X(0)) = Y\left(X(0) + \frac{H}{2}\right) - Y\left(X(0) - \frac{H}{2}\right)$$

Since the value of these two Y points are not known, we are forced to use either

$$Y(X(1)) - Y(X(0))$$

or

$$Y(X(0)) - Y(X(-1))$$

as the first central difference that we will add to $Y(0)$. Choosing the former, we may thus write

$$\delta Y\left(X(0) + \frac{H}{2}\right) = Y(X(1)) - Y(X(0))$$

and the general relation

$$\delta Y\left(X(K) + \frac{H}{2}\right) = Y(X(K+1)) - Y(K)$$

may therefore be used to find first-order central differences. Since the second-order difference which we will add will involve the three points $X(-1)$, $X(0)$, and $X(1)$, it will be symmetrical with respect to $X(0)$ and would be determined from the following general relation

$$\delta^2 Y(K) = \delta Y\left(X(K) + \frac{H}{2}\right) - \delta Y\left(X(K) - \frac{H}{2}\right)$$

Similarly, the third-order central differences are defined by

$$\delta^3 Y\left(X(K) + \frac{H}{2}\right) = \delta^2 Y(X(K+1)) - \delta^2 Y(X(K)), \text{ and so on.}$$

The equal-interval central difference table therefore becomes

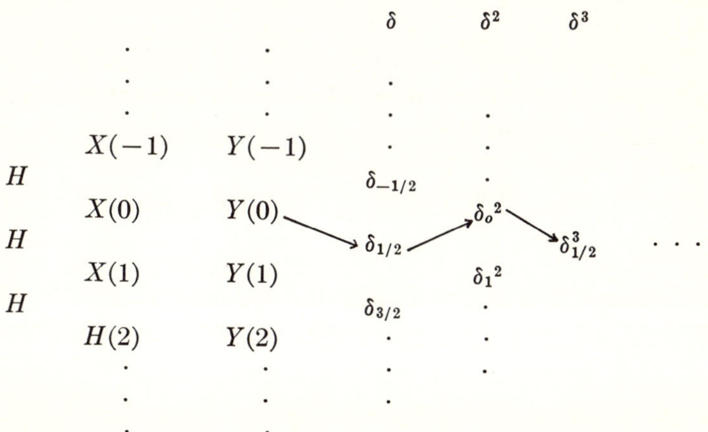

Starting at $Y(0)$ and using the zigzag path shown, we may thus write Gauss' forward interpolation polynomial

$$F_N(X(0) + SH) = Y(0) + S\delta Y\left(X(0) + \frac{H}{2}\right) + S(S-1)\,\frac{\delta^2 Y(X(0))}{2!}$$
$$+ S(S-1)(S+1)\,\frac{\delta^3 Y(X(0) + H/2)}{3!} + \cdots$$
$$= Y(0) + S\delta_{1/2} + S(S-1)\,\frac{\delta_0{}^2}{2!} + S(S-1)(S+1)\,\frac{\delta_{1/2}^3}{3!} + \cdots$$

For a numerical example that will illustrate how this polynomial may be used, assume that we have the following difference table:

I	$X(I)$	$Y(I)$	δ	δ^2
.	.	.		
.	.	.	.	
.	.	.	.	.
-1	2	4	.	.
			4	.
0	3	$8 = Y(0)$ →		→ $3 = \delta_0{}^2$
			$7 = \delta_{1/2}$	
1	4	15		3
			10	.
2	5	25	.	.
.	.	.	.	.
.	.	.	.	
.	.	.		

and we wish to interpolate at $XR = 3.5$. Since the value of S for $XR = 3.5$ and $H = 1.0$ will be equal to

$$S = \frac{XR - X(0)}{H} = \frac{3.5 - 3.0}{1.0} = 0.5$$

we find by Gauss' forward interpolation polynomial that

$$F_2(3.5) = 8 + 0.5*7 + 0.5(0.5 - 1) * \frac{3}{2}$$
$$= 11.125$$

Alternately, we could follow a backward zigzag path through the difference table to obtain Gauss' backward interpolation formula

$$F_N(X) = Y(0) + S\delta Y\left(X(0) - \frac{H}{2}\right) + S(S+1)\,\frac{\delta^2 Y(X(0))}{2!} + S(S-1)(S+1)\,\frac{\delta^3 Y(X(0) - H/2)}{3!} + \cdots$$

The backward zigzag path for the previous numerical example would thus be

I	$X(I)$	$Y(I)$	δ	δ^2
−1	2	4		
			$4 = \delta_{-1/2}$	
0	3	$8 = Y(0)$		$3 = \delta_0{}^2$
			7	
1	4	15		3
			10	
2	5	25		

and for $XR = 3.5$, this would again yield a value of $S = 0.5$; thus,

$$\begin{aligned} F_2(3.5) &= 8 + 0.5*4 + 0.5(0.5+1) * \frac{3}{2} \\ &= 11.125 \end{aligned}$$

At this point the reader may realize that if we now add Gauss' backward difference formula to Gauss' forward difference formula and take one half their sum, the resulting polynomial will be symmetric with respect to $X(0)$ and would be given by

$$\begin{aligned} F_N(X) = {} & Y(0) + \frac{S}{2}\left(\delta Y\left(X(0) + \frac{H}{2}\right) + \delta Y\left(X(0) - \frac{H}{2}\right)\right) \\ & + S^2\,\frac{\delta^2 Y(X(0))}{2!} + \frac{S(S-1)(S+1)}{2}\,\frac{\delta^3 Y\left(X(0) + \frac{H}{2}\right) + \delta^3 Y\left(X(0) - \frac{H}{2}\right)}{3!} \\ & + S^2(S^2 - 1^2)\,\frac{\delta^4 Y(X(0))}{4!} \\ & + \frac{S(S^2 - 1^2)(S^2 - 2^2)}{2}\,\frac{\delta^5 Y\left(X(0) + \frac{H}{2}\right) + \delta^5 Y\left(X(0) - \frac{H}{2}\right)}{5!} \\ & + S^2(S^2 - 1^2)(S^2 - 2^2)\,\frac{\delta^6 Y(X(0))}{6!} + \cdots \end{aligned}$$

```
              C        STIRLING INTERPOLATION.
              C
S.0001                 DIMENSION X(100),Y(100),DY(100,6)
S.0002                 READ(5,1)M,(X(I),Y(I),I=1,M)
S.0003               1 FORMAT(I5,/,(5E15.7))
S.0004                 DO 5 I=1,M
S.0005                 DO 5 J=1,6
S.0006               5 DY(I,J)=0.0
S.0007                 DO 10 I=1,M
S.0008              10 DY(I,1)=Y(I)
S.0009                 LL=M-1
S.0010                 DO 20 I=2,6
S.0011                 DO 30 J=1,LL
S.0012              30 DY(J,I)=(DY(J+1,I-1)-DY(J,I-1))
S.0013              20 LL=LL-1
S.0014                 WRITE(6,2)
S.0015               2 FORMAT('1STIRLING INTERPOLATION',///,7X,'XR',10X,'F(XR)',12X,'N',/
                      1/)
S.0016               3 READ(5,4)XR,N
S.0017               4 FORMAT(E15.7,I10)
S.0018                 DO 40 I=1,M
S.0019                 IF(XR-X(I))50,50,40
S.0020              50 K=I-1
S.0021                 GO TO 60
S.0022              40 CONTINUE
S.0023              60 H=X(2)-X(1)
S.0024                 S=(XR-X(K))/H
S.0025                 A=S
S.0026                 SUM=Y(K)
S.0027                 L=1
S.0028              70 LL=2*L
S.0029                 I=K-L
S.0030                 SUM=SUM+A*(DY(I,LL)+DY(I+1,LL))/2.
S.0031                 IF(L-N)80,90,90
S.0032              80 FLL=LL
S.0033                 A=A/FLL
S.0034                 SUM=SUM+A*S*DY(I,LL+1)
S.0035                 IF(L-N+1)85,90,90
S.0036              85 FL=L
S.0037                 A=A*(S+FL)*(S-FL)/(FLL+1.)
S.0038                 L=L+1
S.0039                 GO TO 70
S.0040              90 WRITE(6,6)XR,SUM,N
S.0041               6 FORMAT(2E15.7,I8)
S.0042                 GO TO 3
S.0043                 END
```

Figure 12.29 Stirling interpolation program

This latter equation is known as Stirling's formula and is widely used in interpolation problems.

Figure 12.29 is a Stirling interpolation program. Initially, the number of data points M and their corresponding X and Y values are read into the computer memory. This is followed by the DO 5, 10, 20, and 30 loops that are used to generate the difference table. Computer statements S.0016 and S.0017 are then used to read in the value at the point of interpolation and the order N of the interpolating polynomial.

The DO 40 loop is now used to determine the central point of interpolation. For example, in the case of the following data:

K	*X(I)*	*Y(I)*
1	1.0	2.0
2	2.0	4.0
3	3.0	8.0
4	4.0	15.0
5	5.0	25.0
6	6.0	40.0

```
STIRLING INTERPOLATION

        XR                 F(XR)              N

 0.35000000E 01   0.10750000E 02              1
 0.35000000E 01   0.11125000E 02              2
 0.35000000E 01   0.11101560E 02              3
 0.35000000E 01   0.11110140E 02              4

IHC2171
```

Figure 12.30 The output results of the program of Figure 12.29

a value of $XR = 3.5$ would result in a value of $K = 3$. After the value of K has been determined, computer statements S.0023 and S.0024 are used to compute the value of H and S. In the case of $K = 3$, computer statement S.0026 would then set SUM equal to $Y(3)$ and computer statements S.0023 through S.0039 would be used to add the necessary additional contributions to $Y(3)$. For example, if $N = 1$, only one additional term is added

$$\text{A*(DY(I,IL)+DY(I+1,IL))/2.} = \text{S*(DY(2,2)+DY(3,2))/2.}$$

After the last term has been added to SUM, control is transferred to statement 90 and the value of XR, SUM, and N are printed before the next interpolation is made.

Figure 12.30 shows the computer results obtained for the above data at $XR = 3.5$ and $N = 1$, 2, 3, and 4. It is suggested that the reader verify these results by executing each of the appropriate program statements.

Aitken-Neville Interpolation Polynomials

Polynomial interpolation may be quite lengthy if one first calculates the polynomial coefficients for a given number of data points and then computes the value of the polynomial for the particular point being interpolated and then repeats this process for a greater number of points until the difference between successive evaluations is less than some specified value. The Aitken-Neville method reduces the calculations by using repeated linear interpolation that is based on the values of previously determined lower-order polynomials.

In order to see how linear interpolation can be repeated to generate higher-order polynomials, consider once again the straight line passing through the points $X(1),Y(1)$ and $X(2),Y(2)$ shown in Figure 12.13. From our earlier discussion, the straight line equation passing through these two points would be given by

$$F(X) = \frac{Y(2) - Y(1)}{X(2) - X(1)} X + \frac{X(2)*Y(1) - X(1)*Y(2)}{X(2) - X(1)}$$

It is important now to realize that the above relation may also be written in determinant form

$$F_{1,2}(X) = \frac{\begin{vmatrix} Y(1) & X(1) - X \\ Y(2) & X(2) - X \end{vmatrix}}{X(2) - X(1)}$$

$$= \frac{Y(1)*(X(2) - X) - Y(2)*(X(1) - X)}{X(2) - X(1)}$$

$$= \frac{Y(2) - Y(1)}{X(2) - X(1)} X + \frac{X(2)*Y(1) - X(1)*Y(2)}{X(2) - X(1)}$$

Note also that

$$F_{1,2}(X(1)) = \frac{\begin{vmatrix} Y(1) & X(1) - X(1) \\ Y(2) & X(2) - X(1) \end{vmatrix}}{X(2) - X(1)} = Y(1)$$

and

$$F_{1,2}(X(2)) = \frac{\begin{vmatrix} Y(1) & X(1) - X(2) \\ Y(2) & X(2) - X(2) \end{vmatrix}}{X(2) - X(1)} = Y(2)$$

and that the subscripts 1,2 of the function $F(X)$ are used to indicate that this polynomial was based on points 1 and 2 and is therefore a linear polynomial.

We now show that we can use the linear interpolation determinant scheme shown above to form a second-order polynomial that passes through the points $X(1),Y(1)$; $X(2),Y(2)$; and $X(3),Y(3)$

$$F_{1,2,3}(X) = \frac{\begin{vmatrix} F_{1,2}(X) & X(1) - X \\ F_{2,3}(X) & X(3) - X \end{vmatrix}}{X(3) - X(1)}$$

Note that this last determinant involves two first-order polynomials that are respectively based on the points 1-2 and 2-3. Since these two polynomials are both multiplied by linear factors in the evaluation of the determinant, the resulting function is therefore a second-order polynomial. It may also be shown by the following that this quadratic function passes through these three points:

$$F_{1,2,3}(X(1)) = \frac{\begin{vmatrix} Y(1) & X(1) - X(1) \\ F_{2,3}(X(1)) & X(3) - X(1) \end{vmatrix}}{X(3) - X(1)} = Y(1)$$

$$F_{1,2,3}(X(2)) = \frac{\begin{vmatrix} Y(2) & X(1) - X(2) \\ Y(2) & X(3) - X(2) \end{vmatrix}}{X(3) - X(1)} = Y(2)$$

$$F_{1,2,3}(X(3)) = \frac{\begin{vmatrix} F_{1,2}(X(3)) & X(1) - X(3) \\ Y(3) & X(3) - X(3) \end{vmatrix}}{X(3) - X(1)} = Y(3)$$

Similarly it can be seen that

$$F_{1,2,3,4}(X) = \frac{\begin{vmatrix} F_{1,2,3}(X) & X(1) - X \\ F_{2,3,4}(X) & X(4) - X \end{vmatrix}}{X(4) - X(1)}$$

is a third-order polynomial that passes through $X(1),Y(1)$; $X(2),Y(2)$; $X(3),Y(3)$; and $X(4),Y(4)$; and

$$F_{1,2,\ldots,N}(X) = \frac{\begin{bmatrix} F_{1,2,\ldots,N-1}(X) & X(1) - X \\ F_{2,3,\ldots,N}(X) & X(N) - X \end{bmatrix}}{X(N) - X(1)}$$

is an $N-1$ order polynomial that passes through $X(1),Y(1)$; $X(2),Y(2)$; . . . ; $X(N),Y(N)$. The divided difference scheme shown below may therefore be used to organize the calculation of higher order polynomials:

			$X(I)$	$Y(I)$	$X(I)-X$	$F(I,1)$	$F(I,2)$	$F(I,3)$	
. . .	$X(3)-X(1)$	$X(2)-X(1)$	$X(1)$	$Y(1)$	$X(1)-X$	$F_{1,2}$	$F_{1,2,3}$	$F_{1,2,3,4}$	. . .
	$X(4)-X(2)$	$X(3)-X(2)$	$X(2)$	$Y(2)$	$X(2)-X$	$F_{2,3}$	$F_{2,3,4}$	$F_{2,3,4,5}$	
	$X(5)-X(3)$	$X(4)-X(3)$	$X(3)$	$Y(3)$	$X(3)-X$	$F_{3,4}$	$F_{3,4,5}$	.	
	.	$X(5)-X(4)$	$X(4)$	$Y(4)$	$X(4)-X$	$F_{4,5}$	.	.	
	.	.	$X(5)$	$Y(5)$	$X(5)-X$	.	.	.	
	.	.	.	.	.	.	.		
		.	.	.	.	.			
			.	.	.				

Figure 12.31 shows an Aitken-Neville interpolation program. Initially, we read into memory the number of data points M (maximum value = 10) and their corresponding X and Y values. The heading and subheadings are now printed and computer statements S.0006 and S.0007 then read in the values of XR (the point of interpolation) and N (the number of points used in obtaining the interpolation polynomial = one less than the degree of the polynomial). The DO 10 loop is then used to evaluate the first-order interpolating polynomials $F_{1,2}(XR)$, $F_{2,3}(XR)$, . . . , $F_{N-1,N}(XR)$. Based on the values of the first-order polynomials, the DO 20

```
                C      AITKEN-NEVILLE INTERPOLATICN.
                C
S.0001                 DIMENSION X(10),Y(10),F(10,10)
S.0002                 READ(5,1)M,(X(I),Y(I),I=1,M)
S.0003               1 FORMAT(I5,/,(5E15.7))
S.0004                 WRITE(6,2)
S.0005               2 FORMAT('1AITKEN-NEVILLE INTERPOLATION',///,6X,'N',12X,'XR',10X,'F(
                      11,N)',//)
S.0006               3 READ(5,4)XR,N
S.0007               4 FORMAT(E15.7,I10)
S.0008                 DO 1C I=1,N
S.0009              1C F(I,1)=(Y(I)*(X(I+1)-XR)-Y(I+1)*(X(I)-XR))/(X(I+1)-X(I))
S.0010                 DO 2C J=2,N
S.0011                 JN=N+1-J
S.0012                 DO 3C K=1,JN
S.0013                 KJ=K+J
S.0014              3C F(K,J)=(F(K,J-1)*(X(KJ)-XR)-F(K+1,J-1)*(X(K)-XR))/(X(KJ)-X(K))
S.0015              2C CONTINUE
S.0016                 WRITE(6,5)N,XR,F(1,N)
S.0017               5 FORMAT(I7,8X,2E15.7)
S.0018                 GO TO 3
S.0019                 END
```

Figure 12.31 An Aitken-Neville interpolation program

and DO 30 loops evaluate the $(N-1)$-th order polynomial, $F_{1,2,\ldots,N}(XR)$. Computer statements S.0016 and S.0017 are then used to print the number of points used in the interpolation (N), the point of interpolation (XR), and the value of the $(N-1)$-th order interpolation polynomial $(F(1,N))$. Finally, control is transferred to statement 3 so that the next interpolation can be made.

Figure 12.32 gives the computer results for $XR = 2.2$ and $N = 1$, 2, and 3 (see Figure 12.13). The reader is urged to hand compute each of the computer statements for the case of $N = 3$ in order to make a comparison between the Aitken-Neville method and the Newton forward method.

Interpolation Error

One obvious cause of an interpolation error is any inaccuracies in the input data values (input or inherent errors). Another source of error is the limitation of the computer that requires all numbers to be rounded off since the capacity of the storage elements is finite (rounding errors). A third source of error is the polynomial itself and the necessity of limiting it to a particular degree (truncation error). The reader should recall, however, that we have already pointed out earlier in this section that the use of a higher-degree polynomial does not necessarily result in a more accurate interpolation.

```
AITKEN-NEVILLE INTERPOLATICN

     N              XR              F(1,N)

     1        0.220CCCOE 01   0.4799999E 01
     2        0.22CCCCOF C1   0.4559999E 01
     3        0.22CCCCOE 01   0.4559998E 01

IHC217I
```

Figure 12.32 The output results of the program of Figure 12.31

Let us first consider the effect of an inaccurate input data value by examining the difference table shown below:

$X(I)$	- $Y(I)$	$\Delta\epsilon$	$\Delta^2\epsilon$	$\Delta^3\epsilon$	$\Delta^4\epsilon$
$X(1)$	$Y(1)$				
		0			
$X(2)$	$Y(2)$		0		
		0		0	
$X(3)$	$Y(3)$		0		$+\epsilon$
		0		$+\epsilon$	
$X(4)$	$Y(4)$		$+\epsilon$		-4ϵ
		$+\epsilon$		-3ϵ	
$X(5)$	$Y(5)+\epsilon$		-2ϵ		$+6\epsilon$
		$-\epsilon$		$+3\epsilon$	
$X(6)$	$Y(6)$		$+\epsilon$		-4ϵ
		0		$-\epsilon$	
$X(7)$	$Y(7)$		0		$+\epsilon$
		0		0	
$X(8)$	$Y(8)$		0		
		0			
$X(9)$	$Y(9)$				

This table introduces an error in the fifth data set ($Y(5) + \epsilon$) to show how a small error in the Y column will result in larger errors in the higher order differences.

In order to see how we may make use of this property, consider the following difference table for the function ϵ^{-X} in which all Y values are listed accurate to four significant figures:

$X(I)$	$Y(I) = \epsilon^{-X(I)}$	Δ	Δ^2	Δ^3	Δ^4
0.0	1.0000				
		−0.0952			
0.1	0.9048		+0.0091		
		−0.0861		−0.0009	
0.2	0.8187		0.0082		+0.0001
		−0.0779		−0.0008	
0.3	0.7408		0.0074		0.0001
		−0.0705		−0.0007	
0.4	0.6703		0.0067		0.0001
		−0.0638		−0.0006	
0.5	0.6065		0.0061		0.0000
		−0.0577		−0.0006	
0.6	0.5488		0.0055		0.0000
		−0.0522		−0.0006	
0.7	0.4966		0.0049		
		−0.0473			
0.8	0.4493				

Note that the original round-off of each of the $Y(I)$ values produces an irregular behavior since the given values of Y were accurate to four significant figures. The reader may easily verify, however, that the effect of round-off will be much greater if the given Y values are specified in terms of fewer significant figures.

Let us now reexamine the above difference table assuming that a reading error has been made in the fifth Y value and we have used 0.6730 instead of 0.6703. The following difference table thus results:

$X(I)$	$Y(I) = \epsilon^{-X(I)}$	Δ	Δ^2	Δ^3	Δ^4
0.0	1.0000				
		−0.0952			
0.1	0.9048		+0.0091		
		−0.0861		−0.0009	
0.2	0.8187		0.0082		+0.0028
		−0.0779		+0.0019	
0.3	0.7408		0.0101		−0.0107
		−0.0678		−0.0088	
0.4	[0.6730]		0.0013		0.0163
		−0.0665		0.0075	
0.5	0.6065		0.0088		−0.0108
		−0.0577		−0.0033	
0.6	0.5488		0.0055		0.0027
		−0.0522		−0.0006	
0.7	0.4966		0.0049		
		−0.0473			
0.8	0.4493				

Note that the ratios of the fourth differences are approximately $+1$, -4, $+6$, -4, and $+1$. The error therefore exists in the fifth entry for Y and is given by

$$6\epsilon = 0.0163$$

or

$$\epsilon = +0.0027$$

Alternatively, we may find the magnitude of the error by taking the sum of the absolute values of the fourth differences

$$|16\epsilon| = 0.0028 + 0.0107 + 0.0163 + 0.0108 + 0.0027$$

or

$$|\epsilon| = 0.0027$$

The reader is referred to Problems 12-31 through 12-33 for a further discussion of the use of difference tables.

Let us now examine the truncation error, assuming that an infinite series is represented by a finite number of terms. Recalling from Sec. 5.5 that if $f(t)$ is approximate by an n-th degree polynomial

$$F(t) = a_1 + a_2t + a_3t^2 + \cdots + a_{n+1}t^n$$

we may write

$$f(t) = F(t) + \frac{f^{n+1}(\epsilon)}{(n+1)!} \prod_{I=1}^{n+1} (t - t_I)$$

where the second term in the right member corresponds to the truncation error that occurs for using an n-th degree polynomial and $f^{n+1}(\epsilon)$ is the $(n+1)$-th derivative of $f(t)$ evaluated over $X_1 \leq \epsilon \leq X_{n+1}$. In this sense, then, the above error equation corresponds to the truncation error that occurs when Lagrange, Newton (forward), or Aitken-Neville interpolation polynomials are used. Note that although the value of ϵ is unknown in the above error formula, it still may be used under certain circumstances to estimate the error bounds.

As an alternative approach, we may use the next higher term of the interpolating polynomial to approximate the truncation error. For example, consider the equal interval Newton forward interpolation formula

$$F_N(S) = Y(1) + S\,\Delta Y(1) + S(S-1)\frac{\Delta^2 Y(1)}{2!} + S(S-1)(S-2)\frac{\Delta^2 Y(1)}{3!} + \cdots$$

and the difference table

X(I)	Y(I)	Δ	Δ²	Δ³
2	4			
		4		
3	8		3	
		7		1
4	15		4	
		11		
5	26			

If we now assume that we are using the first three points to obtain a second-degree interpolating polynomial and the point of interpolation is $XR = 2.2$, it follows, since $H = 1$, that

$$S = \frac{XR - X(1)}{H} = \frac{2.2 - 2.0}{1.0} = 0.2$$

and

$$F_2(2.2) = 4.0 + 0.2*4.0 + 0.2(0.2 - 1.0) * \frac{3}{2}$$

$$= 4.56$$

If we now use the first neglected term to estimate the magnitude of the truncation error, we find

$$S(S-1)(S-2)\,\frac{\Delta^3 Y(1)}{3!} = 0.2(0.2 - 1)(0.2 - 2)\,\frac{1}{6}$$
$$= 0.048$$

Note that if the value of the fourth data point had been (5,30) instead of (5,26), a much greater error value would have been found. This does not necessarily mean that our answer would be more inaccurate for the point (5,30) since the value $XR = 2.2$ is a value very close to the beginning data point. We should also recognize that in prediction or extrapolation problems (interpolations for XR values outside the data X values), the estimated value of the error may be very large since the magnitude of the factors $S(S-1)(S-2)\cdots(S-N+1)$ may be considerable.

12.4 LEAST-SQUARES POLYNOMIAL APPROXIMATION

In Section 12.3 we considered the problem of finding an N-th order interpolation polynomial that passed through $N+1$ points (see Figure 12.33). Since experimental data may often be subject to large errors,

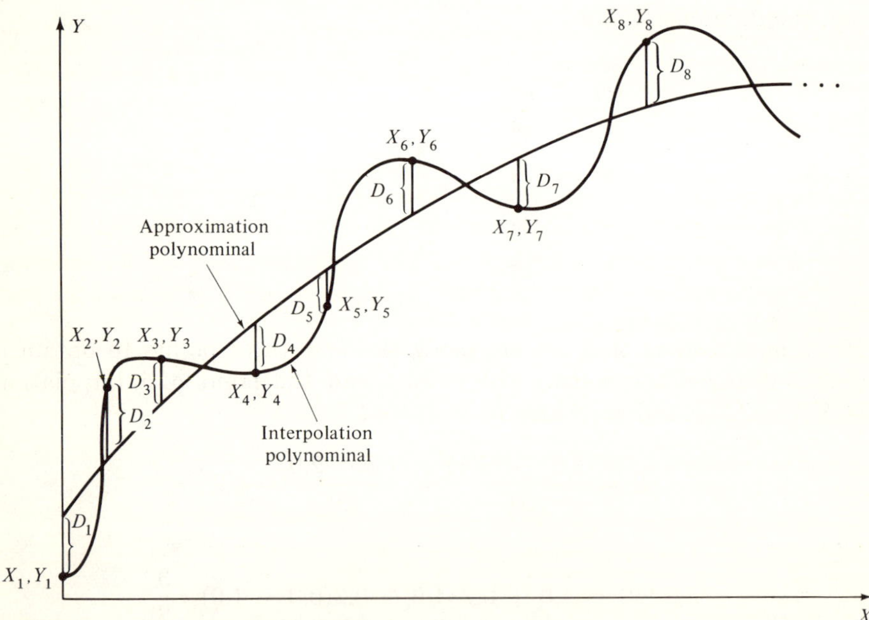

Figure 12.33 Interpolation versus approximation polynomial

it is usually undesirable to approximate experimental data by a polynomial curve that passes exactly through all the given points. Such a functional relation will obviously retain all the experimental errors. In addition, the estimation of derivatives based on an interpolating polynomial will usually lead to very poor results (see Figure 12.33). A much more useful polynomial is one that exhibits the general trend of the data and smooths out the variations due to experimental errors.

In attempting to obtain such a smooth curve as the one shown in Figure 12.33, we might initially try to balance out the differences (D's), or residuals, so that their sum is equal to zero. A little thought will show, however, that the minimization of residuals is really not what we want, since a very large positive residual may compensate for a very large negative residual. A better scheme might minimize the sum of the magnitudes of the residuals. A still better procedure minimizes the sum of the squares of the residuals. This latter scheme is known as the least-squares method and yields a sum that has only a minimum and no maximum.

In order to illustrate how we may find a least-squares polynomial, let us assume that we are given N data points (similar to those shown in Figure 12.33) and we are trying to find a least-squares second-order polynomial approximation

$$P_2(X) = A_1 + A_2X + A_3X^2$$

where A_1, A_2, and A_3 are chosen in order to minimize the sum of the squares of the residuals

$$F = \sum_{I=1}^{N} D_I^2 = \sum_{I=1}^{N} (Y_I - (A_1 + A_2X_I + A_3X_I^2))^2$$

Since we are dealing with the sum of the squares of the deviations, we have a minimum and no maximum. The minimum value of the function F is therefore found by setting the partial derivatives of F with respect to A_1, A_2, and A_3 equal to zero.

$$\frac{\partial F}{\partial A_1} = 0 = \sum_{I=1}^{N} 2(Y_I - (A_1 + A_2X_I + A_3X_I^2))$$

$$\frac{\partial F}{\partial A_2} = 0 = \sum_{I=1}^{N} 2(Y_I - (A_1 + A_2X_I + A_3X_I^2))X_I$$

$$\frac{\partial F}{\partial A_3} = 0 = \sum_{I=1}^{N} 2(Y_I - (A_1 + A_2X_I + A_3X_I^2))X_I^2$$

or

$$\sum_{I=1}^{N} A_1 + \sum_{I=1}^{N} X_I A_2 + \sum_{I=1}^{N} X_I^2 A_3 = \sum_{I=1}^{N} Y_I$$
$$\sum_{I=1}^{N} X_I A_1 + \sum_{I=1}^{N} X_I^2 A_2 + \sum_{I=1}^{N} X_I^3 A_3 = \sum_{I=1}^{N} X_I Y_I$$
$$\sum_{I=1}^{N} X_I^2 A_1 + \sum_{I=1}^{N} X_I^3 A_2 + \sum_{I=1}^{N} X_I^4 A_3 = \sum_{I=1}^{N} X_I^2 Y_I$$

Noting that $\sum_{I=1}^{N} A_1 = A_1 \sum_{I=1}^{N} 1 = A_1 * N$, the augmented matrix thus becomes

$$\begin{bmatrix} \sum_{I=1}^{N} 1 & \sum_{I=1}^{N} X_I & \sum_{I=1}^{N} X_I^2 & \sum_{I=1}^{N} Y_I \\ \sum_{I=1}^{N} X_I & \sum_{I=1}^{N} X_I^2 & \sum_{I=1}^{N} X_I^3 & \sum_{I=1}^{N} X_I Y_I \\ \sum_{I=1}^{N} X_I^2 & \sum_{I=1}^{N} X_I^3 & \sum_{I=1}^{N} X_I^4 & \sum_{I=1}^{N} X_I^2 Y_I \end{bmatrix}$$

and the Gauss-Jordan (or some other elimination method) may be used to solve this linear set of equations for A_1, A_2, and A_3.

As a simple numerical example, let us assume that we are given

I	1	2	3	4
$X(I)$	2	3	4	5
$Y(I)$	4	8	15	25

and we desire a straight line or first-order polynomial approximation that minimizes the sum of the squares of the residuals. Thus

$$F = \sum_{I=1}^{N} D_I^2 = \sum_{I=1}^{N} (Y_I - (A_1 + A_2 X_I))^2$$

and

$$\frac{\partial F}{\partial A_1} = 0 = \sum_{I=1}^{4} 2(Y_I - (A_1 + A_2 X_I))$$
$$\frac{\partial F}{\partial A_2} = 0 = \sum_{I=1}^{4} 2(Y_I - (A_1 + A_2 X_I))X_I$$

or

$$\sum_{I=1}^{4} A_1 + \sum_{I=1}^{4} X_I A_2 = \sum_{I=1}^{4} Y_I$$
$$\sum_{I=1}^{4} X_I A_1 + \sum_{I=1}^{4} X_I^2 A_2 = \sum_{I=1}^{4} X_I Y_I$$

The augmented matrix thus becomes

$$\begin{bmatrix} 4 & 14 & 52 \\ 14 & 54 & 217 \end{bmatrix}$$

Solving this set of equations we find

$$A_1 = -11.5$$
$$A_2 = 7.0$$

and

$$P_1(X) = -11.5 + 7.0X$$

Figure 12.34 is a general Fortran program for obtaining up to a ninth-order least-squares polynomial approximation based on a maximum of 100 data points. Initially, the degree of the polynomial M, the number of data points N, and the corresponding X and Y values are read into memory. The dummy variables LL, JJ, and KK are then respectively set equal to $2*M+1$, $M+2$, and $M+1$. The DO 2 and DO 3 loops are then used to set the initial values of the subscripted variables SUM and C equal to zero. These latter two variables are used to store the coefficient values of the general linear equation set

$$\begin{array}{lllllll}
\mathrm{SUM}(1)\, A_1 & + & \mathrm{SUM}(2)\, A_2 & + \cdots + & \mathrm{SUM}(M+1)\, A_{M+1} & = & C(1) \\
\mathrm{SUM}(2)\, A_1 & + & \mathrm{SUM}(3)\, A_2 & + \cdots + & \mathrm{SUM}(M+2)\, A_{M+1} & = & C(2) \\
\vdots & & \vdots & & \vdots & & \vdots \\
\mathrm{SUM}(M+1)\, A_1 & + & \mathrm{SUM}(M+2)\, A_2 & + \cdots + & \mathrm{SUM}(2M+1)\, A_{M+1} & = & C(M+1)
\end{array}$$

where

$$\mathrm{SUM}(1) = \sum_{I=1}^{N} 1 = N$$
$$\mathrm{SUM}(2) = \sum_{I=1}^{N} X_I$$
$$\vdots \qquad \qquad \vdots$$
$$\mathrm{SUM}(2M+1) = \sum_{I=1}^{N} X_I^{2M}$$

```
C        LEAST SQUARE POLYNOMIAL APPROXIMATION.
C
         DIMENSION Y(100),X(100),A(10,11),SUM(19),C(10)
     100 READ(5,1) M,N,(X(I),Y(I),I=1,N)
       1 FORMAT(2I5,/,(5E15.7))
         WRITE(6,61) (I,X(I),Y(I),I=1,N)
      61 FORMAT('1INPUT DATA',//,8X,'I',10X,'X(I)',10X,'Y(I)',/,(8X,I2,4X,2
        1E15.7))
         LL=2*M+1
         JJ=M+2
         KK=M+1
         DO 2 J=2,LL
       2 SUM(J)=0.0
         SUM(1)=N
         DO 3 J=1,KK
       3 C(J)=0.0
         DO 4 I=1,N
         B=1.0
         C(1)=C(1)+Y(I)
         DO 5 J=2,KK
         B=X(I)*B
         SUM(J)=SUM(J)+B
       5 C(J)=C(J)+Y(I)*B
         DO 4 J=JJ,LL
         B=X(I)*B
       4 SUM(J)=SUM(J)+B
         DO 6 I=1,KK
         DO 6 J=1,KK
         K=I+J
       6 A(J,I)=SUM(K-1)
         DO 7 I=1,KK
       7 A(I,JJ)=C(I)
         WRITE(6,62) ((A(I,J),J=1,JJ),I=1,KK)
      62 FORMAT(////,' COEFFICIENTS OF AUGMENTED MATRIX',//,(5E15.7))
C
C
C        GAUSS-JORDAN SOLUTION.
C
         M=JJ
         N=KK
         KK=0
         JJ=0
         DO 10 K=1,N
         JJ=KK+1
         IL=JJ
         KK=KK+1
      20 IF(ABS(A(JJ,KK))-EPS)21,21,22
      21 JJ=JJ+1
         GO TO 20
      22 IF(LL-JJ) 23,24,23
      23 DO 25 MM=1,7
         ATEMP=A(LL,MM)
         A(LL,MM)=A(JJ,MM)
      25 A(JJ,MM)=ATEMP
C
C        THE EQUATIONS ARE FORCED INTO THE DIAGONAL FORM.
C
      24 DO 30 LJ=1,M
         J=M+1-LJ
      30 A(K,J)=A(K,J)/A(K,K)
         DO 10 I=1,N
         DO 10 LJ=1,M
         J=M+1-LJ
         IF(I-K) 31,10,31
      31 A(I,J)=A(I,J)-A(I,K)*A(K,J)
      10 CONTINUE
         WRITE(6,8)
       8 FORMAT(///,' LEAST SQUARES POLYNOMIAL APPROXIMATION',///,' POLYNOM
        2IAL COEFFICIENTS =',//)
         DO 40 I=1,II
         C(I)=A(I,M)
      40 WRITE(6,9)I,C(I)
       9 FORMAT(' A(',I2,') =',E15.7)
         GO TO 100
         END
```

Figure 12.34 Least-square polynomial approximation program

and

$$
\begin{aligned}
C(1) &= \sum_{I=1}^{N} Y_I \\
C(2) &= \sum_{I=1}^{N} X_I Y_I \\
&\vdots \\
C(M+1) &= \sum_{I=1}^{N} X_I^M Y_I
\end{aligned}
$$

The DO 4 loops are now used to compute the coefficient values. For example, in the case of $M = 2$, the first iteration of the outer DO 4 loop would set $C(1) = Y(1)$ (computer statement S.0016). The DO 5 loop would then be executed two times, yielding

$$
\begin{aligned}
\mathrm{SUM}(2) &= X(1) \\
\mathrm{C}(2) &= Y(1)*X(1)
\end{aligned}
$$

and

$$
\begin{aligned}
\mathrm{SUM}(3) &= X(1)*X(1) \\
\mathrm{C}(3) &= Y(1)*X(1)*X(1)
\end{aligned}
$$

The inner DO 4 loop would then be executed twice, yielding

$$
\begin{aligned}
\mathrm{SUM}(4) &= X(1)*X(1)*X(1) \\
\mathrm{SUM}(5) &= X(1)*X(1)*X(1)*X(1)
\end{aligned}
$$

Control would then be returned to the outer DO 4 loop to add the contribution due to the second data point. This procedure would continue until all N contributions had been added. The DO 6 loops would then form the coefficient part of the augmented matrix and the DO 7 loop would add the right member terms or the $(M+1)$-th column values. This would complete the augmented matrix of the $M+1$ set of linear equations.

Computer statements S.0032 through S.0056 would next be used to solve for the $M+1$ polynomial coefficients. This part of the program is essentially identical to computer statements S.0009 through S.0031 in the Gauss-Jordan program of Figure 6.13. Finally, computer statements S.0057 through S.0062 would be used to print the results. Control is then transferred to statement 100 so that the next set of data may be processed.

In order to test the computer program of Figure 12.34, let us use some of the data that was obtained from the analog voltage-to-pulse-width

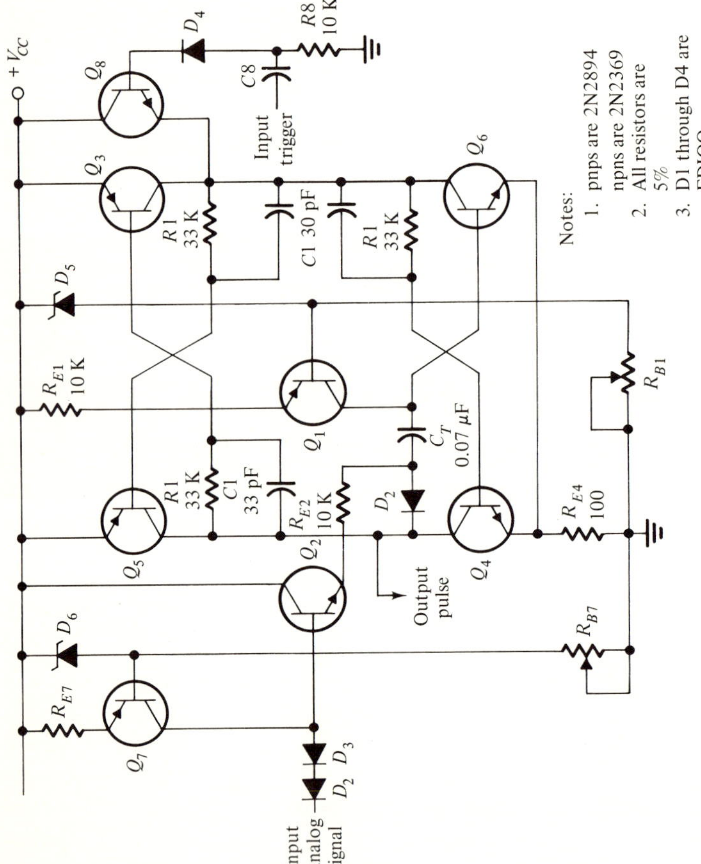

Figure 12.35 D-C voltage-to-pulse-width converter (Courtesy, Electronic Design)

converter shown in Figure 12.35.[4] This type of circuit is very useful in telemetry where analog-to-digital conversion is required. The circuit uses a complementary monostable flip-flop and a constant current generator to obtain an output pulse width that is a function of the input voltage amplitude. The two sets of data shown at bottom of page 720 represent the measured circuit performance for R_{B1} values of 20-K and 80-K ohms.

Figure 12.36 gives the computer output for these two data sets and Figure 12.37 shows the corresponding graph of these results. Note that in the case of R_{B1} = 20K ohms, the Y intercept of the linear approximating polynomial approaches zero microseconds and the output of the converter is essentially linear over the entire input range with a maximum RMS deviation of 0.5 percent.

For another example of the use of the computer program of Figure 12.34 consider the calibration curve of the chromel-to-constantan thermocouple shown in Figure 12.38. In attempting to use this calibration curve

[4] See Warren R. Crockett, "Here's a DC-to-Pulse-Width Converter," *Electronic Design 16*, Vol. 15 (Aug., 1967), pp. 66–69.

```
INPUT DATA

 I          X(I)               Y(I)
 1       0.0                0.125CCCCE C1
 2       C.25CCCCOE 00      0.30CCCCCE C1
 3       C.50CCCCCE C0      0.6250000E 01
 4       0.7500000E 00      0.115CCCCE C2
 5       C.1CCCCCCE 01      0.1650000E 02
 6       0.1250000E 0.1     0.22CCCCOE C2
 7       C.15CCCCCE 01      0.2625000E 02
 8       0.1750000E 01      0.32CC0CCE C2
 9       C.20CCCCCE 01      0.3675000E 02
10       0.2250000E 01      0.42250C0E C2
11       C.25CCCCCE 01      0.4700000E 02
12       0.2750000E 01      0.525CCCCE C2
13       C.30CCCCOE 01      0.5700000E 02
14       0.3250000E 01      0.63CCCCCE C2
15       C.35CCCCCE 01      0.6700000E 02
16       0.3750000E 01      0.72CCCCCE C2
17       C.40CCCCCE 01      0.7700000E 02
18       0.4250000E 01      0.82250CCE C2
19       C.45CCCCCE C1      0.8714999E 02
20       0.4750000E 01      0.9279999E C2
21       0.50CCCCOE 01      0.9725000E 02


COEFFICIENTS OF AUGMENTED MATRIX

 0.2100000E 02  0.5250000E 02  0.9926997E 03  0.5250000E 02  0.179375CE C3
 0.343935CE C4


LEAST SQUARES POLYNOMIAL APPROXIMATION


POLYNOMIAL COEFFICIENTS =


A( 1) = -0.2474075E 01
A( 2) =  0.1989819E 02
```

Figure 12.36 The output results of the program of Figure 12.34 for the circuit of Figure 12.35

```
INPUT DATA

    I         X(I)             Y(I)
    1      0.0              0.0
    2      0.2500000E 00    0.3599999E 01
    3      0.5000000E 00    0.6900000E 01
    4      0.7500000E 00    0.1010000E 02
    5      0.1000000E 01    0.1325000E 02
    6      0.1250000E 01    0.1650000E 02
    7      0.1500000E 01    0.1989999E 02
    8      0.1750000E 01    0.2300000E 02
    9      0.2000000E 01    0.2595000E 02
   10      0.2250000E 01    0.2925000E 02
   11      0.2500000E 01    0.3270000E 02
   12      0.2750000E 01    0.3600000E 02
   13      0.3000000E 01    0.3909999E 02
   14      0.3250000E 01    0.4225000E 02
   15      0.3500000E 01    0.4529999E 02
   16      0.3750000E 01    0.4879999E 02
   17      0.4000000E 01    0.5209999E 02
   18      0.4250000E 01    0.5545000E 02
   19      0.4500000E 01    0.5859999E 02
   20      0.4750000E 01    0.6179999E 02
   21      0.5000000E 01    0.6500000E 02


COEFFICIENTS OF AUGMENTED MATRIX

 0.2100000E 02  0.5250000E 02  0.6855488E 03  0.5250000E 02  0.1793750E 03
 0.2336836E 04


LEAST SQUARES POLYNOMIAL APPROXIMATION

POLYNOMIAL COEFFICIENTS =

A( 1) =  0.2833710E 00
A( 2) =  0.1294472E 02

IHC217I
```

Figure 12.36 Continued

THE X INPUT (VOLTS)	THE Y OUTPUT PULSE WIDTH (μs) $R_{B1} = 20$ K	$R_{B1} = 80$ K
0.00	0.00	1.25
0.25	3.60	3.00
0.50	6.90	6.25
0.75	10.10	11.50
1.00	13.25	16.50
1.25	16.50	22.00
1.50	19.90	26.25
1.75	23.00	32.00
2.00	25.95	36.75
2.25	29.25	42.25
2.50	32.70	47.00
2.75	36.00	52.50
3.00	39.10	57.00
3.25	42.25	63.00
3.50	45.30	67.00
3.75	48.80	72.00
4.00	52.10	77.00
4.25	55.45	82.25
4.50	58.60	87.15
4.75	61.80	92.80
5.00	65.00	97.25

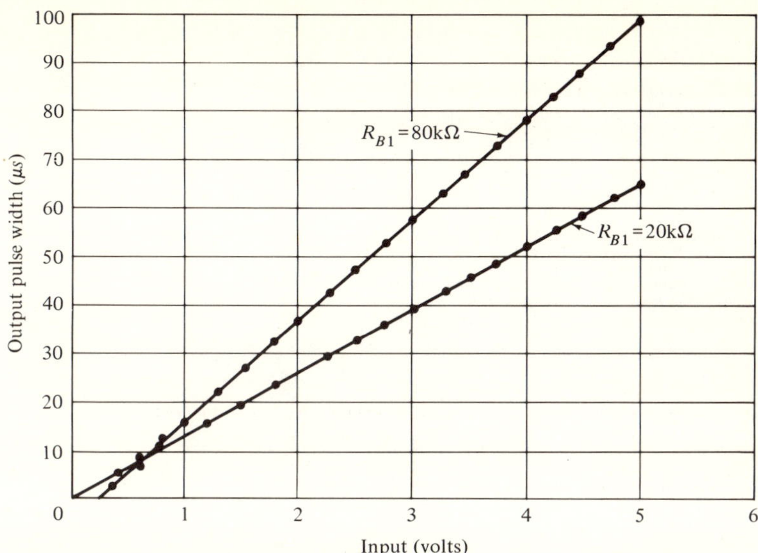

Figure 12.37 The corresponding graph of the output results of Figure 12.36 (Courtesy, Electronic Design)

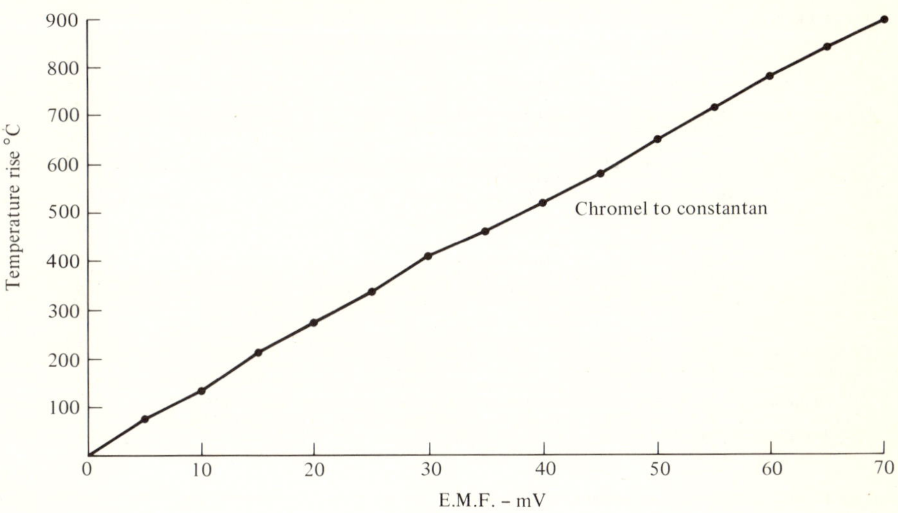

Figure 12.38 Calibration curve of chromel-to-constantan thermocouple

in connection with a large number of temperature measurements and subsequent computer calculations, an approximating polynomial was required. Note that an interpolation polynomial would not be desirable since the subsequent computer calculations required derivative calculations in computing the thermal conductivity.

```
INPUT DATA

   I         X(I)              Y(I)
   1      0.0              0.0
   2      0.4999999E-02    0.7200000E 02
   3      0.9999998E-02    0.1420000E 03
   4      0.1500000E-01    0.2080000E 03
   5      0.2000000E-01    0.2750000E 03
   6      0.2500000E-01    0.3420000E 03
   7      0.3000000E-01    0.4040000E 03
   8      0.3500000E-01    0.4670000E 03
   9      0.4000000E-01    0.5280000E 03
  10      0.4500000E-01    0.5910000E 03
  11      0.5000000E-01    0.6520000E 03
  12      0.5500000E-01    0.7130000E 03
  13      0.6000000E-01    0.7780000E 03
  14      0.6500000E-01    0.8420000E 03
  15      0.6999999E-01    0.9120000E 03


COEFFICIENTS OF AUGMENTED MATRIX

 0.1500000E 02  0.5249997E 00  0.6926000E 04  0.5249997E 00  0.2537498E-01
 0.3321946E 03


LEAST SQUARES POLYNOMIAL APPROXIMATION


POLYNOMIAL COEFFICIENTS =


A( 1) =  0.1280981E 02
A( 2) =  0.1282639E 05


INPUT DATA

   I         X(I)              Y(I)
   1      0.0              0.0
   2      0.4999999E-02    0.7200000E 02
   3      0.9999998E-02    0.1420000E 03
   4      0.1500000E-01    0.2080000E 03
   5      0.2000000E-01    0.2750000E 03
   6      0.2500000E-01    0.3420000E 03
   7      0.3000000E-01    0.4040000E 03
   8      0.3500000E-01    0.4670000E 03
   9      0.4000000E-01    0.5280000E 03
  10      0.4500000E-01    0.5910000E 03
  11      0.5000000E-01    0.6520000E 03
  12      0.5500000E-01    0.7130000E 03
  13      0.6000000E-01    0.7780000E 03
  14      0.6500000E-01    0.8420000E 03
  15      0.6999999E-01    0.9120000E 03


COEFFICIENTS OF AUGMENTED MATRIX

 0.1500000E 02  0.5249997E 00  0.2537498E-01  0.6926000E 04  0.5249997E 00
 0.2537498E-01  0.1378124E-02  0.3321946E 03  0.2537498E-01  0.1378124E-02
 0.7980429E-04  0.1797766E 02


LEAST SQUARES POLYNOMIAL APPROXIMATION


POLYNOMIAL COEFFICIENTS =


A( 1) =  0.5827593E 01
A( 2) =  0.1347090E 05
A( 3) = -0.9207359E 04

IHC217I
```

Figure 12.39 The output results of the program of Figure 12.34 for the thermocouple calibration curve of Figure 12.38

Figure 12.39 gives the resulting linear and second-degree polynomial obtained from the computer program of Figure 12.34. The reader is encouraged to hand calculate the second of these results in order to better understand the method of least squares.

PROBLEMS

12-1. Extend the program of Figure 12.1 so that it is capable of evaluating the arc tangent (X) for $|X| > 1$. Test this program by determining the value of the inverse tangent function for argument values equal to 0.5, 1.0, 1.5, 5.0, and 50.0.

12-2. Write a program, similar to that shown in Figure 12.1, for computing the following functions:

a. $\text{Sin } X = X - \frac{X^3}{3!} + \frac{X^5}{5!} - \frac{X^7}{7!} + \cdots ; |X| < \infty$

b. $\text{Cos } X = 1 - \frac{X^2}{2!} + \frac{X^4}{4!} - \frac{X^6}{6!} + \cdots ; |X| < \infty$

c. $\text{Arcsin } X = X + \frac{X^3}{2 \cdot 3} + \frac{3X^5}{2 \cdot 4 \cdot 5} + \frac{3 \cdot 5X^7}{2 \cdot 4 \cdot 6 \cdot 7} + \cdots ; |X| < 1$

d. $\text{Sinh } X = X + \frac{X^3}{3!} + \frac{X^5}{5!} + \frac{X^7}{7!} + \cdots ; |X| < \infty$

e. $\text{Cosh } X = 1 + \frac{X^2}{2!} + \frac{X^4}{4!} + \frac{X^6}{6!} + \cdots ; |X| < \infty$

f. $\text{Ln } X = (X-1) - \frac{1}{2}(X-1)^2 + \frac{1}{3}(X-1)^3 - \cdots ;$ $|X-1| \leq 1$ and $X \neq 0$

g. $\epsilon^X = 1 + X + \frac{X^2}{2!} + \frac{X^3}{3!} + \cdots$

Test each of these programs by using appropriate argument values.

12-3. Explain why the results of Figure 12.2 for the accuracy values of 1.0E−07 and 1.0E−08, are the same.

12-4. Explain how you would modify the program of Figure 12.1 so that you could compute the function to an accuracy of 1.0E−10.

12-5. Using appropriate argument values, show that the program of Figure 12.5 produces more accurate results than the program of Figure 12.1 for the same specified accuracy.

12-6. Explain why the third, fourth, fifth, and sixth partial sum values in Figure 12.8 are not equal to 0.2500000E−04, 0.1000000E−03, 0.25000000E−02, and 0.1000000E−01, respectively.

12-7. What changes could be made in the program of Figure 12.7 so that it would be more efficient?

12-8. Show how you would incorporate the Wolfe Summing program of Figure 12.7 with a numerical integration routine. Test the resulting program by integrating ϵ^{-X} over a suitable range of X.

12-9. Write a subroutine that uses nesting to evaluate a general n-th degree polynomial and its derivative.

12-10. Use the subroutine of Problem 12-9 to evaluate each of the following polynomials and their derivatives:

a. $P_3(s) = 3.54s^3 + 2.976s^2 - 1.864s + 10.28$ at $s = -1.694$

b. $P_5(s) = 5.12\ 10^4\ s^5 - 6.74\ 10^2\ s^3 + 5.25\ 10^2\ s^2 + 10.65$ at $s = 1.2\ 10^1$

c. $P_8(s) = 18.2s^8 + 16.44s^6 - 22.0s^4 + 9.24s^2 - 8.45$ at $s = -4.2$

12-11. Write a Fortran program that uses nesting to evaluate the following polynomial approximations:

a. $\text{Log}_{10} X \approx A_1Y + A_3Y^3 + A_5Y^5 + A_7Y^7 + A_9Y^9$ for $X = 10.0$, 25.0, and 10000 where

$$Y = (X-1)/(X+1) \qquad A_5 = 0.1775221$$
$$A_1 = 0.8685917 \qquad A_7 = 0.09437648$$
$$A_3 = 0.2893355 \qquad A_9 = 0.1913377$$

b. $10^X \approx (1 + A_1X + A_2X^2 + A_3X^3 + A_4X^4 + A_5X^5 + A_6X^6 + A_7X^7)^2$ for $X = 0.1$, 0.5, and 0.95 where

$$A_1 = 1.151293 \qquad A_4 = 0.07295174 \qquad A_7 = 0.0009326427$$
$$A_2 = 0.6627309 \qquad A_5 = 0.01742112$$
$$A_3 = 0.2543936 \qquad A_6 = 0.002554918$$

c. $\dfrac{\text{Sin } X}{X} \approx 1 + A_2X^2 + A_4X^4 + A_6X^6 + A_8X^8 + A_{10}X^{10}$ for $X = 0.1$, 0.5, and 1.0 where

$$A_2 = -0.1666666 \qquad A_8 = 0.0000027526$$
$$A_4 = 0.008333332 \qquad A_{10} = -0.0000000239$$
$$A_6 = -0.0001984090$$

12-12. Use the program of Figure 12.9 to convert the following numbers to their decimal equivalents: a. $(1220)_3$ b. $(1321)_4$ c. $(8765)_9$

12-13. Modify the program of Figure 12.9 so that it is capable of converting hexadecimal numbers to their decimal equivalent.

12-14. Write a program for converting Roman numbers to their decimal Arabic equivalent. Test your program with the following data: IV, IIII, MXL, and MCV.

12-15. Write a program for converting decimal numbers to their base R equivalent. Hint: Note that the general number expression may be nested.

$$N = (\ldots ((a_nR + a_{n-1})R + a_{n-2})R + \cdots + a_1)R + a_0$$

If we now divide this equation by R, the remainder will be equal to a_0. The first quotient divided by R will yield the remainder a_1, and so on. For example, dividing 36 by 5 will give a quotient of 7 and a remainder of 1, so that $a_0 = 1$. Dividing the first quotient (7) by 5 gives a quotient of 1 and a remainder of 2, so that $a_1 = 2$. Finally, dividing the second quotient by 5 yields a zero quotient and a remainder of 1 or $a_3 = 1$. Thus

$$\begin{array}{l} 0 \leftarrow 1 \leftarrow 7 \leftarrow 36 \;\; \underline{|\div 5} \\ \downarrow \;\; \downarrow \;\; \downarrow \\ 1 \;\; 2 \;\; 1 = \text{Remainders} \end{array}$$

or

$$36)_{10} = 121)_5 = 1*5^2 + 2*5^1 + 1*5^0$$
$$= 25 + 10 + 1 = 36)_{10}$$

12-16. Write a Fortran program for converting whole numbers of one base to their equivalent in any other base. Test your program with several data values.

12-17. In converting decimal numbers that have a fractional part to their base R equivalent, we must treat the fractional part separately.

$$N = a_nR^n + a_{n-1}R^{n-1} + \cdots + a_1R^1 + \underbrace{a_0R^0 \, . + a_{-1}}_{\text{radix point}}R^{-1} + a_{-2}R^{-2} + \cdots$$

The fractional part may be written as

$$R^{-1}(a_{-1} + \cdots + R^{-1}(a_{-(n-2)} + R^{-1}(a_{-(n-1)} + a_{-n}R^{-1})) \cdots)$$

If we now multiply this expression by R, the a_{-1} digit will appear in the units position. Successive multiplications by R will isolate a_{-2}, and so on. For example,

$$\begin{array}{lccccccc} 2*0.075 \rightarrow & 0.150 \rightarrow & 0.300 \rightarrow & 0.600 \rightarrow & 0.200 \rightarrow & 0.400 \rightarrow & 0.800 \rightarrow & 0.600 \ldots \\ \downarrow & \downarrow & \downarrow & \downarrow & \downarrow & \downarrow & \downarrow & \\ \text{units value} = 0 & 0 & 0 & 1 & 0 & 0 & 1 & \end{array}$$

Thus

$$0.075)_{10} = 0.0001001 \ldots)_2$$

Write a Fortran program for converting decimal numbers that contain a fractional part to their base R equivalent.

12-18. Use the program of Figure 12.16 to evaluate the arctangent X from the following table values:

X	ARCTAN X	X	ARCTAN X
0.200	0.1973956	0.205	0.2021986
0.201	0.1983569	0.206	0.2031581
0.202	0.1993179	0.207	0.2041172
0.203	0.2002785	0.208	0.2050759
0.204	0.2012387	0.209	0.2060342
		0.210	0.2069922

for $X = 0.2004$, 0.20009, and 0.20956.

12-19. Modify the program of Figure 12.20 so that the value of W will be computed for specified values of T, B, ϵ_r, and Z_0. Test your program for the following data sets:

	T	B	ϵ_r	Z_0
a.	0.0027	0.125	4.8	50.0
b.	0.0027	0.125	4.8	75.0
c.	0.0027	0.125	4.8	100.0

12-20. The input impedance of a short-circuited lossless line of characteristic impedance Z_0 and length l is given by

$$Z = jX = jZ_0 \tan \frac{\omega l}{v}$$

where ω is the angular frequency and v is the velocity of propagation ($= 3\ 10^{10}/\sqrt{\epsilon_r}$ cm/sec). If ω is much less than $\pi v/(2l)$, the line will be equivalent to a high Q inductance where

$$\omega L = Z_0 \tan \frac{\omega l}{v}$$

or

$$L = \frac{Z_0}{\omega} \tan \frac{\omega l}{v} \text{ henries}$$

Write a Fortran program that reads into memory the values of T, B, W, ϵ_r, ω, and L. Then, using the program of Figure 12.20, have your program compute the value of Z_0 and then, in turn, the length of the required line l. Test your program for values of $T = 0.0027$ inches, $B = 0.125$ inches, $W = 0.062$ inches, $\epsilon_r = 4.8$, $\omega = 400$ MHz, and $L = 0.01$ μHy.

12-21. Using the method employed in the program of Figure 12.20, extend the program of Figure 7.2 so that it may be used to find the operating point for various values of R_B. Test your program for values of R_B = 25 kilohms, 50 kilohms, and 100 kilohms.

12-22. A germanium temperature sensor that is used in cryogenics is specified in terms of the following temperature-resistance values:

TEMP IN DEG KELVIN	*R* MEASURED IN OHMS
1.250	14665.398
1.261	14389.898
1.371	11588.000
1.497	9293.199
1.745	6340.297
1.965	4782.797
2.235	3538.300
2.491	2773.700
2.731	2283.600
2.986	1904.600
3.236	1620.800
3.489	1400.600
3.772	1202.800
3.980	1088.700
4.224	975.200

Use the Lagrangian Interpolation program of Figure 12.23 to determine the temperature for measured sensor resistance values of 1250.36, 3495.5, and 11588.00 ohms. Hint: Inverse interpolation is used to find the argument value corresponding to a given function value. One method of inverse interpolation is based on interchanging the roles of $X(I)$ and $F(X)$.

12-23. One of the difficulties with the method of inverse interpolation suggested in Problem 12-22 is that table values of $F(X)$ (R measured in ohms) may not always be single valued. An alternative method of inverse interpolation that avoids this difficulty is based on obtaining a polynomial approximation of the given function $F(X)$ and then solving this polynomial to obtain the corresponding argument for the specified function value. Using an appropriate polynomial interpolation scheme, write a program that uses this method of inverse interpolation. Test your program with the same data sets that were specified in Problem 12-22.

12-24. Lagrangian interpolation is greatly simplified when the $X(I)$ values are equally spaced. Modify the program of Figure 12.23 so that it will handle only equally spaced arguments.

12-25. Modify the program of Figure 12.25 so that the differences are printed on alternate lines in the usual triangular form.

12-26. Given the following data values:

I	1	2	3	4	5	6
$X(I)$	1.0	2.0	3.0	4.0	5.0	6.0
$Y(I)$	2.0	4.0	8.0	15.0	25.0	40.0

hand compute the value of Y for the value of $X = 2.5$. Use a second-order interpolation polynomial for each of the following methods:

a. Lagrange interpolation
b. Newton's forward interpolation
c. Newton's backward interpolation
d. Gauss's forward interpolation
e. Stirling's interpolation.

Compare your results and explain any differences.

12-27. An alternate form of an Aitken interpolating polynomial is based on the following scheme:

$$F_{1,2}(X) = \frac{\begin{vmatrix} Y(1) & X(1) - X \\ Y(2) & X(2) - X \end{vmatrix}}{X(2) - X(1)}$$

$$F_{1,2,3}(X) = \frac{\begin{vmatrix} F_{1,2}(X) & X(2) - X \\ F_{1,3}(X) & X(3) - X \end{vmatrix}}{X(3) - X(2)}$$

$$F_{1,2,3,4}(X) = \frac{\begin{vmatrix} F_{1,2,3}(X) & X(3) - X \\ F_{1,2,4}(X) & X(4) - X \end{vmatrix}}{X(4) - X(3)}$$

and so on.

Write a Fortran program based on this alternate scheme and compare the results obtained from this program with those shown in Figure 12.32.

12-28. When would there be an advantage in using the program of Problem 12-27 instead of the program shown in Figure 12.31?

12-29. Modify the Aitken-Neville program shown in Figure 12.31 so that the numerical values of the various interpolating polynomials are printed as follows:

$$\begin{matrix} F_{1,2} & F_{1,2,3} & F_{1,2,3,4} & F_{1,2,3,4,5} \\ F_{2,3} & F_{2,3,4} & F_{2,3,4,5} & \cdot \\ F_{3,4} & F_{3,4,5} & \cdot & \cdot \\ F_{4,5} & \cdot & \cdot & \cdot \\ \cdot & \cdot & & \\ \cdot & \cdot & & \\ \cdot & & & \end{matrix}$$

12-30. Using the program of Problem 12-29 and the data for the arctan X function shown in Problem 12-18, evaluate the arctangent function for a value of $X = 0.20956$. Explain the differences in the printed polynomial values.

12-31. Hand compute the difference tables for each of the following functions: a. $Y = X$; b. $Y = X^2$; c. $Y = X^3$; d. $Y = X^4$; $X = 1$, 2, 3, 4, 5, and 6. Note that the first, second, third, and fourth differences of a., b., c., and d., respectively, are equal. This observation may be used to estimate the order of an appropriate interpolating polynomial. For example, if the third differences are all equal, it is quite evident that a third-order polynomial may be used.

12-32. Prediction or extrapolation is estimating the value of Y outside the data range of the X arguments. Using the idea presented in Problem 12-31, estimate the value of Y for $X = 5$ and 6.

X	0	1	2	3	4	5	6
Y	1	−3	−1	13	45	?	?

Hint: Once the degree of the polynomial has been established, it is not necessary to find the polynomial coefficients. The missing values of Y can be found by working back from the equal difference column.

12-33. Using successive differences, determine the error in one of the tabulated values below:

X(I)	Y(I)
0.0	1.0000
0.1	0.9048
0.2	0.8187
0.3	0.7408
0.4	0.6703
0.5	0.6055
0.6	0.5488
0.7	0.4966
0.8	0.4493
0.9	0.4066
1.0	0.3679

12-34. Use the next higher term in the interpolating polynomial to approximate the error in each of the results found in Problem 12-26.

12-35. Correct the single error in the data set of Problem 12-33. Then alternately change each data set value by +0.0001, −0.0001, +0.0001, and so on. Hand compute the difference table, noting that the round-off error essentially doubles for each higher difference.

12-36. Problem 12-35 was used to illustrate how small round-off errors could produce very large errors in higher-order differences. Since round-off errors will seldom vary in exactly this way, let us investigate the effect of a more random variation.

Write a program that uses a random number generator to change the values of the corrected data set of Problem 12-33 by + or −0.0001. For example, if the most significant digit of the random number is a 0, 1, 2, 3, or 4, change the data value by +0.0001. Alternately, change the data value by −0.0001 if the most significant digit of the random number is equal to a 5, 6, 7, 8, or 9. Run the problem several times and examine the values of the fourth differences. What conclusion can you draw?

12-37. Hand compute a second-order least-squares polynomial for the data given in Problem 12-26. Use this polynomial to approximate the value of Y for a value of $X = 2.5$.

12-38. Develop a method for obtaining an approximating polynomial that is based on minimizing the sum of the absolute values of the residuals. Compare this procedure with the least-squares method.

12-39. To estimate the effectiveness of the least-squares approximating polynomial, we may define the RMS error as

$$\text{RMS Error} = \left[\sum_{I=1}^{N} (Y_I - P_N(X(I))^2\right]^{1/2}$$

Modify the program of Figure 12.34 to include the calculation of this RMS error value. Rerun the problem of the voltage-to-pulse-width converter shown in Figure 12.35 and discuss the significance of the two RMS error values you obtain.

CHAPTER 13

The Solution of Ordinary and Partial Differential Equations

13.1 INTRODUCTION

Much of the formal class work taken by electrical engineering students is devoted to the study of the classical solutions of differential equations. This should be expected since the practicing engineer is interested mainly in the behavior of dynamic systems. In other words, the engineer often approaches his problem by obtaining a suitable model that is composed of linear, nonlinear, and/or time-varying elements. Since the equations that describe the model are usually differential equations, it is therefore important that he be well versed in their solution. Unfortunately, the undergraduate is usually exposed, in depth, to only the solution of ordinary constant-coefficient linear differential equations and these are generally limited to a maximum of three degrees of freedom. The main reason for this limitation is that even when such classical techniques as the Laplace transform method are employed, the solution of a three-degree-of-freedom problem is still quite formidable—and when nonlinear differential equations are encountered, the problem is indeed *very* formidable, if not impossible, even for the simpler cases. In fact, the reader should be aware that only 40 years ago, our knowledge of the solution of nonlinear differential equations was approximately equivalent to Newton's knowledge of the solution of linear differential equations. In recent years, however, thanks to the development of the high-speed computer, many advances have been made in the ability of the engineer to solve very complicated systems. This chapter is concerned with the numerical (computer) solution of both linear and nonlinear differential equations; therefore, before we begin our discussion of various numerical techniques that can be employed, let us first discuss the various types of differential equations we might have to solve.

In order to define the various types of differential equations, we must first define the derivative. The derivative of a function of a single variable (the ordinary derivative) may be defined as the limit of the ratio of the increment of the dependent variable to the increment of the independent variable, as the increment of the independent variable approaches zero as a limit. This may be expressed mathematically by assuming y is the dependent variable and is defined in terms of some function of the independent variable x, thus,

$$y = f(x)$$

If x is now changed by an increment Δx, the dependent variable will be changed by an increment Δy and the new value of the function will be given by

$$y + \Delta y = f(x + \Delta x)$$

We may find the increment Δy by subtracting the former equation from the latter equation, thus yielding

$$\Delta y = f(x + \Delta x) - f(x)$$

The derivative may be obtained by dividing both sides of this equation by Δx and taking the limit as Δx approaches zero; thus,

$$\frac{dy}{dx} = \lim_{\Delta x \to 0} \frac{f(x + \Delta x) - f(x)}{\Delta x}$$

or

$$\frac{dy}{dx} = \lim_{\Delta x \to 0} \frac{\Delta y}{\Delta x}$$

We next define the partial derivative of a function w by assuming it is a function of two variables x and y; thus,

$$w = f(x,y)$$

If y is kept constant, the derivative of w with respect to x is called the partial derivative of w with respect to x and is denoted by

$$\begin{aligned}\frac{\partial w}{\partial x} &= \text{partial derivative of } w \text{ with respect to } x\ (y \text{ constant})\\ &= \lim_{\Delta x \to 0} \frac{f(x + \Delta x,\, y) - f(x,y)}{\Delta x}\end{aligned}$$

Similarly,

$$\begin{aligned}\frac{\partial w}{\partial y} &= \text{partial derivative of } w \text{ with respect to } y\ (x \text{ constant})\\ &= \lim_{\Delta y \to 0} \frac{f(x,y + \Delta y) - f(x,y)}{\Delta y}\end{aligned}$$

and the same kind of notation would be used for partial derivatives of functions of three or more variables.

Having defined the ordinary and partial derivatives, we now define a differential equation as any equation that contains one or more derivatives. Differential equations are thus divided into two main classes, ordinary differential equations and partial differential equations; and the order of the differential equation is the order of its highest derivative.

The solution for a single dependent variable will generally be found from the solution of a single differential equation. If there are two or more dependent variables, a set of simultaneous differential equations will have to be solved and the number of equations required will be equal to the number of dependent variables.

Ordinary differential equations whose dependent variable y and its derivatives are of the first degree only, are said to be linear. For example,

$$a\frac{d^2y}{dx^2} + b\frac{dy}{dx} + cy = d$$

and

$$L\frac{d^2q}{dt^2} + R\frac{dq}{dt} + \frac{1}{C}q = V \sin(\omega t + \phi)$$

are linear equations of the second order. Since the superposition principle holds for linear equations, an analytic solution may be found that consists of two parts: a complementary part that contains two arbitrary constants and a particular integral part that depends on the forcing function d or $V \sin(\omega t + \phi)$.

Ordinary differential equations whose coefficients are not constants but depend on the independent variable are also linear equations that obey the superposition principle. For example,

$$\frac{d^2q}{dt^2} + (1 + a \sin \omega t)q = 0$$

is a second order linear equation with nonconstant coefficient $(1 + a \sin \omega t)$, which has the form of the Mathieu equation. This equation will describe an LC circuit in which the capacitance varies sinusoidally with time. Another equation with nonconstant coefficients is the LeGendre's equation

$$(x^2 - 1)\frac{d^2y}{dx^2} + 2x\frac{dy}{dx} - n(n + 1)y = 0$$

This equation often results in boundary-value problems involving spherical coordinates. It should be noted that the LeGendre's equation is linear in the dependent variable y, but it contains coefficients that are a function of the independent variable x. A solution of this equation is called a LeGendre function or a spherical function.

Ordinary differential equations whose dependent variable y and its derivatives are not of the first degree, are said to be nonlinear. For example,

$$\frac{di}{dt} + 0.666i^{1/2} = 0$$

is an example of a first order nonlinear equation that results from a circuit that contains a linear capacitance and a nonlinear resistance. Although this particular nonlinear equation has an explicit analytic solution, the reader will find that this is true for only a few very special cases and a computer solution is almost always required in solving nonlinear equations of second order or higher.

Partial differential equations may be similarly classified. For example, one of the most common forms of a partial differential equation encountered by electrical engineers is the second-order equation that has the following form

$$A\frac{\partial^2 w}{\partial x^2} + 2B\frac{\partial^2 w}{\partial x \partial y} + C\frac{\partial^2 w}{\partial y^2} + D\frac{\partial w}{\partial x} + E\frac{\partial w}{\partial y} + Fw = G$$

When the coefficients A, B, C, D, E, F, and G are a function of x and y only, the equations are linear, The various forms that this particular type of equation may take are called elliptic, parabolic, and hyperbolic, and the form depends respectively on whether $B^2 - AC$ is less than, equal to, or greater than zero.

Since the numerical solution of differential equations and, in particular, partial differential equations is perhaps the most difficult computer task, we will restrict our discussion of this subject to an introduction, and we begin by first considering the solution of ordinary differential equations.

13.2 EULER'S METHOD

Euler's method is the simplest one-step method that can be employed in the solution of initial value problems where the initial values of the dependent variable and its first $n - 1$ derivatives are known.

In order to illustrate Euler's method of solution, let us consider the first-order ordinary differential equation

$$\frac{dy}{dx} = f(x,y)$$

or

$$y' = f(x,y)$$

The solution of this equation is generally regarded as the curve $y = y(x)$, whose slope equals the value of the original differential equation over the required range of x. Euler's method of finding this solution is based on computing the slope at the initial (or starting) point (x_0,y_0) and then moving in this direction to a new point (x_1,y_1), where $x_1 = x_0 + \Delta x$ and y_1 equals the value of y_0 plus Δx times the value of the initial slope. At this new point, the values of x_1 and y_1 are used to compute the slope from the original differential equation and we then move in the direction of this new slope for an increment Δx and thus determine the location of the new point (x_2,y_2). This process is repeated until the solution $y(x)$ is obtained over the required range of x.

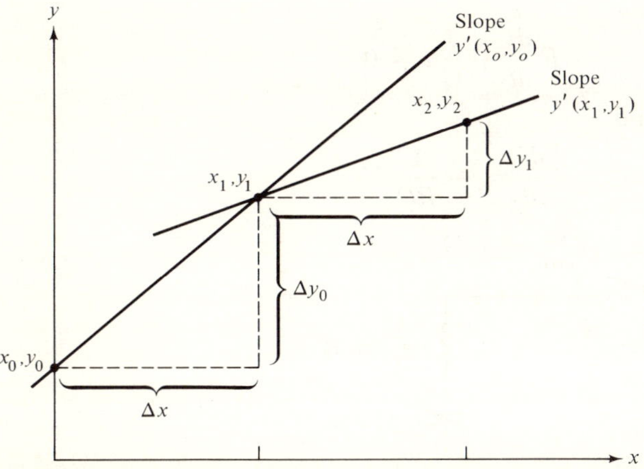

Figure 13.1 The graphical equivalent of Euler's method

Figure 13.1 shows the graphical equivalent of this procedure. Initially we may write

$$dy_0 = \left.\frac{dy}{dx}\right|_{x=0} dx = y'(x_0,y_0)\,dx$$

so that on an increment basis we have

$$\Delta y_0 = y'(x_0,y_0)\,\Delta x$$

Thus

$$x_1 = x_0 + \Delta x$$

and

$$\begin{aligned} y_1 &= y_0 + y'(x_0,y_0)\,\Delta x \\ &= y_0 + \Delta y_0 \end{aligned}$$

Similarly,

$$\begin{aligned} x_2 &= x_0 + 2\Delta x \\ &= x_1 + \Delta x \end{aligned}$$

and

$$\begin{aligned} y_2 &= y_1 + y'(x_1,y_1)\,\Delta x \\ &= y_1 + \Delta y_1 \end{aligned}$$

and so on.

For a simple electrical network that illustrates the use of Euler's method in solving a first-order differential equation, consider the circuit shown in Figure 13.2. Writing Kirchhoff's voltage law for $t > 0$, we find

$$Ri + \frac{1}{C}\int_{t=0} i\,dt = V$$

Since this is an integral equation, rather than a differential equation, we first differentiate obtaining

$$R\frac{di}{dt} + \frac{1}{C}i = 0$$

Solving for di/dt we find

$$\frac{di}{dt} = -\frac{1}{RC}i$$

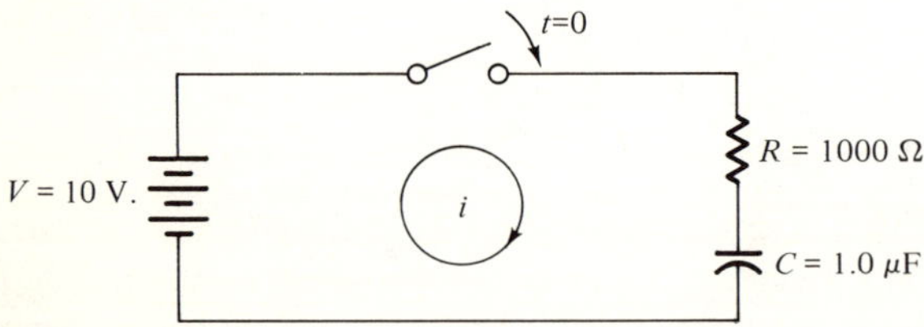

Figure 13.2 A simple RC circuit

```
                C       EULER'S METHOD
                C       SOLUTION OF V-R-C CIRCUIT
                C
S.0001                  READ(5,1)V,R,C
S.0002                1 FORMAT(3E15.7)
S.0003                  WRITE(6,2)V,R,C
S.0004                2 FORMAT('1THE SOLUTION OF THE V-R-C CIRCUIT USING EULER''S METHOD',
                       1//,' V =',E15.7,'   R =',E15.7,'   C =',E15.7,///,'  COMPUTED I',
                       28X,'EXACT I',11X,'TIME',//)
S.0005                  T=0.0
S.0006                  CUR=V/R
S.0007                  DELTAT=1.0E-6
S.0008                  DO 10 I=1,51
S.0009                  CUREXT=V/R*EXP(-T/(R*C))
S.0010                  WRITE(6,3)CUR,CUREXT,T
S.0011                3 FORMAT(3E15.7)
S.0012                  DO 10 J=1,100
S.0013                  T=T+DELTAT
S.0014                  DELCUR=-1./(R*C)*CUR*DELTAT
S.0015               10 CUR=CUR+DELCUR
S.0016                  STOP
S.0017                  END
```

Figure 13.3 A Fortran program that uses Euler's method to compute the transient solution of the current in the circuit of Figure 13.2

We may now start from the initial values, $t_0 = 0$ and $i_0 = V/R$, and use the following Euler iterative relations to compute the current:

$$t_{n+1} = t_n + \Delta t$$

$$\Delta i_n = -\frac{1}{RC} i_n \Delta t$$

$$i_{n+1} = i_n + \Delta i_n$$

Figure 13.3 shows the corresponding Fortran program for finding the solution of the current in the circuit of Figure 13.2. The values of V, R, and C are first read into memory by computer statements S.0001 and S.0002. The heading and the numerical values of V, R, and C are then printed. Computer statements S.0005, S.0006, and S.0007 then specify the initial conditions, $t_0 = 0$ and $i_0 = V/R$, and the time increment DELTAT (Δt = 1.0E-6).

Note that a value of $\Delta t = 10^{-6}$ has been chosen to insure accuracy in the Euler solution. In general, the selection of Δt should be small compared to all natural time constants. Note that in this example the value of Δt has been chosen so that it is smaller than the time constant, $1/RC$, by a factor of 1000.

We now enter the outer DO 10 loop and use computer statement S.0009

$$\text{CUREXT} = \text{V/R*EXP(-T/(R*C))}$$

to compute the exact solution of the current so that we may compare this to the Euler's solution. The exact solution formula shown here has been found from the calculus solution of the original integral equation. We next print the numerical values of the computed current found from Euler's method, the exact current, and the corresponding time.

We next enter the inner DO 10 loop to compute the Euler value of current. Note that computer statements S.0013 through S.0015 are identical to the Euler iterative relations previously discussed. Note also that the solution values are not printed every Δt seconds. Since the value of Δt is equal to 10^{-6}, the two DO 10 loops are used to print the computed results every 100 computations (10^{-4} seconds). This conserves a great deal of paper, as well as eliminating a large number of intermediate irrelevent results, and still preserves the accuracy requirements that dictate a small Δt value.

Figure 13.4 shows the computer results. Note that the Euler results are

THE SOLUTION OF THE V-R-C CIRCUIT USING EULER'S METHOD

V = 0.1000000E 02 R = 0.1000000E 04 C = 0.9999994E-06

COMPUTED I	EXACT I	TIME
0.9999998E-02	0.9999998E-02	0.0
0.9047769E-02	0.9048376E-02	0.9999941E-04
0.8186188E-02	0.8187313E-02	0.1999987E-03
0.7406618E-02	0.7408280E-02	0.2999860E-03
0.6701279E-02	0.6703440E-02	0.3999635E-03
0.6063104E-02	0.6065659E-02	0.4999409E-03
0.5485702E-02	0.5488560E-02	0.5999184E-03
0.4963260E-02	0.4966363E-02	0.6998959E-03
0.4490569E-02	0.4493851E-02	0.7998734E-03
0.4062880E-02	0.4066296E-02	0.8998509E-03
0.3675993E-02	0.3679422E-02	0.9998283E-03
0.3325999E-02	0.3329357E-02	0.1099806E-02
0.3009328E-02	0.3012594E-02	0.1199783E-02
0.2722807E-02	0.2725970E-02	0.1299761E-02
0.2463564E-02	0.2466612E-02	0.1399738E-02
0.2229003E-02	0.2231933E-02	0.1499716E-02
0.2016775E-02	0.2019582E-02	0.1599693E-02
0.1824753E-02	0.1827434E-02	0.1699671E-02
0.1651012E-02	0.1653568E-02	0.1799648E-02
0.1493814E-02	0.1496244E-02	0.1899626E-02
0.1351582E-02	0.1353888E-02	0.1999603E-02
0.1222891E-02	0.1225076E-02	0.2099581E-02
0.1106453E-02	0.1108520E-02	0.2199558E-02
0.1001101E-02	0.1003053E-02	0.2299536E-02
0.9057790E-03	0.9076199E-03	0.2399513E-02
0.8195327E-03	0.8212670E-03	0.2499491E-02
0.7414969E-03	0.7431295E-03	0.2599468E-02
0.6708889E-03	0.6724265E-03	0.2699445E-02
0.6070049E-03	0.6084507E-03	0.2799423E-02
0.5492035E-03	0.5505607E-03	0.2899400E-02
0.4969058E-03	0.4981791E-03	0.2999378E-02
0.4495881E-03	0.4507813E-03	0.3099355E-02
0.4067740E-03	0.4078930E-03	0.3199333E-02
0.3680363E-03	0.3690850E-03	0.3299310E-02
0.3329865E-03	0.3339695E-03	0.3399288E-02
0.3012733E-03	0.3021949E-03	0.3499265E-02
0.2725802E-03	0.2734433E-03	0.3599243E-02
0.2466186E-03	0.2474275E-03	0.3699220E-02
0.2231374E-03	0.2238867E-03	0.3799198E-02
0.2018923E-03	0.2025856E-03	0.3899175E-02
0.1826699E-03	0.1833353E-03	0.3999021E-02
0.1652777E-03	0.1659155E-03	0.4098859E-02
0.1495414E-03	0.1501510E-03	0.4198696E-02
0.1353033E-03	0.1358842E-03	0.4298534E-02
0.1224208E-03	0.1229731E-03	0.4398372E-02
0.1107647E-03	0.1112887E-03	0.4498210E-02
0.1002183E-03	0.1007144E-03	0.4598048E-02
0.9067613E-04	0.9114502E-04	0.4697885E-02
0.8204249E-04	0.8248475E-04	0.4797723E-02
0.7423077E-04	0.7464743E-04	0.4897561E-02
0.6716281E-04	0.6755473E-04	0.4997399E-02

Figure 13.4 The output results of the program of Figure 13.3 for a value of DELTAT $= 10^{-6}$

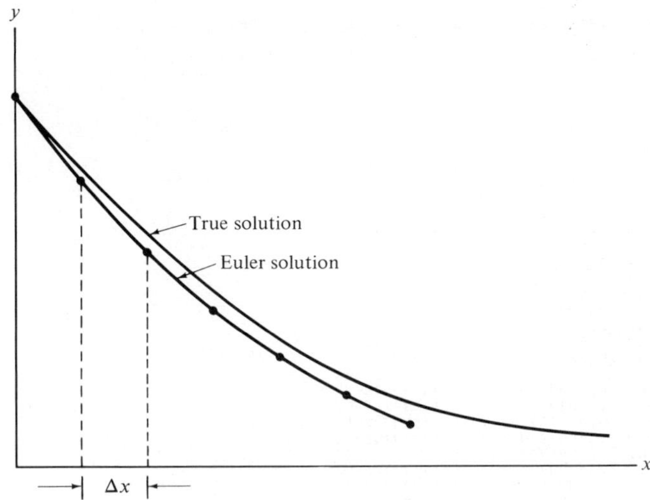

Figure 13.5 A graphical representation that shows increasing successive errors for monotonic solutions

accurate to three or four significant figures and that the accuracy in general, decreases with increasing time. This can be understood by realizing that when the true solution is concave upward, as it is in this example, the computed solution must always be less than the exact solution because it uses the slope of point n to obtain the $n+1$ point value. The deviations will therefore become successively larger for monotonic solutions as shown in Figure 13.5.

In order to show the importance of the choice of the value of Δt on the accuracy of the solution, the program of Figure 13.3 was modified so that

```
C        EULER'S METHOD
C        SOLUTION OF V-R-C CIRCUIT
C
         READ(5,1) V,R,C
       1 FORMAT(3E15.7)
         WRITE(6,2) V,R,C
       2 FORMAT('1THE SOLUTION OF THE V-R-C CIRCUIT USING EULER''S METHOD',
        1//,' V =',E15.7,'   R =',E15.7,'   C =',E15.7,///,'  COMPUTED I',
        28X,'EXACT I',11X,'TIME',//)
         T=0.0
         CUR=V/R
         DELTAT=1.0E-5
         DO 10 I=1,51
         CUREXT=V/R*EXP(-T/(R*C))
         WRITE(6,3)CUR,CUREXT,T
       3 FORMAT(3F15.7)
         DO 10 J=1,10
         T=T+DELTAT
         DELCUR=-1./(R*C)*CUR*DELTAT
      10 CUR=CUR+DELCUR
         STOP
         END
```

Figure 13.6 The program of Figure 13.3 is modified so that the effects of a larger value of DELTAT may be observed

THE SOLUTION OF THE V-R-C CIRCUIT USING EULER'S METHOD

V = 0.1000000E 02   R = 0.1000000E 04   C = 0.9999994E-06

COMPUTED I	EXACT I	TIME
0.9999998E-02	0.9999998E-02	0.0
0.9043805E-02	0.9048369E-02	0.9999990E-04
0.8179039E-02	0.8187305E-02	0.1999998E-03
0.7396959E-02	0.7408187E-02	0.2999988E-03
0.6689660E-02	0.6703213E-02	0.3999972E-03
0.6049991E-02	0.6065328E-02	0.4999957E-03
0.5471490E-02	0.5488142E-02	0.5999941E-03
0.4948299E-02	0.4965886E-02	0.6999925E-03
0.4475139E-02	0.4493326E-02	0.7999910E-03
0.4047222E-02	0.4065733E-02	0.8999894E-03
0.3660230E-02	0.3678835E-02	0.9999878E-03
0.3310245E-02	0.3328753E-02	0.1099986E-02
0.2993725E-02	0.3011985E-02	0.1199985E-02
0.2707471E-02	0.2725361E-02	0.1299983E-02
0.2448587E-02	0.2466012E-02	0.1399982E-02
0.2214457E-02	0.2231343E-02	0.1499980E-02
0.2002714E-02	0.2019006E-02	0.1599978E-02
0.1811218E-02	0.1826875E-02	0.1699977E-02
0.1638032E-02	0.1653028E-02	0.1799975E-02
0.1481406E-02	0.1495723E-02	0.1899974E-02
0.1339756E-02	0.1353389E-02	0.1999972E-02
0.1211651E-02	0.1224599E-02	0.2099971E-02
0.1095794E-02	0.1108064E-02	0.2199969E-02
0.9910157E-03	0.1002620E-02	0.2299967E-02
0.8962560E-03	0.9072092E-03	0.2399966E-02
0.8105570E-03	0.8208780E-03	0.2499964E-02
0.7330521E-03	0.7427623E-03	0.2599963E-02
0.6629582E-03	0.6720803E-03	0.2699961E-02
0.5995664E-03	0.6081241E-03	0.2799960E-02
0.5422358E-03	0.5502542E-03	0.2899958E-02
0.4903874E-03	0.4978913E-03	0.2999956E-02
0.4434967E-03	0.4505115E-03	0.3099955E-02
0.4010897E-03	0.4076401E-03	0.3199953E-02
0.3627376E-03	0.3688487E-03	0.3299952E-02
0.3280526E-03	0.3337485E-03	0.3399950E-02
0.2966835E-03	0.3019886E-03	0.3499949E-02
0.2683147E-03	0.2732507E-03	0.3599947E-02
0.2426583E-03	0.2472480E-03	0.3699946E-02
0.2194558E-03	0.2237196E-03	0.3799944E-02
0.1984718E-03	0.2024301E-03	0.3899942E-02
0.1794943E-03	0.1831688E-03	0.3999930E-02
0.1623313E-03	0.1657401E-03	0.4099917E-02
0.1468095E-03	0.1499698E-03	0.4199903E-02
0.1327718E-03	0.1357000E-03	0.4299890E-02
0.1200764E-03	0.1227881E-03	0.4399877E-02
0.1085948E-03	0.1111048E-03	0.4499864E-02
0.9821117E-04	0.1005330E-03	0.4599851E-02
0.8882034E-04	0.9096724E-04	0.4699837E-02
0.8032746E-04	0.8231164E-04	0.4799824E-02
0.7264664E-04	0.7447963E-04	0.4899811E-02
0.6570027E-04	0.6739286E-04	0.4999798E-02

Figure 13.7 The output results of the program of Figure 13.6

a value of $\Delta t = 10^{-5}$ was used instead of 10^{-6}. Figure 13.6 shows the modified program. Note that in this program the inner DO 10 loop was also modified so that after each ten iterations (rather than 100) the computed values would be printed. This modification allows the computer results shown in Figure 13.7 to be compared on a one-to-one basis with the corresponding results of Figure 13.4. Observe the significant difference in the accuracy of these two solutions. The reader is encouraged to rerun this program with larger and smaller values of Δt so that he may see that too small, as well as too large, a value of DELTAT will produce inaccurate results.

The Solution of Nonlinear Differential Equations

Euler's method may also be used to solve nonlinear differential equations. For example, consider the nonlinear RC circuit shown in Figure 13.8. In this circuit a nonlinear resistor, a diode, is connected in series with a 10-farad capacitor, which has been previously charged to +100 volts. The voltage-current characteristics of the diode may be approximated by

$$v_R = 0.1i^{3/2}$$

and thus for $t > 0$, the application of Kirchhoff's voltage law yields the following nonlinear integral equation:

$$\frac{1}{10}\int_0^t i\,dt - 100 + 0.1i^{3/2} = 0$$

Differentiating this equation, we obtain

$$\frac{1}{10}i + \frac{3}{20}i^{1/2}\frac{di}{dt} = 0$$

and solving for di/dt yields

$$\frac{di}{dt} = -\frac{2}{3}i^{1/2}$$

which is a first-order ordinary nonlinear differential equation.

Since there will be +100 volts across the diode at $t = 0+$, the initial conditions are $t_0 = 0$ and $i_0 = +100$. Starting with these initial conditions, we may use the following Euler iterative relations to compute the current:

$$\Delta i_n = -\frac{2}{3}i_n^{1/2}\,\Delta t$$

$$i_{n+1} = i_n + \Delta i_n$$
$$t_{n+1} = t_n + \Delta t$$

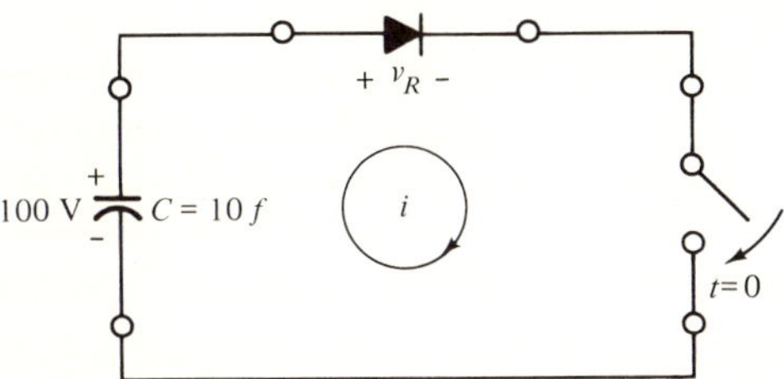

Figure 13.8 A nonlinear RC circuit

```
               C      THE EULER METHOD IS USED TO SOLVE A
               C      NON-LINEAR R-C CIRCUIT
               C
S.0001                WRITE(6, 1)
S.0002              1 FORMAT('1THE EULER SOLUTION OF A NON-LINEAR R-C CIRCUIT',///,7X,'T
                     1IME',8X,'EULERS I',7X,'EXACT I',8X,'ERROR',//)
S.0003                CE=100.
S.0004                T=0.0
S.0005                DELTAT=0.01
S.0006                N=0
S.0007              2 IF(N-N/100*100)4,10,4
S.0008             10 CEXACT=(10.-T/3.)**2
S.0009                ERRORE=CEXACT-CE
S.0010                WRITE(6, 3)T,CE,CEXACT,ERRORE
S.0011              3 FORMAT(4E15.7)
S.0012                IF(CEXACT)5,5,4
S.0013              4 DIDTE=-2./3.*CE**0.5*DELTAT
S.0014                CE=CE+DIDTE
S.0015                T=T+DELTAT
S.0016                N=N+1
S.0017                GO TO 2
S.0018              5 CONTINUE
S.0019                STOP
S.0020                END
```

Figure 13.9 Euler's method is used to solve a nonlinear differential equation

Figure 13.9 is a Fortran program for finding the current. Initially, the heading and subheadings are printed. The initial conditions, $i_0 = 100$ and $t_0 = 0$, and the time increment Δt (0.01) are then specified. The dummy count variable N is then set equal to zero. Computer statement S.0007 next tests to see whether

$$N - N/100*100$$

is equal to zero. Since this expression will be equal to zero during the first iteration (also the 101, 201, and so on), control will be transferred to statement 10 and the value of the exact current will be computed for this value of t ($t = 0$ initially) The exact solution shown here has been found from the calculus solution of the original differential equation. Using the separation of variables technique, we may write

$$i^{-1/2}\, di = -\frac{2}{3}\, dt;\ i \geq 0$$

Integrating both sides of this equation yields

$$\int i^{-1/2}\, di = -\frac{2}{3} \int dt$$

or

$$2i^{1/2} = -\frac{2}{3} t + \text{Const.}$$

Since $i^{1/2}(0) = 10$, we find

$$i^{1/2} = -\frac{1}{3} t + 10$$

or

$$i = \left(-\frac{1}{3}t + 10\right)^2;\ 0 \leq t \leq 30 \text{ seconds}$$

The reader should realize that a direct analytic solution is possible here only because of the simplicity of the nonlinear differential equation.

Returning to the computer program, we next compute the error from

$$\text{ERRORE} = \text{CEXACT} - \text{CE}$$

and then print the numerical values of T, CE, CEXACT, and ERRORE. We then test to see whether CEXACT equals zero. If CEXACT is equal to or less than zero, the solution is obtained and we transfer to statement 5 to terminate the process. If CEXACT is greater than zero, control is transferred to statement 4 and we use the three Euler relations to compute the next step of the solution. We then make $N = N + 1$ and return control to statement 2 to compute the next iteration. Note that the statement

$$\text{IF(N} - \text{N/100*100) 4,10,4}$$

causes the current to be computed every Δt (0.01) seconds, but the results to be printed every 1.0 seconds ($100*\Delta t$).

Figure 13.10 shows the corresponding results. We may observe that these results are initially accurate to almost five significant figures, but

```
THE EULER SOLUTION OF A NON-LINEAR R-C CIRCUIT
```

TIME	EULERS I	EXACT I	ERROR
0.0	0.1000000E 03	0.1000000E 03	0.0
0.9999990E 00	0.9344269E 02	0.9344444E 02	0.1754761E-02
0.1999927E 01	0.8710767E 02	0.8711156E 02	0.3890991E-02
0.2999854E 01	0.8099495E 02	0.8100087E 02	0.5920410E-02
0.3999782E 01	0.7510451E 02	0.7511237E 02	0.7858276E-02
0.4999709E 01	0.6943636E 02	0.6944606E 02	0.9704590E-02
0.5999637E 01	0.6399049E 02	0.6400194E 02	0.1144409E-01
0.6999564E 01	0.5876694E 02	0.5878000E 02	0.1306152E-01
0.7999492E 01	0.5376567E 02	0.5378026E 02	0.1458740E-01
0.8999419E 01	0.4898671E 02	0.4900270E 02	0.1599121E-01
0.9999347E 01	0.4443005E 02	0.4444734E 02	0.1728821E-01
0.1099927E 02	0.4009572E 02	0.4011417E 02	0.1844788E-01
0.1199920E 02	0.3598367E 02	0.3600319E 02	0.1951599E-01
0.1299913E 02	0.3209398E 02	0.3211440E 02	0.2041626E-01
0.1399906E 02	0.2842661E 02	0.2844778E 02	0.2117920E-01
0.1499898E 02	0.2498154E 02	0.2500337E 02	0.2183533E-01
0.1599891E 02	0.2175882E 02	0.2178116E 02	0.2233887E-01
0.1699837E 02	0.1875845E 02	0.1878249E 02	0.2403259E-01
0.1799782E 02	0.1598047E 02	0.1600581E 02	0.2534485E-01
0.1899727E 02	0.1342544E 02	0.1345113E 02	0.2568722E-01
0.1999672E 02	0.1109274E 02	0.1111840E 02	0.2566624E-01
0.2099617E 02	0.8982370E 01	0.9007663E 01	0.2529240E-01
0.2199562E 02	0.7094357E 01	0.7118901E 01	0.2454472E-01
0.2299507E 02	0.5428708E 01	0.5452113E 01	0.2340508E-01
0.2399452E 02	0.3985447E 01	0.4007308E 01	0.2186108E-01
0.2499397E 02	0.2764602E 01	0.2784480E 01	0.1987839E-01
0.2599342E 02	0.1766211E 01	0.1783628E 01	0.1741695E-01
0.2699287E 02	0.9903371E 00	0.1004756E 01	0.1441890E-01
0.2799232E 02	0.4371066E 00	0.4478630E 00	0.1075637E-01
0.2899178E 02	0.1066868E 00	0.1129463E 00	0.6259441E-02

```
IHC253I ALOG-ALOG10 ARG=-0.6649991E-05 LE ZERO
```

Figure 13.10 The output results of the program of Figure 13.9

this accuracy reduces to approximately two significant figures in the vicinity of $t = 27$ seconds.

At this point the reader should recognize that the Euler formula for finding the next point

$$y_1 = y_0 + y'(x_0,y_0)\,\Delta x$$

or

$$y_{n+1} = y_n + y'_n\,\Delta x$$

is equivalent to using the first two terms of the Taylor's expansion for $y(x_0 + \Delta x)$

$$y(x_0 + \Delta x) = y(x_0) + \Delta x\,y'(x_0) + \frac{\Delta x^2}{2!}\,y''(x_0) + \cdots + \frac{\Delta x^n}{n!}\,y^n(x_0) + \frac{\Delta x^{n+1}}{(n+1)!}\,y^{n+1}(\Sigma);\ x_i < \Sigma < x_i + \Delta x$$

It is therefore possible to estimate the truncation error in using a first degree ($n = 1$) Taylor-series approximation; thus,

$$\text{TE}_1 = \frac{\Delta x^2}{2!}\,y''(\Sigma);\ x_i < \Sigma < x_i + \Delta x$$

Since this truncation error is proportional to Δx^2, it follows that a reduction in the step size Δx will reduce the truncation error. This, however, will require more mathematical operations and hence a possible increase in the round-off error. In fact, the reader should realize that both truncation and round-off errors occur at each step and these errors are propagated from one step to the next. From a practical standpoint, then, it is usually good practice to obtain a solution with a relatively coarse step size and then to repeat the process, using half the previous step size. If there is considerable difference between the two results, the process should be repeated again, this time using half the previous step size, and so on. Whenever two successive solutions agree to n significant digits, the procedure may be terminated because it will usually be found that the last solution is accurate to better than n significant figures.

The Solution of Higher-Order Equations

Euler's method may be extended to obtain the solution of higher-order differential equations. For example, consider the RLC circuit shown in Figure 13.11 where the capacitor has been previously charged to +2 coulombs. Writing Kirchhoff's voltage law for $t = 0+$, we obtain

$$\frac{1}{C}\int_{t=0} i\,dt - \frac{Q(0)}{C} + Ri + L\frac{di}{dt} = 0$$

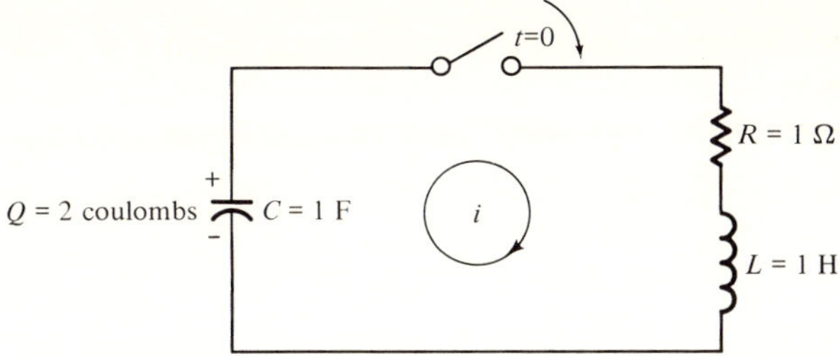

Figure 13.11 An RLC circuit

or

$$\frac{q}{C} - \frac{Q(0)}{C} + R\frac{dq}{dt} + L\frac{d^2q}{dt^2} = 0$$

If we are interested in the solution for the current rather than the charge, we may alternately differentiate the original intergrodifferential equation, obtaining

$$\frac{1}{C}i + R\frac{di}{dt} + L\frac{d^2i}{dt^2} = 0$$

Solving for d^2i/dt^2, we find

$$\frac{d^2i}{dt^2} = -\frac{R}{L}\frac{di}{dt} - \frac{1}{LC}i$$

We may now obtain an Euler solution for the current by starting from the initial conditions, $i_0 = 0$ and $\left.\frac{di}{dt}\right|_0 = \frac{Q(0)}{LC}$, obtaining successive values of each of the variables from the following interative relations:

$$\left.\frac{d^2i}{dt^2}\right|_n = -\frac{R}{L}\left.\frac{di}{dt}\right|_n - \frac{1}{LC}i_n$$

$$t_{n+1} = t_n + \Delta t$$

$$i_{n+1} = i_n + \left.\frac{di}{dt}\right|_n \Delta t$$

$$\left.\frac{di}{dt}\right|_{n+1} = \left.\frac{di}{dt}\right|_n + \left.\frac{d^2i}{dt^2}\right|_n \Delta t$$

Figure 13.12 is the corresponding program for computing the current. First we read in and then we print the numerical values of Q, R, L (called AL in the program because of mode), and C. The initial conditions $t_0 = 0$, $i_0 = 0$, and $\left.\frac{di}{dt}\right|_0$ are then specified along with the value of Δt (DELTAT = 0.001 seconds). We next enter the outer DO 10 loop which

```
C         EULER'S METHOD
C         SOLUTION OF R-L-C CIRCUIT
C
          READ(5,1)Q,R,AL,C
        1 FORMAT(4F1C.6)
          WRITE(6,2)Q,R,AL,C
        2 FORMAT('1THE SOLUTION CF THE R-L-C CIRCUIT USING EULER''S METHOD',
         1//,' Q =',F6.2,3X,'R =',F6.2,3X,'L =',F6.2,3X,'C =',F6.2,///,7X,'T
         2IME',9X,'CLRRENT',9X,'DIDT',//)
          T=C.C
          CUR=C.C
          DIDT=Q/(AL*C)
          DELTAT=C.001
          DO 1C I=1,50
          WRITE(6,3)T,CUR,DIDT
        3 FORMAT(3E15.7)
          DO 1C J=1,10C
          D2IDT2=-R/AL*DIDT-CUR/(AL*C)
          T=T+DELTAT
          CUR=CUR+DIDT*DELTAT
       1C DIDT=DIDT+D2IDT2*DELTAT
          STOP
          END
```

Figure 13.12 The Euler solution of a second-order differential equation

```
THE SOLUTION OF THE R-L-C CIRCUIT USING EULER'S METHOD

Q =  2.00    R =  1.CC    L =  1.CO    C =  1.00


      TIME            CLRRENT            DIDT


0.0              C.0              0.20CCCC0E 01
0.9999937E-C1    C.19C1C34E C0    0.1800275E 01
0.1999981E 0C    C.36C3110E C0    C.1602424E C1
0.2999968E CC    C.51C8972E C0    0.1408145E 01
0.3999955E CC    C.6422953E C0    0.1218950E C1
0.4999942E CC    C.755C8C8E CC    0.1036158E 01
0.5999929E CC    C.8499522E C0    0.8609228E 00
0.6999916E CC    C.9277130E C0    C.6941727E 00
0.7999904E CC    C.9892513E C0    0.5366799E 00
0.8999891E CC    C.1035488E C1    0.3890634E 00
0.9999878E CC    C.1067463E 01    0.2517910E 00
0.1099933E C1    C.1C86237E 01    0.1251847E C0
0.1199878E C1    C.1C92889E C1    0.9435970E-C2
0.1299823E C1    C.1088511E C1   -C.9538627E-01
0.1399768E C1    C.1074193E C1   -0.1893249E 00
0.1499713E C1    C.1C51C12E C1   -0.2725235E 00
0.1599658E C1    C.102CC35E C1   -C.3452120E C0
0.1699603E C1    C.9823118E C0   -0.4C76989E C0
0.1799548E C1    C.9388524E C0   -C.4603628E 00
0.1899493E 01    C.89C5952E C0   -0.5036377E C0
0.1999438E C1    C.8384558E CC   -0.5380C50E C0
0.2099383E C1    C.7832990E CC   -0.5639848E 00
0.2199328E C1    C.7259368E CC   -C.5821279E CC
0.2299273E C1    C.6671249E C0   -0.5930C71E 00
0.2399219E C1    C.60756C6E C0   -C.5972099E 00
0.2499164E C1    C.547882CE C0   -0.5953313E 00
0.2599109E C1    C.4886674E C0   -0.5879679E C0
0.2699054E C1    C.43C4362E CC   -0.5757111E 00
0.2798999E C1    C.3736490E CC   -0.5591429E C0
0.2898944E C1    C.3187C86E CC   -0.5388310E C0
0.2998889E C1    C.2659618E C0   -0.5153249E CC
0.3098834E C1    C.2157C22E C0   -0.4891515E C0
0.3198779E C1    C.1681712E CC   -C.4608141E 00
0.3298724E C1    C.1235611E C0   -C.4307881E CC
0.3398669E C1    C.82C1885E-C1   -0.3995196E 00
0.3498614E C1    C.4364890E-C1   -0.3674244E 00
0.3598559E C1    C.8514769E-C2   -0.3348866E 00
0.3698504E C1   -C.23358C3E-01   -C.3022574E C0
0.379845CE C1   -C.5197661E-C1   -0.2698565E 00
0.3898395E C1   -C.773766CE-C1   -0.2379702E C0
0.399834CE C1   -C.99623C2E-01   -0.2068533E 00
0.4098285E C1   -C.1188C46E C0   -0.1767287E 00
0.419823CE C1   -C.135C315E C0   -C.1477883E C0
0.4298175E C1   -C.1484298E CC   -0.1201950E 00
0.439812CE C1   -C.1591411E C0   -C.9408325E-C1
0.4498C65E C1   -C.1673188E C0   -0.6956C59E-C1
0.4598C1CE C1   -C.1731268E C0   -C.4671047E-C1
0.4697955E C1   -C.1767349E C0   -C.2559042E-01
0.47979CCE C1   -C.1783184E C0   -C.6236073E-02
0.4897845E C1   -C.178C552E CC    0.1133624E-01
```

Figure 13.13 The output results of the program of Figure 13.12

prints the numerical values of T, CUR, and DIDT every 0.1 seconds (100*DELTAT). We then enter the inner DO 10 loop to compute the Euler value of current. Note that computer statements S.0013 through S.0016 are the four Euler relations we discussed above. Note also that we have again used two DO 10 loops so that the computed values will be printed only every 100 computations (every 0.1 seconds).

Figure 13.13 gives the computer results. The reader may verify the accuracy of these results by comparing them to the calculus solution that was graphically printed by the program of Figure 11.6 (see Figure 11.7a).

At this point, the reader should recognize that it is possible to extend the above and use Euler's method to solve a general m-th order differential equation

$$a_m \frac{d^m i}{dt^m} + a_{m-1} \frac{d^{m-1} i}{dt^{m-1}} + \cdots + a_1 \frac{di}{dt} + a_0 i + C = 0$$

The general iterative procedure would require the sequential evaluation of the following $m + 2$ relations:

$$\left.\frac{d^m i}{dt^m}\right|_n = -\left(\frac{a_{m-1}}{a_m} \left.\frac{d^{m-1} i}{dt^{m-1}}\right|_n + \frac{a_{m-2}}{a_m} \left.\frac{d^{m-2} i}{dt^{m-2}}\right|_n + \cdots + \frac{a_0}{a_m} i_n + \frac{C}{a_m}\right)$$

$$t_{n+1} = t_n + \Delta t$$

$$i_{n+1} = i_n + \left.\frac{di}{dt}\right|_n \Delta t$$

$$\left.\frac{di}{dt}\right|_{n+1} = \left.\frac{di}{dt}\right|_n + \left.\frac{d^2 i}{dt^2}\right|_n \Delta t$$

$$\left.\frac{d^2 i}{dt^2}\right|_{n+1} = \left.\frac{d^2 i}{dt^2}\right|_n + \left.\frac{d^3 i}{dt^3}\right|_n \Delta t$$

$$\vdots$$

$$\left.\frac{d^{m-1} i}{dt^{m-1}}\right|_{n+1} = \left.\frac{d^{m-1} i}{dt^{m-1}}\right|_n + \left.\frac{d^m i}{dt^m}\right|_n \Delta t$$

For an alternative method of solving higher-order equations, one should recall that Euler's method was based on using only the first two terms of the Taylor-series expansion. It is therefore possible to use additional terms, as the problem warrants, and thereby increase the accuracy of the solution by reducing the truncation error. For example, consider the previous RLC circuit where the calculation of the current and its derivative was based on using only two terms of the Taylor series. If we modify

```
C          EULER'S METHOD
C          SOLUTION OF R-L-C CIRCUIT
C          CONSTANT ACCELERATICN METHCD
C
           READ(5,1)Q,R,AL,C
         1 FORMAT(4F1C.6)
           WRITE(6,2)Q,R,AL,C
         2 FORMAT('1THE SCLUTICN CF THE R-L-C CIRCUIT USINC EULER''S METHOD',
          1//,' Q =',F6.2,3X,'R =',F6.2,3X,'L =',F6.2,3X,'C =',F6.2,///,6X,'T
          2IME',9X,'CURRENT',8X,'DIDT',//)
           T=C.C
           CUR=C.C
           DIDT=Q/(AL*C)
           DELTAT=C.CC1
           DO 1C I=1,5C
           WRITE(6,3)T,CUR,DIDT
         3 FORMAT(3E15.7)
           DO 1C J=1,1CC
           D2IDT2=-R/AL*DIDT-CUR/(AL*C)
           T=T+DELTAT
           CUR=CUR+DIDT*DELTAT+D2IDT2*DELTAT**2/2.
        1C DIDT=DIDT+D2IDT2*DELTAT
           STOP
           END
```

Figure 13.14 Euler's constant acceleration method is used to solve the RLC circuit

```
THE SOLUTION OF THE R-L-C CIRCUIT USING EULER'S METHCD

Q =  2.CC   R =  1.CC   L -  1.CC   C =  1.00


     TIME            CURRENT           DIDT


 0.0              C.0               0.2CC0000E 01
 0.9999937E-C1    C.19CC026E CO     0.18C0281E 01
 0.1999981E CC    C.36C11C5E CO     0.1602439E 01
 0.2999968E CC    C.5105981E CO     0.1408185E 01
 0.3999955E CC    C.6419C6CE CO     0.1219021E 01
 0.4999942E CC    C.7546C49E CO     C.1036259E 01
 0.5999929E CC    C.8493989E CO     0.8610626E 00
 0.6999916E CC    C.927C9C4E 00     C.6943556E 00
 0.7999904E CC    C.9885681E CO     0.5369077E 00
 0.8999891E CC    C.1034739E C1     0.3893375E 00
 0.9999878E CC    C.1C66651E C1     0.2521138E 00
 0.1099933E C1    C.1C85363E C1     0.1255578E 00
 0.1199878E C1    C.1C91956E 01     0.9859540E-02
 0.1299823E C1    C.1C87527E C1    -0.9491175E-01
 0.1399768E C1    C.1073155E C1    -C.1887997E CO
 0.1499713E C1    C.1C49934E 01    -0.2719476E 00
 0.1599658E C1    C.1C18921E C1    -0.3445865E 00
 0.1699603E C1    C.9811987E CO    -0.4070262E 00
 0.1799548E C1    C.9377789E CO    -C.4596503E 00
 0.1899493E C1    C.8895706E CO    -C.5028934E 00
 0.1999438E C1    C.8374871E CO    -0.5372365E 00
 0.2099383E C1    C.7823921E CC    -0.5632002E 00
 0.2199328E 01    C.725C972E CO    -C.5813348E 00
 0.2299273E 01    C.6663572E CO    -0.5922129E 00
 0.2399219E C1    C.6C68658E 00    -0.5964217E 00
 0.2499164E 01    C.5472651E CO    -0.5945558E 00
 0.2599109E C1    C.4881269E CO    -0.5872111E 00
 0.2699054E C1    C.4299741E CC    -0.5749782E 00
 0.2798999E C1    C.3732649E CO    -0.5584394E 00
 0.2898944E C1    C.3183995E CC    -0.5381613E 00
 0.2998889E C1    C.265726CE CO    -C.5146930E 00
 0.3098834E 01    C.2155397E CO    -0.4885610E 00
 0.3198779E C1    C.168C774E CC    -C.4602677E 00
 0.3298724E C1    C.1235319E CO    -C.4302878E 00
 0.3398669E C1    C.82C4937E-C1    -0.3990672E 00
 0.3498614E C1    C.4373644E-C1    -0.3670204E 00
 0.3598559E C1    C.8656263E-C2    -0.3345318E 00
 0.36985C4E C1   -C.2316719E-01    -0.3019524E 00
 0.379845CE 01   -C.5174133E-C1    -0.2696C10E 00
 0.3898395E 01   -C.771C123E-01    -0.2377637E 00
 0.399834CE C1   -C.9931147E-C1    -0.2066945E 00
 0.4098285E C1   -C.1184617E CC    -0.1766161E 00
 0.419823CE C1   -C.1346617E CO    -0.1477205E 00
 0.4298175E C1   -C.148C373E CO    -0.1201701E 00
 0.439812CE C1   -C.15873C1E CO    -0.9409904E-01
 0.4498065E C1   -C.1668965E CO    -0.6961447E-01
 0.4598010E C1   -C.1726995E CO    -0.4679968E-01
 0.4697955E C1   -C.1763C62E CO    -0.2571189E-01
 0.47979CCE C1   -C.1778911E OC    -0.6386742E-02
 0.4897845E C1   -C.1776315E CO     0.1115937E-01
```

Figure 13.15 The output results of the program of Figure 13.14

the current relation as follows:

$$i_{n+1} = i_n + \left.\frac{di}{dt}\right|_n \Delta t + \left.\frac{d^2i}{dt^2}\right|_n \frac{\Delta t^2}{2}$$

we will find that this additional term will increase the accuracy of our solution. When three terms of the Taylor series are used, the method is sometimes referred to as Euler's constant acceleration method.

Figure 13.14 is a modification of the program of Figure 13.12 and incorporates the use of a third term in the calculation of i_{n+1}. Figure 13.15 gives the corresponding results. The reader should observe the added accuracy that has resulted. The reader should also note that in the solution of a third-order differential equation, the first four terms of the Taylor series may be used in the calculation of i_{n+1} and the first three terms in the calculation of $\left.\frac{di}{dt}\right|_{n+1}$, and so on.

13.3 RUNGE-KUTTA METHODS

There are many varieties of Runge-Kutta methods. All these are one-step methods that do not require the evaluation of higher derivatives of the function, but they nevertheless produce solutions accurate to the same order of magnitude as those produced by using a corresponding number of terms in a Taylor-series expansion.

Second-Order Runge-Kutta Methods

The second-order Runge-Kutta methods are based on finding the value of the next step from the formula

$$y_{n+1} = y_n + y' \Delta x$$

where y' is a weighted slope. These second-order methods are equivalent to using the first three terms of the Taylor series (through Δx^2, hence second order) and thus have a local truncation error that is proportional to Δx^3.

One such second-order formula is based on the weighted average slope

$$y' = \frac{y'(x_n,y_n) + y'(x_n+\Delta x, y_n+y'(x_n,y_n)\,\Delta x)}{2}$$

The geometric interpretation of this weighted slope may be seen from Figure 13.16. The slope $y'(x_n,y_n)$ represents the slope at the starting point. Using this slope, we go one step forward and compute $y'(x_n+\Delta x,$

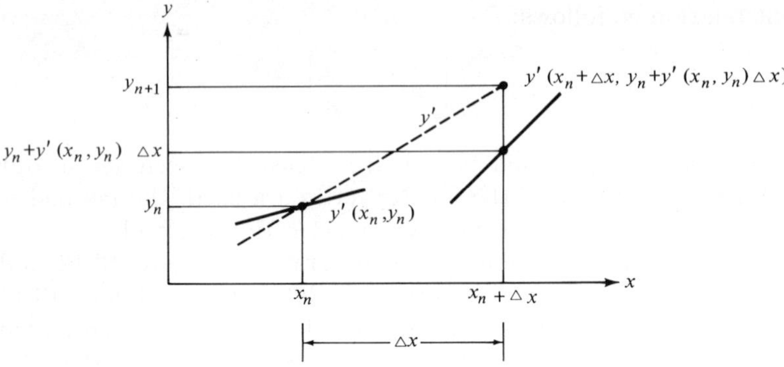

Figure 13.16 The use of weighted slopes

$y_n + y'(x_n, y_n)\ \Delta x)$ which is the slope at the point $x = x_n + \Delta x$, $y = y_n + y'(x_n, y_n)\ \Delta x$. These two slopes are then averaged to obtain the weighted slope. The reader should observe the difference in the computed value of y_{n+1} when the weighted slope is used in contrast with the Euler value of $y_n + y'(x_n, y_n)\ \Delta x$, which uses only the slope at (x_n, y_n). This particular second-order method is sometimes referred to as the improved Euler method.

For a simple numerical example that uses this method, consider the solution of the first-order differential equation

$$\frac{dy}{dx} = x + y$$

using a step size of $\Delta x = 0.1$ and initial conditions of $x_0 = 0$, $y_0 = 1.0$. The following table shows the numerical values obtained for the first four iterations:

ITERATION $(n+1)$	x_{n+1}	$y'(x_n, y_n)$	$y'(x_n + \Delta x, y_n + y'(x_n, y_n)\ \Delta x)$	y'	y_{n+1}
1	0.1	1.0	1.2	1.10	1.110
2	0.2	1.210	1.431	1.321	1.242
3	0.3	1.442	1.686	1.564	1.398
4	0.4	1.698	1.968	1.833	1.581

Another second-order Runge-Kutta formula, which is sometimes called the modified Euler method, is based on the slope

$$y' = y'\left(x_n + \frac{\Delta x}{2}, y_n + y'(x_n, y_n)\ \frac{\Delta x}{2}\right)$$

This is equivalent to starting from the point (x_n,y_n) and moving one-half a step forward on the slope $y'(x_n,y_n)$ and then computing

$$y'\left(x_n + \frac{\Delta x}{2}, y_n + y'(x_n,y_n)\frac{\Delta x}{2}\right)$$

the slope at the point

$$x = x_n + \frac{\Delta x}{2}, y = y_n + y'(x_n,y_n)\frac{\Delta x}{2}$$

This slope is then used to compute the next point y_{n+1}. The following table shows the corresponding results for the solution of the last example:

ITERATION $(n+1)$	x_{n+1}	$y'(x_n,y_n)$	y'	y_{n+1}
1	0.1	1.0	1.10	1.110
2	0.2	1.210	1.321	1.242
3	0.3	1.442	1.564	1.398
4	0.4	1.698	1.833	1.581

Third-Order Runge-Kutta Methods

Third-order Runge-Kutta methods are based on finding the value of the next step from the formula

$$y_{n+1} = y_n + y'\,\Delta x$$

where y' is again a weighted slope. These third-order formulas are equivalent to using the first four terms of the Taylor series (through Δx^3, hence, third order) and thus have a local truncation error that is proportional to Δx^4

One such third-order formula is based on using the weighted average slope

$$y' = \frac{\Delta x}{6}\left(y'(x_n,y_n) + 4y'\left(x_n + \frac{\Delta x}{2}, y_n + y'(x_n,y_n)\frac{\Delta x}{2}\right) + y'\left(x_n + \Delta x, y_n - y'(x_n,y_n)\,\Delta x + 2y'\left(x_n + \frac{\Delta x}{2}, y_n + y'(x_n,y_n)\frac{\Delta x}{2}\right)\Delta x\right)\right)$$

The geometric interpretation of this weighted slope may be seen from an examination of Figure 13.17. The slope $y'(x_n,y_n)$ represents the slope at the starting point which we will call

$$SK1 = y'(x_n,y_n)$$

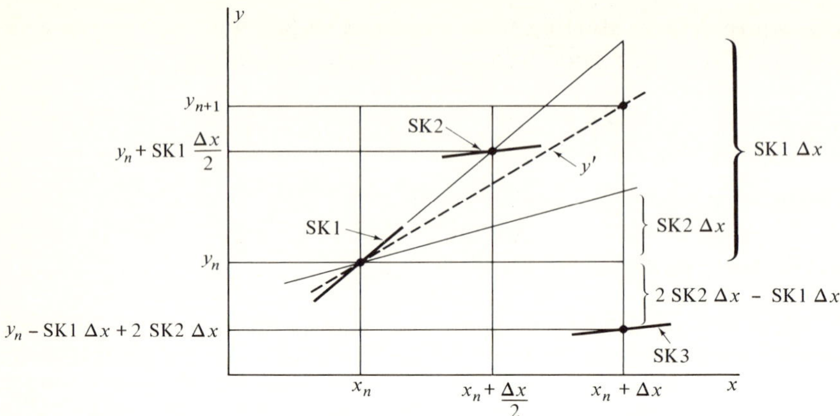

Figure 13.17 Weighted slopes of a third-order Runge-Kutta formula

(Note that this slope is arbitrarily assumed to be positive.) Using this slope, we go one-half step forward and compute the second slope

$$SK2 = y'\left(x_n + \frac{\Delta x}{2}, y_n + SK1 \frac{\Delta x}{2}\right)$$

(Note that this slope is arbitrarily assumed to be positive.) Using the first and second slopes, we now go one step forward and compute the third slope

$$SK3 = y'(x_n + \Delta x, y_n - SK1 \, \Delta x + 2 \, SK2 \, \Delta x)$$

(Note that this slope is arbitrarily assumed to be positive.) These three slopes are now averaged, using weights of (1/6), (4/6), and (1/6), respectively, to obtain y'

$$y' = \frac{1}{6}(SK1 + 4 \, SK2 + SK3)$$

and this slope is then used to compute the next point

$$y_{n+1} = y_n + y' \, \Delta x$$

Another third-order method is based on computing the slopes

$$SK1 = y'(x_n, y_n)$$

$$SK2 = y'\left(x_n + \frac{\Delta x}{3}, y_n + SK1 \frac{\Delta x}{3}\right)$$

$$SK3 = y'\left(x_n + \frac{2\Delta x}{3}, y_n + SK2 \frac{2\Delta x}{3}\right)$$

and then using these to obtain the weighted slope

$$y' = \frac{SK1 + 3\ SK3}{4}$$

This weighted slope is then used, in turn, to compute the next point

$$y_{n+1} = y_n + y'\ \Delta x$$

Fourth-Order Runge-Kutta Methods

Fourth-order Runge-Kutta formulas are equivalent to using the first five terms of the Taylor series (through Δx^4) and thus have a local truncation error that is proportional to Δx^5. One well known fourth-order formula is based on first computing the slopes $SK1$, $SK2$, $SK3$, and $SK4$. These four slopes are then averaged, using weights of 1/6, 2/6, 2/6, and 1/6 respectively; and the weighted slope is then used, in turn, to compute the next point

$$\begin{aligned} y_{n+1} &= y_n + y'\ \Delta x \\ &= y_n + \frac{(SK1 + 2\ SK2 + 2\ SK3 + SK4)}{6}\ \Delta x \end{aligned}$$

The geometric interpretation of this weighted slope may be seen from an examination of Figure 13.18. The slope

$$SK1 = y'(x_n, y_n)$$

represents the slope at the starting point (x_n, y_n) and is arbitrarily assumed to be positive. Using this slope, we go one-half step forward and compute

$$SK2 = y'\left(x_n + \frac{\Delta x}{2}, y_n + SK1\ \frac{\Delta x}{2}\right)$$

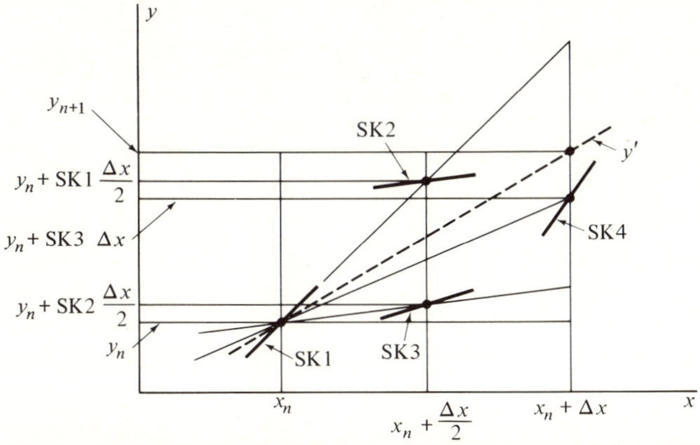

Figure 13.18 Weighted slopes of a fourth-order Runge-Kutta formula

Using this new slope, which is also arbitrarily assumed to be positive, we go one-half step forward from (x_n,y_n) and compute

$$SK3 = y'\left(x_n + \frac{\Delta x}{2}, y_n + SK2\,\frac{\Delta x}{2}\right)$$

Using this new slope, which is also assumed to be positive, we go one full step forward from (x_n,y_n) and compute

$$SK4 = y'(x_n + \Delta x,\, y_n + SK3\,\Delta x)$$

These four slopes are then weighted and averaged to obtain the weighted average slope, y', shown in Figure 13.18.

Figure 13.19 is a Fortran program that uses this four-order Runge-Kutta formula to determine the solution of the current in the nonlinear RC circuit of Figure 13.8. Computer statement S.0001 first defines the arithmetic statement function, DIDT(CRK). We then print the heading and subheadings. This is followed by specifying the initial conditions CE = 100.0 (the initial Euler current), CRK = 100.0 (the initial Runge-Kutta current), and T = 0.0. The value of DELTAT is then specified as 0.5 seconds. Note that this value of DELTAT is 50 times larger than the value used in the program of Figure 13.9. We next compute the exact value of current for this value of T and then we compute the errors in the solution for the current when the respective Euler and Runge-Kutta methods are used. We then print the numerical values of the time, the

```
             C      EULER VERSUS RUNGE-KUTTA SOLUTICN
             C                     OF A
             C      NON-LINEAR R-C CIRCUIT
             C
S.0001              DIDT(CRK)=-2./3.*CRK**0.5
S.0002              WRITE(6,1)
S.0003            1 FORMAT('1THE EULER AND RUNGE KUTTA SOLUTICN CF A NCN-LINEAR R-C CI
                   1RCUIT',///,7X,'TIME',9X,'EULERS I',3X,'RUNGE KUTTA I',4X,'EXACT I'
                   2,8X,'ERROR E',9X,'ERROR R-K',//)
S.0004              CE=100.
S.0005              CRK=100.
S.0006              T=0.0
S.0007              DELTAT=0.5
S.0008            2 CEXACT=(10.-T/3.)**2
S.0009              ERRORE=CEXACT-CE
S.0010              ERRORK=CEXACT-CRK
S.0011              WRITE(6,3)T,CE,CRK,CEXACT,ERRORE,ERRORK
S.0012            3 FORMAT(6E15.7)
S.0013              IF(CEXACT)5,5,4
S.0014            4 DIDTE=-2./3.*CE**0.5*DELTAT
S.0015              CE=CE+DIDTE
S.0016              SK1=DIDT(CRK)
S.0017              SK2=DIDT(CRK+SK1*DELTAT/2.)
S.0018              SK3=DIDT(CRK+SK2*DELTAT/2.)
S.0019              SK4=DIDT(CRK+SK3*DELTAT)
S.0020              CRK=CRK+DELTAT*(SK1+2.*SK2+2.*SK3+SK4)/6.
S.0021              T=T+DELTAT
S.0022              GO TO 2
S.0023            5 CONTINUE
S.0024              STOP
S.0025              END
```

Figure 13.19 A program that compares the Euler and Runge-Kutta solutions of the nonlinear circuit of Figure 13.8

THE EULER AND RUNGE KUTTA SOLUTION OF A NON-LINEAR R-C CIRCUIT

TIME	EULERS I	RUNGE KUTTA I	EXACT I	ERROR E	ERROR R-K
0.0	0.1000000E 03	0.100000 E 03	0.1000000E 03	0.0	0.0
0.5000000E 00	0.9666666E 02	0.966944 E 02	0.9669443E 02	0.2777100E-01	-0.1525879E-04
0.1000000E 01	0.9338934E 02	0.9344444E 02	0.9344443E 02	0.5508423E-01	-0.1525879E-04
0.1500000E 01	0.9016806E 02	0.9025000E 02	0.9025000E 02	0.8193970E-01	0.0
0.2000000E 01	0.8700282E 02	0.8711111E 02	0.8711110E 02	0.1082764E 00	-0.1525879E-04
0.2500000E 01	0.8389365E 02	0.8402779E 02	0.8402776E 02	0.1341095E 00	-0.3051758E-04
0.3000000E 01	0.8084053E 02	0.8100000E 02	0.8100000E 02	0.1594696E 00	0.0
0.3500000E 01	0.7784348E 02	0.7802777E 02	0.7802779E 02	0.1843109E 00	0.1525879E-04
0.4000000E 01	0.7490250E 02	0.7511110E 02	0.7511111E 02	0.2086182E 00	0.1525879E-04
0.4500000E 01	0.7201761E 02	0.7224998E 02	0.7225000E 02	0.2323914E 00	0.1525879E-04
0.5000000E 01	0.6918883E 02	0.6944443E 02	0.6944444E 02	0.2556152E 00	0.1525879E-04
0.5500000E 01	0.6641615E 02	0.6669443E 02	0.6669444E 02	0.2782898E 00	0.1525879E-04
0.6000000E 01	0.6369962E 02	0.6399998E 02	0.6400000E 02	0.3003845E 00	0.1525879E-04
0.6500000E 01	0.6103922E 02	0.6136108E 02	0.6136111E 02	0.3218994E 00	0.3051758E-04
0.7000000E 01	0.5843495E 02	0.587777 E 02	0.5877777E 02	0.3428192E 00	0.3051758E-04
0.7500000E 01	0.5588686E 02	0.562499 E 02	0.5625000E 02	0.3631439E 00	0.4577637E-04
0.8000000E 01	0.5339493E 02	0.537777 E 02	0.5377779E 02	0.3828583E 00	0.6103516E-04
0.8500000E 01	0.5095920E 02	0.5136105E 02	0.5136111E 02	0.4019165E 00	0.6103516E-04
0.9000000E 01	0.4857967E 02	0.4899994E 02	0.4900000E 02	0.4203339E 00	0.6103516E-04
0.9500000E 01	0.4625636E 02	0.4669438E 02	0.4669444E 02	0.4380798E 00	0.6103516E-04
0.1000000E 02	0.4398929E 02	0.4444438E 02	0.4444444E 02	0.4551544E 00	0.6103516E-04
0.1050000E 02	0.4177847E 02	0.4224994E 02	0.4225000E 02	0.4715271E 00	0.6103516E-04
0.1100000E 02	0.3962393E 02	0.4011105E 02	0.4011111E 02	0.4871826E 00	0.6103516E-04
0.1150000E 02	0.3752568E 02	0.3802771E 02	0.3802777E 02	0.5020905E 00	0.6103516E-04
0.1200000E 02	0.3548373E 02	0.3599992E 02	0.3600000E 02	0.5162659E 00	0.7629395E-04
0.1250000E 02	0.3349812E 02	0.3402769E 02	0.3402777E 02	0.5296478E 00	0.7629395E-04
0.1300000E 02	0.3156886E 02	0.3211102E 02	0.3211111E 02	0.5422516E 00	0.9155273E-04
0.1350000E 02	0.2969598E 02	0.3024991E 02	0.3025000E 02	0.5540161E 00	0.9155273E-04
0.1400000E 02	0.2787952E 02	0.2844435E 02	0.2844444E 02	0.5649261E 00	0.9155273E-04
0.1450000E 02	0.2611948E 02	0.2669435E 02	0.2669444E 02	0.5749664E 00	0.9155273E-04
0.1500000E 02	0.2441589E 02	0.2499991E 02	0.2500000E 02	0.5841064E 00	0.9155273E-04
0.1550000E 02	0.2276880E 02	0.2336102E 02	0.2336111E 02	0.5923157E 00	0.9155273E-04
0.1600000E 02	0.2117824E 02	0.2177769E 02	0.2177777E 02	0.5995331E 00	0.7629395E-04
0.1650000E 02	0.1964424E 02	0.202499 E 02	0.2025000E 02	0.6057587E 00	0.9155273E-04
0.1700000E 02	0.1816684E 02	0.187776 E 02	0.1877777E 02	0.6109314E 00	0.9155273E-04
0.1750000E 02	0.1674608E 02	0.173610 E 02	0.1736110E 02	0.6150208E 00	0.9155273E-04
0.1800000E 02	0.1538201E 02	0.159999 E 02	0.1600000E 02	0.6179857E 00	0.9727478E-04
0.1850000E 02	0.1407469E 02	0.146943 E 02	0.1469445E 02	0.6197634E 00	0.9632111E-04
0.1900000E 02	0.1282415E 02	0.1344436E 02	0.1344445E 02	0.6203012E 00	0.8869171E-04
0.1950000E 02	0.1163045E 02	0.1224992E 02	0.1225000E 02	0.6195469E 00	0.8106232E-04
0.2000000E 02	0.1049367E 02	0.1111104E 02	0.1111111E 02	0.6174421E 00	0.7915497E-04
0.2050000E 02	0.9413876E 01	0.1002771E 02	0.1002778E 02	0.6139040E 00	0.7247925E-04
0.2100000E 02	0.8391142E 01	0.8999934E 01	0.9000000E 01	0.6088581E 00	0.6580353E-04
0.2150000E 02	0.7425560E 01	0.8027716E 01	0.8027781E 01	0.6022205E 00	0.6484985E-04
0.2200000E 02	0.6517231E 01	0.7111053E 01	0.7111113E 01	0.5938816E 00	0.5912781E-04
0.2250000E 02	0.5666269E 01	0.6249947E 01	0.6250000E 01	0.5837307E 00	0.5340576E-04
0.2300000E 02	0.4872806E 01	0.5444395E 01	0.5444447E 01	0.5716410E 00	0.5149841E-04
0.2350000E 02	0.4136992E 01	0.4694400E 01	0.4694446E 01	0.5574541E 00	0.4577637E-04
0.2400000E 02	0.3459005E 01	0.3999960E 01	0.4000000E 01	0.5409946E 00	0.4005432E-04
0.2450000E 02	0.2839059E 01	0.3361075E 01	0.3361113E 01	0.5220537E 00	0.3719330E-04
0.2500000E 02	0.2277409E 01	0.2777747E 01	0.2777779E 01	0.5003700E 00	0.3147125E-04
0.2550000E 02	0.1774372E 01	0.2249974E 01	0.2250000E 01	0.4756279E 00	0.2574921E-04
0.2600000E 02	0.1330354E 01	0.1777758E 01	0.1777779E 01	0.4474249E 00	0.2098083E-04
0.2650000E 02	0.9458838E 00	0.1361097E 01	0.1361112E 01	0.4152278E 00	0.1430511E-04
0.2700000E 02	0.6216953E 00	0.9999942E 00	0.1000000E 01	0.3783047E 00	0.5841255E-05
0.2750000E 02	0.3588699E 00	0.6944492E 00	0.6944455E 00	0.3355756E 00	-0.3695488E-05
0.2800000E 02	0.1591841E 00	0.4444659E 00	0.4444448E 00	0.2852607E 00	-0.2104044E-04
0.2850000E 02	0.2619117E-01	0.2500546E 00	0.2500000E 00	0.2238088E 00	-0.5459785E-04
0.2900000E 02	-0.2775446E-01	0.1112601E 00	0.1111115E 00	0.1388659E 00	-0.1485348E-03

IHC253I ALOG-ALOG10 ARG=-0.2775446E-01 LE ZERO

Figure 13.20 The output results of the program of Figure 13.19

Euler current solution, the Runge-Kutta current solution, the exact current, and the respective Euler and Runge-Kutta errors. Computer statement S.0013 then tests to see if CEXACT equals zero. If the exact current is still positive, control is transferred to statement 4 and computer statements S.0014 and S.0015 are used to compute the next Euler value of current. This is followed by computer statements S.0016 through S.0020 which compute the next Runge-Kutta value of current. Computer statement S.0021 then increments T by the value of DELTAT and control is transferred to statement 2 so that the process may be repeated. This procedure is continued until the exact value of current is equal to or less than zero.

Figure 13.20 gives the corresponding results. Note that for this value of DELTAT (0.5), the Euler solution is accurate to one or two significant figures, while the Runge-Kutta solution is accurate to five or six significant figures. The accuracy of this Runge-Kutta solution should also be compared with the accuracy of the Euler solution of Figure 13.10. Note that for the Runge-Kutta formula, a step size 50 times larger can be used and it will still produce more accurate results.

Another fourth-order formula is based on computing the slopes

$$\begin{aligned}
SK1 &= y'(x_n,y_n) \\
SK2 &= y'\left(x_n + \frac{\Delta x}{3}, y_n + SK1\frac{\Delta x}{3}\right) \\
SK3 &= y'\left(x_n + \frac{2\Delta x}{3}, y_n - SK1\frac{\Delta x}{3} + SK2\,\Delta x\right) \\
SK4 &= y'(x_n + \Delta x,\ y_n + SK1\,\Delta x - SK2\,\Delta x + SK3\,\Delta x)
\end{aligned}$$

and then using these to obtain the weighted slope

$$y' = \frac{SK1 + 3\,SK2 + 3\,SK3 + SK4}{8}$$

This weighted slope is used, in turn, to compute the next point

$$y_{n+1} = y_n + y'\,\Delta x$$

Still another fourth-order method is the Runge-Kutta-Gill formula

$$\begin{aligned}
y_{n+1} &= y_n + y'\,\Delta x \\
&= y_n + \frac{\left(SK1 + 2\left(1 - \frac{1}{\sqrt{2}}\right)SK2 + 2\left(1 + \frac{1}{\sqrt{2}}\right)SK3 + SK4\right)}{6}\Delta x
\end{aligned}$$

where

$$\begin{aligned}
SK1 &= y'(x_n,y_n) \\
SK2 &= y'\left(x_n + \frac{\Delta x}{2}, y_n + SK1\frac{\Delta x}{2}\right) \\
SK3 &= y'\left(x_n + \frac{\Delta x}{2}, y_n + \left(-\frac{1}{2} + \frac{1}{\sqrt{2}}\right)SK1\,\Delta x + \left(1 - \frac{1}{\sqrt{2}}\right)SK2\,\Delta x\right) \\
SK4 &= y'\left(x_n + \Delta x,\ y_n - \frac{1}{\sqrt{2}}SK2\,\Delta x + \left(1 + \frac{1}{\sqrt{2}}\right)SK3\,\Delta x\right)
\end{aligned}$$

This version was suggested by Gill around 1950 and was designed to save computer storage. It has since lost some of its importance due to the development of large high-speed storage machines.

The Solution of Higher-Order Differential Equations

Higher-order differential equations may be solved by applying the Runge-Kutta formulas in steps. This is done by first transforming the higher-order equation into a series of first-order differential equations. As an example, let us again consider the solution of the second-order differential equation that describes the network of Figure 13.11

$$\frac{d^2i}{dt^2} = -\frac{R}{L}\frac{di}{dt} - \frac{1}{LC}i$$

We may transform this second-order equation into two first-order equations by introducing the variable p as follows:

$$\frac{di}{dt} = p$$

$$\frac{dp}{dt} = -\frac{R}{L}p - \frac{1}{LC}i$$

where

$$i_0 = 0$$

and

$$p_0 = \frac{Q(0)}{LC}$$

We may now obtain the solution of this set of equations by using one of the Runge-Kutta formulas. For example, using the modified Euler method (a second-order Runge-Kutta formula), we first compute

$$SI1 = i'(t_n,i_n,p_n) = P$$
$$SP1 = p'(t_n,i_n,p_n) = -R*P/L - \mathrm{CUR}/(L*C)$$

and then compute improved values

$$SI2 = i'\left(t_n + \frac{\Delta t}{2}, i_n + SI1\frac{\Delta t}{2}, p_n + SP1\frac{\Delta t}{2}\right) = P + SP1*\mathrm{DELTAT}/2.$$

$$SP2 = p'\left(t_n + \frac{\Delta t}{2}, i_n + SI1\frac{\Delta t}{2}, p_n + SP1\frac{\Delta t}{2}\right)$$
$$= -R*(P+SP1*\mathrm{DELTAT}/2.)/L - (\mathrm{CUR}+\mathrm{SI1}*\mathrm{DELTAT}/2.)/(L*C)$$

These slope values are now used to find the next point values

$$
\begin{aligned}
t_{n+1} &= t_n + \Delta t \\
i_{n+1} &= i_n + SI2\ \Delta t \\
p_{n+1} &= p_n + SP2\ \Delta t
\end{aligned}
$$

Figure 13.21 is the corresponding Fortran program. Initially the numerical values of Q, R, L, and C are read into memory. These values are then printed along with the heading and subheadings. The initial conditions ($T = 0.0$, CUR $= 0.0$, and $P = Q/(AL{*}C)$) and the value of the time increment DELTAT (0.01 seconds) are then specified. Computer statements S.0009 through S.0014 are then used to compute the constants that are required in the solution of the exact current (see Figure 11.6).

We next enter the outer DO 10 loop where we compute and print the values of the exact current, the computed current, and the ERROR for the corresponding value of time. The statements of the inner DO 10 loop are then used to compute the next step Runge-Kutta values. Note that these statements are identical to the relations we described above. Note also that we have again used two DO 10 loops, but this time the computed values are printed every ten iterations (every 0.1 second).

Figure 13.22 gives the computer results. The reader may observe that these results are accurate to four or five significant figures. These results

```
          C       SECOND-ORDER RUNGE-KUTTA SOLUTION
          C                        OF
          C       HIGHER-ORDER DIFFERENTIAL EQUATIONS.
          C
S.CC01            READ(5,1)Q,R,AL,C
S.CC02          1 FORMAT(4F10.6)
S.CC03            WRITE(6,2)Q,R,AL,C
S.CC04          2 FORMAT('1SECOND-ORDER RUNGE-KUTTA SOLUTION OF A SECOND-DEGREE DIFF
                 1ERENTIAL EQUATION',//,' Q =',E15.7,' R =',E15.7,' L =',E15.7,' C =
                 2',E15.7,///,7X,'TIME',7X,'CURRENT RK',4X,'EXACT CURRENT',5X,'ERROR
                 3',/)
S.0C05            T=C.C
S.CC06            CUR=C.C
S.CC07            P=Q/(AL*C)
S.CC08            DELTAT=C.C1
S.CC09            ALPHA=R/(2.*AL)
S.0010            DISC=ALPHA**2-1./(AL*C)
S.0C11            FO=C.15915494/SQRT(AL*C)
S.0C12            BETA=SQRT(-DISC)
S.0013            F=C.15915494*BETA
S.0014            AI=6.2831853*FO**2*Q/F
S.0C15            DO 1C I=1,5C
S.0C16            TI=AI*EXP(-ALPHA*T)*SIN(BETA*T)
S.0C17            ERROR=TI-CUR
S.0C18            WRITE(6,3)T,CUR,TI,ERROR
S.0C19          3 FORMAT(4E15.7)
S.0C20            DO 1C J=1,10
S.0021            SI1=P
S.0022            SP1=-R*P/AL-CUR/(AL*C)
S.0023            SI2=P+SP1*DELTAT/2.
S.0C24            SP2=-R*(P+SP1*DELTAT/2.)/AL-(CUR+SI1*DELTAT/2.)/(AL*C)
S.0C25            T=T+DELTAT
S.0C26            CUR=CUR+SI2*DELTAT
S.0027         1C P=P+SP2*DELTAT
S.0C28            STOP
S.CC29            END
```

Figure 13.21 A Fortran program that uses a second-order Runge-Kutta formula for solving the network of Figure 13.11

```
SECOND-ORDER RUNGE-KUTTA SOLUTION OF A SECOND-DEGREE DIFFERENTIAL EQUATION

Q =  0.2000000E 01 R =  0.1000000E 01 L =  0.1000000E 01 C =  0.1000000E 01

     TIME           CURRENT RK       EXACT CURRENT       ERROR

0.0                0.0              0.0                0.0
0.9999990E-01    0.1900074E 00    0.1900076E 00    0.2384186E-06
0.1999998E 00    0.3601254E 00    0.3601272E 00    0.1788139E-05
0.2999997E 00    0.5106295E 00    0.5106336E 00    0.4172325E-05
0.3999996E 00    0.6419550E 00    0.6419622E 00    0.7152557E-05
0.4999995E 00    0.7546782E 00    0.7546890E 00    0.1078844E-04
0.5999994E 00    0.8494979E 00    0.8495128E 00    0.1484156E-04
0.6999993E 00    0.9272166E 00    0.9272353E 00    0.1865625E-04
0.7999992E 00    0.9887227E 00    0.9887452E 00    0.2241135E-04
0.8999991E 00    0.1034969E 01    0.1035000E 01    0.3051758E-04
0.9999990E 00    0.1066974E 01    0.1067013E 01    0.3910065E-04
0.1099992E 01    0.1085786E 01    0.1085833E 01    0.4673004E-04
0.1199985E 01    0.1092484E 01    0.1092538E 01    0.5435944E-04
0.1299977E 01    0.1088158E 01    0.1088219E 01    0.6103516E-04
0.1399970E 01    0.1073895E 01    0.1073965E 01    0.7057190E-04
0.1499963E 01    0.1050778E 01    0.1050858E 01    0.7915497E-04
0.1599956E 01    0.1019870E 01    0.1019958E 01    0.8773804E-04
0.1699948E 01    0.9822072E 00    0.9823031E 00    0.9596348E-04
0.1799941E 01    0.9387911E 00    0.9388918E 00    0.1006722E-03
0.1899934E 01    0.8905803E 00    0.8906853E 00    0.1050234E-03
0.1999927E 01    0.8384894E 00    0.8385982E 00    0.1087785E-03
0.2099919E 01    0.7833830E 00    0.7834955E 00    0.1125336E-03
0.2199912E 01    0.7260720E 00    0.7261869E 00    0.1148582E-03
0.2299905E 01    0.6673113E 00    0.6674277E 00    0.1164079E-03
0.2399898E 01    0.6077974E 00    0.6079146E 00    0.1171827E-03
0.2499890E 01    0.5481679E 00    0.5482849E 00    0.1170039E-03
0.2599883E 01    0.4890010E 00    0.4891173E 00    0.1162887E-03
0.2699876E 01    0.4308150E 00    0.4309295E 00    0.1144409E-03
0.2799869E 01    0.3740700E 00    0.3741816E 00    0.1115799E-03
0.2899861E 01    0.3191683E 00    0.3192764E 00    0.1081228E-03
0.2999854E 01    0.2664565E 00    0.2665607E 00    0.1041889E-03
0.3099847E 01    0.2162276E 00    0.2163268E 00    0.9918213E-04
0.3199840E 01    0.1687227E 00    0.1688161E 00    0.9340048E-04
0.3299832E 01    0.1241344E 00    0.1242216E 00    0.8714199E-04
0.3399825E 01    0.8260900E-01    0.8268988E-01    0.8088350E-04
0.3499818E 01    0.4425003E-01    0.4432366E-01    0.7362664E-04
0.3599811E 01    0.9120882E-02    0.9186845E-02    0.6596372E-04
0.3699803E 01   -0.2275186E-01   -0.2269386E-01    0.5800277E-04
0.3799796E 01   -0.5137506E-01   -0.5132482E-01    0.5023554E-04
0.3899789E 01   -0.7678562E-01   -0.7674372E-01    0.4190207E-04
0.3999782E 01   -0.9904772E-01   -0.9901428E-01    0.3343821E-04
0.4099774E 01   -0.1182494E 00   -0.1182244E 00    0.2497435E-04
0.4199767E 01   -0.1344993E 00   -0.1344823E 00    0.1704693E-04
0.4299760E 01   -0.1479238E 00   -0.1479148E 00    0.8940697E-05
0.4399753E 01   -0.1586641E 00   -0.1586629E 00    0.1192093E-05
0.4499745E 01   -0.1668737E 00   -0.1668800E 00   -0.6318092E-05
0.4599738E 01   -0.1727152E 00   -0.1727285E 00   -0.1329184E-04
0.4699731E 01   -0.1763585E 00   -0.1763783E 00   -0.1978874E-04
0.4799724E 01   -0.1779782E 00   -0.1780041E 00   -0.2586842E-04
0.4899716E 01   -0.1777514E 00   -0.1777828E 00   -0.3141165E-04
```

Figure 13.22 The output results of the program of Figure 13.21

should also be compared with the Euler results of Figure 13.13 where a value of DELTAT = 0.001 seconds was used (ten times smaller).

This same problem may also be solved by other Runge-Kutta formulas. For example, using the Runge fourth-order Runge-Kutta formula, we start with the two first-order differential equations

$$\frac{di}{dt} = p$$

$$\frac{dp}{dt} = -\frac{R}{L}p - \frac{1}{LC}i$$

and compute the four slopes for these respective equations

$$SP1 = p'(t_n, i_n, p_n) = -R*P/L - \text{CUR}/(L*C)$$

$$SP2 = p'\left(t_n + \frac{\Delta t}{2}, i_n + SI1\,\frac{\Delta t}{2}, p_n + SP1\,\frac{\Delta t}{2}\right)$$

$$= -R*(P+SP1*\text{DELTAT}/2.)/L - (\text{CUR}+SI1*\text{DELTAT}/2.)/(L*C)$$

$$SP3 = p'\left(t_n + \frac{\Delta t}{2}, i_n + SI2\,\frac{\Delta t}{2}, p_n + SP2\,\frac{\Delta t}{2}\right)$$

$$= -R*(P+SP2*\text{DELTAT}/2.)/L - (\text{CUR}+SI2*\text{DELTAT}/2.)/(L*C)$$

$$SP4 = p'(t_n + \Delta t,\ i_n + SI3\ \Delta t,\ p_n + SP3\ \Delta t)$$

$$= -R*(P+SP3*\text{DELTAT})/L - (\text{CUR} + SI3*\text{DELTAT})/(L*C)$$

$$SI1 = i'(p_n) = P$$

$$SI2 = i'\left(p_n + SP1\,\frac{\Delta t}{2}\right) = P + SP1*\text{DELTAT}/2.$$

$$SI3 = i'\left(p_n + SP2\,\frac{\Delta t}{2}\right) = P + SP2*\text{DELTAT}/2.$$

$$SI4 = i'(p_n + SP3\ \Delta t) = P+SP3*\text{DELTAT}$$

```
C      FOURTH-ORDER RUNGE-KUTTA SOLUTION
C                     OF
C      HIGHER-ORDER DIFFERENTIAL EQUATIONS.
C
       READ(5,1)Q,R,AL,C
     1 FORMAT(4F10.6)
       WRITE(6,2)Q,R,AL,C
     2 FORMAT('1FOURTH-ORDER RUNGE-KUTTA SOLUTION OF A SECOND-DEGREE DIFF
      1ERENTIAL EQUATION',//,' Q =',E15.7,' R =',E15.7,' L =',E15.7,' C =
      2',E15.7,///,7X,'TIME',7X,'CURRENT RK',4X,'EXACT CURRENT',5X,'ERROR
      3',/)
       T=0.0
       CUR=0.0
       P=Q/(AL*C)
       DELTAT=0.01
       ALPHA=R/(2.*AL)
       DISC=ALPHA**2-1./(AL*C)
       F0=0.15915494/SQRT(AL*C)
       BETA=SQRT(-DISC)
       F=0.15915494*BETA
       AI=6.2831853*F0**2*Q/F
       DO 10 I=1,50
       TI=AI*EXP(-ALPHA*T)*SIN(BETA*T)
       ERROR=TI-CUR
       WRITE(6,3)T,CUR,TI,ERROR
     3 FORMAT(4F15.7)
       DO 10 J=1,10
       SI1=P
       SP1=-R*P/AL-CUR/(AL*C)
       SI2=P+SP1*DELTAT/2.
       SP2=-R*(P+SP1*DELTAT/2.)/AL-(CUR+SI1*DELTAT/2.)/(AL*C)
       SI3=P+SP2*DELTAT/2.
       SP3=-R*(P+SP2*DELTAT/2.)/AL-(CUR+SI2*DELTAT/2.)/(AL*C)
       SI4=P+SP3*DELTAT
       SP4=-R*(P+SP3*DELTAT)/AL-(CUR+SI3*DELTAT)/(AL*C)
       T=T+DELTAT
       CUR=CUR+(SI1+2.*SI2+2.*SI3+SI4)/6.*DELTAT
    10 P=P+(SP1+2.*SP2+2.*SP3+SP4)/6.*DELTAT
       STOP
       END
```

Figure 13.23 A Fortran program that uses a fourth-order Runge-Kutta formula for solving the network of Figure 13.11

These slope values are then weighted and averaged and the weighted-average slopes are used to compute the next point of the solution for the respective first-order differential equations.

$$\text{CUR} = \text{CUR} + (SI1+2.*SI2+2.*SI3+SI4)/6.*\text{DELTAT}$$
$$P = P + (SP1+2.*SP2+2.*SP3+SP4)/6.*\text{DELTAT}$$

Figure 13.23 is the corresponding Fortran program. Note that this program is essentially the same as the program of Figure 13.21 except that a fourth-order Runge-Kutta formula has been used to obtain the solution.

Figure 13.24 gives the computer results. The reader may observe that these results are accurate to five or six significant figures. This accuracy

```
FOURTH-ORDER RUNGE-KUTTA SOLUTION CF A SECCND-DEGREE CIFFERENTIAL EQUATION

Q =  0.2CCCCCCE C1 R =  0.1CCCCOOE 01 L =  0.1000000E 01 C =  0.1000000E C1


      TIME             CURRENT RK         EXACT CURRENT        ERRCR

 0.0               C.0                0.0                0.0
 0.999999CE-C1     C.19CCC77E  CO     0.1900076E 00     -0.5960464E-07
 0.1999998E  CC    C.36C1267F  CC     0.3601272E 00      0.5364418E-06
 0.2999997E  CC    C.51C6320E  CO     0.5106336E 00      0.1609325E-05
 0.3999996E  CC    C.6419592E  CO     0.6419622E 00      0.2920628E-05
 0.4999995E  CC    C.7546844E  CC     0.7546890E 00      0.4649162E-05
 0.5999994E  CC    C.8495C63E  CC     0.8495128E 00      0.6437302E-05
 0.6999993E  CC    C.9272271E  CO     C.9272353E 00      0.8165836E-05
 0.7999992E  CC    C.9887356E  CC     0.9887452E 00      0.9596348E-05
 0.8999991F  CC    C.1034986E  C1     0.1035C00E 01      0.1335144E-04
 0.999999CE  CC    C.1C66994E  C1     0.1067013E 01      0.1907349E-04
 0.1C99992E  C1    C.1C8581CE  C1     0.1085833E 01      0.2288818E-04
 0.1199985E  C1    C.1C925C9E  C1     0.1092538E 01      0.2861023E-04
 0.1299977E  C1    C.1C88184E  C1     0.1088219E 01      0.3433228E-04
 0.139997CE  C1    C.1C73923E  C1     0.1073965E 01      0.4196167E-04
 0.1499963E  C1    C.1C5C8C7E  C1     0.1050858E 01      0.5054474E-04
 0.1599956E  C1    C.1C199C1E  C1     0.1019958E 01      0.5626678E-04
 0.1699948E  C1    C.9822389E  CO     0.9823031E 00      0.6419420E-04
 0.1799941F  C1    C.938823CE  CC     0.9388918F 00      0.6878376E-04
 C.1899934E  C1    C.89C612CE  CC     C.8906853E 00      0.7331371E-04
 0.1999927E  C1    C.83852C8E  CC     C.8385982E 00      0.7742643E-04
 0.2C99919E  C1    C.7834137E  CC     0.7834955E 00      0.8183718E-04
 0.2199912E  C1    C.7261C19E  CC     0.7261869E 00      0.8493662E-04
 0.2299905E  C1    C.66734C0E  CO     C.6674277E 00      0.8767843E-04
 0.2399898E  C1    C.6C78249F  CC     0.6079146E 00      0.8976460E-04
 C.249989CF  C1    C.5481938E  CO     0.5482849E 00      0.9113550E-04
 0.2599883F  C1    C.489C251E  CO     0.4891173E 00      0.9220839E-04
 0.2699876E  C1    C.43C8371E  CO     0.4309295E 00      0.9238720E-04
 C.2799869E  C1    C.374C9CCE  CO     0.3741816E 00      0.9161234E-04
 C.2899861E  C1    C.3191860E  CC     0.3192764E 00      0.9042025E-04
 0.2999854E  C1    C.2664719E  CC     C.2665607E 00      0.8881092E-04
 C.3C99847F  C1    C.21624C6E  CO     C.2163268E 00      0.8618832E-04
 0.319984CE  C1    C.1687334F  CC     0.1688161E 00      0.8273125E-04
 0.3299832F  C1    C.1241428E  CC     C.1242216E 00      0.7879734E-04
 C.3399825E  C1    C.8261496E-C1      C.8268988E-01      0.7492304E-04
 C.3499818F  C1    C.4425367E-C1      0.4432366E-01      0.6999075E-04
 0.3599811E  C1    C.9122327E-C2      0.9186845E-02      0.6451830E-04
 0.3699803F  C1   -C.2275252E-C1     -0.2269386E-01      0.5866587E-04
 0.3799796F  C1   -C.5137771E-C1     -C.5132482E-01      0.5288795E-04
 0.3899789F  C1   -C.7679C21E-C1     -0.7674372E-01      0.4649162E-04
 C.3999782E  C1   -C.99C54C4E-C1     -C.9901428E-01      0.3975630E-04
 C.4C99774E  C1   -C.1182573E  CC    -0.1182244E 00      0.3296137E-04
 0.4199767F  C1   -C.1345C87E  CC    -0.1344823E 00      0.2640486E-04
 0.429976CE  C1   -C.1479346E  CC    -0.1479148E 00      0.1972914E-04
 C.4399753E  C1   -C.1586761E  CC    -0.1586629E 00      0.1317263E-04
 0.4499745E  C1   -C.1668866E  CC    -0.1668800E 00      0.6616116E-05
 C.4599738C  C1   -C.1727290E  CO    -C.1727285E 00      0.4172325E-06
 0.4699731E  C1   -C.1763728E  CO    -C.1763783E 00     -0.5483627E-05
 0.4799724E  C1   -C.1779930E  CC    -0.1780041E 00     -0.1108646E-04
 0.4899716E  C1   -C.1777666F  CO    -0.1777828E 00     -0.1627207E-04
```

Figure 13.24 The output results of the program of Figure 13.23

should also be compared with the accuracy of the results of Figures 13.13 and 13.22.

13.4 PREDICTOR-CORRECTOR METHODS

The predictor-corrector method of solving a differential equation employs an open formula for prediction and a closed formula for correction. An open formula is one in which we predict or approximate the value at the next step without using any information about the function or its derivatives at the new point. Euler's relation

$$y_{n+1} = y_n + y'_n \, \Delta x$$

is an example of an open formula. A closed formula is one in which we correct the value at the next point by using the value of the function and/or its derivatives at the new point. The trapezoidal formula is an example of one of the simplest closed formulas

$$y_{n+1} = y_n + \frac{y'_{n+1} + y'_n}{2} \, \Delta x$$

Note that this equation is identical to the improved Euler relation of Figure 13.16.

The following four steps are used in carrying out the predictor-corrector process:

1. Predict y_{n+1} by using some open formula.
2. Compute the value of y'_{n+1} from the given differential equation.
3. Using a closed formula and the value of y'_{n+1} found in step 2, compute an improved value for y_{n+1}.
4. If the value of y_{n+1} found in step 3 is essentially the same as the value of y_{n+1} that was computed in step 1, we repeat the process to find y_{n+2}; otherwise, we return to step 2 and repeat steps 2 and 3 until there is no change in two successive corrected values. The added accuracy, produced by the iteration at each integration step, makes predictor-corrector methods very popular.

For a simple numerical example that illustrates this process, consider the solution of

$$\frac{dy}{dx} = x + y$$

using a step size of $\Delta x = 0.1$ and initial conditions $x_0 = 0$, $y_0 = 1.0$.

1. Using Euler's formula we first predict y_{n+1}

$$
\begin{aligned}
y_{n+1} &= y_n + y'_n \, \Delta x \\
y(0.1) &= 1.0 + (0 + 1.0)0.1 = 1.10
\end{aligned}
$$

2. We now find y'_{n+1} from the given differential equation

$$
\begin{aligned}
y' &= x + y \\
y'_{n+1} &= x_{n+1} + y_{n+1} \\
y'(0.1) &= 0.1 + 1.10 = 1.20
\end{aligned}
$$

3. Using the trazepoidal formula, we now compute an improved value for y_{n+1}

$$
y_{n+1} = y_n + \frac{y'_{n+1} + y'_n}{2} \Delta x
$$

$$
y(0.1) = 1.0 + \frac{1.20 + 1.0}{2} 0.1 = 1.110
$$

4. Since the value of y_{n+1} computed in step 3 is different from the value of y_{n+1} we found in step 1, we repeat steps 2 and 3 using $y(0.1) = 1.110$

Step 2:

$$
y'(0.1) = 0.1 + 1.110 = 1.210
$$

Step 3:

$$
y(0.1) = 1.0 + \frac{1.210 + 1.0}{2} 0.1 = 1.1105
$$

Since the value of $y(0.1) = 1.1105$ is different from the previous corrected value ($y(0.1) = 1.110$), we repeat steps 2 and 3 using $y(0.1) = 1.1105$.

Step 2:

$$
y'(0.1) = 0.1 + 1.1105 = 1.2105
$$

Step 3:

$$
y(0.1) = 1.0 + \frac{1.2105 + 1.0}{2} 0.1 = 1.110525
$$

At this point, we see that the difference between two successive corrected values differs in only the sixth place. We therefore terminate the iteration and proceed to step 1, so that the next point, $y(0.2)$, may be computed.

Milne's Multi-Step Predictor-Corrector Method

Multi-step methods require the evaluation of the function and/or its derivatives at one or more additional points. As a consequence, these methods are not self-starting and some other one-step method, such as the Runge-Kutta formulas, must be used to start the procedure.

Milne's method is one of the more popular multi-step methods. It has a local or step truncation error that is proportional to Δx^5 and, unlike the Runge-Kutta formulas, it requires only one derivative evaluation at each integration step. Milne's predictor formula for solving a first-order differential equation is

$$y_{n+1} = y_{n-3} + \frac{4\Delta x}{3}(2y'_n - y'_{n-1} + 2y'_{n-2})$$

This numerical value of y_{n+1} may now be used to compute y'_{n+1} from the given differential equation. Milne's corrector formula may then be used to obtain a corrected value for y_{n+1}

$$y_{n+1} = y_{n-1} + \frac{\Delta x}{3}(y'_{n-1} + 4y'_n + y'_{n+1})$$

The above procedure may also be used to solve higher-order differential equations. For example, consider the solution of the second-order differential equation that describes the network of Figure 13.11

$$\frac{d^2i}{dt^2} = -\frac{R}{L}\frac{di}{dt} - \frac{1}{LC}i$$

Assuming that the numerical values of i, i', and i'' are known at the previous four steps, we may advance the solution to the next step by using the Milne predictor formula to integrate i''; thus,

Step 1:

$$i'_{n+1} = i'_{n-3} + \frac{4\Delta x}{3}(2i''_n - i''_{n-1} + 2i''_{n-2})$$

We now use the corrector formula to integrate i'; thus,

Step 2:

$$i_{n+1} = i_{n-1} + \frac{\Delta x}{3}(i'_{n-1} + 4i'_n + i'_{n+1})$$

Substituting these values in the given differential equation, we obtain

Step 3:

$$i''_{n+1} = -\frac{R}{L}i'_{n+1} - \frac{1}{LC}i_{n+1}$$

We now recompute i'_{n+1} from the corrector equation

Step 4:

$$i'_{n+1} = i'_{n-1} + \frac{\Delta x}{3}\,(i''_{n-1} + 4i''_n + i''_{n+1})$$

If the value of i'_{n+1} obtained in step 4 is essentially the same as the value of i'_{n+1} computed in step 1, we repeat the process to find i_{n+2}; otherwise, we return to step 2 and repeat steps 2, 3, and 4 until no change occurs between successive iterations.

```
C     MILNE'S PREDICTOR-CORRECTOR METHOD IS USED TO SOLVE
C     A SECOND-ORDER DIFFERENTIAL EQUATION.
C
      READ(5,1)Q,R,AL,C
    1 FORMAT(4F10.6)
      WRITE(6,2)Q,R,AL,C
    2 FORMAT('1MILNE''S PREDICTOR-CORRECTOR SOLUTION OF A SECOND-ORDER D
     1IFFERENTIAL EQUATION',//,' Q =',E15.7,' R =',E15.7,' L =',E15.7,'
     2C =',E15.7,///,7X,'TIME',8X,'CURRENT M',4X,'EXACT CURRENT',5X,'ERR
     3OR',/)
      T=0.0
      CUR=0.0
      DELTAT=0.01
      ALPHA=R/(2.*AL)
      DISC=ALPHA**2-1./(AL*C)
      FO=0.15915494/SQRT(AL*C)
      BETA=SQRT(-DISC)
      F=0.15915494*BETA
      AI=6.2831853*FO**2*Q/F
      TI=AI*EXP(-ALPHA*T)*SIN(BETA*T)
      ERROR=TI-CUR
      WRITE(6,3)T,CUR,TI,ERROR
    3 FORMAT(4E15.7)
C
C     STARTING VALUES.
C
      CUR3=0.1719050E+00
      CUR4=0.1900077E+00
      CURP1=0.1860109E+01
      CURP2=0.1840164E+01
      CURP3=0.1820234E+01
      CURP4=0.1800322E+01
      CURPP2=-0.1993767E+01
      CURPP3=-0.1992139E+01
      CURPP4=-0.1990329E+01
      CUR=0.1900077E+00
      T=0.10
      DO 10 I=1,50
      TI=AI*EXP(-ALPHA*T)*SIN(BETA*T)
      ERROR=TI-CUR
      WRITE(6,3)T,CUR,TI,ERROR
      DO 10 J=1,10
      CURP=CURP1+4.*DELTAT*(2.*CURPP4-CURPP3+2.*CURPP2)/3.
   20 CUR=CUR3+DELTAT*(CURP3+4.*CURP4+CURP)/3.
      CURPP=-R*CURP/AL-CUR/(AL*C)
      CCURP=CURP3+DELTAT*(CURPP3+4.*CURPP4+CURPP)/3.
      IF(ABS(CCURP-CURP)-1.E-4)30,30,40
   40 CURP=CCURP
      GO TO 20
   30 CUR3=CUR4
      CUR4=CUR
      CURP1=CURP2
      CURP2=CURP3
      CURP3=CURP4
      CURP4=CURP
      CURPP2=CURPP3
      CURPP3=CURPP4
      CURPP4=CURPP
      T=T+DELTAT
   10 CONTINUE
      STOP
      END
```

Figure 13.25 A Fortran program that uses the Milne predictor-corrector method to solve the network of Figure 13.11

```
       C          RUNGE-KUTTA STARTING VALUES.
       C
S.0001            READ(5,1)Q,R,AL,C
S.0002          1 FORMAT(4F10.6)
S.0003            T=0.0
S.0004            CUR=0.0
S.0005            P=Q/(AL*C)
S.0006            DELTAT=0.01
S.0007            DO 10 I=1,10
S.0008            SI1=P
S.0009            SP1=-R*P/AL-CUR/(AL*C)
S.0010            SI2=P+SP1*DELTAT/2.
S.0011            SP2=-R*(P+SP1*DELTAT/2.)/AL-(CUR+SI1*DELTAT/2.)/(AL*C)
S.0012            SI3=P+SP2*DELTAT/2.
S.0013            SP3=-R*(P+SP2*DELTAT/2.)/AL-(CUR+SI2*DELTAT/2.)/(AL*C)
S.0014            SI4=P+SP3*DELTAT
S.0015            SP4=-R*(P+SP3*DELTAT)/AL-(CUR+SI3*DELTAT)/(AL*C)
S.0016            T=T+DELTAT
S.0017            CUR=CUR+(SI1+2.*SI2+2.*SI3+SI4)/6.*DELTAT
S.0018            P=P+(SP1+2.*SP2+2.*SP3+SP4)/6.*DELTAT
S.0019            PP=-R*P/AL-CUR/(AL*C)
S.0020         10 WRITE(6,20)T,CUR,P,PP
S.0021         20 FORMAT(4E15.7)
S.0022            STOP
S.0023            END
```

Figure 13.26 A program that uses the fourth-order Runge-Kutta method to obtain starting values for the program of Figure 13.25

Figure 13.25 shows a Fortran program that employs the Milne predictor-corrector method to solve the second-order differential equation that describes the network of Figure 13.11. Computer statements S.0001 through S.0017 are essentially the same as those that were used in the program of Figure 13.23. These statements are used in the calculation of the exact value of current.

Computer statements S.0018 through S.0027 are used to provide the necessary starting values. The numerical values of these starting values were obtained from the program of Figure 13.26. This program is very similar to the fourth-order Runge-Kutta program of Figure 13.23, except that only ten iterations have been computed. Figure 13.27 shows these corresponding ten outputs. The reader may readily associate these results with the starting values shown in computer statements S.0018 through S.0027 (Figure 13.25) by recognizing that the time at the next computation step is 0.11 seconds and

$$i'_{n+1} = i'_{n-3} + \frac{4\Delta x}{3}\,(2i''_n - i''_{n-1} + 2i''_{n-2})$$

```
0.9999998E-02  0.1989999E-01  0.1980000E 01  -0.1999899E 01
0.2000000E-01  0.3959998E-01  0.1960001E 01  -0.1999600E 01
0.2999999E-01  0.5910000E-01  0.1940007E 01  -0.1999106E 01
0.3999999E-01  0.7840008E-01  0.1920019E 01  -0.1998419E 01
0.4999999E-01  0.9750032E-01  0.1900039E 01  -0.1997539E 01
0.5999999E-01  0.1164008E 00  0.1880068E 01  -0.1996469E 01
0.6999993E-01  0.1351017E 00  0.1860109E 01  -0.1995211E 01
0.7999992E-01  0.1536030E 00  0.1840164E 01  -0.1993767E 01
0.8999991E-01  0.1719050E 00  0.1820234E 01  -0.1992139E 01
0.9999990E-01  0.1900077E 00  0.1800322E 01  -0.1990329E 01
```

Figure 13.27 The output results of the program of Figure 13.26

is equivalent to

```
CURP = CURP1 + 4.*DELTAT(2.*CURPP4
                          -CURPP3 + 2.*CURPP2)/3.
```

and

$$i_{n+1} = i_{n-1} + \frac{\Delta x}{3} (i'_{n-1} + 4i'_n + i'_{n+1})$$

is equivalent to

```
CUR = CUR3 + DELTAT*(CURP3 + 4.*CURP4 + CURP)/3.
```

Returning now to the program of Figure 13.25, we must realize that since the Milne predictor-corrector method is to be used for computing the solution for $t > 0.10$ seconds, computer statement S.0028 must therefore set $T = 0.1$ seconds before we enter the outer DO 10 loop. An examination of the limits of the outer and inner DO 10 loops shows that results are printed every tenth iteration. Thus for each iteration of the outer DO 10 loop, the exact value of current and the error are computed, and these values are printed along with the Milne value of current and the corresponding time, every 0.1 seconds, starting at 0.1 and ending at 5.0 seconds.

The inner DO 10 loop is used to perform the Milne predictor-corrector calculations. Computer statements S.0034, S.0035, S.0036, and S.0037 correspond respectively to steps 1, 2, 3, and 4 of the procedure. Computer statement S.0038 is used to see whether the value of CCURP (i'_{n+1}) found in computer statement S.0037 is essentially the same as the value of CURP found in computer statement S.0034. If these two values are different, we return to step 2 of the procedure (statement 40) and repeat steps 2, 3, and 4 until no change occurs between successive iterations; otherwise, we go on to statement 30 and redefine the starting values before we compute the next step values.

Figure 13.28 gives the computer solution to the problem. An examination of these results shows that the solution is accurate to approximately five significant figures.

The reader should also make note of the fact that the above procedure may easily be generalized to handle higher-order differential equations. All that is required is a set of starting values, the predictor equation

$$y_{n+1}^{k-1} = y_{n-3}^{k-1} + \frac{4\Delta x}{3} (2y_n{}^k - y_{n-1}^k + 2y_{n-2}^k)$$

the corrector equation

$$y_{n+1}^{k-1} = y_{n-1}^{k-1} + \frac{\Delta x}{3} (y_{n+1}^k + 4y_n{}^k + y_{n-1}^k)$$

MILNE'S PREDICTOR-CORRECTOR SOLUTION OF A SECOND-ORDER DIFFERENTIAL EQUATION

Q = 0.2000000E 01 R = 0.1000000E 01 L = 0.1000000E 01 C = 0.1000000E 01

TIME	CURRENT M	EXACT CURRENT	ERROR
0.0	0.0	0.0	0.0
0.9999996E-01	0.1900077E 00	0.1900076E 00	-0.5960464E-07
0.1999999E 00	0.3601269E 00	0.3601273E 00	0.4768372E-06
0.2999998E 00	0.5106328E 00	0.5106336E 00	0.8940697E-06
0.3999997E 00	0.6419608E 00	0.6419623E 00	0.1490116E-05
0.4999996E 00	0.7546872E 00	0.7546890E 00	0.1788139E-05
0.5999995E 00	0.8495105E 00	0.8495128E 00	0.2264977E-05
0.6999994E 00	0.9272326E 00	0.9272355E 00	0.2861023E-05
0.7999993E 00	0.9887421E 00	0.9887453E 00	0.3159046E-05
0.8999992E 00	0.1034995E 01	0.1035000E 01	0.4768372E-05
0.9999991E 00	0.1067005E 01	0.1067013E 01	0.7629395E-05
0.1099992E 01	0.1085823E 01	0.1085833E 01	0.9536743E-05
0.1199985E 01	0.1092526E 01	0.1092538E 01	0.1144409E-04
0.1299977E 01	0.1088202E 01	0.1088219E 01	0.1621246E-04
0.1399970E 01	0.1073944E 01	0.1073965E 01	0.2098083E-04
0.1499963E 01	0.1050831E 01	0.1050858E 01	0.2670288E-04
0.1599956E 01	0.1019924E 01	0.1019958E 01	0.3337860E-04
0.1699948E 01	0.9822621E 00	0.9823031E 00	0.4106760E-04
0.1799941E 01	0.9388455E 00	0.9388918E 00	0.4631281E-04
0.1899934E 01	0.8906336E 00	0.8906853E 00	0.5173683E-04
0.1999927E 01	0.8385413E 00	0.8385982E 00	0.5692244E-04
0.2099919E 01	0.7834331E 00	0.7834955E 00	0.6240606E-04
0.2199912E 01	0.7261201E 00	0.7261869E 00	0.6675720E-04
0.2299905E 01	0.6673570E 00	0.6674277E 00	0.7075071E-04
0.2399898E 01	0.6078404E 00	0.6079146E 00	0.7426739E-04
0.2499890E 01	0.5482081E 00	0.5482849E 00	0.7683039E-04
0.2599883E 01	0.4890381E 00	0.4891173E 00	0.7921457E-04
0.2699876E 01	0.4308488E 00	0.4309295E 00	0.8064508E-04
0.2799869E 01	0.3741004E 00	0.3741816E 00	0.8118153E-04
0.2899861E 01	0.3191952E 00	0.3192764E 00	0.8118153E-04
0.2999854E 01	0.2664800E 00	0.2665607E 00	0.8076429E-04
0.3099847E 01	0.2162476E 00	0.2163268E 00	0.7915497E-04
0.3199840E 01	0.1687394E 00	0.1688161E 00	0.7677078E-04
0.3299832E 01	0.1241477E 00	0.1242216E 00	0.7385015E-04
0.3399825E 01	0.8261907E-01	0.8268988E-01	0.7081032E-04
0.3499818E 01	0.4425701E-01	0.4432366E-01	0.6664544E-04
0.3599811E 01	0.9124938E-02	0.9186845E-02	0.6190687E-04
0.3699803E 01	-0.2275056E-01	-0.2269386E-01	0.5670637E-04
0.3799796E 01	-0.5137628E-01	-0.5132482E-01	0.5146116E-04
0.3899789E 01	-0.7678920E-01	-0.7674372E-01	0.4547834E-04
0.3999782E 01	-0.9905344E-01	-0.9901428E-01	0.3916025E-04
0.4099774E 01	-0.1182570E 00	-0.1182244E 00	0.3260374E-04
0.4199767E 01	-0.1345084E 00	-0.1344823E 00	0.2610683E-04
0.4299760E 01	-0.1479341E 00	-0.1479148E 00	0.1925230E-04
0.4399753E 01	-0.1586754E 00	-0.1586629E 00	0.1245737E-04
0.4499745E 01	-0.1668856E 00	-0.1668800E 00	0.5602837E-05
0.4599738E 01	-0.1727274E 00	-0.1727285E 00	-0.1132488E-05
0.4699731E 01	-0.1763706E 00	-0.1763783E 00	-0.7748604E-05
0.4799724E 01	-0.1779897E 00	-0.1780041E 00	-0.1436472E-04
0.4899716E 01	-0.1777620E 00	-0.1777828E 00	-0.2086163E-04
0.4999709E 01	-0.1758652E 00	-0.1758926E 00	-0.2741814E-04

Figure 13.28 The output results of the program of Figure 13.25

and the equation for obtaining the highest order derivative value

$$y^k = f(x, y, y', \ . \ . \ . \ , y^{k-1})$$

13.5 THE SOLUTION OF STATE EQUATIONS

This section concerns itself with a brief introduction to the formulation and solution of network state equations. Although this is a relatively new method of solving complex systems, it is not based on any new mathematical concepts. It is, however, a very powerful tool that lends itself to a computer solution in which a large number of variables are involved in various matrix manipulations. This method is also readily applicable to the solution of time-variant and nonlinear network problems and for these reasons it is particularly attractive in the digital computer solution of complex active networks.

Let us first consider the formulation of the network equations. For a linear system, the state equations may be written in matrix form as follows

$$[\dot{X}] = [A][X] + [B][U]$$

where

$[X]$ is a column vector that contains the N state variables and is called the state vector.
$[\dot{X}]$ is the time derivative of $[X]$.
$[A]$ is an N by N matrix called the system matrix.
$[B]$ is an N by M matrix called the distribution matrix.
$[U]$ is a column vector that contains M elements for the input variables.

In general, the state variables are defined in terms of the voltage across the capacitors and the currents through the inductors, and they therefore summarize, at any instant, the energy stored by the network. For example, consider the simple shunt-peaked network shown in Figure 13.29. Making the state variable X_1 equal to the voltage across the capacitor and X_2 equal to the current through the inductor, we may write for Kirchhoff's current law

$$C\frac{dX_1}{dt} = I - X_2$$

and for Kirchhoff's voltage law

$$L\frac{dX_2}{dt} + RX_2 = X_1$$

Rearranging these equations we find

$$\dot{X}_1 = I/C - X_2/C$$
$$\dot{X}_2 = X_1/L - R*X_2/L$$

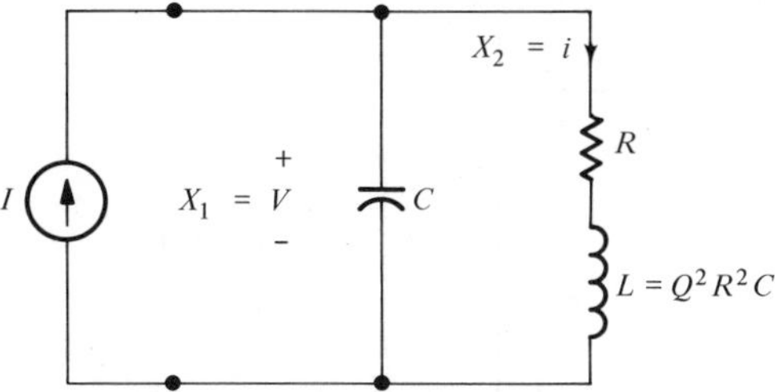

Figure 13.29 Shunt-peaked network

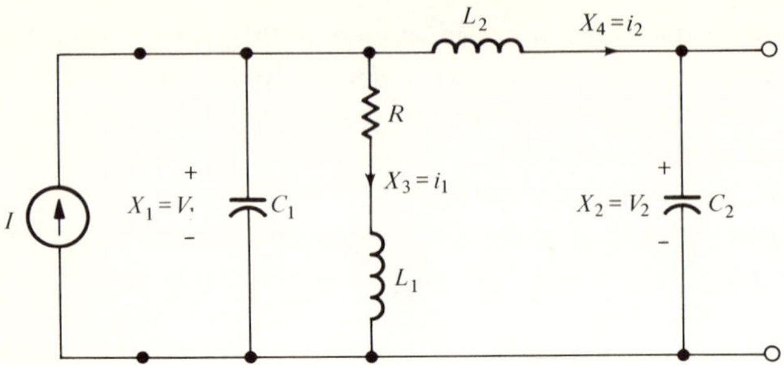

Figure 13.30 The series-shunt-peaked network

so that in matrix form we have

$$\begin{bmatrix} \dot{X}_1 \\ \dot{X}_2 \end{bmatrix} = \begin{bmatrix} 0 & -\dfrac{1}{C} \\ \dfrac{1}{L} & -\dfrac{R}{L} \end{bmatrix} \begin{bmatrix} X_1 \\ X_2 \end{bmatrix} + \begin{bmatrix} \dfrac{1}{C} \\ 0 \end{bmatrix} [I]$$

As a second example, consider the series-shunt-peaked network shown in Figure 13.30. Choosing the state variable X_1 equal to the voltage across capacitor C_1, X_2 equal to the voltage across the capacitor C_2, X_3 equal to the current through L_1, and X_4 equal to the current through L_2, we may apply Kirchhoff's voltage and current laws to obtain the following equations

$$C_1 \frac{dX_1}{dt} = I - X_3 - X_4$$

$$C_2 \frac{dX_2}{dt} = X_4$$

$$L_1 \frac{dX_3}{dt} = X_1 - RX_3$$

$$L_2 \frac{dX_4}{dt} = X_1 - X_2$$

and writing these in matrix form we thus find

$$\begin{bmatrix} \dot{X}_1 \\ \dot{X}_2 \\ \dot{X}_3 \\ \dot{X}_4 \end{bmatrix} = \begin{bmatrix} 0 & 0 & -\dfrac{1}{C_1} & -\dfrac{1}{C_1} \\ 0 & 0 & 0 & \dfrac{1}{C_2} \\ \dfrac{1}{L_1} & 0 & -\dfrac{R}{L_1} & 0 \\ \dfrac{1}{L_2} & -\dfrac{1}{L_2} & 0 & 0 \end{bmatrix} \begin{bmatrix} X_1 \\ X_2 \\ X_3 \\ X_4 \end{bmatrix} + \begin{bmatrix} \dfrac{1}{C_1} \\ 0 \\ 0 \\ 0 \end{bmatrix} [I]$$

Note that these state equations are simultaneous first-order differential equations and that the number of equations is equal to (or less than) the number of reactive elements of the network (the number of degrees of freedom). In the case of networks that contain tie sets of capacitors and cut sets of inductors, the number of degrees of freedom (the number of independent equations) is equal to the total number of inductors and capacitors less the number of tie sets of capacitors and cut sets of inductors.

Let us now turn to the solution of these simultaneous first-order differential equations. A little thought will show us that this problem is essentially the same as that encountered in the solution of a higher-order differential equation (see Section 13.3) in which we transformed the given equation into a set of first-order equations. For example, in the case of the second-order equation

$$y'' = f(t,y,y')$$

we obtained the set

$$\begin{aligned} y' &= p \\ p' &= f(t,y,p) \end{aligned}$$

In the case of the shunt-peaked network above, no transformation is necessary since we actually start with the set

$$\begin{aligned} X_1' &= I/C - X_2/C \\ X_2' &= X_1/L - R*X_2/L \end{aligned}$$

or, in general, we have for this two-degree freedom problem the set

$$\begin{aligned} X_1' &= f_1(t,X_1,X_2) \\ X_2' &= f_2(t,X_1,X_2) \end{aligned}$$

Any of our previous methods for solving first-order equations may now be used to solve this simultaneous set, provided the method is applied in parallel. For example, in the case of the fourth-order Runge-Kutta method of Section 13.3, we would start with the initial conditions $X_1(0)$ and $X_2(0)$, and then use the following relations to obtain the subsequent step values:

$$\begin{aligned} AK1 &= H*f_1(t_n,X1_n,X2_n) \\ AL1 &= H*f_2(t_n,X1_n,X2_n) \\ AK2 &= H*f_1(t_n+H/2,\ X1_n+AK1/2,\ X2_n+AL1/2) \\ AL2 &= H*f_2(t_n+H/2,\ X1_n+AK1/2,\ X2_n+AL1/2) \\ AK3 &= H*f_1(t_n+H/2,\ X1_n+AK2/2,\ X2_n+AL2/2) \\ AL3 &= H*f_2(t_n+H/2,\ X1_n+AK2/2;\ X2_n+AL2/2) \\ AK4 &= H*f_1(t_n+H,\ X1_n+AK3,\ X2_n+AL3) \\ AL4 &= H*f_2(t_n+H,\ X1_n+AK3,\ X2_n+AL3) \end{aligned}$$

$$X1_{n+1} = X1_n + \frac{1}{6}\,(AK1 + 2\;AK2 + 2\;AK3 + AK4)$$

$$X2_{n+1} = X2_n + \frac{1}{6}\,(AL1 + 2\;AL2 + 2\;AL3 + AL4)$$

where H is equal to the time step size. In the case of N sets of simultaneous equations, we would follow a similar procedure using N sets of formulas rather than two.

Figure 13.31 is the corresponding Fortran program for solving the shunt peaked network of Figure 13.29. Initially, we read in the values of R (1000 ohms), Q (0.707), and C (10 pf). The heading and subheadings are then printed, and the value of L (AL) is computed by computer statement S.0005. The value of the step current is then specified as equal to 0.001 amperes. The variables T, $X1$, and $X2$ (initial conditions) are next set equal to zero and the time step is specified as 1.0E-11. The DO 10 loops are then used to compute the Runge-Kutta solutions and the error in the solution of the capacitor voltage.

In each iteration of the outer DO 10 loop (each 1.0E-9 seconds), the exact capacitor voltage is computed from the known solution (for the 1.0 ma step input current)

$$v(t) = \left(1 + 2Q^2\,\frac{\epsilon^{-\frac{t}{2Q^2RC}}}{\sqrt{4Q^2-1}}\,\sin\left(\frac{\sqrt{4Q^2-1}}{2Q^2RC}\,t + \psi\right)\right) * 0.001 * R$$

```
        C      STATE SPACE SOLUTION OF SHUNT-PEAKED COUPLING NETWORK.
        C
S.0001         READ(5,1)R,Q,C
S.0002       1 FORMAT(3F15.7)
S.0003         WRITE(6,2)
S.0004       2 FORMAT('1STATE SPACE SOLUTION OF SHUNT-PEAKED NETWORK',///,4X,'TIM
              1E',13X,'RK-V',12X,'RK-I',6X,'EXACT VOLTAGE',4X,'ERROR V-X1',/)
S.0005         AL=(Q*R)**2*C
S.0006         AI=0.001
S.0007         T=0.0
S.0008         X1=0.0
S.0009         X2=0.0
S.0010         H=1.0E-11
S.0011         DO 10 I=1,50
S.0012         V=(1.+EXP(-T/(R*C))*SIN(T/(R*C)-1.5707963))*AI*R
S.0013         ERROR=V-X1
S.0014         WRITE(6,3)T,X1,X2,V,ERROR
S.0015       3 FORMAT(5E15.7)
S.0016         DO 10 J=1,100
S.0017         AK1=H*( AI-X2)/C
S.0018         AL1=H*(X1-R*X2)/AL
S.0019         AK2=H*( AI-(X2+AL1/2.))/C
S.0020         AL2=H*((X1+AK1/2.)-R*(X2+AL1/2.))/AL
S.0021         AK3=H*( AI-(X2+AL2/2.))/C
S.0022         AL3=H*((X1+AK2/2.)-R*(X2+AL2/2.))/AL
S.0023         AK4=H*( AI-(X2+AL3))/C
S.0024         AL4=H*((X1+AK3)-R*(X2+AL3))/AL
S.0025         X1=X1+(AK1+2.*AK2+2.*AK3+AK4)/6.
S.0026         X2=X2+(AL1+2.*AL2+2.*AL3+AL4)/6.
S.0027      10 T=T+H
S.0028         STOP
S.0029         END
```

Figure 13.31 State-space solution of shunt-peaked coupling network

where

$$\psi = \tan^{-1}\sqrt{4Q^2 - 1} - \tan^{-1}\frac{\sqrt{4Q^2 - 1}}{(-1)}$$

which, in the case of $Q = 0.707$, is equal to

$$v(t) = 1 + \epsilon^{-\frac{t}{RC}} \sin\left(\frac{t}{(RC)} - \frac{\pi}{2}\right)$$

The error is then obtained by computing the difference between the exact voltage and the Runge-Kutta solution for X_1. The inner DO 10 loop is used to compute the Runge-Kutta solution for X_1 and X_2. An examination of the program will show that computer statements S.0017 through

STATE SPACE SOLUTION OF SHUNT-PEAKED NETWORK

TIME	RK-V	RK-I	EXACT VOLTAGE	ERROR V-X1
0.0	0.0	0.0	0.0	0.0
0.9999828E-09	0.9968156E-01	0.9352540E-05	0.9968126E-01	-0.2980232E-06
0.1999961E-08	0.1975843E 00	0.3494075E-04	0.1975856E 00	0.1311302E-05
0.2999939E-08	0.2922601E 00	0.7335779E-04	0.2922637E 00	0.3635883E-05
0.3999848E-08	0.3825797E 00	0.1215823E-03	0.3825810E 00	0.1251698E-05
0.4999581E-08	0.4676991E 00	0.1769632E-03	0.4676848E 00	-0.1424551E-04
0.5999315E-08	0.5470195E 00	0.2372002E-03	0.5469939E 00	-0.2563000E-04
0.6999048E-08	0.6201576E 00	0.3003136E-03	0.6201239E 00	-0.3367662E-04
0.7998782E-08	0.6869105E 00	0.3646475E-03	0.6868720E 00	-0.3844500E-04
0.8998516E-08	0.7472278E 00	0.4288172E-03	0.7471875E 00	-0.4035234E-04
0.9998249E-08	0.8011846E 00	0.4916899E-03	0.8011448E 00	-0.3975630E-04
0.1099798E-07	0.8489571E 00	0.5523642E-03	0.8489205E 00	-0.3653765E-04
0.1199772E-07	0.8908023E 00	0.6101388E-03	0.8907709E 00	-0.3135204E-04
0.1299745E-07	0.9270372E 00	0.6644945E-03	0.9270121E 00	-0.2503395E-04
0.1399718E-07	0.9580234E 00	0.7150667E-03	0.9580061E 00	-0.1728535E-04
0.1499692E-07	0.9841513E 00	0.7616272E-03	0.9841427E 00	-0.8583069E-05
0.1599665E-07	0.1005816E 01	0.8040594E-03	0.1005828E 01	0.1144409E-04
0.1699638E-07	0.1023415E 01	0.8423373E-03	0.1023479E 01	0.6484985E-04
0.1799612E-07	0.1037391E 01	0.8765198E-03	0.1037507E 01	0.1163483E-03
0.1899585E-07	0.1048146E 01	0.9067308E-03	0.1048314E 01	0.1678467E-03
0.1999559E-07	0.1056074E 01	0.9331442E-03	0.1056289E 01	0.2145767E-03
0.2099532E-07	0.1061541E 01	0.9559698E-03	0.1061800E 01	0.2593994E-03
0.2199505E-07	0.1064898E 01	0.9754447E-03	0.1065194E 01	0.2965927E-03
0.2299479E-07	0.1066462E 01	0.9918248E-03	0.1066794E 01	0.3328323E-03
0.2399452E-07	0.1066535E 01	0.1005375E-02	0.1066896E 01	0.3614426E-03
0.2499425E-07	0.1065386E 01	0.1016362E-02	0.1065770E 01	0.3843307E-03
0.2599399E-07	0.1063254E 01	0.1025053E-02	0.1063658E 01	0.4034042E-03
0.2699372E-07	0.1060359E 01	0.1031708E-02	0.1060778E 01	0.4186630E-03
0.2799345E-07	0.1056889E 01	0.1036575E-02	0.1057320E 01	0.4310608E-03
0.2899319E-07	0.1053012E 01	0.1039890E-02	0.1053452E 01	0.4396439E-03
0.2999292E-07	0.1048874E 01	0.1041874E-02	0.1049317E 01	0.4434586E-03
0.3099266E-07	0.1044593E 01	0.1042732E-02	0.1045040E 01	0.4472733E-03
0.3199239E-07	0.1040301E 01	0.1042658E-02	0.1040724E 01	0.4224777E-03
0.3299212E-07	0.1036027E 01	0.1041820E-02	0.1036453E 01	0.4262924E-03
0.3399186E-07	0.1031873E 01	0.1040369E-02	0.1032297E 01	0.4243851E-03
0.3499159E-07	0.1027886E 01	0.1038443E-02	0.1028310E 01	0.4234314E-03
0.3599132E-07	0.1024111E 01	0.1036162E-02	0.1024533E 01	0.4224777E-03
0.3699106E-07	0.1020578E 01	0.1033633E-02	0.1020997E 01	0.4186630E-03
0.3799079E-07	0.1017308E 01	0.1030947E-02	0.1017722E 01	0.4138947E-03
0.3899052E-07	0.1014311E 01	0.1028179E-02	0.1014720E 01	0.4091263E-03
0.3999026E-07	0.1011590E 01	0.1025394E-02	0.1011995E 01	0.4053116E-03
0.4098999E-07	0.1009147E 01	0.1022648E-02	0.1009548E 01	0.4014969E-03
0.4198973E-07	0.1006972E 01	0.1019984E-02	0.1007371E 01	0.3986359E-03
0.4298946E-07	0.1005062E 01	0.1017432E-02	0.1005456E 01	0.3938675E-03
0.4398919E-07	0.1003401E 01	0.1015021E-02	0.1003789E 01	0.3881454E-03
0.4498893E-07	0.1001971E 01	0.1012767E-02	0.1002355E 01	0.3833771E-03
0.4598866E-07	0.1000756E 01	0.1010684E-02	0.1001139E 01	0.3824234E-03
0.4698839E-07	0.9997557E 00	0.1008777E-02	0.1000122E 01	0.3663898E-03
0.4798813E-07	0.9989630E 00	0.1007055E-02	0.9992886E 00	0.3256202E-03
0.4898786E-07	0.9983332E 00	0.1005517E-02	0.9986181E 00	0.2849102E-03

Figure 13.32 The output results of the program of Figure 13.31 for the network of Figure 13.29

S.0026 are the statements that would result from applying the formulas listed above.

Figure 13.32 gives the computer solution for the network of Figure 13.29. Note that the accuracy of these results is approximately four significant figures and that an increase or decrease in the step size H would increase or decrease this error.

Figure 13.33 shows a similar program that can be used to find the unit step current solution to the series-shunt-peaked network shown in Figure 13.30. The element values of data have been supplied for the case of a maximally flat time delay (linear phase) and require for three degrees of correction that

$$C_1 = 0.5C_2,\ L_1 = 0.2C_2R^2,\ \text{and}\ L_2 = 0.7C_2R^2$$

Figure 13.34 gives the computer results. Note the small overshoot that takes place for a linear phase adjustment. The reader is encouraged to extend this investigation and to use the computer program to investigate the effect on the transient response of the network as a function of the

```
        C       STATE SPACE SOLUTION OF SERIES-SHUNT-PEAKED COUPLING NETWORK.
        C
S.0001          WRITE(6,1)
S.0002        1 FORMAT('1STATE-SPACE SOLUTION OF SERIES-SHUNT-PEAKED COUPLING NETW
               1ORK',///,4X,'TIME',14X,'X1',14X,'X2',13X,'X3',13X,'X4',/)
S.0003          R=1000.
S.0004          C2=1.0E-11
S.0005          C1=0.5*C2
S.0006          ALL1=0.2*C2*R**2
S.0007          ALL2=0.7*C2*R**2
S.0008          T=0.0
S.0009          X1=0.0
S.0010          X2=0.0
S.0011          X3=0.0
S.0012          X4=0.0
S.0013          AI=0.001
S.0014          H=1.0E-11
S.0015          DO 10 I=1,50
S.0016          WRITE(6,2)T,X1,X2,X3,X4
S.0017        2 FORMAT(5E15.7)
S.0018          DO 10 J=1,100
S.0019          AK1=H*(AI-X3-X4)/C1
S.0020          AL1=H*(X4/C2)
S.0021          AM1=H*(X1-R*X3)/ALL1
S.0022          AN1=H*(X1-X2)/ALL2
S.0023          AK2=H*(AI-(X3+AM1/2.)-(X4+AN1/2.))/C1
S.0024          AL2=H*(X4+AN1/2.)/C2
S.0025          AM2=H*(X1+AK1/2.-R*(X3+AM1/2.))/ALL1
S.0026          AN2=H*(X1+AK1/2.-(X2+AL1/2.))/ALL2
S.0027          AK3=H*(AI-(X3+AM2/2.)-(X4+AN2/2.))/C1
S.0028          AL3=H*(X4+AN2/2.)/C2
S.0029          AM3=H*(X1+AK2/2.-R*(X3+AM2/2.))/ALL1
S.0030          AN3=H*(X1+AK2/2.-(X2+AL2/2.))/ALL2
S.0031          AK4=H*(AI-(X3+AM3)-(X4+AN3))/C1
S.0032          AL4=H*(X4+AN3)/C2
S.0033          AM4=H*(X1+AK3-R*(X3+AM3))/ALL1
S.0034          AN4=H*(X1+AK3-(X2+AL3))/ALL2
S.0035          X1=X1+(AK1+2.*AK2+2.*AK3+AK4)/6.
S.0036          X2=X2+(AL1+2.*AL2+2.*AL3+AL4)/6.
S.0037          X3=X3+(AM1+2.*AM2+2.*AM3+AM4)/6.
S.0038          X4=X4+(AN1+2.*AN2+2.*AN3+AN4)/6.
S.0039       10 T=T+H
S.0040          STOP
S.0041          END
```

Figure 13.33 State-space solution of the series-shunt peaked coupling network

STATE-SPACE SOLUTION OF SERIES-SHUNT-PEAKED COUPLING NETWORK

TIME	X1	X2	X3	X4
0.0	0.0	0.0	0.0	0.0
0.9999828E-09	0.1315996E 00	0.3183046E-03	0.2833683E-04	0.9496176E-05
0.1999961E-08	0.2540819E 00	0.2503293E-02	0.9559908E-04	0.3707038E-04
0.2999939E-08	0.3614053E 00	0.8263722E-02	0.1814934E-03	0.8055987E-04
0.3999848E-08	0.4501536E 00	0.1904415E-01	0.2719150E-03	0.1369092E-03
0.4999581E-08	0.5189156E 00	0.3595486E-01	0.3574130E-03	0.2025421E-03
0.5999315E-08	0.5677894E 00	0.5973495E-01	0.4320464E-03	0.2736568E-03
0.6999048E-08	0.5979920E 00	0.9074003E-01	0.4924922E-03	0.3464769E-03
0.7998782E-08	0.6115360E 00	0.1289623E 00	0.5373962E-03	0.4174414E-03
0.8998516E-08	0.6109682E 00	0.1740525E 00	0.5668516E-03	0.4833371E-03
0.9998249E-08	0.5991474E 00	0.2253622E 00	0.5819884E-03	0.5414141E-03
0.1099798E-07	0.5790636E 00	0.2819940E 00	0.5846510E-03	0.5894508E-03
0.1199772E-07	0.5536863E 00	0.3428565E 00	0.5771376E-03	0.6257973E-03
0.1299745E-07	0.5258424E 00	0.4067217E 00	0.5619775E-03	0.6493849E-03
0.1399718E-07	0.4981198E 00	0.4722847E 00	0.5417736E-03	0.6597135E-03
0.1499692E-07	0.4727924E 00	0.5382169E 00	0.5190556E-03	0.6568213E-03
0.1599665E-07	0.4517701E 00	0.6032189E 00	0.4961775E-03	0.6412338E-03
0.1699638E-07	0.4365659E 00	0.6660653E 00	0.4752316E-03	0.6138976E-03
0.1799612E-07	0.4282832E 00	0.7256440E 00	0.4579972E-03	0.5761206E-03
0.1899585E-07	0.4276178E 00	0.7809880E 00	0.4458996E-03	0.5294802E-03
0.1999559E-07	0.4348758E 00	0.8312978E 00	0.4399922E-03	0.4757578E-03
0.2099532E-07	0.4500015E 00	0.8759605E 00	0.4409568E-03	0.4168483E-03
0.2199505E-07	0.4726157E 00	0.9145539E 00	0.4491124E-03	0.3546951E-03
0.2299479E-07	0.5020612E 00	0.9468492E 00	0.4644450E-03	0.2912050E-03
0.2399452E-07	0.5374552E 00	0.9728052E 00	0.4866354E-03	0.2282033E-03
0.2499425E-07	0.5777396E 00	0.9925559E 00	0.5151045E-03	0.1673670E-03
0.2599399E-07	0.6217385E 00	0.1006371E 01	0.5490575E-03	0.1101728E-03
0.2699372E-07	0.6682104E 00	0.1014681E 01	0.5875335E-03	0.5788282E-04
0.2799345E-07	0.7158985E 00	0.1018050E 01	0.6294574E-03	0.1150695E-04
0.2899319E-07	0.7635772E 00	0.1017116E 01	0.6736862E-03	-0.2820784E-04
0.2999292E-07	0.8100929E 00	0.1012562E 01	0.7190593E-03	-0.6078089E-04
0.3099266E-07	0.8543993E 00	0.1005120E 01	0.7644410E-03	-0.8599013E-04
0.3199239E-07	0.8955840E 00	0.9955434E 00	0.8087584E-03	-0.1038590E-03
0.3299212E-07	0.9328912E 00	0.9845587E 00	0.8510342E-03	-0.1146382E-03
0.3399186E-07	0.9657332E 00	0.9728323E 00	0.8904166E-03	-0.1187724E-03
0.3499159E-07	0.9936973E 00	0.9610004E 00	0.9261917E-03	-0.1168712E-03
0.3599132E-07	0.1016513E 01	0.9496298E 00	0.9577975E-03	-0.1096787E-03
0.3699106E-07	0.1034139E 01	0.9392083E 00	0.9848336E-03	-0.9803830E-04
0.3799079E-07	0.1046677E 01	0.9301354E 00	0.1007071E-02	-0.8285046E-04
0.3899052E-07	0.1054366E 01	0.9227201E 00	0.1024445E-02	-0.6504217E-04
0.3999026E-07	0.1057556E 01	0.9171783E 00	0.1037030E-02	-0.4553597E-04
0.4098999E-07	0.1056718E 01	0.9136347E 00	0.1045041E-02	-0.2521927E-04
0.4198973E-07	0.1052379E 01	0.9121285E 00	0.1048802E-02	-0.4920027E-05
0.4298946E-07	0.1045138E 01	0.9126197E 00	0.1048736E-02	0.1461496E-04
0.4398919E-07	0.1035616E 01	0.9149985E 00	0.1045338E-02	0.3273698E-04
0.4498893E-07	0.1024445E 01	0.9190959E 00	0.1039156E-02	0.4890771E-04
0.4598866E-07	0.1012247E 01	0.9246947E 00	0.1030773E-02	0.6270634E-04
0.4698839E-07	0.9996125E 00	0.9315419E 00	0.1020776E-02	0.7383221E-04
0.4798813E-07	0.9871342E 00	0.9393601E 00	0.1009758E-02	0.8210717E-04
0.4898786E-07	0.9752524E 00	0.9478603E 00	0.9982772E-03	0.8746791E-04

Figure 13.34 The output results of the program of Figure 13.33 for the network of Figure 13.30

tolerance of the various circuit elements. The reader may also be interested in pursuing a more challenging extension of this program that would involve the formulation as well as the equation solution. The resulting program would thus be very similar to the steady-state a-c circuit analysis program of Section 9.4, except that in this case we would be looking for the transient solution of the state variable equations. The reader will benefit from reading C. Pottle's excellent discussion of "State-space Techniques for General Active Network Analysis," which is found in Chapter 3 of *System Analysis by Digital Computer*.[1]

[1] Franklin F. Kuo and James F. Kaiser, *System Analysis by Digital Computer* (New York: John Wiley and Sons, Inc., 1966), pp. 59–98.

13.6 INVERSE LAPLACE TRANSFORM

The reader is undoubtedly familiar with the Laplace transform method of solution of linear integrodifferential equations.[2] The reader will recall that this method of solution converts the problem from the time domain into the frequency domain. The main advantage of using this frequency approach is that the frequency domain solution can be easily found by simple algebraic operations. Once the frequency domain solution has been found, the inverse Laplace transformation is needed to find the corresponding time response.

This section concerns itself with the computer determination of the inverse transformation. The program is based on our assuming that the frequency function is expressed as a proper fraction having the following general form

$$F(s) = \frac{A_1 + A_2 s + A_3 s^2 + \cdots + A_M s^{M-1}}{(s - s_{p1})(s - s_{p2}) \cdots (s - s_{pN})}$$

where the A's are real coefficients and the distinct poles are arbitrarily limited to a maximum number of 20 and are expressed in general as

$$s_{pI} = \text{ROOTR(I)} + j\ \text{ROOTI(I)}$$

Note that this form of the frequency function assumes that we already know the roots (poles) of the denominator polynomial. If this is not the case, the reader is referred to Section 8.5 (SUBROUTINE POLRT) or Chapter 4. Note also that this treatment assumes that the roots of the denominator are distinct, but not necessarily real. The reader will find that, in practice, this is not a severe limitation. In other cases where multiple roots occur, the reader will find that it is a relatively simple matter to extend the following program if the repeated roots are read into memory in groups and are limited to a maximum multiplicity of two or three.

The program is based on expanding the frequency function into its partial fraction expansion

$$F(s) = \frac{C_1}{s - s_{p1}} + \frac{C_2}{s - s_{p2}} + \cdots + \frac{C_N}{s - s_{pN}}$$

and then taking the inverse Laplace transform of each of these terms.

[2] For a more complete discussion see George R. Cooper and Clare D. McGillem, *Methods of Signal and System Analysis* (New York: Holt, Rinehart, and Winston, Inc., 1967), Chs. 6 and 7.

The solution is thus the sum of the residues evaluated at each pole

$$f(t) = \sum_{I=1}^{N} \operatorname*{Lim}_{s \to s_{pI}} (s - s_{pI}) \frac{A_1 + A_2 s + \cdots + A_M s^{M-1}}{(s - s_{p1})(s - s_{p2}) \cdots (s - s_{pN})} \epsilon^{s_{pI} * t}$$

In order to allow complex roots (conjugate pairs) to be present, additional Fortran statements are needed if the program is to be run on a computer that does not perform complex arithmetic. Let us assume that this will be the case and therefore treat each root as if it were a complex number

$$s_{pI} = \text{ROOTR}(I) + j\ \text{ROOTI}(I)$$

Note that a real root would therefore have a zero imaginary part (ROOTI(I) = 0.0).

Figure 13.35 shows the corresponding Fortran program. After the numerator coefficients and the denominator roots have been read into the computer memory, we see that computer statements S.0013 through S.0029 are used to compute the numerator part of each of the residues

$$\left[A_1 + A_2 s + \cdots + A_M s^{M-1}\right]_{s = s_{pI}} : I = 1, 2, \ldots, N$$

```
          C        THE INVERSE LAPLACE TRANSFCRM
          C
          C        ROOTS MUST BE DISTINCT.
          C
S.CC01             DIMENSION RCOTR(20),RCCTI(20),A(20),SNR(20),SNI(20),CR(2C),CI(2C),
                  1SDR(2C),SDI(2C)
S.CC02         2CC READ(5,1)M,(A(I),I=1,M)
S.CC03           1 FORMAT(I1C,/,(5E15.7))
S.CC04             READ(5,2)N,(RCCTR(I),RCCTI(I),I=1,N)
S.CC05           2 FORMAT(I1C,/,(5E15.7))
S.CC06             WRITE(6,3)
S.CC07           3 FORMAT('1THE INVERSE LAPLACE TRANSFCRM',///,' THE NUMERATOR POLYNO
                  1MIAL COEFFICIENTS',//)
S.CC08             CO 1C J=1,M
S.CC09          1C WRITE(6,4)J,A(J)
S.CC10           4 FORMAT(' A(',I2,') =',E15.7)
S.0011             WRITE(6,5)(I,RCCTR(I),RCCTI(I),I=1,N)
S.CC12           5 FORMAT('-THE RCCTS OF THE DENCMINATCR ',///,2X,'ROOT', 7X,'REAL',8
                  1X,'IMAGINARY',/,(I5,2X,2E15.7))
          C
          C        CALCULATION OF THE NUMERATCR PART CF THE RESICUE.
          C
S.0C13             CO 2C I=1,N
S.CC14             SR=ROOTR(I)
S.0C15             SI=ROOTI(I)
S.0C16             PR=1.C
S.0C17             PI=C.C
S.0C18             SUMNR=A(1)
S.0C19             SUMNI=C.0
S.CC20             IF(M-1)1CC,21,22
S.0C21          22 CO 3C J=2,M
S.0C22             B=PR*SR-PI*SI
S.0C23             C=PI*SR+PR*SI
S.CC24             PR=B
S.CC25             PI=C
S.0C26             SUMNR=SUMNR+A(J)*PR
S.0C27          3C SUMNI=SUMNI+A(J)*PI
S.0C28          21 SNR(I)=SUMNR
S.0C29          2C SNI(I)=SUMNI
```

Figure 13.35 A program for finding the inverse Laplace transform; the roots must be distinct

```
         C
         C      CALCLLATION OF THE DENCMINATCR PART CF THE RESICUE.
         C
S.CC30          DO 4C J=1,N
S.0C31          PR=1.C
S.0C32          PI=C.C
S.0C33          DO 5C I=1,N
S.0034          IF(J-I)6C,5C,60
S.CC35       6C SUMDR=ROOTR(J)-ROCTR(I)
S.CC36          SUMDI=ROOTI(J)-ROCTI(I)
S.0C37          B=PR*SLMDR-PI*SUMDI
S.0C38          C=PR*SLMDI+PI*SUMDR
S.CC39          PR=B
S.0C40          PI=C
S.0C41       5C CONTINLE
S.0C42          SDR(J)=PR
S.0C43       4C SDI(J)=PI
         C
         C      CALCLLATION OF THE CCNSTANT PART CF THE RESICUE.
         C
S.0C44          DO 7C I=1,N
S.0C45          CR(I)=(SNR(I)*SDR(I)+SNI(I)*SDI(I))/(SDR(I)*SDR(I)+SDI(I)*SDI(I))
S.0C46       7C CI(I)=(SNI(I)*SDR(I)-SNR(I)*SDI(I))/(SDR(I)*SDR(I)+SDI(I)*SDI(I))
S.0C47          WRITE(6,6)(CR(I),CI(I),RCCTR(I),RCCTI(I),I=1,N)
S.0048        6 FORMAT(///,' F(T)  = +  (',E15.7,' + J ',E15.7,')  EXP(',E15.7,' + J
               1 'E15.7,')T',(/,7X,' +  (',E15.7,' + J ',E15.7,')  EXP(',E15.7,' +
               2J ',E15.7,')T'))
         C
         C      CALCLLATION OF THE TIME RESPCNSE.
         C
S.0C49          READ(5,7)TINIT,TINCR,NUMB
S.C050        7 FORMAT(2E15.7,I10)
S.0C51          T=TINIT-TINCR
S.CC52          WRITE(6,8)
S.0053        8 FORMAT(///,' TOTAL RESIDUE AS A FUNCTICN CF TIME',//,7X,'REAL',9X,
               1 'IMAGINARY',7X,'TIME',/)
S.0054          DO 8C I=1,NUMB
S.CC55          FTR=C.C
S.CC56          FTI=C.C
S.0C57          T=T+TINCR
S.CC58          DO 9C J=1,N
S.0C59          FTR=EXP(ROOTR(J)*T)*(CR(J)*CCS(RCCTI(J)*T)-CI(J)*SIN(ROOTI(J)*T))+
               1FTR
S.CC60       9C FTI=EXP(ROOTR(J)*T)*(CI(J)*CCS(RCCTI(J)*T)+CR(J)*SIN(ROOTI(J)*T))+
               1FTI
S.0C61       8C WRITE(6,9)FTR,FTI,T
S.0C62        9 FORMAT(3E15.7)
S.0C63          GO TO 2CC
S.CC64      1CC STOP
S.0C65          END
```

Figure 13.35 Continued

Note that this is done for each root by first setting SR and SI equal to the real and imaginary part of the root. We then set the dummy variables $PR = 1.0$ and $PI = 0.0$. This is followed by computer statements S.0018 and S.0019 which set SUMNR $= A(1)$ and SUMNI $= 0.0$. We now test to see whether M is equal to 1. If M is less than 1, we transfer to STOP since this is obviously a data error. If M is equal to 1, we transfer to statement 21 and thus bypass the DO 30 loop. If M is greater than 1, we transfer to the DO 30 loop. The DO 30 loop is then used to evaluate the numerator polynomial and computer statements S.0028 and S.0029 are used to store the real and imaginary parts of the result for each root.

The calculation of the denominator part of the residue is now performed by computer statements S.0030 through S.0043. For each iteration of the DO 40 loop, the DO 50 loop computes

$$\text{SDR}(J) = \prod_{I=1}^{N} (s - s_{pI}) \text{ for } s = s_{pJ},\ J = 1, 2, \ldots, N;\ J \neq N$$

and the real and imaginary parts of this expression are stored respectively in SDR(J) and SDI(J). The ratio of the numerator and denominator parts of the residues are then computed by the DO 70 loop, and the corresponding general time response is printed by computer statements S.0047 and S.0048.

Finally, computer statements S.0049 through S.0061 are used to compute the time step responses. This is accomplished by specifying the initial value of time TINIT, the increment in the time step TINCR, and the number of time calculations NUMB. The DO 90 loop is then used to compute the function value for each time step and computer statements S.0061 and S.0062 are used to print the results. After all the time steps have been computed, control is transferred to statement 200 so that the next frequency function can be evaluated.

Figure 13.36 shows the results obtained for several input examples:

a. $F(s) = \dfrac{1}{s+1}$

b. $F(s) = \dfrac{1}{(s+1+j1)(s+1-j1)}$

c. $F(s) = \dfrac{1}{(s+2+j2)(s+2-j2)}$

d. $F(s) = \dfrac{-2+s}{s(s+1)(s+3)}$

e. $F(s) = \dfrac{1}{s(s+1)}$

f. $F(s) = \dfrac{5+s}{\left(\begin{array}{l} s(s+1.078756+j2.107114)(s+1.078756-j2.107114) \\ \quad (s+1.421242+j0.7276187)(s+1.421242-j0.7276187) \end{array}\right)}$

The first four examples were arbitrarily chosen and correspond to the time solutions:

A. $f(t) = \epsilon^{-t}$

B. $f(t) = \epsilon^{-t} \sin t$

C. $f(t) = 0.5\epsilon^{-2t} \sin 2t$

D. $f(t) = -\dfrac{2}{3} + 1.5\epsilon^{-t} - \dfrac{5}{6}\epsilon^{-3t}.$

These computed time functions and the corresponding computed time responses may be readily verified by the reader.

The fifth and sixth examples correspond to practical circuits we have previously considered. Example five represents the normalized step response of the single-stage amplifier we considered in Section 3.3. The

```
THE INVERSE LAPLACE TRANSFORM

THE NUMERATOR POLYNOMIAL COEFFICIENTS

A( 1) =   C.1CCCCCCE C1

THE ROOTS OF THE DENOMINATOR

 ROOT        REAL          IMAGINARY
   1    -0.1CCCCCCE C1   C.C


F(T) = + (   C.1CCCOCCE C1 + J    0.0              )  EXP( -0.1000000E 01 + J    0.C              )T
       + (


TOTAL RESIDUE AS A FUNCTION OF TIME

      REAL            IMAGINARY          TIME

 0.1CCOCCCE C1   C.0                  C.0
 0.60653C7E CC   C.0                  C.5CCCCCCE CC
 0.3678795E CC   C.0                  C.1CCOCOOE 01
 0.22313C2E CC   C.0                  0.15OCCOOE 01
 0.1353353E CC   C.0                  0.2CCCCCOE 01
 0.82085C1E-C1   C.0                  0.25CCCCCE C1
 0.4978707E-C1   C.0                  0.3OCOCOOF 01
 0.3C19738E-C1   C.0                  0.35CCCCCE C1
 C.1831564E-C1   C.0                  C.4COOCOOE 01
 0.11109CCE-C1   C.0                  0.45CCCCOE C1



THE INVERSE LAPLACE TRANSFORM

THE NUMERATOR POLYNOMIAL COEFFICIENTS

A( 1) =   C.1CCCCCCE C1

THE ROOTS OF THE DENOMINATOR

 ROOT        REAL          IMAGINARY
   1    -C.1CCCCCCE C1 -C.1CCCCCCOE 01
   2    -C.1CCCCCCE C1  C.1CCCCCCE 01


F(T) = + (   C.C              + J   0.5COCOOOE 00)  EXP( -0.1000C000E 01 + J  -0.1CCCCCCE C1)T
       + (   C.C              + J  -0.5COCOOOE 00)  EXP( -0.1000C00E 01 + J   C.1CCCCCCE C1)T


TOTAL RESIDUE AS A FUNCTION OF TIME

      REAL            IMAGINARY          TIME

 0.0             C.0                  0.0
 0.2907861E CC   C.0                  C.5CCCCCOE 00
 C.3C95598E CC   C.0                  C.1CCCCCOE 01
 0.2225711E CC   C.0                  0.15COCCOE 01
 C.1230600E CC   C.0                  0.2CCCCCCE C1
 C.4912559E-C1   C.0                  C.25C00COE 01
 C.7C25946E-C2   C.0                  C.3CCCCCCE C1
-0.1C59273E-C1   C.0                  0.35CCCCOE C1
-C.1386132E-C1   C.0                  C.4CCCCCOE C1
-C.1C85938E-C1   C.C                  C.45OCCCOE 01
```

Figure 13.36 The output results of the program of Figure 13.35

```
THE INVERSE LAPLACE TRANSFORM

THE NUMERATOR POLYNOMIAL COEFFICIENTS

A( 1) =  C.1CCCCCCE C1

THE ROOTS OF THE DENOMINATOR

 ROOT        REAL            IMAGINARY
   1    -C.2CCCCCCE C1 -C.2CCCCCCE 01
   2    -C.2CCCCCCE C1  C.2CCCCCCE 01


F(T) = + (  C.C              + J   0.2500C000E 00)  EXP( -0.2000000E 01 + J  -C.2CCCCCCE C1)T
       + (  C.C              + J  -C.250C000E 00)  EXP( -0.2000000E 01 + J   C.2CCCCCCE C1)T


TOTAL RESIDUE AS A FUNCTION OF TIME

      REAL             IMAGINARY          TIME

 C.0               C.0               C.0
 0.1547799E CC     C.0               C.5C00C00E 00
 0.6153CCCE-C1     C.0               0.1CCCCCCE 01
 0.3512976E-C2     C.0               C.15C0000E 01
-0.6930660E-C2     C.0               0.20C0C00E 01
-0.3230590E-C2     C.0               0.2500C00E 01
-0.3463C07E-C3     C.0               C.30CCCCCE C1
 0.2995469E-C3     C.0               0.35C0000E 01
 0.1659463E-C3     C.0               0.4C00C00E 01
 0.2542972E-C4     C.0               C.45CCCCOE 01

THE INVERSE LAPLACE TRANSFORM

THE NUMERATOR POLYNOMIAL COEFFICIENTS

A( 1) = -C.2CCCCCCE C1
A( 2) =  C.1CCCCCCE C1

THE ROOTS OF THE DENOMINATOR

 ROOT        REAL            IMAGINARY
   1     C.C                C.0
   2    -0.1CCCCCCE C1  C.C
   3    -C.3CCCCCCE C1  C.0


F(T) = + ( -0.6666666E C0 + J   0.0            )  EXP(  0.0            + J   0.C          )T
       + (  C.15CCCCCE C1 + J   0.0            )  EXP( -0.1000000E 01 + J   0.C          )T

       + ( -C.8333333E C0 + J   0.C            )  EXP( -0.3000000E 01 + J   C.C          )T


TOTAL RESIDUE AS A FUNCTION OF TIME

      REAL             IMAGINARY          TIME

 0.5960464E-C7    C.0               0.0
 0.5718762E-C1    C.0               C.5CCOCC0E C0
-0.1563367E CC    C.0               C.1C00C00E 01
-0.3412289E CC    C.0               C.15C0C00E 01
-0.4657294E CC    C.0               0.2CCOCOOE 01
-0.5440CCCE CC    C.0               C.25C0CC0E C1
-0.5920889E CC    C.0               0.3CC0000E 01
-0.6213934E CC    C.0               0.350CC0E 01
-0.6391982E CC    C.0               0.4C0C000E 01
-0.6500C42E CC    C.0               0.450C000E 01
```

Figure 13.36 Continued

```
THE INVERSE LAPLACE TRANSFORM

THE NUMERATOR POLYNOMIAL COEFFICIENTS

A( 1) =  C.1CCCCCCE C1

THE ROOTS OF THE DENOMINATOR

 ROOT          REAL            IMAGINARY
   1     C.C                 C.C
   2    -C.1CCCCCCE C1       C.0


F(T) = + (  C.1CCCOCCE C1 + J    0.0             )  EXP(  0.0            + J    0.0         )T
         + ( -C.1CCCCCCE C1 + J    0.0             )  EXP( -0.1000000E 01 + J    0.0         )T


TOTAL RESIDUE AS A FUNCTION OF TIME

       REAL              IMAGINARY         TIME

 0.0                C.0                0.0
 0.39346693E CC     C.0                0.5COCCOOE 00
 0.63212C5E CC      C.0                0.1CCOCCOE 01
 0.7768698E CC      C.0                0.15CCCCOE 01
 0.8646647E CC      C.0                0.2CCCCCCE C1
 0.917915CE CC      C.0                0.25COOCOOE 01
 0.95C213CE CC      C.0                0.3CCOCCOE C1
 0.9698C26E CC      C.0                0.35CCCCOF C1
 0.9816844E CC      C.0                0.4CCOCOOE 01
 0.988891CE CC      C.0                0.45COCOOE 01


THE INVERSE LAPLACE TRANSFORM

THE NUMERATOR POLYNOMIAL COEFFICIENTS

A( 1) =  C.5CCCCCCE C1
A( 2) =  C.1CCCCCCE C1

THE ROOTS OF THE DENOMINATOR

 ROOT          REAL            IMAGINARY
   1     C.C                 C.C
   2    -C.1C78756E C1 -C.21C7114E 01
   3    -C.1C78756E C1  C.2107114E 01
   4    -C.1421242E C1 -C.7276187E CC
   5    -C.1421242E C1  C.7276187E CC


F(T) = + (  C.35CCCC7E CC + J   0.C             )  EXP(  0.0            + J   C.C          )T
         + (  C.1C19618E CC + J  -0.4113927E-01)  EXP( -0.1078756E 01 + J  -0.21C7114E C1)T

         + (  C.1C19619E CC + J   C.4113926E-01)  EXP( -0.1078756E 01 + J   0.21C7114E C1)T

         + ( -C.2765622E CC + J  -0.2706819E 00)  EXP( -0.1421242E 01 + J  -0.7276187E CC)T

         + ( -C.276962CE CC + J   C.2706822E 00)  EXP( -0.1421242E 01 + J   0.7276187E CC)T
```

Figure 13.36 Continued

TOTAL RESIDUE AS A FUNCTION OF TIME

REAL	IMAGINARY	TIME
0.1788139E-C6	C.3576279E-C6	C.0
0.1657C09E-C3	C.3576279E-C6	0.9999996E-01
0.1299620E-C2	C.2384186E-06	0.1999999E 00
C.4258633E-C2	C.298C232E-C6	0.2999999E 00
0.9711087E-C2	C.2384186F-C6	C.3999999E C0
0.1809716E-C1	C.2384186E-C6	C.4999998E CC
0.2961755E-C1	C.2C4891CE-C6	C.5999998E 00
0.4424542F-C1	C.16C1875E-C6	C.6999997E C0
0.6175393E-C1	C.141561CE-C6	0.7999997E 00
0.8175462E-C1	C.1378357E-C6	0.8999997E CC
0.103739CF CC	C.1229346E-C6	0.9999996E 00
0.1271241F CC	C.1C59379E-C6	0.1099999E 01
0.1512938E CC	C.745C581E-C7	C.1199999E C1
0.1756368E CC	C.6332994E-C7	C.1299998E 01
0.1995788E CC	C.596C464E-C7	0.1399998E 01
0.2226061E CC	C.3352761E-C7	C.1499997E C1
0.2442845E CC	C.372529CF-C7	0.1599997E 01
0.2642698E CC	C.26C7703E-07	0.1699996E 01
0.2823118E CC	C.1862645E-C7	0.1799995E 01
0.2982541E CC	C.745C581E-C8	0.1899995E 01
0.3120282E CC	C.745C581F-C8	C.1999994E 01
C.3236445E CC	C.372529CE-C8	0.2099994E 01
0.333181CE CC	C.0	0.2199993E 01
0.34077C4E CC	-C.372529CE-C8	0.2299993E 01
0.3465866E CC	C.0	0.2399992E 01
0.3508324E CC	-C.3725290E-C8	C.2499991E 01
0.3537259E CC	-C.745C581F-C8	0.2599991E 01
0.35549C5E CC	C.0	0.2699990E 01
0.3563448E CC	-C.3725290E-C8	0.2799990E 01
C.3564956E CC	-C.372529CE-C8	0.2899989E 01
0.3561314E CC	-C.745C581E-C8	C.2999989E 01
0.355419CE CC	-C.3725290E-C8	0.3099988E 01
0.3545CC6E CC	C.0	0.3199987E 01
0.3534932E CC	C.6984919E-C9	C.3299987E 01
0.3524888E CC	C.23283C6E-C9	0.3399986E 01
0.3515556E CC	C.23283C6E-C9	0.3499986E 01
0.35074C8E CC	-C.4656613E-C9	0.3599985E 01
0.35C0724E CC	C.0	C.3699985E 01
0.3495626E CC	-C.23283C6E-C9	0.3799984E 01
0.3492107E CC	-C.1164153E-C8	0.3899983E 01

IHC217I

Figure 13.36 Continued

sixth example represents the normalized step response of the series-shunt-peaked coupling network we considered in Sections 8.5 (Subroutine POLRT) and 13.5. The reader may readily verify that the normalized transfer function of this network would be given by

$$V_2(s) = \frac{s+5}{\left(\begin{array}{l} s(s+1.078756+j2.107114)(s+1.078756-j2.107114) \\ \qquad (s+1.421242+j0.7276187)(s+1.421242-j0.7276187) \end{array}\right)}$$

and would yield the normalized time response shown (t/RC_2).

13.7 PARTIAL DIFFERENTIAL EQUATIONS

The Elliptic Equation

Elliptic partial differential equations often occur in steady-state boundary-value problems. A typical electrical example of an elliptic equation would be the solution of Laplace's equation in a two-dimensional region subject to specific conditions at the boundary of the region. For

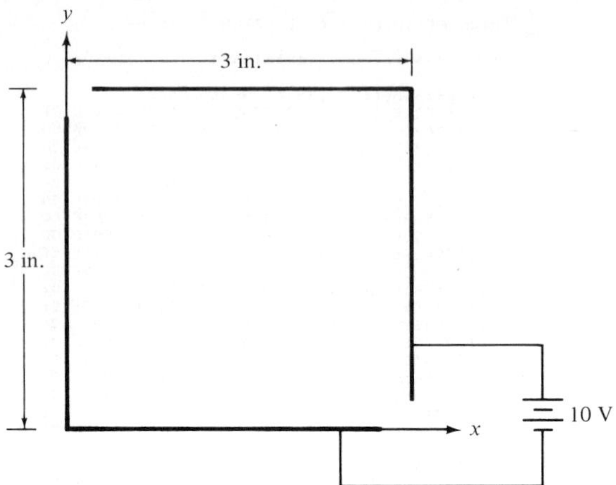

Figure 13.37 A two-dimensional field created by two right-angle metal brackets

example, let it be required to determine the potential distribution within a region bounded by two right-angle metal brackets having a difference of potential of ten volts, as shown in Figure 13.37. Assuming that the brackets are infinitely long in the z direction, the potential distribution in the x-y plane may be found from the solution of the two-dimensional Laplace equation

$$\frac{\partial^2 V}{\partial x^2} + \frac{\partial^2 V}{\partial y^2} = 0$$

In most problems of this type, it is not practical to obtain an analytic solution and one is usually forced to employ some numerical technique to obtain an approximate solution. The method of finite differences is perhaps the most widely used method. This scheme is based on superimposing a grid of mesh size Δx by Δy over the region of interest (see Figure 13.38) and determining the approximate values of the potential at each of these points.

The potentials may be computed from a partial difference equation that approximates the partial differential equation. We may derive the appropriate difference equation by first assuming $\Delta x = \Delta y = h =$ the grid size. We next determine the forward rate of change in potential with respect to x, from the point x, y in Figure 13.38. Thus,

$$\frac{\Delta V}{\Delta x} = \frac{V(x + h,\, y) - V(x,y)}{h}$$

or

$$= \frac{V_1 - V_0}{h}$$

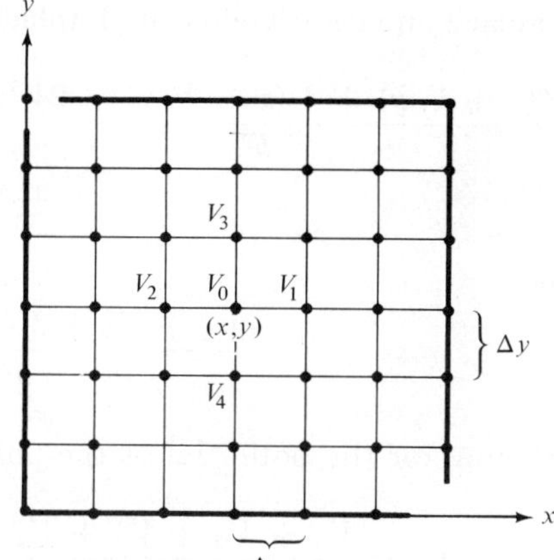

Figure 13.38 The superimposed grid

We then find the backward rate of change

$$\frac{\Delta V}{\Delta x} = \frac{V(x - h,\, y) - V(x,y)}{-h}$$

or

$$= \frac{V_0 - V_2}{h}$$

and from these two rates we compute the rate of change of the rate of change of potential in the x direction

$$\frac{\Delta \dfrac{\Delta V}{\Delta x}}{\Delta x} = \frac{V(x + h,\, y) + V(x - h,\, y) - 2V(x,y)}{h^2}$$

$$= \frac{V_1 + V_2 - 2V_0}{h^2}$$

Similarly, we compute the rate of change of the rate of change of potential in the y direction

$$\frac{\Delta \dfrac{\Delta V}{\Delta y}}{\Delta y} = \frac{V(x,\, y + h) + V(x,\, y - h) - 2V(x,y)}{h^2}$$

$$= \frac{V_3 + V_4 - 2V_0}{h^2}$$

Substitution of these finite difference expressions in the two-dimensional

Laplace equation yields the Laplacian difference equation

$$\frac{V(x+h,\,y)+V(x-h,\,y)-2V(x,y)}{h^2} + \frac{V(x,\,y+h)+V(x,\,y-h)-2V(x,y)}{h^2} = 0$$

or

$$\frac{V_1+V_2+V_3+V_4-4V_0}{h^2}=0$$

Solving for the potential at the point x, y we thus find that[3]

$$V_0 = \frac{V_1+V_2+V_3+V_4}{4}$$

$=$ the average of the potentials surrounding the point.

Similar algebraic equations may also be written for each of the N interior mesh points and the solution of the difference equation thus involves the solution of a set of N linear simultaneous algebraic equations.

[3] This method may also be extended to cover the case of the three-dimensional Laplace equation

$$\frac{\partial^2 V}{\partial x^2}+\frac{\partial^2 V}{\partial y^2}+\frac{\partial^2 V}{\partial z^2}=0$$

and the average of the potentials surrounding the point would be given by

$$V_0 = \frac{V_1+V_2+V_3+V_4+V_5+V_6}{6}$$

where V_5 and V_6 are the grid potentials at the points $(x,\,y,\,z+h)$ and $(x,\,y,\,z-h)$ respectively.

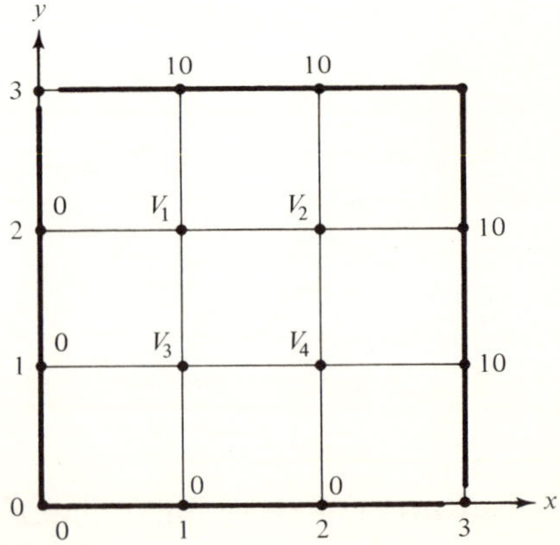

Figure 13.39 A one-inch mesh size

In order to obtain an accurate solution, it is desirable to make the mesh size h as small as possible so that the difference equation will more nearly approximate the differential equation. This will result in a large number of mesh points and a corresponding large number of equations. Since we have already pointed out in Chapter 6 that the use of determinants (as in Cramer's rule), the Gauss elimination method, or the Gauss-Jordan method can not be used for large N, we turn to the Gauss-Seidel method for an iterative solution of these equations.

For a simple numerical illustration that shows how we may use the Gauss-Seidel method, consider a one-inch mesh size as shown in Figure 13.39. Writing algebraic equations that correspond to the average of the potentials that surround each of the interior mesh points, we have

$$\begin{aligned} V_2 + 0 + 10 + V_3 - 4V_1 &= 0 \\ 10 + V_1 + 10 + V_4 - 4V_2 &= 0 \\ V_4 + 0 + V_1 + 0 - 4V_3 &= 0 \\ 10 + V_3 + V_2 + 0 - 4V_4 &= 0 \end{aligned}$$

or

$$\begin{bmatrix} 4 & -1 & -1 & 0 \\ -1 & 4 & 0 & -1 \\ -1 & 0 & 4 & -1 \\ 0 & -1 & -1 & 4 \end{bmatrix} \begin{bmatrix} V_1 \\ V_2 \\ V_3 \\ V_4 \end{bmatrix} = \begin{bmatrix} 10 \\ 20 \\ 0 \\ 10 \end{bmatrix}$$

The Gauss-Seidel method may now be used to solve these equations by assuming an initial approximation of zero volts for each of the interior mesh points. We thus first compute the potential at the point (1,2) by finding the average of the potentials surrounding this point, thus

$$V_1 = \frac{0 + 0 + 10 + 0}{4} = 2.5 \text{ volts}$$

We next find the average of the potentials surrounding the point (2,2)

$$V_2 = \frac{10 + 2.5 + 10 + 0}{4} = 5.625 \text{ volts}$$

We now find the potential at point (1,1)

$$V_3 = \frac{0 + 0 + 2.5 + 0}{4} = 0.625 \text{ volts}$$

and then the potential at point (2,1)

$$V_4 = \frac{10 + 0.625 + 5.625 + 0}{4} = 4.0625 \text{ volts}$$

At this point we have completed one iteration of the Gauss-Seidel method, and we continue the process until we obtain the solution to the desired degree of accuracy. The following table shows the results of the first four iterations:

	ITERATION	V_1	DIFFERENCE	ITERATION	V_2	DIFFERENCE
	0	0.0		0	0.0	
	1	2.5	2.5	1	5.625	5.625
Y = 2	2	4.0625	1.5625	2	7.03125	1.40625
	3	4.765625	0.703125	3	7.3828125	0.3515625
	4	4.94140625	0.17578125	4	7.470703125	0.087890625
	ITERATION	V_3	DIFFERENCE	ITERATION	V_4	DIFFERENCE
	0	0.0		0	0.0	
	1	0.625	0.625	1	4.0625	4.0625
Y = 1	2	2.03125	1.96875	2	4.765625	0.703125
	3	2.3828125	0.3515625	3	4.94140625	0.17578125
	4	2.470703125	0.087890625	4	4.9853515625	0.0439453125
		X = 1			X = 2	

An examination of the table will show that convergence to the exact answers is quite rapid. Successive differences show a ratio of approximately four-to-one. A general formula for computing the limiting λ is given by[4]

$$\lambda = \cos^2 \frac{\pi}{N}$$

$$\approx 1 - \left(\frac{\pi}{N}\right)^2$$

An examination of this relation shows that for a value of $N = 20$, $\lambda = 0.975$. This means that the error in the mesh point potentials will be reduced by approximately 2.5 percent each iteration. Since the number of iterations, I, required to reduce the error to a specified accuracy, EPS, can be approximated by

$$\lambda^I = \text{EPS}$$

it follows that

$$I \approx \left(\frac{N}{\pi^2}\right)^2 (-\log \text{EPS})$$

[4] See David Young, *Iterative Methods for Solving Partial Differential Equations of Elliptic Type*, (1954), Vol. 76, Transcript of the American Mathematics Society pp. 92–111.

or the number of iterations varies as N^2. We may thus expect very slow convergence as the number of mesh points is increased.

We should also note that the Jacobi method or the "method of simultaneous displacements" is quite similar to the Gauss-Seidel method. The Jacobi method, however, does not use new potential values in calculating the potentials of other mesh points, during a given iteration, until after the iteration is completed. As a consequence, the Jacobi method requires approximately twice the number of iterations (as the Gauss-Seidel method) to obtain a specified accuracy.

In order to investigate the rate of convergence of the Gauss-Seidel method, let us return to the problem of Figure 13.39 and reduce the mesh size to one-quarter of an inch, as shown in Figure 13.40. This size mesh will therefore lead to a total of 169 mesh points and the potentials at each of these points may be designated as $V_1, V_2, \ldots, V_{169}$.

Figure 13.41 is a Fortran program for computing the first ten iterations of a Gauss-Seidel solution of this field problem. Initially the DO 10 loop sets all 169 mesh voltages equal to zero. The DO 20 and DO 30 loops then set the potentials of the right-hand boundary points equal to ten volts. The DO 40 loop is then executed ten times, with each iteration corresponding to one iteration of the Gauss-Seidel method. During each

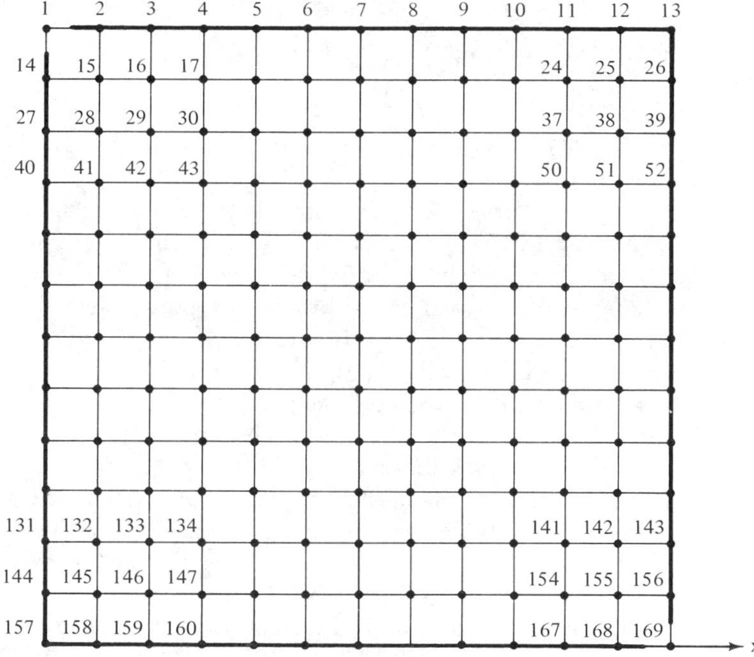

Figure 13.40 A quarter-inch mesh size

```
               C
               C        TWO DIMENSION LAPLACES EQUATION SOLUTION
               C
S.0001                  DIMENSION V(169)
S.0002                  DO 10 J=1, 169
S.0003               10 V(J)=0.0
S.0004                  DO 20 J=2, 12
S.0005               20 V(J)=10.0
S.0006                  DO 30 J=13, 156, 13
S.0007               30 V(J)=10.0
S.0008                  DO 40 I=1, 10
S.0009                  DO 50 K=1,131,13
S.0010                  L=14+K
S.0011                  M=24+K
S.0012                  DO 50 J=L,M
S.0013               50 V(J)=(V(J-1)+V(J+1)+V(J-13)+V(J+13))/4.
S.0014                  WRITE(6,60) I
S.0015               60 FORMAT(/////,' ITERATION =',I3//)
S.0016                  WRITE(6,70) (V(J),J=1,169)
S.0017               70 FORMAT(13F8.2)
S.0018               40 CONTINUE
S.0019                  RETURN
S.0020                  END
```

Figure 13.41 A Fortran program for solving Laplace's equation in a two-dimensional region

iteration of the DO 40 loop, the two DO 50 loops are used to compute the average potential of the points surrounding each of the interior mesh points. For example, when K is equal to 1, $L = 15$ and $M = 25$. The inner DO 50 loop then sets $J = 15$ and computer statement S.0013 computes the average of the potentials surrounding V_{15}, thus

$$V(J) = (V(J-1)+V(J+1)+V(J-13)+V(J+13))/4$$

or

$$V_{15} = \frac{V_{14} + V_{16} + V_2 + V_{28}}{4}$$

and the execution of this loop is continued until $J > M$ $(J > 25)$. At this point, the average potentials of $V_{15}, V_{16}, \ldots, V_{25}$ have been computed and control is returned to the outer DO 50 loop (S.0009) where K is incremented by 13 $(K = 14)$. The values of L and M are now made equal to 28 and 38, respectively, and the execution of the inner DO 50 loop thus computes the average potentials of points $V_{28}, V_{29}, \ldots, V_{38}$. This iteration of the DO 50 loops is continued until the average potential of all interior points has been found. At this point, computer statements S.0014 through S.0017 are used to print the number of the iteration and the potentials at each of the 169 mesh points.

Figure 13.42 shows the corresponding results of the first ten iterations. Note that the physical location of the mesh point voltages corresponds to the physical location of the actual mesh points. Note also that several iterations are required before the lower left potentials change in value from their initial value of zero volts. We may also observe that even after ten iterations the solution is still very inaccurate, since an exact solution would require all the potentials along the main diagonal (upper left to lower right) to be equal to five volts.

```
ITERATION =   1

 0.0   10.00   10.00   10.00   10.00   10.00   10.00   10.00   10.00   10.00   10.00   10.00   10.00
 0.0    2.50    3.12    3.28    3.32    3.33    3.33    3.33    3.33    3.33    3.33    5.83   10.00
 0.0    0.63    0.94    1.05    1.09    1.11    1.11    1.11    1.11    1.11    1.11    4.24   10.00
 0.0    0.16    0.27    0.33    0.36    0.37    0.37    0.37    0.37    0.37    0.37    3.65   10.00
 0.0    0.04    0.08    0.10    0.11    0.12    0.12    0.12    0.12    0.12    0.12    3.44   10.00
 0.0    0.01    0.02    0.03    0.04    0.04    0.04    0.04    0.04    0.04    0.04    3.37   10.00
 0.0    0.00    0.01    0.01    0.01    0.01    0.01    0.01    0.01    0.01    0.01    3.35   10.00
 0.0    0.00    0.00    0.00    0.00    0.00    0.00    0.00    0.00    0.00    0.00    3.34   10.00
 0.0    0.00    0.00    0.00    0.00    0.00    0.00    0.00    0.00    0.00    0.00    3.33   10.00
 0.0    0.00    0.00    0.00    0.00    0.00    0.00    0.00    0.00    0.00    0.00    3.33   10.00
 0.0    0.00    0.00    0.00    0.00    0.00    0.00    0.00    0.00    0.00    0.00    3.33   10.00
 0.0    0.00    0.00    0.00    0.00    0.00    0.00    0.00    0.00    0.00    0.00    3.33   10.00
 0.0    0.0     0.0     0.0     0.0     0.0     0.0     0.0     0.0     0.0     0.0     0.0     0.0
```

```
ITERATION =   2

 0.0   10.00   10.00   10.00   10.00   10.00   10.00   10.00   10.00   10.00   10.00   10.00   10.00
 0.0    3.44    4.41    4.70    4.78    4.80    4.81    4.81    4.81    4.81    5.44    7.42   10.00
 0.0    1.13    1.72    1.96    2.05    2.08    2.09    2.10    2.10    2.10    3.04    6.03   10.00
 0.0    0.36    0.62    0.76    0.82    0.85    0.86    0.86    0.86    0.86    1.92    5.35   10.00
 0.0    0.11    0.21    0.28    0.31    0.33    0.34    0.34    0.34    0.34    1.44    5.04   10.00
 0.0    0.03    0.07    0.10    0.12    0.13    0.13    0.13    0.13    0.13    1.24    4.91   10.00
 0.0    0.01    0.02    0.03    0.04    0.05    0.05    0.05    0.05    0.05    1.16    4.85   10.00
 0.0    0.00    0.01    0.01    0.01    0.02    0.02    0.02    0.02    0.02    1.13    4.83   10.00
 0.0    0.00    0.00    0.00    0.00    0.01    0.01    0.01    0.01    0.01    1.12    4.82   10.00
 0.0    0.00    0.00    0.00    0.00    0.00    0.00    0.00    0.00    0.00    1.11    4.82   10.00
 0.0    0.00    0.00    0.00    0.00    0.00    0.00    0.00    0.00    0.00    1.11    4.82   10.00
 0.0    0.00    0.00    0.00    0.00    0.00    0.00    0.00    0.00    0.00    1.11    3.98   10.00
 0.0    0.0     0.0     0.0     0.0     0.0     0.0     0.0     0.0     0.0     0.0     0.0     0.0
```

Figure 13.42 The output results of the program of Figure 13.41

```
ITERATION =  3

 0.0   10.00  10.00  10.00  10.00  10.00  10.00  10.00  10.00  10.00  10.00  10.00  10.00
 0.0    3.89   5.08   5.45   5.58   5.62   5.63   5.64   5.64   5.79   6.56   8.15  10.00
 0.0    1.49   2.29   2.64   2.78   2.84   2.86   2.86   2.87   3.14   4.41   6.98  10.00
 0.0    0.56   0.95   1.17   1.28   1.33   1.35   1.35   1.36   1.69   3.22   6.31  10.00
 0.0    0.20   0.38   0.49   0.55   0.59   0.60   0.61   0.61   0.97   2.62   5.96  10.00
 0.0    0.07   0.14   0.20   0.23   0.25   0.26   0.26   0.26   0.63   2.33   5.78  10.00
 0.0    0.02   0.05   0.08   0.09   0.10   0.11   0.11   0.11   0.48   2.20   5.70  10.00
 0.0    0.01   0.02   0.03   0.04   0.04   0.04   0.04   0.04   0.42   2.14   5.67  10.00
 0.0    0.00   0.01   0.01   0.01   0.02   0.02   0.02   0.02   0.39   2.12   5.65  10.00
 0.0    0.00   0.00   0.00   0.00   0.01   0.01   0.01   0.01   0.38   2.11   5.64  10.00
 0.0    0.00   0.00   0.00   0.00   0.00   0.00   0.00   0.00   0.37   2.10   5.43  10.00
 0.0    0.00   0.00   0.00   0.00   0.00   0.00   0.00   0.00   0.37   1.61   4.26  10.00
 0.0    0.0    0.0    0.0    0.0    0.0    0.0    0.0    0.0    0.0    0.0    0.0    0.0
```

```
ITERATION =  4

 0.0   10.00  10.00  10.00  10.00  10.00  10.00  10.00  10.00  10.00  10.00  10.00  10.00
 0.0    4.14   5.47   5.92   6.08   6.14   6.16   6.16   6.21   6.48   7.26   8.56  10.00
 0.0    1.75   2.70   3.14   3.33   3.41   3.44   3.46   3.54   4.03   5.37   7.56  10.00
 0.0    0.73   1.24   1.54   1.69   1.76   1.79   1.80   1.91   2.53   4.21   6.93  10.00
 0.0    0.29   0.54   0.71   0.80   0.85   0.88   0.89   1.01   1.70   3.55   6.57  10.00
 0.0    0.12   0.23   0.31   0.36   0.39   0.41   0.42   0.54   1.26   3.20   6.37  10.00
 0.0    0.04   0.09   0.13   0.16   0.17   0.18   0.19   0.31   1.05   3.02   6.26  10.00
 0.0    0.02   0.04   0.05   0.07   0.07   0.08   0.08   0.21   0.95   2.94   6.21  10.00
 0.0    0.01   0.01   0.02   0.03   0.03   0.03   0.04   0.16   0.90   2.90   6.19  10.00
 0.0    0.00   0.00   0.01   0.01   0.01   0.01   0.01   0.14   0.88   2.88   6.12  10.00
 0.0    0.00   0.00   0.00   0.00   0.00   0.01   0.01   0.13   0.87   2.70   5.77  10.00
 0.0    0.00   0.00   0.00   0.00   0.00   0.00   0.00   0.13   0.65   1.90   4.42  10.00
 0.0    0.0    0.0    0.0    0.0    0.0    0.0    0.0    0.0    0.0    0.0    0.0    0.0
```

Figure 13.42 Continued

```
ITERATION =  5

 0.0   10.00  10.00  10.00  10.00  10.00  10.00  10.00  10.00  10.00  10.00  10.00  10.00
 0.0    4.30   5.73   6.24   6.43   6.50   6.53   6.55   6.64   6.98   7.73   8.82  10.00
 0.0    1.93   3.01   3.53   3.77   3.87   3.91   3.95   4.13   4.75   6.06   7.95  10.00
 0.0    0.87   1.49   1.86   2.05   2.14   2.18   2.23   2.48   3.28   4.96   7.37  10.00
 0.0    0.38   0.70   0.92   1.04   1.11   1.15   1.20   1.48   2.39   4.28   7.00  10.00
 0.0    0.16   0.32   0.43   0.51   0.55   0.57   0.63   0.92   1.89   3.89   6.79  10.00
 0.0    0.07   0.14   0.19   0.24   0.26   0.28   0.32   0.62   1.62   3.68   6.67  10.00
 0.0    0.03   0.06   0.08   0.11   0.12   0.13   0.17   0.48   1.48   3.57   6.61  10.00
 0.0    0.01   0.02   0.04   0.05   0.05   0.06   0.10   0.40   1.42   3.51   6.56  10.00
 0.0    0.00   0.01   0.01   0.02   0.02   0.03   0.07   0.37   1.38   3.43   6.44  10.00
 0.0    0.00   0.00   0.01   0.01   0.01   0.01   0.05   0.35   1.27   3.09   5.99  10.00
 0.0    0.00   0.00   0.00   0.00   0.00   0.00   0.05   0.26   0.86   2.09   4.52  10.00
 0.0    0.0    0.0    0.0    0.0    0.0    0.0    0.0    0.0    0.0    0.0    0.0    0.0
```

```
ITERATION =  6

 0.0   10.00  10.00  10.00  10.00  10.00  10.00  10.00  10.00  10.00  10.00  10.00  10.00
 0.0    4.42   5.92   6.47   6.63   6.77   6.81   6.85   6.99   7.37   8.06   9.00  10.00
 0.0    2.07   3.25   3.84   4.11   4.23   4.29   4.38   4.65   5.34   6.58   8.24  10.00
 0.0    0.99   1.70   2.12   2.35   2.47   2.54   2.65   3.01   3.93   5.54   7.69  10.00
 0.0    0.46   0.85   1.11   1.27   1.36   1.42   1.54   1.97   3.02   4.86   7.34  10.00
 0.0    0.21   0.41   0.56   0.65   0.71   0.76   0.89   1.34   2.47   4.45   7.11  10.00
 0.0    0.09   0.19   0.27   0.32   0.36   0.39   0.52   0.99   2.15   4.21   6.98  10.00
 0.0    0.04   0.08   0.12   0.15   0.17   0.20   0.32   0.80   1.98   4.08   6.91  10.00
 0.0    0.02   0.04   0.05   0.07   0.08   0.10   0.22   0.70   1.90   3.99   6.83  10.00
 0.0    0.01   0.02   0.02   0.03   0.04   0.05   0.18   0.65   1.81   3.83   6.66  10.00
 0.0    0.00   0.01   0.01   0.01   0.02   0.03   0.15   0.59   1.59   3.38   6.14  10.00
 0.0    0.00   0.00   0.00   0.01   0.01   0.02   0.11   0.39   1.02   2.23   4.59  10.00
 0.0    0.0    0.0    0.0    0.0    0.0    0.0    0.0    0.0    0.0    0.0    0.0    0.0
```

Figure 13.42 Continued

```
ITERATION =  7

 0.0   10.00  10.00  10.00  10.00  10.00  10.00  10.00  10.00  10.00  10.00  10.00  10.00
 0.0    4.50   6.06   6.64   6.88   6.98   7.03   7.10   7.28   7.67   8.31   9.14  10.00
 0.0    2.18   3.44   4.08   4.39   4.53   4.62   4.75   5.10   5.82   6.98   8.45  10.00
 0.0    1.09   1.88   2.36   2.62   2.76   2.86   3.04   3.51   4.47   6.00   7.95  10.00
 0.0    0.54   0.98   1.29   1.48   1.59   1.69   1.90   2.44   3.56   5.34   7.60  10.00
 0.0    0.26   0.50   0.68   0.80   0.88   0.96   1.18   1.77   2.98   4.91   7.37  10.00
 0.0    0.12   0.24   0.34   0.41   0.46   0.54   0.76   1.37   2.64   4.65   7.23  10.00
 0.0    0.06   0.11   0.17   0.21   0.24   0.30   0.52   1.14   2.44   4.50   7.14  10.00
 0.0    0.02   0.05   0.08   0.10   0.12   0.17   0.39   1.02   2.32   4.37   7.04  10.00
 0.0    0.01   0.02   0.04   0.05   0.06   0.11   0.33   0.94   2.17   4.15   6.83  10.00
 0.0    0.00   0.01   0.02   0.02   0.03   0.08   0.27   0.80   1.84   3.59   6.25  10.00
 0.0    0.00   0.00   0.01   0.01   0.01   0.05   0.18   0.50   1.14   2.33   4.65  10.00
 0.0    0.0    0.0    0.0    0.0    0.0    0.0    0.0    0.0    0.0    0.0    0.0    0.0

ITERATION =  8

 0.0   10.00  10.00  10.00  10.00  10.00  10.00  10.00  10.00  10.00  10.00  10.00  10.00
 0.0    4.56   6.16   6.78   7.04   7.15   7.22   7.31   7.52   7.91   8.51   9.24  10.00
 0.0    2.27   3.60   4.28   4.62   4.79   4.90   5.09   5.48   6.21   7.29   8.62  10.00
 0.0    1.17   2.03   2.55   2.85   3.02   3.17   3.41   3.95   4.93   6.38   8.15  10.00
 0.0    0.60   1.10   1.45   1.68   1.82   1.96   2.25   2.88   4.03   5.73   7.81  10.00
 0.0    0.31   0.58   0.79   0.94   1.05   1.18   1.49   2.18   3.44   5.30   7.59  10.00
 0.0    0.15   0.30   0.42   0.51   0.58   0.70   1.02   1.75   3.07   5.02   7.44  10.00
 0.0    0.07   0.15   0.21   0.26   0.32   0.43   0.75   1.49   2.84   4.84   7.33  10.00
 0.0    0.03   0.07   0.10   0.13   0.17   0.28   0.59   1.33   2.68   4.68   7.21  10.00
 0.0    0.02   0.03   0.05   0.07   0.09   0.19   0.50   1.20   2.47   4.39   6.96  10.00
 0.0    0.01   0.01   0.02   0.03   0.05   0.14   0.40   0.99   2.05   3.76   6.34  10.00
 0.0    0.00   0.01   0.01   0.01   0.03   0.09   0.25   0.59   1.24   2.41   4.69  10.00
 0.0    0.0    0.0    0.0    0.0    0.0    0.0    0.0    0.0    0.0    0.0    0.0    0.0
```

Figure 13.42 Continued

```
ITERATION =  9

 0.0   10.00  10.00  10.00  10.00  10.00  10.00  10.00  10.00  10.00  10.00  10.00  10.00
 0.0    4.61   6.25   6.89   7.16   7.29   7.38   7.50   7.72   8.11   8.66   9.32  10.00
 0.0    2.34   3.72   4.45   4.81   5.01   5.16   5.39   5.82   6.54   7.55   8.75  10.00
 0.0    1.24   2.16   2.73   3.06   3.26   3.45   3.76   4.35   5.32   6.69   8.31  10.00
 0.0    0.66   1.21   1.60   1.85   2.03   2.23   2.59   3.29   4.45   6.06   7.99  10.00
 0.0    0.35   0.66   0.91   1.08   1.22   1.41   1.80   2.57   3.85   5.63   7.76  10.00
 0.0    0.18   0.35   0.49   0.60   0.71   0.89   1.30   2.11   3.45   5.34   7.61  10.00
 0.0    0.09   0.18   0.26   0.33   0.41   0.58   0.99   1.82   3.20   5.14   7.49  10.00
 0.0    0.04   0.09   0.13   0.17   0.24   0.40   0.81   1.63   2.99   4.93   7.35  10.00
 0.0    0.02   0.04   0.07   0.09   0.14   0.30   0.68   1.44   2.72   4.59   7.07  10.00
 0.0    0.01   0.02   0.03   0.05   0.09   0.22   0.53   1.15   2.22   3.89   6.41  10.00
 0.0    0.00   0.01   0.01   0.02   0.05   0.13   0.31   0.68   1.33   2.48   4.72  10.00
 0.0    0.0    0.0    0.0    0.0    0.0    0.0    0.0    0.0    0.0    0.0    0.0    0.0


ITERATION =  10

 0.0   10.00  10.00  10.00  10.00  10.00  10.00  10.00  10.00  10.00  10.00  10.00  10.00
 0.0    4.65   6.32   6.98   7.27   7.41   7.52   7.66   7.90   8.27   8.79   9.39  10.00
 0.0    2.40   3.83   4.59   4.98   5.20   5.39   5.66   6.11   6.81   7.76   8.86  10.00
 0.0    1.31   2.27   2.88   3.24   3.48   3.71   4.08   4.70   5.66   6.95   8.45  10.00
 0.0    0.72   1.31   1.74   2.02   2.24   2.49   2.91   3.66   4.81   6.34   8.14  10.00
 0.0    0.39   0.74   1.01   1.21   1.39   1.64   2.11   2.93   4.20   5.91   7.92  10.00
 0.0    0.21   0.41   0.57   0.71   0.85   1.09   1.57   2.44   3.80   5.61   7.75  10.00
 0.0    0.11   0.22   0.31   0.40   0.52   0.75   1.24   2.13   3.51   5.39   7.62  10.00
 0.0    0.05   0.11   0.17   0.22   0.32   0.54   1.02   1.89   3.26   5.15   7.46  10.00
 0.0    0.03   0.06   0.09   0.13   0.21   0.41   0.85   1.65   2.93   4.76   7.16  10.00
 0.0    0.01   0.03   0.04   0.07   0.14   0.30   0.66   1.30   2.36   4.00   6.47  10.00
 0.0    0.01   0.01   0.02   0.04   0.08   0.17   0.38   0.75   1.40   2.53   4.75  10.00
 0.0    0.0    0.0    0.0    0.0    0.0    0.0    0.0    0.0    0.0    0.0    0.0    0.0
```

Figure 13.42 Continued

```
                C
                C          AN INVESTIGATION OF THE SPEED UP FACTOR IN THE SOLUTION OF LAPLACES
                C          FIELD EQUATION.
                C
S.0001                     DIMENSION V(169)
S.0002                   1 READ(5,2)C
S.0003                   2 FORMAT(F3.1)
S.0004                     DO 10 J=1,169
S.0005                  10 V(J)=0.0
S.0006                     DO 20 J=2,12
S.0007                  20 V(J)=10.0
S.0008                     DO 30 J=13,156,13
S.0009                  30 V(J)=10.0
S.0010                     DO 40 I=1,75
S.0011                     SWITCH=0.0
S.0012                     DO 50 K=1,131,13
S.0013                     L=14+K
S.0014                     M=24+K
S.0015                     DO 50 J=L,M
S.0016                     VNEW=(V(J-1)+V(J+1)+V(J-13)+V(J+13))/4.
S.0017                     IF(ABS(VNEW-V(J))-0.001)50,50,60
S.0018                  60 SWITCH=1.0
S.0019                  50 V(J)=VNEW+C*(VNEW-V(J))
S.0020                     IF(SWITCH)40,70,40
S.0021                  40 CONTINUE
S.0022                  70 WRITE(6,80)I,C
S.0023                  80 FORMAT( '1ITERATION =',I3,5X,'CONSTANT (C) =',F4.1,///)
S.0024                     WRITE(6,90)(V(J),J=1,169)
S.0025                  90 FORMAT(13F8.2)
S.0026                     GO TO 1
S.0027                     RETURN
S.0028                     END
```

Figure 13.43 A program that investigates a simple speed-up factor used in the Gauss-Seidel solution of Laplace's equation

Figure 13.43 is a Fortran program that uses a simple scheme for speeding up the rate of convergence of a Gauss-Seidel solution of Laplace's equation. Computer statements S.0002 and S.0003 first read in the numerical value of the constant C that will be used as a speed-up factor. The DO 10 loop then sets all 169 potentials to zero volts and the DO 20 and DO 30 loops set the right-hand boundary points to ten volts as before. The DO 40 loop now permits a maximum of 75 iterations in the calculation of the solution for each value of C read into memory. During each iteration of the DO 40 loop, computer statement S.0011 first sets the dummy variable SWITCH equal to zero. The outer DO 50 loop is then entered and the value of K is set equal to one. This causes the value of L to be equal to 15 and the value of M to be equal to 25. The inner DO 50 loop is then entered and J is set equal to L ($J = 15$ the first iteration). The new potential of this point, VNEW, is then computed as before by taking the average of the potentials surrounding V_{15}. We then test to see whether the absolute value of the difference between the new value of V_{15} and its previous value is greater than 0.001 volts. If the difference is more than 0.001 volts, SWITCH is made equal to one. Computer statement S.0019

$$V(J) = V\text{NEW} + C*(V\text{NEW} - V(J))$$

is then used to compute the new value of V_{15}. Note that if C were equal to zero, V_{15} would have the same value that the previous program would compute. Since C does not equal zero, the new value of V_{15} will be larger (or smaller) than the average of the surrounding potentials (VNEW) by the factor C times the change in the potential of this point. Note that if VNEW $-$ $V(J)$ is quite large, this added factor will also be quite large and this will tend to speed up the rate of convergence. Note also that VNEW $-$ $V(J)$ may be positive or negative and therefore the factor may add or subtract from VNEW. The potential of each point is thus computed by using this speed-up factor and during each iteration of the DO 50 loops the variable SWITCH is thrown to 1 (made equal to 1) if any one of the corrections are greater than 0.001 volts.

After the iteration of the DO 50 loops is completed, computer statement S.0020 tests to see whether SWITCH is equal to zero. If SWITCH is equal to zero, this means that the correction of every one of the interior points was less than 0.001 volts and therefore control is transferred to statement 70 so that the results may be printed. On the other hand, if the value of SWITCH were found equal to 1, this would mean that at least one of the potentials had a correction greater than 0.001 volts and therefore the DO 50 loops would be executed again until either 75 iterations were reached or until all corrections in a given iteration were less than 0.001 volts.

Figure 13.44 shows the computer output for seven different values of C. Note that in each case the final results are identical except for the case of $C = 1$. In the case of $C = 1$, we see that the potentials actually diverge rather than converge on the exact solutions. An examination of the other choices of C reveal that a value of $C = 0.6$ is close to optimum since only 25 iterations are required to obtain the correct solutions to the specified accuracy.

The reader should be aware of the importance of the initial approximations to the mesh potentials as far as they relate to the speed of convergence of the solutions. Although it is impractical to make an intelligent guess of the potential for each point and then read these in the form of data, it is possible to use an initial potential other than zero volts. This usually greatly improves the speed of convergence. For example, Figure 13.45 is a modification of the program of Figure 13.43. This new program differs only in the DO 10 loop where the initial values of the interior mesh points are set at five volts rather than zero volts. Figure 13.46 shows the corresponding results. Note that only 17 iterations are required for a value of $C = 0.4$. The reader should also observe that fewer iterations are required for all values of C except for $C = 1$ where the process again diverges. The reader should recognize that the choice of the initial approximations can be easily extended. For example, in this problem it would be a fairly simple matter to set the initial values of the mesh points close to the main diagonal at five volts, those above the diagonal at

```
ITERATION = 64      CONSTANT (C) = 0.2

0.0   10.00  10.00  10.00  10.00  10.00  10.00  10.00  10.00  10.00  10.00  10.00  10.00
0.0    5.00   6.98   7.91   8.43   8.77   9.01   9.21   9.38   9.54   9.69   9.85  10.00
0.0    3.02   5.00   6.24   7.05   7.63   8.07   8.44   8.77   9.08   9.39   9.69  10.00
0.0    2.08   3.75   4.99   5.92   6.63   7.21   7.71   8.17   8.63   9.08   9.54  10.00
0.0    1.56   2.93   4.07   4.99   5.76   6.42   7.02   7.60   8.18   8.77   9.38  10.00
0.0    1.23   2.36   3.36   4.23   4.99   5.69   6.35   7.02   7.71   8.44   9.21  10.00
0.0    0.98   1.92   2.78   3.57   4.29   4.99   5.69   6.42   7.21   8.07   9.01  10.00
0.0    0.79   1.55   2.28   2.96   3.63   4.29   4.99   5.76   6.63   7.63   8.77  10.00
0.0    0.62   1.22   1.81   2.39   2.97   3.57   4.23   4.99   5.92   7.06   8.44  10.00
0.0    0.46   0.91   1.36   1.81   2.28   2.78   3.36   4.07   5.00   6.24   7.91  10.00
0.0    0.31   0.61   0.91   1.22   1.55   1.92   2.36   2.94   3.75   5.00   6.98  10.00
0.0    0.15   0.31   0.46   0.62   0.79   0.98   1.23   1.56   2.08   3.02   5.00  10.00
0.0    0.0    0.0    0.0    0.0    0.0    0.0    0.0    0.0    0.0    0.0    0.0    0.0


ITERATION = 42      CONSTANT (C) = 0.4

0.0   10.00  10.00  10.00  10.00  10.00  10.00  10.00  10.00  10.00  10.00  10.00  10.00
0.0    5.00   6.98   7.91   8.43   8.77   9.01   9.21   9.38   9.54   9.69   9.85  10.00
0.0    3.02   5.00   6.24   7.06   7.63   8.07   8.44   8.77   9.08   9.39   9.69  10.00
0.0    2.08   3.75   4.99   5.92   6.63   7.21   7.71   8.18   8.63   9.08   9.54  10.00
0.0    1.56   2.94   4.07   4.99   5.76   6.42   7.02   7.60   8.18   8.77   9.38  10.00
0.0    1.23   2.36   3.36   4.23   4.99   5.69   6.36   7.02   7.71   8.44   9.21  10.00
0.0    0.98   1.92   2.78   3.57   4.30   4.99   5.69   6.42   7.21   8.07   9.01  10.00
0.0    0.79   1.55   2.28   2.97   3.63   4.30   5.00   5.76   6.63   7.63   8.77  10.00
0.0    0.62   1.23   1.82   2.39   2.97   3.57   4.23   5.00   5.92   7.06   8.44  10.00
0.0    0.46   0.92   1.37   1.82   2.28   2.78   3.36   4.07   5.00   6.24   7.91  10.00
0.0    0.31   0.61   0.92   1.23   1.55   1.92   2.36   2.94   3.76   5.00   6.98  10.00
0.0    0.15   0.31   0.46   0.62   0.79   0.98   1.23   1.56   2.08   3.02   5.00  10.00
0.0    0.0    0.0    0.0    0.0    0.0    0.0    0.0    0.0    0.0    0.0    0.0    0.0
```

Figure 13.44 The output results of the program of Figure 13.43

ITERATION = 32 CONSTANT (C) = 0.5

0.0	10.00	10.00	10.00	10.00	10.00	10.00	10.00	10.00	10.00	10.00	10.00	10.00
0.0	5.00	6.96	7.91	8.43	8.77	9.01	9.21	9.38	9.54	9.69	9.85	10.00
0.0	3.02	5.00	6.24	7.06	7.63	8.07	8.44	8.77	9.08	9.39	9.69	10.00
0.0	2.08	3.75	5.00	5.92	6.63	7.21	7.71	8.18	8.63	9.08	9.54	10.00
0.0	1.56	2.94	4.07	4.99	5.76	6.42	7.02	7.60	8.18	8.77	9.38	10.00
0.0	1.23	2.36	3.36	4.23	4.99	5.69	6.36	7.02	7.71	8.44	9.21	10.00
0.0	0.98	1.92	2.78	3.57	4.30	5.00	5.69	6.42	7.21	8.08	9.01	10.00
0.0	0.79	1.55	2.28	2.97	3.63	4.30	5.00	5.76	6.63	7.63	8.77	10.00
0.0	0.62	1.23	1.82	2.39	2.97	3.57	4.23	5.00	5.92	7.06	8.44	10.00
0.0	0.46	0.92	1.37	1.82	2.28	2.78	3.36	4.08	5.00	6.24	7.91	10.00
0.0	0.31	0.61	0.92	1.23	1.56	1.92	2.36	2.94	3.76	5.00	6.98	10.00
0.0	0.15	0.31	0.46	0.62	0.79	0.98	1.23	1.56	2.08	3.02	5.00	10.00
0.0	0.0	0.0	0.0	0.0	0.0	0.0	0.0	0.0	0.0	0.0	0.0	0.0

ITERATION = 25 CONSTANT (C) = 0.6

0.0	10.00	10.00	10.00	10.00	10.00	10.00	10.00	10.00	10.00	10.00	10.00	10.00
0.0	5.00	6.98	7.91	8.44	8.77	9.01	9.21	9.38	9.54	9.69	9.85	10.00
0.0	3.02	5.00	6.24	7.06	7.64	8.08	8.44	8.77	9.08	9.39	9.69	10.00
0.0	2.09	3.76	5.00	5.92	6.63	7.21	7.72	8.18	8.63	9.08	9.54	10.00
0.0	1.56	2.94	4.03	5.00	5.76	6.43	7.03	7.60	8.18	8.77	9.38	10.00
0.0	1.23	2.36	3.37	4.24	5.00	5.70	6.36	7.03	7.72	8.44	9.21	10.00
0.0	0.99	1.92	2.79	3.57	4.30	5.00	5.70	6.43	7.21	8.08	9.01	10.00
0.0	0.79	1.56	2.28	2.97	3.64	4.30	5.00	5.76	6.63	7.64	8.77	10.00
0.0	0.62	1.23	1.82	2.40	2.97	3.57	4.24	5.00	5.92	7.06	8.44	10.00
0.0	0.46	0.92	1.37	1.82	2.28	2.79	3.37	4.08	5.00	6.24	7.91	10.00
0.0	0.31	0.61	0.92	1.23	1.56	1.92	2.36	2.94	3.76	5.00	6.98	10.00
0.0	0.15	0.31	0.46	0.62	0.79	0.99	1.23	1.56	2.09	3.02	5.00	10.00
0.0	0.0	0.0	0.0	0.0	0.0	0.0	0.0	0.0	0.0	0.0	0.0	0.0

Figure 13.44 Continued

ITERATION = 27 CONSTANT (C) = 0.7

0.0	10.00	10.00	10.00	10.00	10.00	10.00	10.00	10.00	10.00	10.00	10.00	10.00
0.0	5.00	6.98	7.91	8.44	8.77	9.01	9.21	9.38	9.54	9.69	9.85	10.00
0.0	3.02	5.00	6.24	7.06	7.63	8.08	8.44	8.77	9.08	9.39	9.69	10.00
0.0	2.08	3.76	5.00	5.92	6.63	7.21	7.72	8.18	8.63	9.08	9.54	10.00
0.0	1.56	2.94	4.08	5.00	5.76	6.43	7.03	7.60	8.18	8.77	9.38	10.00
0.0	1.23	2.36	3.36	4.23	5.00	5.70	6.36	7.03	7.72	8.44	9.21	10.00
0.0	0.99	1.92	2.79	3.57	4.30	5.00	5.70	6.43	7.21	8.08	9.01	10.00
0.0	0.79	1.56	2.28	2.97	3.64	4.30	5.00	5.76	6.63	7.64	8.77	10.00
0.0	0.62	1.23	1.82	2.40	2.97	3.58	4.24	5.00	5.92	7.06	8.44	10.00
0.0	0.46	0.92	1.37	1.82	2.28	2.79	3.37	4.08	5.00	6.24	7.91	10.00
0.0	0.31	0.61	0.92	1.23	1.56	1.92	2.36	2.94	3.76	5.00	6.98	10.00
0.0	0.15	0.31	0.46	0.62	0.79	0.99	1.23	1.56	2.09	3.02	5.00	10.00
0.0	0.0	0.0	0.0	0.0	0.0	0.0	0.0	0.0	0.0	0.0	0.0	0.0

ITERATION = 48 CONSTANT (C) = 0.8

0.0	10.00	10.00	10.00	10.00	10.00	10.00	10.00	10.00	10.00	10.00	10.00	10.00
0.0	5.00	6.98	7.91	8.44	8.77	9.01	9.21	9.38	9.54	9.69	9.85	10.00
0.0	3.02	5.00	6.24	7.06	7.64	8.08	8.44	8.77	9.08	9.39	9.69	10.00
0.0	2.09	3.76	5.00	5.92	6.63	7.21	7.72	8.18	8.63	9.08	9.54	10.00
0.0	1.56	2.94	4.08	5.00	5.76	6.43	7.03	7.60	8.18	8.77	9.38	10.00
0.0	1.23	2.36	3.37	4.23	5.00	5.70	6.36	7.03	7.72	8.44	9.21	10.00
0.0	0.99	1.92	2.79	3.57	4.30	5.00	5.70	6.43	7.21	8.08	9.01	10.00
0.0	0.79	1.56	2.28	2.97	3.64	4.30	5.00	5.76	6.63	7.64	8.77	10.00
0.0	0.62	1.23	1.82	2.40	2.97	3.57	4.23	5.00	5.92	7.06	8.44	10.00
0.0	0.46	0.92	1.37	1.82	2.28	2.79	3.37	4.08	5.00	6.24	7.91	10.00
0.0	0.31	0.61	0.92	1.23	1.56	1.92	2.36	2.94	3.76	5.00	6.98	10.00
0.0	0.15	0.31	0.46	0.62	0.79	0.99	1.23	1.56	2.09	3.02	5.00	10.00
0.0	0.0	0.0	0.0	0.0	0.0	0.0	0.0	0.0	0.0	0.0	0.0	0.0

Figure 13.44 Continued

ITERATION = 75 CONSTANT (C) = 1.0

0.0	10.00	10.00	10.00	10.00	10.00	10.00	10.00	10.00	10.00	10.00	10.00	10.00
0.0	4.28	8.14	8.54	10.56	10.86	10.94	11.27	8.28	14.22	8.59	8.97	10.00
0.0	2.34	3.96	6.09	7.12	7.80	8.54	6.23	9.10	13.51	6.69	11.18	10.00
0.0	0.99	2.43	4.67	5.20	5.77	4.66	5.68	9.21	10.94	7.71	10.60	10.00
0.0	-0.68	0.95	2.40	2.76	2.41	3.80	7.67	7.23	14.15	7.18	10.93	10.00
0.0	-1.03	0.44	2.72	1.51	2.08	5.83	5.74	7.69	11.73	4.73	10.32	10.00
0.0	-0.92	0.28	1.15	-0.39	3.52	5.00	5.70	6.55	10.11	4.63	10.54	10.00
0.0	-1.87	-1.51	-1.10	1.02	4.26	3.29	4.90	4.48	10.19	4.50	11.06	10.00
0.0	-3.35	-4.03	0.48	2.77	1.59	2.19	3.03	4.41	8.67	3.54	10.46	10.00
0.0	-2.90	0.69	4.06	1.28	1.53	3.12	3.94	4.40	9.61	3.46	12.06	10.00
0.0	1.53	3.31	1.08	0.68	1.75	0.62	0.70	0.85	5.37	0.22	3.35	10.00
0.0	1.02	-1.63	-1.84	-0.59	-2.22	-1.64	-1.39	-0.82	3.80	-4.95	7.43	10.00
0.0	0.0	0.0	0.0	0.0	0.0	0.0	0.0	0.0	0.0	0.0	0.0	0.0

Figure 13.44 Continued

```
      C
      C     AN INVESTIGATION OF THE SPEED UP FACTOR IN THE SOLUTION OF LAPLACES
      C     FIELD EQUATION.
      C
            DIMENSION V(169)
          1 READ(5,2)C
          2 FORMAT(F3.1)
            DO 10 J=1,169
         10 V(J)=5.0
            DO 20 J=2,12
         20 V(J)=10.0
            DO 30 J=13,156,13
         30 V(J)=10.0
            DO 31 J=14,144,13
         31 V(J)=0.0
            DO 32 J=157,168
         32 V(J)=0.0
            DO 40 I=1,75
            SWITCH=0.0
            DO 50 K=1,131,13
            L=14+K
            M=24+K
            DO 50 J=L,M
            VNEW=(V(J-1)+V(J+1)+V(J-13)+V(J+13))/4.
            IF(ABS(VNEW-V(J))-0.001)50,50,60
         60 SWITCH=1.0
         50 V(J)=VNEW+C*(VNEW-V(J))
            IF(SWITCH)40,70,40
         40 CONTINUE
         70 WRITE(6,80) I,C
         80 FORMAT('1ITERATION =',I3,5X,'CONSTANT (C) =',F4.1,///)
            WRITE(6,90) (V(J),J=1,169)
         90 FORMAT(13F8.2)
            GO TO 1
            RETURN
            END
```

Figure 13.45 A modification of the program of Figure 13.43 in which the initial values of the interior mesh points are set at 5 volts instead of 0 volts

7.5 volts, and those below the diagonal at 2.5 volts. Thus, by using a very coarse mesh size, the programmer can estimate the approximate potentials in various regions nad make use of DO loops to set the initial values accordingly.

Another method for speeding up the convergence of the Gauss-Seidel method is called the "successive overrelaxation method" by Young[5] and the "extrapolated Liebmann method" by Frankel.[6] This method calculates the new potential of a point by using the relaxation parameter W in the following expression

$$\text{VNEW} = \text{W}*(\text{V}(\text{J}+1) + \text{V}(\text{J}-1) + \text{V}(\text{J} + 13) + \text{V}(\text{J}-13))/4. + (1. - \text{W})*\text{V}(\text{J})$$

For the Gauss-Seidel method, overrelaxation must be used to speed up the convergence and values of W between 1 and 2 should be used. The actual optimum value of W will depend on the number of mesh points.

Figure 13.47 is a Fortran program that employs the overrelaxation technique to speed up convergence. This program is identical to that of

[5] David Young, *Iterative Methods for Solving Partial Differential Equations.*

[6] S. Frankel, *Convergence Rates of Iterative Treatments of Partial Differential Equations,* Math Tables Aids Comp., Vol. 4 (1950), pp. 65–75.

```
ITERATION = 27        CONSTANT (C) = 0.2

 5.00  10.00  10.00  10.00  10.00  10.00  10.00  10.00  10.00  10.00  10.00  10.00  10.00
 0.0    5.00   6.98   7.91   8.44   8.77   9.01   9.21   9.38   9.54   9.69   9.85  10.00
 0.0    3.02   5.00   6.24   7.06   7.63   8.07   8.44   8.77   9.08   9.39   9.69  10.00
 0.0    2.09   3.76   5.00   5.92   6.63   7.21   7.71   8.18   8.63   9.08   9.54  10.00
 0.0    1.56   2.94   4.08   5.00   5.76   6.42   7.02   7.60   8.18   8.77   9.38  10.00
 0.0    1.23   2.37   3.37   4.24   5.00   5.70   6.36   7.02   7.71   8.44   9.21  10.00
 0.0    0.99   1.93   2.79   3.58   4.30   5.00   5.70   6.42   7.21   8.08   9.01  10.00
 0.0    0.79   1.56   2.29   2.98   3.64   4.30   5.00   5.76   6.63   7.63   8.77  10.00
 0.0    0.62   1.23   1.82   2.40   2.98   3.58   4.24   5.00   5.92   7.06   8.44  10.00
 0.0    0.46   0.92   1.37   1.82   2.29   2.79   3.37   4.08   5.00   6.24   7.91  10.00
 0.0    0.31   0.61   0.92   1.23   1.56   1.92   2.36   2.94   3.76   5.00   6.98  10.00
 0.0    0.15   0.31   0.46   0.62   0.79   0.99   1.23   1.56   2.09   3.02   5.00  10.00
 0.0    0.0    0.0    0.0    0.0    0.0    0.0    0.0    0.0    0.0    0.0    0.0    5.00


ITERATION = 17        CONSTANT (C) = 0.4

 5.00  10.00  10.00  10.00  10.00  10.00  10.00  10.00  10.00  10.00  10.00  10.00  10.00
 0.0    5.00   6.98   7.91   8.44   8.77   9.01   9.21   9.38   9.54   9.69   9.85  10.00
 0.0    3.02   5.00   6.24   7.06   7.63   8.08   8.44   8.77   9.08   9.39   9.69  10.00
 0.0    2.09   3.76   5.00   5.92   6.63   7.21   7.71   8.18   8.63   9.08   9.54  10.00
 0.0    1.56   2.94   4.08   5.00   5.76   6.42   7.03   7.60   8.18   8.77   9.38  10.00
 0.0    1.23   2.37   3.37   4.24   5.00   5.70   6.36   7.03   7.72   8.44   9.21  10.00
 0.0    0.99   1.92   2.79   3.58   4.30   5.00   5.70   6.43   7.21   8.08   9.01  10.00
 0.0    0.79   1.56   2.29   2.97   3.64   4.30   5.00   5.76   6.63   7.64   8.77  10.00
 0.0    0.62   1.23   1.82   2.40   2.97   3.57   4.24   5.00   5.92   7.06   8.44  10.00
 0.0    0.46   0.92   1.37   1.82   2.28   2.79   3.37   4.08   5.00   6.24   7.91  10.00
 0.0    0.31   0.61   0.92   1.23   1.56   1.92   2.36   2.94   3.76   5.00   6.98  10.00
 0.0    0.15   0.31   0.46   0.62   0.79   0.99   1.23   1.56   2.09   3.02   5.00  10.00
 0.0    0.0    0.0    0.0    0.0    0.0    0.0    0.0    0.0    0.0    0.0    0.0    5.00
```

Figure 13.46 The output results of the program of Figure 13.45

```
ITERATION = 20      CONSTANT (C) = 0.5

 5.00   10.00   10.00   10.00   10.00   10.00   10.00   10.00   10.00   10.00   10.00   10.00   10.00
 0.0     5.00    6.98    7.92    8.44    8.77    9.01    9.21    9.38    9.54    9.69    9.85   10.00
 0.0     3.02    5.00    6.24    7.06    7.64    8.08    8.44    8.77    9.08    9.39    9.69   10.00
 0.0     2.08    3.76    5.00    5.92    6.63    7.21    7.72    8.18    8.63    9.08    9.54   10.00
 0.0     1.56    2.94    4.08    5.00    5.76    6.43    7.03    7.60    8.18    8.77    9.38   10.00
 0.0     1.23    2.36    3.37    4.24    5.00    5.70    6.36    7.03    7.72    8.44    9.21   10.00
 0.0     0.99    1.92    2.79    3.57    4.30    5.00    5.70    6.43    7.21    8.08    9.01   10.00
 0.0     0.79    1.56    2.28    2.97    3.64    4.30    5.00    5.76    6.63    7.64    8.77   10.00
 0.0     0.62    1.23    1.82    2.40    2.97    3.57    4.23    5.00    5.92    7.06    8.44   10.00
 0.0     0.46    0.92    1.37    1.82    2.28    2.79    3.37    4.08    5.00    6.24    7.91   10.00
 0.0     0.31    0.61    0.92    1.23    1.56    1.92    2.36    2.94    3.76    5.00    6.98   10.00
 0.0     0.15    0.31    0.46    0.62    0.79    0.99    1.23    1.56    2.09    3.02    5.00   10.00
 0.0     0.0     0.0     0.0     0.0     0.0     0.0     0.0     0.0     0.0     0.0     0.0     5.00


ITERATION = 24      CONSTANT (C) = 0.6

 5.00   10.00   10.00   10.00   10.00   10.00   10.00   10.00   10.00   10.00   10.00   10.00   10.00
 0.0     5.00    6.98    7.91    8.44    8.77    9.01    9.21    9.38    9.54    9.69    9.85   10.00
 0.0     3.02    5.00    6.24    7.06    7.64    8.08    8.44    8.77    9.08    9.39    9.69   10.00
 0.0     2.09    3.76    5.00    5.92    6.63    7.21    7.72    8.18    8.63    9.08    9.54   10.00
 0.0     1.56    2.94    4.08    5.00    5.76    6.43    7.03    7.60    8.18    8.77    9.38   10.00
 0.0     1.23    2.36    3.37    4.24    5.00    5.70    6.36    7.03    7.72    8.44    9.21   10.00
 0.0     0.99    1.92    2.79    3.57    4.30    5.00    5.70    6.43    7.21    8.08    9.01   10.00
 0.0     0.79    1.56    2.28    2.97    3.64    4.30    5.00    5.76    6.63    7.64    8.77   10.00
 0.0     0.62    1.23    1.82    2.40    2.97    3.57    4.24    5.00    5.92    7.06    8.44   10.00
 0.0     0.46    0.92    1.37    1.82    2.28    2.79    3.37    4.08    5.00    6.24    7.91   10.00
 0.0     0.31    0.61    0.92    1.23    1.56    1.92    2.36    2.94    3.76    5.00    6.98   10.00
 0.0     0.15    0.31    0.46    0.62    0.79    0.99    1.23    1.56    2.09    3.02    5.00   10.00
 0.0     0.0     0.0     0.0     0.0     0.0     0.0     0.0     0.0     0.0     0.0     0.0     5.00
```

Figure 13.46 Continued

```
ITERATION = 27      CONSTANT (C) = 0.7

 5.00  10.00  10.00  10.00  10.00  10.00  10.00  10.00  10.00  10.00  10.00  10.00  10.00
 0.0    5.00   6.98   7.91   8.44   8.77   9.01   9.21   9.38   9.54   9.69   9.85  10.00
 0.0    3.02   5.00   6.24   7.06   7.64   8.08   8.44   8.77   9.08   9.39   9.69  10.00
 0.0    2.09   3.76   5.00   5.92   6.63   7.21   7.72   8.18   8.63   9.08   9.54  10.00
 0.0    1.56   2.94   4.08   5.00   5.76   6.43   7.03   7.60   8.18   8.77   9.38  10.00
 0.0    1.23   2.36   3.37   4.23   5.00   5.70   6.36   7.03   7.72   8.44   9.21  10.00
 0.0    0.99   1.92   2.79   3.57   4.30   5.00   5.70   6.43   7.21   8.08   9.01  10.00
 0.0    0.79   1.56   2.28   2.97   3.64   4.30   5.00   5.76   6.63   7.64   8.77  10.00
 0.0    0.62   1.23   1.82   2.40   2.97   3.57   4.24   5.00   5.92   7.06   8.44  10.00
 0.0    0.46   0.92   1.37   1.82   2.28   2.79   3.37   4.08   5.00   6.24   7.91  10.00
 0.0    0.31   0.61   0.92   1.23   1.56   1.92   2.36   2.94   3.76   5.00   6.98  10.00
 0.0    0.15   0.31   0.46   0.62   0.79   0.99   1.23   1.56   2.09   3.02   5.00  10.00
 0.0    0.0    0.0    0.0    0.0    0.0    0.0    0.0    0.0    0.0    0.0    0.0    5.00
```

```
ITERATION = 40      CONSTANT (C) = 0.8

 5.00  10.00  10.00  10.00  10.00  10.00  10.00  10.00  10.00  10.00  10.00  10.00  10.00
 0.0    5.00   6.98   7.91   8.44   8.77   9.01   9.21   9.38   9.54   9.69   9.85  10.00
 0.0    3.02   5.00   6.24   7.06   7.64   8.08   8.44   8.77   9.08   9.39   9.69  10.00
 0.0    2.09   3.76   5.00   5.92   6.63   7.21   7.72   8.18   8.63   9.08   9.54  10.00
 0.0    1.56   2.94   4.08   5.00   5.76   6.43   7.03   7.60   8.18   8.77   9.38  10.00
 0.0    1.23   2.36   3.37   4.24   5.00   5.70   6.36   7.03   7.72   8.44   9.21  10.00
 0.0    0.99   1.92   2.79   3.57   4.30   5.00   5.70   6.42   7.21   8.08   9.01  10.00
 0.0    0.79   1.56   2.28   2.97   3.64   4.30   5.00   5.76   6.63   7.64   8.77  10.00
 0.0    0.62   1.23   1.82   2.40   2.97   3.57   4.24   5.00   5.92   7.06   8.44  10.00
 0.0    0.46   0.92   1.37   1.82   2.28   2.79   3.37   4.08   5.00   6.24   7.91  10.00
 0.0    0.31   0.61   0.92   1.23   1.56   1.92   2.36   2.94   3.76   5.00   6.98  10.00
 0.0    0.15   0.31   0.46   0.62   0.79   0.99   1.23   1.56   2.09   3.02   5.00  10.00
 0.0    0.0    0.0    0.0    0.0    0.0    0.0    0.0    0.0    0.0    0.0    0.0    5.00
```

Figure 13.46 Continued

```
ITERATION = 75     CONSTANT (C) = 1.0

 5.00  10.00  10.00  10.00  10.00  10.00  10.00  10.00  10.00  10.00  10.00  10.00  10.00
 0.0    5.00   7.90   8.78  10.62  10.95  10.93  11.57  10.81  13.56   8.53   8.97  10.00
 0.0    2.10   5.00   6.83   8.08   8.68   9.13   8.87  11.56  11.41   6.69  11.40  10.00
 0.0    1.22   3.17   5.00   6.40   6.52   6.75   8.39   9.37   8.44   8.32  11.22  10.00
 0.0   -0.62   1.92   3.60   5.00   5.45   7.10   8.32   7.23  11.43   8.25  10.76  10.00
 0.0   -0.95   1.32   3.48   4.55   5.00   6.15   5.74   8.05  10.10   6.49  11.27  10.00
 0.0   -0.93   0.87   3.25   2.90   3.85   5.00   6.20   7.18   8.50   7.01  11.09  10.CO
 0.0   -1.57   1.13   1.61   1.68   4.26   3.80   5.00   5.72   8.12   6.9C  11.22  10.00
 0.0   -0.81  -1.56   0.63   2.77   1.95   2.82   4.28   5.00   7.14   6.35  10.64  10.00
 0.0   -3.56  -1.41   1.56  -1.43  -0.10   1.50   1.88   2.86   5.00   5.04   9.13  10.00
 0.0    1.47   3.31   1.68   1.75   3.51   2.99   3.10   3.65   4.96   5.00   9.15  10.00
 0.0    1.02  -1.40  -1.22  -0.76  -1.27  -1.09  -1.22  -0.64   0.87   0.85   5.00  10.00
 0.0    0.0    0.0    0.0    0.0    0.0    0.0    0.0    0.0    0.0    0.0    0.0    5.00

IHC217I
```

Figure 13.46 Continued

```
C       AN INVESTIGATION OF THE SPEED UP FACTOR IN THE SOLUTION OF LAPLACES
C       FIELD EQUATION. USING THE RELAXATION METHOD.
C
        DIMENSION V(169)
      1 READ(5,2)W
      2 FORMAT(E15.7)
        DO 10 J=1,169
     10 V(J)=0.0
        DO 20 J=2,12
     20 V(J)=10.0
        DO 30 J=13,156,13
     30 V(J)=10.0
        DO 40 I=1,75
        SWITCH=0.0
        DO 50 K=1,131,13
        L=14+K
        M=24+K
        DO 50 J=L.M
        VNEW=W*(V(J-1)+V(J+1)+V(J-13)+V(J+13))/4.+(1.-W)*V(J)
        IF(ABS(VNEW-V(J))-0.001)50,50,60
     60 SWITCH=1.0
     50 V(J)=VNEW
        IF(SWITCH)40,70,40
     40 CONTINUE
     70 WRITE(6,80)I,W
     80 FORMAT( '1ITERATION =',I3,5X,'CONSTANT (W) =',E15.7,///)
        WRITE(6,90)(V(J),J=1,169)
     90 FORMAT(13F8.2)
        GO TO 1
        RETURN
        END
```

Figure 13.47 The overrelaxation technique is used to speed up convergence

Figure 13.43 except for the overrelaxation statement above. Figure 13.48 shows the corresponding computer output. Note that a choice of $W = 1.6$ requires only 26 iterations to achieve the same accuracy that was specified for the program of Figure 13.43.

Before leaving this problem, the reader should realize that the knowledge of the potentials at the regular boundaries of Figure 13.37 avoided certain other difficulties. Since the potentials of the two boundaries were specified, only the potentials of interior mesh points were unknown. If some of the boundaries had been insulated (like the two sides), the mesh potentials of points on these boundaries would have to be computed. This may be done in a way similar to the method employed for interior points. For example, Figure 13.49a shows a mesh point on an insulated boundary that is not at a corner. The potentials of points similar to this one would be found from

$$V(I,J) = (2.*V(I,J-1) + V(I+1,J) + V(I-1,J))/4.$$

Similarly, for points at insulated exterior corners, such as the one shown in Figure 13.49b, the potential would be found from

$$V(I,J) = (V(I,J-1) + V(I+1,J))/2.$$

and for points at reentrant corners, such as the one shown in Figure 13.49c, the potential would be determined from

$$V(I,J) = (V(I,J+1) + 2.*V(I,J-1) + 2.*(I+1,J) + V(I-1,J))/6.$$

```
ITERATION = 66     CONSTANT (W) =  0.1200000E 01

 0.0   10.00  10.00  10.00  10.00  10.00  10.00  10.00  10.00  10.00  10.00  10.00  10.00
 0.0    5.00   6.98   7.91   8.43   8.77   9.01   9.21   9.38   9.54   9.69   9.85  10.00
 0.0    3.02   5.00   6.24   7.06   7.63   8.07   8.44   8.77   9.08   9.39   9.69  10.00
 0.0    2.08   3.75   4.99   5.92   6.63   7.21   7.71   8.18   8.63   9.08   9.54  10.00
 0.0    1.56   2.94   4.07   4.99   5.76   6.42   7.02   7.60   8.18   8.77   9.38  10.00
 0.0    1.23   2.36   3.36   4.23   4.99   5.69   6.35   7.02   7.71   8.44   9.21  10.00
 0.0    0.98   1.92   2.78   3.57   4.30   4.99   5.69   6.42   7.21   8.07   9.01  10.00
 0.0    0.79   1.55   2.28   2.97   3.63   4.30   4.99   5.76   6.63   7.63   8.77  10.00
 0.0    0.62   1.22   1.81   2.39   2.97   3.57   4.23   5.00   5.92   7.06   8.44  10.00
 0.0    0.46   0.91   1.36   1.81   2.28   2.78   3.36   4.07   5.00   6.24   7.91  10.00
 0.0    0.31   0.61   0.91   1.23   1.55   1.92   2.36   2.94   3.75   5.00   6.98  10.00
 0.0    0.15   0.31   0.46   0.62   0.79   0.98   1.23   1.56   2.08   3.02   5.00  10.00
 0.0    0.0    0.0    0.0    0.0    0.0    0.0    0.0    0.0    0.0    0.0    0.0    0.0


ITERATION = 54     CONSTANT (W) =  0.1299999E 01

 0.0   10.00  10.00  10.00  10.00  10.00  10.00  10.00  10.00  10.00  10.00  10.00  10.00
 0.0    5.00   6.98   7.91   8.43   8.77   9.01   9.21   9.38   9.54   9.69   9.85  10.00
 0.0    3.02   5.00   6.24   7.06   7.63   8.07   8.44   8.77   9.08   9.39   9.69  10.00
 0.0    2.08   3.75   5.00   5.92   6.63   7.21   7.71   8.18   8.63   9.08   9.54  10.00
 0.0    1.56   2.94   4.07   4.99   5.76   6.42   7.02   7.60   8.18   8.77   9.38  10.00
 0.0    1.23   2.36   3.36   4.23   4.99   5.69   6.36   7.02   7.71   8.44   9.21  10.00
 0.0    0.98   1.92   2.78   3.57   4.30   4.99   5.69   6.42   7.21   8.07   9.01  10.00
 0.0    0.79   1.55   2.28   2.97   3.63   4.30   4.99   5.76   6.63   7.63   8.77  10.00
 0.0    0.62   1.23   1.81   2.39   2.97   3.57   4.23   5.00   5.92   7.06   8.44  10.00
 0.0    0.46   0.91   1.37   1.82   2.28   2.78   3.36   4.07   5.00   6.24   7.91  10.00
 0.0    0.31   0.61   0.92   1.23   1.55   1.92   2.36   2.94   3.75   5.00   6.98  10.00
 0.0    0.15   0.31   0.46   0.62   0.79   0.98   1.23   1.56   2.08   3.02   5.00  10.00
 0.0    0.0    0.0    0.0    0.0    0.0    0.0    0.0    0.0    0.0    0.0    0.0    0.0
```

Figure 13.48 The output results of the program of Figure 13.47

```
ITERATION = 44      CONSTANT (W) =  0.1400000E 01

 0.0   10.00  10.00  10.00  10.00  10.00  10.00  10.00  10.00  10.00  10.00  10.00  10.00
 0.0    5.00   6.98   7.91   8.43   8.77   9.01   9.21   9.38   9.54   9.69   9.85  10.00
 0.0    3.02   5.00   6.24   7.06   7.63   8.07   8.44   8.77   9.08   9.39   9.69  10.00
 0.0    2.08   3.75   5.00   5.92   6.63   7.21   7.71   8.18   8.63   9.08   9.54  10.00
 0.0    1.56   2.94   4.07   5.00   5.76   6.42   7.02   7.60   8.18   8.77   9.38  10.00
 0.0    1.23   2.36   3.36   4.23   5.00   5.69   6.36   7.02   7.71   8.44   9.21  10.00
 0.0    0.98   1.92   2.78   3.57   4.30   5.00   5.69   6.42   7.21   8.08   9.01  10.00
 0.0    0.79   1.55   2.28   2.97   3.63   4.30   5.00   5.76   6.63   7.63   8.77  10.00
 0.0    0.62   1.23   1.82   2.39   2.97   3.57   4.23   5.00   5.92   7.06   8.44  10.00
 0.0    0.46   0.92   1.37   1.82   2.28   2.78   3.36   4.07   5.00   6.24   7.91  10.00
 0.0    0.31   0.61   0.92   1.23   1.55   1.92   2.36   2.94   3.76   5.00   6.98  10.00
 0.0    0.15   0.31   0.46   0.62   0.79   0.98   1.23   1.56   2.08   3.02   5.00  10.00
 0.0    0.0    0.0    0.0    0.0    0.0    0.0    0.0    0.0    0.0    0.0    0.0    0.0


ITERATION = 34      CONSTANT (W) =  0.1500000E 01

 0.0   10.00  10.00  10.00  10.00  10.00  10.00  10.00  10.00  10.00  10.00  10.00  10.00
 0.0    5.00   6.98   7.91   8.44   8.77   9.01   9.21   9.38   9.54   9.69   9.85  10.00
 0.0    3.02   5.00   6.24   7.06   7.63   8.07   8.44   8.77   9.08   9.39   9.69  10.00
 0.0    2.08   3.75   5.00   5.92   6.63   7.21   7.71   8.18   8.63   9.08   9.54  10.00
 0.0    1.56   2.94   4.07   5.00   5.76   6.42   7.02   7.60   8.18   8.77   9.38  10.00
 0.0    1.23   2.36   3.36   4.23   5.00   5.69   6.36   7.02   7.71   8.44   9.21  10.00
 0.0    0.98   1.92   2.78   3.57   4.30   5.00   5.69   6.42   7.21   8.08   9.01  10.00
 0.0    0.79   1.55   2.28   2.97   3.64   4.30   5.00   5.76   6.63   7.63   8.77  10.00
 0.0    0.62   1.23   1.82   2.39   2.97   3.57   4.23   5.00   5.92   7.06   8.44  10.00
 0.0    0.46   0.92   1.37   1.82   2.28   2.79   3.36   4.08   5.00   6.24   7.91  10.00
 0.0    0.31   0.61   0.92   1.23   1.56   1.92   2.36   2.94   3.76   5.00   6.98  10.00
 0.0    0.15   0.31   0.46   0.62   0.79   0.99   1.23   1.56   2.09   3.02   5.00  10.00
 0.0    0.0    0.0    0.0    0.0    0.0    0.0    0.0    0.0    0.0    0.0    0.0    0.0
```

Figure 13.48 Continued

```
ITERATION = 29      CONSTANT (W) =  0.1549999E 01

 0.0   10.00  10.00  10.00  10.00  10.00  10.00  10.00  10.00  10.00  10.00  10.00  10.00
 0.0    5.00   6.98   7.91   8.44   8.77   9.01   9.21   9.38   9.54   9.69   9.85  10.00
 0.0    3.02   5.00   6.24   7.06   7.63   8.08   8.44   8.77   9.08   9.39   9.69  10.00
 0.0    2.08   3.75   5.00   5.92   6.63   7.21   7.71   8.18   8.63   9.08   9.54  10.00
 0.0    1.56   2.94   4.07   5.00   5.76   6.42   7.03   7.60   8.18   8.77   9.38  10.00
 0.0    1.23   2.36   3.36   4.23   5.00   5.69   6.36   7.03   7.72   8.44   9.21  10.00
 0.0    0.98   1.92   2.79   3.57   4.30   5.00   5.70   6.42   7.21   8.08   9.01  10.00
 0.0    0.79   1.56   2.28   2.97   3.64   4.30   5.00   5.76   6.63   7.64   8.77  10.00
 0.0    0.62   1.23   1.82   2.39   2.97   3.57   4.23   5.00   5.92   7.06   8.44  10.00
 0.0    0.46   0.92   1.37   1.82   2.28   2.79   3.37   4.08   5.00   6.24   7.91  10.00
 0.0    0.31   0.61   0.92   1.23   1.56   1.92   2.36   2.94   3.76   5.00   6.98  10.00
 0.0    0.15   0.31   0.46   0.62   0.79   0.99   1.23   1.56   2.09   3.02   5.00  10.00
 0.0    0.0    0.0    0.0    0.0    0.0    0.0    0.0    0.0    0.0    0.0    0.0    0.0


ITERATION = 26      CONSTANT (W) =  0.1599999E 01

 0.0   10.00  10.00  10.00  10.00  10.00  10.00  10.00  10.00  10.00  10.00  10.00  10.00
 0.0    5.00   6.98   7.91   8.44   8.77   9.01   9.21   9.38   9.54   9.69   9.85  10.00
 0.0    3.02   5.00   6.24   7.06   7.64   8.08   8.44   8.77   9.08   9.39   9.69  10.00
 0.0    2.09   3.76   5.00   5.92   6.63   7.21   7.72   8.18   8.63   9.08   9.54  10.00
 0.0    1.56   2.94   4.08   5.00   5.76   6.43   7.03   7.60   8.18   8.77   9.38  10.00
 0.0    1.23   2.36   3.37   4.24   5.00   5.70   6.36   7.03   7.72   8.44   9.21  10.00
 0.0    0.99   1.92   2.79   3.57   4.30   5.00   5.70   6.43   7.21   8.08   9.01  10.00
 0.0    0.79   1.56   2.28   2.97   3.64   4.30   5.00   5.76   6.63   7.64   8.77  10.00
 0.0    0.62   1.23   1.82   2.40   2.97   3.57   4.24   5.00   5.92   7.06   8.44  10.00
 0.0    0.46   0.92   1.37   1.82   2.28   2.79   3.37   4.08   5.00   6.24   7.91  10.00
 0.0    0.31   0.61   0.92   1.23   1.56   1.92   2.36   2.94   3.76   5.00   6.98  10.00
 0.0    0.15   0.31   0.46   0.62   0.79   0.99   1.23   1.56   2.09   3.02   5.00  10.00
 0.0    0.0    0.0    0.0    0.0    0.0    0.0    0.0    0.0    0.0    0.0    0.0    0.0
```

Figure 13.48 Continued

```
ITERATION = 26       CONSTANT (W) =  0.1650000E 01

  0.0    10.00   10.00   10.00   10.00   10.00   10.00   10.00   10.00   10.00   10.00   10.00   10.00
  0.0     5.00    6.98    7.91    8.44    8.77    9.01    9.21    9.38    9.54    9.69    9.85   10.00
  0.0     3.02    5.00    6.24    7.06    7.64    8.08    8.44    8.77    9.08    9.39    9.69   10.00
  0.0     2.09    3.76    5.00    5.92    6.63    7.21    7.72    8.18    8.63    9.08    9.54   10.00
  0.0     1.56    2.94    4.08    5.00    5.76    6.42    7.03    7.60    8.18    8.77    9.38   10.00
  0.0     1.23    2.36    3.37    4.23    5.00    5.70    6.36    7.03    7.72    8.44    9.21   10.00
  0.0     0.99    1.92    2.79    3.57    4.30    5.00    5.70    6.43    7.21    8.08    9.01   10.00
  0.0     0.79    1.56    2.28    2.97    3.64    4.30    5.00    5.76    6.63    7.64    8.77   10.00
  0.0     0.62    1.23    1.82    2.40    2.97    3.57    4.23    5.00    5.92    7.06    8.44   10.00
  0.0     0.46    0.92    1.37    1.82    2.28    2.79    3.37    4.08    5.00    6.24    7.91   10.00
  0.0     0.31    0.61    0.92    1.23    1.56    1.92    2.36    2.94    3.76    5.00    6.98   10.00
  0.0     0.15    0.31    0.46    0.62    0.79    0.99    1.23    1.56    2.09    3.02    5.00   10.00
  0.0     0.0     0.0     0.0     0.0     0.0     0.0     0.0     0.0     0.0     0.0     0.0     0.0
```

```
ITERATION = 28       CONSTANT (W) =  0.1700000E 01

  0.0    10.00   10.00   10.00   10.00   10.00   10.00   10.00   10.00   10.00   10.00   10.00   10.00
  0.0     5.00    6.98    7.91    8.44    8.77    9.01    9.21    9.38    9.54    9.69    9.85   10.00
  0.0     3.02    5.00    6.24    7.06    7.64    8.08    8.44    8.77    9.08    9.39    9.69   10.00
  0.0     2.08    3.76    5.00    5.92    6.63    7.21    7.72    8.18    8.63    9.08    9.54   10.00
  0.0     1.56    2.94    4.08    5.00    5.77    6.43    7.03    7.60    8.18    8.77    9.38   10.00
  0.0     1.23    2.36    3.37    4.24    5.00    5.70    6.36    7.03    7.72    8.44    9.21   10.00
  0.0     0.99    1.92    2.79    3.58    4.30    5.00    5.70    6.43    7.21    8.08    9.01   10.00
  0.0     0.79    1.56    2.28    2.97    3.64    4.30    5.00    5.76    6.63    7.64    8.77   10.00
  0.0     0.62    1.23    1.82    2.40    2.97    3.57    4.23    5.00    5.92    7.06    8.44   10.00
  0.0     0.46    0.92    1.37    1.82    2.28    2.79    3.37    4.08    5.00    6.24    7.91   10.00
  0.0     0.31    0.61    0.92    1.23    1.56    1.92    2.36    2.94    3.76    5.00    6.98   10.00
  0.0     0.15    0.31    0.46    0.62    0.79    0.99    1.23    1.56    2.09    3.02    5.00   10.00
  0.0     0.0     0.0     0.0     0.0     0.0     0.0     0.0     0.0     0.0     0.0     0.0     0.0
```

Figure 13.48 Continued

```
ITERATION = 35      CONSTANT (W) =  0.1750000E 01

  0.0   10.00  10.00  10.00  10.00  10.00  10.00  10.00  10.00  10.00  10.00  10.00  10.00
  0.0    5.00   6.98   7.91   8.44   8.77   9.01   9.21   9.38   9.54   9.69   9.85  10.00
  0.0    3.02   5.00   6.24   7.06   7.63   8.08   8.44   8.77   9.08   9.39   9.69  10.00
  0.0    2.08   3.76   5.00   5.92   6.63   7.21   7.72   8.18   8.63   9.08   9.54  10.00
  0.0    1.56   2.94   4.08   5.00   5.76   6.42   7.03   7.60   8.18   8.77   9.38  10.00
  0.0    1.23   2.36   3.37   4.23   5.00   5.70   6.36   7.03   7.72   8.44   9.21  10.00
  0.0    0.99   1.92   2.79   3.57   4.30   5.00   5.70   6.43   7.21   8.08   9.01  10.00
  0.0    0.79   1.56   2.28   2.97   3.64   4.30   5.00   5.76   6.63   7.64   8.77  10.00
  0.0    0.62   1.23   1.82   2.40   2.97   3.57   4.23   5.00   5.92   7.06   8.44  10.00
  0.0    0.46   0.92   1.37   1.82   2.28   2.79   3.37   4.08   5.00   6.24   7.91  10.00
  0.0    0.31   0.61   0.92   1.23   1.56   1.92   2.36   2.94   3.76   5.00   6.98  10.00
  0.0    0.15   0.31   0.46   0.62   0.79   0.99   1.23   1.56   2.09   3.02   5.00  10.00
  0.0    0.0    0.0    0.0    0.0    0.0    0.0    0.0    0.0    0.0    0.0    0.0    0.0

ITERATION = 49      CONSTANT (W) =  0.1799999E 01

  0.0   10.00  10.00  10.00  10.00  10.00  10.00  10.00  10.00  10.00  10.00  10.00  10.00
  0.0    5.00   6.98   7.91   8.44   8.77   9.01   9.21   9.38   9.54   9.69   9.85  10.00
  0.0    3.02   5.00   6.24   7.06   7.64   8.08   8.44   8.77   9.08   9.39   9.69  10.00
  0.0    2.09   3.76   5.00   5.92   6.63   7.21   7.72   8.18   8.63   9.08   9.54  10.00
  0.0    1.56   2.94   4.08   5.00   5.76   6.43   7.03   7.60   8.18   8.77   9.38  10.00
  0.0    1.23   2.36   3.37   4.23   5.00   5.70   6.36   7.03   7.72   8.44   9.21  10.00
  0.0    0.99   1.92   2.79   3.57   4.30   5.00   5.70   6.43   7.21   8.08   9.01  10.00
  0.0    0.79   1.56   2.28   2.97   3.64   4.30   5.00   5.76   6.63   7.64   8.77  10.00
  0.0    0.62   1.23   1.82   2.40   2.97   3.57   4.24   5.00   5.92   7.06   8.44  10.00
  0.0    0.46   0.92   1.37   1.82   2.28   2.79   3.37   4.08   5.00   6.24   7.91  10.00
  0.0    0.31   0.61   0.92   1.23   1.56   1.92   2.36   2.94   3.76   5.00   6.98  10.00
  0.0    0.15   0.31   0.46   0.62   0.79   0.99   1.23   1.56   2.09   3.02   5.00  10.00
  0.0    0.0    0.0    0.0    0.0    0.0    0.0    0.0    0.0    0.0    0.0    0.0    0.0
```

Figure 13.48 Continued

```
ITERATION = 61      CONSTANT (W) =  0.1849999E 01
```

0.0	10.00	10.00	10.00	10.00	10.00	10.00	10.00	10.00	10.00	10.00	10.00	10.00
0.0	5.00	6.98	7.92	8.44	8.77	9.01	9.21	9.38	9.54	9.69	9.85	10.00
0.0	3.02	5.00	6.24	7.06	7.64	8.08	8.44	8.77	9.08	9.39	9.69	10.00
0.0	2.09	3.76	5.00	5.92	6.63	7.21	7.72	8.18	8.63	9.08	9.54	10.00
0.0	1.56	2.94	4.08	5.00	5.76	6.43	7.03	7.60	8.18	8.77	9.38	10.00
0.0	1.23	2.36	3.37	4.24	5.00	5.70	6.36	7.03	7.72	8.44	9.21	10.00
0.0	0.99	1.92	2.79	3.57	4.30	5.00	5.70	6.43	7.21	8.08	9.01	10.00
0.0	0.79	1.56	2.28	2.97	3.64	4.30	5.00	5.76	6.63	7.64	8.77	10.00
0.0	0.62	1.23	1.82	2.40	2.97	3.57	4.23	5.00	5.92	7.06	8.44	10.00
0.0	0.46	0.92	1.37	1.82	2.28	2.79	3.37	4.08	5.00	6.24	7.91	10.00
0.0	0.31	0.61	0.92	1.23	1.56	1.92	2.36	2.94	3.76	5.00	6.98	10.00
0.0	0.15	0.31	0.46	0.62	0.79	0.99	1.23	1.56	2.09	3.02	5.00	10.00
0.0	0.0	0.0	0.0	0.0	0.0	0.0	0.0	0.0	0.0	0.0	0.0	0.0

```
ITERATION = 75      CONSTANT (W) =  0.1900000E 01
```

0.0	10.00	10.00	10.00	10.00	10.00	10.00	10.00	10.00	10.00	10.00	10.00	10.00
0.0	5.00	6.98	7.92	8.44	8.77	9.02	9.21	9.38	9.54	9.69	9.85	10.00
0.0	3.02	5.00	6.24	7.06	7.64	8.08	8.44	8.77	9.08	9.39	9.69	10.00
0.0	2.09	3.76	5.00	5.92	6.64	7.21	7.72	8.18	8.63	9.08	9.54	10.00
0.0	1.56	2.94	4.08	5.00	5.76	6.42	7.03	7.60	8.18	8.77	9.38	10.00
0.0	1.23	2.36	3.37	4.23	5.00	5.70	6.36	7.03	7.72	8.44	9.21	10.00
0.0	0.99	1.92	2.79	3.57	4.30	5.00	5.70	6.42	7.21	8.08	9.01	10.00
0.0	0.79	1.56	2.28	2.97	3.64	4.30	5.00	5.76	6.64	7.63	8.77	10.00
0.0	0.62	1.23	1.82	2.40	2.97	3.57	4.23	5.00	5.92	7.06	8.44	10.00
0.0	0.46	0.92	1.37	1.82	2.28	2.79	3.37	4.08	5.00	6.24	7.92	10.00
0.0	0.31	0.61	0.92	1.23	1.56	1.92	2.36	2.94	3.76	5.00	6.98	10.00
0.0	0.15	0.31	0.46	0.62	0.79	0.98	1.23	1.56	2.09	3.02	5.00	10.00
0.0	0.0	0.0	0.0	0.0	0.0	0.0	0.0	0.0	0.0	0.0	0.0	0.0

```
IHC217I
```

Figure 13.48 Continued

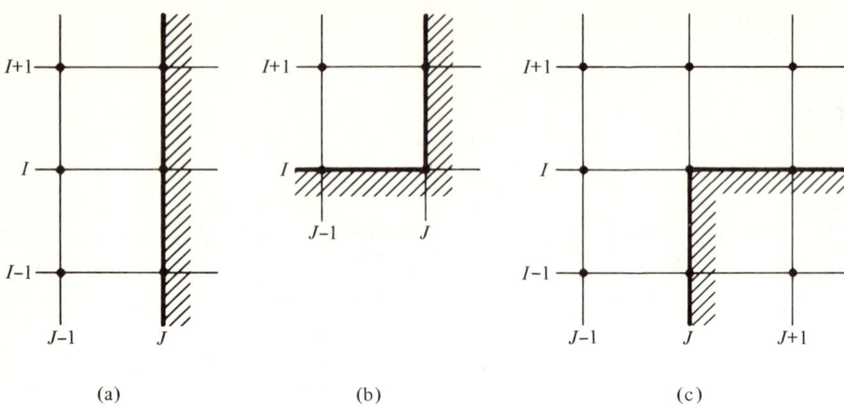

Figure 13.49 The calculation of points on insulated boundaries

Another problem that must be considered is the calculation of the potentials of interior mesh points close to irregular boundaries. Figure 13.50 shows such a point (x,y) located close to an irregular boundary. Since the boundary points 1 and 3 do not lie a distance h away from the point 0, we can not calculate the potential of V_0 as the average of V_1, V_2, V_3, and V_4. One solution to this type of problem is to use linear interpolation. In other words, the potential V_0 may be computed from

$$V_0 = \frac{V_2 + \dfrac{1}{1+a}(V_1 - V_2) + V_4 + \dfrac{1}{1+b}(V_3 - V_4)}{4}$$

It is left as an exercise for the reader to show how linear interpolation may be incorporated into the computer program.

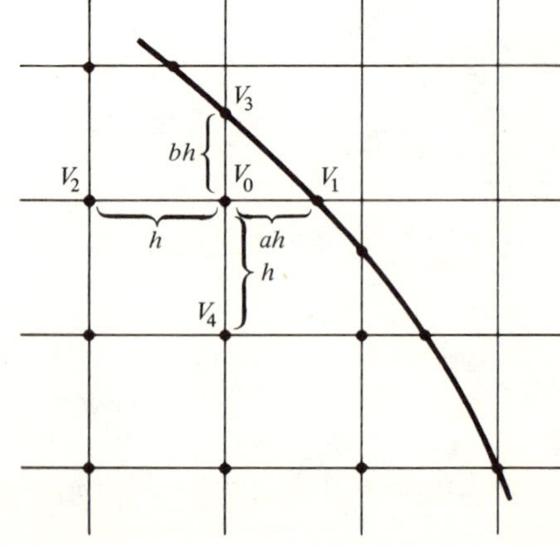

Figure 13.50 The calculation of interior mesh potentials close to irregular boundaries

Parabolic and Hyperbolic Equations

For a simple electrical example that gives rise to a parabolic or hyperbolic partial differential equation, consider the flow of electricity in a transmission line. Representing the transmission line in terms of a series of cascaded incremental T-sections having a series resistance R and inductance L per unit length and a shunt conductance G and capacitance C per unit length, we first write the voltage and current equations for the arbitrary section shown in Figure 13.51. From Kirchhoff's voltage law we find the incremental change in voltage for the length Δx is given by

$$\Delta v = -R\,\Delta x i - L\,\Delta x\,\frac{\partial i}{\partial t}$$

and from the current law, the incremental change in current is given by

$$\Delta i = -G\,\Delta x v - C\,\Delta x\,\frac{\partial v}{\partial t}$$

Rewriting these equations for an infinitesimal section, we obtain

$$\frac{\partial v}{\partial x} + Ri + L\,\frac{\partial i}{\partial t} = 0$$

$$\frac{\partial i}{\partial x} + Gv + C\,\frac{\partial v}{\partial t} = 0$$

Differentiating the first equation with respect to x and the second equation with respect to t and solving the two resulting equations for v, we find

$$\frac{\partial^2 v}{\partial x^2} = LC\,\frac{\partial^2 v}{\partial t^2} + (RC + GL)\,\frac{\partial v}{\partial t} + RG\,v$$

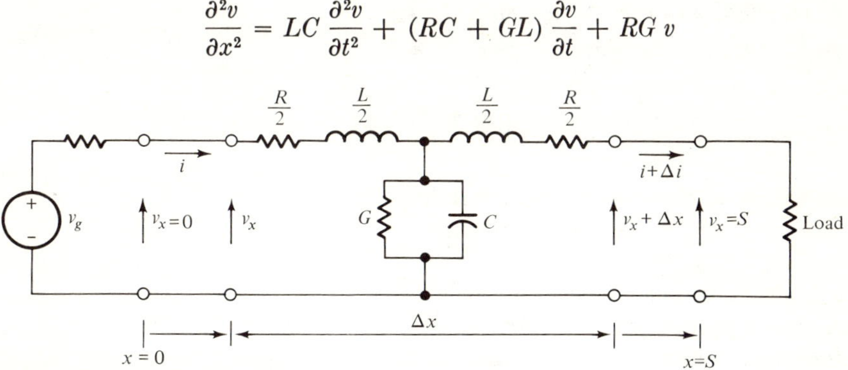

Figure 13.51 A transmission represented in terms of a cascade of infinitesimal T-sections

This is a one-dimensional electric field equation and is often referred to as the telegraph equation because it defines the voltage distribution along telegraph transmission lines.

Similarly, we obtain the equation for the current i

$$\frac{\partial^2 i}{\partial x^2} = LC \frac{\partial^2 i}{\partial t^2} + (RC + GL) \frac{\partial i}{\partial t} + RG\, i$$

Since these two equations have the same mathematical form, the general equation may be written as

$$\frac{\partial^2 u}{\partial x^2} = LC \frac{\partial^2 u}{\partial t^2} + (RC + GL) \frac{\partial u}{\partial t} + RG\, u$$

where $u(x,t)$ is replaced by either $v(x,t)$ or $i(x,t)$.

Frequently certain transmission line constants may be neglected, and this neglect leads to special cases. For example,

1. $L = C = 0$ yields

$$\frac{\partial^2 u}{\partial x^2} = K$$

   which is the one-dimensional elliptic partial differential equation.

2. $G = L = 0$ or $R = G = 0$ yields

$$\frac{\partial^2 u}{\partial x^2} = K \frac{\partial u}{\partial t}$$

   which is the one-dimensional parabolic partial differential equation called the diffusion equation.

3. $R = G = 0$ yields

$$\frac{\partial^2 u}{\partial x^2} = K \frac{\partial^2 u}{\partial t^2}$$

   which is the one-dimensional hyperbolic partial differential equation called the wave equation.

We see that under certain circumstances the one-dimensional transmission line equation may be an elliptic, parabolic, or hyperbolic partial differential equation. The complete solution for these three cases would therefore require a knowledge of the *boundary conditions* ($u(0,t)$ and $u(S,t)$) and the *initial conditions* ($u(x,0)$ and $\partial u(x,0)/\partial t$).

The general form of these second-order partial differential equations would employ the Laplacian operator ∇^2, which is equal to

$$\nabla^2 = \frac{\partial^2}{\partial x^2} + \frac{\partial^2}{\partial y^2} + \frac{\partial^2}{\partial z^2}$$

in the rectangular cartesian coordinate system. The elliptic partial differential equation known as Laplace's equation would therefore be written as

$$\nabla^2 u = 0$$

The general diffusion equation would be written as

$$\nabla^2 u = K \frac{\partial u}{\partial t}$$

and the general wave equation as

$$\nabla^2 u = K \frac{\partial^2 u}{\partial t^2}$$

The finite difference method may also be employed in the solution of parabolic and hyperbolic partial differential equations. For example, in the case of the transmission line one-dimensional diffusion equation we would let

$$\frac{\partial u}{\partial t} \approx \frac{u(x, t + \Delta t) - u(x,t)}{\Delta t}$$

and

$$\frac{\partial^2 u}{\partial x^2} \approx \frac{u(x + h, t) - 2u(x,t) + u(x - h, t)}{h^2}$$

so that

$$u(x, t + \Delta t) = u(x,t) + \frac{\Delta t}{h^2} RC(u(x + h, t) - 2u(x,t) + u(x - h, t))$$

This equation can therefore be used to determine the voltage (current) at a particular mesh point $(x, t + \Delta t)$ from the knowledge of the voltage at the mesh points $(x - h, t)$, (x,t), and $(x + h, t)$. A graphical interpretation of this procedure is shown in Figure 13.52.

Note that in the case of the solution of the elliptic partial differential equation, the value of the solution at any mesh point depended on the values at all the boundary points. Note also that we were required to obtain the solution at all mesh points before we could obtain the solution at any particular mesh point. In the case of the parabolic differential

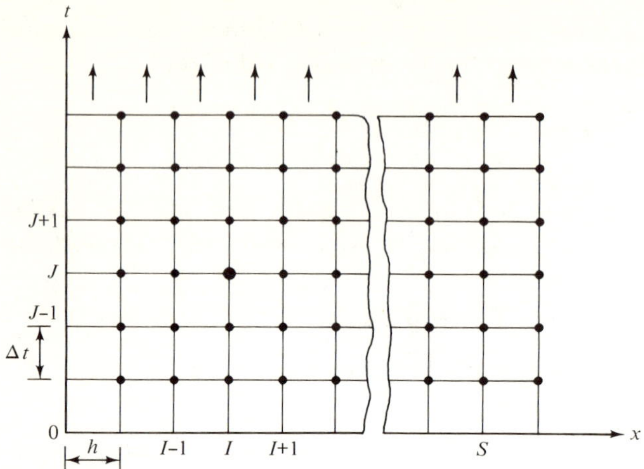

Figure 13.52 A graphical interpretation of the finite difference method of solution of the one-dimensional parabolic partial differential equation

equation, the solution at any mesh point depends on the initial and boundary conditions. The solution of the parabolic equation therefore requires a sort of "marching" process to compute the solution at a given x for subsequent values of time. For example, knowing the voltage at position I and time J (see Figure 13.52), we can find the voltage at point I and time $J + 1$ from the relation

$$u(I, J + 1) = u(I,J) + \frac{\Delta t}{h^2} RC(u(I + 1, J) - 2u(I,J) + u(I - 1, J))$$

The above procedure is often limited because of the problem of stability. In many cases a very small value of Δt is required to obtain a stable solution and, as a consequence, a large number of time steps is needed to obtain the solution for a given value of time. In general, stable operation will result for

$$RC \frac{\Delta t}{h^2} \leq 0.25$$

In the case of the solution of the transmission line one-dimensional hyperbolic partial differential equation, the finite difference method requires that we let

$$\frac{\partial^2 u}{\partial t^2} = \frac{u(x, t + \Delta t) - 2u(x,t) + u(x, t - \Delta t)}{\Delta t^2}$$

and

$$\frac{\partial^2 u}{\partial x^2} = \frac{u(x+h,\,t) - 2u(x,t) + u(x-h,\,t)}{h^2}$$

so that

$$u(x,\,t+\Delta t) = 2u(x,t) - u(x,\,t-\Delta t) + \frac{\Delta t^2}{h^2}\,LC(u(x+h,\,t) - 2u(x,t) + u(x-h,\,t))$$

This method is again subject to the problem of stability and for a stable solution we require

$$LC\,\frac{\Delta t^2}{h^2} \leq 1.0$$

PROBLEMS

13-1. Using the adjectives first, second, and so on; ordinary or partial; linear or nonlinear; constant coefficient or time-varying coefficient; describe each of the following differential equations:

a. $\dfrac{d^2Y}{dX^2} - \epsilon\left(1 - \dfrac{1}{3}\left(\dfrac{dY}{dX}\right)^2\right)\dfrac{dY}{dX} + X = 0$

b. $\dfrac{dY}{dX} + \dfrac{1}{2X}\,Y + X^2 = 0$

c. $3X\,\dfrac{d^3Y}{dX^3} + 2\,\dfrac{dY}{dX} + 7Y = 0$

d. $\dfrac{\partial F}{\partial t} = \dfrac{\partial^2 F}{\partial X^2}$

e. $\dfrac{\partial^2 Y}{\partial t^2} + k\,\dfrac{\partial Y}{\partial t} = a^2\,\dfrac{\partial^2 Y}{\partial X^2}$

f. $\dfrac{\partial^2 U}{\partial t^2} = k^2\,\dfrac{\partial^2 U}{\partial X^2}$

13-2. Modify the Euler method program of Figure 13.3 and solve for the current in the network of Figure 13.2 when the capacitor C is replaced by a 0.5 henry inductance. Use a value of DELTA = 1.0E-6 and print your results for time values of 0.0 to 0.005 seconds, in 0.0001 second steps.

13-3. Rerun the program of Figure 13.6 with DELTAT equal to 1.0E-4, 1.0E-5, 1.0E-6, and 1.0E-7. In each case, print the solution as well as the error at 1.0E-4 second intervals. Plot the corresponding errors as a function of time.

13-4. Repeat Problem 13-3 for the network of Problem 13-2.

13-5. Modify the program of Figure 13.9 and obtain the solution to the following nonlinear networks:

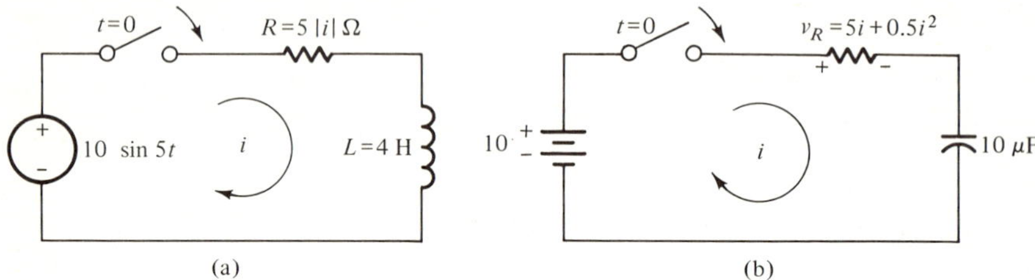

13-6. Modify the program of Figure 13.12 and determine the solution for the following network:

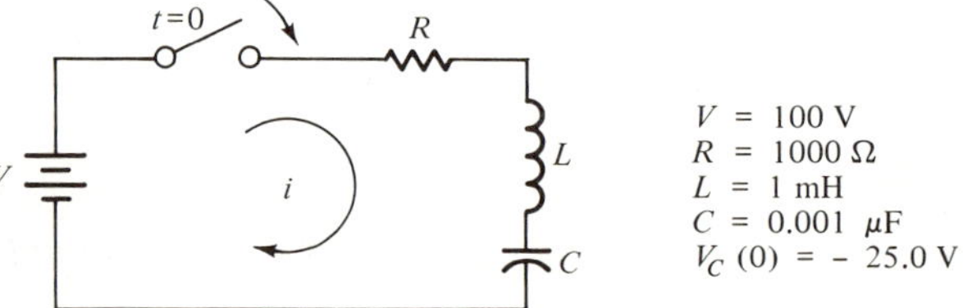

13-7. Assume that the switch is in position 1 and the circuit is in the steady state. At $t = 0$ the switch is thrown to position 2. Formulate the differential equation for $v_2(t)$ and modify the program of Figure 13.12 so that it can be used to obtain the solution for $v_2(t)$.

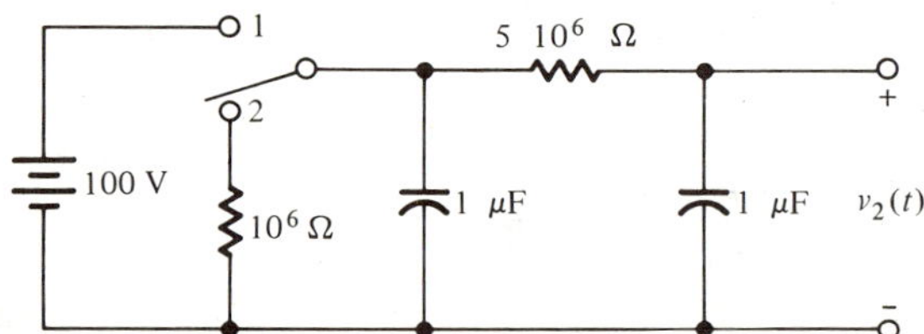

13-8. Assuming zero initial conditions, use Euler's method to solve for the currents in each of the following networks:

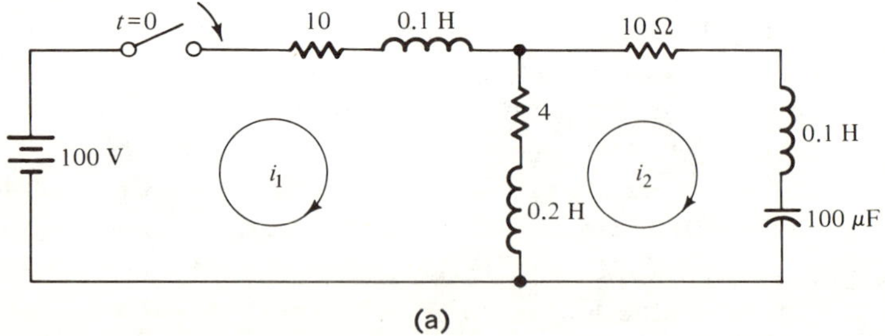

(a)

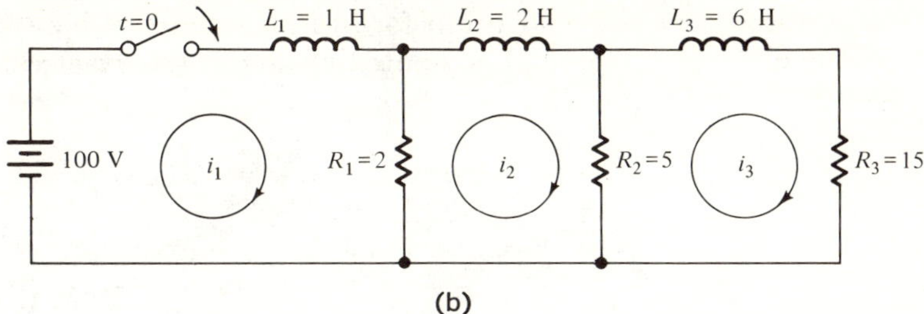

(b)

13-9. Modify the program of Figure 13.14 and use it to solve for the current of the network of Problem 13-2.

13-10. Using one of the second-order Runge-Kutta formulas, hand compute the first four iterations of the solution of the following first-order differential equations. Use $Y(0) = 1$ and $\Delta X = 0.1$.
a. $Y' = XY$ b. $Y' = X + Y^{1/2}$ c. $Y' = X^2 + Y^2$

13-11. Repeat Problem 13-10 using one of the Runge-Kutta third-order formulas.

13-12. Modify the program of Figure 13.21 and use it to obtain the solution for the networks of:
a. Problem 13-7 b. Problem 13-8.

13-13. Repeat Problem 13-12 using the program of Figure 13.23.

13-14. Repeat Problem 13-10 using Milne's multi-step predictor-corrector method.

13-15. Determine the solution for q, i, and di/dt at $t = 0.0$, 0.1, 0.2, 0.3, 0.4, and 0.5 seconds using the following methods:
a. The constant acceleration method

$$q_{n+1} = q_n + h*q_n' + \frac{h^2}{2}q_n''$$

$$q_{n+1}' = q_n' + h*q_n''$$

b. The linear acceleration method

$$q_{n+1} = q_n + h*q_n' + \frac{h^2}{6}(2q_n'' + q_{n+1}'')$$

$$q_{n+1}' = q_n' + \frac{h}{2}(q_n'' + q_{n+1}'')$$

c. The Adams-Stormer method

$$q_{n+1} = 2q_n - q_{n-1} + h^2*q_n''$$

Note that b. is a predictor-corrector method and will require a starting value. Note also that the Adams-Stormer method requires a solution for $t = 0$ and 0.1 before the method can be applied. Compare your three hand computed results to the exact calculus solution for this problem.

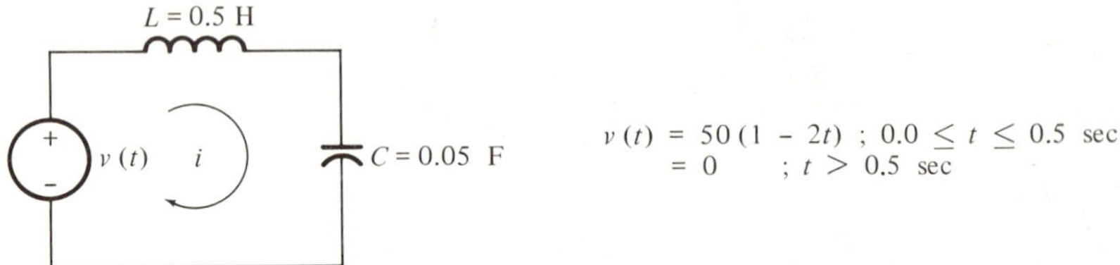

13-16. Repeat Problem 13-12 using the program of Figure 13.25.

13-17. Choosing approprate state variables, write the matrix equations for the following networks:

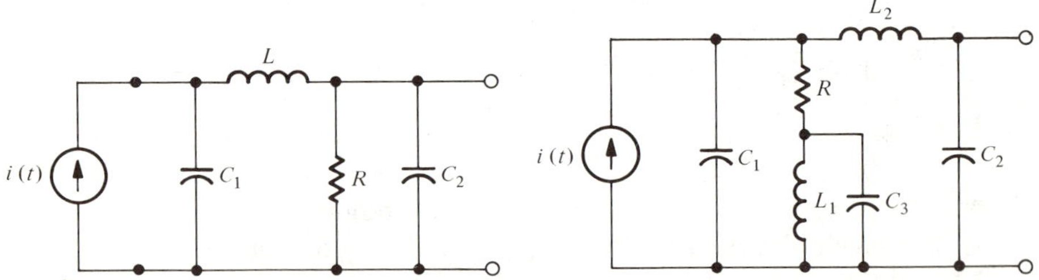

13-18. Write appropriate Fortran programs (see Figures 13.31 and 13.33) for solving for the output voltage of the networks of Problem 13-17. In network a. let $i(t) = 0.001u(t)$ and $C = C_1 + C_2$, $N = C_1/C_2 = 5$, and $L = KR^2C = 0.48R^2C$. In network b. let $i(t) = 0.001u(t)$ and $C_1 = 0.5C_2$, $C_3 = 0.22C_2$, $L_1 = 0.2R^2C_2$, and $L_2 = 0.7R^2C_2$.

13-19. Modify the program of Figure 13.31 and use it to solve for the output voltage response, due to a unit step current input, of the following networks shown in:

a. Figure 9.14.
b. Figure 9.17.
c. Figure 9.20.

13-20. Using the state variable method, write a general program for obtaining the transient response of a network. Model your program after the a-c analysis program of Section 9.4.

13-21. Improve the program of Problem 13-20 by incorporating a plot routine.

13-22. Explain why the total residue as a function of time has a zero imaginary value for all examples in Figure 13.36 except Figure 13.36f.

13-23. Use the program of Figure 13.35 to obtain the inverse Laplace transform of the following transfer functions. Verify each of the solutions by hand computing the inverse transforms of the partial fraction expansions.

a. $F(s) = \dfrac{5}{(s+1)(s+2)(s+5)}$

b. $F(s) = \dfrac{s^2 + 6s + 5}{s(s^2 + 4s + 5)}$

c. $F(s) = \dfrac{4s^4 + 20s^3 + 37s^2 + 29s + 10}{s(s^2 + 2s + 2)(s^2 + 4s + 5)}$

13-24. Modify the program of Figure 13.35 so that it will be able to obtain the inverse transform when repeated real roots are present. Assume that the maximum repetition will be equal to three and the repeated roots will be read in as separate data. Note that the program should be capable of handling several sets of repeated roots.

13-25. Modify the program of Problem 13-24 so that it will handle repeated complex roots (maximum repetition = 2).

13-26. Extend the one-inch mesh size two-dimensional problem of Figure 13.39 to the three-dimensional problem shown below. Assume that the cube is cut along the heavy lines shown and the isolate front part (front face, bottom face, and left-side face) has a potential of +10 volts with respect to the rear part (rear face, top face, and right-side face). Starting with zero initial values for the internal points, use the Gauss-Seidel method and hand compute the first five iterations.

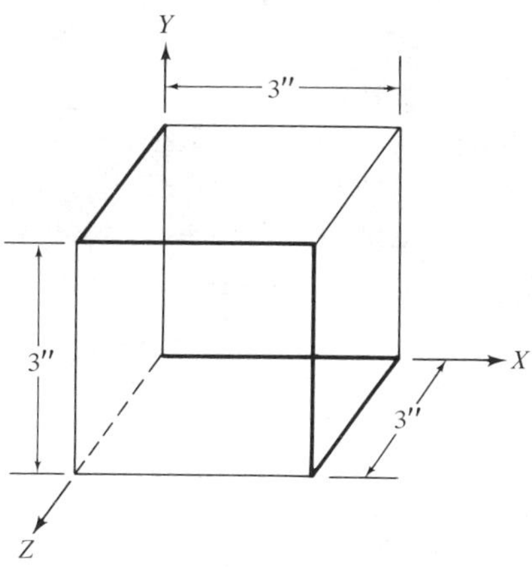

13-27. Repeat Problem 13-26 using the Jacobi method. Compare the two results.

13-28. Modify the program of Figure 13.41 so that the Jacobi method is used instead of the Gauss-Seidel method. Compare your results with those of Figure 13.42.

13-29. Modify the program of Figure 13.41 and solve the three-dimensional problem of Problem 13-26 (use a 1/4 in. mesh size).

13-30. Rerun the program of Figure 13.45 with the tri-diagonal interior points initially set at 5 volts, the points below the diagonal set at 2.5 volts, and the points above the diagonal set at 7.5 volts. Compare your results with those previously obtained.

13-31. Rerun the program of Figure 13.47 and obtain the optimum value of W.

13-32. Repeat Problem 13-31 using a mesh size of 1/8 in. Discuss your results.

13-33. Modify the program of Figure 13.45 and solve for the potential distribution within the T-shaped block. Assume that the top edge has a potential of +10 volts with respect to the bottom edge, and all other sides are perfectly insulated. Hint: See Figure 13.39.

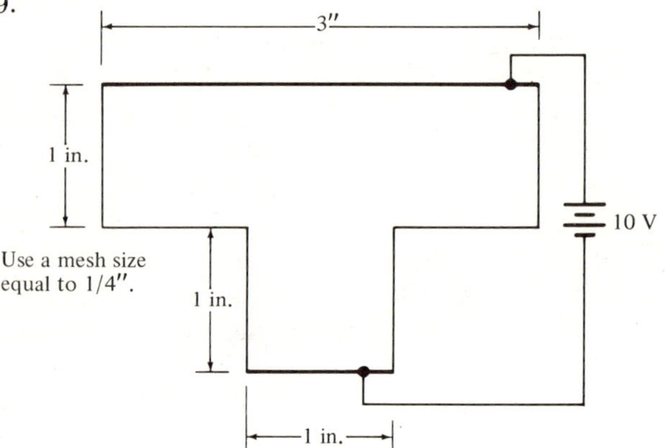

13-34. Modify the program of Figure 13.45 and solve for the interior potentials in the following figure. Hint: See the method discussed in connection with Figure 13.50 and use a 1/4 in. mesh size.

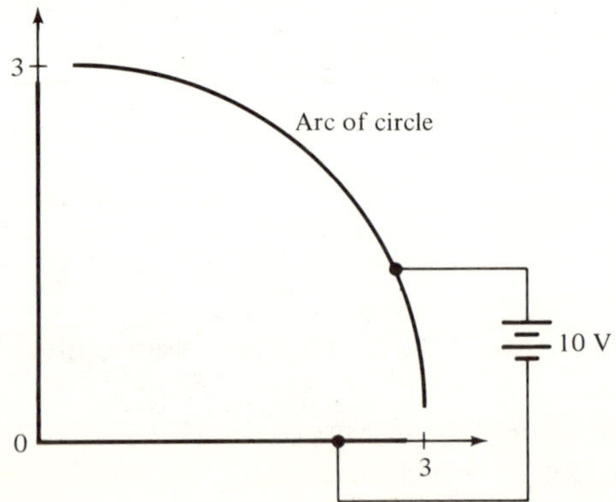

13-35. One Monte Carlo method of solution of the two-dimensional Laplace's equation for Figure 13.39 is based on a "random walk" from a given point to a point on the boundary. The path taken in the "random walk" is based on comparing the most significant digit of two successive random numbers and would follow the following scheme:

FIRST RANDOM NUMBER MOST SIGNIFICANT DIGIT	SECOND RANDOM NUMBER MOST SIGNIFICANT DIGIT	NEW X VALUE	NEW Y VALUE
0, 1, 2, 3, or 4	0, 1, 2, 3, or 4	X	$Y+h$
5, 6, 7, 8, or 9	5, 6, 7, 8, or 9	X	$Y-h$
0, 1, 2, 3, or 4	5, 6, 7, 8, or 9	$X+h$	Y
5, 6, 7, 8, or 9	0, 1, 2, 3, or 4	$X-h$	Y

For example, the most significant digits of the random number generator of Figure 10.16 are 9, 1, 3, 8, 8, 0, 0, etc. Starting at point V_1 or (1,2), the first two random numbers, 9 and 1, would take us to the boundary point (0,2). Returning to the starting point, the next two numbers, 3 and 8, would take us to the point (2,2). The next two numbers would take us back to the starting point (1,2). The next two numbers, 0 and 0, would take us to the boundary point (1,3). If we now took the sum of the boundary potentials that were reached and divided this sum by the number of walks, we would approximate the potential of V_1 as $^{10}/_2 = 5$ volts.

Write a Fortran program that uses this method to compute the potentials at V_{30}, V_{50}, V_{89}, and V_{123} in the figure of Figure 13.40.

13-36. Given a one-mile open-circuited transmission line with

$$R = 2.5 \text{ ohms/mile and } C = 0.004\ \mu\text{F/mile}$$

Write a Fortran program that will compute the line voltage as a function of time. Assume an input unit step voltage. Also discuss your choices of Δt and h.

APPENDIX A

The 026 Keypunch and 82 Sorter

THE IBM 026 KEYPUNCH

The IBM 026 printing card punch, or key punch, is shown in Figure A.1.[1] This machine can be used to punch the Fortran coded program and data cards that will later be processed by the computer. Let us first discuss some of the more general features of the unit.

I. General

a. The *power switch* is located at the upper left. It should be thrown to the *down* position to turn the machine on and a 60-second warm-up period is required.

b. The drum or *program control lever* is a Y-shaped switch located below the program drum and above and to the left of the back space switch. When the switch is rotated counterclockwise, the program sensing arms are lowered to the program drum so that they may read the codes that have been punched on the drum control card for automatic control operation. When the switch is rotated clockwise, the machine must be manually controlled. This switch must be rotated clockwise to the *off* position when removing or inserting the program drum.

c. The *column indicator* is located at the base of the drum holder (under the window) and indicates the number of the next column to be punched.

d. The *back space* switch is located at the bottom of the machine just below the reading station. The machine will back space as long as this switch is held down. Back spacing of cards located at either the punching station or reading station, as well as the program drum control card, is controlled by this switch.

[1] Only the IBM 026 keypunch is described here. The operation of the IBM 29 key punch is very similar to the 26 machine and the reader will find that once he has learned how to use the 26 machine it will be a simple matter for him to operate the 29 machine.

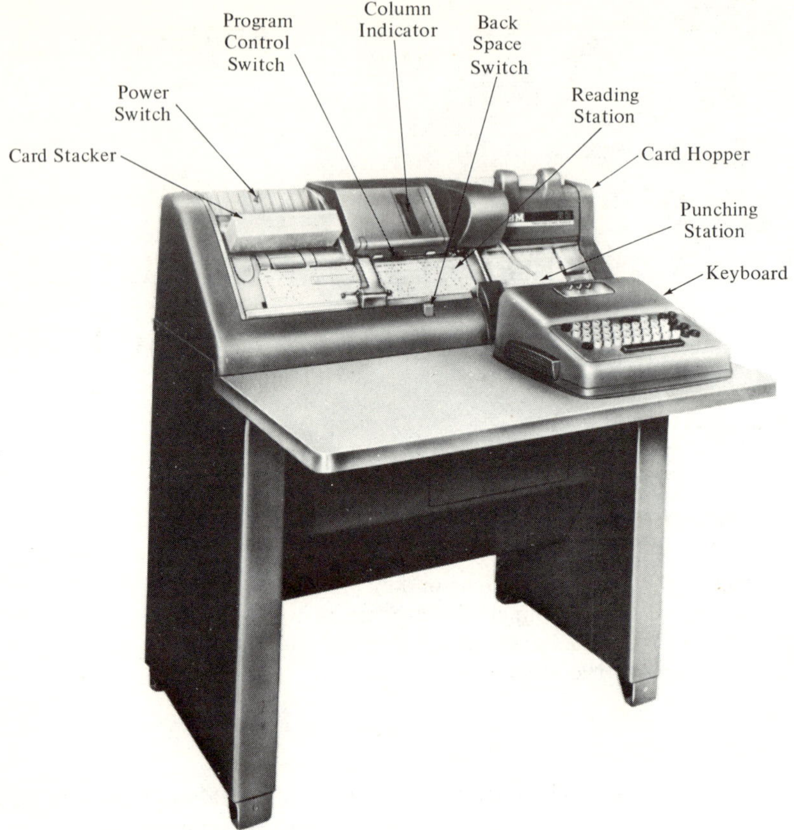

Figure A.1 IBM 26 printing card punch (Courtesy, IBM)

e. The *reading station* is located at the middle of the machine just above the back space switch and just below the program control lever. Normally cards feed from right to left, passing from the punching station through the reading station and ultimately left and up to the card stacker. Direct manual insertion of cards into the read station is reserved only for cards that are to be corrected or duplicated. This feature will be explained later.

f. The *punching station* is located on the right, next to the read station. Normally, cards feed from the card hopper down into the punching station, and after they have been punched, they are transferred from right to left through the read station and then out and up to the *card stacker*, which is located at the upper left just below the power switch.

g. The *card hopper* is located at the upper right. The spring loaded pressure plate must first be drawn back before cards can be inserted. Cards must be inserted face forward and with the 9-edge down. The hopper has a capacity for approximately 500 cards.

h. In this picture, the *keyboard* is shown at the right, but it is possible to move the keyboard any place on the reading board for the greater convenience of the user.

A more detailed view of the keyboard is shown in Figure A.2, a and b. An examination of Figure A.2a shows that there are three on–off control switches located at the top of the keyboard. When the right-hand Print switch is thrown to the *on* position, striking a key will cause a character to be printed, as well as punched, in a particular column. The printed character will appear at the top of the column. If the switch is thrown to the *off* position, no printing will result during key punch operation. In normal operation the Print switch should be left in the *on* position. The middle switch is the Automatic Skip and Automatic Duplication switch. It is used in connection with the program drum control

(a)

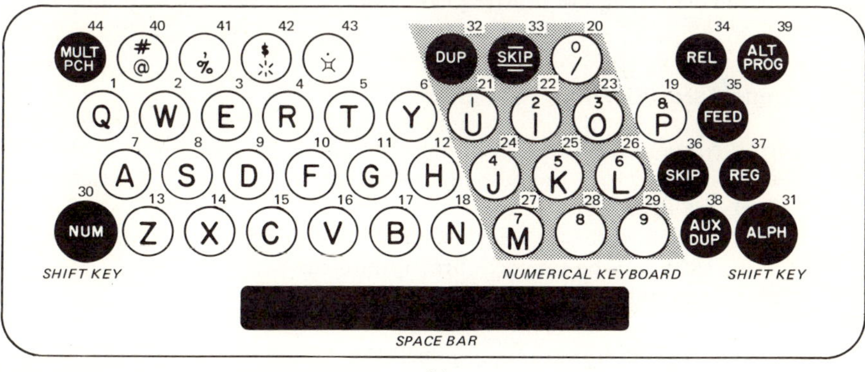

(b)

Figure A.2 (a) Combination alphabetic and numerical keyboard; (b) keyboard chart (Courtesy, IBM)

card and will be explained later. The left-hand switch is the Automatic Feed switch. This switch should be set to the *on* position when more than one card is to be punched.

Figure A.2b shows a more detailed view of the keyboard itself. Keys numbered 1 through 18 are alphabetic keys containing one character per key. Keys numbered 19 through 29 and 40 through 43 contain two characters per key. These keys contain the necessary alphabetic, numerical, and special characters that are used in the punching of a Fortran coded program. Since the use of all the above keys is quite obvious, the following discussion will be limited to the use of some of the remaining special purpose keys.

Key number 30 is the NUM key. In effect it is similar to the shift key on a typewriter. It is depressed to punch the top symbols on any of the two character keys. It must therefore be used when numbers are to be punched. The machine is automatically locked when any one of the keys 1 through 18 is depressed at the same time that the numeric key is depressed. The character will, however, be punched once the numeric key has been released.

Key number 31 is the ALPHA key. It is used in conjunction with the drum control card and causes the machine to punch alphabetic rather than numeric symbols when it is depressed.

Key number 36 is the SKIP key. This key is also used in conjunction with the drum control card and causes the punch to skip the remainder of the field (columns) marked with + punches on the drum card.

Key number 33 is the -SKIP-key. This key causes a − sign to be punched if the numeric key is not depressed, but if in numeric shift, the pressing of this key results in the remainder of the current field being skipped.

Key number 44 is the MULT PUNCH key. It permits successive punching in the same column with more than one key. The operation of this key should be avoided by all those who are not thoroughly familiar with this procedure in order to avoid an unusable punch hole pattern.

The following is a discussion of some of the ways in which the key punch machine may be used. In each case it is assumed that the power switch has been turned on and that a supply of cards has been inserted into the card hopper.

II. Single Card Punching Procedure

a. Turn the program control lever clockwise to the *off* position, turn the automatic feed switch off, turn the automatic skip and automatic duplication switch off, and turn the Print switch on.

b. Press key number 35, the FEED key, to bring down a card from the hopper. Alternately a card may be inserted directly into the punch station.

c. Press key number 37, the REG (Register) key, to position column 1 under the punching head.

d. By using the appropriate keys, punch the desired symbols on this card.

e. To retrieve this card from the machine, press key number 34, the REL (Release) key. This will cause the card to move into the *read* area. Now press key number 37, the REG key, and then the REL key to release the card.

III. Correcting or Duplicating Procedure

a. Same conditions as in item a, under II.

b. Manually insert the card that is to be corrected or duplicated into the read station. Insert a blank card into the punch station. Alternately, if a supply of cards is in the hopper, press the FEED key and then insert the card that is to be duplicated into the read station.

c. Hold down the REG key to register the card that is to be duplicated with the blank card.

d. Press key number 32, the DUP (Duplicate) key. This punches into the card located in the punching station the characters in the corresponding column of the card located in the reading station. If a correction is to be made in a particular column, release the duplicate key so that the machine stops duplicating at the desired column and punch the correct symbol in that column. Use the column indicator to determine what column is being duplicated, and as you near the column that is to be corrected, use a short striking motion on the DUP key to duplicate one column at a time.

e. To retrieve the old as well as the new corrected card, alternately press the REG and REL keys.

Note that the correction of a card does not require the card to be completely repunched, and in addition, that it is possible to make more than one correction at a given time.

IV. Normal Program Card-Punching Procedure

a. Turn the program control lever clockwise to the *off* position, turn the automatic feed switch on, turn the automatic skip and automatic duplicate switch off, and turn the Print switch on.

b. Press the FEED key to bring down a card from the hopper.

c. Press the FEED key again. This will register or position the first card so that its column 1 will be under the punching head. It will also bring down a second card from the hopper.

d. By using the appropriate keys, punch the desired symbols on the first card.

e. Now press the REL (Release) key. This moves the punched card into the read station, registers the next card, and brings down another card.

f. When the program has been completely punched, turn the AUTO FEED switch to the *off* position and alternately press the REL (Release) and REG (Register) keys until the cards have been cleared from the machine.

V. The Program Drum Card

When the program control lever is rotated counterclockwise to the *on* position, the program drum card will automatically control the operation of the key punch. The program control card is shown fastened to the program drum in Figure A.3.

To change a program control card, it is first necessary to raise the cover plate to expose the program drum. The program control lever must next be turned

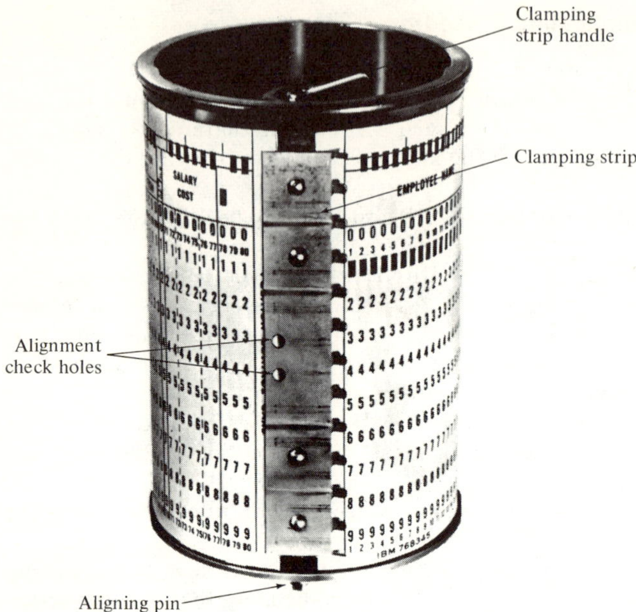

Figure A.3 Program drum (Courtesy, IBM)

clockwise to the *off* position. This raises the star wheels, which are used to sense the punches in the drum card, and permits the program card to be removed from the machine.

To remove the program control card, rotate the clamping strip handle, shown in Figure A.3, counterclockwise as far as it will go. This will loosen the clamping strip and permit the drum card to be removed from the drum. A new drum card may now be attached by inserting the column 80 edge of the card under the left edge of the clamping strip. The two alignment check holes shown in the figure may now be used to insure that the 80-edge is flush with the clamping strip. This edge may now be firmly clamped by rotating the clamping strip handle clockwise to its center position. When the clamping strip handle is in its center position, the left edge of the clamping strip is tightened and the right or toothed edge is loosened. The drum card may now be wrapped tightly around the drum and the column one edge inserted under the toothed edge of the clamping strip. Turning the handle fully clockwise completes the fastening of the drum card. The program drum may now be returned to the machine by placing it on the drum spindle. The drum should then be rotated until the aligning pin falls in the aligning hole in the column indicator. This completes the alignment of column one on the drum control card with column one of the column indicator. The cover plate can now be lowered and the program control lever rotated to the counterclockwise or *on* position so that the program control card will automatically control the operation of the machine.

A blank drum card will cause all numeric punching, making the use of the NUM key unnecessary. The absence of a drum card will cause all alphabetic punching, making the use of the ALPHA key unnecessary.

The following symbols may be used on drum cards for control purposes:

a. A + or 12-row punch in each column in the field defines the field.

b. A − or 11-row punch in a column followed by + symbols in succeeding columns will cause the machine to skip automatically to the column following the last + symbol.

c. A 0 punch in the drum card causes the machine to duplicate what has been punched in the corresponding column of the card in the read station.

d. A 1 punch causes the machine to shift to alphabetic characters for this column.

e. A 2 punch causes the keypunch to print leading zeros and certain other characters that normally would not be printed.

f. A 3 punch in a given column will prevent character printing even though the Print switch is in the *on* position.

VI. Program Drum Card Operation

To illustrate the usefulness of a program drum control card, consider the Fortran statement card shown in Figure A.4. It may be observed that columns 1 through 5 are used for the statement label or number, columns 7 through 72 are used for the actual Fortran statement, and columns 73 through 80 are not interpreted by the Fortran complier and may be used for identification purposes.

Figure A.5 is an example of a typical Fortran drum control card. The operation of the machine under this drum control card would be as follows:

a. Turn the program control lever counterclockwise to the *on* position, turn the automatic feed switch on, turn the automatic skip and automatic duplicate switch on, and turn the print switch on.

b. Press the FEED key to bring down a Fortran statement card from the hopper.

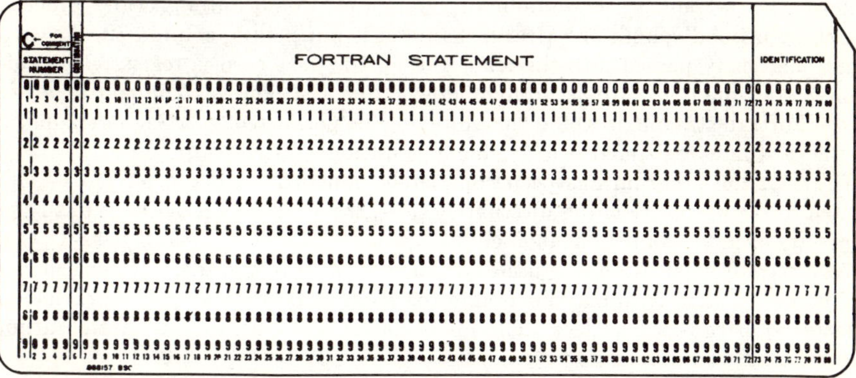

Figure A.4 The Fortran statement punch card

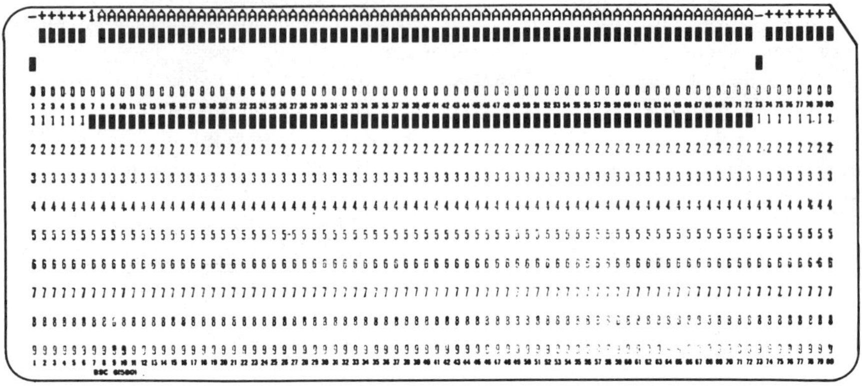

Figure A.5 An example of a Fortran drum card

c. Press the FEED key again. This will bring down a second Fortran statement card from the hopper and will register or position the first card according to the punches on the program drum control card. The − punch in column 1 of the drum control card and the + punches in columns 2 through 6 will automatically cause the first Fortran statement card to be positioned in column 7. This will permit normal punching of Fortran statements that must begin in column 7. If a Fortran statement requires a number or label, the backspace switch should be pressed until the appropriate column is reached.

d. By using the appropriate keys, punch the desired Fortran statement.

e. If the Fortran statement card punched in item d did not run beyond column 72, the next Fortran statement may be punched by pressing the REL key. This will move the punched card into the read station and register the next Fortran statement card in column 7 and bring down another Fortran statement card from the hopper. Alternately, if the Fortran statement punched in item d went beyond column 72, the − punch on the drum control card in column 73 followed by + punches in column 74 through 80 would automatically prevent punching beyond colunn 72. After a character is punched in column 72, the card is automatically passed into the read station, the next card is registered in column 7, and another card is brought down from the hopper. The continuation of the Fortran statement would therefore now be permitted, starting in column 7 of the card now registered in the punching station.

f. When the program has been completely punched, turn the AUTO FEED switch to the *off* position and alternately press the REL and REG keys until the cards are cleared from the machine.

As a second illustration of a useful drum control card, consider the card shown in Figure A.6. This card may be used in the sequential numbering of a previously punched Fortran source deck. The operation of the machine under this drum control card would be as follows:

a. Load the hopper with the Fortran source deck that is to be numbered sequentially in columns 78 through 80.

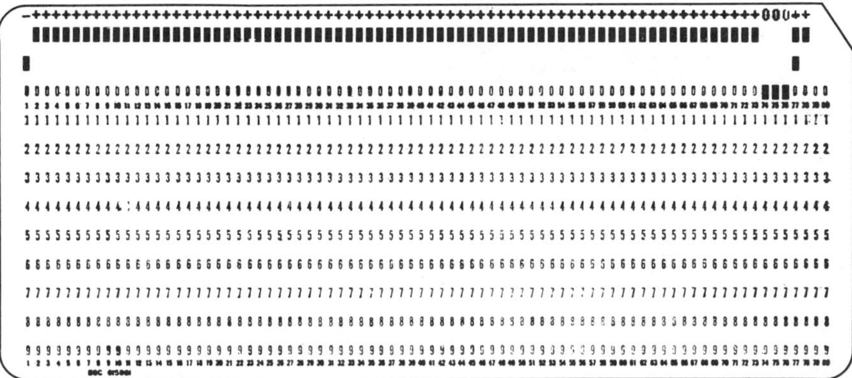

Figure A.6 A drum control card that can be used in the identification of Fortran source cards

b. Turn the program control lever counterclockwise to the *on* position, turn the automatic feed switch on, turn the automatic skip and automatic duplicate switch on, and turn the Print switch on.

c. Press the FEED key to bring down the first source card from the hopper.

d. Press the FEED key again. This will bring down the second source card from the hopper and will register or position the first card according to the punches on the program drum control card. The − punch in column 1 of the drum control card and the + punches in columns 2 through 73 will automatically cause the first source card to be positioned in column 74. For the first card, turn the drum control switch off and type the letter *N* in column 74, 0 in column 75, and a period in column 76. The drum control switch should now be turned on. This can be followed by two blank spaces and the number 01 printed in columns 79 and 80.

e. The punching of the 1 in column 80 will automatically move this card into the read station, bring down the third source card, and register the second source card in column 78 with the NO. symbols being duplicated in columns 74, 75, and 76 respectively, due to the 0 punches in columns 74, 75, and 76 on the drum control card. The numbers 0 and 2 may now be punched in columns 79 and 80 and this process may be repeated until all the cards have been numbered.

The numbering system described above will prove to be extremely useful in resorting the deck if it should be accidentally dropped. The resorting may be done by hand or automatically with the card sorter shown in Figure A.7.

The IBM Card Sorter operates at a speed of 650 cards per minute and sorts one column at a time. In order to illustrate how it may be used to numerically sequence a deck of cards that might have been accidentally dropped, consider the random number arrangement of 50 cards shown in Figure A.8. Card 1 is considered the first card in the deck and will correspond to the first card that will be sorted, card 2 the second card, and card 50 the last card to be sorted.

The cards are first sorted according to their column 80 number. The result of this sorting is shown in Figure A.9 and indicates the cards that have been sorted

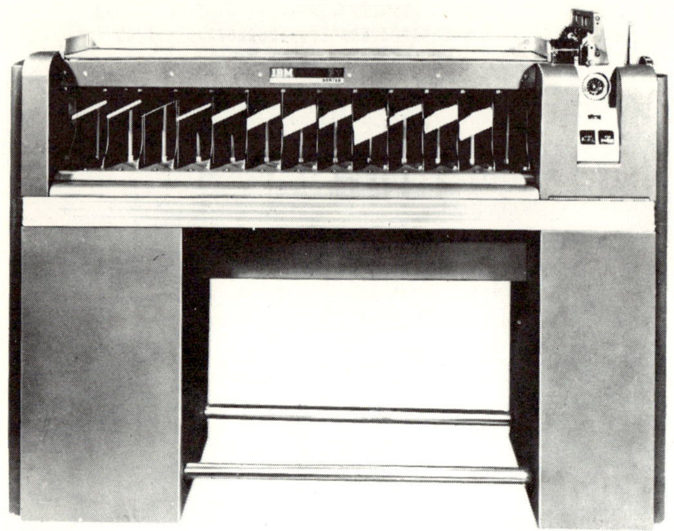

Figure A.7 The IBM 82 Sorter (Courtesy, IBM)

into bins 0 through 9. The bottom number represents the first card that was found to have the corresponding bin number in its column 80 and the top number corresponds to the last card found with this bin number.

The cards are now stacked, bin below bin, with the 0 bin stack being first and the 9 bin stack last. The result is shown in Figure A.10 with card 1 being the first card and card 50 the last.

CARD	NUMBER	CARD	NUMBER	CARD	NUMBER	CARD	NUMBER	CARD	NUMBER
1	45	11	01	21	08	31	18	41	37
2	36	12	03	22	10	32	12	42	48
3	07	13	23	23	02	33	05	43	31
4	50	14	32	24	33	34	44	44	21
5	25	15	47	25	46	35	22	45	04
6	16	16	41	26	14	36	27	46	15
7	30	17	06	27	19	37	43	47	20
8	28	18	26	28	35	38	09	48	38
9	34	19	39	29	42	39	13	49	29
10	49	20	40	30	17	40	24	50	11

Figure A.8 A Random Number Arrangement of 50 Numbers.

BIN	0	1	2	3	4	5	6	7	8	9
	20	11	22	13	04	15	46	37	38	29
	10	21	12	43	24	05	26	27	48	09
	40	31	42	33	44	35	06	17	18	19
	30	41	02	23	14	25	16	47	08	39
	50	01	32	03	34	45	36	07	28	49

Figure A.9 The Bin Distribution of 50 Random Numbers Numerically Sorted on Column 80.

The cards are now sorted numerically for the second time, but this time according to column 79. The bin results of this sorting are shown in Figure A.11.

The cards are now stacked, bin below bin, with the 0 bin first. The results of the column 79 sort are shown in Figure A.12 and it is apparent that the accidentally dropped deck is now back in its numerically correct sequence.

Another example of the need for a kind of drum control card different from the two cards described above would be where a large number of data cards would have to be punched. A drum control card could similarly be punched to control automatically the operation of the machine as defined by the required data fields. In general, drum control cards will prove useful whenever a large number of cards with identical fields are to be punched.

CARD	NUMBER	CARD	NUMBER	CARD	NUMBER	CARD	NUMBER	CARD	NUMBER
1	50	11	32	21	34	31	36	41	28
2	30	12	02	22	14	32	16	42	08
3	40	13	42	23	44	33	06	43	18
4	10	14	12	24	24	34	26	44	48
5	20	15	22	25	04	35	46	45	38
6	01	16	03	26	45	36	07	46	49
7	41	17	23	27	25	37	47	47	39
8	31	18	33	28	35	38	17	48	19
9	21	19	43	29	05	39	27	49	09
10	11	20	13	30	15	40	37	50	29

Figure A.10 50 Random Numbers Rearranged After Column 80 Sort.

BIN	0	1	2	3	4	5
		19	29	39	49	
	09	18	28	38	48	
	08	17	27	37	47	
	07	16	26	36	46	
	06	15	25	35	45	
	05	14	24	34	44	
	04	13	23	33	43	
	03	12	22	32	42	
	02	11	21	31	41	
	01	10	20	30	40	50

Figure A.11 The Bin Distribution of 50 Random Numbers Sorted on Columns 80 and 79.

CARD	NUMBER	CARD	NUMBER	CARD	NUMBER	CARD	NUMBER	CARD	NUMBER
1	01	11	11	21	21	31	31	41	41
2	02	12	12	22	22	32	32	42	42
3	03	13	13	23	23	33	33	43	43
4	04	14	14	24	24	34	34	44	44
5	05	15	15	25	25	35	35	45	45
6	06	16	16	26	26	36	36	46	46
7	07	17	17	27	27	37	37	47	47
8	08	18	18	28	28	38	38	48	48
9	09	19	19	29	29	39	39	49	49
10	10	20	20	30	30	40	40	50	50

Figure A.12 50 Random Numbers Rearranged After Column 79 Sort.

APPENDIX B

Computer Arithmetic

It is important to realize that the basic operation of addition or subtraction is the same mathematical operation when we deal with signed numbers. For example, in the addition of $A + B$ the following four possibilities exist:

1. If both A and B are positive, we add directly and put a + sign on the result.
2. If A is positive and B is negative, we take the complement of B and add it to A. If the result is positive, we affix a + sign; if not, we recomplement and affix a − sign.
3. If A is negative and B is positive, we take the complement of A and add it to B. If the result is positive, we affix a + sign; if not, we recomplement and affix a − sign.
4. If both A and B are negative, we add directly and put a − sign on the result.

Correspondingly in the case of the subtraction of $A - B$, the following four possibilities exist:

1. If both a and b are positive, we take the complement of b and add it to a. If the result is positive we affix a + sign; if not, we recomplement and affix a − sign.
2. If a is positive and b is negative, we add directly and put a + sign on the result.
3. If A is negative and B is positive, we add directly and put a − sign on the result.
4. If both A and B are negative, we take the complement of A and add it to B. If the result is positive, we affix a + sign; if not, we recomplement and affix a − sign.

An examination of the above two sets of possibilities show that the subtraction of signed numbers is identical to the addition of signed numbers if possibility 1 is interchanged with 2 and 3 is interchanged with 4.

The true or radix complement is formed by subtracting each digit of the number from the radix (base) minus one of the number system and then adding one to the least significant digit. The true complement in the decimal system is referred to as the 10's complement and in the binary system as the 2's complement. The following numerical examples will help to clarify the above:

$$A - B = 89 - 23 = ?$$

Since this may be considered as the addition of a positive A and a negative B, we first take the complement of B. This is accomplished by subtracting each digit from the radix minus 1 so that in the decimal system this would correspond to subtracting each digit from 9 and then adding *one* to the least significant digit.

$$\begin{array}{r} 99 \\ -23 \\ \hline 76 \\ +\ 1 \\ \hline 77 \end{array} \quad \text{the 10's complement}$$

Note that the equivalent of this operation is to add 100 to −23, so that the complement 77 is larger than −23 by exactly 100. If we now add A to the complement of B, the result should be high by exactly 100.

$$\begin{array}{r} 89 \\ +77 \\ \hline \underline{1}/66 \end{array}$$

In this numerical example, the addition of 89 and 77 produced a carry into the third place. Since this carry represents the exact excess of the sum of A and the complement B, it should be disregarded. Electronically it is equivalent to an overflow from the accumulator register, so that the digits remaining in the register would represent the signed addition of $A + B$.

It is instructive to reverse the numerical values of A and B and repeat the example.

$$A - B = 23 - 89 = ?$$

Since this again may be considered as the addition of a positive A and a negative B, we first find the complement of B.

$$\begin{array}{r} 99 \\ -89 \\ \hline 10 \\ +\ 1 \\ \hline 11 \end{array} \quad \text{the 10's complement}$$

The equivalent of this operation is to add 100 to −89 so that the complement 11 is larger than −89 by exactly 100. We now add A to the complement of B, realizing that the result should be high by exactly 100.

$$\begin{array}{r} 23 \\ +11 \\ \hline \_/34 \end{array}$$

In this numerical example, the addition of 23 and 11 produces no carry into the third place. As a consequence, the lack of an overflow from the accumulator indicates that the sum is actually negative and the digits remaining in the register are the 10's complement of the result. We now recomplement or take the 10's complement of 34 and affix a − sign to the result.

$$\begin{array}{r} 99 \\ -34 \\ \hline 65 \\ +\ 1 \\ \hline -66 \end{array} = \text{the signed addition of } 23 - 89$$

Before we consider these same arithmetic operations in terms of binary numbers, let us first recall that the decimal number 23 is actually a contraction and really means

$$2*10 + 3*1 = 2*10^1 + 3*10^0 = 23$$

Thus in taking advantage of the symbol position we disregard the power of ten and write only 23.

In general we may represent any number N as

$$N = a_nR^n + a_{n-1}R^{n-1} + \cdots + a_1R^1 + \underbrace{a_0R^0 \,.\, +a_{-1}}_{\text{radix point}}R^{-1} + a_{-2}R^{-2} + \cdots$$

where R is called the base or radix of the number system. In the base ten or decimal system, $R = 10$ and ten different symbols (a's) are used

$$a = 0, 1, 2, 3, 4, 5, 6, 7, 8, \text{ or } 9$$

In the base two or binary system, $R = 2$ and two different symbols are defined

$$a = 0 \quad \text{or} \quad 1$$

The decimal number $23)_{10}$ would thus correspond to the binary number $10111)_2$ or

$$\begin{aligned} 23)_{10} &= 1*2^4 + 0*2^3 + 1*2^2 + 1*2^1 + 1*2^0 = 10111)_2 \\ &= 16 + 0 + 4 + 2 + 1 = 23)_{10} \end{aligned}$$

Similarly the decimal number $89)_{10}$ would correspond to the binary number $1011001_2)$ or

$$\begin{aligned} 89)_{10} &= 1*2^6 + 0*2^5 + 1*2^4 + 1*2^3 + 0*2^2 + 0*2^1 + 1*2^0 = 1011001)_2 \\ &= 64 + 0 + 16 + 8 + 0 + 0 + 1 = 89)_{10} \end{aligned}$$

Now using binary representation, we repeat the above examples.

$$A)_2 - B)_2 = 1011001 - 0010111 = ?$$

We first find the 2's complement of B.

$$\begin{array}{r} 1111111 \\ -0010111 \\ \hline 1101000 \\ +\qquad\quad 1 \\ \hline 1101001 \end{array} \quad \text{the 2's complement}$$

The equivalent of this operation is to add 10000000 to −0010111, so that the complement 1101001 is larger than −0010111 by exactly 10000000 (the decimal equivalent is 128_{10}). Note that taking binary complements is a relatively simple electronic procedure; ones are changed to zeros and zeros are changed to ones and then one is added to the least significant digit. We now add A to the complement of B, realizing that the result should be high by exactly 10000000.

$$\begin{array}{r} 1011001 \\ +1101001 \\ \hline \underline{1}/1000010 \end{array}$$

In this numerical example, the addition of A and the complement of B produced a carry into the eighth place. Since this carry represents the exact excess of the sum of A and the complement of B, it should be disregarded and considered as overflow. The binary digits remaining in the register are equivalent to the decimal number $66)_{10}$.

The corresponding binary addition of 23 − 89 requires the 2's complement of the binary equivalent of 89.

$$\begin{array}{rl} 1111111 & \\ -1011001 & \\ \hline 0100110 & \\ + \quad\quad 1 & \\ \hline 0100111 & \text{the 2's complement} \end{array}$$

We now add A to the complement of B, realizing that the result should be high by 10000000.

$$\begin{array}{r} 0010111 \\ +0100111 \\ \hline \_/0111110 \end{array}$$

Since there is no carry into the eighth place, the sum is negative and the digits remaining in the register are the two complements of the result. Taking the 2's complement of 0111110 and affixing a − sign yields,

$$\begin{array}{rl} 1111111 & \\ -0111110 & \\ \hline 1000001 & \\ + \quad\quad 1 & \\ \hline -1000010 & = \text{the signed addition of } 0010111 - 1011001 \end{array}$$

and is equivalent to the decimal number $-66)_{10}$.

Consider now the arithmetic operation of multiplication. In general, the multiplication of an N digit number by an N digit number results in a number $2N$ digits long. The result of an arithmetic multiplication operation is therefore usually stored in a double length register and the programmer has the option of using the upper or lower N digits. In general, no overflow is possible unless negative numbers are represented in their signed 2's complement form.

The basic multiplication operation can be regarded as essentially repeated addition. For example, consider one scheme described by the four-digit accumulator register and the two-digit distributor register shown in Figure B.1.

	Overflow	Accumulator				Distributor	
Step 1.		2	3	0	0	8	9
Step 2.	2	3	0	0	0		
Step 3.		3	1	7	8		
Step 4.	3	1	7	8	0		
Step 5.		2	0	4	7		

Figure B.1 The multiplication of two two-digit numbers.

In step 1, a multiplicand having a decimal value of 89 is shown in the distributor register, and a multiplier having a decimal value of 23 is shown in the upper part of the accumulator. In step 2, the contents of the accumulator are shifted left by one digit. The digit that overflows is examined and determines the number of times the contents of the distributor register (the multiplicand) will be added into the lower end of the accumulator. This result is shown in step 3. The contents of the accumulator are again shifted left by one digit in step 4. Since the overflow digit is this time equal to 3, the multiplicand value of 89 is added three times into the lower end of the accumulator. The final result is shown in step 5 and indicates that the most significant digits are located in the upper part of the accumulator and the least significant digits are in the lower part.

Similarly, the process of division may be regarded as repeated subtraction or the inverse of multiplication. For a numerical example, let us use binary representation and registers that are 8 bits long. The value of each bit may be represented by the state of a flip-flop. Since subtraction is involved, the 2's complement will be used and the extreme left flip-flop or bit will denote the sign of the number. A bit value of 1 will represent a negative number and a bit value of 0 will correspond to a positive number. The binary point will be assumed to be between the first and second bit, or the number will be assumed to be stored in its floating-point form. A floating-point number is written in two parts. The first part of the number contains the most significant digits of the number and may be regarded as a fraction having a magnitude between 0 and 1. The second part of the number is the exponent for the base of the number system and may be regarded as the number of places the radix point has been shifted. The sign of the exponent will be plus when the radix point is shifted to the left and minus when it is shifted to the right. For example, the decimal number 89 would be represented by

.89	+2
fractional part	exponent part

Note that this is similar to scientific notation, differing only by a factor of 10.

For the actual numerical example let 23 be divided by 89. In binary notation this would correspond to

$$23)_{10} = 0010111)_2$$

and

$$89)_{10} = 1011001)_2$$

In floating-point notation we would have

+0010111. =	0.0010111	111	= the dividend
and			
+1011001. =	0.1011001	111	= the divisor
	fractional part	exponent part	

It should now be obvious that the position of the binary point is unimportant as long as the exponent parts of the two numbers are the same.

One of the division algorithms proceeds by subtracting the divisor from the dividend. This is done by first taking the 2's complement of the divisor and adding this to the dividend.

Step 1.

```
  1.1111111
 -0.1011001
 ----------
  1.0100110
+  .0000001
 ----------
  1.0100111   the 2's complement of the divisor
  0.0010111   the dividend
 ----------
  1.0111110
```

Since the sign bit shows that the result is negative, the first quotient bit is 0, and we restore by adding back the divisor.

```
  1.0111110
  0.1011001
 ----------
  0.0010111
```

Step 2.

We then shift the divisor right one digit and take the 2's complement.

```
  1.1111111
 -0.0101100
 ----------
  1.1010011
+  .0000001
 ----------
  1.1010100   the 2's complement of the divisor shifted
              one digit to the right
```

We now add the dividend and the 2's complement of the shifted divisor.

```
 0.0010111
 1.1010100
 ---------
 1.1101011
```

Since the sign bit shows that the result is negative, the second quotient is also zero and we restore by adding back the shifted divisor.

```
 1.1101011
+0.0101100
 ---------
 0.0010111
```

Step 3.

We now shift the divisor another position to the right and take the 2's complement.

```
 1.1111111
-0.0010110
 ---------
 1.1101001
+ .0000001
 ---------
 1.1101010   the 2's complement of the divisor shifted
             two digits to the right
```

We now add the dividend and the 2's complement of the shifted divisor.

```
 0.0010111
 1.1101010
 ---------
 0.0000001
```

Since the sign bit shows that the sign bit is positive, the third quotient bit is a one. The divisor is now shifted another position to the right and its 2's complement is added to the previous result.

Step 4.

```
 1.1111111
-0.0001011
 ---------
 1.1110100
+ .0000001
 ---------
 1.1110101   the 2's complement of the divisor shifted
             three digits to the right

 1.1110101
 0.0000001
 ---------
 1.1110110
```

Since the sign bit shows that the result is negative, the fourth quotient bit is zero and we restore by adding back the shifted divisor.

```
 1.1110110
 0.0001011
 ---------
 0.0000001
```

Step 5.

We now shift the divisor another position to the right and take its 2's complement.

$$\begin{array}{rl} 1.1111111 & \\ -0.0000101 & \\ \hline 1.1111010 & \\ +\ .0000001 & \\ \hline 1.1111011 & \text{the 2's complement of the divisor shifted four digits to the right} \end{array}$$

The results of step four are now added to the 2's complement of the shifted divisor.

$$\begin{array}{r} 1.1111011 \\ 0.0000001 \\ \hline 1.1111100 \end{array}$$

Since the sign bit shows that the result is negative, the fifth quotient bit is zero, and we restore by adding back the shifted divisor. This process can be continued until 15 digits are found (the length of the double register). Thus

$$23)_{10}/89)_{10} = 0.25842+)_{10}$$

or

$$.0010111)_2/.1011001)_2 = 0.0100+)_2 = 0.25+)_{10}$$

Note that the above process is identical to the familiar process of long division. The divisor is first subtracted from the dividend and then it is shifted one digit to the right and subtracted from the new dividend, and so on. Each subtraction that results in a positive remainder yields a 1 in the corresponding quotient bit. For each subtraction that results in a negative remainder, the divisor is added back before the next shift and subtraction, and a 0 results in the corresponding quotient bit.

Selected Bibliography

FORTRAN

Anderson, Decima M., *Computer Programming Fortran IV*. New York: Appleton-Century-Crofts, 1966.

Dimitry, Donald and Mott, Thomas H. Jr., *Introduction to Fortran IV Programming*. New York: Holt, Rinehart and Winston, Inc., 1966.

Golde, Hellmut, *Fortran II and IV for Engineers and Scientists*. New York: The Macmillan Company, 1966.

Golden, James T., *Fortran IV Programming and Computing*. Englewood Cliffs, N.J.: Prentice-Hall Inc., 1965.

Harris, L. Dale, *Fortran Programming*. Columbus, Ohio: Charles E. Merrill Books, Inc., 1964.

Ledley, Robert Steven, *Fortran IV Programming*. New York: McGraw-Hill Book Company, 1966.

Lee, R. M., *Fortran IV Programming*. New York: McGraw-Hill Book Company, 1967.

McCracken, Daniel D., *A Guide to Fortran IV Programming*. New York: John Wiley and Sons, Inc., 1965.

McCracken, Daniel D., *Fortran with Engineering Applications*. New York: John Wiley and Sons, Inc., 1967.

Organick, Elliott I., *A Fortran IV Primer*. Reading, Mass.: Addison-Wesley Publishing Co., 1966.

Plumb, S. C., *Introduction to Fortran*. New York: McGraw-Hill Book Company, 1964.

Prager, William, *Introduction to Basic Fortran Programming and Numerical Methods*. New York: Blaisdell Publishing Company, 1965.

Price, Wilson T., *Elements of Basic Fortran IV Programming*. New York: Holt, Rinehart and Winston, Inc., 1969.

Wimmert, Robert J., *Computer Programming Techniques Volume II Fortran Language*. New York: Holt, Rinehart and Winston, Inc., 1969.

NUMERICAL METHODS

Abramowitz, Milton and Stegun, Irene, *Handbook of Mathematical Functions*, National Bureau of Standards, Applied Mathematics Series. Washington, D.C.: U.S. Government Printing Office, 1964.

Arden, Bruce W., *An Introduction to Digital Computing*. Reading, Mass.: Addison-Wesley Publishing Company, 1963.

Carnahan, Brice, Luther, H. A., and Wilkes, James O., *Applied Numerical Methods*. New York: John Wiley and Sons, Inc., 1969.

Hamming, Richard W., *Numerical Methods for Scientists and Engineers*. New York: McGraw-Hill Book Company, 1962.

Hastings, Cecil, Jr., *Approximations for Digital Computers*. Princeton, N.J.: Princeton University Press, 1955.

Hildebrand, F. B., *Introduction to Numerical Analysis*. New York: McGraw-Hill Book Company, 1956.

Householder, A. S., *Principles of Numerical Analysis*. New York: McGraw-Hill Book Company, 1953.

James, M. L., Smith, G. M., and Wolford, J. C., *Applied Numerical Methods for Digital Computation with Fortran*. Scranton, Pa.: International Textbook Co., 1967.

Kunz, K. S., *Numerical Analysis*. New York: McGraw-Hill Book Company, 1957.

Kuo, Franklin F. and Kaiser, James F., *System Analysis by Digital Computer*. New York: John Wiley and Sons, Inc., 1966.

Kuo, Shan S., *Numerical Methods and Computers*. Reading, Mass.: Addison-Wesley Publishing Company, 1965.

McCracken, Daniel D. and Dorn, William S., *Numerical Methods and Fortran Programming*. New York: John Wiley and Sons, Inc., 1964.

McCalla, Thomas Richard, *Introduction to Numerical Methods and Fortran Programming*. New York: John Wiley and Sons, Inc., 1967.

McCormick, John M. and Salvadori, Mario G., *Numerical Methods in Fortran*. Englewood Cliffs, N.J.: Prentice-Hall Inc., 1964.

Pennington, Ralph H., *Introductory Computer Methods and Numerical Analysis*. New York: The Macmillan Company, 1965.

Scheid, Francis, *Theory and Problems of Numerical Analysis*. New York: McGraw-Hill Book Company, 1968.

Index